Vehicle and engine technology

Vehicle and Engine Technology

Second Edition

Heinz Heisler
MSc, BSc, FIMI, MIRTE, MCIT
Formerly Principal Lecturer and Head of Transport Studies,
College of North West London,
Willesden Centre,
London, UK

A member of the Hodder Headline Group
LONDON • SYDNEY • AUCKLAND

First edition published in Great Britain 1985
Second edition first published in Great Britain in 1999 by
Arnold, a member of the Hodder Headline Group,
338 Euston Road, London NW1 3BH

http://www.arnoldpublishers.com

British Library Cataloguing in Publication Data
A catalogue record for this book is available from the British Library

ISBN 0 340 69186 7

1 2 3 4 5 6 7 8 9 10

Commissioning Editor: Sian Jones
Production Editor: Liz Gooster
Production Controller: Sarah Kett

Typeset by J&L Composition Ltd, Filey, North Yorkshire
Printed and bound in Great Britain by The Bath Press, Bath

To my long-suffering wife, who has provided support and understanding throughout the preparation of both editions

Contents

Preface to the first edition

Although much technical literature and many books are available dealing with the design, construction, and maintenance of various parts of the motor vehicle, the craft or technician student seldom has the time to search through several sources of information for a solution to a problem or question. This book, with the second volume, therefore aims to present a unified, precise, and clear description and explanation of the fundamentals of all the essential components of the motor vehicle, making extensive use of illustrations alongside the written material.

This first volume covers the part-1 technology content of the City and Guilds of London Institute motor-vehicle craft studies course 381 and the objectives of the Business & Technician Education Council standard units Vehicle technology I (U82/005) and Engine technology I (U82/878). However, in places the content goes beyond the requirements of these syllabuses, to provide a better foundation for subsequent studies, and the fullness of information given should also make the book useful as a reference work for practising craft and technician personnel as well as to students.

Acknowledgements
The author would like to express his thanks to the many friends and colleagues who provided suggestions and encouragement in the preparation of this book; to Bob Davenport of Edward Arnold (Publishers) Ltd who scrutinised and edited the text; and to his daughter Glenda who typed the manuscript.

Preface to the second edition

Although much technical literature and many books are available dealing with the design, construction and maintenance of various parts of the motor vehicle, the craft or technician student seldom has the time to search through several sources of information for a solution to a problem or question. This book presents a unified, precise and clear description and explanation of the fundamentals of the essential components of the motor vehicle, making extensive use of illustrations alongside the written material.

It is now more than twelve years since the first edition of this book was published, and in this time span the demands which have been imposed on the motor vehicle have compelled its design and construction to become much more complex and sophisticated. This enlarged text with more than 300 new illustrations drawn by the author brings into focus advancements in motor vehicle technology which include mechanical refinements, electrical applications and electronically controlled systems. The basic fundamentals of the motor vehicle are covered. In addition the text explores the following more recent developments and improvements which have taken place.

Utilisation of electronically controlled units, electrical sensors, solenoid valve-controlled petrol and diesel injection and electronic ignition, dual stage air intakes, supercharging and detailed engine construction changes have resulted in considerable improvements in engine performance, such as higher maximum speeds, better acceleration response, fuel economy, lower levels of mechanical and combustion noise, refined engine balance, smoothness of running, reduction in exhaust emission, reliability and prolonged life expectancy. Automatic transmission and power steering are now commonly used to reduce driver fatigue and four wheel drive is a popular option available to improve ground wheel grip on poor road surfaces. Varied suspension arrangements are now used for both light and heavy vehicles to improve the steering, handling response, control of body-roll, dive, squat and ride comfort. Vehicle braking safety has been upgraded by adopting split brake line systems, servo-brake units, load sensing valves, and antilock brake systems. A section on commercial vehicle braking has also been included. An extensive coverage of electrically operated equipment has been added to this edition such as light and heavy duty starter motors, alternator charging circuits, central door locks, electrical window winders, two speed and intermittent windscreen wipers, thermal and electronically controlled signal light systems, fuel and temperature sensor transmitters and gauges and air conditioning. Minor changes and additions have been made in other topic areas of this updated text.

Heinz Heisler MSc, BSc, FIMI, MIRTE, MCIT
Formerly Principal Lecturer and Head of Transport Studies
College of North West London
Willesden Centre
London, UK

1 Vehicle body and chassis layout

1.1 The motor car

A motor car is designed to carry passengers in the sitting position facing the direction of motion and to accommodate their luggage. Additional space must be provided for the engine which produces the propelling force, for the transmission system which conveys the power from the engine to the driving road-wheels, for the steering which enables the vehicle to be manoeuvred, for the suspension layout which partially isolates the body from road-wheel impact shocks, and for the braking system which enables the moving vehicle to be slowed down or brought to rest.

While all these functions must be satisfied, consideration must also be given to the styling of the body to suit various aesthetic tastes and application requirements.

Light motor vehicles which are built to carry passengers and sometimes goods may be broadly classified as follows:

a) saloon car
b) coupé
c) convertible
d) estate car
e) pick-up.

1.1.1 Saloon car (figs 1.1 and 1.2)

Most saloon cars have an enclosed compartment within which are situated a row of front and a row of rear seats (fig. 1.1). There is no partition between the driver and rear-passenger seats, but separate luggage space will be provided either at the front or the rear, depending on at which end the engine is mounted. Normally there are one or two doors on each side. If the car is a hatchback, a door replaces the luggage-space boot (fig. 1.2)

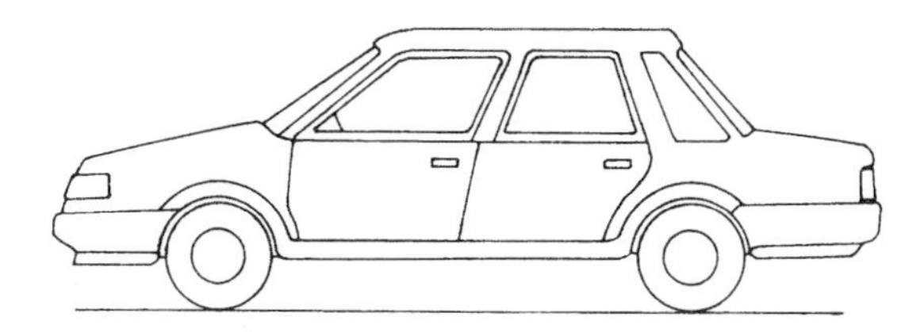

Fig. 1.1 Saloon car

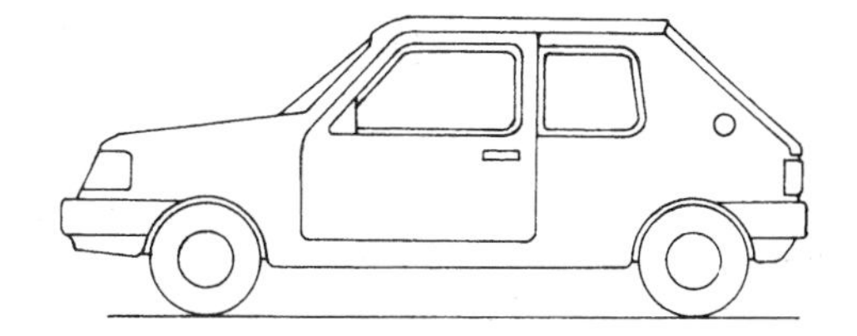

Fig. 1.2 Hatchback

1.1.2 Coupé (fig. 1.3)

A variation in saloon-car design is the coupé, which has two doors, two front seats, and a hard roof. Sometimes two additional small seats are arranged at the rear, such layouts being commonly known as 'two-plus-two'.

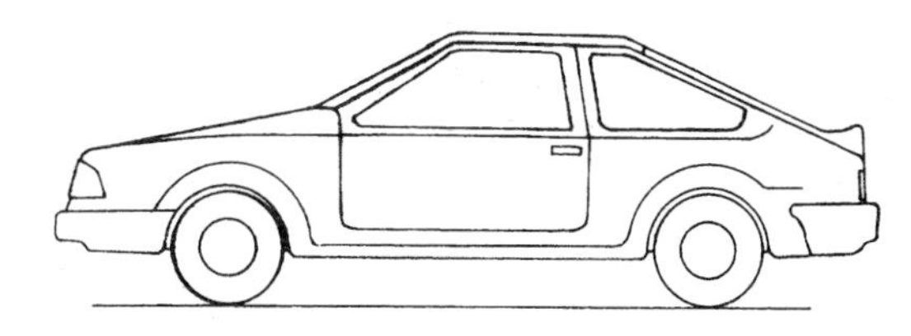

Fig. 1.3 Coupé

1.1.3 Convertible (fig. 1.4)

Cars of this type normally have two doors and two seats, but occasionally two extra seats are included. A soft folding roof and wind-up windows are usually provided so that the driver-and-passenger compartment may be either open or enclosed.

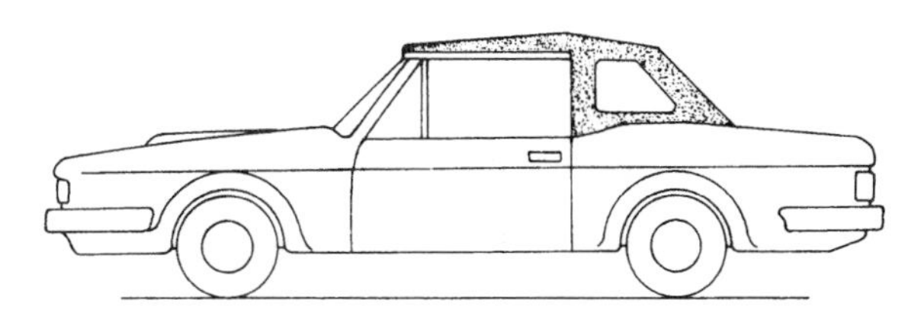

Fig. 1.4 Convertible

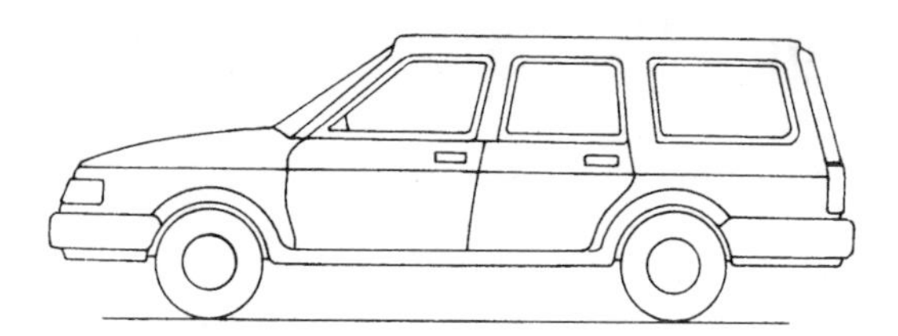

Fig. 1.5 Estate car

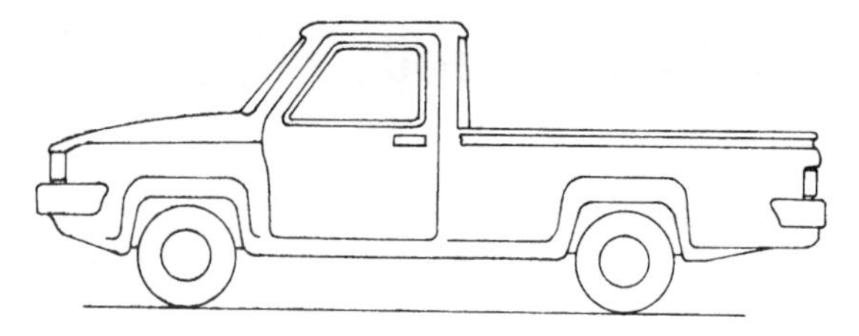

Fig. 1.6 Pick-up

1.1.4 Estate car (fig. 1.5)
To increase the utility of the saloon car, estate versions enlarge the rear space by extending the passenger roof completely to the back end. Access for loading is accommodated by a rear door, and sometimes the rear seats are made to collapse so that additional space is provided for carrying goods.

1.1.5 Pick-up (fig. 1.6)
This class of vehicle may be described as a two-door front-seating van with an open back (with or without a canvas roof) designed to carry an assortment of goods.

1.1.6 Integral body construction
Unlike commercial vehicles which have a separate cab attached to a vehicle chassis, car bodies are now mostly of frameless monobox construction, usually known as an integral construction (fig. 1.7). These body shells are made up from pillars, rails, sills, and panels all welded together, and – to replace the chassis – a reinforcing channel-section underframe with an extended subframe at the front is provided.

Integral body construction has the following advantages over the conventional chassis and attached body:

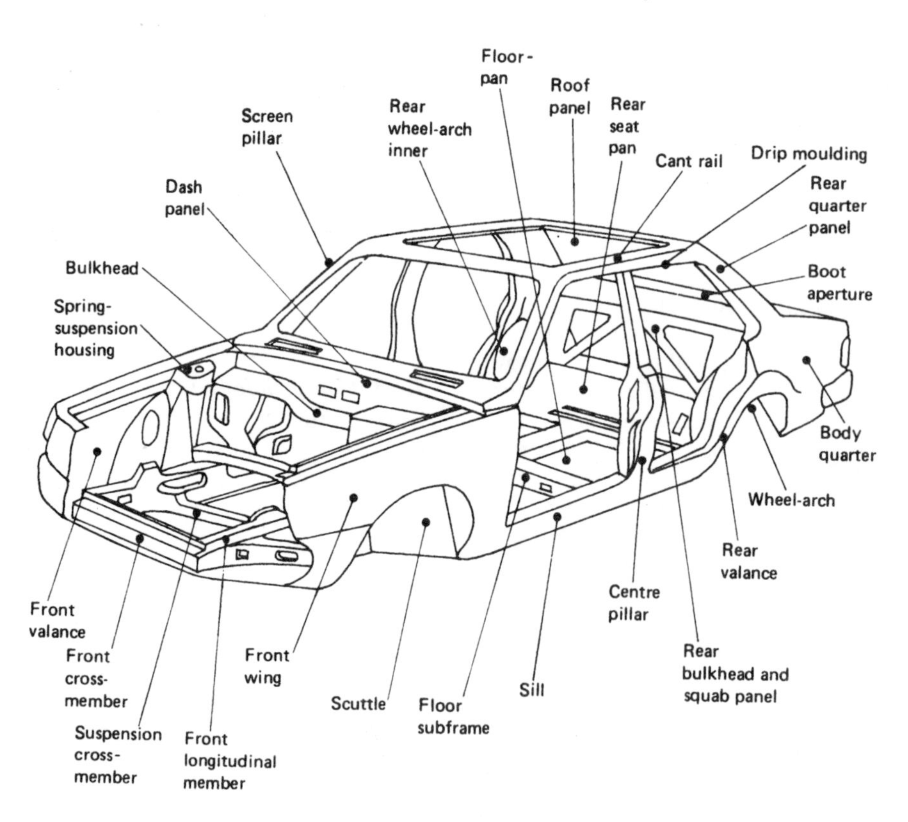

Fig. 1.7 Integral body construction

a) it is much more rigid when subjected to twisting and vertical loading, since the whole body is designed to contribute to the distribution of these forces;
b) the interior flooring can be deep and broad and the outside body can be mounted much lower and be styled to reduce air drag since there are no chassis constraints;
c) engine and transmission components can be more conveniently positioned, and greater support can be provided for all-round independent suspension which has been a weakness with chassis layouts;
d) compared to its chassis predecessor, the combined body construction is overall generally lighter and more adaptable and is preferred for large-scale production.

1.2 Commercial vehicles

Figure 1.8 shows a typical commercial vehicle constructed with a cab and a load-carrying compartment attached to a separate chassis.

The chassis resembles a ladder, having two side-members made from C-channel-section high-tensile steel with pressed-steel cross-members riveted, bolted, or welded to them.

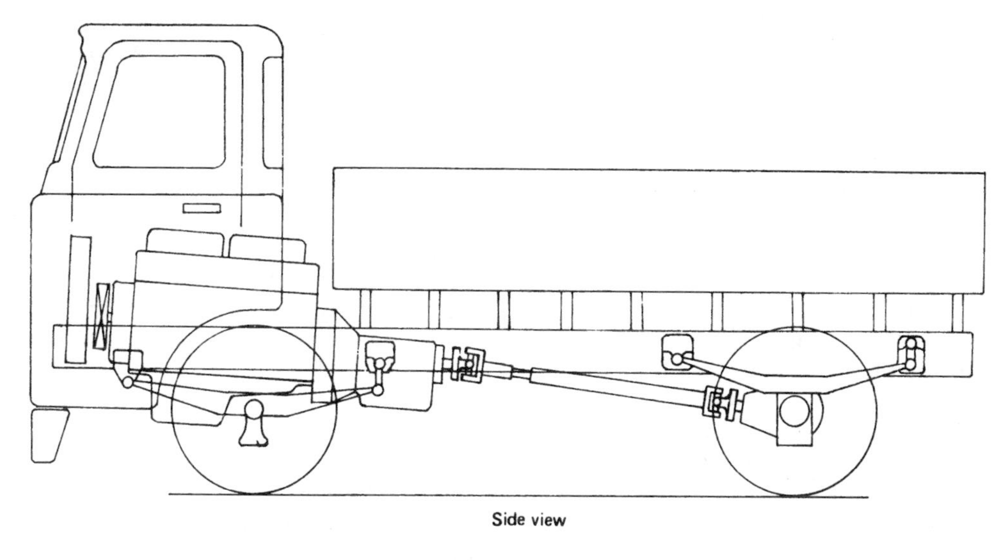

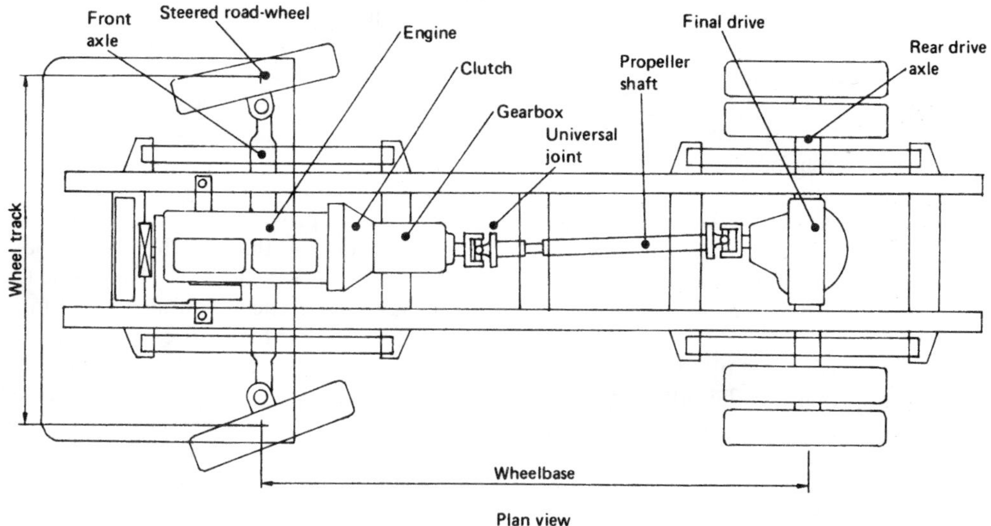

Fig. 1.8 Rigid 4 × 2 truck

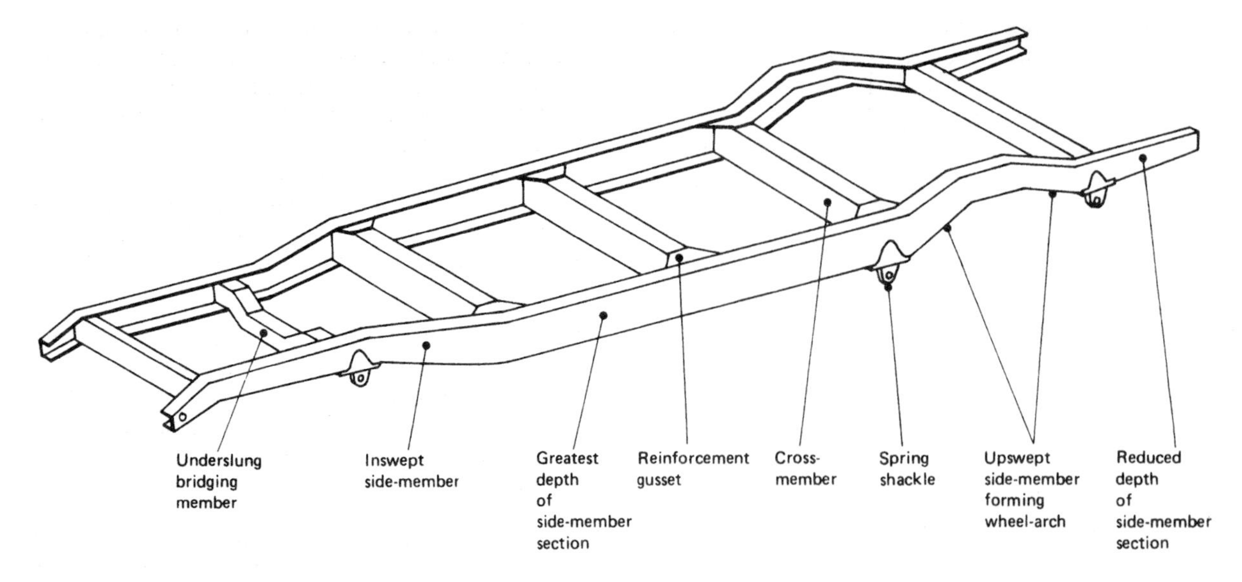

Fig. 1.9 Light-commercial-vehicle chassis

Mostly the side-members are straight parallel constant-depth sections (fig. 1.8), but they may sometimes be upswept at the rear to form a wheel-arch to accommodate the axle movement relative to the chassis and they may be inswept at the front to provide adequate wheel-turning clearance when the vehicle is being steered (fig. 1.9). In addition, the depth of the channelling may be reduced towards each end to reduce the weight. For heavy-duty applications, certain sections of the side-members may be reinforced with flitch double-channel sections (see fig. 1.22(i)).

Sturdy outboard shackle-hangers are bolted to the chassis side-members to support the semi-elliptic leaf springs. The cross-members are normally situated adjacent to these points to provide adequate resistance to torsional flexing.

A solid axle-beam is attached to the centres of both front springs. This supports the pivoting stub-axles and wheels, and also provides a degree of rigidity in the centre region of the leaf-springs. Only single wheels are used at the front, to minimise tyre scrub and wear.

The rear-drive axle-casing is bolted across the mid-span of the leaf springs. At the outer ends of the axle-casing are wheel hubs which support twin road-wheels – these considerably increase the load-carrying capacity of the vehicle. The load-carrying distribution between front and rear axles when fully loaded is usually of the order of one-third at the front to two-thirds at the rear.

The engine unit, clutch bell-housing, and gear-box casing are all bolted together to form one rigid assembly. This whole package is then positioned between the chassis side-members and over the front axle-beam. The whole assembly is usually supported on rubber mountings at three different points – one on each side of the engine on to the chassis side-members and one underneath the gearbox directly on to a cross-member.

Power from the gearbox output shaft is transmitted to the back-axle final drive, via the universal joints and the propeller shaft. The final drive then redirects and splits the drive between the rear road-wheels to provide the necessary tractive effort at ground level.

Forward control (fig. 1.8) Most commercial-vehicle cabs provide forward-control driver's accommodation – that is, the cab is positioned partially over and ahead of the front axle (fig. 1.8).

This layout has the advantage that it maximises the length available for load-carrying and provides superior driving visibility; but the disadvantages are that the engine is positioned in the centre of the cab, so there is less passenger space, and the cab usually has to be tilted about 60° forwards to gain access to the engine unit for maintenance.

The body does not rest directly on the chassis members but is supported on longitudinal runners and cross-bearers or by a separate subframe. This enables a raised flat loading platform to be used, which makes loading easier.

Vehicle dimensions (fig. 1.8) To describe the size of any vehicle, it is necessary to provide two fundamental dimensions: the wheel track and the wheelbase. These are defined as follows (fig. 1.8):

a) *Wheel track* – the transverse distance between the off-side and near-side wheel-to-ground contact centres.
b) *Wheelbase* – the longitudinal distance between the front- and rear-wheel axle centres.

1.2.1 Lorries

Heavy goods commercial vehicles are in general referred to as 'lorries', but these vehicles may be grouped into two categories: rigid trucks and articulated vehicles.

Rigid trucks (figs 1.8 to 1.11) These are vehicles designed to carry goods and – unlike articulated vehicles – they are built so that all the axles are attached to a single chassis frame.

A simple truck has two axles and four wheels, which rather limits its load-carrying capacity. It is now standard practice to use twin wheels on all non-steered axles (except where giant wide balloon tyres are fitted) – the payload on the axle will thus be subdivided between four tyres. By increasing the number of axles, the load can be shared between more axles – this reduces the load per axle and protects the tyres from overloading and the road surface from damage.

Wheel axles may be described as 'live' or 'dead' – that is, drive or non-drive respectively. A live axle supports the payload and provides driving tractive effort, but a dead axle just supports the load.

Classification of a rigid truck The different sizes of rigid trucks can be described and classified by the number of wheel hubs (that is, the number of wheel locations) and the number of drive axle-hubs.

Examples of rigid-truck classification are as follows:

a) A four-wheeler – '4 × 2' indicates four wheel locations of which two are for driving wheels (fig. 1.8).
b) A six-wheeler – '6 × 4' indicates six wheel locations of which four are for driving wheels (fig. 1.10).
c) A six-wheeler – '6 × 2' indicates six wheel locations of which only two are for driving wheels.
d) An eight-wheeler – '8 × 4' indicates eight wheel locations of which four are for driving wheels (fig. 1.11).

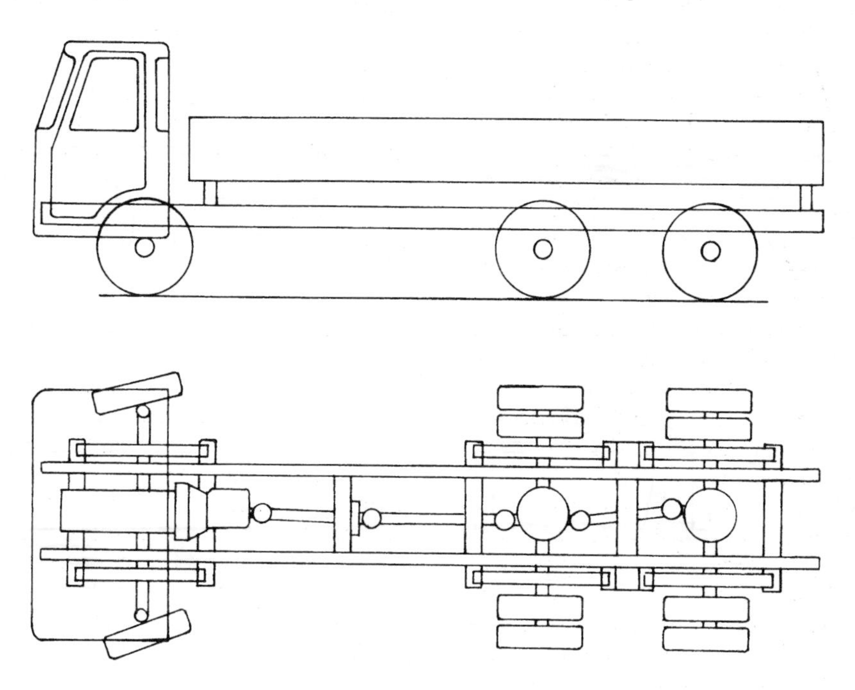

Fig. 1.10 Rigid 6 × 4 truck

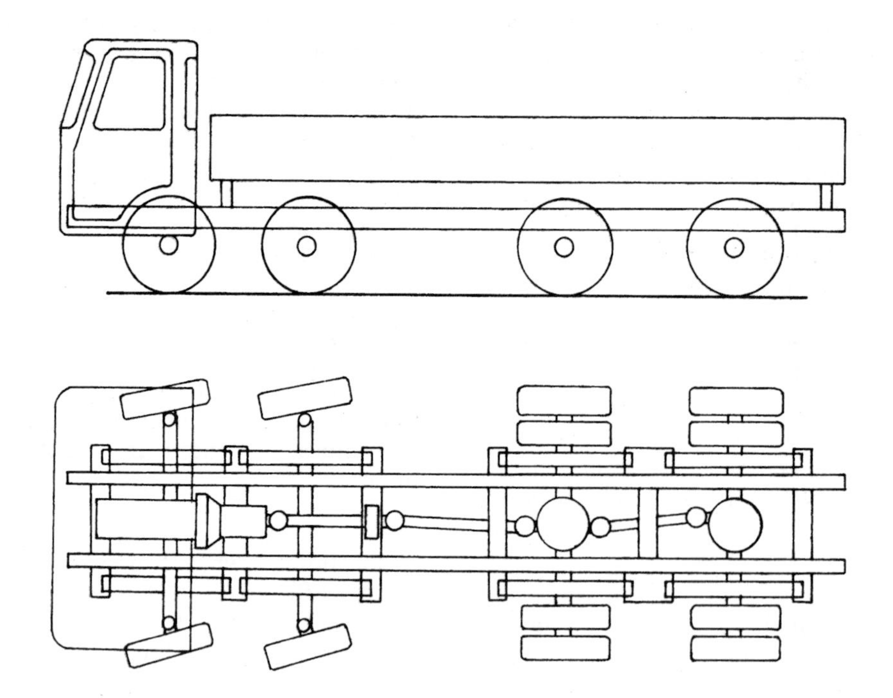

Fig. 1.11 Rigid 8 × 4 truck

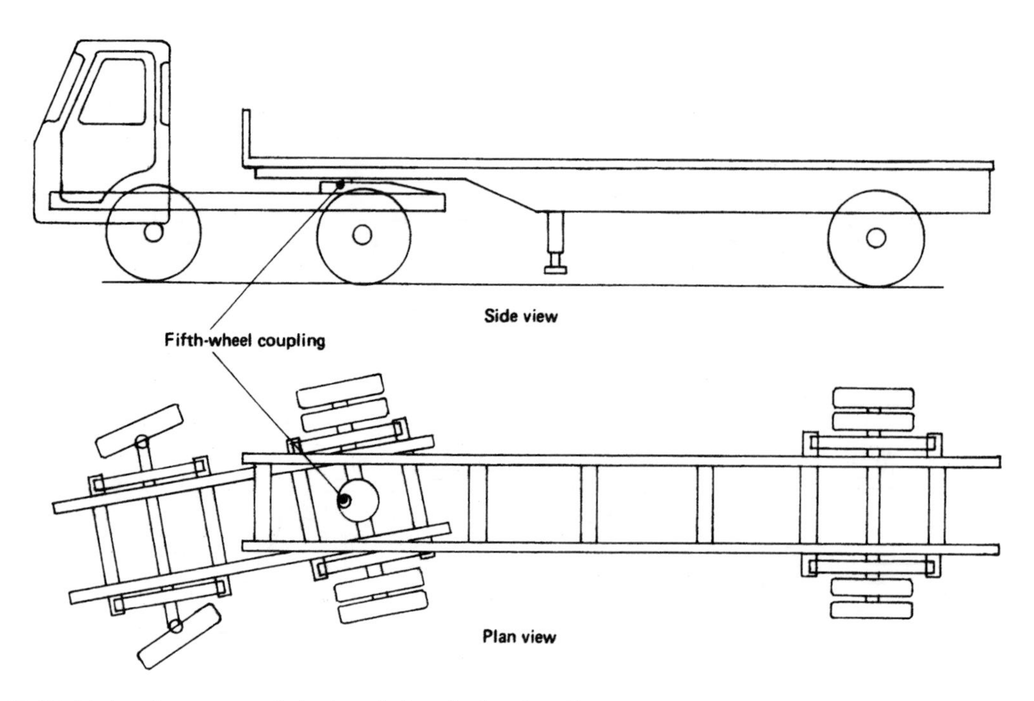

Fig. 1.12 Rigid 4 × 2 tractor and single-axle 2 articulated trailer

Articulated tractor and semi-trailer (figs 1.12 to 1.14) Articulated vehicles consist of a tractor unit which supplies the propulsive power and a semi-trailer which carries the payload.

The tractor unit has a short rigid chassis and may have two or three axles. The front axle supports the steered road-wheels, and the rear axle is a driving one. The middle axle may either be a second drive axle or provide the second axle for dual steering.

The semi-trailer has a long rigid chassis with a single-axle, tandem-axle, or tri-axle layout at the rear end, which supports the majority of the payload. The front end of the trailer chassis is supported on the rear of the tractor chassis, where it is free to swivel about a pivot known as the fifth-wheel coupling. It should be observed that all the trailer axles are non-drive dead axles.

Fifth-wheel coupling (fig. 1.12) This refers to the swivel mechanism by which the payload-carrying trailer is attached to the tractor unit which supplies the propelling tractive effort (fig. 1.12). It consists of a turntable mounted on the rear of the tractor unit to support the underside front end of the tractor, with a pivot joint consisting of a king-pin which pivots between two half jaws surrounding it.

Hitching and unhitching of the trailer and the tractor are achieved by moving the half jaws either together to secure the king-pin or apart to release it.

Classification of articulated vehicles An articulated tractor and trailer can be classified as follows:

a) Four-wheeler and two-wheel trailer – 'rigid 4 × 2 tractor and single-axle 2 articulated trailer' (fig. 1.12).
b) Six-wheeler tandem-drive-axle tractor and four-wheel trailer – 'rigid 6 × 4 tractor and tandem-axle 4 articulated trailer' (fig. 1.13).
c) Six-wheeler dual-steer-axle tractor and six-wheel trailer – 'rigid 6 × 2 tractor and tri-axle 6 articulated trailer' (fig. 1.14).

Advantages and disadvantages of articulated vehicles compared with rigid trucks

Advantages

a) The trailer is designed to be detachable so that the tractor can be immediately coupled to another loaded trailer unit.
b) Articulated vehicles have much smaller turning circles than rigid trucks of the same length

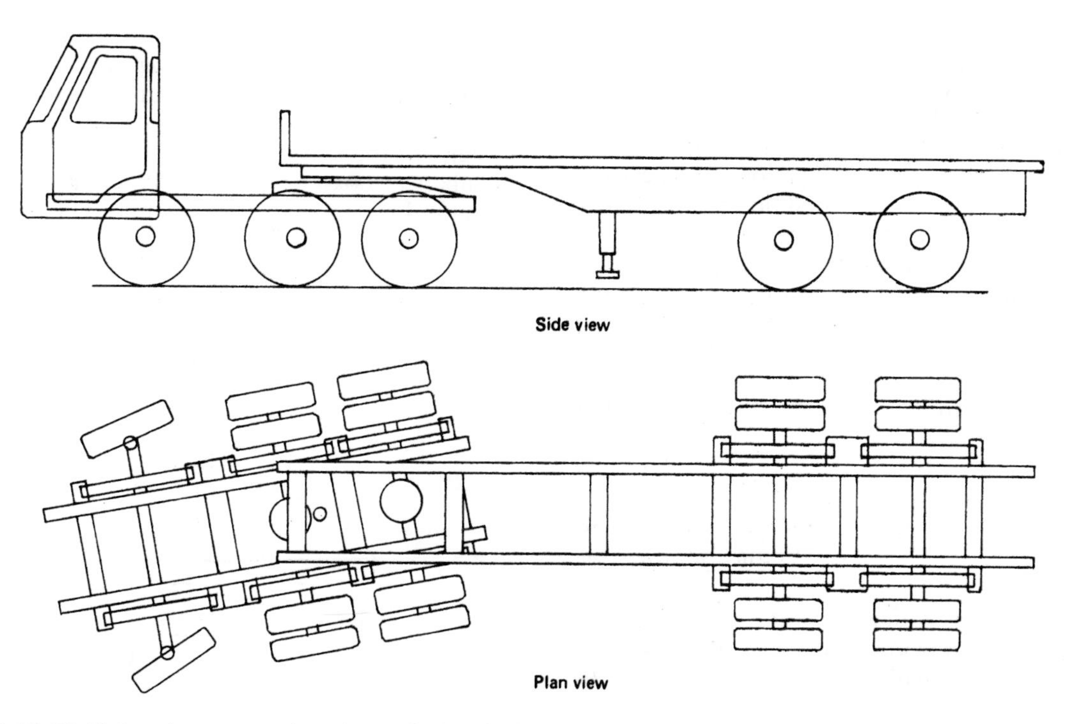

Fig. 1.13 Rigid 6 × 4 tractor and tandem-axle 4 articulated trailer

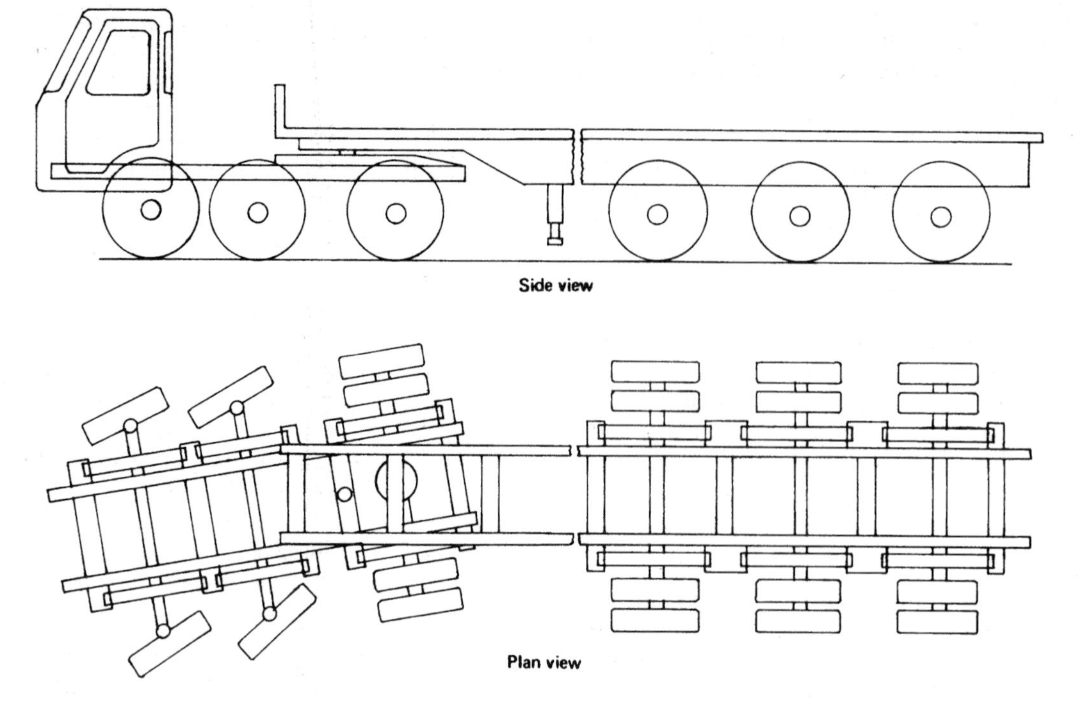

Fig. 1.14 Rigid 6 × 2 tractor and tri-axle 6 articulated trailer

and are easier to manoeuvre backwards and forwards.

Disadvantages

a) Because only the front end of the semi-trailer is supported by the tractor, there is less traction available compared with the rigid truck.
b) Tractor-and-trailer combinations under certain steered and braking conditions have a tendency to jack-knife about the fifth wheel.
c) The tractor and trailer are capable of only a small degree of pivoting in the vertical plane and tend to be unstable over rough ground.
d) Articulated-trailer wheels do not follow the same path as the tractor wheels, but tend to cut in or across the road when turning about a corner.

1.2.2 Vans (fig. 1.15)

Vans can be described as light goods vehicles. They may be used for long-distance journeys or door-to-door delivery.

The basic vehicle has seats in the front for the driver and for only one or maybe two passengers. The engine is usually mounted over or just in front of the front axle, and the cab could be described as a semi-forward control.

There are doors opposite the seats on each side which may be hinged or of the sliding type – the choice depends upon the service application. There are double doors at the rear of the van which open outwards for loading purposes.

Small vans combine the cab and body and are of integral or monobox construction. With this design the body shell is reinforced with frame members of box-sectioning, and there are usually structural members either side of deep-channel or 'top-hat' section.

Large vans sometimes separate the cab and the body, which would be mounted on an independent chassis frame. The rear axle may have twin road-wheels secured to each hub to increase the load-carrying capacity of the vehicle.

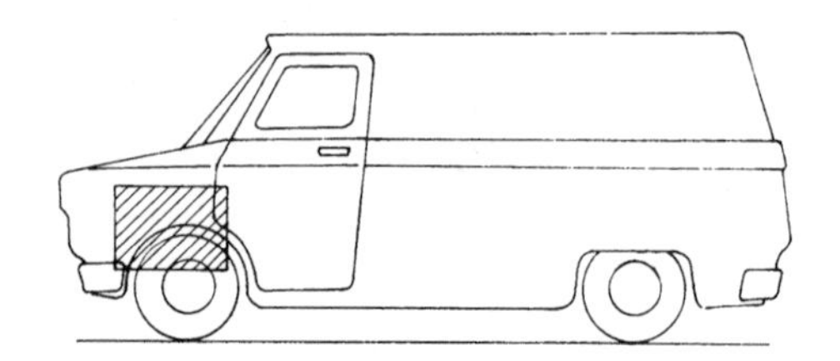

Fig. 1.15 Medium-sized van

1.2.3 Coaches (fig. 1.16)

Coaches are used for carrying passengers on journeys of considerable distance, therefore the interior is of luxury quality to provide the best possible comfort and to minimise fatigue.

Seats are positioned facing the front, so that passengers have the benefit of looking ahead as they are moving forward. Passenger visibility is of prime importance, so large panelled windows on either side extend the full length of the vehicle and across the back seats. There is a door adjacent to the driver, and the passenger's door is opposite the driver's seat. An emergency door is usually situated either at the back or on one side towards the rear.

Most coaches are of the two-axle arrangement, but sometimes an extra axle is employed at the front to provide dual steering as a safety feature.

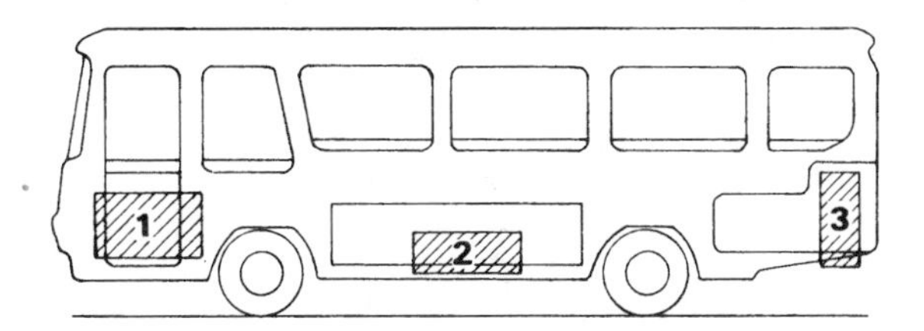

Fig. 1.16 Coach

Engines may be mounted longitudinally in the front (1), or in the mid-position horizontally (2), or at the rear transversely (3). Note there are other suitable engine-positioning configurations which can be used. The positioning of the engine and transmission depends greatly on the length of the coach, the number of passenger seats required, the luggage space to be provided, and high or low floorboard and seat-mounting requirements.

1.2.4 Double-decker bus (fig. 1.17)

Buses are used to transport large numbers of people having very little luggage over short journeys, usually in high-density traffic. The obvious solution so that the vehicle occupies the minimum amount of road space but carries the maximum number of passengers is to build a second floor. An inherent problem with these buses is that stair space is necessary for people to climb up to the upper deck.

The ground floor of the bus is divided into seat and standing space. The size and quality of seats can be minimal, as people will be seated for only a short time. Visibility within the bus must be sufficient for passengers to be able to see where they are and where to get off. Most modern buses have

Fig. 1.17 Double-decker bus

two sets of doors – passengers starting their journey will enter through the front side door and pay their fare, and people can disembark by the second door which is further to the rear.

The engine is normally mounted transversely across the back of the bus, but sometimes the engine is located longitudinally to one side at the back.

1.3 Chassis construction

To appreciate the design and construction of a vehicle's chassis, an understanding of the operating environment is necessary. Once the operating conditions are known, a comparison of the different available chassis-member cross-setion shapes can be made. The completion of chassis design then involves the reinforcement of the chassis side- and cross-member joints, and the various methods of fastening them together. The following sections examine and illustrate the basic requirements of the chassis.

1.3.1 Chassis operating conditions (figs 1.18 to 1.21)

To appreciate the design of a vehicle's chassis, it is first necessary to examine the kind of conditions it is likely to meet on the road. There are four major loading situations which the chassis will experience, as follows:

i) vertical bending
ii) longitudinal torsion
iii) lateral bending
iv) horizontal lozenging.

Vertical bending (fig. 1.18) If a chassis frame is supported at its ends (such as by the wheel axles) and a weight equivalent to the vehicle's equipment, passengers, and luggage is concentrated across the middle of its wheelbase, the side-members will be subjected to vertical bending making them sag in the centre region.

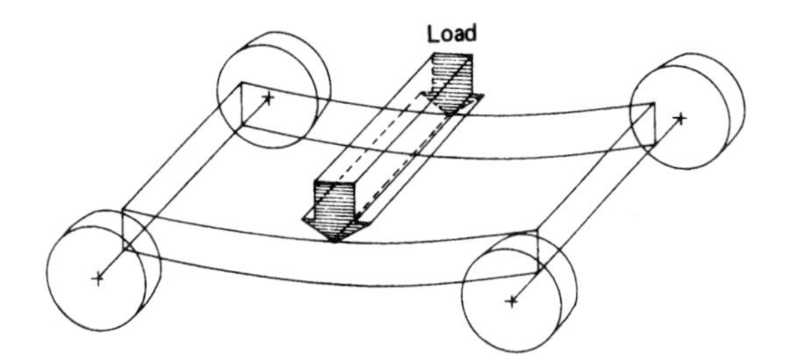

Fig. 1.18 Vertical bending

Longitudinal torsion (fig. 1.19) When front and rear diagonally opposite road-wheels roll over bumps simultaneously, the two ends of the chassis will be twisted in opposite directions. Both the side- and the cross-members will thus be subjected to longitudinal torsion which distorts the chassis.

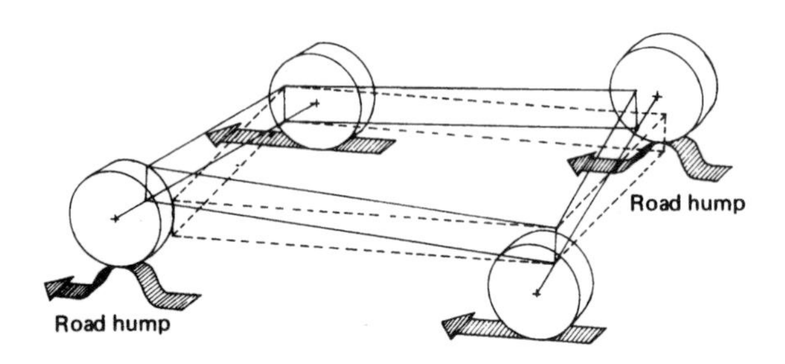

Fig. 1.19 Longitudinal torsion

Lateral bending (fig. 1.20) Under certain conditions, the chassis may be exposed to lateral (side) forces – due possibly to the camber of the road, side wind, centrifugal force as when turning a corner, or collision with some object. The adhesion reaction of the road-wheel tyres will oppose these lateral forces, with the net result that the chassis side-members will be subjected to a bending moment which tends to bow the chassis in the direction of the force.

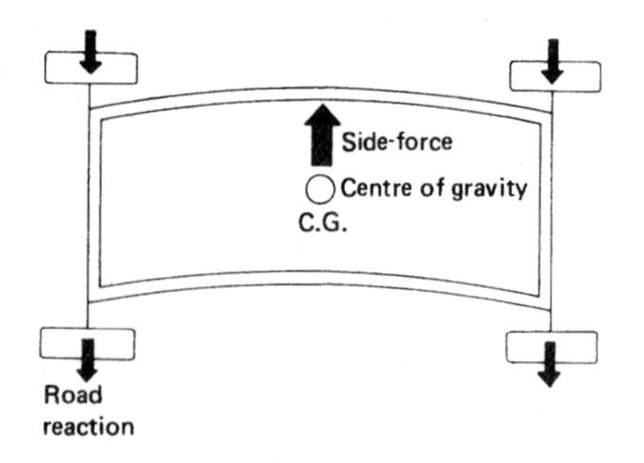

Fig. 1.20 Lateral bending

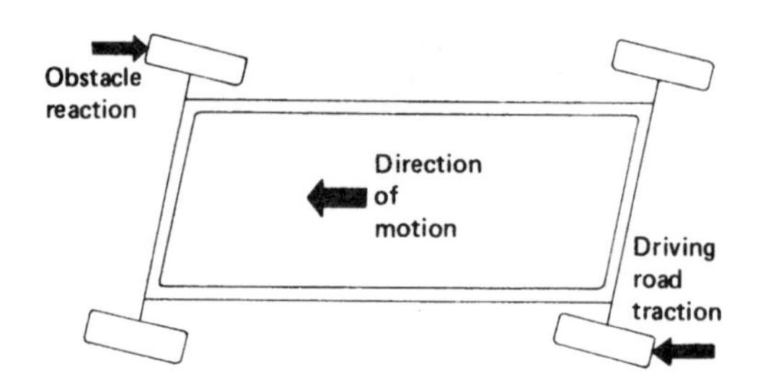

Fig. 1.21 Lozenging

Horizontal lozenging (fig. 1.21) A chassis frame driven forward or backwards will continuously be exposed to wheel impact with road obstacles such as pot-holes, road joints, surface humps, and curbs while other wheels will be providing the propelling thrust. Under such conditions the rectangular chassis will distort to a parallelogram shape. This is known as 'lozenging'.

1.3.2 Chassis sections

A chassis frame supporting the body and the wheel axles when travelling over normal road surfaces will be subjected to both bending and torsional distortion. The various chassis-member cross-section shapes which have application are as follows:

a) solid round or rectangular cross-sections
b) enclosed thin-wall hollow round or rectangular box-sections
c) open thin-wall rectangular channelling such as 'C', 'I', or 'top-hat' sections.

The different sections' ability to resist bending and twisting will now be examined.

Side-member bending resistance (fig. 1.22) The chassis side-members span the wheelbase between the front and rear axles and thus must take the majority of the sprung weight, i.e. the weight of that part of the vehicle which is supported by the suspension system. The natural tendency for these members to sag must be resisted by their bending resistance.

If steel beams of solid square or round cross-section were used as side-members, the weight of these members would be out of all proportion to the bending resistance they would contribute. The maximum possible bending stiffness of chassis members relative to their weight can be achieved by the use of either pressed-out open-channel sections or enclosed thin-wall hollow round or rectangular box-sections.

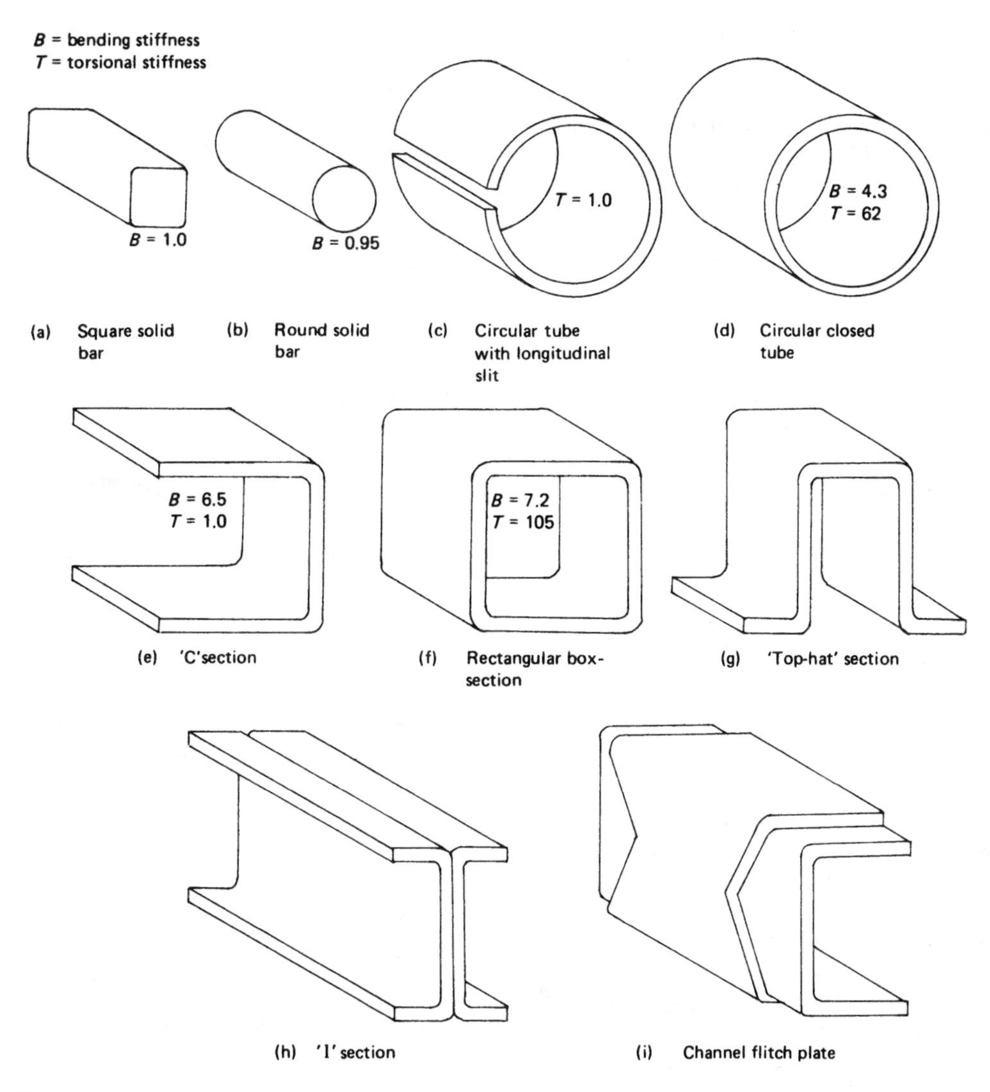

Fig. 1.22 Chassis-member sections

A comparison of the bending stiffnesses of different cross-sections is shown in figs. 1.22(a) to (f). All these have the same cross-sectional area and wall thickness. Taking the solid square section as having a stiffness of 1, the relative bending stiffnesses are as follows:

Square bar	1.0
Round bar	0.95
Round hollow tube	4.3
Rectangular C-channel	6.5
Square hollow section	7.2

This comparison shows that the further the metal is distributed from the centre of each section in the vertical plane, the more effective it is in opposing bending for a given amount of steel used. In actual practice, chassis side-members have a ratio of channel web depth to flange width of about 3:1 – for a 4 mm thick C-section channel, this would provide a bending resistance 15 times as great as for a solid square section with the same cross-sectional area.

For special heavy-duty applications, two C-section channels may be placed back-to-back to form a rigid I-section load-supporting member (fig. 1.22(h)).

To provide additional strength and support for an existing chassis over a highly loaded region – for example, part of the side-member spanning a rear tandem-axle suspension – the side-members may have a double-section channel. This second skin is known as a flitch frame or plate (fig. 1.22(i)).

Side- and cross-member torsional resistance (fig. 1.22) It has been established that open-channel sections have excellent resistance to bending, but they have very little resistance to twist. Chassis side- and cross-members must both be designed to increase their twist rigidity to prevent torsional distortion along their length.

An illustration of the relative torsional stiffness between open-channel sections and closed thin-wall box-sections can be seen in figs 1.22(c) to (f). There are two comparisons: first between the open and closed circular sections and secondly between the rectangular sections, the open section being treated as having a resistance of 1 in each case.

Longitudinal split tube	1.0
Enclosed hollow tube	62
Open rectangular C-channel	1.0
Closed rectangular box-section	105

This clearly focuses the advantages of enclosing channel sections to increase their torsional stiffness. These examples show that just closing the slit in the circular tube results in a resistance to twist 62 times as great. Adding a fourth side to a 'top-hat' section fulfills the boxing-in requirements, as when its brim is spot-welded to the underside of the floor-pan to form the chassis frame structure for both vans and cars (fig. 1.22(g)).

An important factor that must not be overlooked is that the chassis frame is not designed to be completely rigid, but to combine both strength and flexibility to some degree.

1.3.3 Chassis side- and cross-member joints (fig. 1.23)

Cross- and side-members are joined together to form a rectangular one-piece frame. Most cross-members are of open-channel section, but tube sections are sometimes used for special purposes. The individual channel members do not have much resistance to twist, but when joined together they form a relatively rigid structure which can withstand both bending and torsional loading.

The attachment of the cross-members to the side channels is of considerable importance, since both bending and torsional stresses will be concentrated at the junctions.

Commercial-vehicle side-members are generally made from flat strip steel pressed into C-channel section. The middle section is known as the web and is designed to resist any vertical bending. The bent-over flats are known as the top and bottom flanges and are there to prevent the web from buckling along its length and to provide additional resistance to both bending and torsional stresses.

The most heavily stressed parts of the channel are the flanges or the outer regions of the web, so any attachment should preferably be in the web

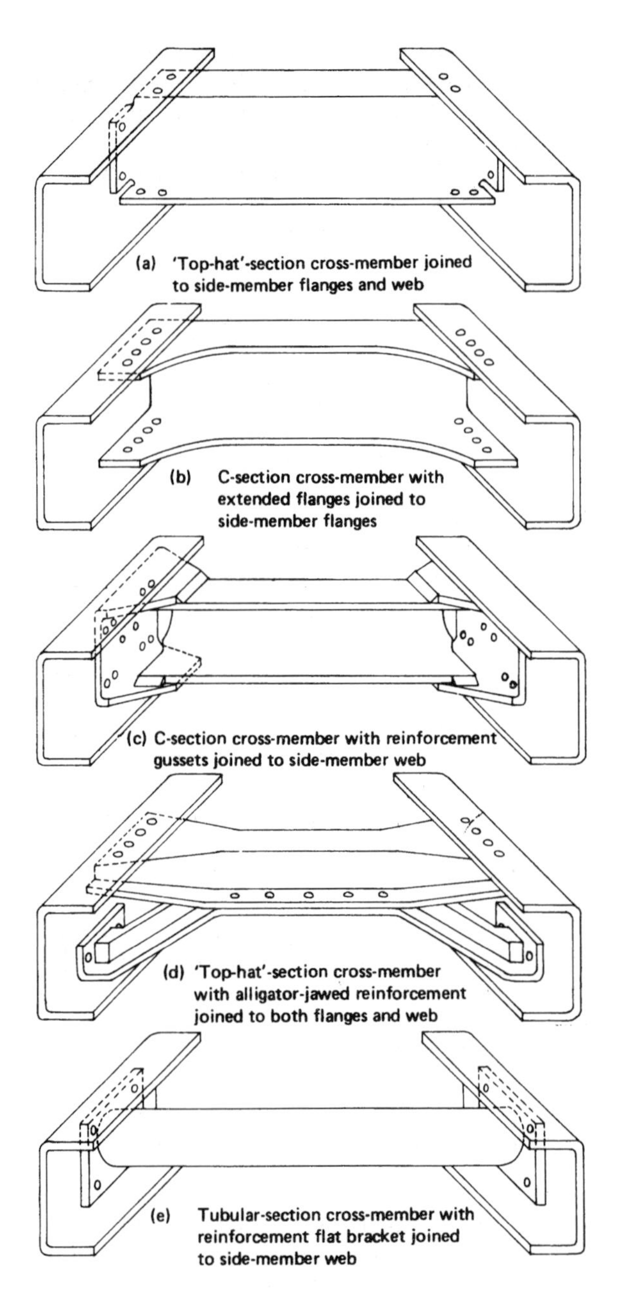

Fig. 1.23 Chassis side- and cross-member reinforcement joints

just in from the top or bottom flange. In practice, for convenience, joints may have to be made between flanges or a combination of both web and flange joints.

Figure 1.23(a) shows a cross-member of 'top-hat' section joined between the web and both flanges. Sometimes just the web alone will be joined, or alternatively the upper and lower flanges form the attachments. These joints are mostly used for light- and medium-duty work.

Figure 1.23(b) shows a pure channel-section flange joint, but the cross-member flanges have been widened to provide reinforcement to the joint. This joint would be suitable for medium-duty work.

Figure 1.23(c) shows a widely used side- and cross-member joint suitable for heavy-duty trucks. Here the cross-member has a lap-welded end gusset (triangular) bracket which is joined to the side-channel web only. This method of joint reinforcement relieves the flange of holes, which are normally a weakness in joint design.

Figure 1.23(d) shows a pressed-out two-piece cross-member which opens up at the ends to form an alligator-jaw flange-and-web reinforcement joint. This form of cross-member and attachment is often used when it is necessary to have an underslung bridging member to clear the engine's sump-pan.

Figure 1.23(e) shows a round tube-section cross-member with a fillet-welded rectangular end bracket joined directly to the side-member web. Tubular-section cross-members are particularly suitable to withstand both bending and torsional distortion at concentrated points, such as spring shackle-hangers and tandem-axle suspension pivoting supports.

1.3.4 Chassis side- and cross-member fastening (fig. 1.24)

The durability of a chassis structure will depend to some extent on the way in which the various members are joined together. The three different methods available are riveting, bolting, and lap-welding.

Riveted joints (fig. 1.24(a)) Cold-riveted joints are the most popular method of providing a permanent joint between two chassis members. The unformed rivet has a shank and a set head. It is fastened by passing its shank through a pair of holes punched or drilled in the members, after which the protruding shank is upset with a forming punch under the impact of a hammer to form a second closing head. In the process of forging

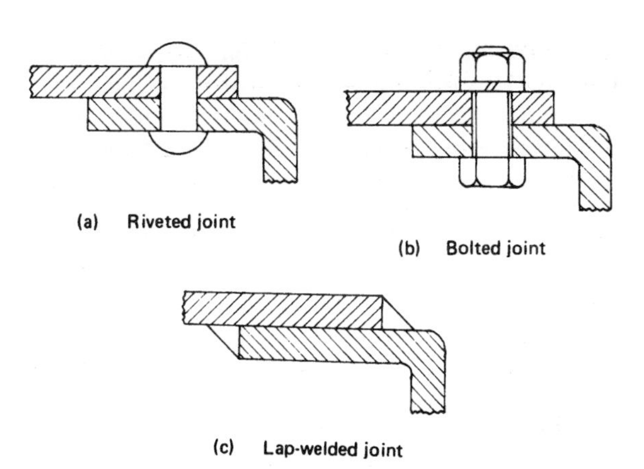

Fig. 1.24 Chassis-member joint fastening

the second head, the shank, if cold, will spread out and occupy any clearance existing in the hole.

These joints can provide a moderately large compressive force between the plates and thus prevent relative movement.

Bolted joints (fig. 1.24(b)) For heavy-duty applications, the use of nuts and bolts may be justified as a fastening device, particularly if additional components are to be attached. These joints rely on the tightened nuts and bolts setting up compressive forces between the plates, so that the corresponding friction forces generated prevent relative movement. If the nuts are insufficiently tightened, or if they work loose due to continuous flexing and vibration of the chassis, relative movement between the joined plates due to any clearance between the bolt and the hole may lead to fretting, noise, corrosion, and finally fatigue failure.

Welded joints (fig. 1.24(c)) Very rarely are chassis side- and cross-members welded together directly, but subsections are frequently joined by lap-welding. The problem with welded joints is that they produce thermal distortion and, if the frame is very rigid, high concetrated stresses will exist at the joints which may eventually crack. In addition, welding will destroy any previous heat treatment around the joint, which may make some parts soft and others brittle. The consequence of this could be to weaken the structure. Elaborate precautions can be taken to prevent these problems, but they are expensive.

1.4 Vehicle components and their methods of attachment and location

It is intended that the reader with no background knowledge in motor-vehicle engineering will read this general section, form an overall appreciation of the functions of each main part of the vehicle, and know how and where they are mounted and located.

This section is divided into three parts as follows:

i) functions of the main vehicle components;
ii) engine, clutch, gearbox, and final-drive mounting;
iii) location and mounting with various car layouts.

1.4.1 Functions of the main vehicle components

This section will briefly explain the functions of the following essential vehicle components:

a) engine
b) gearbox
c) clutch
d) propeller shaft and universal joints
e) drive shafts
f) final drive
g) steering
h) brakes.

Engine power unit The engine is a machine which takes in a combustible mixture of air and fuel, burns it, and converts the released heat energy into mechanical energy which is then harnessed to a rotating crankshaft. The output power developed may be described as the rate of doing useful rotational work – that is, the product of the crankshaft rotational speed and its turning effort.

Gearbox This is a machine which takes in the engine's power from the crankshaft and, through a relay of gearwheels, alters the output turning-effort and speed to suit the vehicle's propelling-force and road-speed requirements. A number of different gear-ratios can be selected to cover the speed range and accelerating conditions expected from the vehicle.

Clutch This is a device which is designed to separate the engine's power unit, when running, from the final drive. It thus enables the vehicle to be brought to a standstill without stalling the engine. It provides smooth engagement of the engine's power to the drive wheels from rest and enables different gear-ratios to be selected and engaged when the vehicle is in motion.

Propeller shaft and universal joints The propeller shaft is a hollow shaft whose purpose is to join the gearbox output drive to the final drive. Propeller shafts operate in conjunction with couplings known as universal joints whose function is to accommodate any angular misalignment which may occur between the gearbox and the final drive under operating conditions.

Drive shafts These are usually solid shafts which join the final drive to the road-wheel live stub-axles through the medium of universal or constant velocity joints.

Final drive This consists of two dissimilar sized meshing gears which change the direction of the drive through a right angle and provide a permanent gear-reduction. As a result, the connecting shafts going to the road-wheels will be slowed down, and the turning-effort (usually referred to as torque) is multiplied to match the road-condition requirements.

It should be noted that, when the final drive and gearbox are combined to form one single unit, this is sometimes referred to as a trans-axle unit.

Steering The steering provides the means for the driver to alter the vehicle's direction when it is moving. The steering system is designed to convert rotary movement of the driver's steering-wheel to a linear to-and-fro transverse movement of the steering link-rods. This movement is changed back to an angular movement of the stub-axles about their swivel pins, thus turning and pointing the front wheels along the path the vehicle is expected to follow.

Suspension The road-wheels are attached indirectly to the vehicle chassis or body structure through a sprung and hinged linkage arrangement known as the suspension. This method of mounting and supporting the body prevents bumps caused by road-surface irregularities being transferred in the form of noise and vibrations to the body or chassis, passengers, and any goods being transported.

Brakes The braking system is designed to slow down or bring to a standstill a moving vehicle either progressively or rapidly. A brake-drum and shoes or a brake-disc and pads are attached to each road-wheel and, when the foot-brake or hand-brake is applied, friction will be generated between the drum and the shoes or the disc and the pads. The friction produced converts the vehicle's kinetic energy of motion into mechanical and heat energy, resulting in a reduction in vehicle speed.

1.4.2 Engine, clutch, gearbox, and final-drive support mounting

It is usual for the engine, clutch, and gearbox combination and (with front-wheel drive layouts) the final drives to be supported on a three-point mounting system. This configuration allows the best possible freedom of movement about an imaginary roll-centre axis, but provides a degree of rigidity for withstanding the torque reaction when the engine is developing power.

Rubber blocks bonded on to steel plates are usually used to mount the various components. Sometimes these mounts are positioned at 45° to the horizontal so that the rubber is subjected to a combination of both compression and shear elastic distortion. This method of loading the rubber provides a flexible mount which increases in stiffness with increased driving torque.

Mountings are necessary for the following reasons:

a) to absorb the torque reaction when power is being transmitted
b) to cushion the rocking movement caused by the engine's out-of-balance forces
c) to prevent engine and transmission vibrations being transmitted to the body structure
d) to accommodate any misalignment between the engine or transmission units relative to the body frame.

1.4.3 Location and mounting with various car layouts

Front-mounted engine and rear-wheel drive (fig. 1.25) With this type of car layout, the engine, clutch, and gearbox are bolted together in series, and the whole assembly is then supported between both halves of the front-wheel suspension by a three-point mounting system. The propeller shaft and universal joints join the gearbox to the final drive.

If the final drive is unsprung, as with the rigid rear axle (fig. 1.25(a)), the drive is transferred within the axle-casing to the road-wheels. If the final drive is sprung and mounted underneath the body structure by a three-point rubber mounting (fig. 1.25(b)), the drive then goes through each drive shaft and universal joint to the independently suspended road-wheel.

Front-mounted engine and front-wheel drive (fig. 1.26) Car layouts of this kind have the engine, clutch, gearbox, and final drive built together to form a single integral assembly.

The transverse engine arrangement (fig. 1.26(a)) has the engine, clutch, and gearbox bolted together in series in that order, with the final drive forming part of the clutch bell-housing and gearbox casing. The drive shafts and their respective universal joints are situated on each side of the final-drive housing, so that they can convey the propelling power to each front drive stub-axle and road-wheel.

In the case of the longitudinal-mounted engine (fig. 1.26(b)), the four components – engine, clutch, final drive, and gearbox – are bolted together in that order. The power flows from the final drive to each drive shaft and through its universal joint to the road-wheel.

Both layouts use a three-point rubber mounting system to support the whole power- and transmission-unit assembly.

Rear-mounted engine and rear-wheel drive (fig. 1.27) Figure 1.27(a) shows a horizontally opposed four-cylinder engine which is connected in series with the clutch, final drive, and gearbox in that order. The power output is then split at the final-drive housing and transferred by the drive shaft and coupling joints to the rear road-wheels.

An alternative transverse in-line four-cylinder engine (fig. 1.27(b)) has the power flow in series with the engine, clutch, gearbox, and final drive, where it is then divided and transferred to each rear driving road-wheel through the respective couplings and drive shafts.

As in the previous examples, the power and transmission assembly is supported on the classic three-point mounting.

1.5 Comparison of major component layouts

The private motor car may be designed to have the engine, clutch, gearbox, and final-drive components in one of a number of possible combinations and configurations, each having certain merits and limitations. The different power- and transmission-unit layout options available, and their advantages and disadvantages, can be broadly grouped under the following three headings:

a) front-mounted engine and rear-wheel drive
b) front-mounted engine and front-wheel drive
c) rear-mounted engine and rear-wheel drive.

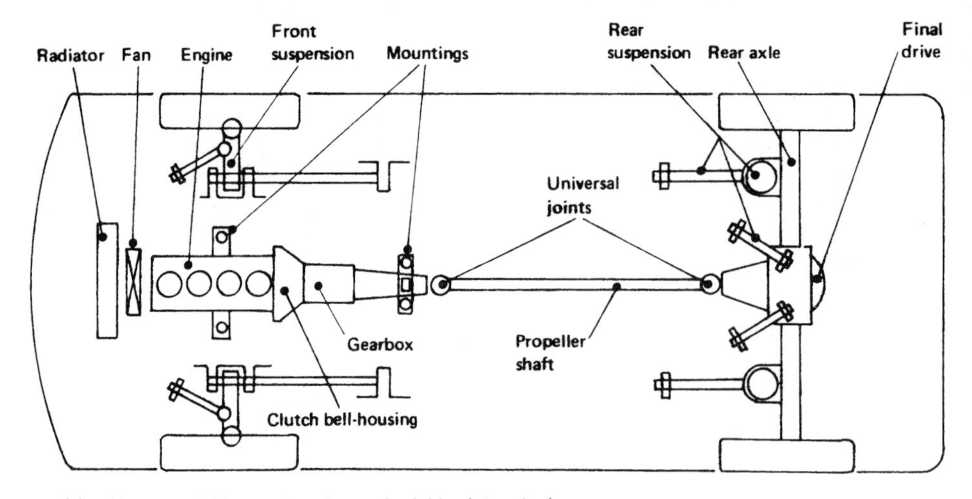

(a) Unsprung rigid rear axle and rear-wheel drive (plan view)

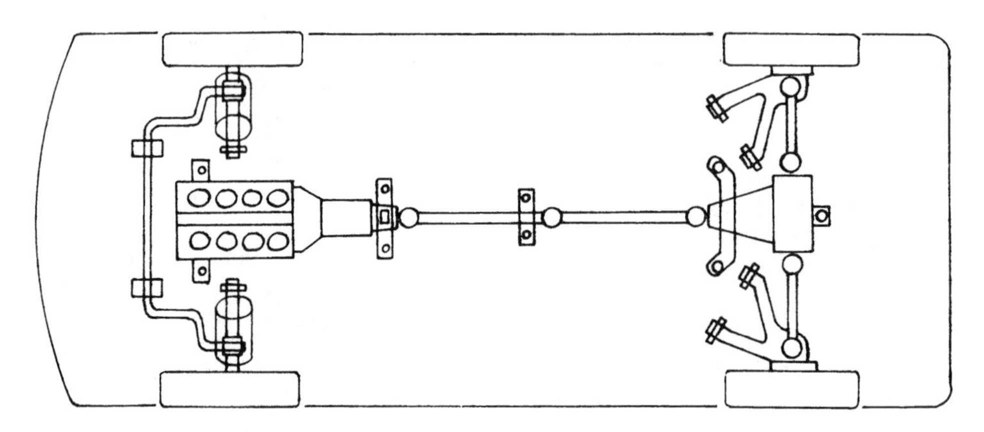

(b) Independent rear suspension and rear-wheel drive (plan view)

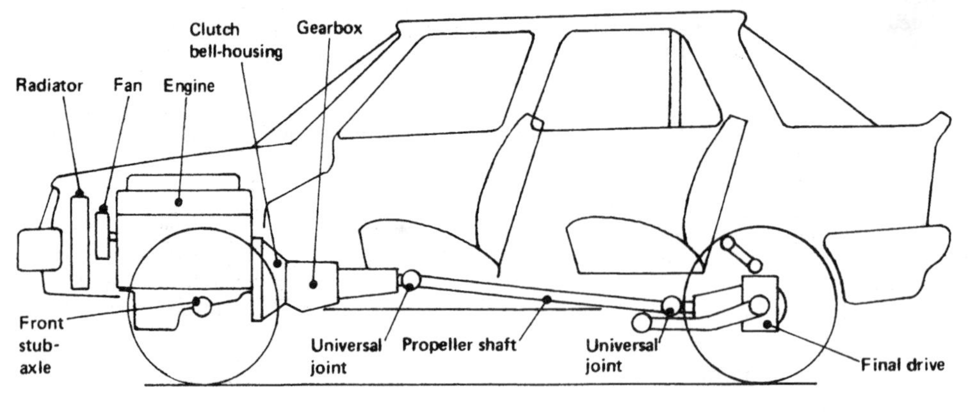

(c) Side view

Fig. 1.25 Front-mounted engine and rear-wheel drive

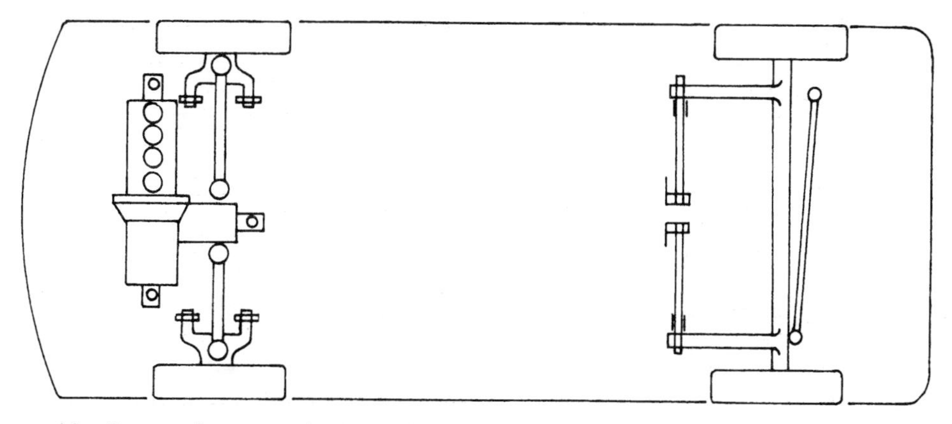

(a) Transverse front-mounted engine and front-wheel drive (plan view)

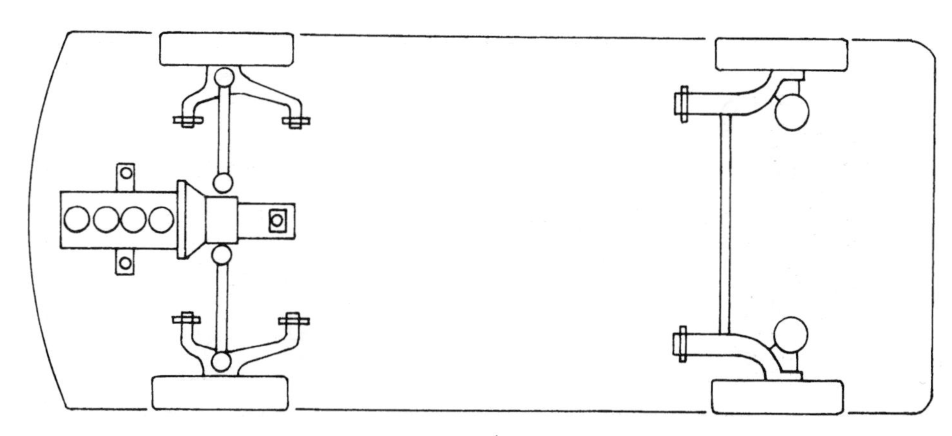

(b) Longitudinal front-mounted engine and front-wheel drive (plan view)

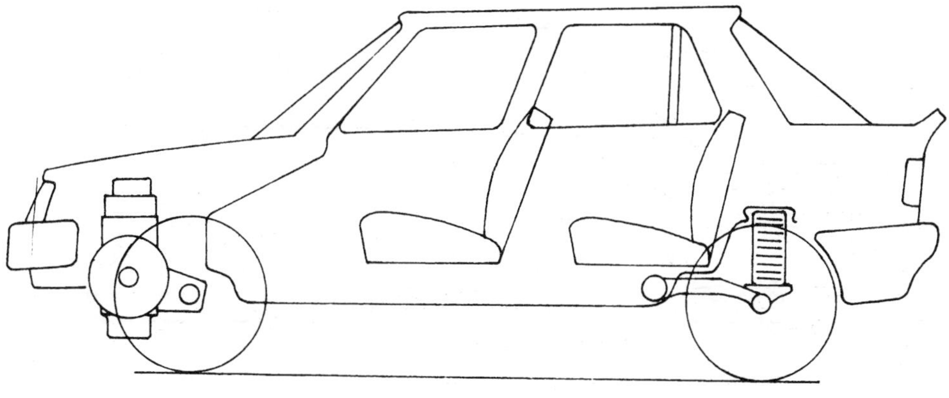

(c) Side view

Fig. 1.26 Front-mounted engine and front-wheel drive

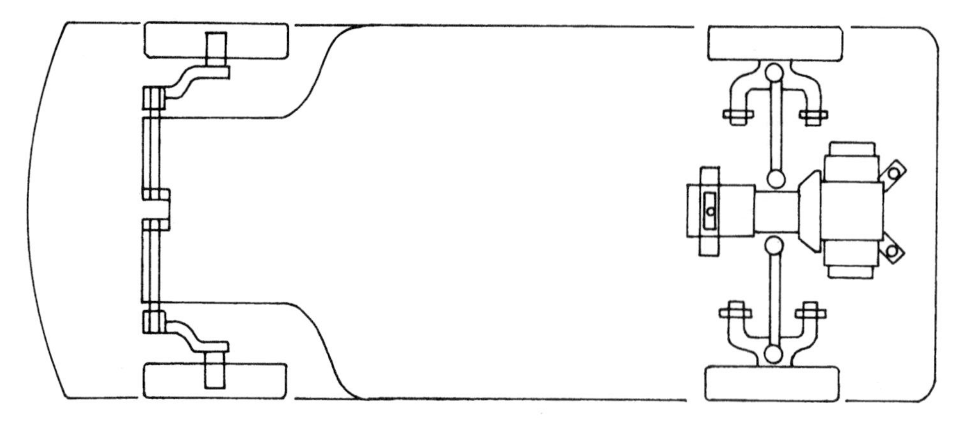

(a) Longitudinal rear-mounted engine and rear-wheel drive (plan view)

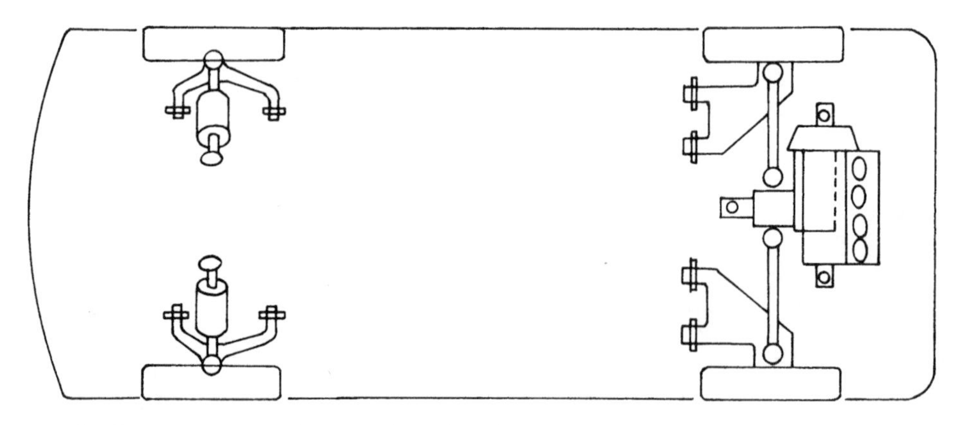

(b) Transverse rear-mounted engine and rear-wheel drive (plan view)

(c) Side view

Fig. 1.27 Rear-mounted engine and rear-wheel drive

1.5.1 Front-mounted engine and rear-wheel drive (fig 1.25)

Advantages

a) A front-mounted engine establishes a forward centre of gravity. This tends to stabilise the car handling at speed.
b) There is a degree of crash safety with the engine mounted in front of the driver and passengers.
c) The radiator can be situated in the front so that it utilises the air-stream ramming effect as the car moves forward.
d) The slope of the propeller shaft between the gearbox and the final drive is small, so simple Hookes' universal joints may be used (see chapter 4).
e) When the car climbs a steep slope, there will be a partial weight transfer to the rear wheels which improves tyre-to-road grip.
f) The interconnecting linkages from the various driver's controls to the clutch, gearbox, and engine can be simple and direct.
g) Front longitudinal-mounted engines are generally easily accessible for routine maintenance.
h) Tyre wear on wheels which only steer is marginally less than on wheels which both steer and drive.

Disadvantages

a) A single or split propeller shaft with universal joints and supporting bearings is necessary between the front-mounted gearbox and the rear axle, and under certain operating conditions this may generate vibration, drumming, howl, and other noises.
b) Some form of floor tunnel is essential to provide clearance for the propeller-shaft system to operate. This may interfere with passenger leg-room, particularly if there is a third passenger in the rear seat.
c) If a rigid casing is employed for the axle and final drive, there will be more weight not supported by the suspension system and so the quality of the suspension ride may be reduced.
d) Independent rear suspension necessitates additional universal joints and drive shafts.
e) When stuck in mud, a rear-wheel-drive vehicle will tend to plough further into the ground when attempts are made to drive away.

1.5.2 Front-mounted engine and front-wheel drive (fig. 1.26)

Advantages

a) The concentraton of engine, transmission, and final-drive components on the front driving wheels improves road adhesion and acceleration.
b) The propeller shaft is eliminated, so a low floor profile can be used and in some cases the vehicle's centre of gravity may be lowered.
c) The engine, gearbox, clutch, and final drive can be built into a compact single assembly which may be installed or removed as a whole.
d) Front-wheel-drive steered wheels are capable of driving out of pot-holes, ditches, loose soil, and boggy ground.
e) Simplified rigid or independent rear suspension can be incorporated which requires the minimum of servicing.
f) A transverse-mounted engine provides more passenger room for a given size of car.
g) Attaching the final drive to the power and transmission unit reduces the unsprung weight and so improves the quality of ride.
h) Combined steering and driving road-wheels improve wheel traction and road-holding on bends.
i) With a forward centre of gravity, handling characteristics such as oversteer and understeer tend towards the more desirable understeer response.
j) Engine, gearbox, and clutch-control actuating linkages may be simplified.

Disadvantages

a) Constant-velocity universal joints (see chapter 4) have to be built into the front suspension and steering system to drive the live stub-axles.
b) Initial costs are generally higher and the periods between component replacements are usually shorter compared with the conventional rear-wheel drive cars.
c) Hill-climbing will move the centre of gravity of the car backwards slightly, so less weight will then act on the front driving wheels and reduction of tyre traction will result.
d) Having all the power and transmission components in front may subject the driver and passengers to more noise, heat, and fumes.
e) The increased concentration of weight at the front tends to produce slightly heavier steering.

1.5.3 Rear-mounted engine and rear-wheel drive (fig. 1.27)

Advantages

a) Greater weight concentration on the rear wheels improves driving traction when climbing hills.
b) The rear wheels can be designed to take a larger proportion of braking, due to the increased weight distribution at the rear end.
c) Passengers will not experience excessive noise, heat, and fumes, as these will be left behind when the car moves in the forward direction.
d) With non-drive front wheels, the steering and suspension can be simplified and there will be no steering interference due to worn transmission components.
e) Generally, because there is less weight on the front wheels, steering is relatively lighter than for other arrangements.
f) The exhaust-pipe and silencer system does not span the length of the car but can be short, direct, and compact.

Disadvantages

a) Engine, gearbox, and clutch-control linkages will have to be extended to the driver's position.
b) The luggage-boot width must be reduced to provide sufficient side-clearance for the front steered wheels.
c) The relative large proportion of weight towards the rear will tend to make the car unstable at speed. Also, handling characteristics differ when driving empty, with just additional luggage in the boot, or with passengers in the rear seat.
d) A lighter front end tends to make the car oversteer and very sensitive to cross-winds.
e) Accommodation for the cooling-system radiator and its effective air supply can be difficult to arrange. Adequate interior heating for the driver and front passenger may be more complicated.
f) Access to the power and transmission units is more difficult, so servicing and repairs may take longer.
g) The most convenient location for the petrol tank is in the front, which could be a safety hazard in a collision.

1.6 Seat location and securing (fig. 1.28)

Seating accommodation must be comfortable. There must be sufficient leg-room for passengers, and the driver must be positioned so that his legs and feet can control the accelerator and clutch and provide effective leverage to the brake-pedal for emergency. In addition, the seats must be designed to give adequate support to the lower part of the body and back and against any lateral forces which might be experienced. Consideration must also be given to the positioning of the seats for good visibility towards the front and sides and to the driver's natural arm movement so that he can steer and change gear with the minimum of effort.

It should be appreciated that, the lower the seating position relative to the floor, the more the legs will be straightened out so that increased leg-room is needed. Conversely, a high-mounted seat will bend the legs so that far less leg-room is necessary in front.

1.6.1 Car front-seat mounting (fig. 1.28(a) and (b))

Car seats are generally supported on a steel channelling of 'top-hat' section, the rims of which are bolted to the floor-pan (fig. 1.28(a) and (b)). The front seats are mounted on steel male and female runners which provide fore and aft movement adjustment.

The fixed-runner member has ladder-like slots cut out on its top face, and a release-lever and pawl pivoting on the moving runner engages one of the fixed-runner ladder slots (fig. 1.28(b)). If adjustment is necessary, the release-lever is pushed down. This frees the sliding runner and seat so that it can be moved fore or aft to a selected position. The lever is then released so that the spring-loaded pawl snaps into one of the fixed runner slots.

A slightly different method of engaging the slots in the runner is shown in fig. 1.28(c). Here there are slotted runners on each side, and short spring-loaded plungers engage the slots. To move the seat position, the lever is pushed over so that the plungers are pulled out of their slots. The seat is then slid fore or aft as required, and the lever is then released so that the plungers can engage their respective slots. Being able to lock both sides of the seat provides a more secure adjustment mechanism for heavier cars.

1.6.2 Truck seat mounting (fig. 1.28(d))

Commercial-vehicle seating adjustments can take many forms which may include the following:

a) fore and aft seat setting
b) vertical seat-height setting
c) seat squab tilt
d) seat cushion rake.

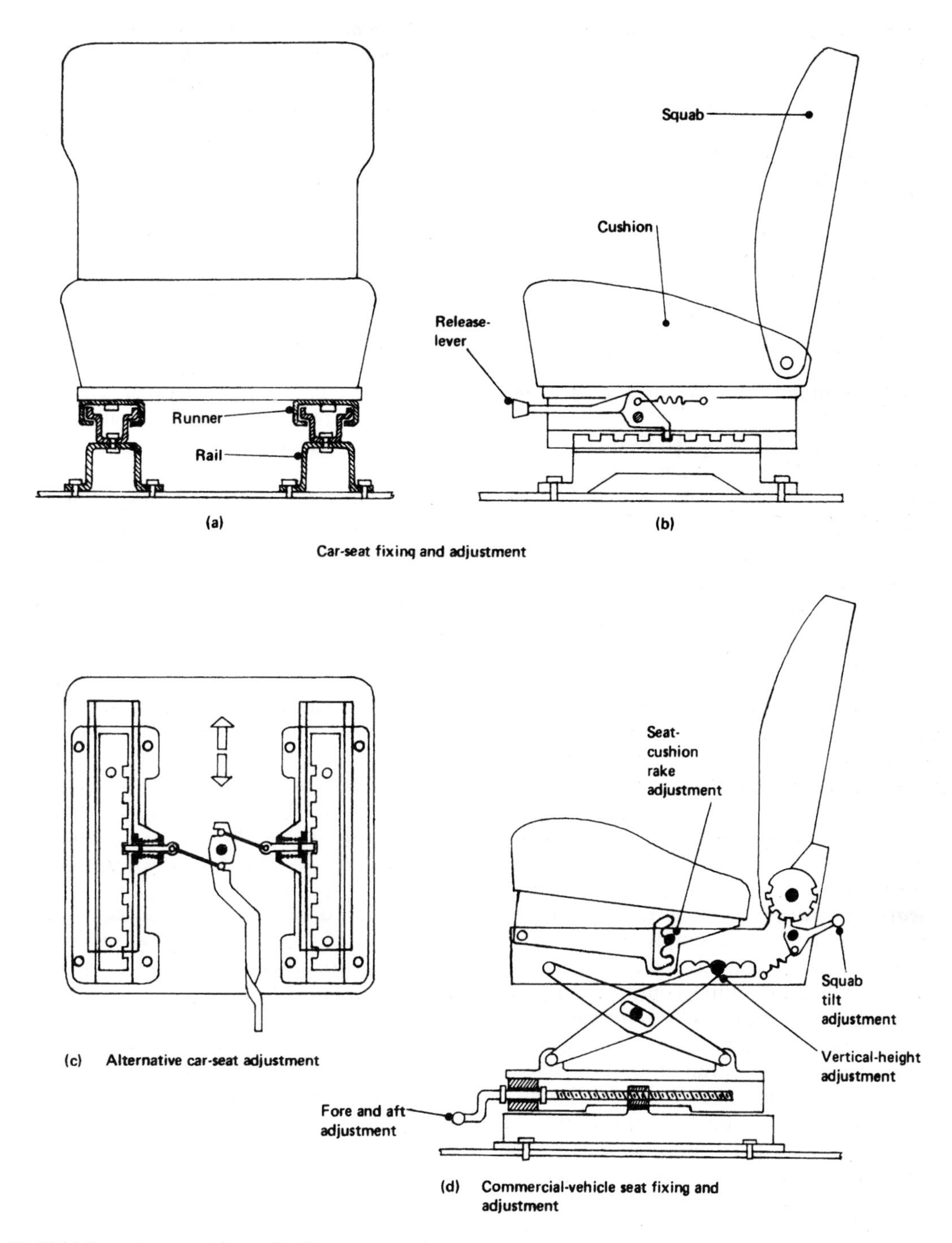

Fig. 1.28 Vehicle seat mounting and adjustment

Figure 1.28(d) shows one seat arrangement with the various adjustments, and it can be seen that the seat is bolted to the floor structure in a similar way to the car seat.

Fore and aft movement in this example is obtained by a screw-and-nut action in which the nut is attached to the lower part of the seat. When rotated, the screw pulls or pushes the seat backwards or forwards depending on the direction of rotation. Height can be altered by changing the scissor angle between the cross-arms by moving the rear upper cross-arm engagement pin into one of the elongated location notches.

Seat squab or back-rest angle can be altered by raising the pawl-lever and tilting the squab the desired amount. Releasing the lever will lock the pawl in one of the toothed-disc gaps. Finally the seat cushion rake can be changed by pushing forwards the adjusting pin, altering the cushion slope, and pulling back the pin to locate one of the three vertical notches.

1.7 Seat-belt location, fitting points and operation (fig. 1.29)

Consideration must now be given to the safety of the driver and passengers in their seats when the vehicle is rapidly braked or involved in an accident. During quick stopping, the seated driver and passengers tend to be jerked forward relative to their seats and, under extreme impact, may even be thrown through the windscreen.

To reduce discomfort when braking and the chance of serious injury or even death, seat belts are installed as standard interior features for the front seats. The objective of seat belts is to restrain and prevent the occupants of the front seats being flung dangerously towards the front during emergency stops etc. but to cause the minimum of interference to the driver and passengers and not restrict their movements during normal driving conditions.

1.7.1 Seat-belt arrangements (fig. 1.29(a) and (b))

Some early seat belts were simply strapped and secured at either side to the sill and tunnel at floor level, but this did not support and prevent the person in the seat being tilted forward and doubling up under heavy braking.

An improvement in belt strap support came with the fitting of the shoulder-strap two-point anchorage (fig. 1.29(a)) in which one end of the belt was attached to the centre pillar and the other to the centre floor-pan or tunnel, but this did not prevent the body slipping forward relative to the seat.

The present preferred arrangement is to have a combined lap-and-shoulder strap with a three-point anchorage layout (fig. 1.29(b)). Here the lap is held firm against, the seat, to prevent the lower part of the body sliding out of the seat, and at the same time the shoulder strap resists the upper part of the body being jerked forward relative to the back seat support.

1.7.2 Seat-belt anchor fixture (fig. 1.29(c))

The safety-belt body attachments consist of steel anchorage brackets which have bolt holes and an elongated slot to take the strap. The anchor-bracket bolt hole is enlarged to locate a short spacer sleeve so that the bracket can still swivel freely when the anchor nut and bolt are screwed up tight to the body structure. In addition, a large flat steel washer is sometimes placed on the underside of the floor, next to the nut, so that the bolt and nut will not pull through the floor when the strap is subjected to shock loads. Generally the belt body-attachment points are reinforced with short steel strips as an extra precaution. A wavy spring washer is slipped between the bolt and the bracket, so that the attachment will not rattle with vehicle movement.

1.7.3 Automatic inertia-controlled belt reel (fig. 1.29(d) to (f))

The major disadvantage of wearing seat belts is that it restricts the movement of the body and limbs. This problem can be minimised by having the centre-pillar end of the belt wound on to a spring-loaded reel so that the body can move freely relative to the seat under normal driving conditions (fig. 1.29(d)). Built into the reel is a mechanism which senses the rate of deceleration of the vehicle and locks the wound part of the belt in the reel when the vehicle is rapidly retarded, so that the exposed length of belt which fits across the lap and diagonally over the chest prevents the body being propelled forward (fig. 1.29(e) and (f)).

This automatic locking device works on the principle of inertia, and relies on the mass of a steel ball to actuate the jamming of the toothed belt reel. This toothed reel houses both a spiral spring and a rolled-up length of belt. The inertia device consists of a steel ball resting between two shallow saucer-shaped members. The lower member is fixed to the reel casing, and the upper member forms part of a hinged toggle-arm which is in contact with a toothed pawl.

When the vehicle is braked, it rapidly slows down but the inertia of the ball continues to carry the ball forward. This forces the saucer members apart, so that the toggle-arm pushes the toothed pawl into engagement with the toothed reel (fig. 1.29(f)). The instant this happens, the reel locks so that the belt will not wind out any more, thus restraining the body from being thrown towards the front. Immediately the vehicle speed steadies, the ball rolls back between the saucer-shaped

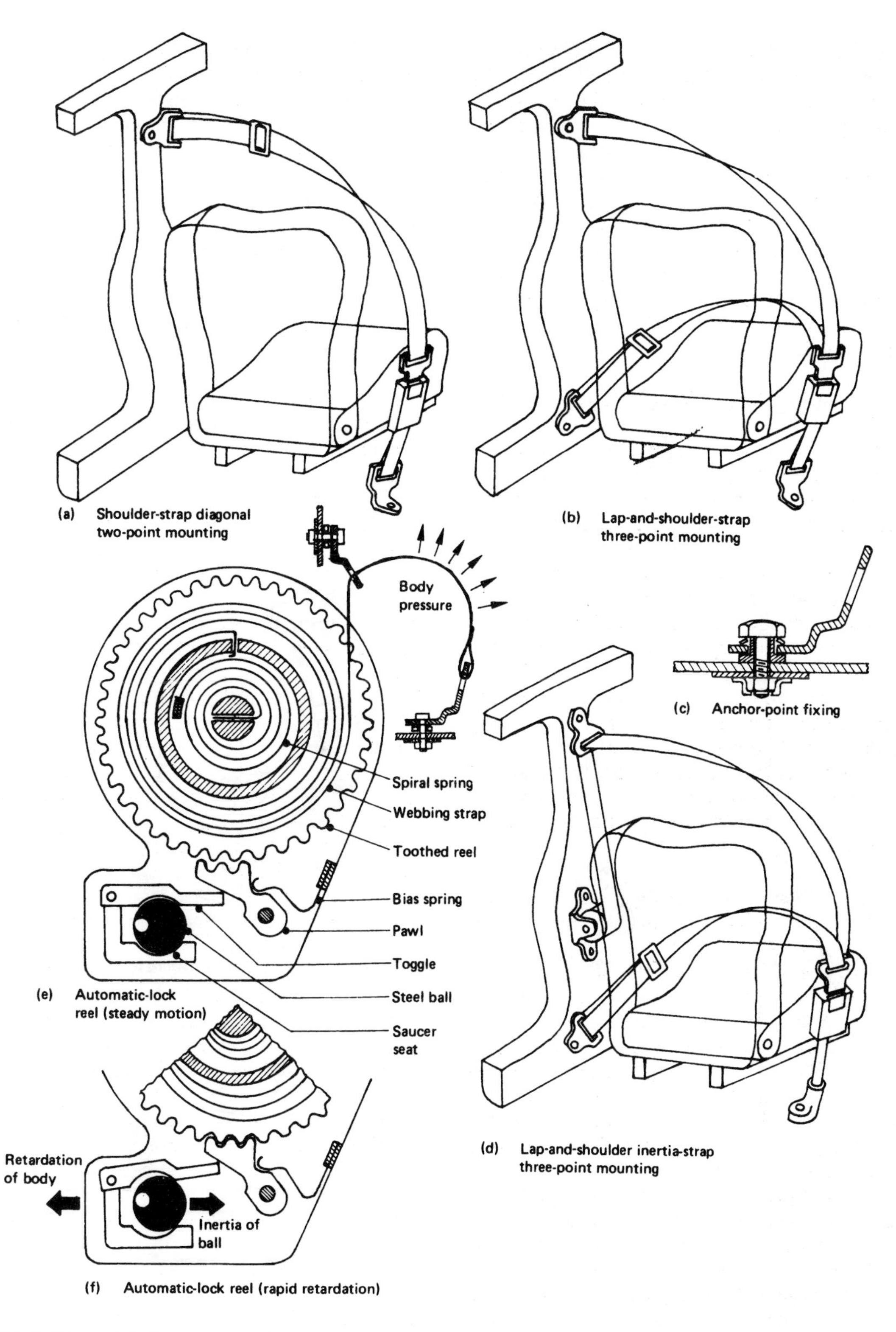

Fig. 1.29 Safety seat-belt arrangements

cavities, thus allowing the spring-loaded pawl to disengage itself from the toothed reel again (fig. 1.29(e)).

1.8 Jacks and jacking points (fig. 1.30)

Jacks are used to raise some part of the vehicle higher from the ground under emergency conditions, such as for changing wheels which have punc-

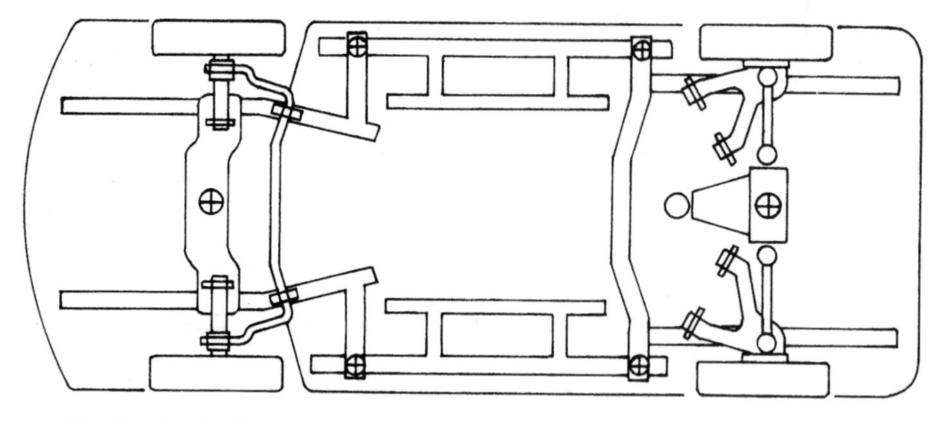

(a) Plan view of car jacking points

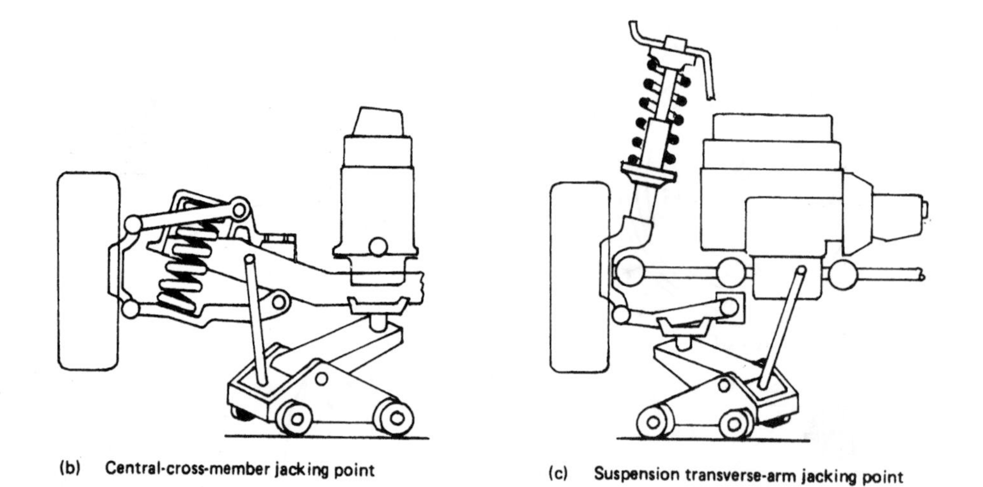

(b) Central-cross-member jacking point

(c) Suspension transverse-arm jacking point

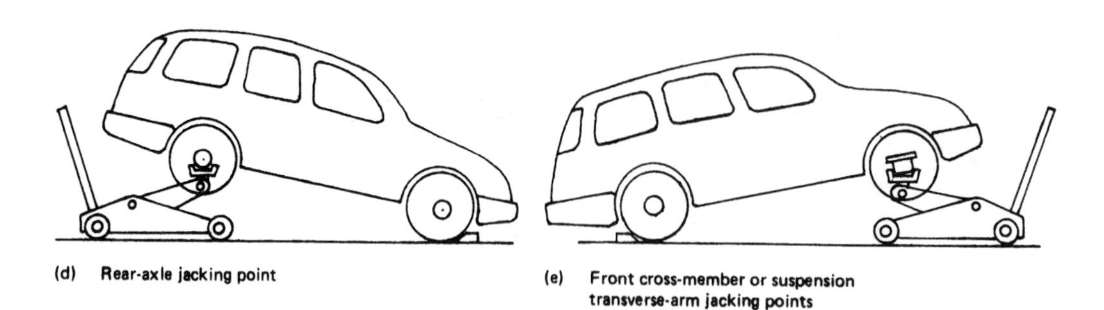

(d) Rear-axle jacking point

(e) Front cross-member or suspension transverse-arm jacking points

Fig. 1.30 Car jacking positions

tured tyres, or for routine maintenance procedures where a mechanic may need to crawl underneath the vehicle to examine or service a component.

For simple emergency checks or repairs, the vehicle is equipped with its own jack and it has reinforced jacking points built on to the existing body structure to withstand the concentrated load created when the vehicle is lifted at one point. The reinforcement to the body structure usually takes the form of a box-section welded to various parts of the underside of the body. The locations chosen are usually on deep-channel strengthened sections designed to withstand both torsional and bending stresses.

Figure 1.30(a) shows a plan view of a car with the jacking points marked with a circle and cross. The side jacking points are designed to be used with the car's own jack, but the front-cross-member and the rear-axle points require the use of a more substantial bottle jack or trolley jack.

When jacking-up rear-wheel-drive cars, there is usually a front cross-member where the jack may be located centrally (fig. 1.30(b)). Unfortunately, with front-wheel-drive cars the cross-members are usually situated much further back, so that one of the lower transverse suspension arms provides the most suitable jacking location (fig. 1.30(c)). Figure 1.30(d) and (e) show side views of the most suitable jacking locations for raising either the rear or the front half of the car. Note the wooden chocks placed next to the wheels to prevent the car from rolling away from the jack. At no time should anybody crawl under a vehicle which has been jacked up without first placing an axle-stand underneath a suitable jacking point.

1.8.1 Screw-and-nut jacks (fig. 1.31)

Figure 1.31 shows three types of portable car jacks suitable for light emergency repairs only.

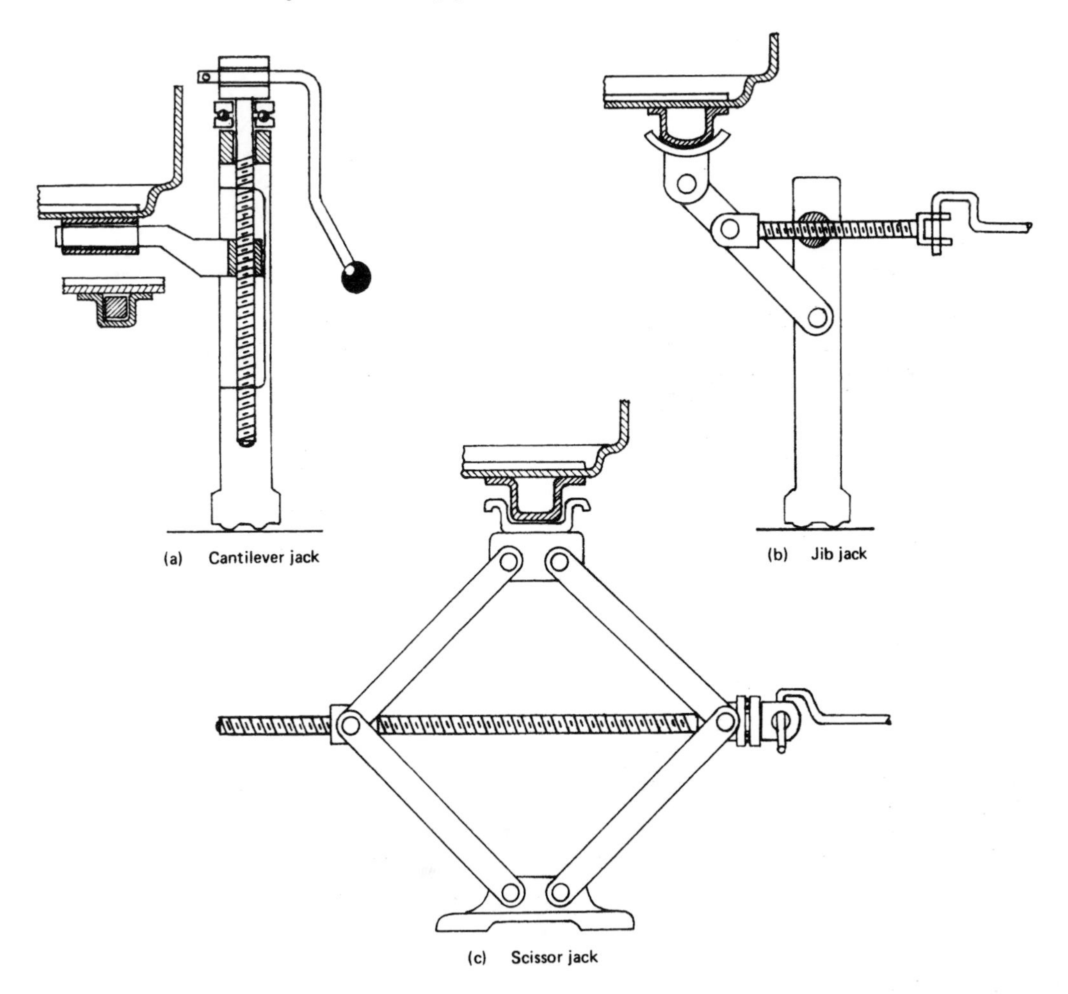

Fig. 1.31 Screw-and-nut jacks

All three are designed to lift from the side and to clear only individual road-wheels off the ground.

Each of these jacks operates on the screw-and-nut principle in which the nut is prevented from turning and the screw is rotated with a handle so that it moves axially towards or away from the nut. Whatever position the screw is in relative to the nut, there will always be sufficient friction between the meshing threads to prevent the screw turning on its own when the handle is released.

With the cantilever jack (fig. 1.31(a)), the nut is raised or lowered when the screw is twisted, so a thrust ball-race is necessary to reduce the friction and to support the weight of the part of the car being lifted. Conversely, with the jib jack (fig. 1.31(b)) the nut is supported by the vertical post, and it is the screw which moves relative to the stationary nut when it is rotated, raising or lowering the jib arm accordingly.

Alternatively, a four-bar diamond-shaped frame known as a scissor jack (fig. 1.31(c)) may be preferred. In this arrangement the screw is supported horizontally at mid-bar joint level by the nut and a short sleeve. When the screw is rotated, the nut

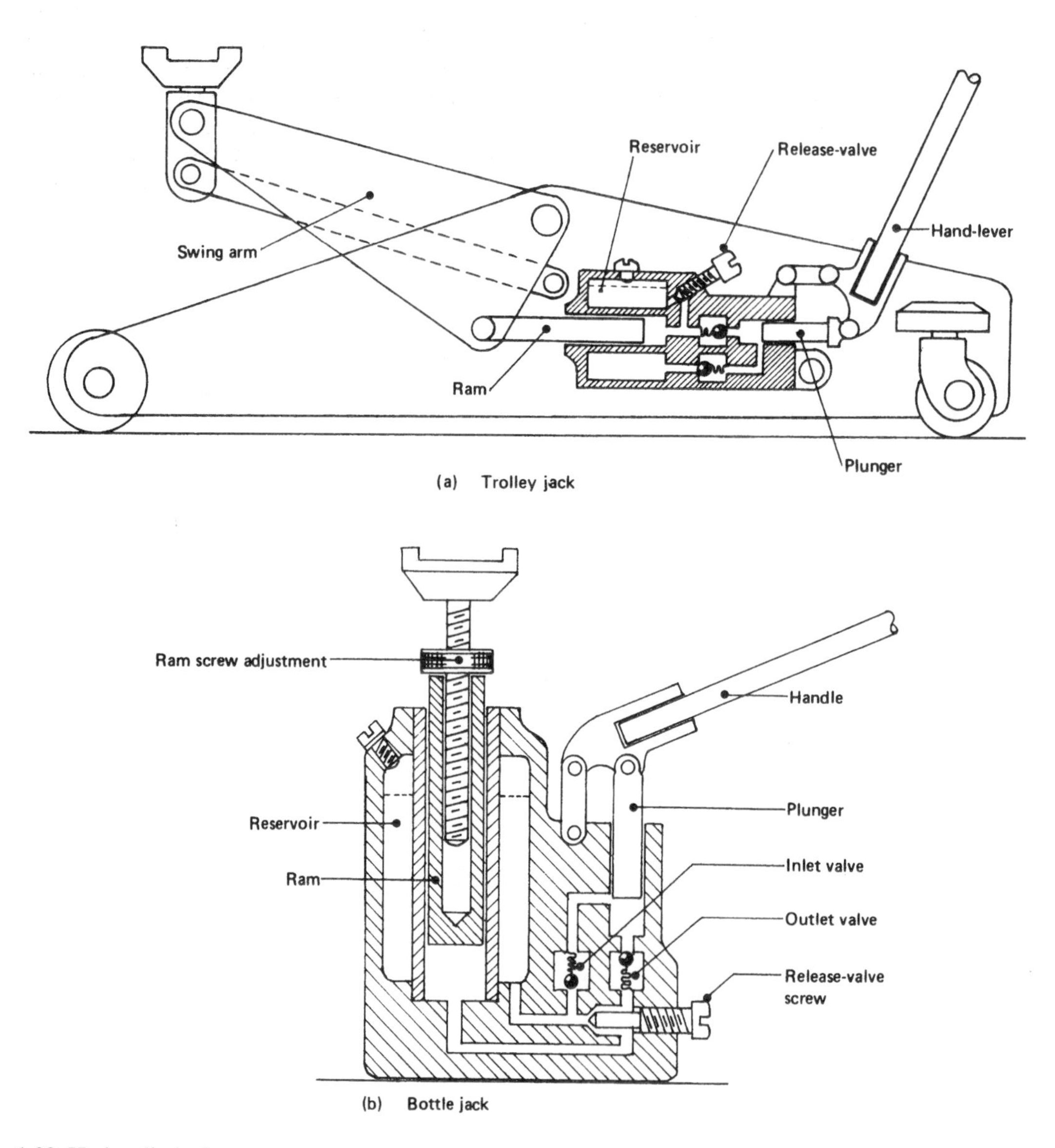

Fig. 1.32 Hydraulic jacks

moves either towards or away from the sleeve forming the other bar joint, so that the bars are forced vertically to extend or contract respectively.

1.8.2 Hydraulic jacks (fig. 1.32)
For routine maintenance servicing, more substantial jacks such as the hydraulic trolley jack or bottle jack are used. In these, a small-diameter input-effort plunger and cylinder displace fluid into a larger-diameter output-load cylinder chamber with ram. The ratio of the different cross-sectional areas of the two cylinders is equal to the force ratio of the jack.

The basic operation of the bottle and trolley jacks are identical (fig. 1.32). When the hand-lever is moved upwards, the small plunger moves clear of the small inlet drilling about half-way along its stroke. A slight vacuum is thus created in the plunger cylinder so that the outlet valve will close and the inlet valve opens. Fluid from the reservoir will then fill the plunger cylinder chamber. When the hand-lever is pushed downwards, the plunger moves beyond the inlet hole so that the fluid is trapped and is then pressurised. The inlet valve will now close and the outlet valve is forced to open. Pressurised fluid now passes into the output cylinder chamber. The repeated pumping action of the handle will displace more fluid from the reservoir into the plunger chamber, and the plunger in turn will force more fluid into the output chamber so that the ram will eventually move outwards.

In the case of the bottle jack, the ram pushes directly against the underside of the vehicle to raise it. With the trolley jack, the ram pushes against the lower part of the swing arm so that this rapidly rises and contacts the jacking point selected. It should be noted that initially the bottle-jack central screw must be adjusted until the ram touches the jacking point – only then should the hydraulic pump be operated.

To lower either of the jacks, the release-valve screw is unscrewed so that pressurised fluid can escape back to the reservoir.

2 The friction clutch

2.1 The purpose of a clutch

The name transmission *clutch* has become established simply due to its meaning of grasp or grip tight. It is a device which enables one rotary drive shaft to be coupled to another shaft, either when both shafts are stationary or when there is relative motion between them.

The need for the clutch stems mainly from the characteristics of the turning-effort developed by the engine over its lower speed range. When idling, the engine develops insufficient torque for the transmission to be positively engaged – for a car, the engine speed has to be increased to about 800 to 1200 rev/min before complete coupling without slip may be made. The vehicle also has to be moving at about 5 to 8 km/h (depending on the transmission ratios) if no snatch is to take place. To obtain a smooth engagement, the clutch has to be progressively engaged to take up the drive until the torque transmitted from the engine equals that required to propel the vehicle.

2.1.1 *Clutch fundamentals*

The clutches which will be described in this chapter depend on friction for their ability to transmit motion.

Consider a block of material which is made to slide over a flat-surface board (fig. 2.1). Frictional force (F) is created which tends to oppose this movement. If the pressure between the stationary and sliding surfaces is increased by placing a weight (W) on the block, the force required to pull the block over the surface will increase directly with the magnitude of this load. The ratio of the frictional force (F) to load (W) applied between the friction faces is known as the coefficient of friction, symbol μ (pronounced 'mu'),

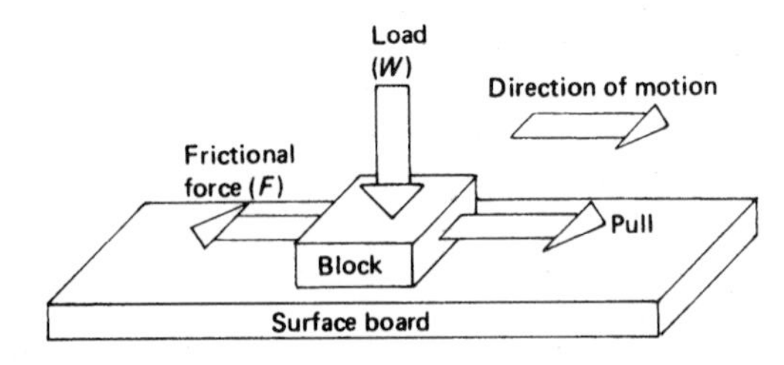

Fig. 2.1 Board-and-block friction generator

i.e. $\text{coefficient of friction} = \dfrac{\text{frictional force}}{\text{applied load}}$

or

$$\mu = \frac{F}{W}$$

$\therefore \text{frictional force} = \text{coefficient of friction} \times \text{applied load}$

or $F = \mu W$

Since μ is a ratio of like quantities, it has no units.

The friction clutch (fig. 2.2) is a gradual-engagement clutch which allows a pair of discs (a flywheel and a driven- and pressure-plate) attached to shafts to come together slowly. The friction generated between the discs will transmit the turning effort from one shaft to the other. If the pressure which holds the discs together is varied, the rate of slip will change and in a similar manner so will the degree of power transfer. Note that the rubbing pair will never run out of sliding surface as in the case of a sliding block and board.

The frictional force created between the two surfaces drags round the output disc and shaft, and the greater the grip (W) between the two discs provided by the thrust springs, the greater will be the frictional drag force and the turning-effort that can be transmitted before the clutch slips.

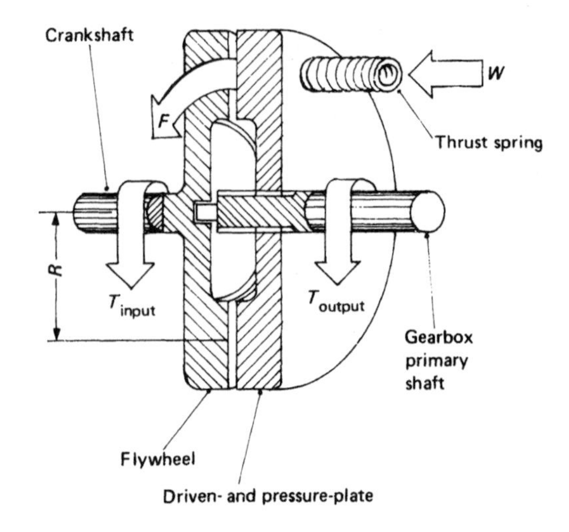

Fig. 2.2 Single-rubbing-pair friction clutch

This turning-effort is measured as frictional torque and is expressed as the product of the frictional force and the mean radius of the annular friction discs:

$$\text{frictional torque (N m)} = \text{friction force (N)} \times \text{mean radius (m)}$$

i.e. $T = FR$

Example An engine develops a torque of 108 N m. The clutch discs have a mean radius of 90 mm, and the coefficient of friction between the rubbing pair is 0.4. If there are six thrust springs, what clamping force must each spring provide to transmit this torque?

$$T = FR$$

but $F = \mu W$

$$\therefore \quad T = \mu W \times R$$

$$\therefore \quad W = \frac{T}{\mu R} = \frac{108 \text{ N m}}{0.4 \times 0.09 \text{ m}} = 3000 \text{ N}$$

$$\therefore \quad \text{force applied per spring on the disc} = \frac{3000 \text{ N}}{6} = 500 \text{ N}$$

For practical purposes, modern clutch units have a centre driven-plate lined on both sides with friction material and sandwiched between a flywheel and a pressure-plate made of steel or cast iron (fig. 2.3). With this design, there is a rubbing pair of surfaces on each side of the driven-plate so that twice as much frictional force is created for the same spring thrust. This means that this clutch can transmit twice as much torque as the simple single-rubbing-pair clutch,

i.e. $T = 2FR$

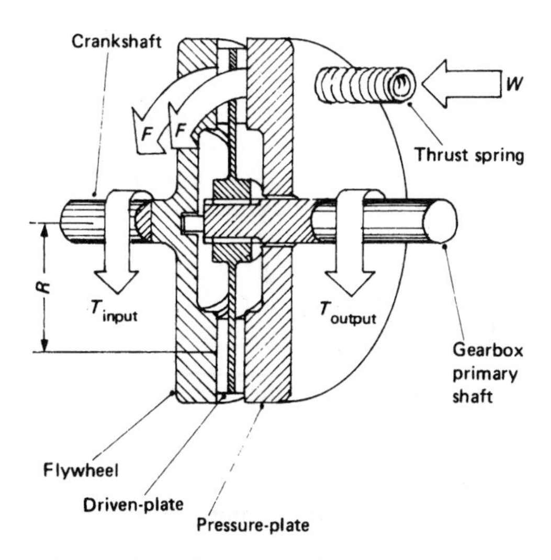

Fig. 2.3 Double-rubbing-pair friction clutch

Example A double-rubbing-pair clutch has a mean radius of 90 mm. The total spring thrust amounts to 1500 N, and the coefficient of friction between the lining and the steel faces is 0.4. What is the maximum torque which can be transmitted?

$$T = 2FR$$

but $F = \mu W$

$$\therefore \quad T = 2(\mu W)R$$

$$= 2 \times 0.4 \times 1500 \text{ N} \times 0.09 \text{ m}$$

$$= 108 \text{ N m}$$

2.1.2 Summary of the purpose of a clutch

a) To enable the rotary motion of one shaft (the engine crankshaft) to be transmitted to a stationary or slowly revolving output shaft (the gearbox first-motion shaft) smoothly and without snatch.
b) To provide a positive linkage, capable of transmitting the maximum engine torque when the road speed of the vehicle is high enough for the engine to be directly coupled to the transmission without it stalling.
c) To enable the engine to be rapidly separated and re-engaged from the transmission while one or both are in motion, for gear-changing and emergency stops.

2.1.3 Desired clutch features

a) The force required to separate the drive must not be excessive.
b) A reasonable coefficient of friction must be maintained under all working conditions.
c) Correctly matched rubbing surfaces must be hard enough to resist wear but not so hard that they become scored.
d) Adequate surface area and mass of rubbing surfaces to absorb the heat generated.
e) Adequate cooling or ventilation to dissipate the heat generated.
f) Reasonable thermal conductivity to dissipate the heat and so avoid distortion of the flywheel and pressure-plate.
g) A friction material which will not crush at high temperatures and clamping loads.

2.2 Description and operation of a multi-coil spring clutch unit (fig. 2.4)

The principal components of this type of clutch unit are a flywheel, a cover-pressing, a pressure-plate, a driven-plate, thrust springs, and a release-lever mechanism.

2.2.1 Input flywheel member

The flywheel forms the input member of the clutch and has four main functions:

i) it stores kinetic energy during the engine power stroke and releases it during the three idle periods,
ii) it provides a location for the ring gear used for transmitting the starter-pinion drive to the crankshaft,
iii) it provides a rigid base to which to bolt the clutch assembly,
iv) it also forms the frictional faceplate for one side of the driven centre-plate.

Cover-pressing This is a steel pressing which houses the pressure-plate assembly. It is bolted to the flywheel, so it forms one part of the driving member of the clutch. This cover is made to take the reaction of the thrust coil springs and it also provides a pivot point for the release levers. It must have ventilation holes for heat dissipation and yet be sufficiently rigid to support the loaded thrust springs.

Pressure-plate This plate is an annular cast-iron disc providing a smooth rubbing surface for one side of the driven centre-plate. It must be sufficiently thick to resist distortion and to dissipate the generated heat while the clutch is slipped. Cast on the back of the plate are four protruding lugs which are machined so they can locate and support the release-lever and strut assemblies.

Drive straps The cover-pressing and the pressure-plate are held together by four laminated spring-steel straps arranged tangentially around the pressure-plate. One end of each strap is riveted to the cover-pressing, while the other end is bolted to the pressure-plate. These straps deflect during clutch operation without disturbing the concentricity of the cover-pressing and the pressure-plate and transmit the drive from the cover-pressing to the pressure-plate. This design eliminates friction between the cover and the pressure-plate.

Thrust springs A ring of helical-coil thrust springs is assembled between the pressure-plate and the cover in a state of compression. When the clutch is engaged, these load the pressure-plate against the driven-plate, so transmitting the drive from the cover-pressing (which is bolted to the flywheel) to the driven friction-plate. The drive is then passed through the splined hub to the gearbox primary shaft.

Release-lever plate This is an annular cast-iron disc carried on the inner ends of the release-levers. It is retained in position by specially formed springs, and its function is to act as the frictional faceplate for the release or withdrawal bearing.

Solid driven-plate This is a disc plate sandwiched between the flywheel and the pressure-plate. It consists of a spring-steel disc mounted on a splined hub, and moulded asbestos friction facing is riveted to each side of the disc. The drive is passed from the flywheel and pressure-plate assembly to the friction discs. It is then transmitted to the splined hub and hence to the splined gearbox primary shaft. The splined hub allows the driven-plate to float between the flywheel and the pressure-plate while the clutch is being engaged or disengaged.

Engagement-and-disengagement mechanism Four case-hardened pressed-steel release-levers are provided for clutch release. A knife-edge strut is located in a groove formed in the outer end of each lever and is retained in position by two extended ears which fit into grooves machined in the bosses of the pressure-plate. The levers pivot on loose pins which fit into recesses in the levers and are free to move in eyebolts. The threaded ends of the eyebolts pass through the clutch cover and are retained by nuts. The plain ends of the eyebolts are located in clearance holes in the pressure-plate. The release-levers are fitted with anti-rattle springs located between the levers and the clutch cover-pressing.

Clutch operation Pressing the clutch pedal conveys movement through the push-rod to the clutch fork. This pivots on the cross-shaft, and the forked end moves the withdrawal bearing along the gearbox primary-shaft bearing cover and tubular extension support which acts as a guide until the bearing contacts the inner tips of the release-levers.

Further movement pivots the release-levers on the eyebolt pins and, operating through the strut, moves the clutch pressure-plate, relieving the clutch driven-plate friction-disc facings of driving

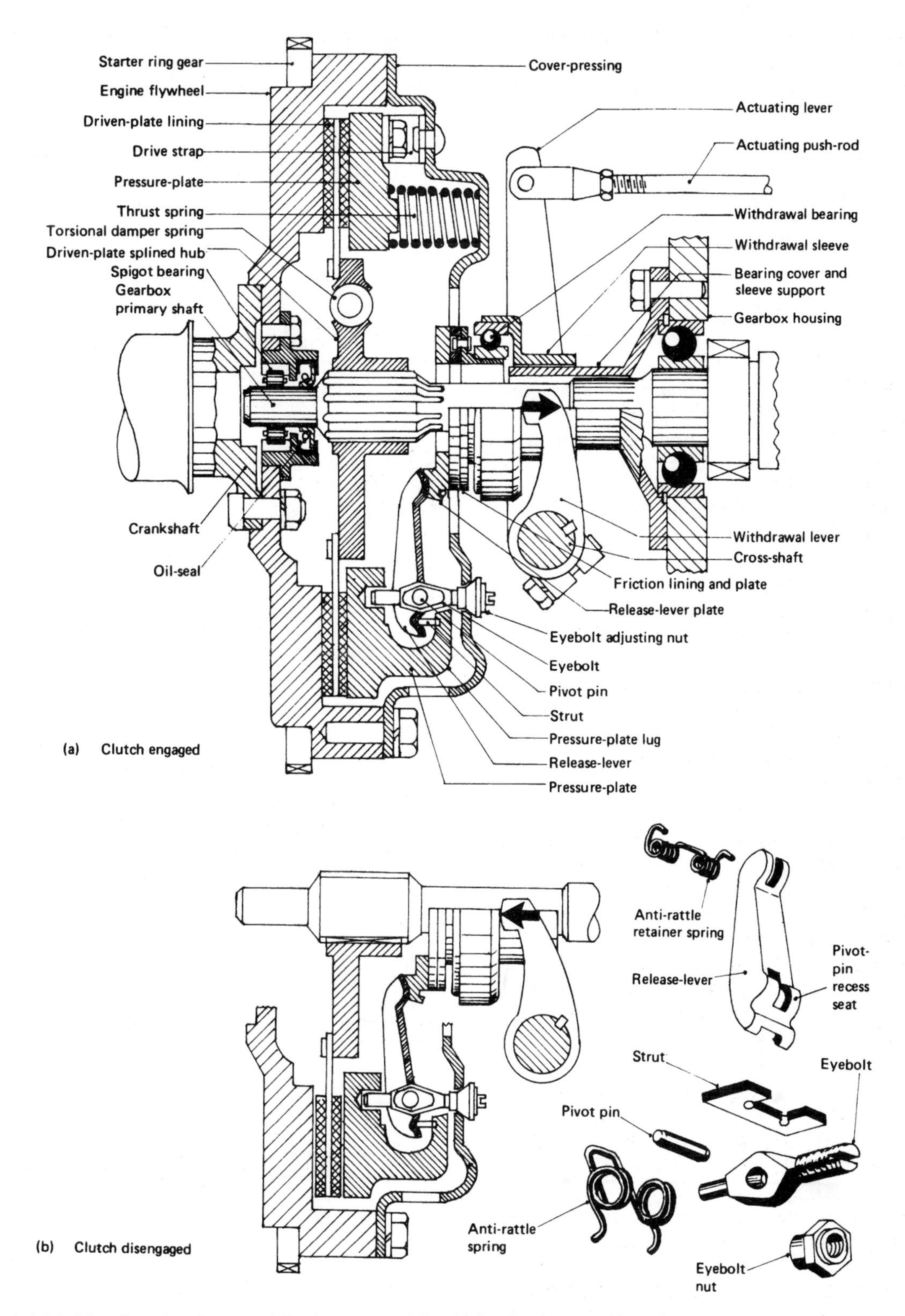

Fig. 2.4 Multi-coil spring Borg and Beck commercial-vehicle clutch-assembly unit

pressure. When the clutch pressure is released, the movements are reversed and the driving pressure is re-applied to the disc faces.

Application Multi-coil spring clutch units are not normally used for cars, but they are used on heavy commercial vehicles where it is difficult to provide sufficient clamping thrust by a single diaphragm spring (see section 2.2.2). Further research and development will no doubt eventually produce heavy-duty diaphragm-type pressure-plate units.

2.2.2 Description and operation of a diaphragm-spring clutch unit (figs 2.5 to 2.8)

Diaphragm-spring clutch units are similar in construction to the multi-coil spring clutch, but a single dished diaphragm-type spring is used to apply the clamping thrust, and this also serves as part of the release mechanism. There are two types shown in figs 2.5 and 2.6.

Diaphragm spring (fig. 2.5) The diaphragm spring resembles a dished washer and has been developed from the Belleville washer (named after Julian François Belleville, a French civil engineer who was granted a British patent for the device in 1866). This diaphragm spring is a steel disc with a hole through the centre, the inner portion of the disc being radially slotted to form a number of actuating (release-lever) fingers. The outer ends of these slots have enlarged blunting holes formed, to spread the concentrated stresses created at the ends of the slots over an enlarged area when the fingers are being deflected, thus preventing fatigue cracking, and also providing a means of locating the shouldered rivets which restrain the fulcrum rings.

Spring and fulcrum-ring action (figs 2.5 and 2.6) The diaphragm spring supplies the thrust load and is positioned between the pressure-plate and the cover-pressing. The outer edge of the dished spring bears against the pressure-plate, and

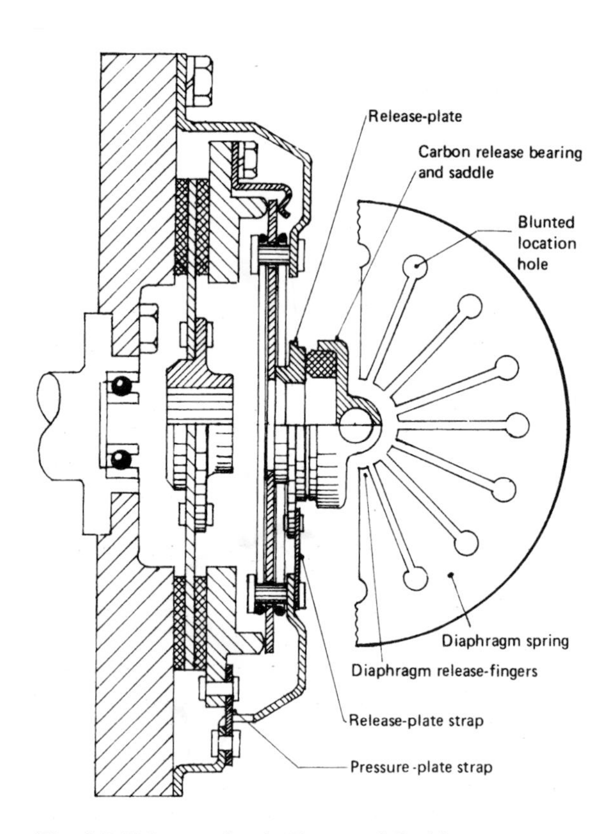

Fig. 2.5 DS-type clutch (Borg and Beck)

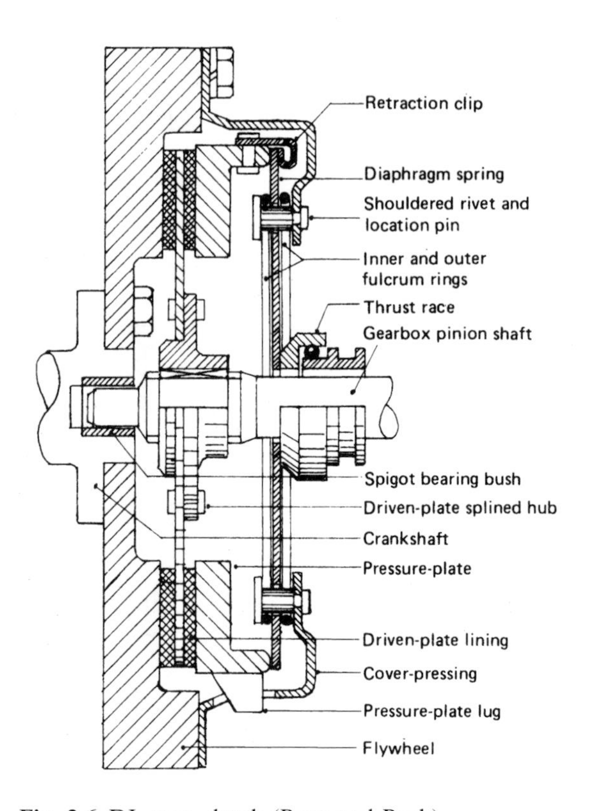

Fig. 2.6 DL-type clutch (Borg and Beck)

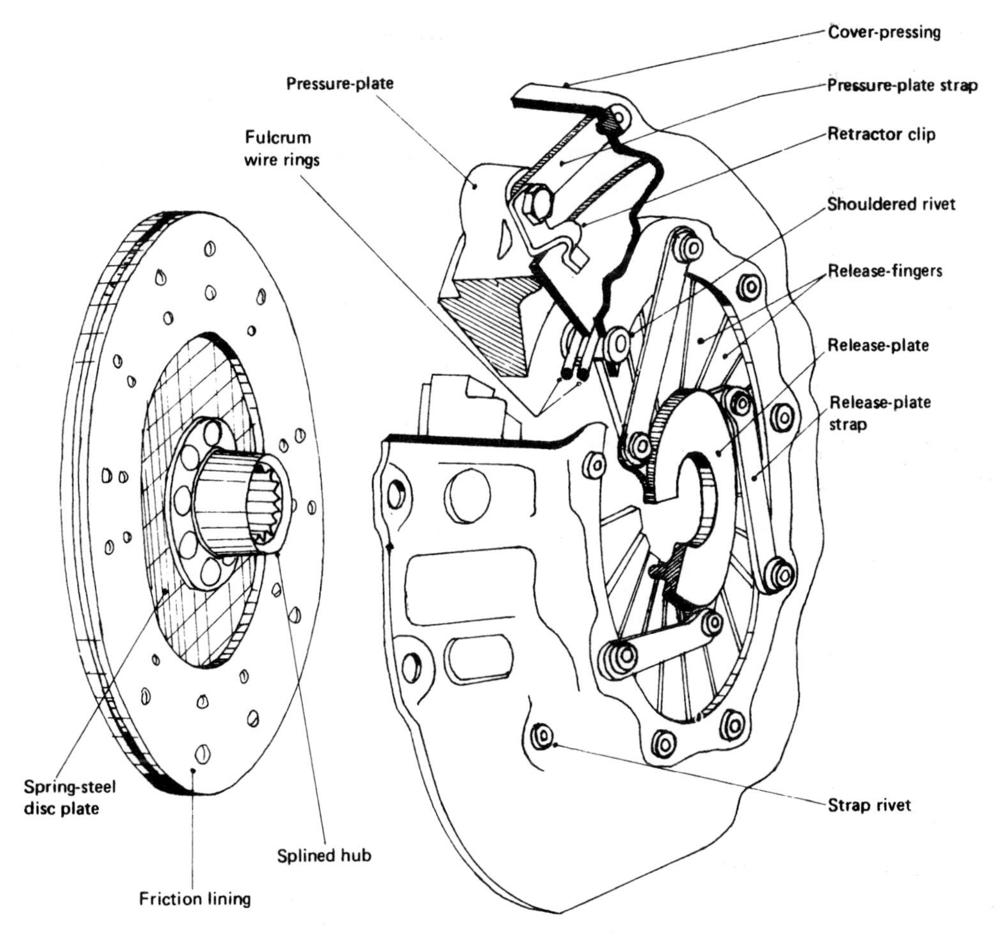

Fig. 2.7 Rigid driven-plate and diaphragm-spring clutch assembly

a short distaince in from its outer edge are situated two round-section wire rings, one on each side of the dished spring. These two rings are located and held in position by shouldered rivets which are themselves supported by the cover-pressing (see the pictorial view of the clutch assembly – fig. 2.7).

When the cover-pressing is bolted to the fly-wheel, the dished spring will be slightly flattened. This loads the pressure-plate against the driven friction discs and, at the same time, the spring reaction will be taken through the outer ring to the cover-pressing. The inner ring acts as a pivot or fulcrum point for all the individual release-lever fingers, so that they can roll on this ring with the minimum of friction.

The position of the fulcrum rings will control the force ratio for releasing the clutch driven-plate, and they are therefore positioned near the periphery of the diaphragm spring to increase the leverage.

Clutch operation (fig. 2.8) To disengage the clutch, the release bearing is pressed against the release-fingers by means of the clutch actuating linkage (see section 2.3). These pivot and tilt about the inner ring, so that the spring's cone angle is increased, and the spring thrust load is relaxed and even withdrawn from the driven-plate friction faces. The drive between the engine and gearbox is thus interrupted. Retractor clips are provided so that the diaphragm spring can with-draw the pressure-plate from the flywheel.

As the driven-plate friction discs wear, the spring will automatically expand towards the fly-wheel to compensate. This reduces the cone angle (fig. 2.10(b)) and increases the dishing of the spring, so that the back of the spring will convey a rolling reaction against the outer fulcrum ring with the least friction. With a new driven-plate, the dished spring will be almost flat and so cannot exert its maximum thrust load against the

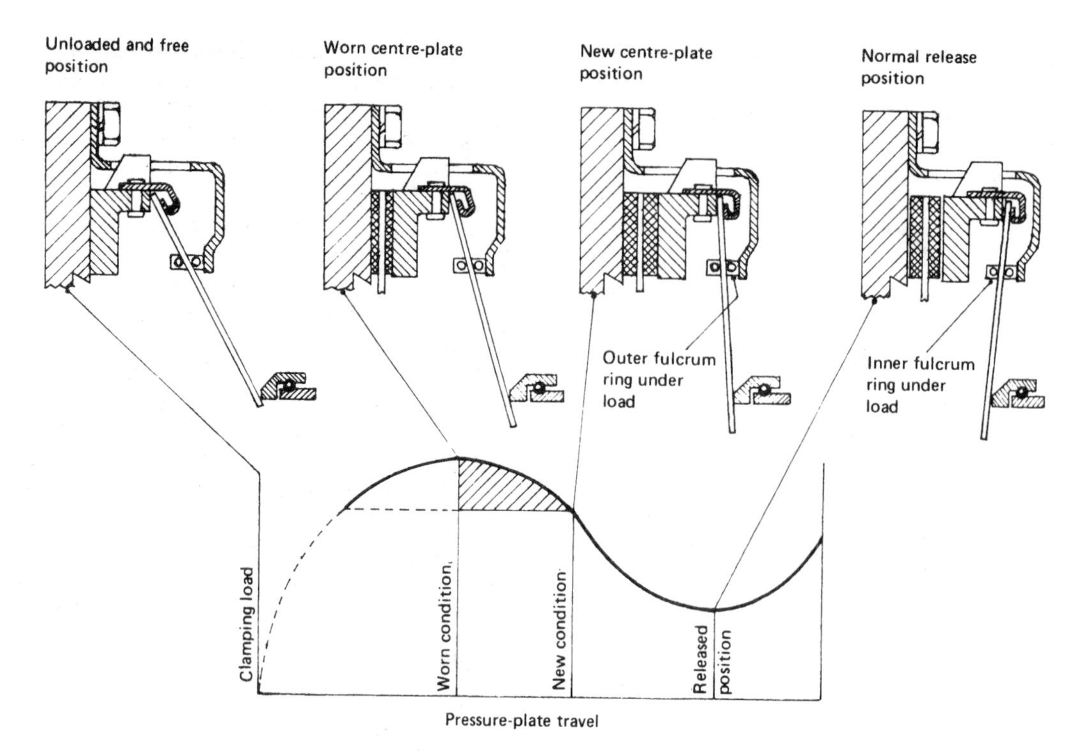

Fig. 2.8 Diaphragm-spring Borg and Beck car and van clutch-assembly units

pressure-plate. However, as wear occurs, the cone angle is reduced, enabling initially a greater clamping thrust to be exerted. With further wear, the distortion will result in the release-fingers becoming over-relaxed – that is, their force on the pressure-plate will decrease – which will reduce the clamping thrust.

Drive from the cover-pressing to the pressure-plate The drive from the cover-pressing may be transferred to the pressure-plate by two methods:

i) small-torque-capacity clutches use three lugs cast as part of the pressure-plate which engage slots aligned in the cover-pressing (fig. 2.6);
ii) larger clutches tend to have three straps, one end of each being attached to the cover-pressing and the other end to the pressure-plate (fig. 2.5).

Release-bearing assembly This provides a means whereby the clutch foot-pedal linkage – which is stationary – transmits the disengagement leverage to a rotating clutch assembly. There are two basic types:

i) a release-plate is attached to the release-fingers and a carbon saddle-type release bearing is pressed against this plate to disengage the clutch (fig. 2.5);
ii) a ball thrust-race assembly is made to act directly against the release-fingers when releasing the driven-plate (fig. 2.6).

2.2.3 Comparison of multi-coil and diaphragm-spring clamping characteristics

The relationship between load and deflection of a multi-coil spring clutch is linear, as shown in fig. 2.9. As the springs are compressed (deflected), the compressive clamping load will increase propor-

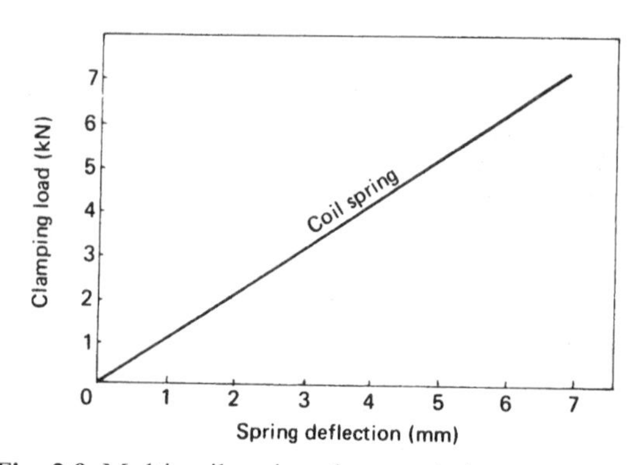

Fig. 2.9 Multi-coil spring characteristics

tionally, producing the straight-line characteristic curve shown.

The relationship between load and deflection of a diaphragm-spring clutch, on the other hand, is a non-linear one. The shape of the load–deflection curve mainly depends on the ratio of the dish height (h) in the free state to the thickness (t) of the diaphragm spring for a given spring size, i.e. h/t – see fig. 2.10(a) and (b).

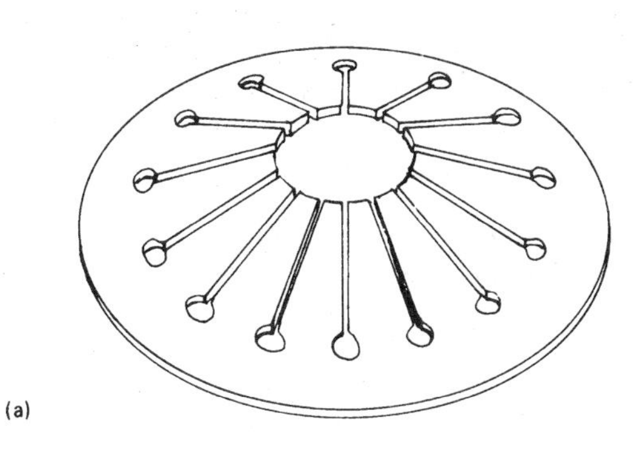

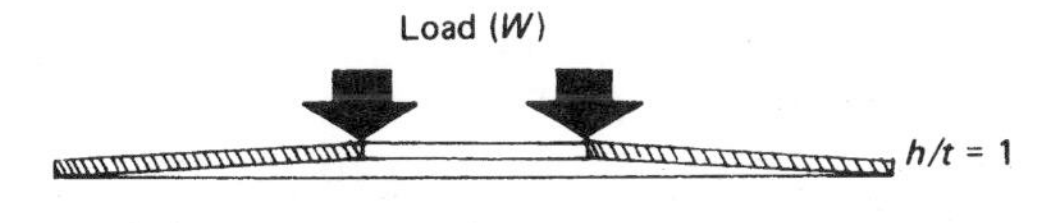

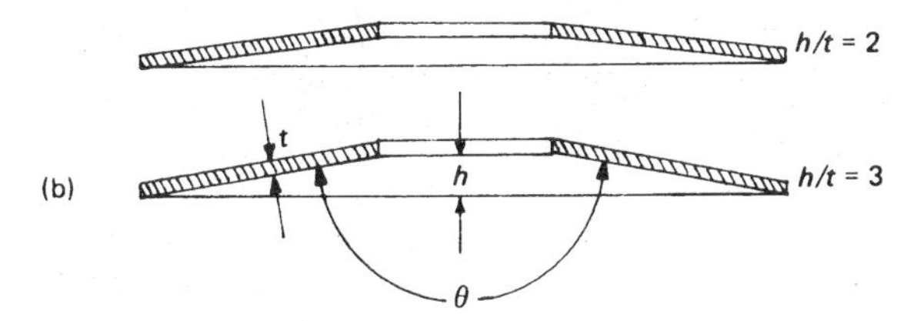

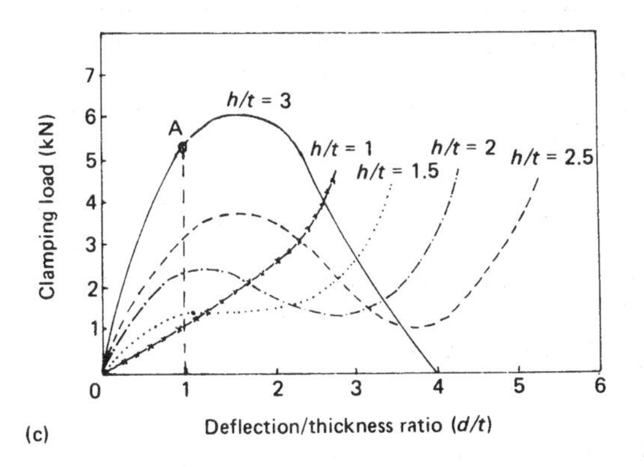

Fig. 2.10 Diaphragm spring characteristics: W = diaphragm load; t = diaphragm thickness; h = diaphragm dish height; θ = diaphragm cone-angle

The load–deflection characteristics are shown in fig. 2.10(c), which shows the clamping load plotted against the ratio of spring deflection (d) to diaphragm thickness (t), i.e. d/t, for five different ratios of h/t. It will be seen that the spring load increases nearly linearly up to a deflection equal to the thickness of the spring – point A. After this, the characteristics differ for different ratios of h/t. It can be seen that for $h/t = 1$ there is a tendency for the spring load to increase at a greater rate; conversely, for $h/t = 3$ the compressive load will progressively drop with further deflection. Between these two extremes – i.e. for $h/t = 1.5$, 2.0, and 2.5 – the clamping load tends to decline with more spring distortion, but with a tendency to bottom-out and even to increase again as it nears its full movement. This means that the designer can alter the ratio h/t so that the load–deflection characteristic of the spring can be varied to suit the application, a typical value for clutches being 1.7.

2.2.4 *Relationship between driven-plate wear and clamping thrust*

Coil springs apply their maximum compressive clamping load when the driven-plate is new and at its thickest – see fig. 2.11. However, as the lining wears, the coil springs have to extend to take up the distance between the pressure-plate and the cover-pressing, and this means that the spring will progressively loose its compressive thrust as the driven-plate wears.

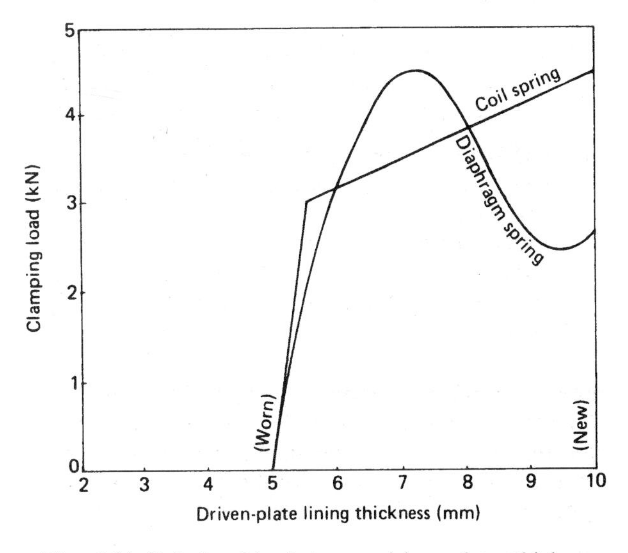

Fig. 2.11 Relationship between driven-plate thickness and clamping load for coil and diaphragm springs

In contrast, the diaphragm spring is practically flattened when the driven-plate is new, so it cannot then exert its maximum clamping load on to the pressure-plate. However, the dishing of the diaphragm increases as the lining wears, which effectively increases the load acting on the pressure-plate. Eventually, though, this increased diaphragm thrust due to the dishing of the spring will be counteracted by the gap between the pressure-plate and the cover-pressing enlarging to such an extent that the compressive force of the spring will become over relaxed, hence the pressure-plate loading will decrease with further driven-plate wear.

2.2.5 Merits of the diaphragm spring over multi-coil springs

a) Compact design enables the use of a shallow clutch bell-housing which encloses the clutch unit.
b) Fewer moving parts help to eliminate squeaks, rattles, and wear.
c) No initial adjustment is needed for the pressure-plate unit – unlike the multi-coil spring clutch units, where a small clearance must be set between the release-lever plate and the thrust bearing.
d) Accurate balance of the clutch assembly is maintained under all operating conditions.
e) The diaphragm acts as both clamping spring and release-fingers.
f) Spring axial load is self-compensating as the driven-plate wears.
g) Clamping load is unaffected by engine speed, whereas coil springs tend to bow – that is, bend along their length – and lose their thrust at high speeds.
h) Disengagement pressures are reduced with increased pedal movement.

2.3 Cushioned driven-plate with torsional damper (fig. 2.12)

For a clutch to operate smoothly and effectively the rigid driven-plate is designed to incorporate two additional features which are as follows:

1) spring cushioned friction linings
2) spring vibration damper hub.

2.3.1 Driven-plate axial cushioned friction lining (fig. 2.12)

The incompressibility of the solid annular friction lining material of the driven-plate (see fig 2.7), when sandwiched between the flywheel and pressure-plate contact faces, prevents a gradual transfer of torque from the engine to the gearbox, that is, the clutch is either engaged or disengaged with very little slip take-up in between.

However if two annular friction linings are separated by a series of flexible crimped leaf spring segments, then the compressive load/central gap contraction characteristics between the two linings provides a progressive parabolic clamp force take-up. Accordingly torque will be smoothly transferred from the engine to the transmission instead of a jerky shudder slip-stick engagement which tends to be generated by a solid driven-plate.

2.3.2 Driven-plate torsional damping (fig. 2.12)

Crankshafts are subjected at various speeds to torsional vibration in which the crankshaft, instead of rotating at a uniform speed, alternately accelerates and decelerates in rapid succession. If this motion was to be transmitted to the gearbox, the input shaft gear teeth would oscillate between adjacent meshing teeth on the driven gears, thus causing a clattering noise and correspondingly tooth-profile wear. To prevent the crankshaft oscillation being transferred to the transmission gears, the output splined-hub is separated from the driven double-sided annular friction linings by means of six coil-springs positioned around the splined-hub. Thus every time the input torque fluctuates, the springs will compress and rebound in sympathy with the torsional vibration. A pair of friction washers assembled on either side of the hub-flange and held in compression by a Belleville dished-washer spring absorbs the crankshaft vibration at the flywheel end. This energy absorption is achieved by the rubbing friction generated between the side-member plate walls, friction washers and hub-flange whenever there is any angular to and fro movement of the driven-plate side-members, relative to the steady splined-hub and gearbox input shaft rotation. A Belleville dished-washer spring provides a roughly constant compressive thrust, as the angle of the dished-washer increases to compensate for the friction washer wear; it therefore maintains the frictional damping characteristic throughout the lifetime of the driven-plate.

2.4 Clutch operating linkage

The clutch unit is bolted to the flywheel and therefore rotates as part of the crankshaft assembly – it is thus subjected to the same engine and trans-

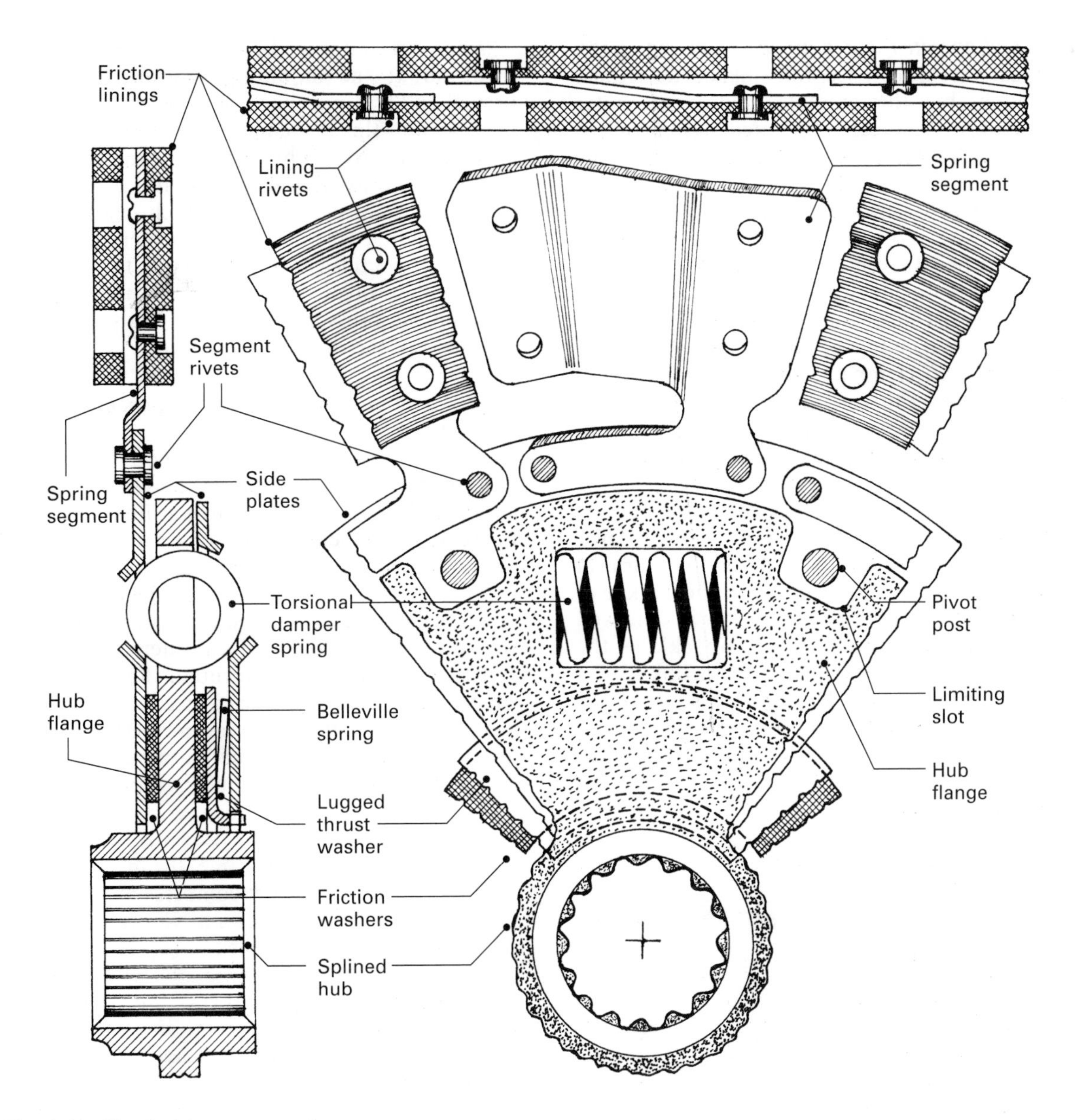

Fig. 2.12 Clutch driven centre plate

mission vibrations such as rocking, shaking, and pitching. These reactions cannot be eliminated, but they are practically isolated from the body by supporting the engine and gearbox on flexible rubber mountings. In contrast, the clutch foot-pedal assembly is attached to the body structure which may be subjected to slight bump and rebound movements which the suspension has not been able to absorb.

For the clutch to be operated smoothly without any jerks and consistently in whatever stage of engagement or disengagement it happens to be, some sort of flexible linkage system must be employed which will not interfere with the clutch adjustment, is compact, needs very little maintenance, has a long working life, is simple, and is reasonably cheap.

There are three main methods of transmitting movement and force to the clutch from the foot-pedal:

i) rod linkage,
ii) cable linkage,
iii) hydraulic pipelines.

*2.4.1 **Rod-operated clutch arrangement*** (fig. 2.13)
Most early vehicles used rod linkage to actuate the clutch, but the present trend is to prefer cable and hydraulically operated systems, and only a few commercial vehicles still utilise the rod linkage.

Two foot-pedal arrangements are shown in fig. 2.13: a chassis-mounted pedal and a bulkhead-positioned pedal lever, the latter nowadays being more suitable for tilt-cab installation. The engine end of the clutch operating mechanism is enclosed in the clutch bell-housing and consists of a cross-shaft clamped to which is a fork-arm which supports a thrust bearing and saddle. One end of the cross-shaft extends outside the bell-housing, and a withdrawal lever is clamped to it. This lever is connected to the foot-pedal either directly by means of the withdrawal rod for the chassis-mounted foot-pedal or by a relay lever which transforms the horizontal withdrawal movement into a vertical one for the bulkhead pedal layout. Adjustment of both is achieved either by a screw-nut or by a turnbuckle as the clutch driven-plate wears. With both linkages, a pedal adjustment stop is provided so that the pedal is positioned suitably for the driver's foot and, in the case of the chassis-mounted pedal, clears the floorboards.

When the clutch pedal is pressed down, it will pivot about the mounting bracket. The withdrawal rod will pull back or forward as the case may be, and this movement will be relayed to the withdrawal lever which rotates the fork-arm about the cross-shaft and pushes the thrust-bearing saddle against the clutch release-plate, taking up the free-play. Any further movement will press the release-levers towards the flywheel, so that they pivot about the eyebolt pins, withdrawing the pressure-plate from the driven-plate to interrupt the drive.

2.4.2 Cable-operated clutch arrangement
(fig. 2.14)
Cable linkage is a popular method of relaying movement from the pedal to the clutch. Cables are very effective in conveying linear motion between components which have a small amount of movement between them, such as an engine rocking on its mounts relative to the body.

The cable assembly consists of an inner multi-strand steel-wire core and an outer cable sheath consisting of a spirally wound flexible sleeve which may have nylon end-pieces (this plastic has very good rubbing wear properties and in general does not require lubricating). Sometimes the inside of the sleeve is lined with anti-friction

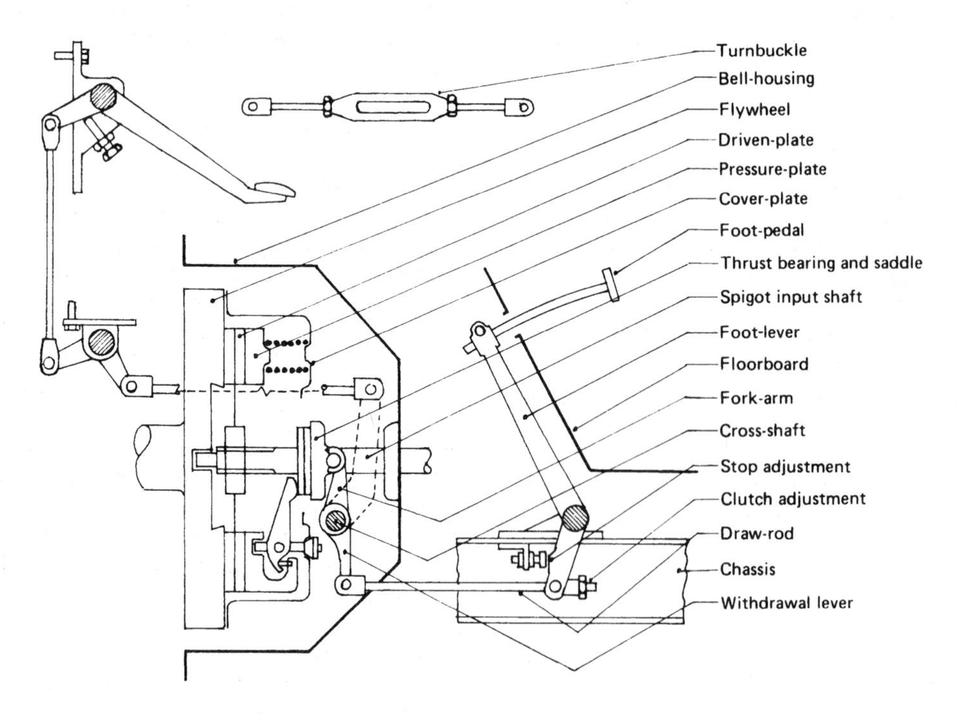

Fig. 2.13 Rod-operated clutch arrangement

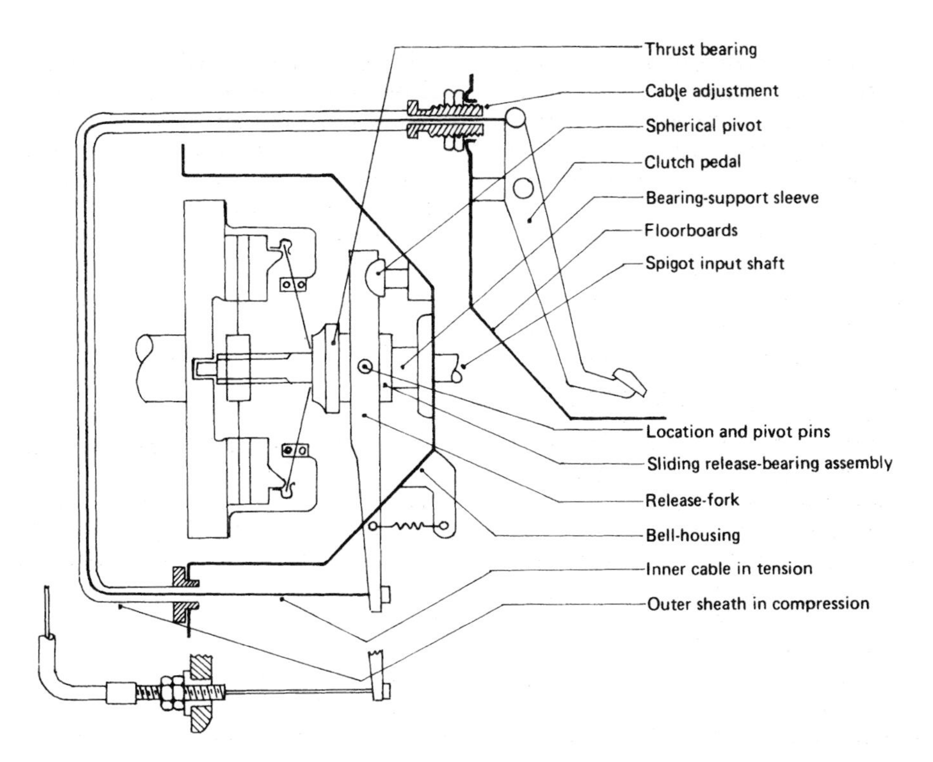

Fig. 2.14 Cable-operated clutch arrangement

material such as polytetrafluorethylene (PTFE) plastic to minimise any 'slip–stick' friction.

The cable is made to span between the clutch bell-housing and the car bulkhead, where the pedal is generally situated. A screw adjusment is provided at either the pedal or the bell-housing end as shown – it acts by altering the length of the outer cable sleeve, which has the effect of increasing or decreasing the free-play of the inner cable.

Most manufacturers nowadays prefer to support the thrust bearing on an extended tube which fits over the gearbox primary input shaft – this means that it can slide freely and perpendicularly against the clutch release-fingers. This method ensures that equal pressure will be applied to all the release-fingers and avoids any uneven loading of the diaphragm spring and pressure-plate.

A pressed-steel release-fork lever is used to provide the leverage from the cable to the thrust bearing. The end of the lever is supported and pivoted on a spherical-headed bolt, and two adjacent pins locate the thrust bearings at about one third the length of the lever from the pivot. The outer end of the lever extends outside the bell-housing and is connected to the inner cable. When the clutch is disengaged, the inner cable will be subjected to tension and the equal and opposite reaction to this will put the outer sleeve into compression. The fork-lever will then tilt about its pivot so that it will force the release bearing against the release-fingers and so disengage the drive.

2.4.3 Hydraulically operated clutch arrangement (fig. 2.15)

An alternative and convenient method of transmitting force and movement is by forcing fluid through a flexible plastic pipeline running between the foot-pedal and the clutch bell-housing.

Controlled clutch action is achieved by having a master-cylinder bolted to the bulkhead and a push-rod connecting the clutch-pedal movement to the sliding piston. (More details on master-cylinder operation are given in chapter 9 on brakes.) A second cylinder and piston – known as the slave-cylinder unit – are located and supported on an extension formed on the bell-housing flange. The piston inside this cylinder conveys the slightest movement to the fork-lever through the slave push-rod. The fork-lever has the thrust-bearing assembly attached to one end, and a spherical pivot is situated slightly in from this end.

When the clutch pedal is depressed, the master-cylinder piston moves forwards and pushes a

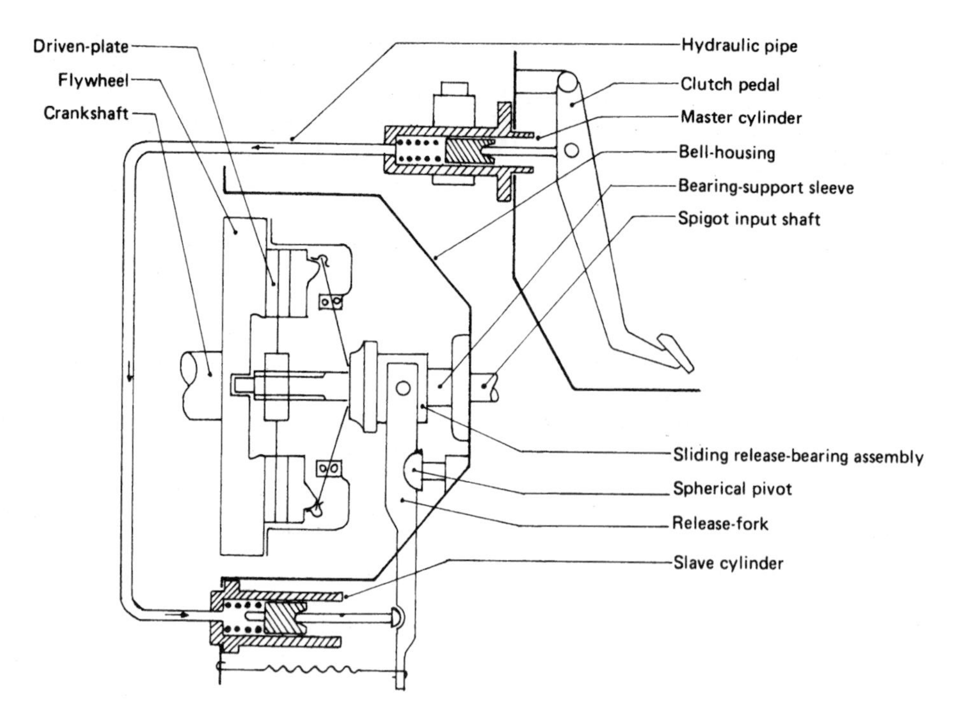

Fig. 2.15 Hydraulically-operated clutch arrangement

continuous column of fluid through the pipeline. Consequently, an equal volume of fluid must be displaced into the slave-cylinder so that the piston moves out and tilts the fork-lever. The net result is that the thrust bearing defects the release-fingers so that the driven-plate will slip.

Engagement of the clutch occurs when the pedal is released – this allows the fluid to return to the master-cylinder and its reservoir. The return-spring in the slave-cylinder will then maintain a slight pressure on the fork-lever, so that the thrust bearing will always be in contact with the release-fingers.

Driven-plate wear will be compensated for by the slave return-spring and piston automatically moving out to take up the increased fork-lever tilt.

When subjected to large leverage forces, hydraulic actuating mechanisms do not suffer from frictional wear as do cables. This makes them particularly suitable for heavy-duty applications on large trucks.

2.5 Clutch faults, causes and remedies

The four main clutch faults – slip, judder, drag, and fierceness – will now be considered briefly.

2.5.1 Slip

This is the condition when a gear ratio has been selected, the clutch pedal has been released and the accelerator pedal has been pushed down, and yet no power is transmitted to the driving wheels although the engine accelerates.

The causes and remedies for slippage are as follows:

a) Glazing of the surfaces, caused by oil or grease on the friction faces due to leakage from the engine crankcase or the gearbox or to excessive lubrication of the spigot shaft and its support bearing during assembly. Clean the components and renew the driven-plate.
b) Clutch mechanism binding. Check the linkage and rectify interference.
c) Incorrectly adjusted clutch pedal. Adjust the linkage to provide correct clearance.
d) Worn-out driven-plate. Renew the driven-plate and pressure-plate assembly.

2.5.2 Judder

This is the condition when the clutch is re-engaged by the gradual release of the clutch pedal but, instead of a smooth take-up, the combined engine-and-gearbox unit shudders and vibrates.

The causes and remedies for judder are as follows:

a) Worn-out rubber engine and gearbox mountings. Replace the mountings.

b) Body-to-engine tie-bar out of alignment. Readjust the tie-bar.
c) The pressure-plate is out of parallel with the flywheel face. Reposition the pressure-plate, making sure that the location dowels are fitted and that the set-bolts are evenly tightened.
d) Distorted driven-plate or bent splined shaft which permits uneven friction contact around the contact faces. Check the alignment of the shaft and renew the driven-plate.
e) Misalignment of the engine to the gearbox, due to incorrect assembly of the bell-housing to the engine. Check the bell-housing location dowels and observe if the housing has been squarely drawn up to the engine by the set-bolts.
f) Oil, grease, or other contamination causing uneven engagement. Clean the flywheel and pressure-plate friction faces thoroughly and renew the driven-plate.

*2.5.3 **Drag or spin***

This condition exists when it is difficult to engage a gear either when the vehicle is stationary or in motion, due to the driven-plate being dragged around when the clutch pedal has supposedly released it.

The causes and remedies for drag are as follows:

a) Oil or grease on the driven-plate faces, due either to leakage from the crankcase or gear-box or to excessive greasing of the spigot shaft and support bearing during assembly. Wash the components, lubricate the splined shaft and spigot bearing, and renew the driven-plate.
b) Misalignment between the engine and the splined clutch-shaft. Check the alignment of the engine-to-gearbox housing joint.
c) Insufficient clutch linkage movement preventing the release bearing from completely separating the driven-plate from the flywheel and pressure-plate faces. Adjust the linkage to obtain correct free movement.
d) Damaged or distorted pressure-plate or clutch cover. Renew the pressure-plate and driven-plate.
e) Distorted driven-plate, due to the weight of the gearbox being allowed to hang on the clutch spigot shaft and driven-plate while being aligned to the engine during assembly. Renew the driven-plate, and the pressure-plate if the release-fingers or diaphragm spring appear to have been strained.
f) Driven-plate hub binding on the splined shaft. Wash the driven-plate hub and splined shaft and lubricate with high-melting-point grease.
g) Splined-spigot-shaft support bush or ball-bearing binding. Examine the bearing and either lubricate or renew.
h) Broken driven-plate or friction lining. Renew the driven-plate.
i) Dirt or foreign particles embedded in the driven-plate faces. Renew the driven-plate.

*2.5.4 **Fierceness or snatch***

This condition occurs when the clutch pedal is released to couple the engine to the transmission drive and, instead of a gradual smooth take-up of the drive, a snatch or jerky engagement is experienced.

The causes and remedies for fierceness or snatch are as follows:

a) Oil or grease on the driven-plate faces causes uneven and irregular gripping of the driven-plate by the pressure-plate while being engaged. Wash all components, lubricate the splined shaft, and renew the driven-plate.
b) Clutch-pedal linkage binding. Check the cause of interference and rectify.
c) Worn-out driven-plate faces causing variation in friction grip while being engaged. Renew the driven-plate and possibly the pressure-plate.
d) Misalignment of the pressure-plate to the flywheel face. Check the pressure-plate location dowels and realign the pressure-plate to the flywheel and evenly tighten the set-bolts.
e) Driven-plate damper springs loose or broken, so that their cushioning action during clutch engagement is not effective. Renew the driven-plate.

3 Gearbox construction and operation

3.1 The purpose of the gearbox

To appreciate the reasons for using a gearbox, an understanding of the engine's performance characteristics and of how a pair of gear wheels forms a leverage system is required.

3.1.1 Engine performance characteristics (fig. 3.1)

The performance characteristics of an engine show the relationship between engine power (that is, the rate of doing useful work), the turning-effort or torque developed, and the engine rotary speed.

It has been found that power is proportional to the product of torque and speed,

i.e. $P \propto TN$

where P = brake power (kW)

T = torque (N m)

and N = speed (rev/min)

This means that, if the power remains constant, the torque and speed are inversely proportional to each other

i.e. $T \propto \frac{P}{N}$

so that if the torque is doubled then the speed will be halved and vice versa (fig. 3.1).

3.1.2 The need for gear reduction

At low engine speed, a reciprocating-piston engine does not develop sufficient turning-effort or torque to propel a vehicle forward from a standstill. Even the greater torque produced at higher engine speed would be insufficient to accelerate the vehicle at a reasonable rate. Therefore means must be provided to multiply the engine's torque and to increase the rotary engine speed relative to the driving road-wheels – that is, to produce a speed reduction from the engine.

The gearbox provides a way of varying the engine's output torque and speed to match the vehicle's speed and load, which will include factors such as the passengers, luggage, or goods being carried and the road conditions, which involve such things as road surface finish, wind resistance, road gradients, and the degree of acceleration desired.

3.1.3 Understanding gear ratios (fig. 3.2)

The two circles in fig. 3.2 represent a pair of gear-wheel blanks of different diameter. Superimposed

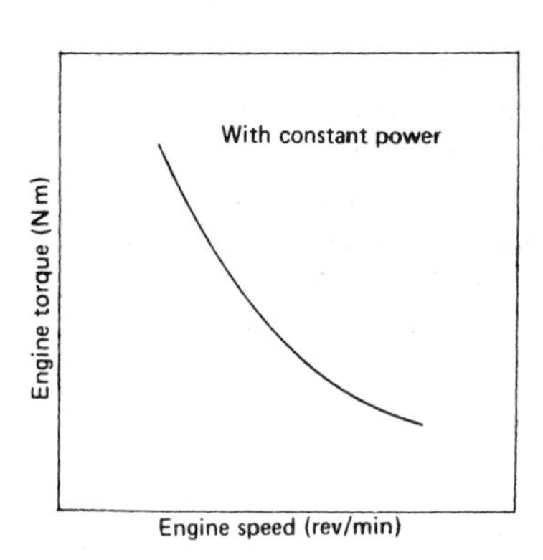

Fig. 3.1 Relationship between engine torque and speed for a constant power output

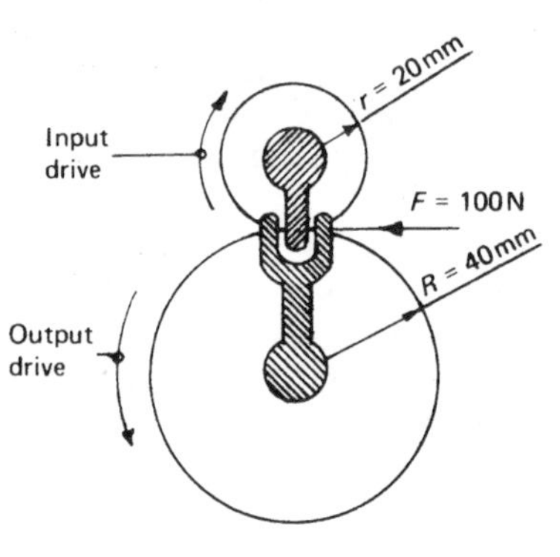

Fig. 3.2 Leverage relationship between meshing gears of different diameter

on the wheels are a lever-arm and a fork-arm interlocked to represent gear teeth in mesh. If the larger (output) wheel has a radius of 40 mm, the smaller (input) wheel has a radius of 20 mm, and a force of 100 N is applied perpendicular to the arms at the point of contact, then

input torque = 100 N × 0.02 m = 2 N m

and

output torque = 100 N × 0.04 m = 4 N m

This shows that with a leverage ratio of 2:1, the output torque is double the input torque.

Obviously, if two meshing single lever arms revolve, they will very soon be misaligned and will cease to contact each other. A continuous supply of levers is therefore necessary to produce an uninterrupted torque multiplication and rotary drive. This difficulty is simply overcome by forming spur teeth on the periphery of the blank discs, so that a never-ending lever system is created.

If the diameter of a gearwheel is doubled, then the circumference and the number of teeth of the same pitch on the circumference are also doubled. It follows that, for one revolution of the input wheel in fig. 3.2, the meshing output wheel with twice the number of teeth will have only revolved half a turn; or, for two revolutions of the input shaft, the output shaft will have completed one revolution. The result is that, with a gear-reduction ratio of 2:1, the output speed has halved.

From the previous two relationships discovered between a pair of meshing gearwheels of different size, the output torque is inversely proportional to the speed or gear ratio. So, if the output torque is doubled, the output speed will be halved, and vice versa. Thus gearwheels provide a method of transmitting motion from one shaft to another and at the same time enabling a change in torque and speed to be achieved.

3.1.4 Gear train

If two or more gearwheels are meshed in series – either in the same plane or in different planes – the gearwheels are said to form a gear train.

Simple gear train When the gearwheels are supported on separate shafts and are in the same plane, the gear train is known as a simple gear train (fig. 3.3). Such arrangements are used for engine timing gears. If only two gearwheels are involved then a 'single-stage' simple gear train is formed (fig. 3.4). This arrangement is used with most front-wheel-drive gearboxes.

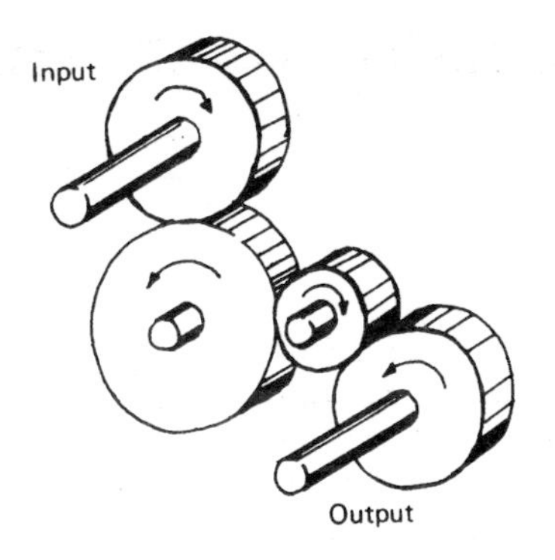

Fig. 3.3 Simple gear train

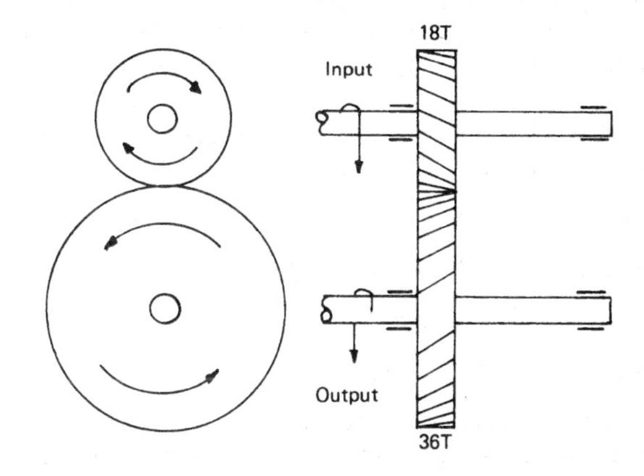

Fig. 3.4 Single-stage gear train

The gear ratio of a gear train may be defined as the input speed divided by the output speed, or it may be derived by using the formula

$$\text{gear ratio} = \frac{\text{product of teeth on driven gears}}{\text{product of teeth on driver gears}}$$

$$= \frac{\text{driven}}{\text{driver}}$$

Example In a given gearbox, the primary-shaft second gearwheel has 18 teeth and the secondary output-shaft gearwheel has 36 teeth. Calculate the gear ratio.

$$\text{Gear ratio} = \frac{\text{driven}}{\text{driver}} = \frac{36}{18} = 2{:}1$$

Compound or multi-stage gear train If two or more pairs of gearwheels are joined in series and the driven gearwheel of one gear train is connected by a common shaft to the driver of the

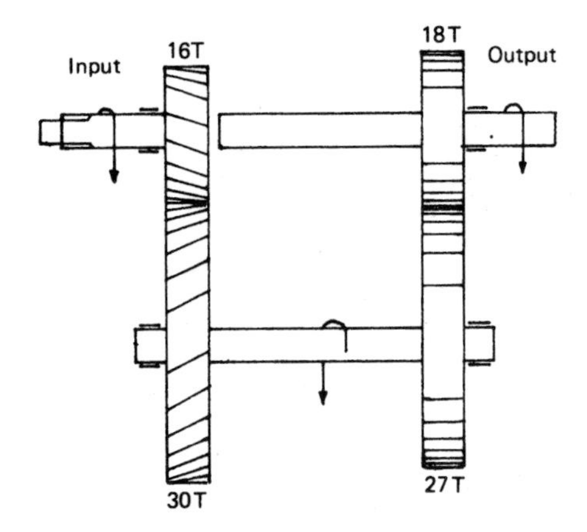

Fig. 3.5 Compound (double-stage) gear train

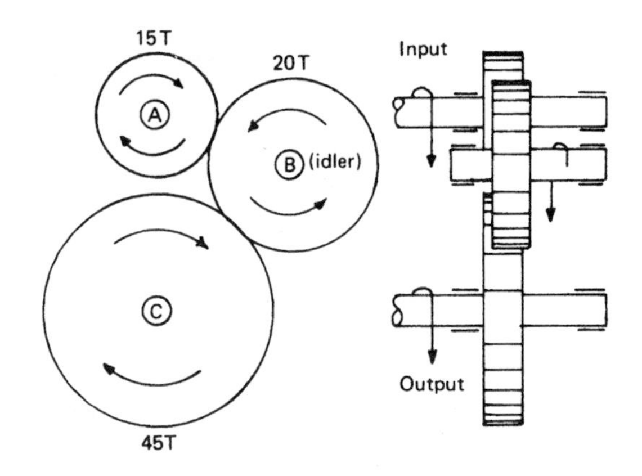

Fig. 3.6 Simple reverse gear train

next gear train, these gears are said to form a compound gear train (fig. 3.5). When two gearwheels are joined together by a single shaft, a 'double-stage' compound gear train is formed (fig. 3.5). This layout is found in gearboxes used in vehicles with front-mounted engines and having a rear-wheel drive.

Example In a gearbox, the first-stage constant-mesh gearwheels have 16 teeth on the primary shaft and 30 teeth on the layshaft. The third-gear output main-shaft wheel has 18 teeth, and the meshing lay-shaft wheel has 27 teeth. Calculate the third-gear ratio.

$$\text{Gear ratio} = \frac{\text{driven}}{\text{driver}} \times \frac{\text{driven}}{\text{driver}}$$

$$= \frac{30}{16} \times \frac{18}{27} = 1.25{:}1$$

Reverse gear train When two gearwheels are meshed together by a third or middle gearwheel, the additional gearwheel does not affect the overall gear ratio so it is known as an idler gear. However, the idler gear does change the direction of rotation (fig. 3.6).

$$\text{Gear ratio} = \frac{\text{driven}}{\text{driver}} \times \frac{\text{driven}}{\text{driver}}$$

$$= \frac{\text{B} \times \text{C}}{\text{A} \times \text{B}} = \frac{\text{C}}{\text{A}}$$

Example Find the reverse gear ratio if the primary input shaft gearwheel has 15 teeth, the idler gear has 20 teeth, and the output secondary shaft has 45 teeth.

$$\text{Gear ratio} = \frac{\text{driven}}{\text{driver}} \times \frac{\text{driven}}{\text{driver}}$$

$$= \frac{20}{15} \times \frac{45}{20} = 3{:}1$$

3.1.5 Epicyclic gears (fig. 3.7)

Essentially an epicyclic single-stage gear train consists of an internally toothed annular (ring) gear (A in fig. 3.7) with a band brake encircling it. In the centre of this gear is a sun gear (S) forming part of the input shaft. Bridging the space between the sun gear and the annular gear are a number of planet (pinion) gears (P) which engage both these gears and are mounted on a carrier (C) which is integral with the output shaft.

For torque to be transmitted, either the sun gear, the carrier, or the annular gear must be held stationary, to provide a reaction. In the following description, only the annular-gear reaction situation will be considered.

When the annular-gear band brake is tightened to provide the reaction and the input sun-gear shaft is driven, the planet gears rotate around their own axes and at the same time are forced to revolve and 'walk' around the inside of the annular gear. Consequently the carrier and output shaft which support the planet-gear axes will also be made to rotate, but at a speed lower than that of the input shaft.

Epicyclic gear trains are used for automatic transmission, overdrives, and final drives.

The gear ratio of a single-stage epicyclic gear train may be calculated by the following formula:

$$\text{Gear ratio} = \frac{S + A}{S} = 1 + \frac{A}{S}$$

also $$A = S + 2P$$

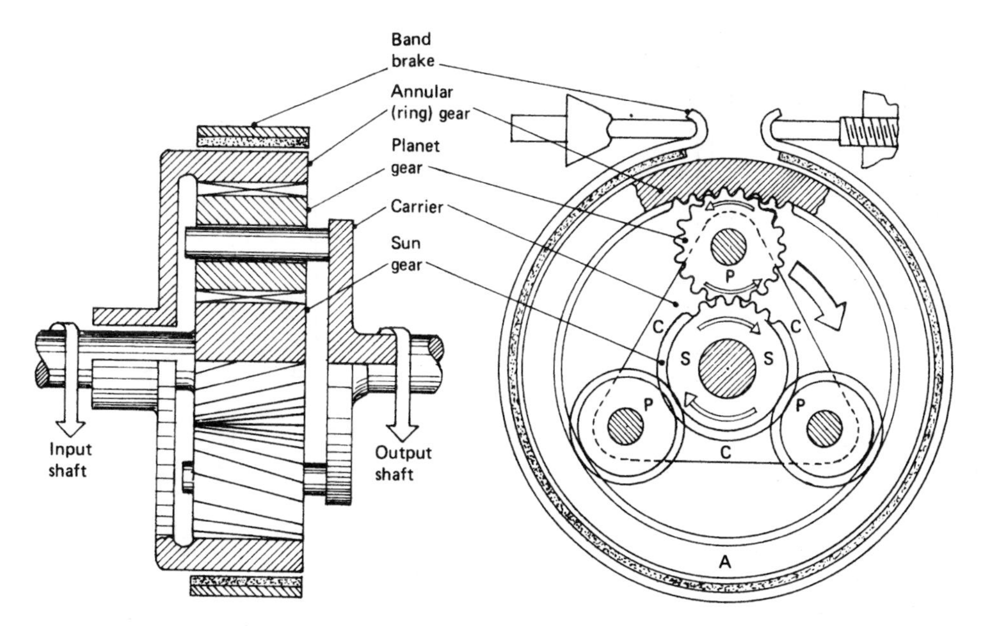

Fig. 3.7 Epicyclic gear train

where A = number of annular-gear teeth
S = number of sun-gear teeth
P = number of planet-gear teeth

Example An epicyclic gear train has sun and annular gears with 28 and 56 teeth respectively. If the input speed from the engine drives the sun shaft at 3000 rev/min, determine (a) the overall gear ratio, (b) the number of planet-gear teeth, (c) the carrier-shaft output speed.

a) $$\text{Gear ratio} = 1 + \frac{A}{S} = 1 + \frac{56}{28} = 1 + 2 = 3:1$$

b) $$A = S + 2P$$

$$\therefore \quad P = \frac{A - S}{2} = \frac{56 - 28}{2} = \frac{28}{2} = 14 \text{ teeth}$$

c) $$\text{Output speed} = \frac{3000 \text{ rev/min}}{3} = 1000 \text{ rev/min}$$

3.1.6 Types of spur gearwheels

Tooth gearing is used for positive transmission of rotary motion from one shaft to another. When the shafts are parallel, the gearwheels are called 'spur' gears and are cylindrical discs with teeth formed on their circumference. The materials used are usually low-alloy nickel–chromium–molybdenum steels. Gear-tooth profile terminology is shown in fig. 3.8.

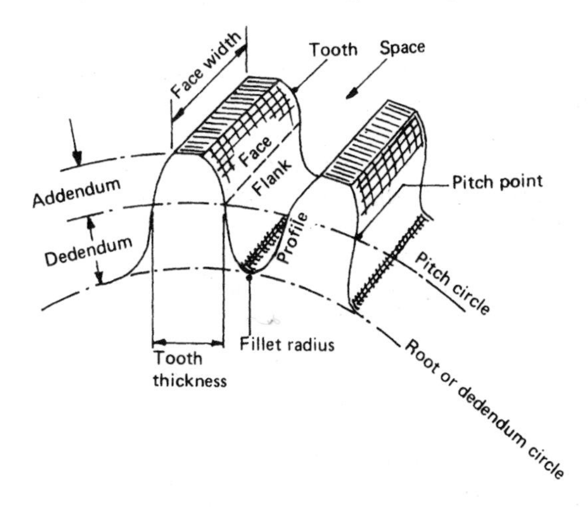

Fig. 3.8 Gear-tooth profile terminology

Straight-tool spur (fig. 3.9(a) and (b)) With straight-tooth spur gears, teeth are cut at right angles to the face and parallel to the axis of the gear-wheel. Where the teeth profiles contact, the relative motion is a rolling action at the pitch point, changing to a sliding one when contact occurs on either the face or flank of the tooth.

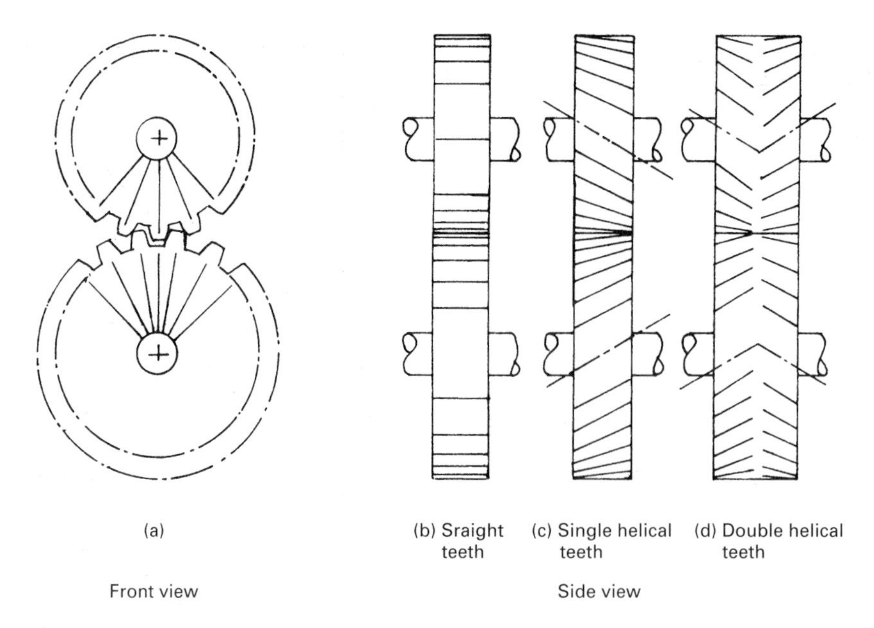

Fig. 3.9 Spur gearing

Radial forces between the teeth in contact tend to separate the gears and must be absorbed by radial-type bearings.

Helical-tooth spur (fig. 3.9(c) and (d)) With helical-tooth spur gears, teeth are cut at an angle both to the face and to the axis of the gearwheel. This means that contact between teeth in mesh takes place along a diagonal line across the faces and flanks of the teeth. Thus one pair of meshing teeth remains in contact until the following pair engages, and so the load on the teeth is distributed over a larger area. This reduces tooth loading and promotes smoother and quieter running. Axial- or end-thrust will be produced on the shafts and must be absorbed by bearings. Double helical gears are sometimes used to counteract side-thrust.

3.1.7 Characteristics of vehicle resistance to motion–tractive effort and power

The characteristics of a vehicle's resistance to motion and the engine's torque or tractive-effort performance at the drive wheels are shown in fig. 3.10(a).

The force opposing the vehicle's motion is seen to increase proportionally with speed; but, at the upper end of the speed range, wind drag causes a more noticeable rise in resistance.

Compared with the vehicle's resistance to motion, the engine's torque or tractive effort developed at the wheels in top gear initially rises at a greater rate, reaching its rather flat peak at about mid-speed, and then drops steadily. At the point where the tractive effort intersects the road resistance, the vehicle has reached its maximum speed – in this case 164 km/h.

The gap between the tractive effort and road resistance at any point over the vehicle's speed range indicates the surplus tractive effort available for the vehicle's acceleration.

If the gear is changed from a 1:1 to a 2:1 gear ratio, the tractive-effort characteristics over the engine's speed range are now compressed into half the vehicle's speed range, but the peak tractive effort will have doubled, with a very rapid rise and fall of tractive effort either side of the peak performance. Thus, over a very narrow speed range in a low gear, the large tractive effort in excess of the road resistance will provide a lively acceleration response.

Another way of illustrating vehicle performance is to show the relationship between the road resistance power on a level road which the engine has to overcome to maintain the vehicle at any given speed and the surplus power, that is, the amount of power developed by the engine above that of the power consumed by the road resistance (see fig. 3.10(b)). The surplus power is obtained by taking the vertical distance between the engine power and road resistance power curves at a number of different vehicle speeds and plotting them against vehicle speed, thus producing a

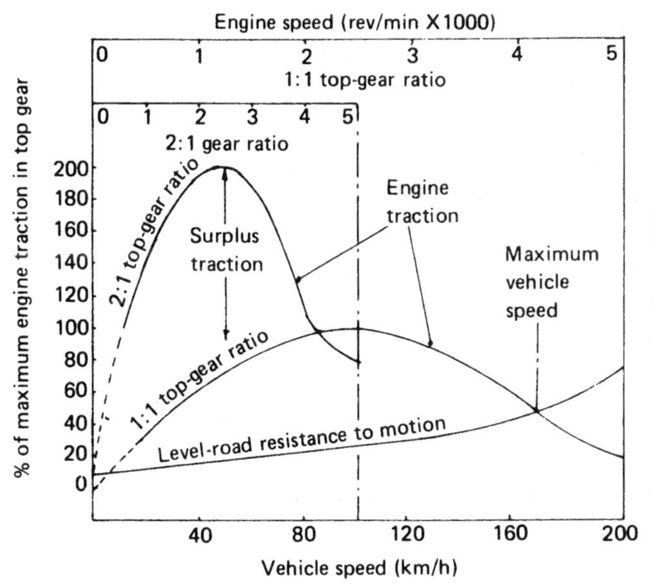

Fig. 3.10(a) Relationship between engine speed, traction, vehicle speed, and road resistance for 1:1 and 2:1 gear ratios

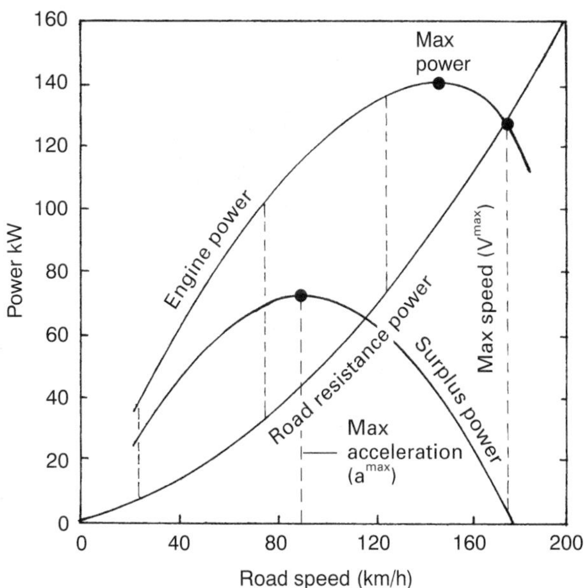

Fig. 3.10(b) Relationship between road power and engine power over vehicle's speed range

separate surplus power curve as shown in fig. 3.10(b). This surplus power, that is, the power for acceleration, is used to increase the vehicle's speed: if the surplus power is small the vehicle's speed increases slowly, conversely if there is a large amount of surplus power the vehicle's speed can increase to a higher speed in a much shorter time. In other words the vehicle's acceleration at any given speed increases as the excess power above the road resistance power becomes greater.

3.1.8 Multi-speed-ratio gearbox

The gearbox is a machine which takes in power from the engine at various speeds and, through a gear train, gives this same power out again to the final drive, but at a reduced speed. The actual ratio of input to output speed will depend on which of the fixed gear-reduction ratios is selected. The object of having a range of gear ratios is to match the various parts of the engine's speed range to the speed of the road-wheels under various road conditions, such as accelerating from a standstill, cruising on the level, climbing hills, or coasting down slopes.

3.2 Four-speed sliding-mesh and constant-mesh gearboxes

The construction and operation of a four-speed gearbox and the individual component details and action will now be considered in some depth. First the sliding-mesh crash gearbox will be considered, followed by the single- and double-stage constant-mesh sliding-dog-clutch gear-engagement arrangements.

3.2.1 Four-speed-and-reverse double-stage sliding-mesh gearbox (fig. 3.11)

A sliding-mesh gearbox is very similar to a constant-mesh gearbox, but differs in the method of engaging the individual gear ratio. The individual components will be examined in more detail in section 3.2.2 on constant-mesh gearboxes.

With the sliding-mesh gearbox, the individual gear ratio chosen is obtained by sliding the actual gearwheel selected axially along the splined main output shaft until it meshes fully with the corresponding layshaft gear cluster. The sliding main-shaft gearwheels and their corresponding layshaft gearwheel clusters are of the spur straight-tooth form. Straight-cut teeth are necessary so that when engaged there will be no side-thrust which would tend to force the meshing teeth apart under load, as would be the case with helical-cut teeth.

The difficulty experienced by the driver with this method of gear engagement is that, unless the speeds of the input and output shafts are skillfully matched first, the sliding teeth will not align and will crash into each other when attempting a gear change.

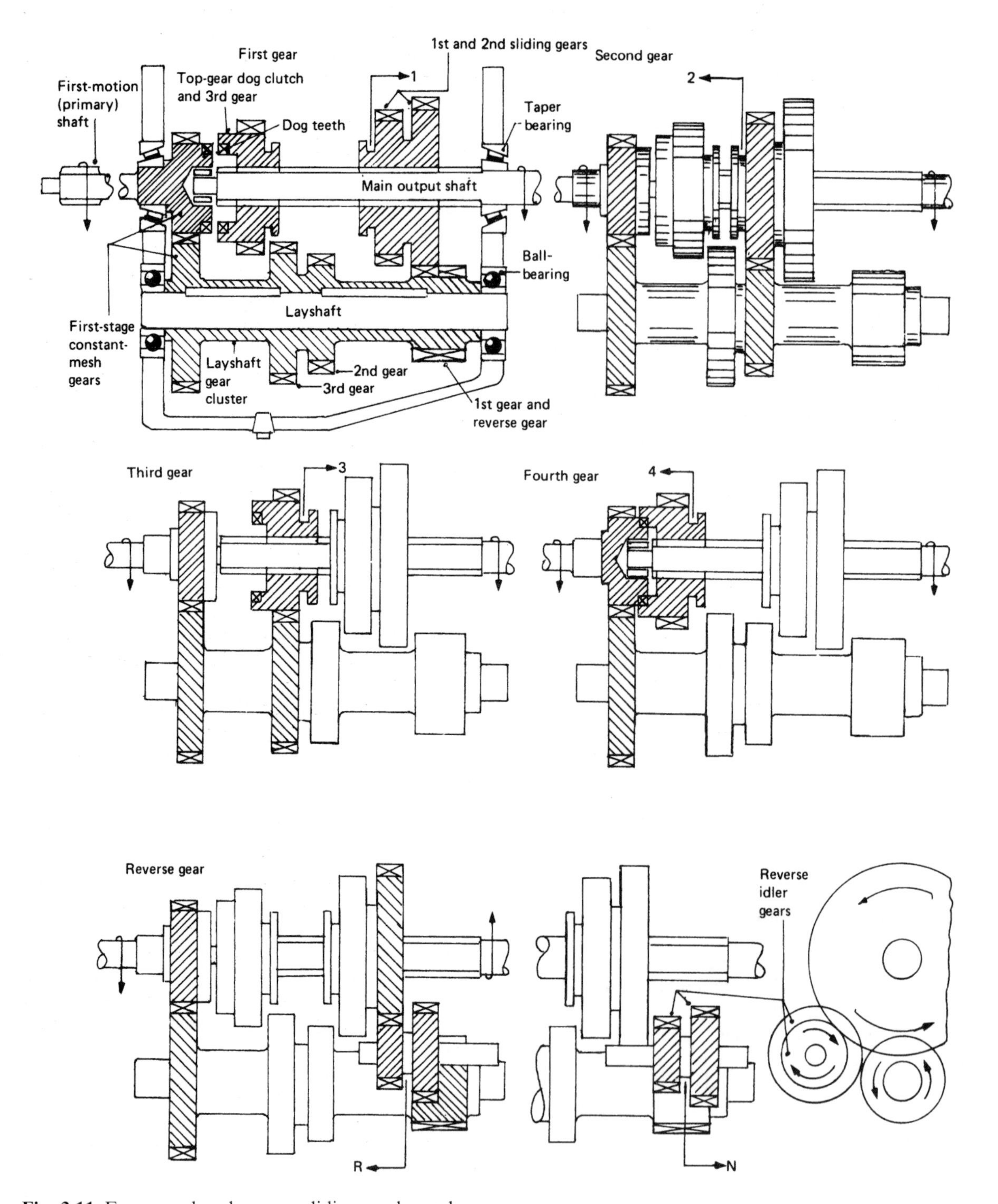

Fig. 3.11 Four-speed-and-reverse sliding-mesh gearbox

Transmission shafts and gears are generally made from low-alloy nickel–chromium–molybdenum steels.

The application of this type of gearbox is now restricted to certain commercial-vehicle boxes where a large number of close gear ratios are required in a compact form.

3.2.2 Four-speed double-stage constant-mesh gearbox (fig. 3.12)

First-motion (primary) input shaft This shaft is splined at the flywheel end and has a first-stage constant-mesh helical gearwheel and a fourth-gear toothed dog clutch formed on it at the gearbox end. It is supported at the flywheel end by a small bush or ball-bearing housed inside a recess machined in the crankshaft end-flange and at the gearbox end by a ball-bearing or taper-roller bearing.

Layshaft and layshaft cluster gears The layshaft cluster gears are held rigidly together. The gears may be cast or forged as a one-piece unit for small and medium-sized gearboxes or, for larger heavy-duty gearboxes, the gearwheels may be machined separately and then held together on a common splined layshaft.

With one-piece laycluster gears, the supporting layshaft may be a force fit at its ends in the gearbox housing and supports the laycluster gears on needle-roller bearings recessed in the ends of the gear-cluster. Thrust washers are provided between the gearcluster and the gearbox housing to absorb any side-thrust generated by the helical gears and during engagement. With large heavy-duty gearboxes, the splined layshaft will have ball- or taper bearings fitted at its ends, and these will be located within the gear-housing itself.

Main or output shaft This shaft has stepped-diameter sections, some having a plain polished surface finish to allow various gears to revolve relative to this shaft, while other sections are splined to provide the drive path from the gears to the constant-mesh sliding-dog-clutch inner hubs.

This shaft supports the first, second, third, and reverse final output-reduction gearwheels. These gearwheels are free to revolve relative to this shaft and are in constant mesh with the laycluster gearwheels. In addition this shaft carries the first/second-, and third/fourth-gear sliding-dog-clutch inner hubs which are fixed to the shaft by splines.

To support the assembly of main shaft, output gears, and hub, one end of the shaft has a reduced-diameter-spigot plain bearing surface machined on it which fits into a recess in the first-motion-shaft gear end with a needle-roller bearing between them. The other (output) end of the shaft is supported by either ball- or taper bearings located in the gearbox housing.

Sliding dog clutch This is a positive locking device which enables the driver to connect the power flow from the first-motion shaft to the output shaft when the friction clutch has disengaged the gearbox from the engine.

This dog clutch consists of an inner and outer hub. The inner hub is fixed by splines to the output main shaft and has external splines machined around it. The outer hub has a single groove formed round the outside to locate a selector fork and is internally splined to mesh with the exterior splines of the inner hub.

When a gear is selected, the speed of the input shaft and that of the output shaft are initially unified – that is, brought to a common speed or equalised – either by allowing the engine speed to drop when changing up or by revving the engine slightly when changing into a lower gear. The outer hub is then pushed over the dog teeth cut on the side of the particular gear chosen. This method of engagement provides a positive means of transmitting power through the compound gear train to the output main shaft. (An alternative version of a sliding dog clutch is shown in figs 3.20 and 3.21.)

Power path for first-, second-, and third-gear selection The power flow passes from the input first-motion shaft to the laycluster gear via the first-stage constant-mesh gear. The power path then divides three ways to the main shaft, via first, second, and third output gearwheels. In the neutral position these three output gearwheels will be driven by the meshing laycluster gear; but they will only be able to revolve about their own axis relative to the main shaft, and the main shaft itself will not revolve.

Each individual gear ratio may be selected by sliding the outer dog-clutch hub towards and over the dog teeth forming part of the particular gearwheel required. This engages and locks the selected output gearwheel to the output main shaft, so completing the power-flow path.

Top-gear selection Top or fourth gear does not provide a gear reduction but gives a direct

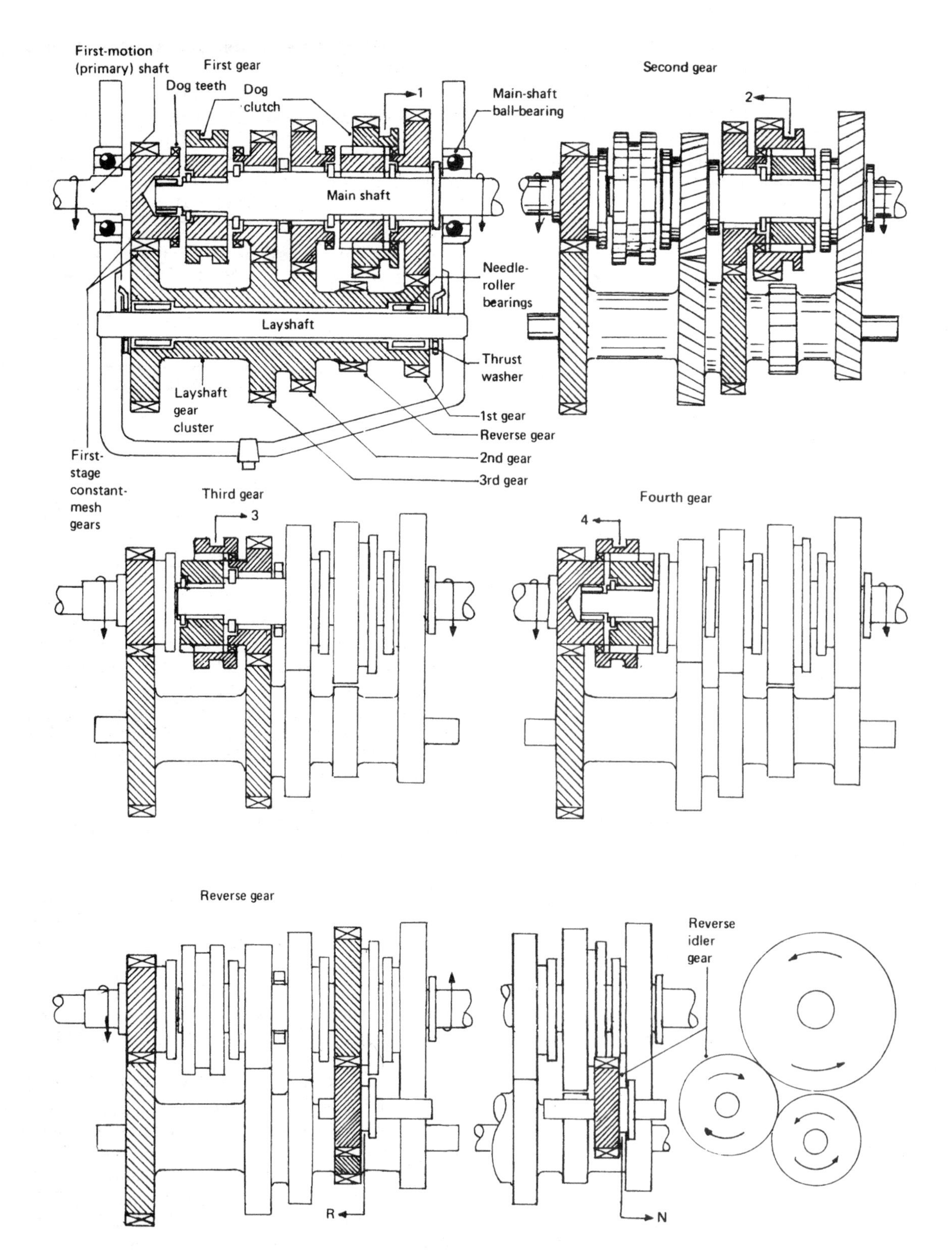

Fig. 3.12 Four-speed-and-reverse double-stage constant-mesh gearbox

power-flow path from the input to the output shaft. When top gear is selected, the third/fourth-gear dog-clutch hub is pushed over the dog teeth cut on the input first-motion shaft, so allowing the power to flow directly from the input to the output main shaft. All the other constant-mesh gearwheels supported on the main shaft will still be able to revolve about their axis at their own speeds relative to the main shaft, as they are not engaged.

Reverse-gear selection Power flows from the input shaft to the first pair of constant-mesh gearwheels and to the laycluster gears, producing the first-stage gear reduction. When the reverse sliding-mesh idler gear is slid into mesh, it transmits motion from the laycluster reverse gearwheel to the reverse idler and to the reverse output gear which forms part of the first/second-gear dog-clutch outer hub. Thus this provides a second-stage gear reduction. In addition to producing the second-gear reduction train, the idler wheel changes the direction of rotation and so provides a reverse gear train.

3.2.3 Five-speed-and-reverse single-stage constant mesh gearbox (fig. 3.13(a)–(f))

This arrangement has only a primary input shaft and a secondary output shaft and all gear reduction takes place in a single stage – that is, between only one pair of gearwheels (fig. 3.13(a)–(f)).

The primary shaft supports a cluster of gears – the shaft and gears may be a single casting or forging, or the gears may be splined or keyed on to the shaft. The output secondary shaft supports constant-mesh gearwheels which are free to revolve relative to it when not engaged.

Power path for first-, second-, third-, fourth- and fifth-gear selection The power flow passes from the input primary shaft, and the power path then divides five ways to the secondary shaft via first, second, third, fourth and fifth output gearwheels. In the neutral position, these four output gearwheels will be driven by the meshing primary-shaft gear cluster, but they will only be able to revolve about their own axis relative to the output secondary shaft.

Each individual gear ratio may be selected by sliding the appropriate outer dog-clutch hub towards and over the dog teeth forming part of the particular gearwheel required. This engages and locks the selected output gearwheel to the output secondary shaft, so completing the power-flow path.

Reverse-gear selection Power flows from the input primary shaft to the reverse gearwheel attached to it. When the reverse sliding-mesh idler gear is slid into mesh, motion is transmitted from the primary reverse wheel to the reverse idler and to the reverse output gear which forms part of the first/second-gear dog-clutch outer hub. This produces a gear reduction, the idler wheel changing the direction of rotation, and so provides a reverse gear drive.

Note: in the forward gears, the output-gearwheel direction of rotation is reversed relative to the input shaft, but with the reverse gear the direction of the output reverse gear is the same as that of the input gearwheel.

The difficulty of unifying the speeds of the input and output shafts when changing gear with a constant-mesh gearbox has been overcome by the development of the synchromesh gearbox.

Overdrive gearing The first, second and third gear ratios in fig. 3.13(a)–(c) provide an under-drive (UD) speed reduction at the output final drive pinion since in each case the input primary shaft gears are smaller than the output secondary shaft gears. Fourth gear in fig. 3.13(d) provides a direct-drive (DD) gear ratio of 1 to 1 since both the primary and secondary shaft fourth gears are the same size (same number of teeth on gear wheels). Fifth gear ratio in fig. 3.13(e) provides an over-drive (OD) speed increase at the output pinion since the input primary shaft gear is larger than the output secondary shaft gear. This means that the output speed from the gearbox is higher than the crankshaft speed. In other words the engine's revolutions per minute at a given vehicle speed in overdrive fifth gear will be reduced compared with when direct-drive 1 to 1 gear ratio in fourth gear is engaged.

The advantage of the overdrive is that the vehicle can be driven at a higher speed with the engine's maximum rotational speed (rev/min) reduced by anything between 10% and 30% while the overdrive gearing is selected (see graph in fig. 3.13(g)). Thus overdrive reduces engine wear, noise, improves fuel consumption and permits the vehicle's maximum speed to be raised when travelling on the level or on a down gradient without the engine rotational speed exceeding its upper safe limit. Note that the maximum engine power developed is the same in each gear ratio.

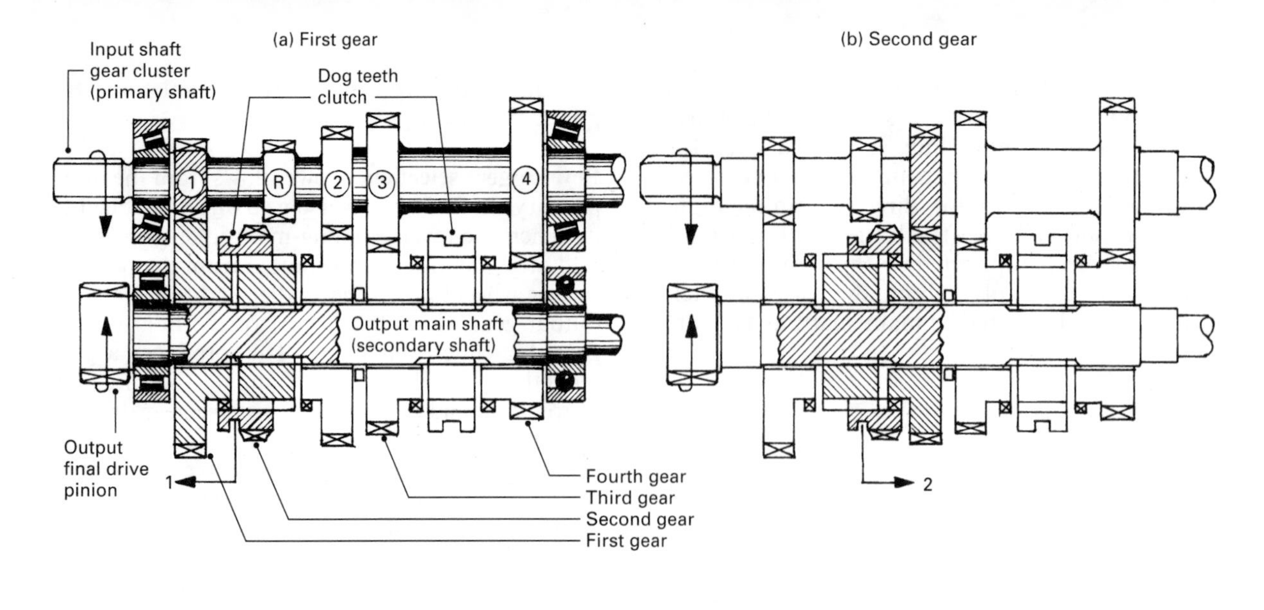

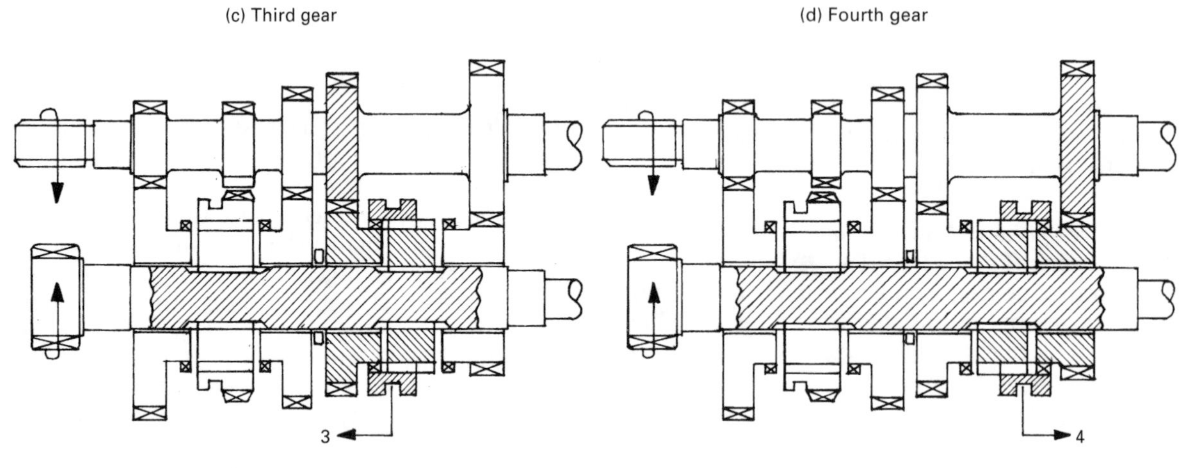

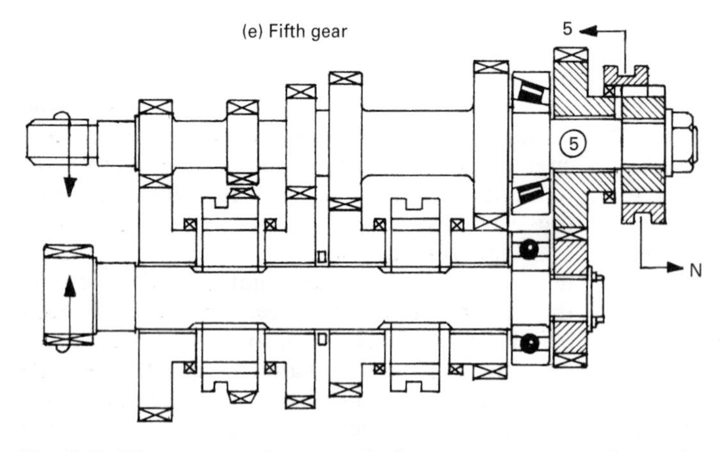

Fig. 3.13 Five-speed-and-reverse single-stage constant-mesh gearbox

(f) Reverse gear

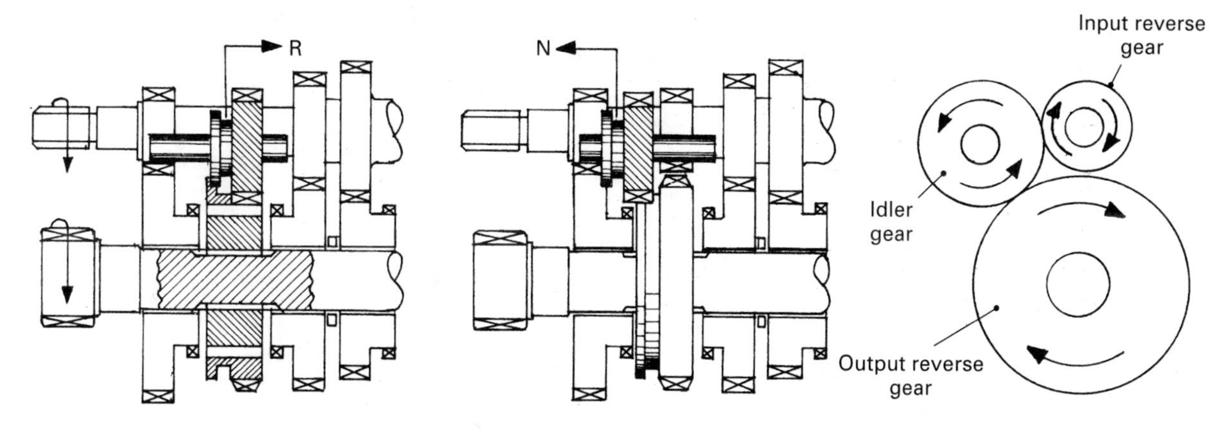

(g) Relationship between under and over gearing on vehicle performance

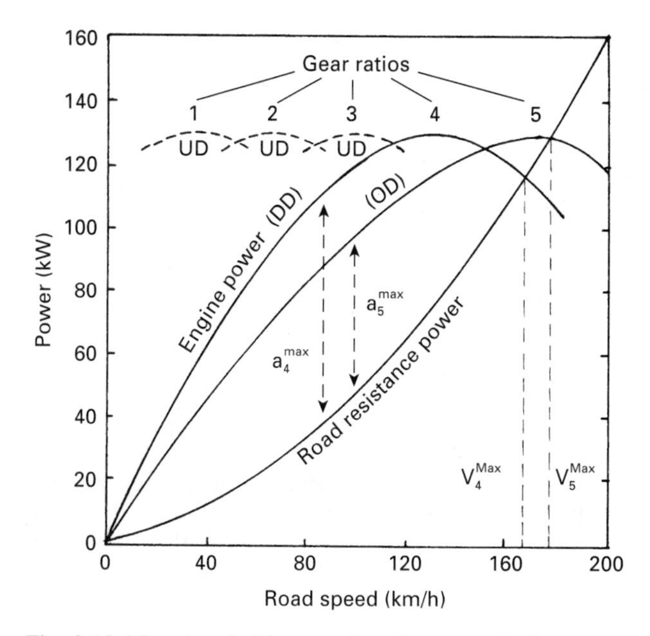

Fig. 3.13 (*Continued*) Five-speed-and-reverse single-stage constant-mesh gearbox

The disadvantage of the overdrive compared with the direct-drive gear ratio is that there is less surplus power (see graph fig. 3.13(g)) between the power developed by the engine and the road resistance power curves so that the power available for maximum acceleration will generally be far less.

3.3 Rod-and-fork gear selectors (fig. 3.14)

A gear-selector mechanism is a device which allows the driver to select and engage a particular gear ratio. The selector engagement is achieved via three guiding selector rods supported at their ends either in the actual gearbox housing or in the selector cover housing. Sliding with or over these rods are selector forks which fit over saddle-like grooves machined on the outer sliding-dog-clutch hub. Pushing a selector fork towards one of the constant-mesh gearwheels will move the hub of the particular dog-clutch hub over the exposed ring of dog teeth machined on the gearwheel. When the hub and gearwheel dog teeth are engaged, the drive is coupled from the layshaft cluster gear to the main output shaft, thus completing the power-flow path through the gearbox.

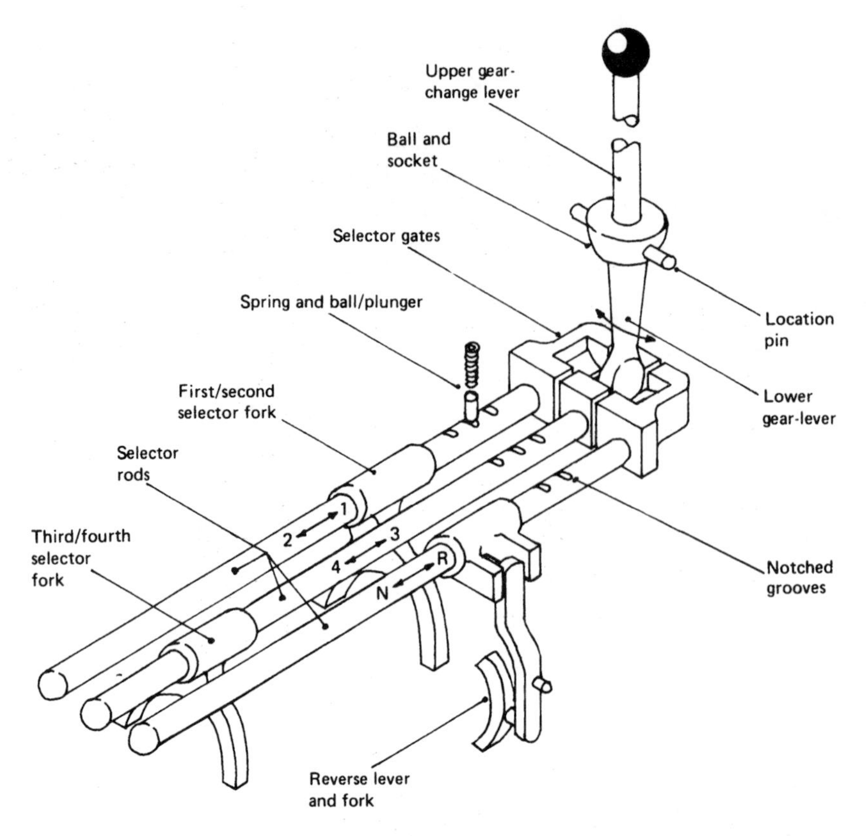

Fig. 3.14 Rod-and-fork gear-selector arrangement

To engage neutral, the selector forks are shifted so that the outer dog-clutch hubs are directly over the inner hubs, so that each gearwheel on the main shaft is free to revolve independently relative to the main shaft.

3.3.1 Gear-change-lever action (fig. 3.14)
The gear-change lever is a forged steel rod or stick with a control knob screwed on its upper end. The lower end has a spherical shape with flats machined on each side. A swivel ball is formed on the lever at about one fifth of its length from the lower end, to provide a pivot point for the lever gear-selection action.

The lower spherical end fits in the selector-rod gate with the machined flats lying parallel with the rods. The swivel-ball portion of the lever sits in a socket machined in the gear-change-lever cover casting. Fitted over the lever swivel ball are a spring-loaded dust-cover and a spherical washer which holds the lever ball in position (fig. 3.15).

To select and engage a gear ratio, the gear-change lever is tilted at right angles to the rods so that the spherical lever end sweeps across the three selector gates and aligns with the selected rod. The gear-change lever knob is then pushed parallel to the selector rods and in a direction which, via the selector fork, shifts the dog-clutch outer hub over the dog teeth formed on the side of the selected gearwheel.

3.4 Spring-loaded ball or plunger and selector-rod grooves (fig. 3.15)
The selector rods or forks are located and held in their selected position by spring-loaded balls or plungers and grooves (notches). The action of the spring load forces the balls into their respective grooves when they align.

Three grooves or notches are cut on each of the two forward-gear selector rods and two on the reverse selector rod. When the selector rods and their rigidly attached forks are shifted one way or the other, these grooves will eventually align with spring-loaded balls or plungers located in the gearbox housing assembly. On the forward-gear selector rods the central groove provides the

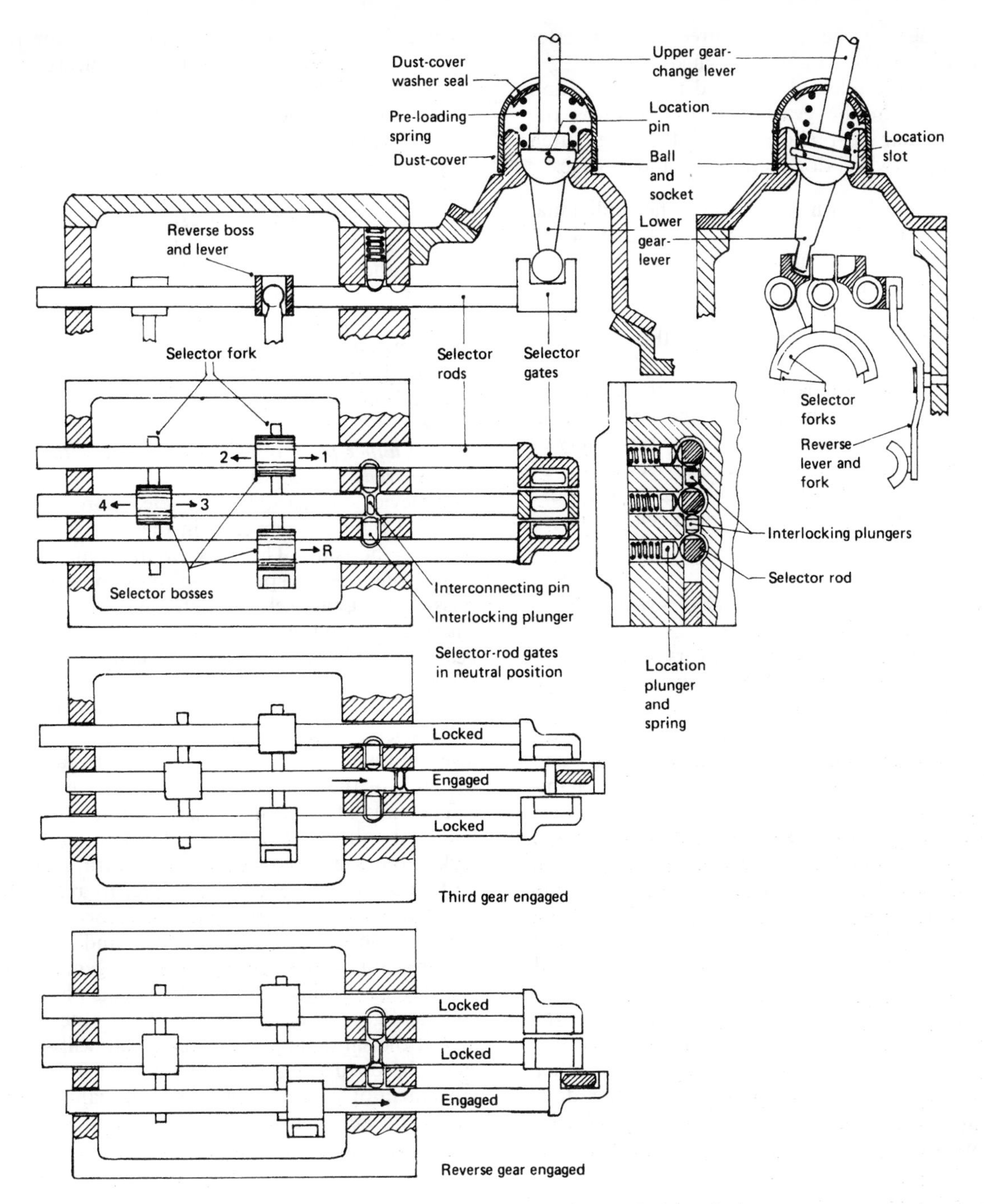

Fig. 3.15 Three-rod gear-selector mechanism with plunger-and-pin interlocking device

neutral position and the grooves on either side will be for the engaged position for one of the gear ratios, i.e. first or second or third or fourth gear. It should be noted that the reverse selector has only a neutral and a reverse position.

3.5 The need for a gear interlocking device

The action of changing the gear ratio consists of two separate operations:

i) selection of the particular selector-rod gate by swinging the gear-lever lower end across the channel formed by the three selector gates until the desired gate is aligned,

ii) sliding the selected gate and rod axially and parallel to the gearbox shafts towards the particular gear chosen until the dog-clutch and gearwheel dog teeth mesh and engage.

Unfortunately, when selecting the individual gates it is possible for the rectangular-sectioned lower lever with its semi-rounded tip to be placed or aligned between two selector gates and then for the lever to force both gates to engage two different gears simultaneously. Obviously this situation would provide two power-flow paths which would jam the whole gear pack and, with the vehicle in motion, would smash and strip the weakest gear teeth from their roots. To prevent the possibility of this occurring, all gearboxes employ some sort of safety interlocking device which prevents more than one gear-ratio power-flow path being engaged at any one time.

3.5.1 Plunger-and-pin interlocking device with three selector rods (fig. 3.15)

This arrangement uses two plungers and an interconnecting pin to allow only one gear ratio to be engaged at any one time.

The mechanism consists of three selector rods. The central rod has a radial hole drilled through it, the mouth of each side of the hole being chamfered, and fitted inside this hole is an interconnecting pin. The selector rods on each side of this central rod have single grooves cut facing the middle rod. Holes drilled in the gearbox casing or cover housing at right angles to the selector rods communicate between the rods and hold two interlocking balls or spherical-ended plungers.

When a gear ratio involving one of the outer selector rods is selected, the particular gear-ratio selector rod will be moved one way or the other from the neutral position to engage the dog-clutch outer hub with the dog teeth of the selected gearwheel. As this selector rod moves, the contacting interlocking plunger with its spherical ends is pushed outwards and away from the rod as the groove is misaligned with the plunger, so that the plunger end now contacts the straight shank of the rod. This movement of the interlocking plunger at right angles to the selector rod pushes the other end of this plunger into the chamfered interconnecting-pin hole and relays this movement to the pin and to the other interlocking plunger which now will be partially located in the groove in the third selector rod. The net result will be that the two selector rods not in use will be wedged in the neutral position by the plungers and pin while the first rod is free to be shifted to and fro unrestricted.

Similarly, if a selected gear ratio requires the central selector rod to be shifted, both interlocking plungers will be pushed outwards from the central rod as its straight shank is made to contact the plunger spherical ends; consequently the other ends of both plungers will now be positioned and wedged against the grooves of the outer selector rods. This again allows only one rod to be moved, while locking the other two rods in their neutral position.

3.5.2 Caliper-plate interlocking device with three selector rods (fig. 3.16)

This device consists of three selector rods which are held rigid at their ends in the gearbox casing or cover housing. The selector forks are free to slide along their respective selector rods through bosses which have selector gates machined on top. These gates are there to locate and align both the gear-lever spherical end and the caliper plate (or 'swing plate').

When assembled in the neutral position, all three selector gates will be aligned, and the gear-lever end will be situated in the middle gate with the ends of the caliper plate on each side of it.

The ends of the caliper plate will fit across the first and third gate so that these two gates and forks cannot be engaged, but the central gate will be free to shift to and fro to select its respective gear ratio involving the middle selector rod.

When one of the outer gates and forks is required for gear selection and engagement, the gear-lever is pushed over or tilted. This places the spherical gear-lever end in one of the outer gates, and at the same time the caliper plate will swivel, since the open ends of the plate must follow the movement of the gear-lever. Thus the caliper plate will again be jammed across two of the selector gates, preventing there being any accidental shifting and engagement of other gear ratios.

3.5.3 Caliper-plate interlocking device with single selector rod (figs 3.17 and 3.18)

With this layout there is only one selector rod or rail, and fixed to it is a selector boss and striking pin. Unlike the conventional three-selector-rod arrangement, where selection is obtained with a forward or backward movement of the selector fork, this single-rod method adopts both

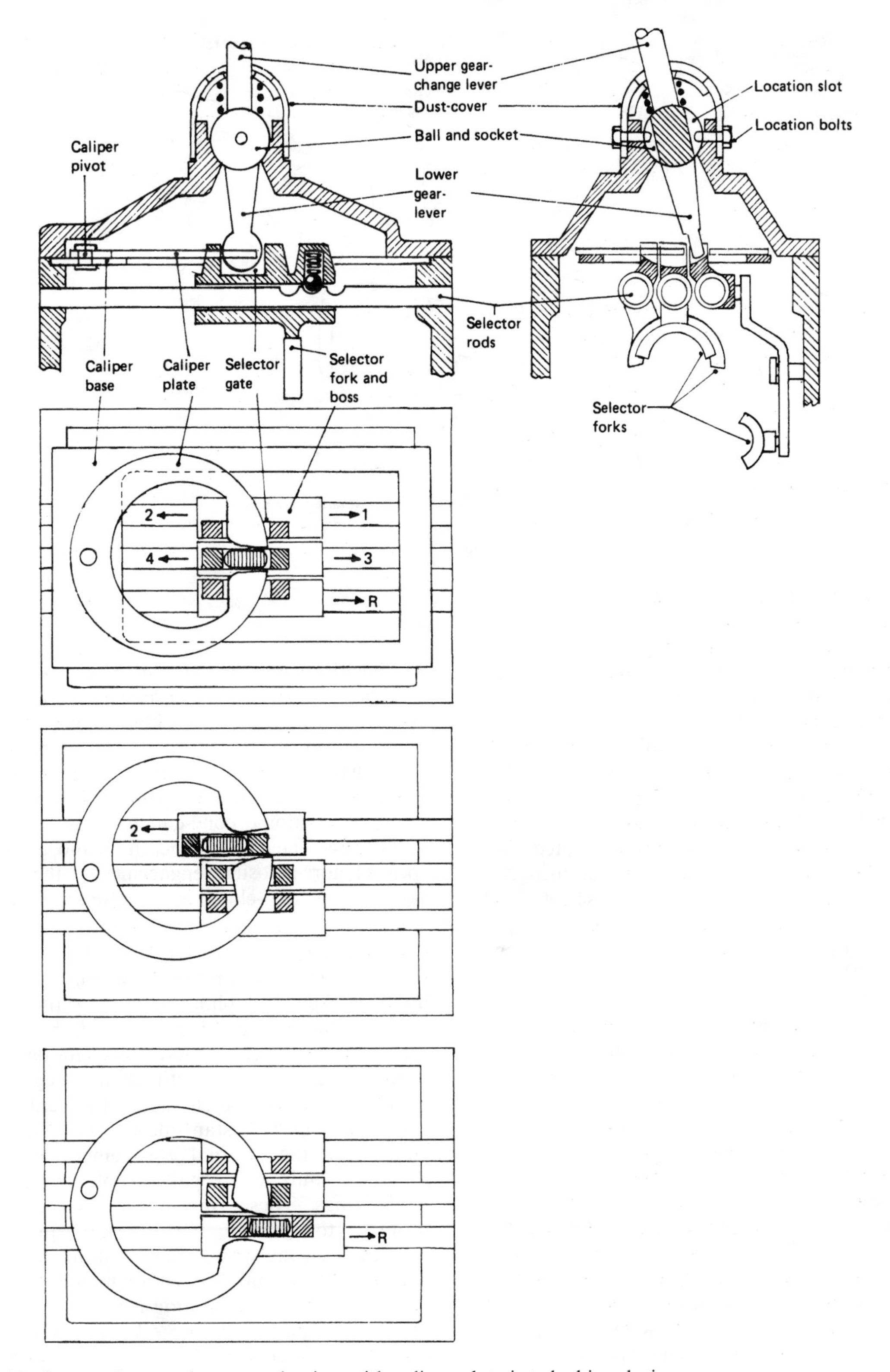

Fig. 3.16 Three-rod gear-selector mechanism with caliper-plate interlocking device

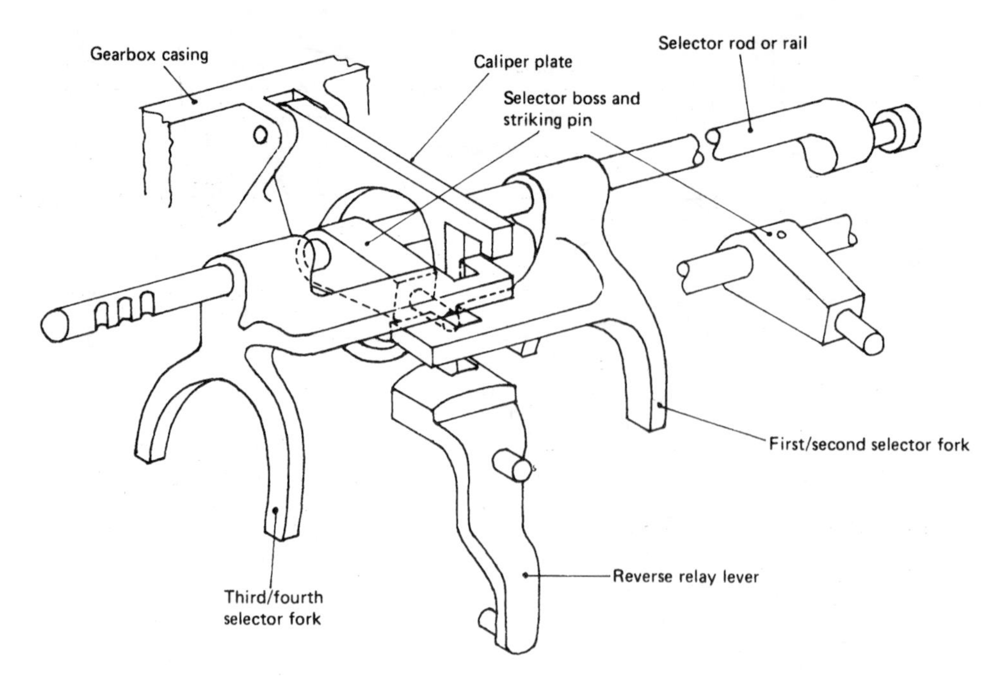

Fig. 3.17 Single-rail gear-selector arrangement

a to-and-fro movement and a rotary twisting motion for gear-ratio selection and engagement.

There are two selector-fork-and-gate bosses which are free to slide on the rail, the third selector gate being provided by forming part of the reverse relay lever.

Situated at right angles and half-way along this rail is the caliper plate, which is pivoted and supported at one end by the gearbox housing.

To select a particular gear ratio, the selector rail is twisted or rotated slightly until the striking pin aligns with the selector gate required. At the same time the caliper plate is compelled to follow this movement since its free ends are adjacent to and on each side of this pin. The thickness of the caliper plate is such as to allow it to just slip across the selector gates so that, whatever gate the striking pin is aligned with, the other two gates will have a portion of the caliper plate jamming them and hence the other selector fork or the reverse relay lever is prevented from moving along the rail.

To change to another gear ratio, the selector rail is moved backwards or forwards until the striking pin is in the neutral position – that is, with the three gates level with each other. The rail is then rotated until it aligns with the selected gate, followed by the pin pushing one of the forks or relay levers into an engaged position.

3.6 Positive baulk-ring synchromesh unit

The purpose of the synchromesh gearbox is to enable the gear-ratios to be changed smoothly and quietly with very little skill on the part of the driver when the vehicle is in motion. A synchromesh gearbox is a constant-mesh gearbox which utilises a frictional cone-clutch device to initially unify the selected gear-ratio shaft and gear-wheel speeds before a positive engagement with a conventional sliding dog-clutch is permitted to take place.

Construction (fig. 3.19(a)–(c)) This type of synchromesh unit comprises an inner and outer hub attached by splines to the gearbox main output-shaft. Either side of the hub assembly are constant-mesh gear-wheels mounted on needle-roller bearings with their integral dog-teeth and conical-clutch facing the central hub. Separating the constant-mesh gear-wheel dog-teeth from the internal engagement splines of the outer hub are a pair of phosphor-bronze baulk rings. These rings have conical internal profiles to match the conical projections machined on the side of the constant-mesh gear-wheels and external dog-teeth which in turn have similar profiles to the dog-teeth and internal-splines of the constant-mesh gear-wheels and outer hub, respectively. Three equally spaced slots are formed around the inner hub to locate and

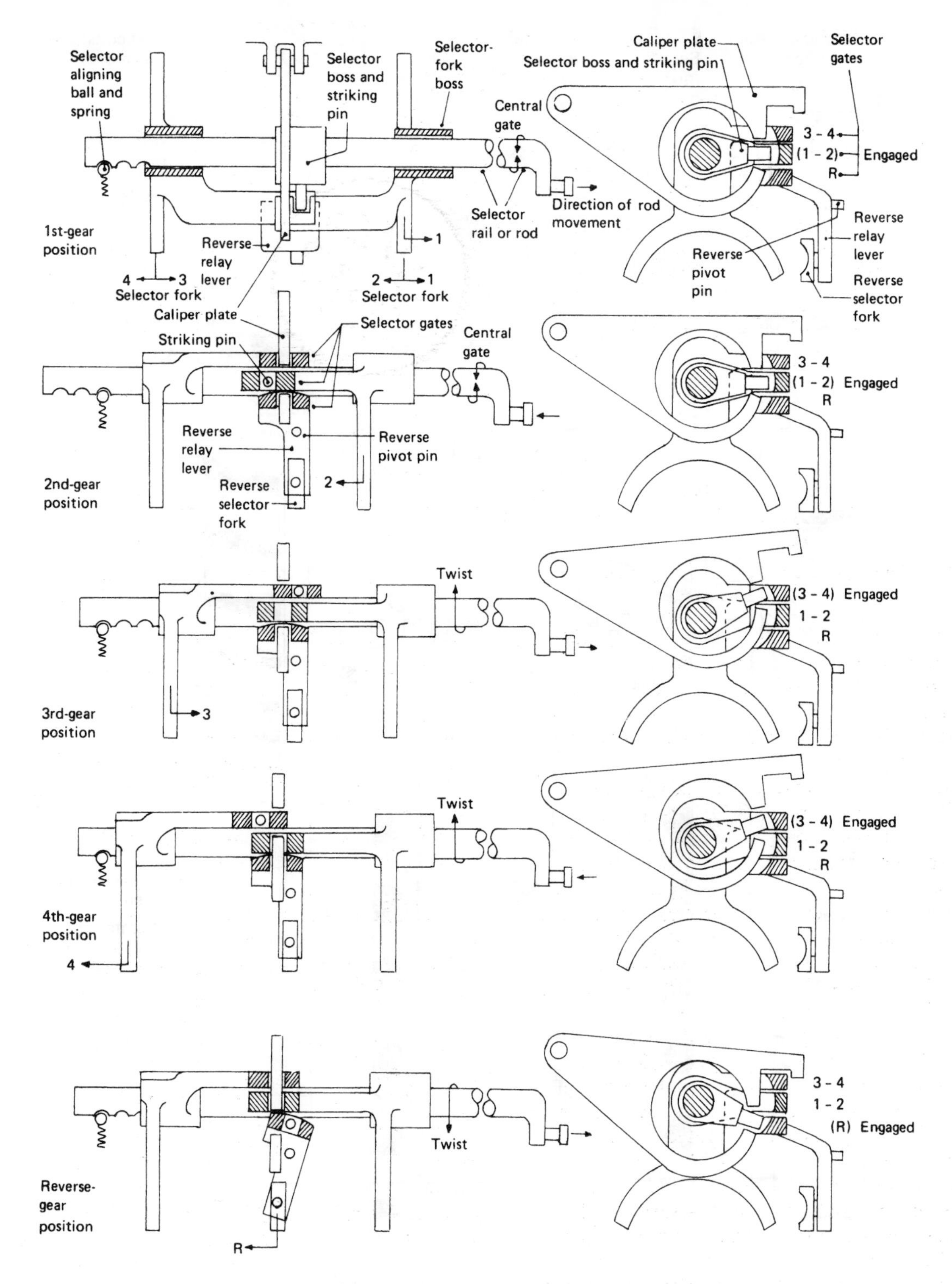

Fig. 3.18 Gear-selection mechanism with single-rod and caliper interlocking device

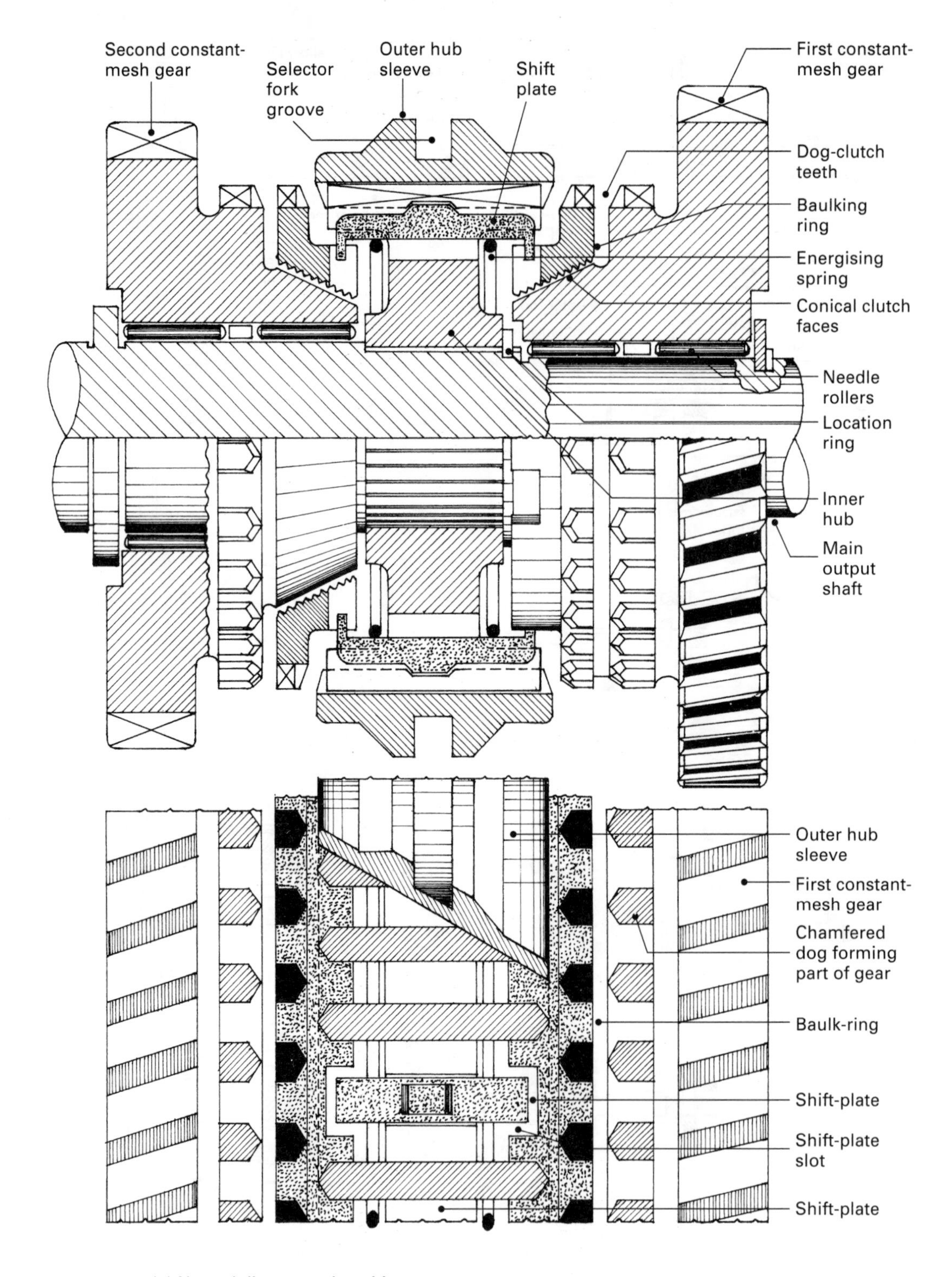

(a) Neutral disengaged position

Fig. 3.19 Positive baulk-ring syncromesh unit

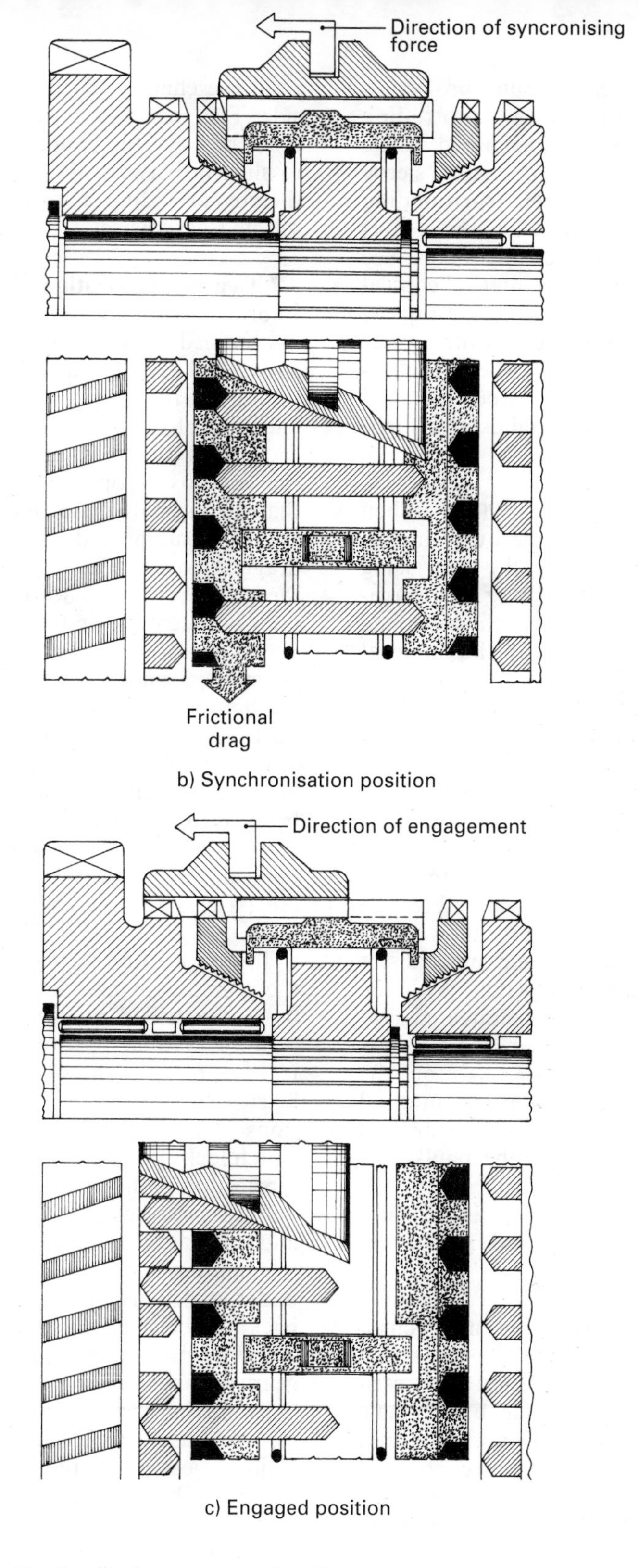

Fig. 3.19 (*Continued*) Positive baulk-ring syncromesh unit

restrain the three shift-plates; the outer projection of these shift-plates is formed with a protruding hump which fits into a groove machined inside and around the periphery of the outer sleeve hub. In addition to the dog-teeth surrounding the baulk rings, each ring end-face has three elongated slots which align with the shift-plates. Vee grooves are also machined on the internal conical surface of the baulk rings; these grooves rapidly scrape the oil film from the tapered profile projection on the side of the gear-wheel when synchronisation is to take place. A pair of circular shaped wire energising springs support and exert a slight radial thrust onto the shift-plates.

Neutral disengaged position (fig. 3.19(a)) When the gear selector is in neutral, the outer hub is held in its central position by the energising spring's radial outward thrust pushing the protrusion (hump) on the outer face of each shift-plate into the circular groove cut into the internal splines of the outer sleeve hub.

Synchronisation (fig. 3.19(b)) When the driver selects the desired gear with the vehicle in motion, the selector-fork forces the outer sleeve-hub to the left (see fig. 3.19(b)); at the same time the shift-plate pushes the baulk-ring against the conical profile of the constant-mesh gear-wheel. Immediately the frictional torque of the male and female conical clutch drags the baulk-ring round with the gear-wheel until the shoulder of the elongated slot in the baulk-ring hits the shift-plates. With the shift-plate offset to the baulk-ring elongated slot, any further movement of the outer hub-sleeve is prevented by the chamfered dog-teeth on the baulk-ring aligning with the chamfered internal splines of the outer hub. However, the driver's force on the selector fork pushes the baulk-ring female cone further into contact with the constant-mesh gear male-cone, and as a result, rapidly unifies the main-shaft and gear-wheel speeds.

Gear engagement (fig. 3.19(c)) Once the synchronisation of the main-shaft and constant-mesh gear speeds occurs, due to the frictional resistance generated between the male and female members, the unified speeds of the shaft and gear remove the frictional drag (since no drag can occur if both rotating members are moving at the same speed). Hence the force applied to the selector-fork to move it to the left is now sufficient for the chamfered splines of the hub to push the chamfered baulk-ring dog-teeth to one side. This instantly permits the sliding splines to move between the baulk-ring dog-teeth, and then to slide in between the constant-mesh dog-teeth, thereby providing a positive engagement of the selected gear-wheel.

3.7 Gearbox lubrication (figs 3.20 and 3.21)
Lubrication for the moving parts in the gearbox is obtained by partially filling the box with the correct oil through a level plughole situated on the side of the casing, until the oil starts to drain back out of the hole. The plug is then screwed in to prevent the remaining oil spilling out when the gearbox is operating. The oil level is such that the layshaft or secondary-shaft cluster gears are just submerged, so that the oil will be dragged round with the gearwheel teeth when the gears revolve. The oil will then spread and flow between the individual gearwheels, output main shaft and primary shaft, dog-clutch assemblies, and support bearings. The selector mechanisms will be lubricated by oil splashing up from the gear teeth (fig. 3.20). A drain-hole and screw plug are usually included at the lowest point in the oil-bath casing, so that old oil can be drained off. Overfilling the gearbox will create a pumping action which will generate a pressure build-up within the box and will eventually force oil past the input-shaft and output-shaft oil seals.

To help oil find its way between the shafts and the gears revolving relative to them, two or three holes are drilled radially in each gearwheel (fig. 3.20). It should be observed that there is only relative rotation between the gears and shafts when they are disengaged under no-load conditions – when a particular gear is engaged, the dog clutch locks the gear to the shaft so that very little lubrication is then necessary.

Bush wear between the main-shaft gearwheels and the main shaft is considered to be mainly due to minute relative motion between the two while under load. This is explained by the slight angular backlash between the meshing dog-clutch teeth and the initial diametrical clearance between the gearwheel bushing and the shaft.

In special cases, heavy-duty commercial vehicles may use a forced-feed lubrication system in which a gear pump pressurises oil along an axial drilling in both the primary and main shafts (fig. 3.21). Intersecting this central drilling are radial holes which feed oil outwards between the shaft and the gears, both when engaged and when

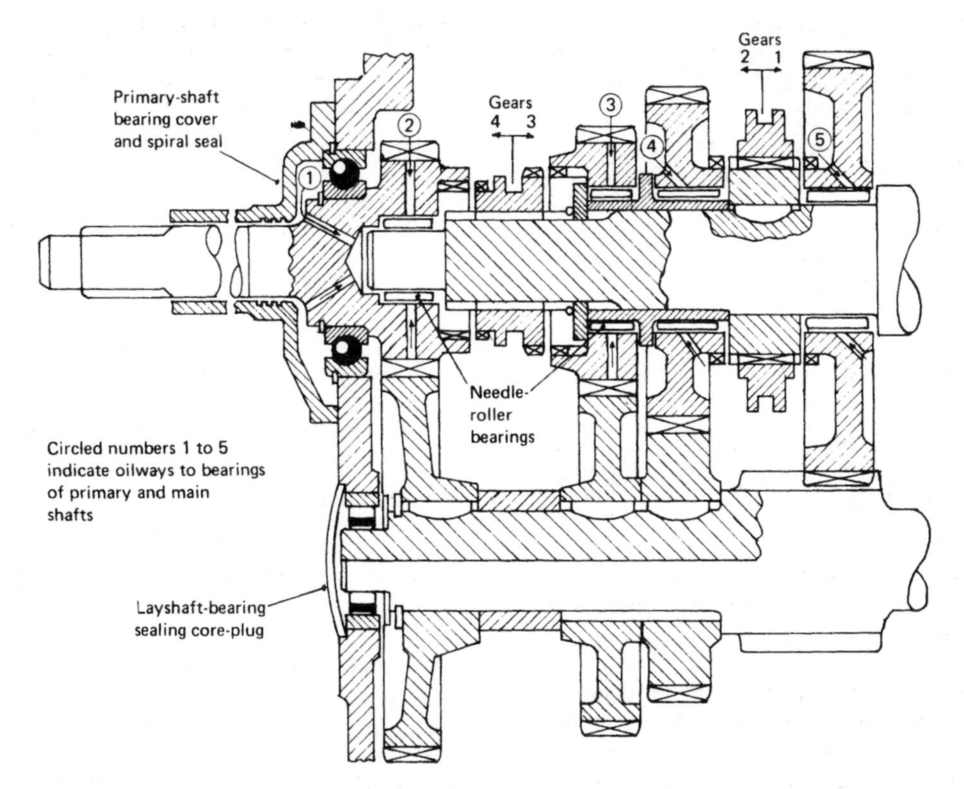

Fig. 3.20 Splash-feed lubrication system for gearbox

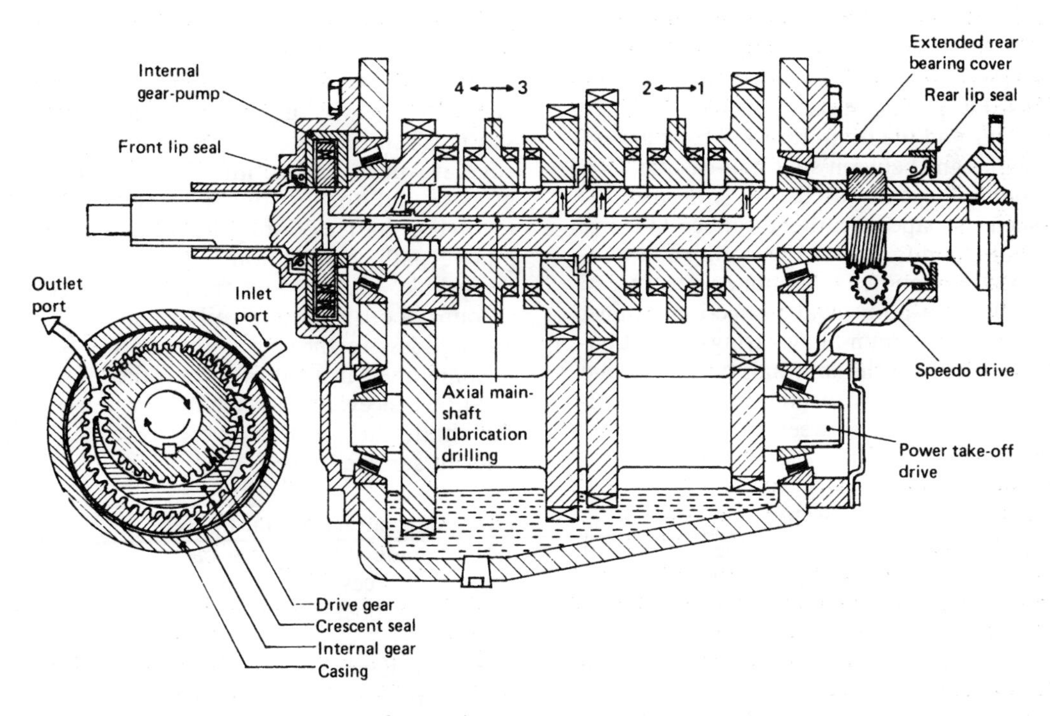

Fig. 3.21 Forced-feed lubrication system for gearbox

disengaged. Generally there is no need to pressurise the layshaft and gear cluster.

With front-wheel-drive cars, a single oil supply for both the gearbox and the final drive is generally provided. The lubricant used must be the one recommended, and its viscosity may depend on whether it is to be used as a common oil for both the gearbox and the final-drive crown wheel and pinion or just for straight or helical gear teeth. In the former case a higher-viscosity oil may be selected; conversely, with synchromesh gearboxes, a thinner oil is generally preferred to give good clean quiet gear changes.

3.8 Gearbox oil-leakage prevention

A gearbox is assembled from many components and must be externally leakproof. The sealing of these components may be divided into two groups.

i) Components which are joined together by machined pressure-face joints. These mating faces use static seals.
ii) Components which have rotary motion relative to the gearbox casing or housing and have an external drive. These use dynamic seals.

3.8.1 Static seals

A static seal is sandwiched between two pressure faces so that when they are clamped together it deforms to take up exactly their surface roughness, waviness, and misalignment within set limits. These seals are used to prevent leakage between the main gearbox casing and the front, top, and back covers, or in some designs between split half-casings.

The common seals used are non-metallic gaskets such as paper, synthetic-rubber O-rings, and joint compounds of both the setting and non-setting types. Sometimes compounds are spread on gaskets to give additional protection agianst leakage. With aluminium-alloy casings – particularly if the casing is of the split-half type – the two halves are bolted directly together with only sealing compound as an in-between packing.

3.8.2 Dynamic seals

To prevent oil leaking out between a rotating shaft and its bearing housing, some method of sealing is necessary. This can take the form of either a spiral-thread clearance seal or a contact radial-lip seal.

Spiral-thread clearance seal (fig. 3.22) With this form of seal, a parallel portion of a revolving shaft fits inside an extended sleeve housing, and an internal spiral (or, more accurately, helical) groove is machined inside the sleeve bore (fig. 3.22), there being a small clearance between the shaft and the sleeve.

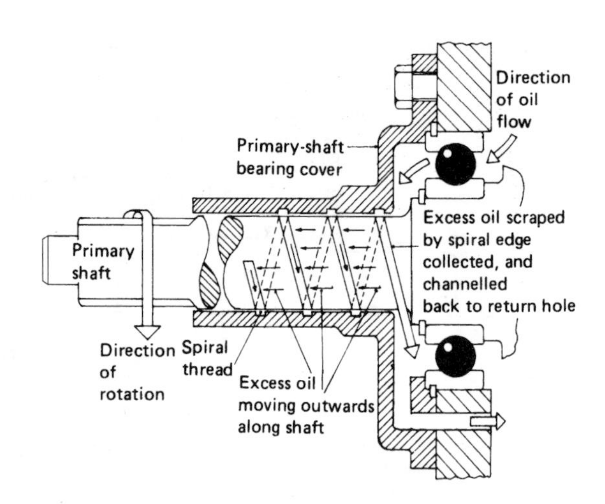

Fig. 3.22 Dynamic spiral-thread seal action

The primary-shaft bearing is lubricated by splashing of oil against the side of the ball-bearing. Some of this oil will creep between the inner and outer bearing race tracks and will drip into the outer bearing housing on to the primary shaft. When sufficient oil is trapped between the shaft and the housing-sleeve bore, the revolving shaft will generate a pumping action which will further move the oil along the shaft. When the oil encounters the helix or spiral edge, it will be pushed into the spiral rectangular groove, and the existing oil then covering the revolving shaft will drag the oil along the spiral-groove path, back towards the bearing and the drain-hole. Each groove turn will collect a given amount of oil, but excess-oil overspill will become less and less further along the shaft.

Sometimes, instead of having the spiral groove inside the sleeve, it is machined on the outside of the shaft, the inside sleeve bore then having a smooth surface finish.

Dynamic contact radial-lip seal (fig 3.23) The seal is made up of three parts: (i) a short steel cylindrical sleeve flanged at one end, (ii) a flexible synthetic-rubber lip seal which is bonded to the steel flanged sleeve, and (iii) a circumferential garter or band spring which is located at the back of the rubber lip (fig. 3.23).

The steel sleeve is a force fit in its housing bore,

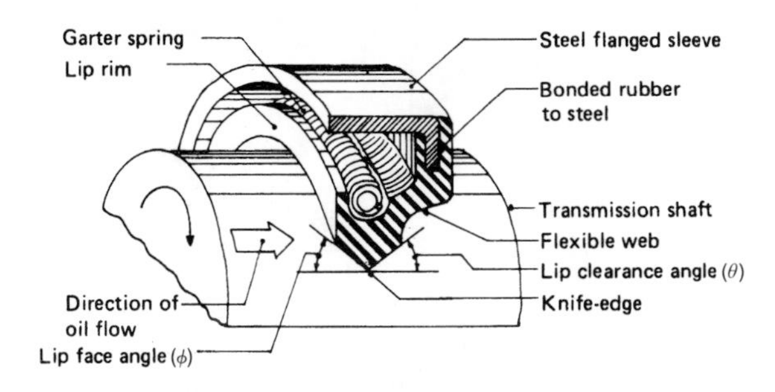

Fig. 3.23 Dynamic lip-seal action

its job being to locate and support the flexible seal. The rubber seal consists of a flexible web which stays concentric with the shaft at all times, even when the shaft is deflected slightly to one side under load. The seal lip has a knife-edge contact at the sealing point and applies light pressure on the shaft. The large included angle between the face angle and the clearance angle (ϕ and θ respectively) stiffens the sealing point against local distortion. To improve the contact between the shaft and the knife-edge rubber lip, a circumferential spring bears on the back of the seal knife-edge rubber and it is located by a moulded deep rim to allow for the contraction action of the spring.

The knife-edge rubbing contact with any oil film formed on the shaft surface tends to create a viscous oil joint which will then reject any further outward leakage.

4 Automatic transmission

Automatic transmission replaces the conventional clutch and manual synchromesh gearbox by a torque converter and a compound epicyclic gear train. The torque converter provides a smooth and automatic take-up of the drive and at the same time multiplies the output torque from the engine, whereas the compound epicyclic gear train gears are permanently in mesh, relying on a series of wet clutch and brake-bands or brake multiplate discs to actuate the engagement and disengagement of the various gear ratios. Automatic selection of gear ratios is achieved by a hydraulic control unit which monitors the driving load and speed conditions; it then directs the various hydraulic servo-units to actuate the appropriate clutch and brake-bands or multiplate discs, so that the necessary gear ratio changes take place to match the engine's output capability and the road resistance at any one time.

4.1 Hydrokinetic fluid coupling (fig. 4.1)

A fluid coupling uses hydrokinetic energy as a way of transferring power from the engine to the transmission so that the drive is taken up smoothly and positively without undue slip. Hydrokinetic fluid couplings, commonly known as fluid flywheels or just fluid couplings, consist of two saucer-shaped discs, an input impeller-member, also known as the pump, and an output turbine-member sometimes referred to as the runner (fig. 4.1). These aluminium alloy members are cast with flat radial vanes (blades) for controlling the flow path of the fluid (fig. 4.2). A core guide-ring is cast with the vanes on the impeller-wheel to establish the circulation of fluid between adjacent pairs of cells; originally a core was also cast on the turbine-wheel, but this produced an excessive amount of turbine-wheel fluid drag at low speed. It was therefore dispensed with in later fluid-coupling designs. There will always be some relative slip between the input impeller-wheel and the output turbine-wheel so that the fluid flow-path established between adjacent pairs of cells will be continuously aligning and misaligning with different cells. Therefore the impeller and turbine wheels are nearly always manufactured with a different number of cells in the two half-members so that two adjacent cells never completely align; this prevents a repetition of jolts as cells break their alignment and new cell passage-ways are formed. A typical figure would be a 51-cell impeller-wheel matching a 45-cell turbine-wheel.

4.1.1 Principle of operation (fig. 4.2)

When the engine is rotating the impeller (pump), this causes the reservoir of fluid stored to about three-quarters of the height of the enclosed casing also to rotate with the impeller-wheel and the slower rotating turbine-wheel since half the fluid is basically confined to the impeller-wheel cells. This subjects the fluid trapped in the impeller cells to centrifugal force which pressurises it and forces it to flow radially outwards from the entrance at A to its exit at B. As the fluid moves further outwards from the centre of rotation it is subjected to a rising centrifugal force, and eventually, when the fluid reaches the outer edge of the impeller at B, it will have acquired a certain amount of fluid kinetic energy, that is hydrokinetic energy. This energy of motion propels the fluid across the flow path where the impeller and turbine cells meet, from the impeller's exit B to the turbine's entrance at C, thus causing it to impinge on the cell walls of the turbine-wheel and in so doing impart its kinetic energy to the turbine-wheel. Consequently the turbine-wheel will now be compelled to rotate though at a much reduced speed. Hence the centrifugal force acting on the fluid within the much slower rotating turbine-wheel cells will be far less, so that the large centrifugal drive force will transfer hydrokinetic energy from the impeller to the turbine wheel, which will push the fluid in the turbine cells inward until it reaches its innermost radii. The fluid at this point then readily flows from the turbine exit at point D to the impeller entrance at point A to repeat the circulation cycle of events.

The greater the difference in speed between the impeller and turbine-wheels, the greater will be the circulation vortex velocity and consequently the greater the hydrokinetic energy imparted to

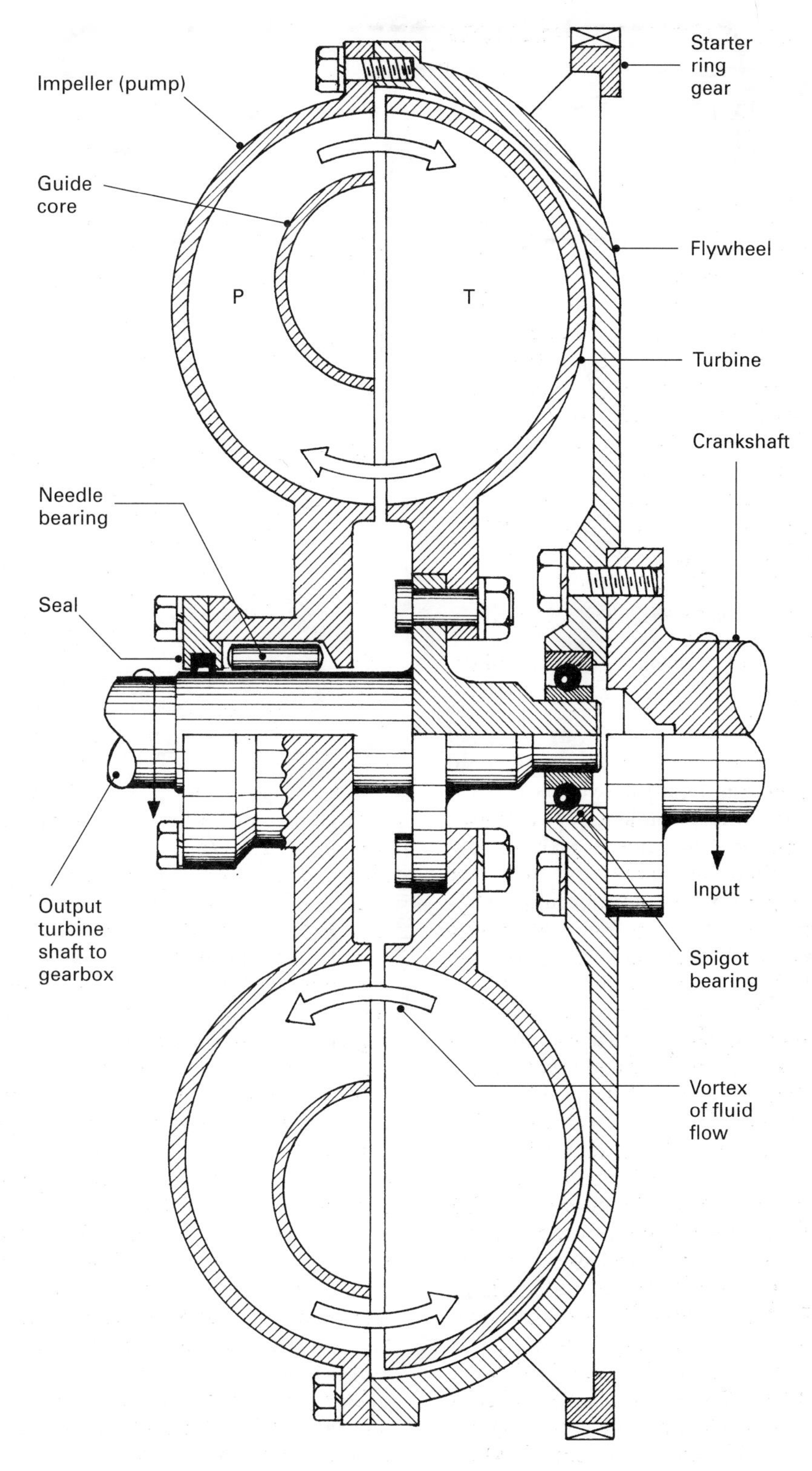

Fig. 4.1 Fluid coupling

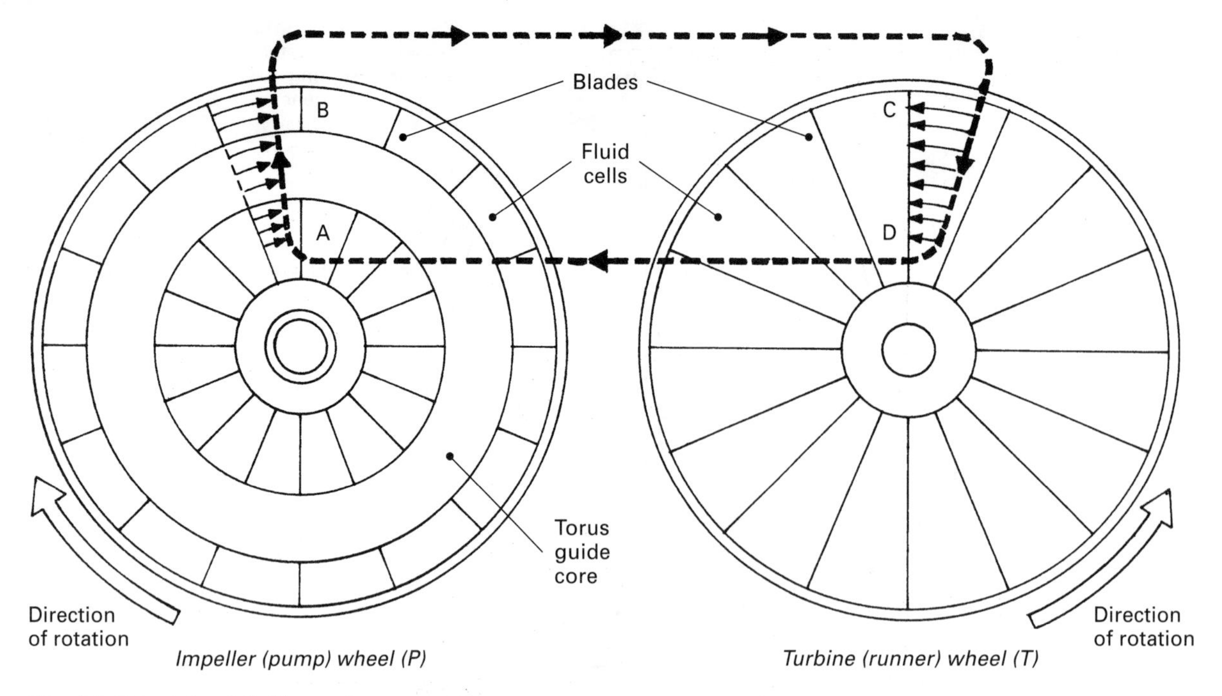

Fig. 4.2 Principle of fluid coupling

the output turbine-wheel. Conversely, as the speed of the vehicle increases the difference in speed between the impeller and turbine-wheels become less (the turbine-wheel is catching up with the impeller-wheel), the vortex velocity reduces and the transfer of hydrokinetic energy from the impeller to the turbine falls considerably; as a result less torque is transmitted from the engine to the transmission system.

Since it is inherent that a vortex circulation velocity is essential to transmit torque to the turbine-wheel, there must always be a minimum of 2–3% slip between the impeller and turbine-wheel rotational speeds when cruising. This slip increases to something like 20% or more in the low speed range when the vehicle is accelerating. On overrun, the transmission drives the engine so that the roles of impeller and turbine are reversed, that is, the turbine becomes the input wheel and the impeller the output wheel; thus overrun engine braking is conveyed through the coupling to the transmission as in a conventional friction clutch.

4.1.2 Fluid coupling performance characteristics (fig. 4.3)

The slip efficiency curves (fig. 4.3) show the slip at 600 rev/min to be zero, that is, the turbine-wheel is

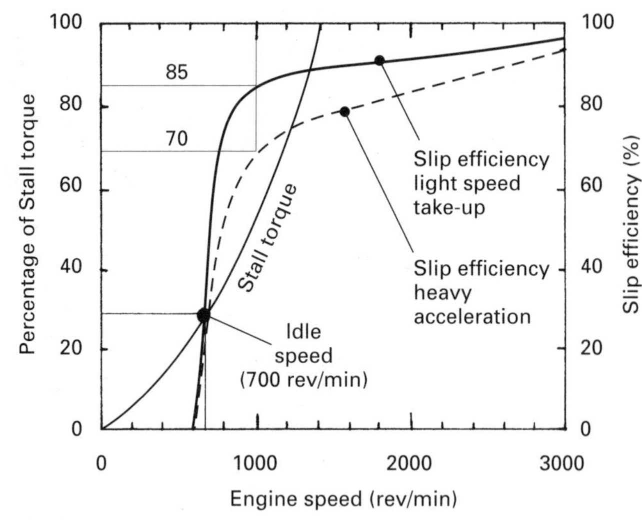

Fig. 4.3 Performance curves for a two-element fluid coupling

stationary, and at idle engine speed of 700 rev/min the slip efficiency has risen to 30%. As the engine speed increases to 1000 rev/min the coupling efficiency rapidly rises with a light load acceleration to 85% or with a heavy acceleration to 70%, after which the efficiency curve bends and therefore increases at a much slower rate until at cruising

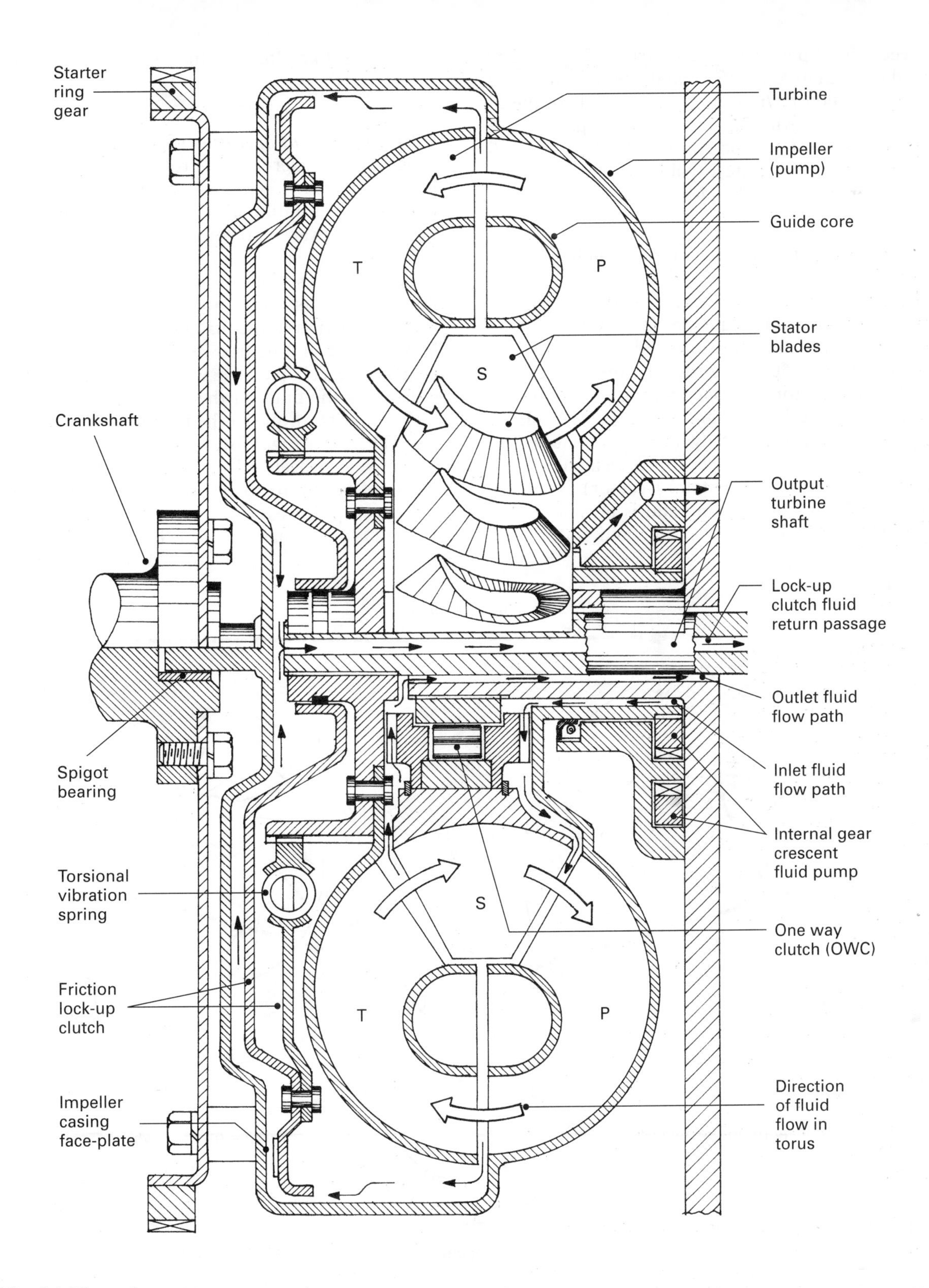

Fig. 4.4 Three element torque converter

speeds the slip efficiency may reach 97–99%. The stall torque curve shows a rapid rise from no drive torque at 600 rev/min to 100% of stall torque at 1400 rev/min whereas the drive torque at idle speed amounts to something like 30% of stall torque, which is quite capable of making a car creep forward if the hand-brake is not applied.

4.2 Hydrokinetic torque converter-coupling (fig. 4.4)

A torque converter uses hydrokinetic energy as a means of smoothly transferring power from the engine to the transmission and simultaneously multiplies the output torque in the process. A three-element torque converter coupling consists of an input steel pressing impeller-wheel encasing the output turbine-wheel which is also fabricated from a steel pressing. The third element of the converter, known as the stator (reactor), is usually an aluminium-alloy casting and is normally held stationary. The impeller, turbine and stator have something like 26, 23 and 15 blades (vanes) respectively.

4.2.1 Principle of operation (fig. 4.5)

When the impeller-wheel is rotated by the engine, it acts as a centrifugal pump drawing in fluid near the centre of rotation at the entrance A, forcing it radially outwards through the cell passages formed by the vanes and torus guide core to the impeller-wheel's peripheral exit at B. It is then flung due to its momentum across the impeller/turbine junction to the turbine-wheel cell entrance at C, where it imparts its hydrokinetic energy against the cell passage walls, hence the thrust exerted by the fluid on the drive side of the vane walls imparts torque to the turbine-wheel (see fig. 4.5). The centrifugal force created by the fluid in the impeller-wheel cell passage-ways now pushes the fluid in the relatively slow rotating turbine-wheel cell passages inwards to the turbine-wheel exit at D (this is made possible because the oppos-

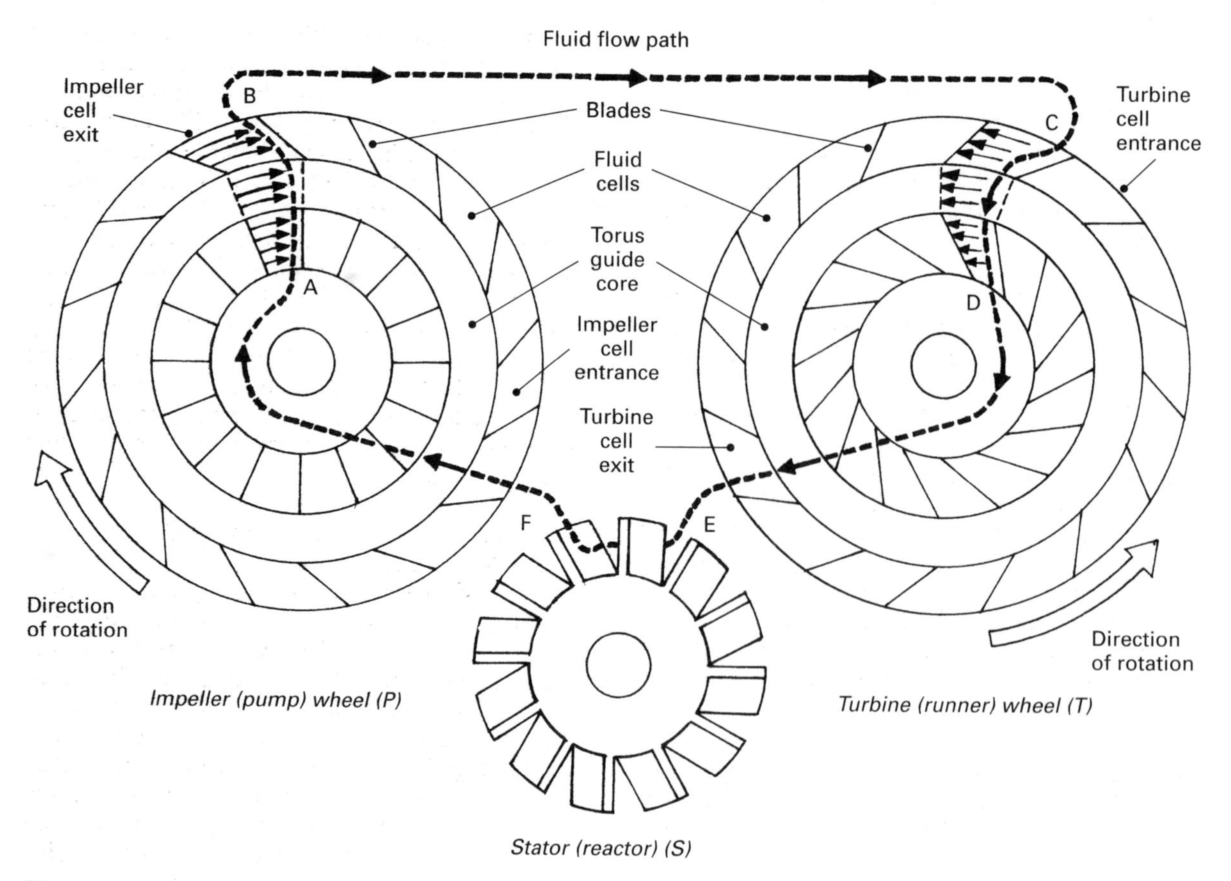

Fig. 4.5 Principle of three element torque converter

ing centrifugal force created by the fluid trapped in the turbine cell passages will be far less than that produced by the faster rotating impeller-wheel). The fluid is then compelled to flow between the fixed stator blades from its entrance at E to its exit at F (see fig. 4.5). The reaction of the fluid's momentum as it glides over the curved surfaces of the blades is absorbed by the fixed casing to which the stator is held, and in so doing is redirected towards the impeller entrance at A. Fluid from the stator blade passage-ways again now enters the passages shaped by the impeller-wheel vanes. As this fluid acts on the drive side of the impeller vanes, it imparts a thrust equal to the stator reaction in the direction of rotation (see fig. 4.5).

It therefore follows that the engine torque T_e delivered to the impeller and the reaction torque T_r transferred by the fluid to the impeller are both transmitted to the output turbine as the turbine torque T_t through the media of the fluid.

i.e.

Engine torque + reaction torque
= output turbine torque

that is

$$T_e + T_r = T_t$$

4.2.2 Torque converter-couple performance characteristics and the need for a one-way clutch (fig. 4.6(a)–(c))

Stall point (fig. 4.6(a)) One of the limitations of a torque converter is that, though the fluid is guided by the cell vane passages, at different speed conditions the actual direction the fluid wants to go in will vary considerably. At stall conditions, when the impeller is being rotated by the engine and the turbine-wheel is stationary or only just starting to rotate, the direction of the fluid leaving the impeller will be deep into the concave side of the stator blades where it reacts and is made to flow towards the entrance of the impeller in a direction which provides the maximum thrust see (fig. 4.6(a)). Under these conditions maximum torque multiplication takes place but converter efficiency is at its lowest (see fig. 4.7).

Normal conversion speed range (fig. 4.6(b)) Once the turbine begins to rotate, the fluid's direction of flow changes somewhat, so that it now flows nearly parallel to the impeller-wheel blades. This means the impingement of the fluid on the impeller blades will not be so effective (see fig. 4.6(b)). Hence the torque multiplication is reducing, whereas the impeller/turbine slip efficiency is rising (see fig. 4.7) until it reaches its peak of about 85% at roughly 1600 rev/min (in other words there is a 15% relative rotational slip between the input and output members). As the turbine-wheel speed of rotation rises even higher, the direction of the fluid flow will change even more, so that it now acts on the concave (back) side of the stator blades (see fig. 4.6(c)). Above the critical speed known as the coupling point, where the fluid's thrust changes from the concave to the convex side of the stator blades, the stator reactor torque will now act in the opposite sense and redirects the fluid. Thus its new direction towards the impeller entry passages will hold back instead of assisting the impeller-wheel motion. The result of this would be in effect to cancel out some of the engine input torque with further speed increases. Therefore under these circumstances there is very little torque multiplication and the slip efficiency commences to fall (see fig. 4.7).

Coupling speed range (fig. 4.6(c)) The inherent speed limitation of a torque converter is overcome by building into the stator hub a one-way-clutch (free-wheel) device (see fig. 4.6(c)). Therefore, when the direction of fluid flow changes sufficiently for it to impinge onto the back of the stator blades, the stator hub is released, thus allowing it to spin freely between the input and output members. The free-wheeling of the stator causes very little fluid interference; thus the three-element converter now becomes a two element fluid coupling. This condition prevents the decrease in turbine torque at higher output speeds, and at the same time produces a sharp rise in the slip efficiency at turbine-wheel output speeds above the coupling point (see fig. 4.7).

4.3 Sprag-type free-wheel one-way clutch (fig. 4.8(a) and (b))

This compact heavy load capacity one-way free-wheeling-type clutch consists of an inner and outer ring race member which has cylindrical external and internal tracks respectively. Interconnecting the inner and outer ring members are short struts, known as sprags, which are stacked closely together in a circular row so that each sprag shares a portion of the interlocking load. The contact ends of the sprags form inclined planes which are semicircular with their radius of curvature offset to each other; it thus makes

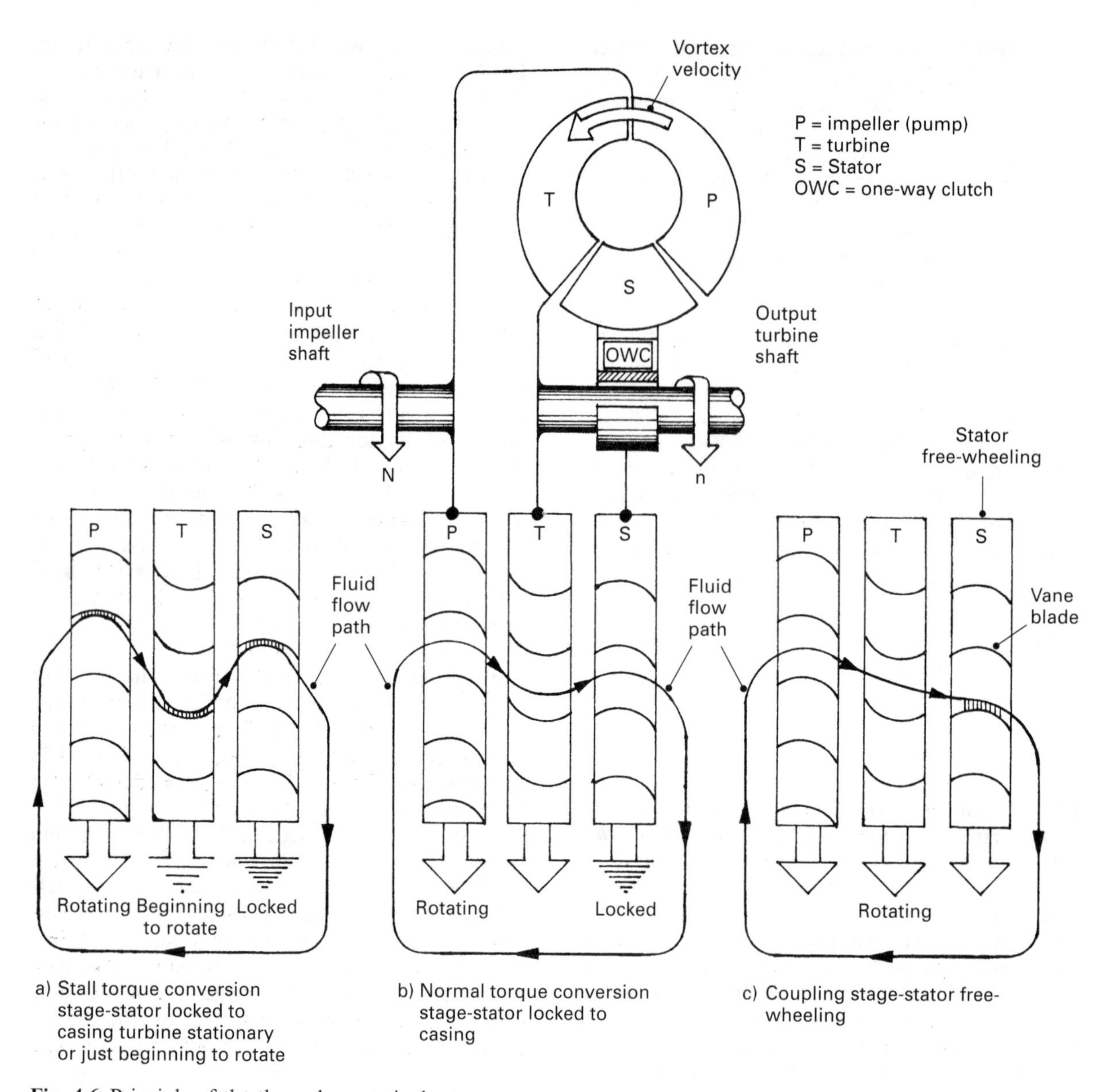

Fig. 4.6 Principle of the three-element single stage torque converter

the sprags look slightly lopsided. A pair of cages are installed between the inner and outer race members; these cages have rectangular slots formed to equally space and locate the sprags around the inner and outer tracks. In between the cages is a ribbon-type spring which twists the sprags into light contact with their respective tracks when the clutch is expected to overrun or free-wheel.

When the inner race member is fixed to an extension to the casing, and the direction of thrust by the working fluid is on the concave side of the stator blades, the outer race member is pushed towards the left-hand side (see fig. 4.8(a)), and at the same time the ribbon spring tension lightly presses the sprag inclined planes against their respective track. The sprags are thus forced to tilt counterclockwise, thereby wedging their inclined planes hard against the inner and outer tracks; hence the inner and outer race members become jammed by the many wedged sprag struts.

As the rotational speed of the turbine-wheel increases (see fig. 4.6(c)) the resultant direction of

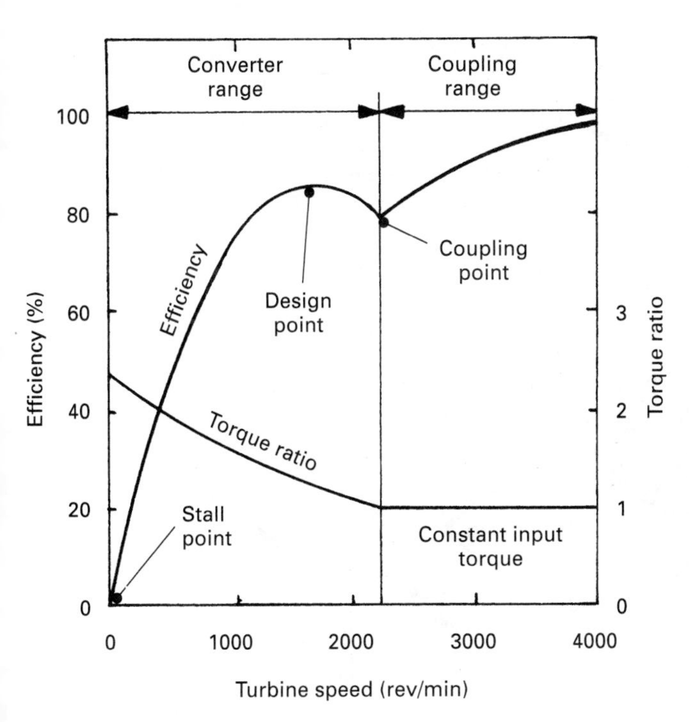

Fig. 4.7 Performance curves for a three-element converter-coupling

the working fluid changes to a point where it now imparts its thrust against the convex side of the stator blades (see fig. 4.8(b)). Accordingly the sprag struts will now be tilted clockwise; this tends to roll the sprags so that their point of contact with their respective tracks is at a lower position on their inclined planes. As a result the outer race member is released from the stationary inner race member so that the stator member is now permitted to 'free-wheel'.

4.4 Torque converter lock-up clutch (fig. 4.4)
A drawback to both fluid couplings and particularly the torque converter is that there is always a small percentage of slip between the impeller and turbine-wheels.This slip is necessary to provide a vortex circulation of working fluid so that hydrokinetic energy can be transferred from the impeller to the turbine-wheel. Thus, whenever the turbine-wheel rotational speed catches up with the impeller-wheel speed, the vortex circulation stops and therefore there can be no fluid energy imparted to the output turbine-wheel. Consequently under these conditions slip occurs, that is, the turbine-wheel having no input torque slows down; thus drive power will be lost and more fuel will be consumed than is necessary.

To overcome this limitation of the torque converter-coupling under cruising conditions, a friction lock-up clutch is incorporated inside the impeller-wheel casing. This clutch is similar to a conventional friction single-plate clutch and is mounted on a splined projection attached to the turbine-wheel hub. When not engaged, pressurised fluid from the torque converter circulates on both sides of the drive-plate disc. Above a certain turbine-wheel rotational speed set by the automatic transmissions hydraulic control, fluid will be exhausted from the left-hand impeller casing side of the drive-plate through a central passage in the turbine shaft to the transmission sump via a lock-up control valve (not shown). Under these conditions, the fluid pressure acting on the right-hand side of the drive-plate will force the drive-plate to the left, and thereby engages the impeller casing faceplate, thus locking the turbine-wheel to the impeller casing.

When the turbine-wheel rotational speed falls to some predetermined level, the lock-up control valve commences to supply fluid from the transmission's internal gear crescent pump to the space between the impeller casing and the left-hand side of the drive-plate. As a result the lock-up drive-plate will be pushed to the right against the lower fluid pressure supplied through the torque converter to release the impeller/turbine lock-up.

4.5 Epicycle gear train (figs 4.9(a)–(f) and 4.10(a)–(f))
Essentially an epicyclic single-stage gear train (fig. 4.9(a)) consists of an internally toothed annulus (ring) gear (A) with a band encircling it, and in the centre of this gear is a sun-gear (S). Bridging the space between the sun-gear and the annulus-gear are planet (pinion) gears (P) which engage both these gears, and are mounted on a planet-carrier (Cp). For torque to be transmitted, either the annulus-gear, sun-gear or planet-carrier must be held stationary (fixed) to provide a reaction torque while the other two members become either the input or output drives.

When the brake-band grips the annulus-gear (fig. 4.9(a)) the annulus-gear is held stationary to provide the reaction, and the input sun-gear is driven. Then the planet-gears rotate around their own axes and at the same time are forced to revolve and 'walk' around the inside of the annulus-gear. Consequently the planet-carrier and the output shaft which supports the planet-gear axes (pins) will also be made to rotate, but at a speed lower than that of the input shaft.

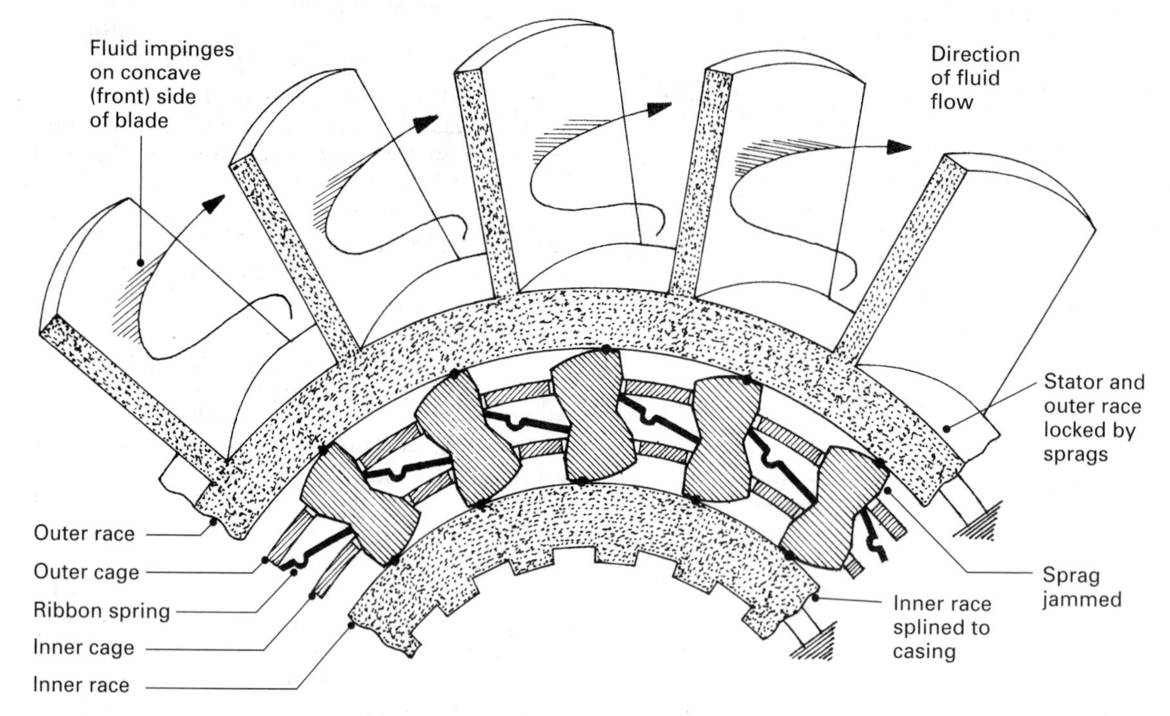

a) Stator hub and blades locked to inner race

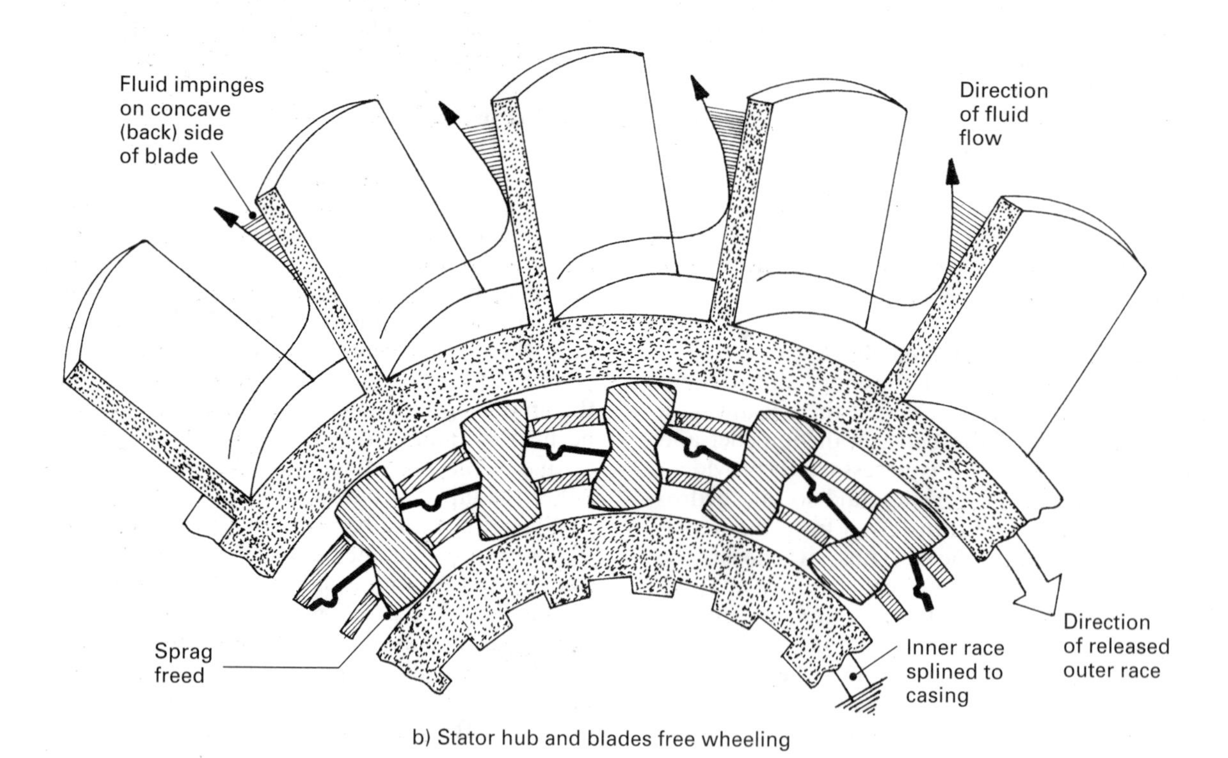

b) Stator hub and blades free wheeling

Fig. 4.8 Overrun free wheel (one way clutch) sprag type clutch

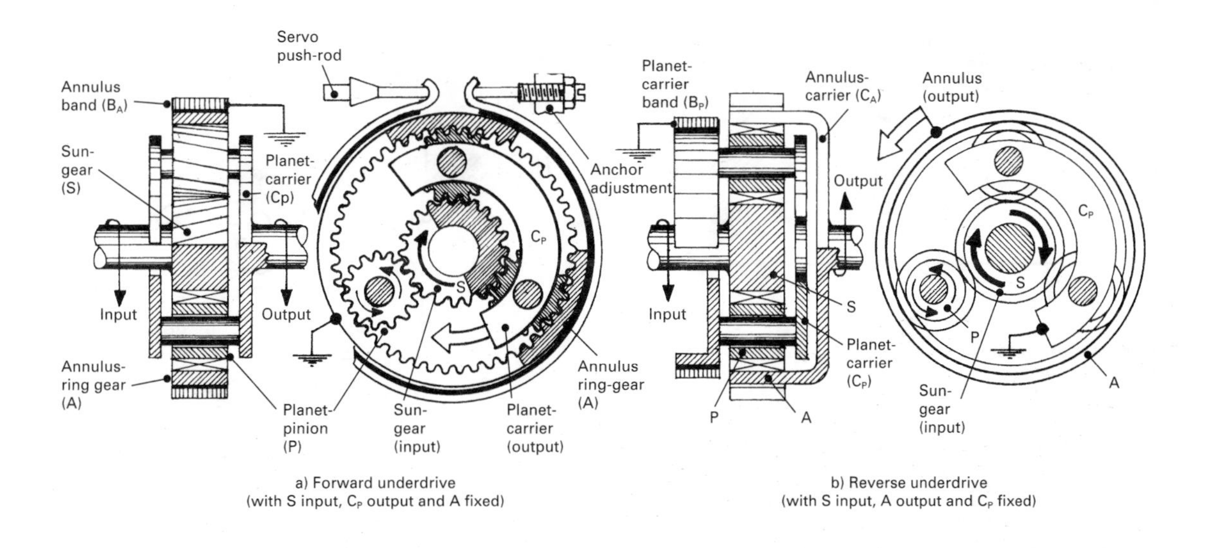

a) Forward underdrive
(with S input, C_P output and A fixed)

b) Reverse underdrive
(with S input, A output and C_P fixed)

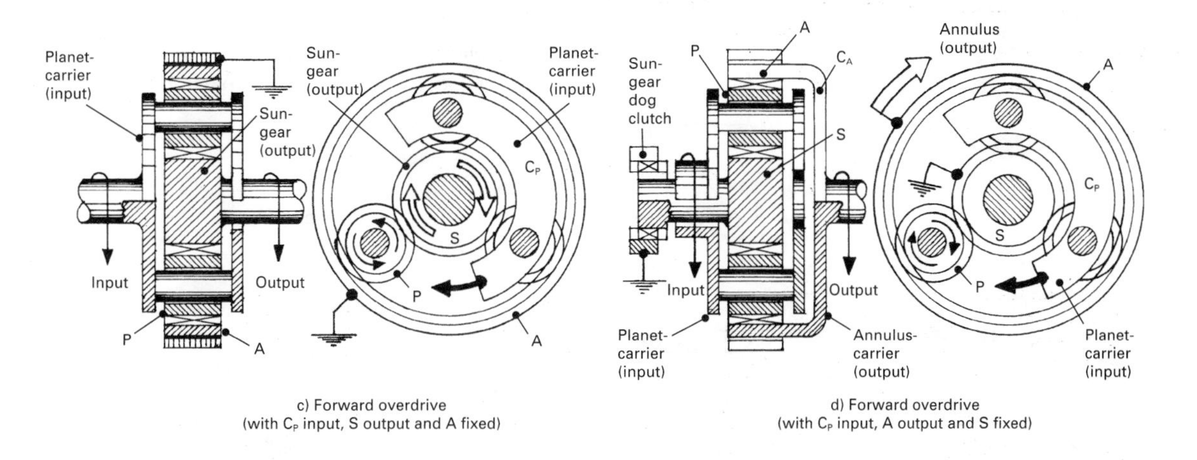

c) Forward overdrive
(with C_P input, S output and A fixed)

d) Forward overdrive
(with C_P input, A output and S fixed)

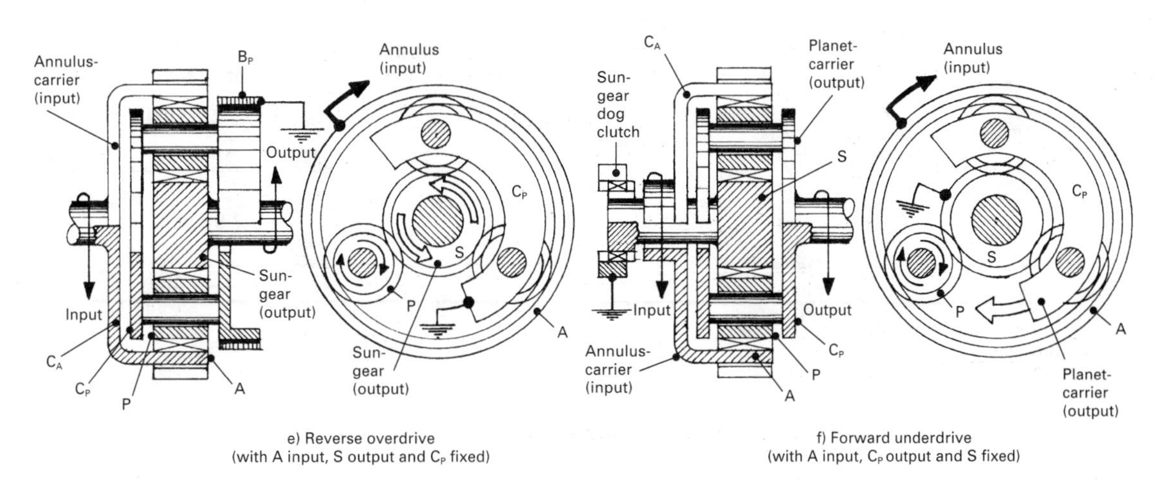

e) Reverse overdrive
(with A input, S output and C_P fixed)

f) Forward underdrive
(with A input, C_P output and S fixed)

Fig. 4.9 Single stage epicyclic gear train arrangements

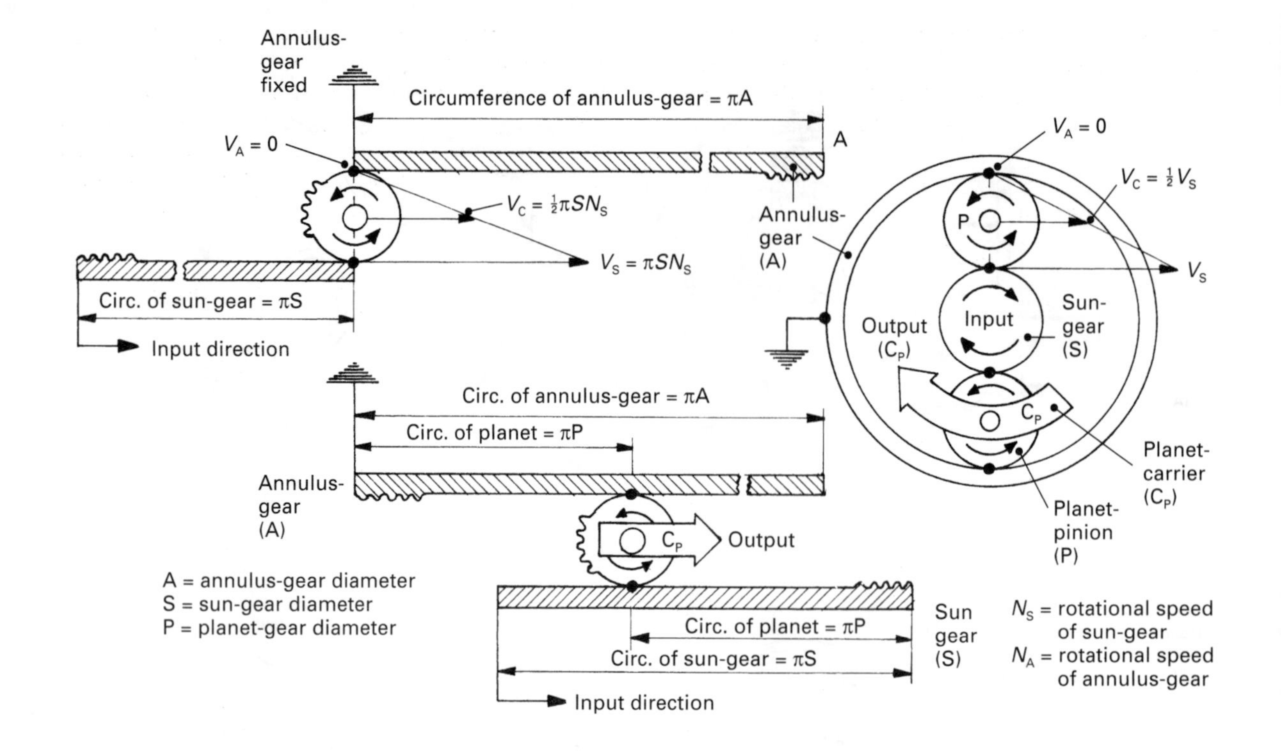

a) Forward underdrive (with 'S' input, 'C_P' output and 'A' fixed)

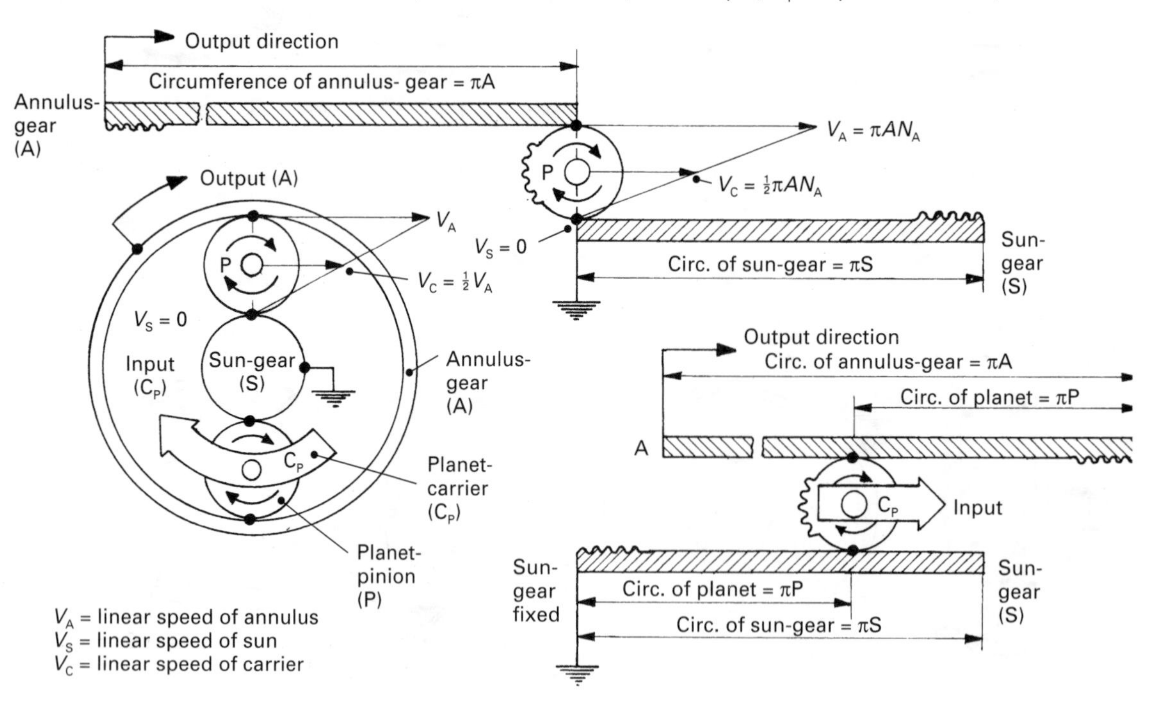

b) Forward overdrive (with 'C_P' input, 'A' output and 'S' fixed)

Fig. 4.10 Principle of the epicyclic gears

Consider the toothed gears replaced by rollers which rotate without slipping and whose diameters are proportional to the number of teeth (see fig. 4.10(a)). Let S = sun-gear diameter, P = planet-gear diameter, A = annulus-gear diameter = $S + 2P$, N_s = rotational speed of the sun-gear, N_a = rotational speed of the annulus-gear, N_c = rotational speed of the carrier and C_p = pitch circle diameter of carrier. The internal and external toothed annulus and sun-gears respectively can be considered as straight tooth tracks interconnected by a toothed planet-gear (see fig. 4.10(a)). If the annulus-gear (S) is held stationary (fixed), and the sun-gear (S) and planet-carrier (Cp) are made the input and output drive members respectively, then rotating the input sun-gear (S) once is equivalent of advancing the sun-gear's external straightened track a distance equal to its circumference πS (or number of sun-gear teeth). Correspondingly the planet-gear will be forced to roll around the annulus-gear's straightened internal track a similar distance. Hence the relative linear (peripheral) speed (fig. 4.10(a)) of the planet-gear at its furthest point from the fixed annulus-gear at the point of contact between the planet and annulus-gear teeth is at a maximum, that is, $V_s = \pi S N_s$, while the point of contact between the stationary annulus and the planet-gear has a zero relative speed, that is, $V_a = 0$. Therefore, by similar triangles, at the mid-point of the planet-gear at its axis of rotation it has a linear (peripheral) speed of half that of the maximum linear speed, that is, $V_c = 1/2\pi\, SN_s$. Similarly if the sun-gear is held stationary (fig. 4.10(b)) and the annulus-gear forms the input member, then the maximum linear speed will be at the point of contact between the planet-gear and annulus-gear, whereas the contact point of the planet-gear with the fixed sun-gear is zero, and likewise the mid-point at the centre of the planet-carrier will have a linear speed of half the maximum linear speed.

Forward underdrive with S input, C_p output and A fixed (figs 4.9(a) and 4.10(a)) If the sun-gear S rotates at N_s rev/min the linear (peripheral) speed V_s of the contact point between S and P is $\pi S\, N_s$ i.e.

$$V_s = \pi S\, N_s$$

However it can be seen by similar triangles that the gear P at the point of contact with the sun-gear S moves forwards at twice the speed of the planet carrier Cp. This means that the linear speed V_c of the planet carrier Cp is half that of the gear P at the point of contact with the sun-gear S, that is, at a linear speed of $1/2\ \pi S\, N_s$.

Now the linear speed V_c of the planet-carrier is equal to the pitch circle circumference $\pi\,(P + S)$ of the planet-carrier, times the rotational speed N_c of the planet-carrier, that is

linear speed of planet-carrier
= pitch circle circumference of carrier
× rotational speed of carrier

$$V_c = \pi\,(P + S)\, N_c$$

but

$$V_c = 1/2\ \pi S\, N_s$$

therefore

$$1/2\ \pi N_s = \pi\,(P + S)\, N_c$$

or

$$\frac{N_s}{N_c} = \frac{2(S + P)}{S}$$

$$\text{But GR} = \frac{\text{input}}{\text{output}} = \frac{N_s}{N_c}$$

therefore

$$\text{GR} = \frac{2(S + P)}{S}$$

Reverse underdrive with S input, A output and C_p fixed (fig. 4.9(b))

$$\text{gear ratio (GR)} = \frac{\text{input}}{\text{output}} \times \frac{\text{input}}{\text{output}}$$

$$= \frac{\pi S\, N_s}{\pi P\, N_p} \times \frac{\pi P\, N_p}{\pi A\, N_a} = \frac{S\, N_s}{A\, N_a}$$

or

$$\frac{N_s}{N_a} = \frac{A}{S}$$

But

$$A = S + 2P$$

$$\text{therefore GR} = \frac{S + 2P}{S}$$

Forward overdrive with C_p input, A output and S fixed (figs 4.9(d) and 4.10(b))

linear speed of carrier
= pitch circle circumference of carrier
× rotational speed of carrier

$$V_c = (\pi C) N_c$$

But

$$V_c = 1/2\ \pi A N_a \quad \text{and} \quad C = S + P$$

Therefore

$$1/2\ \pi A N_a = \pi (S + P) N_c$$

or
$$\frac{N_c}{N_a} = \frac{1/2\ \pi A N_c}{\pi (S + P)} = \frac{A}{2 (S + P)}$$

But
$$A = S + 2P$$

or
$$\frac{N_c}{N_a} = \frac{S + 2P}{2 (S + P)}$$

But GR
$$= \frac{\text{input}}{\text{output}} = \frac{N_c}{N_a}$$

therefore GR
$$= \frac{S + 2P}{2(S + P)}$$

Forward overdrive with C_p input, S output and A fixed (fig. 4.9(c))

linear speed of carrier
= pitch circle circumference
× rotational speed of carrier

$$V_c = \pi C N_c$$

But

$$V_c = 1/2\ \pi S N_s \quad \text{and} \quad C = S + P$$

therefore

$$1/2\ \pi S N_s = \pi(S + P) N_c$$

or
$$\frac{N_c}{N_s} = \frac{S}{2 (S + P)}$$

But
$$\text{GR} = \frac{\text{input}}{\text{output}} = \frac{N_c}{N_s}$$

therefore
$$\text{GR} = \frac{S}{2(S + P)}$$

Reverse overdrive with A input, S output and C fixed (fig. 4.9(e))

$$\text{gear ratio (GR)} = \frac{\text{input}}{\text{output}} \times \frac{\text{input}}{\text{output}}$$

$$= \frac{\pi A N_s}{\pi P N_p} \times \frac{\pi P N_p}{\pi S N_a} = \frac{A N_a}{S N_s}$$

or
$$\text{GR} = \frac{N_a}{N_s} = \frac{S}{A}$$

But
$$A = S + 2P$$

therefore
$$\text{GR} = \frac{S}{S + 2P}$$

Forward underdrive with A input, C_p output and S fixed (fig. 4.9(f))

linear speed of carrier
= pitch circle circumference
× rotational speed of carrier

$$V_c = \pi C N_c \quad V_c = 1/2\ \pi A N_a \quad \text{and} \quad C = S + P$$

therefore

$$1/2\ \pi A N_a = \pi(S + P) N_c$$

or
$$\frac{N_a}{N_c} = \frac{2(S + P)}{A}$$

But
$$\text{GR} = \frac{\text{input}}{\text{output}} = \frac{N_a}{N_c}$$

therefore
$$\text{GR} = \frac{2(S + P)}{S + 2P}$$

Note
$$A = S + 2P$$

Example An epicyclic gear train has sun and annulus gears with 28 and 56 teeth respectively. If the input speed from the engine drives the sun shaft at 3000 rev/min, determine (a) the number of planet-gear teeth, (b) the overall gear ratio, and (c) the pinion carrier-shaft output speed.

a)
$$A = S + 2\ P$$

therefore

$$P = \frac{A + S}{2}$$

$$= \frac{56 + 28}{2} = \frac{28}{2} = 14 \text{ teeth}$$

b) $\text{GR} = \frac{2(S + P)}{S} = \frac{2(28 + 14)}{28} = \frac{2 \times 42}{28} = 3{:}1$

c) $\text{GR} = \frac{N_s}{N_c}$

therefore

$$N_c = \frac{N_s}{\text{GR}} = \frac{3000}{3} = 1000 \text{ rev/min}$$

4.5.1 Single-stage twin-planet pinion epicyclic gear train (reverse underdrive) (fig. 4.11)
With the annulus-gear braked, the input drive passes from the sun-gear to the inner and outer planet pinions in turn being in mesh with the internal teeth of the annulus-gear. Since the annulus-ring-gear is held stationary, the outer planet-gear absorbs the torque reaction and in the process is compelled to revolve around the internal teeth of the annulus-gear. As a result the planet-pinions are forced to revolve on their own axes; this therefore compels the pinion-carrier to revolve at a reduced speed relative to that of the input. Twin planet-pinion epicyclic gear trains are used when the direction of rotation of the output planet-carrier must be in the opposite direction to that of the input of the input sun-gear, whereas if a single planet-pinion epicyclic gear train is used, the planet-carrier will rotate in the same direction as the sun-gear. If both the inner and outer planet-pinions have the same number of teeth, then the overall gear ratio will be the same as the single planet-pinion epicyclic gear train which has the same number of planet-pinion teeth.

4.5.2 Single-stage compound epicyclic gear train (forward underdrive) (fig. 4.12)
With the annulus-gear braked, the input drive passes from the sun-gear to the integral large and small planet-pinions; it then flows to the planet-carrier to the output shaft. At the same time, the small planet-pinion absorbs the drive torque reaction and in the process is made to revolve around the internal teeth of the fixed annulus-gear. The underdrive (gear reduction) condition is created by the small planet-pinion being compelled to roll 'walk' about the internal teeth of the fixed annulus-gear; the pinions therefore are forced to revolve on their own axes. As a result, the large planet-pinion, also revolving on the same carrier-pin as the small planet-gear, drives forward the pinion-carrier at a reduced speed relative to that of the input shaft. The overall gear ratio step-down is achieved by having two stages of meshing gear teeth, one between the large pinion and sun-gear and the other between the small pinion and annulus-ring-gear. By using this compound epicyclic gear train, a relatively large step-down (gear reduction) gear ratio can be obtained for a

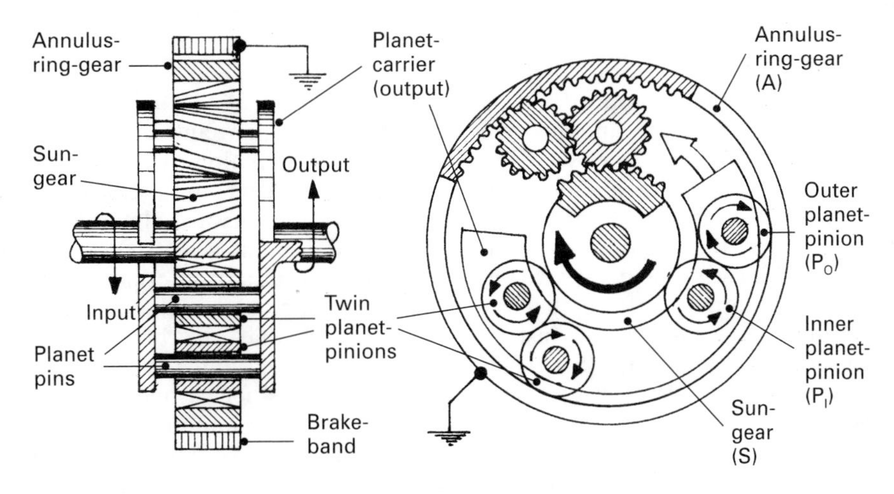

Fig. 4.11 Single stage twin planet pinion epicyclic gear train (reverse underdrive)

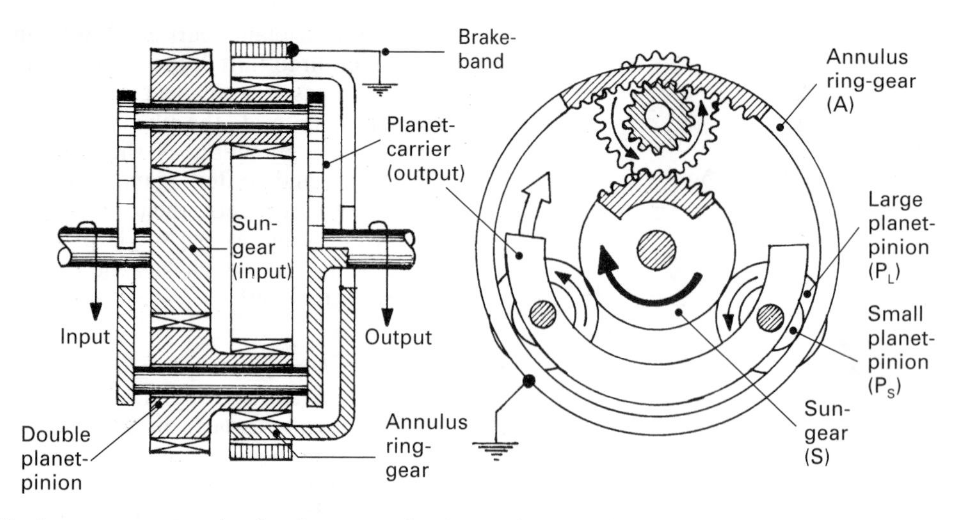

Fig. 4.12 Single stage compound epicyclic gear train (forward underdrive)

given diameter of annulus-ring-gear compared to a simple single-stage epicyclic gear train.

Conversely, if the sun-gear is held stationary and the input drive passes to the pinion-carrier, while the annulus-gear becomes the output drive, then a compact step-up overdrive epicyclic gear train is obtained.

4.5.3 Simpson compound epicyclic three speed and reverse gear train power flow (fig. 4.13(a)–(d))

A three speed and reverse epicyclic gear arrangement designed by Howard W. Simpson using two planetary gear train sets was originally incorporated in the 1955 Chrysler Torque-Flite automatic transmission. Ever since, it has successively been employed by many other car manufacturers and therefore is worth some consideration.

First gear (fig. 4.13(a)) With the forward-clutch (FC) and one-way-clutch (OWC) applied, power flows from the input shaft to the annulus-gear (Af) of the forward planetary-gear set. The clockwise rotation of the forward annulus-gear causes the forward planet-gears also to rotate clockwise, thus driving the double sun-gear counterclockwise. The forward planetary-carrier (Cf) is attached to the output shaft so that the planet-gears (Pr) drive the sun-gear (Sf) instead of 'walking' around the sun-gear (Sf). This counterclockwise rotation of the sun-gear (Sr) causes the reverse planet-gears (Pr) to rotate clockwise. With the one-way clutch (OWC) holding the reverse planet-carrier (Crp), the reverse planetary-gears (Pr) turn the reverse annulus-gear (Ar) and output shaft clockwise at a reduced speed.

Second gear (fig 4.13(b)) With the forward clutch (FC) and second gear brake-band (2GB) applied, power flows from the input shaft to the forward annulus-gear (Af). With the double sun-gear held by the applied second gear brake brake-band (2GB), the clockwise rotation of the forward annulus-gear (Af) compels the planet-gears (Pf) to rotate on their own axes and roll 'walk' around the stationary sun-gear (Sf) in a clockwise direction. Because the forward planet-gear pins are mounted on the forward planet-carrier (Cf), which is itself attached to the output shaft, the output shaft will be driven clockwise at a reduced speed and at a speed slightly higher than that of the first gear.

Third gear (fig. 4.13(c)) With the forward clutch (FC), and high and reverse clutch ((H+R)C) applied, power flows from the input shaft to the double sun-gear in a clockwise direction and similarly the forward clutch (FC) will rotate the forward annulus-gear (Af) clockwise. This causes both external and internal gears on the forward gear set to revolve in the same direction at similar speeds so that the bridging planet-gears (Pf) become locked and the whole gear set therefore revolves together as one. The output shaft drive via the reverse annulus-gear carrier (Cra) therefore turns clockwise with no relative speed reduction to the input shaft, that is, as a direct drive ratio 1:1.

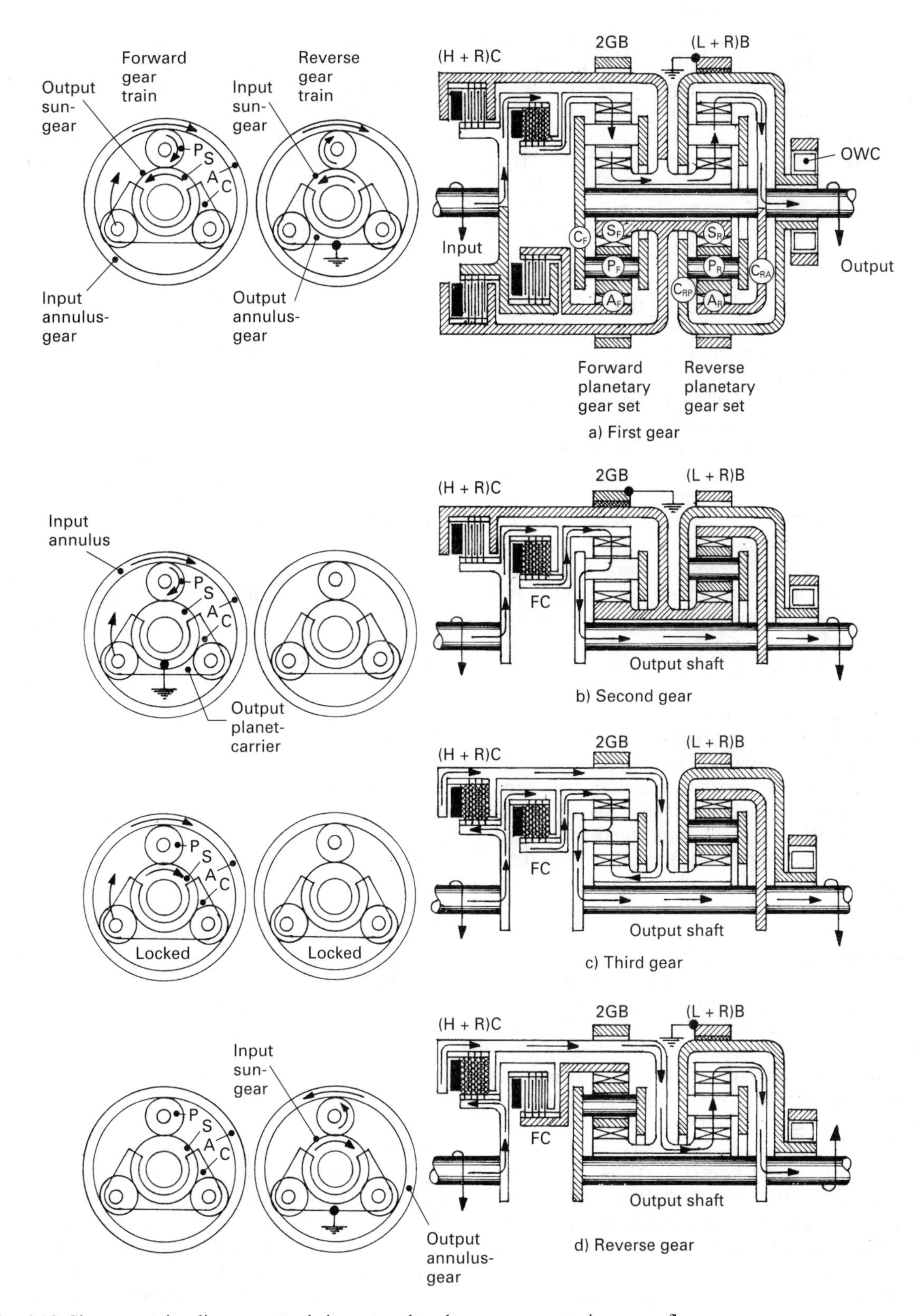

Fig. 4.13 Simpson epicyclic compound three speed and reverse gear train power flow

Reverse gear (fig. 4.13(d)) With the low and reverse brake-band ((L+R)B), and high and reverse clutch ((H+R)C) applied, power flows from the input shaft via the high and reverse clutch ((H+R)C) to the reverse sun-gear (Sr) in a clockwise direction. Because the reverse planet-carrier (Crp) is held by the low and reverse brake-band ((L+R)B), the planet-gears (Pr) are forced to rotate counterclockwise on their axes, and in doing so compel the reverse annulus-gear (Ar) to also rotate counterclockwise. As a result, the output shaft, which is attached to the reverse annulus-gear (Ar), rotates counterclockwise, that is, in the reverse direction at a reduced speed.

4.5.4 Wilson four-speed and reverse compound epicyclic gear train power flow (fig. 4.14(a)–(e))
Major W.G. Wilson designed a four-speed and reverse gearbox using epicyclic gears operated by a friction clutch and brake-bands to obtain smooth gear changes. This gearbox was originally used in the 1898 Wilson–Pilcher motor car; later it was incorporated in armoured cars and tanks built in the First World War. Since then, it has been used for semi-automatic transmission for both cars and buses, and in more recent years it has been adapted for heavy duty automatic transmission.

First gear (fig. 4.14(a)) With the drive-clutch (DC) and first gear brake-band (1GB) applied, power flows from the input shaft to the first gear sun-gear (S1) via the drive clutch (DC). This drives the planet-gears (P1) which are compelled to revolve around the annulus-gear (A1) taking with them the planet-carrier (C1), and hence the output shaft at a much slower speed.

Second gear (fig. 4.14(b)) With the drive-clutch and second-gear brake-band (2GB) applied, power flows from the input shaft to the (sun-gears (S1and S2) via the drive clutch (DC). This drives the planet-gears (P2) which are compelled to revolve around the annulus (A2) taking with them the planet-carrier (C2) and the annulus (A1). Hence not only is sun-gear (S1) driving forward the planet gears (P1), but the annulus-gear (A1) is also driving forward the planet-gear (P2). Thus the planet-carrier (C1) and the output shaft will be driven forward at a slightly higher speed than when only the first epicyclic gear train is engaged.

Third gear (fig. 4.14(c)) With the drive-clutch (DC) and third gear brake-band (3GB) applied, power flows from the input shaft to the sun-gears (S1and S2) via the drive-clutch (DC) as when in second gear, but this time power also flows from the annulus (A2) and carrier (C3) to the planet-gear (P3). Since the sun-gear (S3) is held stationary, the planet-gears (P3) roll forward and so drive the annulus (A3); this in turn drives the planet-carrier (C2) and planet-gears (P2). Again the forward moving planet-carrier (C2) drives the annulus-gear (A1) so that both sun-gears (S1 and S2) impart motion to the planet gears (P1); it thereby drives the planet-carrier (C1) at a slightly higher speed than when only the first and second planetary gear sets are engaged.

Fourth gear (fig. 4.14(d)) With the drive-clutch (DC) and fourth gear clutch (4GC) applied, power flow will be from the input shaft to the first and second sun-gears (S1 and S2) respectively, and at the same time power will be diverted via the fourth gear clutch (4GC) to the third sun-gear (S3). Consequently the three sets of planetary-gears cannot revolve at their own individual speeds so that all the planetary-gear train sets lock, that is, they jam together, causing the whole assembly to rotate as one solid unit. Hence the fourth top gear provides a straight-through direct drive.

Reverse gear (fig. 4.14(e)) With the drive-clutch (DC) and reverse brake-band (RGB) applied, power flows from the input shaft to the first sun-gear (S1); this rotates the first planet-gear (P1) counterclockwise and the first annulus-gear (A1) in the opposite direction to the first sun-gear (S1). This is made use of by joining the first annulus-gear (A1) to the reverse sun-gear (Sr). The reverse sun-gear (Sr) is therefore made to rotate in the same direction as the annulus-gear (A1) that is, in the opposite direction to the first sun-gear (S1). Consequently the reverse planet-gear (Pr) and planet-carrier (Cr) will revolve in the reverse direction to the input shaft; a reverse motion at a reduced speed is therefore imparted to the output shaft.

4.6 Four-speed and reverse transaxle automatic transmission mechanical power flow (Nissan Primera) (fig. 4.15)
A description of the power flow through the various gear ratios for a typical four-speed and reverse transaxle automatic transmission now follows with the aid of a list of key components and abbreviations used in the automatic transmission layout

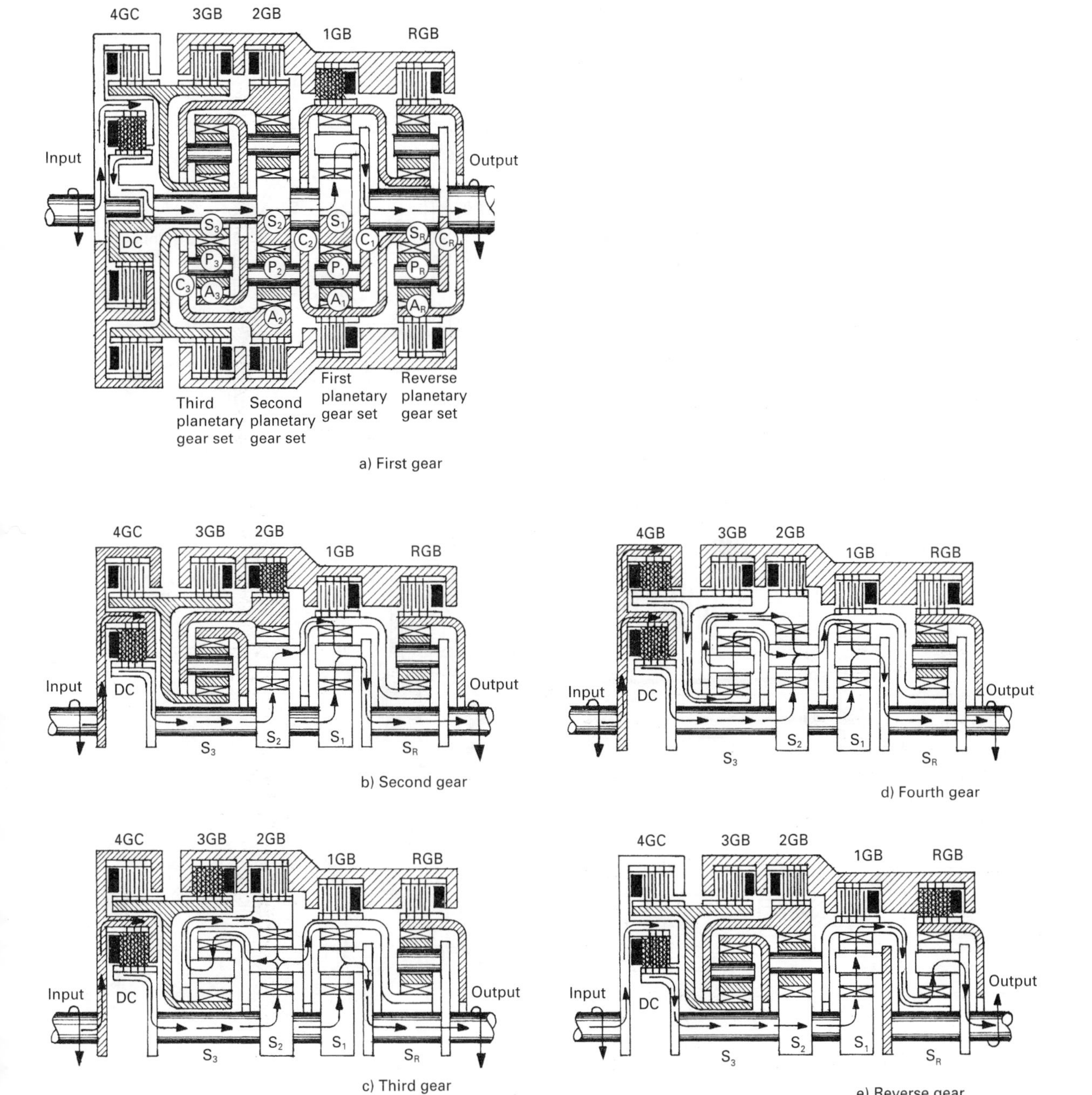

Fig. 4.14 Wilson epicyclic compound four speed and reverse gear train power flow

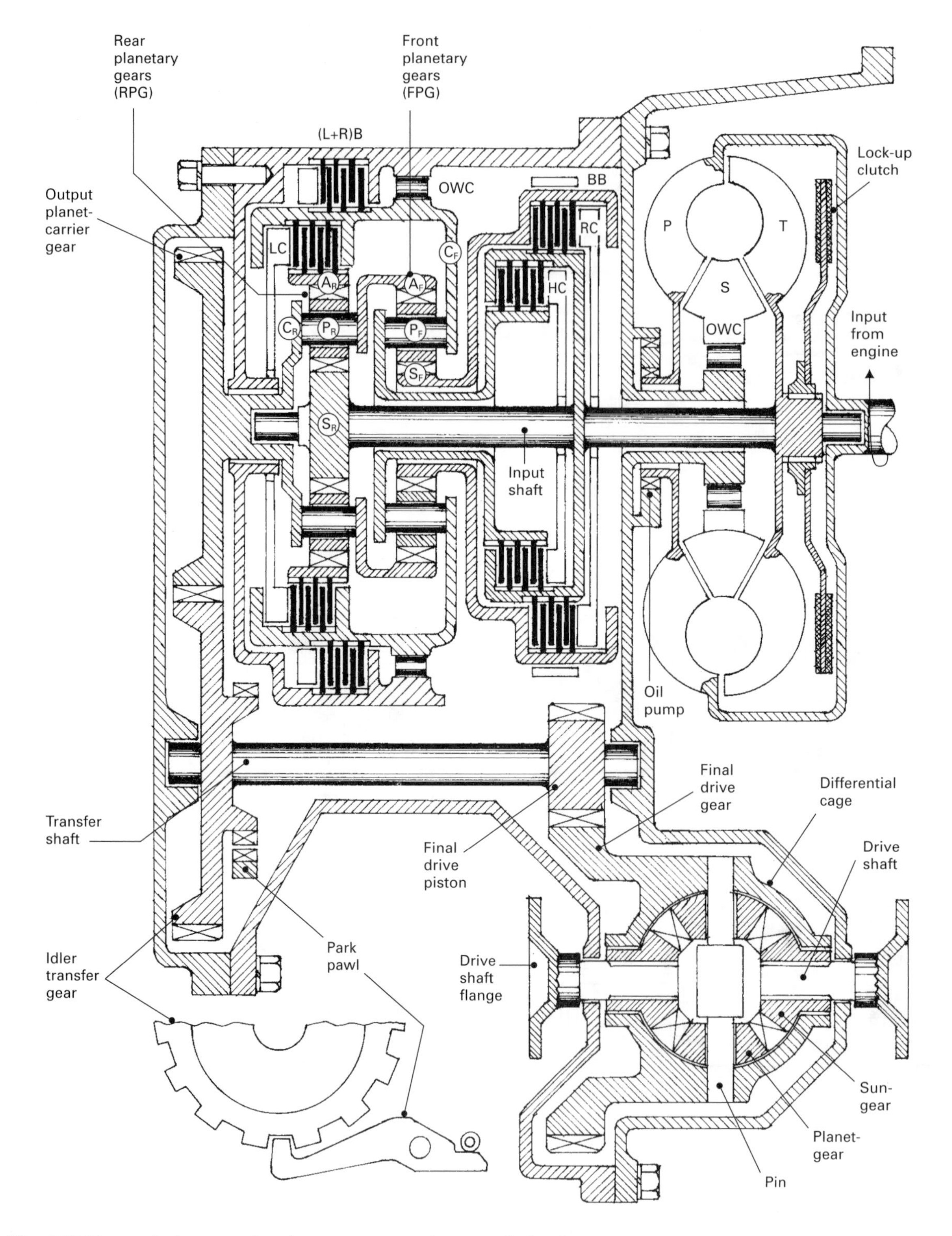

Fig. 4.15 Transaxle four speed and reverse automatic transmission layout

shown in fig. 4.15 and a clutch and brake engagement sequence shown in Table 4.1.

1) front annulus-gear Af
2) front planet-gear Pf
3) front sun-gear Sf
4) front planet-carrier Cf
5) reverse annulus-gear Ar
6) reverse planet-gear Pr
7) rear sun-gear Sr
8) rear planet-carrier Cr
9) impeller (pump) wheel P
10) turbine-wheel T
11) stator S
12) reverse clutch RC
13) low clutch LC
14) low and reverse clutch (L+R)C
15) high clutch HC
16) brake-band BB
17) one-way-clutch OWC
18) front planetary gear set FPG
19) rear planetary gear set RPG

Neutral (N) With the selector lever in N, all the clutches and brake are disengaged so that the torque-converter drive is separated from the planetary gear train output.

Parking-pawl (P) (fig. 4.15) With the selector lever in P, the parking-pawl engages the external teeth of the parking ring-gear. The engaged parking-pawl prevents the vehicle creeping forward when the transmission is in neutral due to torque converter fluid drag.

D drive range first gear (fig. 4.16(a)) With the selector lever in D drive range, the one-way-clutch (OWC) holds the front planet-carrier and the outer member of the low clutch while the low clutch (LC) is applied; this enables the rear annulus-gear to be held stationary. Power flows from the torque converter to the rear sun-gear via the input shaft. The rear sun-gear rotates clockwise thus driving the rear planet-gears counterclockwise and attempts to turn the rear annulus-gear (Ar) also counterclockwise. However, since the counterclockwise rotation of the annulus-gear (Ar) is prevented by the one-way-clutch (OWC), the rear planet-carrier is compelled to rotate clockwise at a reduced speed. If the 'L' range first gear is selected, both the low clutch (LC) and the low and reverse brake ((L+R)B) are applied. As a result, instead of the one-way-clutch (OWC) allowing the vehicle to free-wheel on overrun, the low and

Table 4.1 Clutch and brake engagement sequence for each gear ratio.

Clutch and brake engagement sequence									
Range		Low clutch LC	Low & reverse clutch (L+R)C	Brake-band BB	High clutch HC	One-way clutch OWC	Reverse clutch RC	Lock-up	Parking-pawl
N									
P									applied
Drive	D1	applied				applied			
	D2	applied		applied					
	D3	applied			applied			applied	
	D4			applied	applied			applied	
	L1	applied	applied			applied			
	L2	applied		applied					
Reverse			applied				applied		

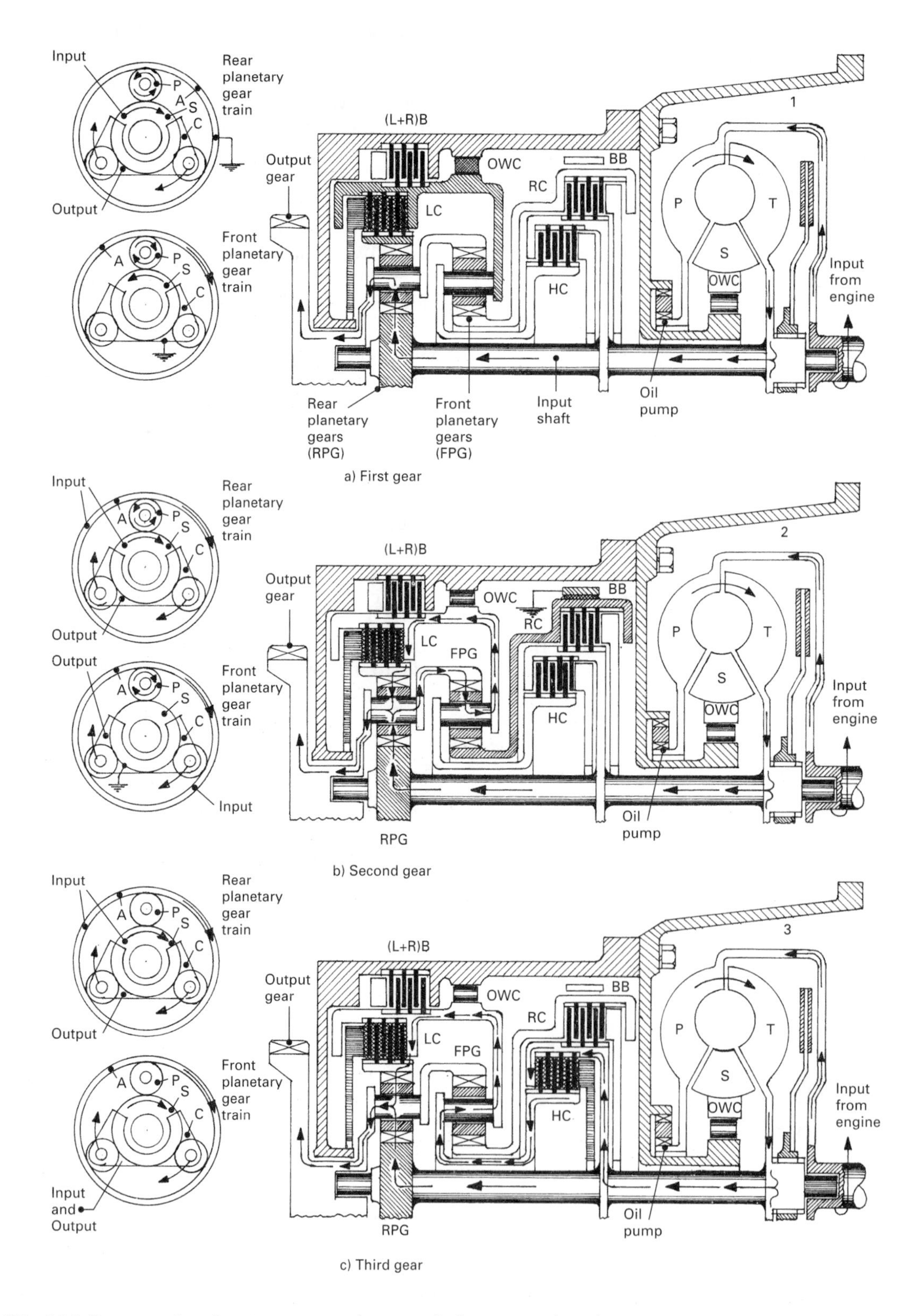

Fig. 4.16 Four speed and reverse automatic transmission transaxle unit

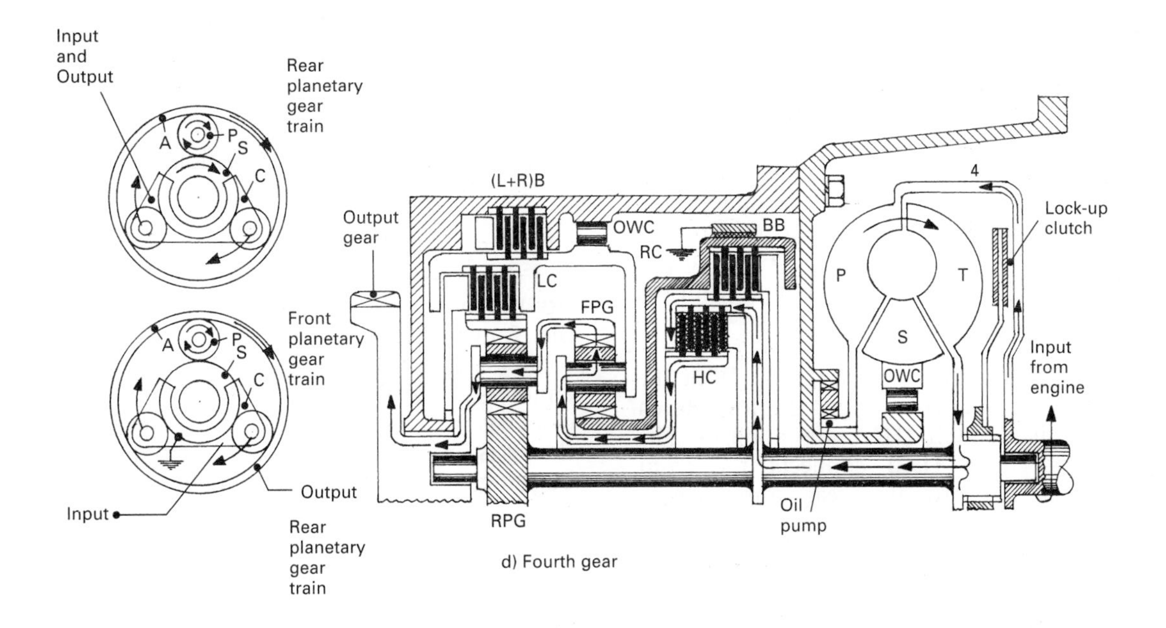

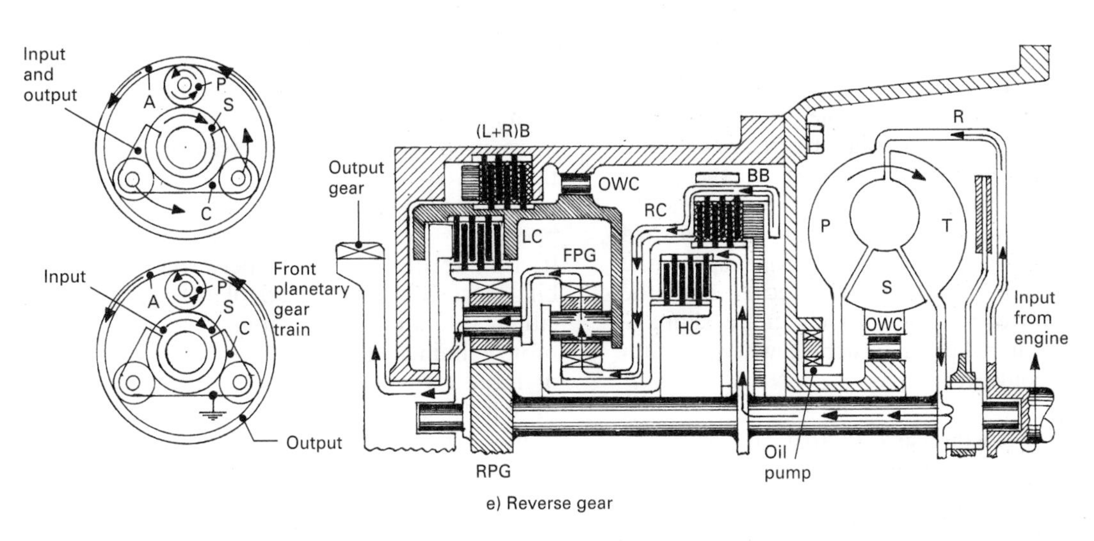

Fig. 4.16 (*Continued*) Four speed and reverse automatic transmission transaxle unit

reverse brake ((L+R)B) locks the low clutch outer member and the rear annulus-gear Ar. Consequently a positive drive exists between the engine and transmission on both drive and overrun; it thus enables engine braking to be applied to the transmission when the transmission is overrunning the engine.

D drive range second gear (fig. 4.16(b)) With the selector lever in D drive range, the low clutch (LC) and brake-band (BB) are applied. The applied low clutch (LC) connects the front planet-carrier (Cf) to the rear annulus-gear (Ar) while the applied brake-band (BB) holds the front sun-gear (Sf) stationary. Power flows from the torque-converter to the rear sun-gear (Sr) via the input shaft. The rear sun-gear (Sr) rotates clockwise and the rear annulus-gear (Ar) which is joined to the front planet-carrier (Cf) rotates in the same direction at the same speed.

The output rear planet-carrier (Cr) is joined to the front annulus-gear (Af); therefore the front annulus-gear (Af) drives the front planet-gears (Pf) in a clockwise direction; this compels the planet-gears to roll around the fixed front sun-gear (Sf) also in a clockwise direction. As a result, the front planet-carrier (Cf) will drive the rear annulus-gear (Af) also in a clockwise direction. Accordingly the clockwise rotation of the rear sun-gear, with the additional clockwise rotation of the rear annulus-gear (Ar) in the same direction, slows down the rear planet-gears (Pr) and speeds up the rear planet-carrier (Cr) and output gear at a slightly higher speed compared with the first gear ratio.

D drive range third gear (fig. 4.16(c)) With the selector lever in D drive range, the low clutch (LC) and high clutch (HC) are applied. The applied low clutch (LC) connects the front planet-carrier (Cf) to the rear annulus-gear (Ar), while the applied high clutch (HC) connects the input shaft to the front planet-carrier (Pf). Power flows from the torque-converter to the rear sun-gear (Sr) via the input shaft in a clockwise rotation. At the same time the applied high and low clutches provide a power flow path to the rear annulus-gear (Ar) via the front planet-carrier (Cf). Since the rear sun and rear annulus gears rotate in the same direction and at the same speed, the rear planet-carrier (Cr) and output gear will also be driven at the same speed. Third gear therefore provides a straight through one-to-one gear ratio in the forward direction. Above a programmed road speed the lock-up clutch automatically locks out the torque-converter in order to eliminate slippage between the impeller and turbine wheels within the torque-converter.

D drive range fourth gear (fig. 4.16(d)) With the selector lever in D drive range, the high clutch (HC) and brake-band (BB) are applied. The brake-band holds the front sun-gear (Sf) stationary, whereas the high clutch (HC) provides the flow path to the front planet-carrier (Cf). As a result the front planet-gears (Pf) are compelled to revolve around the fixed front sun-gear (Sf) thereby driving the front annulus-gear (Af) at an increased speed in the same clockwise direction. The power flow path then passes via the rear planet-carrier (Cr) to the output gear at a higher speed than that of the input shaft. Fourth gear therefore provides an overdrive gear ratio in the forward direction. As with the third gear, the lock-up clutch, at some predetermined vehicle speed, automatically joins the engine's power flow from the flywheel directly to the planetary gear train sets, thus by-passing the torque-converter which inherently suffers from slippage.

R drive range reverse gear (fig. 4.16(e)) With the selector lever in R reverse position, the low and reverse brake ((L+R)B) and the reverse clutch (RC) are applied. The low and reverse brake ((L+R)B) holds the front planet-carrier stationary and the power now flows though the rear clutch (RC) to the front annulus-gear (Af). Consequently the front planet-gears (Pf) are now compelled to rotate on their own axes, thereby driving the front annulus-gear (Af) at a reduced speed in a counter-clockwise (reverse) direction. The power flow path then passes via the rear planet-carrier (Cr) to the output gear at a lower speed than that of the input shaft. Reverse gear therefore provides an under-drive gear ratio in the reverse direction.

4.6.1 Multi-plate hydraulic actuated clutch (fig. 4.17)

Automatic transmission clutches comprise a pack of annular discs or plates, alternative plates being internally and externally circumferentially grooved to match up with the input and output splined drive members respectively (see fig. 4.17). When these plates are pressed together, torque will be transmitted from the input to the output members by way of these splines and grooves and the friction torque generated between pairs of rubbing surfaces. These steel plates are faced with either resinated paper linings or with sintered bronze linings depending on whether moderate or large torques are to be transmitted. Because the whole gear cluster assembly will be submerged in fluid, these linings are designed to be operated wet, that is in transmission fluid. These clutches are hydraulically operated by servo-pistons via a dished annular spring which provides a progressive squeeze load take-up.

The hydraulic fluid is supplied under pressure through axial and radial passages drilled in the input shaft and the torque-converter casing. To transmit pressurised fluid from one member to another where there is relative angular movement between components, the casing support sleeve has machined grooves on either side of all the radial supply passages. Square sectioned nylon sealing-rings are then pressed into these grooves so that when the casing sleeve is in position, these rings expand and seal lengthways a portion of the casing sleeve with the corresponding bore formed in the outer rotating sleeve member.

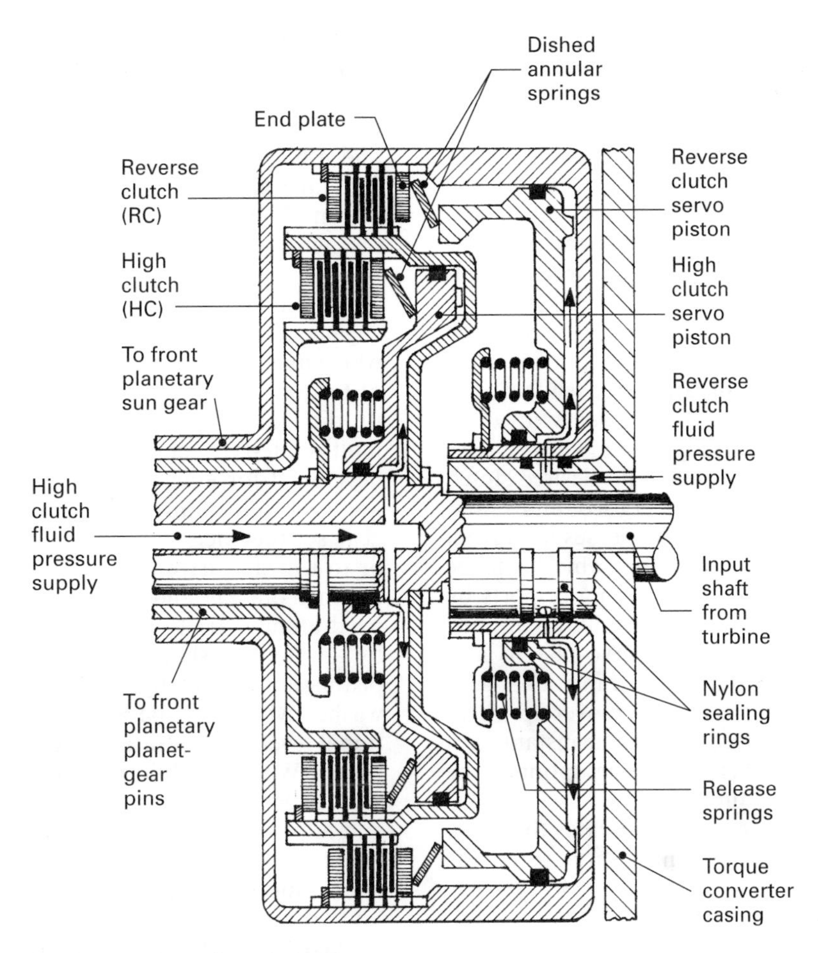

Fig. 4.17 Multiplate hydraulically activated clutches

High clutch (HC) When pressurised fluid is supplied to the high clutch servo-piston chamber, the piston will move over to the left-hand side and through the dish-shaped annular spring will smoothly clamp the drive plates together. The front planetary planet gear-pins will now be locked to the input turbine-shaft; therefore the power flow path, in addition to going to the rear sun-gear via the input turbine-shaft, also flows to the front planet gear-carrier (see fig. 4.16(c) and (d)). Releasing the pressurised fluid from the high clutch servo-chamber separates the input turbine-shaft from the front planet gear-carrier so that either the third or fourth gear is disengaged.

Reverse clutch When pressurised fluid is supplied to the reverse clutch servo-chamber, the servo-piston will be forced via the dished annular spring against the end-plate of the reverse multi-plate pack. This compresses the dished annular spring and so sandwiches the drive and driven plates together; the input turbine-shaft is now connected to the front sun-gear (see fig. 4.16(e)). Releasing the fluid pressure supply instantly disengages the input turbine-shaft from the front sun-gear due to the release springs and thus puts the transmission drive into neutral.

4.7 Hydraulic automatic gear change control system

Gear up and down shift is achieved by applying and releasing the various clutches and brakes at an appropriate time so that the predetermined power flow path for a particular gear ratio selected is complete. A hydraulic control circuit is used to actuate the gear changes and this is sensitive to vehicle speed and load conditions and to the acceleration pedal demands dictated by the driver.

The hydraulic circuitry operates by a fluid pump driven by the transmission's output shaft which generates pressure and directs fluid via a pressure regulator valve to the torque converter and the various clutch and brake servo-cylinders via a network of passages and valves.

Vehicle speed and engine load sensing valves (governor-valve and throttle-valve, respectively) monitor the operating conditions, and signal to a valve (known as a shift-valve) which switches the hydraulic circuitry when an appropriate gear change should take place. There is one shift-valve to actuate an up or down gear shift between each consecutive gear ratio; thus for a four-forward-speed transmission there are three shift-valves. The appropriate shift-valve for a given gear up or down shift then moves to a position which allows pressurised fluid to be directed to the desired clutch or brake servo which is to be energised, and simultaneously opens passages which releases fluid from the clutch or brake servo-chamber which is to be de-energised. The opening and closing of ports by the shift-valves are so designed that fluid is applied to and released from individual servo-pistons in a precise manner, so that during the transitional gear change period there is no hydraulic overlap where two gear ratios are engaged at the same instant, or a hydraulic delay where there is an interruption to the power flow for a short time.

The basic hydraulic circuitry operates though a number of components which are listed as follows:

1) fluid pump (P)
2) pressure regulator valve (PRV)
3) manual valve (MV)
4) torque converter relief valve (TCRV)
5) governor valve (GV)
6) throttle valve (TV)
7) detent valve (DV)
8) throttle and kickdown cam (TKC)
9) low and reverse ball valve ((L + R)V)
10) 1–2 shift-valve (1–2 SV)
11) 2–3 shift-valve (2–3 SV)
12) 3–4 shift-valve (3–4 SV)
13) multi-plate clutch servo chamber/piston units (LCS), (HCS) and (RCS)
14) multi-plate brake servo chamber/piston unit ((L + R)BS)
15) band-brake servo chamber/piston unit (BBS)

4.7.1 Clutch and brake servo-units with epicycle gearing

Smooth and silent gear change synchronisation is achieved by engaging or locking out various members of the planetary gear sets with virtually uninterrupted delivery of engine power. Rapid and precise gear change is made possible by incorporating multi-plate clutches, a multi-plate brake and a band-brake. Up or down gear shifts therefore take place with virtually simultaneous application of one clutch or brake and a corresponding disengagement of another clutch or brake.

Multi-plate clutches and brakes (fig. 4.18) To engage and disengage input and output drives which may be transmitting relatively large amounts of torque and at the same time have the capacity to dissipate the heat generated, compact wet multi-plate-type clutches and brakes have been developed for epicyclic gear transmissions. They are used to lock any two members of a planetary gear set together, to hold one member stationary or to transfer drive from one shaft or member to another quickly and smoothly. The rotating and fixed annular friction plates can also be engaged by an annular shaped hydraulic operated servo-piston either directly or indirectly by a dished washer which can be arranged to provide additional leverage to multiply the operating clamping thrust. When the fluid pressure is released, return springs separate the pairs of rubbing faces. Wear and adjustment of the friction plate pack is automatically compensated for by the piston being free to move further forward.

Brake-band and servo-units (fig. 4.18) Brake-bands form a compact and effective method of engaging or disengaging a part of an epicyclic gear train which has to be held stationary to produce a torque reaction before torque can be transmitted. The brake resembles a split friction spring band wrapped around an external drum; every time the band is tightly drawn together, the drum, which forms part of the planetary gears, is brought to a standstill. The braking of the drum is achieved through a push-rod actuated by a combined large stepped double-acting second gear servo-piston and a smaller single-acting fourth gear servo-piston. Fluid line-pressure is introduced either to the underside of the annular face of the large piston or to the underside of the small piston to energise the band-brake when shifting into second or fourth gear respectively. To release the band-brake, line-pressure is directed to the upper spring-chamber side of the large piston. Applying and releasing the band-brake via these different sized servo-pistons provides an extended and progressive energising and de-energising

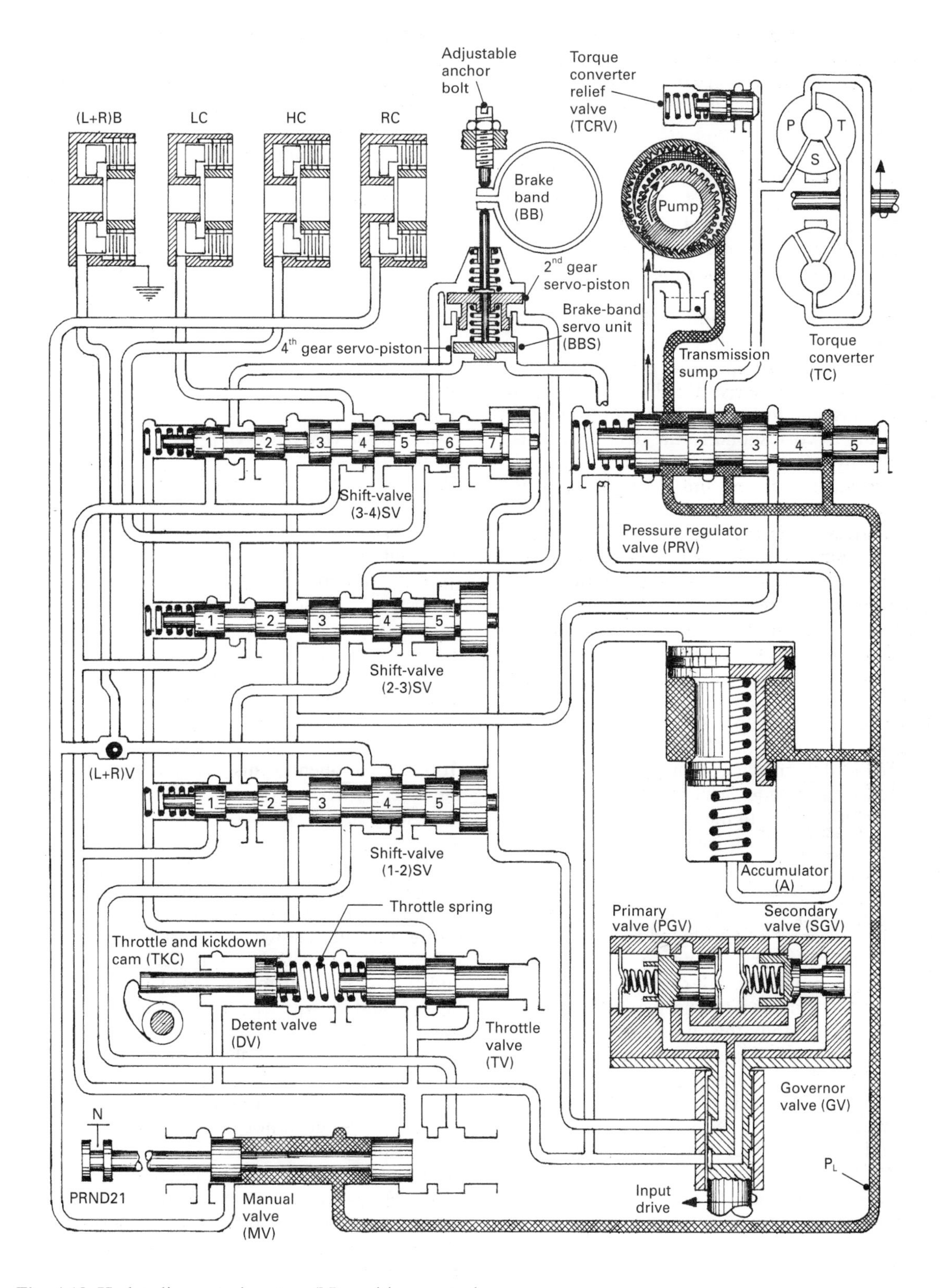

Fig. 4.18 Hydraulic control system (N) position neutral

action which matches the overlap between releasing one member and then holding another member of the planetary gear sets. Band wear slackness can be taken up by externally adjusting the anchor screw.

4.7.2 Pressure regulator valve (PRV) (figs 4.18–4.22)

Fluid pressure is generated by the pump driven by the engine. Output pressure from the pump will increase approximately in proportion to the engine speed. This pressure build-up does not always match the brake-band and clutch servo-piston clamping load requirements for different vehicle speed and load operating conditions.

The pressure regulator valve (PRV) is incorporated to adjust the fluid pump's output flow rate and pressure to correspond to the operating conditions at any one time. The pressure is adjusted to provide maximum pressure clamping load at low speed and heavy acceleration, when high torque loads are transmitted, and to reduce this pressure for smaller loads and light acceleration conditions so that smooth and gentle gear changes take place instead of fierce clutch and brake engagements. A reduction in pump output pressure also reduces the power consumption, wear and level of noise generated by the hydraulic pump.

Most pressure regulator valves (PRV) are of the spring loaded spool type, that is, a plunger with several reduced diameter sections along its length, positioned in a cylinder which has a number of passages interacting with the cylinder walls. When the engine speed and corresponding pump pressure is low, fluid flows via the inlet ports around the wasted section of the plunger and out unrestricted along a passage leading to the manual valve (MV), where it is distributed to the various control valves and operating servo-pistons. As the pump pressure builds up with rising engine speed, line-pressure is directed to the rear face of the plunger's fourth land; this progressively moves the plunger forward against the control-spring, causing the land (2) to uncover the port leading to the torque-converter; it thereby circulates fluid under pressure around the impeller, turbine and stator blades. With further pump pressure rise, the plunger's first land will uncover the port leading to the intake side of the hydraulic pump, so that a portion of the output from the pressure is circulated back to the suction side of the pump, thus preventing an excess in output pressure. It therefore regulates the output fluid pressure known as line-pressure according to the control spring stiffness.

To increase the torque transmitting capacity during kickdown driving conditions, a kickdown pressure is introduced to the plunger between lands 4 and 5; this assists the line-pressure in moving the plunger against the control-spring so that the effective port orifice leading to the torque-converter is enlarged, it thus subjects the torque-converter to higher fluid pressure.

Most automatic transmissions also subject the pressure-regulator-valve plunger to throttle-pressure in such a way that line-pressure is made to rise with the severity of load and acceleration to increase the engagement damping load on the brake and clutch piston servos.

4.7.3 Torque converter relief valve (TCRV) (fig. 4.19)

This valve acts as a safety valve to relieve torque-converter pressure if it should become excessively high. When the pressure rises to the preset maximum determined by the spring stiffness, the spring is compressed and the plunger moves back to open the exit port; it thereby relieves the excess fluid pressure.

4.7.4 Manual valve (MV) (fig. 4.19)

The manual valve directs the line-pressure generated by the pump and controlled by the pressure-regulator-valve to the various hydraulic controlled components, this being dependent upon the selector lever position (PRND21) which is chosen by the driver.

4.7.5 Governor valve (GV) (figs 4.18–4.22)

The governor valve senses vehicle speed and converts an input line-pressure to a reduced variable output pressure known as governor-pressure (GP) which increases with a rise in vehicle road speed and then conveys it to the shift-valves to control the up and down gear shifts. The governor is driven by a gear which meshes with a gear attached to the output shaft.

The governor valve consists of a drive spindle with a disc-shaped flange at one end; mounted on this disc is the governor body which incorporates both primary and secondary spring loaded valves. Line-pressure enters the governor unit via a circumferential groove formed in the drive spindle; it then flows to the wasted portions of the secondary and primary valves, respectively. When the vehicle is stationary the outer land on the primary valve blocks the exit from the governor body. However

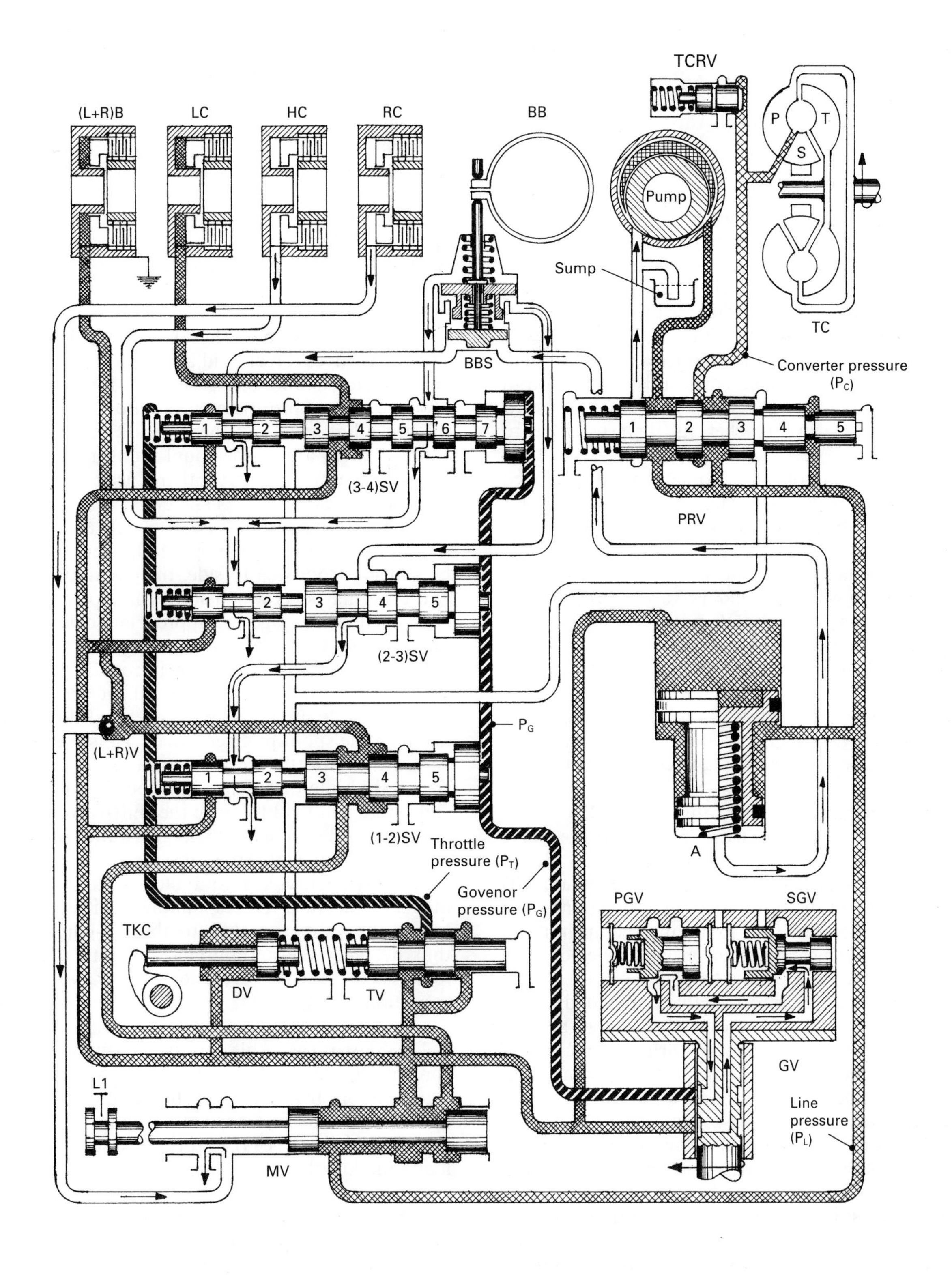

Fig. 4.19 Hydraulic control system (L) range first gear (similar to (D) range except (L+R)B is applied)

when the vehicle starts to move and the governor drive spindle rotates, centrifugal force acts on the primary valve mass, moving it outwards against the resisting governor spring. The outer land of the primary valve now partially uncovers the governor exit port passage. This permits fluid to flow to the right-hand side of the three shift valves at a reduced pressure known as governor-pressure; it thus provides the first stage of governor pressure control. As the vehicle speed and hence the governor spindle drive rotational speed increases, so does the centrifugal force acting on the primary-valves mass, and further out will move the primary-valve; this reduces the exit port/land restriction which thus raises the output governor pressure. At still higher vehicle speeds the primary-valve reaches its limit to outward movement. When this occurs, the increased line-pressure acting on the secondary-valve's stepped annular land face will cause the secondary-valve to be pushed inwards against the secondary spring and the centrifugal force trying to force the secondary-valve outwards. As a result, the inward moving secondary-valve produces a second stage inlet port/land restriction output pressure control. Thus it can be seen that this type of governor valve provides two distinct phases of regulation, the first being sensitive to low vehicle speed control, and the second phase providing governor pressure control over the much wider medium to high vehicle speed range.

4.7.6 Throttle valve (TV) (figs 4.18–4.22)
The throttle valve senses engine throttle valve opening (engine load) and provides a reduced variable pressure known as throttle pressure (Pt) to one end of the shift-valves in opposition to the governor-pressure (Pg) to modify the upshift and downshift vehicle speed according to the demands of the driver and vehicle load.

The spool-type throttle valve is actuated by the throttle and kickdown cam through a throttle-spring which applies an end thrust on the spool-valve in proportion to the throttle opening, that is, according to the downward movement of the accelerator pedal. Line-pressure acting on the right-hand end annular land face opposes the throttle spring thrust, so that the throttle-spool-valve is continuously finding a new equilibrium position as operating conditions change. When the accelerator pedal is only lightly pressed down, the throttle-spool-valve right-hand land blocks the passage leading to the spring end of all three shift-valves. As the accelerator pedal is pushed down more, the increased throttle spring thrust moves the throttle-spool-valve to the right-hand side against the opposing line-pressure. Consequently the right-hand land inner edge uncovers the port which permits line-pressure to pass to the spring end of each of the three shift-valves. Throttle pressure will now assist the shift-valve spring to oppose governor-pressure and therefore delays the respective gear-ratio upshift until a higher governor-pressure is delivered to the shift-valves.

4.7.7 Shift-valves (SV) (figs 4.18–4.22)
The shift-valves direct fluid at line-pressure to the various clutch and brake servo-piston units to produce the appropriate gear-ratio change for the prevailing load and speed conditions.

Shift-valves are of the spool plunger type; the plunger length has several reduced diameter sections which act as passageways for fluid to pass to and from the various clutch and brake servo-piston units. These spool-plungers operate by shifting from side to side, thereby uncovering and covering ports which supply line-pressure to engage and disengage the appropriate clutch and brake servo-units. Shift-valves act as buffers between the throttle-pressure introduced at the spring end of the spool-valve and the governor-pressure to dominate over the eventual direction in which the spool-plunger moves. Thus if there is very little throttle-pressure at a given vehicle speed, the state of balance between the opposing end forces will tend to move the shift-valve to the left-hand side to actuate a gear-ratio upshift. However if the throttle-pressure is high, the governor-pressure has to build up to a much higher value (and corresponding vehicle speed) before the valve will shift to the left-hand side, thus delaying the appropriate gear-ratio upshift for that particular vehicle speed and load condition.

1–2 Shift-valve This valve automatically controls the hydraulic circuitry which shifts the transmission from first up to second or from second down to first depending upon governor-pressure (vehicle speed), throttle-pressure (engine load) and kickdown-pressure (rapid acceleration).

2–3 Shift-valve This valve automatically controls the hydraulic circuitry which shifts the transmission from second up to third or from third down to second gear according to the state of balance between governor-pressure, throttle-pressure and kickdown-pressure.

3–4 Shift-valve This valve automatically controls the hydraulic circuitry which shifts the transmission from third up to fourth or from fourth down to third gear according to the imposing signals supplied by governor-pressure, throttle-pressure and kickdown-pressure.

4.7.8 Accumulator (figs 4.18–4.22)
An accumulator is basically a spring loaded piston encased in a cylinder and is connected through line-pressure passages to the low clutch, high clutch and band-brake piston servo-units. The accumulator shown in fig. 4.18 has a large and small diameter cylinder which meets end on, matching corresponding large and small diameter pistons joined together by a hollow wasted section, and accommodated within the piston and small cylinder is the accumulator spring.

When any of these clutch or brake servo-pistons are applied, the inertia of line-pressure introduced to any of the servo-piston chambers is liable to cause a harsh or shock engagement. To ensure a smooth and progressive clutch or brake engagement, the line-pressure about to actuate a piston servo-unit is at the same time also directed against the spring-loaded accumulator piston. Thus immediately fluid enters the servo/accumulator circuit, the accumulator piston is pushed back to a point of equilibrium to absorb and damp the surge of fluid which has suddenly been released to an appropriate servo-piston chamber during an upshift or downshift of gears. The annular space behind the piston is also subjected to line-pressure so that the amount of expansion of the accumulator cylinder chamber and hence its damping response caused by the piston compressing the spring also influence the magnitude of a line-pressure fed to the middle reduced waste region of the piston which assists the resisting spring thrust at any one time when a gear change takes place.

It should also be observed that the band-brake servo-end piston chamber has an additional passage that connects it to the small diameter accumulator chamber which contains the spring. Thus when the band-brake is applied to engage fourth gear, the shock-absorbing small diameter expansion chamber piston now has to act with the line-pressure in the middle region of the piston, and the spring thrust against the opposing line-pressure coming from the large end of the accumulator piston. The accumulator in this instant is double ended.

4.7.9 Supplementary valves
Additional valves not shown on the hydraulic layout are generally incorporated in the hydraulic system to improve the quality of the time sequence for the up and down gear shifts so that smoother transitional gear changes take place. One such valve is a 2–3 timing-valve which provides a restriction orifice to slow the application fluid to the high clutch while the slower reacting band is being released. It thus prevents a hard 2–3 or 3–2 down shift.

4.8 A four-speed and reverse simplified hydraulic control system
A step-by-step explanation of how the hydraulic control system provides the basic automatic up and down gear shift now follows.

Neutral position (fig. 4.18) With the selector lever in the N neutral position, the manual-valve (MV) spool plunger blocks off the line-pressure (Pl) supplied by the fluid pump via the pressure-regulator-valve (PRV) from the rest of the passages leading to the various control valves, clutches, and brake servos, and at the same time uncovers its two exhaust ports so that all line-pressure in the hydraulic control system is released. Consequently all the clutches and brake are in their respective disengaged position.

First gear (fig. 4.19) With the sector lever in D drive range, fluid is delivered from the oil sump to the pressure-regulator-valve. It then divides, some being delivered to the torque-converter, the remainder passing out to the manual-valve (MV) as regulated pressure, more commonly referred to as line-pressure (Pl).

The low clutch (LC) is applied by line-pressure from the manual-valve (MV) relayed via the 3–4 land groove of the 3–4 shift-valve; it thus causes the front planet-carrier to hold. If the sector lever is in L drive range, the low and reverse brake ((L + R)B) will be also applied by line-pressure from the manual-valve via the 3–4 land groove of the 1–2 shift-valve and the low and reverse ball-valve ((L + R)V). At the same time, line-pressure (Pl) from the manual-valve passes to the throttle-valve, detent-valve and the governor-valve. The reduced pressure output from the governor-valve which is known as governor-pressure (Pg) is directed to the end faces of the three shift-valves, whereas the output pressure from the throttle-valve (TV), known as throttle-pressure (Pt), is conveyed to the spring ends of the (1–2), (2–3) and (3–4) shift-valves.

Whilst the transmission is in D drive first gear, the one-way-clutch will engage, so preventing the front planetary gear set planet-carrier from rotating, that is, holding (not shown in hydraulic system).

Second gear (fig. 4.20) With the manual-valve (MV) still in D drive range, hydraulic conditions will be similar to first gear, that is, line-pressure (Pl) from the manual-valve (MV) is directed via the 3–4 land groove of the (3–4) shift-valve to engage the low clutch (LC), except that rising vehicle speed increases the governor-pressure sufficiently to push the 1–2 shift-valve to the left-hand side against both spring and throttle-pressure. Line-pressure is now directed via the groove between lands 1–2 of the 1–2 shift-valve and the 3–4 land groove of the 2–3 shift-valve to the band-brake piston servo on the applied side, thus engaging the band-brake and causing the front sun-gear to hold.

If there is a reduction in vehicle speed or if the engine load is increased sufficiently, the resulting imbalance between the spring and throttle pressure load as opposed to governor-pressure acting on the 1–2 shift-valve at opposite ends causes the shift-valve to move against the governor-pressure to the right-hand side. Consequently the hydraulic circuitry will switch back to first gear conditions, causing the transmission to shift down from second to first gear again.

Third gear (fig. 4.21) As for second gear, line-pressure (Pl) passes from the manual-valve (MV) to the low clutch (LC) via the 3–4 shift-valve 3–4 land groove to continue the engagement of the low clutch, thus causing the front planet-carrier and rear annular-gear to be connected. At even higher road speeds in D drive range, the governor-pressure (Pg) will have risen to a point where it is able to overcome the spring and throttle-pressure thrust of the 2–3 shift-valve. This causes the spool-valve to shift over to the left-hand side. Line-pressure will now pass through the 2–3 shift-valve between lands 1–2 to the release side of the band-servo via the 3–4 shift-valve 5–6 land groove to disengage the band-brake, and at an instant later, this line-pressure will be directed to the high clutch (HC) servo-piston thus engaging the high clutch; it thus joins the input shaft to the front planet-carrier. Consequently both front and rear planetary gear sets lock up permitting the input drive from the torque-converter to be transmitted directly through the transmission's output gear. The actual vehicle speed at which the 2–3 shift-valve switches over will be influenced by the throttle opening (throttle-pressure). A light engine load (low throttle-pressure) will cause an early gear upshift whereas a high engine load (high throttle-pressure) will raise the upshift speed.

Fourth gear (fig. 4.22) As for third gear, the high clutch (HC) is still applied by line-pressure passing from the manual-valve (MV) to the high clutch piston-servo via the land groove 1–2 of the 2–3 shift-valve so that the input-shaft and front planet-carrier will be connected. With still higher road speeds in D drive range, the increased governor-pressure (Pg) will actuate the 3–4 shift-valve, forcing it to shift across so that it closes the passage leading to the low clutch (LC) between lands 3–4 and then uncovers the exhaust or drain port between the land grooves 4–5 to release the line-pressure being applied against the low clutch servo-piston. Immediately after the low clutch (LC) is disengaged, the 3–4 shift-valve uncovers the line-pressure passage between the grooved lands 1–2 so that line pressure passes from the manual-valve (MV) to the applied side of the band-servo small piston; this energises the band-brake, thereby causing the first sun-gear to hold.

Reverse gear (fig. 4.23) With the manual-valve in R reverse gear position the low clutch (LC), high clutch (HC) and band-brake servos have all been exhausted of line-pressure. Line-pressure from the manual-valve (MV) is now directed directly to the reverse-clutch (RC) and to the low and reverse brake ((L + R)B) via the low and reverse ball valve ((L + R)V). This causes the input-shaft to be connected to the front sun-gear and the front planet-carrier to be held respectively. It should be observed that the low and reverse ball valve ((L + R)V) is pushed to the right-hand side to block the 1–2 shift-valve 3–4 land groove passage when the reverse-gear is selected.

D Drive range kickdown (fig. 4.24) A forced downshift from fourth to third, third to second or from second to first gear when a rapid acceleration response is required can be achieved when the accelerator-pedal is fully depressed. This is accomplished by the throttle and kickdown-cam

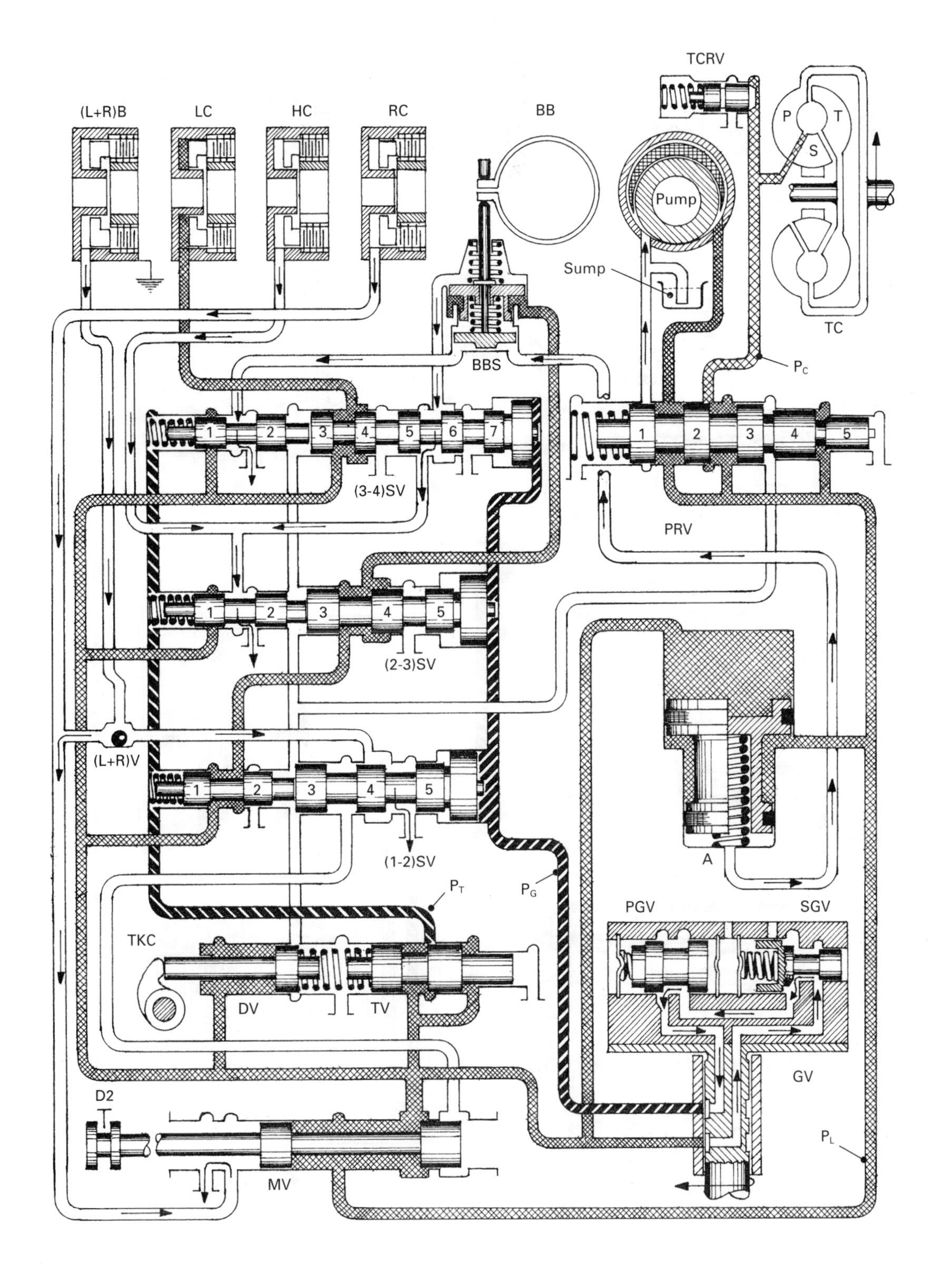

Fig. 4.20 Hydraulic control system (D) range second gear

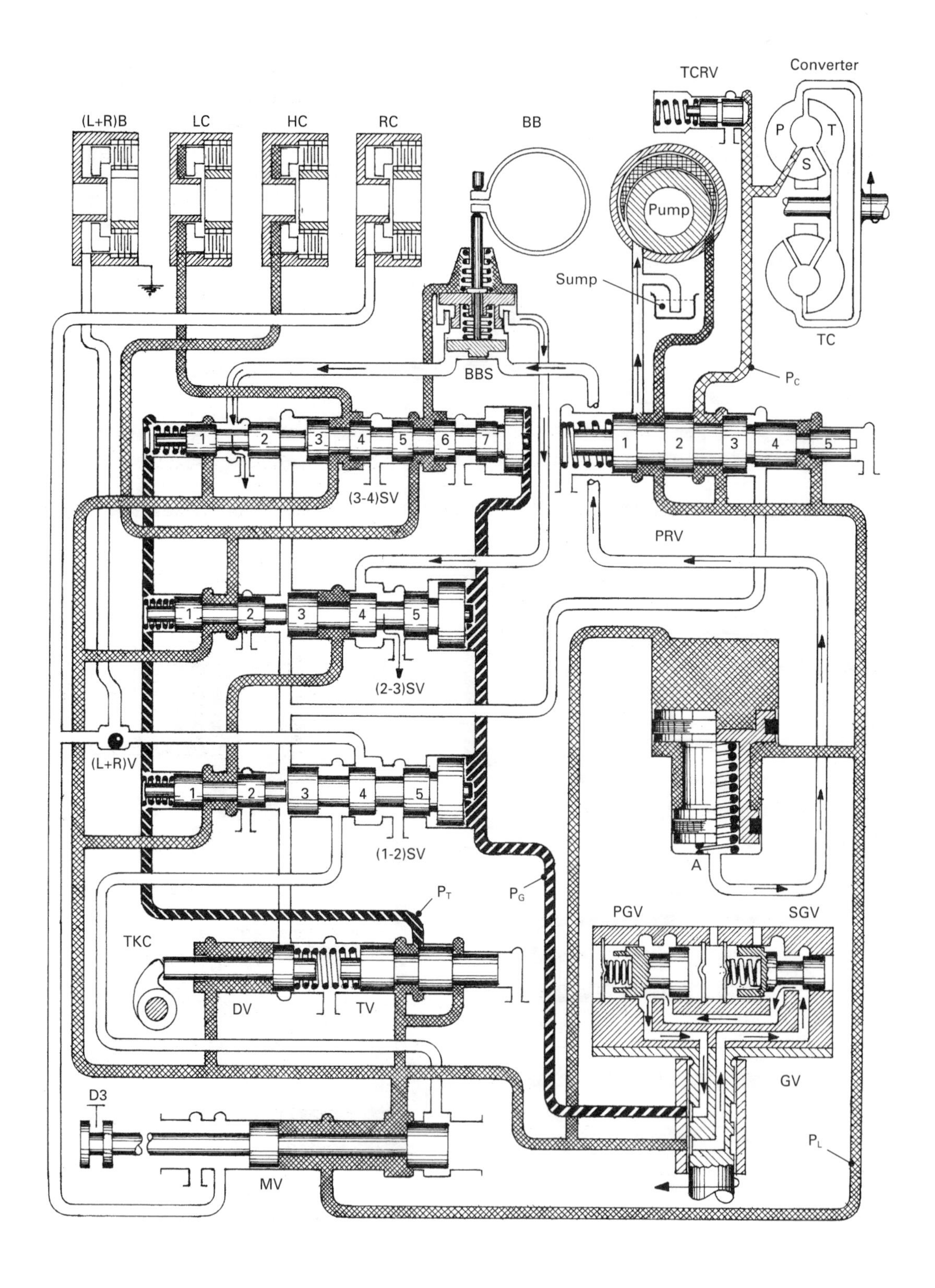

Fig. 4.21 Hydraulic control system (D) range third gear

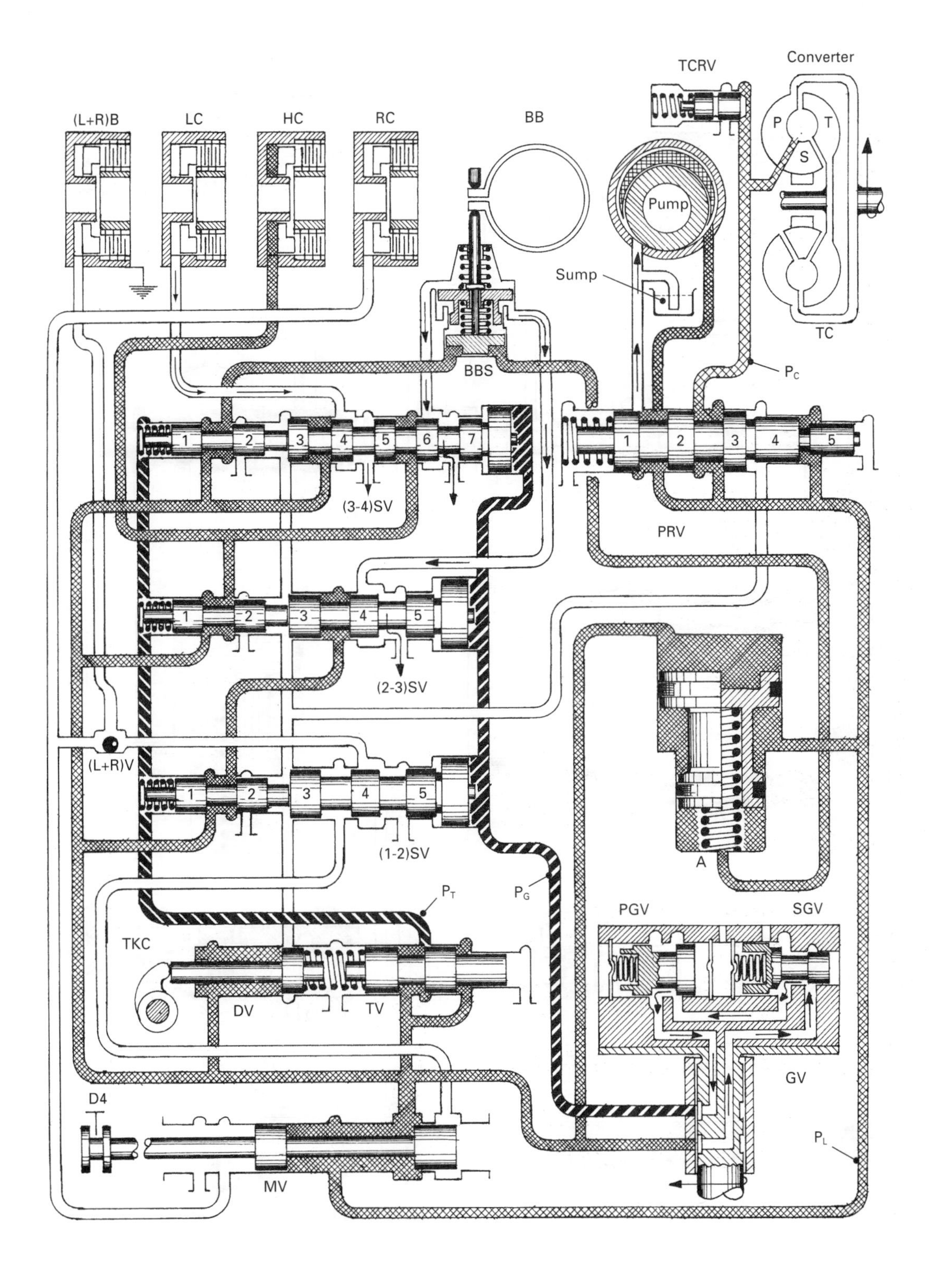

Fig. 4.22 Hydraulic control system (D) range fourth gear

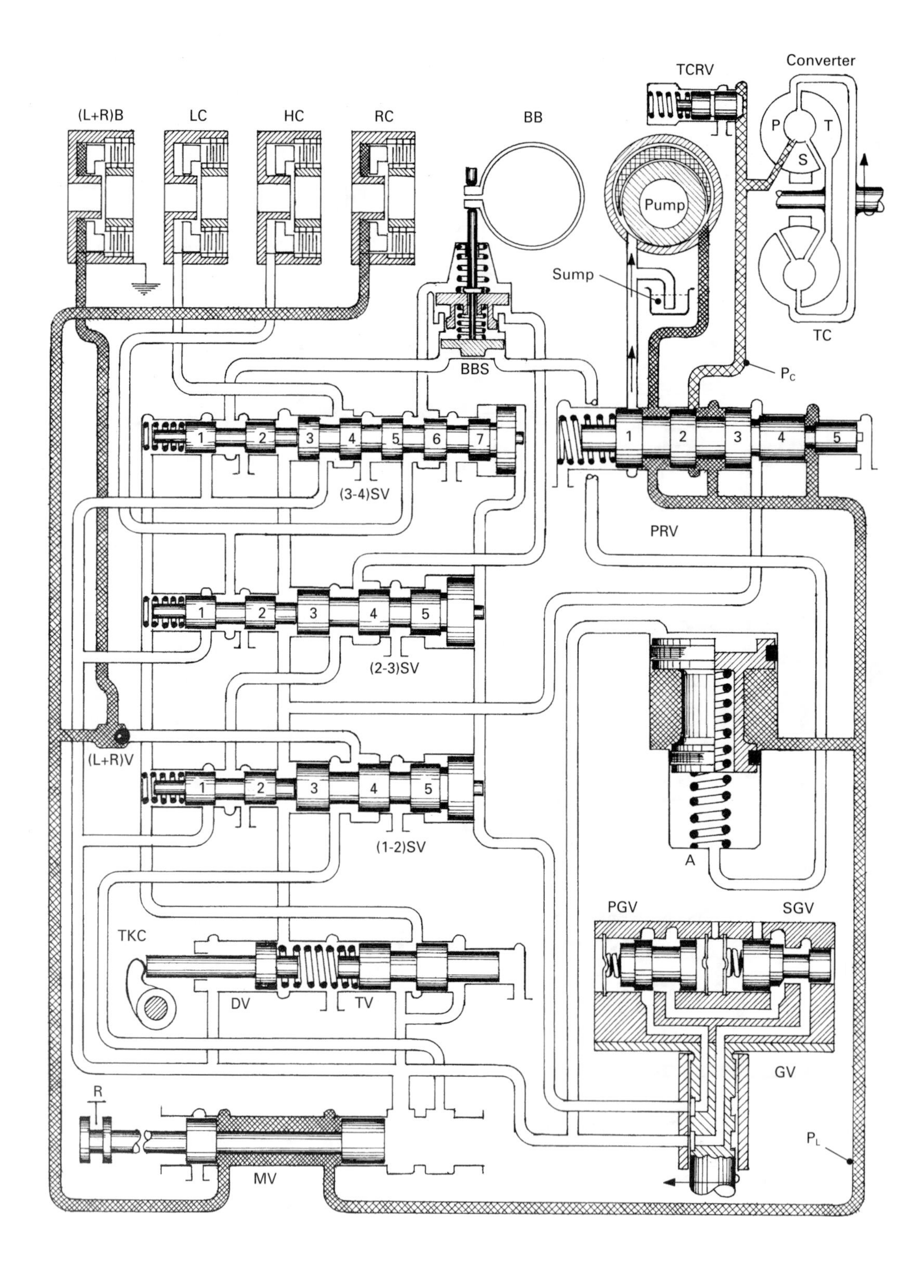

Fig. 4.23 Hydraulic control system (R) position reverse gear

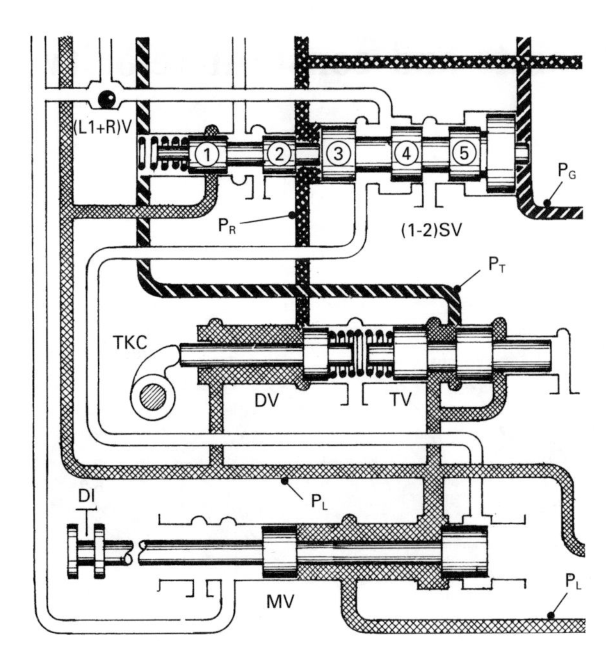

Fig. 4.24 Fluid flow in drive range – first gear forced kickdown

(TKC) which is linked to the accelerator-pedal pushing over the detent-valve (DV) until it uncovers the passage connecting the manual-valve (MV) to the 2–3 land grooves of each of the three shift-valves. Accordingly, since the exposed annular area on land 3 is larger than that of land 2, there will be an additional kickdown-pressure (P_k) opposing the governor-pressure (P_g). Hence under full throttle acceleration, the spring force F_s, throttle-pressure thrust F_t and the additional kickdown-pressure thrust F_k opposes and overcomes the governor-pressure thrust F_g, that is, $F_s + F_t + F_k > F_g$. As a result the respective shift-valve operating at a particular vehicle speed will move to a downshift position (shuttle spool valve shifts to the right-hand side) and so actuates a gear shift to a lower gear-ratio, or if the transmission is in first gear, it delays the upshift until the vehicle is moving at a much higher speed. When the accelerator-pedal is released, the anti-clockwise rotation of the throttle and kickdown-cam (TKC) allows the detent-valve to move to the left-hand side and in the process blocks off the line-pressure leading to the shift-valves and then opens the exhaust port. The collapsed kickdown-pressure (Pk) now permits the shift-valves to operate in a normal manner between the opposing throttle and governor pressures.

5 Propeller shafts, universal joints and constant velocity joints

5.1 Propeller shaft (figs 5.1 and 5.2)

The propeller shaft transfers the power drive from the output shaft of the front body-mounted gearbox to the unsprung rear-axle final drive, so providing the means of propelling the vehicle forward or in reverse.

The propeller shaft consists of a low-carbon-steel tube, either formed from rolled steel sheet which is then butt-welded along its seam or made from seamless drawn tubing. The hollow shaft ends are made a force fit over solid cylindrical recesses turned down on both universal-joint yoke bosses supporting the shaft, or alternatively between one universal-joint yoke boss and a sliding-joint splined stub shaft. The ends of the tube are then butt-welded to their respective yoke-boss or stub-shaft shoulder to make a rigid drive structure which is concentric to the input and output support shafts (fig. 5.1).

The shaft is made hollow to apply its mass in the most effective position to resist twisting, longitudinal sagging, and rotational whirl.

The torsional stiffness and resistance to bending of a shaft increase directly in proportion to the cube of the shaft's diameter. This means that only a small increase in diameter will considerably stiffen the shaft, but the centre portion of the shaft contributes little to the torsional stiffness of the shaft.

A comparison of weight and torque-capacity properties of solid and hollow shafts can be illustrated by the example in fig. 5.2. Taking a typical size for a tubular propeller shaft of 60 mm outside diameter and 2 mm wall thickness, this can be shown to be equivalent to a 21.5 mm diameter solid shaft of the same weight. Similarly it can be proved that the torque-carrying capacity for the hollow shaft is approximately 5.2 times as large as that for the solid shaft. Conversely, a solid shaft having an equivalent torsional stiffness to the 60 mm diameter hollow shaft would have to have its diameter increased to 37.3 mm, but this would make it about 3 times as heavy as the hollow shaft.

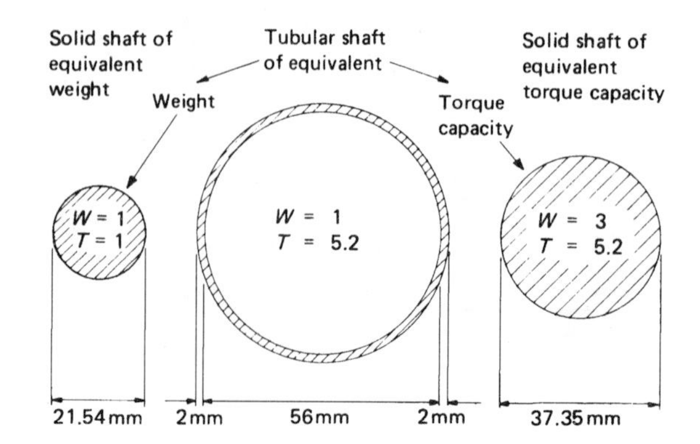

Fig. 5.2 Comparison of weight (W) and torque-carrying capacity (T) of solid and hollow propeller shafts (three shafts drawn in proportion)

5.2 Hooke's universal joint (fig. 5.3)

The first universal joint was known to have been designed by the sixteenth-century physicist and

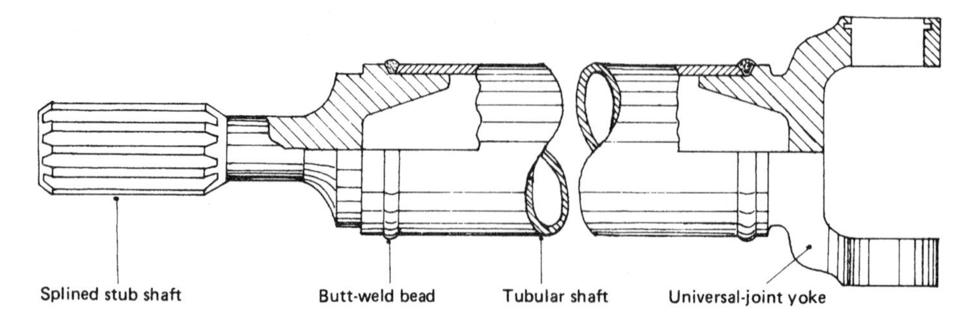

Fig. 5.1 Typical propeller-shaft assembly

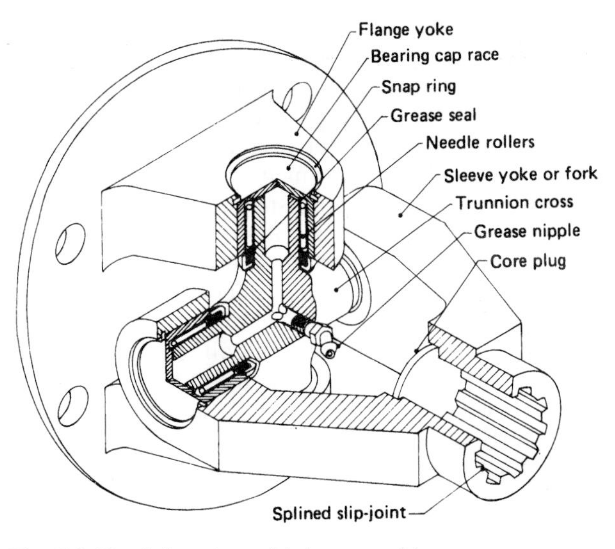

Fig. 5.3 Hooke's universal-joint assembly

mathematician Jerome Cardan, but it is to Robert Hooke, living about 100 years later, that the first practical universal joint is usually attributed.

The Hooke's universal joint consists of two forged-steel yokes or forks joined to the two shafts being coupled and situated at right angles to each other (fig. 5.3). A trunnion cross or 'spider' hinges these two yokes together. Since the arms of the trunnion cross are at right angles, there will be four extreme positions during each revolution when the entire angular movement is being taken by only one half of the joint. This means that the trunnion-cross arm rocks backwards and forwards between these extremes and, if the input rotational speed is uniform, the output-shaft speed will therefore increase or decrease twice in one revolution, the speed fluctuation becoming greater as the inclination between the two shafts increases and vice versa.

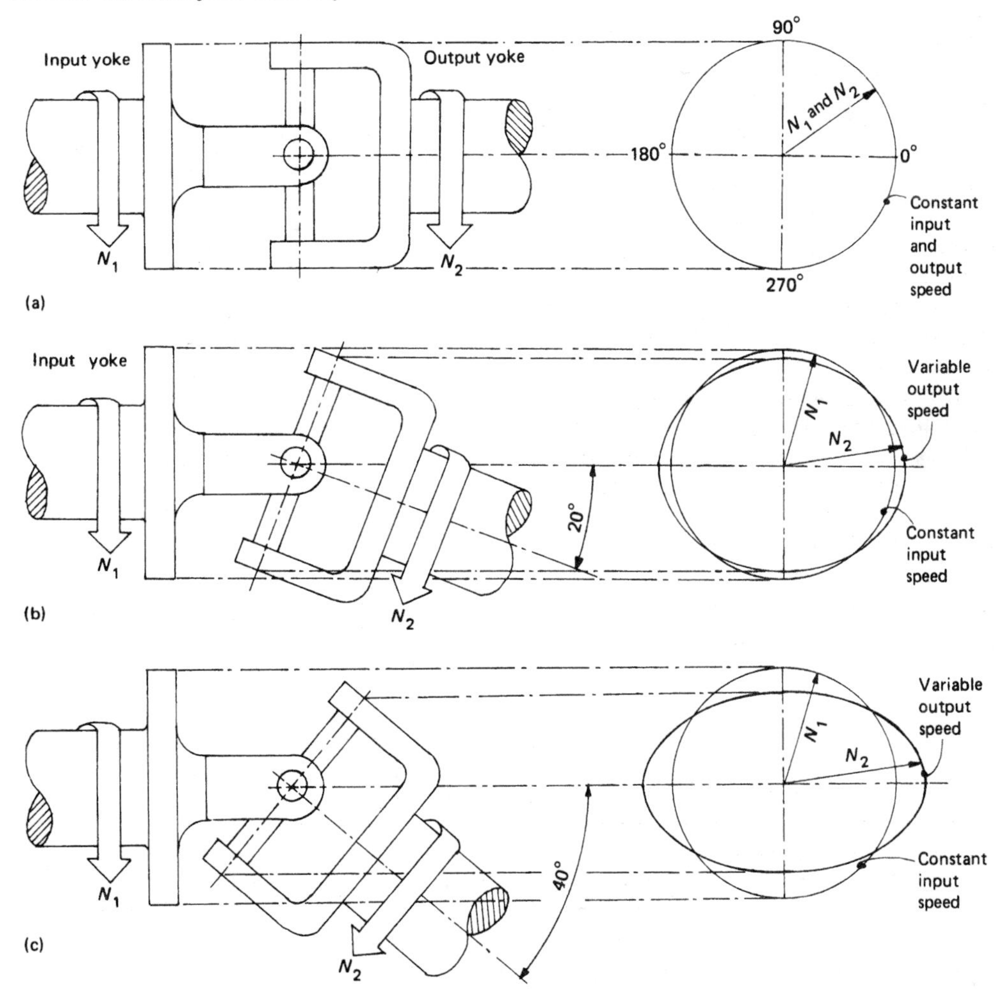

Fig. 5.4 Universal-joint speed variation with different working angles

Friction due to rubbing between the trunnion arms and the yoke bores is minimised by incorporating needle-roller bearings between the hardened trunnion-arm journals and hardened bearing caps pressed into the yoke bores.

5.2.1 Speed variation with angle of inclination (fig. 5.4)

Figure 5.4(a) shows the two shafts aligned, and the radius of the circle represents the magnitude of the constant input speed. In this case, with a zero working angle between the shafts, the output speed is also uniform throughout one complete revolution.

When the output shaft is inclined at an angle of 20° to the input shaft (fig. 5.4(b)), it can be seen that the projected lines from the tips of the output trunnion arms lie inside the input-speed circle, but if the shaft is rotated through 90° the projections would lie outside the circle. A complete projection plot would produce the elliptical figure shown, which would then represent the variation in output speed over one cycle, compared to the constant input speed shown by the circle. The lengths of the radius arms N_1 and N_2 represent the magnitudes of the input and output speeds respectively.

With still greater inclination of the output shaft to 40° (fig. 5.4(c)), the pivoting of the universal joint will produce a much greater increase and decrease in speed relative to the constant input-shaft speed as shown.

The graph in fig. 5.5 illustrates the percentage speed fluctuation over a possible working-angle range for a Hooke's universal joint; however, for practical reasons the limiting angles would depend upon shaft speed and would therefore very rarely exceed 20°.

Large joint angles, with their inherent output-speed fluctuation, may at certain speeds produce propeller-shaft inertia excitation – that is, torsional oscillations – which will result in excessive noise, unpleasant transmission vibration, and consequent rapid wear of the needle-roller-bearing and splined-slip-joint parts.

A point which is not generally appreciated is that it is preferable to have a small operating angle, rather than none at all, with a Hooke's joint incorporating needle-roller bearings. If the angle is greater than about 1°, the needles will roll slightly and thus prevent the highly stressed contact areas being in the same position at all times. Joints which are operated with no joint angle under heavy loads may squeeze away the boundary grease, and the needles then 'bed' or 'brinell' into the cap race and trunnion journal.

5.3 Universal-joint needle-bearing lubrication (fig. 5.6)

The flange and sleeve yokes are assembled on the trunnion journals with four needle-roller-bearing assemblies. The needle bearings are self-contained, being assembled in a hardened-steel cap race, and operate directly on to hardened trunnion-arm journals (fig. 5.6). The bearing cap race is located and retained either by a snap ring in a groove situated in the yoke bore or by a lockplate and set-screws.

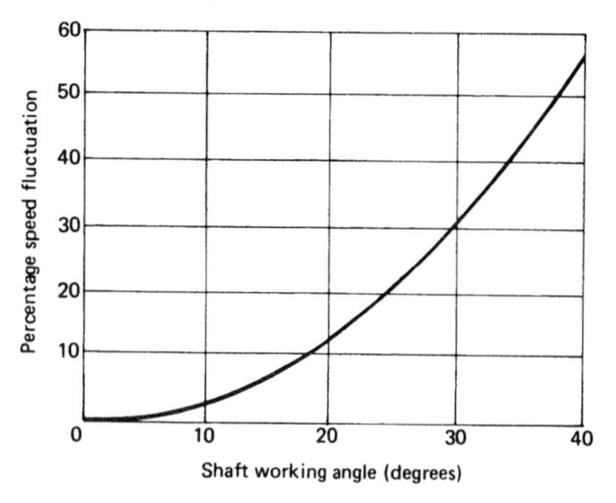

Fig. 5.5 Relationship between universal-joint angle and output-speed variation

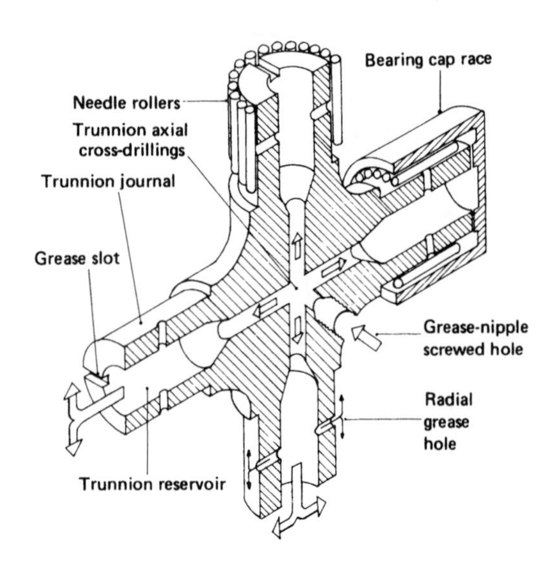

Fig. 5.6 Universal-joint trunnion-cross and needle-race lubrication

The trunnion cross or spider is a steel forging with four journal arms protruding from the central core. Axial holes drilled through the middle of each arm intersect in the centre of the trunnion core. A grease nipple or plughole, internally threaded, is drilled between two adjacent trunnion arms and intersects the cross-drillings at the centre. Larger holes drilled at the outer ends of the trunnion journal arms form grease reservoirs. Some large universal joints used for heavy-duty conditions have small radial drillings in each arm, about midway along the journal. A slot is machined across the tip of each trunnion arm so that grease can flow between the arm ends and the bearing cap.

Lubrication is achieved by pumping grease through the grease nipple. The grease is forced outwards to fill up each arm reservoir, and further greasing will push grease through the arm-tip slots, around and between the cap race and the trunnion journal to fill up the spaces between the needle rollers.

During operation – that is, when the propeller shaft is revolving – the centrifugal force acting on the grease mass in the trunnion-arm reservoir will force grease outwards from the centre to replace

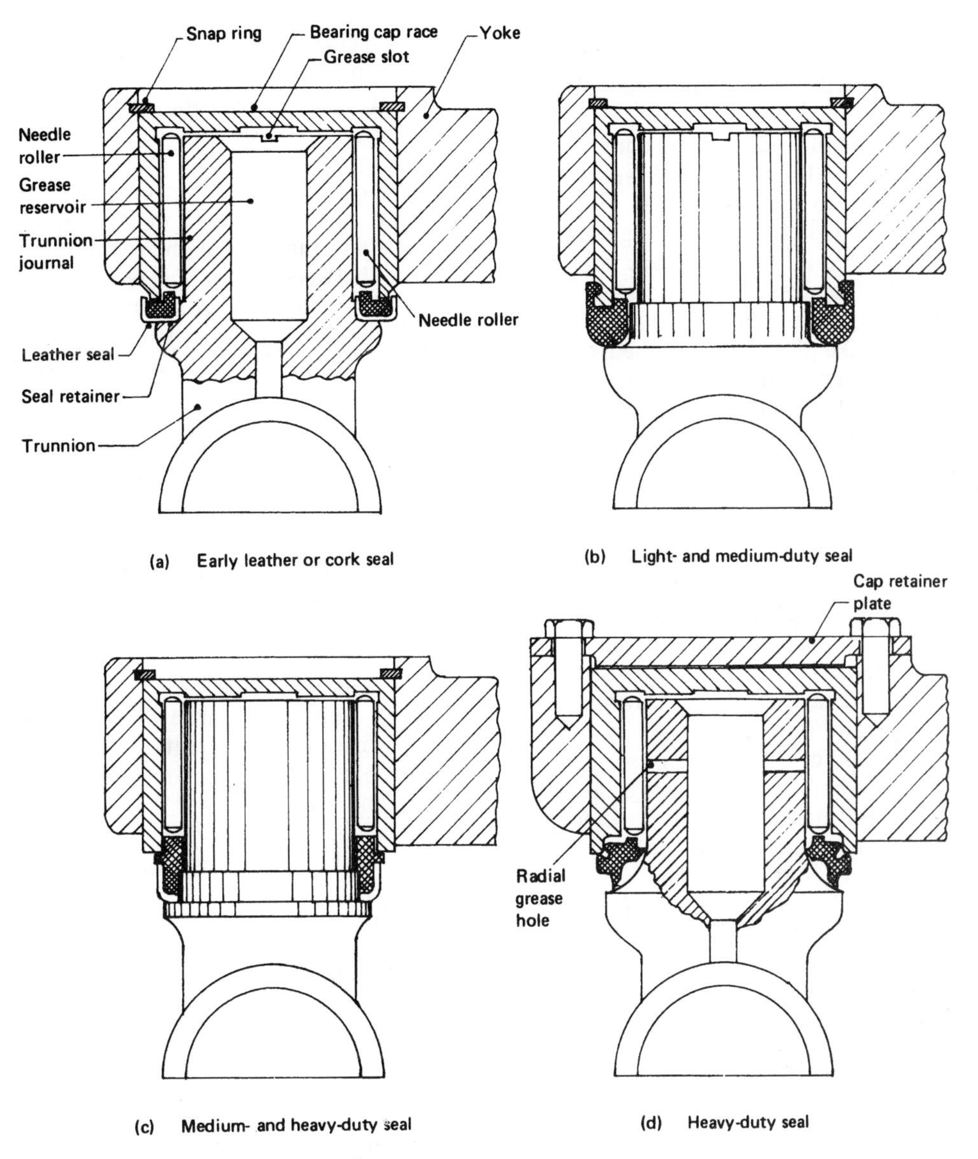

Fig. 5.7 Needle-roller-bearing grease retention

the existing grease which may have disintegrated due to continuous use.

On some cars, the universal joints are prepacked with grease for life through a grease nipple which is then removed and replaced by a plug to seal the screw hole.

Heavy-duty universal joints have a relief valve positioned between two of the trunnion arms to prevent damage to the seals through overlubrication or the use of high-pressure grease equipment. This serves as an indicator to show when the joint is completely filled.

5.3.1 Grease retention in needle race (fig. 5.7)

Grease seals are situated between the trunnion journal and the bearing cap race to retain the grease which is forced between the needle-roller bearings.

Early seals for both cars and commercial vehicles were made of leather or cork supported in an annular steel pressing (fig. 5.7(a)). Here the steel seal dust-cover is lightly pressed against the trunnion shoulder and a leather or cork ring seal sits in the recess in the pressing. The seal assembly will then move as part of the trunnion, so that wiping action will take place between the ring seal and the bearing-cap circumferential edge.

Modern seals used for light and medium duty are made of synthetic moulded rubber and do not use a dust-cover (fig. 5.7(b)). The sealing action is slightly different from that of the previous leather and dust-cover arrangement, as the rubber seal is so shaped as to fit over a recessed groove machined on to the outside of the bearing cap. The seal grips the bearing cap so that the wiping action is between the rubber seal and the recess formed on the inner part of the trunnion journal.

For medium- to heavy-duty applications, the rubber seal is given additional support by a steel dust-pressing (fig. 5.7(c)). The seal is so shaped that it fills the space between the bearing cap race and the trunnion journal. This helps to keep the needle bearings in position and prevents the grease being squeezed out. In addition to this, the rubber moulding has a shoulder flange which is sandwiched between the dust-cover and the bearing-cap end face, its purpose being to protect the grease-retention part of the seal from dust and dirt. The seal assembly is made to grip the trunnion journal – this means that, when the joint pivots, the relative movement takes place between the seal and the cap.

A comparatively new type of moulded-synthetic rubber seal suitable for heavy-duty use has both a radial and an axial lip (fig. 5.7(d)). It fits tightly against a recess formed in the inside of the bearing cap race. Grease retention is obtained by the radial lip, and dust exclusion is provided by the axial lip. One of the problems with high-pressure lubrication is that the excess grease forced into the bearing must find a way out, and it usually does so by damaging a section of the seal. With this design, overfilling the needle-bearing chamber with grease will collapse the back of the rubber seal which presses against the cap inner recess and the surplus grease will then just squeeze past without permanently damaging the seal.

5.4 Propeller-shaft slip-joint (figs 5.8 and 5.9)

The propeller shaft is provided with a universal joint at both the sprung gearbox end and the unsprung rear-axle final drive. The axle is restrained in its vertical movement from bump to rebound by the leaf springs. The front end of the leaf spring is held in position by a fixed shackle and pin and, as the rear end of the spring deflects between bump and rebound, the axle will pivot about the fixed shackle-pin through an arc of radius equal to approximately half the spring length. This suspension and transmission layout is known as the Hotchkiss drive (fig. 5.8(a) and (b)).

Vertical lines drawn through the centre of the axle under unloaded and loaded conditions will show that the horizontal distance between the front universal joint and the centre of the rear axle changes (fig. 5.8(b)).

In addition to the deflection of the spring changing the distance between the universal-joint centres, the torque reaction on the axle when the vehicle is accelerated or braked will distort the spring and twist the axle about its centre, so the rear universal joint will constantly alter its mean position, thus creating further changes in length between the universal-joint centres (fig. 5.8(c) and (d)).

A variable-length or telescopic propeller-shaft assembly is thus necessary, and this is simply obtained by using a splined slip-joint. These slip-joints tend to take two forms:

i) a sliding joint forming part of the propeller shaft,
ii) a sliding joint forming part of the gearbox main shaft.

i) Some cars and most commercial vehicles use a slip-joint which has an externally splined stub shaft which is an extension of the propeller

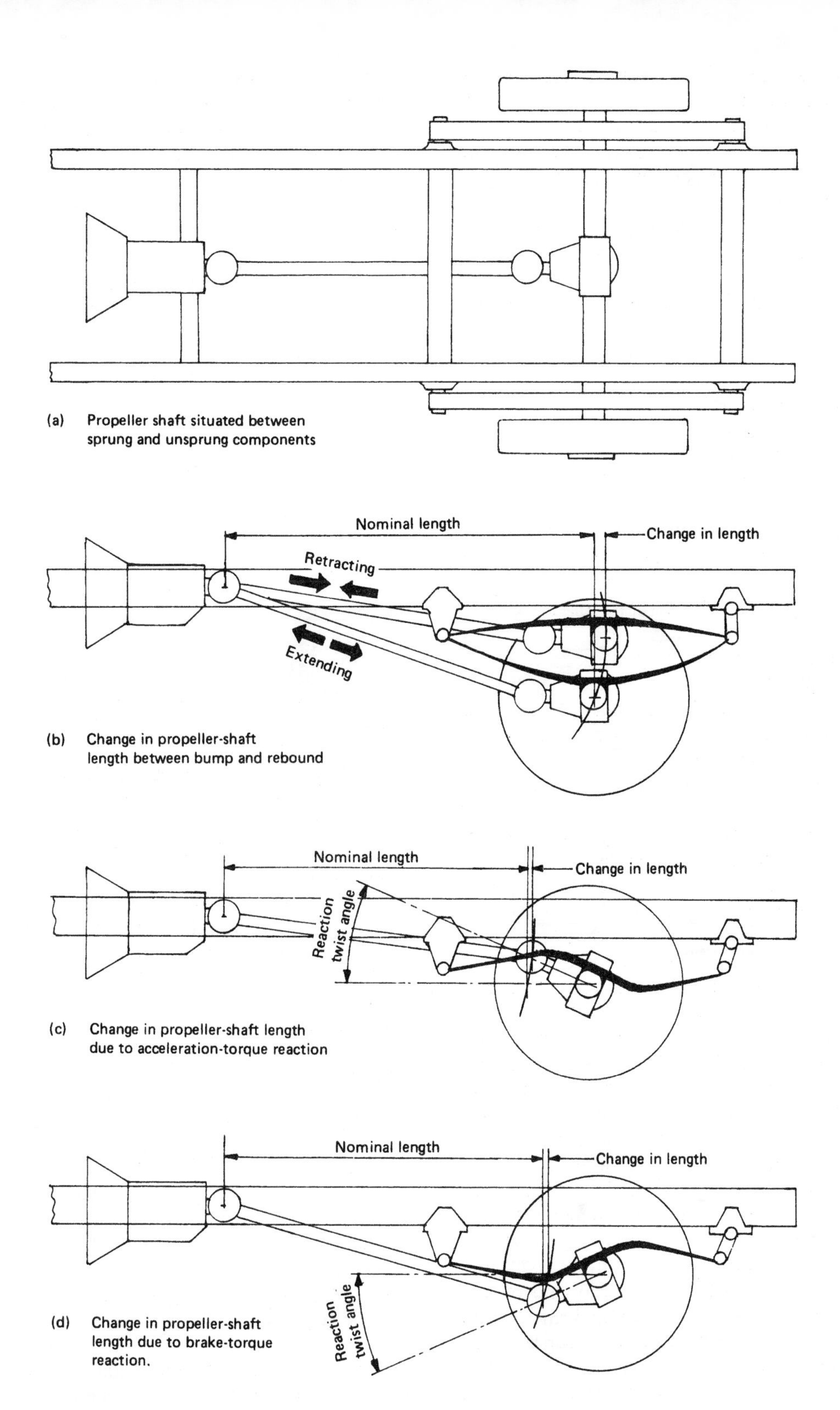

Fig. 5.8 Hotchkiss drive and suspension

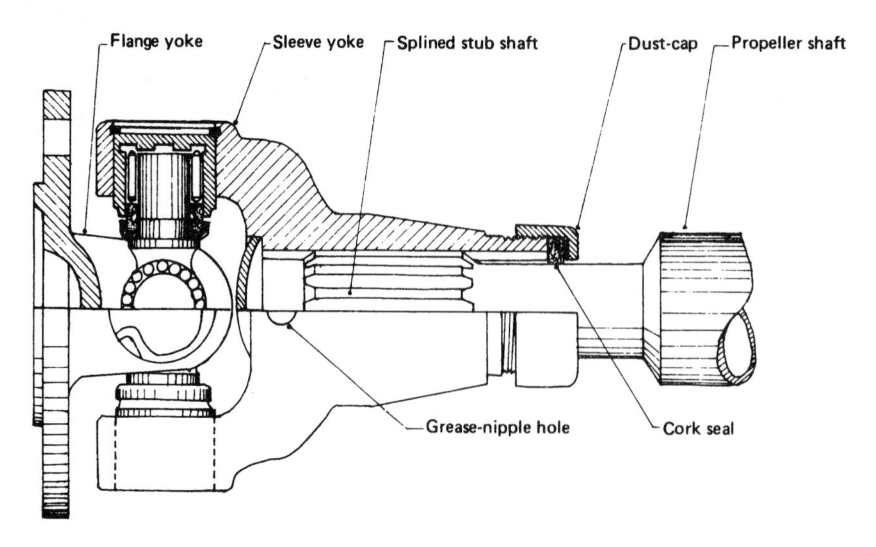

Fig. 5.9 Splined slip-joint between sleeve yoke and propeller shaft

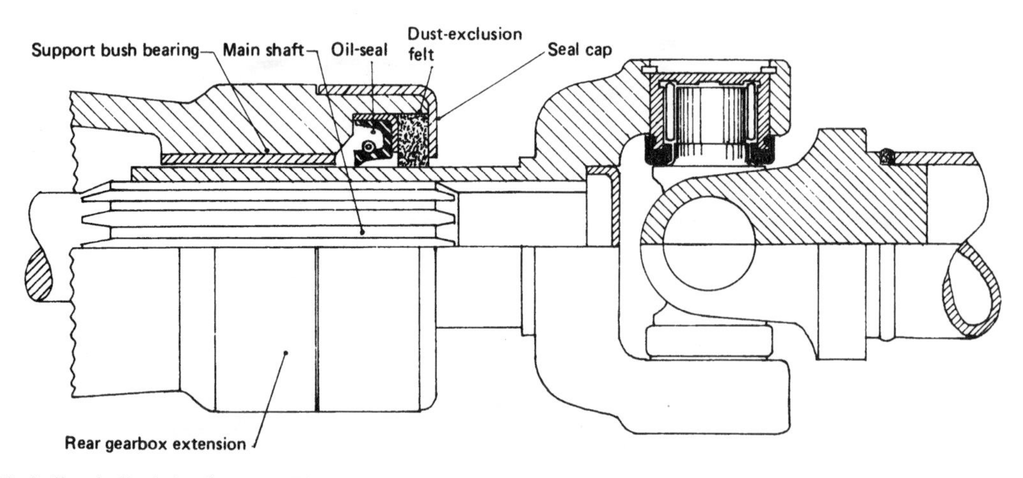

Fig. 5.10 Splined slip-joint between sleeve yoke and gearbox main shaft

shaft, and this is made to engage the internally splined coupling sleeve which is an integral part of one of the universal-joint yokes (fig. 5.9). A grease-nipple lubricator is provided on the slip-joint for spline lubrication, leakage being prevented by a cork washer and a dust-cap. A relief valve is not fitted.

ii) A popular slip-joint used on cars and vans has an internally splined extended coupling-sleeve yoke forming part of the front universal joint. This yoke sleeve engages the rear end of the gearbox externally splined main shaft and is supported in the gearbox rear-cover-housing bush bearing (fig. 5.10). An oil-seal and a dust-exclusion felt operate directly on the sliding sleeve.

5.5 Series-coupled universal joints (fig. 5.11)

The purpose of using a pair of universal joints in the Hotchkiss-type drive is to enable power to be transmitted from the gearbox output shaft to the final-drive pinion shaft. These shafts are usually approximately parallel but do not lie on the same axis – that is, they are offset to each other.

If only one universal joint is used at the gearbox end, the propeller shaft and the final-drive pinion shaft will form one long rigid drive-line. The problem with this arrangement comes when the vehicle accelerates or is braked – the axle will tend to twist due to torque reaction which will be absorbed by the propeller shaft and universal joint, and this will impose considerable strain on the transmission system. By using a universal

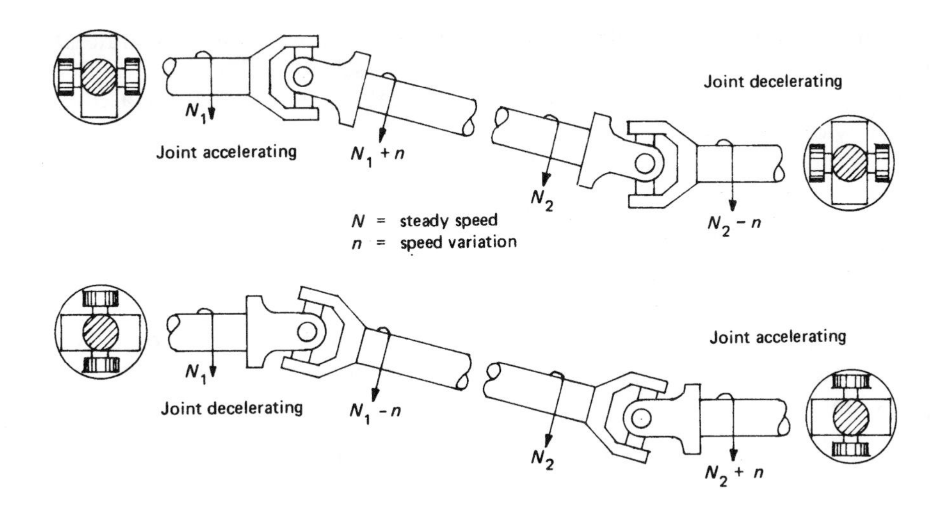

Fig. 5.11 Two universal joints connected in series

joint at each end of the propeller shaft, any torque reaction tending to rotate the axle-casing will be taken by the springs and tie-rods (if fitted) alone. This leaves the propeller shaft and universal joints to concentrate on the transmission of power between sprung and unsprung shafts which are continuously changing their offset.

A second problem when using only one single Hooke's-type universal joint is that the output speed from the joint will change four times in one revolution. This may lead to resonant torsional vibrations in the transmission drive, but fortunately this can be rectified by incorporating a second universal joint in series with the first (fig. 5.11). If the second universal joint is angularly positioned relative to the first joint so that both the yokes on the propeller shaft are in line, then the two joints will be 90° out of phase with each other. This means that the rear output joint will be speeding up when the output speed of the front joint is slowing down, so the speed variations at the two joints over one 360° cycle for any angular displacement will be changing in the opposite sense and so cancel each other.

A summary of what is achieved by using two joints is that, within the normal operating angles of both joints (which under extreme conditions could be up to 15°) and with an input drive revolving at constant velocity, a constant-velocity output will exist at the final drive, rather than the periodic angular-velocity fluctuations which would be experienced with a single joint.

5.6 Propeller-shaft vibration

Small cars and short vans and trucks are able to use a single propeller shaft with a slip-joint at the front end without experiencing any undue vibration. However, with vehicles of longer wheelbase, the longer propeller shaft required would tend to sag and under certain operating conditions would tend to whirl (fig. 5.12) and then set up sympathetic resonant vibrations in the body of the vehicle – that is, cause the body to 'drum' or vibrate as the shaft whirls.

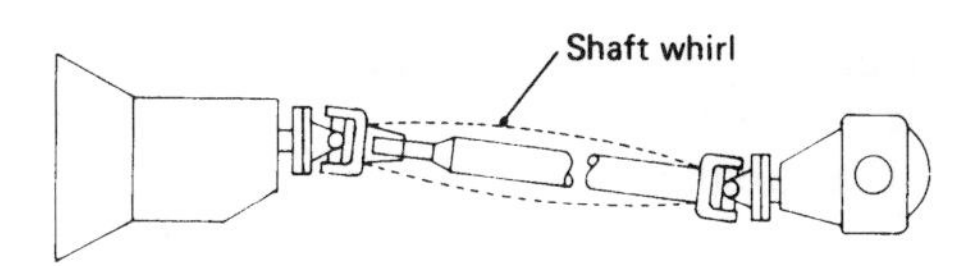

Fig. 5.12 Simple one-piece propeller shaft with one slip-joint and two universal joints

The main factors which affect the resonant frequency of the propeller shaft which is creating the vibration may be divided into two groups:

i) propeller-shaft factors:

   a) the diameter and length of the shaft,
   b) the degree of balance of the assembled shaft and joints,
   c) the bending resistance of the shaft;

ii) vehicle-body factors:

   a) the type and shape of body structure, reinforcement box-sections, etc.,

b) the position of components within the body structure,
c) drive-line vibration-clamping qualities provided by engine and transmission mounts, spring bushes, panel insulation, etc.

5.6.1 An insight into propeller-shaft whirl

Whirling of a rotating shaft happens when the centre of gravity of the shaft's mass is eccentric and so is acted upon by a centrifugal force which tends to bend or bow the shaft so that it orbits about the shaft's longitudinal axis like a rotating skipping rope. As the speed rises, the eccentric deflection of the shaft increases, with the result that the centrifugal force will also increase. The effect is therefore cumulative and will continue until the whirling becomes critical, at which point the shaft will vibrate violently.

It is in the nature of things that a circular-sectioned shaft horizontally supported between bearings should have its centre of gravity displaced to one side of the central axis. The reasons for this are as follows:

a) the shaft will sag between centres, due to its own mass;
b) if a tubular seamless-drawn propeller shaft is used, the wall thickness may not be uniform all the way round;
c) if a tubular shaft is rolled up from flat sheet stock and butt-welded, the amount of weld metal may not be equivalent to the mass on the opposite side;
d) if the tubular shaft is forced on to universal-joint stub-shaft recesses which have been turned between loose centres, then the shaft could be eccentric to the axis of rotation;
e) if universal joints are attached to the ends of the shaft which are then supported on bearings, the joint yokes and trunnion arms may be assembled very slightly to one side;
f) if a slip-joint coupling is used at one end of the shaft, the clearance between the male and female splines will allow the shaft to be moved over to a limited extent.

From the theory of whirling, it has been found that the critical whirling speed of a shaft is inversely proportional to the square of the shaft's length. If, therefore, a shaft having, for example, a critical whirling speed of 6000 rev/min is doubled in length, the critical whirling speed of the new shaft will be reduced to a quarter of this, i.e. the shaft will now begin to vibrate at 1500 rev/min. Conversely, by halving the length of the shaft, the critical speed would be increased fourfold to a new speed of 24 000 rev/min.

This example shows that doubling the length of the original shaft brings the whirling speed well within the operating range of a normal propeller shaft, but halving the length puts the critical speed considerably above the maximum propeller-shaft speed for a vehicle.

A common approach to increasing the rigidity of the propeller shaft is to extend either the rear end of the gearbox main shaft and housing (fig. 5.13) or the final-drive pinion shaft and housing (fig. 5.14). The former is preferred for medium-sized cars, and the latter has had some success on larger cars which use rear coil-spring suspension with trailing arms and tie-rod stabilisers. A slip-joint is usually provided at the gearbox end of the propeller shaft, so that the propeller shaft automatically adjusts its length in accordance with suspension-deflection changes.

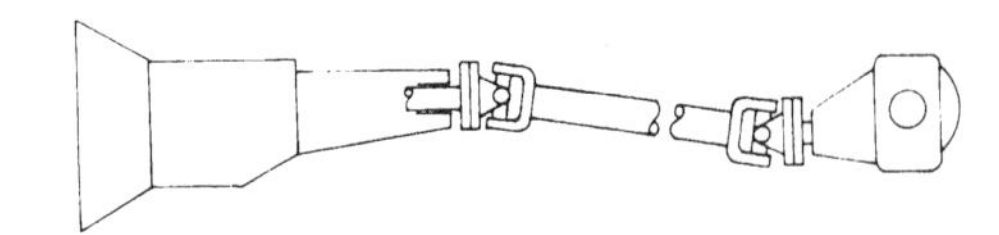

Fig. 5.13 One-piece drive-line with extended gearbox housing

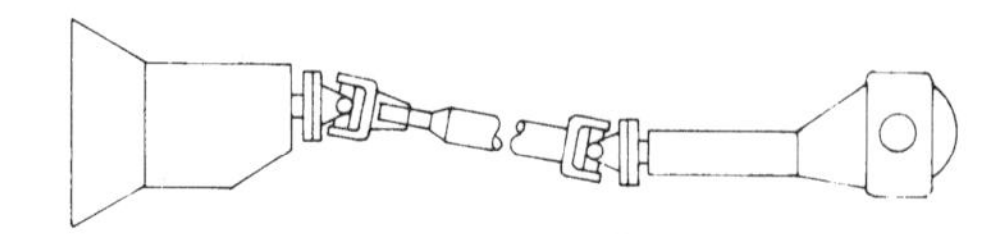

Fig. 5.14 One-piece drive-line with extended differential housing

The vibration problem could also be solved by increasing the diameter of the shaft, but this would increase its strength beyond its torque-carrying requirements and at the same time increase its inertia, which would oppose the vehicle's acceleration and deceleration.

An alternative solution frequently adopted by car, van, and commercial-vehicle manufacturers is the use of divided propeller shafts supported by intermediate or centre bearings. This arrangement has been employed in the past on large cars to achieve a lowering of the transmission drive from the front-mounted gearbox to the rear axle, thus reducing the floorboard tunnel height, but, used to avoid vibration, it avoids the disadvantages of a thicker shaft and a further point in its favour when used on commercial vehicles is that large offsets between gearbox centre lines and the final-drive pinion centre line can be achieved in two or three stages.

5.7 Divided propeller shafts and their support

Two-piece drive-lines, with two shafts and an intermediate support bearing (fig. 5.15), are generally used on trucks with wheelbases from 3.4 to 4.8 m, but there is some overlap depending on the vehicle's work role.

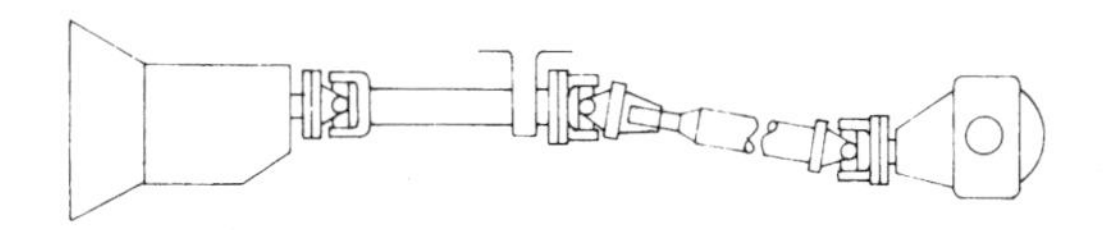

Fig. 5.15 Two-piece drive-line with single intermediate support bearing

The two-piece propeller shaft has three universal joints, and the primary propeller shaft is of the fixed-joints-and-tube-assembly type, but the secondary propeller shaft has a slip-joint at the support-bearing end to accommodate any elongation due to suspension movement. Usually the primary shaft is in line with the gearbox mainshaft axis, but the secondary propeller shaft is inclined slightly so that it intersects the rear-axle final-drive pinion shaft. For high-chassis-mounted vehicles, both shafts may be sloped to reduce the effective shaft inclination angle. Sometimes when the primary shaft is in line with the output shaft of the gearbox, rubber-type universal couplings are employed which tend to damp out torsional transmission vibration better than conventional steel joints (see section 5.10).

For vehicles with wheelbases over 4.8 m, a three-piece drive-line with two intermediate support bearings may be necessary (fig. 5.16). There are four universal-joints, and it can be seen that the intermediate shaft lies parallel to the output

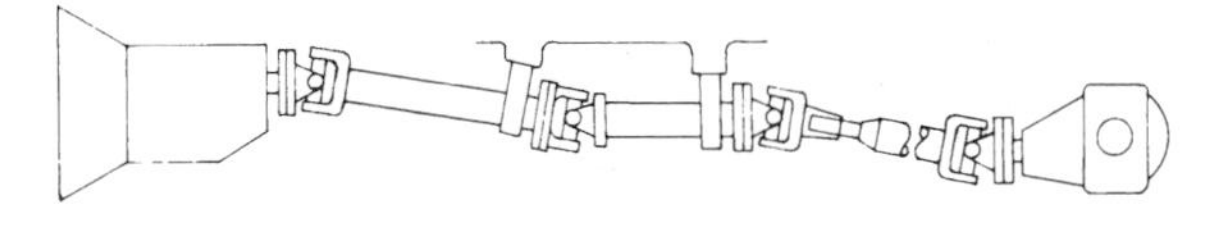

Fig. 5.16 Three-piece drive-line with two intermediate support bearings

shaft of the gearbox. Again only the rear propeller shaft incorporates a slip-joint to compensate for shaft length change.

5.7.1 Intermediate propeller-shaft support bearings

Intermediate propeller-shaft bearing-and-mount assemblies are provided to position and support the divided or split propeller shafts. The support-bearing assemblies are of two basic types:

i) self-aligning bearing supports,
ii) flexible-mounted bearing supports.

Self-aligning intermediate-bearing supports have mostly been used on heavy-duty trucks. One type of self-aligning bearing support is a double-row ball-bearing with a deep-grooved inner race and an internally semicircular outer race (fig. 5.17(a)). With this arrangement, any shaft deflection is accommodated by the inner race and balls tilting about the fixed outer-race spherical seat.

A second method of providing self-alignment for the split shafts is achieved by a single-row deep-grooved ball-bearing with a spherical profile on the periphery of the outer races. The ball race is then received in a steel support ring which has an internal profile to match the outside of the bearing (fig. 5.17(b)). If there is any misalignment in service, relative movement of the bearing and the ring can take place without imposing strains on the bearing assembly. Both arrangements discussed need to be lubricated periodically, and oil-seals are provided to retain the grease and keep dirt out of the bearing tracks.

Flexible-mounted intermediate-bearing supports for divided shafts may be used for both light and heavy vehicles. All these types tend to have a single-row deep-grooved ball-bearing which fits directly over one of the divided shafts, and surrounding this bearing is a rubber member which is enclosed in a steel frame. The intermediate shafts are then given support by bolting this assembly to the chassis or body shell. This rubber mounting provides a flexible support for the bearing so that a slight tilt of the shaft can be accommodated. In addition, the flexible rubber acts as a vibration damper and isolates any propeller-shaft vibrations from the body members.

For extra-heavy duty, a solid rubber ring block fitting over a bearing hub, may be used (fig. 5.17(c)). It can be seen that the inner bearing race is held in position by the universal-joint

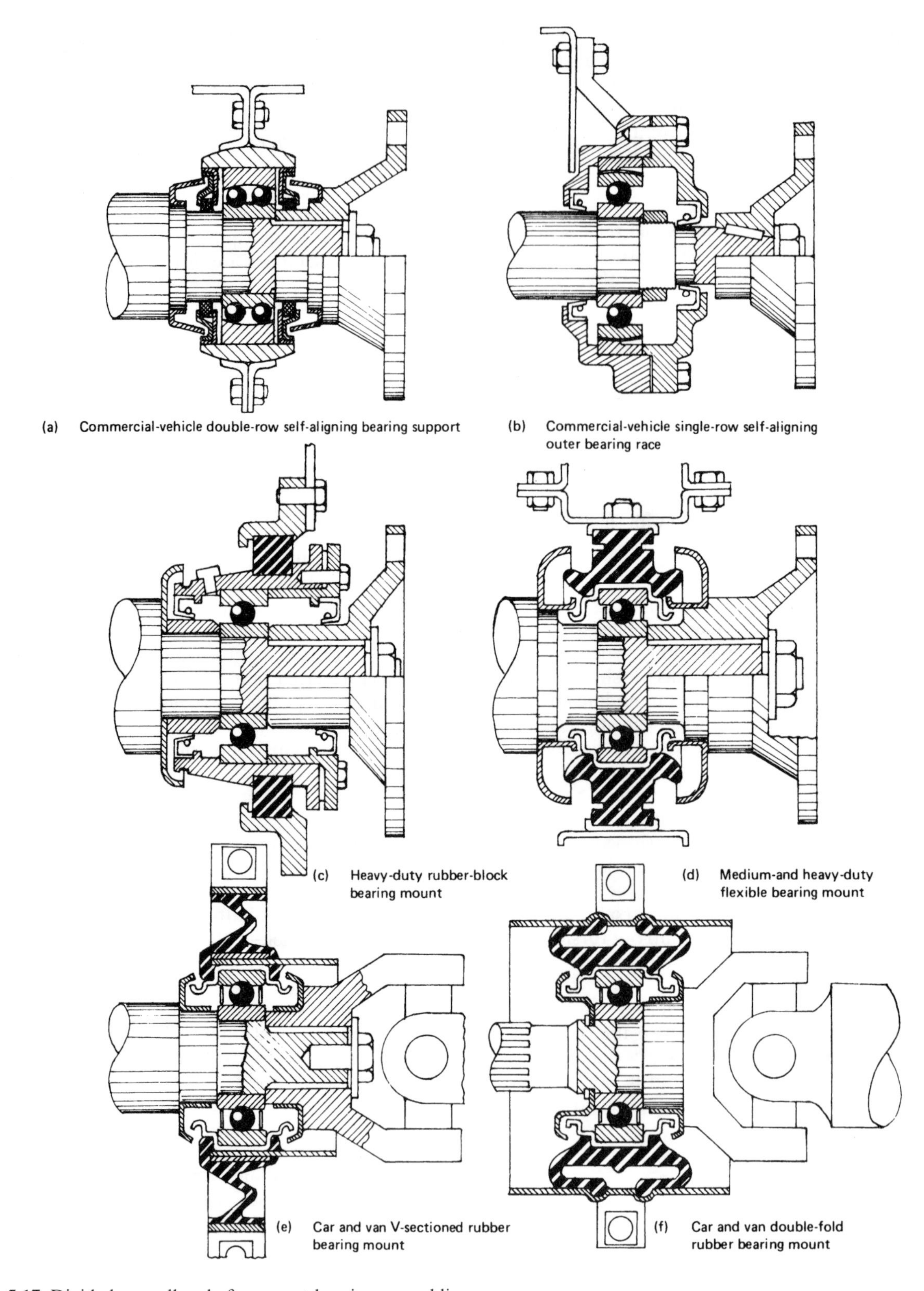

Fig. 5.17 Divided-propeller-shaft support-bearing assemblies

flange and that the outer bearing race is located by a shimmed sleeve. Regular lubrication is necessary with this assembly.

Most light- and heavy-duty intermediate-bearing assemblies now use pre-greased and sealed-for-life deep-groove bearings. The rubber element is bonded both to the external steel casing and to the outer bearing-race steel pressing during the rubber moulding and curing (vulcanisation) process. Dust-baffles mounted on the shaft protect the bearing against grit and wet weather.

Figure 5.17(d) shows a bearing arrangement used for commercial vehicles, and it can be seen that a slot is formed on each side of the rubber moulding to improve flexibility.

Figure 5.17(e) shows a bearing layout suitable for cars and vans. It can be seen that the rubber element has a V-shaped section so that it can fold and move about its mean position more readily as conditions demand. This also improves the vibration-damping properties of the rubber assembly.

Figure 5.17(f) shows an alternative layout for light vehicles, but here the moulded rubber section forms double link-arms which give greater rigidity for the bearing but at the same time will easily tilt and still give excellent damping properties.

5.8 Universal- and slip-joint alignment

The correct alignment of universal joints connected in series is essential, otherwise the synchronising of one joint's angular acceleration with the equivalent angular deceleration of the other joint every quarter of a revolution will not be obtained. In fact, instead of smoothing out the speed fluctuations, both joints if not correctly assembled can produce a mismatch such that they both speed up or down simultaneously. This then results in compounding of the speed variation so that unpleasant vibrations are likely to become more pronounced.

Propeller shafts and universal joints are balanced by the manufacturer after assembly. Therefore most commercial-vehicle propeller-shaft assemblies are marked so that, when they are dismantled and new needle rollers and cap races are fitted, the flange yoke and the propeller-shaft yoke or the sleeve yoke can be put together the same way as before they were dismantled.

Marks such as arrows or letters or both (fig. 5.18) on the two end flange yokes when aligned with other marks on the propeller-shaft and sleeve yokes ensure the following:

a) each pair of yokes forming one universal joint is assembled as it was originally when balanced,
b) the two outer flange yokes are opposite to each other, as are the inner sleeve and propeller-shaft yokes, which makes the flange yokes 90° out of phase with the other yokes as required with universal joints connected in series.

Divided propeller shafts should always be fitted so that the flange yokes are in the positions shown in figs 5.15 and 5.16, with the flange yoke at the gearbox at 90° to the flange yoke on the second propeller shaft and the third-propeller-shaft flange yoke at 90° to the second. The first and third propeller shafts should therefore always be in alignment with each other.

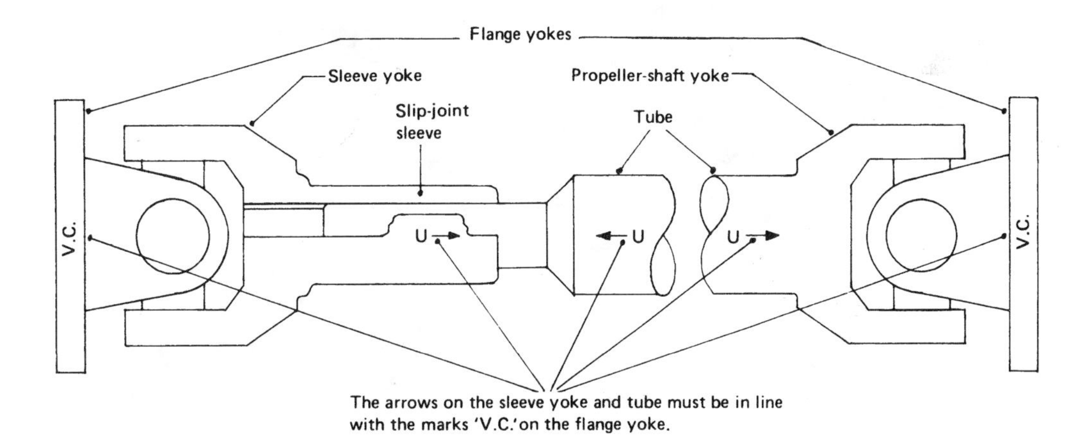

Fig. 5.18 Universal-joint, slip-joint, and propeller-shaft alignment

5.9 Constant velocity joints

Universal joints are used to couple shafts together which are inclined to each other; their angle of inclination may be fixed or may vary constantly. However, with the normal Hooke's joint the inclined output shaft speed will increase and then decrease relative to the constant input speed four times every revolution; the greater the articulation angle, the greater will be this speed variation every quarter of a revolution.

Constant-velocity-type joints are designed to transmit torque from the input to the output shafts with no angular speed fluctuation every revolution in the output drive as is the case with the Hooke's double-pronged trunnion-cross-joint.

Constant-velocity-type joints are essential when the axes of the input and output members are to be subjected to two forms of shaft joint articulation simultaneously. These variations in the input to output shaft angle of inclination are caused by two movements:

1) the vertical up and down movement of the vehicle body relative to the front road-wheels which are supported and restrained by the independent suspension;
2) the horizontal movement of the stub-axles about the swivel-pins when the vehicle is being steered. Universal joints which do suffer from rotational speed fluctuation could and would compound the two sources of speed fluctuation; thus speed variation would then be transferred to the steering/drive wheels when the vehicle is being steered, and at certain critical speeds could create and feed vibration into the steering system.

5.9.1 Birfield constant-velocity joint based on the Rzeppa principle (fig. 5.19)

These joints basically resemble a ball and socket coupling consisting of an outer socket or cup member and an inner ball or spherical shaped member. Bridging these two members are six steel

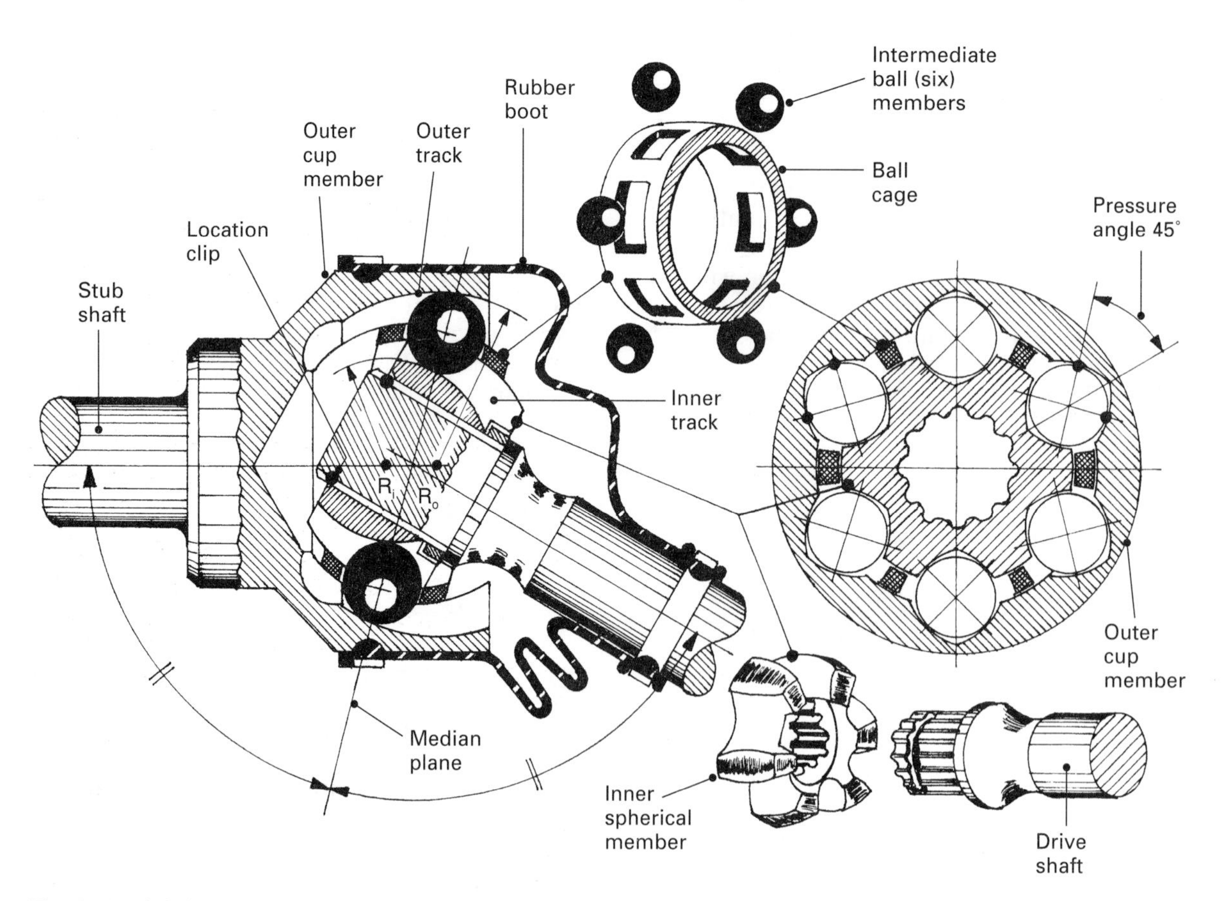

Fig. 5.19 Birfield Rzeppa type constant-velocity joint

balls which fit into curved track grooves formed in both outer cup and inner ball members. It is the function of these balls to transfer the torque load from the input inner-ball member to the output outer-socket or cup member with the minimum amount of friction.

Principle of operation (fig. 5.19) The basic principle in providing constant velocity between input and output shafts in which the intersecting angle of the two shafts may vary at any instant is to maintain the contact points between both halves of the joint in a plane which bisects the driving and driven shaft angle, this being known as the 'median plane'.

Ball and groove tracks (fig. 5.19) The six balls are moved into the median plane by the ball and socket grooves which have their geometric centres offset to each other and to the true centre of intersection of the drive and driven half-members. The curvature mismatch of the ball and socket grooves are such that the distance between the groove tracks of the inner and outer members along the median plane is always equal to the ball diameters, whereas the distance becomes less on either side of the median plane. Thus at all times the converging tracks push the top and bottom balls towards the median plane no matter what the articulated angle of the joint.

Ball cage (fig. 5.19) Although the converging track grooves of the inner and outer joint members are the main source for keeping the six balls aligned to the median plane, a spherical steel cage sandwiched between the external and internal surfaces of the inner and outer half-member is used to initially assemble and then to hold these balls in position during articulation of the joint.

Track groove wall profile (fig. 5.19) Load transfer from the inner ball member to the outer cup member is by way of the six balls. There is a large circular arc profile on each side wall of the track grooves, which permits the balls to contact the tracks near their mid-wall position (see fig. 5.19). Thus edge contact overloading is avoided and the balls are now subjected to a less severe compressive stress, as opposed to shear stress, if a single circular arc track is used. Grease is pre-packed for life into and around the inner and outer ball tracks. A rubber boot protects the working joint from abrasive grit and water exposure which would otherwise rapidly wear and erode both balls and their respective tracks.

5.10 Rubber universal couplings

To improve the smoothness and quietness of the transmission system, some manufacturers incorporate universal couplings made from metal and rubber elements, usually adhesive-bonded together. These are sometimes referred to as 'metalastic' couplings.

The advantages of rubber couplings over metal Hooke's-type universal joints are as follows:

a) Impact loads such as sudden changes in transmitted torque are absorbed and limited by the torsional flexibility of the rubber joints.
b) Rubber joints provide a break in the path of transmitted vibrations, so the rubber acts as an insulator to noise transfer.
c) Backlash in transmission gears is cushioned by the rubber elements. This reduces the impact or shock loads on bearings, gear teeth, splines, and other components – consequently there will be less noise, and fatigue life will be prolonged.
d) Rubber joints have axial flexibility which can accommodate considerable change in effective shaft length, so end-thrust is considerably cushioned.
e) No maintenance is necessary over the whole life-span of the rubber coupling.

5.10.1 Doughnut-type ('Rotoflex') rubber coupling (fig. 5.20)

This type of coupling consists of a regular polygon of square-section rubber (fig. 5.20(a)). One-piece steel tubular sleeves and inter-leaf plates bisect each corner angle of the polygon-shaped rubber elements that are bonded to the rubber elements. In most of the couplings there are six metal inserts, and three-arm yokes or spiders are used to support them and transfer the drive from the input to the output shaft. To ensure maximum life and torque capacity, the rubber is pre-compressed with an encircling metal band at the factory (fig. 5.20(c)). This band must be removed after the coupling has been attached to both yokes.

In action, the rubber elements between the interleaf plates attached to the propeller-shaft (input) yoke and the plates attached to the drive-shaft (output) yoke are compressed to transfer the drive from input to output.

The angle between the two shafts being coupled should not be more than 3° for normal loading,

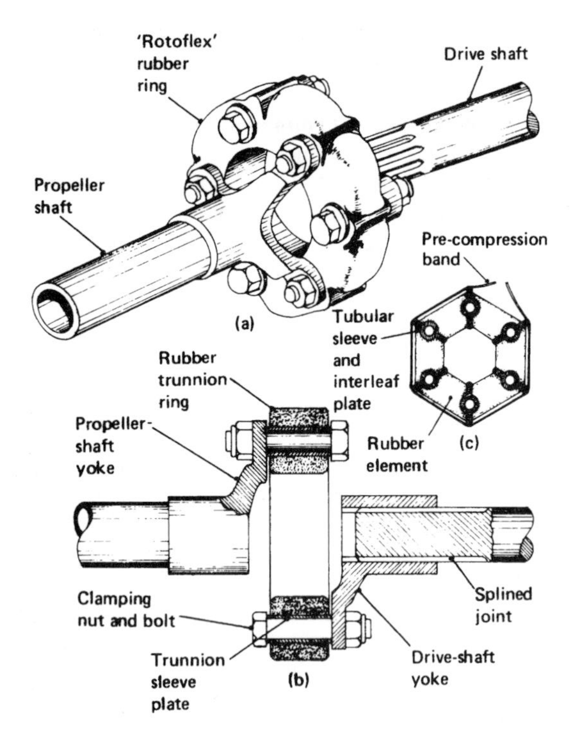

Fig. 5.20 Doughnut-type ('Metalastik rotoflex') coupling

but angular misalignment caused by bump and rebound or other shock conditions is acceptable because it is infrequent. If the coupling is to be used for low-speed half-shaft drives, no centring device is required; however, for high-speed propeller-shaft application a centring device is necessary to compensate for the low radial stiffness of the coupling.

The 'Rotoflex' coupling eliminates torsional irregularities in the torque transmitted and at the same time accommodates relatively large shaft operating angles. It provides not only torsional flexibility but also axial flexibility – this allows the suspension members to take the side-loads from the drive wheels and hence protects the gearbox shaft bearings.

5.10.2 Bush-type ('Metalastik') rubber coupling (fig. 5.21)

Essentially this coupling consists of four rubber ball-joints housed in a circular pressed-steel casing (fig. 5.21(a) and (b)). The ball-joints are equally spaced round a circle and are alternately connected to the driving and driven members. The connecting bolts pass through steel trunnion sleeves having a spherical external form to which the internal surfaces of the rubber elements are bonded, and the external surfaces of the rubber elements are bonded to the hemispherical housings formed in the case pressing. The casing is made from two steel-half pressings, one being slightly larger in diameter than the other so that the overlap of the larger pressing can be rolled over the edge of the smaller one. Pre-compression of the rubber prolongs the life and increases the load capacity of the coupling.

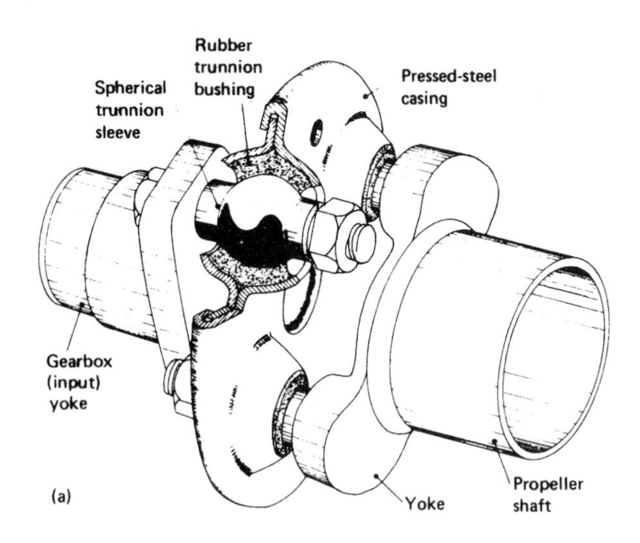

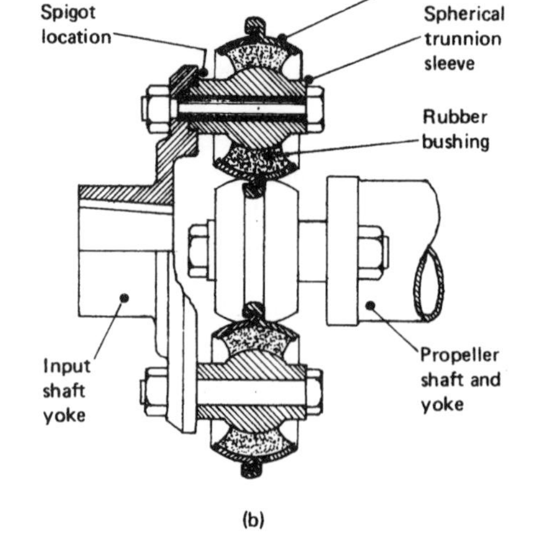

Fig. 5.21 'Layrub' bush-type ('Metalastik') coupling

The coupling is located in relation to the yokes by means of a spigot on one end of each spherical sleeve, so that there are two spigots on each side of the coupling. The spigots enter counterbores in the yokes but do not transmit any torque – all the torque being instead transmitted through the clamped interfaces between the sleeves and the yokes (fig. 5.21(b).

The spherical trunnion sleeves provide an appreciable degree of axial and torsional flexibility without significant loss of radial stiffness, and a higher torque capacity than would be obtained with cylindrical sleeves. The assembly is well balanced because of its symmetrical design, so replacement couplings will not result in vibrational difficulties.

For continuous running, a maximum conical deflection of 5° is acceptable; while 12° to 15° can be accommodated on full bump and rebound. These couplings are suitable for high-speed operation such as is encountered on propeller-shaft drives.

6 The final drive and four-wheel drive

6.1 The purpose of the final-drive gears

The final-drive gears may be directly or indirectly driven from the output gearing of the gearbox.

Directly driven final drives are used when the engine and transmission units are combined together to form an integral construction (fig. 6.1(a) to (c)). In fig. 6.1(a) and (c) it can be seen that the final-drive gears consist of a pair of different-sized bevel gears positioned between the engine and the gearbox. The size difference between the two gears will make the large bevel gear rotate much slower than the small input gear. An alternative arrangement (fig. 6.1(b)) shows the transmission layout with all shafts situated parallel but in different planes to each other – this final drive does not use bevel gears.

Indirectly driven final drives are used at the rear of the vehicle, being either sprung and attached to the body structure or unsprung and incorporated in the rear-axle casing as in fig. 6.1(d). Again, bevel gears are generally used, but sometimes worm and worm-wheel final drives are preferred for special heavy-duty applications.

The final-drive gears are used in the transmission system for the following reasons:

a) to redirect the drive from the gearbox or propeller shaft through 90°,
b) to provide a permanent gear reduction between the engine and the driving road-wheels.

6.1.1 Redirected right-angle final drive

For vehicles with engines mounted longitudinally – that is, parallel with the vehicle's wheelbase – the engine and gearbox rotating axes are at right angles to the vehicle's driving-wheel axles. Therefore, to transmit the power from the engine to the wheel axles, the drive has to be turned through 90°, and to achieve this a pair of bevel-type gear-wheels is necessary.

6.1.2 Permanent gear reduction

When top or direct-drive gear is selected and engaged, the crankshaft rotational motion is transmitted straight through to the final drive and then to the wheels by way of the drive shafts or half-shafts. If the final drive were not put in series with the engine, the road-wheels would then revolve at the same speed as the engine, which would not suit the vehicle's road-wheel speed and traction requirements. This miss-match of speeds may be illustrated by the following example.

Consider a vehicle with road-wheels of 0.6 m effective diameter and an engine speed range from 800 to 5000 rev/min. The minimum and maximum road speed of the vehicle can be determined as follows:

Distance moved per wheel revolution = rolling circumference of wheels

$$= \pi d$$

$$= 3.142 \times 0.6 \text{ m}$$

$$= 1.885 \text{ m}$$

Minimum vehicle speed $= \pi dN_1$

$$= 1.885 \text{ m/rev} \times 800 \text{ rev/min}$$

$$= 1508 \text{ m/min}$$

$$= \frac{1508 \times 60}{1000} \text{ km/h}$$

$$= 90.5 \text{ km/h}$$

Maximum vehicle speed $= \pi dN_2$

$$= 1.885 \text{ m/rev} \times 5000 \text{ rev/min}$$

$$= 9425 \text{ m/min}$$

$$= \frac{9425 \times 60}{1000} \text{ km/h}$$

$$= 565.5 \text{ km/h}$$

As can be seen, both these speeds are far too high, and the simplest remedy is to place in series a constant or permanent gear reduction between the gearbox and the road-wheel axles, a suitable and typical gear-reduction ratio being something of the order of 4:1. This then will provide a lower vehicle speed range as shown:

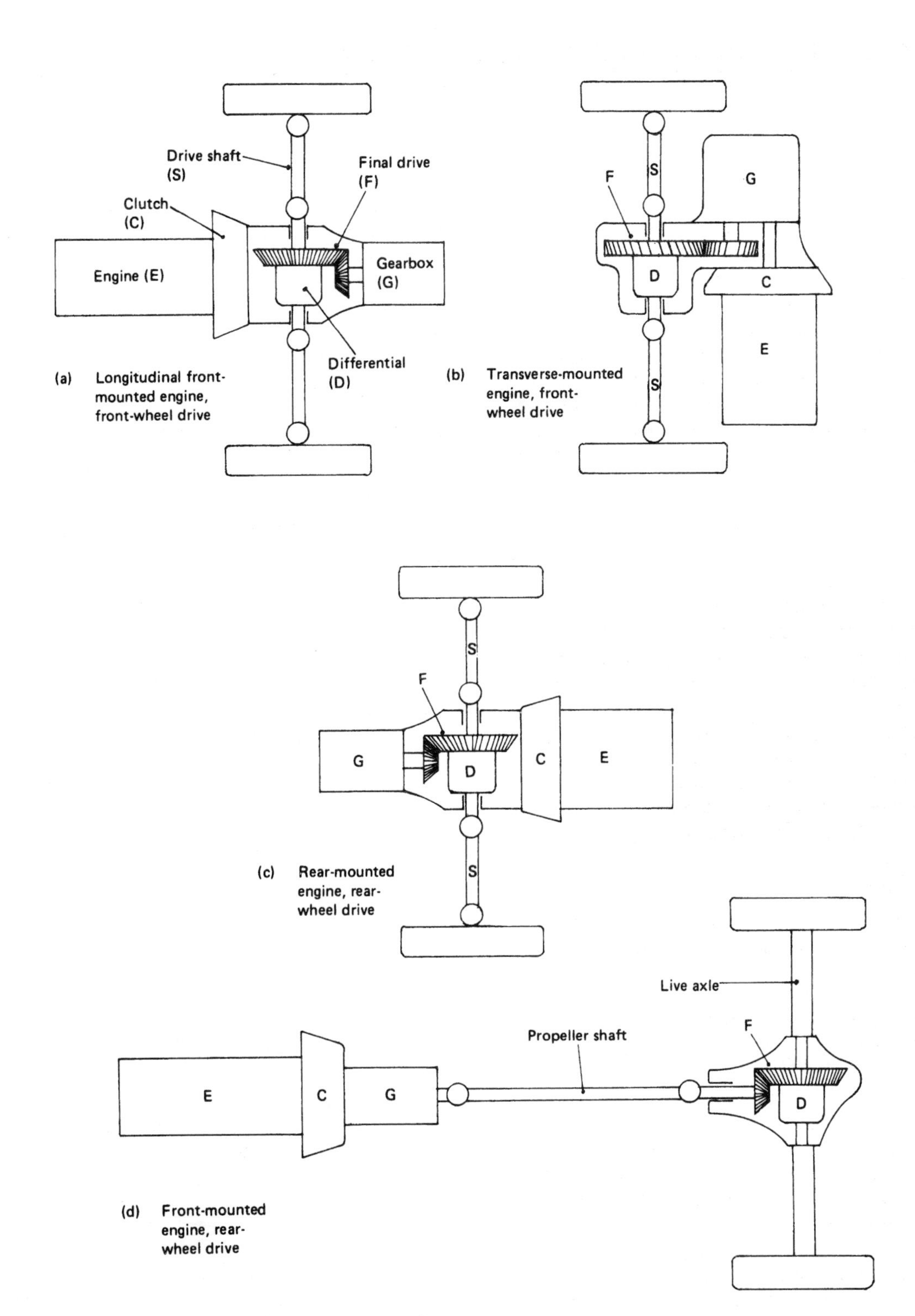

Fig. 6.1 Engine gearbox and final-drive configurations

$$\text{New minimum speed} = \frac{90.5 \text{ km/h}}{4} = 22.6 \text{ km/h}$$

$$\text{New maximum speed} = \frac{565.5 \text{ km/h}}{4} = 141.4 \text{ km/h}$$

This is thus a much more realistic match of engine to road-wheel speeds.

The more the road-wheel speed is reduced by the introduction of the final drive, the more the output torque will be increased. This ensures that even at low engine speeds there is sufficient turning-effort to propel the vehicle when in direct gear – and it also contributes to increasing the torque in all gears, so that the driver does not have to change gear ratio so frequently.

6.2 Types of final-drive gearing

Final-drive gears which are inclined at right angles to each other mainly take the form of a pair of bevel-shaped meshing gears, one being a large gear known as the crown wheel, since when held horizontally it resembles a royal crown, and the other being a small gear commonly referred to as the bevel pinion gear. Sometimes for special applications worm and worm-wheel gearing is employed.

Final-drive gear ratios generally vary from 3.8:1 to 4.2:1 for cars and up to 6:1 or even more for heavy vehicles. A typical number of teeth on the crown wheel for cars is between 35 and 45. The bevel pinion gear usually has an odd number of teeth, which ensures that each tooth on the pinion will periodically engage every tooth on the crown wheel, a very common size of pinion being one with nine teeth. The final drive and the axle half-shafts are enclosed in a casing which resembles a double-ended banjo – hence a rear-axle housing is commonly known as a banjo casing. This casing also supports the vehicle's weight.

The following types of gearing will now be studied in greater depth:

a) straight bevel-tooth gears,
b) spiral bevel-tooth gears,
c) hypoid bevel-tooth gears,
d) worm and worm-wheel gears.

6.2.1 Straight bevel-tooth gears (figs 6.2 and 6.3)

Straight bevel-tooth gears are used for connecting shafts at right angles in the same plane.

A bevel gearwheel may be considered to resemble a cone with its top removed, this shape being known as a conical frustum.

If two conical frustums are in rubbing contact as shown in fig. 6.2, friction between them will enable a pure rolling action to transmit the same motion as bevel wheels whose apexes coincide with the point of intersection of the axes of their shafts. The apexes of the 'complete cones' obviously coincide with the point of intersection of the axes of the two shafts. If the two conical frustums roll together without slipping, the peripheral speeds are equal at any particular point on the line of contact. The angular speed of each shaft will be inversely proportional to the diameter of the conical frustum on that shaft. Note that the conical surfaces are known as pitch surfaces.

In order to provide a positive drive between the two shafts, it is necessary to have interlocking

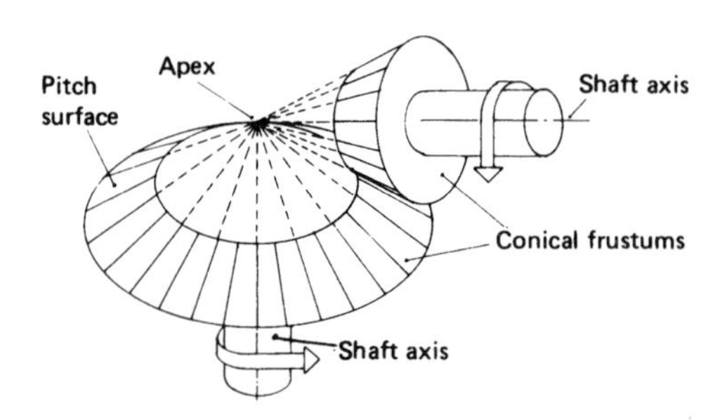

Fig. 6.2 Conical development of bevel gearing

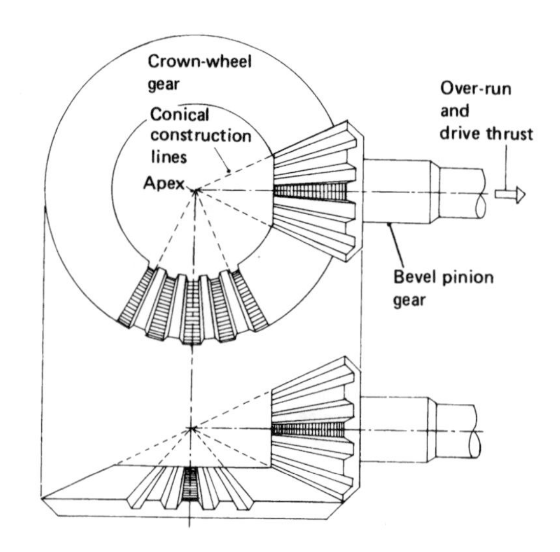

Fig. 6.3 Geometry of straight-tooth bevel-gear crown wheel and pinion

projections or teeth on the conical frustums (fig. 6.3). Straight bevel gears have teeth which radiate from the point of intersection of the shafts and which, while they are of geometrically similar shape at all points along their length, diminish dimensionally closer to the apex. The profile of the teeth must be shaped as in the case of spur gears and is usually of the involute type.

With involute tooth profiles, sliding takes place along the tooth profile between the teeth in contact at all times except the instant when they touch at the pitch point, when the direction of sliding is reversed. The degree of sliding increases rapidly as tooth contact moves further away from the pitch point.

The elementary straight bevel gear is not designed for continuous heavy-duty and high speed use, as is necessary for the crown-wheel and pinion gears, but it does form the basis for modified bevel gearing suitable for final drives. One special application of straight bevel gears is their adoption for the differential gear train – here only low relative speeds are experienced.

6.2.2 Spiral bevel-tooth gears (figs 6.4 and 6.5)

An understanding of the design of the crown wheel and pinion will become clearer if the crown wheel is described as a disc with gear teeth projecting from one side. Now, if this disc could be split and stretched out straight, it would resemble a toothed rack. Now consider the crown wheel and pinion being constructed from two gearwheels in mesh – if the radius of one of these gearwheels is increased indefinitely, the pitch circle becomes a straight line and the wheel becomes a rack meshing with a smaller gear known as the pinion (fig. 6.4). Therefore, because of the straight pitch line, the side profile of the teeth on involute crown-wheel teeth will also be straight; conversely, these straight teeth will mesh with curved involute teeth of the pinion.

With straight bevel gears, only one pair of teeth will be in contact at any one instant, and each pair will come into mesh over the whole length of the teeth at the same time. This means that tooth-profile contact will continuously change from one side to the other of the pitch line of the pinion tooth, and the net result is an uneven transmission of motion from the pinion to the crown wheel as the load is transferred from one pair of teeth to the next. Thus these gears tend to be noisy, subject to high rates of wear, and possibly may suffer from early fatigue failure if used for final-drive applications.

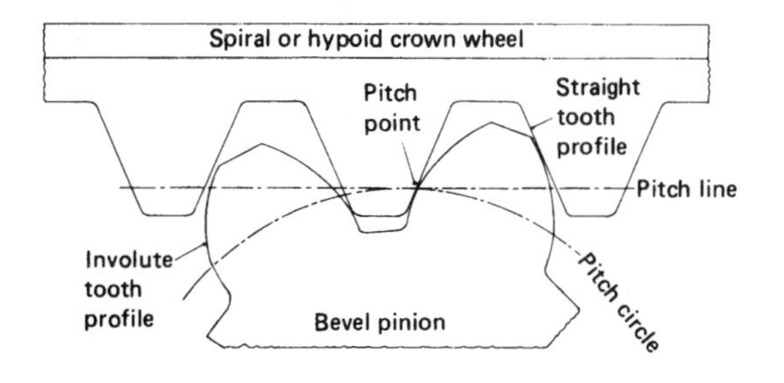

Fig. 6.4 Enlarged section of a crown wheel and pinion

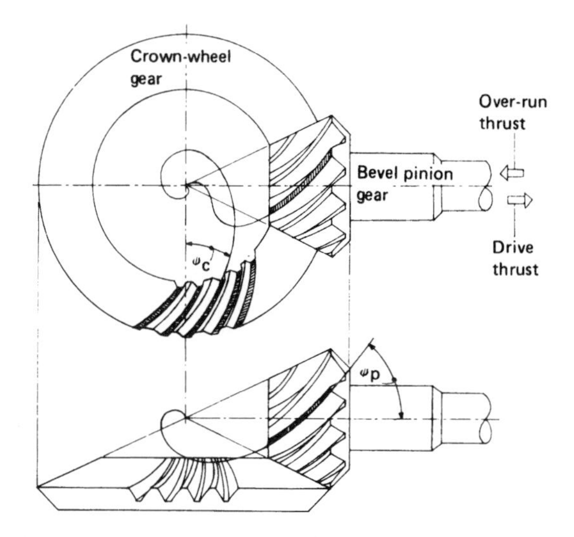

Fig. 6.5 Geometry of spiral-tooth bevel-gear crown wheel and pinion

The characteristics of spiral bevel gears (fig. 6.5) are that more than one pair of teeth come into contact at any instant and the meshing teeth engage gradually so that there will be contact at some point of the pitch line of a tooth at all times. Because of the continuous pitch-line contact and the progressive tooth engagement and disengagement, these gears are inherently quiet and tend to operate smoothly.

A further feature of these spiral gears is that the tooth-profile surface areas in contact at any time will be considerably larger than for the straight-cut teeth, so that higher operating loads can be sustained.

6.2.3 Hypoid bevel-tooth gears (figs 6.6 and 6.7)

Hypoid bevel gears are used for connecting shafts at right angles but not lying in the same plane.

The crown-wheel and bevel-pinion gear-tooth shapes are very similar to those of the spiral

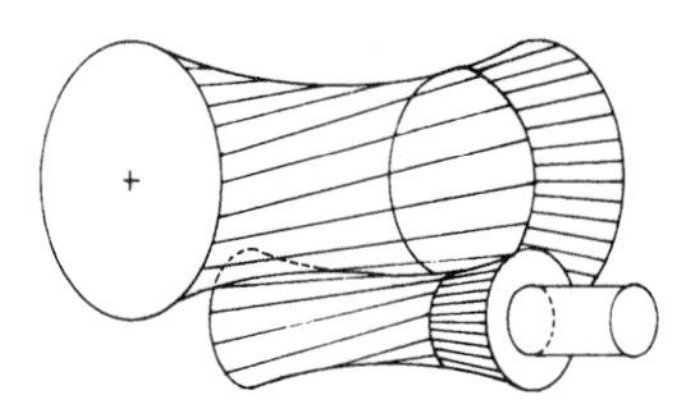

Fig. 6.6 Hyperboloid (hypoid) gearing

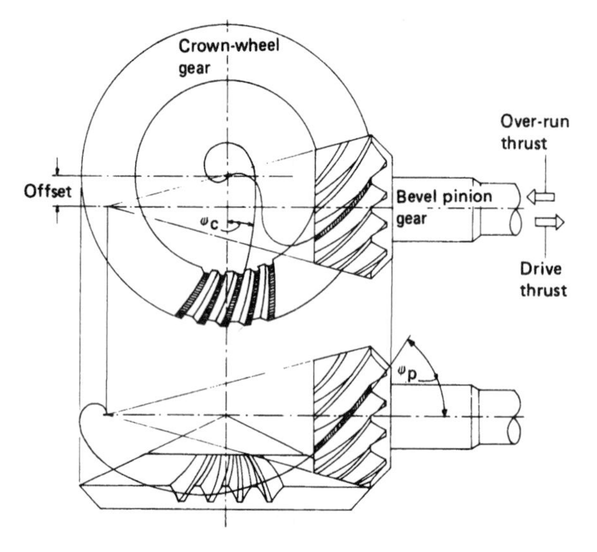

Fig. 6.7 Geometry of hypoid-tooth bevel-gear crown wheel and pinion

bevel-tooth gears, but the pinion is offset below the crown-wheel centre line.

The name 'hypoid' is an abbreviation of 'hyperboloid' of revolution, as the basic surfaces on which the teeth are cut are 'hyperboloids' instead of cones – see fig. 6.6. A hyperboloid is the solid formed by rotating a hyperbola about an offset axis, such that each point on the hyperbola remains at a constant distance from the axis. Hyperboloids have the distinctive property, common to the cylinder and cone, that a number of straight lines may be drawn upon their surfaces. If one hyperboloid is brought into contact with another, the axes may be so arranged that contact is over a straight line. Hence hyperboloidal surfaces are suitable for the transmission of motion by rolling contact. For an actual bevel gearwheel, only a frustum of the hyperboloid (a small width) is used, and teeth are formed partly above and partly beneath its surface.

The original attraction of hypoid gearing was the lowering of the pinion-gear axis below the centre axis of the crown wheel, so that the propeller-shaft tunnel formed as part of the floorboards between the passenger seats would not protrude so much. This then was followed by the designers' lowering the transmission components and engine, a feature which enabled the centre-of-gravity height from the ground to be reduced.

The adoption of hypoid gears also provides other advantages: it allows the pitch diameter of the pinion to increase with the offset, so that a larger pinion for a given size of crown wheel can be used. This increases pinion tooth strength by about 20 to 30% and gives a greater pitch overlap, resulting in almost silent running when the two gears are correctly meshed (fig. 6.7).

The geometry of an offset pinion alters the relative tooth contact action; hence, in addition to the sliding motion along the tooth profile, there is a sliding action in the direction of the length of the tooth. In order to limit this relative sliding velocity across the width of the teeth, a maximum offset of 0.2 times the crown-wheel diameter for passenger cars and 0.1 times the diameter for trucks should not be exceeded. Even so, lubrication problems were initially experienced as the two-dimensional rubbing generated additional friction and heat. To counteract this, extreme-pressure lubricants have been developed for hypoid final-drive axles – these oils provide a boundary film which is sufficiently stable to prevent tooth scuffing.

It is of interest to observe that the spiral-tooth angles of the crown wheel and pinion are equal ($\psi_c = \psi_p = 35°$ to $45°$) – see fig. 6.5. In contrast to spiral bevel gears, with hypoid bevel gears the crown-wheel spiral angle is smaller than that of the bevel pinion gear ($\psi_c = 10°$ and $\psi_p = 50°$) – see fig 6.7. This larger spiral angle of the hypoid pinion has the effect of increasing the pinion end-thrust by something of the order of 10 to 15%.

For most bevel types of gearing both the crown-wheel and the pinion shaft are at present made from nickel–chrome alloy steel and are carburised after machining and then case-hardened by an oil quench.

6.2.4 Worm and worm-wheel gears (figs 6.8 to 6.10)

A worm and worm-wheel are used to connect two non-intersecting shafts which are at right angles and are offset to each other.

In appearance (see figs 6.9 and 6.10) the worm resembles a multi-thread screw and the worm-

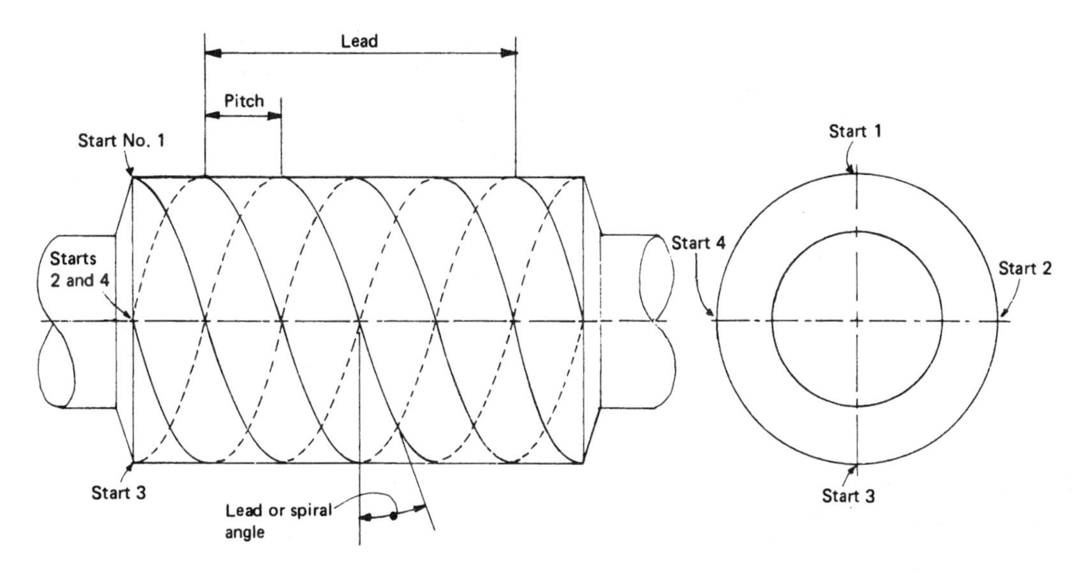

Fig. 6.8 Geometry of a four-start worm

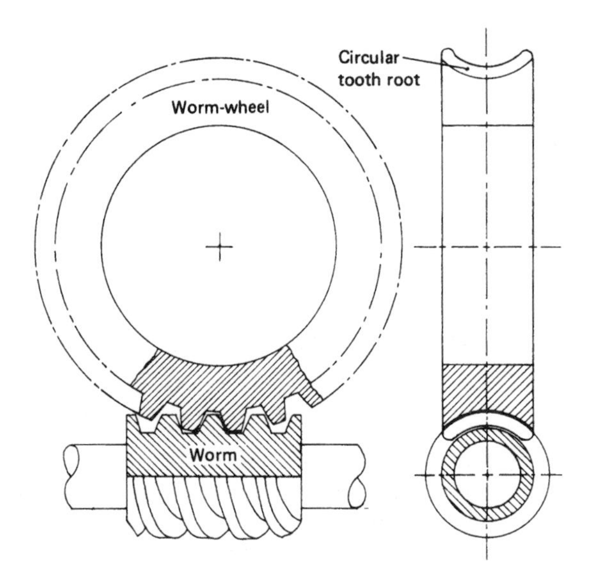

Fig. 6.9 Parallel worm and worm-wheel final drive

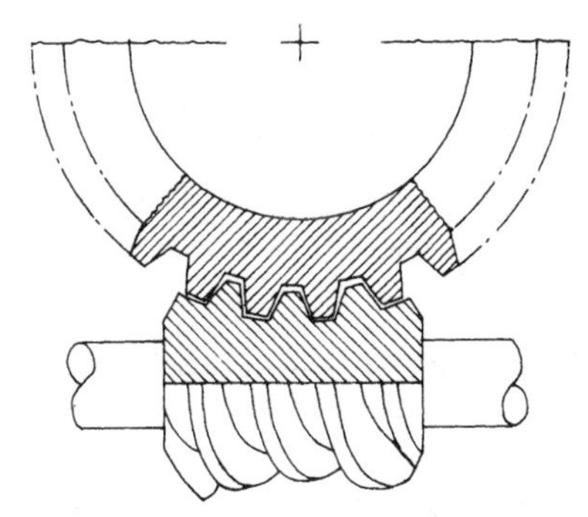

Fig. 6.10 Hourglass worm and worm-wheel final drive

wheel rim resembles a toothed rack bent round to form a ring. When the worm is rotated, the wedge profiles of the spiral-like screw threads slide across and push against the worm-wheel helical-cut teeth, thus providing a wedge action which causes the worm-wheel to rotate on its axis. The most effective spiral or lead angle is about 45°, which under good lubrication conditions may provide a mechanical efficiency of 97% for meshing pairs with highly polished surfaces.

The worms used are the multi-start type. This means that two or more threads (teeth) form helices parallel to each other along the length of the worm (fig. 6.8). The number of separate threads or teeth observed at one end of the worm is the number of starts. When a single-thread worm revolves once, it advances the worm-wheel a distance equivalent to the axial pitch of the worm. A four-start worm in one revolution will advance the worm-wheel a distance equivalent to the lead of the worm – which is four thread pitches (4 × pitch) and not one thread pitch as is the case for the single-start worm.

Worms are cut with multi-starts in order to decrease their gear ratios. Final-drive worms may have anything between four to eight starts, depending on the gear reduction required. The number of teeth on a typical worm-wheel can range between 26 and 38.

The gear ratio of a worm and worm-wheel combination is given by

$$\text{gear ratio} = \frac{\text{number of teeth on worm-wheel}}{\text{number of worm starts}}$$

Example Calculate the final-drive gear ratio of a worm and worm-wheel if the worm has five starts and there are 28 teeth on the worm-wheel.

$$\text{Gear ratio} = \frac{28}{5} = 5.6{:}1$$

The worm is usually made of case-hardened nickel steel and the worm-wheel of phosphor bronze, which includes about 12% tin. If a high working efficiency is to be maintained, both gears must be highly polished after being machined.

Worm drives operate best with large gear ratios of more than 6:1. They have a high load-carrying capacity, run smoothly, and are quiet in operation; but – due to the nature of the meshing-teeth rubbing action – their mechanical efficiency is lower than that of bevel types of gearing for a single-stage final drive. Materials and manufacturing costs are higher than for spiral and hypoid gearing, so they are not in common use. Because of their ability to provide a large gear reduction, they are sometimes preferred to double-reduction final drives as they can reduce the axle speed in a single stage and also because the overall mechanical efficiency of a double-reduction axle would not be better than that of the worm drive.

The worms may be underslung when used in cars or overslung for commercial-vehicle application.

There are two basic types of worm gears: the parallel (fig. 6.9) and the hourglass (fig. 6.10). For normal light and medium duty, the parallel type is suitable, but for heavy-duty work the hourglass type is recommended. It is claimed, that the hourglass-shaped worm provides greater bearing-surface tooth contact, but it is more difficult to machine and to align to the worm-wheel. It should be noted that for both types the root of the worm-wheel tooth is circular in cross-section instead of straight, so that a greater load-carrying capacity is obtained.

6.3 The need for final-drive differential gearing (fig. 6.11)

A given turning circle of a vehicle may be defined as the average circumferential distance moved by the vehicle with the driver's steering-wheel held at some point to one side of the central straight-ahead position. In actual fact, the paths followed by the inner and outer road-wheels will be slightly different to each other – the inner wheels will have a smaller turning radius than the outer wheels about some common centre (fig. 6.11). This implies that the outer wheels will travel a greater distance than the inner ones for the vehicle to complete a circular path. The only way this can happen is if the two sets of wheels revolve at slightly different speeds, with the outer wheels rotating faster than the inner ones and so completing more revolutions than the inner wheels for a given turning circle.

The speed variation between the inner and outer wheels may be illustrated by the following example – see fig. 6.11.

Example A vehicle has its steering set to provide a turning-circle radius of 5 m with a wheel-track width of 1.4 m. If the effective road-wheel rolling diameter is 0.6 m, determine the number of complete revolutions made by the inner and outer wheels for one turning circle.

Mean turning radius = 5 m

Inner-wheel turning radius, $R_i = 5\ \text{m} - \dfrac{1.4\ \text{m}}{2}$

$= 4.3$ m

Outer-wheel turning radius, $R_o = 5\ \text{m} + \dfrac{1.4\ \text{m}}{2}$

$= 5.7$ m

Rolling circumference of road-wheel $= \pi d$

$= 3.142 \times 0.6$ m

$= 1.885$ m

For one complete turning circle,

distance moved by inner wheel $= 2\pi R_i$

$= 2 \times 3.142 \times 4.3$ m

$= 27$ m

distance moved by outer wheel $= 2\pi R_o$

$= 2 \times 3.142 \times 5.7$ m

$= 35.8$ m

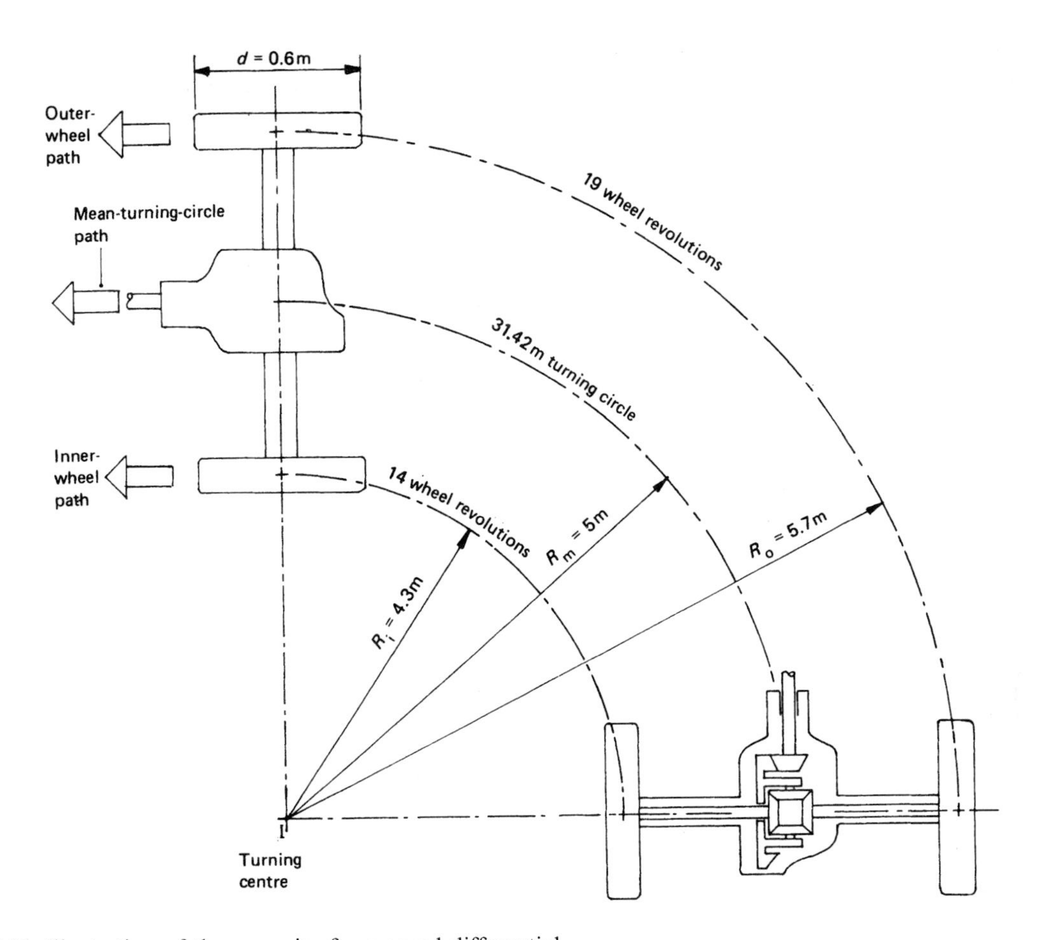

Fig. 6.11 Illustration of the necessity for a speed differential

$$\therefore \text{ revolutions completed by inner wheels} = \frac{\text{distance moved by inner wheels}}{\text{road-wheel circumference}}$$

$$= \frac{27 \text{ m}}{1.885 \text{ m}}$$

$$= 14 \text{ revolutions}$$

$$\text{and revolutions completed by outer wheels} = \frac{\text{distance moved by outer wheels}}{\text{road-wheel circumference}}$$

$$= \frac{35.8 \text{ m}}{1.885 \text{ m}}$$

$$= 19 \text{ revolutions}$$

The difference of 19 − 14 = 5 revolutions between the inner and outer wheels for one complete turning circle of the wheels must be accommodated either by tyre scrub or by some mechanical device known as the final-drive differential.

With non-driving wheels – that is, wheels which do not provide driving tractive effort to propel the vehicle forward, or backward for that matter – the difference in wheel speed is not a problem as the wheels are independent of each other and so can adjust their speeds to whatever is demanded by the driver and road conditions.

There are, however, difficulties with live axles and their road-wheels – that is, axle-shafts and wheels which transmit the tractive effort to the road to propel the vehicle along. In its simplest form, the propeller shaft or gearbox output shaft imparts power to the final drive, which then reduces the output speed and divides the driving motion between the two halves of the axle. The difficulty is to transmit the driving torque to the two half-shafts but at the same time to enable both shafts to adjust their speeds independently and automatically to suit the road demands.

The problem was solved as early as 1827 by Onesiphore Pecqueur, a master mechanic in Paris, France, who conceived the idea while making a model steam vehicle. The device invented enabled equal torque to be transmitted to each axle and wheel, while supplying a wheel-speed differential to match the road requirements. Thus the mechanism now incorporated within the final drive and between the half-shafts or drive shafts is commonly known as bevel differential gearing.

6.3.1 Construction and operation of bevel differential gearing

Differential construction The differential mechanism is built into the centre portion of the final-drive crown wheel. It usually consists of a cast malleable-iron cage housing which incorporates the bevel differential gears (fig. 6.12). The bevel-gear train may be considered to form a closed loop or ring in which there are two bevel side-gears generally called the sun gears and sandwiched between them are two or four bevel pinion gears known as planet gears which indirectly link the sun gears together.

The sun gears have extended sleeve shoulders which are externally machined and polished to provide bearing surfaces to support these gears in bosses formed in the differential cage. The hollow centres of these gears are splined so that they can engage and support the inner ends of the axle half-shafts.

Situated in the centre of the differential cage and at right angles to the sun-gear axis is a cross-pin which provides the axis for the planet gears. This pin is a force fit in the cage casting, but is additionally secured by a small location pin. Note: if for heavy-duty application there are four planet gears instead of two, this cross-pin is then referred to as the 'spider'. The differential cage is generally made in two halves so that the gears can be assembled, but sometimes a single-piece housing may be used for light-duty purposes. The inside of the cage is spherical-shaped to provide an effective bearing reaction surface for the sun-and-planet-gear end-thrust. Externally the cage has a machined flange to locate and support the crown wheel.

Drive division When a vehicle is travelling in a straight path (fig. 6.13(a)), both driving road-wheels will be revolving at the same speed, so the bevel planet gears just act as a bridge coupling the two bevel sun gears together. Power will thus be transmitted from the crown wheel and differential cage to the planet cross-pin (spider). The planet gears will then orbit about the sun-gear axis, pulling both sun gears around with them at the same rotational speed.

Speed division Now consider the vehicle driving on a curved track (fig. 6.13(b) and (c)). The inner wheel will tend to reduce speed, and the outer wheel will increase its speed to compensate for the different turning circles of the two sets of road-wheels. To enable this to happen without tyre scrub, the inner drive wheel will reduce its rotational speed to something less than the speed of the crown wheel and differential cage and the outer drive wheel will proportionally increase its speed above this latter speed. This means that the planet gears will have to rotate on their cross-pin axis to allow relative movement between the planet and sun gears to take place. So that this relative movement between gears can occur, the meshing planet gears will be made to roll ('walk') around the slowed down but still revolving inner sun gear, due to the interlocking gear teeth. Hence the rotational motion about the slow-moving sun gear will be imparted to the outer sun gear by the revolving planet gears in addition to the normal crown wheel rotational speed; consequently the outer wheel rotates faster by the amount that the inner-wheel speed is reduced. Taking this concept to the limit, if one of the sun gears is stopped from rotating completely, the forced rotation of the planet gear on its own axis will double the speed of the other sun gear, so that it will be rotating at twice the speed of the input crown wheel.

From this, it is evident that whichever side the curve in the road is, the differential sun and planet gears will automatically adjust their relative movements independently of the crown wheel so that they can accommodate the continuously changing road conditions.

Consolidating the principle of the differential gears, it can be seen that the speed lost by the inner sun gear will be the speed gained by the outer sun gear, so the speed of the crown wheel and cage will be equal to the mean speed of the two drive wheels,

i.e. $$N_c = \frac{N_i + N_o}{2}$$

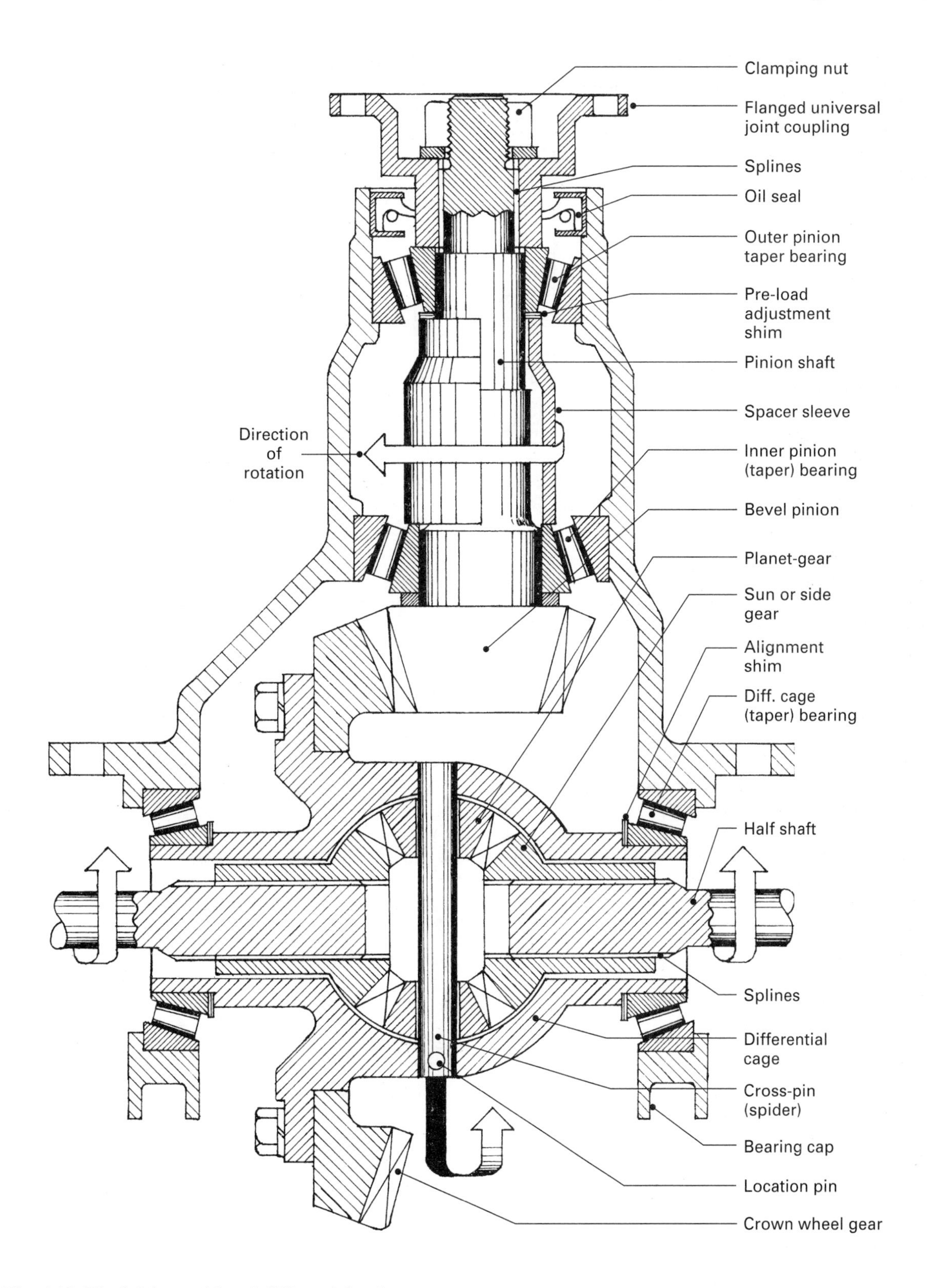

Fig. 6.12 Final-drive and bevel-differential unit

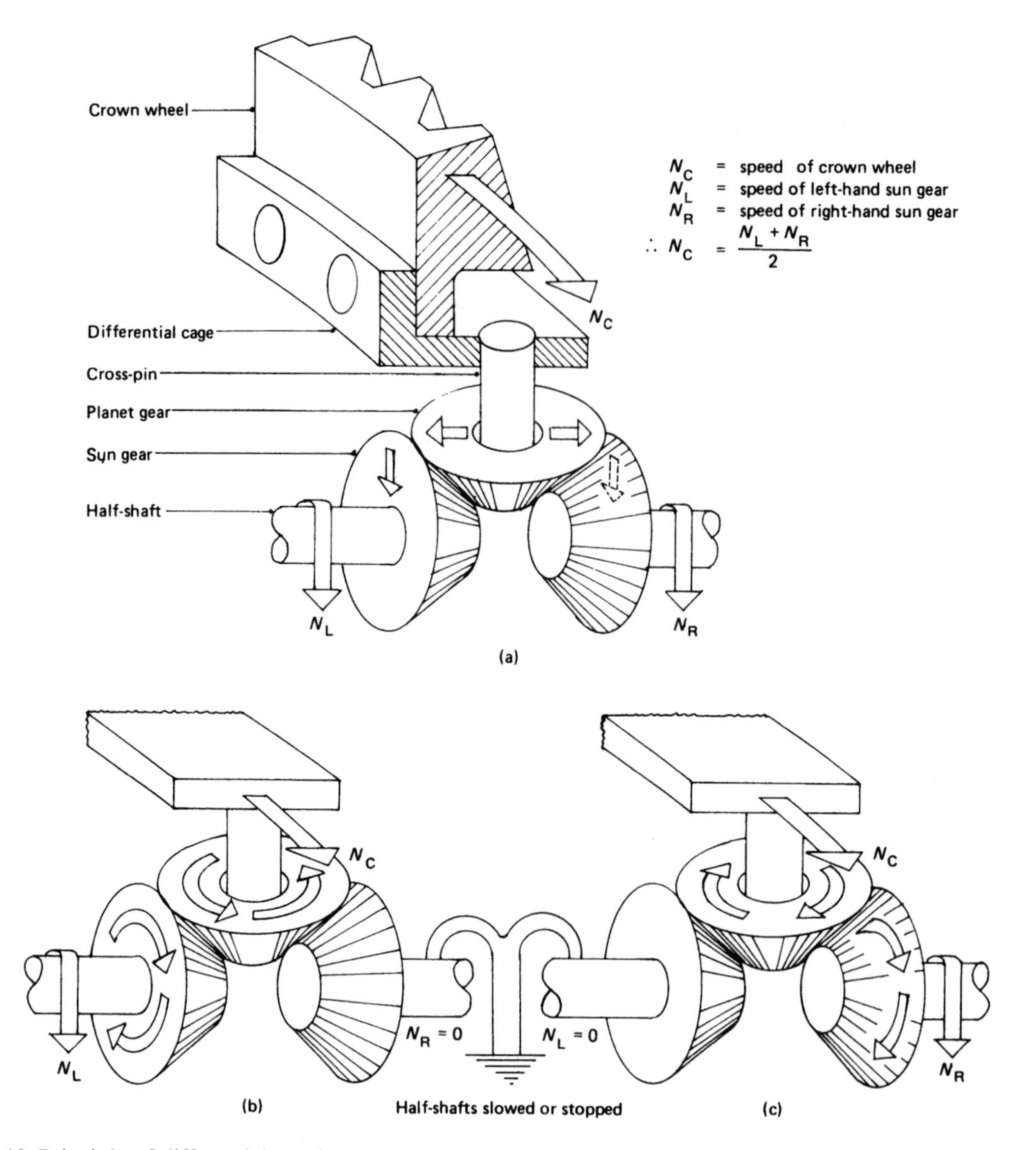

Fig. 6.13 Principle of differential gearing

where N_c = crown-wheel speed
N_i = inner-wheel speed
and N_o = outer-wheel speed

Example 1 If the inner and outer wheels of a vehicle rounding a bend rotate at 80 and 120 rev/min respectively, determine the crown-wheel speed.

$$N_c = \frac{N_i + N_o}{2} = \frac{80 + 120}{2}$$
$$= 100 \text{ rev/min}$$

Example 2 Find the speed of the inner driving wheel of a vehicle if the crown wheel and the outer wheel rotate at 200 and 250 rev/min respectively.

$$N_c = \frac{N_i + N_o}{2}$$
$$200 = \frac{N_i + 250}{2}$$
$$\therefore \quad N_i = (200 \times 2) - 250$$
$$= 150 \text{ rev/min}$$

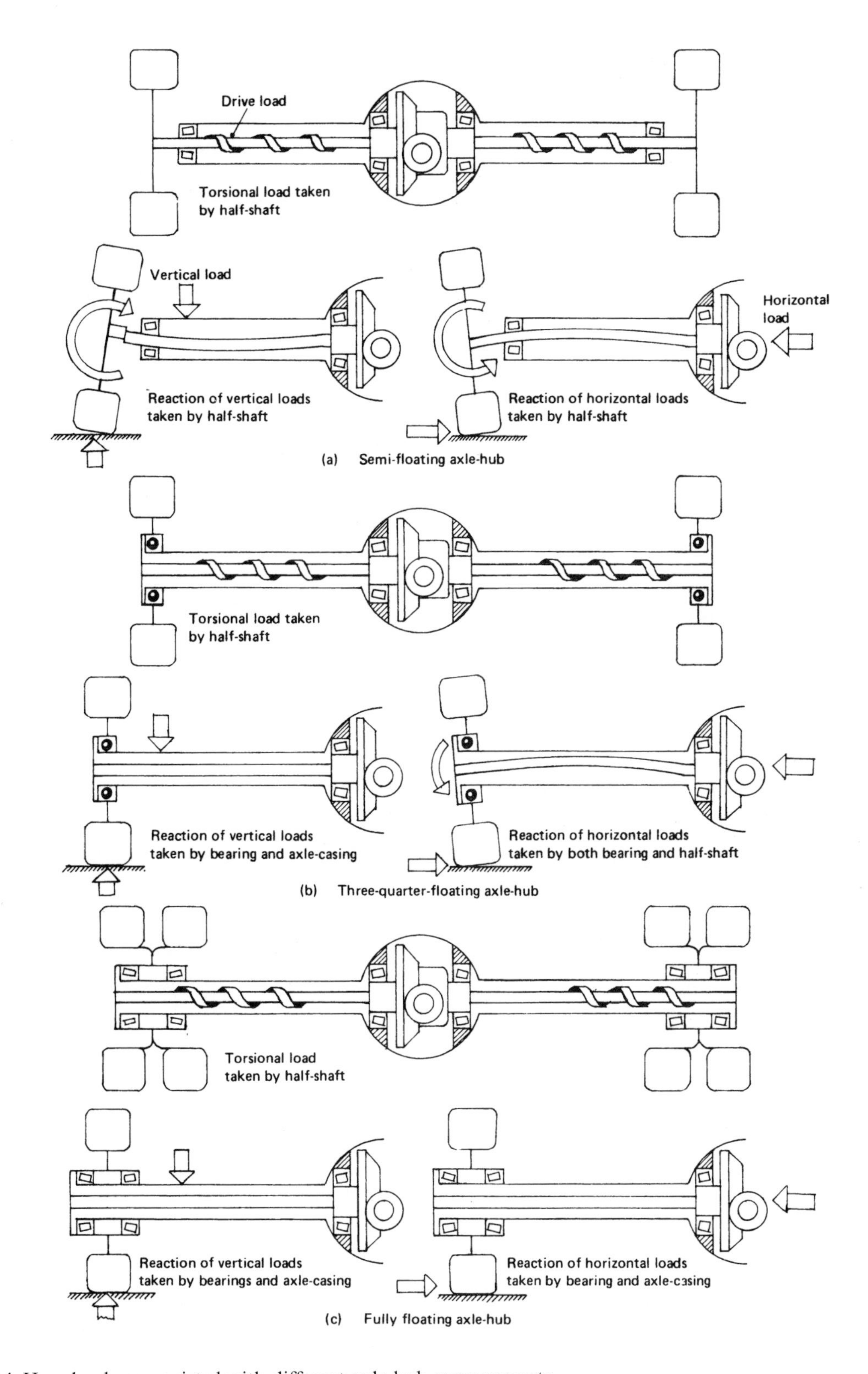

Fig. 6.14 How loads are resisted with different axle-hub arrangements

Torque division The driving torque applied to any one driving wheel must have an equal and opposite torque reaction which is applied through the planet gears by the opposite sun gears and driving wheel. It follows that torque supplied to the planet gears by the cage and cross-pin will be equally divided between the sun gears, so that any traction lost by one of the driving wheels due to skidding will be equally subtracted from the traction of the other wheel. The resulting reduced traction is known as the limiting torque or traction applied to both driving wheels. Thus, if a wheel on one side of a vehicle cannot grip the ground due to a slippery surface, then the wheel on the other side will not be able to produce any tractive effort because the skidding wheel cannot provide a torque reaction to support and hold still or even slow down its own sun gear. The differential can therefore be called a torque equaliser.

6.4 Rear-axle half-shaft and hub arrangements (fig. 6.14)

Axles are designed to transfer the vehicle's vertical weight to the road-wheels and in some cases provide the means to transmit driving effort to the wheels. Axles can be classified as either live axles, which support the vehicle and its load and which transmit power to the road-wheels, or dead axles which only carry the vehicle's weight.

The arrangements for supporting the road-wheels on live axles and providing the driving traction involve using an axle-hub mounted on to the axle-casing and supported by ball- or roller bearings.

Axle half-shafts are situated on each side of the final drive, to convey motion to the road-wheels.

There are basically three different methods of supporting axle wheel hubs on the rear-axle casing. These are classified as follows:

a) semi-floating axle-hub,
b) three-quarter-floating axle-hub,
c) fully floating axle-hub.

6.4.1 Semi-floating axle-hub (fig. 6.14(a))

Description The axle-hub which has the road-wheel attached to it is an extension of the axle half-shaft. The outer end of the shaft is supported by a single bearing inside the tubular axle-casing. The inner end of the shaft is splined and supported by the final-drive unit, which is itself mounted on bearings within the axle-casing.

Shaft hub and bearing loading The semi-floating axle with its overhanging hub is subjected to the driving torque and to both vertical and horizontal loads. The vertical vehicle weight establishes a shearing force, and the distance between the wheel and the suspension-spring seat on the axle produces a bending moment, the reaction of the latter being shared between the axle bearing and the final-drive-unit bearings. The horizontal load due to tilting of the vehicle, cornering centrifugal force, or side wind creates both side-thrust and a bending moment. The latter may act with the vertical bending moment or in the opposite sense, depending on the direction in which the side-force is applied.

Construction Figure 6.15 shows a semi-floating axle suitable for small and medium-sized cars.

The axle half-shaft and flanged hub are forged from a single piece of nickel–chrome steel. The hub end of the shaft has a larger diameter than the rest of its length, to cope with the vertical and horizontal loads it is subjected to in service. The outer face of the flanged hub is shouldered so that the brake-drum can be accurately centralised, and wheel studs are situated in holes evenly spaced around this machined flange.

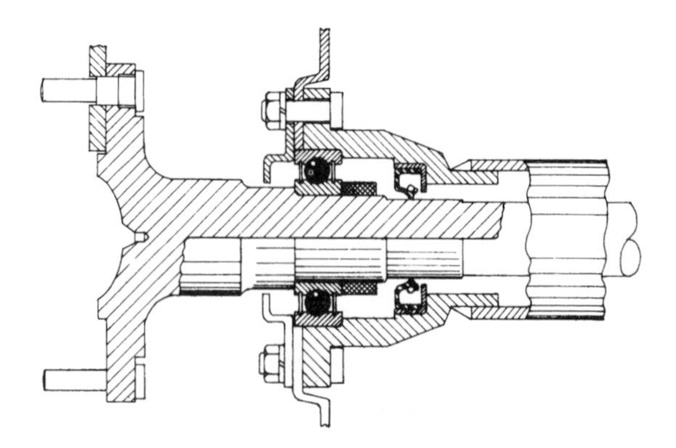

Fig. 6.15 Semi-floating ball-race-bearing axle

A deep-grooved ball-race bearing – pre-greased and sealed – is pressed over and along the shaft up to its shoulder. An interference-fit retaining collar secures the bearing position. The half-shaft outer ball-race track is made a light force fit inside the mouth of the tubular axle-casing. The bearing is sandwiched on one side by the axle-casing and on the other by the brake backplate, and the retaining plate by four nuts and bolts. To prevent excess oil finding its way to the end of the axle-casing, a radial-lip oil-seal is pressed into a recess in the casing. When positioned correctly, the seal lip should align and rub against a machined and polished surface. In this particular axle it can be seen that the enlarged mouthpiece of the axle-casing fits into the tubular casing and that the two halves are then permanently lap-welded.

Figure 6.16 shows a semi-floating axle using a taper-roller bearing, which is preferred for larger and higher-performance cars due to its greater load-carrying capacity. With this arrangement, a separate hub is wedged on to a keyed and tapered half-shaft and is held in position by a castellated nut. The taper-roller-bearing inner cone is a light force fit inside the mouth of the casing. The exact position of the bearing in the casing is determined by shims packed between the casing flange and the brake backplate. Outward thrust on either road-wheel is taken by the adjacent hub bearing, while inward thrust is transmitted to the opposite bearing through the axle half-shafts and a slotted axle-shaft spacer (not shown). Thus each hub bearing takes thrust in one direction only. Increasing the thickness of the shims on one side and decreasing it on the other shifts both half-shafts further to one side relative to the axle-casing. Removing shims from one side without replacing them with a similar thickness on the other side reduces shaft end-float; reversing the process will increase end-float.

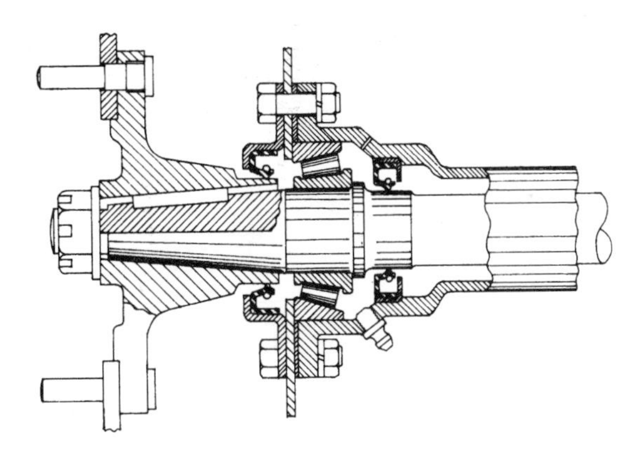

Fig. 6.16 Semi-floating taper-roller-bearing axle

The bearing is lubricated by pre-packing it with grease and periodically pumping in additional grease. Two oil-seals are fitted: the inner one prevents oil escaping from the final drive, and the outer oil-seal prevents excess grease pumped into the bearing finding its way on to the brake-shoes and drum.

6.4.2 Three-quarter-floating axle-hub (fig. 6.14(b))

Description As before, the road-wheel is bolted to the hub which forms part of the axle-shaft. The outer end of the shaft and hub is supported by a bearing positioned over the axle-casing, so that the bearing is between the hub and the casing instead of between the axle and the casing as in the semi-floating layout. The inner end of the half-shaft receives its support by being splined to the final-drive assembly, as in the case of the semi-floating half-shaft.

Shaft hub and bearing loading With the three-quarter-floating axle and hub layout, the shaft transmits the driving torque but the vertical-load shear force and bending moment are taken by the tubular axle-casing through the medium of the hub bearing, provided the road-wheel and the hub bearing lie in the same vertical plane. In practice, a slight offset of wheel and bearing centres can exist which tends to tilt the hub relative to the axle-casing, and this will be opposed by the bearing or, if this offset is large, the half-shaft will supply the additional resistance. Horizontal loads, which create end-thrust, will be restricted by the hub bearing and casing, but the side-forces will create a bending moment which will tend to twist the wheel relative to the axle-casing. This tilting tendency will be resisted by both the hub bearing and the axle-shaft.

Construction Figure 6.17 shows a three-quarter-floating axle which was at one time popular for cars and light commercial vehicles when semi-floating half-shafts frequently fractured, particularly in cold weather. However, improvements in

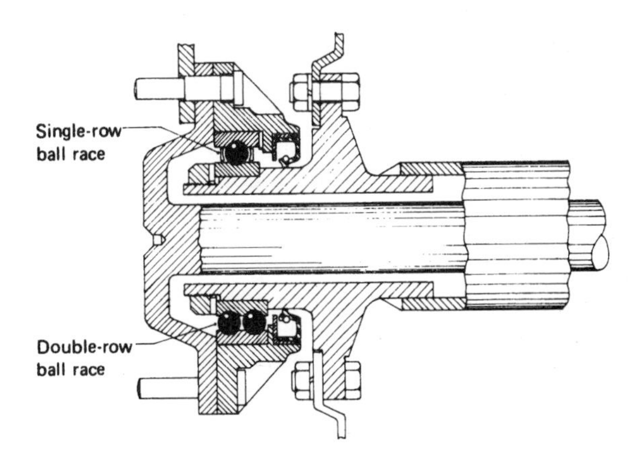

Fig. 6.17 Three-quarter-floating axle

design detail, manufacturing, and materials have made the compact and cheap and now reliable semi-floating axle is a better proposition than the three-quarter-floating arrangement.

As can be seen, the half-shaft has an upset-forged flange outer end which is clamped to the bearing hub by the wheel studs. A rather large-diameter single-row or a double-row ball-race bearing can be employed (fig. 6.17), depending upon light- or heavy-duty applications. This bearing is positioned over the axle-casing and is held in position by a large nut. The outer bearing track supports the hub, and at the back of the hub is an oil-seal which prevents excess oil (coming from the final drive after lubricating the bearing) escaping from the hub to the brakes. Due to the design discussed earlier, the bending moment due to side-thrust will be accommodated partly by the half-shaft but mostly by the bearing, so that an unacceptably large tilting force will tend to overload the bearing if this is not adequately proportioned to sustain the stresses imposed.

6.4.3 Fully floating axle-hub (fig. 6.14(c))

Description This hub layout takes the form of a flanged sleeve which is mounted over the axle-casing, the flange being provided to accommodate the road-wheel or wheels. Two spaced-out bearings are positioned between the hub and the casing to support the hub assembly.

Shaft hub and bearing loading An improvement on the first two types of hub support is provided by the fully floating axle arrangement. Here the axle-shaft takes only the turning-effort or torque, and both the vertical and horizontal load reactions are resisted by a pair of widely spaced taper-roller bearings mounted on the axle-casing. Thus the axle half-shaft is relieved of all the various loads except the torsional drive to the wheel.

Construction Figure 6.18 shows a fully floating axle-hub which is based on a concept similar to that of the three-quarter-floating axle but takes the design one stage further by not only relieving the half-shaft of vertical shear forces and bending moments but also insulating the shaft from all horizontal-load effects. This is achieved by mounting two taper-roller bearings over the axle-casing some distance apart, so that any side-thrust or tilting loads on the hub can easily be restrained by the bearings in any direction.

As can be seen in fig. 6.18, the hub revolves on the axle-casing completely independently of the half-shaft, so the latter can be removed for replacement if necessary without interfering with the load-supporting role of the hub. The vehicle can thus be towed with a broken half-shaft. This construction is larger and more expensive than both the other arrangements, but it is particularly suitable for all truck and heavy-duty vehicles employing live axles and for trailers using dead axles. Single or twin road-wheels may be fitted, depending upon the application.

It should be observed that the inner bearing is slightly larger in diameter than the outer bearing, so that they fit over the tubular casing in step fashion, and the sets of bearing rollers taper towards each other. These axle-hubs will give good service provided that the bearings are correctly assembled and pre-loaded. The correct tightness of the bearings is achieved by adjusting the inner nut at the end of the axle-casing until the required pre-load torque is obtained. The two nuts are then locked together. On some hub arrangements of this type, a spacer tube is placed between the inner bearing cones; also, shims are placed between one of the outer bearing cones and the adjacent hub shoulder for initial bearing adjustments.

Lubrication is provided from the final-drive oil supply so that oil is forced to flow out to the hub and between the bearings, and a radial-lip oil-seal prevents any oil escaping from the rear of the hub. A gasket is sometimes placed between the flanged half-shaft and the outer face of the hub, to prevent any leakage of oil.

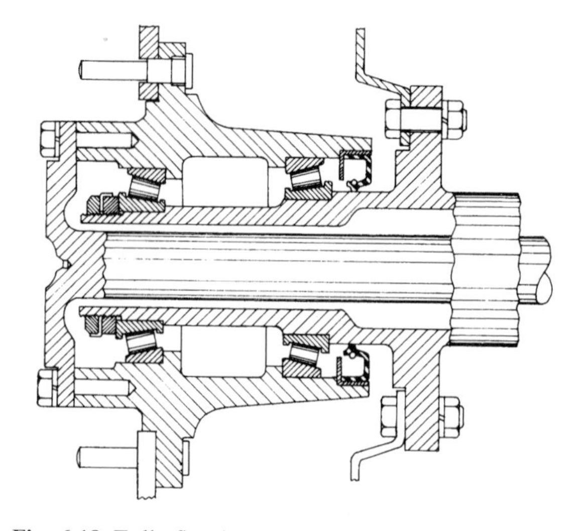

Fig. 6.18 Fully floating axle

6.4.4 Summary of loads imposed on rear-axle wheel-hub assemblies

The forces imparted to the half-shaft, hub, axle-casing, and bearing or bearings with the three different axle-hub arrangements are compared below.

a) Semi-floating axle-hub:

   i) a driving torque;
   ii) a shearing force due to the vertical vehicle load;
   iii) a bending moment due to the offset of the vehicle spring-seat load and the road wheel;
   iv) an end-thrust due to side-forces created by road camber, cornering, side wind, etc.;
   v) a bending moment due to the resistance offered by the tyre on the ground when end-thrust exists.

b) Three-quarter-floating axle-hub:

   i) a driving torque;
   ii) the shear force due to vertical loading is absorbed by the axle casing;
   iii) the offset-vertical-load bending moment is absorbed by the hub bearing and the axle-casing;
   iv) the end-thrust load is absorbed by the hub bearing and the axle-casing;
   v) the end-thrust bending-moment reaction is shared between the hub bearing and the half-shaft, the proportion sustained by the shaft depending on the hub-bearing arrangment.

c) Fully floating axle-hub:

   i) a driving torque;
   ii) the shear force due to vertical loading is absorbed by the axle-casing;
   iii) the offset-vertical-load bending moment is absorbed by the axle-casing;
   iv) the end-thrust load is absorbed by the hub bearing and the axle-casing;
   v) the side-thrust bending-moment reaction is spread out between the hub bearings and is absorbed by the axle-casing.

6.5 Differential lock (fig. 6.19)

Vehicles which are frequently used to drive over rough, sandy, wet, boggy or snowy countryside must deliver traction to both drive wheels even when one of the drive wheels has lost its grip. Conventional final drive differentials enable the drive wheels to rotate at different speeds when the vehicle is negotiating a curved track and at the same time to divide the drive torque equally between these wheels. However, if one wheel should lose traction, the other wheel will likewise lose an equal amount of road surface grip. Consequently the complete loss of grip on one wheel means there will be no drive in the other wheel even though its tyre-to-surface adhesion is good.

The purpose of a differential lock is to rigidly join the half-shafts or drive-shafts of either wheel together. This is normally achieved by locking together any two convenient components of the differential which have relative movement. In the example shown in fig. 6.19, one of the bevel sun-gears has been extended so that dog-teeth can be formed around it. Similarly the differential-cage has been extended and eight equally spaced slots have been machined to mount and support a sliding-ring dog-clutch.

Engagement of the differential lock is obtained which the control-fork operates by a reduction geared control motor via a cable (not shown) that pushes the sliding-ring dog-clutch into mesh with the bevel sun-gear dog-clutch. The locking together of the differential cage with one of the sun-gears is sufficient to jam solid the differential bevel-gearing, hence in effect the two drive-shafts or half-shafts are now rigidly joined together. There is now no speed differentiation between the drive wheels; however, torque will be delivered to both wheels at all times, even if one road wheel loses its grip. This differential lock is designed so that it can be engaged up to 15 km/h (10 mph). Above 25 km/h (15 mph) the differential lock electrical actuating mechanism is made to automatically disengage the dog-clutch lock, thereby preventing undue transmission wind-up and tyre slip/scrub.

6.6 Four-wheel drive (fig. 6.20)

Conventional cars, vans and trucks normally only have one live axle so that traction is shared entirely by the tyres of the two road wheels. With the usual bevel-gear differential, if one of these wheels should lose traction, then the removal of a reaction torque on this spinning wheel prevents torque being transferred to the other wheel which has maintained its grip on the ground. By having both axles live, drive is now transmitted to all four road wheels, hence the terms four or all-wheel drive (FWD or AWD).

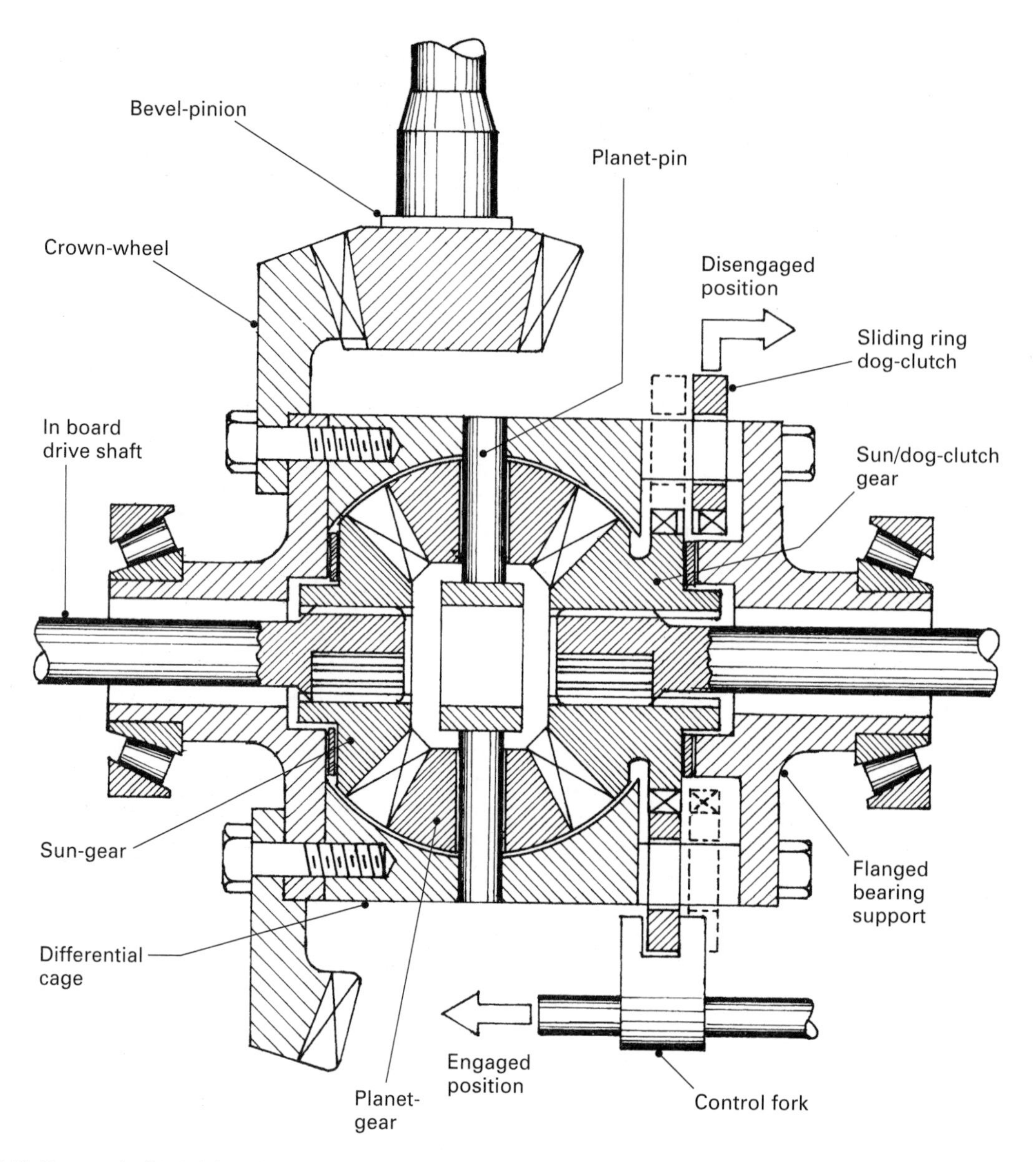

Fig. 6.19 Rear-axle final drive differential lock mechanism

The benefits of a four-wheel drive vehicle compared to a two-wheel drive are as follows:

1) Four-wheel drives are particularly suitable for off-road vehicles which have to travel over rough, rugged, sandy, or boggy terrain and on snow or icy ground.
2) The traction of a four-wheel drive is spread four ways so that the chances of losing complete traction are much smaller; thus slippage when pulling away from a standstill or accelerating when already in motion is more unlikely.
3) If the static load distribution of the vehicle is uneven, then the axle and wheels taking more load will accordingly provide the greater traction; it therefore provides a greater degree of driving stability.
4) If the vehicle is climbing, additional load will be transferred to the rear-drive-axle and wheels, whereas if the vehicle descends a slope more weight will be transferred to the front

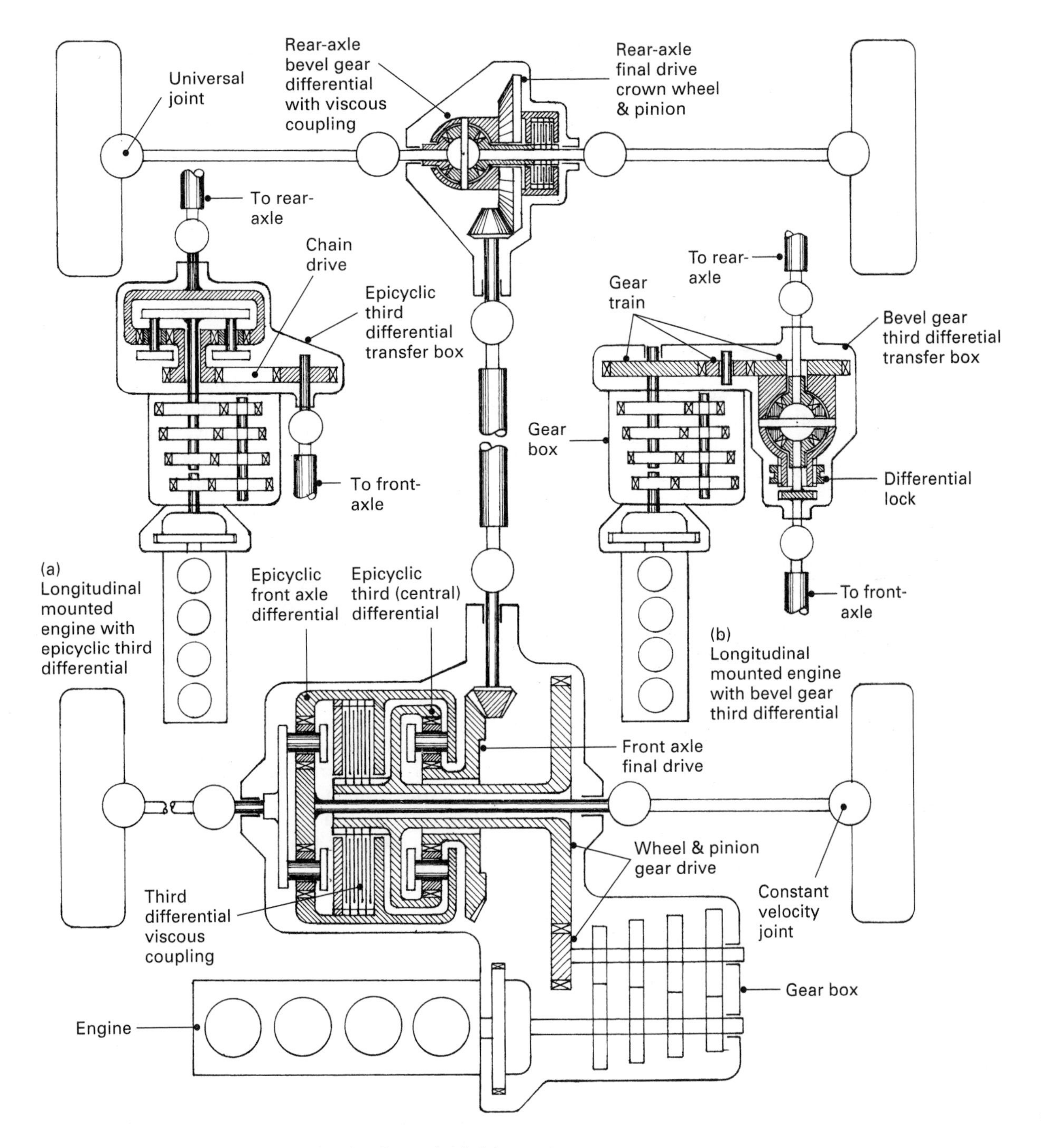

Fig. 6.20 Transversely mounted engine four-wheel drive system

driving wheels; it therefore enables a relatively high degree of traction to be maintained overall at all times.

5) With a conventional rear-wheel drive, if the front axle gets bogged down, the front wheels do not rotate and consequently the forward propelling tractive effort of the rear wheels just digs the front wheels further into the ground. However with a four-wheel drive, if the front wheels get stuck, the propelling force

from behind and the driving torque at the front wheels enables the rotating front wheels to climb out of a rut.

The limitations of a four-wheel drive system are as follows:

1) Under most driving conditions it is unlikely that there is any need for an additional drive axle, since the traction by two wheels is quite adequate and the extra weight and expense of the transmission system may not be justified.
2) If a four-wheel drive vehicle is to travel at speed, then a third differential is essential to prevent torsional transmission wind-up and excessive tyre squeal and wear.
3) With conventional differentials, if one wheel slips then no torque is transmitted to the other road wheels unless a differential-lock, limited-slip differential, or a viscous-coupling is incorporated in the final drive.
4) If the front or rear wheels should lose their traction, then the drive will also be lost on the good drive wheels unless the third differential is locked out by means of a differential-lock or a viscous-coupling built into the third differential.
5) The additional transmission gearing and moving parts will provide more wearing components, cost more to repair, and will very slightly increase the vehicle's fuel consumption.
6) For highway use a good traction control system can be nearly as effective without the economy, weight and cost penalties of a four-wheel drive.

6.6.1 Four-wheel drive arrangements (fig. 6.20)
A typical transverse mounted engine four-wheel drive layout is shown in fig. 6.20. Power from the engine is transferred through the gearbox to the reduction wheel and pinion gear drive to the transfer-box. Here the drive is conveyed to the third differential annulus-carrier and ring-gear via the hollow input-shaft. The drive is then split: one half transfers to the planet-pinion carrier which forms the input to the front final-drive differential, and the other half is transferred to the sun-gear and the attached crownwheel and pinion; the latter in turn transmits the power flow via the propeller-shaft to the rear-axle final-drive.

Thus it can be seen that the transfer-box houses two flat epicyclic differentials with a viscous-coupling in between. The flat epicyclic gear differentials are usually preferred as they provide a more compact transfer-box than if bevel-gear-type differentials were used. The third differential divides the drive between the front and rear axles; it also incorporates a viscous-coupling whose purpose is to prevent any sudden excessive speed difference occurring between the front and rear final drives, whereas the front differential splits the input drive from the third differential into two, so that it can transmit equal torque to both front road wheels and at the same time enables both wheels to rotate at their own speed dictated by the road conditions. The rear-axle final-drive crownwheel and pinion cancels out the gear reduction of the transfer-box crownwheel and pinion, since the wheel and pinion drive already provides the necessary final-drive gear and speed reduction. Built onto the rear-axle crownwheel and pinion is a bevel-gear differential which divides the driving torque between each rear road wheel and simultaneously enables each road wheel to rotate at its own speed. However, if one wheel should slip, the attached viscous-coupling automatically partially jams the differential bevel-gears; this then prevents the slipping wheel from spinning, and simultaneously provides a reaction torque into the differential so that torque will be transmitted to the road wheel which is still maintaining ground traction.

Two insert drawings (a) and (b) are shown in fig. 6.20. These drawings show typical transfer-box layouts which incorporate third differentials for longitudinal mounted engines. Figure 6.20(a) shows an epicyclic third differential which is used to provide an unequal torque division between the front and rear axles, while fig. 6.20(b) shows a bevel-gear differential and lock used when a 50/50 torque distribution is acceptable.

Epicyclic third differential transfer box (fig. 6.20(a)) Power from the gearbox is transferred to the planet-pinion carrier; the planet-pinions then divide the power flow between the sun-gear and the annulus ring-gear. The sun-gear directs the drive via the chain-drive to the front final-drive, whereas the annulus ring-gear and carrier forms the output via the propeller-shaft to the rear final-drive. When both front and rear drives rotate at the same speed the planet-pinions are held stationary so that the whole epicyclic cluster of gears revolves as one solid unit. As soon as there is a difference in speed between the front and rear propeller-shaft drives, then the planet-pinions commence to revolve on their own axes.

This permits the sun and planet-gears to rotate freely at their respective speeds and still maintain the designed torque distribution between the two axles.

Bevel gear third differential/lock transfer box (fig. 6.20(b)) Power from the gearbox is transferred via a train of gears to the third differential cage. The bevel pinion-gears then divide the power between the front and rear bevel sun-gears. The front sun-gear relays power to the front-axle final-drive by way of the front propeller-shaft. This drive also incorporates a differential lock which should only be engaged when negotiating rough,very soft or slippery ground, whereas the rear bevel sun-gear transmits power to the rear-axle final-drive by way of the rear propeller-shaft. When both propeller-shafts are rotating at similar speeds the bevel planet-gears are held still on their pins; therefore there is no relative movement of the bevel-gears within the differential cage and consequently the differential cluster of gears rotates as a single rigid unit. However, when there is a variation in speed between the front and rear propeller-shafts the bevel planet-gears commence to rotate on their pins. This enables the bevel sun-gears to rotate at different speeds to each other and to the input differential cage, while nevertheless still supplying equal torque to both front and rear final-drives.

6.6.2 Third (central) differential (fig. 6.20)

With four-wheel drive vehicles the front and rear drive wheels may not necessarily rotate at the same speed due to a number of factors:

1) There will be a speed variation between axles when a vehicle is being driven round a curved track, since each axle will have a slightly different rolling radius about some instantaneous centre of rotation.
2) Wheels rolling on a flat surface and those rolling on a hump or in a dip will have to rotate at different speeds at any one time to each other if the vehicle is to maintain its mean speed.
3) New and worn tyres on the same vehicle will produce different effective rolling radii to each other.
4) Since the payload distribution between axles will usually be uneven, it therefore causes a corresponding change in the effective rolling radius of each road wheel.

If the front and rear pair of final drives were to be directly coupled together, then the slightest difference in mean speed between the front and rear axles would cause the drive-shafts and propeller-shaft to be torsionally wound up, this torsional strain energy only being released when one or more of the wheels lose their grip, slip or bounce off the ground. To avoid excessive transmission wind-up, four-wheel drives incorporate a third differential (sometimes known as the central differential). Torque from the gearbox is diverted to the third differential where it is then divided between the front and rear final-drives and at the same time the third differential permits both final-drive inputs to rotate at whatever speed the road conditions dictate. Both final-drive and third differentials can be either of the bevel-gear or epicyclic-gear type. Bevel-gear differentials always split the output torque 50/50 whereas the epicyclic-gear differential, known as a flat gear differential, can divide the output torque with other combinations, such as 34/66 or 53/47 front and rear respectively, depending upon the weight distribution and the driving requirements of the vehicle.

6.6.3 Transfer box incorporating an epicyclic third (central) differential / lock and a front axle bevel gear differential (fig. 6.21)

Power from the gearbox (see fig. 6.21) is transferred to the third differential via the annulus ring-gear hollow shaft and carrier; the annulus ring-gear internal teeth mesh with the outer planet-pinions while the external teeth of the sun-gear meshes with the inner planet-pinions. Output to the front-axle differential is taken from the twin planet-carrier, while the rear-axle final drive takes its drive via the sun-gear and crownwheel and bevel-pinion. The twin planet-pinions are used instead of single planet-pinions to make the sun-gear rotate in the same direction of rotation as that of the annulus ring-gear.

Power supplied to the front bevel-gear differential cage from the third differential is then split between the right and left-hand bevel sun-gears via the bevel planet-gears which are free to revolve on their own axes (pins); the bevel planet-gears therefore act as the bridging speed divider.

Action of epicyclic third differential (fig. 6.21) When both front and rear final-drives are rotating at the same speed there will be no relative internal rotation of the twin planet-pinions so that the whole epicyclic unit rotates as one solid member. Should the front axle begin to rotate faster than the rear-axle, then the twin planet-pinions will

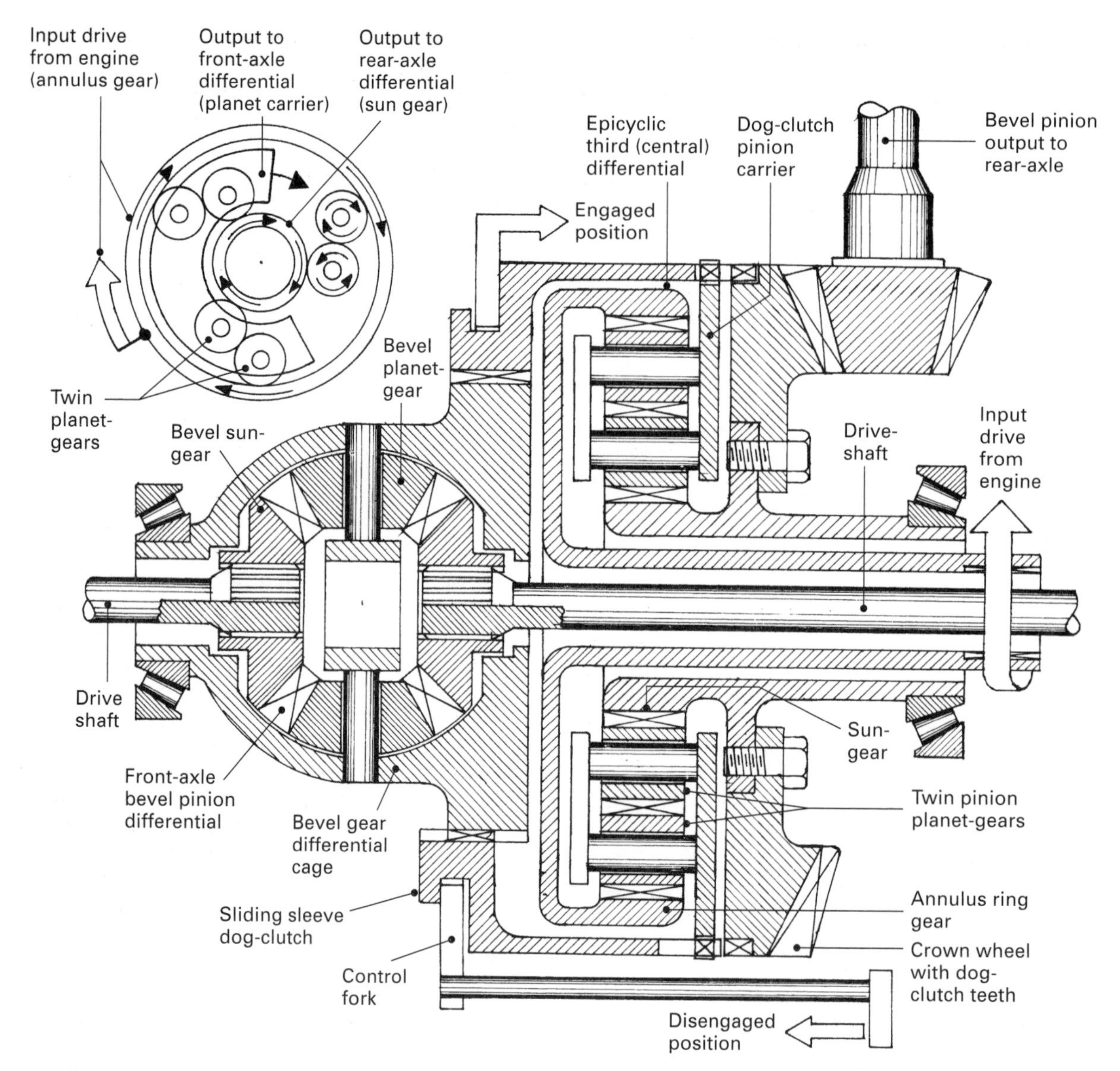

Fig. 6.21 Transfer box incorporating an epicyclic third (central) differential/lock and a front-axle bevel gear differential

commence to rotate on their axes and revolve around the internal and external teeth of the sun and annulus-gears without altering the speed of the rear final-drive. Conversely, should the rear final-drive speed up and rotate faster than the front axle, then again the otherwise stationary twin planet-pinions commence to rotate on their axes and revolve about the internal and external teeth of the sun and planet-pinions respectively without interfering with the speed of the front axle.

Third differential lock (fig. 6.21) The third differential lock simply consists of a sleeve internally splined to the front axle differential cage on the left-hand side with extended dog-teeth at the other end which are in permanent mesh with the twin planet-pinion carrier. When the third differential is locked out a reduction motor and cable pushes the sliding sleeve dog-clutch to the right until they align and engage corresponding dog-teeth spaces machined on the crownwheel. The drive between the front differential and crownwheel and pinion leading to the rear axle final-drive is now bridged; consequently epicyclic gears are jammed and neutralised so that equal torque is delivered to both front and rear axles.

6.6.4 Transfer box incorporating an epicyclic third (central) differential with viscous coupling and a front axle epicyclic differential (fig. 6.22)

Power from the gearbox is transferred to the third differential annulus ring-gear via its hollow shaft and carrier; the planet-pinion gears mesh with both the internal teeth of the annulus ring-gear and the external teeth of the sun-gear. Output to the front-axle differential is taken from the planet-pinion carrier, while the rear-axle final-drive takes its drive via the sun-gear and crownwheel and pinion. Power supplied to the front epicyclic differential annulus ring-gear is then equally divided between the planet-pinion carrier and the left-hand drive-shaft and the sun-gear and the right-hand drive-shaft.

Action of front final-drive epicyclic differential (fig. 6.22) When both front road wheels are rotating at the same speed there will be no relative internal rotation of the planet-pinion and sun-gear so that the whole unit rotates as one solid member. Should the left-hand road wheel and planet-carrier speed up, then the planet-pinion will commence to revolve around the external and internal teeth of the sun and annulus-gears respectively without altering the speed of the sun-gear and its corresponding right-hand road wheel. Conversely, should the right-hand road wheel and sun-gear speed up, the planet-pinions will again commence to rotate on their axes and to revolve about the external and internal teeth of the annulus and sun-gear respectively without changing the speed of the planet-carrier and its corresponding left-hand road wheel.

6.6.5 Viscous coupling (fig. 6.20 and 6.22)

The shortcoming of a differential lock used for final-drives or third differentials is that it is either disengaged or fully engaged and has to be manually selected. However, combining a viscous coupling with a differential provides an automatic way of neutralising the differential action whenever there is a relative speed difference between the front and rear final-drives or when one of the wheels loses its grip and commences to spin. Viscous coupling therefore can be incorporated with either a third differential to transmit more torque to the final-drive whose axle retains good road wheel traction, or when it is incorporated into a final-drive differential to transfer an increasing amount of torque to the wheel which has traction whenever the wheel on the opposite side loses its grip and spins.

A viscous coupling incorporated into a transfer box (see fig. 6.22) is combined with the third differential so that its multiplate clutch can progressively and automatically lock out the third differential every time there is any sudden relative speed fluctuation between the front and rear final drives.

A viscous coupling resembles a multi-plate clutch; it has two sets of steel disc plates: one set is internally splined to an extension to the input hollow shaft and annulus carrier, while the other set is externally splined to a drum attached to the planet-pinion carrier. The external splined plates have radial slots, whereas the internal splined plates have a series of perforated circular holes. These periodic interruptions to the contact surfaces of the annular-plates create additional eddy currents which increase the shear rates and hence the resistance torque. The drum chamber housing the disc-plate pack is sealed and filled with silicone fluid which takes up most of the chamber volume. This fluid is a silicon and oxygen compound with methyl groups attached. These silicon compounds have their atomic structure branch chain lengths arranged to match the viscous shear torque requirements, to drag together the two sets of disc plates and in the extreme to lock the multi-plate clutch. These compounds, known as silicone fluid, are stable at high temperatures and over a long period of time. Although its viscosity does decline somewhat with rising temperature, it rapidly recovers as the temperature falls.

Under normal driving conditions there will be very little relative speed difference between the two sets of disc-plates so that the viscous drag will be minimal. However, if there is an increase in relative speed, be it gradual or sudden, then the viscous shear resistance offered by the layer of silicone fluid separating the plates increases sharply. In addition, any prolonged slippage between the two sets of plates causes the silicone fluid to heat up and expand. This forces the small amount of air trapped in the chamber to emulsify and form bubbles. As the temperature rises even more with continuous slippage, the bubbles are compressed by the expanding fluid, and the pressure in the multi-plate chamber is accordingly raised and consequently increases

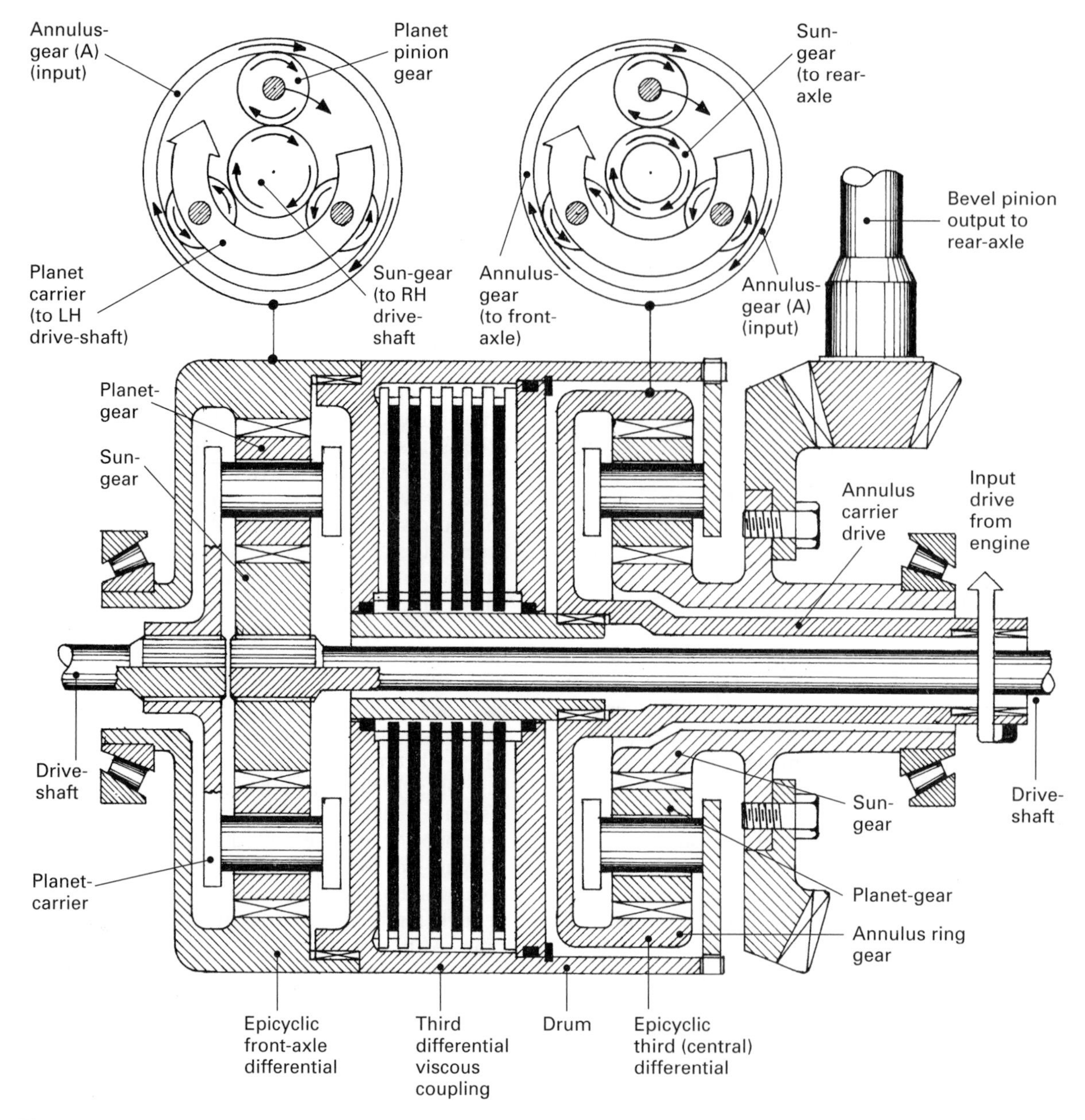

Fig. 6.22 Transfer box incorporating a epicyclic third (central) differential with viscus coupling and a front-axle epicyclic differential

the axial-thrust clamping load, holding the two sets of disc-plates together. Under extreme operating conditions the viscous shear drag is quite sufficient to lock together the plates. However, when the relative movement between the plates has been largely eliminated, the silicone fluid cools and slip can again take place if the road conditions warrant it.

7 Suspension, road-wheels and tyres

7.1 Semi-elliptic leaf springs

To appreciate the construction and operation of springs and suspensions, an understanding of what they do is necessary.

The functions of a sprung suspension There are several functions of a sprung suspension which may be listed as follows:

a) to satisfactorily absorb large and small road impacts and so protect the vehicle occupants from shocks;
b) to maintain the vehicle body on an even keel when travelling over rough ground or when cornering, so that pitch and roll are minimised;
c) to reduce the unsprung mass – i.e. that part of the total mass which is not suspended – to a minimum so that the wheels will follow the contour of the road surface more closely;
d) to reduce the impact stress on the vehicle's various mechanisms.

The function of suspension springs Springs are interposed between the wheels and the body so that the body is partially isolated from the axle. This allows the axle to move independently of the body. When a vehicle rides over rough ground, the wheels will rise when rolling over a bump and will deflect the springs. The energy created due to this movement is momentarily stored in the spring; it is then released again, due to the elasticity of the spring material, and in expending this energy the spring will rebound. The spring will then oscillate at its natural frequency, causing the vehicle to bounce many times before equilibrium is restored. If the body were rigidly connected to the axle, the kinetic energy created by the bump would be imparted directly to the body, creating high impact stresses to the chassis and panelling and discomfort to passengers.

The forces acting on a chassis wheel suspension The suspension should be able to sustain the following forces:

a) The static and dynamic vertical loading of the vehicle. These are absorbed by the elastic compression, shear, bending, or twisting action of the spring medium, which could be either rubber, elastic, or a single- or multi-leaf, helical-coil, or torsion-bar spring.
b) The twisting reaction due to driving and braking torque. These reactions are usually absorbed by the stiffness of the leaf spring alone, by stabiliser arms, or by triangular wishbone arms.
c) The driving and braking thrusts which must be transmitted between the sprung body and the road-wheels. These forces may be conveyed and carried directly by the rigidity of the leaf springs, by wishbone arms, or by the addition of tie-rods.
d) Any side-thrust due to centrifugal force, cross-winds, cambering of the road, going over a bump or pot-hole, etc. These forces are usually absorbed by the rigidity of the leaf spring or the hinge linkage arms of the suspension attachments between the wheel stub-axle and the chassis. Additional sideways support may be obtained by fitting radius and Panhard rods.

7.1.1 Leaf-spring classification (fig. 7.1)

Laminated leaf-spring configurations may be classified by the proportion of an ellipse which they provide – i.e. fully, three-quarter, half, or quarter elliptic.

The original fully elliptic spring (fig. 7.1(a)) goes back to the days of coaches, but it is now used only for commercial-vehicle driver-cab suspension.

The three-quarter-elliptic spring (fig. 7.1(b)) provided a soft but more rigidly supported version. This form of suspension, in a modified version, provides a progressive dual-rate fixed cantilever spring which increases in stiffness the more it is loaded and so is used on some heavy recovery vehicles where there is a considerable weight difference between unladen and laden conditions.

The half- or semi-elliptic spring (fig. 7.1(c)) is today by far the most commonly used leaf spring.

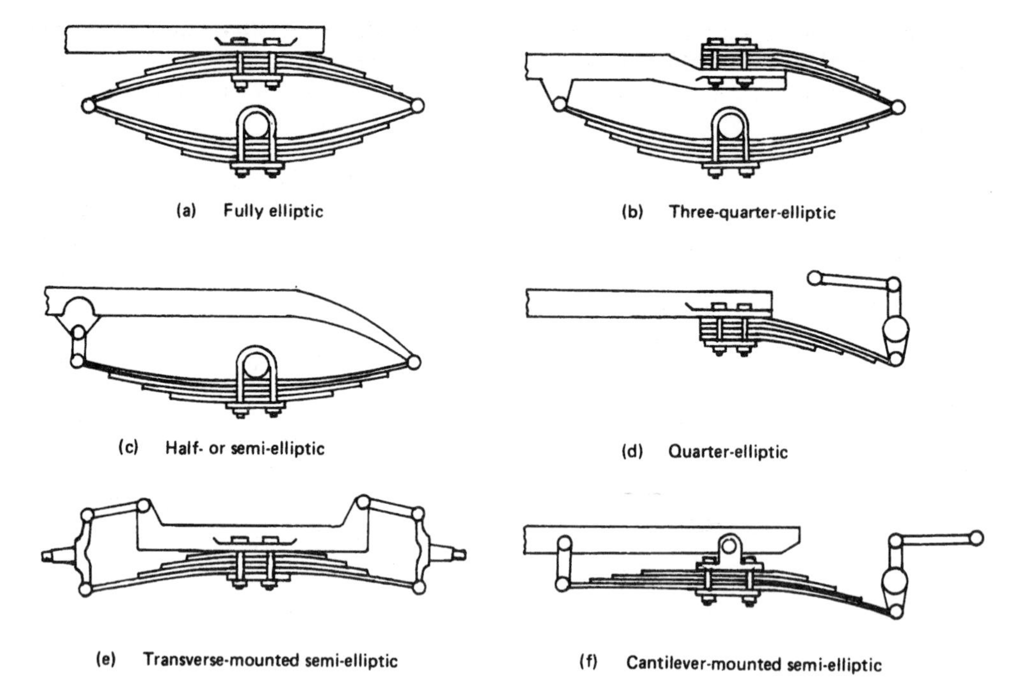

Fig. 7.1 Laminated-spring suspension configurations

It is used for car rear suspension and for both front and rear van and lorry applications.

The quarter-elliptic spring (fig. 7.1(d)) finds its use on small sports cars where a compact short spring is desirable.

Transverse semi-elliptic springs (fig. 7.1(e)) have been used for many years to form bottom, top, or both transverse link-arms for independent suspension for both front and rear suspensions.

Finally, the cantilever-mounted semi-elliptic spring (fig. 7.1(f)) has been used in such cars as the Jaguar for the rear suspension. The central pivot extends the effective spring length and, since it lies parallel and very close to the chassis, provides a compact and effective suspension.

7.1.2 Leaf-spring design features

When the springs are manufactured, each leaf or blade is curved – that is, given a camber set. The greatest set is given to the smallest leaf, and the set is progressively reduced as the span increases, so that the main leaf will have the least (fig. 7.2).

A centre bolt aligns and clamps the various leaves together. In addition, so that the leaves stay together along their span, steel clamps (sometimes rubber-lined) are fitted about halfway between the centre bolt and the spring eyes.

When the multi-leaf spring is deflected, the

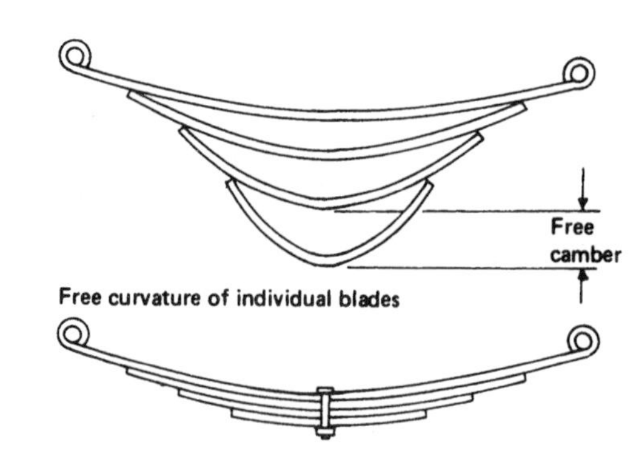

Fig. 7.2 Combined curvature of clamped blades

interleaf action is for the upper side of each leaf tip to slide or rub against the under-side of the blade above it. This creates friction. Under certain conditions this may be useful, as it reduces the amount of bounce – that is, it provides a degree of spring damping when the spring bounces – but usually it does not match the ride characteristics needed and it may make the suspension too stiff, so that the ride will be harsh over light road irregularities.

Inter-leaf rubbing in the presence of dampness or moisture will create fretting corrosion which reduces the fatigue strength and hence the oscillating life of the spring. Phosphate paint between the blades can to some extent reduce this problem.

Fitting a thin layer of lead or an anti-friction disc between the blades as shown in fig. 7.3 reduces the friction interference and reduces fretting.

Spring-blade life can be prolonged if the top surface of each leaf is shot-peened or work-hardened. This changes the stresses on the upper side of the blade from a normal tensile to a compressive state and greatly reduces the risk of fatigue failure. Rounding the edges of the blades also improves the fatigue life of the blades.

A further method of distributing the stresses within the blades more evenly along each blade span and prolonging spring life is to change from a straight cropping of the blade ends (fig. 7.3) to a tapering of the leaves near their ends (fig. 7.4). This reduces the peaking of stress levels.

At present, leaf springs are mostly made from silicon–manganese steels.

7.1.3 Constant-rate and progressive-rate semi-elliptic springing

There are two basic methods of mounting a semi-elliptic spring to the chassis, as follows:

Constant-rate swing springs (fig. 7.5) With this method, the forward end of the spring is directly pinned to the front spring-hanger and the rear end to a swing shackle. When the spring is deflected between the unloaded and the loaded position, the spring camber will be reduced and the spring length will increase. To allow this to take place, the swing shackle will pivot about the upper fixed shackle-pin. The driving thrust can then be transmitted through the forward half of the spring directly to the fixed spring-hanger.

There will be very little change in the spring stiffness as the spring straightens out, hence this is known as a constant-rate suspension spring.

Progressive-rate slipper spring (fig. 7.6) With this method of supporting the spring ends, the

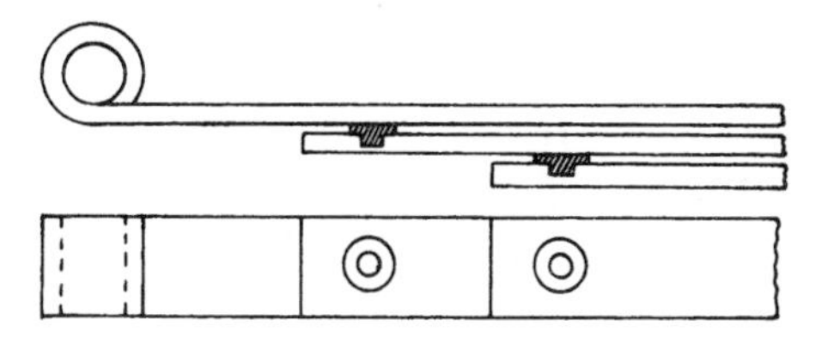

Fig. 7.3 Straight-cropped blade ends with disc inserts

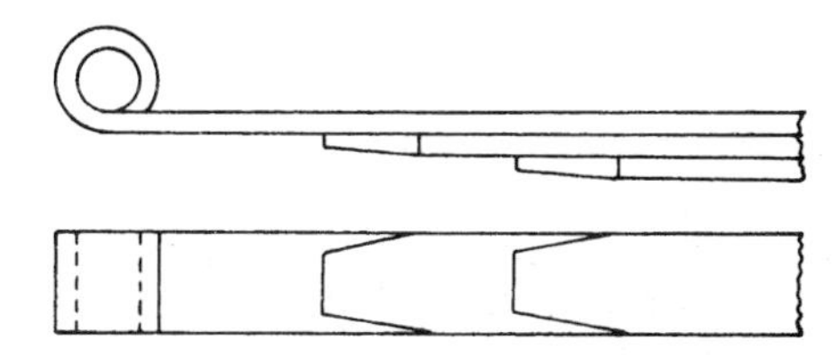

Fig. 7.4 Taper-cropped blade ends

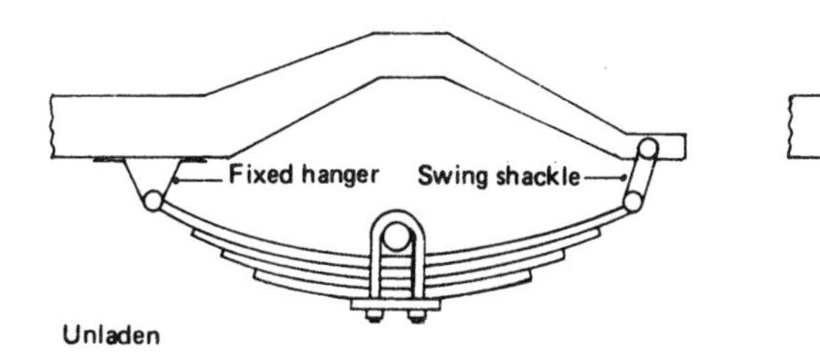

Laden

Fig. 7.5 Constant-rate fixed- and swing-shackled leaf springing with overslung axle-beam

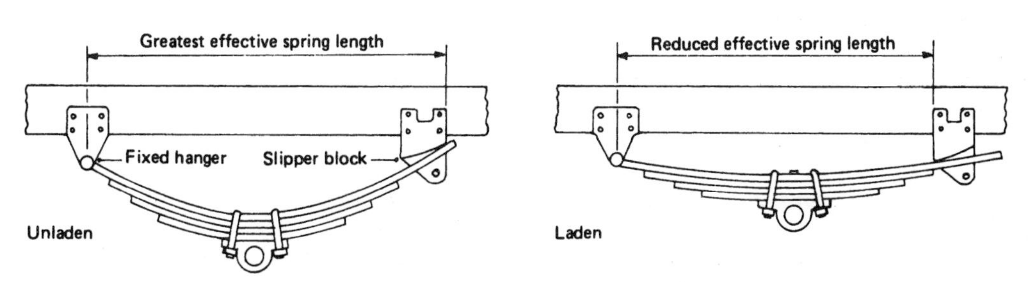

Fig. 7.6 Progressive-rate slipper-block-contact leaf-spring and fixed shackle with underslung axle-beam

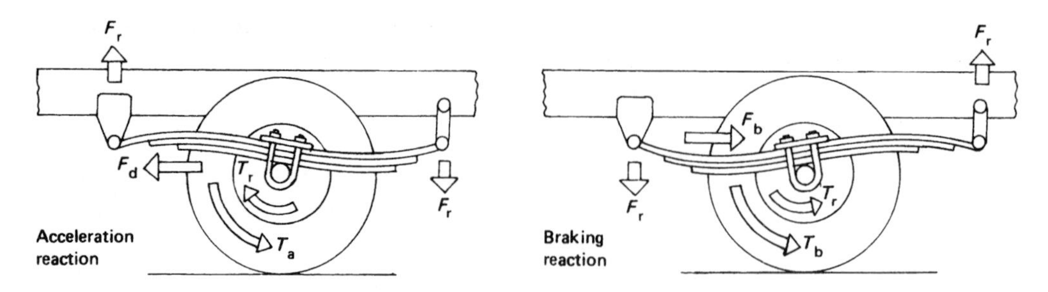

Fig. 7.7 Acceleration and braking reaction forces acting on the spring shackles

forward end is attached directly to the spring-hanger as before, but the rear end has no eye but just rests on a curved slipper block or pad. Initially, when the spring is unloaded, the contact point will be on the outside position of the slipper face, but the straightening of the spring as the load is increased will roll the main-leaf end around the slipper profile from the outer to the inner position. This effectively shortens the spring length. This is equivalent to stiffening the spring, or increasing the spring rate, which will therefore offer a progressively increased resistance to the vehicle payload.

7.1.4 Forces acting on a semi-elliptic suspension (fig. 7.7)

With the semi-elliptic leaf-spring suspension layout, static and dynamic vertical loads are absorbed by the chassis mounts, spring, and axle. The driving thrust (F_d) and the braking retardation (F_b) are taken directly between the axle and the front hanger mounting through the forward half of the spring span. The acceleration torque (T_a) and the braking torque (T_b) are absorbed by distortion of the leaf spring. When accelerating, the reaction torque (T_r) will tend to twist the axle clockwise and the vertical reaction forces (F_r) will tend to lift the front end of the chassis and push the rear end into the ground. When braking, the reverse takes place – the torque reaction on the axle will tend to rotate it anticlockwise, pushing the front end down and raising the rear end.

All side-forces are absorbed by the stiffness or rigidity of the spring blades, shackles, and chassis mountings.

7.1.5 Semi-elliptic single parallel-section spring (fig. 7.8)

Semi-elliptic springs are the most popular method of suspension support for vans and commercial vehicles and are still widely used for the rear suspension on cars. In their simplest form, the main leaf spring has its ends rolled into eyeholes, being attached at one end to a mounting bracket on the chassis and at the other end to a swing shackle which allows the length of the spring span to alter when the spring is deflected. The axle-beam is then bolted rigidly across and at the centre of the springs. The weight of the sprung body acts downwards at the extreme ends of the spring, and the reaction force acts upwards through the wheel to the centre of the spring.

The load (W) pressing downwards on the spring creates a bending moment (M) about the spring eyes. This is a minimum at the spring eyes but progressively rises to a maximum at the centre axle-beam attachment. This is shown by the shaded triangles either side of the axle in Fig. 7.8. The magnitudes of the bending moment at distances x_1, x_2, x_3 from the eye will be equal to $M_1 = Wx_1$, $M_2 = Wx_2$, $M_3 = Wx_3$.

Bending moments produce bending stresses (σ_b) which increase from a minimum at the spring ends to a maximum just either side of the axle-beam. Thus the least stressed section of the spring

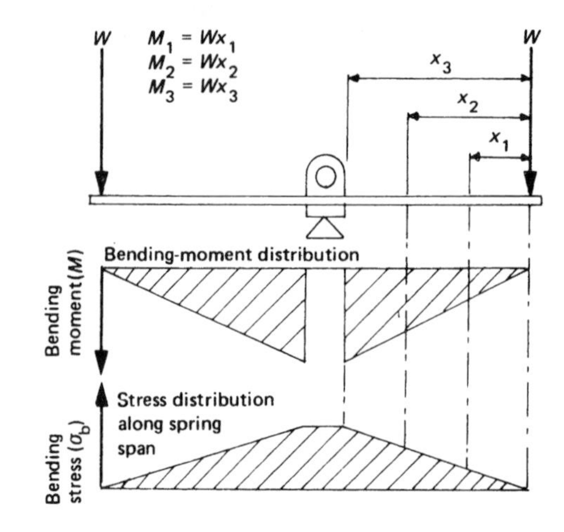

Fig. 7.8 Single parallel-section leaf spring

is at the spring eyes and the most stressed is near the spring centre, which is therefore the weakest point of a uniform-section single-leaf span. This makes the parallel-section leaf-spring unsuitable for a suspension, as it would fail prematurely.

7.1.6 Semi-elliptic multi-leaf spring (fig. 7.9)
To compensate for the uneven stress along the spring span, a multi-leaf spring has several leaves which if laid out side by side would form a triangle or, if half of each leaf could be placed on each side of the main leaf, would form a diamond shape as shown in fig. 7.9. The length of each leaf is chosen so that the total cross-sectional area of leaves at any one point is proportional to the bending stress at that point. This then produces a constant stress across any section along the full spring span.

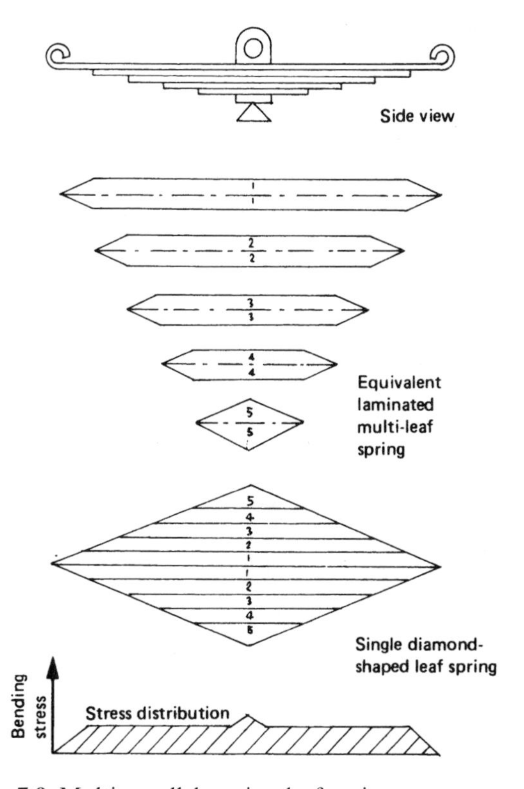

Fig. 7.9 Multi-parallel-section leaf-spring

A centre bolt passing through a hole drilled through each leaf clamps the leaves together and reduces the spring-leaf cross-section at the mid-point. This results in a slight central stress peak on the stress-distribution graph. Similarly the main leaf eyes remain square in section and parallel and do not have a triangular reduced section, so the stress near the eyes will be seen to taper off.

7.1.7 Single trapezium-shaped leaf-spring (fig. 7.10)
Another approach to maintaining an approximately constant stress distribution along the spring span is to have a single spring blade of uniform thickness but increasing in width from its ends towards the mid-span. A plan view shows a trapezium shape (fig. 7.10).

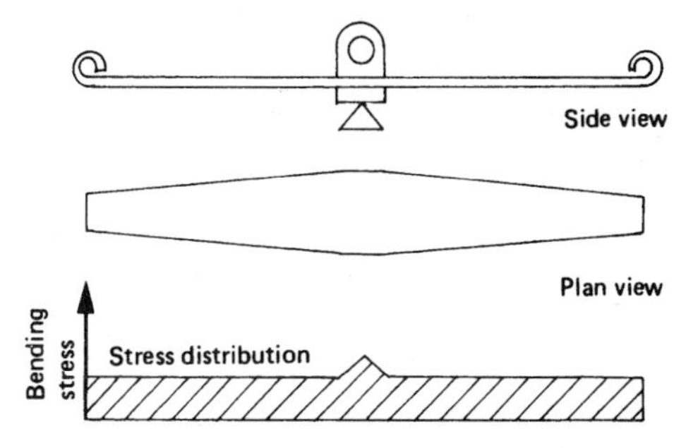

Fig. 7.10 Single trapezium-shaped leaf-spring

The increase in cross-sectional area towards the middle of the spring blade counteracts the increase in bending moment created by the body weight, so that the spring remains uniformly stressed along its length.

7.1.8 Single tapered leaf-spring (fig. 7.11)
A more popular approach using a single-leaf spring is to have the blade of constant width but to taper its thickness from a maximum in the mid-span position to a minimum at its ends as shown in fig. 7.11. With this shape, the increased bending moment from the spring ends of the axle centre will be resisted by the proportionally enlarged cross-sectional area of the blade.

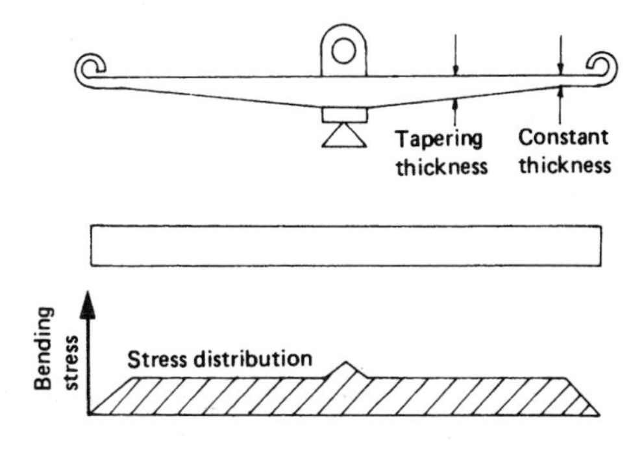

Fig. 7.11 Single tapered leaf-spring (parabolic)

The taper leaf seems to be preferred to the trapezium shape as it is more compact and easier to clamp on to the axle-beam.

7.1.9 Multi-taper-leaf springs (fig. 7.12)

For heavy-duty large tractors or trucks, two or three taper-leaf springs may be used together. Liners may be used between the pressure points at the mid-spring seat position, so that the springs do not touch at any point between the middle seat section and the load-bearing end points (fig. 7.12).

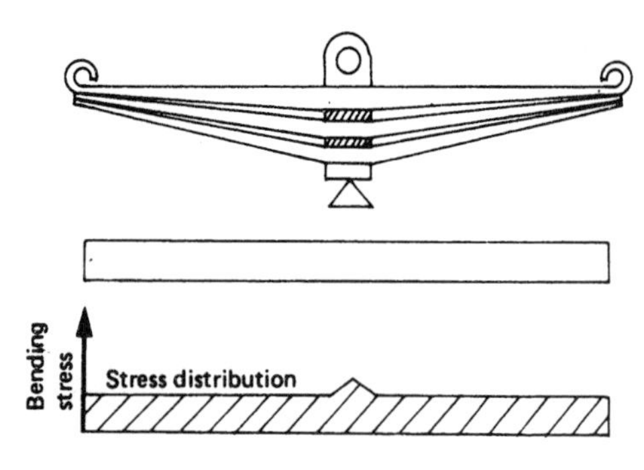

Fig. 7.12 Multi-taper leaf-spring

7.1.10 Advantages of taper-leaf over multi-leaf springs

The advantages of taper-leaf over multi-leaf springs are as follows:

a) The variable-cross-section single-leaf spring is only about half the weight of a multi-leaf spring used for the same payload.
b) There is no inter-leaf friction with the single taper blade. Where the taper-leaf application has more than one leaf, inter-leaf friction is reduced because fewer leaves are required and because these leaves bear upon each other only at the ends. This provides a more sensitive springing for light road shocks and so gives a better ride.
c) The taper-leaf spring stresses are more uniform and lower overall than with the multi-leaf design. Taper-leaf spring life is therefore longer.
d) With the single-taper-leaf spring, there is no inter-leaf collection of moisture and trapped dirt which would promote fretting corrosion and fatigue failure.

7.1.11 Leaf-spring shackle arrangements

To obtain an efficient suspension, the vehicle weight must be transmitted to the leaf spring by means of a fixed spring-mount or hanger at the front end of the spring and usually a swinging shackle at the rear end. The spring is attached or hinged at each end by shackle-pins passing through the spring eye and the mounting or shackle-plate. These pins provide a joint which can rotate or pivot in rubber or metal bushes but at the same time be firmly held together. This reduces wear and noise and does not alter the suspension and steering geometry as the spring deflects and the various forces act on the system.

Rubber bushes are generally used for cars and vans, but metal phosphor-bronze bushes are provided on heavy-duty commercial vehicles.

There are two types of rubber bushing commonly used: flanged rubber half-bushes and silent block bushes.

Flanged rubber half-bushes (fig. 7.13(a)) Flanged rubber half-bushes are pushed into holes with finger pressure and are then clamped together until the shackle side-plates are firmly against the shackle-pin shoulders. Any twisting motion of the spring will be absorbed as a torsional straining within the rubber itself. Frictional grip is provided by the inside and outside rubber being in compression, so there is no relative slip between the rubber and metal.

Silent block rubber bushes (figs 7.13(b) and 7.14(a)) The silent-block-type rubber bush has an outer and an inner steel cylindrical case and the rubber is pre-stressed when manufactured. With this design there is no relative slip between the rubber and the inner and outer steel casings, and any relative twist will be taken up as torsional elastic strain.

The outer casing is a force fit in the spring eye or chassis hole, but the shackle-pin is a slip fit when being assembled, and, to prevent the inner casing or tube from rubbing relative to the pin, the shackle side-plates must be tightened up until the shoulders of the tube are hard against the plates.

Generally a silent block bush is used at the fixed-hanger end, and a split flanged bush forms the bearing for the swinging shackle for small and medium-sized cars. Large cars and vans use a silent block joint at each end of the leaf spring.

Metal bushes (figs 7.14(b) and 7.15) For heavier-duty applications, metal bushes are used. These can either be plain or screw-profiled, and both types are a force fit in the spring eye or spring

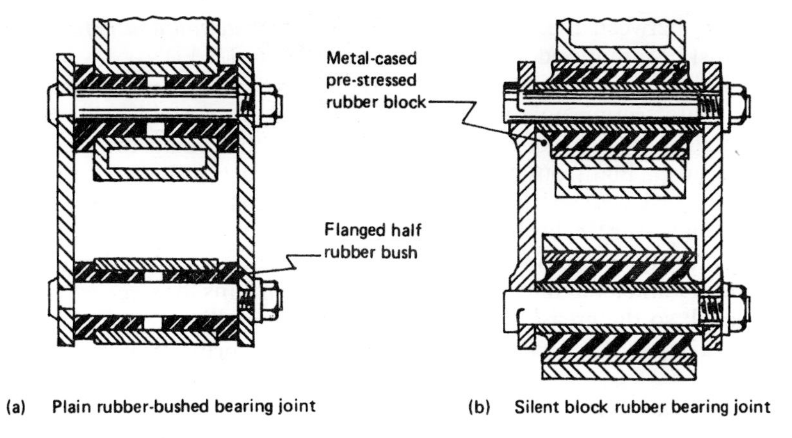

(a) Plain rubber-bushed bearing joint (b) Silent block rubber bearing joint

Fig. 7.13 Swinging-shackle arrangements with rubber bushes

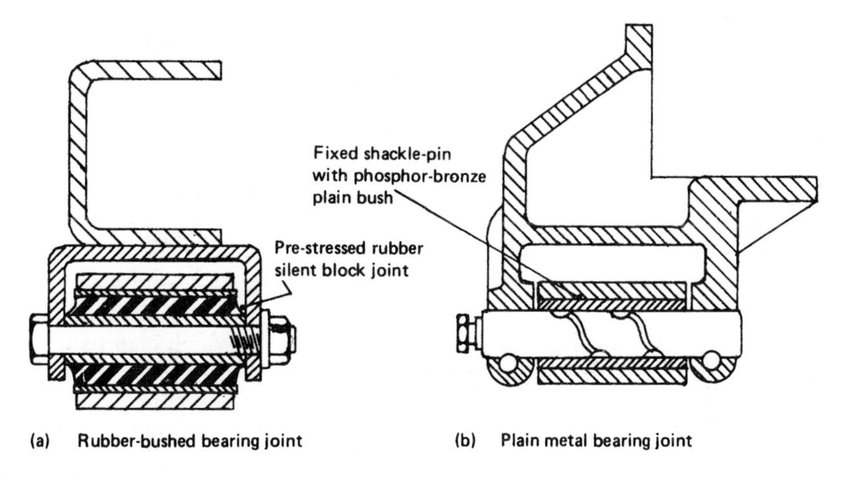

(a) Rubber-bushed bearing joint (b) Plain metal bearing joint

Fig. 7.14 Fixed-shackle arrangements

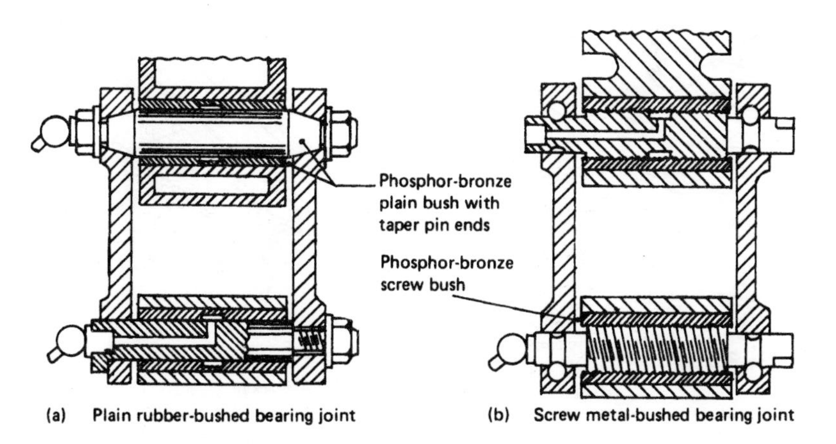

(a) Plain rubber-bushed bearing joint (b) Screw metal-bushed bearing joint

Fig. 7.15 Swinging-shackle arrangements with metal bushes

mounting hanger. Rubbing between the shackle-pin and the metal bush must be minimised, so they are always lubricated by holes drilled axially along the shackle-pin. A radial intersecting hole in the middle of the pin permits the passage of grease between the pin and the bush. Plain bushes are usually helically internally grooved so that the grease will spread more evenly over the bush bearing surface. With the screw-type pin-and-bush joint there will be both rotary and axial movement when the spring is deflected, so the grease should readily spread over the bearing-surface pair.

Heavy-duty shackle-plates and pins (fig. 7.15)
With heavy-duty suspensions, more robust shackle-pin and plate clamping must be provided. The two most common methods are (a) the taper shackle-pin which, when clamped to the shackle-plate, wedges the two parts firmly together and (b) the slotted-shackle-plate and bolt with an alignment groove machined at each end of the pin.

7.2 Rigid-axle-beam suspension

The beam-type axle is the oldest and simplest of sprung axle arrangements, but it still has certain merits which may justify its use in preference to more sophisticated suspension designs.

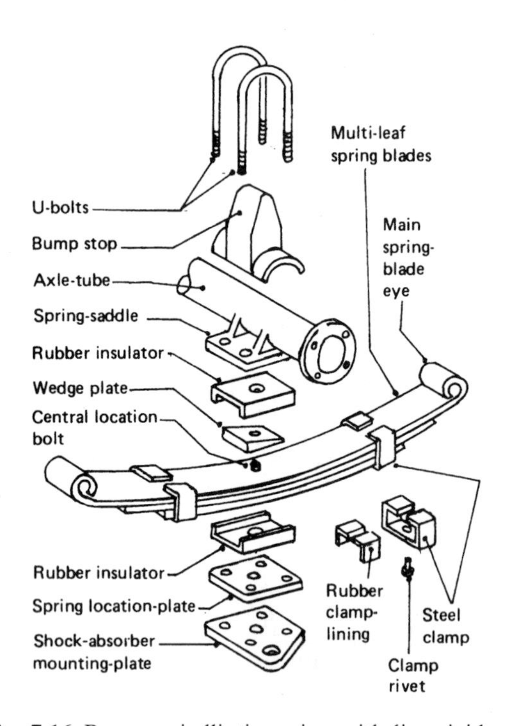

Fig. 7.16 Rear semi-elliptic spring with live rigid-axle suspension

7.2.1 Construction and action of beam-type axles (figs 7.16 and 7.17)

With this design the road-wheels are interconnected by a rigid solid forged axle-beam at the front, this being the non-drive axle, or by a rigid tube-section beam at the rear with the drive half-shafts passing through the centre to transmit the drive to the back wheels. Stub-axles are hinged or pivoted on to each end of the front axle-beam to provide a means for steering the front road-wheels.

The semi-elliptic springs fit on flat spring-saddles or spring-beds specially formed on the underface of the axle-beam (fig. 7.18). In the centre of each spring-saddle is a countersink hole which provides a location point for the spring centre-bolt head. The springs and the axle are clamped together with U-shaped bolts passing through four holes drilled through these spring-saddles.

Sometimes on car rear axles, rubber insulator pads are sandwiched on each side of the spring-blade pack to damp out any small vibrations and noise coming from the unsprung axle (fig. 7.18).

With large trucks, the axle and saddle are clamped directly on to the steel main leaf (fig. 7.19), and axle clamp blocks and separate clamp bolts replace the U-bolts for greater grip. In addition, large springs have their spring leaves dimpled in the centre to eliminate sliding movement between the leaves and to reduce shear force on the centre bolt which clamps the leaves together.

The front axle-beam span between the spring-saddles is of 'I' section, which gives the greatest

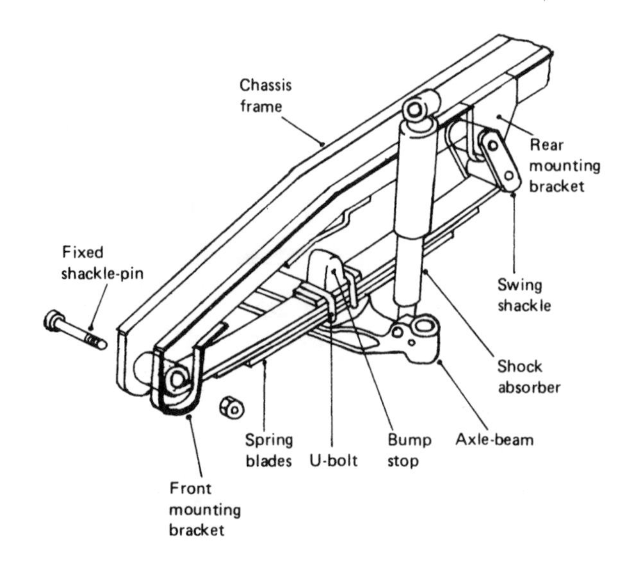

Fig. 7.17 Front semi-elliptic spring with rigid-axle-beam suspension

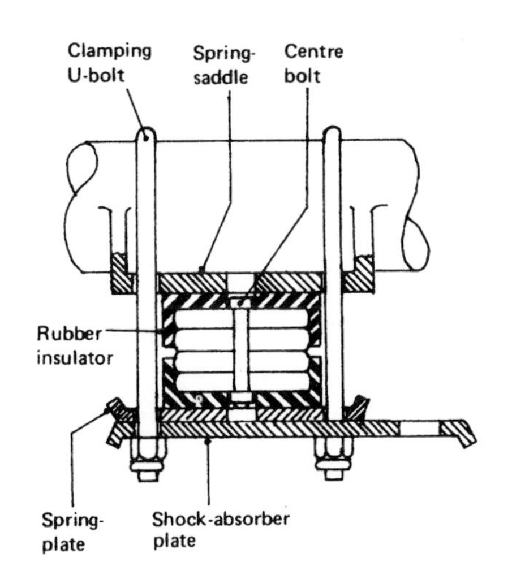

Fig. 7.18 Car and van axle and leaf-spring location and clamping

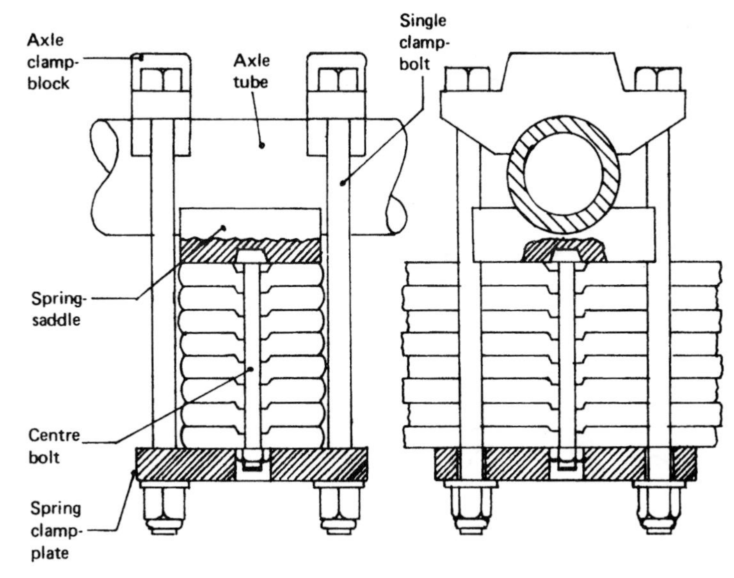

Fig. 7.19 Heavy-commercial-vehicle axle and leaf-spring location and clamping

resistance against bending due to static and dynamic vertical loads. Between the spring-saddles and the stub-axles, the section shape changes to a round section – this is more suitable for resistance to twisting which is created due to acceleration- and braking-torque reaction. The vehicle's weight and payload is passed from the sprung chassis to the front and rear mounting brackets, shackle-pins, leaf springs, and to the unsprung axle-beam. The axle-beam then divides the vertical forces between the two road-wheels.

During body roll, there will be no wheel camber roll since the axle-beam will keep the road-wheels perpendicular to the ground under normal driving conditions (fig. 7.20(a)(i)).

If one of the wheels goes over an obstacle or into a pot-hole, the road-wheel will move vertically up and down following the road surface contour without raising or lowering the average sprung body height (fig. 7.20(a)(ii)). Unfortunately, however, the vertical motion of the wheel will tilt the axle-beam and transmit some of the bump or rebound movement from one end of the axle-beam to the other. This deflection can set up an oscillating tilting motion of the axle-beam – known as axle 'tramp' – which will be experienced as an unpleasant tremor or vibration on the driver's steering-wheel.

Blocks of rubber known as bump stops are usually clamped on top of the axle or spring by the U-bolts. These stiffen the suspension if bouncing becomes too violent or body roll is excessive.

7.2.2 Reasons for using beam-type axles

The reasons for adopting axle-beam suspension for certain applications are as follows:

a) The axle-beam spreads across the mid-span of the two semi-elliptic springs which are attached to the chassis side-member on each side. The axle gives rigid support for each spring, both in the vertical plane when the sprung body bounces and in the horizontal plane when there are side-forces involved.

   Having the springs interconnected is particularly desirable where the vehicle is built on a separate chassis, with side-members spanning the full length of the vehicle and only simple cross-members provided to hold them together, as is the practice with most commercial vehicles at present. The alignment of both road-wheels to the road then relies not on the chassis and body stiffness but on the rigidity of the axle-beam which supports both stub-axles and wheels.

b) Since the road-wheels are attached to the axle-beam by means of the stub-axles, the wheels will remain approximately perpendicular to the road at all times, so that tyre tread contact with the road will be at its best. This will

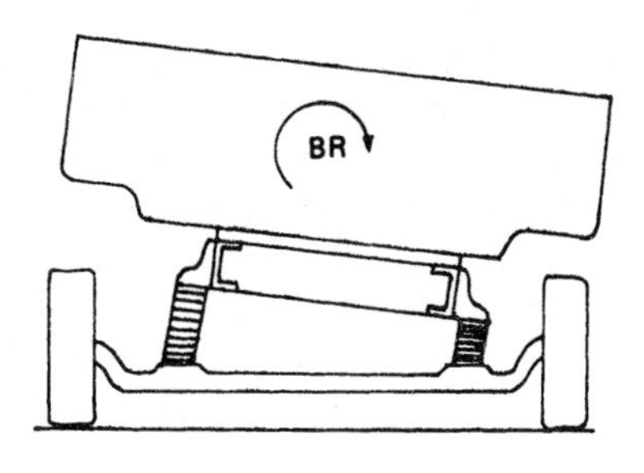

(i) Body rolls; wheels remain upright

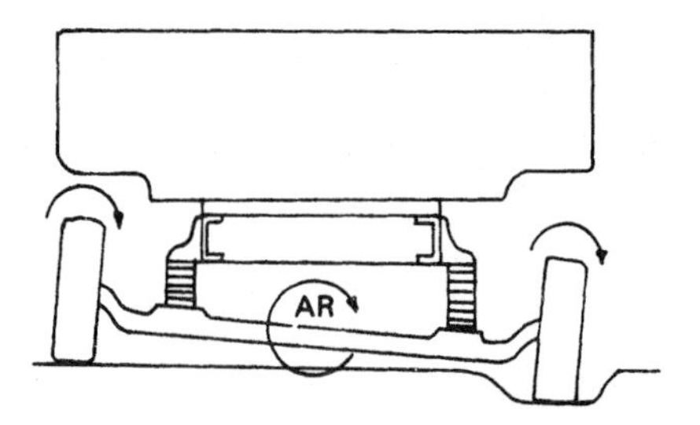

(ii) Wheel enters pothole; axle and wheels tilt

(a) Rigid axle-beam with semi-elliptic springs

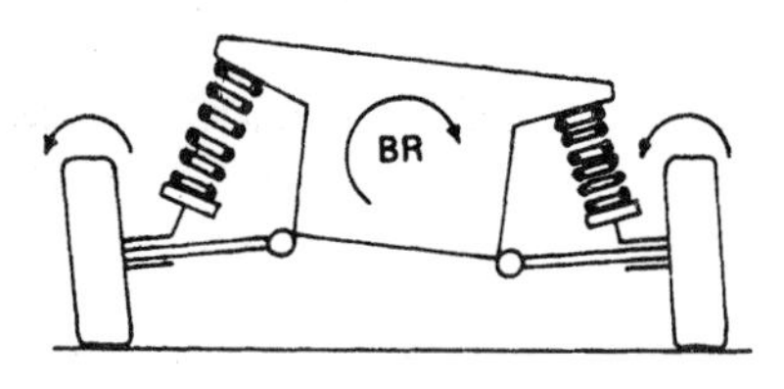

(i) Body rolls; both wheels tilt inwards

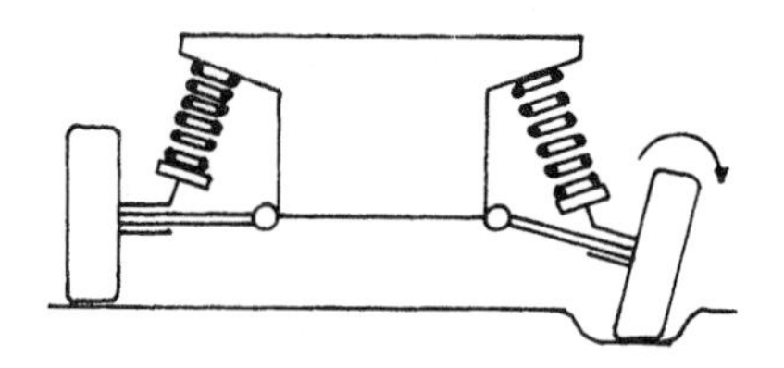

(ii) Wheel enters pothole and tilts outwards; body remains upright

(b) Short swing arm

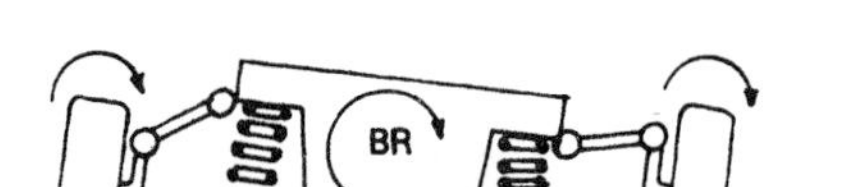

(i) Body rolls; both wheels tilt outwards

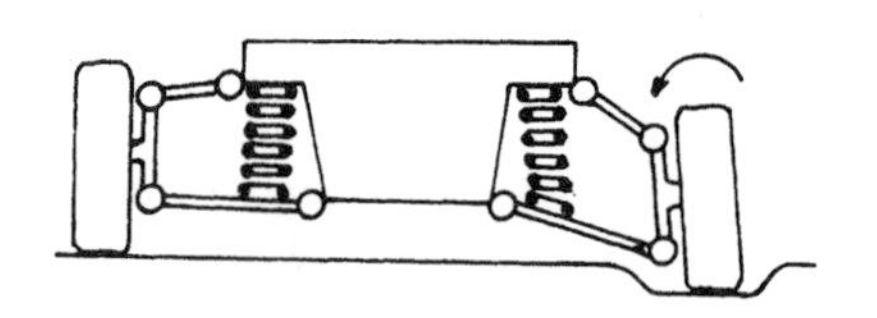

(ii) Wheel enters pothole and tilts inwards; body remains upright

(c) Transverse double wishbone

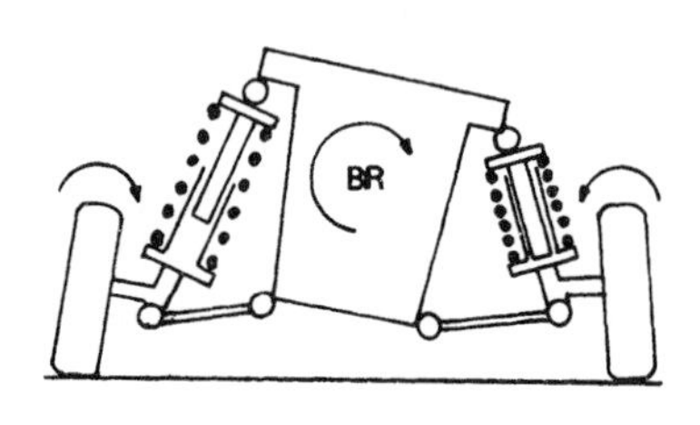

(i) Body rolls; wheels tilt towards the centre

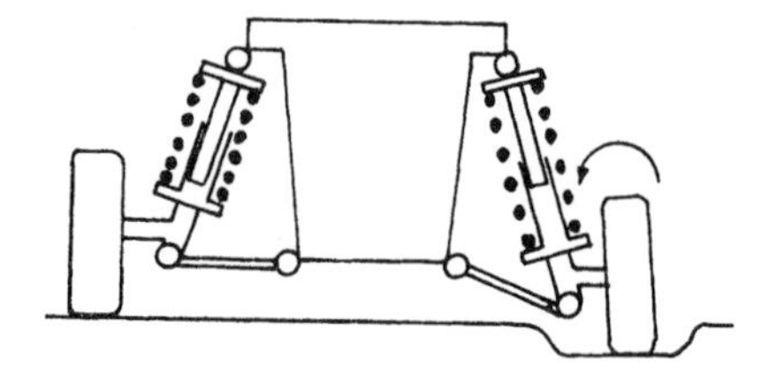

(ii) Wheel enters pothole and tilts inwards; body remains upright

(d) MacPherson leg-strut

Fig. 7.20 Effects of body roll and irregular road surfaces on suspension geometry

provide very good road grip and prolong tyre life.

c) The height of the axle-beam above the road will not alter between the unladen and laden state or when the body rolls or goes over a bump or pothole, so ground clearance remains constant and predictable, which is desirable if the vehicle is to be used over rough ground.

d) The axle-beam stub-axles and semi-elliptic springs form a simple, compact, uncomplicated suspension-and-steering system which is built-up of few parts. It can take considerable amounts of rough handling, it is reliable and easily serviced, and it possibly costs less than other types of front suspension.

7.2.3 Helper springs (fig. 7.21(a) and (b))

It is very difficult for leaf springs to cope with a large variation in load ranging from unladen to laden without the quality of the ride and the ability to adequately support the load to suffer. Thus if the springs have a low spring stiffness to absorb the many bumps and dips in the road when the vehicle is unladen, when fully laden the springs would sink down too far, possibly causing the chassis to bottom onto the axle and for the wheels to scrape the underside of the wheel arches. Conversely if high stiffness springs are chosen, under full load operating conditions there would be sufficient rigidity for the springs to support and maintain the necessary bounce/rebound clearance between the chassis and axle. However this would be at the expense of the ride becoming very harsh when part or all of the load is removed, that is, when the vehicle is driven empty.

To effectively broaden the no-load to full load spring stiffness range, small helper springs are sometimes clamped by U bolts mid-way along and on top of the bowed main semi-elliptic multi-leaf parallel or tapered section spring. Abutment brackets over the ends of the helper leaf spring give additional support to the main spring when the vehicle is laden and the main leaf spring curvature has been flattened or even reversed its camber. Conversely, as weight is removed the main leaf spring camber (curvature) returns. This raises the chassis well above the axle so that the abutment bracket now clears the ends of the helper springs. Only the main spring contributes to the support of the chassis now, so that the much reduced spring stiffness provides a relatively sensitive response to wheel bounce and rebound as the wheels roll over the irregularities on the road surface.

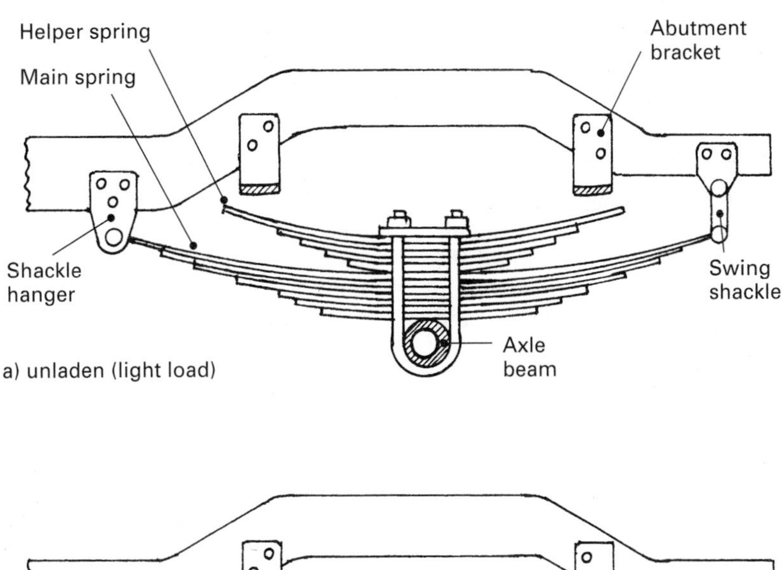

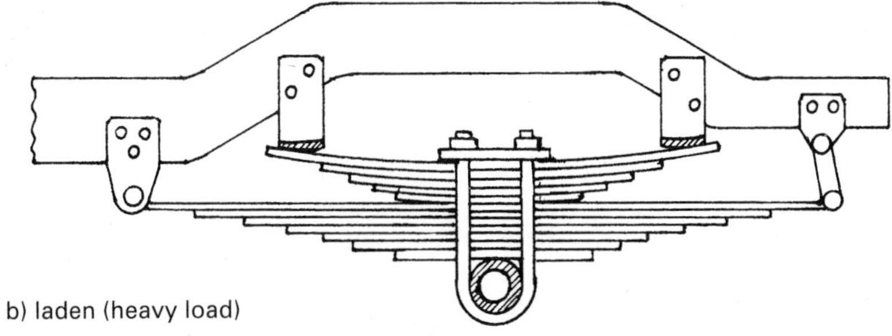

Fig. 7.21 Helper leaf-springs

7.3 Independent front suspension

7.3.1 Suspension geometry

An appreciation of how and why the suspension geometry influences the steering characteristics is essential for intelligent investigation of steering faults and their correction. It is therefore proposed to examine basic suspension concepts and their application to steering requirements.

Swivel-pin axis offset (fig. 7.22) Each front road-wheel is mounted on swivel-pins via a stub-axle carrier. In its basic form the vertical swivel-pin axis is offset to the centre of the tyre contact patch by a distance R (see fig. 7.22). When the vehicle moves forward the propelling force acts through the swivel-pins, whereas the drag resisting force between the tyre and ground reacts in the opposite sense through the centre of the contact patch. Consequently the propelling force and the opposing drag force offset causes the wheel assembly to twist about the swivel-pins in an anticlockwise direction for the left-hand wheel and in a clockwise direction for the right-hand wheel. If both wheel drag forces are the same the left-hand and right-hand anticlockwise and clockwise twisting moment of their respective wheels will cancel each other out through the interconnecting track-rod so that the road-wheels move straight-ahead. However, should the left-hand wheel hit a bump or a pot hole it will considerably increase the resisting drag of that tyre. As a result the imbalance between the left and right-hand turning effort will momentarily turn the steering to the left. Likewise, should the right-hand wheel ride over a small dip or hump the increased drag resistance between the tyre and ground will cause the steering to twitch to the right.

Wheel camber angle (fig. 7.23) Sideways inclination or lateral tilt of a road-wheel from the vertical is known as camber and the angle between the centre line of the wheel and the vertical is called the camber angle. A wheel leaning away from the body has positive camber, whereas an inward-leaning wheel is said to have negative camber. Early road construction adopted a heavily cambered contour across the road to disperse rain water to the side of the curb where it would then flow parallel to the road and drain away. Horse-drawn vehicles had wooden spoked wheels encircled by a flat shrunk-on steel tyre. For the tyre to roll flat against the road surface the road-wheels were laterally inclined at an angle such that the wheel stood approximately perpendicular to the camber of the road. Modern roads have very little or no camber and vehicles these days have pneumatic rubber tyres which can comply to the contour of most road irregularities, hence the need for a large wheel camber ceases to exist.

Wheel camber is achieved by inclining the stub-axle to the horizontal. Early vehicles had positive camber angles in the region of 8°. However modern cars have much reduced camber angles which may range from 0 to 2°.

Nevertheless a small amount of wheel camber makes the wheel and tyre form a frustum of a cone with its apex pointing away from the body. Thus if a frustum of a cone is permitted to roll on a flat surface it will move in a circular path about its apex. Hence with positive wheel camber the left-hand wheel will roll in an anticlockwise direction while the right-hand wheel rotates in a clockwise direction. It follows that if both wheels are linked by a track-rod and the vehicle is moving forwards, that there will be a tendency for both wheels to roll in the opposite direction to each other, that is, away from each other at the front. However this will be prevented by the balanced preload imposed on the track-rod, provided the camber remains the same for both wheels, this track-rod preload being compressive if the track-rod arms are behind the stub-axle and tensile in nature if they are in front while the vehicle is moving forwards. This track-rod preload helps to avoid twitching of the front wheels about their swivel-pins and therefore provides a small amount of straight-ahead steering stability.

Wheel camber to a limited extent assists in reducing the scrub radius between the centre of the contact patch and the swivel-pin axis line projected to ground level. It therefore contributes in reducing the road wheel kickback reaction whenever there is an imbalance between the outward turning moment of the front wheels about the swivel-pin pivots.

Swivel (king)-pin inclination (fig. 7.24) The front road steering wheels pivot on either swivel-pins or king-pins depending upon the design of suspension used. The axes of these swivel or king pins are laterally tilted so that swivel (king)-pin inclination is the lateral inward slope to body from the lower to the upper swivel (king)-pin joints to the vertical. Thus a line drawn through the swivel-pin joints intersects the ground nearer to the centre of

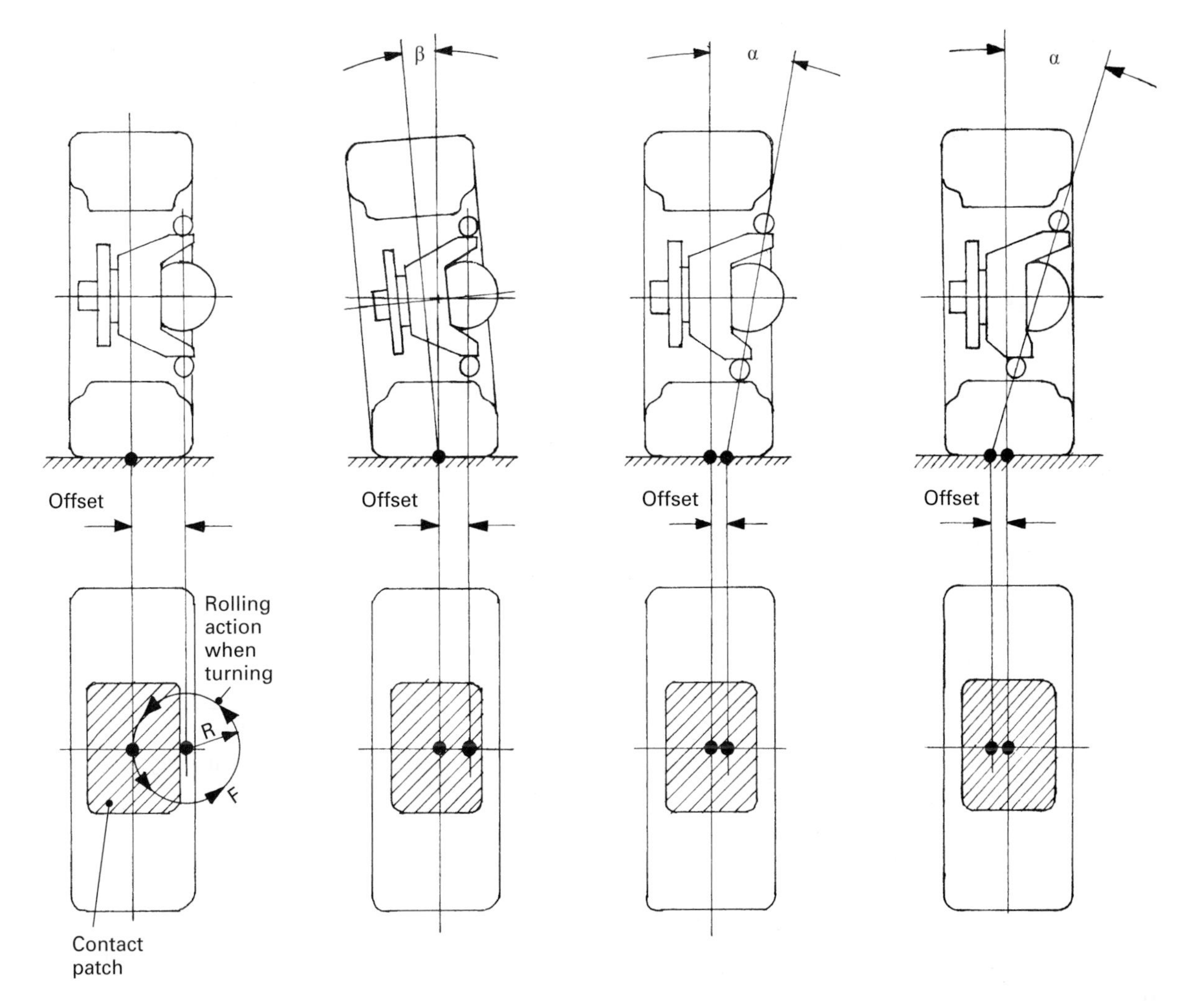

Fig. 7.22 Swivel (king) pin vertical axis offset

Fig. 7.23 Wheel camber angle

Fig. 7.24 Swivel (king) pin inclination with positive offset

Fig. 7.25 Swivel (king) pin inclination with negative off-set

the tyre contact patch than if there was no inclination of the pivot joints. The much reduced offset between the centre line of the wheel and that of the swivel (king)-pins at ground level has several benefits which will be discussed as follows:

1) It reduces the scrub radius between the centre of the tyre contact patch and the pivot axis at ground level. It thus minimises the scrub turning effort, that is, it reduces the drag torque necessary to turn the front wheels so that the manoeuvring of the steering wheel becomes considerably lighter.
2) A reduction in wheel to pivot ground level centre line offset reduces the imbalance twitch when one of the wheels rolls over a bump/hole or when the vehicle is braked and the surface grip between the ground and each wheel is uneven. It therefore helps to minimise road shock transmission to the steering wheel.
3) The geometry of laterally inclining the swivel-pin has the secondary effect of lifting the front suspension and body in proportion to the amount the steering wheels are turned.

In actual fact when the stub-axles are swivelled the road wheels are lowered, but since they cannot normally dip into the ground, the alternative is for the suspension and body to rise in proportion to the amount the steering wheels are

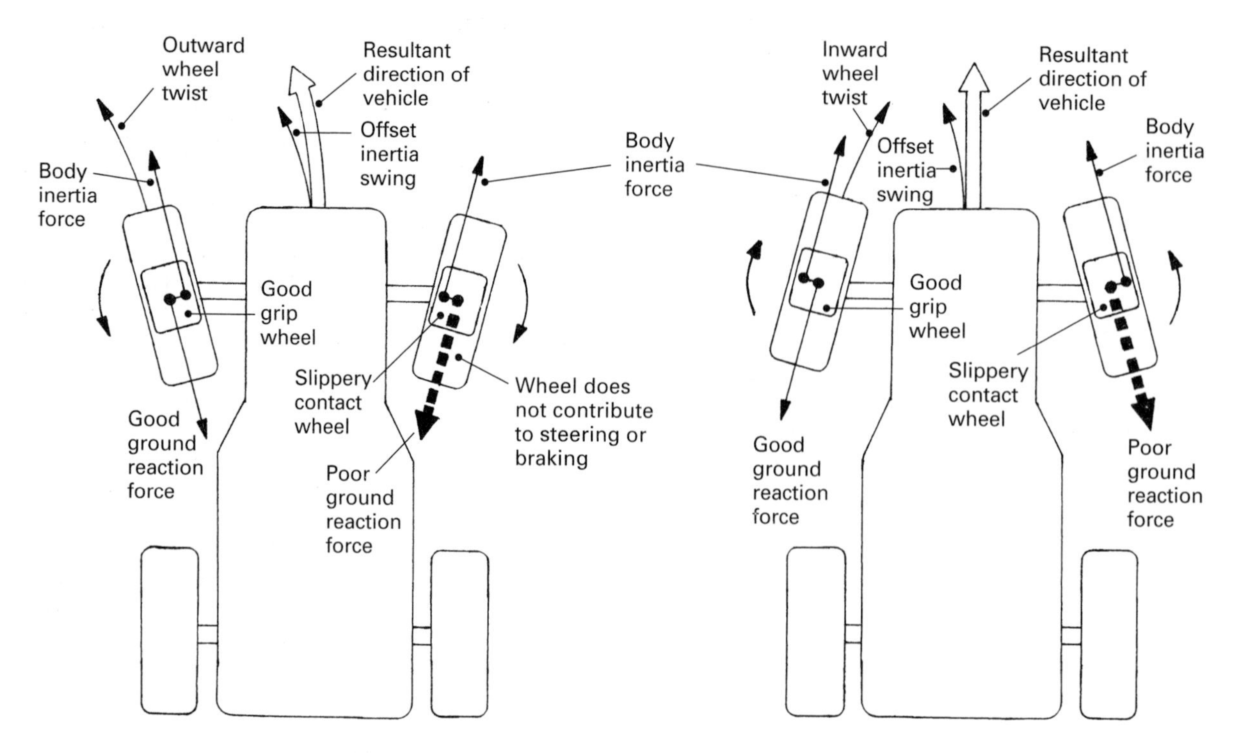

Fig. 7.26 Directional instability with positive off-set when one wheel skids whilst being braked

Fig. 7.27 Directional stability with negative off-set when one wheel skids whilst being braked

turned. This inherent lift of the front end of the vehicle when pivoting the stub-axle assembly and road wheels away from the straight-ahead position provides a degree of instability, thus once the steering wheels have turned, the weight of the body acting down on the suspension swivel-pins tries to push (swing) the stub-axle assembly to its lowest position. Conveniently this provides a desirable self-righting effect on the steering road wheels so that when the driver releases the steering wheel, the road wheels will tend to straighten out to the dead-ahead driving position on their own. However if the swivel-pin-inclination is large or the weight on the front suspension is considerable the steering will be heavy and the self-righting effort will be severe in itself and dangerous. This is the reason why the driver should always keep one hand on the steering wheel when returning the steering wheel to the straight-ahead position.

Swivel (king)-pin inclination commonly varies between 5 and 15° a typical value being something like 8°.

Swivel-pin inclination with positive or negative off-set (figs 7.24 and 7.27) Positive swivel-pin offset is when the swivel-pin axis line intersects the ground on the inside (body side) of the tyre contact patch (see fig. 7.24), whereas negative swivel-pin offset is when the swivel-pin axis line is made to intersect the ground on the outside (away from the body) of the tyre contact patch (see fig 7.25).

When a vehicle is being braked the inertia force of the body acts through the pivots and tries to drag the swivel-pins and wheels forwards, while the tyre to ground reaction grip acts in the opposite sense through the centre of the contact patch to counteract the vehicle's inertia (see figs 7.26 and 7.27). Because these opposing forces are offset, the left and right-hand wheels will attempt to twist in an anticlockwise and clockwise direction respectively for the positive offset and in a clockwise and anticlockwise direction respectively for the negative offset. Thus with a positive swivel-pin offset, both front wheels will tend to swing away from each other at the front of the stub-axle (see fig. 7.26), whereas with a negative swivel-pin off-

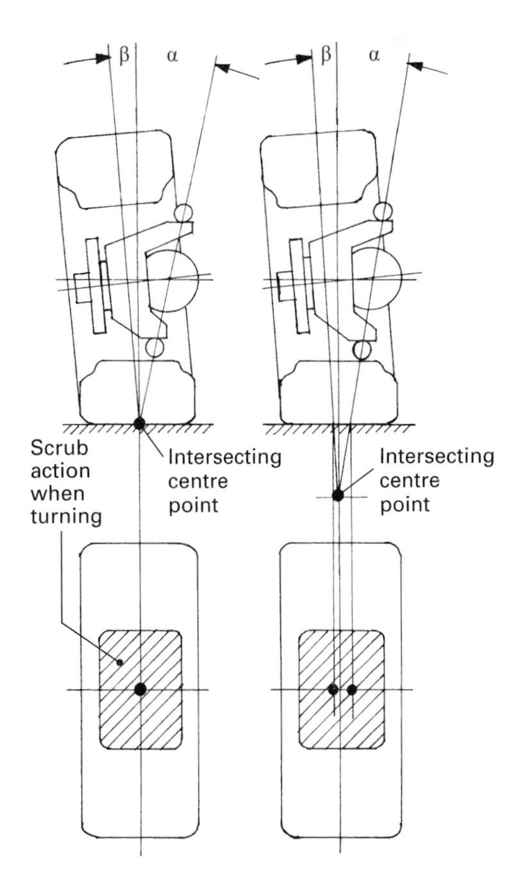

Fig. 7.28 Camber and swivel pin inclination centre point steering

Fig. 7.29 Camber and swivel pin inclination producing semi-centre point steering

set the wheels will try to pivot towards each other at the front (see fig. 7.27).

With a positive swivel-pin offset, if the right-hand front wheel should lose its grip when braking, then due to the vehicle's inertia, the vehicle will tend to swing about the wheel with the good road grip, that is, to the left aided by the swivel-pin offset turning moment of the good wheel (see fig. 7.26). Conversely, with a negative swivel-pin offset, if the right-hand front wheel should lose its grip, that is, it slips when braking, then again the vehicle's inertia will tend to swing the body about the wheel which has the good tyre to ground grip, that is, the left-hand wheel. However the left-hand swivel-pin negative offset produces a clockwise turning moment which partially counteracts the anticlockwise swing tendency of the vehicle's body (see fig. 7.27). As a result the negative swivel-pin offset helps to maintain the direction of motion of a braking vehicle in the straight-ahead direction even if one or the other of the front wheels tends to skid.

Centre point and semi-centre point steering (figs 7.28 and 7.29) True centre point steering is when a centre line through the wheel and a line through the swivel-pin axis meet at ground level. This condition eliminates the scrub radius and greatly reduces the twisting effort on the swivel-pin necessary to steer the vehicle. It also minimises tyre to road impact reaction feedback to the steering wheel (see fig. 7.28). Nevertheless it has been found that a semi-centre point steering where there is a small offset between the centre of the road wheel and the projected swivel-pin axis line is desirable (see fig. 7.29). A small offset makes the wheels roll about a circular path when pivoting about the swivel-pins. This distorts the contact area of the tyre on the ground to make the tyre follow the direction the wheels are being steered. Without this offset the wheel and tyre swivels about a single pivot or apex point at ground level in a slip-jerk fashion and at the same time causes excessive tyre scrub.

Swivel-pin castor angle (figs 7.30 and 7.31) The tilting of the swivel-pin axis or king-pin axis in the longitudinal direction of the vehicle will make the axis of the swivel (king)-pin when projected to the ground level be either in front of or behind the centre of the tyre contact patch. The angle of inclination of the swivel-pin in the fore and aft direction is known as the castor angle (see fig. 7.30(a)). When the pivot axis projected line intersects the ground ahead of the contact patch centre, the distance between these two points is known as 'trail' since the contact patch centre is trailing behind the pivot point (see fig. 7.30(a)) and is known as positive castor, whereas when the pivot axis projected line is to the rear of the contact patch centre, the distance between these two points is known as 'lead' because the wheel centre now leads the swivel point (see fig. 7.30(c)) and is known as negative castor.

Positive castor angle used with rear-wheel drive (fig. 7.31) Turning the steering with positive castor angle geometry on a rear-wheel drive vehicle when in motion introduces an unbalanced force-couple (turning moment) at ground level. This couple is produced by the driving thrust of the rear wheels pushing forwards against the projected swivel-pin axis and the offset of the opposing ground reaction force at the centre of

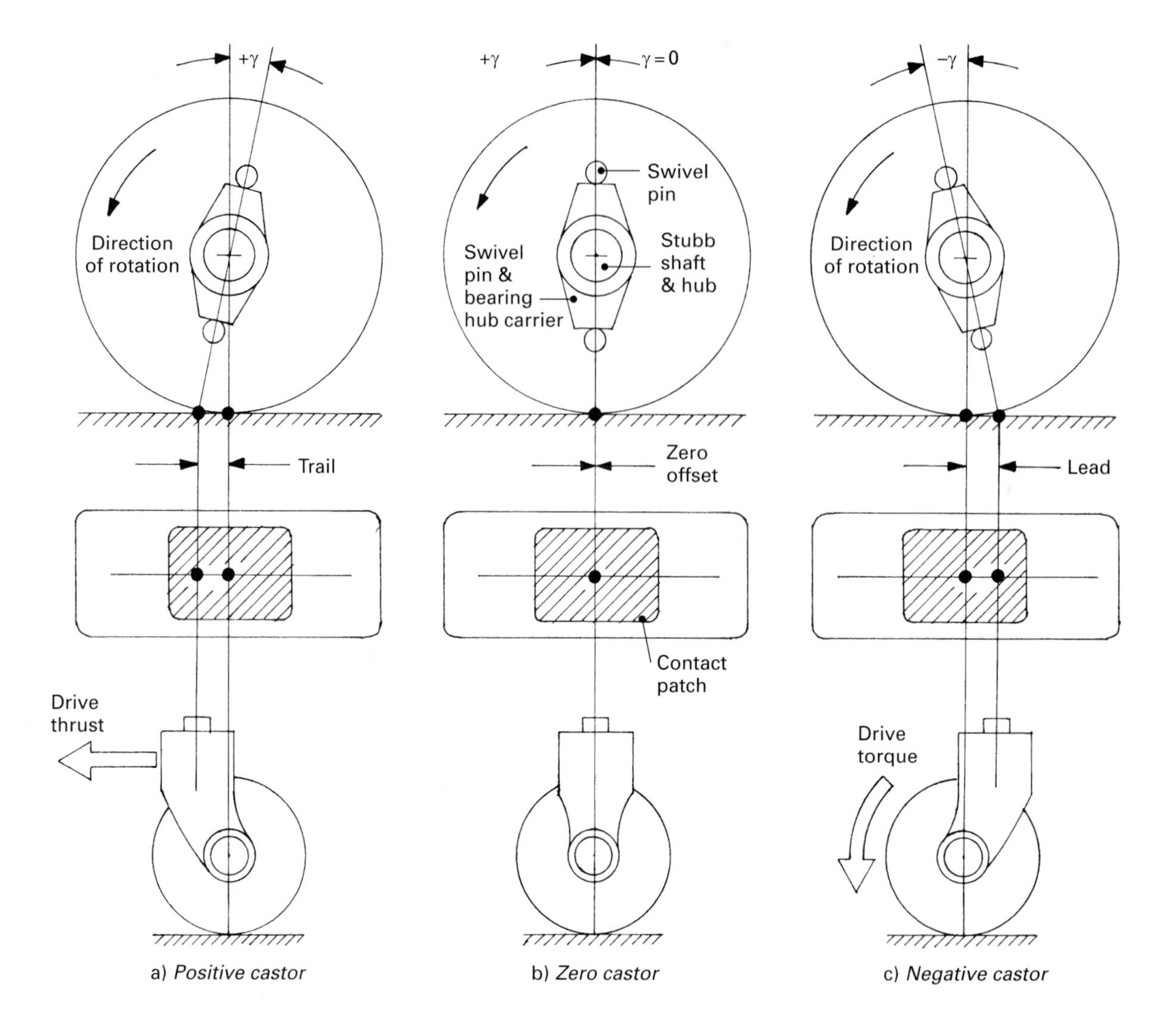

Fig. 7.30 Castor angle steering geometry

the contact patch. Thus when the steering turns to move the front wheels towards one lock, the offset opposing forces will tend to align themselves by twisting the front wheels back to the straight-ahead position. Thus castor angle provides a degree of self-righting after a turn.

Negative castor angle used with front-wheel drive (fig. 7.32) With negative castor geometry on a front-wheel drive vehicle being driven around a bend there will be an unbalanced force-couple (turning moment) at ground level. This couple is created by the tractive effort or thrust on the front wheels propelling the vehicle forwards and the offset of the opposing rolling resistance forces (pulling the body along) at the projected swivel-pin axis. When the applied effort is removed from the steering-wheel, the tractive effort thrust and the offset opposing rolling resistance forces which are trying to move apart tend to eliminate the offset by bringing the front wheels back to the straight-ahead position.

With front-wheel drive in theory negative castor geometry will provide a self-righting tendency for the steering. Nevertheless most front-wheel drive cars seem to prefer a small amount of positive castor angle going onto zero castor.

The castor angle for rear-wheel drive vehicles may range from +0.5 to +5.0°, whereas the castor angle for front-wheel drive cars can vary from −2.0 to +3.0° but is more likely to be between −0.5 and +1.5°.

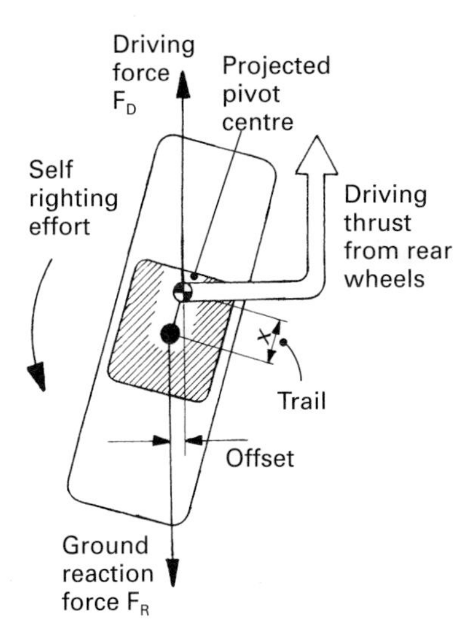

Fig. 7.31 Positive castor rear-wheel drive

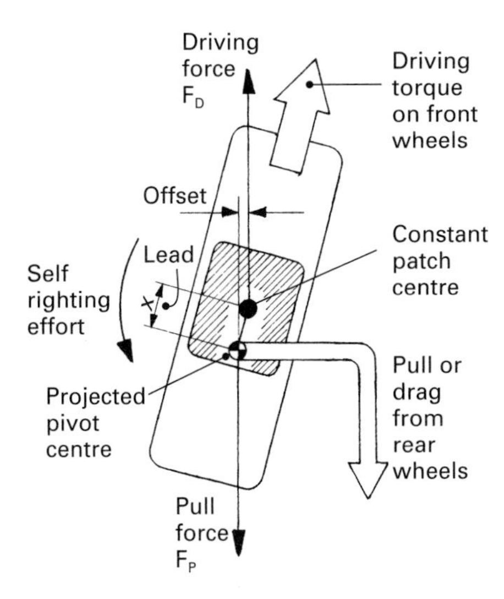

Fig. 7.32 Negative castor front-wheel drive

7.3.2 Roll centres and roll axis (fig. 7.33)

Road wheel suspensions are designed to suspend and support the body through link-arms, struts and the media of springs so that the body is isolated and protected from road wheel impact reaction and wheel bounce when the vehicle travels over irregular road surfaces. To achieve this objective each wheel axle must be allowed to rise and fall relative to the mean body height from the ground by one of two methods:

1) By semi-elliptic leaf springs mounted longitudinally on each side of the chassis. The springs are attached by shackle-pin joints at their extreme ends to the chassis, while the spring's mid-position is rigidly bolted to an axle-beam which spans both wheels. Thus when one wheel lifts, the springs torsionally twist along their length between the shackled ends of the main leaf and the mid-axle position permits the axle-beam to tilt.
2) By transverse or trailing arms which connect the body to the stub-axles or an axle beam respectively via pivot joints, the vertical support normally being provided by coil or torsion-bar springs.

With both methods the body is always subjected to side forces caused by a centrifugal force acting through the centre of gravity of the body and load mass whenever the vehicle is cornering. However this side force is resisted by the tyre grip on the road surface. Now if the body was rigidly attached to the wheel axles the whole vehicle would tend to tilt outwards from the turn, but in fact the body is suspended and supported by either flexible restraining leaf springs or hinge-jointed link arms. Thus instead of the whole vehicle attempting to tip about the outer wheels, the body alone will tend to roll about an imaginary roll-axis created by the front and rear suspension linkage configuration. A small amount of body roll is advantageous as it provides a sense of danger to the driver when negotiating a bend or driving along a curving road; conversely a large degree of body roll is likely to give an uncomfortable ride.

Roll centre height (fig. 7.33) The roll centre height is the vertical distance from the ground at either the front or rear axle at which the vehicle's body tends to roll about when it experiences centrifugal side force. Suspension type, length and pivoting position of linkage determines the location about which the body will roll when it is subjected to side forces. Therefore since the body is supported at the front and rear by different suspension layouts, there will be a separate roll centre height for both front and rear suspensions.

The roll centre height is usually fixed by the suspension design somewhere between ground and axle height level. Independent front suspension normally has very low roll centre heights fixed just above ground level, whereas rigid and independent rear suspension has higher roll centre heights usually set slightly below axle level.

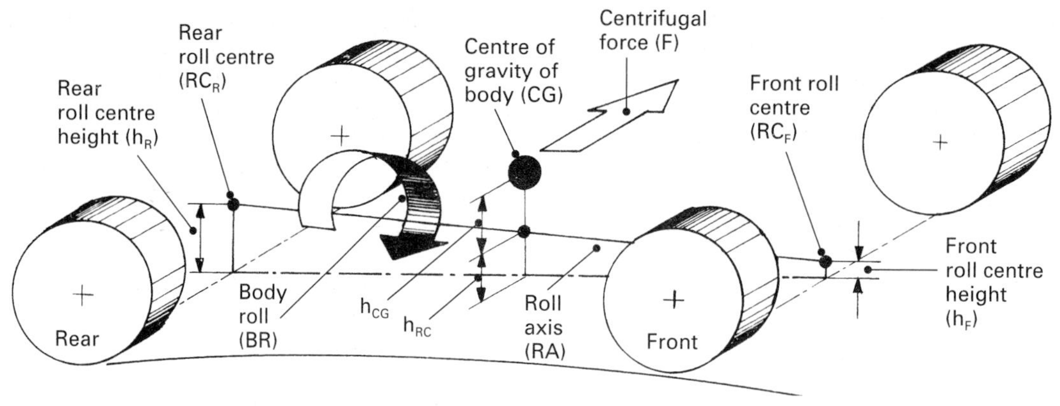

Fig. 7.33 Roll axis and roll centre heights

Roll axis (fig. 7.33) A roll axis is an imaginary line drawn between the front and rear suspension roll centre heights about which the body is assumed to roll about. Since the body is suspended only at these two extremes, then nothing else but the type and layout geometry of the suspension chosen influences the position of the roll axis. The roll axis is normally made to slope down from just below axle height at the rear to near ground level at the front.

7.4 Benefits of limitations of independent front suspension

The reasons for adopting independent front suspension for passenger cars and light vans are as follows:

a) One of the problems with sprung vehicle bodies is that the centrifugal force created when cornering tends to form a roll couple which tilts or rolls the body outwards.

To counteract the body roll, a resisting couple must exist which is made up from the product of the springs' reaction forces and their effective distance apart.

If the springs are situated further apart, the necessary reaction stiffness of the springs will be smaller. If the springs are placed closer, the spring stiffness will have to be greater to resist the roll couple. In fact the roll angle is inversely proportional to the effective spring-base width squared.

With the beam axle, the greatest distance between the springs is determined by the width of the chassis which supports the shackles attached. With the independent suspension with transverse-arm linkage, the effective width between springs becomes the vehicle's wheel track. This means that, compared to the axle-beam type, softer springs can be used with independent suspension without increasing body roll. The advantage of soft springs is that they will respond and deflect with the smallest road ripple without transmitting the accompanying shocks to the body and passengers, thus improving ride comfort.

b) For a given spring weight, the elastic strain energy stored in a coil or torsion-bar spring is greater than for a semi-elliptic multi-leaf spring, so lighter springs may be used with independent wishbone suspension.

c) With independent wishbone suspension, the spring has only to support vertical loads and to absorb shocks – driving, braking, and lateral forces are sustained by the suspension linkage alone.

d) The unsprung mass is reduced with independent suspension, so road-wheels will follow the contour of the road ripples and irregularities at up to higher speeds than for the heavy rigid-axle-beam suspension. Effectively this will reduce tyre scrub and wear.

e) If an anti-roll bar is used in conjunction with the independent suspension, even softer springs may be employed for normal vertical loads but, when the vehicle is cornering, the anti-roll-bar action will provide the necessary resisting stiffness to oppose body roll.

f) Using a separate or independent suspension for each side of the car will reduce any interaction between opposite road-wheels, so there is less chance of vibrational resonance causing wheel wobble.

g) With the wishbone hinging on the subframe structure, the arcs followed by the unsprung stub-axle swivel-joints relative to the sprung body structure when the suspension bounces will produce a precise and predictable wheel-path in the vertical plane which is essential for consistent steering geometry.

h) The engine and chassis structure can be lowered so that the centre of the car may also be lowered and the engine may be brought forward to provide more room for the passengers.
i) Independent suspension usually lowers the roll centres, so that the body will roll before the wheels break away from the road, which gives a warning to the driver.

There are some disadvantages to independent suspension systems compared with axle-beam suspensions, as follows:

a) There is reduced cornering power, due to wheel cambering with body roll.
b) When one wheel bounces there will be a slight change in wheel track, causing tyre scrub.
c) A more rigid chassis or subframe structure is required with the independent linkage.
d) Independent suspension requires more complicated suspension and steering linkage and pivot joints, which makes the suspension more expensive and subject to more wear.
e) Unbalanced-wheel-assembly effects will be transmitted to the steering-wheel more easily and will be more pronounced.
f) Steering-geometry alignment is more critical and needs more frequent attention.

7.5 Construction and action of independent front suspension

With an understanding of the merits and shortcomings of independent suspension, the construction and operation of the various arrangements will now become clear.

7.5.1 Independent short swing arm (fig. 7.20(b))

This is the simplest type of independent suspension. It has only one transverse arm – referred to as a wishbone member – held rigidly through a kingpin steering joint to the wheel stub-axle, but pivoting on the body subframe.

The wishbone-arm forks are spread wide apart at the pivot points, so that any driving- and braking-reaction torque can be absorbed solely by the swing-arm member. The spring is interposed between the body structure and the swing arm and its only function is to support the vehicle's weight – side-forces and reactions are taken completely by the swing arm and the supporting subframe pivot points.

The body will roll when on a curved track, and this will make both wheels lean inwards towards the centre of the circular path the car is following (fig. 7.20(b)(i)), thus producing camber roll.

If one wheel goes into a dip or over an obstacle, its deflection will be confined to one side of the car only and, due to its elasticity, the spring will be compressed without disturbing the body height to any great extent (fig. 7.20(b)(ii)).

With this suspension, the slightest swing of the arm will considerably alter the uprightness of the wheel to the ground.

7.5.2 Independent double-transverse-wishbone front suspension (fig. 7.34)

Independent double-transverse-wishbone suspension uses a pair of triangular forked wishbone members. These arms are hinged to the front subframe or cross-member so that they pivot outwards towards the wheels (fig. 7.34).

The upper and lower ends of these members are connected to a vertical stub-axle post by ball-and-socket swivel-joints. This forms a four-bar chain which can swing about the body hinge and which pivots as the wheel deflects relative to the body position.

The ball-and-socket joints accommodate vertical relative movement of the stub-axle and the wishbone members and also provide the pivot centres for the stub-axle to swivel about for steering purposes.

In between the lower wishbone and the underside of the cross-member is a helical-coil spring which provides the elastic resistance to the body weight and road shocks experienced. The vehicle weight and payload is passed from the sprung body and cross-member to the top of the coil spring. The base of this spring bears against the lower wishbone member, which reacts and supports this weight.

A shock-absorber damper is placed inside the coil spring and is attached by rubber bushes to the underside of the fixed cross-member and to the lower wishbone member.

The fork arms of the lower wishbone members are spaced as far apart as possible, to absorb the accelerating- and braking-torque reactions.

Any side-thrust will be resisted by the stiffness of the wishbone members and the swivel-joints and pivots.

When the car is cornering, the body will roll. Both wheels will then lean outwards from the turning circle, producing a small amount of camber roll (fig. 7.20(c)(i)).

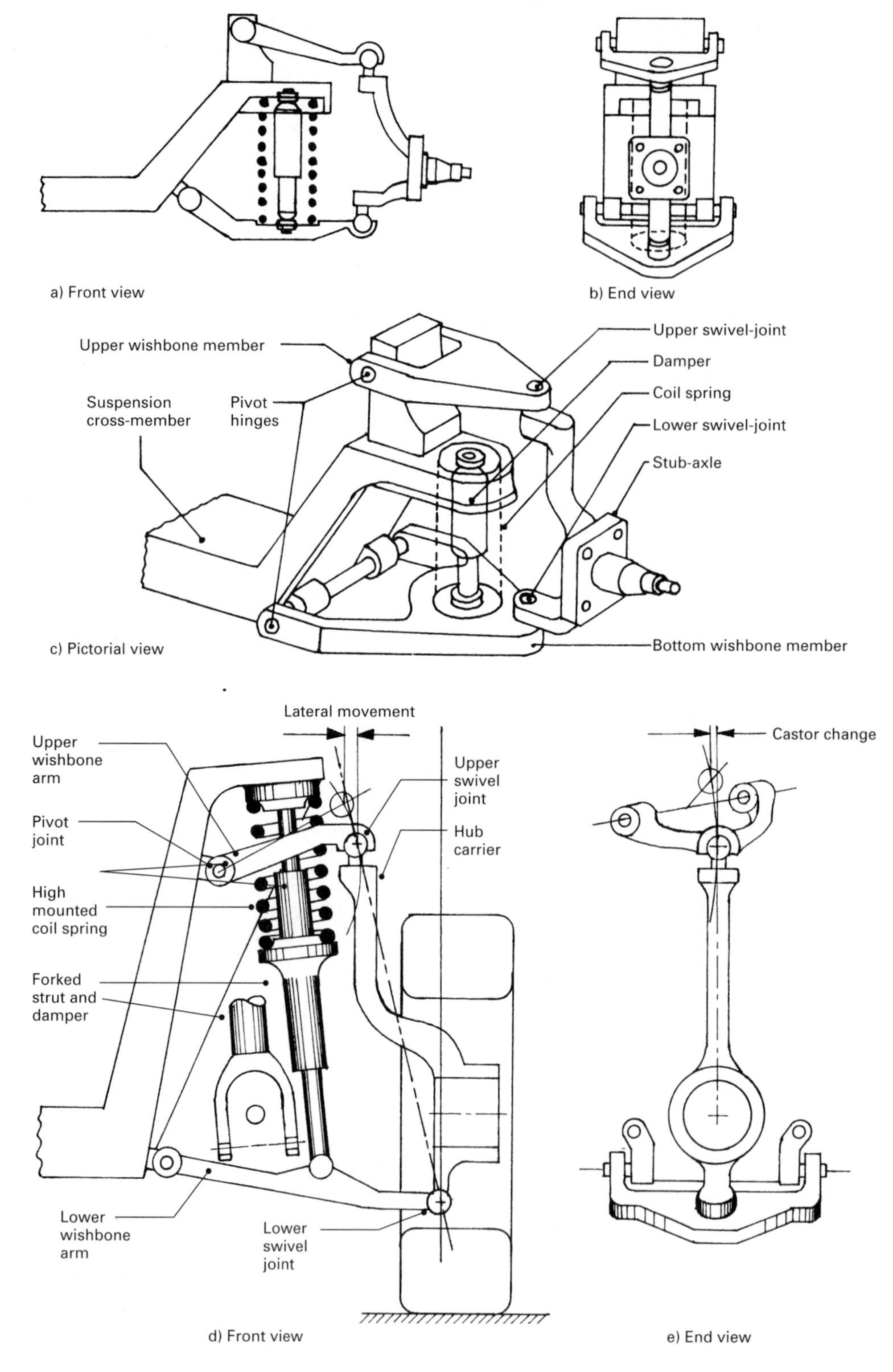

Fig. 7.34 Coil-spring double-transverse-wishbone independent front suspension

When travelling straight ahead, if one of the wheels goes over a bump or a dip in the road, only the individual suspension linkage will be momentarily deflected up or down without altering the average height of the sprung body (fig. 7.20(c)(ii)). This thus provides a completely independent suspension for each wheel with no reaction vibrations transmitted from one side to the other.

To relieve the longitudinal and lateral forces acting on the ball swivel joints of the unequal double wishbone suspension, the hub-carrier is provided with an extended vertical forked strut which is made to curve around the inside of the upper wheel rim and tyre (see Fig 7.34(d)). Thus the distance between the lower and upper ball joints is increased, which considerably reduces the horizontal loads absorbed by both pivot joints. The upper wishbone arm is located in a skewed position so that it not only swings transversely but also in a partial longitudinal direction. Consequently when the wheel and hub-carrier moves towards the bump position, the upper wishbone arm swings such that the upper link arm not only moves inwards towards the body but it also tilts slightly rearwards (see Fig 7.34(e)). It therefore compensates for rubber-bush compliance of the lower wishbone and upper link arm pivots and simultaneously maintains or even increases the effective castor-angle and thus the steering stability when the vehicle is braked and the body tends to dive.

In addition a high mounted coil spring with a central strut type telescope damper unit is used (see Fig 7.34(e)). so that room is made for the horizontally positioned drive-shafts which span the distance between the transaxle's final drive and the hub carriers.

7.5.3 Constant-track unequal double-transverse-wishbone independent front suspension (fig. 7.35)

If the upper and lower transverse wishbone arms are made the same length (fig. 7.35(a)), then, if one wheel deflects, the transverse arms will keep the wheel perpendicular to the ground but the track width will be slightly reduced. This will cause excessive tyre wear.

Constant wheel track can be approached by having short upper and long lower wishbone arms (fig. 7.35(b)). When the wheel deflects, such as will happen if it strikes a road projection, both arms will tend to draw the wheel inwards, but at the same time the greater arc moved by the shorter upper arm will tilt the wheel inward at the top and outwards at road level. Consequently, the expected total inward movement of the wheel is compensated for by the outward tilt action at road level created by the unequal-length transverse arms.

7.5.4 Torsion-bar double-transverse-arm independent front suspension (fig. 7.36(a) and (b))

An alternative version of the transverse-double-wishbone suspension adopts a torsion-bar spring instead of a helical coil spring to provide the elastic resistance to changing suspension vertical load (fig. 7.36(a)). With this arrangement, the torsion bar is situated parallel to the underframe longitudinal members on each side of the vehicle.

The torsion bar is splined at each end. One end is spined to a reaction lever which is bolted to the under-side of the body, while the other end is splined to the lower suspension arm supported by the eyehole pivot, this itself being attached to the longitudinal side-member.

When the suspension is deflected, the lower suspension arm pivots and tends to twist the

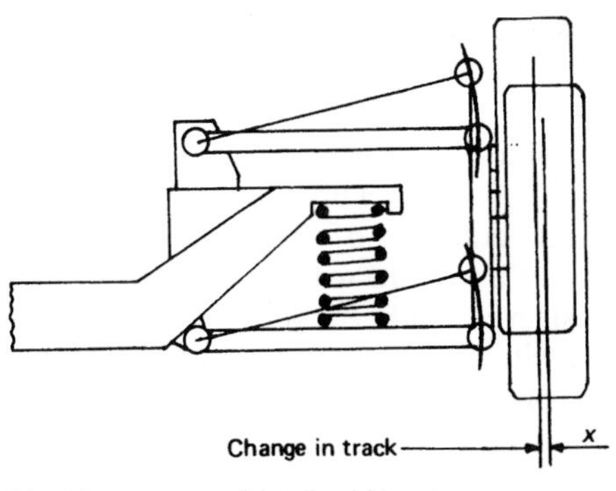

(a) Transverse equal-length wishbones

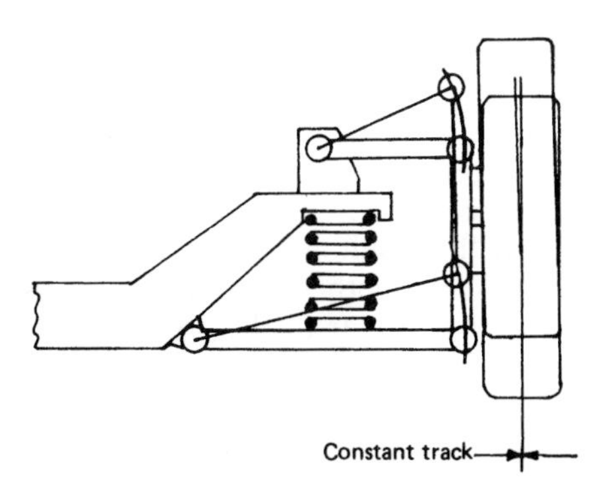

(b) Transverse unequal-length wishbones

Fig. 7.35 Constant-track unequal double-transverse-wishbone independent front suspension

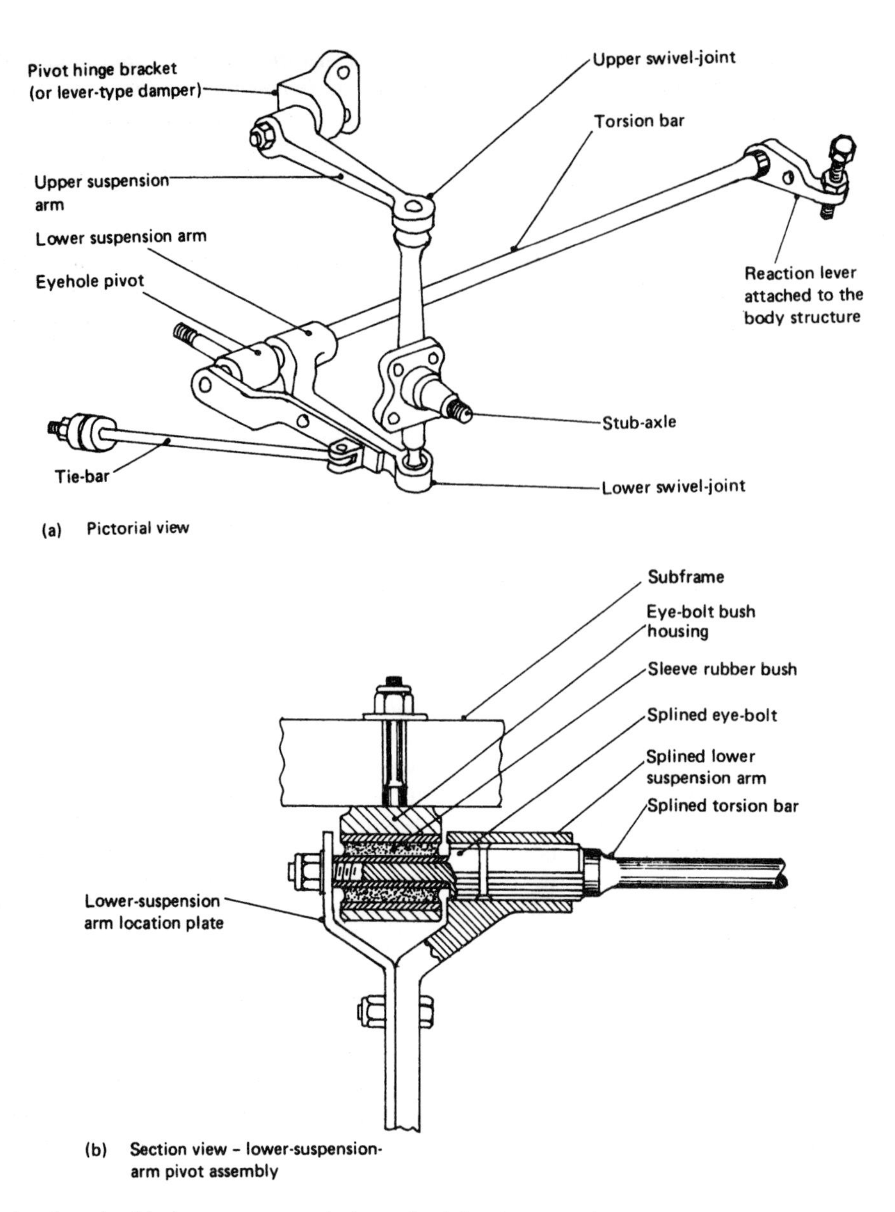

Fig. 7.36 Torsion-bar double-transverse-arm independent front suspension

torsion bar, this being resisted by the rigidly held reaction lever at the far end of the bar. The upper suspension arm completes the four-bar-chain geometry so that the resultant stub-axle vertical movement will keep both front road-wheels approximately perpendicular to the ground at all times. Sometimes the upper suspension-arm pivot forms part of a lever-type shock-absorber (damper) unit – if this is not the case, a separate telescopic damper must be attached between the subframe and the lower suspension arm.

The eyehole pivot assembly allows the lower suspension arm to pivot and simultaneously transfers torsion-bar resilience to the suspension (fig. 7.36(b)). This lower suspension pivot or hinge consists of a one-piece circular eyehole casing and stud which is bolted to the subframe. A sleeved rubber bush is pressed into the casing, and situated in the centre of the bush is a splined eye-bolt, one end of which locates with internal splines formed in the lower suspension arm, while the other end clamps the lower suspension-arm location plate to the protruding inner sleeve of the rubber bush. Thus, when the lower suspension arm partially rotates, the angular movement will be taken up entirely by the torsional distortion of the rubber so that no frictional sliding takes place between the eye-bolt and the inner bush sleeve.

Transfer of static and dynamic elastic resistance is provided simply by the torsion bar's end splines locating in the same extended internally splined hole in the lower suspension arm as the splined eye-bolt. The torsion bar therefore acts only as a spring and does not play any part in the pivoting of the lower suspension arm.

Support of the lower suspension arm is provided by a tie-bar which prevents horizontal twisting of this arm while the vehicle is accelerating or braking.

Alteration of the trim height of the vehicle may be achieved by screwing in or out the torsion-bar reaction-lever adjustment set-bolts.

Body-roll and bump or dip deflection of the suspension will produce wheel tilt similar to that shown in fig. 7.20(c) – that is, similar to the coil-spring double-transverse-wishbone suspension.

The overall simplicity and compactness of the torsion-bar type of springing makes it an attractive proposition for car suspensions.

7.5.5 MacPherson leg-strut independent suspension (fig. 7.37)

This independent suspension consists of a single transverse link – referred to as the track-control arm – hinged on to the lower part of the body subframe or cross-member. A stub-axle and strut assembly is pivoted to the outer end of the track-control arm by a ball and socket at its bottom end, the upper end of the strut forming part of the shock-absorber-damper outer tube or cylinder case. Inside this tube is a sliding piston and rod, the upper end of the rod being pushed on to a ball thrust bearing which is housed in a rubber mounting bolted to the body structure. The thrust bearing allows rotation of the strut, while the rubber mounting accommodates angular movement and carries the body weight. A coil spring is fitted between spring-seat flanges. The upper seat is attached to the end of the piston rod and the lower bearing race track, and the lower seat forms part of the leg-strut. The ball-and-socket joint at the bottom serves as a pivot for the steering strut and the stub-axle in the horizontal plane and as a suspension joint for relative movement between the track-control arm and the stub-axle in the vertical plane.

A bump stop situated at the top of the piston rod serves to stiffen the suspension springing when the coil spring is almost fully compressed. Driving and braking thrust reactions are absorbed by the triangular track-control arm.

During cornering, the body will roll and – depending on the initial angular set of the transverse track-control arm – the inner wheel will lean outwards and the outer wheel will lean inwards. Both wheels then generate camber roll (fig. 7.20(d)(i)). Each wheel suspension will be completely independent of the other side, so that the suspended body will not be influenced by small wheel deflections while the car is being driven (fig. 7.20(d)(ii)).

7.6 Antiroll bars

An antiroll bar stiffens the suspension springing when the body rolls or one wheel goes over a bump or dip in the road. The antiroll bar therefore permits softer suspension springing to be used when the vehicle is moving straight ahead, as this responds far better than hard springing to small ripples or irregularities in the road, hence it improves the ride comfort of the passengers.

The antiroll bar is a circular sectioned torsion-bar mounted transversely on spaced out rubber bush swivel bearings to the body and having cranked arms, that is, bent extended ends which are attached to each swing-arm via short vertical link-rods by means of rubber-bush joints (see fig. 7.38(a))

Antiroll bar action

Equal wheel lift on both sides (fig. 7.38(b)) If both wheels bounce and rebound together the antiroll bar contributes nothing to the suspension spring stiffness since both of the cranked arms swing in the same direction.

Single-wheel lift on one side (fig. 7.38(c)) If one wheel (right-hand) rides over a bump in the road the control-arm swings upwards and in so doing twists the right-hand cranked antiroll bar arm; however, this will be resisted by the left-hand cranked arm which is attached to its respective control-arm. Subsequently the torsional wind-up distorts the antiroll bar in such a way as to exert an upward thrust on the left-hand side of the body via the rubber-bush bearings on that side. In so doing, it lifts the body on the opposite side to the raised wheel, thereby equalising the lift of the body on both sides. The vehicle will therefore continue to travel with the body in an upright

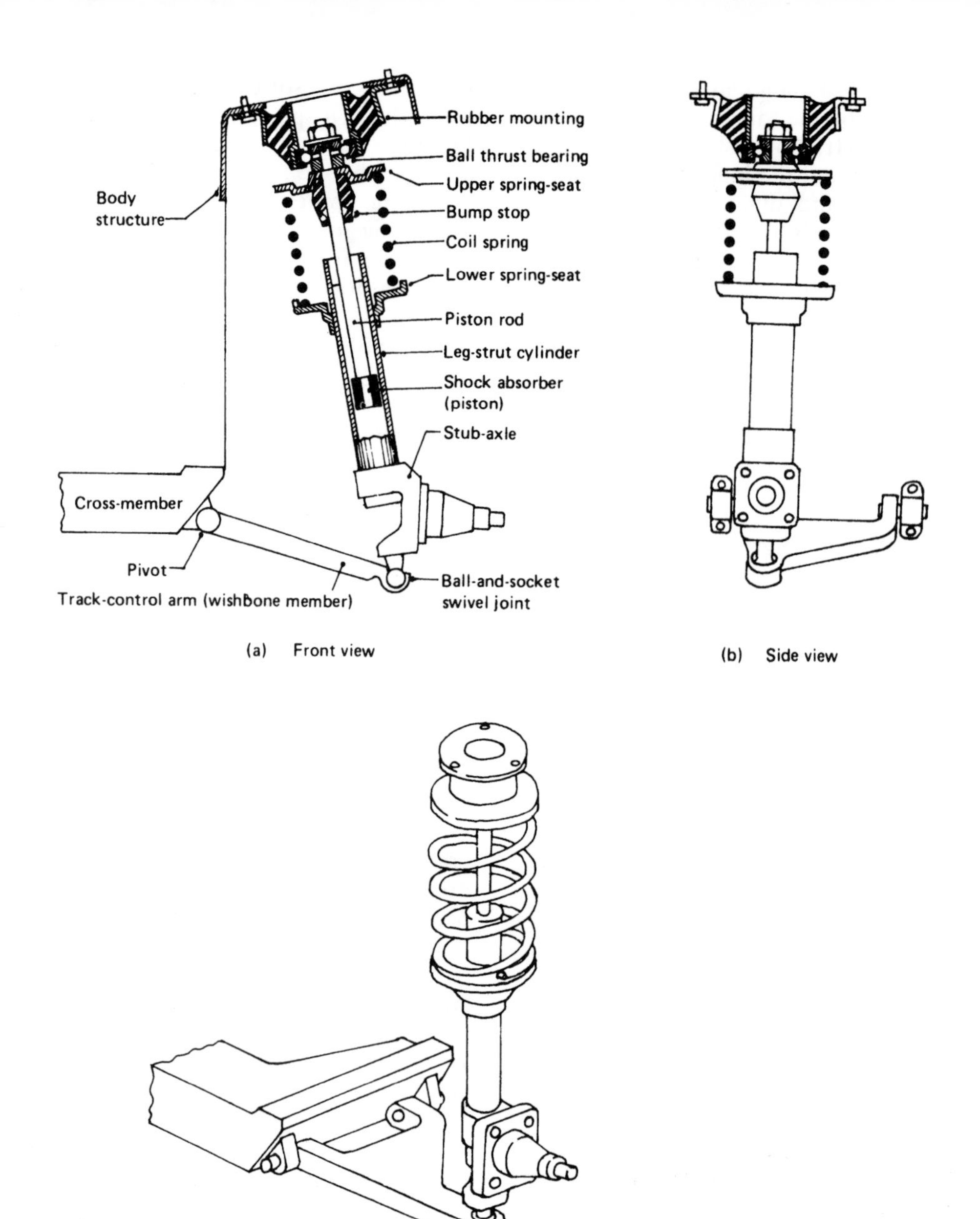

Fig. 7.37 MacPherson leg-strut independent front suspension

position, that is, on an even keel. Thus whenever one or the other wheel should dip or lift the anti-roll bar compensates for changes in body height by raising or lowering the body on the opposite side.

Body roll (fig. 7.38(d)) As the body commences to roll due to the centrifugal force acting through the centre of gravity of the suspended body when the vehicle is being driven round a bend or on a curved track, the body on the inner wheel side will commence to lift, whereas the body on the outer wheel side will tend to dip. Since the antiroll bar is compelled to roll with the body, the cranked arm on the inner wheel side will twist downwards while the cranked arm on the outer wheel side will twist upwards in the opposite direction. Subsequently the reaction to the torsional wind-up of the antiroll bar will be to pull down on the inner wheel side of the body and push up the body

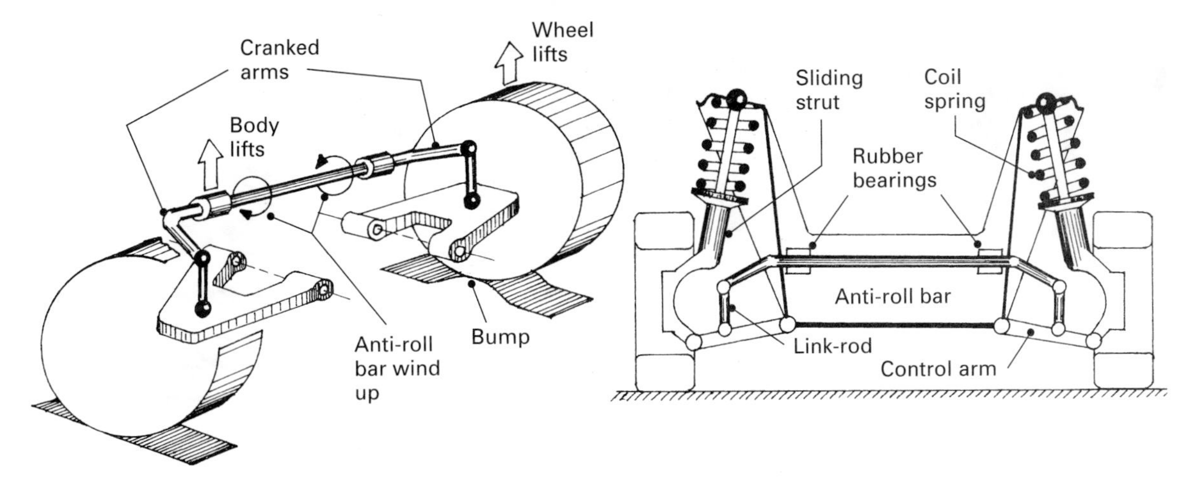

a) Pictorial view of anti-roll bar in action

b) Equal wheel lift causes anti-roll bar to be inactive

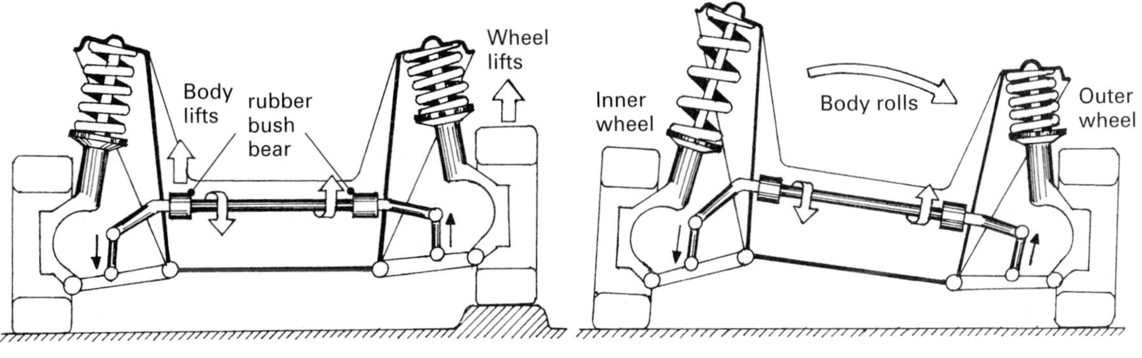

c) Wheel lift causes anti-roll bar torsional twist

d) Body roll causes anti-roll bar torsional twist

Fig. 7.38 Front MacPherson strut independent suspension with anti-roll bar

on the outer wheel side. This results in the reaction to the distorted configuration of the cranked antiroll bar opposing the outward roll of the body. Hence the greater the degree of body roll, the greater will be the opposing resistance offered by the antiroll bar.

Thus the function of the antiroll bar is to relieve the main suspension springs of some of their load every time the body rolls. It thus enables softer springs to be used so that the road wheels are able to follow more closely the contour of the road surface and for the suspension to be more sensitive to small humps and dips in the road. In the extreme, some antiroll bars are tuned to take as much as 30–40% of the total vertical load imposed on the suspension when subjected to severe body roll.

7.6.1 Bump stops and spring aids (fig. 7.39(a)–(c))
Under normal driving and load operating conditions the suspension springs and antiroll bar adequately cope with wheel bounce, rebound and body roll. However, if the body is heavily laden or there is a large amount of body roll, then it is desirable to stiffen the suspension springing under these extreme conditions. This is normally achieved by imposing progressive rubber springs known as bump stops or spring aids between the body and the wheel suspension, usually in the form of a hollow cylindrical convoluted rubber moulding (see fig. 7.39(a) and (b)).

With normal deflection of the load-bearing suspension arm, the distance between the under side of the body and the suspension arm or leg-strut (see fig. 7.39(a) and (b)) is much too wide from the progressive bump stop to be active. However, with large static or dynamic loads the gap between the body and the suspension load-bearing member will be close to a point where the hollow rubber moulding commences to be compressed in a concertina fashion. Accordingly as the ribs of the walls fold the effective thickening

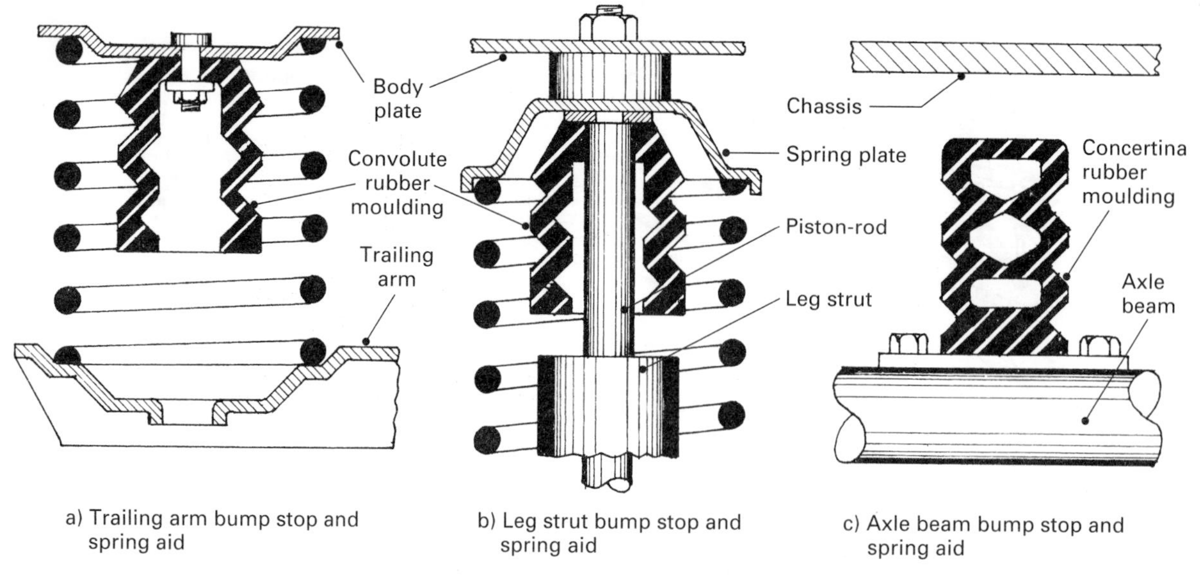

Fig. 7.39 Bump stops and spring aids

of the cylindrical wall progressively stiffens the moulding's resistance to further compression. Thus the bump stop provides an additional stiffening of the suspension springs only as the suspension pivoting arm or arms approach the maximum permitted degree of deflection.

Bump stops are normally located inside the coil springs to the upper body spring seat-plate (see fig. 7.39(a) and (b)) or to the axle-beam/leaf spring plate (see fig. 7.39(c)).

7.7 Telescopic-shock-absorber damper (fig. 7.40)

The function of a telescopic damper The damper or shock absorber cannot support weight and has no resilience. If a damper is compressed, it will not rebound and the energy (shock) imparted to it will not be stored but will be dissipated as heat. When dampers are installed in spring assemblies, part of the energy created due to vertical movement of the axle will be absorbed and changed into heat, leaving insufficient energy in the spring for violent rebounds. This reduces the amplitude and frequency of oscillation of the spring.

Damper-unit construction The main components of a telescopic damper are a cylinder and a piston attached to a rod. The cylinder is placed in an outer tube so that a fluid-reservoir space is created around it. There are passages in the piston to enable fluid to move from one side to the other. In the base of the cylinder, a passage allows fluid to be transferred between the cylinder and the reservoir.

When the piston moves, fluid can pass from one end of the cylinder chamber to the other through the small piston-head passages, and from the lower cylinder chamber through passages formed in the base to the reservoir and back. Since the fluid is incompressible, there is considerable resistance to motion when a relatively large volume of fluid is displaced through small orifices in a short time. It is this action – which is commonly known as damping – which generates the opposing resistance to spring deflection between bump and rebound. Generally a greater resistance is necessary for rebound than for bump, so separate bump and rebound valves are built into the piston and the cylinder-chamber base to provide a differential fluid-resistance control, a typical valve having a rebound-to-bump resistance ratio of 3:1.

The reservoir has two functions. First it absorbs the excess fluid displaced from the lower chamber, as the upper chamber has a smaller capacity due to the space the rod occupies. Its second very important function is to allow air trapped by the rapidly moving fluid to rise and escape – this reduces aeration and frothing of the fluid so that the viscosity of the fluid remains consistent.

Bump or compression stroke (fig. 7.40(a)) With the piston descending, the central rebound valve in the piston head closes and the piston bump

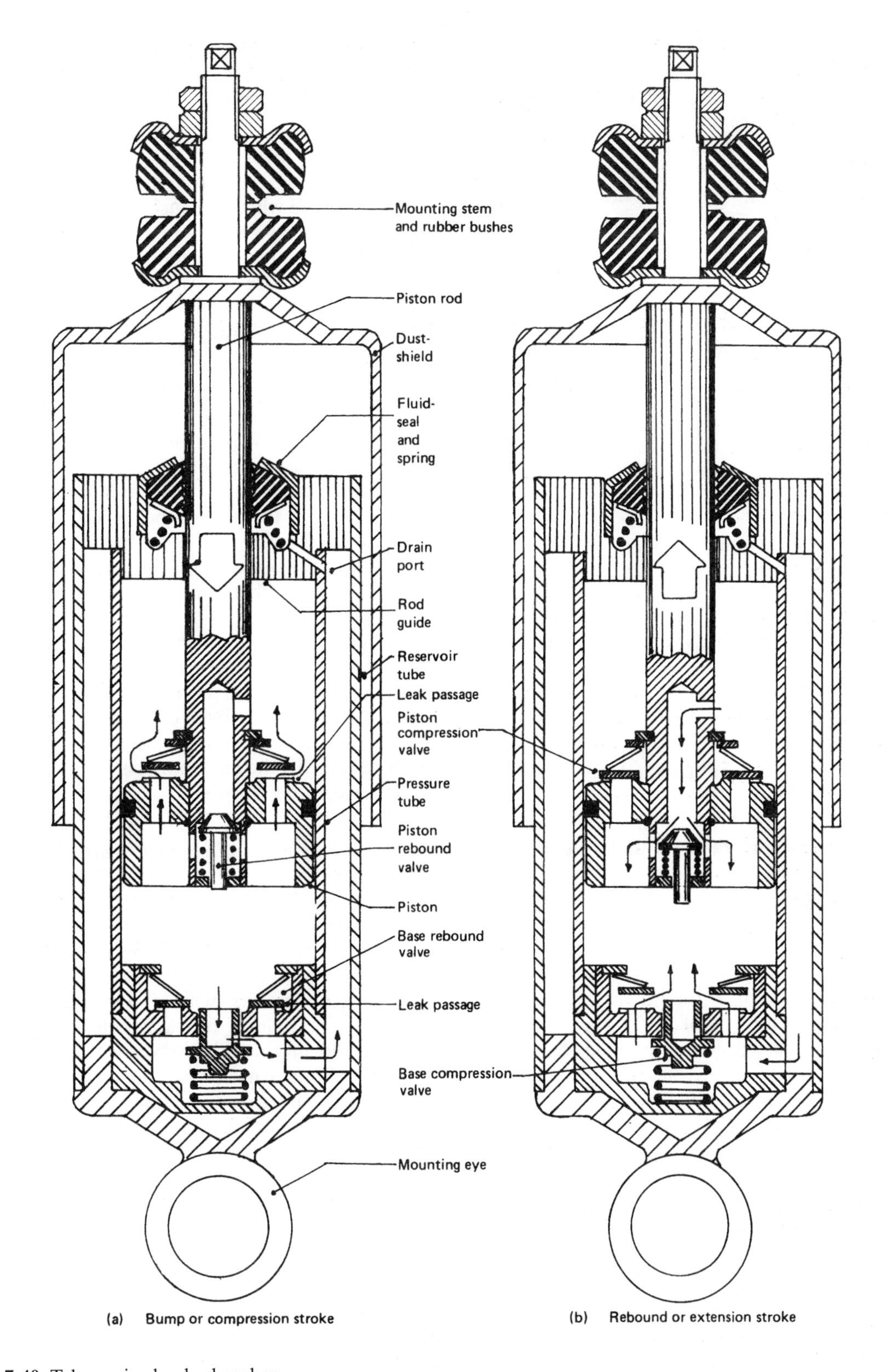

Fig. 7.40 Telescopic shock absorber

valve opens so that fluid is transferred from the lower to the upper cylinder chamber. At the same time, the outer base rebound valve closes and the central base bump valve opens, displacing a quantity of fluid to the outer reservoir.

At high speed, resistance is created by the fluid being forced through the spring-loaded valves, and at low speed by fluid passing through leak passages.

Rebound or extension stroke (fig. 7.40(b)) When the piston moves upwards, the bump valve in the piston head closes and the central piston rebound valve opens so that fluid is transferred from the upper cylinder-and-rod chamber to the enlarged lower chamber. At the same time, the central base bump valve closes and the outer base rebound valve opens. Fluid from the unpressurised reservoir will now be drawn into the lower cylinder chamber to compensate for the reduced volume of fluid transferred from the cylinder-rod upper chamber.

As with bump, the spring-loaded blow-off valves control the high-speed resistance, and fluid transfer at low speed will be restricted through leak passages.

7.8 Rear suspension

Rear suspension is required to support the vehicle body on an axle-beam or with independent suspension on stub-axles. Springs are interposed between the axle and body so that the latter is partially isolated from the axle and wheels. The function of the spring is to permit the axle and wheels to move independently of the body due to the deflection of the springs as the wheels drive over bumps, holes and uneven surfaces. However, the springs normally only cope with the vertical loads imposed on them by the weight of the body and the dynamic impact of the wheels on the ground. Thus the suspension is designed to sustain not only the vertical loads, but the fore and aft loads imposed on the axle due to the driving and braking reaction, lateral forces imposed on the axle caused by centrifugal force when cornering, camber or sloping road and gusts of wind.

Methods of controlling body movement relative to the wheels will now be described.

7.9 Axle beam non-drive rear suspension

Axle beam rear suspension is usually constrained by a trailing arm which absorbs the driving and braking longitudinal force (see fig. 7.41(a)–(d)). However, to maintain a constant wheelbase (distance between axles), an upper arm is sometimes attached between the axle and the rear of the body via rubber bush joints.

7.9.1 Trailing arms (fig. 7.41(a)–(d))

Trailing arms are used to connect the body to the axle-beam and wheels in such a way that the body is free to move vertically relative to the axle, but not to move in the fore and aft direction when the axle and wheels are experiencing both driving and braking thrust, or when the body is subjected to the inertia of acceleration or deceleration.

7.9.2 Panhard-rod (fig. 7.41(a))

A Panhard-rod restrains the body from shifting laterally (sideways) relative to the axle-beam and wheels without hampering the up and down movement of the body. The Panhard-rod is positioned parallel to the axle-beam and attached at one end to the side of the body and to the axle-beam at the opposite end, both ends being secured by a rubber bush and steel pin or bolt joints.

As the body moves vertically up or down, the Panhard-rod swings horizontally about its axle-beam pivot joints so that the movement at the body end is in the form of a small arc. Consequently there is a small lateral movement of the body relative to the axle, but it is insignificant and can therefore be ignored.

7.9.3 Leading and trailing arm Watts linkage (fig. 7.41(a))

A longitudinal positioned Watts linkage restrains the body moving fore and aft relative to the axle-beam when the vehicle is being driven. A Watts linkage is positioned on either side of the axle and parallel to the rear wheels. Upper and lower brackets are welded to each side of the axle-beam to form the middle vertical link, a long sturdy trailing arm forms the lower horizontal link and a short rod becomes the upper rear horizontal link. Rubber bush and pin joints attach the links between the body and axle brackets. As the axle moves up and down relative to the body the axle and bracket assembly will twist slightly to accommodate the changing inclination of both the trailing arms and rods as they articulate about their body attachment pivot joints. By these means the axle will be able to maintain approximately the same horizontal centre position between the trailing arm and rod attachment points to the body.

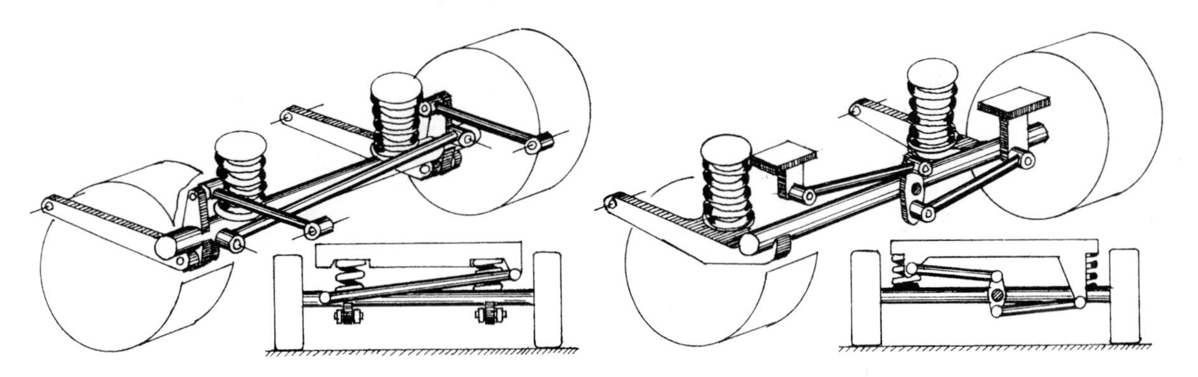

a) Axle beam and leading/trailing arm Watt linkage with Panhard-rod and coil springs

b) Axle beam and trailing arm with transverse Watt linkage and coil spring

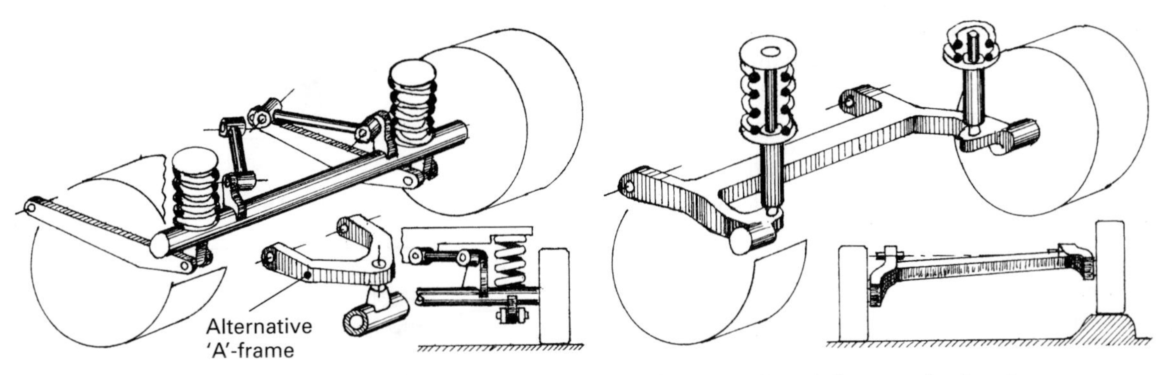

c) Axle beam and trailing arm with diagonal tie-rods and coil springs

d) Trailing arm twist axle beam and coil springs

Fig. 7.41 Axle beam non-drive rear suspension

*7.9.4 **Transverse Watts linkage*** (fig. 7.41(b))
This linkage restrains the body from moving over to one side every time the body is subjected to side forces without interfering with the independent vertical movement of the body relative to the wheel axle which take place under normal driving conditions. With this system a double arm lever centrally pivoted is mounted mid-way along the axle-beam. Horizontal link-rods positioned on either side of the central lever arm are attached at the outer wheel end to the body via high and low mounted brackets; rubber bush joints form end attachments for these rods.

As the body moves up and down relative to the axle-beam, the central lever will twist to take up the changing angular position of the link-rods. Thus the raising or lowering of the body remains unrestricted, whereas lateral (side) movement of the body relative to the axle-beam is prevented.

*7.9.5 **Diagonal twin tie-rods*** (fig. 7.41(c))
These tie-rods, also known as torque-rods, are used to absorb the reaction forces acting on the axle-beam and so prevent it twisting when driving or braking occurs. At the same time the triangular or diagonal layout of the tie-rods which are attached both to the body and axle-beam by rubber bush joints restrain the body moving both longitudinally and sideways (laterally) relative to the axle-beam without interfering with the movement of the body and the deflection of the coil spring under normal driving conditions.

*7.9.6 **'A' Frame*** (fig. 7.41(c))
This is normally a triangular-shaped steel pressing which pivots at two spread-out pivot attachment points on the body and at a single pivot bracket point mid-way along the axle-beam (see insert in fig. 7.41(c)). This simple arm therefore is able to cope with both longitudinal and lateral forces imposed on the body during accelerating, braking, and when driving around bends in the road.

*7.9.7 **Twist axle beam*** (fig. 7.41(d))
This is a semi-independent rear suspension. Each wheel axle is supported by a trailing arm and a

rectangular or circular sectioned beam cross-member which is welded to both trailing arms about a third of the way along their length from the pivoting ends. Lateral or sideways deflection is absorbed by the wide spread trailing arm pivot points and the interconnecting beam which forms a rigid rectangular frame. Body roll and keeping the vehicle on an even keel when one wheel goes over a bump or pothole is partially controlled by the interconnecting beam twisting so that the deflection of the trailing arms on one side is counteracted by the other trailing arm acting as a torsional restraining anchor point.

7.9.8 Scott–Russell linkage (fig. 7.42(a)–(c))

The Scott–Russell two-link mechanism restrains the body from swaying or shifting to one side relative to the wheel and axle beam when the body is subjected to lateral (side) forces. This configuration is formed by lateral link-arms diagonally positioned between the axle-beam and body, and control-rods attached by pivot joints to both lateral link-arms and the axle-beam. The body end of the lateral link-arm is attached by a conventional rubber bush type joint, whereas the axle-beam end is secured by a special large flexible slotted bush joint. The inclination of both the lateral link-arm and control-arm to the horizontal permits the two arms to fold and unfold as the distance between the body and axle changes when the vehicle is being driven over bumpy road surfaces. There is an identical link mechanism incorporated on both sides of the axle-beam. This ensures that there is no side displacement of the body relative to the axle-beam as the axle-beam to body vertical distance varies.

For this mechanism to operate, the distance between the lateral link-arm axle pivot flexible joint and the fixed control-arm axle pivot joint must be allowed to vary. Thus when the vertical spring height contracts, the folding action of the lateral link-arm as it pivots about its body attachment joint will make the lower pivot point connection with the axle move further away from the control-arm axle fixed pivot joint, whereas on rebound the lateral link-arm moves to a more upright position, so that its axle pivot point is compelled to move nearer to the control-arm axle fixed pivot joint. This to and fro horizontal movement is made possible by adapting a large flexible slotted rubber bush joint for the lateral link-arm lower attachment point to the axle-beam. The circular casing of the joint is welded to the lower end of the lateral link-arm, whereas

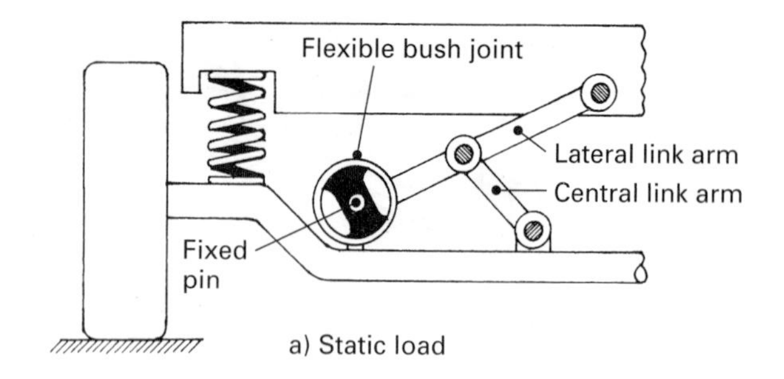

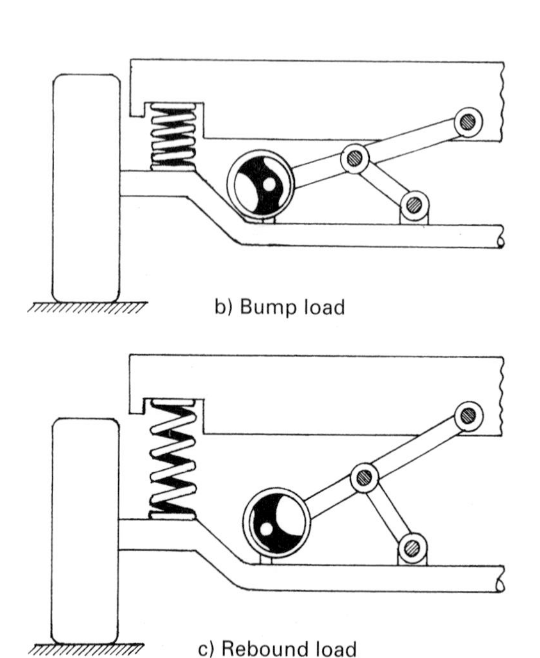

Fig. 7.42 Transverse Scott–Russell linkage

the central pivot-pin of this joint is rigidly anchored to the axle-beam.

The action of this flexible slotted rubber bush joint will now be described:

1) When the vehicle is unladen the pivot-pin and its rubber moulding will be in the centre of the pivot joint circular casing (see fig. 7.42(a)).
2) When the vehicle load is increased, or the wheels are going over a bump and the springs are compressed the lateral link-arm circular joint casing is compelled to move nearer to the pivot-pin by distorting the rubber moulding to the right side of the circular casing (see fig. 7.42(b))
3) Conversely as the vehicle's load is reduced or the springs are experiencing rebound extension conditions, the lateral link-arm moves away from the pivot-pin by distorting the

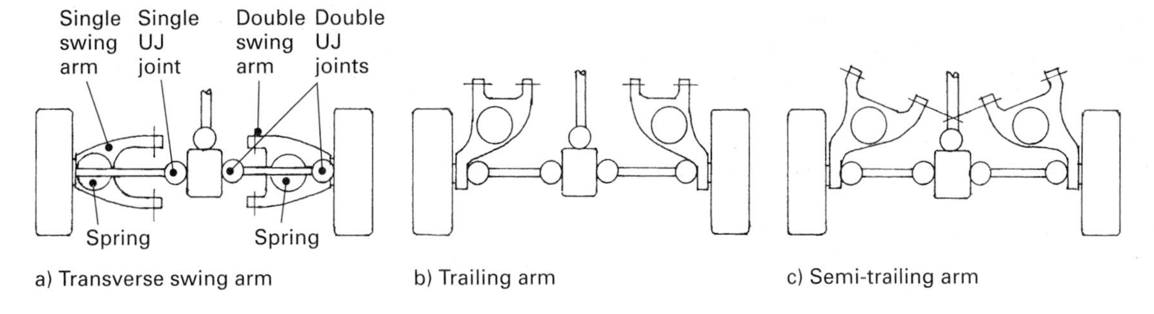

Fig. 7.43 Rear-wheel drive independent suspension

rubber moulding to the left side of the circular casing (see fig. 7.42(c)).

This very compact arrangement ensures that the wheels can only move vertically in relation to the body and at the same time it stabilises any body roll tendencies.

7.10 Rear independent suspension

These suspension arrangements can generally be classified into four categories:

1) transverse swing arm;
2) trailing arm;
3) semi-trailing arm;
4) strut and link-arm.

7.10.1 Transverse swing arm rear-wheel drive rear independent suspension (fig. 7.43(a))

Here the wheel transversely swings on a triangular twin-pronged arm attached by rubber bush pivot joints to the subframe. The pivot joints are positioned as wide apart as possible to absorb the driving and braking thrust. With the single transverse arm the wheel-camber varies considerably as the wheels deflect up and down. Thus if the body rolls, the inner wheel to the turn leans inwards and the outer wheel leans inward, that is the inner wheel camber becomes more positive and the outer wheel moves to a more negative camber position. To counteract this shortcoming it is far better to have an upper and lower transverse arm as this limits the change in wheel-camber. The single swing-arm suspension needs only a single inboard constant-velocity joint (see left-hand suspension in fig. 7.43(a)), but an inner and outer constant-velocity joint is required with the double transverse arm suspension (see right-hand suspension in fig. 7.43(a)). If the single swing arm is used, the coil spring is usually located over the swing arm near the wheel (see left suspension in fig. 7.43(a)) whereas the double swing arm coil spring is more likely to be positioned nearer the pivot joints (see right-hand suspension in fig. 7.43(a)). The track also tends to vary as the load on the springs changes when one wheel is bounced up and down or if the body rolls when cornering.

7.10.2 Trailing arm rear-wheel drive rear independent suspension (fig. 7.43(b))

The trailing arm and coil spring suspension form a compact layout; it has therefore been used on small to medium sized cars. The triangular-shaped trailing arms are hinged transversely across the rear of the body with their pivot joints spaced out as far as possible to sustain the lateral side forces. Transmission is provided via drive-shafts and inboard and outboard constant-velocity joints driven by the body-mounted final-drive.

When the body rolls both wheels tend to lean away from the turn, that is, the inner wheel becomes negatively cambered and the outer wheel becomes more positively cambered; furthermore the track will vary with the individual deflection of the trailing arms.

7.10.3 Semi-trailing arm rear-wheel drive rear independent suspension (fig. 7.43(c))

Transverse swing arm suspension produces positive wheel camber on the inner wheel and negative camber on the outer wheel when cornering, whereas a trailing arm suspension goes into negative and positive wheel-camber on the inner and outer wheel respectively whenever the body rolls.

The semi-trailing arm suspension combines the trailing arm and transverse swing arm movements by skewing the pivot joints to something like 15–25° so that the trailing arm now partially swings sideways. With a well-designed semi-trailing suspension the change in wheel-camber when the body rolls will be minimal. Furthermore,

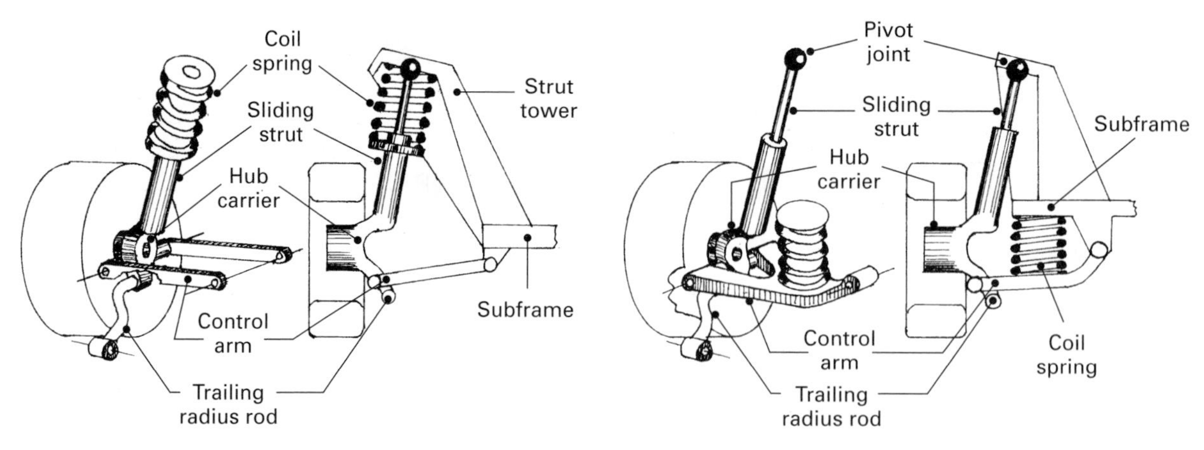

Fig. 7.44 Strut and link non-drive coil spring rear independent suspension

under bump and rebound both the inner and outer wheels toe in and therefore neutralise any roll steer tendency. This form of suspension finds particular favour with medium to large cars.

7.10.4 Strut and link non-drive rear independent suspension (fig. 7.44(a) and (b))

This sliding-strut three-pivot joint suspension uses a piston-rod and cylinder strut (leg) to guide the wheel and hub carrier's vertical motion. A triangular configuration is produced by the upper end of the piston-rod being anchored by a rubber-bush pivot joint to the body structure, whereas at the lower end the cylinder merges with the hub-carrier. Lateral movement between the hub-carrier and subframe is restrained by a transverse control-arm spread apart to withstand longitudinal braking forces.

The sliding-strut vertical movement and the swing of the control-arm combines to maintain an almost constant wheel-camber whenever the wheel and hub-carrier is subjected to bump and rebound. A trailing radius-rod and rubber-bush joint absorbs the dynamic rolling hardness of the tyre in the longitudinal direction. Coil springs can be located between the body tower and the top end of the hub-carrier leg (see fig. 7.44(a)), or it may be sandwiched between the subframe and the lower wishbone-shaped control-arm (see fig. 7.44(b)).

7.10.5 Trailing-arm torsion-bar/tube springs with external anti-roll bar non-drive rear independent suspension (fig. 7.45)

This compact non-drive trailing-arm rear independent suspension which is used on medium-sized cars incorporates a pair of torsion springs transversely located across the body in the form of a compound bar and tube. The relatively short torsion-tubes are welded to the trailing-arms and are supported at their outer ends on a rubber-bush pivot bearing mounts which are themselves bolted onto the body or subframe. At their inner ends the torsion-tubes are splined to centrally located torsion-bars, while the opposite outer ends of these torsion-bars are anchored rigidly to the body via the rubber-bush pivot mount brackets. To provide adequate lateral force constraint, the right-hand torsion-tube is extended and enlarged to encase the left-hand torsion bar/tube assembly. Rubber bushes spaced apart and located between the left-hand torsion tube and the right-hand torsion-tube extension provides the inner pivot support for the trailing-arm longitudinal spring movement and thus prevents wheel-camber and toe-in changes when cornering. An anti-roll bar is located parallel to the torsion bar/tube spring and is securely attached at each end to its respective trailing-arm. The anti-roll bar transfers some of the outer bar/tube torsion resistance to the inner spring whenever one wheel goes over a bump or pothole or when the body rolls when negotiating a bend in the road.

7.10.6 Trailing-arm torsion bar spring with enclosed anti-roll bar non-drive rear independent suspension (fig. 7.46)

This very compact trailing-arm non-drive rear independent suspension comprises a pair of trailing-arms each welded to a transverse pivot tube. Each pivot tube is supported on two sets

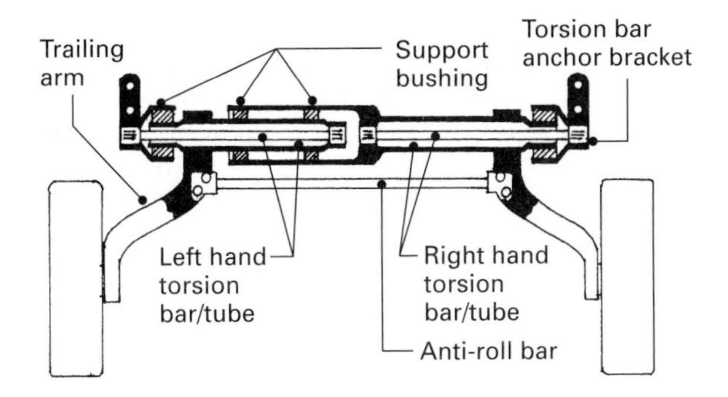

Fig. 7.45 Trailing arm torsion bar/tube spring with external anti-roll bar rear independent suspension

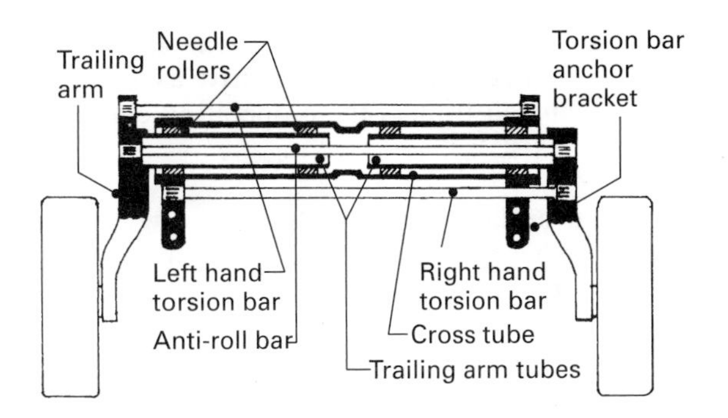

Fig. 7.46 Trailing arm torsion bar spring with enclosed anti-roll bar rear independent suspension

of spaced out needle roller bearings located within the cross tube, which is itself bolted at the left and right-hand ends to the body via each torsion-bar anchor bracket. The left-hand trailing-arm is extended back beyond the pivot tube to form a lever and a splined attachment point for the left-hand torsion-bar. The other end of this torsion-bar is anchored to a similar extension provided on the right-hand torsion-bar anchor bracket. However, the right-hand trailing-arm has its torsion-bar spline attachment point along its arm slightly out from the pivot tube, while the other end of the torsion-bar is gripped by the left-hand torsion-bar anchor bracket just ahead of the attachment bolts.

An anti-roll bar straddling the inside of both pivot tubes is splined at the pivot centres to each trailing-arm. Thus torsion resistance is transferred from one trailing-arm to the other every time one of the arms is deflected and twists about its point tube. This suspension arrangement is credited with precise wheel-camber and toe-in control when the body is subjected to roll.

7.11 Tandem axle leaf spring suspension

7.11.1 Reactive balance beam tandem axle suspension (fig. 7.47(a) and (b))

For carrying heavy loads trucks generally have tandem axle suspension at the rear so that the payload can be shared between both axles. However, to divide the static load equally between axle wheels they must be in contact with the ground at all times. With the conventional fixed hanger bracket and swing shackle semi-elliptic spring, this is not possible, since if one axle should roll over a bump in the road, it will lift the chassis and the other axle wheels clear of the ground, thus subjecting the single active axle momentarily to carry the full load alone.

Staic load equalisation (fig. 7.47(a) and (b)) To equalise axle loading as the wheels travel over road irregularities such as dips and humps, adjacent semi-elliptic spring ends may be interconnected by a short centrally pivoting balance-beam. Therefore, if one axle and its wheels roll into a dip the balance-beam swings to permit the axle still on higher ground to be nearer the chassis. Conversely, if one axle and its wheels climb over a hump the balance-beam tilts, enabling the other axle to move further away from the chassis so that both axle wheels are in contact with the ground at all times. Unfortunately, driving and braking torque distribution between axle and wheels are not equally shared as will be illustrated.

Unequal driving torque distribution (fig. 7.47(a)) When the vehicle is being accelerated either from a standstill or after a gear change, the driving torque reaction T_r on both axles will tend to twist each axle in the opposite direction to that of the drive torque T_d, that is, clockwise, if the vehicle is moving to the left (see fig. 7.47(a)).

As a result the front ends of both axle springs experience an upthrust, while the rear ends are subjected to a downthrust. Therefore, the front and rear axle adjacent spring ends will tend to tilt the balance-beam anticlockwise. Subsequently, it pushes the front axle wheels down hard against the ground whereas the rear axle and its wheels tend to be lifted away from the ground. Thus the front axle will contribute more tractive-effort than the rear axle when being propelled forwards. This is not generally a problem under good road conditions but it can promote tyre scrubbing on the rear axle wheels. However, if both axles are non-drive as when used on articulated trailers these operating conditions will not exist.

Unequal braking torque distribution (fig. 7.47(b)) As the vehicle is being braked, the braking torque reaction on both axles will tend to twist each axle in the same direction to the braking torque, that is, anticlockwise if the vehicle is moving to the left (see fig. 7.47(b)). Consequently the front end of each axle spring will experience a downthrust, while the rear ends of the springs will be subjected to an upthrust. Thus the adjacent spring ends of the front and rear axles will tend to swing the balance-beam in a clockwise direction. Hence the front axle and wheels will tend to lift whereas the rear axle will be forced harder against the ground.

As a result, the rear axle and wheels will contribute more to the braking and retardation of the vehicle than the front axle. It also means that the front axle wheels will lose grip more easily and tyre scrub will be higher on these wheels.

7.11.2 Non-reactive bell-crank lever and rod suspension (fig. 7.48(a) and (b))

The unequal driving and braking torque distribution of the reactive balance-beam tandem axle suspension can be overcome by modifying the front and rear axle and spring support linkage in such a way that the torque reaction when driving or braking of both axles neutralises each other, hence the driving and braking torques are therefore equally distributed between both axles and their respective wheels. Equal load distribution is achieved by having both front spring ends of each axle attached to fixed hanger brackets, whereas the rear spring ends of each axle spring are attached via swing-shackles to a pair of bell-crank levers. These bell-crank levers are themselves pivot-mounted back to back on the side of the chassis just above the rear end of each axle spring and are interlinked by a horizontal connecting-rod.

Static load equalisation (fig. 7.48(a) and (b)) If one axle and wheels should roll over a hump its rear spring shackle will be lifted towards the chassis. This movement is then relayed via the interlinked bell-crank levers which both pivot in the same direction to the other axle spring rear end. This pushes down the second axle spring rear ends so that the axle wheels maintain firm ground contact. Conversely, if one axle and its wheels should roll in a dip, its rear spring shackle will move further away from the chassis. Correspondingly, via the bell-crank lever linkage the other axle spring rear ends will move closer to the chassis, and accordingly both axle wheels will be able to maintain their contact with the road surface.

Equal driving torque distribution (fig. 7.48(a)) When the vehicle is being accelerated the driving torque reaction on both axles will twist them clockwise. Accordingly the front spring ends of both axle springs will experience an upthrust, while the spring rear ends are subjected to a downthrust. This will result in the front and rear bell-crank levers trying to rotate in the opposite direction to each other, that is, anticlockwise and clockwise respectively. However this is prevented by the tensile pull F_t of the interconnecting rod. The outcome will be for both pairs of swing shackles to remain at similar heights from the ground. In this manner the driving torque distribution will be shared equally between axles.

Equal braking torque distribution (fig. 7.48(b)) As the vehicle is being braked, the brake torque reaction on both axles will tend to twist each axle in the opposite direction to the generated braking torque, that is, anticlockwise if the vehicle is moving in the left direction (see fig. 7.48(b)). However, the front end of each axle spring will experience a downthrust while the rear ends of the springs will be subjected to an upthrust. Consequently the front and rear bell-crank levers will attempt to rotate clockwise and anticlockwise respectively, that is, in the opposite direction to each other. Accordingly this is resisted by the compressive push F_c at each end of the interconnecting rod. As a result, both pairs of shackles will remain at the same height from the ground and will thus ensure that the braking torque distribution will be shared equally between axles.

7.11.3 High mounted single inverted semi-elliptic spring with upper and lower torque-rods (fig. 7.49)

This tandem axle suspension comprises both leading and trailing pairs of upper and lower torque-rods. These are attached by rubber-bush joints to the extended side members on each side of the chassis at one end and to the axle's vertical arm at the opposite end. A single inverted multi-leaf semi-elliptic spring is supported at its ends on axle slipper blocks, while the chassis is supported on a pivoting trunnion mounted over the centre of

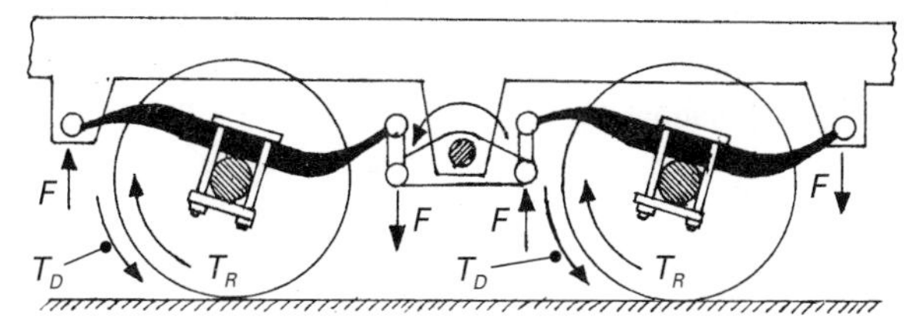

a) Unequal driving torque distribution

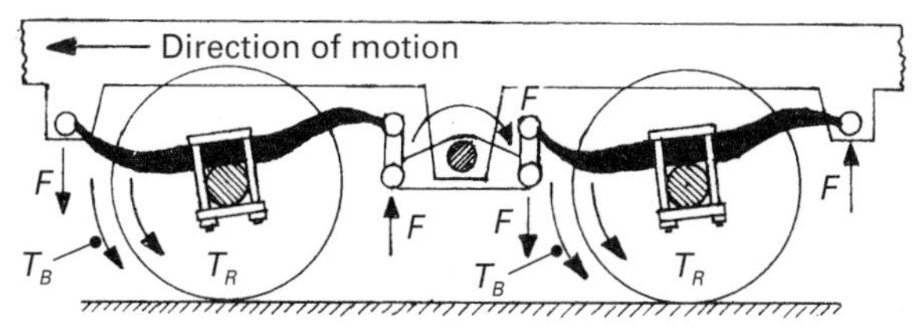

b) Unequal braking torque distribution

Fig. 7.47 Reactive balance beam tandem axle suspension

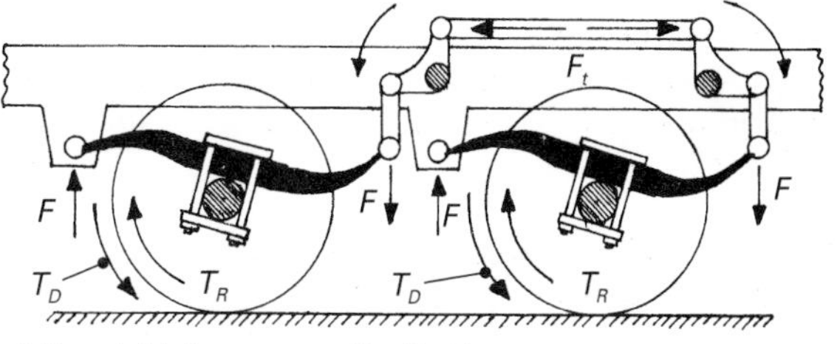

a) Equal driving torque distribution

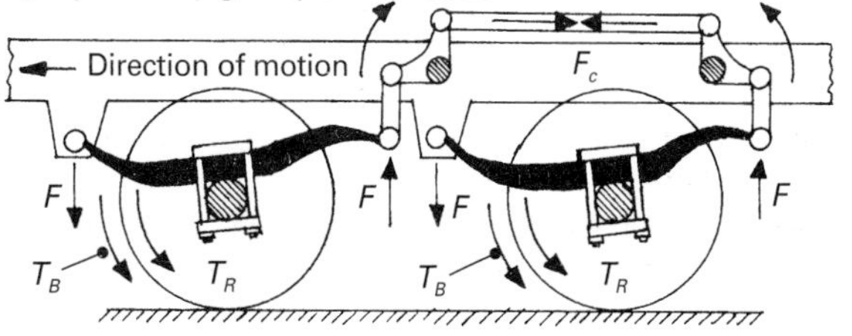

b) Equal braking torque distribution

Fig. 7.48 Non-reactive bell-crank-level and rod suspension

the spring. The upper and lower torque-rods absorb both the longitudinal forces and the driving and braking torque reaction, whereas the upper torque-rods on either side of the axles are mounted diagonally to prevent the lateral (side) forces from moving the chassis sideways relative to the chassis.

Static load equalisation (fig. 7.49) If one axle and its wheels travel over a hump or rut in the road the respective axle wheels are permitted by the swing of the torque-rods to follow the contour of the road surface, so that the axle moves nearer or further from the chassis. Simultaneously the centrally pivoting inverted springs swivel on their trunnion to compensate for the changing height of the axle relative to the chassis. Hence the single leaf span of the springs divides the vehicle's load equally between both axles.

Driving and braking load distribution (fig. 7.49) Both the driving and braking thrust and the reaction torque are absorbed by the upper and lower torque-rods, thus leaving the inverted spring just to deal with the vehicle's vertical payload alone.

7.11.4 Hendrickson long equalisation balance-beam with single semi-elliptic spring suspension (fig. 7.50)

This tandem axle suspension is designed for both on-road and off-road application. It consists of an underslung central pivoting balance-beam which is attached at each end by rubber-bush joints to both the lower axle arms. The upper axle arms are held in their longitudinal position by pairs of diagonally arranged torque-rods attached to the chassis which also restrain the lateral (side) forces, whereas the balance-beam mounted between the axles is attached to both lower axle arms. The weight of the vehicle is transferred to the balance-beam and axles through the semi-elliptic spring which is attached to the chassis by a shackle-pin and a fixed hanger at its front end, while chassis-mounted slipper hangers (one/two) rest on the main leaf at its rear end. Longitudinal thrust is transferred from the axles and balance-beam to the chassis through the front half of the leaf spring.

Static load distribution (fig. 7.50) A trunnion is bolted to the underside of the multi-leaf spring so that the latter is made to pivot at the centre of the balance-beam. It thus divides the static load equally between axles. Every time that one of the axles and its wheels travel over a bump or depression in the road the balance-beam will tilt to permit both sets of axle wheels to closely follow the profile of the road surface. This therefore ensures that the static load distribution is shared equally between axles at all times.

When the vehicle is unladen the extended spring leaf is in contact with the outer slipper bracket. This then provides the suspension with

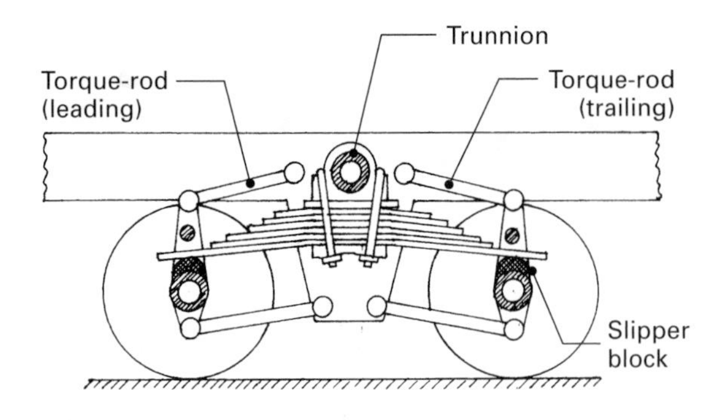

Fig. 7.49 High mounted single inverted semi-elliptic spring with upper and lower torque-rod suspension

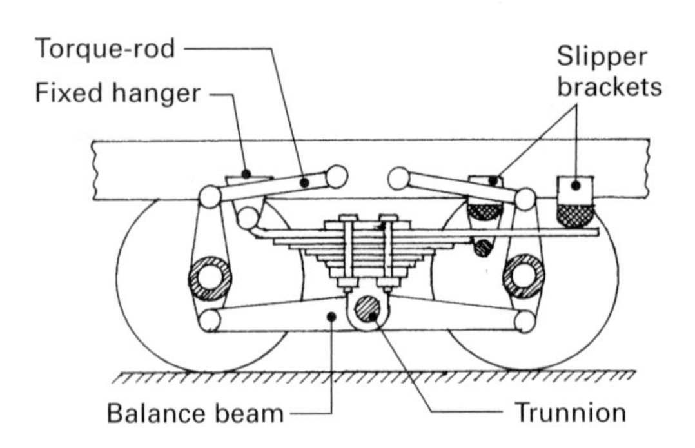

Fig. 7.50 Hendrickson long equalisation balance beam with single semi-elliptic spring suspension

a low spring stiffness, that is, a relatively soft ride. As the vehicle becomes laden, the main leaf deflects until it contacts the inner slipper bracket. This shortens the spring span, raises the spring stiffness, and thus provides a much harder spring to support the heavier loads.

Driving and braking load distribution (fig. 7.50) Driving and braking longitudinal forces are transferred from the axles to the chassis mainly through the front half of the leaf spring and partially via the leading and trailing upper torque-rods, whereas the driving and braking axle torque reaction is sustained by the pivoting balance-beam and the leading and trailing torque-rod chassis attachment joints.

7.12 Rubber spring mounted on a balance-beam with leading and trailing torque-rod suspension (fig. 7.51(a)–(d))

With this suspension, the vehicle's static load is transferred from both sides of the chassis to the mid-region of the underslung balance-beam via pairs of inclined rubber springs. This divides the load equally between the ends of the balance-beam which are themselves attached to the axle's lower arms (see fig. 7.51(a)). Driving and braking longitudinal pull and push is conveyed through the rubber springs to the chassis, whereas the upper leading and trailing torque-rods prevent the axles twisting due to the resisting torque reaction. Lateral forces are accommodated by positioning diagonally the leading and trailing torque-rods. They are attached by rubber-bush joints to both chassis side members at one end but merge to form a single spherical rubber pivot joint at the axle ends. Rubber springs of this kind offer good body roll resistance and stability.

When one of the axles travels over a hump or dip, the balance-beam tilts to permit one or the other of the axles to rise or fall relative to the horizontal chassis, and at the same time the torque-rods swing on their chassis side-member mounted pivots to accommodate the articulation of the balance-beam. The actual spring to balance tilt is catered for by the pivoting load transfer members and the interconnecting link of the rubber springs.

Rubber spring construction and characteristics (fig. 7.51(b) and (c)) The springs are made from a number of circular moulded rubber blocks bonded to and sandwiched between steel supporting plates (fig. 7.51(d)). These steel plates divide the rubber spring into a number of sections. They therefore provide lateral support and prevent bulging of the rubber without interfering with the vertical spring stiffness at any point in their deflection range. The rubber spring pack is held in an inclined position to something like 30° to the vertical between the upper chassis side member and the load transfer member, as this subjects the rubber to both shear and compression strain.

When in the unloaded state the weight on the spring distorts the rubber mainly in shear strain corresponding to a soft low spring stiffness (see fig. 7.51(b)). As the weight on the spring increases, the distortion of the rubber changes from shear to a compressive strain, thus providing a much harder high spring stiffness (see fig. 7.51(c)). Note rubber in shear deflects readily but in compression the rubber becomes more compact and therefore is more difficult to squeeze together. Thus from unladen to fully laden the spring stiffness progres-

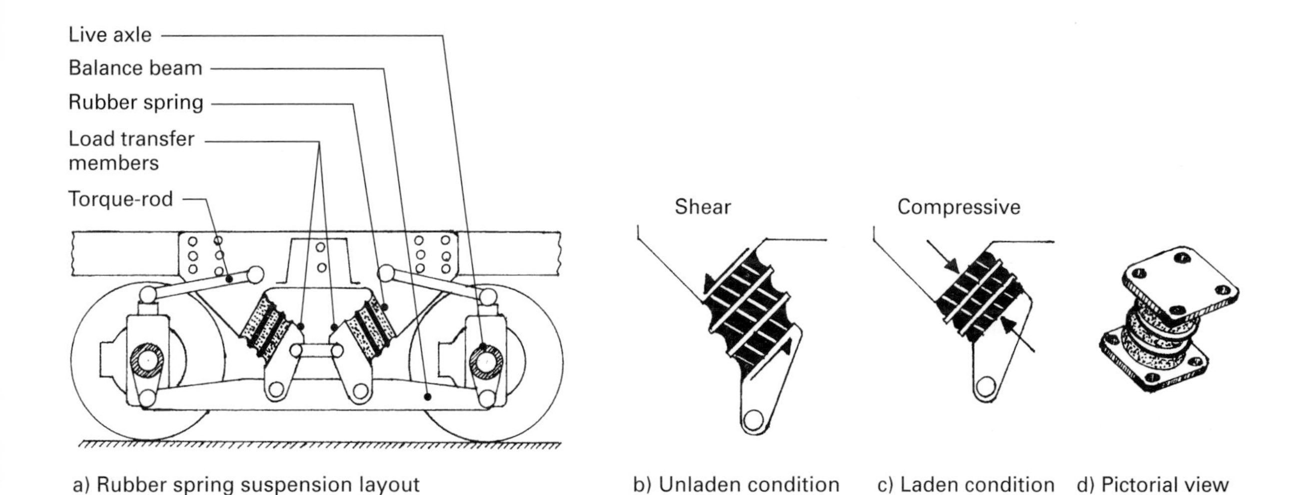

Fig. 7.51 Rubber spring mounted on balance beam with leading and trailing torque arms

sively rises to support the heavier loads. The spring stiffness therefore varies to match the changing load and bounce response as the axles rise and fall in their attempt to follow the contour of the road surface.

Springs of this type usually have only a small vertical deflection from the unloaded to the fully loaded state, but they are lighter than a conventional steel spring layout, are extremely reliable in service, and offer good ride quality.

7.13 Air spring suspension (fig. 7.52)

Air spring suspension have become popular with heavy goods vehicles, coaches and buses in recent years, due to the following reasons:

1) the suspension bounce frequency between unladen and laden operating conditions is much narrower and is much nearer to the ideal frequency for human comfort than conventional steel springs can achieve;
2) it protects fragile loads;
3) it lowers stresses imposed on the vehicle because high peak axle loads are eliminated;
4) constant vehicle ride height irrespective of whether the vehicle is unladen or fully laden;
5) air suspension generates less noise than steel leaf springs in both the laden and particularly in the unladen condition;
6) air spring bag life is as good or even better than for steel leaf springs;
7) air spring suspension prices are only marginally higher than for conventional steel springs;
8) maintenance is minimal and replacement units are easy to fit.

Air springs (fig. 7.52) Air springs, or air-bags as they are commonly called, are usually of the rolling lobe (diaphragm) type, so-called because the flexible rubber walls of the bags roll over its piston. The cylindrical rubber walls are sealed at the upper end (top) by a fixed plate with rolled edges in which the moulded neck of the bag snugly fits, whereas the neck of the air-bag at its lower end (bottom) sits over the inward tapering portion of the piston crown and is made air-tight and held in position by a clamping plate bolted on top of the piston. A bump-stop is installed underneath the upper fixed plate, so if there is a loss of air pressure in the bag, the bump-stop comes to rest on top of the piston, thus allowing the vehicle to continue its journey, though at a much reduced speed. As the vehicle travels over the normal irregularities of the road surface the trapped elastic compressed air spring media contracts and expands to cope with the bounce and rebound caused by the continuous rise and fall of the axle. The air spring is designed to operate with a continuously varying spring bag height by adopting the rolling lobe principle. When the axle drives over a bump, the lower spring member moves up. This makes the flexible rubber walls roll over the skirt of the piston forming the lower spring member. Conversely on rebound the sagging overlap of the rubber walls now rolls up the piston skirt, compensating for the extension of the spring bag height. By these means the air spring is permitted to continuously change its height while subjecting the air-bag walls to the minimum of stress over a very long life span.

Air pressure is supplied to the air-bags from the air braking system which may be as high as 8 bar, but the supply pressure is reduced before it is introduced to the air-bag to something like 5 bar when the vehicle is fully laden, and decreasing to about 2.5 bar in the unladen state. Rolling-lobe air-bags can operate at bounce frequencies as low as 80 cycles per minute fully laden, rising to roughly 100 cycles per minute when unladen.

Axle suspension (fig. 7.53) Trailing arms arranged on either side of the chassis are commonly used to support the air-bag springing. These arms are made to pivot on low-slung hanger brackets by means of metallastic joints. The length of these arms is made to extend beyond the axle to support the air-bags. The axle is mounted on the trailing arms ahead of the air-bag by means of metallastic pivot joints, while the air-bags are positioned as far back as possible from the axle, as this provides the maximum leverage for the resisting upthrust of the air spring.

Driving thrust is transferred from the axle to the chassis via the trailing arm and hanger bracket, whereas the driving and braking axle reaction torque twist is prevented by the upper diagonally positioned torque-rods which also restrain any sideways movement of the chassis when the body is subjected to lateral (side) forces.

Body roll stability (figs 7.52 and 7.53) Conventional antiroll bars are generally incorporated into the trailing arm suspension between the chassis and the axle-beam to increase the body roll resistance when negotiating a bend in the road (not shown in fig. 7.53).

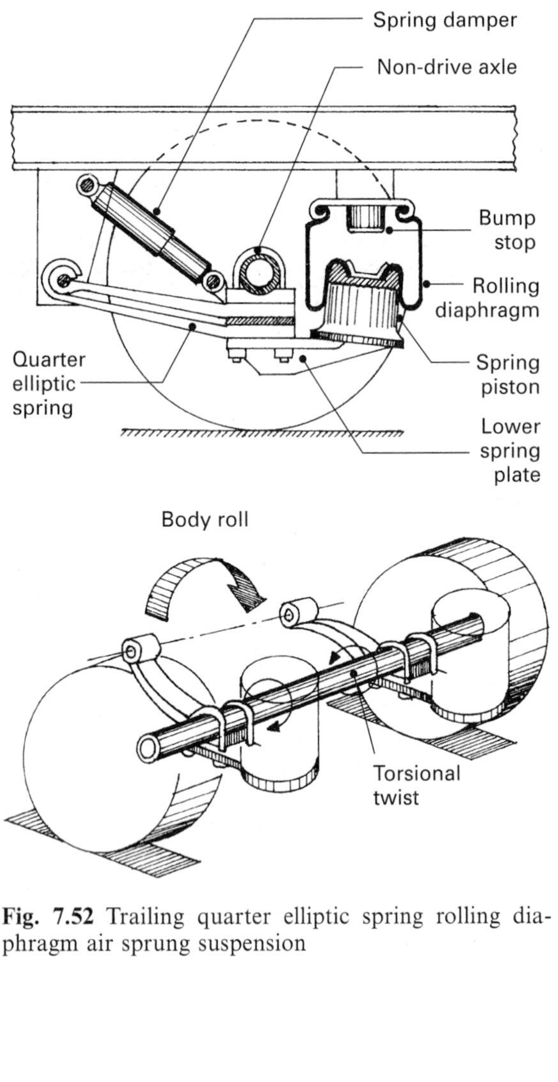

Fig. 7.52 Trailing quarter elliptic spring rolling diaphragm air sprung suspension

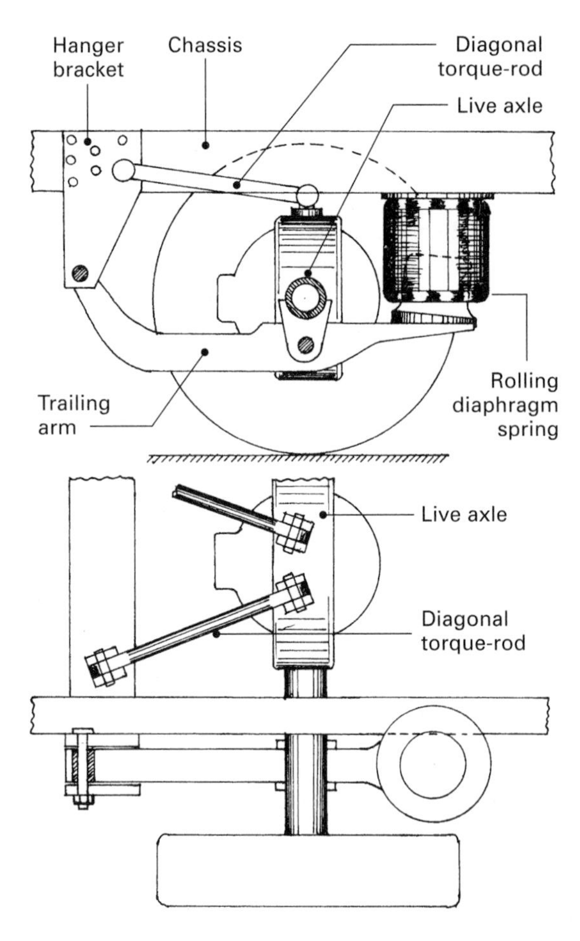

Fig. 7.53 Trailing arm rolling diaphragm air sprung suspension

An alternative non-drive air spring suspension arrangement utilises trailing quarter-elliptic leaf-springs hinged on a low slung hanger bracket, with the tube-sectioned axle-beam bolted rigidly to the swinging ends of the trailing leaf-springs (see fig. 7.52). When the vehicle travels round a bend in the road, the high centre-of-gravity of the body and payload relative to its roll-centre causes it to sway or roll about its roll axis. However this is resisted by the deflection of the quarter-elliptic springs and by the torsional resistance offered by the axle-beam tube. The axle and trailing spring leaves therefore virtually perform the role of an antiroll bar.

Shock absorber damper (fig. 7.52) Pairs of inclined shock-absorbers spanning both sides of the chassis result in low damper piston speed and minimum heat generation and therefore reduce wear. The shock-absorbers ensure that the tyres remain in contact with the road surface and wheel hop is virtually eliminated even when operating in the unladen state.

Compressed air operated air spring system (figs 7.54 and 7.55) Compressed air from the air brake system is supplied through a protection-valve to the air-bags via one or two suspension levelling valves which are mounted on the chassis or hanger-bracket and are actuated by a rod linkage attached to the axle beam (see fig. 7.54). These valves sense the height of the chassis relative to the axle, and automatically adjust the air supply to maintain the set level. On each side of the chassis the air-bags in a multi-axle suspension are interconnected to give static and dynamic axle load equalisation by ensuring the air-bag spring pressure on each side of the vehicle are equal no matter what degree of individual air-bag deflection is experienced (see fig. 7.55).

Brake distribution in proportion to axle load is achieved via a load-sensing valve. This valve is controlled automatically by air-bag pressure which changes with load. Accordingly it varies the braking pressure applied to the brake actuators of each axle.

7.14 Road-wheels

Wheel construction must fulfil a number of objectives as below.

Structure Wheels should be rigid, so that they retain their shape under all operating conditions. If they are subjected to abnormal impact, they must not shatter or collapse but should preferably buckle. The dimensional tolerances of the wheel should be sufficiently accurate to provide wheel alignment for the wheels to be balanced.

Weight Wheels should be made as light as possible, to reduce the unsprung weight. Light wheels and tyres will follow the road surface irregularities more accurately and so minimise wheel bounce, which should result in improved road-holding and reduced tyre wear.

Tyre attachment The wheel must be designed so that the tyre can be fitted easily and be firmly located and secured. This is necessary since the wheel–tyre combination forms the path for the transmission of traction to the road or for steering reaction.

Wheel mounting The wheel attachment must perform the job of locating, securing, and supporting the wheel, and should allow the wheel to be easily fitted or removed from its axle-hub.

Cost Wheels should be made out of materials which are low in cost, which can easily be fabricated, cast, or forged, and which then require the minimum amount of machining to bring the wheels within the necessary dimensional tolerances. The chosen material should not rapidly deteriorate with age and weathering; or, if it is susceptible to corrosion, it must be provided with protection treatment which should also improve its finish and appearance.

7.14.1 Pressed-steel-disc wheels (fig. 7.56)

These wheels consist of a formed disc pressed into a rolled-section well-base rim and retained in position by spot welding.

The basic rim for the car disc-type wheel is formed from flat steel strip by a series of rolling operations. The heavier commercial-vehicle rims are made from bars in which the section has been formed by hot rolling. The steel strip is cut or cropped to the required length and then rolled into a circle before being butt-welded. The circular steel strip is then pressed into the correct profile shape by means of several rolling operations, to emerge as a completed rim. The rim is then expanded to size and the valve hole is punched.

The well in the base of the rim is deep enough to receive the beads of the tyre for mounting or demounting, so that the stiff beads can be passed

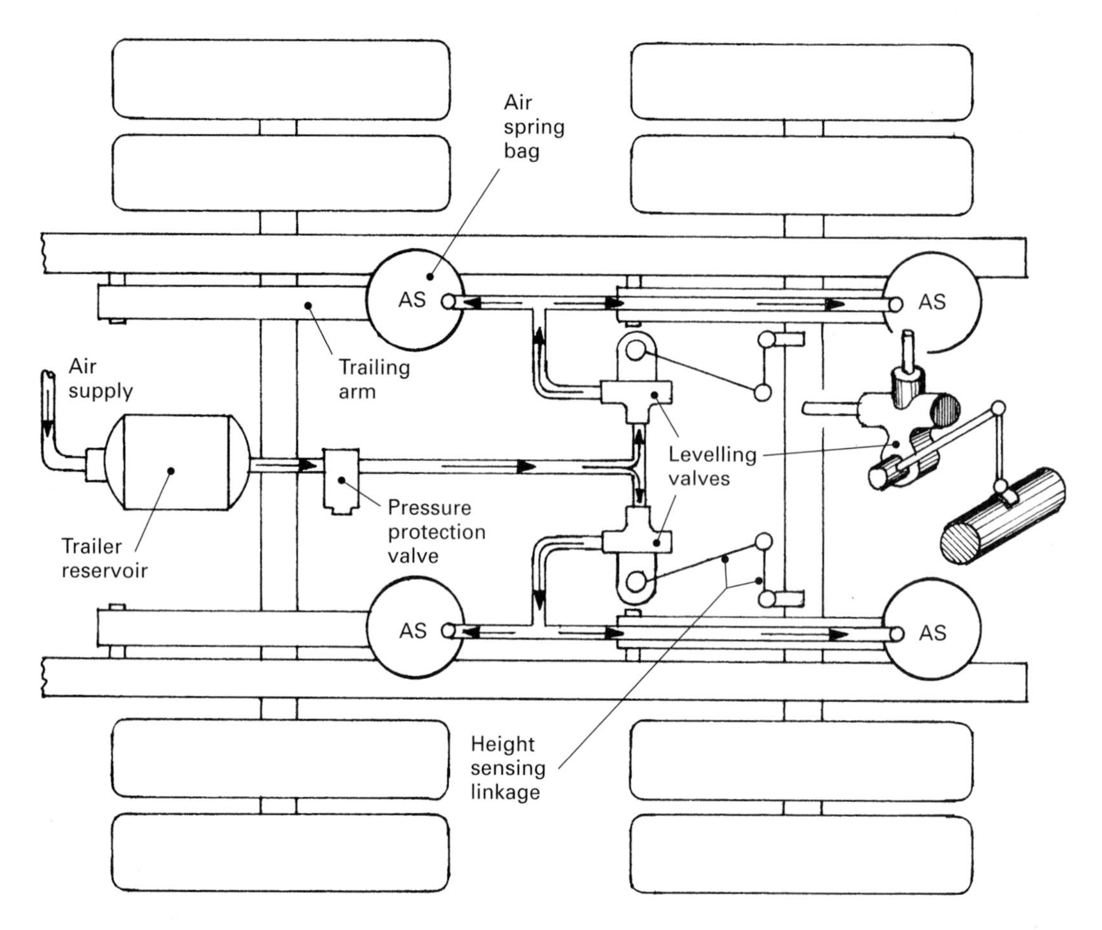

Fig. 7.54 Tandem axle air spring suspension air supply system circuitry

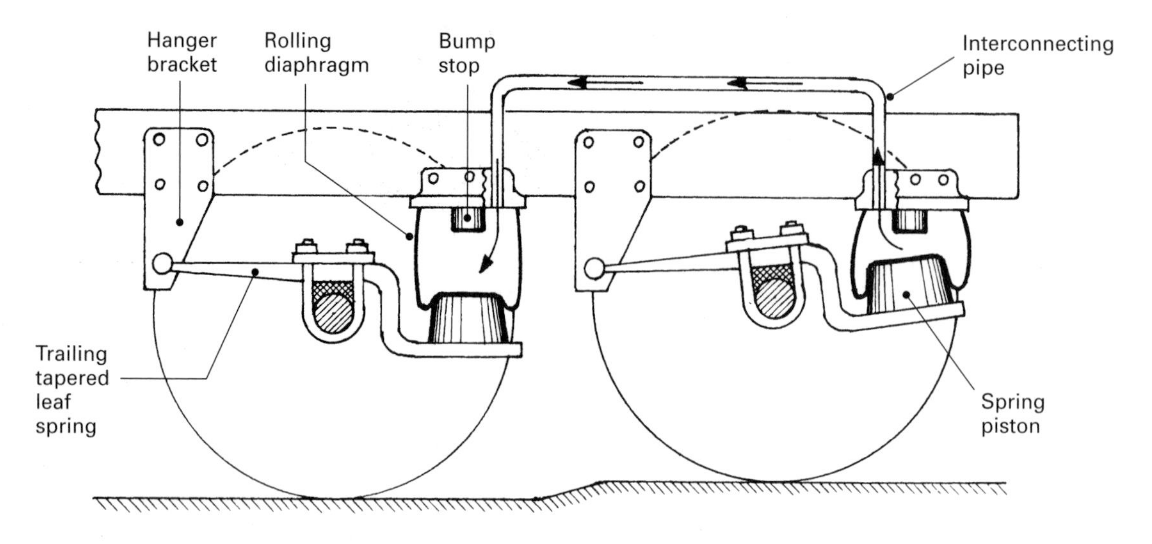

Fig. 7.55 Tandem axle air spring suspension axle load equalisation

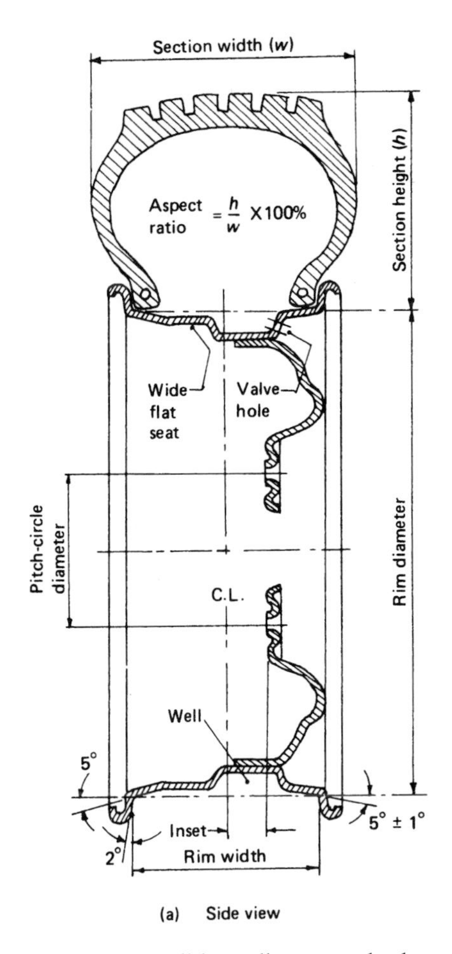

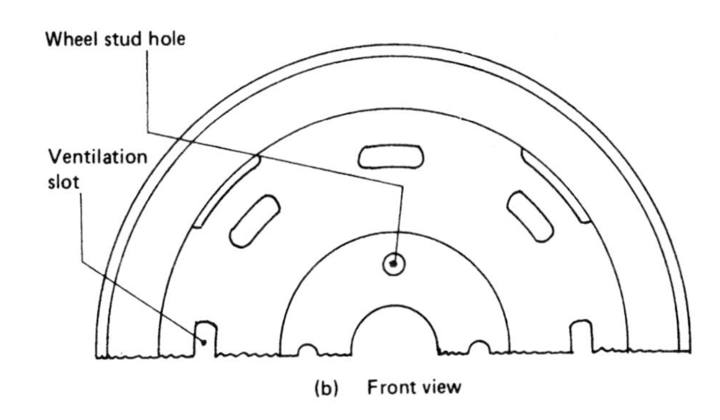

Fig. 7.56 Car well-base disc-type wheel

over the edge of the rim. The seats of the rim upon which the tyre sits have a 5° taper so that, as the tyre is inflated, the beads are forced up the taper to give a wedge fit, and with tubeless tyres a good seal is made.

The disc itself is manufactured from an octagonal blank of even gauge which, when formed, contacts the rim in a number of equal-spaced arcs. The disc pressing has a number of equal-spaced shallow elongated slots immediately under the base of the rim well (figs 7.56(b) and 7.57(a)). These allow the passage of air from one side to the other, which improves brake cooling and lessens the transfer of heat from the brake-drum to the tyre.

The disc is attached to the axle-hub by a series of studs and nuts with conical or spherical seatings which locate in depressions in the disc.

Car well-base rims (figs 7.56(a) and 7.57(a)) Wheel rims are now designed to reduce the possibility of the tyre becoming dislodged from the rim when subjected to heavy cornering.

Originally the rim was tapered from the rim flange to the edge of the well, and it was possible for the tyre bead to be forced back into the well by pressure acting over a short length of the bead. Such a situation could occur at the bottom of the rim, due to the interaction of the cornering side-force acting on the wheel and the reaction of the tyre and the road.

Safety rims of the flat-ledge rim profile (fig. 7.56(a)) help to prevent this, by providing a parallel flat ledge between the outer taper adjacent to the flange and the edge of the well. With a close-fitting tyre bead, only pressure acting all the way round the bead can dislodge the bead into the rim well – if pressure is applied to one narrow section, all that will happen is that the bead will jam and so prevent separation.

A further approach for improving the bead rim seal and joint is to form a hump slightly in from

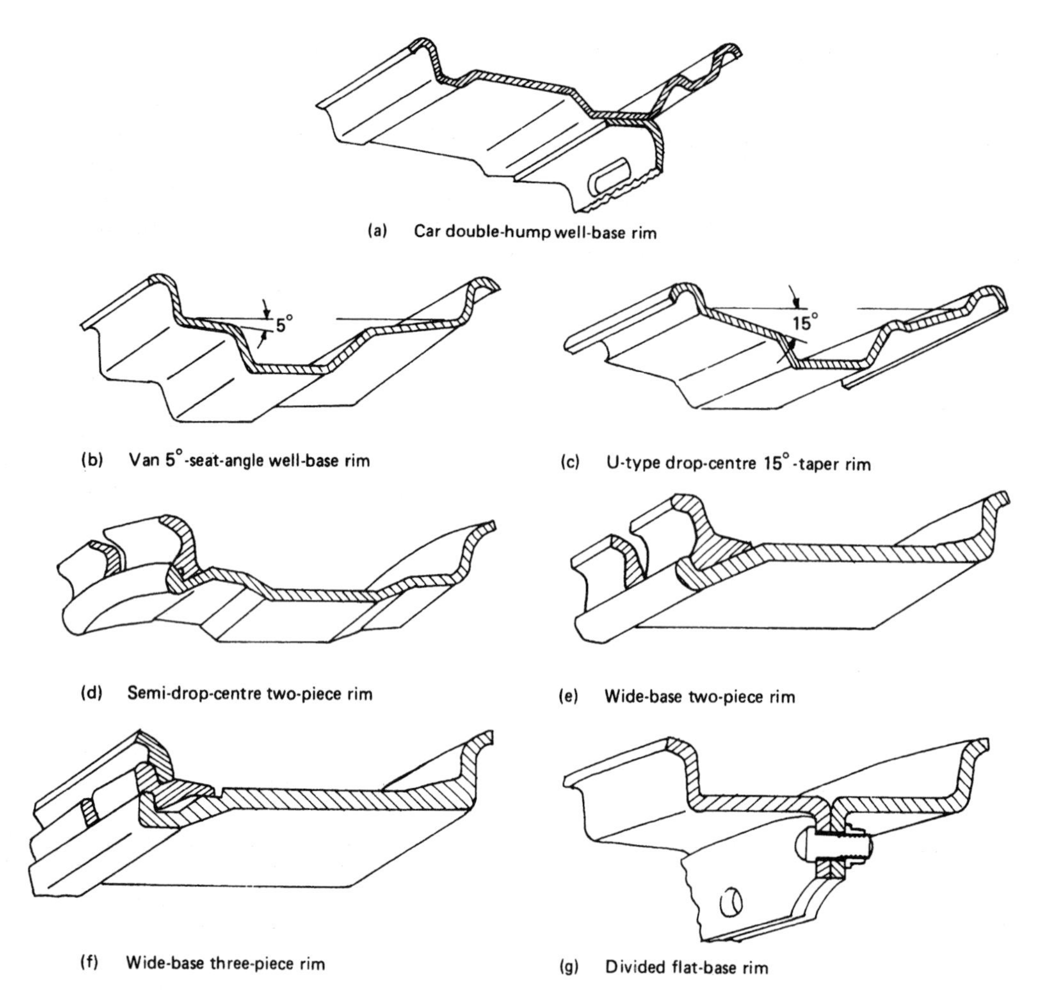

Fig. 7.57 Vehicle wheel rims

the shorter taper next to the rim flange (fig. 7.57(a)). Once the bead has been positioned over the short taper portion adjacent to the rim flange, it will be very difficult for it to slip or climb over the hump, back into the rim well.

Van 5°-seat-angle well-base rim (fig. 7.57(b)) These rims are an extension of the range of car-type rims, having high wall flanges to protect the tyre beads and walls from external interference and damage. These rims are generally used for both tubed and tubeless tyres, but large stiff tyres cannot be mounted over these flanges, so this form of rim is only suitable for vans and small commercial vehicles.

U-type drop-centre 15°-taper rim (fig. 7.57(c)) Tubeless tyres and large-section tyres with relatively flexible beads are increasingly becoming popular for vans, buses, and medium-sized trucks. These applications use a single-piece well-base wheel rim, with relatively shallow flanges at the edge of 15°-taper bead-seats. This profile allows the mounting and dismounting of the tyre from the wheel, and provides a sound joint and seal between the tyre bead and the rim taper.

7.14.2 Detachable-rim wheels

Large tyres of commercial vehicles demand more plies in their casing and are therefore correspondingly bulkier in the bead region. Such tyres are therefore fitted on rims which have one side-flange which is removable, so that the wheel tyres can slide into position and the flange can then be replaced and locked in place.

Semi-drop-centre two-piece rim (fig. 7.57(d)) This type of rim is made up of inner and outer tapered tyre-bead-seating surfaces separated by a shallow central or near central well. The inner flange is removable for tyre-fitting purposes. With this design, the well depth is sufficient only to allow the tyre beads to pass over the top of the outer beadseat taper. The outer detachable flange is retained in its working position by being sprung into a continuous groove formed along the outer edge of the rim base.

This rim is an intermediate between the well-base and wide-base rims. It is used for heavier beads which are too rigid for fitting on the full well-base type. A typical application is for light trucks.

Wide-base two-piece rim (fig. 7.57(e)) This wide-base rim has one fixed flange integral with the rim base and one split detachable flange. The rim has 5°-taper seats for tyre beads, the one on the detachable-flange side being integral with the flange. The outer detachable flange is retained in its working position by being sprung into a continuous groove formed along the outer edge of the rim base. When the tyre is inflated, the bead sits over the detachable-flange taper, holding it in position. The main use of this rim is on medium-size commercial vehicles.

Wide-base three-piece rim (fig. 7.57(f)) This wide-base rim has one fixed flange integral with the rim base, one detachable endless flange, and a separate flange-retaining split lock-ring. The rim has 5°-taper seats for tyre-bead location, the one on the detachable-flange side usually being on an extension of the spring lock-ring. When the tyre is inflated, the bead will sit over the extension of the spring lock-ring, holding it in position. This form of rim is used on large commercial vehicles.

Divided flat-base rim (fig. 7.57(g)) In this design, the rim is integral with the wheel itself. Tyre fitting and removal are carried out by dividing the two halves of the wheel by dismantling the outer ring of bolts which hold the wheel halves together. These rims are used primarily for large trucks used as military vehicles, where run-flat tyres embodying a compressible bead-spacer are employed.

7.14.3 Cast and forged alloy wheels

These wheels are produced as a single-piece rim and disc. Car wheels are generally shaped by casting or extrusion, but forging is preferred for truck wheels.

The major advantage of light-alloy wheels is their reduced weight. Magnesium and aluminium alloys are the two most commonly used – magnesium alloy has a weight saving of 30% over aluminium alloy and 50% over steel for structures of similar strength. Magnesium alloys have very good fatigue properties and excellent resilience, so they resist vibrational and shock loading better than either aluminium alloy or steel, but they are highly susceptible to corrosion and must be given a protective surface coating which must not be damaged. Aluminium alloys do not have such good fatigue properties but they corrode less readily and are easier to cast or forge. The section thicknesses for the rim and disc have to be greater with light alloys than with steel, but this provides a rigid-structured wheel which is desirable.

After casting or forging there is the disadvantage of having to machine the wheel rim and the studhole flange, but this has the benefit of producing close-tolerance wheels. Finally, most light alloys tend to be better conductors of heat than steel, so any heat generated by the tyre or brake is more readily transferred to the wheel-disc where it is dissipated by the air stream.

Light-alloy wheels are more expensive to produce than pressed-steel wheels, and aluminium-alloy wheels are cheaper than those made of magnesium alloy. Light-alloy wheels are used when appearance and weight are of importance. Usually an aluminium alloy is preferred for passenger cars and trucks, but sports and racing cars tend to use a magnesium alloy. The wheel-rim profile used with light-alloy wheels is similar to that employed with steel rims.

7.14.4 Wheel stud and nut fixtures

Car conical-nut mounting (fig. 7.58(a)) Steel wheels used for cars are mostly aligned with conical-taper nuts which fit into the wheel-disc countersink stud holes. The taper centralises the wheel to the hub axis and provides a wedge action to the nuts when they are tightened. The stud and nut then locate, secure, and hold the wheel.

Spigot mounting with conical set-bolt (fig. 7.58(b)) In recent years there has been some trend towards spigot-mounting the wheel on the axle-hub. This accurately locates and partially supports the wheel, while the conical-shaped set-

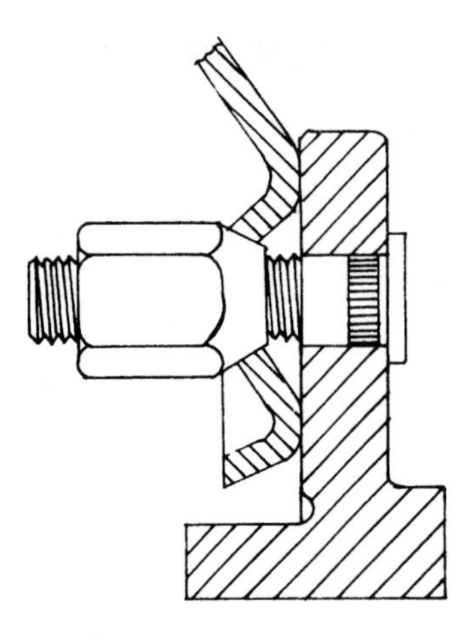

Wheel nuts – conical taper

Stud holes – conical countersink
in steel wheel-pressing

Location – by nuts seating
in stud-hole countersink

(a) Nut with
conical seat

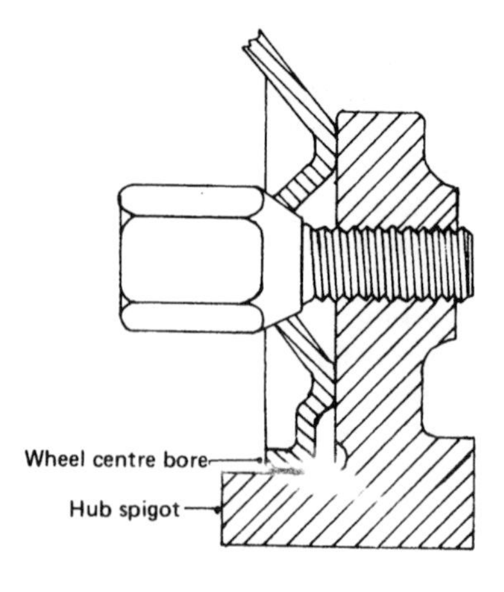

Wheel set-bolt – conical taper

Set-bolt holes – conical countersink
in steel wheel-pressing

Location – close fit of centre-bore
on spigot

(b) Spigot-mounting
with conical set-bolt

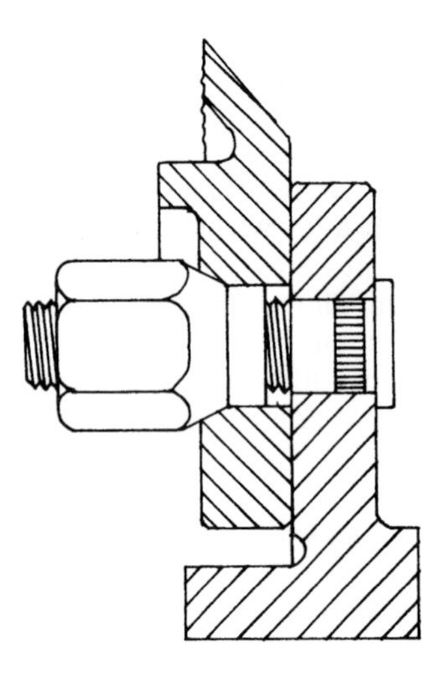

Wheel nuts – conical with parallel fitted shoulder

Stud holes – Conical countersink
with parallel bore in cast light alloy

Location – Close fit of centre-bore on nut shoulder,
wedged by conical-nut seat

(c) Nut with fitted
shoulder and
conical seat

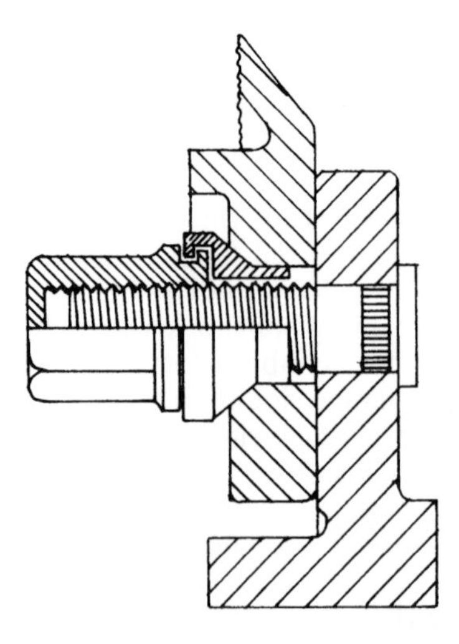

Wheel nuts – plain with conical
and parallel integral sleeve

Stub holes – conical countersink with
parallel bore in cast light alloy

Location – close fit of centre-bore on parallel sleeve.
wedged by conical-sleeve seat

(d) Nut with fitted
sleeve and
conical seat

Fig. 7.58 Passenger-car wheel fixings

bolts secure the disc on to the hub and transmit both the driving and braking torques. This method may be used with both steel and light-alloy wheels and is becoming common on small and medium-sized cars.

Car shouldered-nut mounting (fig. 7.58(c)) With light-alloy car wheels, the softer metal will distort or tear away at the stud holes when the nuts are tightened, and this will tend to enlarge these holes. An improvement in maintaining a concentric wheel is achieved by extending the conical end of the nut with a parallel shoulder or sleeve portion which is a close fit with the stud hole. The nut then locates, secures, and supports the wheel.

Car sleeved-integral-nut mounting (fig. 7.58(d)) A further improvement can be made by incorporating a separate sleeve which has a conical-and-parallel profile. This sleeve is free to rotate relative to the hexagon nut and prevents tearing between the alloy wheel and the stud nut during tightening.

Commercial-vehicle British Standard AU 50 wheel mounting (fig. 7.59(a)) Commercial-vehicle steel wheels can be mounted to the axle-hub flange by conical-taper nuts fitting into countersink stud holes. If twin wheels are employed, the wheel stud-hole flange next to the drum is located by a loose conical-taper-shaped washer placed over the stud or by an integral spherical-shaped seat on the stud. The nuts locate, secure, and support the wheel.

Commercial-vehicle Continental (DIN) standard wheel mounting (fig. 7.59(b)) Another method of attaching the wheel to the hub is by using plain nuts with split spherical-faced washers fitting into spherical countersinks formed in the wheel stud holes. The spherical seats centralise the stud holes, and the relative motion is taken mainly between the washers and the nuts. The centre-bore of the wheel flange does not support the wheel load. When twin wheels are used, a spherical washer or seat supports the inner-wheel flange.

Commercial-vehicle spigot mounting (fig. 7.59(c)) With spigot-mounted fixing, the centre-bore is machined to provide a close-tolerance fit on the hub spigot for location and load-carrying purposes. The wheel nuts have an integral flat-face washer which securely holds the wheel to the brake-drum, and the stud absorbs the acceleration and braking torques. The same mounting arrangement applies when wheels are used in twin rear combinations, the support being between the centre-bore of the wheel stud-hole flange and the hub spigot. This method of wheel mounting is becoming popular for large trucks.

Right- and left-hand-thread studs and nuts A few manufacturers are still producing wheel studs with right- and left-hand threads. When these are used, the 'near-side' wheels have left-hand threads and the nuts are marked 'L'. The 'off-side' wheels have right-hand threads and the nuts are marked 'R'.

It is claimed that, if the wheel nuts were screwed up loosely, the forward-rotating wheel direction would tend to tighten the nuts using the right- and left-hand thread convention. Conversely, when using right-hand threads on the near-side there would be a tendency for the nuts to unscrew.

7.15 Pneumatic-tyre construction

To appreciate the way in which pneumatic tyres have developed and are designed, an understanding of the requirements of the tyre is essential.

The pneumatic tyre performs the following functions:

a) to support the vehicle load,
b) to cushion the wheel against small road shocks,
c) to transmit driving and braking forces between the wheel and the road surface,
d) to convert steering effort into directional movement.

Tyres may be broadly divided into tubed and tubeless, cross-ply and radial-ply constructions.

7.15.1 Inner tubes (fig. 7.60(a))

A tyre inner tube is an endless flexible container tube which when inflated will expand indefinitely like a balloon until it bursts – unless it is restrained from further expansion by a casing known as the tyre. The inner tube then forms a separate air bag which is protected by an outer cover of textile material and rubber. The sole purposes of this inner tube are to take up the shape of the tyre cover when inflated with compressed air and to contain this trapped air over long periods.

The tube is made up from extruded sheets of rubber which are shaped on a former and cured

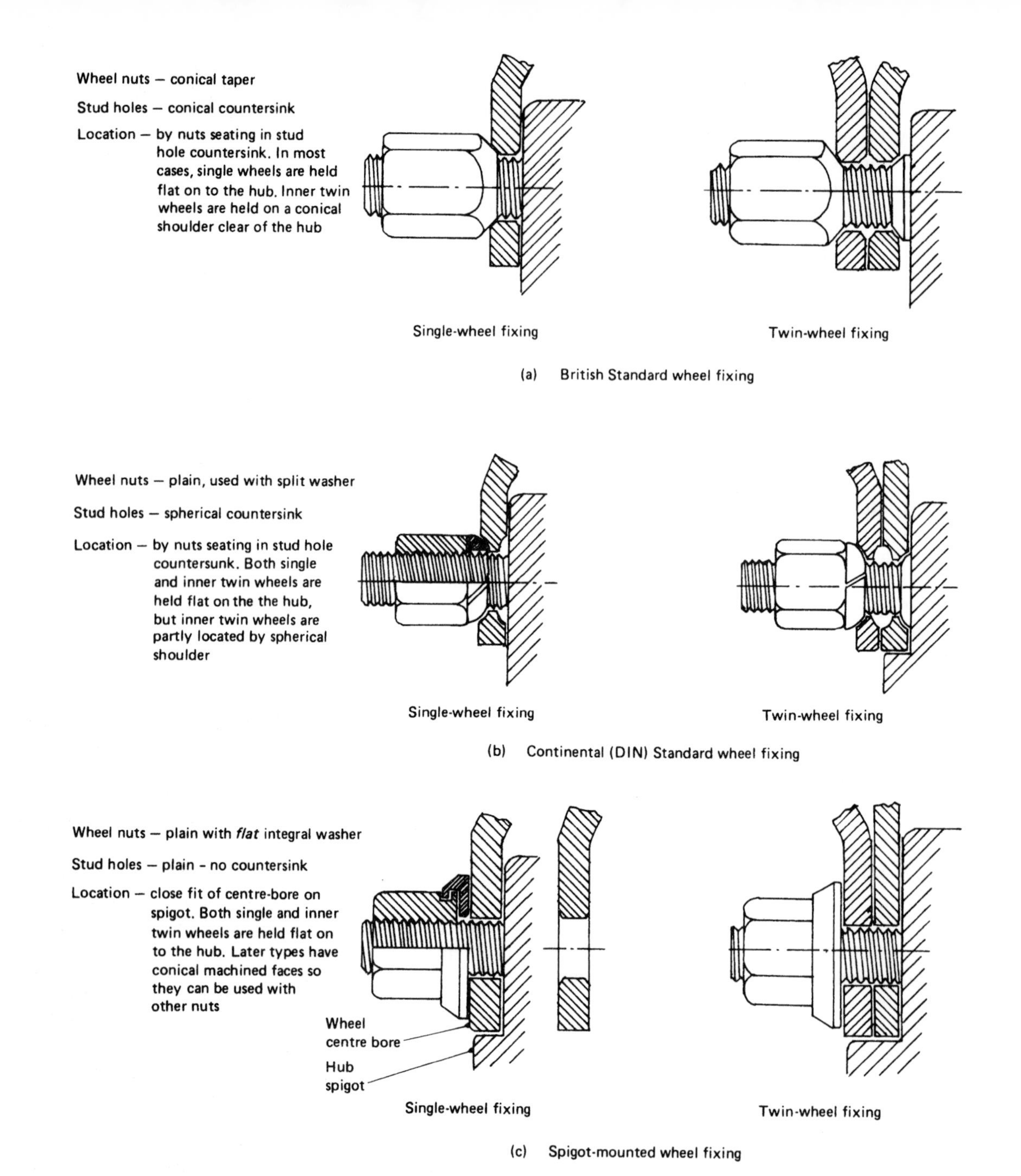

Fig. 7.59 Commercial-vehicle wheel fixings

(vulcanised) in a steam-heated mould to the cover shape, but about 10% undersized. The only outlet or inlet to the tube interior is through a valve stem sealed to the inside circumference during vulcanisation.

The use of inner tubes is now restricted to a few large heavy vehicles.

7.15.2 Tubeless tyres (fig. 7.60(b))

Essentially the tubeless tyre is an outer cover in which the inner tube is replaced by an unstretched rubber lining bonded to the inside of the cover wall and continuing under the beads or edges of the cover to join up with the outer side-wall rubber. The bead portion of the tyre is so constructed

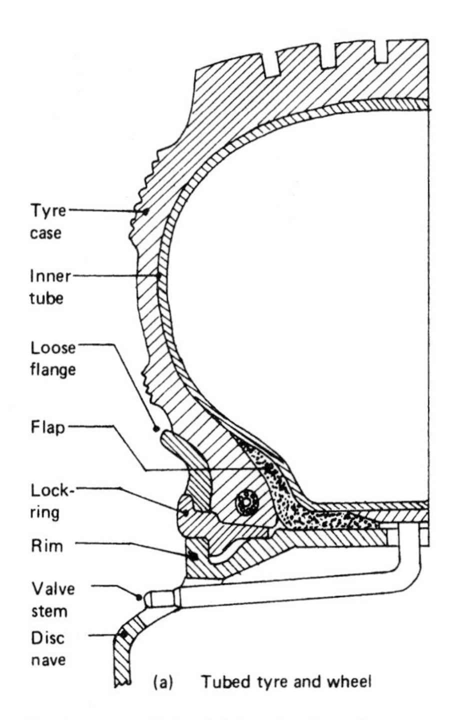

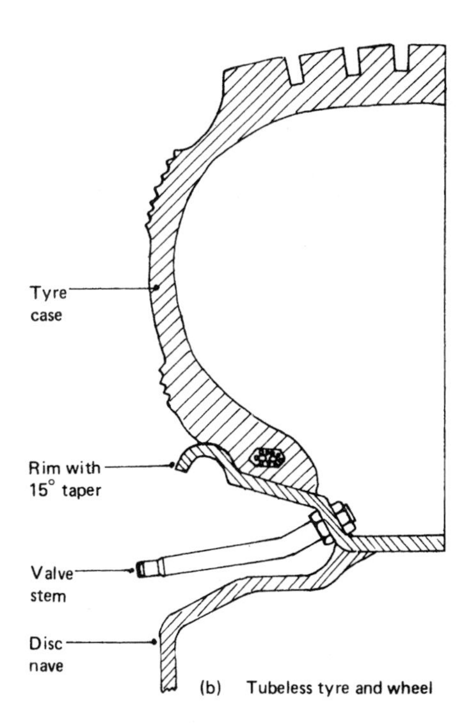

Fig. 7.60 Commercial-vehicle wheel and tyre construction

that it makes an effective air-tight seal between the tyre and the rim. Initial sealing between the bead and the wheel rim when inflating the tyre is assisted by the taper rim seats on each side of the well base and adjacent to the rim flanges. Car and commercial vehicle rims tend to have taper angles of 5° and 15° respectively.

With tubeless arrangements, the tyre valve is usually fitted to a hole formed in the well base of the rim. Car valves are sealed by rubber grommets, but commercial vehicles tend to use all-metal threaded valves secured to the rim by nuts.

7.15.3 Merits of tubeless over tubed tyres

The benefits of using tubeless rather than tubed tyres can be summarised as follows.

a) *Improved air retention* With ordinary inner tubes the rubber is stretched when in use, and this stretching permits the air to leak through the wall more quickly. With the tubeless tyre, the bonded inner lining which replaces the inner tube is in no way stretched, and the air also has to pass through the cover wall itself, so that air loss is very much slower. The seal provided between the tyre bead and the rim itself also allows no leakage of air around the tyre rim.
b) *Improved safety* With tubed tyres, many bursts are caused by the tube chafing against a fabric bruise, which can suddenly puncture without warning. With a tubeless tyre, a slow loss of air develops if a tyre is ruptured – giving warning in time to avoid serious trouble.
c) *Improved ride comfort* The tubeless tyre is lighter than the tubed combination, so the reduced unsprung-mass reaction on the vehicle's suspension system reduces wheel bounce.
d) *Improved tyre cooling* Without an inner tube, the heat generated in the compressed air can be dissipated directly through the metal rim. With an inner tube, the heat passes through the rubber inner tube, which is a relatively poor conductor of heat, thus there will be a greater temperature build-up.
e) *Improved wheel-assembly balance* A commercial-vehicle tubeless tyre does not use a tube flap, loose flange, and lock-ring, but instead the tyre bead fits concentric to the taper wheel rim. This provides a better balanced assembly.
f) *Improved assembly of tyre to wheel* With a tubeless tyre, only the cover has to be fitted over the wheel rim – there is no danger of loose components flying off during inflation due to distortion or carelessness, or of the inner tube being nipped or punctured during assembly.

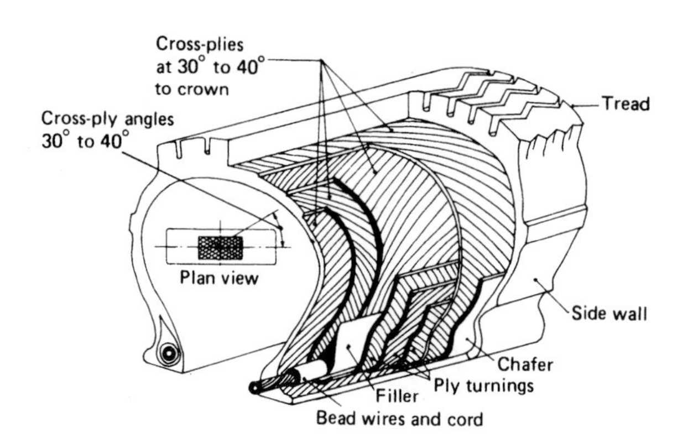

Fig. 7.61 Cross-ply tread construction

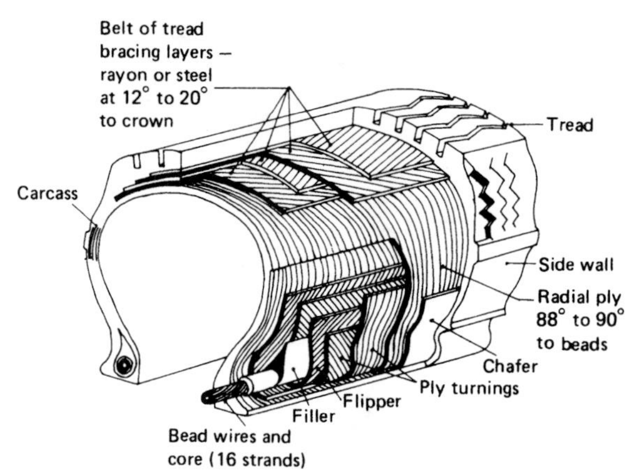

Fig. 7.62 Radial-ply tread construction

*7.15.4 **Tyre cover construction*** (figs 7.61 and 7.62)
The major components of a tyre cover are the carcass, beads, side walls, and tread.

Carcass This consists of the inner horseshoe-shaped lining made up of a number of layers of textile cord plies. The carcass forms the backbone of the tyre construction – the tread, bead, and walls all being moulded on to these cord plies.

Bead This is the inner edge of the tyre, and its purpose is to locate and centralise the cover on to the wheel rim. It also provides the rigidity and strength which is necessary to support the carcass. This is assisted by having an endless wire core moulded circumferentially through the bead.

Side wall This is the rubber outside covering of the carcass between the bead and the tyre thread. Its thickness will determine the degree of protection it can provide to the carcass and the suppleness or stiffness of the tyre when it is deflected.

Tread This is the rubber compound which contacts the road surface when the wheel rolls, and its pattern design considerably influences the tyre's gripping and road-holding ability and its working life.

*7.15.5 **Cross-ply tyre construction*** (fig. 7.61)
Cross-ply tyres are usually constructed of plies of rayon or nylon cord fabric and rubber laid at opposing angles to each other at approximately 30° to 40° to the tread circumference. The section height of the tyre is almost equal to the width, and when inflated the section is horseshoe-shaped.

The cords of a cross-ply tyre are laid from opposite sides of the tyre at an angle of 30° to 40° to the crown of the tyre, and there are usually two or four layers. When the portion of a tyre in contact with the road is deflected, the cords, which are inextensible, move rather than stretch or contract. This deflection causes the angle between them to change. Apart from tending to shear the layer of rubber between the plies so that heat is generated, this movement can cause the outer ribs of the tread pattern to move laterally and shuffle as they roll in contact with the road. At high speeds, the greater friction causes the rolling resistance to increase, the temperature of the tread to rise, and the outer ribs to wear faster than those at the centre. During cornering, of course, this shuffling, combined with the lateral distortion, causes the area of the contact patch to be progressively reduced as the lateral acceleration increases.

*7.15.6 **Radial-ply tyre construction*** (fig. 7.62)
This construction consists of one, two, three, or more carcass plies of textile cords laid at an angle of 88° to 90° to the beads, with a belt of several plies laid circumferentially under the tread. Without the belt, the 90° plies would produce a casing which would greatly increase its section height on inflation. The belt, being inextensible, prevents the casing height increasing, and the tyre retains its flatter profile even when inflated. Obviously the behaviour of such tyres is different from that of cross-ply tyres.

Since the belt is inextensible and its cords lie at an angle of 12° to 20° to the crown, there is little angular movement of the plies as the tyre rolls. Instead, the belt tends to deform as a hoop, while the radial plies give a low vertical deflection rate. The rolling resistance is therefore caused by the longitudinal bending of the cords of the belt, and it is low, while the contact patch is generally longer than that of a cross-ply tyre.

For a given cornering force, radial-ply tyres generate a lower slip angle than cross-ply tyres. The magnitude of the difference depends on the construction of the belt, which is laterally stiff and prevents the tread from shuffling and distorting laterally.

7.15.7 A summary of cross-ply and radial-ply features

The cross-ply tyre gets its strength from the many textile cords, criss-crossed and embedded in the rubber. These cords are arranged in layers, usually referred to as plies. These plies perform two jobs: first they have to make the walls strong enough to contain the air pressure and yet leave them as supple as possible for deflection; secondly they have to support and steady the tread. These two requirements conflict since, to obtain sufficient bracing of the tread, the ply must be reasonably stiff – this then means that the walls will be rigid.

Radial plies have to perform only one job, which is to make the wall of the tyre strong enough to contain the air pressure. They do not have to support the tread. Radial plies do not have to be criss-crossed; instead, they are laid radially – following the natural profile of the tyre. This provides a supple but strong wall, which is desirable. To support the tread, a layer of rayon or steel cords forming a belt lies underneath it. The sole job of this belt is to brace the whole of the tread firmly flat down and open on to the road.

7.15.8 Comparison of cross-ply and radial-ply characteristics

Ride comfort (fig. 7.63(a)) With cross-ply tyres, bending of the walls necessitates a shear action to change the criss-cross ply angle. This makes the walls very stiff, so that the bounce on rough roads is not readily cushioned. Radial-ply tyre construction is such that the natural direction of radial-ply cords provides a supple wall which will bend readily. These tyres can then absorb a great deal of extra bounce and take up a great deal of the jolts and jars of rough and bumpy roads.

The radial-ply tyre is more comfortable at speed, since its shock-absorbing deflection is 25% greater than that of the cross-ply, but at lower speeds the cross-ply tyre tends to ride more smoothly and the steering is lighter, making it more suitable for parking.

Cornering (fig. 7.63(b)) Cross-ply tyres have comparatively stiff walls, so they will not bend very much to take up any sideways strain during cornering. The cross-plies which form part of the wall and tread will therefore begin to pull and lift up one side of the tread from the ground, thus reducing road grip and traction.

Radial-ply tyres under sideways force will bend to absorb this extra strain. The tread is not linked directly to the wall plies but is braced down on to the road by independent layers of rayon or steel cords. This means that the tread will stay firm and flat down, with its whole working area of tread pattern biting into the surface of the road.

Acceleration and braking (fig. 7.63(c)) Conventional cross-ply tyres have no tread bracing – the only support the tread has is just a circular continuation of the wall plies. This means that the tread is affected by every movement of the walls and is not braced and held down on to the ground. Because the tread is not braced and held down on to the road, the tread blocks are able to shift and dance about on the road surface, which will obviously reduce road grip.

With radial-ply tyres, the braced layers of cords act independently of the wall plies. When the wheels revolve, the bracing belt will follow the contour of the road like a caterpillar-track-layer vehicle, and this provides a continuous flat contact-patch area with the road surface. The whole of the tread pattern will then be fully stabilised and so road-wheel acceleration and braking traction will be improved.

Tyre life With a cross-ply tyre, the wall plies are continuously interacting with the plies attached to the tread. When the wheel rolls, the distortion of the walls will tend to pull the tread away from the road surface, scraping the tread blocks as road contact begins and ends.

A radial-ply tyre will provide a flat full-width track-laying action along the wheel-and-road interface (fig. 7.63(c)). This is possible due to the independent action of the bracing belt which continuously flexes and changes its shape from a circular to a flat profile at the contact zone. The full-width tread-pattern contact when cornering and the extended flat zone when driving and stopping reduce wear. This considerably extends tyre life – by as much as 80%.

Fuel consumption The more flexible casing of radial-ply compared to cross-ply tyres reduces

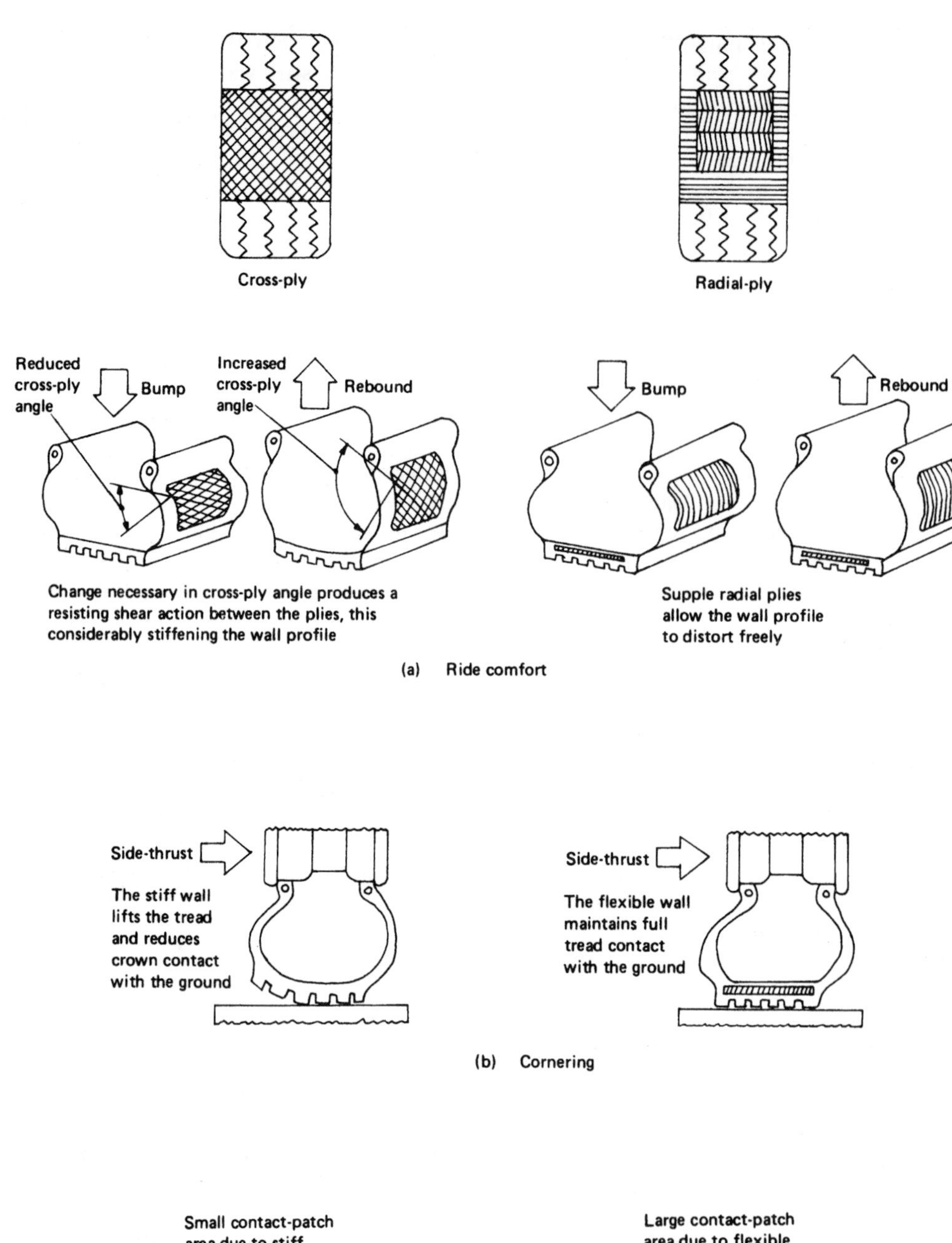

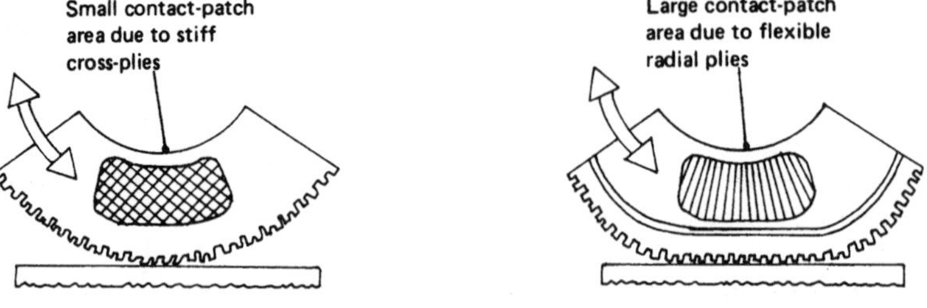

Fig. 7.63 Comparison of cross-ply and radial-ply tyres

the amount of energy consumed while running. Savings in fuel consumption may be of the order of 5%.

First cost The initial cost of cross-ply tyres is about 20% less than for radial-ply tyres.

7.15.9 ***Run-flat car-tyre concept*** (fig. 7.64)
In the inflated state, the conventional tubeless tyre performs the task of containing air and rolling. Once deflation occurs, the bead of the tyre can move across the rim into the well base, allowing the tyre to slide over the rim flange and so causing the tyre to buckle and fold. The rim may then plough into the road, leading to loss of control.

The run-flat tyre developed by Dunlop and known as the 'Denovo II' is designed to operate effectively with or without air, providing acceptable handling qualities when deflated and excellent handling qualities when inflated.

To achieve the run-flat concept, the following must be fulfilled:

a) The tyre bead must be retained against the rim flanges when the tyre deflates.
b) The tyre must be able to support all possible stresses imposed by deflation.
c) The tyre must remain undamaged by internal friction and heat generated when running flat.
d) The tyre should be able to self-seal any punctures, so that the remaining air will be trapped and heat will be generated and so reinflate the tyre within limits.

The design features of the run-flat tyre include the following:

i) The tyre bead and rim are designed so that an enlarged and reinforced toe on the tyre bead engages a small circumferential groove in the bead-set area of the rim. The bead thus remains locked in position under inflated and deflated conditions – even in the event of a high-speed blow-out, the tyre should not be dislodged from the wheel (fig. 7.64(a)).
ii) The tyre has a low-profile radial-ply construction and, to sustain the increased stresses created when running deflated, the tyre side walls are made thicker. The tyre is fitted on to a rim which is narrower than the tyre tread,

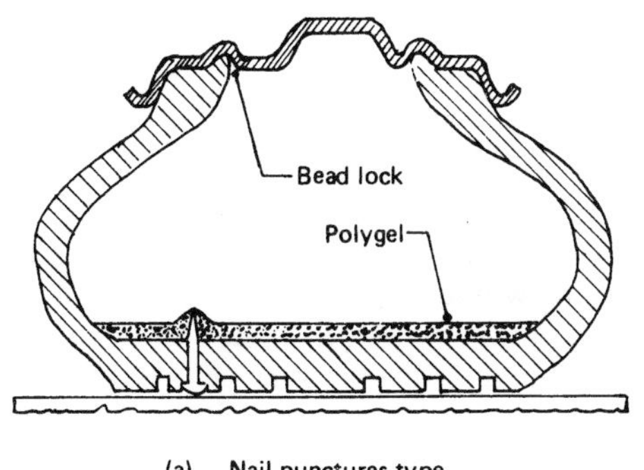

(a) Nail punctures type

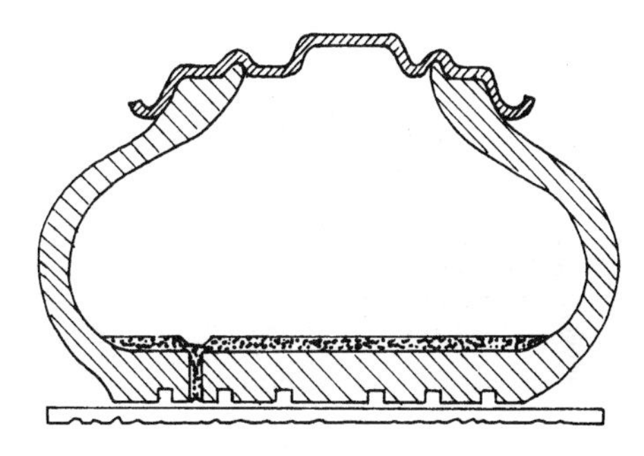

(b) Polygel seals puncture hole

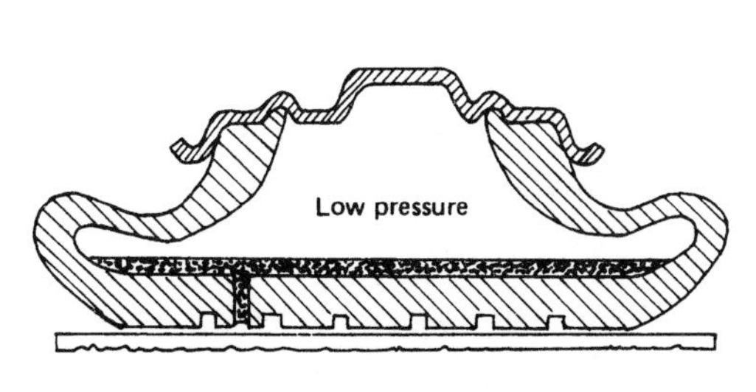

(c) Polygel will seal any other small puncture at low pressure so that the remaining cold air will be trapped.

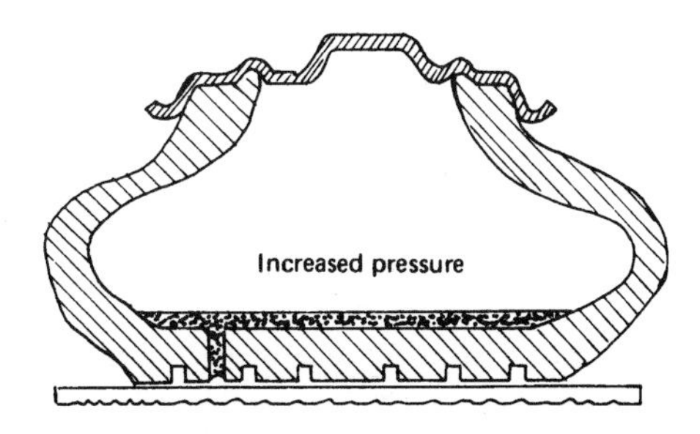

(d) Movement of the wheel and tyre deflection will generate heat. This raises the trapped-air temperature and pressure which re-inflates the tyre

Fig. 7.64 Dunlop 'Denovo II' run-flat car tyre

and the rim flange is shaped to support and not damage the tyre during deflated conditions.

iii) The friction and resultant heat generated are minimised by a one-stage lubricant known as 'Polygel' which is applied to the underside of the tread at high temperature during the final stages of tyre manufacture. Polygel also has self-sealing properties and can automatically seal many puncture penetrations without any great loss of air. This means that in effect the run-flat capability may never be put to the test (figs 7.64(a) to (d)).

7.16 Legal and technical requirements when using cross- and radial-ply tyres

The Motor Vehicle (Construction and Use) Regulations 1978 concerning the fitting of both cross-ply and radial-ply tyres may be interpreted as follows.

The mixing of cross-ply and radial-ply tyres generally is not recommended. If mixing is necessary, the wrong combination of cross-ply and radial-ply tyres can be both illegal and dangerous. When using two differently constructed tyres on the same vehicle, the following rules should be observed:

a) Do not, in any circumstances, have radial-ply tyres on the front with cross-ply on the rear. There is no exception to this – it applies whether the vehicle has front- or rear-wheel drive, or whether the rear tyres are standard-tread cross-ply or winter-tread cross-ply. A quick-check table is shown in Table 7.1.
b) Do not mix cross-ply and radial-ply tyres on the same axle. If fitting the spare wheel results in mixing, drive at low speeds and with extra care and change the combination as soon as possible. It should be noted, however, that such mixing does contravene the law.
c) The best combinations are either cross-ply all round or radial-ply tyres all round. If mixing is essential, it is much safer to have cross-ply tyres on the front and radial-ply on the rear. For high-performance cars, mixed equipment is not recommended.

Table 7.1 Mixing of cross-ply and radial-ply tyres

	Recommended position if mixed		
Tyre mix	Front	Rear	Comments
Radial-ply and cross-ply	Cross-ply	Radial-ply	Essential both technically and legally.
Steel-braced radial-ply and textile-braced radial-ply	Textile-braced radial	Steel-braced radial	Steel-braced radials have slightly higher cornering stiffness than textile-braced radials.

7.16.1 Slip angle (fig. 7.65)

To understand why differently constructed tyres should only be fitted in a certain combination, it is important to know the meaning of the term 'slip angle'.

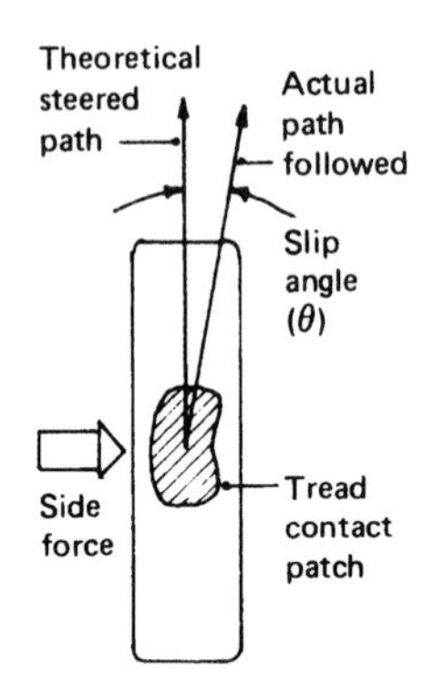

Fig. 7.65 Tyre slip angle

When a road-wheel is pointed in a given direction, it would be expected that it would roll along this path. However, if a side-force is applied to the wheel axle, such as the centrifugal force when cornering, the tyre wall will flex. The bending of the tyre wall will distort the tyre tread, so that the wheel and tyre tread now follow a path at an angle to the plane of the wheel. This angle is known as the slip angle, although this name is misleading since there is no real slip but a drift in a new direction.

7.16.2 Oversteer and understeer characteristics with cross- and radial-ply tyre combinations (fig. 7.66)

Generally, cross-ply tyres have a larger slip angle for a given cornering force than radial-ply tyres. The reduced slip angle created by the radial-ply compared to cross-ply tyres is due to a wider and flatter contact patch with the road surface.

If cross-ply tyres are fitted on the rear axle and radial-ply at the front, the increased slip angle at

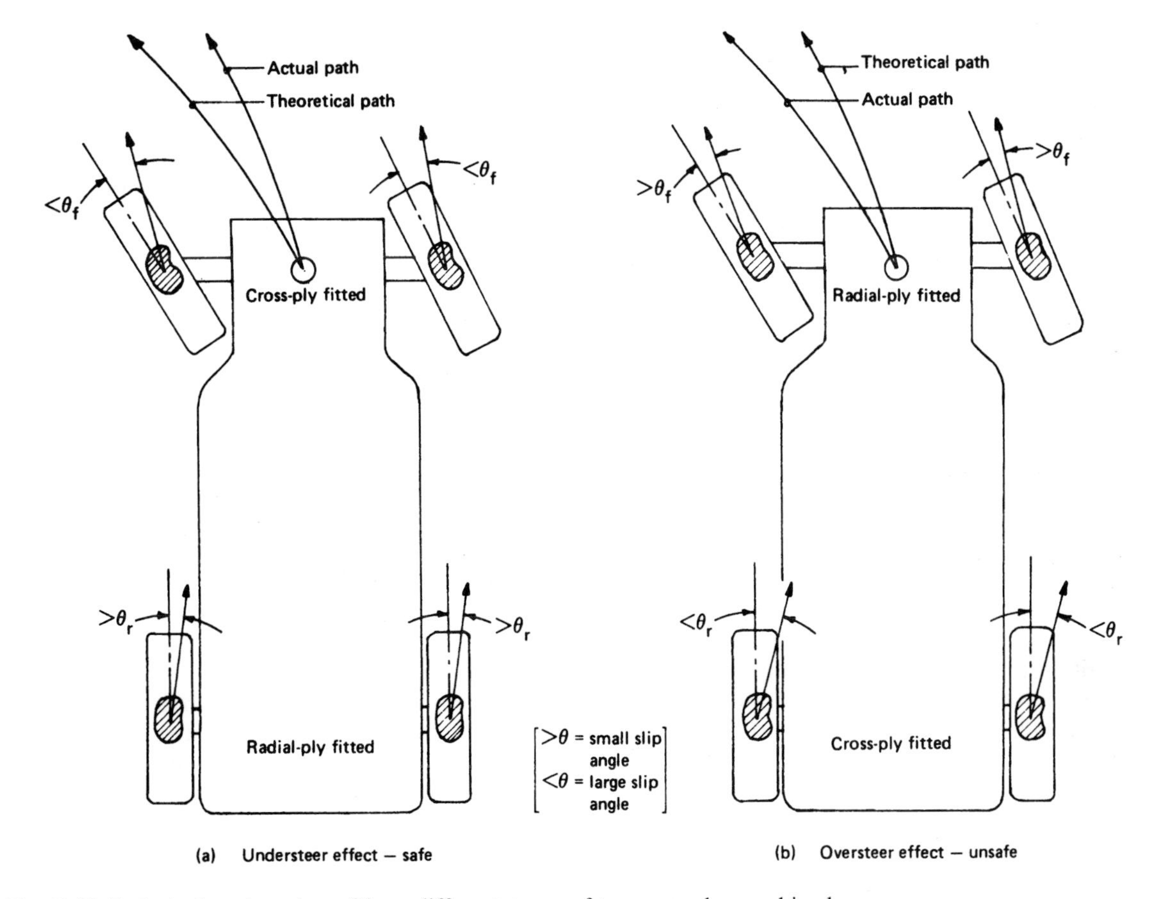

Fig. 7.66 Technical explanation of how different types of tyres may be combined

the rear compared to the front makes the vehicle oversteer, which can be dangerous. Conversely, fitting radial at the rear and cross-ply tyres at the front means that the increased slip angle on the front wheels cancels some of the wheel steering angle, so that the vehicle will tend to understeer – this is a more desirable steering characteristic.

7.16.3 Tread wear

For both types of construction, when the tread has worn down by a certain amount it becomes illegal to use the tyre further. The tread pattern of the tyre (excluding any tie-bars or tread-wear indicators) must be at least one millimetre deep across three quarters of the breadth of the tread and round the entire circumference of the tyre.

7.17 Tyre sizes and designations

At present, tyres are marked in accordance with the standard agreed by the European Tyre and Rim Technical Organisation.

Tyre markings provide up to four pieces of information:

i) construction type,
ii) speed rating,
iii) size,
iv) case profile.

Construction type This relates to the ply or ply-and-brace construction and is indicated by a letter code. Radial-ply tyres use the letter R and bias-belted the letter B, while cross-ply tyres carry no letter identifying construction – i.e. R = radial-ply tyre, B = bias-belted, no marks = cross-ply.

Speed rating This relates to the maximum speed at which a vehicle should be driven and is indicated by a code letter (see Table 7.2). The speed quoted is the maximum speed of which the vehicle is capable, not the speed at which it is normally driven.

Table 7.2 Speed marking of tyres

Cross-ply			
Wall markings	Normal (unmarked)	S	H
10″ (Rim diameter)	Up to 120 km/h (75 mile/h)	Up to 150 km/h (95 mile/h)	Over 150 km/h (95 mile/h)
12″ (Rim diameter)	Up to 135 km/h (85 mile/h)	Up to 160 km/h (100 mile/h)	Over 160 km/h (100 mile/h)
13″ and above	Up to 150 km/h (95 mile/h)	Up to 175 km/h (110 mile/h)	Over 175 km/h (110 mile/h)
Radial-ply			
Wall markings	SR	HR	VR
All rim diameter sizes	Up to 180 km/h (113 mile/h)	Up to 210 km/h (130 mile/h)	Over 210 km/h (130 mile/h)

Size This relates to the tyre nominal section width and wheel-rim diameter (fig. 7.56(a)), which are indicated by the first and second numbers respectively. The section of radial-ply and bias-belted tyres is always quoted in millimetres, and of cross-ply tyres in inches. The rim diameter is in inches for all types of tyre. For example, '5.40–12' indicates a nominal section width of approximately 5.4 inches and a rim diameter of 12 inches. Likewise, '155 SR-14' indicates a nominal section width of 155 mm and a rim diameter of 14 inches.

Casing profile This relates to the ratio of the section height of a casing to the section width (the measurement from bead to crown), this dimension usually being smaller than the first (fig. 7.56(a)). The ratio of these dimensions for earlier tyres was normally 80:100, and a tyre of this form is said to have an 80% aspect ratio. Low-profile tyres with a 60 or 50% aspect ratio are not uncommon with radial-ply tyres, and one or two are now appearing with 40% aspect ratio. If there is a deviation from the normal 80%, then a further number may be added to the side-wall marking; for example, 175/70 HR–13 = 175 HR–13, with a 70% aspect ratio.

7.17.1 Examples of tyre markings (Table 7.2)

a) A recommended tyre for a car is 5.20–12. This means that the tyre will be cross-ply, approximately 5.2 inches wide, and suitable for a 12 inch diameter rim and for up to 135 km/h.
b) A recommended tyre for a car is 155 SR–13. This indicates that the tyre is radial-ply, approximately 155 mm wide, and suitable for a rim 13 inches in diameter and for up to 180 km/h.
c) A recommended tyre for a car is 185/60 HR–14. This indicates that the tyre is radial-ply, approximately 185 mm wide, has a 60% aspect ratio, and is suitable for a rim 14 inches in diameter and for up to 210 km/h.

7.18 Tyre valves

A tyre contains the air which carries the vehicle load. The function of the tyre valve is to permit air under pressure into the tyre chamber formed between the casing and the rim when required, but to release this air for pressure-adjustment purposes or when removal of the tyre is necessary. The valve stem is designed to easily accept a high-pressure air-line adaptor or a pressure-testing gauge.

7.18.1 Valve operation (fig. 7.67(a) and (b))

The valve assembly consists of a core-pin which has a valve set attached. This pin passes through an internally parallel sleeve, the outside of which has a tapered seal attached. When the core assembly is screwed into the core housing, the tapered seal prevents any air leakage between the core housing and the valve sleeve. The return-spring is situated either between the valve-seat washer and the core end-stop, or between the valve sleeve and a crimped wire ring situated on the core-pin. This spring normally holds the valve in the closed position, thus trapping any compressed air which has been pumped into the tyre.

When the tyre needs inflating, a compressed-air pipeline with a valve adaptor is fitted over the valve stem. This has the effect of pushing down the core-pin. Air from the higher-pressure air supply will then be forced through the valve annular core space into the tyre. When sufficient pressure is reached, the adaptor is removed,

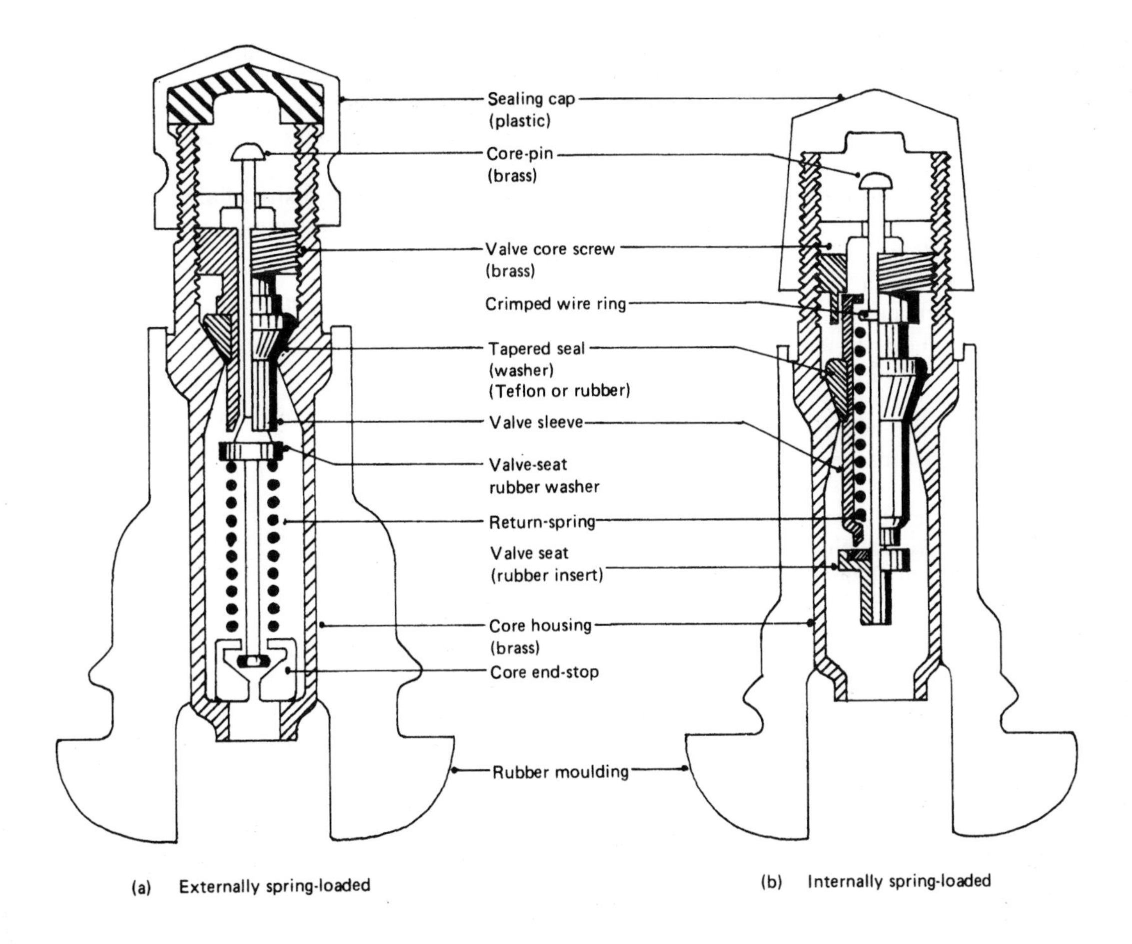

(a) Externally spring-loaded

(b) Internally spring-loaded

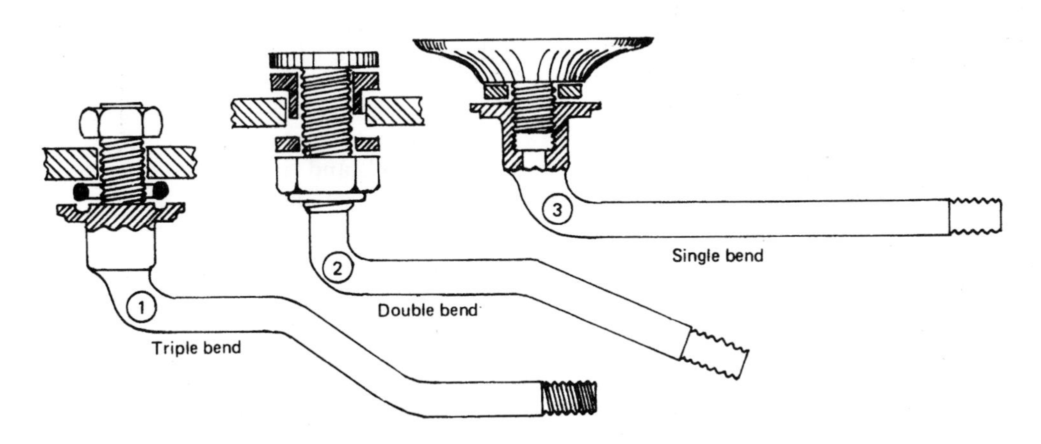

(c) Commercial-vehicle valve shapes and fixings:
1 and 2 – tubeless tyres; 3 – tubed tyres

Fig. 7.67 Tyre valve assemblies

releasing the core-pin which allows the valve to close. If the tyre pressure is too high, the core-pin may be depressed slightly to release the excess air from the tyre.

The outer cap serves to keep out grit and to act as a secondary seal for the valve assembly.

7.18.2 Valve applications

Car tubeless tyres use a snap-in valve which consists of a rubber moulding bonded to the metal stem casing. This rubber moulding has a recess or groove formed at the large-diameter base, so that, when the valve is pulled through the valve hole in the wheel rim, the pliable rubber base will snap into position (fig. 7.67(a) and (b)).

Commercial-vehicle tubeless-tyre valve stems are all-metal. They are attached to the wheel rim by an extended thread formed at the base and secured by a nut. Air-tight sealing is achieved by either an O-ring or a flat and flanged rubber washer (fig. 7.67(c) 1 and 2).

Trucks with tubed tyres have a thread-adaptor tube moulded to a circular rubber patch, this being in turn vulcanised to the inner tube. The valve-stem casing is then screwed on to the tube adaptor (fig. 7.67(c) 3).

Commercial-vehicle valve stems may have a single, double, or triple bend to accommodate different rim profiles and single and twin wheel combinations (fig. 7.67(c)).

7.19 Safety precautions in tyre maintenance

The removing, dismantling, assembly, and fitting of wheels and tyres are potentially dangerous, but reasonable care can considerably reduce accidents, injury, and damage.

The following recommended precautions when servicing wheels and tyres will reduce the risk of incorrect and dangerous practices.

7.19.1 Removing the wheel and tyre

1 Jack the vehicle on a flat hard surface – *never* on soft or uneven ground.
2 Check wheels for damage and ensure that the parking brake is applied and is effective.
3 Always use axle-stands with a jacked-up axle.
4 Personnel should not enter or even lean against a jacked-up vehicle.
5 If a wheel or tyre is damaged, remove the valve core completely to deflate the tyre before removing the wheel from the vehicle or the tyre from the wheel.
6 With divided wheel rims, do not loosen or unscrew the outer clamping nuts.
7 Before removing a tyre from a wheel, make sure that the tyre is fully deflated by first removing the valve core.
8 Take care not to distort flanges and lock-rings when dismounting tyres.

7.19.2 Examination and preparation of the cover and rim

1 Check that the parts of the rim that come into contact with the tyre bead are not distorted or damaged in any way.
2 Remove any accumulation of rust on the flange gutter and rim base to ensure correct seating of components. Clean these surfaces with a wire brush and emery cloth if necessary, wipe clean, and then repaint with epoxy-resin paint if necessary.
3 Examine the tyre case to be fitted inside and outside for any damage or manufacturing defects.
4 Wipe the cover beads with a clean dry rag and lubricate the cover beads, flange, and rim. Do not use talc.
5 Check that the sizes of wheel, tyre, and valve for tubeless, and of the wheel, tyre, flap (if fitted), and valve for tubed assemblies are compatible.
6 Check that stud holes are not damaged or elongated. Correct location of a wheel not only prolongs wheel life but is an essential contribution to satisfactory tyre performance.
7 Make sure that wheels with cracked rims and stud-attachment naves are not replaced. Make sure that split flanges and lock-rings used with truck wheels and rims are not distorted and are the correct size and type.

7.19.3 Reassembly and inflation of wheel and tyres

1 It is recommended that, whenever a tubeless tyre is renewed, the complete valve is renewed too.
2 Place the tyre-and-wheel assembly flat on the ground to enable the cover to take up its correct position on the rim and to obtain a good air-tight seal.
3 Inflate initially with the valve core removed to a pressure of 1 bar.
4 If the tyre will not inflate, keep the air connected and roll the assembly along the ground.

When the tyre beads have seated, disconnect the air line and insert the valve core.

5 Do not use a moist or contaminated air supply.
6 Check that the fitting line at the tyre beads is concentric with the wheel flange, to avoid unnecessary lift or run-out.
7 Place the tyre in a safety cage and ensure that the air line between the valve and the person inflating the tyre is long enough to prevent any damage from flying components in the event of a tyre or wheel burst.
8 If no tyre guard is available, due to breakdown 'on site', lean the assembly against a wall, with the lock-ring/loose flange facing the wall.
9 Establish the correct pressure for the size and application before inflating, and do not over-inflate.
10 Do not stand in front of or lean over the tyre when inflating it.
11 Do not leave a tyre unattended while it is being inflated.
12 Ensure that all pressure gauges and air-metering devices are regularly checked against a master pressure gauge.
13 Do not interrupt a fitter when he is tightening wheel nuts. It is essential that all wheel nuts are fully tightened following the correct sequence, alternating opposite sides of the wheel.
14 Do not undertighten or overtighten wheel nuts and, if a power tool is used, check the correct tightening torque.

7.20 Tyre inflation pressure

Tyre manufacturers recommend tyre inflation pressures to provide the best compromise for load carrying, vehicle handling, and tyre life.

7.20.1 Load capacity

Tyre inflation pressures are usually related to the expected wheel load. Increased inflation pressure provides the tyre casing with more support, so that its load-carrying capacity will become greater (fig. 7.68(a)). The load-carrying capacity will be determined by the carcass ply-cord material, the resilience of the tread grip, the bounce absorption (i.e. the cushioning qualities), uniform wear, and expected life.

Under-inflation of tyres increases the tyre-wall deflection at the base of the wheel. This results in a continuous flexing of the tyre walls which generates heat and reduces the fatigue life of the casing (fig. 7.69(a)).

Over-inflated tyres (fig. 7.69(b)) cause the cord plies and rubber covering to become highly stressed, so that they cannot take the repeated impacts to which the tyre is subjected. A prolonged journey overloaded with over-inflated tyres would eventually promote failure by rupturing of the walls. As a general rule, the harder the tyres are pumped up, the less cushioning there will be and the harder will be the ride.

7.20.2 Vehicle handling

It has been established that the cornering force which can be sustained by the tyre when going round a bend is proportional to the tyre inflation pressure within given limits (fig. 7.68(b)). The consequence of this is that the slip angle for a given cornering force will be reduced with increased air pressure.

A match of inflation pressure between the front and rear wheels can influence the understeer and oversteer characteristics of a vehicle. For example, if the tyre pressures were raised at the front and lowered at the rear, the vehicle would tend to oversteer. Conversely, if the pressure difference were reversed, i.e. reduced at the front and increased at the rear, the vehicle would be inclined to understeer (fig. 7.66(a) and (b)).

7.20.3 Tyre life

Tyre life will depend upon many factors such as inflation pressure, wheel load, and vehicle speed.

The correct tyre pressure will minimise the distortion and straining of the tyre case. Low inflation pressure will rapidly wear the outer edges of the tread, but high inflation pressure will tend to wear the centre of the tread crown. Both extremes, then, reduce tyre life (figs 7.68(c) and 7.69).

Overloading the tyre will highly stress the carcass structure, distort the tread-pattern blocks and increase their scrubbing action, and overheat the tyre. Increasing the wheel load will then reduce tyre life (fig. 7.68(d)).

Finally, with increased wheel speed, the rate of flexing and the rise in carcass temperature shortens tyre life. In addition, the wheels will not closely follow the road surface contour at speed, so that the tread will be in a constant state of scuffing as it contacts and leaves the ground (fig. 7.68(e)).

7.21 Wheel balance

Wheel imbalance is where some section of the circular wheel or tyre, usually tyre, has become

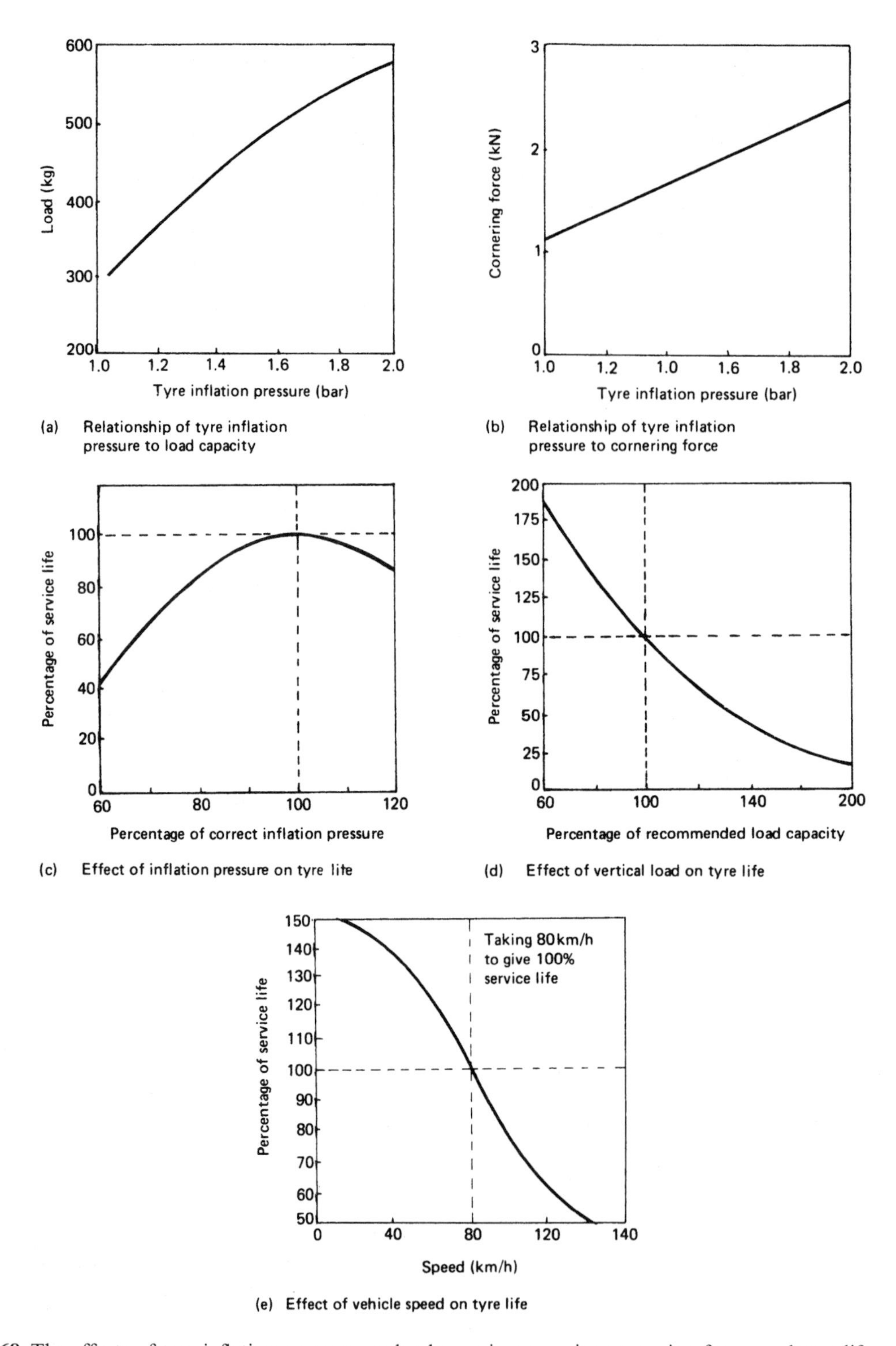

Fig. 7.68 The effects of tyre inflation pressure on load-carrying capacity, cornering force, and tyre life

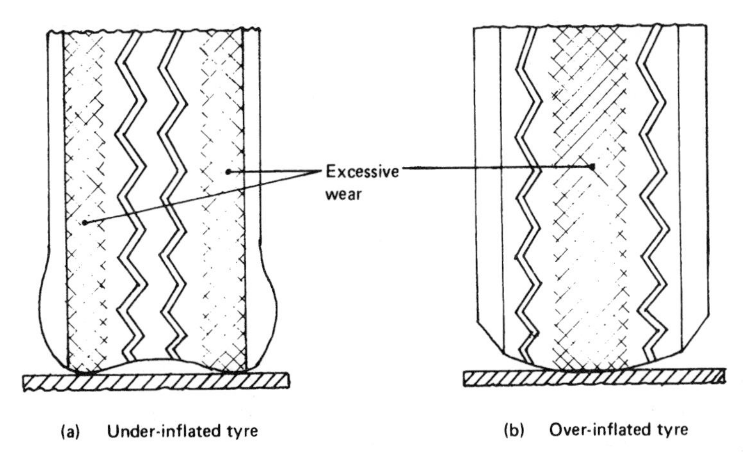

Fig. 7.69 Wear with incorrectly inflated tyres

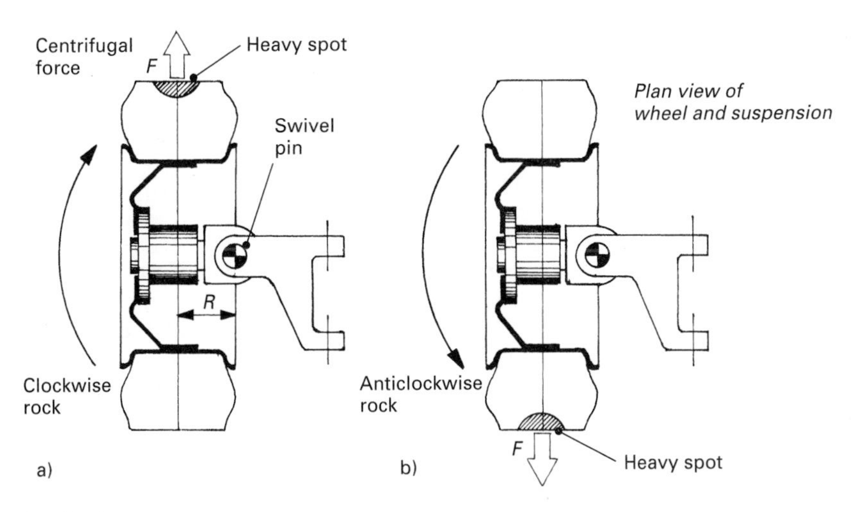

Fig. 7.70 Static wheel imbalance

heavier than a similar section on the opposite side of the wheel and tyre for a number of reasons. The circular mass of wheel and tyre when rotating is continuously subjected to centrifugal force which acts radially outwards. If this mass is uniformly dispersed around the wheel hub, then diametrically opposing centrifugal forces cancel each other and the wheel is said to be in a state of balance.

7.21.1 Static balance (fig. 7.70(a) and (b))

If some of the tread on one side of the tyre has worn, due possibly to the brakes locking, then the opposite side of the wheel assembly where only normal wear has taken place now becomes heavier. Other reasons for wheel imbalance are the mass of the valve-stem assembly fitted onto the well of the wheel, tolerance in manufacturing the wheel, and the assembly of the tyre onto the wheel rim. The difference in weight between diametrically opposed sides of the wheel due to uneven tyre tread wear will cause the unworn tread mass to exert a centrifugal pull which is not opposed. Consequently, when the heavy region or spot is at axle level, the offset between the centre of the heavy spot and the swivel-pin axis produces a moment of force which tends to twist the wheel in the direction of the unbalance force, in this case, clockwise (see fig. 7.70(a)). As the wheel rotates a further half revolution, the heavy spot is now facing in the opposite direction. It therefore produces a moment of force which now tends to twist the wheel in an anticlockwise direction (see fig. 7.70(b)). As can be seen, the revolving wheel will continually rock or twitch clockwise and anticlockwise about the swivel-pin, every 180° of wheel rotation. Hence this can be interpolated as a wobble. The magnitude of this twitch will be small at low vehicle speed but becomes much more prominent as the vehicle speed rises.

Free swing method of correcting static imbalance
Static imbalance can be easily checked by jacking up the axle of the wheel so that the wheel clears the ground, or removing the wheel and mounting it on the spindle of a balancing-machine. Rotate the wheel, say 90° and release it. If the wheel does not move, again rotate it so that it takes up a new position and release it. If it still does not rotate on its own then it can be assumed that the wheel assembly is statically balanced. If, however, when releasing the wheel it swings back and forth until it eventually comes to rest, the wheel is not balanced and the following procedure should be followed:

1) Chalk the top of the wheel rim opposite the heavy side of the tyre which is at the bottom.
2) Choose a weight size (mass) from a selection of magnetic weights and place it on the rim opposite the heavy side of the wheel, that is, in line with the chalk mark.
3) Rotate the wheel about a quarter of a revolution and then release it. If the wheel remains in this position, then the wheel is balanced.
4) If the wheel still swings from side to side but more gently, increase the size (mass) of the magnetic weight slightly and again rotate the wheel a quarter of a revolution to see if there is any swing left.
5) Conversely, when the first magnetic weight is attached, if the wheel swings more violently then before, a smaller weight must be selected. This trial-and-error approach in selecting the correct balance weights is soon mastered.
6) After establishing the correct position and balance weight mass, replace the magnetic weight with a clip-on balance weight.

*7.21.2 **Dynamic balance*** (fig. 7.71(a) and (b))
In practice uneven wear across the tyre tread occurs. This can be due to a number of reasons such as:

1) incorrect steering suspension geometry causing too much wheel camber;
2) wheel track out of alignment either in the straight-ahead position or on turns;
3) excessive scraping or impact against kerbs;
4) overloading of the vehicle;
5) under-inflated tyre pressure;
6) excessive tyre scrub due to fast driving round corners, etc.

The consequence of uneven tread wear across the crown of the tyre is that there can be more than one heavy spot and these are likely to be offset from the centre of the tyre crown, that is, they will be in two different planes. Thus if there are two offset heavy regions opposite each other the centrifugal force generated by these heavy spots will create a force couple which tends to swivel the wheel and tyre about the tyre to ground contact

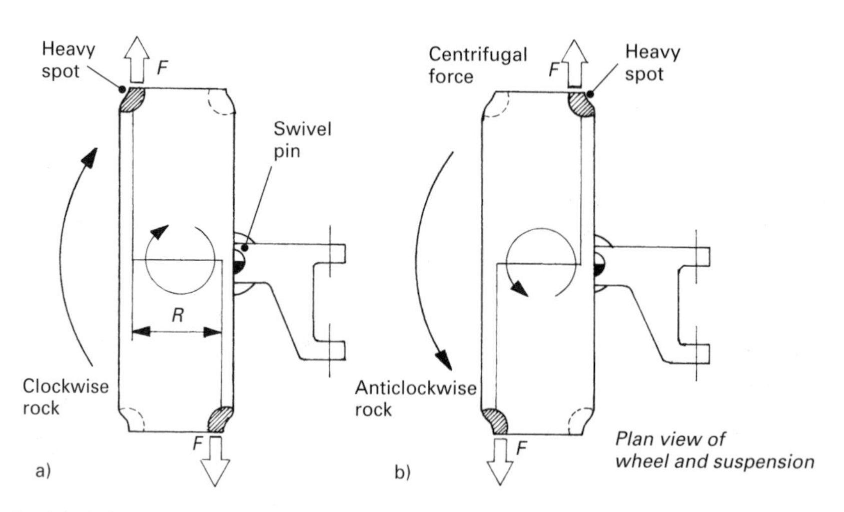

Fig. 7.71 Dynamic wheel imbalance

patch every time the heavy spots are horizontal at axle level (see fig. 7.71(a) and (b)). The direction of twist will alter every half revolution of wheel rotation, so that there will be a continuous tendency for the wheel to repeatedly swivel (rock) clockwise and anticlockwise whenever the vehicle is in motion (see fig. 7.71(a) and (b)).

With dynamic wheel imbalance, driving a wheel (or vehicle) through its speed range will encounter first a narrow critical speed band where resonance sets in and the amplitude of oscillation increases excessively. This critical speed is determined by the magnitude of the out-of-balance force couple. Driving to even higher speeds may encounter a second critical speed band where resonance again occurs, but this time, the amplitude of oscillation is not so large as for the first one. If the speed limit could be extended a third or even a fourth critical speed would be reached. However, the magnitude of the resonant oscillations diminishes as the order of the criticals rise, so that generally only the first or possibly the second is noticeably disturbing. Driving either side of a critical speed band reduces the level of oscillation, so that if a wheel has dynamic imbalance, a temporary measure would be to drive below or above the speed at which steering vibration is experienced. Worn suspension and steering joints can aggravate wheel imbalance oscillation and may result in excessive tyre scrub, uneven tread wear, and vibration fed though the steering-wheel to the driver.

Static and dynamic wheel imbalance Static and dynamic wheel imbalance can be identified by removing the wheel from the axle and mounting it on a wheel-balancing machine hub and spindle. The spindle should then be spun and its speed gradually increased to its upper limit. It should be observed during the speed run up whether the wheel commences to wobble and if it continuously gets worse as the rotational speed rises, then this indicates a state of static imbalance of the wheel and tyre. However, if the wheel wobble builds up to a period of violent oscillation and then steadies out somewhat with further speed increase, then the wheel assembly is more likely to be in a state of dynamic imbalance. Whichever mode of imbalance is established, imagine the effects the magnitude of oscillation can have, that is, tugging backwards and forwards in both vertical and horizontal planes on the suspension swivel-pin joints and on the steering track-rod end ball joints.

7.21.3 Lateral run-out (fig. 7.72)

Before balancing a wheel assembly, the axle or suspension should be jacked up and the wheel spun to see if it wobbles. The wheel rim and tyre should also be checked for lateral run-out which should not exceed 2 mm. The cause of lateral run-out could be due to a buckled wheel rim and unevenly fitted tyre. Buckled wheels should be replaced, whereas poorly fitted tyres should be deflated and repositioned between the tapered sides of the wheel rim.

7.21.4 Radial run-out (fig. 7.73)

This is when the radial distance from the wheel hub to the crown of the tyre varies around its circumference, that is, the tyre has been fitted eccentrically around the rim of the wheel and

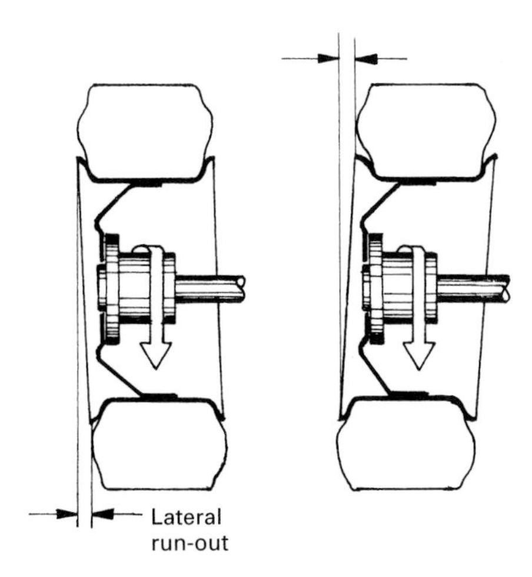

Fig. 7.72 Wheel rim run-out

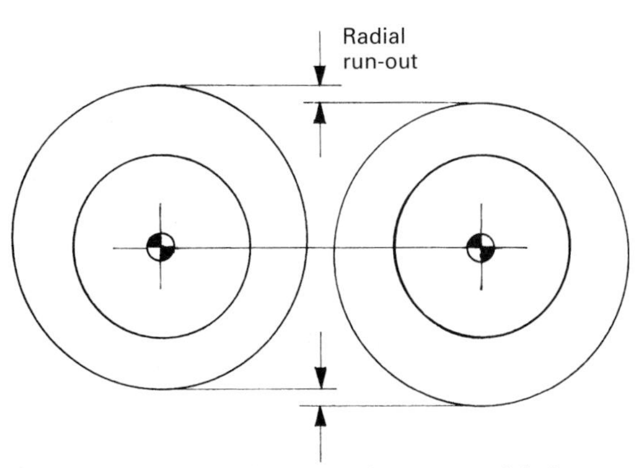

Fig. 7.73 Tyre radial run-out (tyre eccentricity)

therefore will cause wheel and tyre imbalance. Check the radial run-out by spinning the wheel and holding a rigid object above or to one side of the crown of the tyre. If there appears to be a degree of tyre lift and fall, deflate and reposition the tyre on the shoulder of the wheel. Do not try to balance the tyre to compensate for a badly fitted tyre.

8 Steering systems

8.1 Steering linkage arrangements

The steering system is designed to enable the driver to control and continuously adjust the steered path of the vehicle, with a positive response to whatever directional signal the driver may make on the steering-wheel. To achieve these aims, some sort of mechanical linkage has to be incorporated between the front steered road-wheels and the driver's steering-wheel. This mechanism must operate effectively under all normal conditions without interfering with the propelling road-wheel traction or with the suspension movement which copes with road surface irregularities and body roll during cornering.

The layout of any steering arrangement used depends largely on the basic class of vehicle concerned – whether it is a commercial vehicle, which generally has a rigid-axle-beam front suspension, or a private car which relies on independent front suspension. Both the axle-beam-suspension and

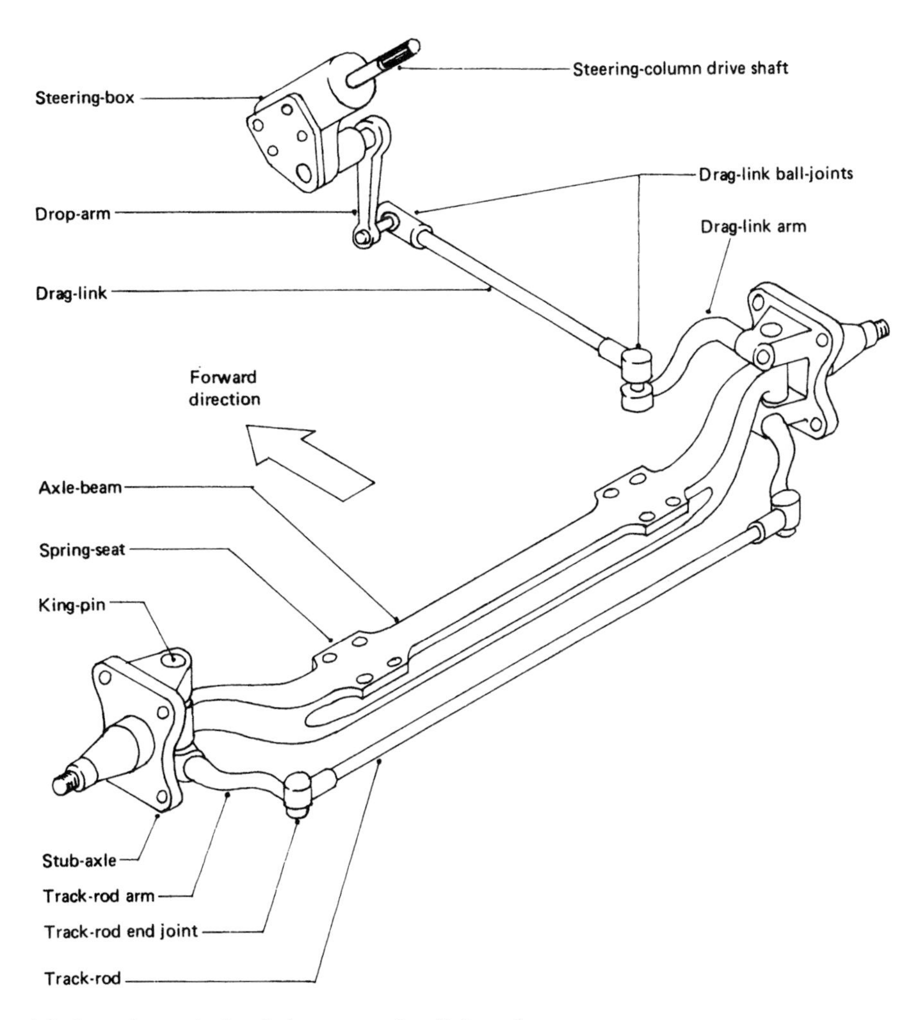

Fig. 8.1 Pictorial view of a typical axle-beam steering-linkage layout

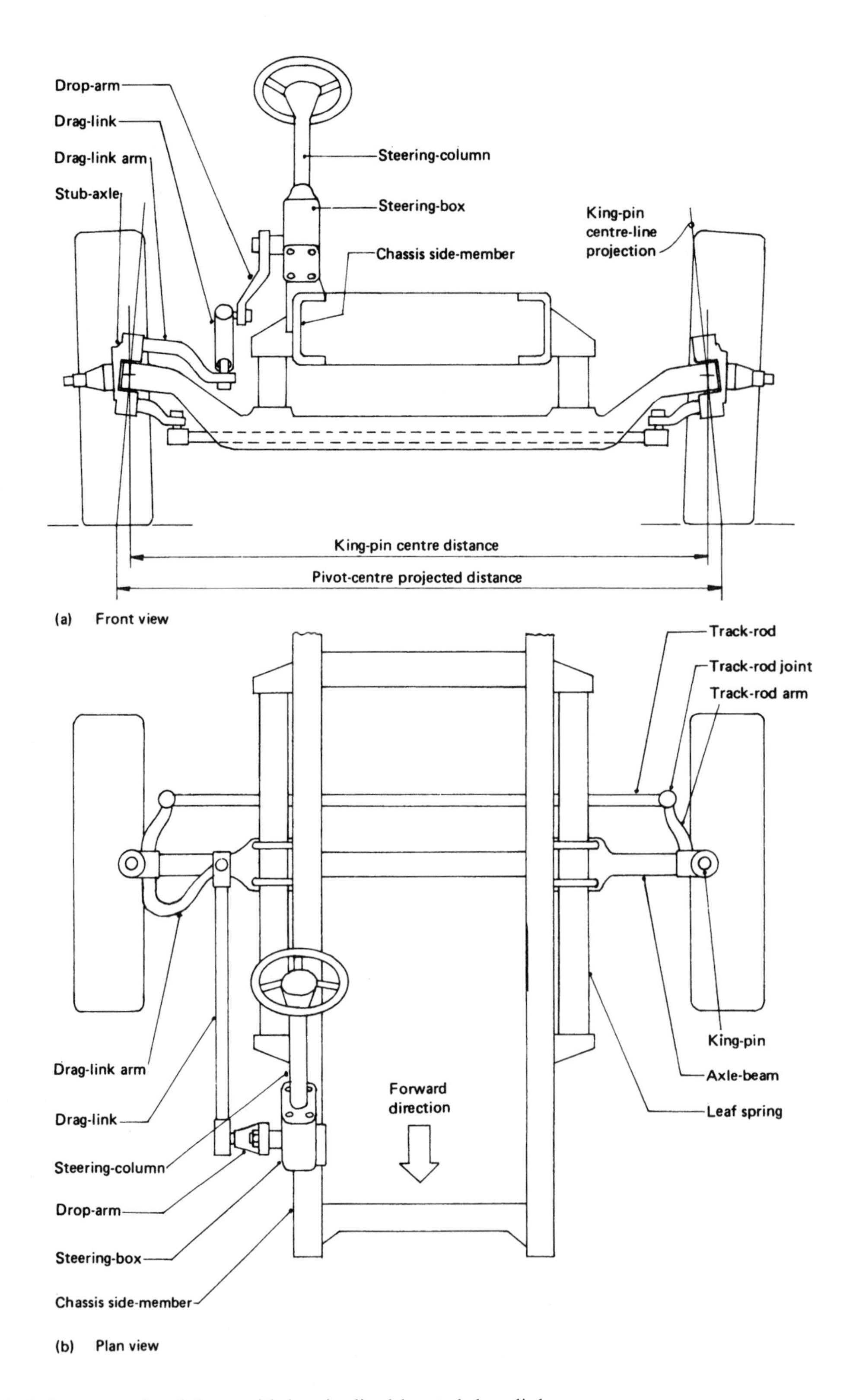

Fig. 8.2 Axle-beam steering linkage with longitudinal-located drag-link

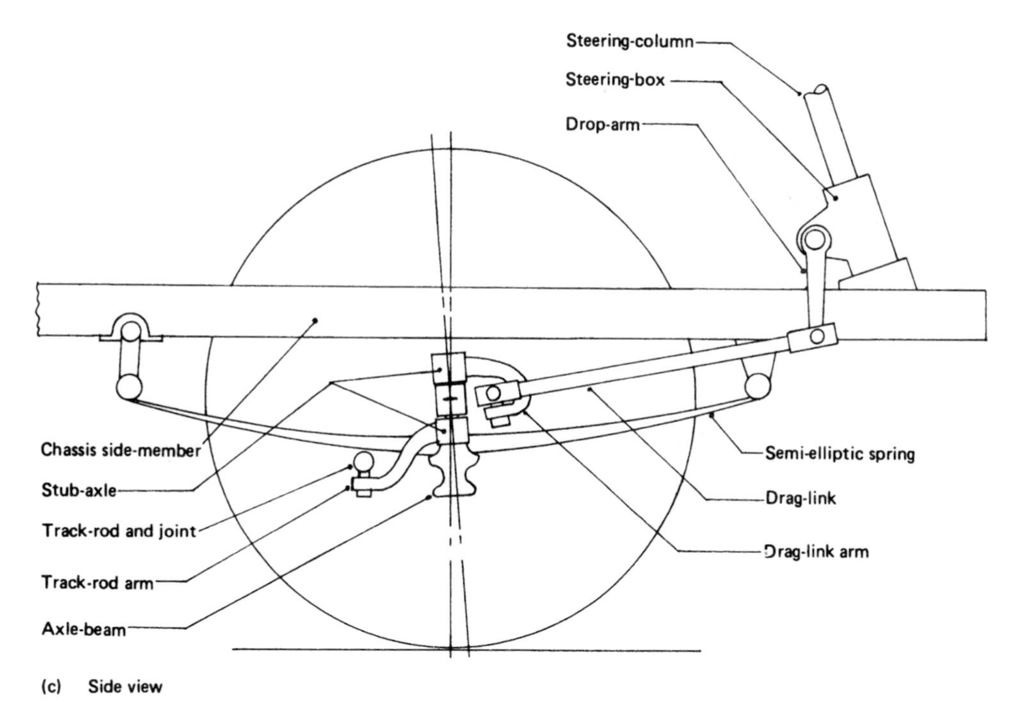

Fig. 8.2 (*continued*) Axle-beam steering linkage with longitudinal-located drag-link

the independent-suspension steering systems and their requirements will now be examined and described.

8.1.1 Axle-beam-suspension steering system (figs 8.1 to 8.3)

This steering system (fig. 8.1) consists of a steering-wheel which imparts motion to the steering-box, and this conveys the steering effort through the drop-arm and drag-link directly to one of the two stub-axles pivoting at the ends of the axle-beam. A track-rod joins both the stub-axles together.

Figure 8.2 shows the axle-beam steering layout in three different views, so that each component and its relative position within the system may be easily studied. The components and their functions are described below.

Steering-box The steering-box provides a gear reduction so that, with only a small effort, a much larger force can be applied to the steering linkage. At the same time, the degree of stub-axle movement will be reduced for a given angular movement of the steering-wheel. This is desirable as it prevents the steering being oversensitive to the driver's touch on the wheel.

Drop-arm This forged lever-arm is bolted on to a tapered steering-box output rocker-shaft. It is usually positioned so that it hangs or drops downwards. Its swing action imparts a circular-arc movement to the drag-link.

Drag-link This tubular rod converts the circular movement of the drop-arm into a linear push or pull motion of the drag-link arm which is rigidly attached to one of the stub-axles. Thus it drags the stub-axle about its pivot. At each end of the rod is a ball-joint which provides relative movement in two planes.

An alternative transverse drag-link layout is shown in fig. 8.3. This is preferred for some vehicles designed for cross-country applications.

Drag-link arm This arm connects the drag-link to one of the stub-axles. Its function is to provide the leverage to convert the linear drag-link movement to an angular movement about the stub-axle king-pin.

Drag-link suspension and steering interaction When the axle-beam moves up and down, it does so in a circular arc pivoting about the fixed front shackle-pins. Similarly, the drag-link will

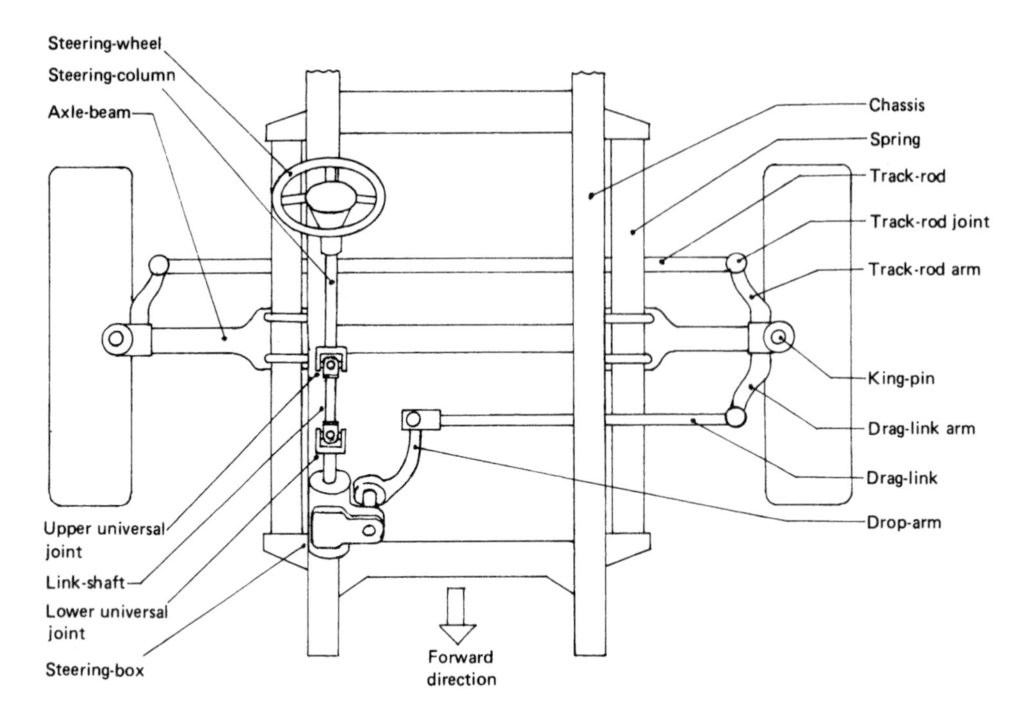

Fig. 8.3 Axle-beam steering linkage with transverse-located drag-link

pivot about the drop-arm ball-joint during any axle vertical movement. If the effective arc radii of the axle movement and the drag-link arm end are approximately equal, then the suspension axle movement relative to the chassis will not interact with the vehicle's steered path. However, if there is a slight difference in the arc radii and movement of the axle and of the drag-link arm when the suspension is deflected, then the drag-link will proportionally increase or decrease the relative angular position of the stub-axle about the king-pin. Consequently the steering will continuously twitch or jerk when travelling over rough surface conditions, unless axle-movement interference with the steering linkage is minimised.

Stub-axles Each steered road-wheel is mounted on a short axle-shaft commonly known as a stub-axle. The stub-axle has two extended horizontal prongs which fit over the ends of the axle-beam. A short circular bar known as the king-pin passes vertically through both prongs and the eye of the axle-beam to form the hinge pivot. The stub-axle thus acts as both the wheel axle and the pivot support member in the horizontal plane for the steered road-wheels.

Track-rod arms Each stub-axle has a forged track-rod arm rigidly bolted approximately at right angles to the wheel axis in the horizontal plane. The length of this arm provides the leverage to rotate the stub-axle about the king-pin, the rotary movement being transferred to the other stub-axle by means of the track-rod.

Track-rod To enable the two stub-axles to pivot together, a tubular track-rod spans the wheel track. The ends of this rod are fitted with ball-joints which are themselves bolted to the track-rod arms which make up part of each stub-axle. These ball-joints are restrained to move in the horizontal plane only. When the drag-link moves, it rotates one of the stub-axles, and this motion is transferred to the other stub-axle through the track-rod. The drag-link movement may be either a pull or a push action.

8.1.2 Independent-suspension steering systems

(figs 8.4 and 8.5)

With the rigid-beam suspension, a stub-axle is pivoted at each end of the axle-beam so that relative movement can take place only in the horizontal plane. Therefore the effective track-rod

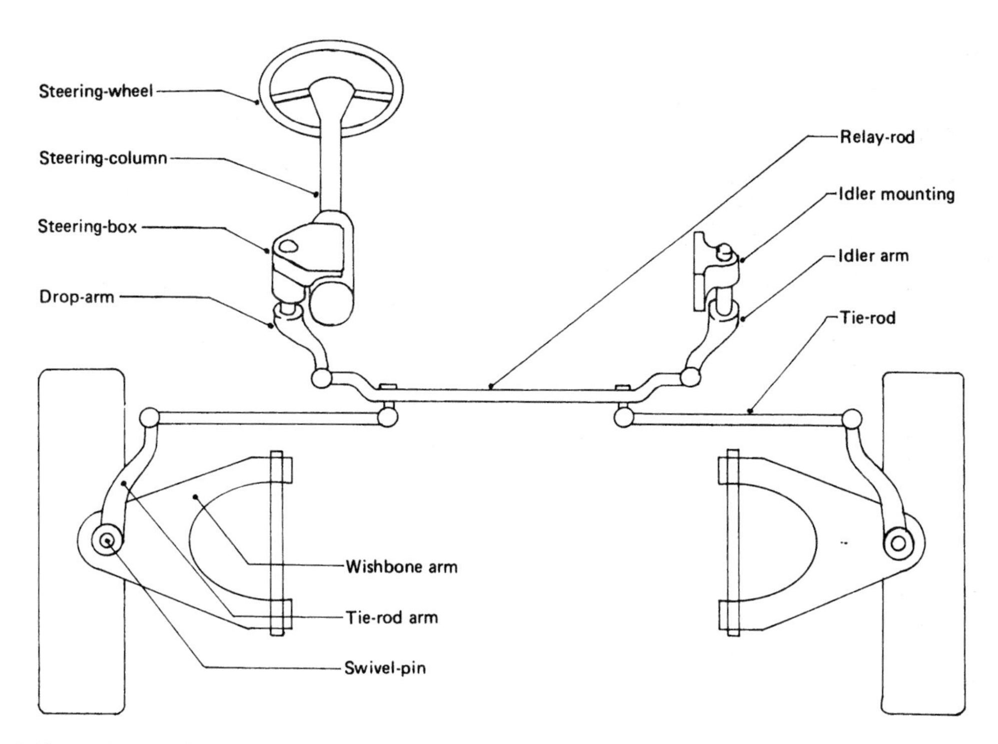

Fig. 8.4 Split track-rod with relay-rod and idler steering-linkage layout

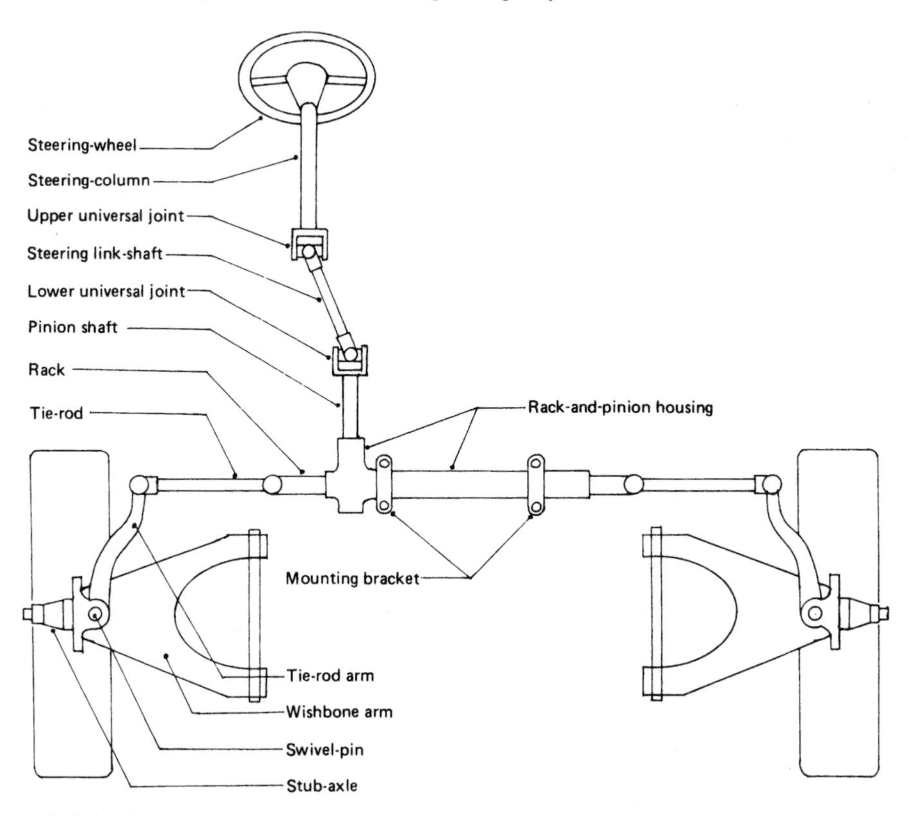

Fig. 8.5 Rack-and-pinion steering-linkage layout

length is not influenced by any vertical suspension deflection.

Independent-suspension steering has to cope with each stub-axle moving up or down independent of the other so that the distance between track-rod-arm ball-joint centres is continually varying. If, therefore, a single track-rod joined the two stub-axles together, the slightest bump or rebound would tend to pull both stub-axle arms at once and would thus interfere with the steering-track toe-in or toe-out.

The difficulty of the changing distance between track-rod-arm ball-joint centres can be overcome by using a three-piece track-rod. The centre portion of the track-rod may either be a relay-rod suspended between the steering-box drop-arm and an idler arm fixed to the body structure (fig. 8.4), or it may form the rack shaft of a rack-and-

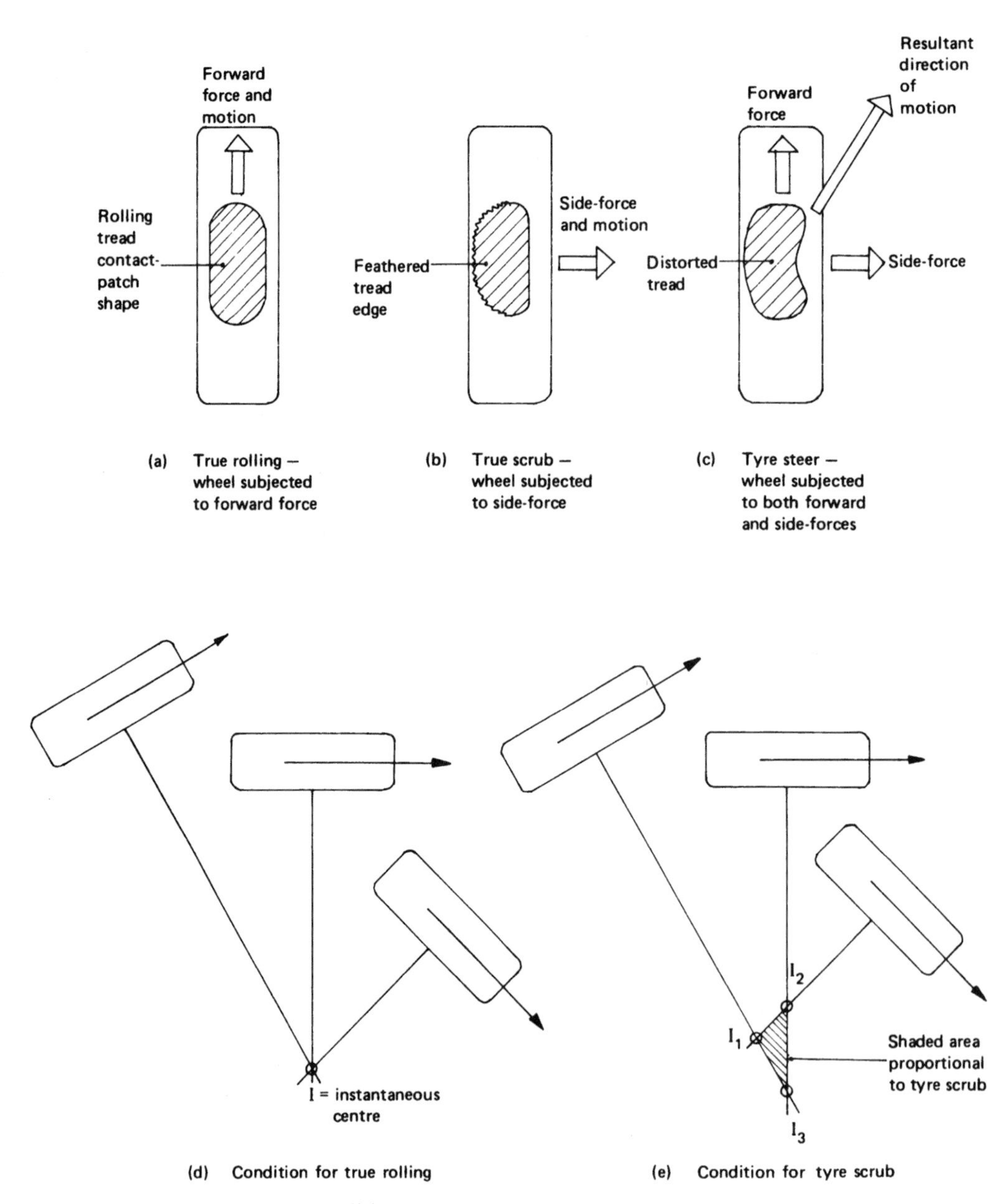

Fig. 8.6 Road-wheel and tyre rolling conditions

pinion steering-box (fig. 8.5). In either case, this part has freedom to move in the horizontal plane but is restrained from moving in the vertical plane. Movement in the vertical plane is provided by the two outer connecting rods, commonly known as tie-rods. The tie-rods swing about the ball-joints situated at the end of the middle track-rod member and describe arcs of radius nearly the same as those of the transverse wishbone members.

Early independent-suspension steering used stub-axles and king-pin pivots similar to those used with the axle-beam, but current trends prefer ball-swivel joints for the stub-axle pivot hinge action as they are spaced further apart and are more reliable than pin-and-bush joints.

Large private cars tend to adopt the system shown in fig. 8.4 When the steering-wheel is moved, the drop-arm will convey movement to the relay-rod which then transmits this motion to both tie-rods and stub-axles. Thus the drop-arm and idler-arm relay joints provide movement in only the horizontal plane, but the tie-rod joints have to provide movement in both the horizontal and vertical planes.

The most popular steering system used for small and medium-sized private cars is the rack-and-pinion layout (fig. 8.5). This type of steering-box has a rack-and-pinion housing bolted along the body cross-member. The driver's steering-wheel angular movement is converted to a linear to-and-fro movement of the rack by means of the pinion and rack teeth meshing together. Each end of the rack shaft is attached to a tie-rod by means of a ball-and-socket joint. The outer tie-rod ends also have ball-joints which are bolted to the stub-axle track-rod arms. The rack shaft thus supplies the transverse steering thrust, with the tie-rod ball-joints allowing pivoting in two planes – that is, suspension vertical deflection and horizontal steering movement.

8.2 The Ackermann principle as applied to steering linkage

Understanding the fundamental principles involved in steering a vehicle about a single centre of rotation is essential for diagnosing steering faults. This section provides an appreciation of how wheel tyres react under forward rolling with and without side-forces. It then surveys the different ways in which a four-wheeled vehicle can have its axles arranged to achieve true rolling with the minimum of tyre scrub. Finally it looks at the evolution of steering geometry and its corresponding interconnecting linkage.

8.2.1 Conditions for true rolling (fig. 8.6)

True rolling occurs only when the direction of motion is perpendicular to the wheel axis (fig. 8.6(a)). A force acting parallel to the wheel axis will produce purely a scrub action (fig. 8.6(b)). When the direction of motion is neither parallel nor perpendicular to the axis of rotation, the movement will be compounded of true rolling and lateral distortion (fig. 8.6(c)) – this is the condition when the wheels are being steered.

For true-rolling conditions on a circular path, the projected axes of several wheels all moving in different curved paths must intersect at a single point which is known as the instantaneous centre (fig. 8.6(d)). If these projected axes do not intersect at a single point, a degree of tyre-scrub must result (fig. 8.6(e)).

8.2.2 Turntable steering (fig. 8.7)

Early horse-drawn carriages used a turntable centrally pivoted steering-axle in which the one-piece axle-beam pivoted about a single pin so that the complete front axle swivelled to steer the front wheels (fig. 8.7).

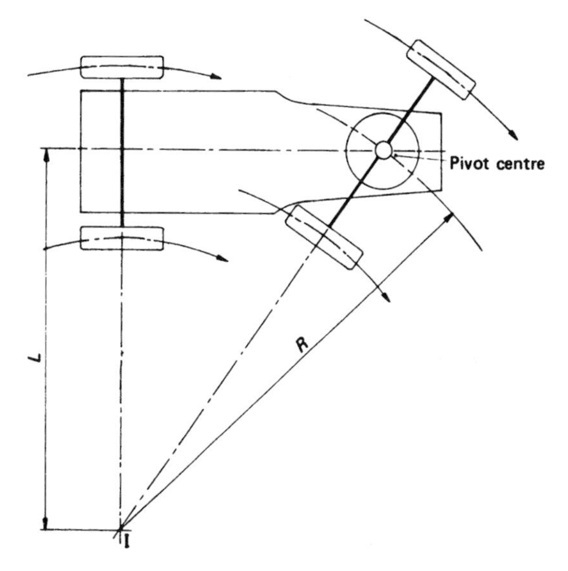

Fig. 8.7 Turntable steering

True rolling is obtained with this arrangement since a line drawn through the swing-axle axis will intersect somewhere along the fixed-rear-axle projected line, the point of intersection then being the instantaneous centre of rotation.

Because the whole axle assembly is pivoting about one point, steering with this layout required a considerable twisting movement about the central pivot. A large-diameter friction turntable is necessary to give body-roll support for the single pivot axle. Considerable clearance space is necessary to allow the axle-beam to swing about its pivot and for the inner road-wheel to clear the side of the body.

Turntable steering is not used for conventional private cars and lorries, but it is used in a modified form for articulated and towed draw-bar trailers.

8.2.3 The Ackermann principle (figs 8.8 and 8.9)
The Ackermann principle is based on the two front steered wheels being pivoted at the ends of an axle-beam. For true or pure rolling conditions when moving on a curved track with a four-wheeled vehicle, lines drawn through each of the four wheel axes must intersect at one common centre of rotation (fig. 8.9). This centre point is termed the instantaneous centre, since its actual position is constantly changing as the front-wheel angular positions alter to correct the steered vehicle's path. Since both rear wheels are fixed on the same axis but the front-wheel axles are independent of each other, the instantaneous centre must lie somewhere along an imaginary extended line drawn through the rear axle.

The original Ackermann linkage had parallel-set track-rod-arms. Both steered wheels then swivelled at equal angles, so the intersecting projection lines did not meet at one point (see fig. 8.8). It can be seen that if both front wheels were free to follow their own natural paths, they would converge and eventually cross each other. Since the vehicle must move along a single mean path, both wheel tracks would be in continuous conflict with each other and this could only be resolved by tyre slip and tread scrub.

The modified linkage has inclined track-rod arms, which makes the inner wheel swivel about its king-pin slightly more than the outer wheel. This enables the projected lines drawn through the stub-axles to converge at a single point somewhere along the rear-axle projection (fig. 8.9).

8.2.4 The Ackermann linkage (fig. 8.10(a))
Almost from the beginning, the self-propelled motor vehicle incorporated the double-pivot wheel-steering system which was invented for horse-drawn vehicles by George Lankensperger, a Munich carriage builder, in 1817. In England, Rudloph Ackermann acted as Lankensperger's agent and a patent was taken out in his own name describing the double-pivot steering arrangement.

With this linkage the track-rod arms are set parallel to each other and a track-rod joins them together. When the steering is positioned straight ahead, the linkage forms a rectangle; but, as the stub-axles are rotated about their king-pins, the steering linkage and axle-beam resembles a parallelogram. This linkage configuration will turn both wheels the same amount – see fig. 8.10(a), which shows the parallel-set linkage positioned to provide both a 20° and a 40° turn for the inner and outer wheels.

8.2.5 The modified Ackermann linkage (fig. 8.10(b) and (c))
An improvement to the Ackermann linkage layout was introduced by Charles Jeantaud in 1878. This modification inclines the track-rod arms so that lines drawn through the centres of the king-pins and the track-rod-arm ball-joints, when projected towards the rear of the vehicle, converge and intersect somewhere near the rear axle (fig. 8.9).

This trapezium linkage configuration enables the inner wheel to rotate about its king-pin pivot by a greater amount than the outer wheel, which is necessary if semi-true rolling is to be achieved (fig. 8.10(b) and (c)). True rolling will be obtained in the straight-ahead position and in one position on the left- and right-hand turns (locks). In between these points, only partial true rolling will take place. The degree of departure from true rolling and hence the amount of tyre scrub which takes place overall will depend mainly on the ratio of track-rod to track-rod-arm lengths and on the track-rod-arm angular inclination or 'set'. If the steering-linkage dimensions and settings are carefully chosen, very little misalignment will occur for angles of turn up to about 15°; however, beyond this turning angle the error increases rapidly.

Because the basic arrangement of the double-pivot linkage is unchanged and the only alteration is the setting of the track-rod arm, this layout is still referred to as the Ackermann linkage.

An important observation when considering the Ackermann-linkage inaccuracy is that the linkage was designed initially to steer rigid steel and solid rubber-tyred wheels. However, in recent years there have been considerable developments and improvements in the performance of the pneumatic tyre, and the deviation from the theoretical true-rolling angles necessary for a given

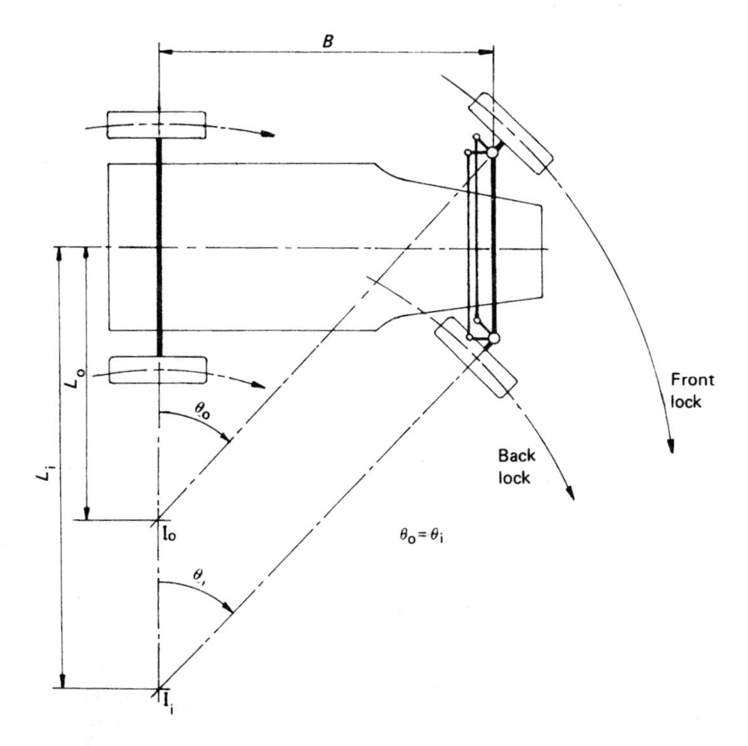

Fig. 8.8 Side-pivot steering with parallel-set track-rod arms

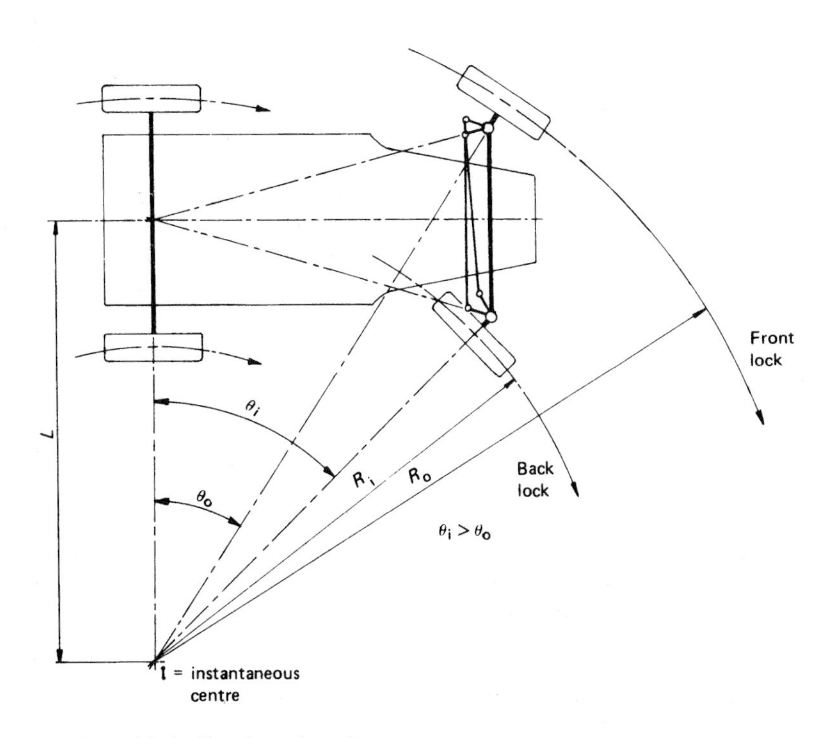

Fig. 8.9 Side-pivot steering with inclined track-rod arms

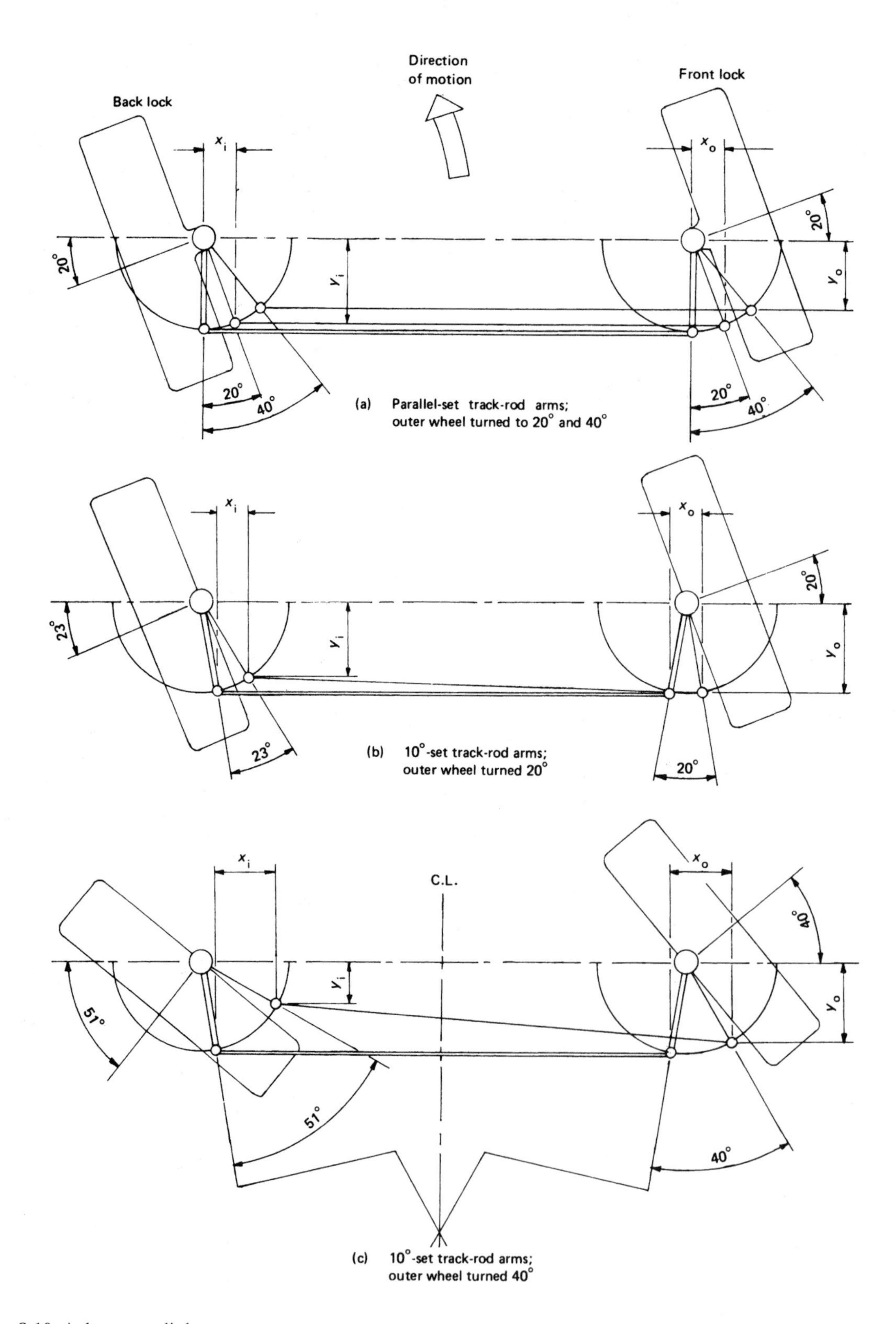

Fig. 8.10 Ackermann linkage geometry

steered circular path can readily be corrected by the tyre's side-wall flexibility and tread distortion, provided the angular error between the steered wheels is not too great.

It should be observed that, since the rear wheels turn on a smaller radius than the front wheels, it is easier to manouevre a vehicle in reverse than in the forward direction when parking.

8.2.6 Ackermann-linkage geometry (fig. 8.10)

With parellel-set steering arms (fig. 8.10(a)), the track-rod dimensions y_i, x_i and y_o, x_o remain equal for all angles of turn. With the inclined arms (fig. 8.10(b) and (c)), the inner-wheel track-rod end dimension y_i is always smaller than the outer-wheel dimension y_o when the wheels are turned. In contrast, there is very little variation between x_i and x_o for small angular movement between wheel steering locks. For small steering angles about the king-pin up to say 10°, there is very little difference between y_i and y_o and between the inner and outer wheel turning angles, but, as the angular turn of the front wheels increases, the transverse positioned track-rod tilts in the horizontal plane and produces a difference between y_i and y_o which becomes more pronounced as the steering angle becomes greater.

Figure 8.10(b) shows the steering linkage with the track-rod arms inclined at 10° and the outer wheel turned 20°. The corresponding inner wheel is shown to rotate 23°. Likewise for 40° outer-wheel turn, the inner wheel rotates 51° (fig. 8.10(c)). In both cases it can be seen that, for a given angular movement of the stub-axles, the inner-wheel track-rod arm and track-rod are more effective than the outer-wheel linkage in turning the steering wheels.

To sum up the relationship between inner- and outer-wheel angular movement between steering locks: it can be said that, for a given amount of transverse track-rod movement with inclined track-rod arms, the least effective angular displacement of a stub-axle pivot occurs in the straight-ahead region, and the most effective angular displacement takes place as the stub-axles move away from the mid-position. In other words, the angular movement of the inner wheel relative to the outer wheel becomes much greater as both wheel turns approach full lock – see fig. 8.11. With modern radial tyres, the difference between front- and back-lock steering angles is sometimes reduced.

8.3 The need for front-wheel alignment (fig. 8.12)

For the road-wheels to roll freely and with the minimum of effort, opposite wheels must be approximately parallel to each other when the vehicle is in motion along a straight path.

Each wheel will want to follow a path perpendicular to its own axis of rotation. This means for example that, if the front wheels are aligned so that they are converging towards the front, then in the forward direction both wheels will be trying to roll closer together. Conversely, if the wheels are set so that they are diverging towards the front, the free-rolling tendency will be for the wheels to move farther apart. In both cases the wheels will be restrained from moving closer or farther apart by the steering linkage, therefore the average path followed will mean that both wheels will have a continuous tendency to either push together or pull apart. Consequently, while rolling forward, each wheel will simultaneously be slipping laterally, so a continuous cross-tread scrub action is established, with the obvious results – excessive tread wear, heavy steering, and probably poor fuel consumption.

Examining a tyre tread subjected to excessive lateral slip or scrub will show a diagonal wear pattern, with the leading side of the tread blocks heavily worn and the trailing side having an extruded feather-like appearance. This may be sensed by rubbing the palm of the hand across the tread and feeling in which direction the hand glides smoothly or encounters surface roughness.

Figure 8.12(a) shows converging wheels with the characteristic inward feathering of the tread blocks. If the hand is moved over the tread and inward, it should feel smooth.

Figure 8.12(b) shows diverging wheels with the

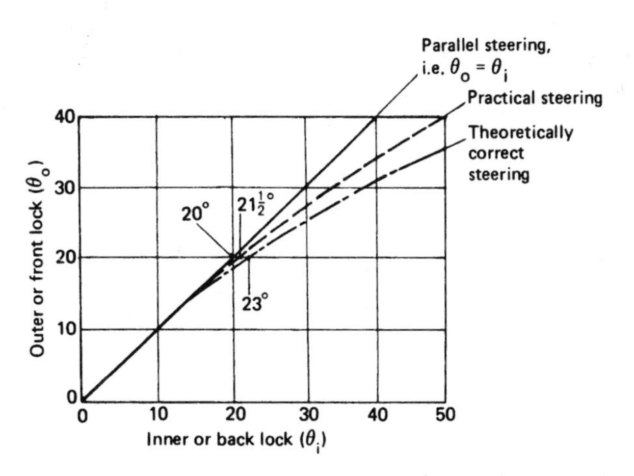

Fig. 8.11 Front- and back-lock steering-angle curves

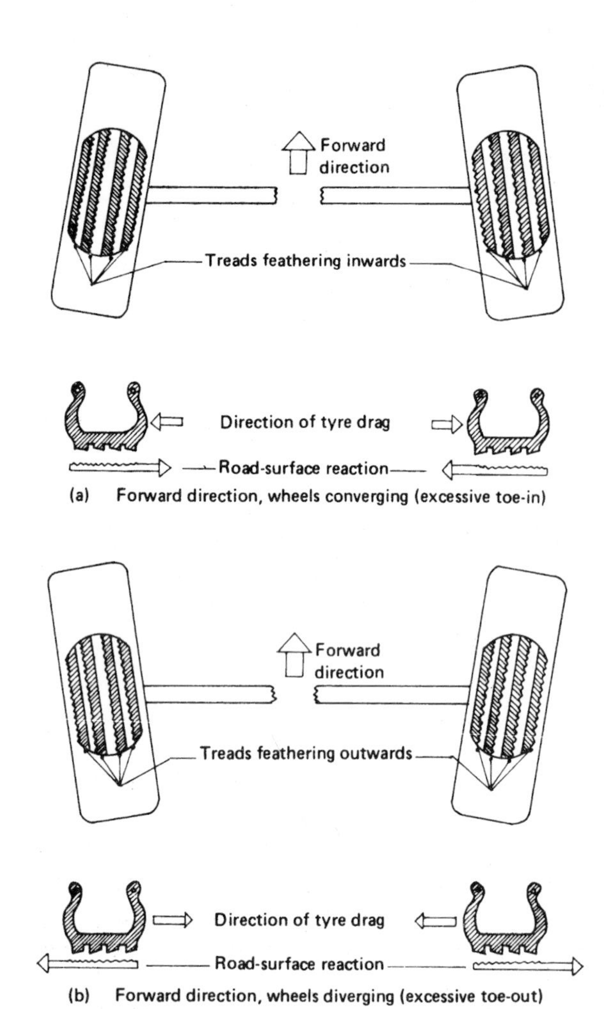

Fig. 8.12 Effects of wheel misalignment

usual outward feathering of the tread blocks and grooves. An inward movement of the hand over the tread will experience a degree of roughness.

Correctly aligned and parallel-rolling wheels will produce an even wear pattern across the tyre tread, so that lateral movement of the hand inwards or outwards across the tread should feel smooth.

8.4 Front-wheel toe-in or toe-out (figs 8.13 to 8.15)

Toe-in is the amount by which the front-wheel rims are set closer together at the front than at the rear with the wheels in a straight-ahead position when the vehicle is stationary (fig. 8.13(a)). Alternatively, toe-out is the amount by which the front-wheels rims are set father apart at the front than at the rear (fig. 8.13(b))

i.e. toe-in = $T_r - T_f$
and toe-out = $T_f - T_r$

Toe-in or toe-out compensates for movement within steering ball-joints, suspension rubber bush-joints, and any slight deflection of the track-rod arms or suspension arms when the vehicle is in motion. The objective of the non-parallel stationary alignment of the front steered wheels is for the toe-in or toe-out to be taken up when the vehicle is moving – this allows both wheels to run parallel under normal driving conditions.

Figure 8.14 shows a rear-wheel-drive steering and suspension layout. It can be seen that the propelling thrust pushes the suspension cross-member and body forwards during drive, which is resisted by the road reaction on the tyres. The resultant outward twisting of both wheels will distort the suspension wishbone rubber bushing so that the front wheels will diverge. To counteract this tendency, the wheels are initially given a toe-in so that under driving conditions the wheels will run parallel.

Figure 8.15 shows a front-wheel-drive steering and suspension layout, and here the driving traction is imparted to the front wheels so that the stub-axle assembly is pulling forward the suspension cross-member and body. The natural reaction of the wheels will then be to twist inwards and so distort the suspension wishbone rubber pivot joints in a horizontal plane so that the front wheels will converge. To correct this converging tendency when in motion, the wheel track is adjusted with a toe-out when the vehicle is

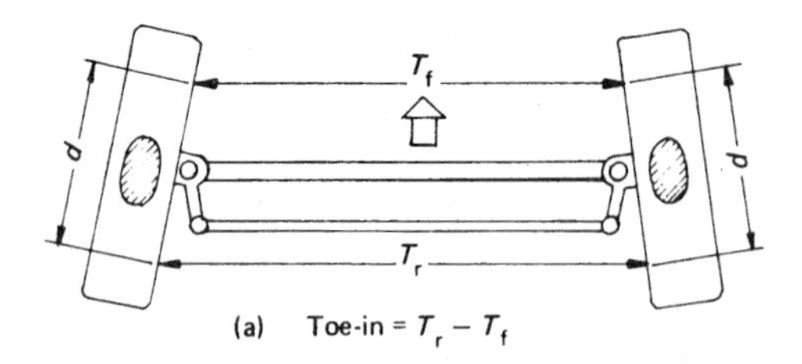

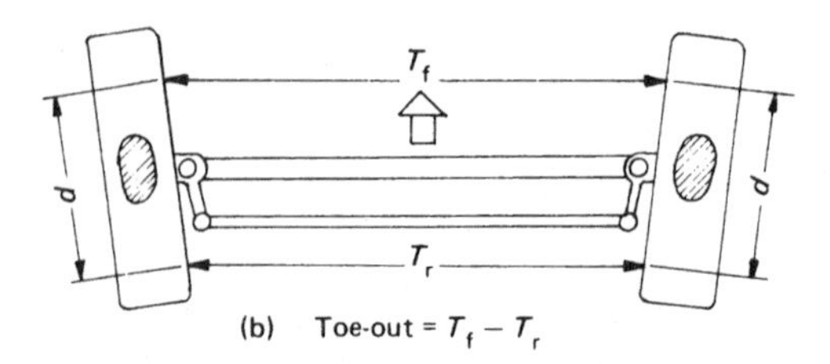

Fig. 8.13 Illustration of steering toe-in and toe-out

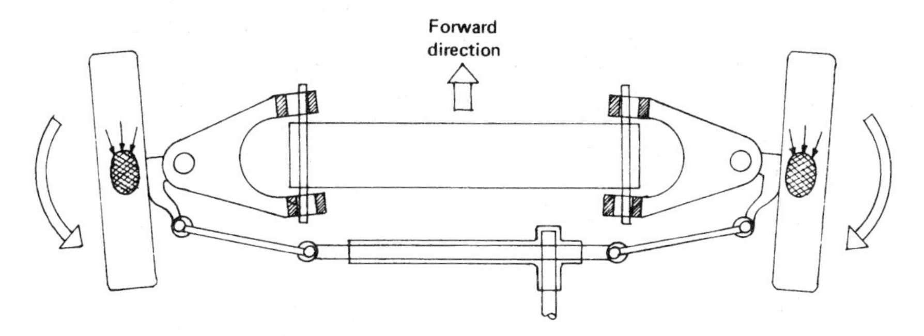

Fig. 8.14 Rear-wheel-drive steered-wheel reaction due to forward motion

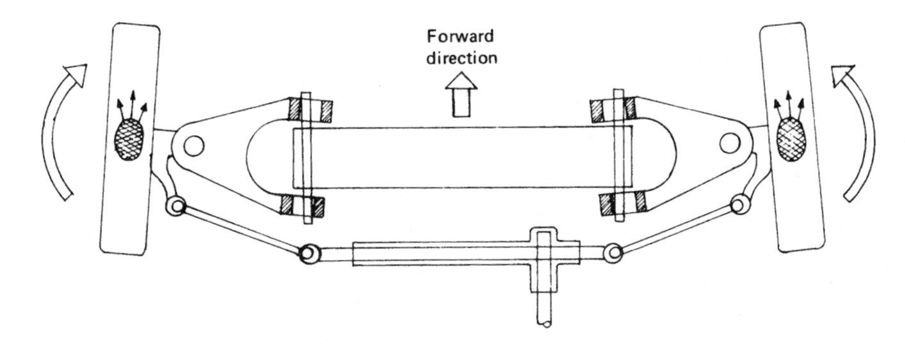

Figure 8.15 Front-wheel-drive steered-wheel reaction due to forward motion

at rest, this being equal to the amount of distortion in the horizontal plane expected when the vehicle is being driven. This explanation given is an oversimplified one, and, due to other variations and combinations in steering and suspension geometry, front-wheel-drive vehicles sometimes have zero toe-out.

8.4.1 Toe-out on turns (fig. 8.16)

Toe-out on turns relates to the diverging in the forward direction of the front wheels when the stub-axles are rotated about their king-pins (fig. 8.16). The difference between the angles of turn imparted to the front steered wheels gives the magnitude of the toe-out in degrees. The amount of toe-out on turns is obtained by the inclination or set of the track-rod arms and will depend upon the track-rod-arm length, its angular set, and the track width.

The measurement of toe-out angle on turns is obtained by rolling the front wheels on to radius or turntable gauges in the straight-ahead position and setting the pointer of each gauge to zero. The outer road-wheel is then turned through 20° as indicated on the turntable scale, and the opposite inner-wheel angle is observed on its scale. This generally should read between 20.5° and 23°. That the toe-out angles on each lock are approximately equal is more important than the actual amount.

A vehicle with excessive tyre wear but a correctly aligned track in the straight-ahead position should be investigated by checking the angular toe-out of the inner wheel with the outer wheel set at 20°. Any error between the readings on both sides may be due to a bent track-rod or arm etc.

8.5 Track alignment and adjustment (fig. 8.17)

Adjustment of the front-wheel parallelism for true rolling must take into account the static toe-in or toe-out recommended by the vehicle manufacturer. The method of adjustment consists of increasing or decreasing the length of the track-rod or tie-rods until the correct track-alignment setting is obtained.

The simplest track-alignment tool is the trammel gauge. This is a horizontal bar that can slide in slots formed in a pair of perpendicular posts that stand on the floor. Each post has a prod pointer attached – one has an adjustable fixed prod and the other has a spring-loaded prod with a reference scale (fig. 8.17).

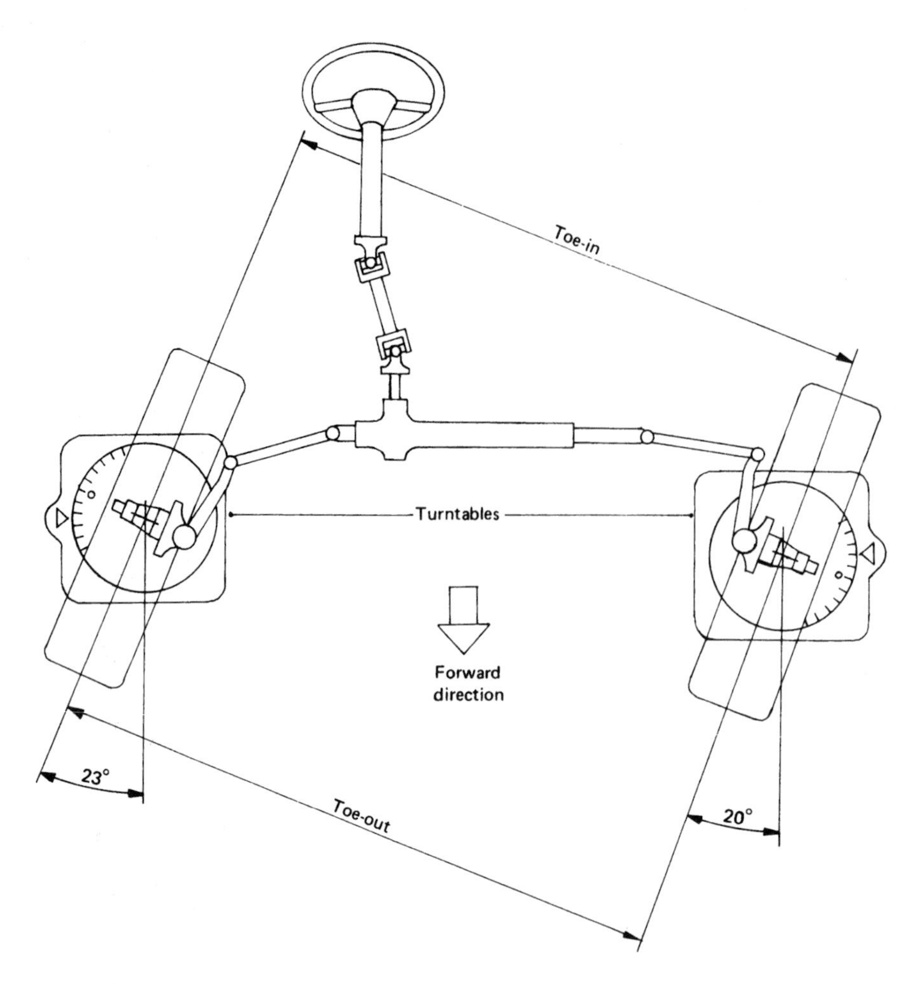

Fig. 8.16 Toe-out angle on turns

Always check track alignment on level ground, otherwise an incorrect reading may be obtained.

To check the track alignment:

1 Place the trammel bar across and in front of the steered wheels, with the posts positioned on the outside of each wheel and the prods pointing towards the wheel rims. Set the fixed prod against the wheel inner rim.
2 Pull and release the spring-loaded prod so that it presses against the other wheel inner rim and observe the scribed scale reading.
3 Mark the points where each prod touches the rim with chalk and roll the vehicle forward so that the wheels turn through exactly 180° so that the chalk marks are now at the rear of the wheels.
4 Now position the trammel bar behind the steered wheels, with the fixed prod just touching the inner rim and the spring-loaded prod free to adjust itself to the inner rim. The difference between the front and rear measurements provides the toe-in or toe-out.

If the alignment is out, carry out the various adjustment procedures as explained in the following sections.

8.5.1 Single-piece axle-beam steering-linkage adjustment (fig. 8.2)

The track-rod assembly used with an axle-beam suspension uses a tubular rod with ball-and-socket track-rod end joints. Each end of the rod tube is internally threaded – one end has a right-hand thread, and the other a left-hand thread. Each ball-and-socket track-rod end has an extended stub which is threaded with either a right- or a left-hand thread. Therefore the right-hand-thread ball-joint end is screwed into the right-hand-thread track-rod end, and likewise

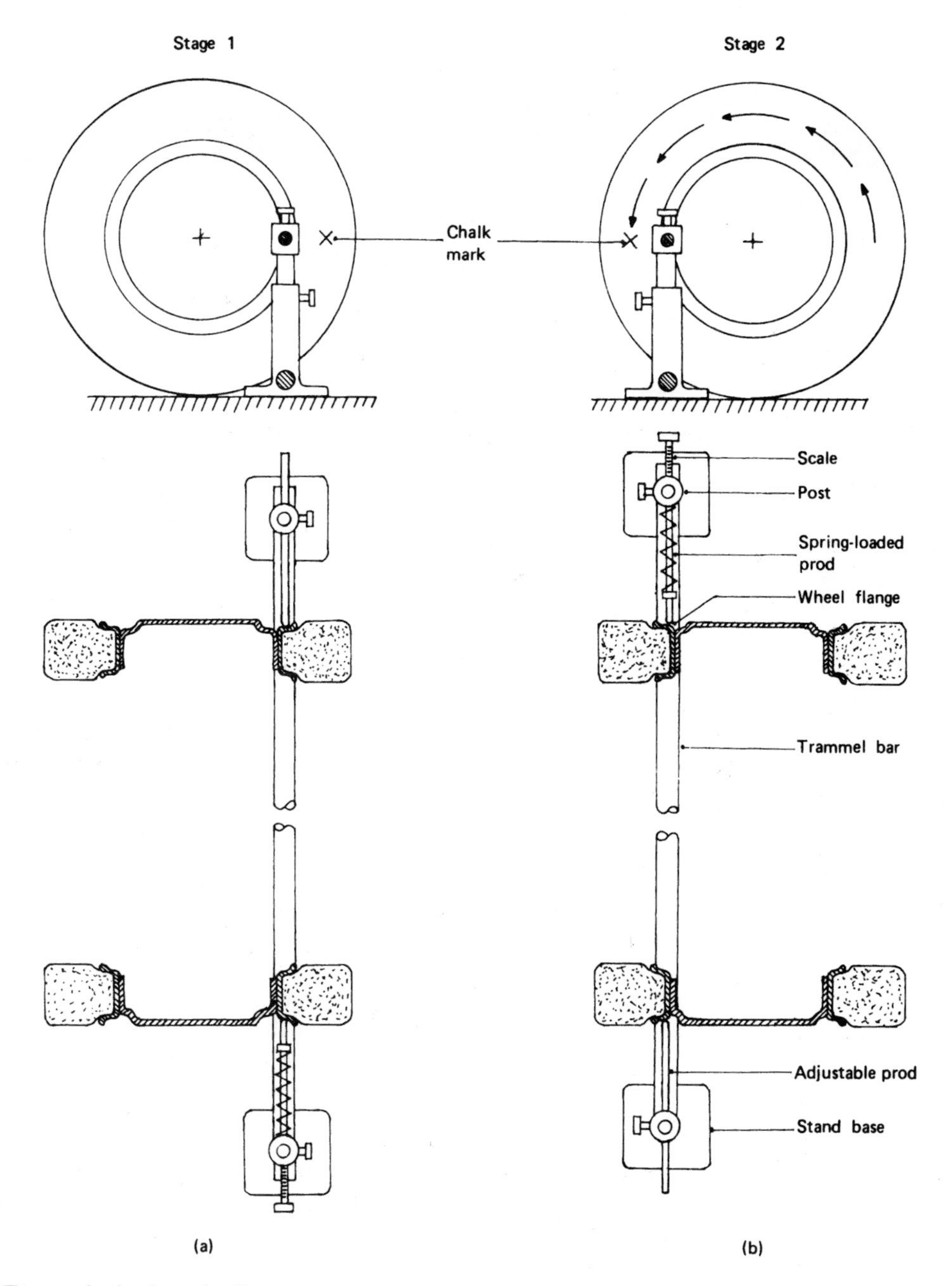

Fig. 8.17 Trammel wheel-track alignment gauge

the left-hand ball-joint and track-rod end are screwed together (fig. 8.18(b) and (c)).

Adjustment of the rod assembly is achieved by slackening off both pinch-clamps and then rotating the rod so that the screwed ends move either together or apart until the desired toe-in is obtained. Finally, both pinch-clamps must be tightened to maintain the required track.

8.5.2 Three-piece steering-box and idler steering linkage (fig. 8.4)

The three-piece steering linkage has a pair of tie-rods, one on each side adjacent to the front wheel. The end of each tie-rod screws on to a right- or left-hand-threaded ball-joint end so that each tie-rod may be individually adjusted in length (fig. 8.18(b)).

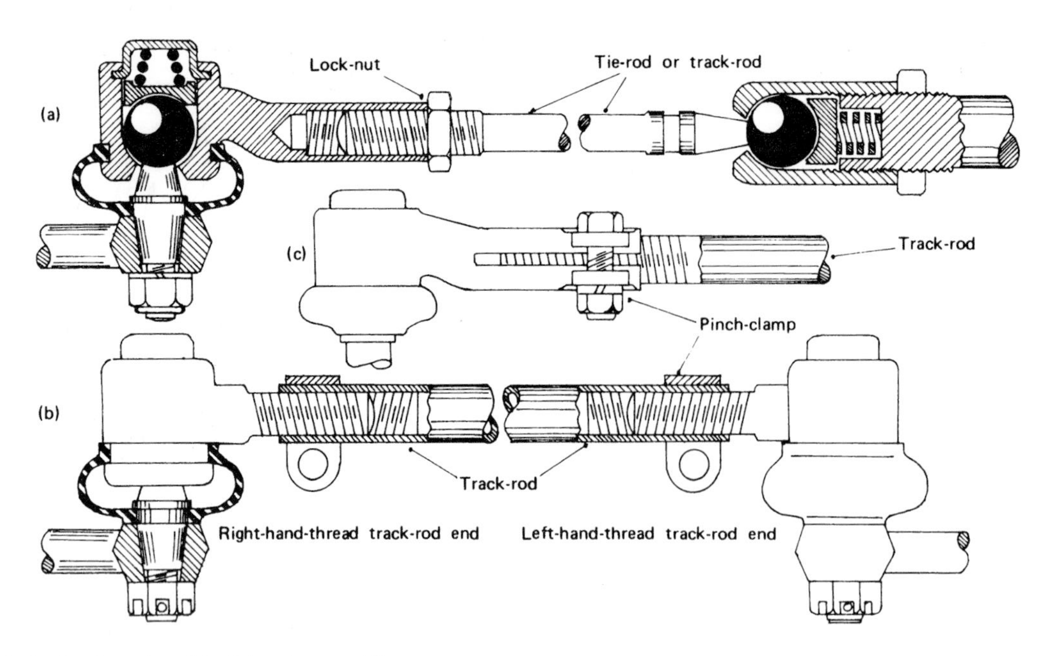

Fig. 8.18 Track-rod-end ball-joint assemblies

The general procedure for checking the track alignment is to set the steering-wheel in the straight-ahead position by dividing the angular rotation of the driver's steering-wheel between the extreme left and right locks. Next observe if the road-wheel on the driver's side is in the straight-ahead position by measuring the horizontal distance from the lower wishbone-arm cross-member pivot centre to the wheel rim, or between two other reference points. This length is usually quoted in the vehicle service manual.

To adjust the tie-rod length on the driver's side, slacken both lock-nuts, rotate the rod tube to give the correct length setting, and then retighten the lock-nuts. Now check the toe-in or toe-out. If it is not within the specified limits, slacken the lock-nuts (or pinch-clamps) on the rod to the opposite side to the driver and adjust the rod as necessary. Secure this setting by retightening the lock-nuts or clamps. Finally, roll the vehicle forward and recheck the track alignment.

8.5.3 Two-piece rack-and-pinion steering linkage (fig. 8.5)

With the two-piece steering linkage there is a tie-rod on each side of the rack shaft. Each tie-rod is a solid rod with a spherical ball on the inboard end. The outer end of each rods has a right-hand thread cut – these screw inside the extended ends of the ball-and-socket joints (fig. 8.18(a)). Protecting the inboard ball-and-socket rack joint is a bellows-type cover which is held in position by a metal clip. This should be loosened if any track adjustment is necessary, to prevent it becoming twisted.

The tracking procedure starts with setting the driver's steering-wheel in the straight-ahead position. Sometimes a dimension is given for the distance from the inner ball-joint socket to the edge of the rack housing. This length should be measured and the rack positioned accordingly.

Now observe if the lengths of the two tie-rods are equal when the steering is in the straight-ahead position. One way of comparing tie-rod lengths is to count the number of exposed threads adjacent to the outer ball-joint assembly on the tie-rod.

If a correction is necessary, release the outer ball-joint lock-nut on each tie-rod while holding the flat or hexagon on the other end of the tie-rod. Preventing the ball-socket rotating while loosening the lock-nut avoids damage to the spring-loaded ball-and-socket seating.

Initially, rotate one of the tie-rods until its length corresponds to the other tie-rod. Next turn both tie-rods by an equal amount until the correct wheel toe-in or toe-out is observed on the gauge. This is achieved by turning both tie-rods in the same direction, otherwise they will become unequal in length. Rotate the rods through only a quarter turn at a time before rechecking the alignment. One complete turn of the rod will alter the alignment by approximately 3 mm.

Note: never grip the threaded part of the tie-rods during adjustment; always release the bellows before adjustment; and, finally, retighten the bellows-cover clip and the tie-rod lock-nuts and recheck the toe-in or toe-out.

8.6 Rack-and-pinion steering assembly (figs 8.19 to 8.21)

The rack-and-pinion steering-box is designed to have the following properties:

a) to give a gear reduction to reduce steering-wheel effort;
b) to reduce the directness of the steering response;
c) to enable the front wheels to turn through an angle of about 70° between steering locks;
d) the gearing should tend to be semi-irreversible – damping out any light wheel wobbles, but allowing direct road-wheel reactions to be

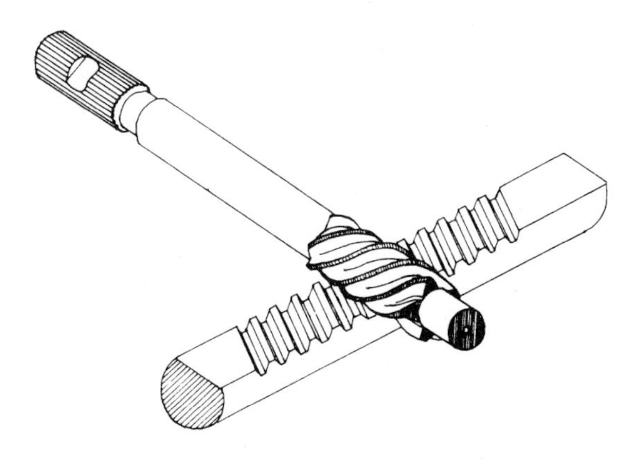

Fig. 8.19 Pictorial view of a rack and pinion

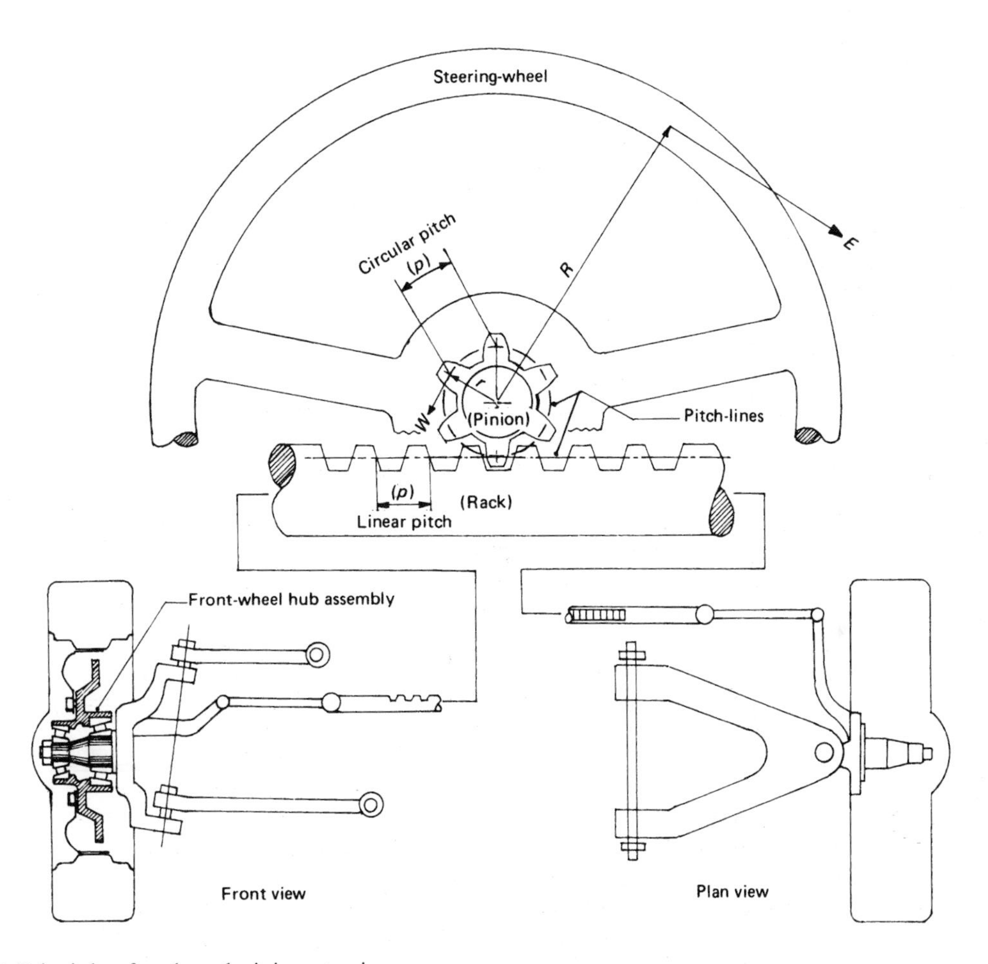

Fig. 8.20 Principle of rack-and-pinion steering

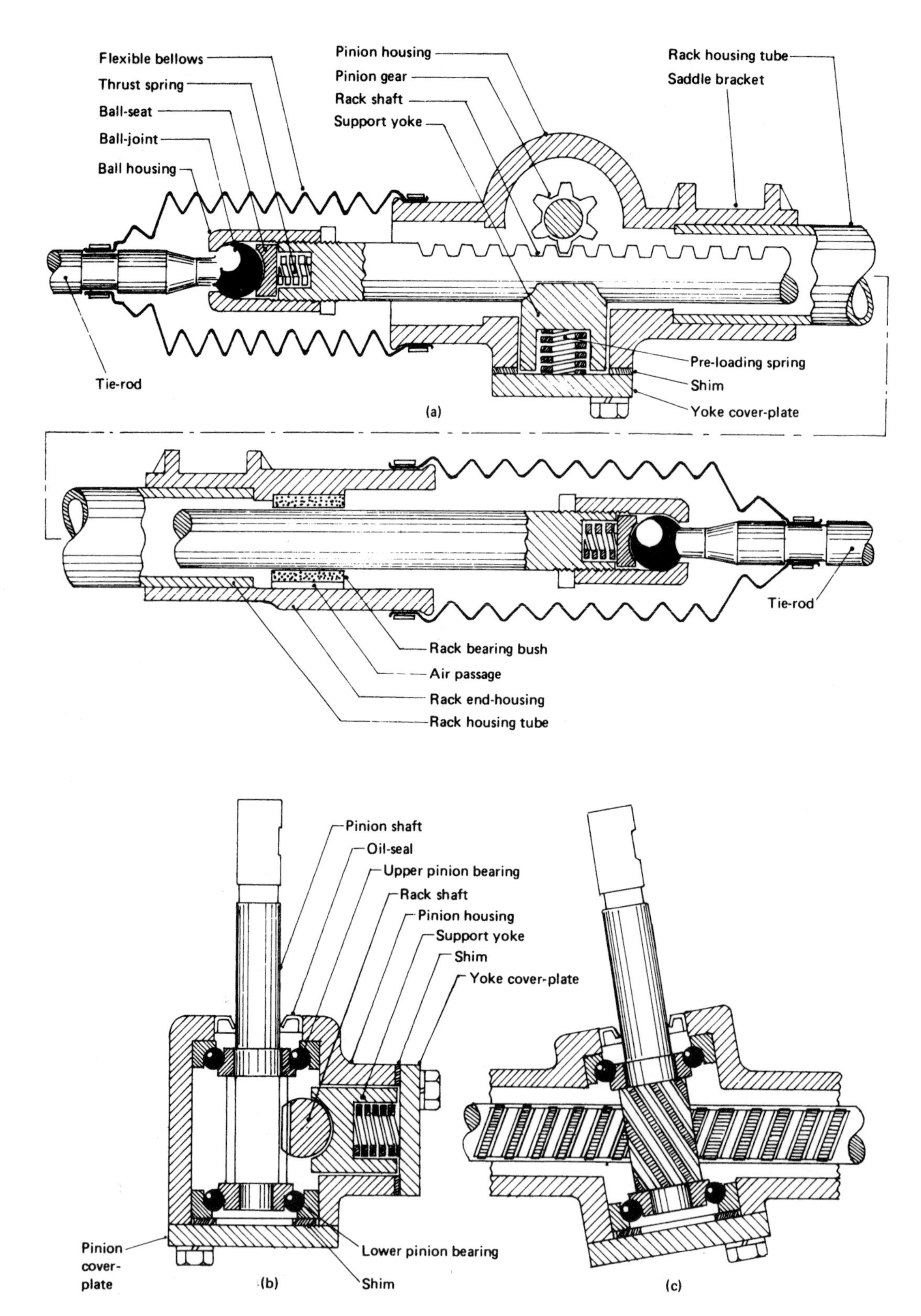

Fig. 8.21 Rack-and-pinion steering-gear

transmitted back to the steering-wheel. If the gear were completely irreversible, road-wheel reaction would strain the steering linkage as a whole and could result in permanent damage.

8.6.1 Principle of operation (figs 8.19 and 8.20)
The rack-and-pinion mechanism is designed to transfer the circular input motion of the pinion into a linear rack output movement so that side-to-side travel of the rack may then be relayed through the tie-rods to the tie-rod steering-arms and the stub-axles (fig. 8.20). The transfer of motion from the pinion to the rack is achieved by the pinion teeth pushing the mating rack teeth along the rack axis in either direction. For correct mesh, the circular pitch of the pinion should equal the linear pitch of the rack.

Let R = steering-wheel radius

r = pinion pitch-circle radius

t = number of pinion teeth

p = linear or circular pitch

E = input steering-wheel effort

and W = output rack load

Now consider the pinion to make one revolution (fig. 8.20);

then input steering-wheel movement $x_i = 2\pi R$

and output rack movement $x_o = 2\pi r$

$= t\,p$

$$\therefore \text{movement ratio MR} = \frac{x_i}{x_o} = \frac{2\pi R}{2\pi r} = \frac{R}{r}$$

$$\text{also MR} = \frac{x_i}{x_o} = \frac{2\pi R}{t\,p}$$

Example 1 A rack-and-pinion steering has a pinion of 16 mm pitch-circle diameter. What effort must be applied on the 320 mm diameter steering-wheel to overcome a resistance of 500 N experienced transversely on the rack?

$$\text{Movement ratio MR} = \frac{R}{r} = \frac{160\text{ mm}}{8\text{ mm}} = 20{:}1$$

$$\text{When there is no friction, MR} = \frac{x_i}{x_o} = \frac{W}{E}$$

$$\therefore \text{input effort } E = \frac{W}{\text{MR}} = \frac{500\text{ N}}{20} = 25\text{ N}$$

Example 2 A rack-and-pinion steering gear has a 5 tooth pinion of 10 mm pitch. If an effort of 15 N is applied by each hand on the 350 mm diameter steering-wheel, determine (a) the movement ratio, (b) the force transmitted to the tie-rods.

a) $$\text{Movement ratio MR} = \frac{2\pi R}{t\,p}$$

$$= \frac{2 \times 22/7 \times 175\text{ mm}}{5 \times 10\text{mm}} = 22{:}1$$

b) Output load $W = E \times \text{MR}$

$= 2 \times 15\text{ N} \times 22 = 660\text{ N}$

8.6.2 Construction and description (fig. 8.21)
The rack-and-pinion steering-box has a pinion-gear shaft connected indirectly (via a coupling and universal joints) to the driver's steering-wheel. Two ball-race bearings straddle and support the pinion gear. Positioned perpendicular to the pinion is a circular-sectioned shaft with teeth cut along one face. This is known as the rack and is aligned to mesh with the pinion. The rack slides in its tubular housing between two bearings: a plain bush bearing at one end of the housing, and an adjustable half-bearing support yoke opposite the pinion gear at the other end of the casing. The half-bearing yoke is made to push the rack teeth into mesh with the pinion teeth – it thus controls the amount of backlash between the rack and pinion teeth.

8.6.3 Design features (figs 8.19 to 8.21)
The rack is a gear of infinite diameter, so its pitch-circle will be a straight line. Both gear-teeth profiles are of the involute form, so the side profile of the pinion teeth will be curved, but, because the rack pitch-circle is a straight line, the sides of the rack teeth will be straight and inclined (fig. 8.20).

Some of the early pinions were of the simple spur-gear form, with straight-cut teeth, but these have been mainly replaced with helical-toothed pinions. Rack-and-pinion straight-cut teeth will mesh with only one pair of teeth contacting at any one moment, and these teeth will come into contact over their whole width at the same time. Thus the tooth-profile contact point will be always moving from one side to the other of the pitch-line of both the rack and the pinion teeth. Uneven movement of the rack will result as the steering load is transferred from one pair of teeth to the next. With helical-cut rack-and-pinion teeth, more than one pair of teeth come into contact at any instant, so the teeth engage progressively and there will be some contact on the pitch-line at all times. Helical pinion teeth (fig. 8.19) are therefore able to take greater loads, they engage smoothly, and they operate quietly.

By tilting the pinion axis to the perpendicular to the rack travel (fig. 8.21(c)), the effective pinion pitch-radius will be increased for a given rack movement so that fewer and stronger pinion teeth may be used. This also enables larger gear-ratio reductions to be employed for a given rack travel.

The helix angle of the teeth and the inclination of the pinion axis to the perpendicular both introduce a sliding action between meshing teeth; consequently friction is increased. Due mainly to the damping effect of this internally generated friction, road-wheel shock impulses which feed back to the steering linkage will thus be prevented from being transmitted back through the rack to the pinion and to the driver's steering-wheel.

The movement ratios of rack-and-pinion steering gears will, for manual steering, depend largely on the weight of the car. Typical values range from 12:1 for small cars up to 22:1 for heavier vehicles.

8.6.4 Pinion-bearing pre-load adjustment (fig. 8.21(b) and (c))

The pinion ball-race bearings must be assembled and adjusted to provide a specified axial pre-load within the limits of a given rotational torque. This initial degree of compression applied to the pinion bearings eliminates any end-float or relative free movement in the bearings when subjected to operating conditions.

Assuming that the rack-and-pinion assembly has been dismantled, the pre-load adjustment procedure is as follows.

1 Press out the old pinion-shaft oil-seal.
2 Fit the upper outer bearing race into the pinion housing.
3 Fit the upper inner bearing race on to the pinion shaft.
4 Refit the rack in the housing.
5 Insert the pinion and centralise the rack.
6 Fit the lower inner and outer bearing assembly.
7 Pack sufficient shimming under the lower bearing so that it stands proud of the pinion housing.
8 Fit the bearing-housing end cover-plate without a gasket and hand-tighten the bolts.
9 Measure with feeler gauges the gap between the pinion housing and the end cover-plate.
10 Remove the bolts and the end cover-plate.
11 Alter the shim pack to obtain between 0.270 mm and 0.33 mm gap between the housing and the cover-plate.
12 Replace the bearing end cover-plate with the correct joint gasket and fully tighten the bolts.
13 Fit a new pinion-shaft oil-seal.

8.6.5 Rack bearing-support-yoke adjustment (fig. 8.21(a) and (b))

The correct meshing of the rack-and-pinion teeth is obtained by moving the bearing support yoke towards or away from the rack. To adjust the relative position of the bearing support yoke to the pinion housing, selected thicknesses of shims are inserted between the housing and the yoke cover-plate.

The procedure for setting the initial depth of mesh for the rack-and-pinion teeth is as follows:

1 Fit the bearing support yoke, cover-plate gasket, and yoke cover-plate without the pre-loading spring and shim pack.
2 Fit the cover-plate bolts and tighten them evenly and progressively while turning the pinion shaft backwards and forward through 180° until it is just possible to rotate the pinion. A pre-load rotational torque of about 0.15 N m is generally specified.
3 Measure with feeler gauges the clearance gap between the cover-plate and the housing and then remove the cover-plate.
4 Select a shim-pack thickness equal to the cover-plate-to-housing gap plus 0.05 mm to 0.15 mm.
5 Reassemble, fitting the pre-loading spring, the prepared shim pack, and the cover-plate.
6 Tighten the bolts securing the yoke cover-plate and check the torque load required to start movement of the pinion. This usually must not exceed 0.25 N m.

8.7 Steering gearbox

Manual steering gearboxes are now only used on a few vans and light trucks and occasionally in conjunction with an externally coupled power cylinder on large commercial vehicles. The majority of cars have adapted the rack and pinion steering layouts which are more suitable for independent suspension. Inevitably power-assisted steering incorporating an integral steering gear, reaction control valve and power cylinder is demanded these days by drivers, as this makes driving easier and reduces physical fatigue.

Nearly all types of steering gearboxes utilise some sort of screw-and-nut mechanism, the screw becoming the input from the steering wheel via a shaft joint linkage, whereas the nut, which is made to move to and fro along the screw, is connected via a rocker-shaft to the drop-arm which thus forms the output member.

8.7.1 Recirculating ball rack and sector steering gearbox (fig. 8.22(a)–(c))

This steering gearbox is basically a manual screw-and-nut-type steering device, which reduces the screwing friction which would be generated between the engaging threads by replacing the screw-and-nut threads by a series of recirculating steel balls which roll between internal and external matching spiral grooves (worms) machined on both the nut and screw-shaft respectively. Within the nut there are two sets of passage-ways with a corresponding guide cage which permits the steel balls to be loaded into the nut during assembly. Thus there are two independent sets of steel balls which circulate as the worm-shaft is made to rotate. It thus spreads the thrust load between the worm-nut and worm-shaft, prevents jamming of the screw mechanism, and minimises ball and track wear.

The gear reduction is obtained in the first stage by the small linear movement of the worm-nut along the worm-shaft for a relatively large angular movement of the worm-shaft, that is, the input drivers steering-wheel. In the second stage the sector (portion of a gear wheel) gear shaft teeth mesh with the rack-teeth formed externally on one side of the worm-nut. It therefore relays the nut movement to the drop-arm by converting the linear movement of the nut back to an angular movement which is imparted to the drop-arm.

The worm-shaft is supported between two ball race bearings which can be adjusted for end-float by screwing in or out of the adjustment sleeve until the correct pre-load is obtained. To minimise wear and to reduce friction even further, the sector-shaft is itself supported between two sets of needle rollers. The rack and sector teeth are tapered slightly so that back-lash between the meshing teeth can be altered by moving the sector shaft up or down relative to the nut, by means of the sector-shaft adjustment screw, until the correct clearance is achieved. For the whole mechanism to operate effectively over a long period of time all the moving components are submerged in oil, the oil being retained by the lip-type oil-seal (see fig. 8.22(a) and (b)). The lubricant level should be periodically checked and topped up to the correct height. Under-filling the box leads to heavy steering; over-filling will promote pressure build-up and may cause a blow-out of the oil seal.

8.8 Front-wheel bearing-hub assembly (fig. 8.23)

This essentially is made up from a stub-axle, an externally cylindrical sleeve hub, a pair of taper-roller bearings, a grease-seal, a castellated adjustment nut and split-pin, a washer, and a dust-cap.

A centrally flanged cylindrical sleeve known as the hub is mounted over small outer and large inner taper-roller bearings supported on the stub-axle. To increase the resistance to tilting loads, the bearings are designed to absorb both radial and axial loads when assembled in the hub. The slackness between the taper rollers and the inner and outer races must be taken up by spinning the hub assembly while at the same time tightening the adjustment nut until all the free play has been taken up. The bearings are then pre-loaded by tightening the nut with a torque wrench to some predetermined torque setting. Alternatively, the castellated adjustment nut may be tightened up fairly tightly, while revolving the hub to ensure that the rollers and their races are in contact, and then slackened by approximately a quarter turn until one of the slots in the nut aligns with the hole in the stub-axle. The split-pin should then be pushed through and bent over to secure the nut in position.

The flanged hub supports the road-wheel and either a brake-drum or a brake-disc.

The whole hub assembly is partially packed with grease and, to prevent this grease from escaping, a radial-lip grease-seal is pressed inside the hub next to the larger bearing. The lip of the seal is positioned to face this bearing and fits over the cylindrically machined surface on the stub-axle.

A dust-cap with a central vent hole encloses the end of the hub. This area of the hub should not be filled with grease.

8.8.1 Bearing lubrication

If the bearing and seal are to be checked, remove the split-pin and the castellated nut on the end of the stub-axle and withdraw the hub assembly. Wash all parts thoroughly in paraffin, and examine the bearing rollers and tracks for wear and damage. Before replacing the hub, pack the inside of the hub and the bearings with grease (see fig. 8.23) – the amount required is about one and a

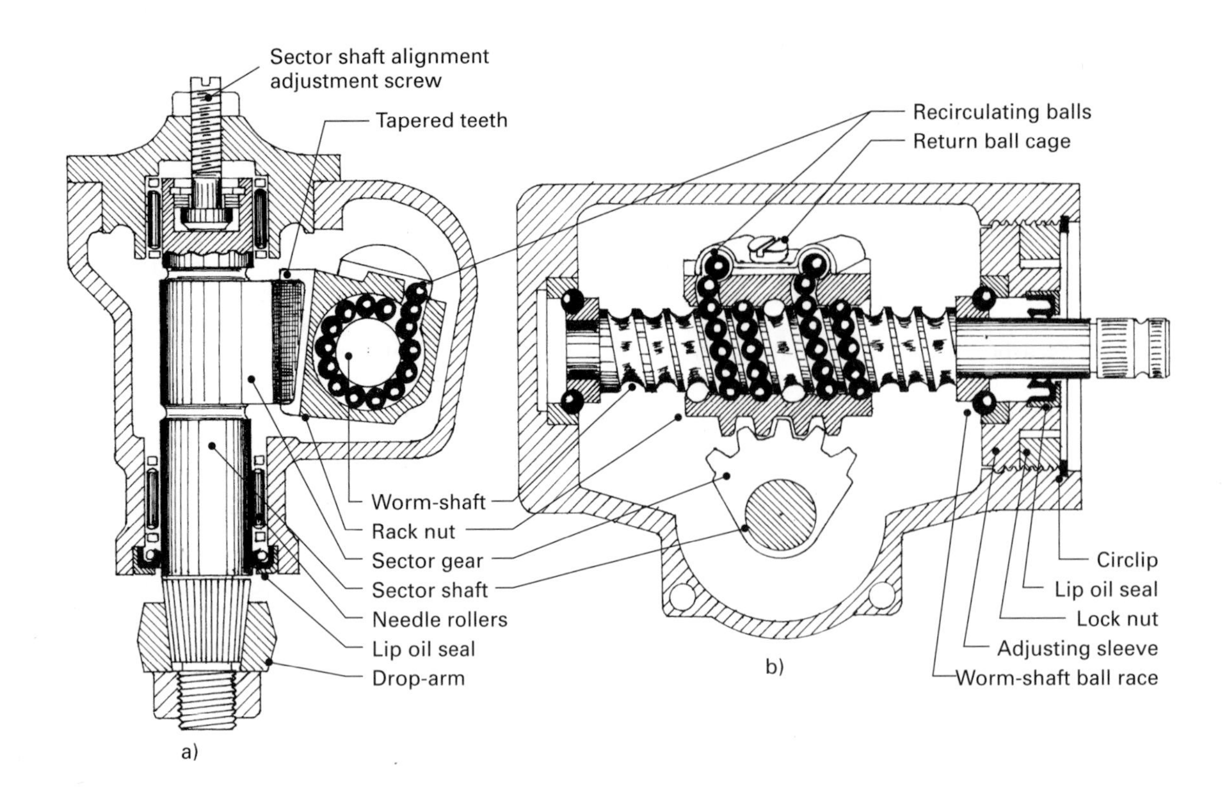

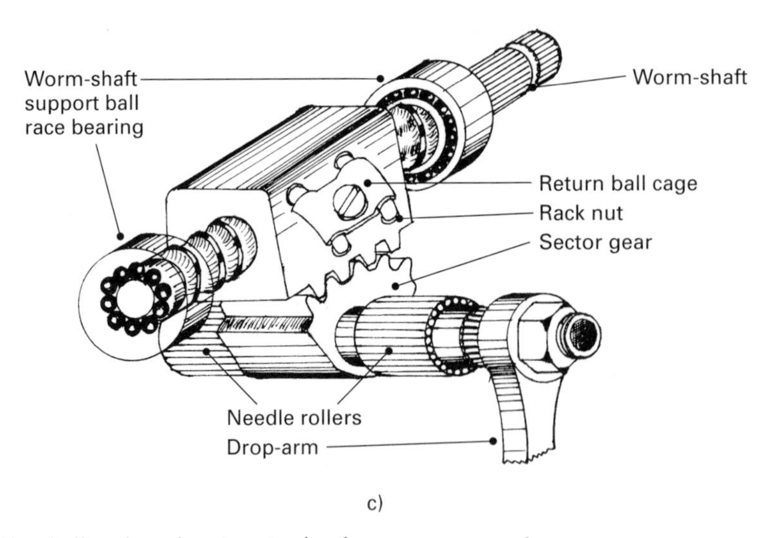

Fig. 8.22 Recirculating ball rack and sector steering box

half dust-capfuls distributed within the hub housing.

Overpacking bearing hubs with grease will cause churning and very high running temperatures. High operating temperatures and excessive mechanical working of the grease between the rollers and their tracks will eventually break down the grease so that it becomes soft. Once the grease has softened, the centrifugal force acting on the grease in the hub will throw it out-

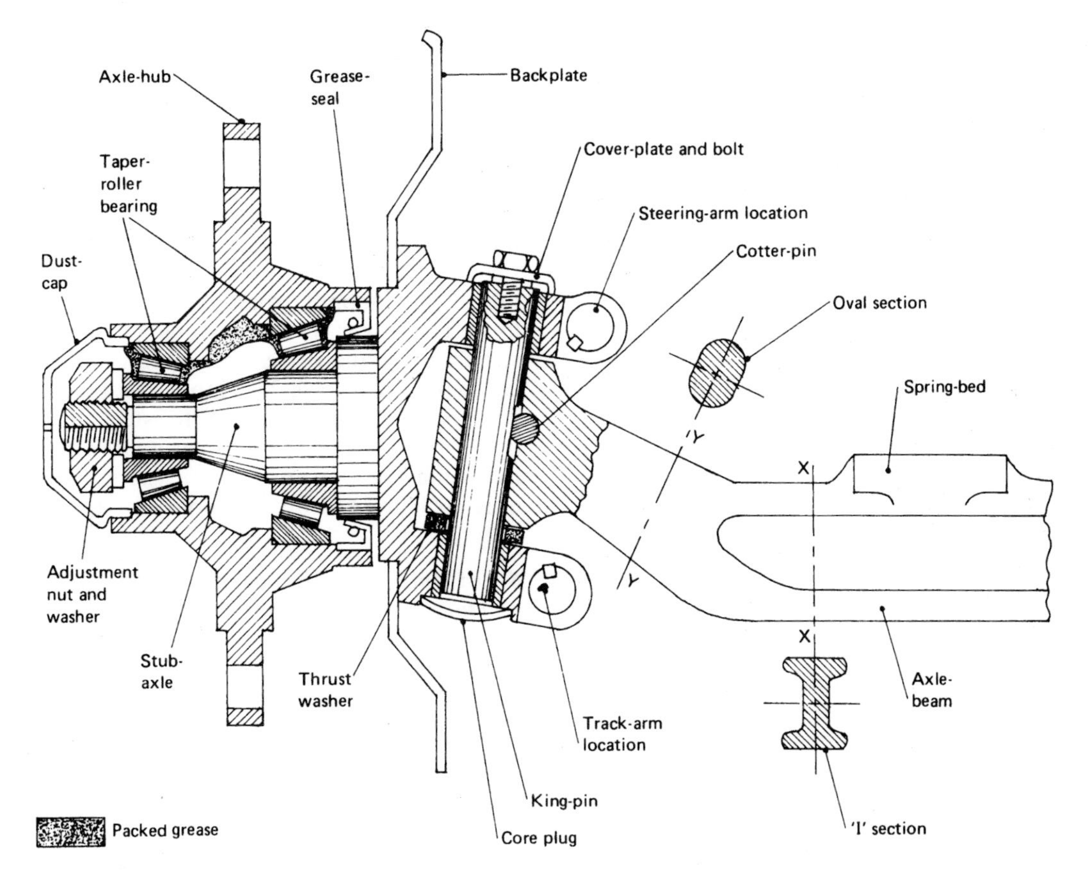

Fig. 8.23 Axle-beam and stub-axle assembly

wards. The grease then flows towards the lip seal and, if the air pressure within the hub is high due to the increased heat, the low-viscosity grease will seep past the seal lips.

Bearing greases are based on mineral oil with soaps of calcium, sodium, or lithium compounds added to thicken the body of the oil.

The consistency of a grease depends both on the viscosity of the base oil and on the structure and properties of the metallic soap added. The three most important properties of a grease are as follows:

a) the melting temperature, at which the grease loses its semi-solid state;
b) its resistance to contamination and dilution by water;
c) its ability to withstand mechanical working before the grease softens.

The relative operating properties of commonly used greases are shown in Table 8.1.

Table 8.1 Operating properties of commonly used greases

	Upper temperature limits			
Type of grease soap	Short durations	Continuous	Water resistance	Mechanical stability
Calcium	80°C	50°C	Good	Fair
Sodium	120°C	80°C	Poor	Good
Lithium	150°C	120°C	Good	Good

8.9 Independent suspension front-wheel drive hub and swivel-pin carrier assembly (fig. 8.24)
The stub-shaft and hub carrier is usually made from forged steel or malleable cast-iron. The carrier basically resembles a cylindrical sleeve which houses the stub-shaft double row ball race, with spaced out upper and lower fork arms supporting the swivel-pin ball and socket joint built into the upper and lower suspension arms. Mounted on the bearings inside the hub carrier is a flanged sleeve wheel hub used to support the stub-shaft, and both the brake disc and road-wheel via a spigot shoulder and conical set bolts. An integral stub-shaft and constant velocity ball joint outer member transfers torque from the drive-shaft to the road-wheel, via the splined stub-shaft/wheel hub joint. The external splines of the stub-shaft mesh with corresponding internal splines of the flanged wheel hub, and they are drawn together by a self-locking nut and washer which pulls the shoulder of both the constant-velocity joint outer member and the flanged wheel hub hard against the inner track end of the double ball race bearing.

The upper and lower swivel-pin ball joints allow the hub-carrier and wheel assembly to be steered by swivelling about these pivots, while simultaneously permitting the transverse suspension arms to swing up and down relative to the body when the wheels are driven over rough road surfaces. To enable the driving torque to be transmitted smoothly from the final-drive to the stub-shaft, hub and road-wheel, the swivel-pin axis line is made to pass through the centre of the constant-velocity joint.

8.10 Power-assisted steering
To reduce the physical effort the driver has to apply to the steering-wheel to manoeuvre a vehicle, it is usual to increase the steering box gear-ratio so that for a given input force on the steering-wheel the output force directed in rotating the steering stub-axle is considerably increased. However, this multiplication of output force is achieved at the expense of having to multiply the distance the input effort has to move (steering-wheel rotation) to bring about the desired angular rotation to the steering road-wheels about their swivel-pins. Power-assisted steering is designed to reduce the effort the driver has to exert on the steering-wheel and to reduce the steering-wheel movement for a given swivel-pin angular turn, that is, to make the input (steering-wheel movement) to output (swivel-pin movement) more direct.

The power assistance provided relieves the driver of a large amount of physical exertion when steering, particularly if the vehicle is large and carries heavy loads, yet there must be sufficient resistance for the driver to overcome when moving the steering-wheel so that he or she feels that they are exerting an input force corresponding to the changing road conditions and vehicle load. The driver is therefore able to experience or sense danger and drives accordingly. Thus a small car may only receive 25% input assistance so that 75% of the effort needed to steer the car is provided by the driver, whereas a large truck fully laden may require up to 90% assistance, that is, only 10% input effort is supplied by the driver.

8.10.1 Integral power-assisted steering box and reaction control valve (figs 8.25, 8.26 and 8.27(a)
The integral power-assisted steering gearbox can be used for rigid front axle suspension commonly used on vans and commercial vehicles (see fig. 8.25), and can also be utilised in conjunction with independent front suspension layouts on large cars and vans (see fig. 8.26).

Description of linear reaction control valve (fig. 8.27(a) and (b)) The rack and pinion recirculating ball steering gear, power cylinder and hydraulic reaction-control-valve are all combined and share a common housing. The stepped rack-piston is permitted to slide in dual diameter cylinders according to the meshing sector-shaft's angular position.

Incorporated within the hollow rack-piston is a reaction-nut supported between ball race bearings mounted within the rack-piston. A worm-shaft supported at the right-hand end by radial and axial needle roller bearings is screwed onto the reaction-nut through a set of recirculating steel balls.

A reaction valve which controls the charging and discharging of the power cylinder is situated opposite the sector-shaft and at right angles to the worm-shaft. The reaction-control-valve is a 'spool'-type valve which has four annular grooves machined along its length. To and fro movement of the spool-valve is controlled by the reaction-nut which has an extended arm or tongue which fits into the second groove from the left-hand end of the spool.

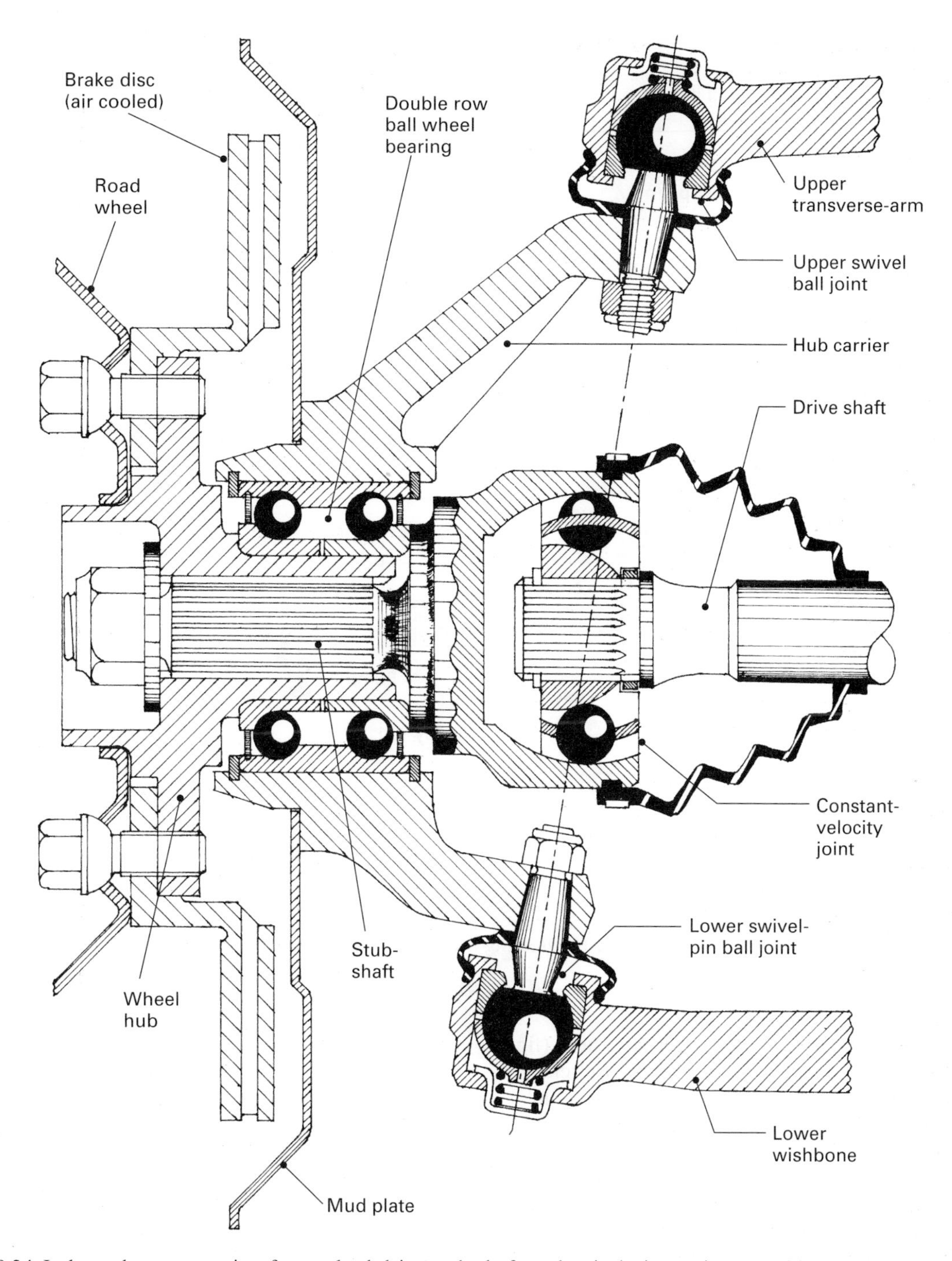

Fig. 8.24 Independent suspension front wheel drive stub-shaft and swivel-pin carrier assembly

Operation of linear reaction control valve (fig. 8.27(a)–(c)

Neutral position (fig. 8.27(c)) With no steering effort, the reaction-control-valve will take up a central position determined by the centring springs. Under these conditions fluid from the hydraulic pump and both ends of the rack-piston is free to flow back to the reservoir tank via the small gaps formed between the spool and bore groove control edges.

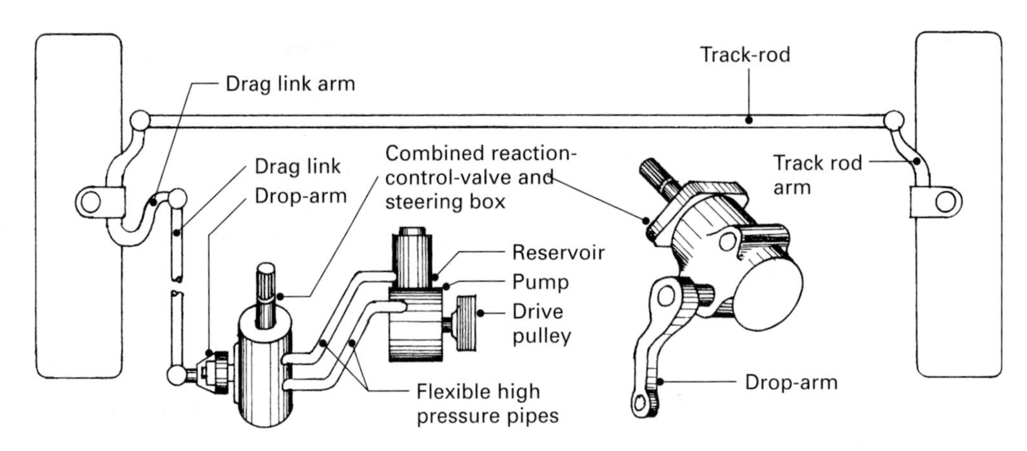

Fig. 8.25 Integral steering box and power assisted steering utilised with rigid axle front suspension (commercial vehicles)

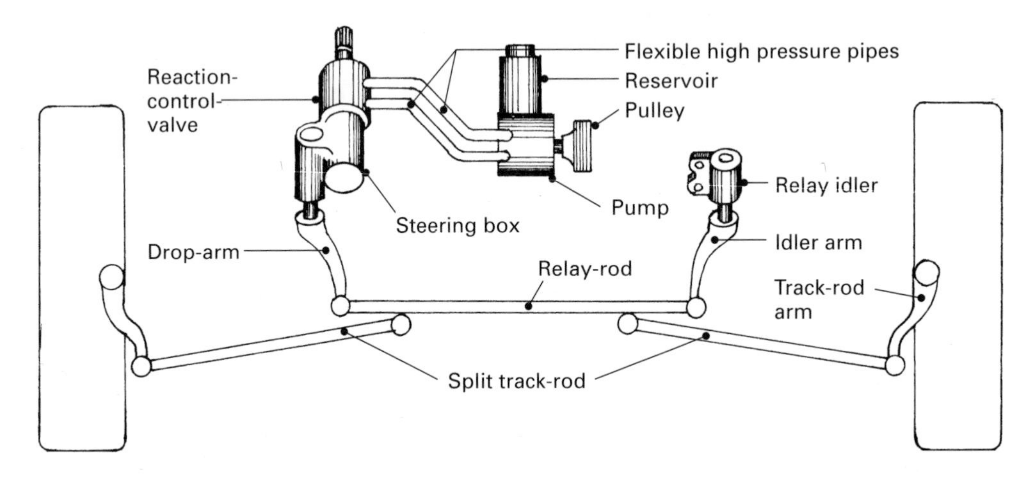

Fig. 8.26 Integral steering box and power-assisted steering utilised with independent front suspension (large cares and light vans)

Anticlockwise rotation of the steering wheel (fig. 8.27(a)) Rotating the steering-wheel anticlockwise screws the rack-nut outwards to the left-hand end, and at the same time it tends to drag the reaction-nut round with the worm-shaft. Immediately the reaction-control-valve will be pushed by the reaction-nut tongue towards the right-hand side so that control land (1) uncovers the fluid exit passage from the left-hand power cylinder to the reservoir. At the same time control land (2) blocks off the pump fluid supply exit to the reservoir, and land (3) uncovers the pump fluid supply passage connecting the pump to the right-hand power cylinder. Fluid pressure now acting on the right-hand end of the rack-piston will thus assist the manual effort applied by the driver on the steering-wheel to rotate the sector-shaft and drop-arm in a clockwise direction.

The greater the effort applied to the steering-wheel by the driver to turn the steering, the greater will be the screwing force between the worm-shaft and reaction-nut. Accordingly as the driver's steering effort increases, so the reaction-nut tilt will. This then enlarges the reaction-valve control edge passage gap feeding the right-hand power cylinder chamber, which therefore raises the pressure acting on the right-hand piston face. Thus it can be seen that the power assistance provided matches the input effort applied by the driver, so that the assistance is progressive and the driver feels that he or she is in control at all times.

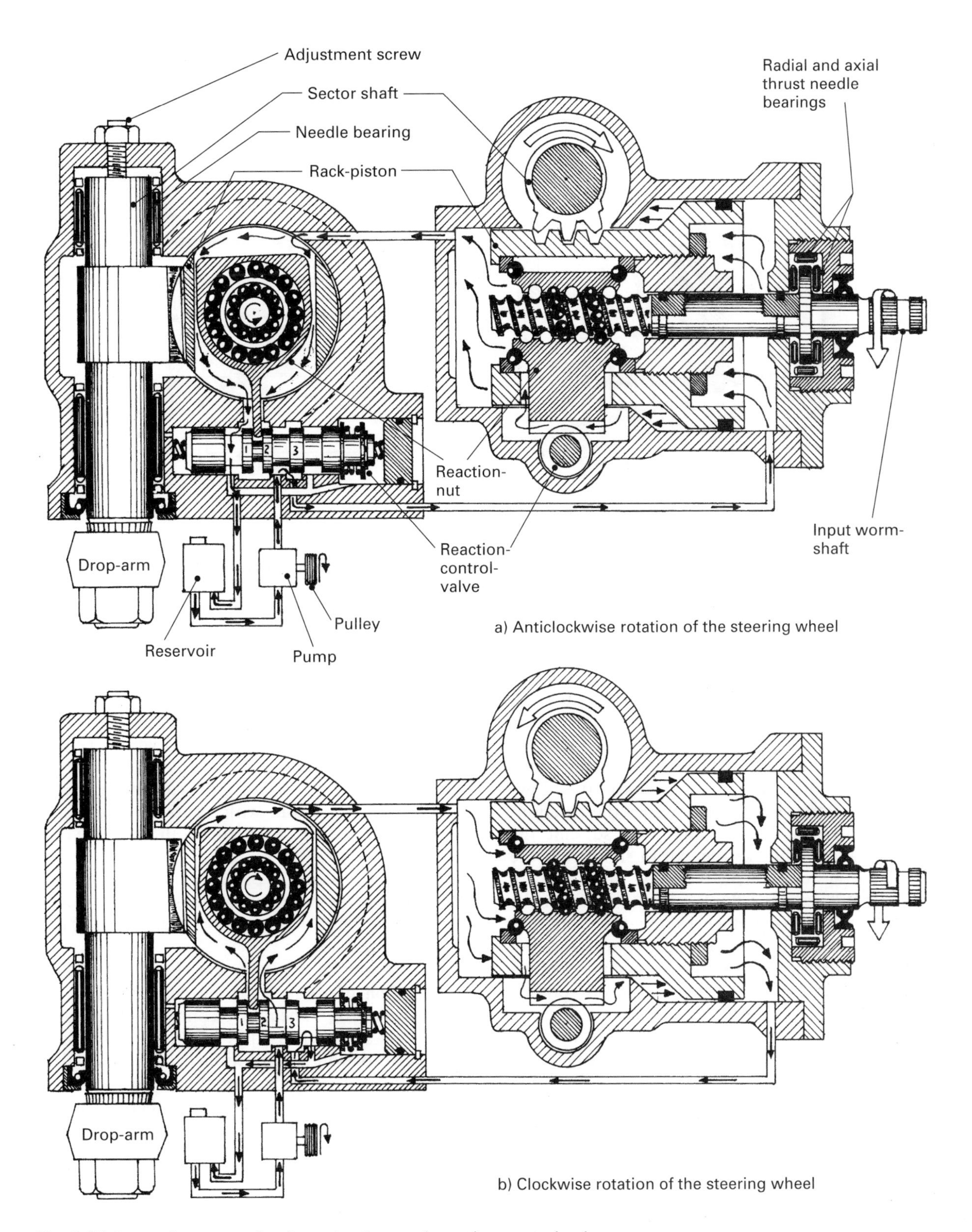

Fig. 8.27 Integral power-assisted steering box and reaction control valve

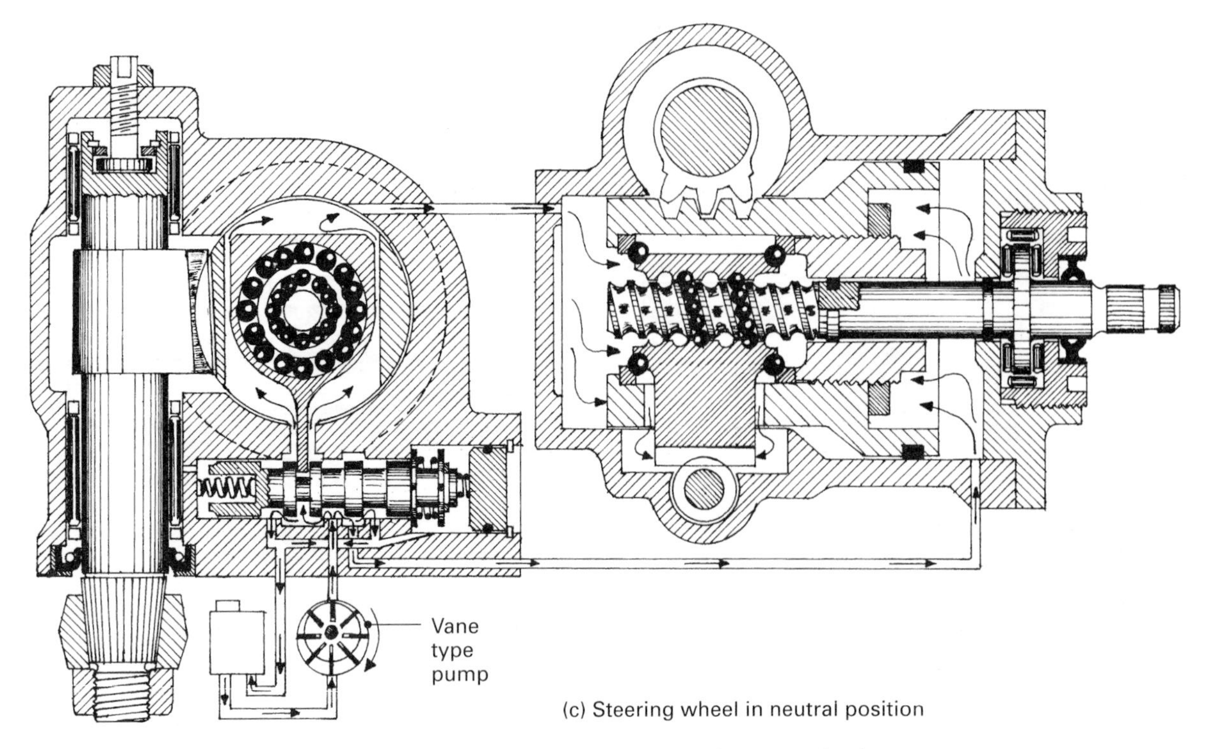

Fig. 8.27 (*Continued*) Integral power-assisted steering box and reaction control valve

Clockwise rotation of the steering wheel (fig. 8.27(b)) Rotating the steering-wheel clockwise screws the reaction-nut and therefore also the rack-piston inwards to the right-hand end. Simultaneously there will be a tendency for the reaction-nut to partially twist round in the same direction as the rotating worm-shaft. It follows that the reaction-nut tongue pushes the reaction-control-valve to the left-hand side so that the control land (1) blocks the exit of fluid from the pump to the reservoir at the same time. Control land (3) uncovers the fluid exit passage from the right-hand power-cylinder to the reservoir and land (2) uncovers the passage connecting the pump fluid supply to the left-hand power-cylinder. Fluid from the right-hand power-cylinder will now return to the reservoir, whereas the fluid supply from the pump will flow into the left-hand power-cylinder. The greater the passageway control land (2) gap, the greater will be the fluid pressure pushing against the left-hand piston side. Thus the amount the sector-shaft and drop-arm are made to rotate is proportional to the tilt of the reaction-nut, that is, it depends upon the reaction-valve control edge passage gap which in turn is determined by the effort the driver applies in rotating the steering-wheel.

8.10.2 Rack and pinion power-assisted steering with rotary control valve (figs 8.28(a) & (b) and 8.29(a) & (b))
The integral rack and pinion power-assisted steering layout can either be arranged with a central double-acting power piston with end linked split track-rods (see fig. 8.28(a)), or with offset double acting power piston with central linked split track-rods (see fig. 8.28(b)). Both these layouts are exclusively used with independent suspension; however, the central piston arrangement finds more favour as in effect the rack-shaft acts as a central track-rod so that shorter end split track-rods can be used.

Description of rotary reaction control valve (fig. 8.29(a) and (b)) The rotary-control-valve consists of a valve-sleeve encased in a cylindrical housing, a valve-rotor which is positioned within the valve-sleeve, and a torsion-bar which is attached at one end to the valve-sleeve and at the other end to the valve-rotor. The valve-sleeve has three external circumferential fluid passageway grooves separated by four teflon ring seals which are positioned on either side of these grooves. Positioned inside the valve-sleeve is the valve-rotor, the splines at its lower end meshing loosely with similar splines

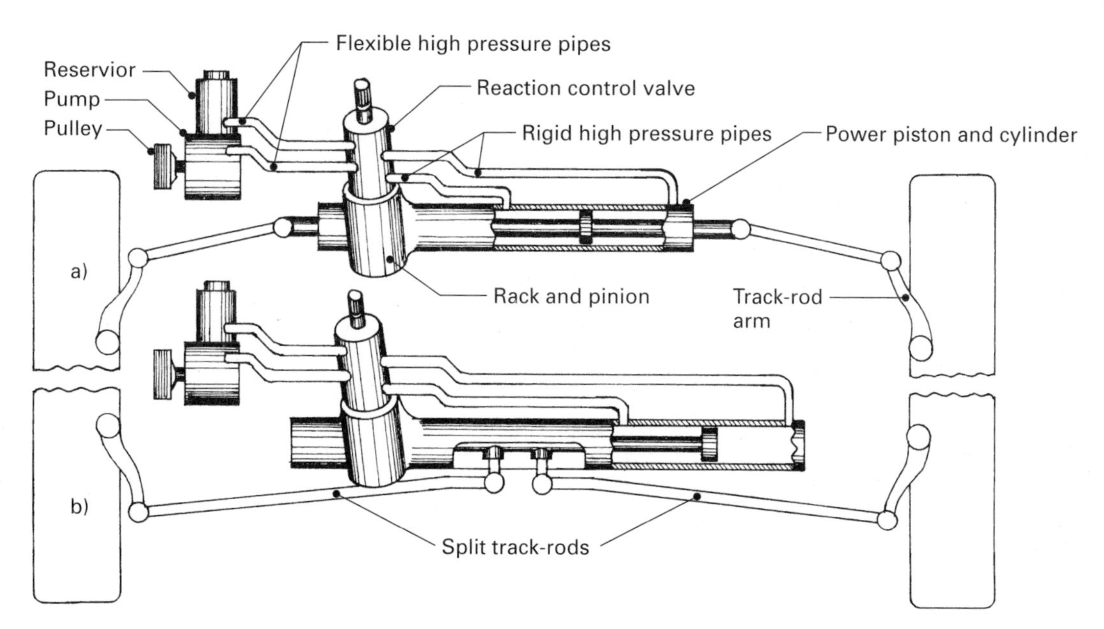

Fig. 8.28 Integral rack and pinion power-assisted steering with track-rods a) end linked b) central linked, utilised with independent suspension

machined inside the valve-sleeve. Six longitudinal slots are machined along and around both the rotor and sleeve. The rotor has three short and three long slots positioned alternately around its circumference, whereas the internal slots inside the sleeve are all the same length. The valve-rotor forms the input from the steering-wheel, while the valve-sleeve is combined with the pinion gear to form the output to the rack-shaft.

Operation of rotary reaction-control-valve (fig. 8.29(a) and (b))

Neutral position With no steering effort the reaction-control-valve will take up a central position determined by the torsion-bar, that is, the longitudinal land of the valve-rotor is aligned mid-way between the longitudinal slots in the valve-sleeve. Under these conditions fluid from the hydraulic pump and both ends of the power-piston is free to flow to the reservoir-tank via the small gaps formed between the rotor and sleeve longitudinal slot control edges.

Turning right – clockwise rotation of the steering-wheel (fig. 8.29(a)) Rotating the steering-wheel clockwise twists the torsion-bar and permits the valve rotor land (3) to block the fluid supply from the pump going to the right-hand power chamber, and at the same time uncovers the passage connecting the right-hand power-cylinder to the reservoir-tank. Similarly, the valve-rotor land (2) blocks the passage connecting the left-hand power chamber to the reservoir and uncovers the fluid supply passage from the pump going to the left-hand power-chamber. Fluid pressure acting on the left-hand side of the power-piston will then assist the manual effort applied by the driver on the steering-wheel to rotate the pinion-gear and hence moves the rack-shaft to the right. The more the torsion-bar is twisted, the greater will be the misalignment of the rotor and sleeve longitudinal slot land edges and the greater will be the valve gap passageway permitting higher fluid pressure to act on the left side of the power-cylinder. Immediately the driver relaxes the force on the steering-wheel, the torsion-bar unwinds permitting the valve-rotor to take up its central position, that is, fluid is permitted to flow on either side of each of the rotor lands so that it can move freely into or out of both sides of the power-cylinder to the reservoir. Consequently the steering system reverts back to manual control.

Turning left – anticlockwise rotation of the steering-wheel (fig. 8.29(b)) Rotating the steering-wheel anticlockwise twists the torsion-bar and permits the valve-rotor to move relative to the valve-sleeve. The rotor land (2) now blocks the fluid supply from the pump going to the

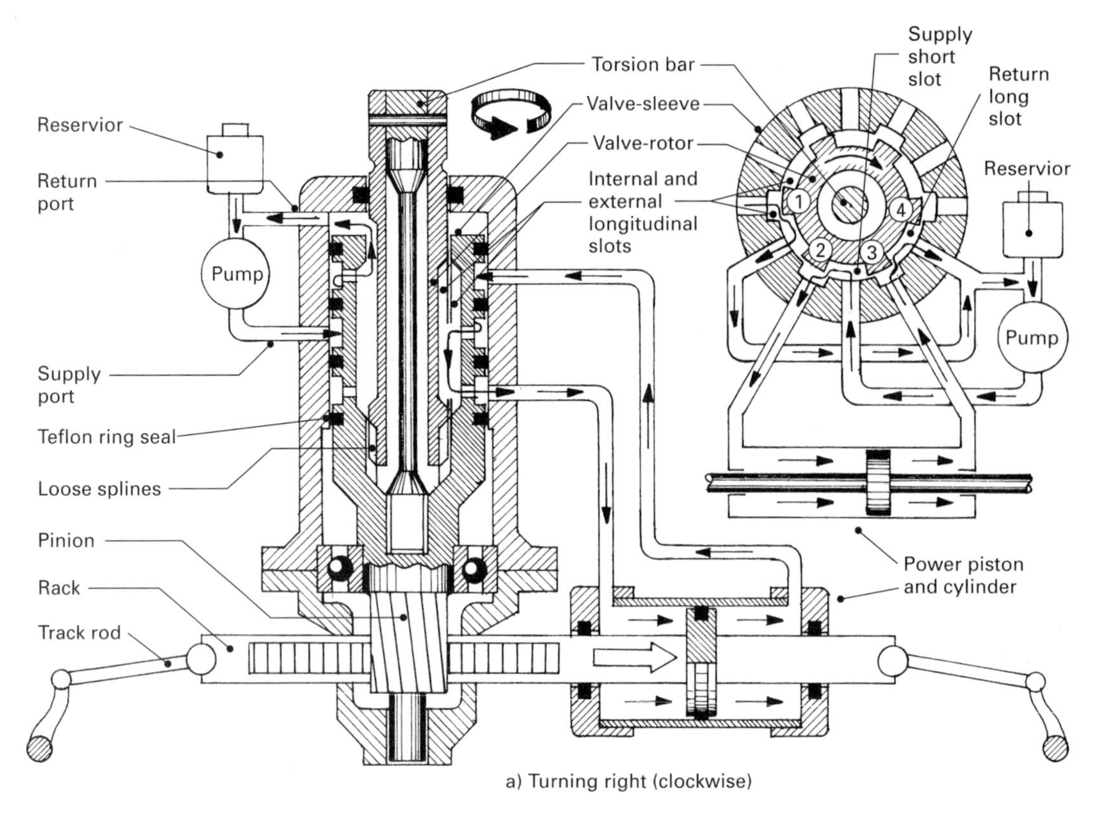

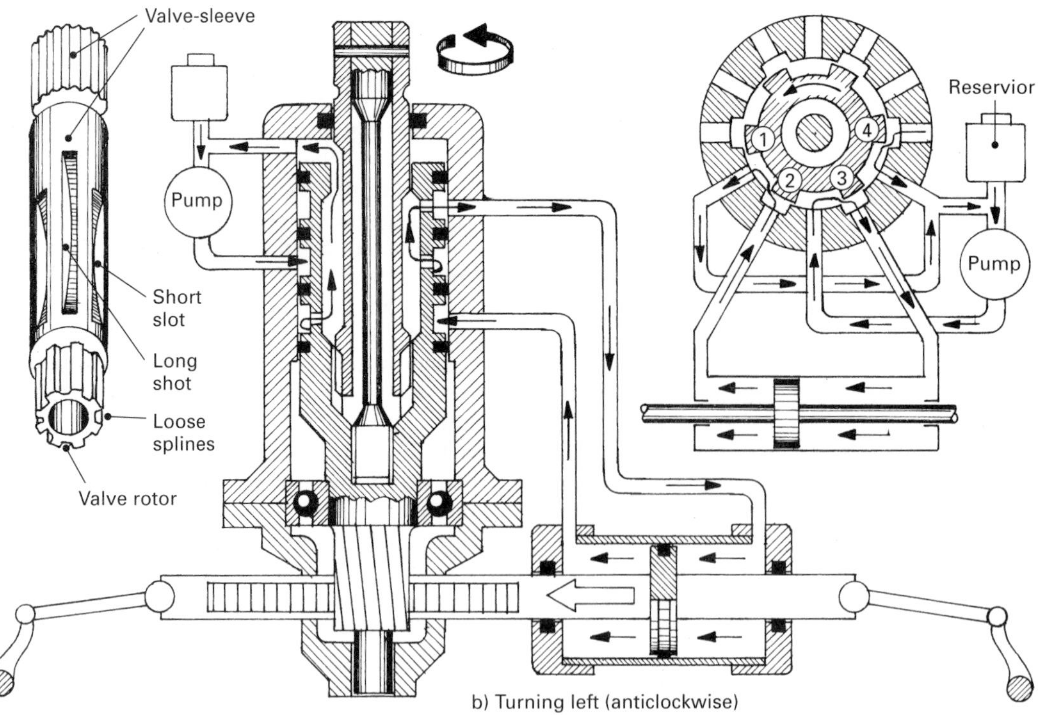

Fig. 8.29 Rack and pinion power-assisted steering with rotary reaction control value

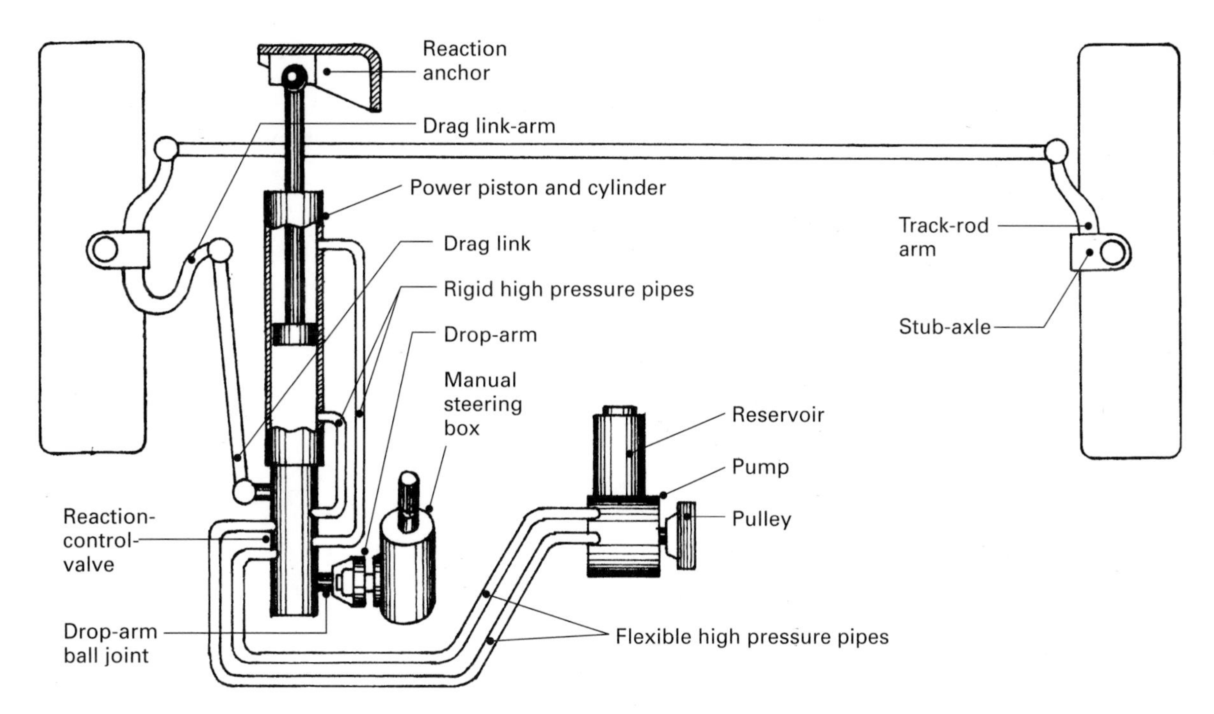

Fig. 8.30 Steering box with external direct coupled power-assisted steering utilised with rigid axle front suspension (commercial vehicles)

left-hand power chamber and simultaneously uncovers the passage connecting the left-hand power chamber to the reservoir. Similarly the valve-rotor land (3) blocks the passage connecting the right-hand power chamber to the reservoir and uncovers the fluid supply passage from the pump going to the right-hand power-chamber. Fluid can now be released from the left-hand chamber and at the same time fluid will be applied to the right-hand power-piston face to assist the manual effort of the driver to rotate the pinion gear and hence move the rack-shaft to the left.

Limit of power assistance Valve control takes place between the relative angular slackness of the rotor and sleeve meshing splines which amounts to something like 9°. If the relative angular deflection of the torsion-bar exceeds 41/2° movement from the central or mid-position, then the power assistance will be at its maximum. Any further rotation of the steering-wheel is only achieved by the increased manual effort of the driver rotating the rotor, sleeve and pinion gear directly via the meshing splines.

8.10.3 Steering gearbox with external direct coupled power cylinder and reaction control valve utilised with rigid axle front suspension (fig. 8.30)

This hydraulic power-assisted steering system is used mainly for large commercial vehicle application and has a combined power cylinder and reaction-control-valve unit which is separate from the manual steering gearbox. The linear reaction-control-valve is similar to that used in the integral power-assisted steering box (see fig. 8.27(a) and (b)), but instead of the reaction-nut tongue moving the spool valve, it is now moved by the drop-arm ball joint. The power-assisted cylinder and reaction-valve unit is supported at the power-cylinder end by the double acting piston's ram-rod which is anchored to the chassis via a ball and socket joint, while the opposite end which houses the reaction-control-valve is attached by a ball joint to the supporting steering box drop-arm. The steering drop-arm swing movement is conveyed to the steering stub-axle via the reaction-control-valve casing, drag-link and drag-link steering-arm. When the steering-wheel is turned the drop-arm moves the linear reaction control-valve spool either forwards or backwards. Automatically the reaction spool valve senses either a right or left-hand steering movement and instantly connects the pump fluid supply to the side of the power-cylinder which needs to expand, whereas the power-cylinder chamber on the other side which is required to contract at the same time is opened to the reservoir tank so that fluid can be removed from the reducing chamber.

9 Brake systems

9.1 Single-line hydraulic braking system

A hydraulic braking system is a compact method of transmitting the driver's foot-pedal effort to the individual road-wheel brakes by conveying pressurised fluid from one position to another and then converting the fluid pressure into useful work at the wheels to apply the brakes and so retard or stop the rotation of the wheels. The characteristics of a hydraulic braking system are that the fluid is incompressible, the pressure generated is uniform throughout, and there is very little fluid movement.

9.1.1 The objectives of a braking system

The purpose of the brakes is to retard or stop the vehicle in motion by converting the kinetic energy of the moving vehicle into rotational frictional torque at the brake-shoes or brake-pads. This energy will then change into heat, the magnitude of the resulting temperature rise depending upon whether the braking application is, for example, a steady hill descent or an emergency stop. Hence the brakes should be able to dissipate large quantities of heat without undue temperature rise and should be rigid enough to absorb the large braking torques encountered.

9.1.2 Construction and operation of a single-line hydraulic system (fig. 9.1)

The functions of the various components of the braking system are as follows.

Brake pipes These provide a continuous fluid circuit between the master-cylinder and the wheel-cylinders. The brake pedal relays the driver's foot effort to the master-cylinder piston which compresses the brake fluid. This pressure is equally transmitted throughout the fluid to the

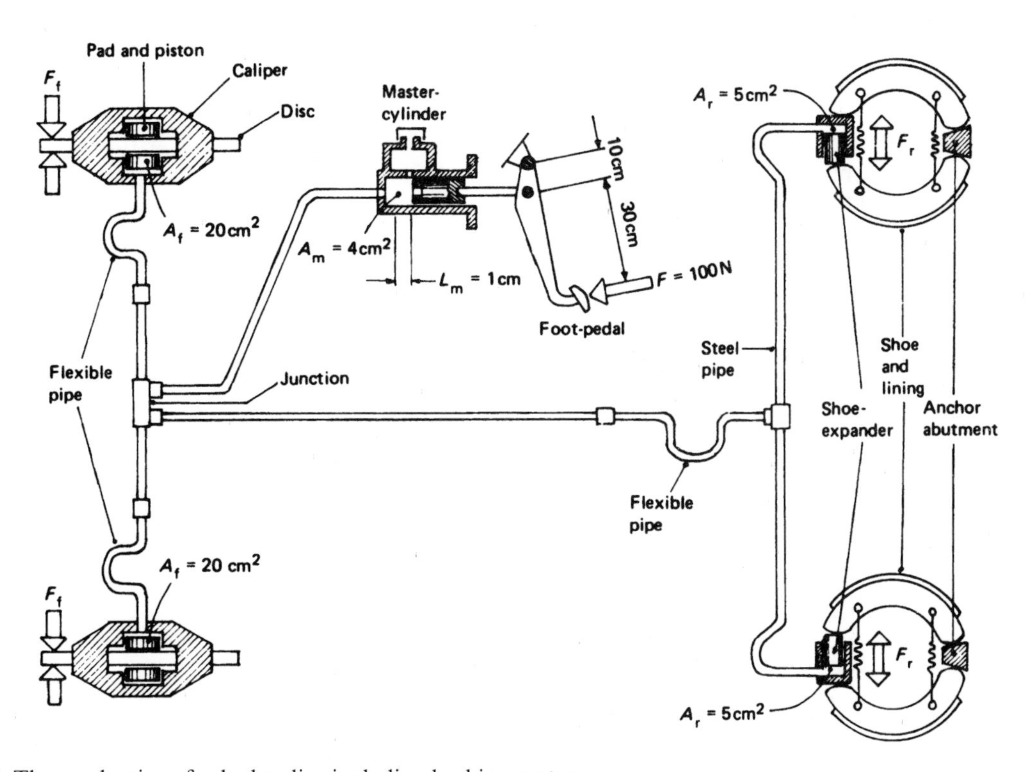

Fig. 9.1 The mechanics of a hydraulic single-line braking system

front disc-caliper pistons and to the rear wheel-cylinder pistons. Steel pipes convey the fluid along the body structure and rigid axle members, but flexible hoses connect the sprung body pipes to the unsprung axle wheel-brake units, to allow for relative movement (fig. 9.1).

Master-cylinder With the assistance of the foot-pedal leverage, this converts foot-pedal force to hydraulic pressure within the fluid system by means of the cylinder and piston every time the driver's foot applies a load on the pedal (fig. 9.1).

Disc-brake This consists of a disc bolted to the wheel hub and sandwiched between two pistons and friction pads supported in a caliper fixed to the stub-axle (figs 9.1 and 9.25). When the brakes are applied, the pistons clamp the friction pads against the two side faces of the disc.

Drum-brake This consists of two brake-shoes and linings supported on a backplate bolted to the axle-casing. These shoes pivot at one end on anchor pins or abutments fixed to the backplate (figs 9.1 and 9.5 to 9.7). The other free ends of the both shoes are forced apart when the brakes are applied, thus the shoes expand radially against a brake-drum located concentrically on the wheel hub.

Wheel-cylinders These convert the transmitted hydraulic line pressure into braking effort as it acts on the cross-sectional area of the disc and drum cylinder pistons (fig. 9.1). This braking effort either squeezes the friction pads against the side faces of the disc or forces the shoe friction linings against the inside of the drum.

9.1.3 The mechanics of a hydraulic braking system
To appreciate the mechanics of the hydraulic braking system, a simple calculation will be used to show how a suitable force ratio is achieved between the foot-pedal and the wheel-cylinder pistons.

Consider the braking system shown in fig. 9.1.

Let F = force on foot-pedal

F_f = force on each front cylinder piston

F_r = force on rear cylinder piston

A_m = cross-sectional area of master-cylinder (4 cm^2)

A_f = cross-sectional area of front pistons (20 cm^2)

A_r = cross-sectional area of rear piston (5 cm^2)

L_m = distance moved by effort (1 cm)

and L_w = distance moved by output

Force on foot-pedal = 100 N

and pedal leverage ratio = 4

$\therefore$ force on master-cylinder piston = 100 N × 4

= 400 N

$\therefore$ pressure in brake pipelines (p) $= \dfrac{F}{A_m}$

$$= \frac{400\text{ N}}{4\text{ cm}^2} = 100\text{ N/cm}^2$$

Force on each front cylinder piston (F_f) $= pA_f$

$= 100\text{ N/cm}^2 \times 20\text{ cm}^2$

= 2000 N

$\therefore$ total force on front brakes = 2000 N × 4 = 8000 N

Force on each rear cylinder piston (F_f) $= pA_r$

$= 100\text{ N/cm}^2 \times 5\text{ cm}^2$

= 500 N

$\therefore$ total force on rear brakes = 500 N × 4 = 2000 N

$$\text{Front-to-rear brake ratio} = \frac{8000\text{ N}}{2000\text{ N}} = 4:1$$

Percentage of front braking = 75%

Percentage of rear braking = 25%

$$\text{Total force ratio} = \frac{\text{output force}}{\text{input effort}} = \frac{10\,000\text{ N}}{100\text{ N}}$$

= 100:1

Consider the master-cylinder stroke (L_m) to be 1 cm, then

volume forced into wheel cylinders = volume displaced in master cylinder

$$(4A_f + 4A_r)\,L_w = A_m L_m$$

$$\therefore \begin{array}{l}\text{wheel-cylinder}\\ \text{piston movement } (L_w)\end{array} = \frac{A_m L_m}{4(A_f + A_r)}$$

$$= \frac{4\ \text{cm}^2 \times 1\ \text{cm}}{4(20 + 5)\ \text{cm}^2}$$

$$= \frac{4}{100}\ \text{cm or } 0.4\ \text{mm}$$

$$\therefore \begin{array}{l}\text{cylinder}\\ \text{movement ratio}\end{array} = \frac{L_m}{L_w} = \frac{1\ \text{cm}}{4/100} = \frac{100}{4}$$

$$= 25{:}1$$

$$\begin{array}{l}\text{Total}\\ \text{movement ratio}\end{array} = \begin{array}{l}\text{cylinder}\\ \text{ratio}\end{array} \times \begin{array}{l}\text{pedal}\\ \text{leverage}\end{array}$$

$$= 25 \times 4 = 100{:}1$$

9.1.4 Characteristics of a hydraulic braking system

A hydraulic braking system is a method of transmitting brake-pedal force to the wheel brakes by conveying pressurised fluid from one position to another and then converting the fluid pressure into useful work at the wheels. The features and characteristics of a hydraulic braking system may be summarised as follows:

a) Pipelines are relatively small-bore and occupy very little space.
b) Pipelines can be bent and shaped to follow the contour of the under-side of the body structure with no mechanical difficulties.
c) Pipelines are installed along the under-side of the body structure and there is no relative pipe-body movement when the brakes are being operated.
d) Pipelines are a convenient means of transferring movement from the brake pedal to the brake wheel-cylinders.
e) Pipelines are made to transmit large forces from the master-cylinder to the brake-wheel cylinders
f) Pipelines are a means of communicating movement from the foot-pedal to the wheel disc- or drum-brakes.
g) Pipeline fluid pressure is uniform throughout the hydraulic circuit and remains consistent even when the axle and wheels are moving between bump and rebound.
h) Braking force ratio in a hydraulic braking system is a direct function of the ratio of the master-cylinder cross-sectional area to the disc- or drum-brake wheel-cylinder cross-sectional areas, so these may be chosen to produce the braking effect desired.
i) The wheel-cylinder cross-sectional areas of the front and rear disc- and drum-brakes respectively may be chosen to produce the best front-to-rear braking ratio.
j) Master-cylinder piston movement is inversely proportional to the force ratio, so the pedal movement will become larger as the braking effort increases.
k) Hydraulic fluid provides minimum response time in transferring force from the pedal to the brake wheel-cylinders.
l) Relative movement between the sprung body and the unsprung axle-and-wheel assembly does not interfere with the transmission of fluid pressure to the brake wheel-cylinders.
m) Hydraulic fluid is incompressible provided there is no trapped air in the system, but if air is somewhere in the braking circuit the foot-brake movement will be spongy.
n) Pipeline systems do not require lubricating, and the cylinder–piston seals are lubricated by the brake fluid.
o) The only internal friction in a hydraulic system is between the cylinder pistons and seals, caused by the fluid pressure squeezing the seal lips against the cylinder walls as the piston moves along its stroke.
p) The only components subjected to wear and which will normally be replaced are flexible hoses and piston and cylinder seals.
q) A hole or joint or seal leakage in the hydraulic circuit will very quickly empty part or all of the system of fluid and the brakes will become ineffective.
r) A hydraulic braking system is suitable only for intermittent braking applications, and a separate mechanical linkage must be used for parking brakes.

9.2 Leading-and-trailing-shoe layout (figs 9.2 and 9.3)

This configuration consists of a pair of shoes pivoted at a common anchor point with a double-piston/cylinder expander which forces out the free ends of both shoes radially against the inside of the brake-drum.

The leading shoe is the one whose expander piston moves outwards in the direction of rotation of the drum. Frictional drag between the shoe and the drum will thus tend to assist the expander piston in wedging the shoe hard against the drum, hence more braking force will be

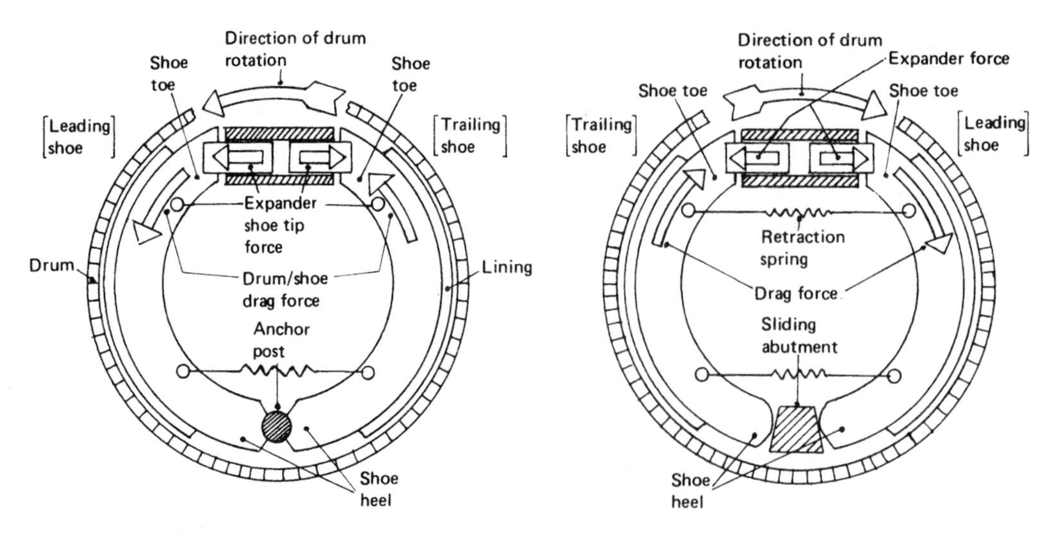

Fig. 9.2 Leading- and trailing-shoe drum-brake arrangements

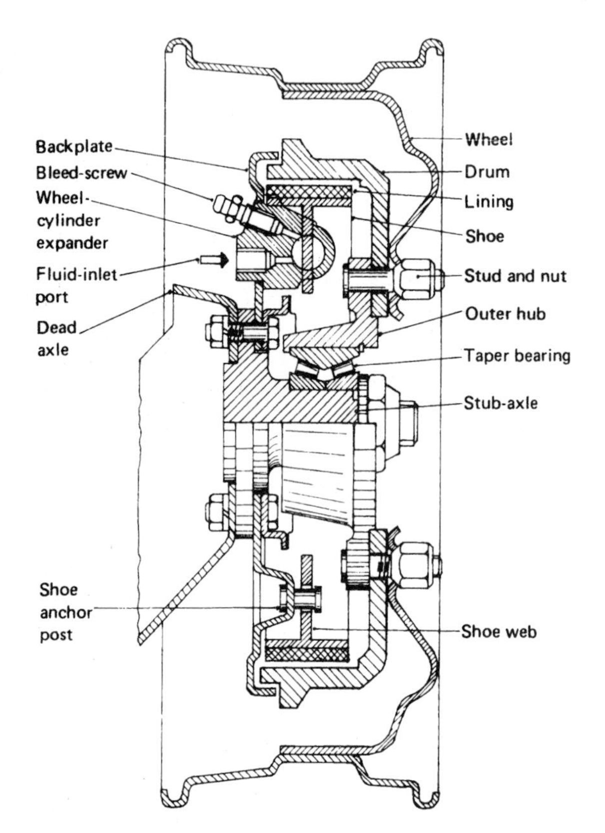

Fig. 9.3 Non-drive rear stub-axle and brake arrangement

obtained for a given actuating load on the expander. This is referred to as the self-servo action of the shoe.

The trailing shoe is the one whose expander piston moves outwards in a direction opposite to the direction of rotation of the drum. Frictional grip between the shoe and the revolving drum will thus tend to oppose the expander piston and to push it back into its cylinder body. Hence the trailing shoe provides less braking force than the leading shoe for a given actuating force.

Wheel-cylinder shoe-expander A leading-and-trailing-shoe layout employs a double-piston/cylinder expander which applies an equal actuating load on each shoe, due to the trapped fluid pushing both pistons outwards simultaneously. This provides a floating expander effect, since the fluid is insensitive to the positioning of the pistons and responds only to the reaction of the pistons against the shoes. Therefore, if one shoe lining is worn more than the other, both pistons will move over to one side, but they will still share equally the actuating load. To prevent the fluid escaping, rubber seals are fitted either in an annular groove in the piston skirt or against the piston crown.

Anchor abutments To prevent the shoes from rotating with the drum when the brakes are applied, an anchor abutment is fixed to the backplate on the opposite side to the expander unit. This anchor may be a single- or double-pin post or a sliding abutment block. The pin post provides a hinge for the shoes, the single and double varieties being used for light and heavy duty respectively. Sometimes for light- and medium-duty applications the sliding abutment is preferred, as it is self-centralising and to some extent compensates for uneven lining wear.

Retraction springs These coil springs – sometimes known as pull-off springs – are used to pull the shoes away from the drum when the hydraulic pressure is released and so prevent any possible shoe drag. At the same time, retraction of the shoes pushes the wheel-cylinder expander pistons together and displaces brake fluid back to the master-cylinder.

Backplate The backplate is usually a pressed-steel construction which is ribbed to increase its rigidity and to provide support for the cylinder expander, anchor abutment, and brake-shoes. The plate is bolted to the axle flange, as it must absorb the entire braking torque reaction of the shoes (because of this, commercial-vehicle manufacturers refer to this plate as the 'torque-plate'). Its other function is to form a dust-shield to protect the drum-and-shoe assembly from dirt and mud.

Brake-shoes These shoes are curved rectangular steel strips shaped to match the inside curvature of the drum. To provide rigidity, a steel web is welded behind and in the middle of the shoe sole so that the shoe assembly forms a 'T' section. The expander end of the shoe is known as the 'toe' and the anchor end as the 'heel'. The toe and heel of the shoes may be flat, curved, or have a semi-circular groove formed to locate the expander or anchor abutments.

Brake linings These materials are usually either a woven or moulded asbestos impregnated with natural resins, bitumens, and drying oils, or with synthetic resins. These materials are chosen because of their ability to withstand high operating temperatures without their frictional properties being greatly affected. The linings for cars and light-duty applications are attached to the shoes by adhesives which require a heat treatment to complete the bond, but commercial vehicles still prefer to have the linings riveted to the shoes.

Brake-drums Brake-drums are generally made of cast or malleable iron. The drum is basically a shallow cylinder with a flange at one end which is designed to be bolted concentrically to the axle-hub. One or two ribs are cast circumferentially on the outside of the drum near its mouth, to provide support against the radial expanding forces and to improve the heat dissipation. Internally the circumference of the drum is machined to present a smooth surface to the brake lining.

Steady-post and rest (fig. 9.4) The job of the steady-post is to keep the brake-shoe square with the drum. The web of the shoe bears against the steady-post on the backplate, thereby counteracting the offset pull of the shoe retraction springs (fig. 9.4 (a) to (c)).

Another form of steady-rest relies on either a coil spring or a leaf spring which presses the shoe on to the pressed-out or riveted-block platform on the backplate (fig. 9.4 (d) to (f)). The coil spring is retained in position on the pin by means of a slotted washer pressed down and turned under the cross-piece on the head of the pin. Alternatively a hook spring may be used which is hooked in place by depressing, turning, and releasing it. The leaf-spring method uses a wavy spring clip which is slotted so that it snaps into position when it is compressed and pushed between the cross-piece of the pin and the shoe web.

9.3 Cam-operated drum-brakes

Originally most drum-brakes were cam-operated, the cam's function being to force the shoes against the drum, but foot-brake shoe-expanders for cars are now mostly of the hydraulic type. Only with large trucks, and particularly articulated trailers, is there a demand for the cam-type expander today. Cams have however found some favour in parking-brake mechanisms for both drum and disc arrangements.

A cam-type brake-shoe expander uses disc-type cams. These consist of a plate rotating about an axis perpendicular to its plane. The profile of the plate is such as to impart a reciprocating motion to its follower – the shoe – which bears against the cam edge.

The object of any shoe-expander is to provide a suitable force ratio between the input effort and the output brake-shoe load. When the brakes are initially applied, only the tension of the pull-off springs and the friction in the mechanism must be overcome, so that a quick-movement device having only a low force ratio is needed during the first part of braking. However, when the shoes are actually pressing hard against the drum and further braking is necessary, a progressively increasing force ratio would be desirable.

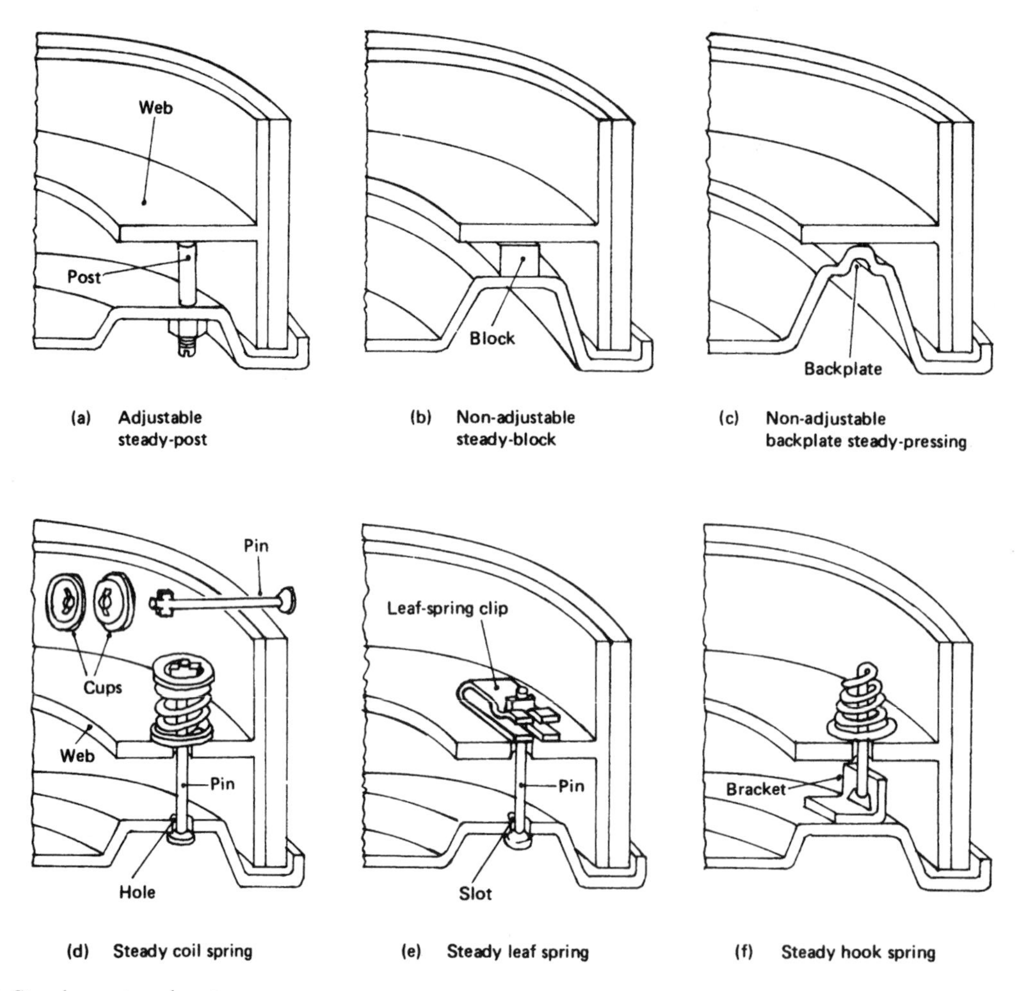

Fig. 9.4 Steady-post and rest

9.3.1 Flat and ovaled cams (fig. 9.5)
To meet these needs, variable-movement-ratio cams have been designed. One of these is the flat-faced cam. The force exerted by the cam perpendicular to the shoe at their contact point is at a maximum distance from the cam's axis when the cam and shoe flat faces are parallel before braking. As the cam rotates, the contact point moves towards the centre of the shoe face and therefore this distance decreases, so the force ratio will become greater.

For some applications the force ratio can be raised even further if an ovaled cam profile contacts the flat shoe faces.

9.3.2 'S' cams (fig. 9.6)
Ovaled cam profiles are suitable only for normal thicknesses of friction linings where the point of contact on the shoe follower face does not change appreciably with wear. For heavy-duty thick linings (12 to 20 mm) used on commercial vehicles, constant-movement-ratio cams acting on rollers are used. These cams are two-lobe cams known as 'S' cams, and their profile is developed from either an Archimedean spiral or an involute curve. This shape provides a constant rate of shoe movement – that is, the shoes are displaced in direct proportion to the cam-lever angular movement from the unworn to the fully worn lining condition.

9.3.3 Fixed and floating cams
Leading shoes work harder than trailing shoes, so the latter wear less. With a fixed cam correctly adjusted, the leading shoe will wear more than the trailing shoe, this then reduces the shoe tip force on the leading shoe until the trailing shoe has worn down by the same amount. Thus brak-

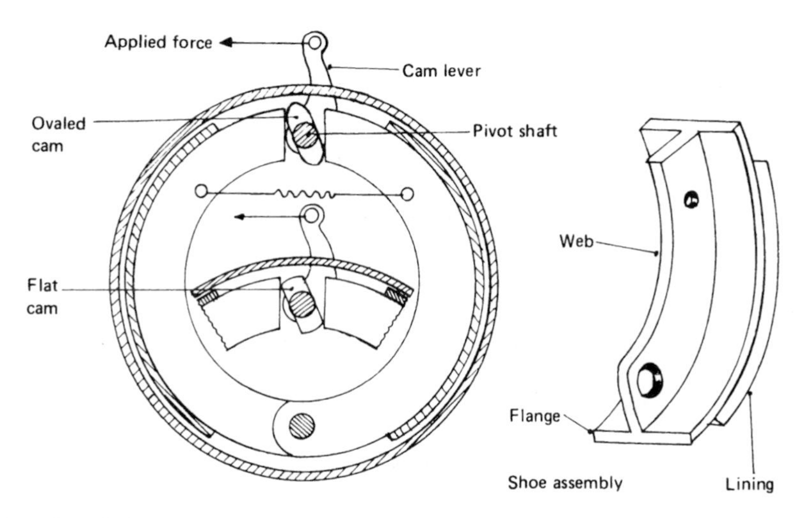

Fig. 9.5 Fixed flat and ovaled cams

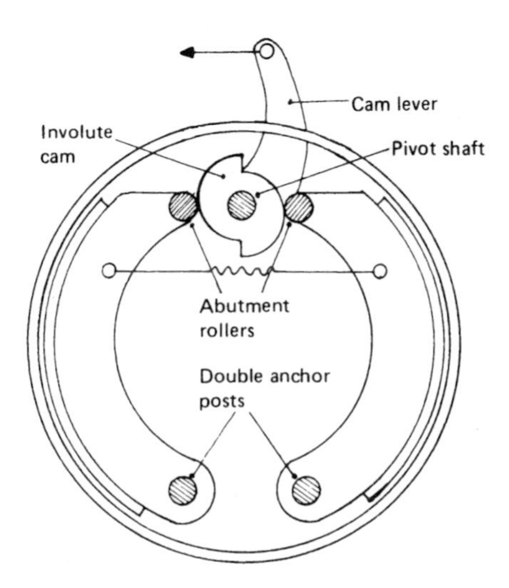

Fig. 9.6 Fixed involute 'S' cam

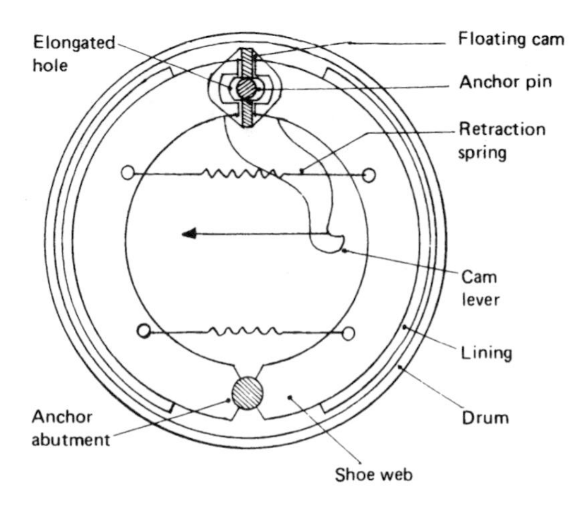

Fig. 9.7 Floating flat cam

ing will be carried out in phases of leading mode followed by trailing mode to equalise the shoe wear. This is known as the equal-work condition.

Equal shoe tip force can only be applied if a different amount of cam lift is provided for each shoe. This can be achieved by having a floating cam which will automatically adjust its position with the state of lining wear (fig. 9.7). Sometimes the cam assembly is allowed to slide on the back-plate, or the cam may have an elongated hole in its middle and be supported on an anchor post. Generally, braking will be more effective with a floating cam, but the total lining wear between the two shoes will be very uneven.

9.4 Wheel-cylinder shoe-expanders

Wheel-cylinder shoe-expander units are incorporated into hydraulic braking systems which have drum-brakes at the rear. The wheel-cylinders transmit the hydraulic pressure to the brake-shoes. These units take two basic forms: the single-piston type, which is common on front-drum-brake vehicles, and the double-piston units which find their application with rear drum-brakes.

9.4.1 Double-piston wheel-cylinder shoe-expanders (rear-wheels) (fig. 9.8)

These units have a cylinder body, two pistons, seals, seal-spreaders and a retainer-spring (if

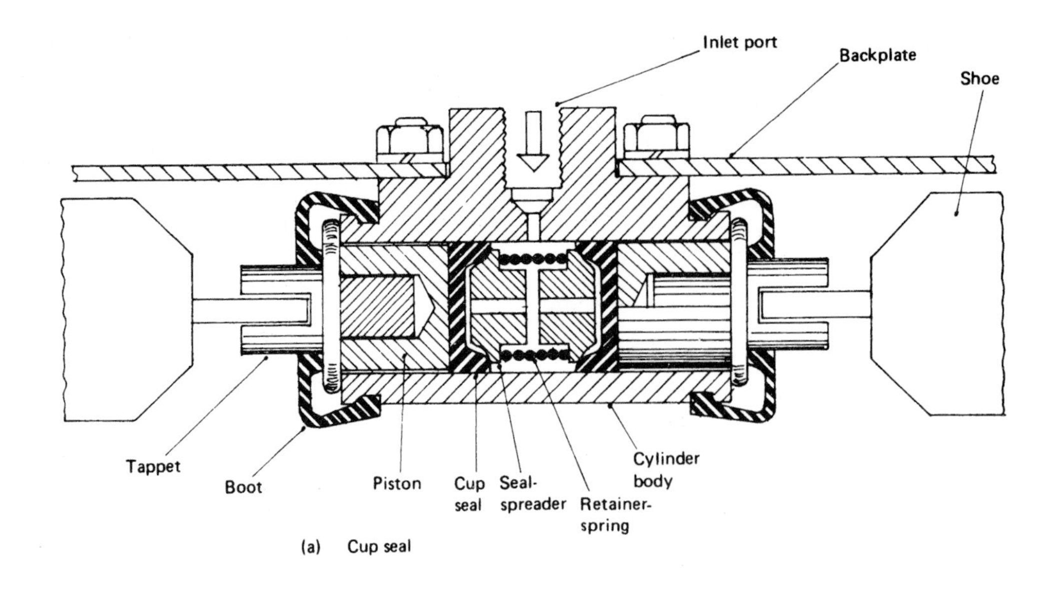

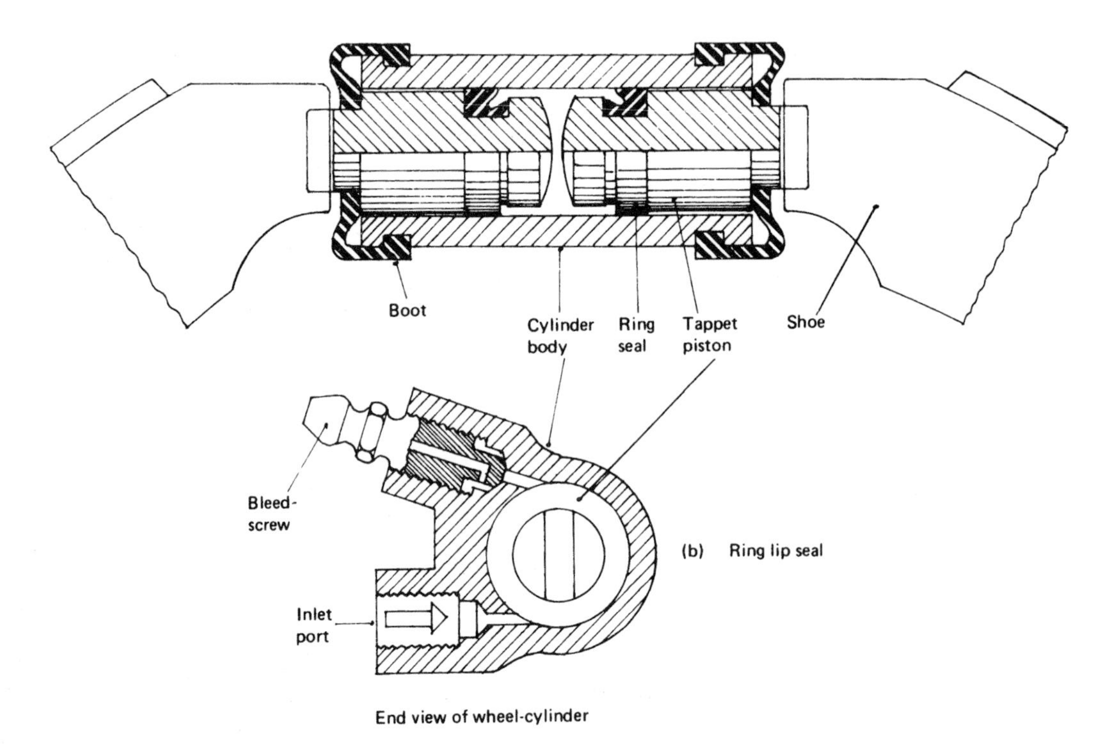

Fig. 9.8 Double-piston wheel-cylinders

cup-type seals are used), rubber dust-boots, and sometimes separate expander tappets.

Wheel-cylinder body The cast-iron wheel-cylinder body has an extended spigot portion which fits into a location hole in the backplate. The body is secured to the plate by either a conical-shaped washer clip or by two studs. The attachment of the wheel-cylinder body to the backplate must be sufficiently substantial to absorb some or all of the braking-torque reaction when the brakes are applied.

A cylindrical hole is bored right through the body to receive the two pistons, seals, and seal-spreaders and a retainer-spring (if fitted). Annular grooves are machined at both ends of the cylinder to position the rubber dust-boots. A bleed-screw valve hole and thread is provided in the centre of the cylinder, usually at the highest point, to assist in the purging of air from the chamber.

Piston, tappet and abutments The two pistons in the wheel-cylinder convert the hydraulic pressure into brake-shoe tip load. Their diameter is so chosen as to provide the correct brake load between the front and rear brakes. The piston outer end may be slotted to receive the shoe toe web so that it acts directly against the shoes. Sometimes, however, the shoes are pushed outwards directly by push-rods or screw tappets or abutments situated between the pistons and the shoe tips.

Retainer-spring and seal-spreader With cup seals, a retainer-spring between two seal-spreaders pushes the cup seals against the piston heads and against the cylinder walls. When the brakes are released, the fluid will not then seep past the piston and air will not enter the wheel-cylinder. With ring lip seals, which fit in grooves formed around the pistons, the natural elasticity of the rubber preloads the seal lip radially against the bore.

A rubber boot or cap fits over each piston's exposed end, to exclude brake-lining dust and dirt from the cylinder walls, which might otherwise become scored.

9.4.2 Single-piston wheel-cylinder shoe-expander (front wheels) (fig. 9.9)

Single-piston wheel-cylinder units are generally used only on front drum-brakes, to increase braking efficiency, and two single-piston units are incorporated diametrically opposite each other. When the outward movement of both the single-piston units is in the direction of forward rotation of the drum, the combination is known as a two-leading-shoe brake.

These single-piston units perform the dual function of expanding one shoe against the drum and acting as the anchor abutment for the other shoe. As for the double-piston units, they are bolted to the backplate.

The action of these single-piston shoe-expanders is similar to that of the double-piston ones, except that the cylinder has a blind end which forms the anchor abutment for the other shoe. The piston seal may be either the ring-seal type or the cup type with a seal-spreader and retainer-spring.

9.4.3 Combined hydraulic / lever rear-wheel shoe-expander (fig. 9.10)

The unit construction (fig. 9.10(a)) consists of a body which is bored to support inner and outer pistons. The other piston has a pressed-steel dust-cover welded to it and is grooved to carry a rectangular-cross-section rubber dust-seal. The inner piston has a cup seal with a seal-spreader and a retainer-spring to pre-load the cup seal against the cylinder wall when the brakes are in the released position.

A triangular slot is formed in each piston to house the tapered end of the cranked lever. This lever is located and pivoted on a pin in the body.

Foot-brake applied (fig. 9.10(b)) When the foot-brake is applied, the fluid pressure will push the inner and outer pistons away from the cylinder's blind enclosed end until the leading shoe is forced against the drum. The hydraulic reaction of the fluid now forces the cylinder body to slide in its slot on the backplate in the opposite direction, until the trailing shoe is applied. In other words, the cylinder body and pistons will float between both shoes and will so apply an equal shoe tip load to each shoe. The movement of the pistons relative to the cylinder body will not interfere with the enclosed end of the cranked lever, as the slots in the pistons provide ample clearance.

Hand-brake applied (fig. 9.10(c)) When the hand-brake is applied, the cable will tend to pull the cranked-lever end away from the backplate, with the result that the lever will rotate about the pivot pin mounted in the cylinder body until its tapered end contacts the outer-piston tapered face and pushes this piston against the leading shoe. Any further cable pull will now produce an equal and opposite reaction at the pivot point, and the cylinder body will now slide on the backplate away from the outer piston and hard against the trailing shoe and drum. Again, equal expanding force will be applied to both shoes, without disturbing the inner hydraulic piston and seal.

A very similar arrangement exists in fig. 9.11, where again the cylinder and piston are both designed to create the foot-brake shoe-expander movement. This time, however, a bell-crank lever engages a rectangular hole in the leading-shoe web. Applying the hand-brake pivots the lever

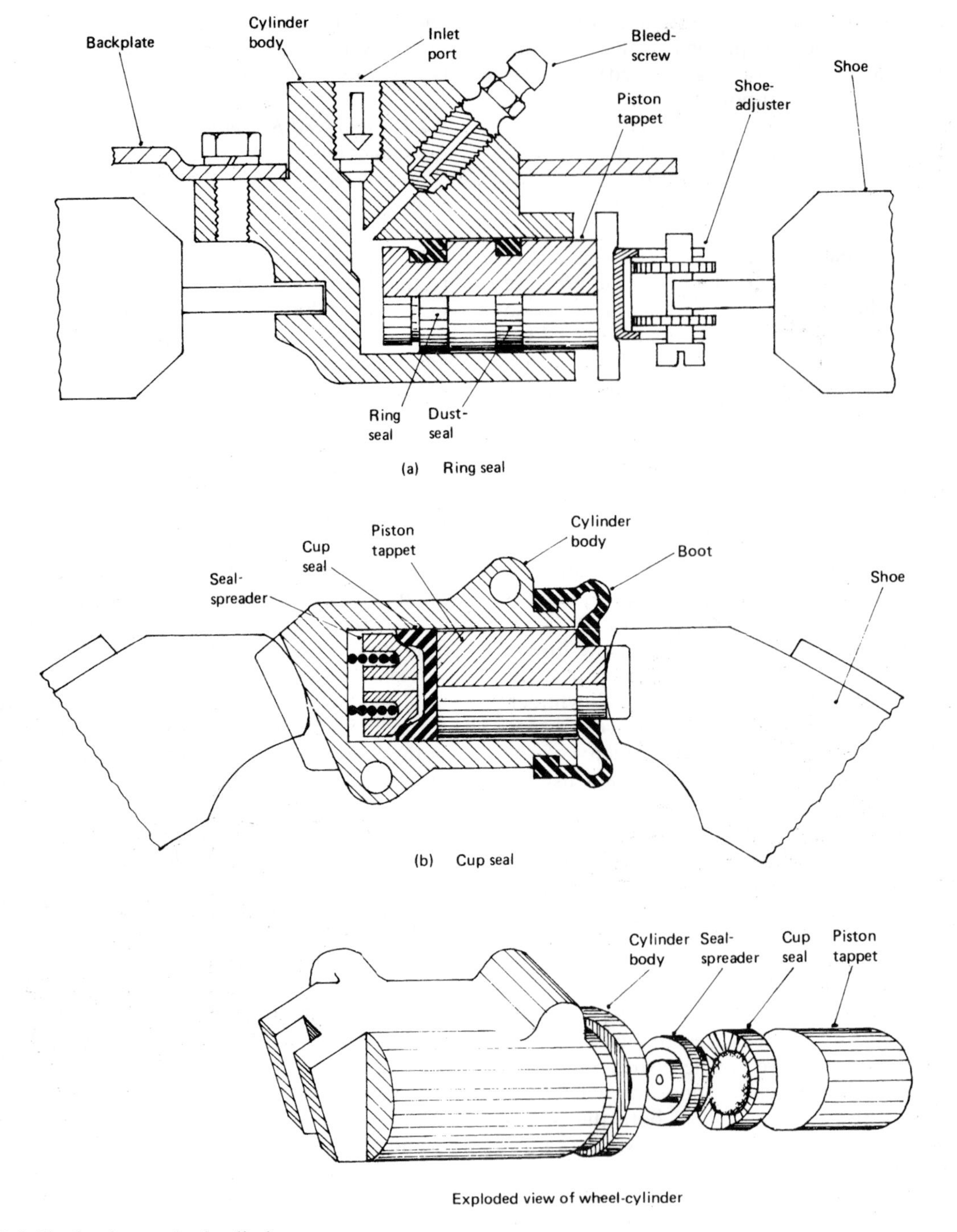

Fig. 9.9 Single-piston wheel-cylinders

so that its short end forces out the leading shoe, and the equal and opposite reaction thrust on the pivot pin now moves the cylinder body in its slot in the backplate to engage the trailing shoe.

9.4.4 Separate rear-wheel parking-brake mechanism (fig. 9.12)

With this parking-brake shoe-expander, the hydraulic foot-brake cylinder body is bolted to

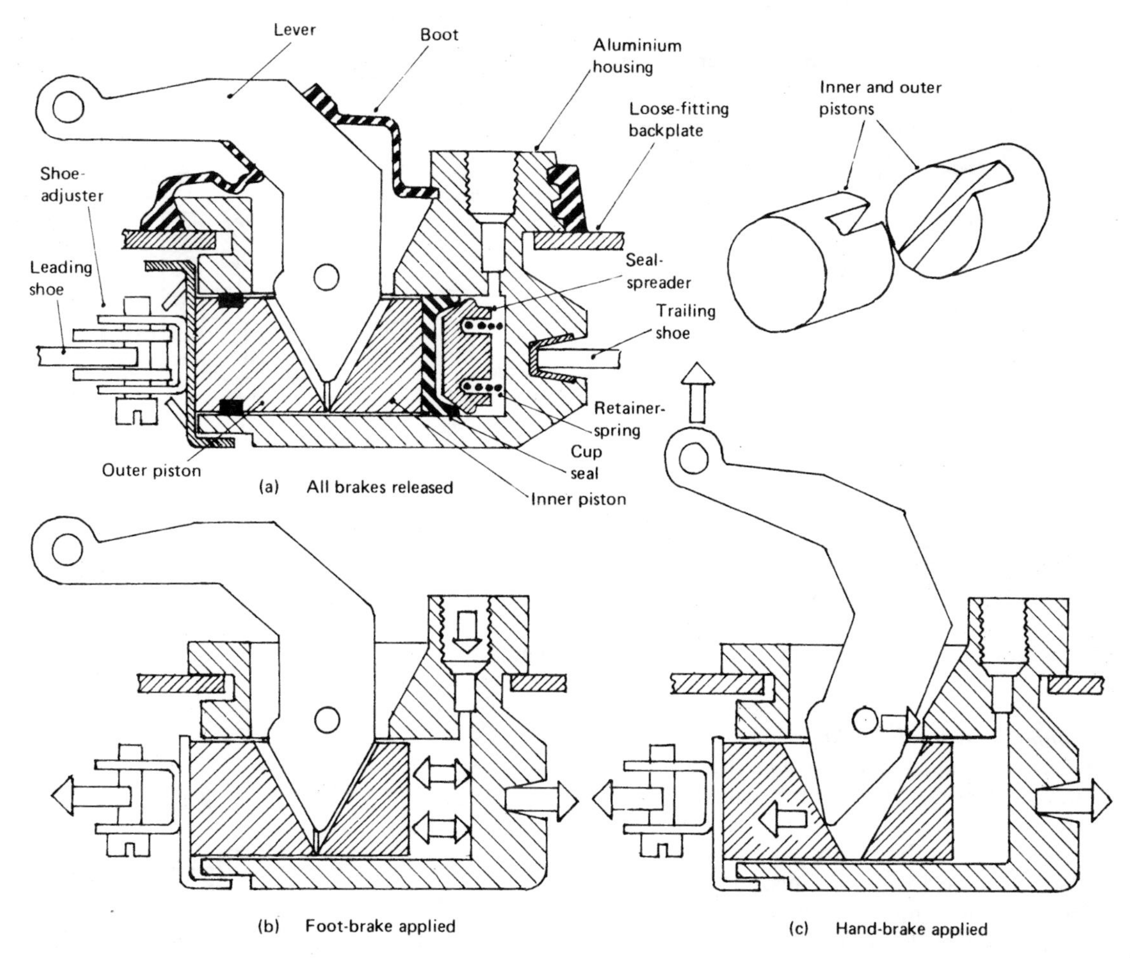

Fig. 9.10 Combined hydraulic/lever rear-wheel shoe-expander (small car)

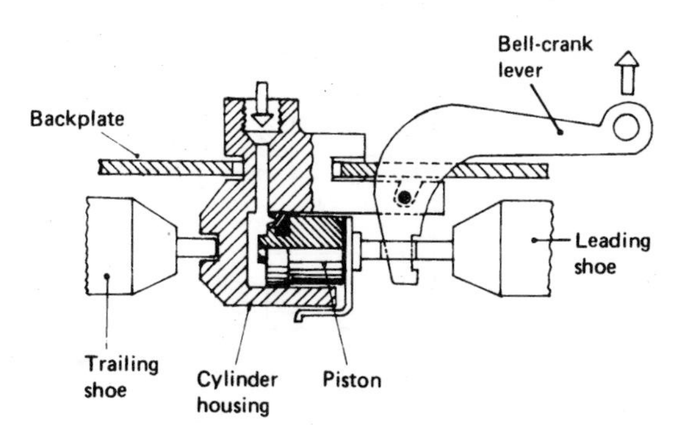

Fig. 9.11 Combined hydraulic/level rear-wheel shoe-expander (large car)

the backplate, and a piston at each end actuates the shoes.

Bridging the two shoes is a link-strut, one end fitting against one shoe web and the other end acting as the pivot point for the parking-brake lever attached to the other shoe. Two alternative lever layouts are shown: one is perpendicular to the shoe (fig. 9.12(a)) and the other is parallel to the shoe (fig. 9.12(b)). The cable is coupled to the free end of the lever and, when the hand-brake is applied, the cable pull pivots the lever so that it pushes the strut one way, actuating the leading shoe and levering the trailing shoe in the opposite direction. Since the link-strut floats between the two shoes, the expanding force will be shared equally between them.

9.5 Wedge-operated shoe-expander units

A wedge is a highly efficient means of producing a force ratio between the input effort at the expander and the output shoe tip load.

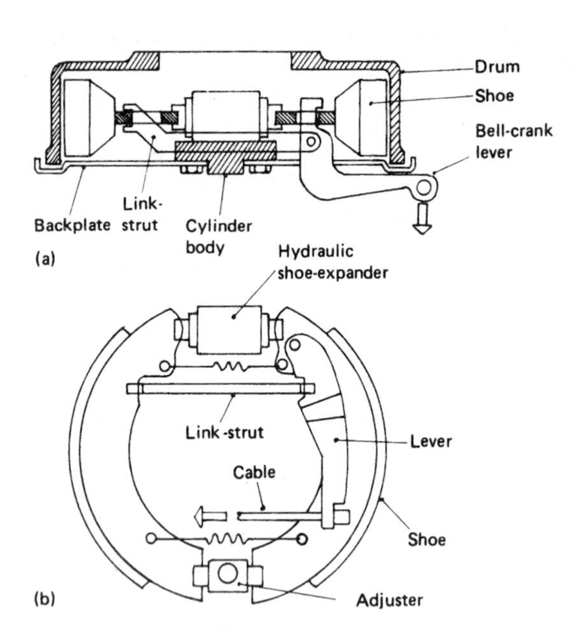

Fig. 9.12 Separate rear-wheel hand-brake lever

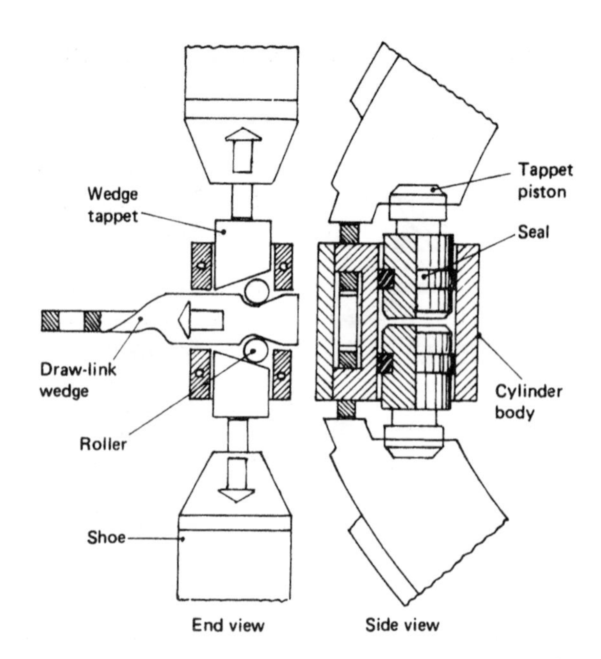

Fig. 9.13 Wedge-and-roller hand-brake shoe-expander

9.5.1 Combined hydraulic-cylinder and mechanical-wedge shoe-expander unit (fig. 9.13)
The combined hydraulic and mechanical expander is enclosed in a cast-iron body, one part of which consists of a bored-through cylinder to house the hydraulic foot-brake pistons while the other part has slots maintained to support and guide the parking-brake wedge mechanism.

The wedge shoe-expander consists of a draw-link of flat rectangular steel section with a taper triangular piece cut out from each side to form the double wedge. Each wedge tappet is a flat parallel rectangular section with a square shoe-contact outer end but with an inner end which is inclined to form a taper ramp, so that it provides half the wedge action.

To reduce friction, rollers are positioned between the central draw-link double-sided wedge ramps and the wedge-tappet ramps.

In the brakes-released position (fig. 9.13), the rollers occupy and contact points between the narrowest width of the central draw-link wedge and the shortest side of the tappets.

When the parking brake is applied, the draw-link will tend to move perpendicular to the axis of the restrained tappets. To accommodate this movement, the rollers will simultaneously roll along and up the draw-link ramps and across and up the tappet ramps. The net result of this double action will be to produce a tappet outward expansion movement equal to the combined lifts of the drawlink and the tappet ramps.

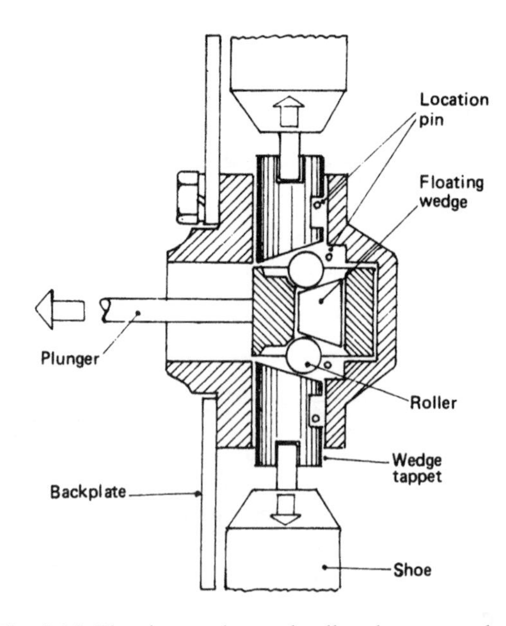

Fig. 9.14 Floating-wedge and roller shoe-expander

9.5.2 Floating-wedge shoe-expander unit (fig. 9.14)
A further improvement in wedge-expander effectiveness is achieved by incorporating a flat wedge in a slot machined at right angles to the plunger axis.

In the brakes-released position, the retraction springs acting on the shoes will push the cylindrical wedge tappets towards the floating wedge, which will be centralised in the plunger slot.

When the brakes are applied, the plunger will be pulled out, so that the rollers will move up the floating-wedge and tappet ramps and so start to push out the two tappets to expand the shoes against the drum. If the leading shoe is worn more than the trailing shoe, then the outward movement of the trailing shoe will contact the drum first. Any further pull on the plunger will continue the expansion between the tappets, due to roller movement; but, to take up the shoe–drum clearance on the trailing shoe, the floating wedge will slide or float across the plunger to take up the slack until equal shoe tip load is applied to both shoes.

9.5.3 Pull-type wedge shoe-expander (fig. 9.15)

The wedge part of this expander is similar to that of the floating-wedge expander described in section 9.5.2, but this unit differs in that the wedge plunger is connected to a hydraulically operated flanged sleeve and yoke by means of a draw-link which has its ends screwed into both the plunger and the yoke. The flanged sleeve slides within one end of the hydraulic cylinder. There are two seals: one rests against the housing face and the other bears against the moving-sleeve flanged end. This expander unit provides both foot-brake and parking-brake actuation, and so is designed for rear-wheel brake assemblies.

When the foot-brake is applied, fluid from the brake-lines will tend to push the seals apart, so the flanged sleeve will move outwards, pushing the yoke ahead of it. As a result of the yoke being forced out, the draw-link will pull out the wedge plunger so that the rollers move up both the floating-wedge and the wedge-tappet ramps to expand the shoes apart.

Releasing the foot-brake and applying the parking brake will pull the brake rod and yoke outwards. This will again move the wedge plunger perpendicular to the tappet axis. As a result, the rollers again move up both the floating-wedge and the wedge-tappet ramps, thus producing the necessary shoe tip braking load. It should be observed that the large space in the centre of the yoke allows the foot-brake to be applied without interfering with the parking-brake mechanical linkage.

9.5.4 Push-type wedge shoe-expander (fig. 9.16)

Both pull- and push-type wedge shoe-expanders operate by forcing rollers between inclined double-wedge and wedge-tappet profiles; but, in the push-type mechanism, instead of a draw-link

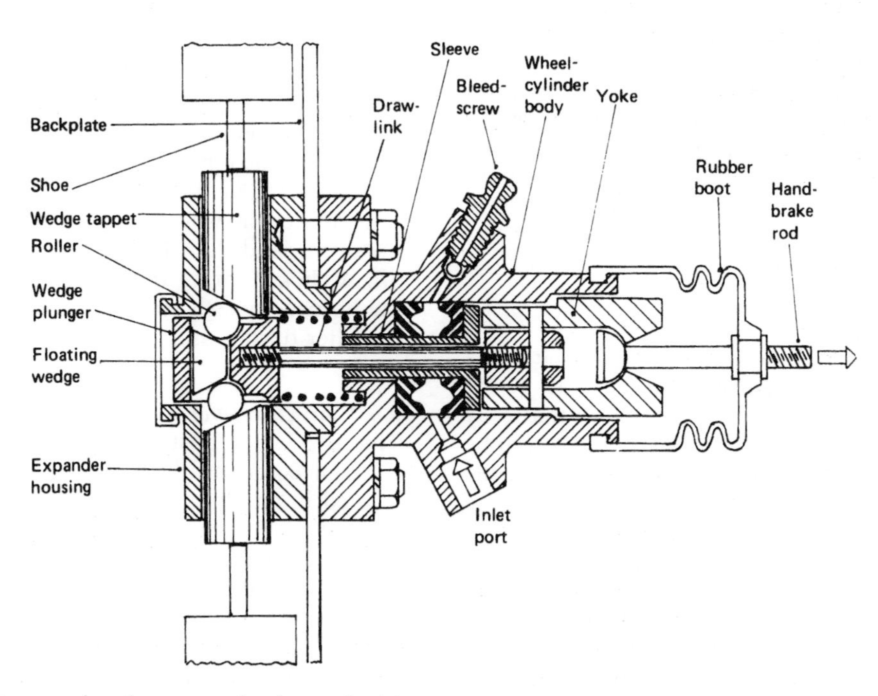

Fig. 9.15 Pull-type wedge shoe-expander (rear wheels)

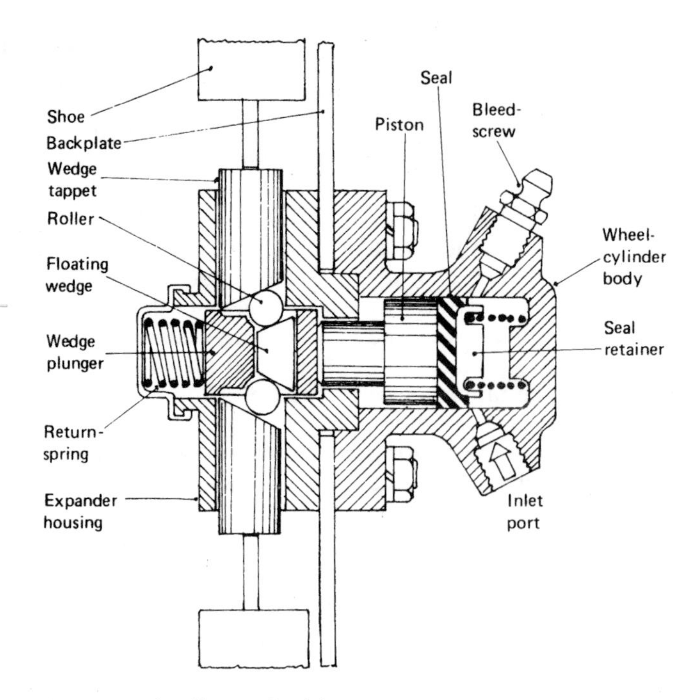

Fig. 9.16 Push-type wedge shoe-expander (front wheels)

which pulls the plunger, a direct-acting piston pushes the plunger and floating wedge across the inclined wedge-tappet faces to expand the shoes against the drum.

When the foot-brake is applied, hydraulic line pressure will be transmitted to the cylinder and will then be converted to force as it operates on the cross-sectional area of the piston. The piston will then move along its stroke and so will force the plungers and shoes against the drum to provide the necessary friction drag to stop the vehicle.

Note that the push-type expander does not have parking-brake provision, and is suitable only for front-wheel brakes.

9.6 Shoe-adjusters

Shoe-adjusters are provided to take up the clearance between the brake-drum and the shoes as the linings wear down. This is necessary so that only the minimum volume of additional fluid has to be displaced into the wheel-cylinders each time the brakes are applied. If the linings are worn and the shoes in the released position are not adjusted, then the movement of the wheel-cylinder pistons will be considerable and a large quantity of fluid will need to be displaced by the master-cylinder piston. The net result will be excessive brake-pedal and master-cylinder-piston movement. Since braking time consists of the reaction thinking time and the time taken for the pedal to move and take up the free-play, it can be seen that adjustment of the wheel brakes is essential.

9.6.1 Wheel-cylinder ratchet shoe-adjuster (fig. 9.17)

A screw tappet and notched adjuster wheel are situated between the wheel-cylinder piston and the shoe. When adjustment is necessary, a screwdriver is pushed into a hole in the side of the drum and is levered against the notched adjuster wheel. This rotates the wheel until the shoe binds against the drum, and the shoe is then slightly slackened off. A clicker-spring bears on the ratchet teeth of the notched wheel, so that the screw setting does not alter during use.

9.6.2 Eccentric-peg shoe-adjuster (fig. 9.18)

This is a pivot post supported in the backplate with an eccentrically mounted peg. This peg is located in an elongated hole in the shoe web. To adjust the shoe, the pivot post is rotated by fitting a suitable spanner to the square end of the post on the outside of the backplate. The eccentric angular movement of the peg will force the shoe against the drum and so take up the free clearance. Friction due to the tight fit of the pivot post in the backplate maintains the adjuster setting.

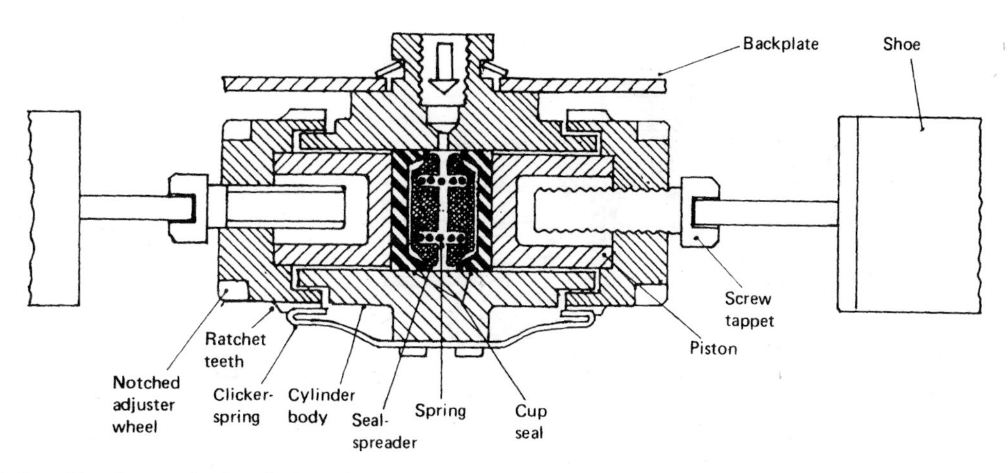

Fig. 9.17 Double-piston wheel-cylinder with ratchet shoe-adjuster

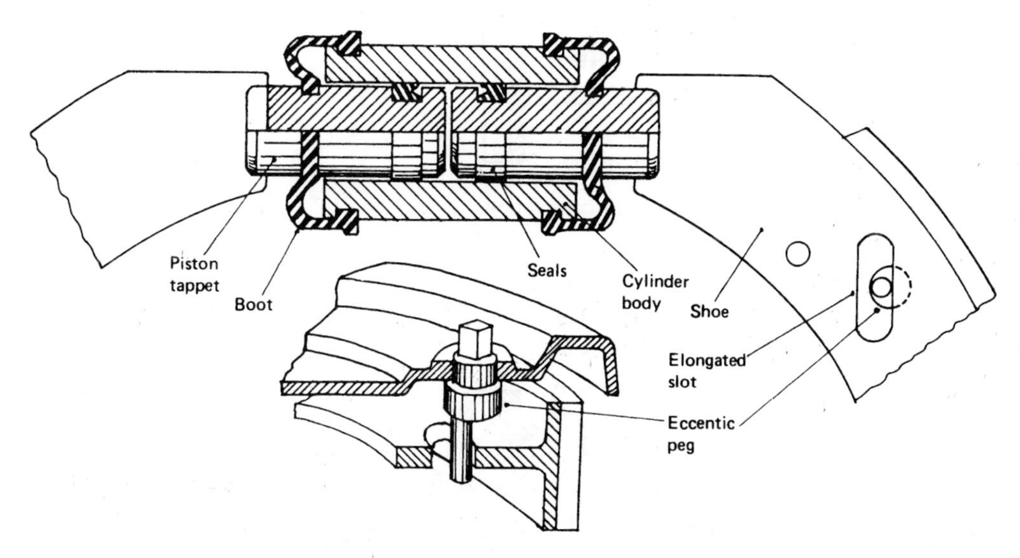

Fig. 9.18 Double-piston wheel-cylinder with eccentric-peg shoe-adjuster

9.6.3 Snail-cam shoe-adjuster (fig. 9.19)
The serrated snail-cam pivots on a hexagonal-headed post located in a hole in the backplate. Rotational friction is provided by a double-coil spring washer fitted between the hexagonal head and the backplate (this is not shown in the diagram). The snail-cam profile bears against a peg fitted in the back of the shoe web. When adjusted, the cam will rotate and push the peg outwards, so that the shoe is forced nearer the drum. Each serration on the cam profile will match up with the peg in turn as the adjuster is rotated from behind the backplate.

9.6.4 'Micran' shoe-adjuster (fig. 9.20)
A cam and holder are positioned between the piston abutment cap and the shoe toe. The double cam has its serrated profile adjacent to the holder notch. When the shoe clearance is to be adjusted, the brake-drum is rotated until a hole in its side aligns with the cam axis. A screwdriver is then placed in the cam-pivot side slot and is rotated until the shoe clearance has been taken up, each position being secured by the cam serration clicking over the fixed holder notch.

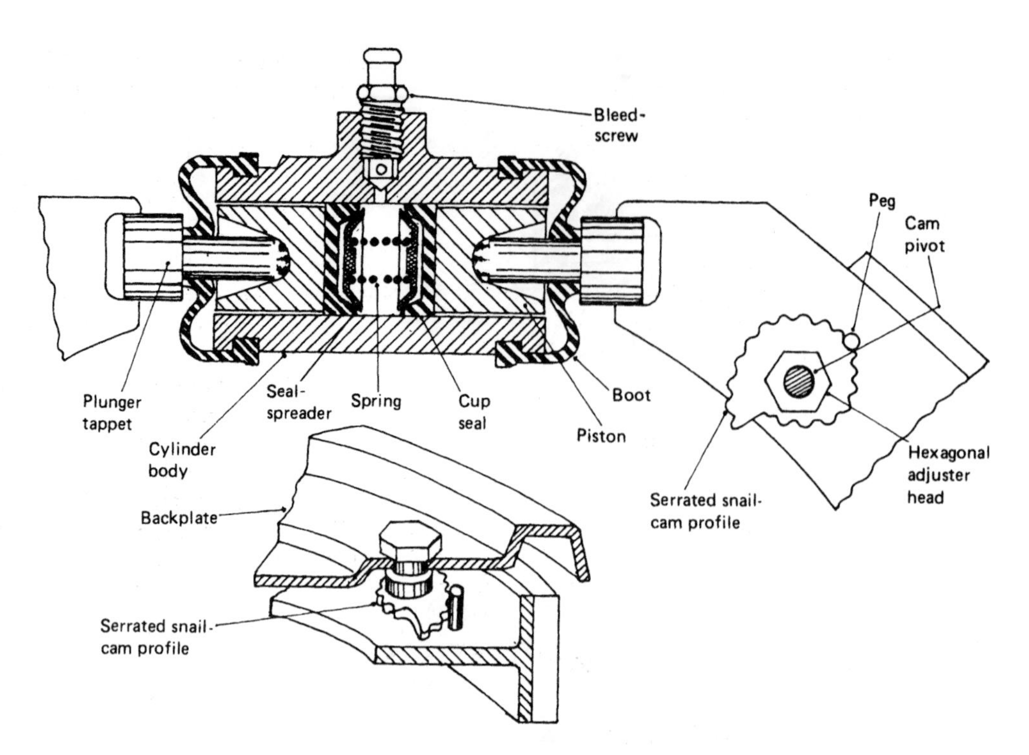

Fig. 9.19 Double-piston wheel-cylinder with snail-cam shoe-adjuster

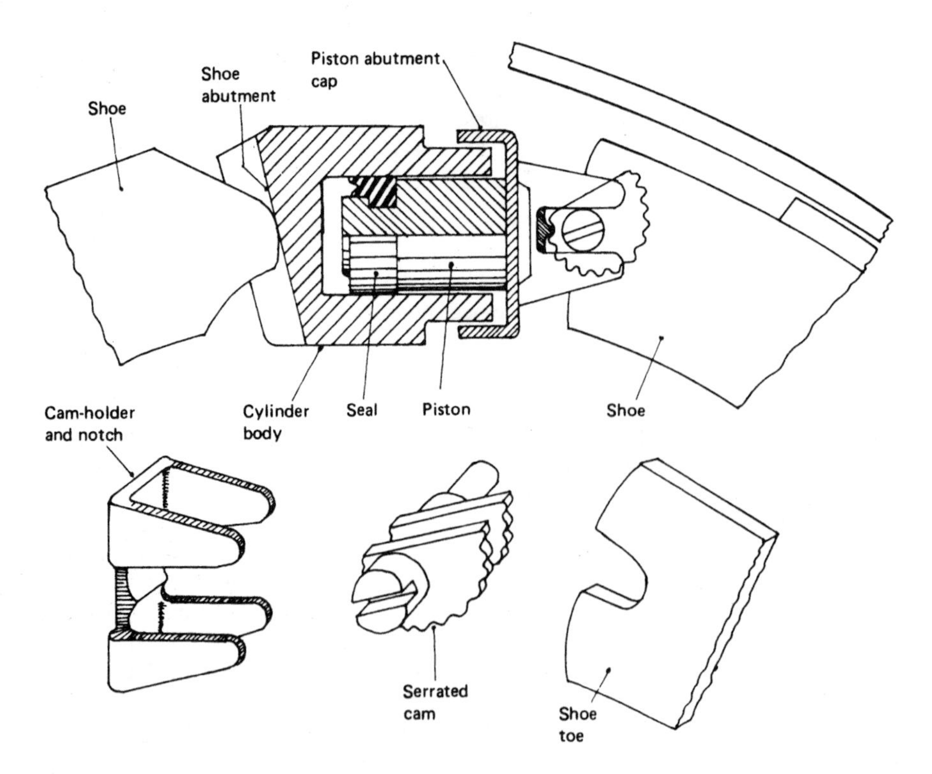

Fig. 9.20 Single-piston wheel-cylinder with 'Micran' shoe-adjuster

9.6.5 Screw-wedge shoe-adjuster (fig. 9.21)
This approach provides the brake-shoes with an anchor point and a means for setting the shoe-to-drum clearance. It consists of two tappets which are tapered at one end and have slots to support and align the shoe toes at the other. They are supported in a housing which is bolted to the backplate. Screwed into this housing at right angles to the tappets is an octagonal wedge with an adjuster screw. When the wedge is screwed in, the tappets are forced apart and so expand the shoes against the drum. The octagonal-wedge flats provide a notch locating effect which is created by the shoe retraction springs pulling the tappets and wedge flats together.

9.6.6 Floating-abutment shoe-adjuster (fig. 9.22)
Some brake arrangements which rely on the self-energising effects of the shoes employ a floating-abutment adjuster. This consists of a screw-and-sleeve tappet. The screw tappet is screwed into a sleeve which has a star-wheel formed on its exposed end, and placed over this inner sleeve is the sleeve tappet. To expand the shoes against the drum, a screwdriver is pushed through a hole in the back-plate and is then made to rotate the star-wheel and inner sleeve. Adjustment is achieved by the resulting unscrewing of the screw tappet, which has the effect of pushing both shoes towards the drum. When adjustment is complete, the retraction spring will brush against the star-wheel and prevent further rotation of the inner sleeve.

9.7 Hand-brake linkage and mechanisms
The foot-brake system is provided to retard or stop a moving vehicle. In contrast, the parking hand-brake system prevents the vehicle rolling or moving once the vehicle has come to rest and the foot-pedal has been released. The hand-brake system may be in operation for very long periods of time, so a mechanical linkage and mechanism separate from that of the hydraulic foot-brake circuit must be incoporated.

9.7.1 Hand-brake linkage layouts (fig. 9.23)
Parking brakes are generally provided only on the rear road-wheels. Either cable or rod linkage may be used to link the various components together, but the general trend these days is for cars and vans to use cable, and only heavy-duty commercial vehicles use rods. The braking effort in both cases is transmitted from the hand-brake lever to the rear wheels by tension, whereas a hydraulic foot-brake circuit is in compression.

The following layouts incorporate a range of hand-levers – these have been shown only as examples and are interchangeable to suit the car's design.

Sliding-equaliser hand-brake cable layout (fig. 9.23(a)) This arrangement uses a single primary cable to connect the hand-lever movement to a sliding equaliser. The middle of the secondary cable is looped around the curved equaliser, and each half of this second cable is connected on one side to the wheel parking-brake lever.

To support and guide the secondary cable, an outer cable cover is provided – this is usually attached to the car's underbody structure at one end. When the hand-brake lever is applied, the primary cable will pull from the middle of the

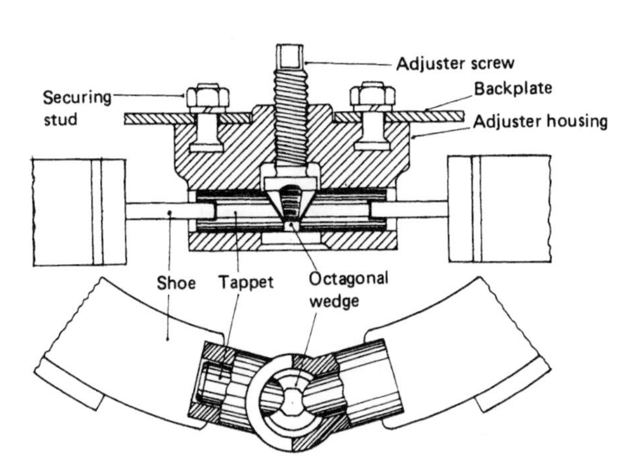

Fig. 9.21 Screw-wedge shoe-adjuster

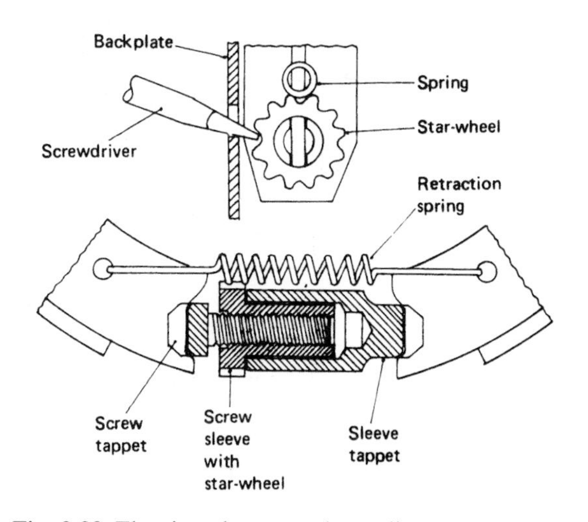

Fig. 9.22 Floating-abutment shoe-adjuster

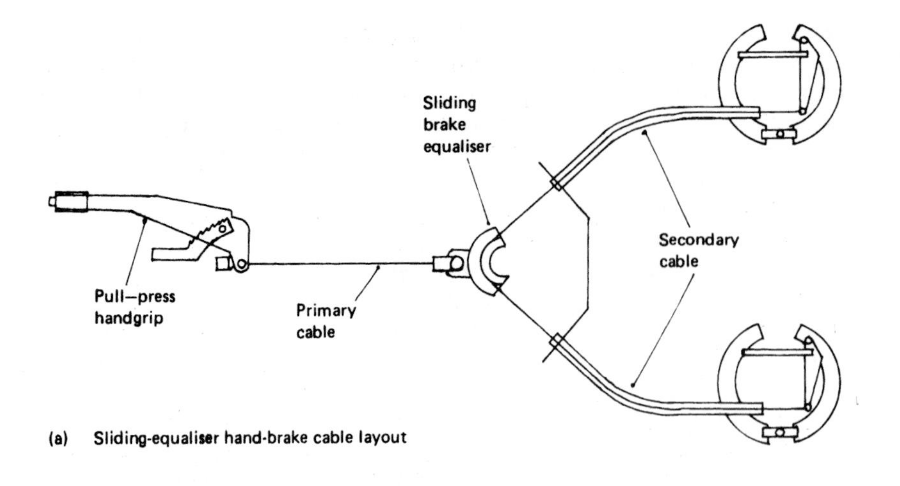

(a) Sliding-equaliser hand-brake cable layout

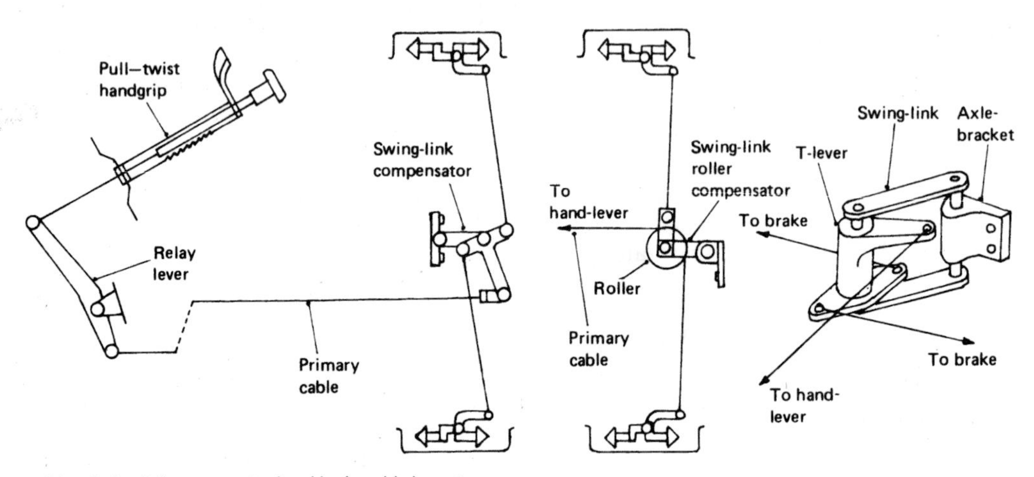

(b) Swing-link-compensator hand-brake cable layout

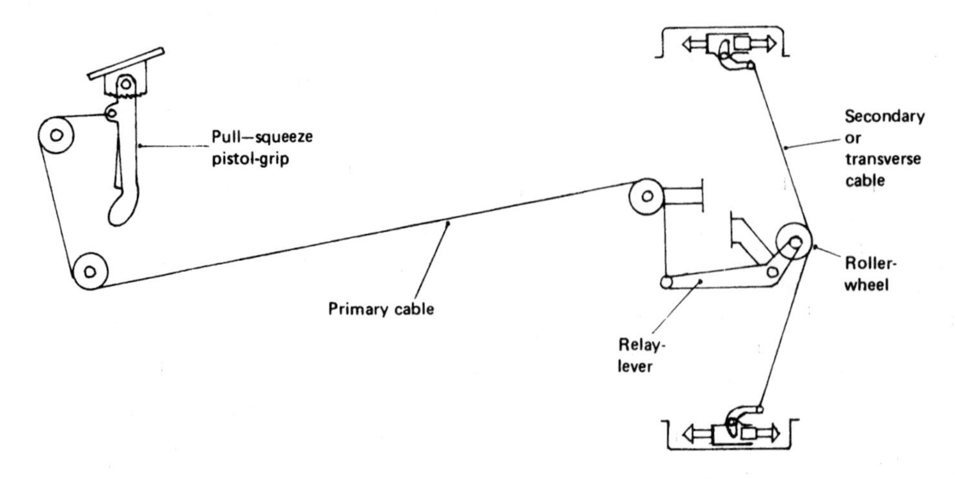

(c) Roller and relay-lever-compensator hand-brake cable layout

Fig. 9.23 Hand-brake cable linkage layout

secondary cable, and any difference in length or slackness between the two halves of the secondary cable will be equalised by the cable sliding relative to the equaliser around a specially formed semi-circular groove. This then will provide equal tension in both halves of the secondary cable and to each rear brake lever.

Adjustment is obtained at the hand-brake lever by a nut screwed on the threaded end of the primary cable.

Swing-link-compensator hand-brake cable layout Figure 9.23(b) shows a pull–twist handgrip which, when applied, relays leverage to the primary cable. The rear end of this cable is connected to a T-shaped lever. The pull in the primary cable will tend to rotate the T-lever so that the two short perpendicular arms will tension the two transverse cables connected to the rear-wheel parking-brake levers. If there is uneven slack in the two transverse cables, the swing-link will compensate by swivelling about its axle-bracket pivot, equalising both slackness and tension in the opposing cables.

An alternative compensator employs a roller attached to the swing-link. One short transverse cable is connected between the swing-link and one of the wheel parking-brake levers, and the other long cable spans from the opposite wheel parking-brake lever to the swing-link roller, turns at right angles, and then runs parallel with the wheelbase to the hand-brake lever. Applying the hand-lever pulls the long primary cable. This tends to shorten its cable span between the swing-link roller and the wheel lever until the slack in the opposite separate cable has been taken up. Any further tensioning by the hand-lever will apply equal braking force in both transverse cables and their parking-brake levers.

Roller-and-relay-lever-compensator hand-brake cable layout (fig. 9.23(c)) A pull–squeeze pistol-grip applies tension to the primary cable, which is guided to the rear of the car by rollers attached to the body structure. The cranked relay-lever with the roller-wheel attached at one end takes up the slackness of the single secondary transverse cable by levering this cable to one side at some point near its centre span. This provides an equal pull to both wheel parking-brake levers. Adjustment is achieved by altering the length of one or both cables.

9.7.2 Hand-brake lever mechanisms

Hand-brake lever assemblies are provided for the driver to apply or release the parking brakes. They may or may not be designed to supply additional leverage to the system.

Pull–squeeze palmgrip hand-lever (fig. 9.24(a)) This consists of a rectangular box-section tube bolted to the dashboard. Within this tube is a drawbar which incorporates a spring-loaded pawl and an actuator strut. The pawl engages a ratchet formed on the underside of the box tube. When the driver pulls the handgrip and squeezes the release lever, the push-rod will force the actuating strut to clear the pawl from the ratchet teeth, and the hand palm-grip is then moved forward to release the brakes. To apply the parking brake, the driver just pulls the handgrip lever out and releases his grip – the pawl then slots into the nearest adjacent ratchet-tooth space and holds the cable in tension.

Pull–twist handgrip lever (fig. 9.24(b)) This construction has a circular draw-bar with ratchet teeth cut on one side. Two pawls are made to engage these teeth. A peg is inserted near the 'lower' end of the bar and limits the free twist movement to about 90°. To apply the parking-brake, the handgrip is twisted. This moves the ratchet teeth away from the pawls, so that the pawl edges are now in contact with the smooth-surface section of the draw-bar. Pulling out the bar and then twisting it back to its original position places the ratchet teeth adjacent to the pawls so that they are free to engage the nearest tooth and so hold the brake cable in tension.

Pull–press-button hand-lever (light-duty) (fig. 9.24(c)) This arrangement seems to be the most popular type at present. It consists of a hand-lever which has one end connected directly to the brake cable while the other end forms the handgrip, with the release press-button incorporated. This lever is pivoted on a ratchet bracket, and, when the driver pulls up the lever arm, the spring-loaded pawl slides over the ratchet teeth until the maximum desired cable tension is achieved. At this point the driver releases his hand from the lever, and the vehicle is parked. To release the parking brake, the driver must pull up the hand-lever slightly to relax the tension on the pawl, press in the press-button with the thumb, which will rotate the pawl teeth clear of

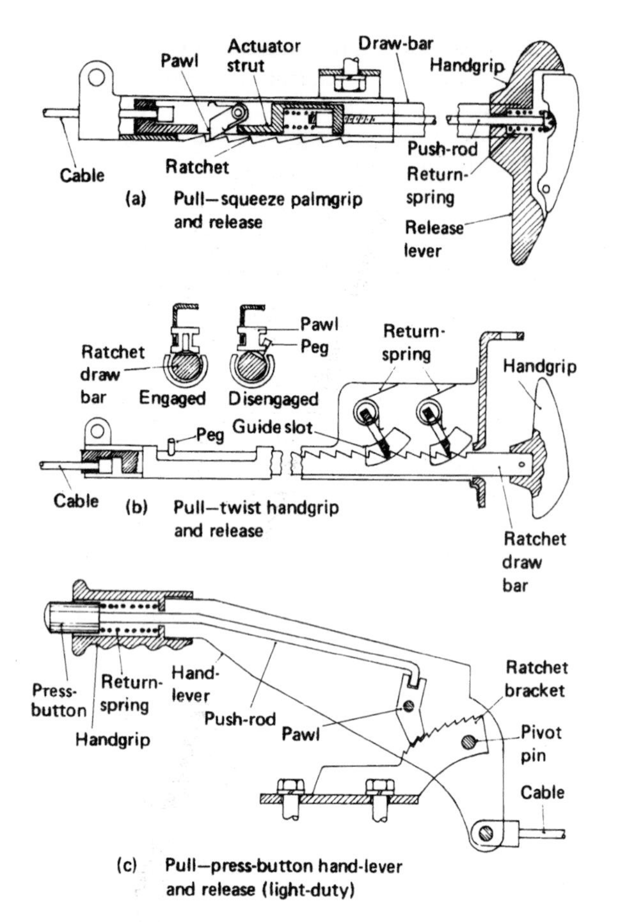

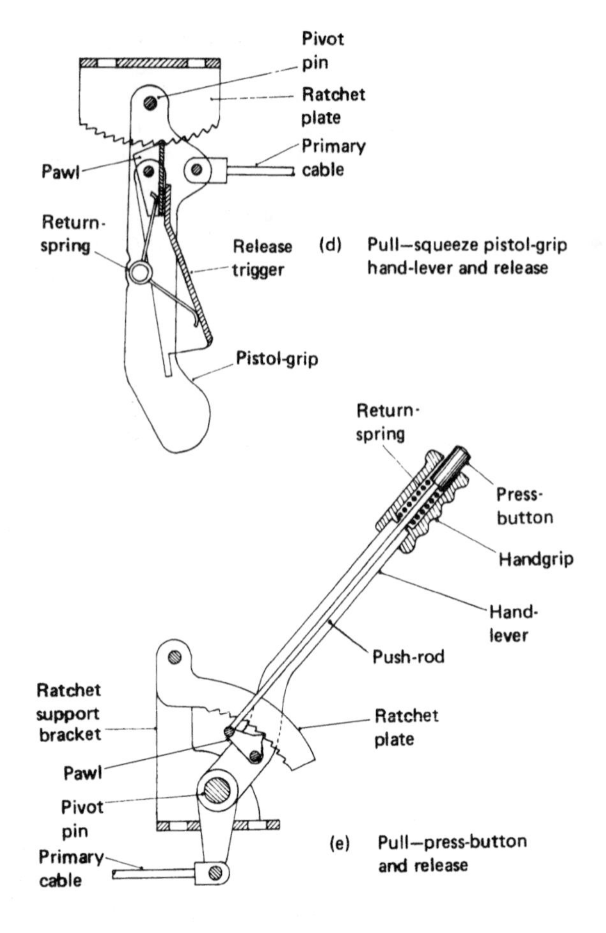

Fig. 9.24 Hand-brake levers

the ratchet teeth, and then return the hand-lever to the fully down and released position.

Pull–squeeze pistol-grip hand-lever (fig. 9.24(d)) The pistol-grip hand-lever is pivoted on the ratchet plate, and leverage is obtained by having the cable connection relatively close to this pivot point. When the lever is pulled, the spring-loaded pawl just slides over the backs of the ratchet teeth until the cable has reached the required tension. At this point the pistol-grip is released, and the slight forward movement will push the spring-loaded pawl into engagement with the nearest adjacent ratchet tooth. To release the parking brake, the release trigger is squeezed and the hand-lever is drawn back slightly. This releases the load from the pawl so that it can rotate and clear the ratchet teeth. The lever is then pushed forward fully to the released position.

Pull–press-button hand-lever (heavy-duty) (fig. 9.24(e)) With this arrangement, the upright lever arm may be longer than for the light-duty type so that increased leverage will be obtained. The ratchet plate is pivoted at one end and is free to swing down and bear against the pawl. When the hand-lever is applied, the pawl just lifts the ratchet plate up and allows the pawl to slide underneath the teeth until the parking position has been reached. The hand-lever will then be released, and the pawl will immediately engage the nearest ratchet-plate tooth. To release the brakes, the lever is initially pulled back, the press-button is depressed to move the pawl away from the ratchet teeth, and then the lever is moved forward towards the 'off' position.

9.8 Disc and pad brakes (fig. 9.25)
A disc-brake consists of a rotating cast-iron disc which is bolted to the wheel hub and a stationary

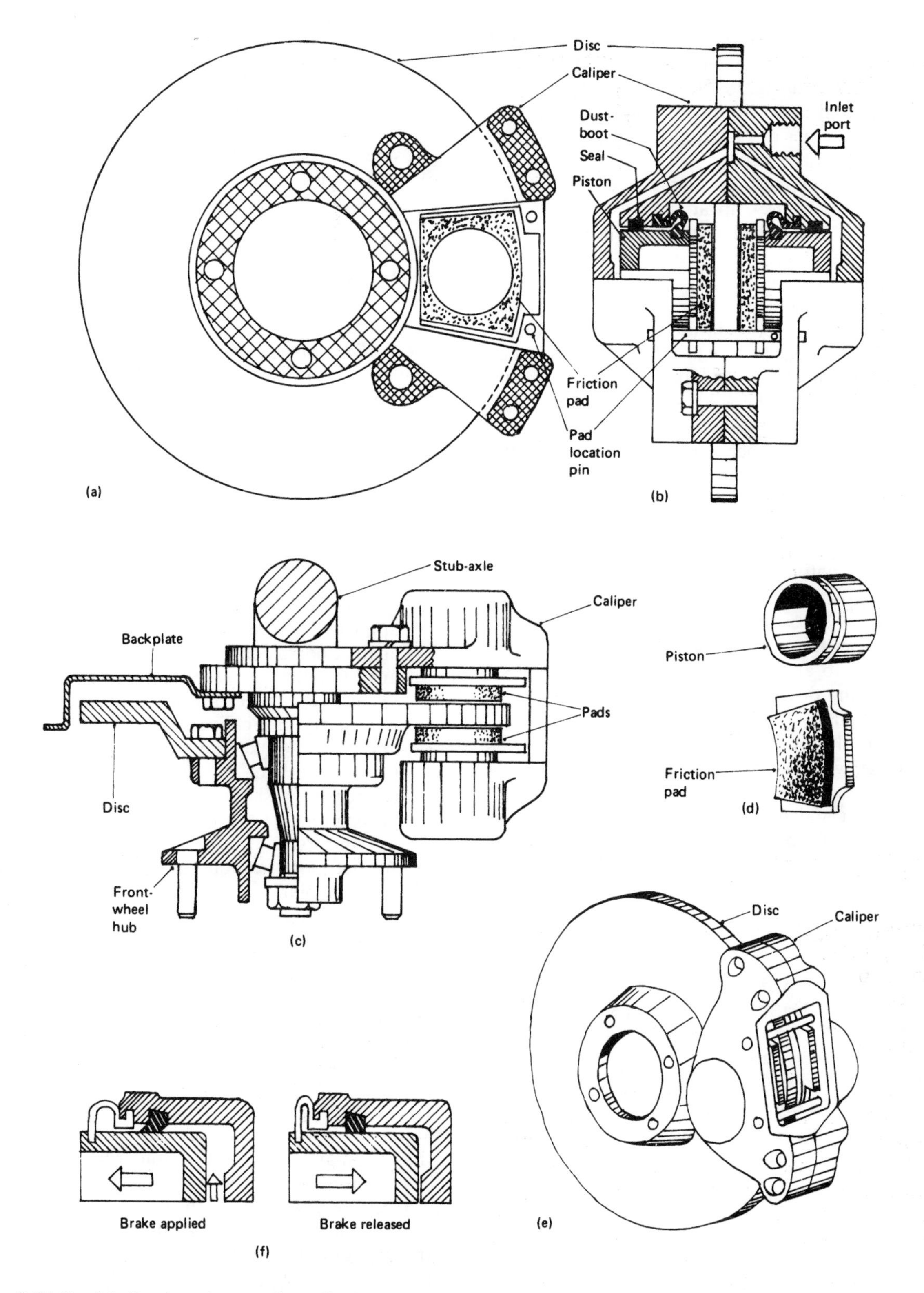

Fig. 9.25 Double floating-piston-caliper disc-brake assembly

caliper unit which straddles the disc and is bolted to the stub-axle or swivel-post flange (fig. 9.25(c)). The caliper unit (fig. 9.25(a) and (b)) is made of cast iron and is in two halves. Each half forms a separate cylinder block with the cylinder axis perpendicular to the disc.

The two cylinders are connected together by drillings which meet up at the pressure faces of the two caliper halves somewhere near the inlet port (fig. 9.25(b)). A bleed-screw drilling also intersects at this junction.

Each cylinder contains a rubber sealing ring positioned in a groove in the body and a hollow piston protected by a dust-cover. Sandwiched between each piston and the disc face on that side is a segment-shaped friction pad bonded to a steel plate. These pads fit into slots formed in each half of the caliper housing (fig. 9.25(e)), and their function is to transfer the braking reaction torque from the disc to the caliper and the stationary stub-axle. They are held in position by locating or retaining pins (fig. 9.25(a) and (b)), or sometimes spring plates.

When the brake pedal is depressed, hydraulic pressure will be generated in the fluid pipelines. This causes the opposing caliper pistons and friction pads to apply equal and opposite forces on the rotating disc in direct proportion to the foot effort applied.

When the brakes are released, the hydraulic pressure will collapse and the distorted rubber seal housed in the cylinder groove will retract the piston and pad sufficiently to clear the disc faces from frictional contact. As the pads wear, the pistons will be pushed out further to compensate, but the retraction movement due to the elasticity of the rubber seals will remain constant, thus pad wear adjustment is automatic (fig. 9.25(f)).

The friction-pad segments operate on only a small sweep area at any one time, so most of the surface subjected to frictional contact will be exposed to the atmosphere. This greatly improves the dissipation of heat compared with the drum-and-shoe braking system.

The advantages of disc-brakes may be summarised as follows.

a) The lack of any self-servo action produces consistent braking.
b) Good air ventilation of the disc and the friction pads provides a low average disc temperature which reduces pad friction fade.
c) The flat friction contact action between the disc and pads produces uniform pad wear.
d) Uniform hydraulic pressure on each side of the disc floats the pad pistons, which thus provide equal grip on the disc, so eliminating side-thrust from the disc to the hub.
e) Pad-to-disc clearance is designed to be automatically taken up as the pads wear.
f) Disc-brakes are simple in design, with very few parts to wear or to malfunction.
g) The friction pads of disc-brakes are easily removed and replaced.

9.8.1 Slide-pin-type disc brake caliper (fig. 9.26)
This design of disc brake comprises the following components:

1) A carrier bracket bolted to the suspension hub carrier which absorbs the brake torque reaction.
2) A caliper and cylinder housing which clamps the friction-pads to the disc.
3) A pair of guide-pins which support the caliper and housing and also permit it to align itself transversely to the disc when the brakes are applied. The pre-greased guide pins are sealed against corrosion by the dust covers to avoid any stickiness or seizure of the pins in their respective carrier bracket bores. These pins only carry the caliper and cylinder housing and do not take the twisting braking load. One or both of the guide pins may have a rubber bush sleeve to prevent noise and to absorb relative movement between the sliding pin and its bore.
4) A pair of friction-pads bonded to steel support plates. These pads press against the annular side faces of the disc when the brakes are applied to create the retardation friction. One of the two lugs formed at the ends of each steel support plate is dragged against a corresponding abutment machined on the inside of the carrier bracket when the brakes are applied to absorb the braking torque.

Operation (fig. 9.26) When the foot brake is applied, the hydraulic pressure created pushes the piston and caliper and cylinder housing apart. The piston forces the inboard pad against its adjacent disc face and at the same time the caliper and cylinder housing slides on the guide-pins in the opposite direction to move the outboard pad also onto its adjacent disc face. By these means the friction-pads transversely align themselves so

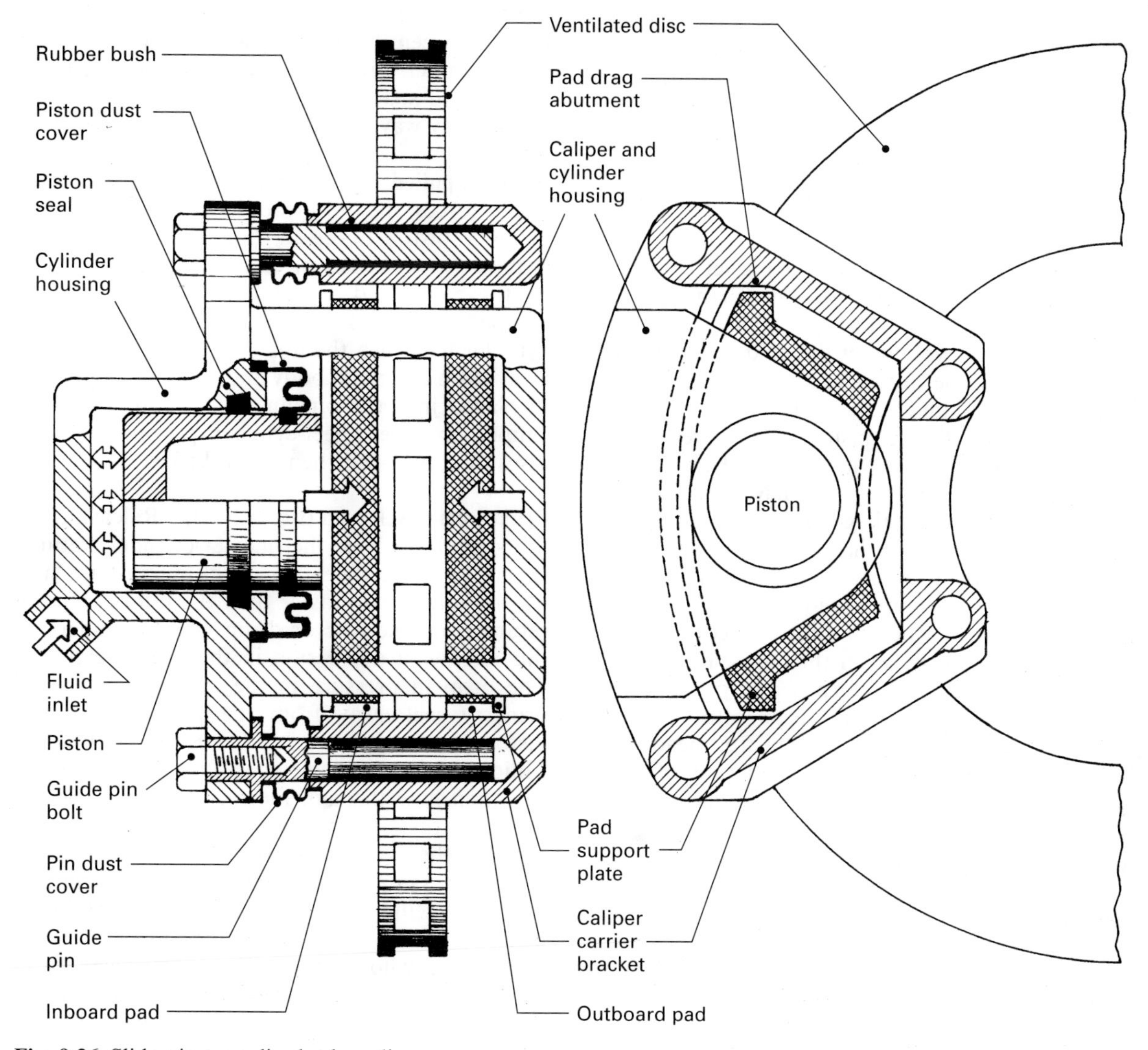

Fig. 9.26 Slide-pin type disc brake caliper

that the clamping load on each side of the disc are equal.

Free clearance between the pads and disc is obtained by the pressure seals which are positioned inside recesses in the cylinder walls and grip the piston. When hydraulic pressure forces the piston outwards, it causes a portion of the seal to distort and move with the piston. Conversely when the brake is released the hydraulic pressure collapses, permitting the strain energy of the elastic rubber to pull back the piston until the seal has been restored to its original shape.

Servicing Check the pads for wear every 10000 km (6000 miles) and fit new pads when the lining thickness has worn to 3 mm. If electrical wear indicators are incorporated the examination should be unnecessary.

9.9 Brake master-cylinders

The brake master-cylinder consists of a cylinder and a piston whose purpose is to generate pipeline hydraulic pressure. This pressure is then converted to force to actuate the wheel-cylinder disc-pads or shoe-expanders.

re are basically two classifications of master-
·r:

i) residual-pressure,
ii) non-residual-pressure.

9.9.1 Residual-pressure master-cylinder. (Lockheed) (fig. 9.27)

Master-cylinder construction The master-cylinder has two chambers: a cylinder pressure chamber and a reservoir chamber. The reservoir compensates for any change in the volume of the fluid in the brake pipeline system due to temperature variations and for a limited amount of fluid leakage.

The master-cylinder piston has a reduced-diameter middle-region which is at all times full of fluid. Both ends of the piston have a rubber lip seal fitted to prevent leakage of fluid. The return-spring end of the piston has a high-pressure cup seal known as the primary seal, and the push-rod piston end has a low-pressure ring seal known as the secondary seal which slips into a recess groove around the piston. Between the cup seal and the piston, a thin washer is interposed to prevent the cup being drawn into the recuperation holes which are drilled around the piston head. A rubber boot encloses the push-rod end of the cylinder to exclude dirt from the cylinder bore.

Residual-pressure check-valve Drum-brakes have a residual-pressure check-valve at the end of the pressure cylinder opposite the push-rod. This creates a low line pressure (0.5 to 1.0 bar) when the brakes are released.

This pressure serves the following purposes:

a) it keeps pedal free-travel to a minimum by opposing the brake-shoe retraction springs;
b) it keeps the wheel-cylinder seal lips in light contact with the cylinder bore, to avoid entry of air;
c) it prevents the re-entry into the master-cylinder of fluid pumped into the line during the bleeding operation (see section 9.10) – this ensures a fresh charge of fluid at each stroke of the brake pedal and a complete purge of air from the system.

Brakes in the 'off' position (fig. 9.27(a)) When the brake pedal is released, and the return-spring has moved the piston fully back against the stop washer and circlip, the bypass or compensation port which is 0.7 mm diameter will be exposed to the pressure chamber and fluid will be able to flow freely between the reservoir and the cylinder pressure chamber. Fluid will at all times be able to fill the annular space formed between the piston and cylinder by way of the large feed port.

Brakes applied (fig. 9.27(b)) Applying the foot-pedal forces the push-rod to push the master-cylinder piston along its bore. Immediately the bypass or compensation port will be sealed 'off', so that any fluid ahead of the piston will be trapped. The pressure generated in the master-cylinder will push the lips of the check-valve cup away from the metal body and so allow fluid to be displaced into the pipelines. This forces out the caliper or shoe wheel-cylinder pistons, and the discs or drums will then be braked.

Brakes being released (fig. 9.27(c)) Releasing the foot-pedal enables the master-cylinder return-spring to move the piston back against its stop washer and circlip faster than the fluid is able to return from the disc or drum wheel-cylinders. It thus creates a depression in the master-cylinder. The drop in pressure in the cylinder draws the primary seal away from the piston head and distorts it, thus uncovering the recuperation holes. Fluid from the annular space around the piston will then flow through the recuperation holes to remove the temporary pressure difference between the two sides of the piston head.

At the same time, fluid returning from the brakes – being under load from the disc-brake piston seals or drum-brake retraction springs – pushes the whole check-valve body away from its rubber seat and so flows back into the master-cylinder. The fully returned piston will then uncover the compensation port connecting the pressure chamber and reservoir and so releases any excess fluid created by the expansion of the heated fluid.

Compensation for excessive pad and lining clearance Excess pad-to-disc or shoe-to-drum clearance will produce a very long master-cylinder piston stroke and pedal movement. A quick release of the pedal will draw additional fluid into the cylinder from the reservoir, and a second quick stroke of the pedal will displace this additional fluid into the pipelines. This procedure will reduce the free pedal movement and will temporarily restore the braking effectiveness until such time as the brakes can be adjusted or the pads and linings are renewed.

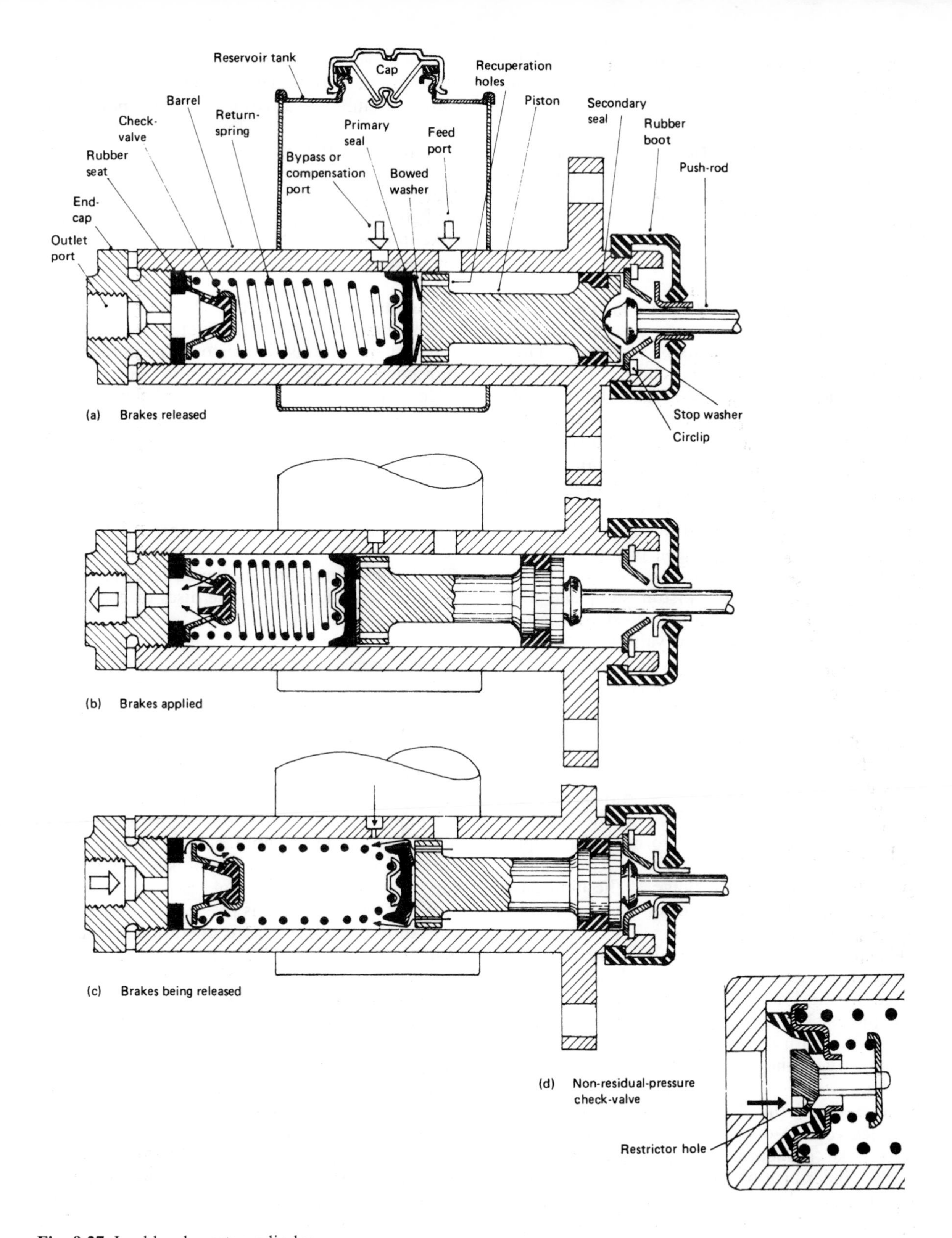

Fig. 9.27 Lockheed master-cylinder

Non-residual-pressure check-valve (fig. 9.27(d)) Unlike drum-brakes, disc-brakes must have no residual pressure in the pipeline, otherwise a complete release of the pads from the disc is not obtained, resulting in overheated discs and rapid wear. This problem may be overcome by having a small restrictor hole in a conical check-valve. A complete pressure release is thus available, yet the system can still be cleared by fairly rapid pumping of the pedal during bleeding.

9.9.2 Non-residual-pressure master-cylinder (Girling) (fig. 9.28)

This master-cylinder consists of two main chambers: the pressure chamber in which the piston operates and an end fluid reservoir. The reservoir permits additional fluid to enter into or return from the pipelines, to maintain a constant volume when the fluid is affected by expansion or contraction due to changes in temperature. In addition, any seepage of fluid due to wear at sealed points is compensated for by the reservoir.

The master-cylinder cast-iron piston is shaped like a cylindrical plunger with a hollow stem at one end. Fitted over the piston stem end and clipped in place is a thimble-shaped steel pressing forming a spring-retainer. The valve stem with its enlarged head rests in the hollow piston. The valve itself is housed on the valve spacer which rests in the slightly reduced bore of the cylinder near the reservoir inlet port.

A rubber ring which acts as a lip seal is provided at each end of the piston. The rubber cup near the return-spring is known as the primary seal and it is subjected to the full line pressure. Its function is to form a fluid-tight piston end. At the push-rod end is a secondary seal which prevents any leakage of the fluid past the primary seal escaping out of the rear end of the piston.

The master-cylinder is usually bolted to the bulkhead of the car, and the push-rod is connected to the brake foot-pedal. A rubber boot over the back end of the master-cylinder and around the push-rod prevents dirt from coming in contact with the cylinder wall.

Brakes released (fig. 9.28(a)) When the brakes are released, the disc-brake piston seals or the drum-brake retraction springs will retract the wheel-cylinder pistons and so displace fluid back to the master-cylinder. The master-cylinder piston return-spring moves the piston to its outermost position, but, just before the piston reaches the end of its stroke, the spring-retainer clipped to the piston stem will catch and pull the valve stem and valve assembly away from the inlet port. Fluid will then be able to flow freely between the reservoir and the pressure chamber. Any expansion or contraction of the fluid in the pipelines will be compensated for by the reservoir, which will give out fluid to the system for contraction or take in more if the fluid in the pipelines becomes very heated and expands.

Brakes applied (fig. 9.28(b)) To apply the brakes, the driver pushes down the foot-pedal and forces the push-rod against the piston. The initial piston movement towards the inlet port will push the edge of the spring-retainer around the mouth of the piston-stem central hole away from the valve-stem head. At the same time, fluid trapped in the hollow piston stem will be momentarily pressurised and so will push the valve-stem assembly towards the inlet port. The valve assembly and seal will then close the inlet port off from the reservoir and any further movement of the piston will displace fluid through the outlet port into the pipeline system to clamp the discs or expand the shoes against the drums.

9.9.3 Compression-barrel master-cylinder (Girling) (fig. 9.29)

A compression-barrel master-cylinder has a stationary primary recuperation seal held in the body, with the plunger moving through the middle to displace and apply pressure to the fluid. There are four small radial compensation ports in the plunger which bypass the recuperation seal to allow movement of fluid between the reservoir and the cylinder when the brakes are released (fig. 9.29(a)).

When the foot-pedal is pressed, the radial compensation ports pass the recuperation seal and so trap the fluid in the pressure half of the barrel. The brake pipeline will then be pressurised (fig. 9.29(b)).

The recuperation-seal shim allows the fluid to flow freely between the horizontal recuperation ports in the body and the back of the recuperation seal when the brakes are released, and prevents the seal being pressed into the recuperation ports when under pressure.

The recuperation-seal support holds the seal in place and limits its travel when the pressure is released. The secondary seal at the push-rod end of the plunger is a wiper seal to prevent any fluid seeping out of the cylinder.

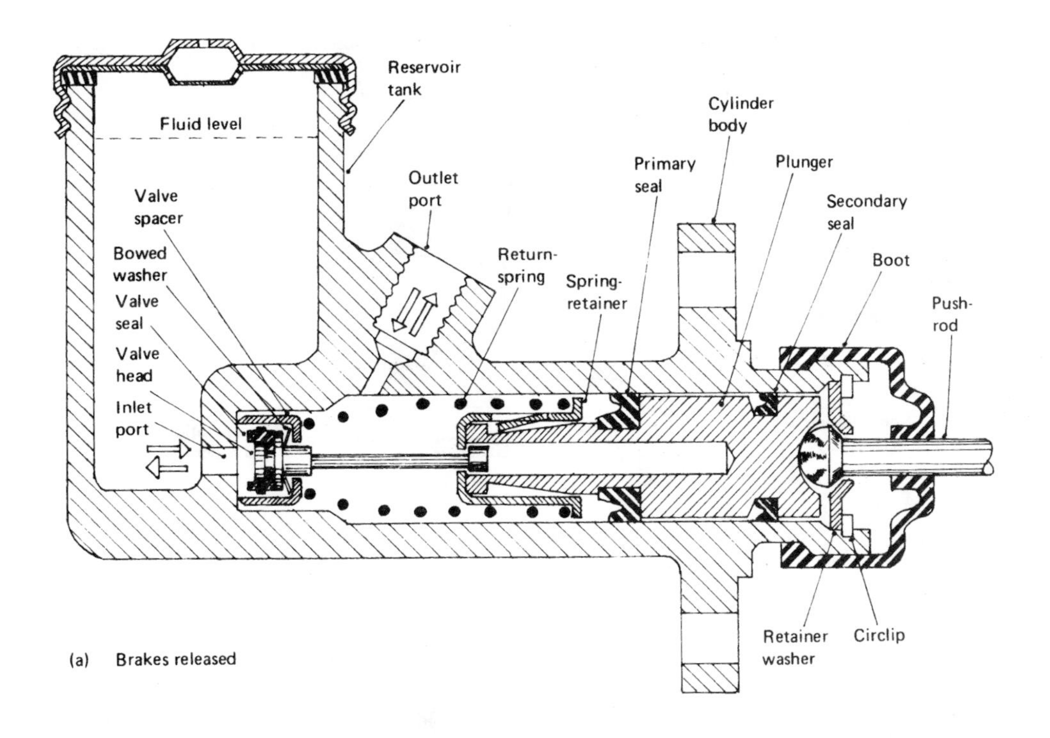

(a) Brakes released

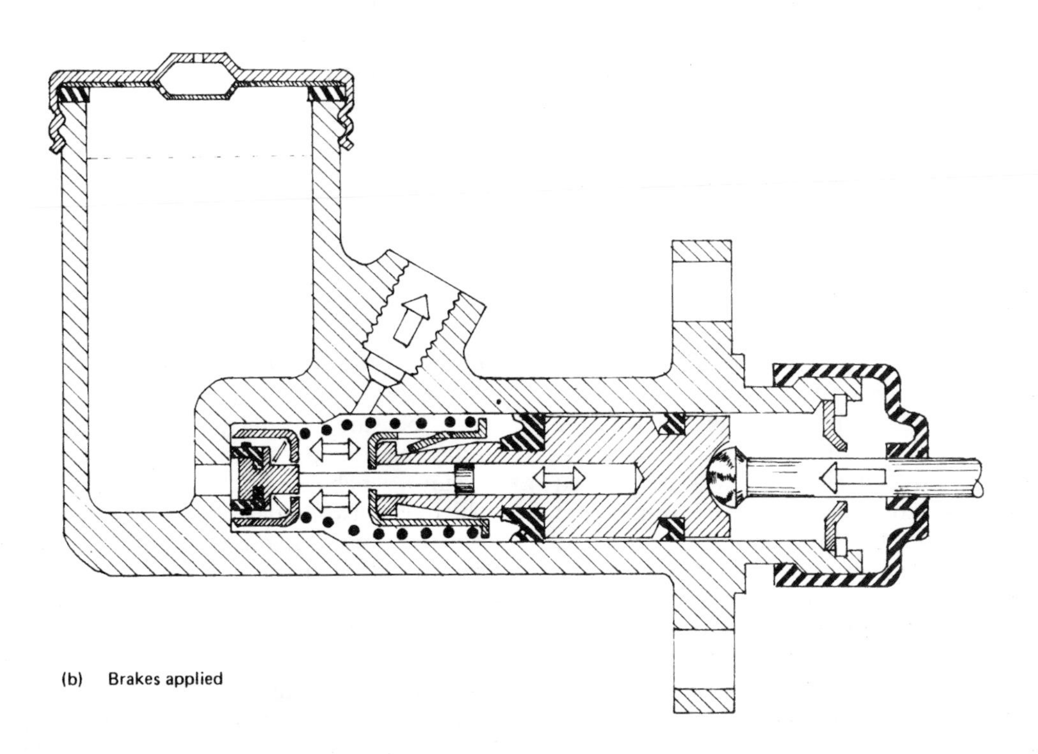
(b) Brakes applied

Fig. 9.28 Girling master-cylinder

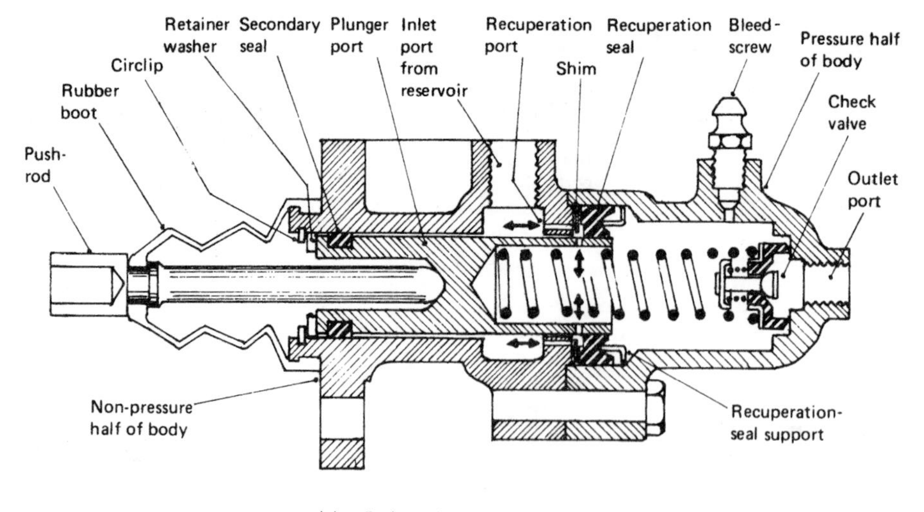

(a) Brakes released

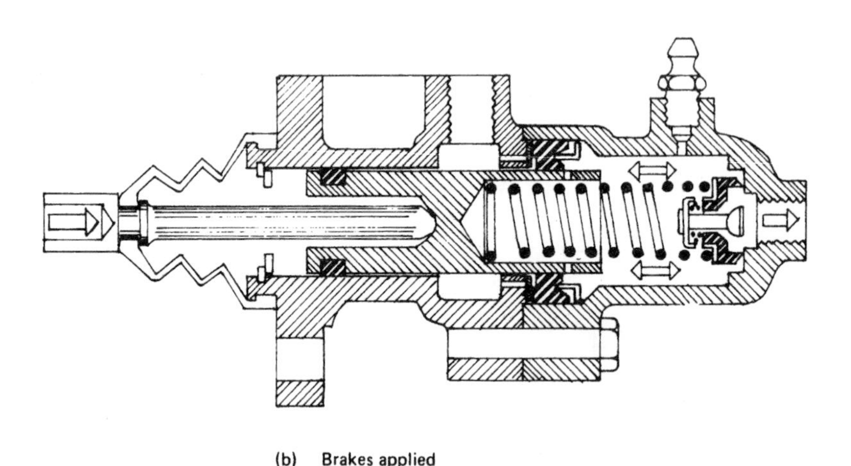

(b) Brakes applied

Fig. 9.29 Compression-barrel master-cylinder (Girling)

Usually for drum-brakes a residual-pressure check-valve is incorporated at the outlet port to provide a small line pressure when the brakes are off. Under generated braking pressure, the central conical valve will be pushed open, and additional fluid will be transferred past the valve to the pipe-lines. Releasing the brakes reverses the process, but this time the central valve will be closed and the whole valve body will be pushed away from the outlet-port face to allow the fluid to escape back into the master-cylinder chamber. The stiffness of the plunger return-spring will determine the minimum pipeline pressure at which the check-valve will close.

9.9.4 Dual or split line braking systems (fig. 9.30(a)–(c))

As a measure of safety, hydraulic brake systems today have two separate hydraulic brake pipe circuits which operate from a master-cylinder which has two independent pressure chambers each transmitting pressure to its own pipeline circuit. Subsequently, should a fault occur in one of the hydraulic circuits, the other one still remains operative and effective.

Front to rear brake line split (fig. 9.30(a)) This arrangement splits the two circuits so that the front brakes are operated by a primary hydraulic circuit and the rear brakes are activated by a

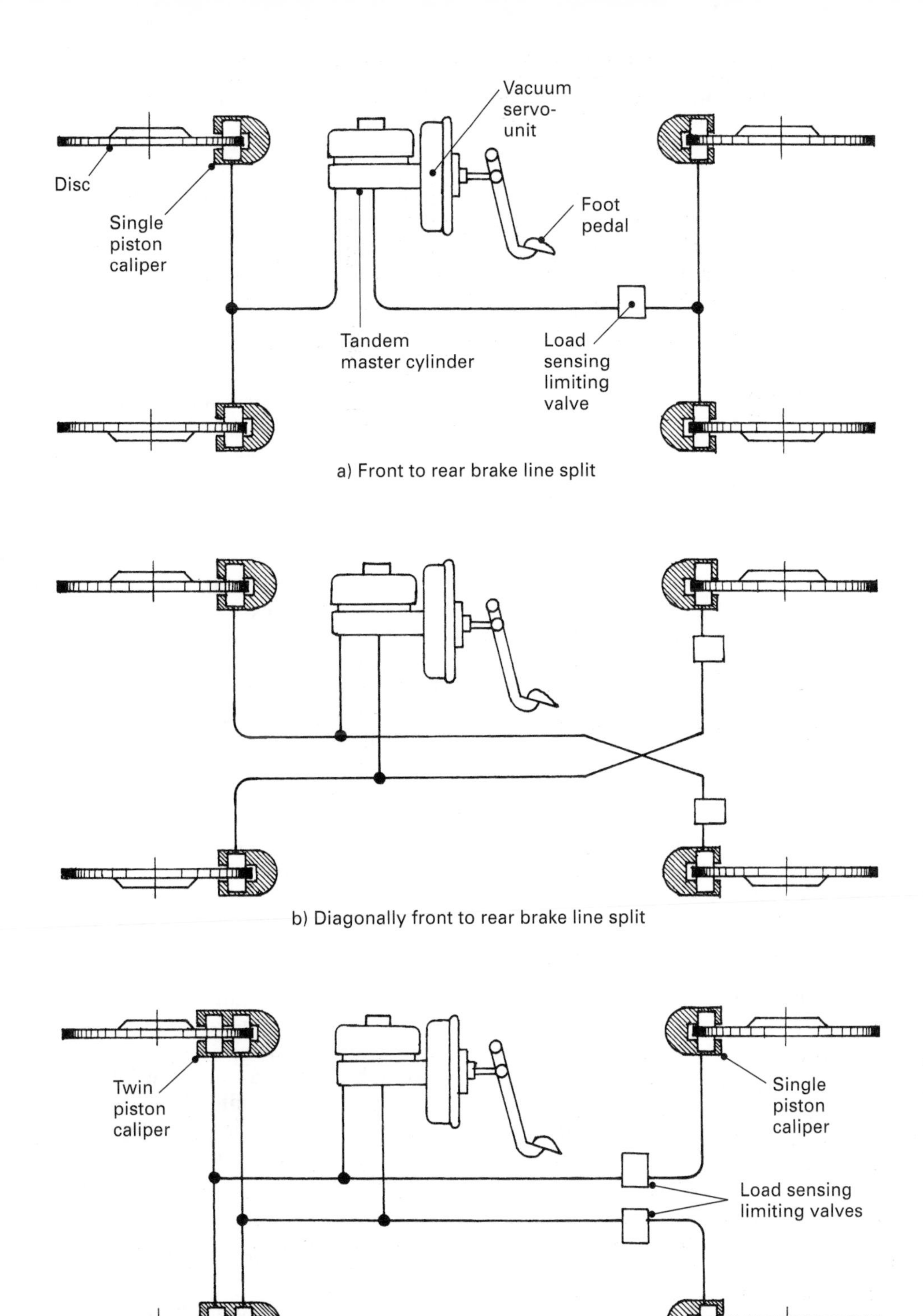

Fig. 9.30 Dual or split-line braking systems

secondary circuit. However the system has its limitations since roughly two-thirds of the total braking is designed to be provided by the front and only one-third by the rear brakes. Therefore, if the rear secondary circuit brakes should fail the front primary brake circuit will still be able to supply up to two-thirds of the total brake which could be considered as reasonable, but if the front primary brake circuit should malfunction the rear secondary brake circuit would only be capable of providing about one-third of the total braking.

Diagonally front to rear brake split (fig. 9.30(b)) The disadvantages of the front to rear brake split can largely be overcome by splitting the hydraulic line diagonally so that one front brake on one side and one rear brake on the other side make up the primary hydraulic brake circuit, and similarly the other front and diagonally opposite rear brake form the secondary hydraulic brake circuit. Thus if one of the hydraulic brake circuits should malfunction there will still be one front and one rear brake operating.

Triangular front to rear brake split (fig. 9.30(c)) For larger, more powerful cars the front brakes sometimes have twin piston disc calipers operating hydraulics independent of each other. With this arrangement the primary and secondary hydraulic split circuits are formed by joining one of the front twin caliper cylinders on each side and one rear caliper cylinder together to form a primary circuit and similarly connecting the other front twin cylinders from either side to the other rear caliper cylinder to make up the secondary circuit. By these means braking will be achieved on both front calipers all the time, even if one of the other hydraulic brake circuits should become faulty.

9.9.5 Tandem master-cylinder with pressure differential warning switch (fig. 9.31(a)–(d))

Tandem master-cylinders are adopted to operate dual or split line hydraulic braking systems. The tandem master-cylinder has two pistons operating from a single cylinder. This enables two separate pressure chambers to be created, each activating one of the hydraulic circuits.

Brakes off (fig. 9.31(a)) With the brakes off both the primary and secondary pistons are pushed outward against their respective stops by the primary and secondary return springs. Fluid is free to circulate between the pressure chamber, reservoir and the wasted region of each piston via the small compensating port and the much larger feed port for both primary and secondary circuits.

Brakes applied (fig. 9.31(b)) Pressing down on the foot pedal pushes the primary piston inwards, simultaneously compressing both the primary and secondary return springs, thus causing the secondary piston also to move further inwards. The front part of both piston movements causes the recuperating seal lips to slide over each compensating port, thus causing the fluid in the cylinder to be sealed off from the reservoir. Thus with increased piston travel, fluid will be pressurised in both primary and secondary pressure chambers and in their respective hydraulic pipeline circuits, thereby operating the appropriate disc or drum brake wheel-cylinder. At the same time fluid from the reservoir gravitates through the feed holes around the annular recesses formed along both pistons.

Brakes released (fig. 9.31(a)) Releasing the foot brake pedal permits the primary and secondary return springs to rapidly stretch out to their original full length in the cylinder and for both pistons to move back against their stops. However, the initial speed at which the swept volume of the pressure chamber increases will far exceed the rate the fluid returns from the wheel cylinder and pipe line. As a result a partial vacuum will be momentarily created within both primary and secondary pressure chambers; consequently the vacuum created in each pressure chamber causes both recuperating seal lips to momentarily collapse. Fluid is now permitted to flow from the reservoir via the annular recessed region of the piston through the horizontal holes in the piston head and around the inwardly distorted recuperating seals into their respective pressure chambers. The additional fluid entering both pressure chambers compensates for any fluid loss within the brake pipeline circuits or for excessive pad to disc or shoe to drum clearance. Should too much fluid be induced into the pressure chambers via the collapsed recuperating seals, then when the return springs have fully retracted both pistons, the compensating ports will again be uncovered, thus allowing the excess fluid to pass back to the fluid reservoir.

Failure in the primary or secondary hydraulic circuit (fig. 9.31(c) and (d)) Should a failure

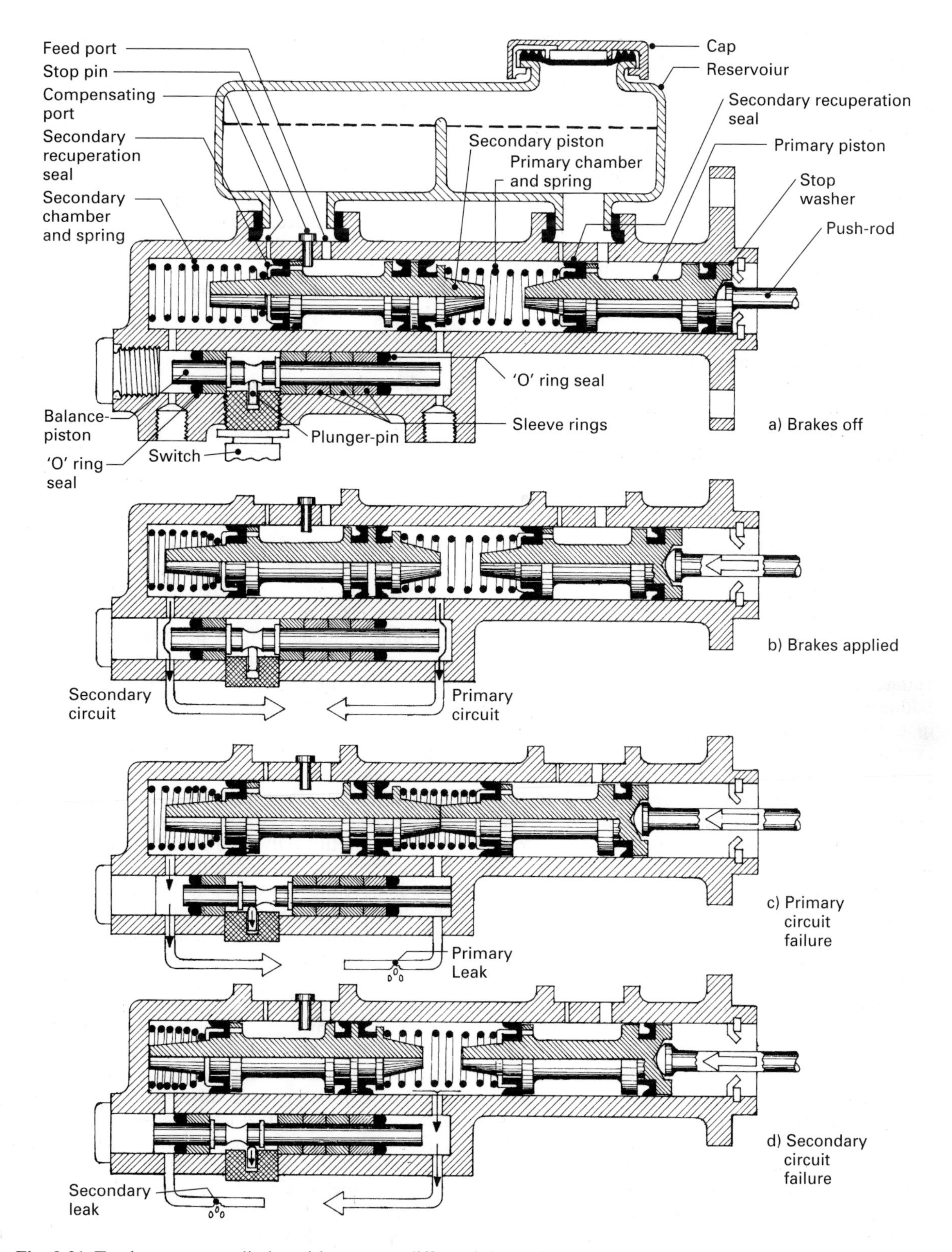

Fig. 9.31 Tandem master-cylinder with pressure differential warning actuator

(leak) occur, either the primary or secondary hydraulic circuit's brake pedal travel will increase and the resultant imbalance of pressure would cause the 'balance-piston' to move towards the system in which the failure and loss of pressure had occurred. As the balance-piston pushes the plunger-pin into the body of the electrical switch it completes the electrical circuit to a warning light on the dashboard. The other hydraulic circuit will still operate effectively and once the fault is traced and rectified, the balance-piston automatically returns to its neutral position.

Failure in the primary circuit (fig. 9.31(c)) If failure in the form of a leak and loss of fluid occurs in the primary hydraulic circuit, there will be no hydraulic pressure created in the pressure chamber. This means the primary piston and push-rod will move further inwards until the primary piston makes contact with the secondary piston. Any more pedal travel will now cause the secondary piston recuperating seal to cover the compensating port. Hence any further increase in pedal effort will pressurise the fluid in the secondary hydraulic pipeline circuit, thereby applying the respective wheel-cylinders controlled by the secondary hydraulic circuit.

Failure in the secondary circuit (fig. 9.31(d)) Should a leak and the loss of fluid cause a failure in the secondary circuit, then there will be no hydraulic pressure generated in the secondary pressure chamber. As a result both the primary and secondary pistons will move further into the cylinder until the secondary piston abuts the blank inner end of the bore. Further pedal effort will now squeeze the fluid trapped between the primary and secondary pistons so that pressurised fluid will operate the primary circuit wheel-cylinders to be applied.

Pressure differential warning actuator (fig. 9.31(a)–(d)) Should either the primary or the secondary hydraulic circuit fail due to loss of fluid, brake-pedal travel will increase, and because of the imbalance of pressure in the two hydraulic circuits, it will cause the balance-piston to move towards the hydraulic circuit in which the failure and loss of pressure has taken place. The effect of the balance-piston position shifting to the right or to the left-hand side will be to push the plunger-pin down into the body of the electrical switch, thus causing the electrical switch contacts to close. Consequently the completed electrical circuit will illuminate the warning light situated on the dashboard display to make the driver aware of a partial hydraulic circuit brake failure. However, the dual hydraulic circuit of the braking system does provide a safety net, where, if one of the hydraulic circuits should fail, the other one will still remain operative. Thus once the fault has been rectified, the balance-piston returns to its central position. This breaks the electrical circuit so that the warning light on the dashboard is extinguished.

9.9.6 Load sensing progressive pressure limiting valve (fig. 9.32(a)–(e))

Rear wheel lock-up and skidding of these wheels can be reduced if the maximum fluid pressure reaching the rear brakes is automatically adjusted to match the weight bearing on the rear suspension. Load sensing progressive pressure limiting valves are incorporated into the rear brake pipeline to limit and adjust the fluid pressure reaching the rear brakes relative to that applied to the front brakes according to the load carried by the rear wheels (see fig. 9.32(c)). However under relatively light braking there is no restriction by the valve on the pressure transferred from the master-cylinder to the rear brakes. When the rear wheels are lightly laden the fluid pressure reaching the rear brakes will be relatively low, but this is allowed to rise proportionally as the weight on the rear wheels increases.

Operation Once the master-cylinder generated fluid pressure has exceeded some minimum pre-determined value corresponding to the laden load on the rear wheels, the load sensing progressive pressure limiting valve commences to regulate the relayed pressure arriving at the rear brakes. As the load imposed on the rear wheels increases, so will the fluid pressure transmitted to the rear brakes, but at a much reduced rate compared to the fluid pressure transferred to the front brakes.

Brakes applied lightly (fig. 9.32(c)) Applying the foot-pedal lightly causes the master-cylinder to generate fluid pressure which is then transferred though the centre of the stepped reaction piston via the open ball cut-off valve to the rear brakes (see fig. 9.32(c)). A further increase in foot-pedal effort also raises the master-cylinder fluid pressure, and at this stage there is no restriction on fluid pressure passing to the rear brakes.

Brakes applied heavily (fig. 9.32(b)) Raising the foot-pedal effort still further increases the

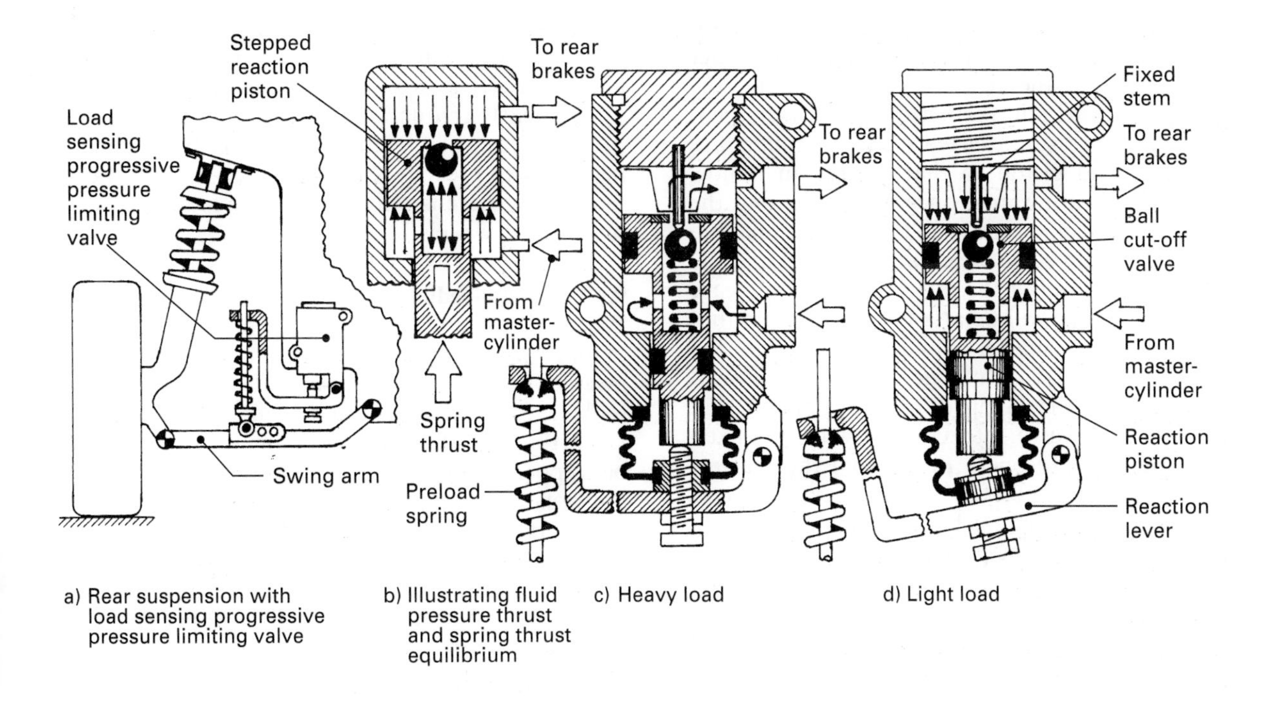

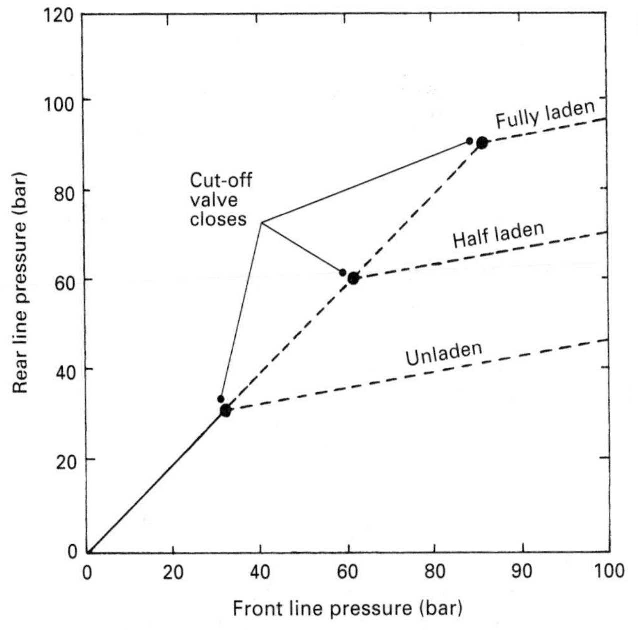

Fig. 9.32 Load sensing progressive pressure limiting valve

generated fluid pressure passing though to the rear brakes until the pressure imposed on the large face area of the stepped reaction piston produces an outward end thrust which eventually will equal and then exceed the opposing preload spring thrust and the fluid pressure thrust imposed on the annular face of the stepped reaction piston (see fig. 9.32(b)). Just above this state of equilibrium, the stepped reaction piston will be forced to move outwards until the ball cut-off valve clears the fixed stem. At this point the ball-valve closes, thus cutting off the master-cylinder fluid supply to the rear brakes. Increasing the foot-pedal effort still further also raises the generated fluid pressure exerted on the annular face of the stepped piston so that the piston is now forced to move inwards, once again opening the ball-valve. This increases still further the pressure of the fluid trapped between the ball-valve and the rear brakes. However the rate of pressure increase to the rear brake will be at a lower rate than that generated by the master-cylinder and transmitted to the front brakes. But once again the inward movement of the stepped reaction piston will bring the ball-valve into contact with the fixed stem, forcing it to open. Fluid from the master-cylinder will immediately communicate directly with the fluid in the rear brake pipelines allowing the rear brake pipeline pressure to increase. Hence the pressure exerted on the large end face of the stepped reaction piston immediately becomes greater than the spring thrust and hydraulic thrust pushing on the annular piston face. This again causes the stepped reaction piston to move outward, thus repeating the closure of the ball-valve.

Progressive pressure increase control (fig. 9.32(e)) Every time the valve is opened with rising master-cylinder generated fluid pressure, the rear pipeline pressure increases in relation to the previous closing of the valve. Over a heavy braking pressure rise phase the stepped reaction piston oscillates around a position of balance, causing a succession of valve openings and closings. It subsequently produces a smaller pressure rise in the rear brake pipeline than with the directly connected front brake pipelines (see fig. 9.32(e)). The ratio of the stepped piston areas determines the degree of rear brake pipeline increase with respect to the front brake pipelines.

Load sensing control (fig. 9.32(a), (c) and (d)) The cut-off or change point depends on the tension of the preload spring which varies with the rear suspension swing-arm deflection (see fig. 9.32(a)). Thus under heavily laden rear suspension conditions, the preload spring thrust against the reaction piston will be high (see fig. 9.32(c)). This means that the closure of the ball-valve occurs relatively late in the master-cylinder generated fluid pressure build-up without the wheels locking. Conversely, under lightly laden rear suspension operating conditions, the preload spring thrust will be low, so that the ball cut-off valve closure will occur early in the generated fluid pressure rise (see fig. 9.32(d)).

This early reduction in the generated fluid pressure rise rate relative to the front brakes is necessary to prevent rear wheel lock-up and skidding when there is only a moderate load imposed on the rear wheels.

9.10 Brake bleeding (fig. 9.33)

When a hydraulic braking system has been dismantled and reassembled, fresh fluid must be poured into the master-cylinder reservoir to replace the original fluid which has been lost. Unfortunately the existing air in the system will be trapped and will then be compressed every time the brakes are applied. The fluid in the pipeline will therefore have to be considerably displaced to take up the sponginess of the air before pressure is actually transmitted from the master-cylinder to the wheel shoe-expanders and disc-caliper pistons.

Bleeding is the process of pumping fresh fluid through the pipeline system until all the air has been expelled (fig. 9.33).

In its simplest form, bleeding is carried out as follows:

1. Connect a rubber tube to the bleed-screw valve nipple situated on either the shoe-expander or the disc caliper, and submerge the free end of the tube in brake fluid contained in a glass jar.
2. Using a spanner, unscrew the bleed valve about half a turn; then, with the aid of an assistant, pump the brake pedal several times until all the air bubbles have disappeared and on the last down-stroke tighten up the bleed valve.
3. After each bleed, top up the reservoir tank with fresh fluid and repeat this procedure at each wheel.

9.11 Vacuum-assisted brake servo-unit (fig. 9.34(a)–(c))

Vacuum-assisted brake servo-units have come into common usage to reduce the brake pedal effort the driver would otherwise have to apply to retard the speed of the vehicle. A vacuum servo-unit converts the induction-manifold vacuum or vacuum-pump vacuum energy into mechanical energy to assist the brake-pedal effort in pressurising the brake fluid in the tandem master-cylinder and pipelines. Note that the air intake in a diesel engine is not throttled so that very little vacuum is created in its induction-manifold. Consequently a separate source of gen-

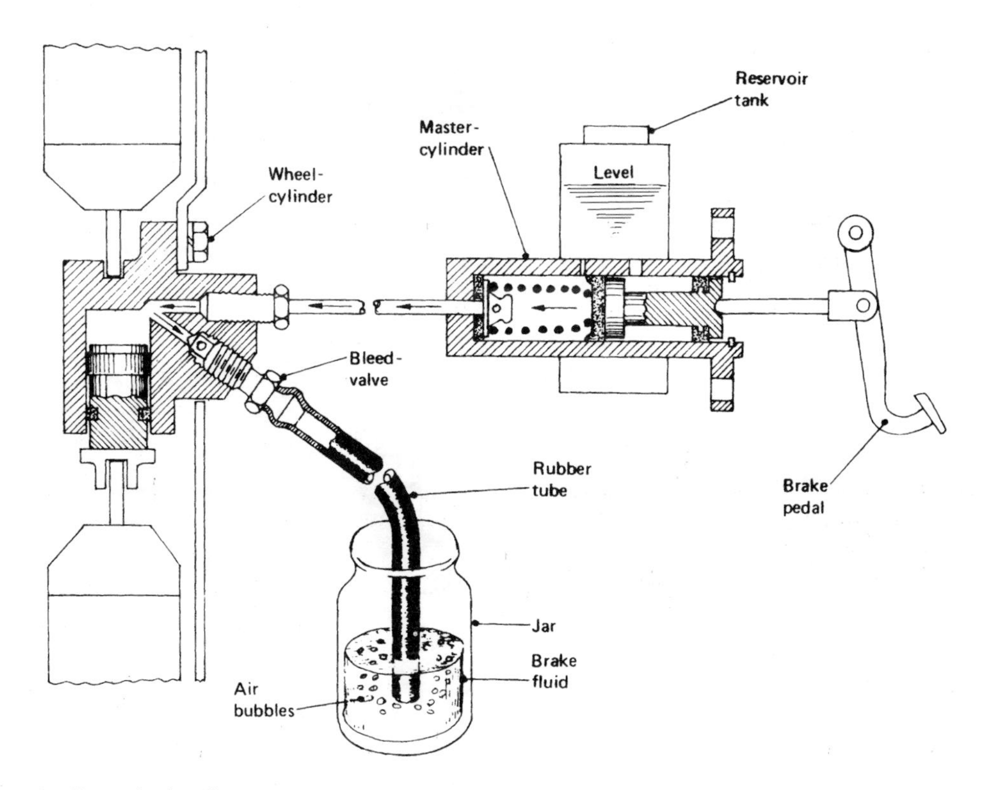

Fig. 9.33 Bleeding a hydraulic system

erated vacuum is necessary. This usually takes the form of a two or four vane vacuum-pump driven off the camshaft.

Direct-acting suspension vacuum-assisted brake servo-unit (fig. 9.34(a)) These units consist of two chambers separated by a rolling diaphragm and a power-piston. The power-piston is linked to the master-cylinder outer primary piston by a power push-rod. The foot-pedal is connected though a pedal push-rod indirectly to the power-piston via a relay-piston, rubber reaction-pad and reaction-piston. With the brakes in the off position, both sides of the power-piston assembly are subjected to manifold or vacuum-pump vacuum. When the brakes are applied, the vacuum in the front left-hand chamber remains undisturbed, whilst the vacuum in the rear right-hand chamber is replaced by atmospheric air. This is accomplished by closing the vacuum supply passage, followed by opening the air inlet passage to the rear chamber. Accordingly the difference of pressure across the power-piston causes it to move towards the lower pressure chamber, that is, the direction of the master-cylinder, so that the thrust imposed on the power push-rod generates fluid pressure in the tandem master-cylinder primary and secondary hydraulic brake lines.

Operation of the reaction control valve

Brakes off (fig. 9.34(a)) With the foot-pedal fully released, the large return spring in the vacuum chamber pushes the rolling diaphragm and power-piston towards and against the air/vacuum chamber stepped steel pressing. With the engine running, the vacuum from the induction-manifold or vacuum-pump draws the non-return valve away from its seat, thereby subjecting the whole left-hand chamber to a similar vacuum existing in the manifold or pump. When the brake-pedal is fully released, the outer spring surrounding the pedal push-rod pulls it and the relay-piston back against the valve retaining-plate. The inlet-valve formed on the end of the relay piston closes against the air/vacuum diaphragm face and at the same time pushes the air/vacuum diaphragm away from the vacuum chamber valve. Vacuum from the left-hand vacuum chamber therefore passes though the inclined passage in the power-piston around the seat of the open vacuum-valve where it then occupies the existing space formed

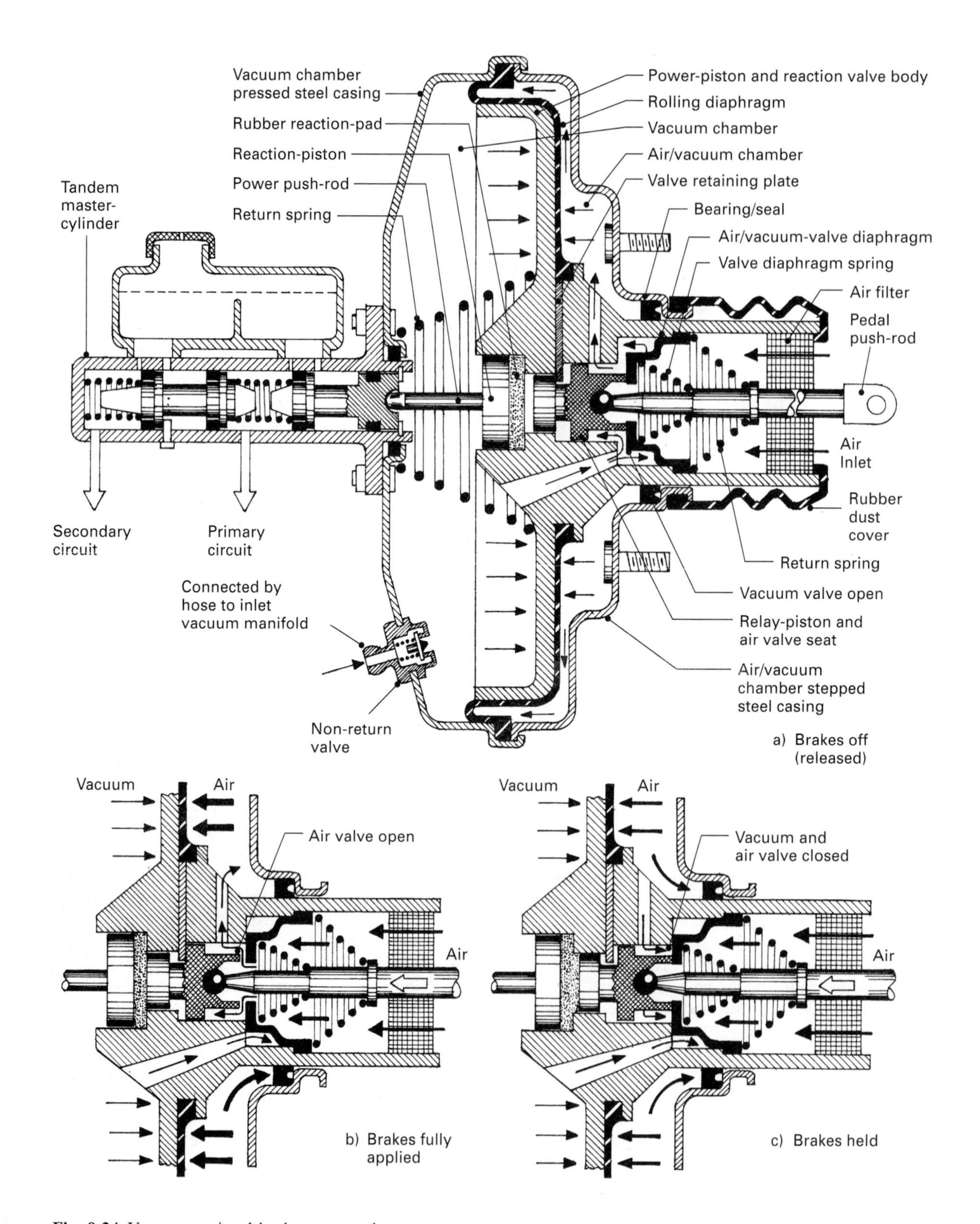

Fig. 9.34 Vacuum-assisted brake servo-unit

in the air/vacuum chamber on the right-hand rear side of the rolling diaphragm. Hence with the air-valve closed and the vacuum-valve open, both sides of the power-piston are suspended in vacuum.

Brakes applied (fig. 9.34(b)) When the foot-pedal is applied the pedal push-rod moves towards the diaphragm power-piston hard against the valve retaining-plate. First of all the air/vacuum diaphragm closes against the vacuum-valve's seat and with further inward push-rod movement the relay-piston inlet-seat separates from the air/vacuum diaphragm face. The right-hand air/vacuum chamber is now cut off from the vacuum supply, and atmospheric air is free to pass though the air-filter between the relay-piston air-valve seat and air/vacuum diaphragm, to replace the vacuum in the right-hand air/vacuum chamber. The difference in pressure between the low pressure vacuum-chamber and the higher pressure air/vacuum chamber causes the power-piston and the reaction-valve body assembly and power push-rod to move forwards against the master-cylinder primary piston so that higher fluid pressure is generated in both brake circuits to actuate the front and rear brakes.

Brakes held (fig. 9.34(c)) Holding the brake-pedal down momentarily continues to move the power-piston with the reaction-valve body forwards under the influence of the greater air pressure in the right-hand air/vacuum chamber, until the rubber reaction-pad is compressed by the shoulder of the power-piston against the opposing reaction of the power push-rod. As a result of squeezing the outer rubber rim of the reaction pad, the rubber distorts and extrudes towards the centre and backwards in the relay piston bore. Subsequently, only the power-piston and valve-body move forward whilst the relay-piston and pedal push-rod remain approximately in the same position until the air-valve seat closes against the air/vacuum diaphragm face. No more atmospheric air can enter the right-hand air/vacuum chamber so that there is no further increase in servo-power assistance. In other words the brakes are on hold. The reaction-pad action therefore provides a progressive servo-assistance in relation to the foot-pedal effort which would not be possible if only a simple reaction spring were positioned between the reaction-piston and the relay-piston.

If a greater brake pedal effort is applied for a given hold position, then the relay piston will again move forward and compress the centre region of the reaction-pad to open the air/vacuum valve. The extra air permitted to enter the air/vacuum chamber will further raise the servo-assistance proportionally. The cycle of increasing or decreasing the degree of braking provides new states of hold which are progressive and correspond to the manual input effort.

Brake released (fig. 9.34(a)) Releasing the brake-pedal allows the pedal push-rod and relay-piston to move outwards; first closing the air-valve and secondly by opening the vacuum-valve. The existing air in the right-hand air/vacuum chamber will then be drawn to the left-hand vacuum chamber via the open vacuum-valve, the power-piston's inclined passage, and finally it is withdrawn to the induction-manifold or vacuum-pump. As in the brake's 'off' position, both sides of the power-piston are suspended in vacuum, thus preparing the servo-unit for the next brake application.

9.12 Antilocking brake system (ABS) (Alfred–Teves) (fig. 9.35(a)–(c))

Maximum effective braking can only occur when the retarding force on the wheel brake matches the grip imposed between the tyre and road surface. Road grip in turn is dependent upon the vertical weight imposed on the wheel and the frictional resistance generated between the tyre tread and road surface. Should the retarding force be too great, the weight on the wheel will be insufficient to keep the wheel rolling. It will thus cause the wheel to lock and skid. If the retarding force is below the maximum to bring the wheel to the point of lock, then the retardation of the wheel will be much slower than its possible maximum. It therefore adds to the minimum possible stopping distance of the vehicle.

It is difficult for the driver to judge the optimum amount of brake-pedal force to bring a wheel to rest in the minimum time on a good road surface, and it is practically impossible on a wet, muddy, or icy surface, as the minimum wheel grip on each wheel will be continuously varying. To prevent the individual wheel from locking and consequently skidding when braking, the driver should rapidly apply and release the brake-pedal repeatedly until the vehicle speed has been reduced to a safe driving speed relative to road conditions. Even so the braking of individual wheels will not necessarily match the changing road surface conditions.

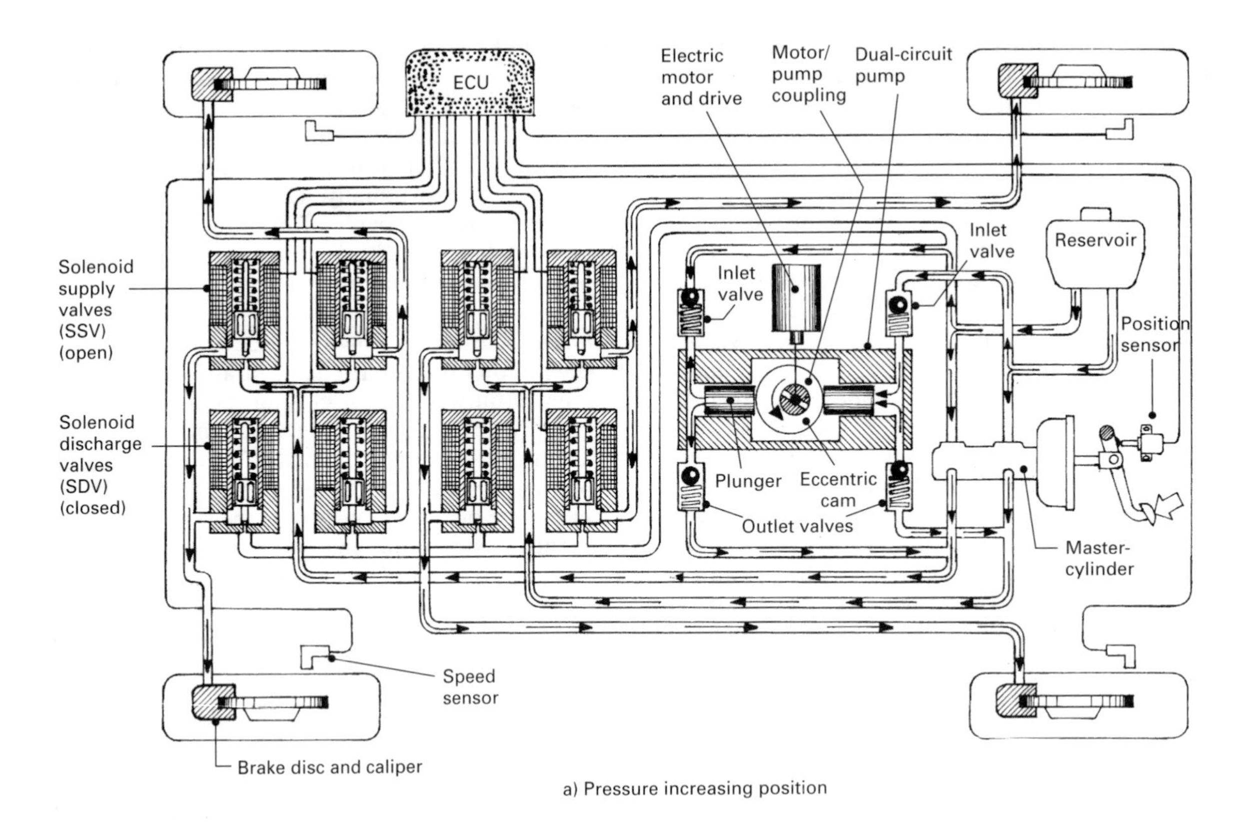

a) Pressure increasing position

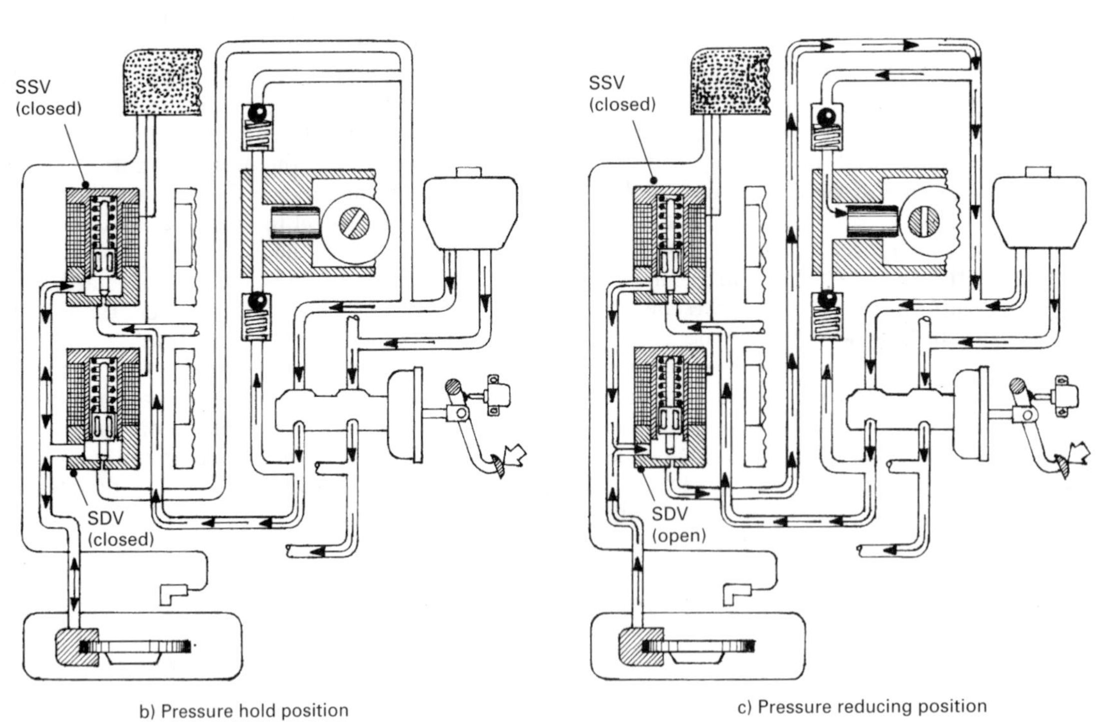

b) Pressure hold position

c) Pressure reducing position

Fig. 9.35 Antilocking braking system (ABS) for cars

Wheel lock is avoided with antilock brake systems (ABS) as the system senses the speed retardation (the onset of wheel lock-up) of individual wheels and automatically delivers a pipeline pressure increase or decrease to each wheel brake to prevent any wheel skid tendency. Simultaneously antilock brake systems provide the optimum fluctuating pipeline pressure to retard the vehicle in the minimum time if so desired. The net gain when braking on slippery surfaces with an antilock brake system is that the vehicle will slow down practically along a straight path instead of swinging around as would be usual if only conventional brakes are fitted to the vehicle.

The antilock brake system (ABS) wheel sensor monitors the rate at which each wheel slows down, as this is a measure of approaching wheel lock-up. It simultaneously sends commands via the electronic control unit (ECU) to the solenoid valve module to increase hold or reduce the build-up of wheel brake pressure. The electronic control unit is programmed to maintain as far as possible the tyre to ground slip ratio within the range 10–30% since this provides the maximum wheel grip and the minimum stopping distance with a high degree of lateral resistance necessary for effective steering control see (fig. 9.36).

Note the slip ratio as a percentage is

$$\frac{(\text{vehicle speed} - \text{wheel speed}) \times 100}{\text{vehicle speed}}$$

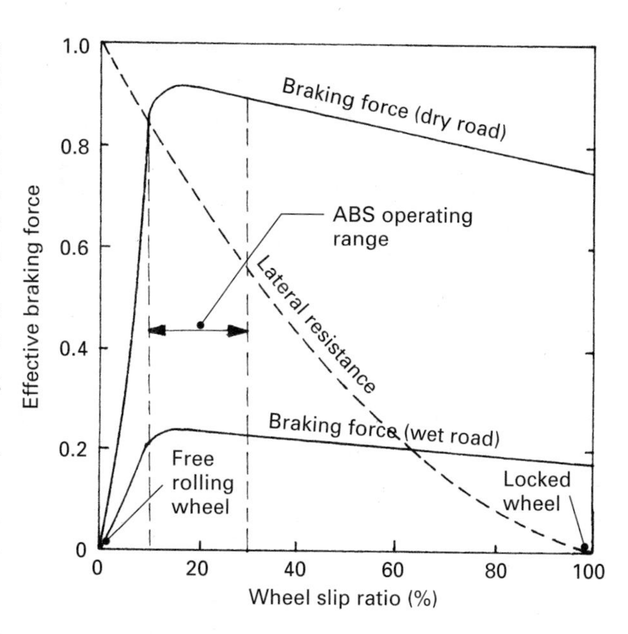

Fig. 9.36 Longitudinal and lateral braking force to slip ratio characteristics

Wheel-excitor and speed-sensor (fig. 9.37) The wheel-excitor and speed-sensor consists of a gear-wheel-shaped (tooth ring) rotor and a sensor element which contains a bar permanent magnet around which a coil is wound. The excitor tooth ring made from low carbon steel is attached to the rotary wheel hub whereas the magnetic sensor element is mounted on the stationary hub carrier or back-plate.

The sensor's permanent magnet produces a magnetic flux which overlaps a portion of the rotating excitor ring (see fig. 9.37). As the wheel and excitor tooth ring revolve, the teeth and gaps of the excitor pass though the fixed magnetic field of the sensor. This continuous movement of the spaced-out teeth across the magnetic field induces an alternating voltage into the stationary coil whose frequency is proportional to the speed of the rotating road wheel. This sensor voltage is produced whenever the excitor rotates and is then relayed directly into the electronic control unit (note the voltage generated is completely independent of the brake system). The increase or decrease in the generated alternating voltage signalled to the electronic control unit is a measure of the wheels' acceleration or deceleration respectively, that is, it relates to a potential on-coming state of wheel lock-up.

Electronic control unit (fig. 9.35(a)) The electronic control unit computes the rotating mean speed of all the wheels and the rate of change of speed of individual wheels by taking in the generated signal current sent from each of the four wheel sensors. An amplified and processed response is then transmitted in the form of a direct current to the solenoid valve module where it actuates and regulates the opening and closing of individual supply and discharge valves. It also controls the on-off operation of the solenoid valve relay and pump motor relay. If any malfunction occurs, it causes a warning light to switch on and the antilock brake system will be cutout, permitting the braking system to revert to a conventional braking arrangement.

Electrically driven dual plunger flow pump (fig. 9.35(a)) The electric motor driven dual plunger flow pump with its inlet and outlet ball valves is automatically actuated when the brakes are applied. Each plunger draws fluid from the

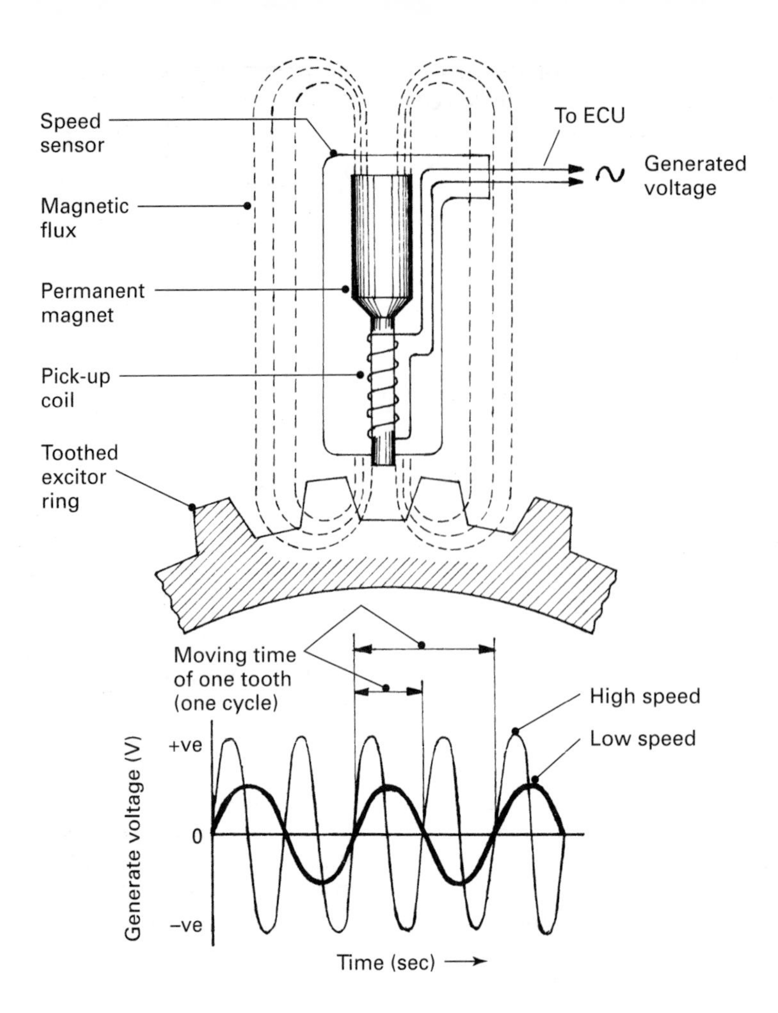

Fig. 9.37 Magnetic speed sensor and excitor

reservoir and pumps it to its respective master-cylinder primary or secondary pipeline circuit. This source of pressurised flow supply compensates for the expulsion and return of fluid to the reservoir from the wheel caliper cylinder pipeline circuit whenever one or more solenoid controlled discharge valves opens. Without this extra source of pressurised fluid pumped into the master-cylinder primary and secondary circuits when incipient wheel lock conditions prevail, the master-cylinder primary and secondary piston would in a short space of time move inwards to the end of the cylinder during each prolonged period of braking until no further braking is possible. A position-sensor switches on and off the pump's electric motor to provide the necessary circulation of fluid for the intermittent supply and discharge of fluid to and from the wheel brakes, and to ensure that sufficient fluid is pumped to the master-cylinder pressure chamber to minimise the brake-pedal travel. The valve module consists of an aluminium valve block with eight solenoid-operated valves, there being one supply and one discharge valve for each of the four wheel brake pipelines.

Operation

Normal breaking conditions (fig. 9.35(a)) When braking with good road to tyre conditions, the position-sensor contact with the brake-pedal switches on the dual-piston-pump electric motor thereby generating a source of high fluid pressure. Under these conditions the four solenoid supply-valves are opened by energising current relayed from the electronic control unit to each of the solenoids, whereas the solenoid discharge-valves remain de-energised and closed. Fluid therefore flows unrestricted from the master-cylinder to the wheel caliper cylinder via the open supply-valves to produce the pressure build-up necessary to slow down the car.

Pressure hold (fig. 9.35(b)) When the rate of speed reduction of one or more wheels exceeds some predetermined value, the speed-sensor or sensors from the wheel or wheels signals to the electronic control unit the danger of wheel lock-up. Immediately current flow to the corresponding solenoid supply-valve or valves is interrupted. This causes the supply-valve or valves to close. Fluid cannot now pass from the master-cylinder to the caliper cylinder or cylinders of the wheel or wheels which are on the point of skidding, so that the fluid pressure trapped between the closed solenoid supply-valve or valves and their respective caliper cylinder or cylinders are therefore held constant see (fig. 9.38).

Pressure reducing (fig. 9.35(c)) Should the wheel speed-sensor still signal an abnormally rapid speed reduction likely to cause the wheel or wheels to lock (see fig. 9.35(c)), then a current will be sent by the electronic control unit to the corresponding solenoid discharge-valve or valves thus opening the valve or valves. Immediately the fluid hold-pressure in this pipeline or lines will collapse, since fluid is now able to escape back to the fluid reservoir. As a result the reduction in line-pressure permits the wheel or wheels to increase speed once again and to re-establish its grip or their grip with the road (see fig. 9.38).

Pressure increasing (fig. 9.35(a)) As soon as the respective wheel speed has changed from a deceleration back to acceleration (see fig. 9.35(a)), the individual wheel speed-sensor signals to the electronic control unit to switch on, and to close the appropriate solenoid discharge-valve, and at the same time to switch off and to open the corresponding solenoid supply-valve. Thus once again fluid will be free to flow from the master-cylinder to the respective wheel caliper cylinder to enable the individual brake to be re-applied (see fig. 9.38). The sensitivity and response time of the solenoids is such that the pulsating regularity takes place at least four to ten times per second.

9.12.1 Traction control system (TCS)

The traction control system is usually incorporated into the antilock brake system to prevent the drive-wheels from spinning when accelerating over slippery road surfaces. With front-wheel drive, spinning front wheels produce a loss of steerability corresponding to the front wheels locking up when braking. Spinning rear wheels with rear-wheel drive causes a loss of driving stability, similar to what happens when the rear wheels lock up when braking, that is, the vehicle tends to swerve.

With a traction control system, the rotary behaviour of all four wheels is monitored continuously by the four wheel-sensors. The signals received by the combined ABS/TCS electronic control unit are then processed. The electronic control unit distinguishes between traction slip at the drive-wheels (tending to cause wheel spin) and brake slip (tending to cause wheel lock-up). If one of the drive-wheels starts spinning due to the frictional resistance between the tyre and the road surface being exceeded, the electronic control unit signals for a brake line pressure to be applied to the wheel spinning as a means to slow down the wheel until it rotates at the optimum speed (brake control). Note that this automatic braking occurs without the driver operating the brake foot-pedal. If the second drive wheel should start to spin, then the engine speed is reduced by a second control circuit which is attached to the acceleration pedal (driving torque control). This speed reduction continues until the engine/driving torque has reached an optimum value to match the existing wheel torque to road grip. The time response for

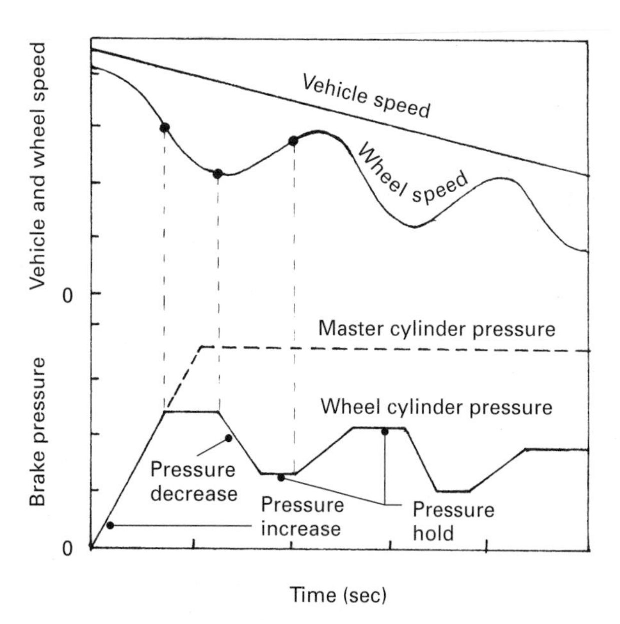

Fig. 9.38 ABS control cycle characteristics

the engine speed to adjust to the road condition is only a matter of milliseconds. Automatic traction control by combining individual wheel braking and engine speed reduction produces an optimum vehicle propelling thrust with differing frictional resistance for the individual drive wheels. When attempting to drive away from a standstill with one wheel tending to spin, traction control momentarily brakes the spinning wheel, hence all the drive is then exerted by the other wheel which is still able to maintain its traction with the road.

9.13 Air-operated power brakes

For large heavy vehicles such as trucks, buses, coaches, etc., vacuum-assisted servos with a manual contribution would not provide adequate braking, particularly for an emergency vehicle stop. Engine-driven compressor-generated compressed air operating at an upper limit of between 7 and 8 bar has a power factor advantage of about 8 times that of vacuum servos which operate at best with a depression of 0.8 to 0.9 bar below atmospheric pressure. With air-operated power brakes the air pipelines may operate at pressures as high as 8 bar. This replaces the hydraulic pipelines which must withstand fluid pressures of up to 80 bar or more. Thus fully air-operated power brake systems function without manual assistance and operate at considerably lower pipeline pressures and yet provide more than sufficient air pressure boost to apply either drum and shoe or disc and pad wheel braking systems.

9.13.1 Truck air over hydraulic brake system (fig. 9.39)

Compressed air supply (fig. 9.39) Atmospheric air is drawn into the reciprocating piston com-

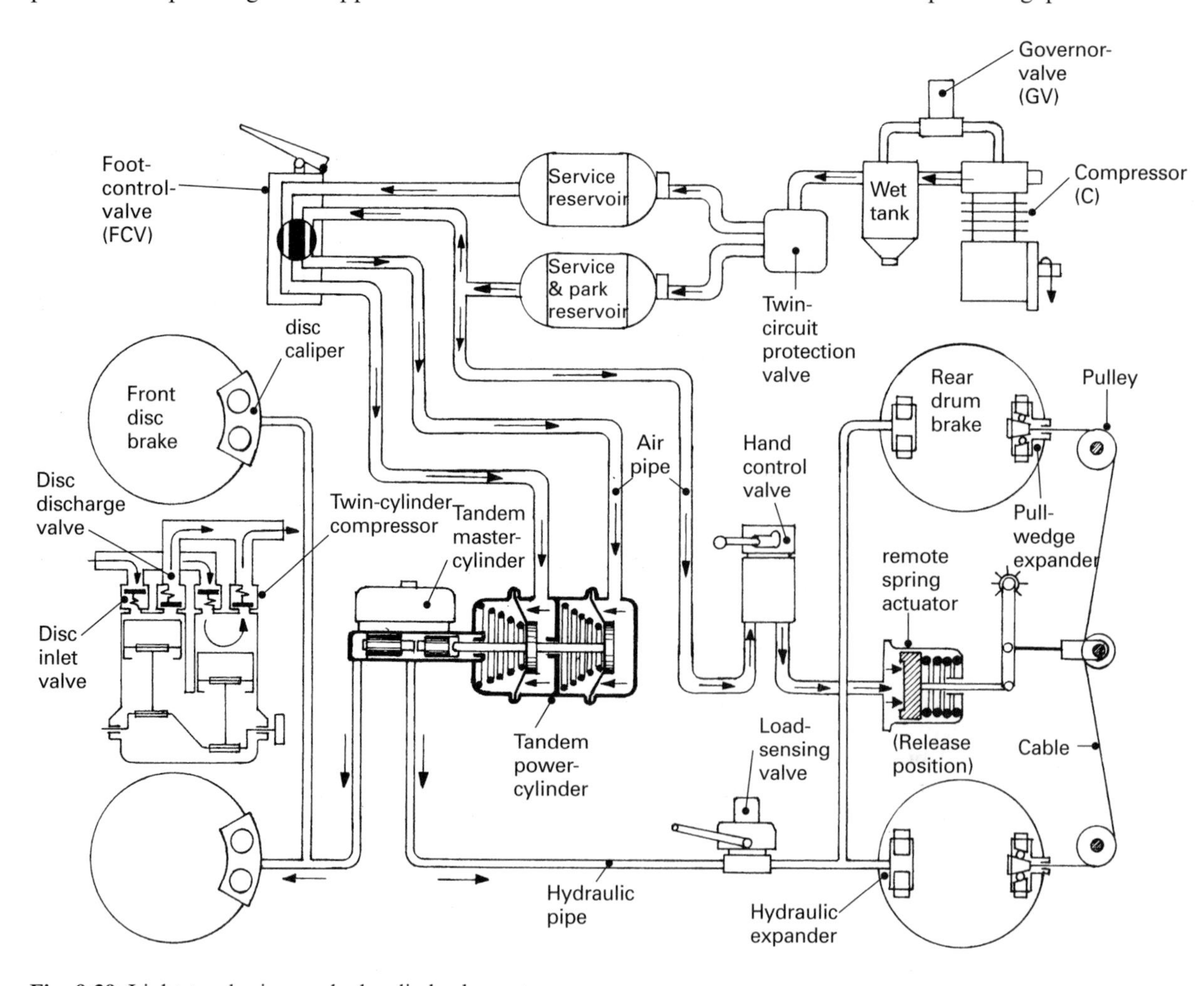

Fig. 9.39 Light truck air over hydraulic brake system

pressor where it is compressed and discharged to the reservoir tank via the wet tank which removes a large proportion of moisture contained in the compressed air supply before passing though the twin circuit pressure-protection valve to the reservoir tanks. A governor-valve senses the air pressure build-up and sends a signal pressure to the compressor to unload the air in the cylinder to the atmosphere once an upper pressure limit has been reached, and to make the compressor operational again when the reservoir pressure falls to a lower pressure limit setting.

Service line circuit (fig. 9.39) Pressurised air is supplied to a tandem power-cylinder via a dual foot-control-valve and two separate service reservoirs; thus if one service line is at fault, the other service line air supply will continue to deliver air to the other tandem power-cylinder chamber. The air pressure delivered to the tandem power-cylinder is converted first to push-rod thrust and secondly into hydraulic fluid pressure as the tandem master-cylinder piston's inward movement squeezes the trapped fluid. The split hydraulic pipeline then provides a separate braking circuit for the front and rear brakes. In this example disc and pad brakes are used on the front wheels whereas the rear wheels use shoe and drum brakes with a hydraulic twin-plunger-expander and a load-sensing limiting-valve to match the proportion of rear braking to the load carried by the rear axle.

Secondary/park line circuit (fig. 9.39) Parking is achieved by an air-controlled remote-spring actuator operating pull-wedge expanders via a rod and cable linkage. Air supply from one of the service reservoirs holds the remote-spring actuator spring in compression the whole time the vehicle is in motion. This fully releases the rear drum brake wedge expanders. When the vehicle has stopped, the hand-control-valve lever is moved to the applied position. This blocks the air supply from the service reservoir and exhausts the air from the remote spring actuator. The power spring is now permitted to expand and in the process tensions the cable which in turn pulls out the wedge expanders to apply the shoe and drum parking brakes.

9.13.2 Truck three-line brake system (fig. 9.40)

Compressed air supply (fig. 9.40) Air is drawn in, compressed, and discharged from the reciprocating piston single or twin-cylinder air compressor to the reservoir tanks by way of the condenser and drain-valve, the unloader-valve and the pressure-protection-valves. The condenser and drain-valve, sometimes referred to as the wet tank, removes by condensation a large proportion of moisture accumulated in the air during the compression process. The condensed water is then expelled to the atmosphere via the automatic drain-valve. The unloader-valve permits compressed air from the compressor to charge the reservoir tanks to an upper pressure limit, at which point the unloader-valve blocks off the passage leading to the reservoirs and then diverts the air delivery from the compressor to the atmosphere until the pressure of the air stored in the reservoirs drops to some predetermined lower limit. By these means the compressor is able to run light for most of a vehicle's journey time.

Service line circuit (fig. 9.40) The service line circuit operates the foot controlled brakes. Compressed air from the two service reservoir tanks flows though the dual service-line to the foot-control-valve. One line directly supplies the front service brake actuators and the other service-line provides a signal pressure to the service relay-valve. When the foot-pedal is applied, the foot-control-valve opens. Compressed air will then be supplied to the front brake actuators via the quick-release-valve to apply the front brakes. Simultaneously a signal pressure from the foot-control-valve opens the relay-valve inlet port and closes its exhaust port. This permits air to be discharged directly from the rear service reservoir to both rear-axle wheel actuators to apply the rear service line brakes. When the foot-pedal is released the foot-control-valve blocks any further air supply to the service-line and exhausts pressurised air that might still be in the service-line. At the same time the service relay-valve closes and its exhaust port expels pressurised air held in the service-line.

Secondary line circuit (fig. 9.40) When the hand-control-valve lever is moved to the applied position, air from the secondary-reservoir tank is delivered to all the wheel-actuator secondary pressure chambers via the hand-control-valve which regulates the air pressure being applied to these pressure chambers. When the hand-control-valve lever is moved to the released position it blocks off the air supply from the secondary

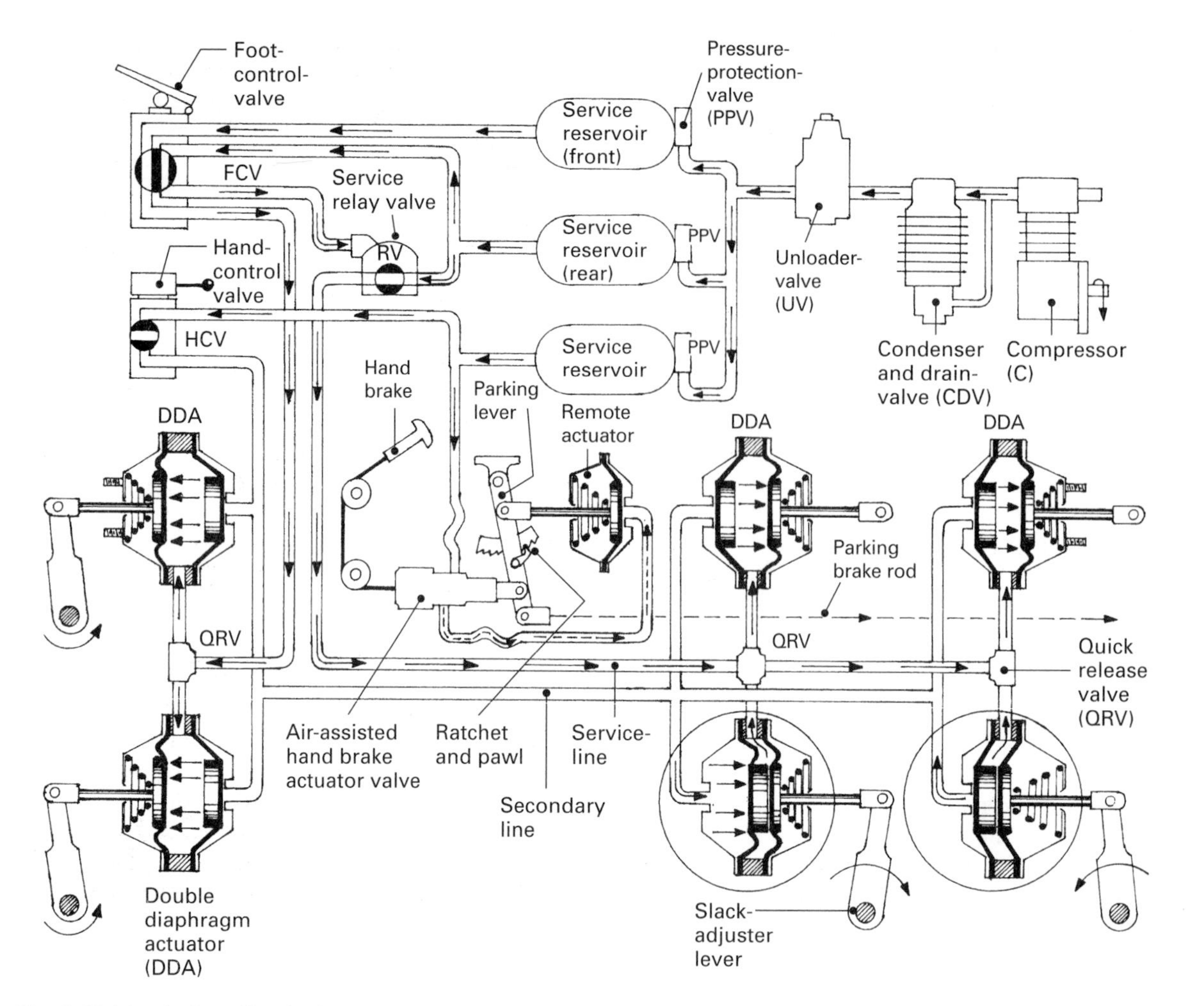

Fig. 9.40 Truck three-line brake system

reservoir to the wheel-actuator and at the same time opens its exhaust port. Air instantly escapes from the wheel-actuator pressure chambers to the atmosphere via the hand-control-valve opened exhaust port, thus releasing the secondary brakes.

Air-assisted hand parking brake (fig. 9.40) The secondary reservoir also supplies the air-assisted mechanical parking brake mechanism. When the parking brake handle is pulled 'out' (brakes applied) the air-assisted parking brake actuator-valve opens so that air from the secondary reservoir now discharges into the remote diaphragm actuator. This pushes forwards the parking lever and pulls the parking brake-rod towards the parking position. Automatically the ratchet and pawl will engage to hold the parking mechanism in the applied position. When the parking brake handle is pushed 'in' (brakes released) the air-assisted parking brake actuator-valve blocks off the air supply from the secondary reservoir and opens its exhaust-valve to allow the pressurised air in the remote actuator chamber to be expelled. Simultaneously the spring-loaded pawl will be released from the ratchet so that the parking lever and rod can move towards the released position.

9.13.3 Truck two-line brake system (fig. 9.41)

Compressed air supply (fig. 9.41) The reciprocating compressor, which is driven by the engine at camshaft speed, draws in fresh air, compresses it to between 6 and 8 bar, and then discharges it into the air-dryer chamber where water and moisture are removed. The dry compressed air

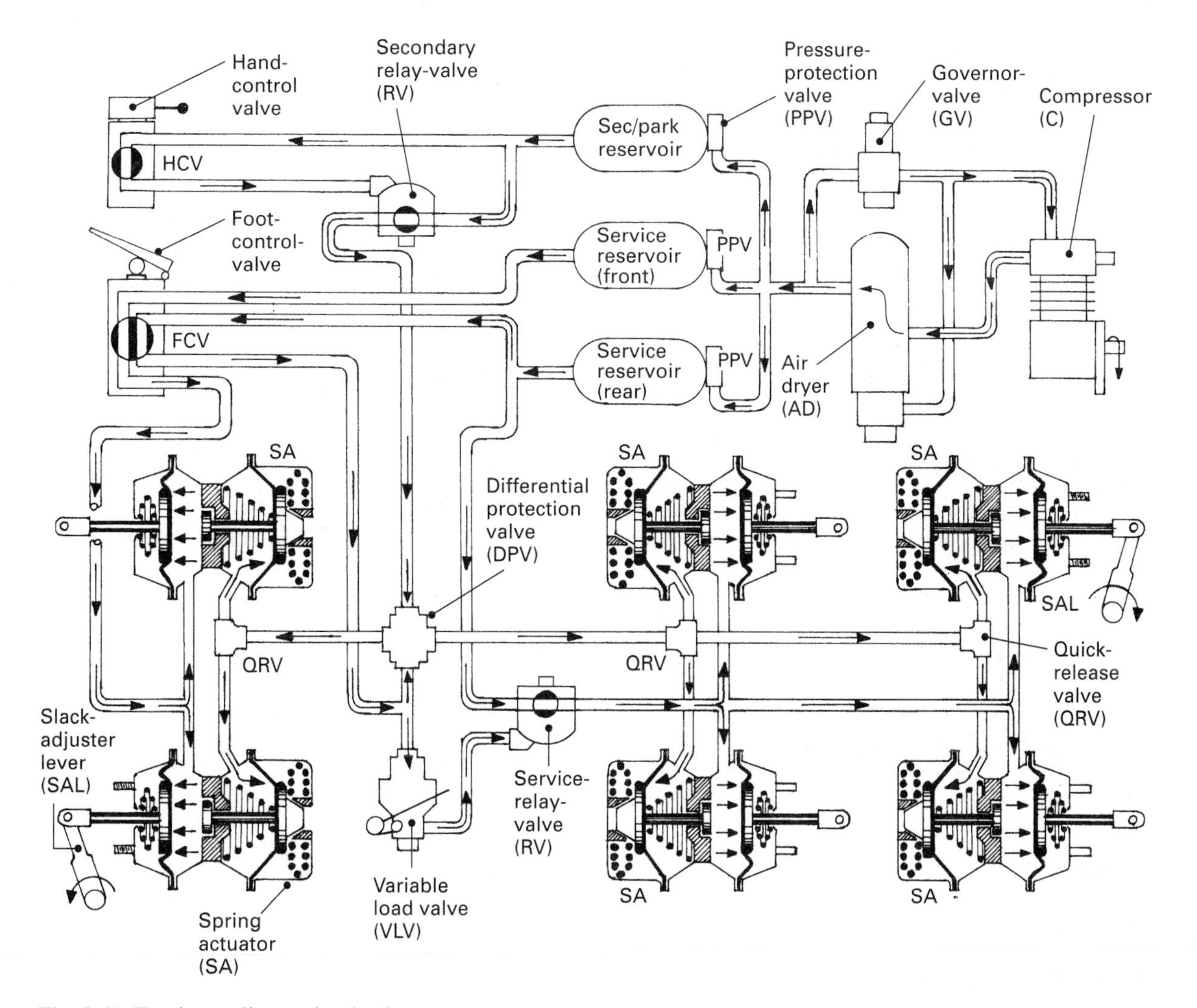

Fig. 9.41 Truck two-line spring brake system

then passes to the three reservoir air storage tanks. An unloader-valve is incorporated in the compressor head which is controlled by the pressure sensing governor-valve. Whenever the line pressure reaches some predetermined upper limit the governor-valve signals to the unloader-valve. The unloader-valve then holds open the compressor's inlet-valve, thereby compelling the air drawn into the compressor's cylinder to be returned to the atmosphere on each of the pistons upstrokes. As the air in the reservoir tanks is used by the brake system, its pressure drops to some minimum set limit. The governor-valve again signals to the unloader-valve to permit the compressor's inlet disc-valve to open and close as normal so that compressed air will once more be delivered to the reservoir tanks. Thus the compressor only works under load for relatively short time spans; the majority of the vehicle's journey time will be spent idling, that is, the piston reciprocates without generating any air pressure.

Service line circuit (fig. 9.41) When the foot-pedal is applied, air flows along each of the dual service lines from their respective reservoir tanks (front and rear) to the foot-control-valve. Metered air is then permitted to flow directly through the front wheel service-lines to operate the front axle brake actuators and along the rear wheel service-line, through the variable-load-valve to the service-line relay-valve, signalling it to open. Immediately the relay-valve opens, air from the rear service reservoir passes via the variable-load-valve to both rear-axle brake-actuators to apply the rear brakes. When the foot-pedal is released the air supply to the brake actuators will be blocked and air in the service-pipelines will exhaust from both the foot-control-valve and the service-relay-valve, thus releasing the brakes controlled by the service-line.

Secondary line circuit (fig. 9.41) With the hand-control-valve in the 'off position' air passes though the valve to transmit a signal pressure to the secondary-relay-valve. Accordingly the relay-valve inlet port opens and its exhaust port closes. This permits air from the secondary/park reservoir tank to be supplied to each of the wheel spring-brake-actuators, power-spring pressure-chambers via the differential-protection-valve, and each of the quick-release-valves, to compress each of the power-springs and thus release the brakes controlled by the secondary line circuit. When the hand-control-valve moves to the 'applied position' its inlet-valve closes and its exhaust-valve opens, thus blocking the signal pressure to the secondary-relay-valve. The relay-valve will therefore close and its exhaust port opens. Air will therefore be prevented from reaching the spring-brake-actuators, and at the same time, it permits air trapped in the secondary pipeline and in the actuator's power-spring pressure chambers to escape though both the quick-release-valve and though the secondary-relay-valve's exhaust port. Consequently, the power-springs will be free to expand and to apply the secondary line brakes.

9.14 Foundation brake and single-diaphragm wheel brake actuator (fig. 9.42)

Foundation brakes refer to the total assembly of the drum, shoes, expander, anchor, retraction springs and torque (back) plate attached to each wheel axle. Air-power-operated brakes in their simplest form incorporate a wheel brake actuator consisting of a diaphragm sandwiched between a return spring exposed to the atmosphere and an air-tight pressure chamber subjected to a supply of pressurised air of varying intensity. A diaphragm-plate and push-rod conveys the applied air pressure thrust to a slack-adjuster lever, which then converts the linear movement of the push-rod to a rotary twist on the 'S'-shaped (involute profile) cam expander shaft. Twin short struts convert the rotary motion of the cam back to a linear expanding outward movement to opposing tappet plungers. The inclined tappet abutments enable the shoe-lining profiles to centralise themselves to the circular shape of the drum as the linings wear. When the air power brakes are applied, the rotating 'S' cam forces the tappets apart, thus resulting in both brake shoes being pushed hard against the drum. The generated frictional drag then causes the rotating drum and wheel to reduce speed or even come to a standstill. Stiff retraction-springs hold both shoes in position and return the shoes back to the 'off' position whenever the foot brake-pedal is released. Slack-adjuster-levers can be set to provide the optimum leverage when the brakes are fully applied. This is achieved by winding a worm and worm-wheel encased at the camshaft end of the lever (see fig. 9.43), until the lever is at its most effective position, that is, perpendicular to the push-rod when all the slack movement has been taken up. Note that the slack movement takes into account the shoe-to-drum clearance and the elastic give of the expander and lever components when under load.

9.14.1 Spring brake actuator (fig. 9.44(a)–(d))

These actuators operate the foundation brakes and are controlled by both the service-line (foot brake) circuit and the secondary/park-line (hand-brake) circuit. These units combine the service-line foot brake diaphragm pressure chamber which operates a push-rod and slack adjuster lever, and a secondary/park-line hand brake diaphragm pressure chamber whose function is to compress the power-spring.

Normal driving (brakes released) (fig. 9.44(a)) As the foot-pedal is released and the hand-control-valve lever is moved to the 'released position', the service-pressure-chamber is exhausted of compressed air, whereas secondary-line compressed air is delivered to the secondary-pressure-chambers to fully compress the power-springs.

Service line applied (fig. 9.44(b)) When the foot-pedal is applied with the hand-control-valve lever in the 'released position', the secondary-line pressure holds the power-springs in each wheel-actuator in the disengaged compressed position. Conversely the service-line air pressure controlled by the foot-control-valve expands the service pressure chamber by deflecting its diaphragm and forcing its push-rod outwards. This causes the slack-adjuster-lever to rotate in the brake applied direction.

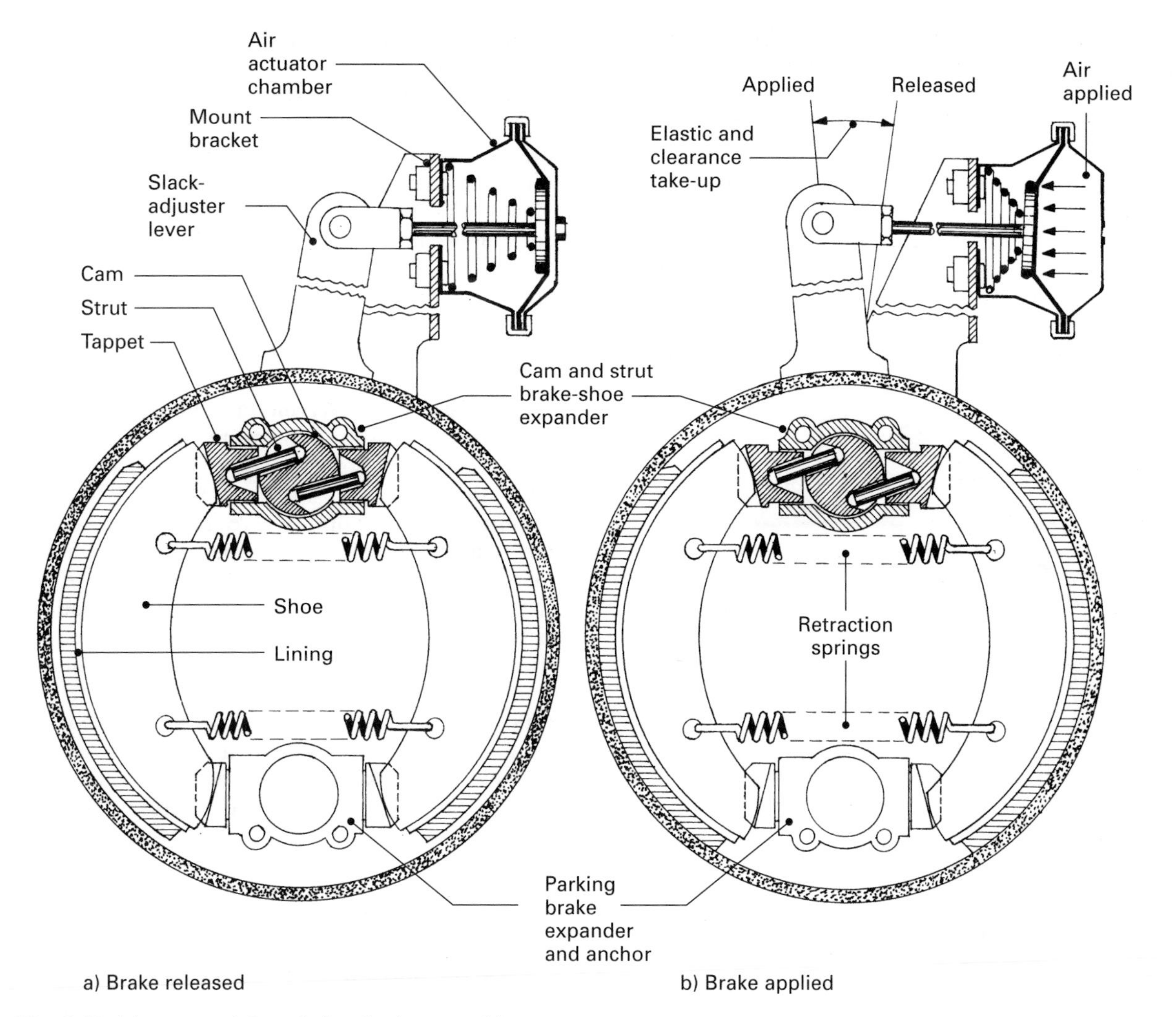

Fig. 9.42 Air-operated foundation brake assembly

Secondary/park-line applied (fig. 9.44(c)) With the hand-controlled-valve lever moved into the 'applied position' and the foot-pedal released, service-line pressure will be exhausted from the service-pressure-chamber. Similarly the secondary/park-line air pressure, controlled by the foot-controlled-valve, is steadily released from the secondary/park-pressure-chamber thus permitting the power-springs to expand. Consequently the power-spring, push-rod and the service-line diaphragm push-rod will both be pushed outwards to rotate the slack-adjuster-lever in the applied direction.

Manual release (fig. 9.44(d)) If the vehicle is to be towed and there is an air supply failure, then it is essential before moving the vehicle to pump air into the power-spring pressure chambers by another source such as the recovery truck emergency air supply, or for each individual power-spring to be compressed manually. Manual releasing of the power-spring brakes is achieved with a 'wind-off screw adapter' housed in a pocket hole formed in the casing separating the two pressure-chambers.

The power-spring is compressed by removing the flap plug and screwing the 'wind-off screw' into the power-spring diaphragm-plate at the rear end of the actuator until it becomes tight. The nut is then wound along the screw thread with a spanner until the power-spring is fully compressed, thus releasing the wheel brake.

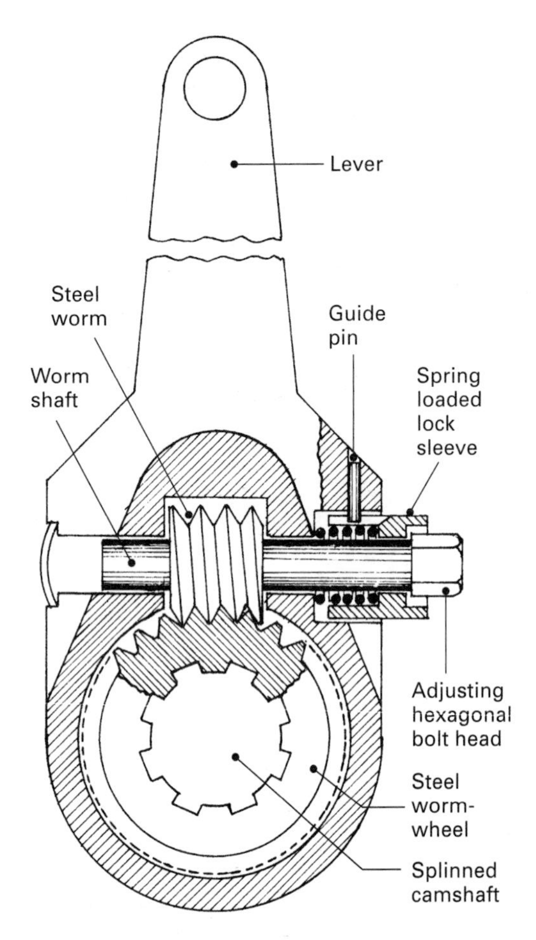

Fig. 9.43 Manual slack-adjuster-lever

9.14.2 Unloader-valve (pressure regulator) (fig. 9.45(a) and (b))
The unloader-valve regulates the air pressure supply charging the braking system's reservoir tanks. When the air pressure is below the lower limit the unloader-valve directs air from the compressor to the reservoir tanks to replenish the air pressure. Above some predetermined upper pressure limit the unloader-valve blocks off the air supply to the reservoir tanks and diverts the air pumped from the compressor back into the atmosphere. This enables the compressor, which is driven at half engine speed from the timing gears, to continue to operate without generating air pressure, that is, for the compressor to idle.

Low air pressure reservoir tank-filling phase (fig. 9.45(a)) When the air pressure drops below the lower pressure setting, the control-spring thrust is able to push down the diaphragm and thereby close and open the control-valve inlet and exhaust respectively. Air will therefore escape from above the control-piston thus permitting the piston and stem to rise to a point when the idle valve closes. Air from the compressor will now flow up and though the annular-shaped gauze air filter and out via the now open check-valve to charge the reservoir-tank. Note the idle-valve cannot open since the pressure around the valve's annular passageway also reacts on the enlarged piston-stem, thus neutralising the air thrust.

High air pressure reservoir tank-idling phase (fig. 9.45(b)) When the air pressure rises to its upper set limit, the air pressure rise from the compressor's air supply is sufficient to push up the diaphragm to a point where the exhaust-valve closes and the inlet-valve opens. Air from the control passage now enters the space above the piston. It therefore pushes down the reaction-piston against the resisting spring upthrust to open the idle-valve. Air supplied by the reservoir is immediately diverted via the idle-valve to the atmosphere. Simultaneously the interruption of the air supply to the reservoir tank reverses the air flow. This makes the check-valve snap back on its seat; air in the reservoir is thus prevented from escaping back though the unloader-valve to the atmosphere.

Once the air pressure in the reservoir-tank is reduced to the lower air pressure setting, the control-spring overcomes the upthrust of the reduced air pressure on the underside of the diaphragm. It thus causes the control-valve inlet to close and its exhaust passage to open. Air pressure on top of the reaction-piston will therefore collapse. It thus makes the piston rise and the idle-valve close. Again the cycle for recharging the reservoir will recommence.

9.14.3 Dual concentric foot control valve (fig. 9.46(a) and (b))
The split-line foot control valve meters out the air supply delivered to the front and rear wheel actuators. The valve unit consists of a foot-treadle, inner and outer plunger, an upper-piston valve-carrier with its inlet/exhaust valve operating the front brakes, and a lower-piston with its inlet/ exhaust valve leading to the rear brakes.

Foot-brake applied (fig. 9.46(a)) With the foot-treadle applied, the upper-piston, valve-carrier and lower-piston are all pushed downwards. Air

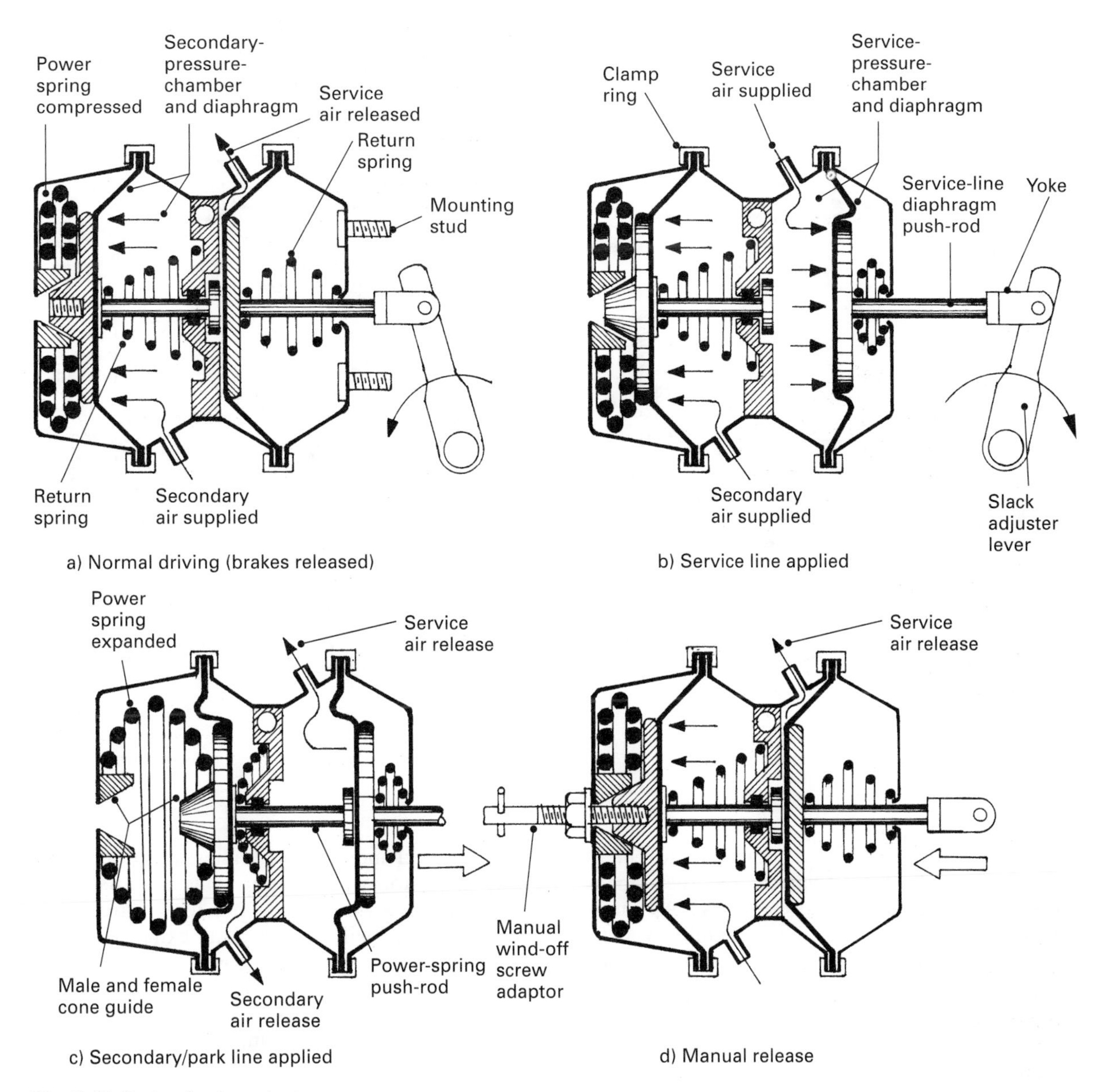

Fig. 9.44 Spring brake actuator

flows from the reservoir-tanks to the front and rear supply-ports. It then passes though the front and rear (upper and lower) open inlet-valves to the chamber underneath the upper and lower reaction-pistons and to their respective wheel-actuator pressure-chambers to apply the brakes. However, as the air passes underneath the upper and lower reaction-pistons it applies an upthrust which tends to lift the pistons against the spring downthrust. As a result the inlet-valve commences to close. This requires further downward pressure on the foot-control-valve to maintain or increase the quantity of air reaching the wheel-actuators. By these means a progressive increase in braking pressure is obtained.

Foot brake released (fig. 9.46(b)) Releasing the foot-treadle permits the lower-piston return-spring to expand and to lift the lower-piston, valve-carrier, and upper-piston to the highest

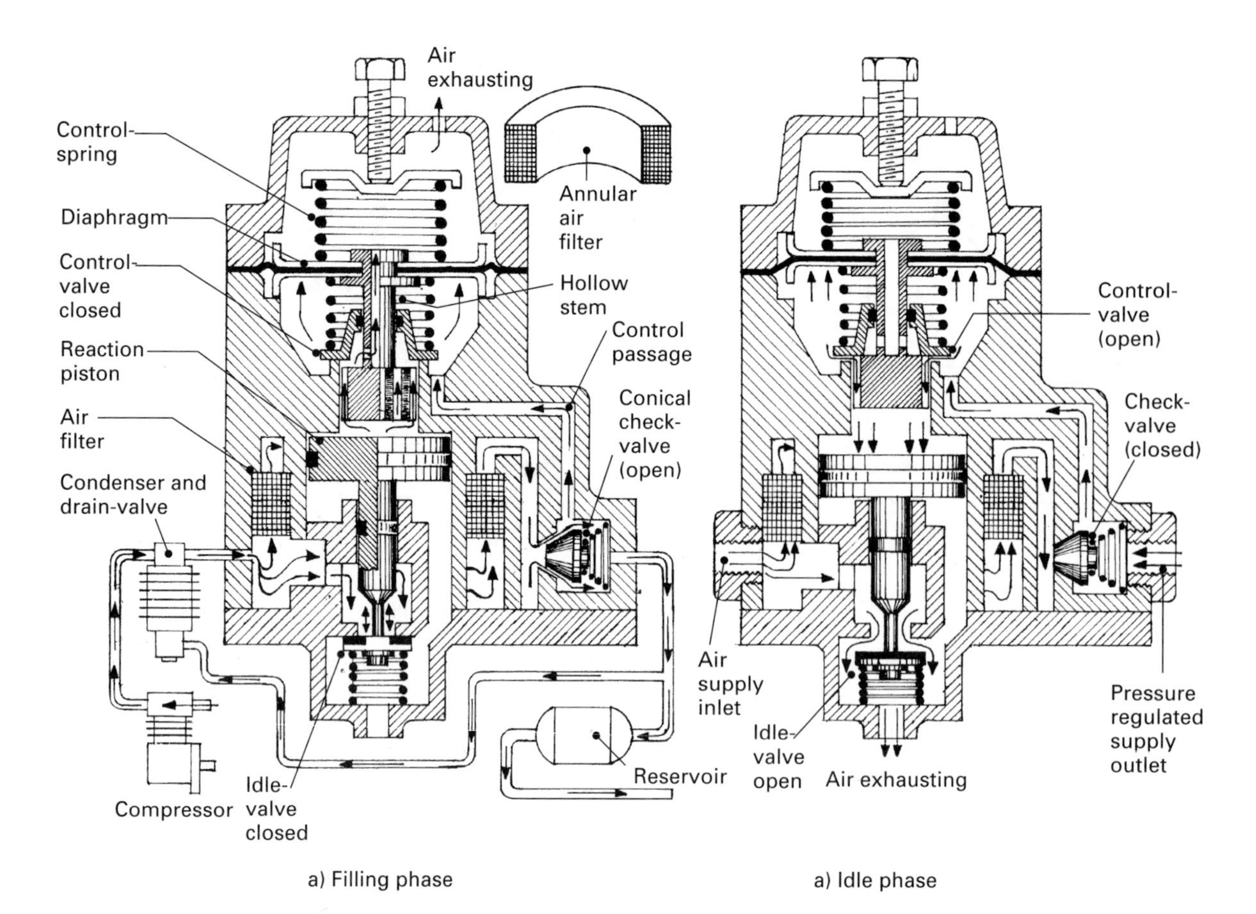

Fig. 9.45 Pressure regulator (unloader-valve)

position. Consequently the upward movement initially closes both inlet-valves, and then opens both the central exhaust passages. Air now escapes from the wheel-actuator chamber via the open exhaust-valve, down the central exhaust port exit to the atmosphere, thereby releasing the brakes.

9.14.4 Hand control valve (fig. 9.47(a) and (b)) The hand control valve regulates the signal pressure to open or close the relay-valve, hence to release or apply the secondary/parking brakes. The valve unit comprises a lever-operated cam with a cam-plate supported and guided by a plunger exhaust-valve. A piston-valve carrier in its cylinder contains the inlet/exhaust-valve-stem, and a graduating-spring reacts with the air supply to control the signal pressure.

Secondary/park brake applied (fig. 9.47(a)) When the hand-control-valve lever is moved to the 'applied position', the cam-plate and exhaust-valve plunger move to the highest position. Similarly the piston-valve-carrier will rise to its upper position. The signal air flow previously directed to the relay-valve will now steadily exhaust via the open exhaust-valve central passage to the atmosphere. Due to the reducing signal pressure, the relay-valve closes and its exhaust passage opens. This permits the air compressing the power-springs to escape to the atmosphere via the relay-valve exhaust passage. Correspondingly the power-springs expand and exert their thrust to apply the brakes.

Secondary/park brake released (fig. 9.47(b)) As the hand-control-valve is moved to the 'released position', the cam-plate and exhaust-valve plunger are pushed downwards. This first closes the exhaust-valve passage and secondly with further downward movement opens the inlet-valve passage. Air will now enter the piston-valve-carrier, pass though the inlet-valve and continue its flow as a signal pressure to the relay-valve. At

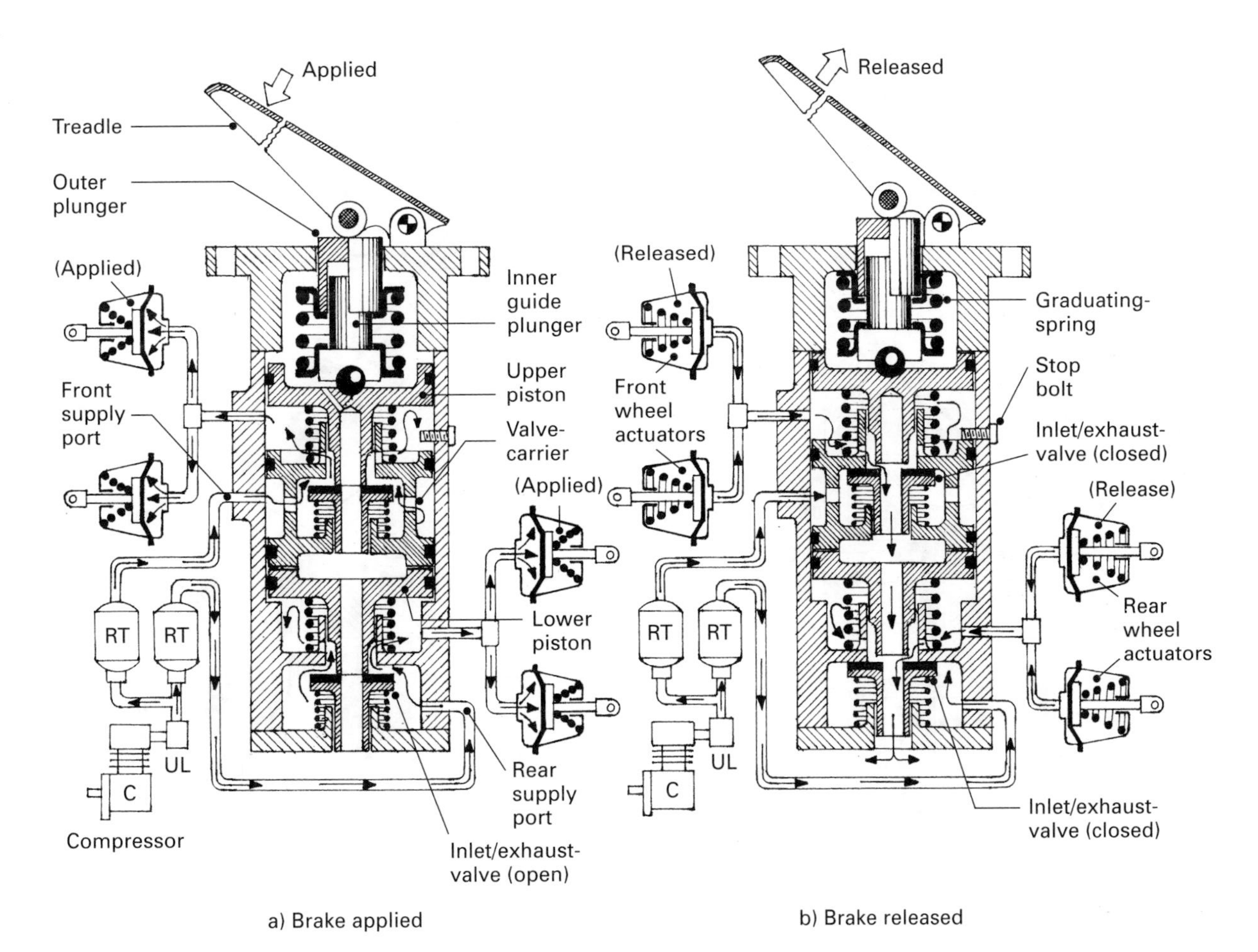

Fig. 9.46 Dual concentric foot control valve

the same time the air pressure above the piston will tend to react and force the piston down to partially close the inlet-valve. Therefore more downward effort is required to again open the inlet-valve to the same level of braking. Thus this air pressure reacts against the piston-carrier-valve, providing a degree of progressive application of the brakes which is essential for braking stability. This results in the relay-valve opening to permit air from the reservoir tank to pass directly to the actuator pressure-chamber. It therefore compresses and releases the brake. Note that while the vehicle is in motion, the power-springs always remain in the compressed state.

9.14.5 Relay-valve (fig. 9.48(a) and (b))
The relay-valve is incorporated in brake-line systems when the hand or foot control valve is situated at a considerable distance from the axle and wheels to speed up the delivery of air from the reservoir tank to the wheel actuators. The relay-valve comprises a large control-piston enclosed in an upper cylinder chamber housing, and an inlet and exhaust hollow-stem-valve contained in a lower junction housing.

Brakes released (fig. 9.48(a)) When the hand-control-valve lever is moved to the 'released position', air above the piston is exhausted via the hand-control-valve central exhaust exit. This permits the control-piston and its solid-ended central stem to move up to its highest position. Consequently the hollow valve-stem will also rise to close off the inlet-valve. In this position, the inlet-valve is closed and there is an exhaust-valve opening between the upper and lower solid and hollow stems. Trapped air in the wheel-actuator chambers now exhausts to the atmosphere via the central passage of the hollow valve-stem and the flap-valve to release the brakes. Note

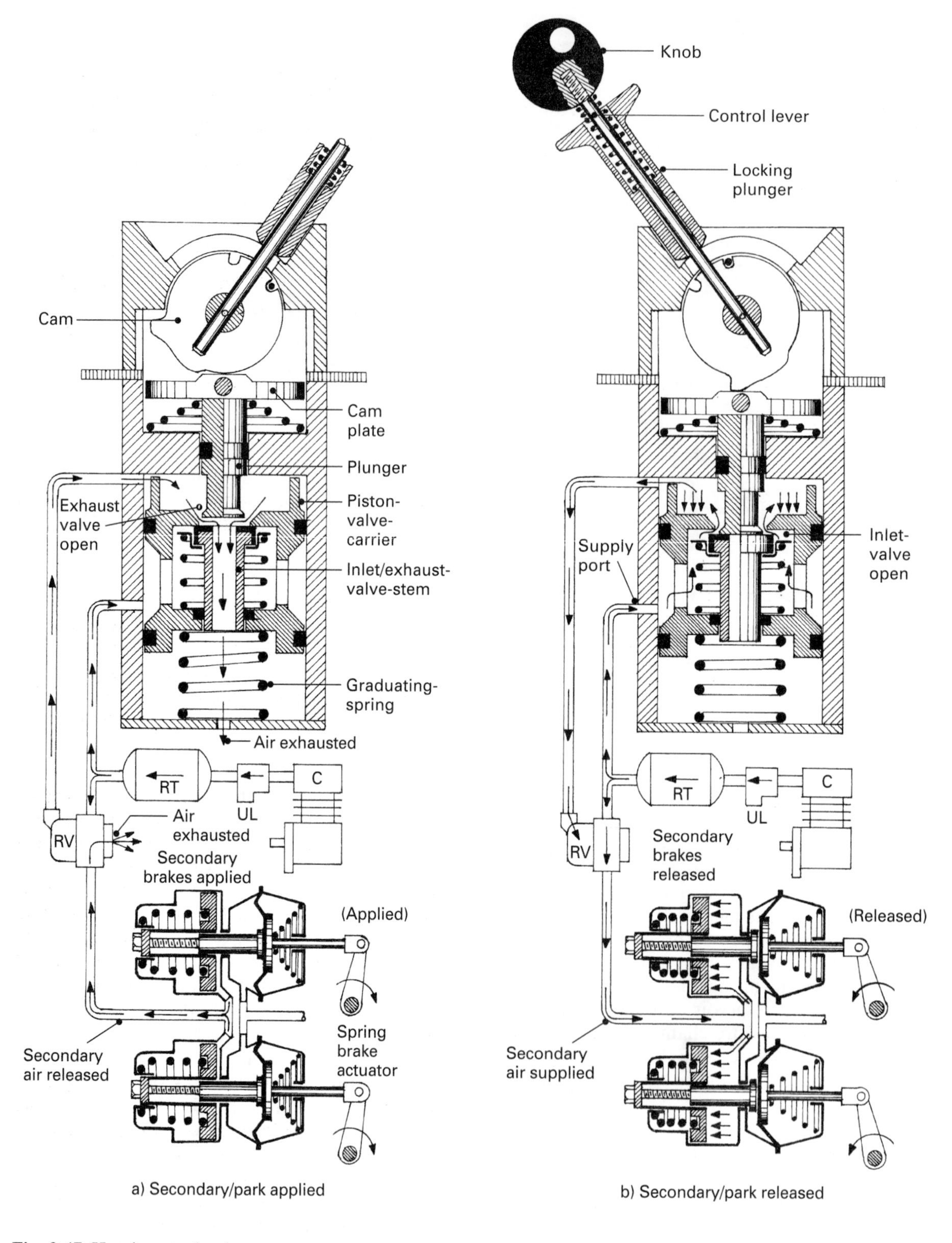

Fig. 9.47 Hand control valve

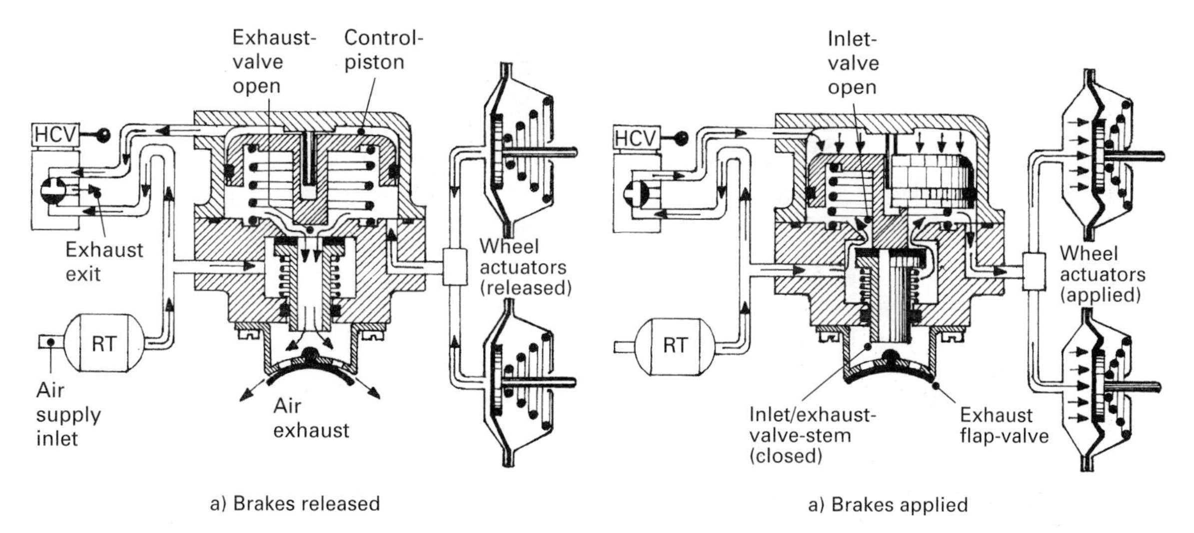

Fig. 9.48 Relay valve

that the flap-valve is there to prevent dust entering the relay-valve.

Brakes applied (fig. 9.48(b)) Moving the hand-control-valve lever to the 'applied position', sends a command signal in the way of air pressure from the reservoir tank to the large control-piston chamber via the hand-control-valve opening. Immediately the control-piston is pushed downwards, first closing the central exhaust passage, and secondly with further downward movement opening the inlet-valve passage. Air now flows directly from the reservoir tank though the open inlet-valve to the chamber under the piston and out to the wheel actuator-chamber to apply the brakes. The air underneath the control-piston tends to oppose and lift the piston so that there will be a continuous state of balance for the valve opening, as higher signal pressures are passed to the upper piston chamber.

9.14.6 Quick-release valve (fig. 9.49(a)–(c))

This valve is designed to speed up the expulsion of compressed air from the wheel actuator chamber when the brakes are released. The quick-release valve is usually incorporated in the pipelines when the wheel brake actuators are far removed from the foot-control-valve. The valve is there to provide a shorter route for the air to exhaust, as opposed to relying on the air to return all the way back to the foot-control-valve before it can escape to the atmosphere.

The quick-release valve is essentially a flexible rubber disc contained between an upper supply housing and a lower junction housing.

Applied position (fig. 9.49(a)) When the foot-pedal is pressed down, air from the foot-control-valve flows into the chamber above the disc. This pushes the rubber disc against the exhaust port thus blocking its exit. Air, however, is permitted to flow around the disc, between the support-ribs and into the wheel actuator's pressure-chambers to apply the brakes.

Hold position (fig. 9.49(b)) After the foot-pedal has been pressed, the brakes are then held in one position. The air supply then pushes down the central region of the rubber disc valve against the exhaust port, and at the same time, the air trapped in the wheel actuator power-chambers pushes up the outer rim of the disc against the upper supply housing. Air is therefore prevented from entering or leaving the wheel actuators, and the brakes are therefore said to be in a state of 'hold'.

Release position (fig. 9.49(c)) Releasing the foot-pedal closes the foot-control-valve and opens its exhaust exit. Air will now be expelled from the pipeline leading to the quick-release-valve. Instantly air trapped in the wheel actuator pressure chambers will flow back and push up the rubber disc valve against the upper supply housing. This closes the supply port and opens the

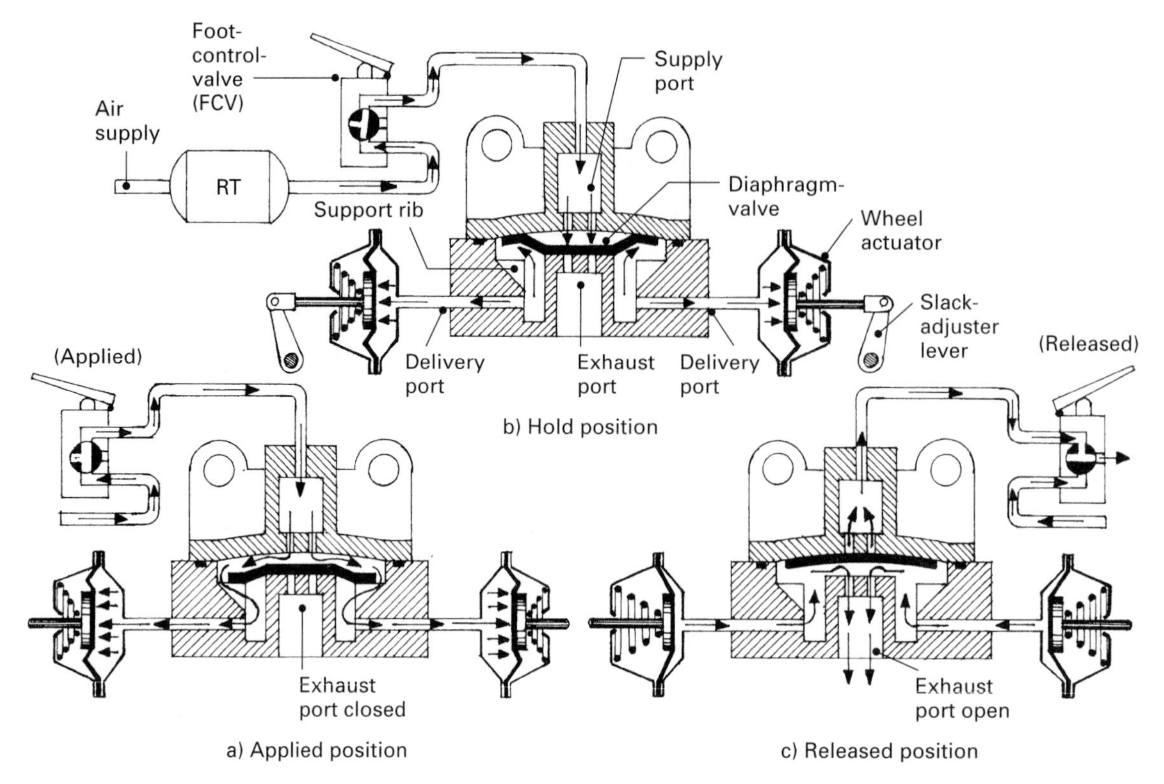

Fig. 9.49 Quick release valve

exhaust port so that air in the wheel actuators can rapidly escape.

9.14.7 Differential-protection-valve (fig. 9.50(a) and (b))

The differential-protection-valve is designed to prevent both service and secondary brake lines being applied at the same time, as this would otherwise cause compounding of the full power-spring push-rod force and service-line push-rod force. The consequence of both systems operating together would be to severely overload the slack adjuster lever, camshaft, and the foundation brake.

Service and secondary brakes released With both the foot-pedal released and the hand-control-valve lever in the 'release position', air is exhausted from the service-line via the quick-release valve whereas air in the secondary-line passes though the open differential-protection-valve coming out at the delivery ports *en route* to the power-spring pressure-chamber where it compresses and holds the power-springs in the 'off position'.

Secondary-line (hand-controlled) brake applied (fig. 9.50(a)) Moving the hand-control-valve (secondary-line) to the spring-brake 'applied position' after a foot brake service-line application which is held, causes the air to exhaust from the secondary-line to the atmosphere via the hand-control-valve. As the pressure in the secondary-line (hand-controlled) falls, the pressure in the service-line (foot-controlled) which is still being held becomes greater than that in the secondary-line and is therefore able to push the inner piston up to close off the secondary port. This is followed by the service-line pressure lifting the outer piston off its seat, thus allowing service-line air to pass though the discharge-port to the power-spring pressure-chamber. As the foot-pedal is released, the reducing service-line pressure permits the return-spring to push down first the outer and then the inner piston, thus closing and opening the service-ports and secondary-ports respectively. As a result the trapped air in the power-spring pressure-chambers is allowed to exhaust, hence the power-spring secondary brakes are applied simultaneously when the foot-operated service-line brakes are released.

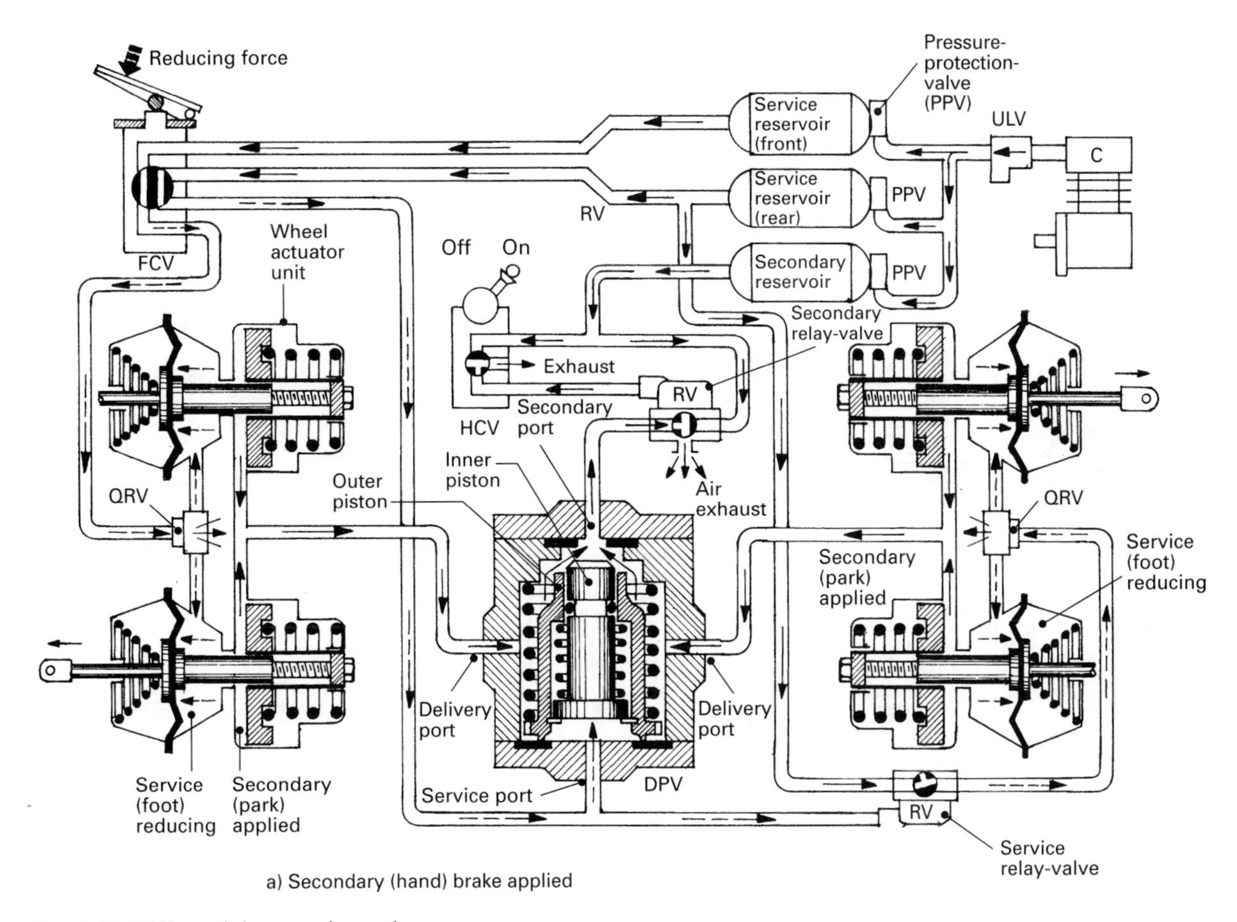

Fig. 9.50 Differential-protection-valve

Service-line (foot-controlled) brake applied (fig. 9.50(b)) When the foot-pedal is pressed down after a secondary-line spring brake application, the secondary-line is exhausted via the relay-valve and hand-control-valve. This enables the pressure in the service-line to push up the inner-piston to close the secondary-port. A further rise in service-line pressure causes the outer-piston also to lift. Service-line air now flows from the service-port, through the differential protection-valve and out from the delivery-ports to the power-spring pressure-chamber. This holds the spring brake in the 'released position', and therefore prevents the compounding of the service and secondary-line brake forces. Reducing the foot-pedal downward movement reduces the service-line air pressure. This enables the return spring thrust to move the outer and inner pistons back, with the result that the service-port is now closed and the secondary-port is open. Consequently the power-spring pressure-chamber is now allowed to exhaust. This again permits the expanding power-springs to operate the brakes.

9.15 Exhaust compression brake retarder

The exhaust compression brake retarder provides an additional method of reducing a vehicle's speed which is particularly useful on long or steep down gradients; however this method of retarding a vehicle is not powerful enough to be used for emergency stops. One major benefit of exhaust compression engine brake retardation is that it reduces the work done by the conventional wheel frictional brakes so that the shoes or pads run cooler and last much longer.

It operates only after the fuel supply to the injector pump has been cut off so that only air is discharged to the exhaust manifold from the cylinders on each of the exhaust-strokes. Its principle of operation relies on a butterfly-valve positioned inside the exhaust down-pipe being

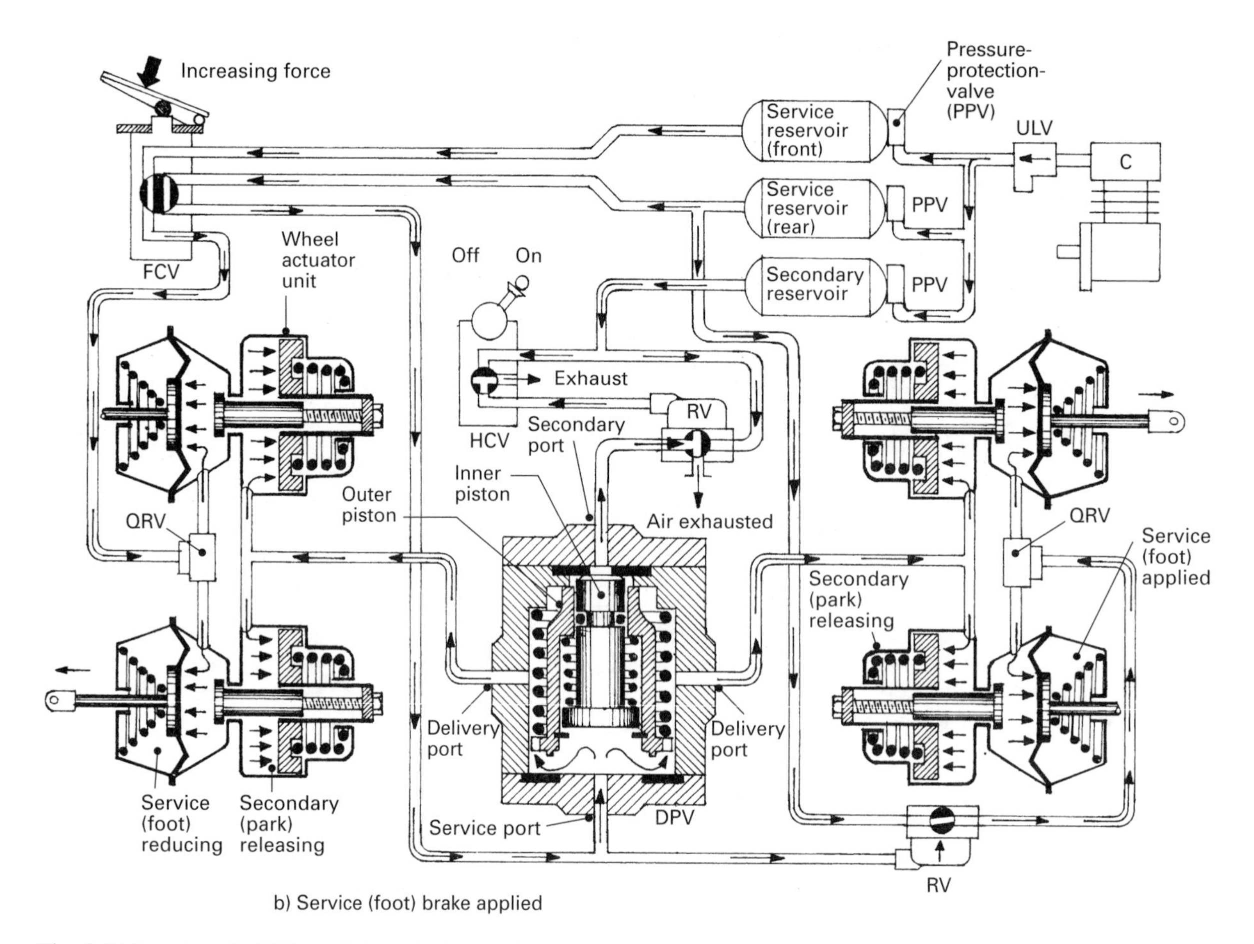

Fig. 9.50 (*continued*) Differential-protection-valve

made to 'close-off' the exit so that the discharged air (not exhaust gas) from each cylinder is pumped continuously into the exhaust manifold against the blocked exit of the exhaust down-pipe. As a result of this accumulating air in the exhaust manifold, there is a build up of back pressure which opposes the up-stroke of each piston on its exhaust-stroke. Consequently this produces a mean resisting torque to the rotating crankshaft which is considerably multiplied at the road drive wheels by the final drive and gearbox gear ratios. The lower the gear ratio selected the greater will be the retarding effort on the vehicle motion.

The generated resisting torque is far greater than that produced by the trapped air in the cylinders when the pistons come up on their compression-strokes since the compressed air at the end of the compression automatically uses this same stored energy in the compressed air to push down the piston on the next non-firing power-stroke. Only a small proportion of the energy generated by the compressed air is absorbed by piston-ring and journal bearing friction and pumping losses, however it is this absorbed energy transformed to resisting crank-shaft torque which produces the vehicle's overrun engine retardation effort.

9.5.1 Operation

Retarder inoperative (Fig 9.51(a)) Under normal driving conditions the exhaust butterfly-valve is maintained in the fully open position by solenoid actuator return-spring. When the exhaust butterfly-valve is fully open the electrical earth-return circuit to the electronic control unit ECU is completed through the closed fuel cut-off switch, thus permitting the ECU to operate the unit injector pump so that fuel is injected into the cylinders in their respective firing sequence.

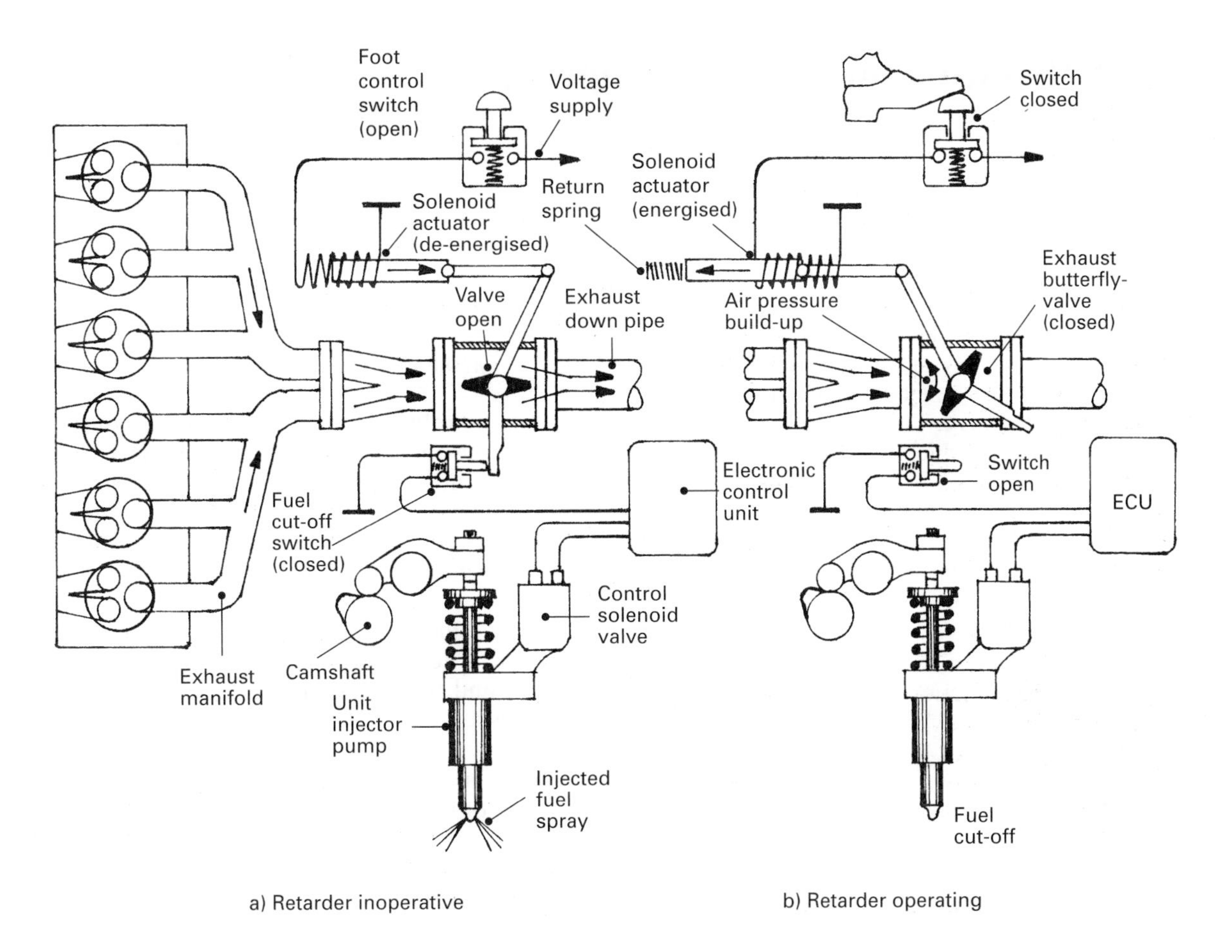

Fig. 9.51 Exhaust compression brake retarder

Retarder operating (Fig 9.51(b)) When the driver takes his foot off the clutch pedal and with the same left foot presses the foot-control-switch, this closes the contacts causing the solenoid butterfly-valve actuator to become energised. The initial anti-clockwise twist to the butterfly-valve spindle opens the fuel cut-off switch contacts, and this is signalled instantly to the ECU which then shuts down all of the individual injector pump units so that fuel injection into the cylinders ceases. This is followed by further movement of the solenoid armature plunger thus causing the butterfly-valve and its spindle to rotate to the fully closed position. Immediately discharging air from the cylinders will be prevented from escaping to the atmosphere by the blocking action of the closed exhaust butterfly-valve. In a very short space of time the back-pressure build up provides an opposing force to the moving pistons and corresponding resisting torque to the rotating crankshaft, thus providing the exhaust compression resisting torque to the drive-axle which is used to control the retardation of the vehicle.

10 Piston-engine cycles of operation

10.1 The internal-combustion engine

The piston engine is known as an internal-combustion heat-engine. The concept of the piston engine is that a supply of air-and-fuel mixture is fed to the inside of the cylinder where it is compressed and then burnt. This internal combustion releases heat energy which is then converted into useful mechanical work as the high gas pressures generated force the piston to move along its stroke in the cylinder. It can be said, therefore, that a heat-engine is merely an energy transformer.

To enable the piston movement to be harnessed, the driving thrust on the piston is transmitted by means of a connecting-rod to a crankshaft whose function is to convert the linear piston motion in the cylinder to a rotary crankshaft movement (fig. 10.1). The piston can thus be made to repeat its movement to and fro, due to the constraints of the crankshaft crankpin's circular path and the guiding cylinder.

The backward-and-forward displacement of the piston is generally referred to as the *reciprocating* motion of the piston, so these power units are also known as reciprocating engines.

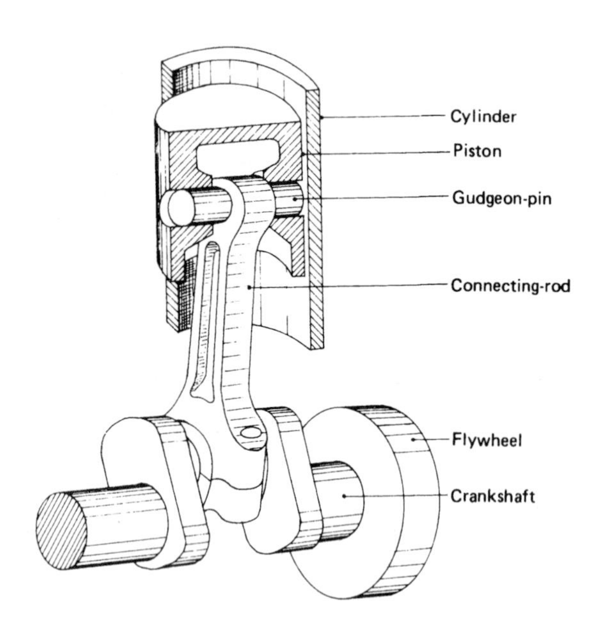

Fig. 10.1 Pictorial view of the basic engine

10.1.1 Engine components and terms (figs 10.1 and 10.2)

The main problem in understanding the construction of the reciprocating piston engine is being able to identify and name the various parts making up the power unit. To this end, the following briefly describes the major components and the names given to them.

Cylinder block This is a cast structure with cylindrical holes bored to guide and support the pistons and to harness the working gases. It also provides a jacket to contain a liquid coolant.

Cylinder head This casting encloses the combustion end of the cylinder block and houses both the inlet and exhaust poppet-valves and their ports to admit air–fuel mixture and to exhaust the combustion products.

Crankcase This is a cast rigid structure which supports and houses the crankshaft and bearings. It is usually cast as a mono-construction with the cylinder block.

Sump This is a pressed-steel or cast-aluminium-alloy container which encloses the bottom of the crankcase and provides a reservoir for the engine's lubricant.

Piston This is a pressure-tight cylindrical plunger which is subjected to the expanding gas pressure. Its function is to convert the gas pressure from combustion into a concentrated driving thrust along the connecting-rod. It must therefore also act as a guide for the small-end of the connecting-rod.

Piston rings These are circular rings which seal the gaps made between the piston and the cylinder, their object being to prevent gas escaping and to control the amount of lubricant which is allowed to reach the top of the cylinder.

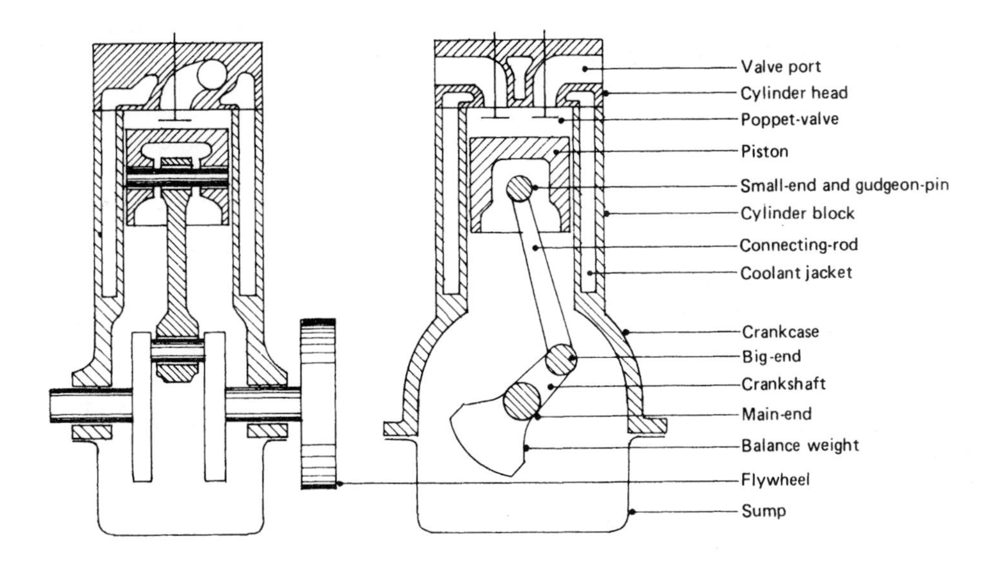

Fig. 10.2 Sectional view of the basic engine

Gudgeon-pin This pin transfers the thrust from the piston to the connecting-rod small-end while permitting the rod to rock to and fro as the crankshaft rotates.

Connecting-rod This acts as both a strut and a tie link-rod. It transmits the linear pressure impulses acting on the piston to the crankshaft big-end journal, where they are converted into turning-effort.

Crankshaft A simple crankshaft consists of a circular-sectioned shaft which is bent or cranked to form two perpendicular crank-arms and an offset big-end journal. The unbent part of the shaft provides the main journals. The crankshaft is indirectly linked by the connecting-rod to the piston – this enables the straight-line motion of the piston to be transformed into a rotary motion at the crankshaft about the main-journal axis.

Crankshaft journals These are highly finished cylindrical pins machined parallel on both the centre axes and the offset axes of the crankshaft. When assembled, these journals rotate in plain bush-type bearings mounted in the crankcase (the main journals) and in one end of the connecting-rod (the big-end journal).

Small-end This refers to the hinged joint made by the gudgeon-pin between the piston and the connecting-rod so that the connecting-rod is free to oscillate relative to the cylinder axis as it moves to and fro in the cylinder.

Big-end This refers to the joint between the connecting-rod and the crankshaft big-end journal which provides the relative angular movement between the two components as the engine rotates.

Main-ends This refers to the rubbing pairs formed between the crankshaft main journals and their respective plain bearings mounted in the crankcase.

Line of stroke The centre path the piston is forced to follow due to the constraints of the cylinder is known as the line of stroke.

Inner and outer dead centres When the crankarm and the connecting-rod are aligned along the line of stroke, the piston will be in either one of its two extreme positions. If the piston is at its closest position to the cylinder head, the crank and piston are said to be at inner dead centre (IDC) or top dead centre (TDC). With the piston at its furthest position from the cylinder head, the crank and piston are said to be at outer dead centre (ODC) or bottom dead centre (BDC). These reference points are of considerable importance for valve-to-crankshaft timing and for either ignition or injection settings.

Clearance volume The space between the cylinder head and the piston crown at TDC is known as the clearance volume or the combustion-chamber space.

Crank-throw The distance from the centre of the crankshaft main journal to the centre of the big-end journal is known as the crank-throw. This radial length influences the leverage the gas pressure acting on the piston can apply in rotating the crankshaft.

Piston stroke The piston movement from inner to outer dead centre is known as the piston stroke and corresponds to the crankshaft rotating half a revolution or 180°. It is also equal to twice the crank-throw.

i.e. $L = 2R$
where L = piston stroke
and R = crank-throw

Thus a long or short stroke will enable a large or small turning-effort to be applied to the crankshaft respectively.

Cylinder bore The cylinder block is initially cast with sand cores occupying the cylinder spaces. After the sand cores have been removed, the rough holes are machined with a single-point cutting tool attached radially at the end of a rotating bar. The removal of the unwanted metal in the hole is commonly known as boring the cylinder to size. Thus the finished cylindrical hole is known as the cylinder bore, and its internal diameter simply as the bore or bore size.

10.1.2 The four-stroke-cycle spark-ignition (petrol) engine (fig. 10.3)

The first internal-combustion engine to operate successfully on the four-stroke cycle used gas as a fuel and was built in 1876 by Nicolaus August Otto, a self-taught German engineer at the Gasmotoreufabrik Deutz factory near Cologne, for many years the largest manufacturer of internal-combustion engines in the world. It was one of Otto's associates – Gottlieb Daimler – who later developed an engine to run on petrol which was described in patent number 4315 of 1885. He also pioneered its application to the motor vehicle.

Petrol engines take in a flammable mixture of air and petrol which is ignited by a timed spark when the charge is compressed. These engines are therefore sometimes called spark-ignition (S.I.) engines.

These engines require four piston strokes to complete one cycle: an air-and-fuel intake stroke moving outward from the cylinder head, an inward movement towards the cylinder head compressing the charge, an outward power stroke, and an inward exhaust stroke.

Induction stroke (fig. 10.3(a)) The inlet valve is opened and the exhaust valve is closed. The piston descends, moving away from the cylinder head (fig. 10.3(a)). The speed of the piston moving along the cylinder creates a pressure reduction or depression which reaches a maximum of about 0.3 bar below atmospheric pressure at one third from the beginning of the stroke. The depression actually generated will depend on the speed and load experienced by the engine, but a typical average value might be 0.12 bar below atmospheric pressure. This depression induces (sucks in) a fresh charge of air and atomised petrol in proportions ranging from 10 to 17 parts of air to one part of petrol by weight.

An engine which induces fresh charge by means of a depression in the cylinder is said to be 'normally aspirated' or 'naturally aspirated'.

Compression stroke (fig. 10.3(b)) Both the inlet and the exhaust valves are closed. The piston begins to ascend towards the cylinder head (fig. 10.3(b)). The induced air-and-petrol charge is progressively compressed to something of the order of one eighth to one tenth of the cylinder's original volume at the piston's innermost position. This compression squeezes the air and atomised-petrol molecules closer together and not only increases the charge pressure in the cylinder but also raises the temperature. Typical maximum cylinder compression pressures will range between 8 and 14 bar with the throttle open and the engine running under load.

Power stroke (fig. 10.3(c)) Both the inlet and the exhaust valves are closed and, just before the piston approaches the top of its stroke during compression, a spark-plug ignites the dense combustible charge (fig. 10.3(c)). By the time the piston reaches the innermost point of its stroke, the charge mixture begins to burn, generates heat, and rapidly raises the pressure in the cylinder until the gas forces exceed the resisting load. The burning gases then expand and so change the piston's direction of motion and push it to

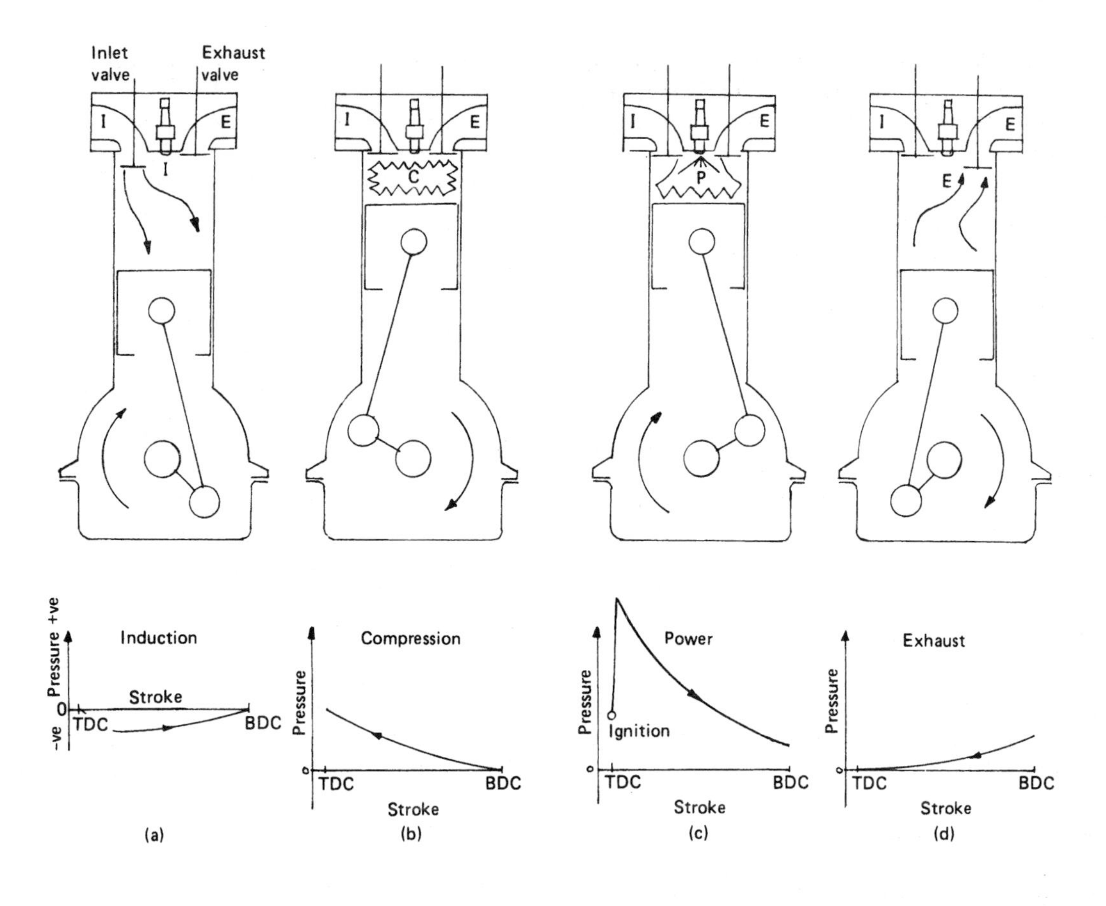

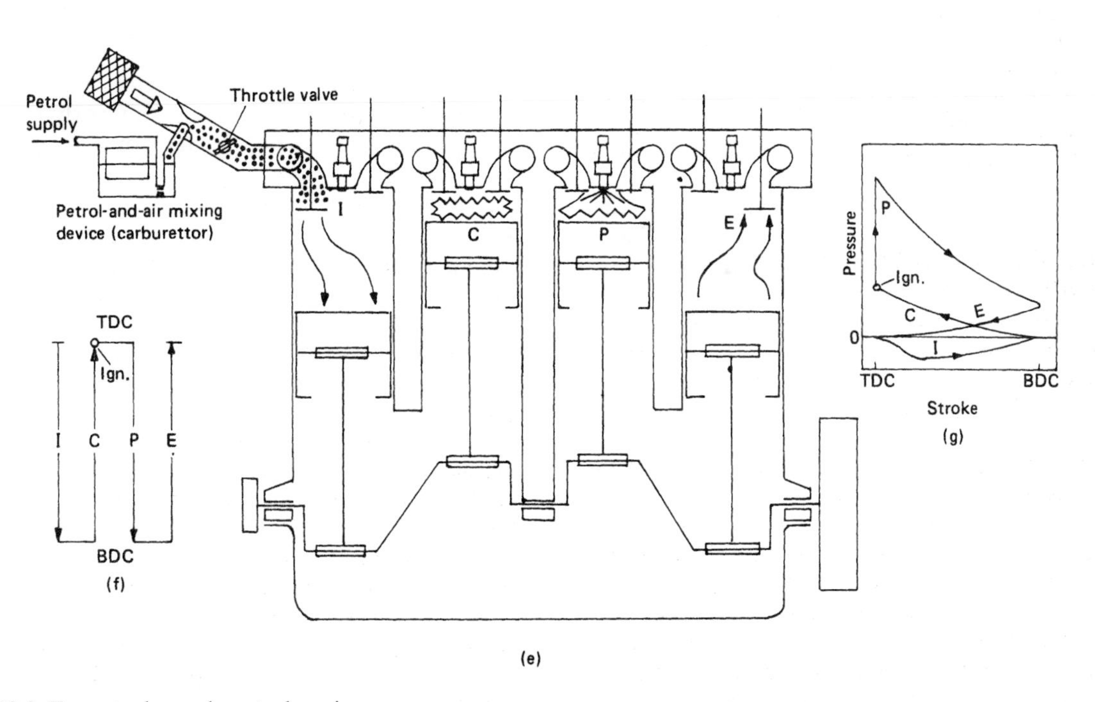

Fig. 10.3 Four-stroke-cycle petrol engine

its outermost position. The cylinder pressure then drops from a peak value of about 60 bar under full load down to maybe 4 bar near the outermost movement of the piston.

Exhaust stroke (fig. 10.3(d)) At the end of the power stroke the inlet valve remains closed but the exhaust valve is opened. The piston changes its direction of motion and now moves from the outermost to the innermost position (fig. 10.3(d)). Most of the burnt gases will be expelled by the existing pressure energy of the gas, but the returning piston will push the last of the spent gases out of the cylinder through the exhaust-valve port and to the atmosphere.

During the exhaust stroke, the gas pressure in the cylinder will fall from the exhaust-valve opening pressure (which may vary from 2 to 5 bar, depending on the engine speed and the throttle-opening position) to atmospheric pressure or even less as the piston nears the innermost position towards the cylinder head.

Cycle of events in a four-cylinder engine (fig. 10.3(e), (f), and (g)) Figure 10.3(e) illustrates how the cycle of events – induction, compression, power, and exhaust – is phased in a four-cylinder engine. The relationship between cylinder pressure and piston stroke position over the four strokes is clearly shown in fig. 10.3(f) and (g) and, by following the arrows, it can be seen that a figure of eight is repeatedly being traced.

10.1.3 Valve timing diagrams

In practice, the events of the four-stroke cycle do not start and finish exactly at the two ends of the strokes – to improve the breathing and exhausting, the inlet valve is arranged to open before TDC and to close after BDC and the exhaust valve opens before BDC and closes after TDC. These early and late opening and closing events can be shown on a valve timing diagram such as fig. 10.4.

Valve lead This is where a valve opens so many degrees of crankshaft rotation before either TDC or BDC.

Valve lag This is where a valve closes so many degrees of crankshaft rotation after TDC or BDC.

Valve overlap This is the condition when both the inlet and the exhaust valves are open at the same time during so many degrees of crankshaft rotation.

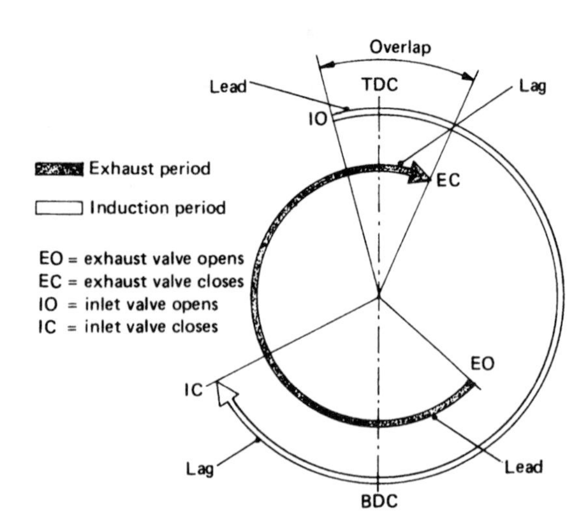

Fig. 10.4 Valve timing diagram

10.2 The two-stroke-cycle petrol engine (fig. 10.5)

The first successful design of a three-port two-stroke engine was patented in 1889 by Joseph Day & Son of Bath. This employed the underside of the piston in conjunction with a sealed crankcase to form a scavenge pump ('scavenging' being the pushing-out of exhaust gas by the induction of fresh charge).

This engine completes the cycle of events – induction, compression, power, and exhaust – in one revolution of the crankshaft or two complete piston strokes.

Crankcase-to-cylinder mixture transfer (fig. 10.5(a)) The piston moves down the cylinder and initially uncovers the exhaust port (E), releasing the burnt exhaust gases to the atmosphere. Simultaneously the downward movement of the underside of the piston compresses the previously filled mixture of air and atomised petrol in the crankcase (fig. 10.5(a)). Further outward movement of the piston will uncover the transfer port (T), and the compressed mixture in the crankcase will then be transferred to the combustion-chamber side of the cylinder. The situation in the cylinder will then be such that the fresh charge entering the cylinder will push out any remaining burnt products of combustion – this process is generally referred to as cross-flow scavenging.

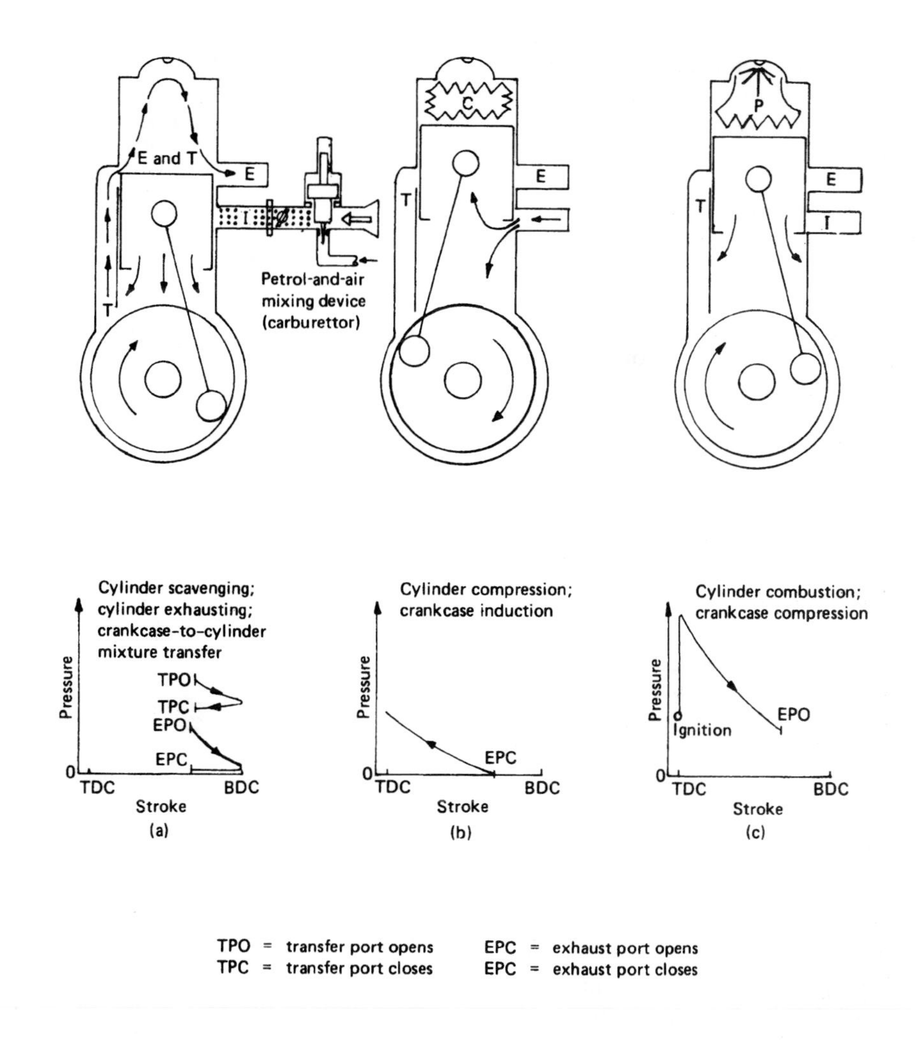

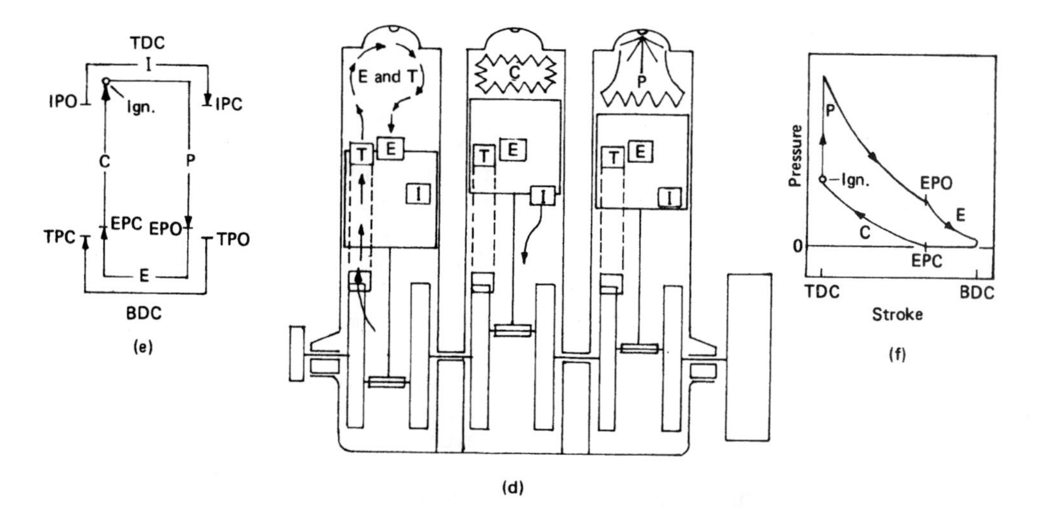

Fig. 10.5 Two-stroke-cycle petrol engine

Cylinder compression and crankcase induction (fig. 10.5(b)) The crankshaft rotates, moving the piston in the direction of the cylinder head. Initially the piston seals off the transfer port, and then a short time later the exhaust port will be completely closed. Further inward movement of the piston will compress the mixture of air and atomised petrol to about one seventh to one eighth of its original volume (fig. 10.5(b)).

At the same time as the fresh charge is being compressed between the combustion chamber and the piston head, the inward movement of the piston increases the total volume in the crankcase so that a depression is created in this space. About half-way up the cylinder stroke, the lower part of the piston skirt will uncover the inlet port (I), and a fresh mixture of air and petrol prepared by the carburettor will be induced into the crankcase chamber (fig. 10.5(b)).

Cylinder combustion and crankcase compression (fig. 10.5(c)) Just before the piston reaches the top of its stroke, a spark-plug situated in the centre of the cylinder head will be timed to spark and ignite the dense mixture. The burning rate of the charge will rapidly raise the gas pressure to a maximum of about 50 bar under full load. The burning mixture then expands, forcing the piston back along its stroke with a corresponding reduction in cylinder pressure (fig. 10.5(c)).

Considering the condition underneath the piston in the crankcase, with the piston initially at the top of its stroke fresh mixture will have entered the crankcase through the inlet port. As the piston moves down its stroke, the piston skirt will cover the inlet port, and any further downward movement will compress the mixture in the crankcase in preparation for the next charge transfer into the cylinder and combustion-chamber space (fig. 10.5(c)).

The combined cycle of events adapted to a three-cylinder engine is shown in fig. 10.5(d). Figure 10.5(e) and (f) show the complete cycle in terms of opening and closing events and cylinder volume and pressure changes respectively.

10.2.1 Reverse-flow (Schnuerle) scavenging (fig. 10.6)

To improve scavenging efficiency, a loop-scavenging system which became known as the reverse-flow or (after its inventor, Dr E. Schnuerle) as the Schnuerle scavenging system was developed (fig.

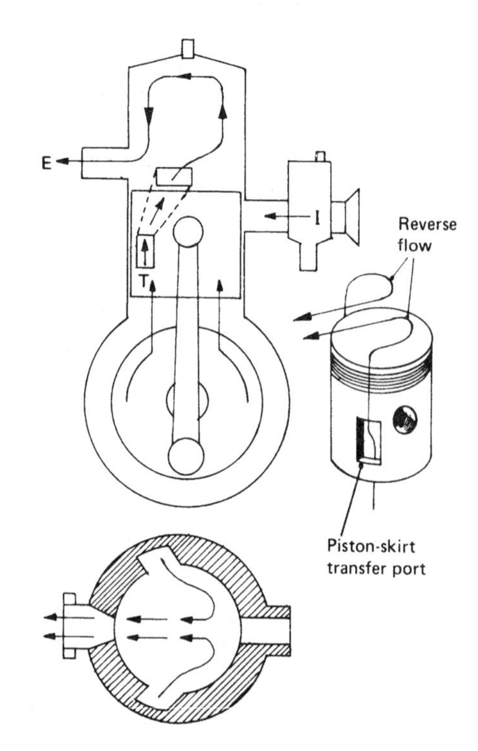

Fig. 10.6 Reverse flow or Schnuerle scavenging

10.6). This layout has a transfer port on each side of the exhaust port, and these direct the scavenging charge mixture in a practically tangential direction towards the opposite cylinder wall. The two separate columns of the scavenging mixture meet and merge together at this wall to form one inward rising flow which turns under the cylinder head and then flows down on the entry side, thus forming a complete loop. With this form of porting, turbulence and intermixing of fresh fuel mixture with residual burnt gases will be minimal over a wide range of piston speeds.

Note that in this particular design the charge mixture is transferred through ports formed in the piston skirt. Alternatively, extended transfer passages may be preferred so that the piston skirt plays no part in the timed transfer.

10.2.2 Crankcase disc-valve and reed-valve inlet charge control (fig. 10.7)

An alternative to the piston-operated crankcase inlet port is to use a disc-valve attached to and driven by the crankshaft (fig. 10.7(a)). This disc-valve is timed to open and close so that the fresh charge is induced to enter the crankcase as early as possible, and only at the point when the charge is about to be transferred into the cylinder is it

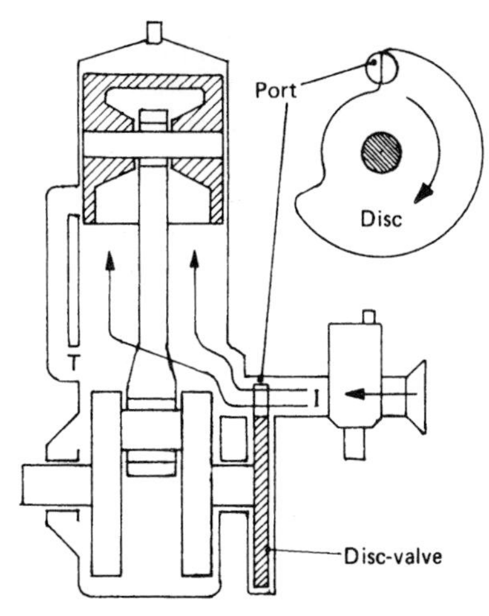

(a) Crankcase disc-valve induction

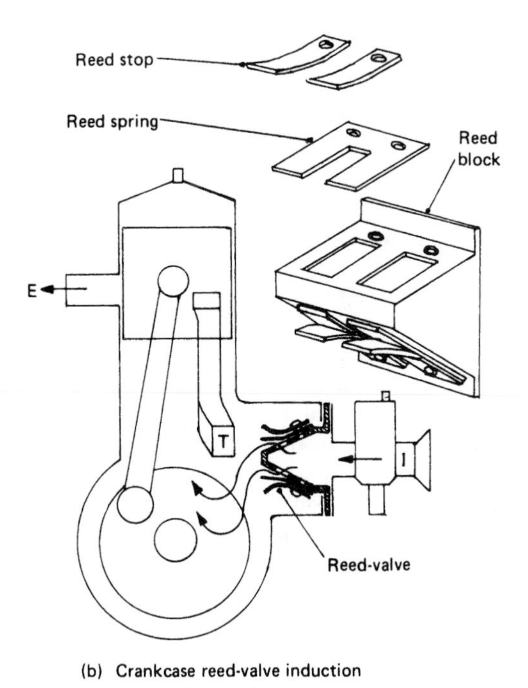

(b) Crankcase reed-valve induction

Fig. 10.7 Crankcase disc-valve and reed-valve induction

closed. This method of controlling crankcase induction does not depend upon the piston displacement to uncover the port – it can therefore be so phased as to extend the filling period.

A further method of improving crankcase filling is the use of reed-valves (fig. 10.7(b)). These valves are not timed to open and close, but operate automatically when the pressure difference between the crankcase and the air intake is sufficient to deflect the reed-spring. In other words, these valves sense the requirements of the crankcase and so adjust their opening and closing frequencies to match the demands of the engine.

10.2.3 Comparison of two- and four-stroke-cycle petrol engines

The following remarks compare the main points regarding the effectiveness of both engine cycles.

a) The two-stroke engine completes one cycle of events for every revolution of the crankshaft, compared with the two revolutions required for the four-stroke engine cycle.
b) Theoretically, the two-stroke engine should develop twice the power compared to a four-stroke engine of the same cylinder capacity.
c) In practice, the two-stroke engine's expelling of the exhaust gases and filling of the cylinder with fresh mixture brought in through the crankcase is far less effective than having separate exhaust and induction strokes. Thus the mean effective cylinder pressures in two-stroke units are far lower than in equivalent four-stroke engines.
d) With a power stroke every revolution instead of every second revolution, the two-stroke engine will run smoother than the four-stroke power unit for the same size of flywheel.
e) Unlike the four-stroke engine, the two-stroke engine does not have the luxury of separate exhaust and induction strokes to cool both the cylinder and the piston between power strokes. There is therefore a tendency for the piston and small-end to overheat under heavy driving conditions.
f) Due to its inferior scavenging process, the two-stroke engine can suffer from the following:
   i) inadequate transfer of fresh mixture into the cylinder,
   ii) excessively large amounts of residual exhaust gas remaining in the cylinder,
   iii) direct expulsion of fresh charge through the exhaust port.

   These undesirable conditions may occur under different speed and load situations, which greatly influences both power and fuel consumption.
g) Far less maintenance is expected with the two-stroke engine compared with the four-stroke engine, but there can be a problem with the products of combustion carburising at the inlet, transfer, and exhaust ports.

h) Lubrication of the two-stroke engine is achieved by mixing small quantities of oil with petrol in proportions anywhere between 1: 16 and 1: 24 so that, when crankcase induction takes place, the various rotating and reciprocating components will be lubricated by a petroil-mixture mist. Clearly a continuous proportion of oil will be burnt in the cylinder and expelled into the atmosphere to add to unwanted exhaust emission.
i) There are fewer working parts in a two-stroke engine than in a four-stroke engine, so two-stroke engines are generally cheaper to manufacture.

10.3 Four-stroke-cycle compression ignition (diesel) engine (fig. 10.8)

Compression-ignition (C.I.) engines burn fuel oil which is injected into the combustion chamber when the air charge is fully compressed. Burning occurs when the compression temperature of the air is high enough to spontaneously ignite the finely atomised liquid fuel. In other words, burning is initiated by the self-generated heat of compression.

Engines adopting this method of introducing and mixing the liquid fuel followed by self-ignition are also referred to as 'oil engines', due to the class of fuel burnt, or as 'diesel engines' after Rudolf Diesel, one of the many inventors and pioneers of the early compression-ignition engine. Note: in the United Kingdom fuel oil is known as 'DERV', which is the abbreviation of 'diesel-engine road vehicle'.

Just like the four-stroke-cycle petrol engine, the compression-ignition engine completes one cycle of events in two crankshaft revolutions or four piston strokes. The four phases of these strokes are (i) induction of fresh air, (ii) compression and heating of this air, (iii) injection of fuel and its burning and expansion, and (iv) expulsion of the products of combustion.

Induction stroke (fig. 10.8(a)) With the inlet valve open and the exhaust valve closed, the piston moves away from the cylinder head (fig. 10.8(a)).

The outward movement of the piston will establish a depression in the cylinder, its magnitude depending on the ratio of the cross-sectional areas of the cylinder and the inlet port and on the speed at which the piston is moving. The pressure difference established between the inside and outside of the cylinder will induce air at atmospheric pressure to enter and fill up the cylinder. Unlike the petrol engine, which requires a charge of air-and-petrol mixture to be drawn past a throttle valve, in the diesel-engine inlet system no restriction is necessary and only pure air is induced into the cylinder. A maximum depression of maybe 0.15 bar below atmospheric pressure will occur at about one third of the distance along the piston's outward stroke, while the overall average pressure in the cylinder might be 0.1 bar or even less.

Compression stroke (fig. 10.8(b)) With both the inlet and the exhaust valves closed, the piston moves towards the cylinder head (fig. 10.8(b)).

The air enclosed in the cylinder will be compressed into a much smaller space of anything from 1/12 to 1/24 of its original volume. A typical ratio of maximum to minimum air-charge volume in the cylinder would be 16:1, but this largely depends on engine size and designed speed range.

During the compression stroke, the air charge initially at atmospheric pressure and temperature is reduced in volume until the cylinder pressure is raised to between 30 and 50 bar. This compression of the air generates heat which will increase the charge temperature to at least 600°C under normal running conditions.

Power stroke (fig. 10.8(c)) With both the inlet and the exhaust valves closed and the piston almost at the end of the compression stroke (fig. 10.8(c)), diesel fuel oil is injected into the dense and heated air as a high-pressure spray of fine particles. Provided that they are properly atomised and distributed throughout the air charge, the heat of compression will then quickly vaporise and ignite the tiny droplets of liquid fuel. Within a very short time, the piston will have reached its innermost position and extensive burning then releases heat energy which is rapidly converted into pressure energy. Expansion then follows, pushing the piston away from the cylinder head, and the linear thrust acting on the piston end of the connecting-rod will then be changed to rotary movement of the crankshaft.

Exhaust stroke When the burning of the charge is near completion and the piston has reached the outermost position, the exhaust valve is opened. The piston then reverses its direction of motion and moves towards the cylinder head (fig. 10.8(d)).

The sudden opening of the exhaust valve

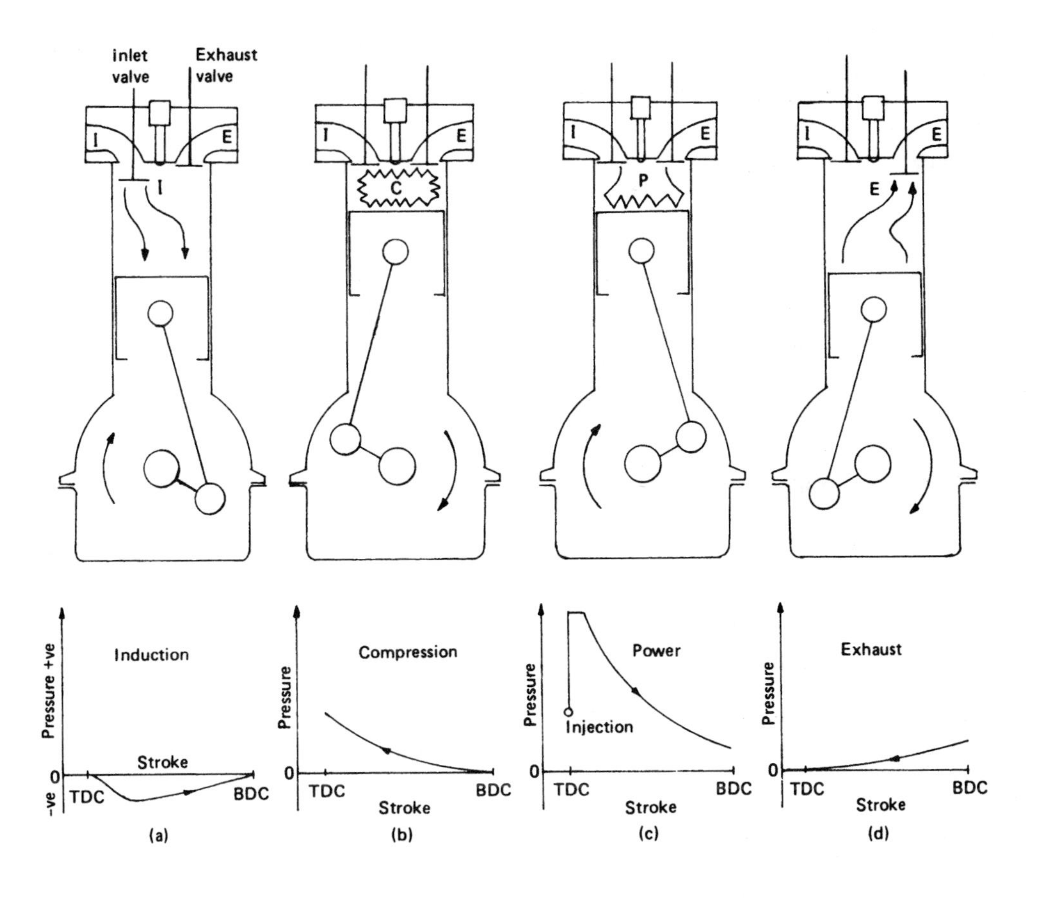

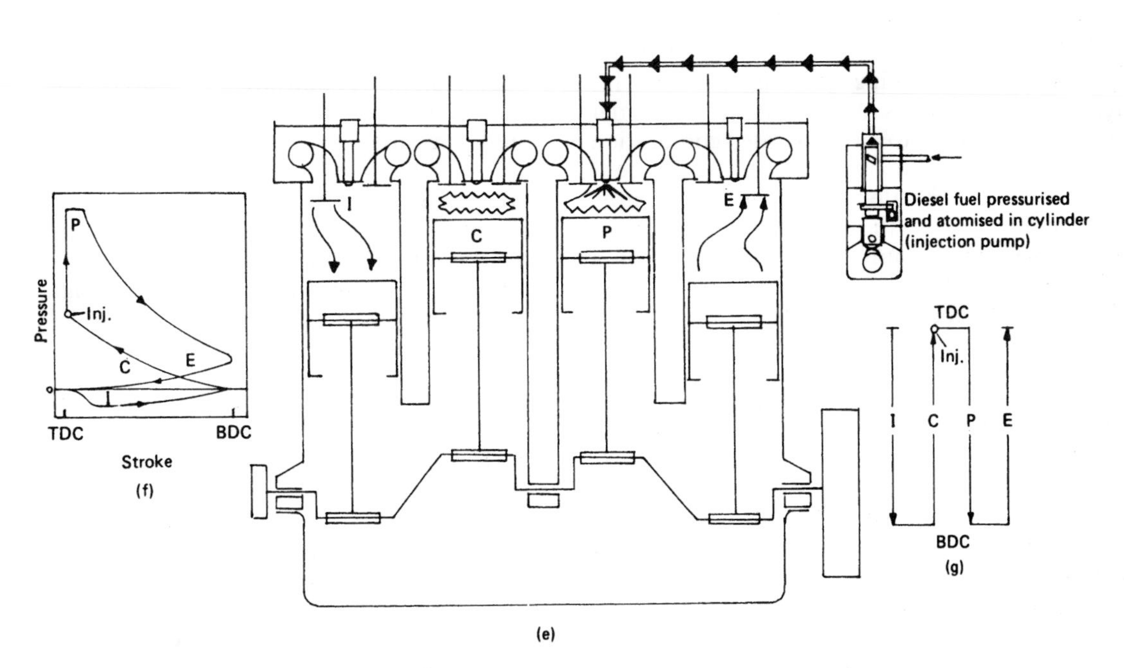

Fig. 10.8 Four-stroke-cycle diesel engine

towards the end of the power stroke will release the still burning products of combustion to the atmosphere. The pressure energy of the gases at this point will accelerate their expulsion from the cylinder, and only towards the end of the piston's return stroke will the piston actually catch up with the tail-end of the outgoing gases.

Figure 10.8(e) illustrates the sequence of the four operating strokes as applied to a four-cylinder engine, and the combined operating events expressed in terms of cylinder pressure and piston displacement are shown in fig. 10.8(f) and (g).

10.3.1 Historical background to the compression-ignition engine

Credit for the origination of the compression-ignition engine is controversial, as eminent engineers cannot agree among themselves as to which of the patents by Herbert Akroyd-Stuart or Rudolf Diesel contributed most to the instigation and evolution of the high-speed compression-ignition engine burning heavy fuel oil. A brief summary of the background and achievements of these two pioneers is as follows.

Herbert Akroyd-Stuart, born 1864, was trained as an engineer in his father's works at Fenny Stratford, England. Between 1885 and 1890 he took out several patents for improvements to oil engines, and later, in conjunction with a Charles R. Binney of London, he took out patent number 7146 of 1890 describing the operation of his engine. Air alone was drawn into the cylinder and compressed into a separate combustion chamber (known as the vaporiser) through a contracted passage or bottle-neck. A liquid fuel spray was then injected into the compressed air near the end of the compression stroke by means of a pump and a spraying nozzle. The combination of the hot chamber and the rise in temperature of the compressed air provided automatic ignition and rapid combustion at nearly constant volume – a feature of the compression-ignition engines of today.

These early engines were of low compression, the explosion taking place mainly due to the heat of the vaporiser chamber itself so that these engines became known as 'hot-bulb' or 'surface-ignition' engines. At starting, the separate combustion chamber was heated externally by an oil-lamp until the temperature attained was sufficient to ignite a few charges by compression. Then the chamber was maintained at a high enough temperature by the heat retained from the explosion together with the heat of the compressed air.

Rudolf Diesel was born in Paris in 1858, of German parents, and was educated at Augsburg and Munich. His works training was with Gebrüder Sulzer in Winterthur. Dr Diesel's first English patent, number 7421, was dated 1892 and was for an engine working on the ideal Carnot cycle and burning all kinds of fuel – solid, liquid, and gas – but the practical difficulties of achieving this thermodynamic cycle proved to be far too much. A reliable diesel oil engine was built in 1897 after four years of experimental work in the Mashinenfabrik Augsburg Nürnberg (MAN) workshops.

In this engine, air was drawn into the cylinder and was compressed to 35 to 40 bar. Towards the end of the compression stroke, an air blast was introduced into the combustion space at a much higher pressure, about 68 to 70 bar, thus causing turbulence in the combustion chamber. A three-stage compressor driven by the engine (and consuming about 10% of the engine's gross power) supplied compressed air which was stored in a reservoir. This compressed air served both for starting the engine and for air-injection into the compressed air already in the cylinder – that is, for blasting air to atomise the oil fuel by forcing it through perforated discs fitted around a fluted needle-valve injector. The resulting finely divided oil mist ignites at once when it contacts the hot compressed cylinder air, and the burning rate then tends to match the increasing cylinder volume as the piston moves outwards – expansion will therefore take place at something approaching constant pressure.

A summary of the combustion processes of Akroyd-Stuart and Diesel is that the former inventor used a low compression-ratio, employed airless liquid-fuel injection, and relied on the hot combustion chamber to vaporise and ignite the fuel; whereas Diesel employed a relatively high compression-ratio, adopted air-injection to atomise the fuel, and made the hot turbulent air initiate burning. It may be said that the modern high-speed compression-ignition engine embraces both approaches in producing sparkless automatic combustion – combustion taking place with a combined process of constant volume and constant pressure known as either the mixed or the dual cycle.

10.4 Two-stroke-cycle diesel engine

The pump scavenge two-stroke-cycle engine designed by Sir Dugald Clerk in 1879 was the first

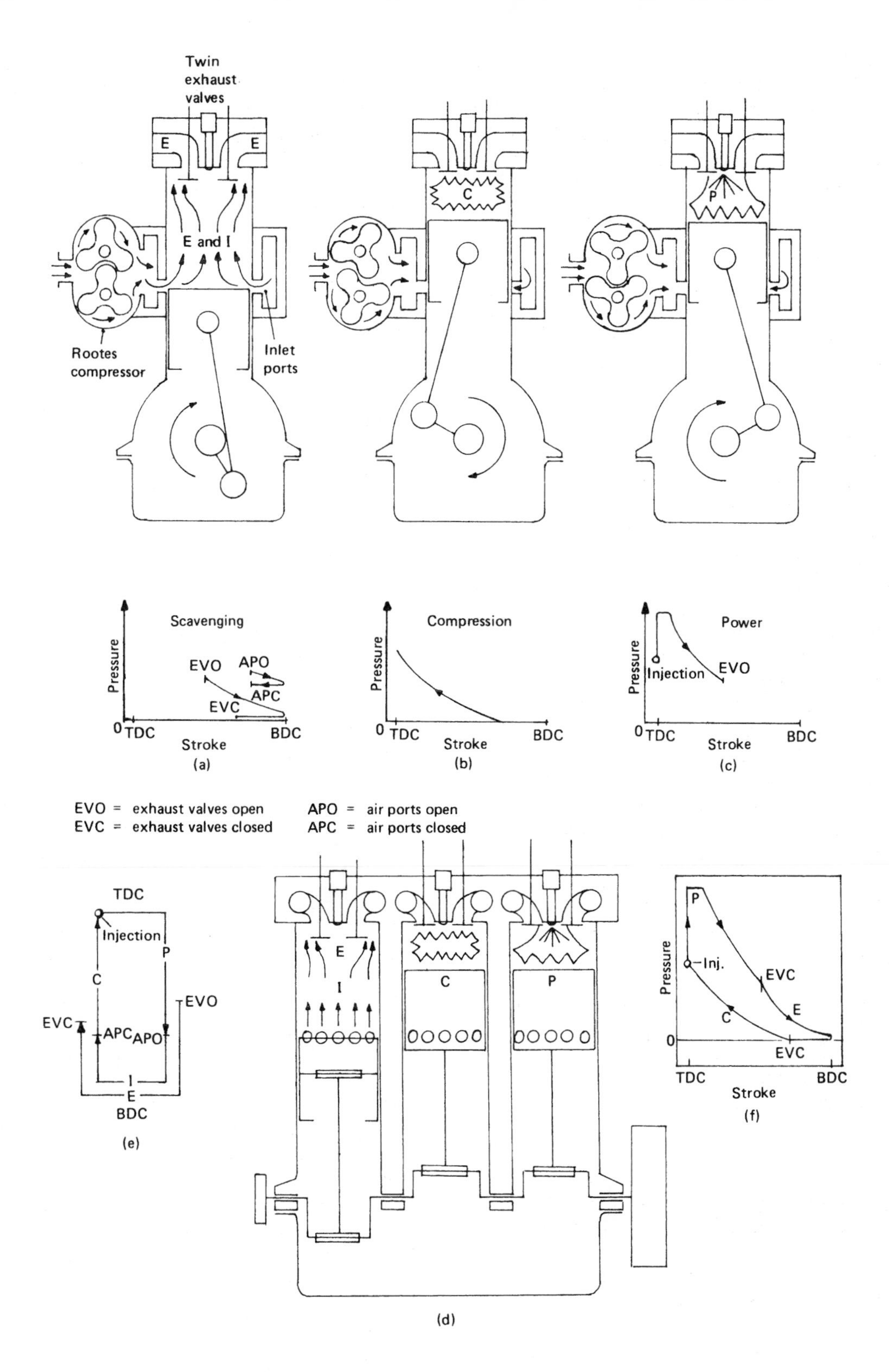

Fig. 10.9 Two-stroke-cycle diesel engine

successful two-stroke engine, thus the two-stroke-cycle engine is sometimes called the Clerk engine. Uniflow scavenging took place – fresh charge entering the combustion chamber above the piston while the exhaust outflow occurred through ports uncovered by the piston at its outermost position.

Low- and medium-speed two-stroke marine diesels still use this system, but high-speed two-stroke diesels reverse the scavenging flow by blowing fresh charge through the bottom inlet ports, sweeping up through the cylinder and out of the exhaust ports in the cylinder head (fig. 10.9(a)).

With the two-stroke-cycle engine, intake and exhaust phases take place during part of the compression and power stroke respectively, so that a cycle of operation is completed in one crankshaft revolution or two piston strokes. Since there are no separate intake and exhaust strokes, a blower is necessary to pump air into the cylinder for expelling the exhaust gases and to supply the cylinder with fresh air for combustion.

Scavenging (induction and exhaust) phase (fig. 10.9(a)) The piston moves away from the cylinder head and, when it is about half-way down its stroke, the exhaust valves open. This allows the burnt gases to escape into the atmosphere. Near the end of the power stroke, a horizontal row of inlet air ports is uncovered by the piston lands (fig. 10.9(a)). These ports admit pressurised air from the blower into the cylinder. The space above the piston is immediately filled with air, which now blows up the cylinder towards the exhaust valves in the cylinder head. The last remaining exhaust gases will thus be forced out of the cylinder into the exhaust system. This process of fresh air coming into the cylinder and pushing out unwanted burnt gas is known as scavenging.

Compression phase (fig. 10.9(b)) Towards the end of the power stroke, the inlet ports will be uncovered. The piston then reaches its outermost position and reverses its direction of motion. The piston now moves upwards so that the piston seals and closes the inlet air ports, and just a little later the exhaust valves close. Any further upward movement will now compress the trapped air (fig. 10.9(b)). This air charge is now reduced to about 1/15 to 1/18 of its original volume as the piston reaches the innermost position. This change in volume corresponds to a maximum cylinder pressure of about 30 to 40 bar.

Power phase (fig. 10.9(c)) Shortly before the piston reaches the innermost position to the cylinder head on its upward compression stroke, highly pressurised liquid fuel is sprayed into the dense intensely heated air charge (fig. 10.9(c)). Within a very short period of time, the injected fuel droplets will vaporise and ignite, and rapid burning will be established by the time the piston is at the top of its stroke. The heat liberated from the charge will be converted mainly into gas-pressure energy which will expand the gas and so do useful work in driving the piston outwards.

An overall view of the various phases of operation in a two-stroke-cycle three-cylinder diesel engine is shown in fig. 10.9(d), and fig. 10.9(e) and (f) show the cycle of events in one crankshaft revolution expressed in terms of piston displacement and cylinder pressure.

10.4.1 Comparison of two- and four-stroke-cycle diesel engines

A brief but critical comparison of the merits and limitations of the two-stroke-cycle diesel engine compared with the four-stroke power unit is made below.

a) Theoretically, almost twice the power can be developed with a two-stroke engine compared with a four-stroke engine.
b) A comparison between a typical 12 litre four-stroke engine and a 7 litre two-stroke engine having the same speed range would show that they would develop similar torque and power ratings. The ratio of engine capacities for equivalent performance for these four-stroke and two-stroke engines would be 1.7:1.
c) In a four-stroke engine, the same parts generate power and empty and fill the cylinders. With the two-stroke engine, the emptying and filling can be carried out by light rotary components.
d) With a two-stroke engine, 40 to 50% more air consumption is necessary for the same power output; therefore the air-pumping work done will be proportionally greater.
e) About 10 to 20% of the upward stroke of a two-stroke engine must be sacrificed to emptying and filling the cylinder.
f) The time available for emptying and filling a cylinder is considerably less in a two-stroke-cycle engine – something like 33% of the completed cycle as compared to 50% in a four-stroke engine. Therefore more power will be needed to force a greater mass of air into the cylinder in a shorter time.

g) Compared with a two-stroke engine, more power is needed by the piston for emptying and filling the cylinder in a four-stroke engine, due to pumping and friction losses at low speeds. At higher engine speeds the situation is reversed, and the two-stroke's Rootes blower will consume proportionally more engine power – this could be up to 15% of the developed power at maximum speed.
h) With reduced engine load for a given speed, a two-stroke engine blower will consume proportionally more of the power developed by the engine.
i) A two-stroke engine runs smoother and relatively quietly, due to the absence of reversals of loading on bearings as compared with a four-stroke engine.

10.5 Comparison of spark-ignition and compression-ignition engines

The pros and cons of petrol and compression-ignition engines are now considered.

Fuel economy The chief comparison to be made between the two types of engine is how effectively each engine can convert the liquid fuel into work energy. Different engines are compared by their thermal efficiencies. Thermal efficiency is the ratio of the useful work produced to the total energy supplied. Petrol engines can have thermal efficiencies ranging between 20 and 30%. The corresponding diesel engines generally have improved efficiencies, between 30 and 40%. Both sets of efficiency values are considerably influenced by the chosen compression-ratio and design.

Power and torque The petrol engine is usually designed with a shorter stroke and operates over a much larger crankshaft-speed range than the diesel engine. This enables more power to be developed towards the upper speed range in the petrol engine, which is necessary for high road speeds; however, a long-stroke diesel engine has improved pulling torque over a relatively narrow speed range, this being essential for the haulage of heavy commercial vehicles.

At the time of writing, there is a trend to incorporate diesel engines into cars. This new generation of engines has different design parameters and therefore does not conform to the above observations.

Reliability Due to their particular process of combustion, diesel engines are built sturdier, tend to run cooler, and have only half the speed range of most petrol engines. These factors make the diesel engine more reliable and considerably extend engine life relative to the petrol engine.

Pollution Diesel engines tend to become noisy and to vibrate on their mountings as the operating load is reduced. The combustion process is quieter in the petrol engine and it runs smoother than the diesel engine. There is no noisy injection equipment used on the petrol engine, unlike that necessary on the diesel engine.

The products of combustion coming out of the exhaust system are more noticeable with diesel engines, particularly if any of the injection equipment components are out of tune. It is questionable which are the more harmful: the relatively invisible exhaust gases from the petrol engine, which include nitrogen dioxide, or the visible smoky diesel exhaust gases.

Safety Unlike petrol, diesel fuels are not flammable at normal operating temperature, so they are not a handling hazard and fire risks due to accidents are minimised.

Cost Due to their heavy construction and injection equipment, diesel engines are more expensive than petrol engines.

10.6 Engine-performance terminology

To enable intelligent comparisons to be made between different engines' ability to pull or operate at various speeds, we shall now consider engine design parameters and their relationship in influencing performance capability.

10.6.1 Piston displacement or swept volume

When the piston moves from one end of the cylinder to the other, it will sweep or displace air equal to the cylinder volume between TDC and BDC. Thus the full stroke movement of the piston is known as either the swept volume or the piston displacement.

The swept or displaced volume may be calculated as follows:

$$V = \frac{\pi d^2 L}{4000}$$

where V = piston displacement (cm^3)
π = 3.142
d = cylinder diameter (mm)
and L = cylinder stroke (mm)

10.6.2 Mean effective pressure (m.e.p.)

The cylinder pressure varies considerably while the gas expands during the power stroke. Peak pressure will occur just after TDC, but this will rapidly drop as the piston moves towards BDC. When quoting cylinder pressure, it is therefore more helpful to refer to the average or mean effective pressure throughout the whole power stroke. The units used for m.e.p. may be either kilonewtons per square metre (kN/m^2) or bars (note: 1 bar = 100 kN/m^2).

10.6.3 Engine torque

This is the turning-effort about the crankshaft's axis of rotation and is equal to the product of the force acting along the connecting-rod and the perpendicular distance between this force and the centre of rotation of the crankshaft. It is expressed in newton metres (N m);

i.e. $T = Fr$
where T = engine torque (N m)
F = force applied to crank (N)
and r = effective crank-arm radius (m)

During the 180° crankshaft movement on the power stroke from TDC to BDC, the effective radius of the crank-arm will increase from zero at the top of its stroke to a maximum in the region of mid-stroke and then decrease to zero again at the end of its downward movement (fig. 10.10). This implies that the torque on the power stroke is continually varying. Also, there will be no useful torque during the idling strokes. In fact some of the torque on the power stroke will be cancelled out in overcoming compression resistance and pumping losses, and the torque quoted by engine manufacturers is always the average value throughout the engine cycle.

The average torque developed will vary over the engine's speed range. It reaches a maximum at about mid-speed and decreases on either side (fig. 10.11).

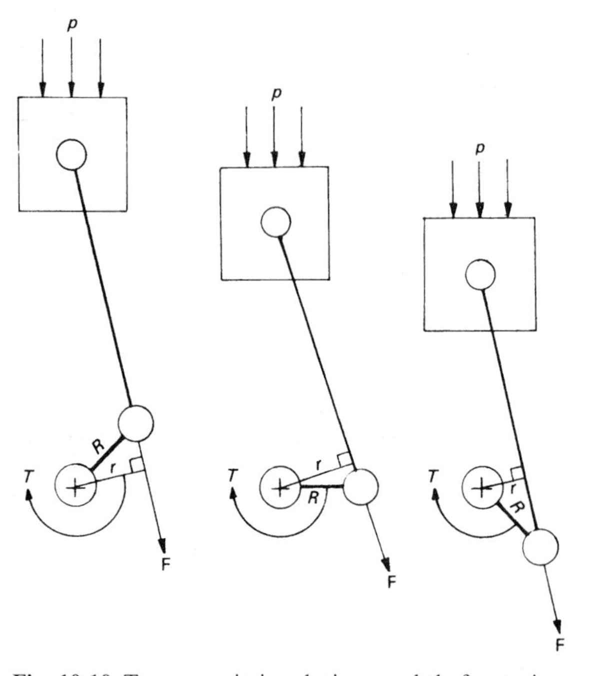

Fig. 10.10 Torque variation during crankshaft rotation (p = cylinder gas pressure; F = connecting-rod thrust; R = crank-throw; r = effective crank radius; T = turning-effort or torque)

10.6.4 Engine power

Power is the rate of doing work. When applied to engines, power ratings may be calculated either on the basis of indicated power (i.p.), that is the power actually developed in the cylinder, or on the basis of brake power (b.p.), which is the output power measured at the crankshaft. The brake power is always less than the indicated power, due to frictional and pumping losses in the cylinders and the reciprocating mechanism of the engine.

Since the rate of doing work increases with piston speed, the engine's power will tend to rise

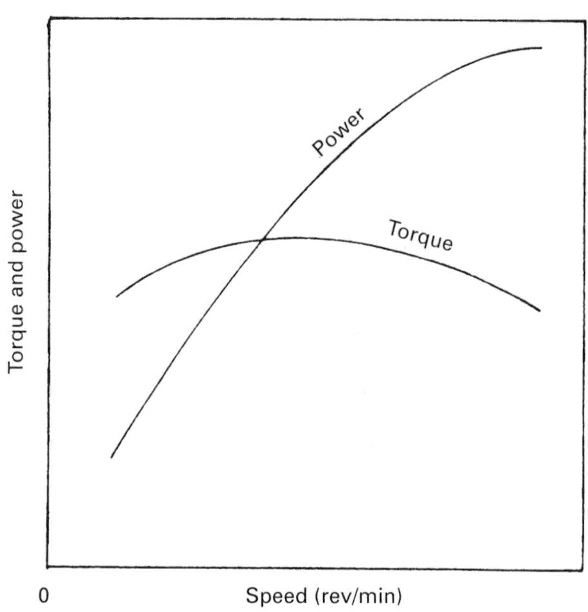

Fig. 10.11 Torque and power variation over engine speed range

with crankshaft speed of rotation, and only after about two thirds of the engine's speed range will the rate of power rise drop off (fig. 10.11).

The slowing down and even decline in power at the upper speed range is mainly due to the very short time available for exhausting and for inducing fresh charge into the cylinders at very high speeds, with a resulting reduction in the cylinders' mean effective pressures.

Different countries have adopted their own standardised test procedures for measuring engine performance, so slight differences in quoted output figures will exist. Quoted performance figures should therefore always state the standard used. The three most important standards are those of the American Society of Automotive Engineers (SAE), the German Deutsch Industrie Normale (DIN), and the Italian Commissione technica di Unificazione nell Automobile (CUNA).

The two methods of calculating power can be expressed as follows:

$$\text{i.p.} = \frac{pLANn}{60\,000}$$

where i.p. = indicated power (kW)
p = effective pressure (kN/m^2)
L = length of stroke (m)
A = cross-sectional area of piston (m^2)
N = crankshaft speed (rev/min)
and n = number of cylinders

$$\text{b.p.} = \frac{2\,\pi\,TN}{60\,000}$$

where b.p. = brake power (kW)
π = 3.142
T = engine torque (N m)
and N = crankshaft speed (rev/min)

The imperial power is quoted in horsepower (hp) and is defined in terms of foot pounds per minute. In imperial units one horsepower is equivalent to 33 000 ft lb per minute or 550 ft lb per second. A metric horsepower is defined in terms of Newton-metres per second and is equal to 0.986 imperial horsepower. In Germany the abbreviation for horsepower is PS derived from the translation of the words 'Pferd-Stärke' meaning horse strength.

The international unit for power is the watt, W or more usually the kilowatt, kW, where 1 kW = 1000 W.

Conversion from watt to hp and visa versa is:

1 kW = 1.35 hp and 1 hp = 0.746 kW

10.6.5 Engine cylinder capacity

Engine sizes are compared on the basis of total cylinder swept volume, which is known as engine cylinder capacity. Thus the engine cylinder capacity is equal to the piston displacement of each cylinder times the number of cylinders,

$$\text{i.e. } V_E = \frac{Vn}{1000}$$

where V_E = engine cylinder capacity (litre)
V = piston displacement (cm^3)
and n = number of cylinders

Piston displacement is derived from the combination of both the cross-sectional area of the piston and its stroke. The relative importance of each of these dimensions can be demonstrated by considering how they affect performance individually.

The cross-sectional area of the piston crown influences the force acting on the connecting-rod, since the product of the piston area and the mean effective cylinder pressure is equal to the total piston thrust;

i.e. $F = pA$
where F = piston thrust (kN)
p = mean effective pressure (kN/m^2)
and A = cross-sectional area of piston (m^2)

The length of the piston stroke influences both the turning-effort and the angular speed of the crankshaft. This is because the crank-throw length determines the leverage on the crankshaft, and the piston speed divided by twice the stroke is equal to the crankshaft speed;

$$\text{i.e. } N = \frac{v}{2L}$$

where N = crankshaft speed (rev/min)
v = piston speed (m/min)
and L = piston stroke (m)

This means that making the stroke twice as long doubles the crankshaft turning-effort and halves the crankshaft angular speed for a given linear piston speed.

The above shows that the engine performance is decided by the ratio of bore to stroke chosen for a given cylinder capacity.

10.7 Compression-ratio (fig. 10.12)

In an engine cylinder, the gas molecules are moving about at considerable speed in the space occupied by the gas, colliding with other molecules

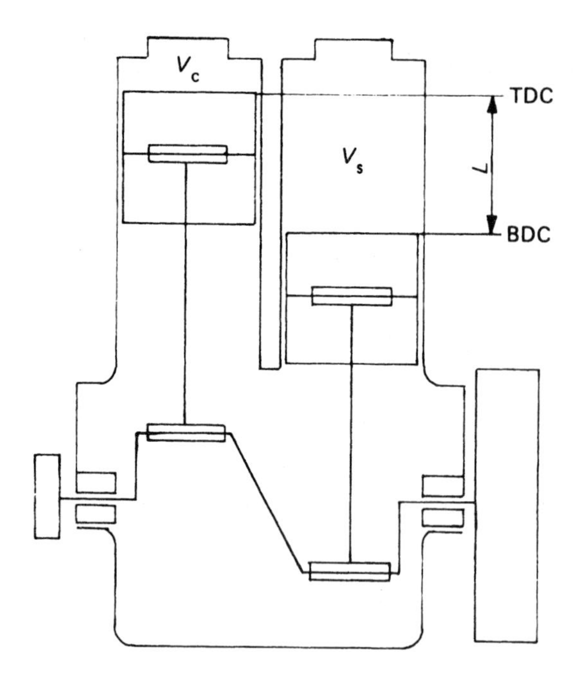

Fig. 10.12 Illustration of compression-ratio

and the boundary surfaces of the cylinder head, the cylinder walls, and the piston crown. The rapid succession of impacts of many millions of molecules on the boundary walls produces a steady continuous force per unit surface which is known as pressure.

When the gas is compressed into a much smaller space, the molecules are brought closer to one another. This raises the temperature and greatly increases the speed of the molecules and hence their kinetic energy, so more violent impulses will impinge on the piston crown. This increased activity of the molecules is experienced as increased opposition to movement of the piston towards the cylinder head.

The process of compressing a constant mass of gas into a much smaller space enables many more molecules to impinge per unit area on to the piston. When burning of the gas occurs, the chemical energy of combustion is rapidly transformed into heat energy which considerably increases the kinetic energy of the closely packed gas molecules. Therefore the extremely large number of molecules squeezed together will thus bombard the piston crown at much higher speeds. This then means that a very large number of repeated blows of considerable magnitude will strike the piston and so push it towards outer dead centre.

This description of compression, burning, and expansion of the gas charge shows the importance of utilising a high degree of compression before burning takes place, to improve the efficiency of combustion. The amount of compression employed in the cylinder is measured by the reduction in volume when the piston moves from BDC to TDC, the actual proportional change in volume being expressed as the compression-ratio.

The compression-ratio may be defined as the ratio of the maximum cylinder volume when the piston is at its outermost position (BDC) to the minimum cylinder volume (the clearance volume) with the piston at its innermost position (TDC) – that is, the sum of the swept and clearance volumes divided by the clearance volume,

i.e. $$CR = \frac{V_s + V_c}{V_c}$$

where CR = compression ratio
V_s = swept volume (cm^3)
V_c = clearance volume (cm^3)

Petrol engines have compression-ratios of the order of 7:1 to 10:1; but, to produce self-ignition of the charge, diesel engines usually double these figures and may have values of between 14:1 and 24:1 for naturally aspirated (depression-induced filling) types, depending on the design.

11 Multi-cylinder engine arrangements

11.1 The need for more than one cylinder

To increase the power and torque produced by an engine, the cylinder capacity has to be enlarged, but to scale up a single-cylinder engine to twice its original size would involve many difficulties.

A comparison will now be made between two single-cylinder engines, one having exactly twice the cylinder diameter and stroke dimensions of the other (fig. 11.1).

The volume of a cylinder is equal to the product of the piston's head or crown area and its stroke;

i.e. $V = AL = \dfrac{\pi d^2 L}{4}$

where V = cylinder volume
A = piston head area
L = piston stroke
and d = cylinder diameter

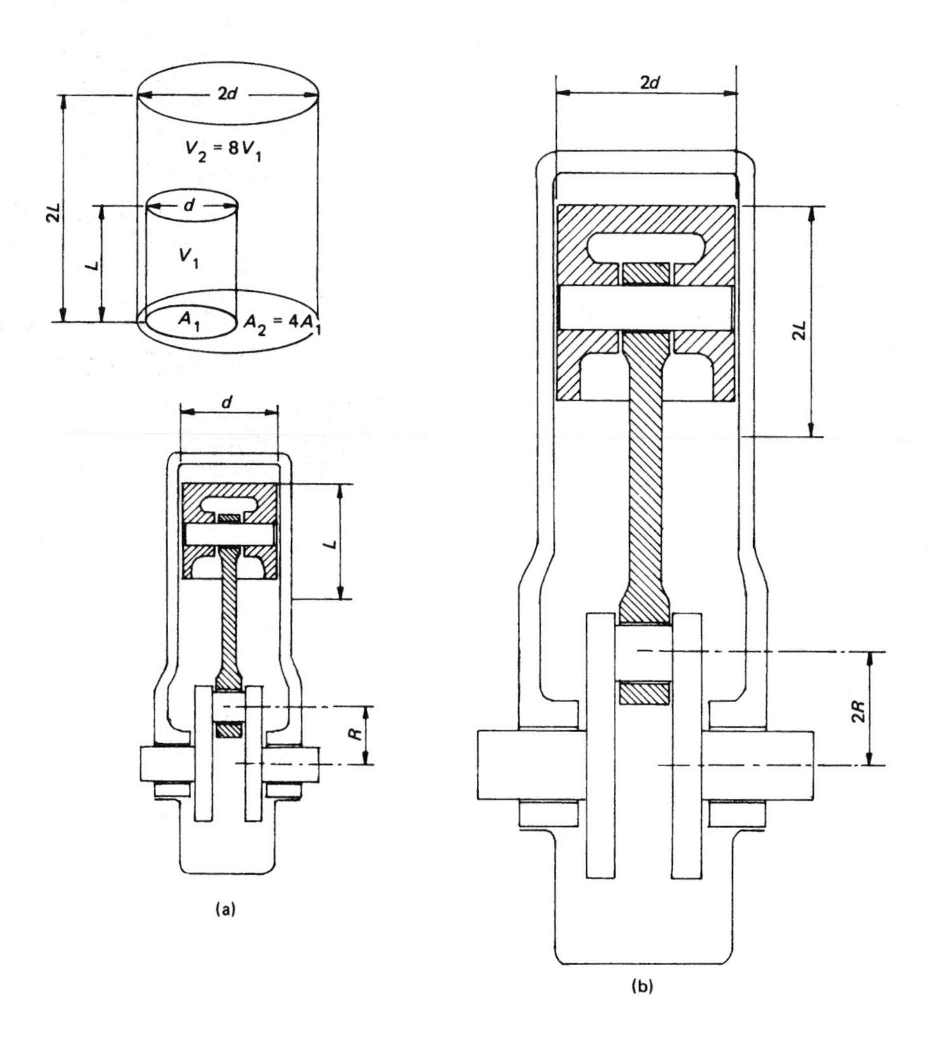

Fig. 11.1 Comparison of cross-sectional areas and volumes of engines (a) and (b)

The piston head area $A = \pi(d/2)^2$; therefore, doubling the diameter will increase the area fourfold. Also, the cylinder's capacity is proportional to the piston stroke, therefore making the piston stroke twice as long doubles the cylinder cubic capacity. Thus the net result of doubling both the cylinder diameter and the piston stroke will be to increase the cylinder's cubic capacity by 4 × 2 or eightfold; i.e., if the small engine has a 1 litre capacity, then the large engine with doubled dimensions will have a capacity of 8 litres.

These relative piston head areas and cylinder cubic capacities of two engines, one of which has twice the linear dimensions of the other, can be demonstrated as follows.

$$\text{Area ratio} = \frac{A_2}{A_1} = \frac{\pi(2d/2)^2}{\pi(d/2)^2} = 4$$

$$\therefore \quad A_2 = 4\ A_1$$

$$\text{Volume ratio} = \frac{V_2}{V_1} = \frac{\pi(2d/2)^2 2L}{\pi(d/2)^2 L} = 8$$

$$\therefore \quad V_2 = 8\ V_1$$

For the same mean effective cylinder gas pressure in both engines, the piston thrust will increase in proportion to the piston head area, therefore doubling the cylinder diameter will increase the piston thrust fourfold.

For a given piston speed and mean effective gas pressure, engine power will increase with the square of the cylinder diameter, therefore doubling the cylinder diameter increases the power fourfold. Conversely the volume and hence mass of the reciprocating components will increase with the cube of their dimensions, so doubling the piston dimensions will increase the mass eightfold, so that maximum piston speed would have to be reduced.

By doubling the piston stroke for a given crankshaft speed, the piston speed will also be doubled; therefore, to maintain the same piston speed for both engines, the crankshaft speed of the larger engine would have to be halved.

Finally, torque is proportional to the piston thrust and the crank-throw length; therefore doubling the piston diameter and stroke increases the piston thrust fourfold and doubles the crank-throw leverage, hence the torque will be increased by 4 × 2 or eightfold.

11.2 Cyclic torque and speed fluctuation

The four-stroke-cycle engine completes one cycle of operations in 720° or two revolutions of the crankshaft, and each of the four strokes corresponds to 180° or half a revolution of the crankshaft.

Out of the four strokes, only the power stroke supplies energy to drive the crankshaft against the various resisting loads, while the other three strokes – induction, compression, and exhaust – absorb some of this energy to overcome the pumping and frictional losses. In addition to these resisting loads, there are reciprocating inertia loads due to the reverse effort necessary to change the direction of motion of the piston assembly every time it reaches its innermost or outermost position. Therefore, because of the variation of useful cylinder pressure throughout the power stroke and the opposing friction, pumping, and inertia loads, there will be a marked fluctuation in crankshaft speed throughout one cycle.

11.2.1 The function of the flywheel

The flywheel attached to the end of the crankshaft performs the task of absorbing excess energy while the crankshaft is accelerating on its 180° of power stroke and of automatically transferring this stored kinetic energy to the crankshaft to overcome the turning resistance experienced over the next 540° during the three idle strokes.

The crankshaft will decelerate as the flywheel gives up energy to drive the crankshaft over the three non-working strokes, but a recovery of speed will follow as combustion accelerates the expansion movement of the piston on its working stroke. The flywheel will thus reduce but never completely eliminate crankshaft speed fluctuations during each cycle of events.

11.2.2 The flywheel effect and the cyclic-torque diagram

The flywheel effect may be understood by considering the actual crank effort or torque produced at any instant over one complete cycle of events. Sometimes the energy imparted to the flywheel and crankshaft exceeds the average resisting load in the engine and other times it can be well below this value, so the flywheel undergoes corresponding speed fluctuations (fig. 11.2).

The average or mean height of the torque diagram can be considered as the torque equivalent to the steady load imposed upon the engine. The shaded area above the average-torque line thus represents the excess energy stored in the flywheel, and the energy below this average line represents the energy which is drawn from the flywheel during one cycle.

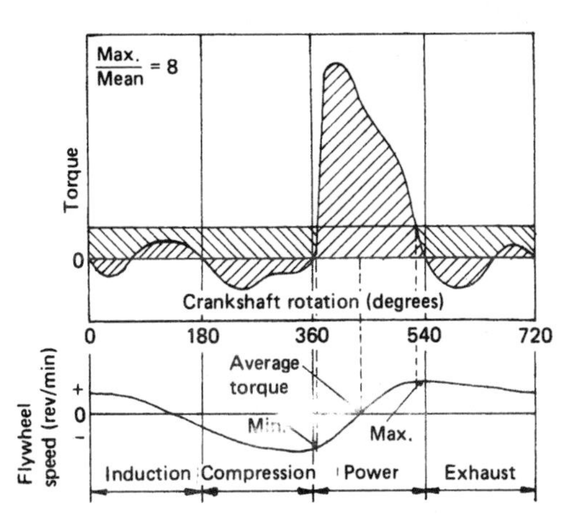

Fig. 11.2 Single-cylinder constant-load flywheel effect

At the beginning of the power impulse, the flywheel will just have stopped supplying energy to the crankshaft and therefore its speed will be at a minimum. Similarly, near the end of the power stroke the flywheel will have just completed receiving surplus energy from the crankshaft and so its speed will be at a maximum. For this cycle of events to continue, the excess and deficiencies of energy must be equal, thus the kinetic energy gained by the flywheel when its speed increases must equal that given back to the system as its speed declines.

The size of the flywheel will determine the degree of speed variation over each cycle. A large flywheel will reduce the speed fluctuation to a minimum, so that the engine will tend to run smoothly at constant speeds. However, a large flywheel mass will oppose any rapid acceleration or deceleration of the engine which might be required – this is because more energy will have to be absorbed, therefore the engine's response will be sluggish. Conversely, a small flywheel will enable the engine to respond to demands of rapid speed changes as might be necessary when driving, but this will be at the expense of uneven and lumpy slow-speed operation.

11.2.3 Multi-cylinder cyclic-torque diagrams

A single-cylinder engine has a power stroke lasting 180° and three idle pumping strokes absorbing power over an interval of 540°, the total angular movement for one cycle of events being 720°. Thus the useful turning-effort produced and the torque absorbed in overcoming the engine load resistances will cause a relatively large fluctuation in crankshaft speed throughout one cycle. The limitations on the size of the flywheel and its ability to smooth out the torque unevenness between cycles to the standard demanded has been largely solved by using multi-cylinder engines in which the single crankshaft and the valve timing are so arranged that the power strokes of the cylinders are phased to take place in sequence, rather than all at the same time.

If the number of cylinders is increased, the intervals between power impulses will be reduced. Consequently, the torque variation throughout the four strokes of the cycle will be smoothed out and there will be insufficient time for the cyclic speed to vary greatly above or below the mean crankshaft speed.

A study of the cyclic-torque curve for the single-cylinder engine (fig. 11.2) will show a firing stroke every 720° with a variation of peak to mean torque over one cycle of around 8:1. By adding a second cylinder, the interval between the firing impulses will be halved to 360°. This reduces the peak to mean torque produced over a cycle by something like half, that is to 4:1 (fig. 11.3(a)).

The addition of a third cylinder brings the interval between the firing impulses to 240°, and there will now be only a 60° interval of crankshaft rotation without a power period. The peak to mean turning-effort will now be considerably smoothed out to something of the order of 2.8:1 (fig. 11.3(b)).

With a fourth cylinder, the firing-impulse intervals are 180° and it can be seen that the power periods are now end-on with each other. This again improves variation of peak to mean torque, a typical value being 2:1 (fig. 11.3(c)).

The five-, six-, and eight-cylinder engines have firing intervals of 144°, 120°, and 90° degrees respectively, with corresponding power periods overlapping by 36°, 60°, and 90°. The respective ratios of peak to mean torque are reduced to 1.7:1, 1.4:1, and 1.1:1 (fig. 11.3(d) to (f)).

11.3 Merits and limitations of single- and multi-cylinder engines

The following is a summary of the major factors which have to be considered when comparing engines of different cubic capacity and various number of cylinders.

a) The shorter the piston stroke, the higher can be the crankshaft speed for a given maximum piston speed.
b) The smaller the cylinder, the lighter will be the

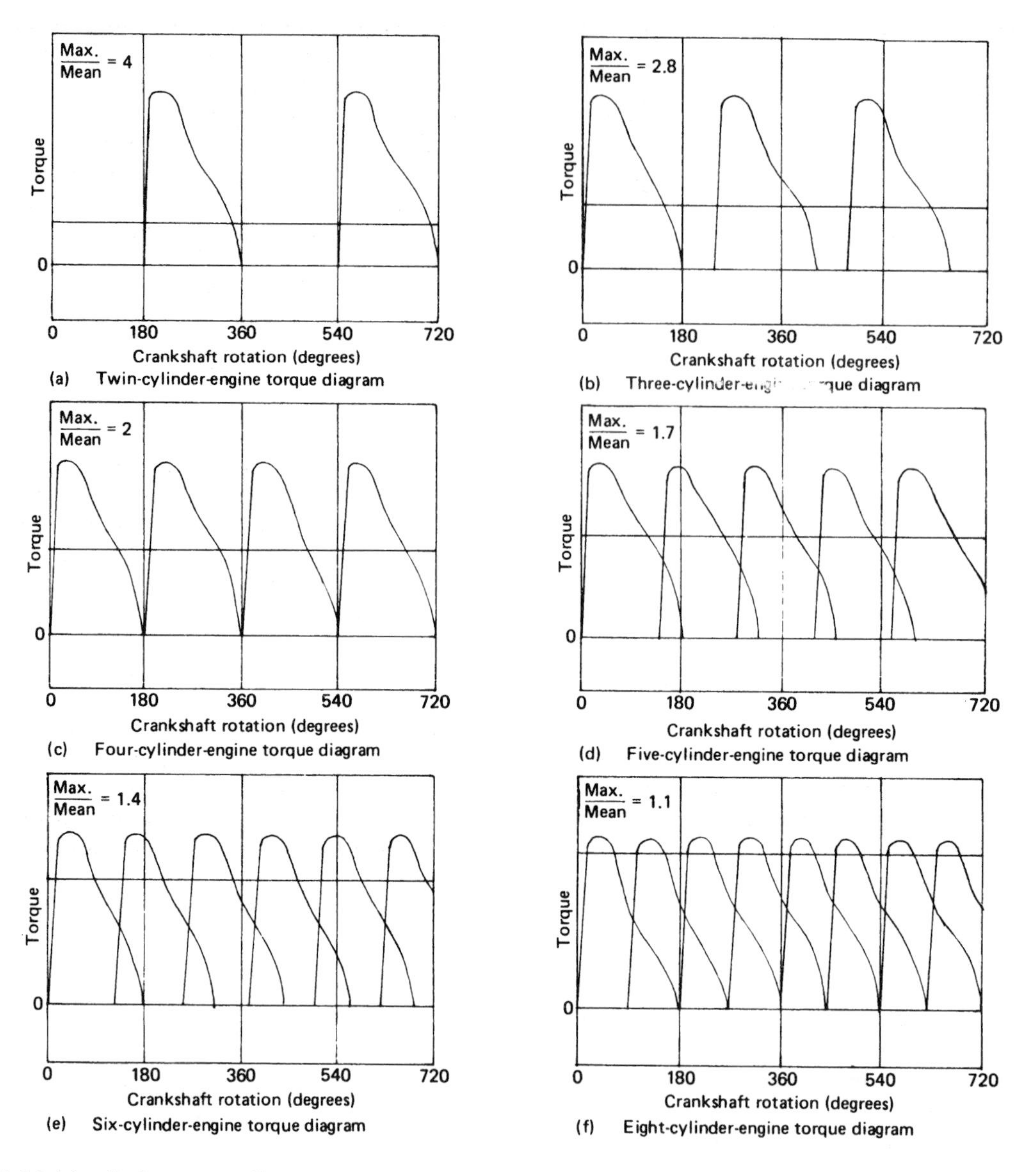

Fig. 11.3 Multi-cylinder torque diagrams

piston in proportion to the cylinder size, so that the limiting inertia forces will permit higher piston speeds to be used.

c) For the same engine cylinder capacity and maximum piston speed, a multi-cylinder engine will produce more power than a single-cylinder engine.
d) A single-cylinder engine having the same piston cross-sectional area as the sum of the piston areas of a multi-cylinder engine will develop a greater torque output.
e) The smaller the cylinder, the greater will be its surface-to-volume ratio, therefore higher compression-ratios are permissible due to the improved cooling of the cylinder, with a subsequent improvement in engine thermal efficiency.
f) Acceleration response improves with the number of cylinders for a given total volume. This is due both to the lighter reciprocating components and to the reduced reliance made on the flywheel so that a smaller flywheel may be adopted.
g) As the number of cylinders and the engine

length increase, torsional vibrations become a problem.

h) As the number of cylinders increases, there will be more power consumed in overcoming rotational and reciprocating drag.

i) As the number of cylinders increases, mixture distribution for carburetted engines becomes more difficult.

j) As the number of cylinders increases, the cost of duplication of components becomes proportionally higher.

k) As the number of cylinders increases, the frequency of power impulses increases too, therefore the power output is more consistent.

11.4 Cylinder firing orders

The sequence in which cylinders are made to fire is chosen to improve the distribution of the fresh charge in the manifold to the cylinders and to aid the release of the exhaust gases, while at the same time suppressing objectionable and dangerous torsional vibrations. These considerations are now discussed as follows.

a) It is best to have successive cylinders firing as far apart as possible. This allows a recovery of charge in the manifold, i.e. complete filling, and will also minimise interference between adjacent or nearby cylinders which may have overlapping induction periods. The usual rule is to arrange that cylinders from opposite ends of the manifold (or from alternate cylinder banks in 'V' engines) draw alternately, but this becomes increasingly more difficult as the number of cylinders is reduced.

b) Similarly on the exhaust side of events: if the exhaust periods overlap with cylinders which are close together, exhaust-gas back-pressure may prevent the products of combustion escaping from the cylinders. Separating or distancing successive cylinders which are exhausting is therefore equally or even more important than for induction.

c) Power impulses tend to wind up the crankshaft and, if the natural torsional oscillations of the shaft happen to coincide with these disturbing impulse frequencies, torsional vibrations may occur. In general, it is better to apply successive power impulses to alternate ends of the crankshaft, but with some crankshafts successive firing of adjacent cylinders prevents relative unwinding, thereby suppressing certain critical torsional vibrations.

11.5 Single-cylinder arrangement (fig. 11.4)

A single-cylinder engine will have a power stroke every 720°/1 = 720° of crankshaft rotation for a four-stroke-cycle engine.

This engine has simply a single-throw crank-arm, and the rotating big-end journal or crankpin is linked to the piston gudgeon-pin by means of a connecting-rod which will have both a linear and an oscillating motion (fig. 11.4).

When the piston is at the top of its stroke, the piston will either be completing its upward (compression) stroke and about to begin its power stroke, or it will be at the end of its exhaust stroke and beginning its downward induction stroke.

Assuming that initially the piston is at TDC on the point of beginning its power stroke, then 180° of crankshaft rotation will move the piston down to the bottom of the cylinder, combustion taking place during this phase. A second 180° of crankshaft rotation moves the piston upward to the top of the cylinder, sweeping the cylinder clear of exhaust products. A third 180° of crankshaft rotation moves the piston downward on its induction stroke, drawing the fresh charge into the cylinder. Finally, a fourth 180° rotation of the crankshaft returns the piston to the top of its stroke, compressing the charge in readiness for the next power stroke. Thus four piston strokes and 720° of crankshaft rotation complete one cycle of events.

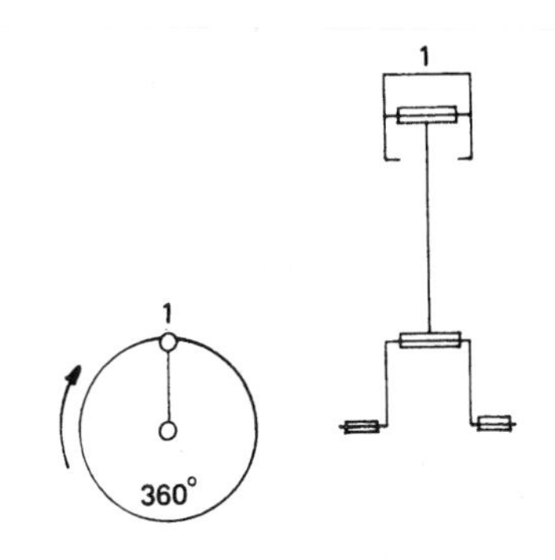

Firing order: 1

Cylinder number	1st revolution		2nd revolution	
	0 — 180°	180° — 360°	360° — 540°	540° — 720°
	TDC — BCD	BCD — TDC	TDC — BDC	BDC — TDC
1	I	C	P	E

Power stroke every 720°

Fig. 11.4 Single-cylinder arrangement

11.6 In-line side-by-side twin-cylinder arrangement (fig. 11.5)

A side-by-side twin-cylinder engine will have a power impulse every 720°/2 = 360° of crankshaft rotation.

The crankshaft consists of a single-throw crank-arm with both pistons and connecting-rods attached to a common big-end journal or crank-pin. Thus the pistons are in phase and positioned parallel and adjacent to each other (fig. 11.5).

When a piston approaches TDC, it is either on the compression or exhaust stroke, therefore if piston 1 is on the top of its compression stroke, piston 2 will be on its exhaust stroke. Crankshaft rotation will cause the pistons to descend. Piston 1 will then be on its power stroke, so piston 2 must be on its induction stroke.

With 180° of crankshaft rotation, both pistons will be at the bottom of their stroke – piston 1 will have completed power and will be about to start its exhaust stroke, and piston 2 will have just finished induction and will be beginning its compression stroke.

A second 180° of crankshaft rotation will bring pistons 1 and 2 to TDC to begin their induction and power strokes respectively.

A third 180° rotation of the crankshaft will move the pistons to TDC. Piston 1 will be completing its exhaust stroke and about to start on its induction stroke, while piston 2 will be at the end of its compression stroke and about to begin its power stroke.

Finally, rotation of the crankshaft through a fourth 180° completes the 720° required for the four-stroke cycle.

11.7 In-line 180°-out-of-phase twin-cylinder arrangement (fig. 11.6)

With this twin-cylinder arrangement there will be uneven intervals between power impulses, which take place every 180° and 540° of crankshaft displacement.

The cylinders are positioned parallel and adjacent to each other (fig. 11.6), and the crankpins are so arranged that when piston 1 is at the top of its stroke, piston 2 is at the bottom of its stroke. The crank-throws are thus 180° out of phase with each other.

With piston 1 at the top of its stroke, piston 2 will be at the bottom. If initially piston 1 is at the end of compression and the beginning of power, piston 2 will therefore be at the end of power and at the beginning of its exhaust stroke.

Rotation of the crankshaft through 180° brings piston 1 to BDC, completing its power stroke and beginning its exhaust stroke. Piston 2 will then be at TDC, at the end of exhaust and about to descend on its induction stroke.

A second 180° of crankshaft rotation will move pistons 1 and 2 to TDC and BDC respectively.

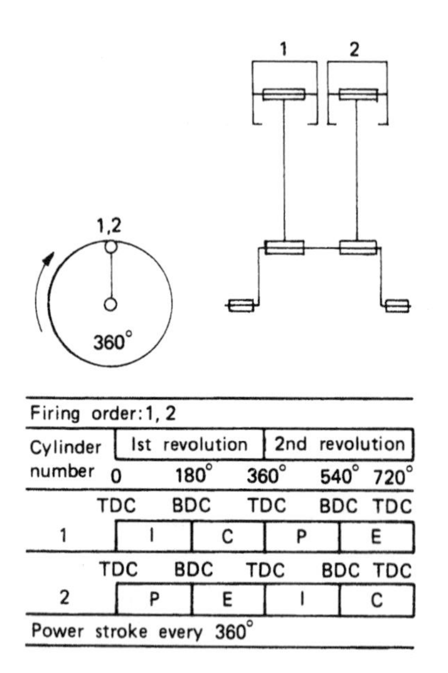

Fig. 11.5 In-line side-by-side twin-cylinder arrangement

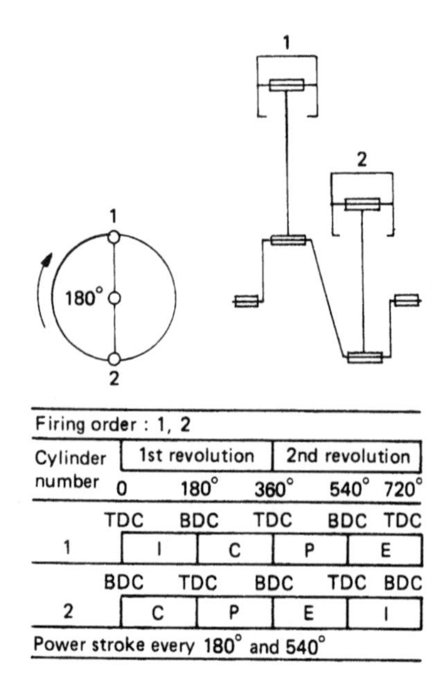

Fig. 11.6 In-line 180°-out-of-phase twin-cylinder arrangement

Piston 1 will then be at the end of exhaust and at the beginning of induction, while piston 2 will just be completing its induction stroke and beginning its compression up-stroke.

The third 180° rotation of the crankshaft will bring piston 1 to BDC, ending induction and starting its compression stroke, while piston 2 will be at the end of compression ready for the next power stroke.

Finally, a fourth 180° of crankshaft rotation moves piston 1 to TDC, at the end of its compression stroke and the beginning of its power stroke again, and piston 2 will move down to BDC to complete its power stroke and hence a complete cycle of events.

11.8 Horizontally opposed twin-cylinder arrangement (fig. 11.7)

This crankshaft-and-cylinder configuration provides power impulses at even intervals of every 360° of crankshaft movement.

With this layout, the crank-throws are 180° out of phase with each other. The connecting-rods and pistons are thus arranged to be positioned on opposite sides of the crankshaft, horizontally opposed, with the cylinder axes offset to each other. Therefore, although the pistons will be moving in opposite directions at all times, they will approach TDC and BDC positions together (fig. 11.7).

Assuming piston 1 is at TDC at the end of its compression stroke and beginning its power stroke, then piston 2 will also be at TDC, finishing exhaust and starting its induction stroke.

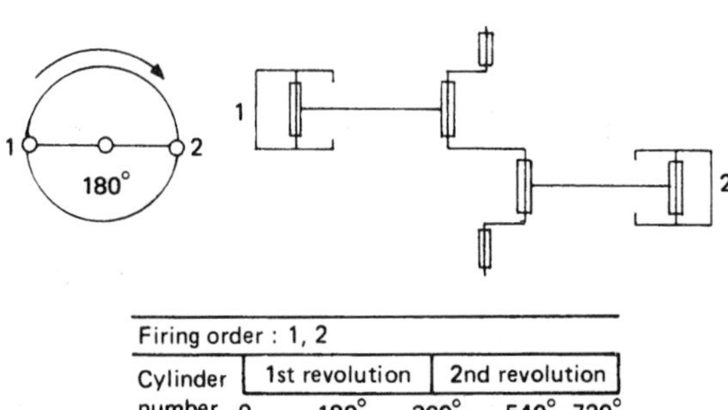

Firing order : 1, 2

Cylinder number	1st revolution		2nd revolution	
	0–180°	180°–360°	360°–540°	540°–720°
	TDC–BDC	BDC–TDC	TDC–BDC	BDC–TDC
1	I	C	P	E
	TDC–BDC	BDC–TDC	TDC–BDC	BDC–TDC
2	P	E	I	C

Power stroke every 360°

Fig. 11.7 Horizontally opposed twin-cylinder arrangement

Rotation of the crankshaft through 180° brings piston 1 to BDC at the end of its power stroke and at the beginning of its exhaust stroke, while piston 2 will also be at BDC having completed induction and about to advance on its compression stroke.

A second 180° of crankshaft rotation will again bring pistons 1 and 2 to TDC position. Piston 1 will then have completed its exhaust phase and will be at the beginning of its induction stroke, while piston 2 will be at the end of compression and just beginning its power stroke.

A third 180° crankshaft movement will move both pistons to BDC position. Piston 1 will then be at the end of its induction stroke and about to begin its compression stroke, and piston 2 will have moved to the end of its power stroke and will be about to begin its exhaust sweeping action.

To complete the cycle of events, the crankshaft will then move through a fourth 180° angular movement to complete two revolutions of the crankshaft.

These engines have been used in small motor cars having cylinder capacities up to about 750 cm^3 and are preferred by some motorcycle manufacturers.

11.9 In-line three-cylinder arrangement (fig. 11.8)

A three-cylinder engine will have a power impulse every 720°/3 = 240° of crankshaft rotation for the four-stroke cycle of operation.

The crank-throws and crankpins are spaced at intervals of 120°, and, to support the crankshaft, four main journals and bearings are provided – one at each end of the crankshaft and two intermediate journals between crankpins 1 and 2 and 2 and 3.

Consider the firing-order of the in-line three-cylinder arrangement shown in fig. 11.8. With piston 1 at the top of the compression stroke and at the start of its power stroke, pistons 2 and 3 will be 60° crank-angle movement from BDC on their induction and exhaust strokes respectively.

Rotation of the crankshaft through 120° brings piston 3 to TDC at the end of its exhaust stroke and at the beginning of its induction stroke, and pistons 1 and 2 will then be 60° from BDC on their power and compression strokes respectively.

A second 120° of crankshaft movement will move piston 2 to TDC, thus completing compression in readiness for its power stroke. Consequently, pistons 1 and 3 will have taken up positions 60° from BDC on their respective exhaust and induction strokes.

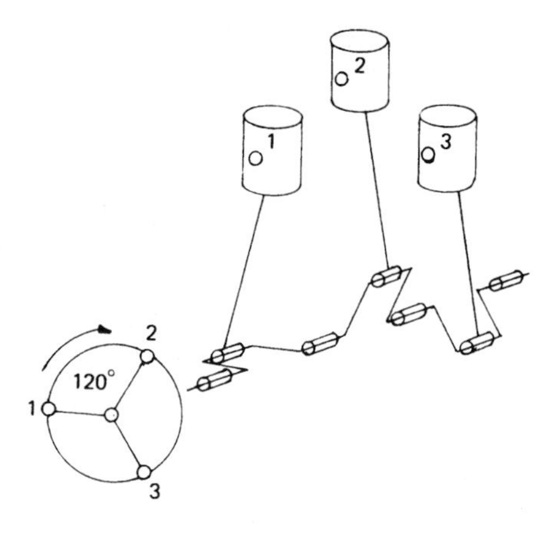

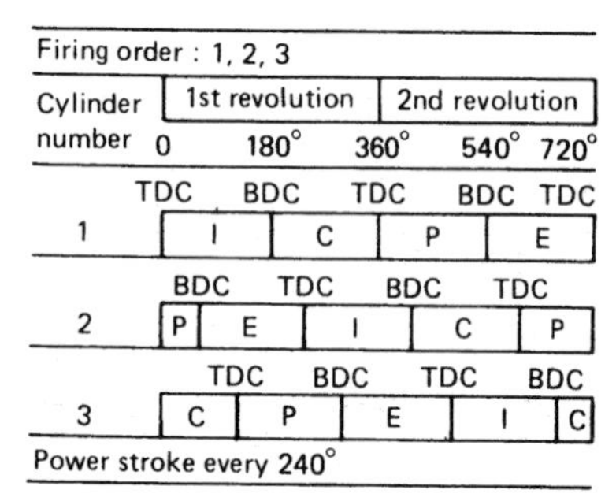

Firing order : 1, 2, 3

Cylinder number	1st revolution (0–180°)	(180°–360°)	2nd revolution (360°–540°)	(540°–720°)
1 (TDC BDC TDC BDC TDC)	I	C	P	E
2 (BDC TDC BDC TDC)	P / E	E / I	I / C	C / P
3 (TDC BDC TDC BDC)	C / P	P / E	E / I	I / C

Power stroke every 240°

Fig. 11.8 In-line three-cylinder arrangement

A further 120° of rotation moves piston 1 to TDC so that it will just be ending exhaust and about to begin its induction stroke. Piston 2 and 3 will now have taken up positions 60° from BDC on their respective power and compression strokes.

Finally a fourth 120° of crank rotation brings piston 3 on its compression stroke to TDC, so that the following downward event would be power. This sequence of events results in a firing order of 1, 2, 3.

These engines are balanced for dynamic shake but there are some unbalanced rocking forces which cannot be eliminated. It is claimed that the extra cylinder smooths out the cyclic torque sufficiently to make the engine a competitor to small engines with the popular four-cylinder configuration. The advantages are savings in weight and length, reduced reciprocating and rotational drag, and (because of the latter) improved fuel consumption.

11.10 In-line four-cylinder arrangement (fig. 11.9)

A four-cylinder in-line engine will have a

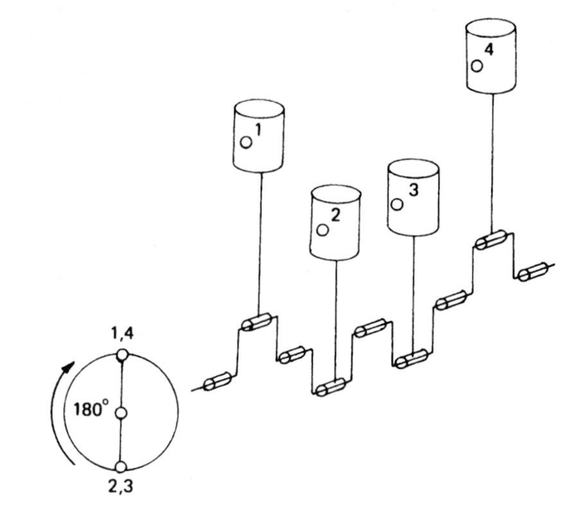

Firing order : 1, 3, 4, 2

Cylinder number	1st revolution (0–180°)	(180°–360°)	2nd revolution (360°–540°)	(540°–720°)
1 (TDC BDC TDC BDC TDC)	I	C	P	E
2 (BDC TDC BDC TDC BDC)	C	P	E	I
3 (BDC TDC BDC TDC BDC)	E	I	C	P
4 (TDC BDC TDC BDC TDC)	P	E	I	C

Power stroke every 180°

Fig. 11.9 In-line four-cylinder arrangement

power impulse every 720°/4 = 180° of crankshaft rotation.

The crankshafts must have crank-throws spaced at intervals of 180° to each other in the order in which it is intended the power impulses should occur (fig. 11.9). With the crankshaft arrangement adopted, it will be seen that all four crank-throws lie in one plane, crankpins 1 and 4 being in phase and at 180° to crankpins 2 and 3.

To support the crankshaft there is a series of either five or three main journals and bearings. If a main journal is provided between all adjacent crankpins, there would thus be five bearings, but sometimes only three main journals and bearings are provided – at each end of the crankshaft and between crankpins 2 and 3.

When considering the firing-order of a four-cylinder in-line engine, it can be seen that when a crankpin and piston are approaching inner dead centre the piston is on either an exhaust stroke or a compression stroke. Therefore assuming crankpin 1 is at the top of a compression stroke, crankpin 4 must be at the top of an exhaust stroke. Further crankshaft rotation will cause them to descend on a power stroke and on an induction stroke respectively.

Rotation of the crankshaft through 180° will position big-ends 1 and 4 at the bottom of their strokes, while big-ends 2 and 3 will be at the top of their strokes after either a compression or an exhaust stroke. It will therefore be obvious that the power stroke can be made to occur in either of these cylinders. Let it be assumed that piston 3 is arranged to be the next to descend on a power stroke, while piston 2 descends on an induction stroke. The order of firing is then 1, 3.

A second 180° of rotation will again bring crankpins and pistons 1 and 4 to the top of a stroke. As piston 1 previously descended on a power stroke and piston 4 on an induction stroke, they will now be at the top of their exhaust and power strokes respectively, so that at this point the order of firing will be 1, 3, 4.

A third crankshaft rotation of 180° will again bring crankpins 2 and 3 to the top of their strokes. As piston 3 previously descended on a power stroke, piston 2 will now provide the power impulse, so that the complete order of firing will be 1, 3, 4, 2.

A final 180° of rotation will complete 720° of crankshaft displacement – that is, two revolutions or four strokes – in readiness for the next cycle of events.

If cylinder 2 was selected instead of cylinder 3 to fire after cylinder 1, then the firing-order would be 1, 2, 4, 3. Both these firing-orders – 1, 3, 4, 2 and 1, 2, 4, 3 – have equal merits and limitations with regard to crankshaft torsional wind-up and the uneven spacing of breathing intervals between adjacent cylinders.

For engine capacities from 0.75 to 2.0 litres, the 'in-line four' is the most popular type of engine. The major components are in dynamic balance, and the number of power impulses provides a smooth torque drive.

11.11 Horizontally opposed flat four-cylinder arrangement (fig. 11.10)

With this arrangement, the crankshaft and cylinders are arranged so that pairs of cylinders are positioned horizontally opposed but offset to each other on each side of the crankshaft. This requires a single-plane crankshaft with crankpins spaced at intervals of 180°, therefore the crank-throws are paired so that crankpins 1 and 4 are diametrically opposite crankpins 2 and 3 (fig. 11.10). There may be three or five main journals and bearings provided to support the crankshaft.

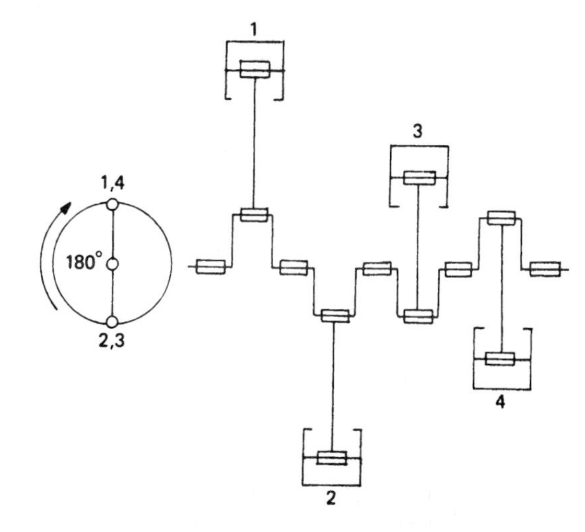

Firing order : 1, 4, 2, 3

Cylinder number	1st revolution (0–180°)	1st revolution (180°–360°)	2nd revolution (360°–540°)	2nd revolution (540°–720°)
1 (TDC BDC TDC BDC TDC)	I	C	P	E
2 (TDC BDC TDC BDC TDC)	P	E	I	C
3 (BDC TDC BDC TDC BDC)	C	P	E	I
4 (BDC TDC BDC TDC BDC)	E	I	C	P

Power stroke every 180°

Fig. 11.10 Horizontally opposed flat four-cylinder arrangement

Considering the firing-order, when pistons 1 and 2 are at TDC, pistons 3 and 4 will be at BDC. Selecting piston 1 to be at the end of its compression stroke and just on the point of firing, then piston 2 will be completing exhaust, while pistons 3 and 4 will be at BDC completing their respective power and induction strokes.

Rotating the crankshaft 180° brings pistons 3 and 4 to TDC at the end of their respective exhaust and compression strokes. Piston 4 will then be on the point of firing, while pistons 1 and 2 will be at BDC completing their respective power and induction strokes. The order of firing is then 1, 4.

A second 180° of rotation (a total of 360°) brings pistons 1 and 2 to TDC, at the end of their respective exhaust and compression strokes, while pistons 3 and 4 will be at BDC completing their respective induction and power strokes. The order of firing is then 1, 4, 2.

A third 180° of rotation (a total of 540°) brings pistons 3 and 4 to TDC at the end of their respective compression and exhaust strokes, while pistons 1 and 2 will be at BDC completing their respective induction and power strokes. The complete order of firing is then 1, 4, 2, 3.

A final 180° of rotation will complete 720° of crankshaft displacement – that is, two revolutions or four strokes – in readiness for the next cycle of events.

The 'flat four' has slightly better dynamic balance than the 'in-line-four' engine, and the smoothness of torque is equal to the latter's. Its low flat shape makes it particularly suitable for rear-mounted engines, but the opposed cylinders allow very little room for cylinder-head servicing.

11.12 In-line five-cylinder arrangement (fig. 11.11)
An in-line five-cylinder engine will have a power impulse every 720°/5 = 144° of crankshaft rotation.

The crank arrangement is such that there are five crank-throws, all in separate planes spaced at intervals of 72° relative to each other.

The crankshaft may be designed with a main journal and bearing at each end and between each pair of crankpins, thus making a six-main-journal crankshaft. Alternatively, the main journals between crankpins 1 and 2 and 4 and 5 may be removed to provide a shorter four-main-journal crankshaft with slightly reduced support.

Consider the firing-order of a five-cylinder engine with the crankshaft arranged as shown in fig. 11.11. With piston 1 at TDC at the end of compression stroke and just about to begin its power stroke, pistons 4 and 5 will be 72° crank-angle movement from TDC on their induction and exhaust strokes respectively, and pistons 2 and 3 will be 36° crankangle movement from BDC on their respective compression and power strokes.

Rotation of the crankshaft through 144° will position piston 2 at the top of its compression stroke and at the beginning of power. In this crank position, pistons 3 and 5 will be 72° from TDC on their respective exhaust and induction strokes. At the same time, pistons 1 and 4 will be 36° from BDC on their respective power and compression strokes.

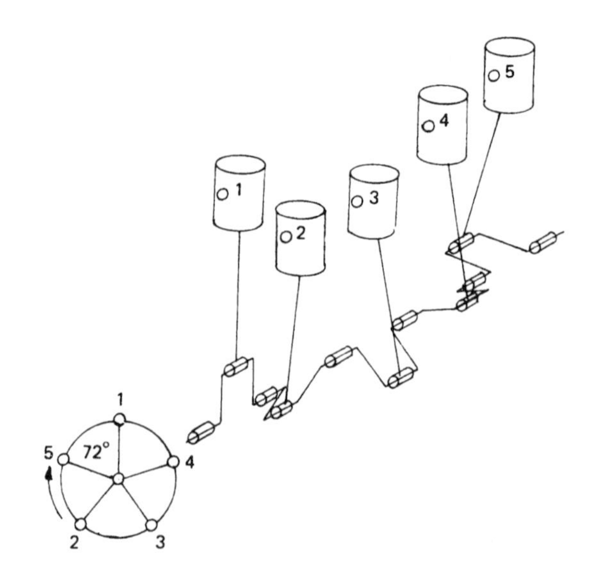

Firing order: 1, 2, 4, 5, 3,

Cylinder number	1st revolution (0 – 180° – 360°) / 2nd revolution (360° – 540° – 720°)
	TDC BDC TDC BDC TDC
1	I C P E
	TDC BDC TDC BDC
2	E I C P E
	BDC TDC BDC TDC
3	I C P E I
	BDC TDC BDC TDC
4	P E I C P
	TDC BDC TDC BDC
5	C P E I C

Power stroke every 144°

Fig. 11.11 In-line five-cylinder arrangement

A second 144° rotation of the crankshaft, brings piston 4 to the top, completing compression and about to begin its power stroke. Pistons 1 and 3 will then be 72° from TDC on their respective exhaust and induction strokes, and pistons 2 and 5 will likewise be 36° from BDC on their respective power and compression strokes.

A third 144° rotation of the crank moves piston 5 to TDC, to the end of compression, and the beginning of its power stroke. At this point, pistons 1 and 2 will be 72° from TDC on their respective induction and exhaust strokes, and pistons 3 and 4 will be 36° from BDC on their respective compression and power strokes.

A fourth 144° rotation of the crankshaft moves piston 3 to maximum compression at the top of

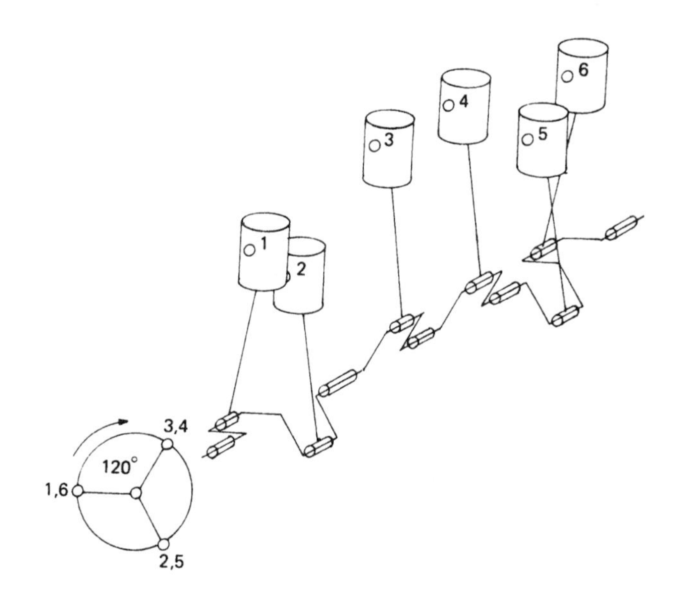

Firing order : 1, 5, 3, 6, 2, 4,

Cylinder number	Dead centres	Strokes (1st revolution 0–360°, 2nd revolution 360°–720°)
1	TDC BDC TDC BDC TDC	I, C, P, E
2	TDC BDC TDC BDC	C, P, E, I, C
3	BDC TDC TDC BDC	P, E, I, C, P
4	BDC TDC BDC TDC	I, C, P, E, I
5	TDC BDC TDC BDC	E, I, C, P, E
6	TDC BDC TDC BDC TDC	P, E, I, C

Power stroke every 120°

Fig. 11.12 In-line six-cylinder arrangement

its stroke in readiness for its downward power stroke. Pistons 2 and 4 will then be on their induction and exhaust strokes respectively, and pistons 1 and 5 will be on their compression and power strokes respectively. This sequence of events provides a firing-order of 1, 2, 4, 5, 3.

A final 144° of rotation will complete 720° of crankshaft displacement – that is, two revolutions or four strokes – ready for the next cycle to begin.

With an odd number of five cylinders, the spacing of the crank-throws ensures that the pistons do not all stop and start together at the top and bottom of each stroke, as they do on a four-cylinder engine. This arrangement therefore provides a very smooth drive.

11.13 In-line six-cylinder arrangement (fig. 11.12)
A six-cylinder in-line engine will have a power impulse every 720°/6 = 120° of crankshaft rotation.

The crankshaft consists of six crank-throws disposed at 120° out of phase with one another. This means that the crank-throws can be arranged in only three planes, therefore the crankpin phasing is arranged in pairs. Thus the axis of any one crankpin if extended will be coincident with the axis of another crankpin further along the shaft (fig. 11.12).

A series of main journals and bearings supports the crankshaft. For heavy-duty diesel engines, journals are provided at each end and between adjacent crankpins – there would thus be seven journals and bearings. For petrol engines it is more usual to dispense with some of the intermediate main journals, so that five or even only four main journals are provided. With the five-main-journal crankshaft, the main journals between crankpins 1 and 2 and 5 and 6 are removed. For the four-main-journal crankshaft the centre main journal between crankpins 3 and 4 is also removed. This removal of main journals reduces the overall length of the crankshaft but imposes additional loading on the remaining main journals.

Consider the firing-order of a six-cylinder engine with the crankshaft arrangement shown in fig. 11.12. With piston 1 at the top of the compression stroke, its opposite piston 6 is at the top of its exhaust stroke.

Rotating the crankshaft through 120° brings pistons 2 and 5 to their TDC position, where it can be arranged for either one of these to have completed a compression stroke. If piston 5 is made to be at the end of compression and at the beginning of its power stroke in the TDC position, then piston 2 must be at TDC on the exhaust stroke.

Rotating the crankshaft another third of a revolution moves it a total of 240° and brings pistons 3 and 4 to the TDC position, so that either one of these can be on the compression stroke. If piston 3 is arranged to be on compression, piston 4 must be on its exhaust stroke.

To complete the first half of the cycle of events, the crankshaft is moved a third 120°. Pistons 1 and 6 are now back again to TDC position. This time round, piston 6 is arranged to be on the compression stroke, so piston 1 must therefore be on its exhaust stroke.

Continuing on the second revolution, the next pair of pistons to reach TDC will be pistons 2 and 5, after the crankshaft has completed 480° of rotation. Piston 2 will now be on compression and piston 5 will be on its exhaust stroke.

Rotating the crankshaft another 120°, that is a total of 600°, brings pistons 3 and 4 to TDC. Piston 4 will then be on compression and piston 3 must therefore be on its exhaust stroke.

Further rotation of the crankshaft through 120° completes the 720° required for the four-stroke cycle and brings the pistons into position for the cycle of operations to be repeated. This cycle of events provides a firing-order of 1, 5, 3, 6, 2, 4.

If the phasings of paired crank-throws 3 and 4 and 2 and 5 are interchanged, then a second equally suitable firing-order of 1, 4, 2, 6, 3, 5 is obtained.

This arrangement has excellent dynamic balance and evenness of torque, and is preferred for engines larger than 2.5 litres if length is not a prime consideration.

11.14 Horizontally opposed flat six-cylinder arrangement (fig. 11.13)

This is a six-cylinder engine with three cylinders arranged in a horizontal plane on each side of the crankshaft. The power impulses are timed as for the in-line six-cylinder arrangement, that is every 120° of crankshaft rotation. The crankshaft has six crankpins spaced at 60° intervals around the crankshaft, and there are usually five main journals and bearings. Pairs of pistons – one from each cylinder bank – will reach TDC and then BDC simultaneously (fig. 11.13).

Consider pistons 1 and 2 at TDC with piston 1 at the end of compression and about to fire and piston 2 at the end of its exhaust stroke. Pistons 3, 4, 5, and 6 will then be at 60° from BDC on their exhaust, compression, induction, and power strokes respectively.

Rotation of the crankshaft by 120° brings pistons 3 and 4 to TDC at the end of their respective exhaust and compression strokes. Pistons 1, 2, 5, and 6 will then be at 60° from BDC on their respective power, induction, compression, and exhaust strokes. The order of firing is now 1, 4.

A second 120° rotation brings pistons 5 and 6 to TDC, completing their respective compression and exhaust strokes. Pistons 1, 2, 3, and 4 will then be at 60° from BDC on their respective exhaust, compression, induction, and power strokes. The order of firing is now 1, 4, 5.

A third 120° rotation brings pistons 1 and 2 to TDC again, completing their respective exhaust and compression strokes. Pistons 3, 4, 5, and 6 will then be at 60° from BDC on their respective compression, exhaust, power, and induction strokes. The order of firing is now 1, 4, 5, 2.

A fourth 120° rotation brings pistons 3 and 4 to

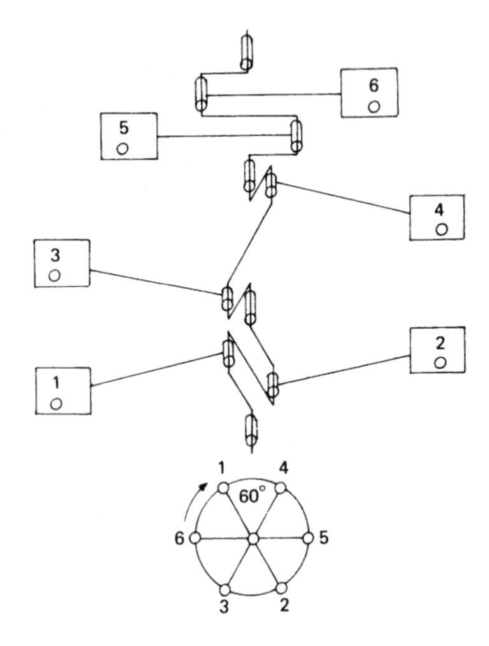

Firing order : 1, 4, 5, 2, 3, 6

Cylinder number	Dead centres (0 – 180° – 360° – 540° – 720°; 1st revolution, 2nd revolution)	Strokes
1	TDC BDC TDC BDC TDC	I C P E
2	TDC BDC TDC BDC TDC	P E I C
3	TDC BDC TDC BDC	C P E I C
4	TDC BDC TDC BDC	E I C P E
5	BDC TDC BDC TDC	P E I C P
6	TDC BDC TDC BDC	I C P E I

Power stroke every 120°

Fig. 11.13 Horizontally opposed flat six-cylinder arrangement

TDC again, completing their respective compression and exhaust strokes. Pistons 1, 2, 5, and 6 will then be at 60° from BDC on their respective induction, power, exhaust, and compression strokes. The order of firing is now 1, 4, 5, 2, 3.

A fifth 120° rotation brings pistons 5 and 6 to TDC again, completing their respective exhaust and compression strokes. Pistons 1, 2, 3, and 4 will then be at 60° from BDC on their respective compression, exhaust, power, and induction strokes. The complete firing order is now 1, 4, 5, 2, 3, 6.

A final rotation of 120° will complete 720° of crankshaft displacement – that is, two revolutions or four strokes – ready for the next cycle to begin.

Like the in-line six-cylinder arrangement, this layout is extremely well balanced; however, its flat wide configuration makes it difficult to install at the front or rear of the car without impairing access to the cylinder assemblies.

11.15 In-line straight eight-cylinder arrangement (fig. 11.14)

'In-line eights' have a power impulse every 720°/8 = 90° of crankshaft rotation.

The crankshafts have crank-throws spaced at intervals of 90° to each other in the order in which it is intended the power impulses should occur (fig. 11.14). This means that there can only be four relative angular positions, therefore the crankpin phasing is arranged in pairs. The crank-throws therefore lie in two planes. There may be either five or nine main journals provided to support the crankshaft. The layout shown resembles a four-cylinder crankshaft in one plane with twin cranks at either end forming a second plane at right angles to the first, so that it is sometimes known as a 'split-four in-line eight'.

Consider pistons 1 and 8 to be at TDC, with piston 1 at the end of compression at the point of firing and piston 8 at the end of its exhaust stroke. Pistons 3 and 6 will then be at mid-stroke on exhaust and compression strokes respectively; pistons 2 and 7 at BDC at the end of induction and power strokes respectively; and pistons 4 and 5 at mid-stroke on power and induction strokes respectively.

Rotating the crankshaft 90° brings pistons 3 and 6 to TDC at the end of their respective exhaust and compression strokes. Pistons 2 and 7 will then be at mid-stroke on compression and exhaust strokes respectively; pistons 4 and 5 at BDC at the end of power and induction strokes respectively; and pistons 1 and 8 at mid-stroke on power and induction strokes respectively. The firing-order after 90° of rotation is 1, 6.

A second 90° rotation brings pistons 2 and 7 to TDC at the end of their respective compression

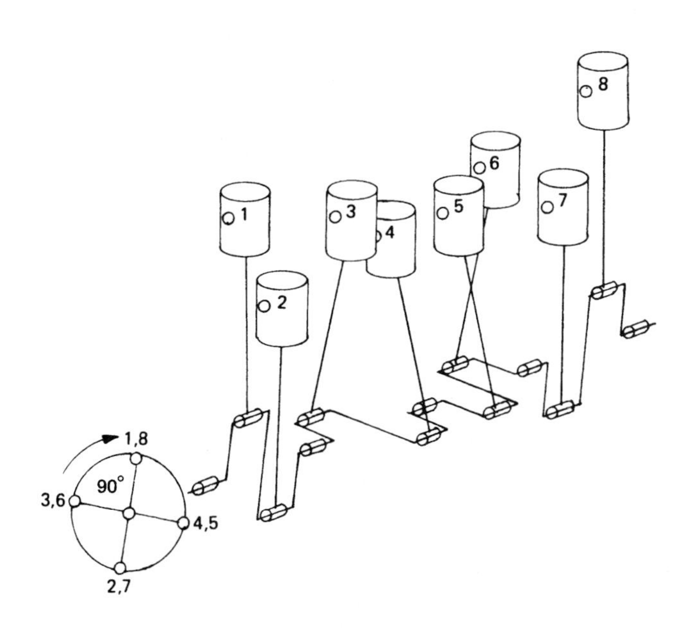

Firing order : 1, 6, 2, 5, 8, 3, 7, 4

Cylinder number	1st revolution (0–180°)	(180°–360°)	2nd revolution (360°–540°)	(540°–720°)
1 (TDC BDC TDC BDC TDC)	I	C	P	E
2 (BDC TDC BDC TDC BDC)	E	I	C	P
3 (TDC BDC TDC BDC)	C / P	P / E	E / I	I / C
4 (BDC TDC BDC TDC)	I / C	C / P	P / E	E / I
5 (BDC TDC BDC TDC)	P / E	E / I	I / C	C / P
6 (TDC BDC TDC BDC)	E / I	I / C	C / P	P / E
7 (BDC TDC BDC TDC BDC)	C	P	E	I
8 (TDC BDC TDC BDC TDC)	P	E	I	C

Power stroke every 90°

Fig. 11.14 In-line straight eight-cylinder arrangement

and exhaust strokes. Pistons 4 and 5 will then be at mid-stroke on exhaust and compression strokes respectively; pistons 1 and 8 at BDC at the end of power and induction strokes respectively; and pistons 3 and 6 at mid-stroke on induction and power strokes respectively. The firing-order after 180° of rotation is 1, 6, 2.

A third 90° rotation brings pistons 4 and 5 to TDC at the end of their respective exhaust and compression strokes. Pistons 1 and 8 will then be at mid-stroke on exhaust and compression strokes respectively; pistons 3 and 6 at BDC at the end of induction and power strokes respectively; and pistons 2 and 7 at mid-stroke on power and induction strokes respectively. The firing-order after 270° of rotation is 1, 6, 2, 5.

A fourth 90° rotation brings pistons 1 and 8 to TDC again, at the end of their respective exhaust and compression strokes. Pistons 3 and 6 will then be at mid-stroke on compression and exhaust strokes respectively; pistons 2 and 7 at BDC at the end of power and induction strokes respectively; and pistons 4 and 5 at mid-stroke on induction and power strokes respectively. The firing-order after 360° of rotation is 1, 6, 2, 5, 8.

A fifth 90° of rotation brings pistons 3 and 6 to TDC again, at the end of their respective compression and exhaust strokes. Pistons 2 and 7 will then be at mid-stroke on exhaust and compression strokes respectively; pistons 4 and 5 at BDC at the end of induction and power strokes respectively; and pistons 1 and 8 at mid-stroke on induction and power strokes respectively. The firing-order after 450° of rotation is 1, 6, 2, 5, 8, 3.

A sixth 90° of rotation brings pistons 2 and 7 to TDC again, at the end of their respective exhaust and compression strokes. Pistons 4 and 5 will then be at mid-stroke on compression and exhaust strokes respectively; pistons 1 and 8 at BDC at the end of induction and power strokes respectively; and pistons 3 and 6 at mid-stroke on power and induction strokes respectively. The firing-order after 540° of rotation is 1, 6, 2, 5, 8, 3, 7.

A seventh 90° of rotation brings pistons 4 and 5 to TDC again, at the end of their respective compression and exhaust strokes. Pistons 1 and 8 will then be at mid-stroke on compression and exhaust strokes respectively; pistons 3 and 6 at BDC at the end of power and induction strokes respectively; and pistons 2 and 7 at mid-stroke on induction and power strokes respectively. The complete firing-order after 630° of rotation will thus be 1, 6, 2, 5, 8, 3, 7, 4.

A further 90° movement, making a total of 720°, will complete two crankshaft revolutions or four strokes in readiness for the next cycle to begin.

By transposing different pairs of crank-throws, other firing orders have been used such as 1, 5, 2, 6, 4, 8, 3, 7 and 1, 7, 3, 8, 4, 6, 2, 5.

When extra cylinder capacity is necessary to pull large loads, but increased piston size and stroke would mean reducing crankshaft speed, the crankshaft may be extended with two more cylinders. This design is dynamically balanced, but there can be some problems with torsional vibrations and the extended length is difficult to accommodate in some trucks.

11.16 90° 'V' twin-cylinder arrangement (fig. 11.15)

This arrangement has the two cylinders positioned at 90° to each other with both big-ends attached to a single crankpin (fig. 11.15). With this configuration there will be uneven intervals between power impulses, which take place every 270° and 450° of crankshaft rotation.

Let piston 1 fire first at the end of its compression stroke. Piston 2 will then be at mid-stroke approaching TDC on either its exhaust or its compression stroke.

Assume then that piston 2 is at mid-stroke on exhaust. Rotation of the crank through 450° will

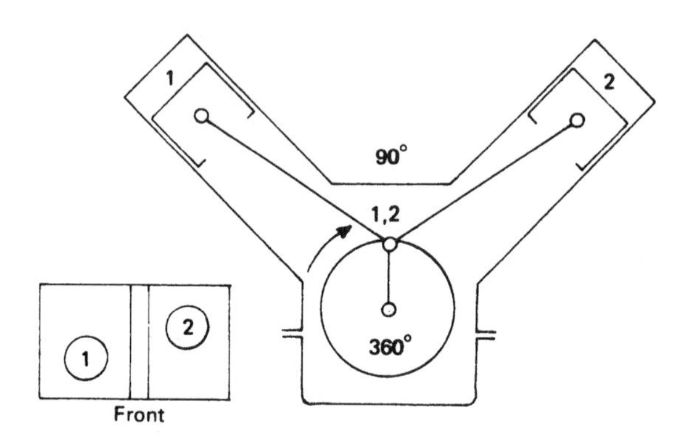

Firing order : 1, 2

Cylinder number	1st revolution		2nd revolution		
	0	180°	360°	540°	720°
	TDC	BDC	TDC	BDC	TDC
1	I	C	P	E	
	TDC	BDC	TDC	BDC	
2	C	P	E	I	C

Power stroke every 270° and 450°

Fig. 11.15 'V' twin-cylinder arrangement

complete its exhaust, induction, and compression strokes in readiness for firing. At this point piston 1 will be at mid-stroke on induction, so rotation of the crank through a further 270° will complete both its induction and its compression strokes. The total crank-angle interval for these two firing events will add up to 450° + 270° = 720°.

Cylinder banks are designated either left-hand or right-hand looking at the engine from the front. Side-by-side connecting rods are used, and thus the two banks of cylinders are offset relative to each other. Either the left- or the right-hand bank may be in the forward position – in the diagram shown, it is the left-hand bank which is staggered ahead of the right-hand bank.

The 'V-twin' has only a moderate degree of dynamic balance, and its uneven filling intervals and insufficient cyclic-torque smoothness make it unsuitable for the motor car. However, it has been included here as it illustrates the basic arrangement of V-banked cylinders with connecting-rods sharing a common crankpin, which is an important engine layout.

11.17 60° 'V' four-cylinder arrangement (fig. 11.16)

The cylinders fire at equally spaced intervals of 180° and are arranged with numbers 1 and 2 in the left-hand bank and numbers 3 and 4 in the right-hand bank.

In the 'V-four' crankshaft arrangement shown in fig. 11.16, the crankpins are unequally spaced at alternate intervals of 60° and 120° and, when viewed from the front, it can be seen that they lie in two planes. Main journals and bearings are provided at each end, with a third journal placed betwen crankpins 2 and 3.

With such an arrangement of cylinder banks and crankshaft, pairs of pistons will be at the top of their strokes but in different cylinder banks. When pistons 1 and 4 are at TDC, either of these may be chosen to be at the end of its compression stroke and about to fire. The other piston would then be at the end of exhaust and just beginning its induction stroke. Let piston 1 be at the end of its compression stroke and about to start its power stroke. Piston 4 will then have completed exhaust and will just be beginning its induction stroke.

Rotation of the crankshaft through 180° will position pistons 2 and 3 at the top of their respective exhaust and compression strokes, so that at this point the order of firing will be 1, 3.

A second 180° rotation (a total of 360°) will bring pistons 1 and 4 to TDC positions again, but this time piston 4 will be at the end of compression and about to begin power, while piston 1 will now have completed exhaust and be about to start its induction stroke. The order of firing up to this point will be 1, 3, 4.

A third 180° of rotation (a total of 540°) will bring pistons 2 and 3 to TDC again, but now piston 2 will be at the end of compression and about to begin its power stroke. The complete order of firing now becomes 1, 3, 4, 2.

Finally, a fourth 180° of rotation will complete 720° of crankshaft movement – that is, two revo-

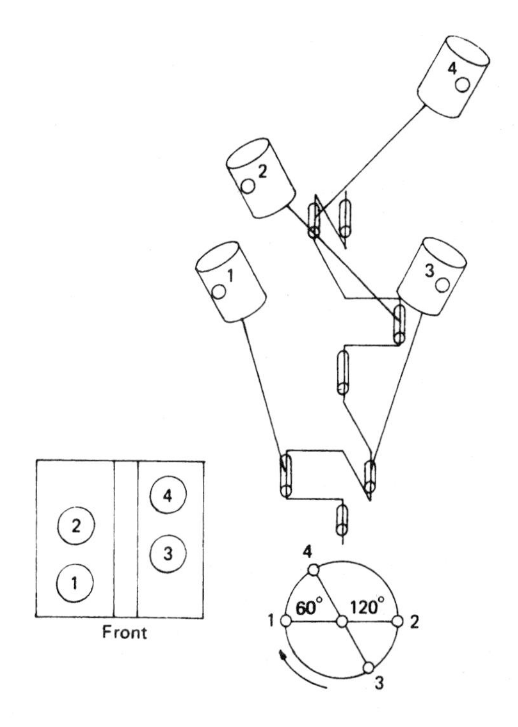

Firing order : 1, 3, 4, 2

Cylinder number	0		180°		360°		540°		720°
	1st revolution				2nd revolution				
	TDC		BDC		TDC		BDC		TDC
1		I		C		P		E	
	BDC		TDC		BDC		TDC		BDC
2		C		P		E		I	
	BDC		TDC		BDC		TDC		BDC
3		E		I		C		P	
	TDC		BDC		TDC		BDC		TDC
4		P		E		I		C	

Power stroke every 180°

Fig. 11.16 'V' four-cylinder arrangement

lutions or four strokes – in readiness for the next cycle of events.

This is an extremely compact engine, but the dynamic balance of this layout is poor, so an additional counterbalance shaft is always necessary to cancel out this imbalance.

11.18 60° 'V' six-cylinder arrangement (fig. 11.17)
The cylinders fire at equally spaced intervals of 120° and are arranged with numbers 1, 2, and 3 in the left-hand bank and numbers 4, 5, and 6 in the right-hand bank. The crankshaft has six crankpins equally spaced at intervals of 60° and arranged to lie in three planes. To support the shaft there are four main journals and bearings placed at each end and between pairs of crankpins, thus providing a relatively short but rigid construction (fig. 11.17).

There are four possible firing-orders, but three of these involve consecutive firing of three cylinders in each bank and only the fourth enables cylinders to be fired alternatively from each bank. The latter firing-order is 1, 4, 2, 5, 3, 6, which probably also offers the best selection from the standpoint of torsional vibration.

With this crankshaft arrangement, pairs of pistons will be at the top of their strokes but in different cylinder banks.

Let pistons 1 and 5 be at TDC after their respective compression and exhaust strokes – that is, piston 1 is about to fire on its power stroke and piston 5 is about to begin its induction stroke.

Rotation of the crankshaft through 120° will position pistons 3 and 4 at the top of their respective exhaust and compression strokes, so that at this point the order of firing will be 1, 4.

A second 120° of rotation (a total of 240°) will bring pistons 2 and 6 to TDC on their respective compression and exhaust strokes. The order of firing will thus be 1, 4, 2.

A third 120° of rotation (a total of 360°) will bring pistons 1 and 5 to TDC on their respective exhaust and compression strokes. The order of firing will thus be 1, 4, 2, 5.

A fourth 120° of rotation (a total of 480°) will bring pistons 3 and 4 to TDC on their respective compression and exhaust strokes. The order of firing wil thus be 1, 4, 2, 5, 3.

A fifth 120° of rotation (a total of 600°) will bring pistons 2 and 6 to TDC on their respective exhaust and compression strokes. The final order of firing will then be 1, 4, 2, 5, 3, 6.

A further 120° of rotation will complete 720° of crankshaft displacement – that is, two revolutions

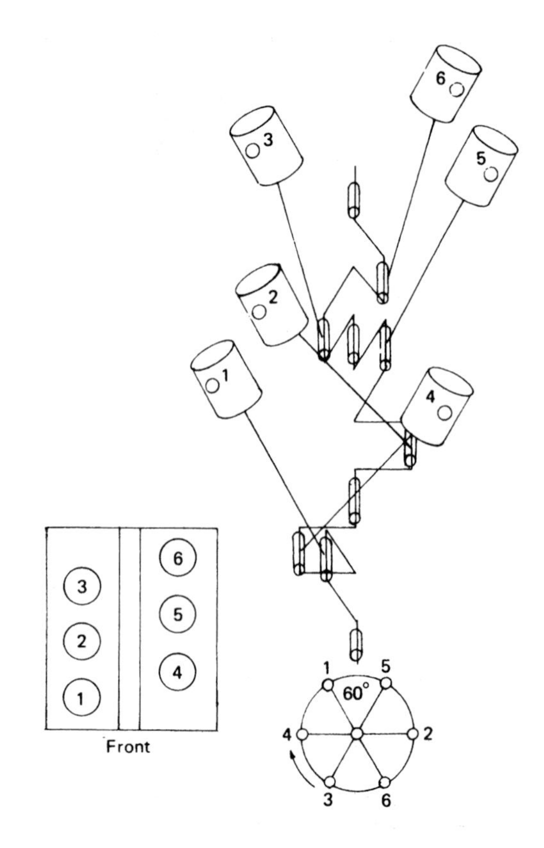

Firing order : 1, 4, 2, 5, 3, 6

Cylinder number	1st revolution (0 – 180° – 360°)	2nd revolution (360° – 540° – 720°)
1	TDC BDC TDC: I, C	TDC BDC TDC: P, E
2	BDC TDC BDC: P, E, I	BDC TDC: I, C, P
3	TDC BDC: C, P, E	TDC BDC: E, I, C
4	TDC BDC: E, I, C	TDC BDC: C, P, E
5	TDC BDC TDC: P, E	TDC BDC TDC: I, C
6	BDC TDC: I, C, P	BDC TDC: P, E, I

Power stroke every 120°

Fig. 11.17 'V' six-cylinder arrangement

or four strokes – in readiness for the next cycle of events.

The dynamic balance is relatively good with this layout, which provides a short compact engine unit compared with the 'in-line six'.

11.19 90° 'V-eight' with single-plane crankshaft

(fig. 11.18)

Like the two-plane crankshaft of the 'in-line eight', the single-plane arrangement used for the 'V-eight' will provide a power impulse every 90° of crank movement.

The single-plane crankshaft has four crankpins paired so that both outer and both inner crankpins are in phase. Each crankpin has two connecting-rod big-ends attached, and generally there are five main journals to support the crankshaft (fig. 11.18).

Consider pistons 1 and 4 to be at TDC, with piston 1 at the end of compression and about to fire, and piston 4 at the end of its exhaust stroke. Pistons 2 and 3 will then be at BDC at the end of their respective power and induction strokes; pistons 5 and 8 will be at mid-stroke on their respective exhaust and compression strokes; and pistons 6 and 7 will be at mid-stroke on their respective induction and power strokes.

Rotation of the crankshaft through 90° brings pistons 5 and 8 to TDC at the end of their respective exhaust and compression strokes. Pistons 1 and 4 will then be at mid-stroke on their respective power and exhaust strokes; pistons 2 and 3 will be at mid-stroke on their respective exhaust and compression strokes; and pistons 6 and 7 will be at BDC at the end of their respective induction and power strokes. The firing-order after 90° of rotation is 1, 8.

A second 90° rotation brings pistons 2 and 3 to TDC at the end of their respective exhaust and compression strokes. Pistons 6 and 7 will then be at mid-stroke on their respective compression and exhaust strokes; pistons 1 and 4 will be at BDC at the end of their respective power and induction strokes; and pistons 5 and 8 will be at mid-stroke on their respective induction and power strokes. The firing-order after 180° of rotation is 1, 8, 3.

A third 90° rotation brings pistons 6 and 7 to TDC at the end of their respective compression and exhaust strokes. Pistons 2 and 3 will then be at mid-stroke on their respective induction and power strokes; pistons 1 and 4 will be at mid-stroke on their respective exhaust and compression strokes; and pistons 5 and 8 will be at BDC at the end of their respective induction and power strokes. The firing-order after 270° of rotation is 1, 8, 3, 6.

A fourth 90° rotation brings pistons 1 and 4 to TDC at the end of their respective exhaust and compression strokes. Pistons 5 and 8 will then be at mid-stroke on their respective power and com-

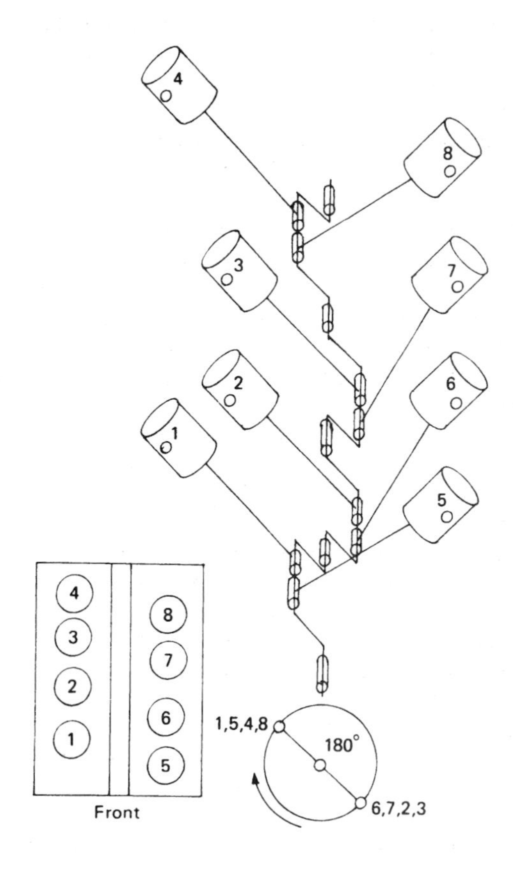

Firing order: 1, 8, 3, 6, 4, 5, 2, 7

Cylinder number	0 – 180° (1st revolution)	180° – 360° (1st revolution)	360° – 540° (2nd revolution)	540° – 720° (2nd revolution)
1 (TDC BDC TDC BDC TDC)	I	C	P	E
2 (BDC TDC BDC TDC BDC)	C	P	E	I
3 (BDC TDC BDC TDC BDC)	E	I	C	P
4 (TDC BDC TDC BDC TDC)	P	E	I	C
5 (TDC BDC TDC BDC)	C / P	P / E	E / I	I / C
6 (BDC TDC BDC TDC)	P / E	E / I	I / C	C / P
7 (BDC TDC BDC TDC)	I / C	C / P	P / E	E / I
8 (TDC BDC TDC BDC)	E / I	I / C	C / P	P / E

Power stroke every 90°

Fig. 11.18 90° 'V-eight' with single-plane crankshaft

pression strokes; pistons 2 and 3 will be at BDC at the end of their respective induction and power strokes; and pistons 6 and 7 will be at mid-stroke on their respective power and induction strokes. The firing-order after 360° of rotation is 1, 8, 3, 6, 4.

A fifth 90° rotation brings pistons 5 and 8 to TDC at the end of their respective compression and exhaust strokes. Pistons 1 and 4 will then be at mid-stroke on their respective induction and power strokes; pistons 2 and 3 will be at mid-stroke on their respective compression and power strokes; and pistons 6 and 7 will be at BDC at the end of their respective power and induction strokes. The firing-order after 450° of rotation is 1, 8, 3, 6, 4, 5.

A sixth 90° rotation brings pistons 2 and 3 to TDC at the end of their respective compression and exhaust strokes. Pistons 1 and 4 will then be at BDC at the end of their respective induction and power strokes; pistons 5 and 8 will be at mid-stroke on their respective power and induction strokes; and piston 6 and 7 will be at mid-stroke on their respective exhaust and compression strokes. The firing-order after 540° of rotation is 1, 8, 3, 6, 4, 5, 2.

A seventh 90° of rotation brings pistons 6 and 7 to TDC at the end of their respective exhaust and compression strokes. Pistons 1 and 4 will then be at mid-stroke on their respective induction and exhaust strokes; pistons 2 and 3 will be at mid-stroke on their respective power and induction strokes; and pistons 5 and 8 will be at BDC at the end of their respective power and induction strokes. The final firing order after 630° of rotation is 1, 8, 3, 6, 4, 5, 2, 7.

An eighth 90° of rotation will complete 720° of crankshaft movement – that is two revolutions or four strokes – in readiness for the next cycle of events.

The single-plane crankshaft, unlike the two-plane V-eight crankshaft, provides at least 180° exhaust-pulse intervals between adjacent cylinders, and with simple manifold modification this can be extended to 360° before pulse interference can occur.

11.20 90° 'V-eight' with two-plane or cruciform crankshaft (fig. 11.19)

The cylinders fire at equally phased intervals of 90° and are arranged with numbers 1, 2, 3, and 4 in the left-hand bank and numbers 5, 6, 7, and 8 in the right-hand bank as shown in fig. 11.19. The two-plane crankshaft has pairs of crank-throws

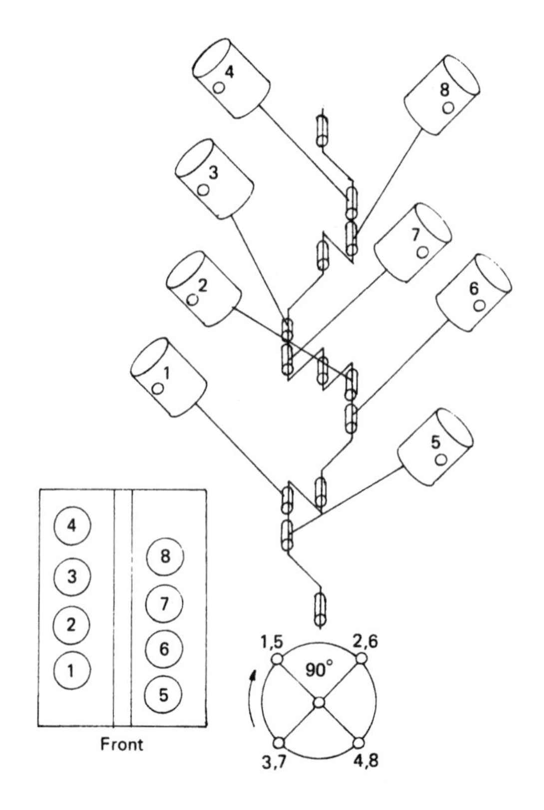

Firing order: 1, 5, 4, 8, 6, 3, 7, 2

Cylinder number (0 – 180° – 360° – 540° – 720°; 1st revolution, 2nd revolution)	Dead centres	Strokes
1	TDC BDC TDC BDC TDC	I C P E
2	BDC TDC BDC TDC	I C P E I
3	TDC BDC TDC BDC	C P E I C
4	BDC TDC BDC TDC BDC	E I C P
5	TDC BDC TDC BDC	E I C P E
6	TDC BDC TDC BDC TDC	P E I C
7	BDC TDC BDC TDC BDC	C P E I
8	BDC TDC BDC TDC	P E I C P

Power stroke every 90°

Fig. 11.19 90° 'V-eight' with two-plane or cruciform crankshaft

phased at intervals of 90°. Each crankpin has two separate connecting-rods attached, these rods being hinged to pistons in different cylinder banks.

With this crankshaft arrangement, there is a main journal and bearing provided at each end and between adjacent crankpins. Since two connecting-rods share a common crankpin, these five-main-journal crankshafts are extremely short and uncomplicated.

Consider the order of cylinder power strokes occurring as the crankshaft rotates as shown in fig. 11.19. With piston 1 at TDC after its compression stroke and at the beginning of power, piston 5 will be at mid-stroke on compression. Pistons 3 and 7 are then at mid-stroke exhaust and at the beginning of exhaust respectively; pistons 4 and 8 will be at the beginning of compression and at mid-stroke on induction respectively; and pistons 2 and 6 will be at mid-stroke power and at the beginning of induction respectively.

Rotation of the crankshaft through 90° will bring piston 5 to TDC on its compression stroke. Piston 1 will then be at mid-stroke power; pistons 3 and 7 at the beginning of induction and at mid-stroke exhaust respectively; pistons 4 and 8 at mid-stroke compression and at the start of compression respectively; and pistons 2 and 6 at the beginning of exhaust and at mid-stroke induction respectively.

A second 90° rotation (a total of 180°) will bring piston 4 to TDC on its compression stroke while piston 8 will be at mid-stroke on compression.

Pistons 2 and 6 will then be at mid-stroke exhaust and at the beginning of compression respectively; pistons 1 and 5 at the beginning of exhaust and at mid-stroke power respectively; and pistons 3 and 7 at mid-stroke induction and at the beginning of induction respectively.

A third 90° of rotation (a total of 270°) will bring piston 8 to TDC on its compression stroke while piston 4 will be at mid-stroke on power. Pistons 2 and 6 will then be at the beginning of induction and at mid-stroke compression respectively; pistons 1 and 5 at mid-stroke exhaust and at the beginning of exhaust respectively; and pistons 3 and 7 at the beginning of compression and at mid-stroke induction respectively.

A fourth 90° rotation (a total of 360°) will bring piston 6 to TDC on its compression stroke while piston 2 will be at mid-stroke on induction. Pistons 1 and 5 will then be at the beginning of induction and at mid-stroke exhaust respectively; pistons 3 and 7 at mid-stroke compression and at the beginning of compression respectively; and pistons 4 and 8 at the beginning of exhaust and at mid-stroke power respectively.

A fifth 90° rotation (a total of 450°) will bring piston 3 to TDC on its compression stroke while piston 7 will be at mid-stroke on compression. Pistons 4 and 8 will then be at mid-stroke exhaust and at the beginning of exhaust respectively; pistons 2 and 6 at the beginning of compression and at mid-stroke power respectively; and pistons 1 and 5 at mid-stroke induction and at the beginning of induction respectively.

A sixth 90° rotation (a total of 540°) will bring piston 7 to TDC on its compression stroke while piston 3 will be at mid-stroke on power. Pistons 4 and 8 will then be at the beginning of induction and at mid-stroke exhaust respectively; pistons 2 and 6 at mid-stroke compression and at the beginning of exhaust respectively; and pistons 1 and 5 at the beginning of compression and at mid-stroke induction respectively.

A seventh 90° rotation (a total of 630°) will bring piston 2 to TDC on its compression stroke while piston 6 will be at mid-stroke on exhaust. Pistons 1 and 5 will then be at mid-stroke compression and at the beginning of compression respectively; pistons 3 and 7 at the beginning of exhaust and at mid-stroke power respectively; and pistons 4 and 8 at mid-stroke induction and at the beginning of induction respectively.

A final eighth 90° of rotation will complete 720° of crankshaft displacement or one cycle of events.

The two-plane crankshaft has a dynamic state of balance far superior to that of the single-plane crankshaft and for this reason is more popular.

12 Balancing of reciprocating components

12.1 Reciprocating motion (fig. 12.1(a))
The mid-crank angle position is not the same as the mid-stroke position; thus considering the piston movement from TDC to BDC, for the first 90° crank angle movement from TDC to mid-crank position, the piston travels further along its stroke than for the second 90° angular displacement from mid-crank position to BDC (fig. 12.1(a)). This means that if the crankshaft is made to rotate at a steady speed, then the piston travels further and therefore faster for the first quarter of a revolution than for the second quarter, where the piston moves slower since it then travels a shorter distance. As a result the piston's downstroke has a high acceleration from TDC to mid-crank position, and a low deceleration from mid-crank position to BDC. On its return upstroke the piston has a low deceleration from BDC to mid-crank, and a high deceleration from mid-crank position to TDC.

Primary force (fig. 12.1(b)) The primary force is the inertia force created by the acceleration or deceleration of the piston assembly mass caused by the rotating crankpin's projected motion along the line of stroke.

Secondary force (fig. 12.1(c)) The secondary force is the inertia force created by the acceleration or deceleration of the piston assembly's mass caused by the rotating crankpin's projected motion perpendicular to the line of stroke. In other words, the secondary force is due to the additional piston acceleration or deceleration produced by the rotating crankpin increasing or decreasing the inclination of the connecting-rod to the line of stroke.

Comparison of cyclic variation for primary and secondary forces (fig. 12.1(d)) The primary and secondary forces are shown plotted adjacent to the crank angle movement (fig. 12.1(d)). As can be seen, the secondary force increases and decreases its magnitude at twice the frequency of the primary force but its maximum values are only about one-quarter of the dominating primary force. At TDC both primary and secondary forces are at a maximum and act together, i.e. $F_p + F_s$. At BDC both primary and secondary forces are at a maximum but act in opposition to each other, i.e. $F_p - F_s$. At 45° and 135° after TDC the secondary forces have decayed to zero, i.e. $F_s = 0$. At mid-crank position 90° after TDC the secondary force is at a maximum, whereas the primary force has declined to zero, i.e. $F_s = \text{max}$ and $F_p = 0$.

12.1.1 In-line three-cylinder engine balance (fig. 12.2(a)–(d))
The in-line three-cylinder crankshaft has three cranks symmetrically spaced at 120° intervals. Both primary and secondary forces are completely balanced, but awkward longitudinal couples are created which change their direction every 180° of crankshaft rotation.

Primary inertia force The reciprocating balance of an in-line three-cylinder engine shows (fig. 12.2(a, b)) that the upward acting primary force of piston No. 1 when at TDC is exactly balanced by the downward acting primary forces of pistons No. 2 and No. 3 with their crankshaft cranks positioned 60° either side of BDC, that is $F_{p1} = F_{p2} + F_{p3}$ (also see graph in fig. 12.2). Likewise when piston No. 1 is at BDC its downward-acting primary force is neutralised by the upward-acting force of pistons No. 2 and No. 3 (fig. 12.2(d)). However, the upward primary force of piston No. 1 and the combination but offset downward primary force of pistons No. 2 and No. 3 (fig. 12.2(b)) produce a clockwise pitching couple which changes its sense every 180°. Thus when piston No. 1 is at BDC the longitudinal couple will now have an anticlockwise sense (fig. 12.2(d)). It will also be observed that when piston No. 2 is at TDC the opposing primary force of pistons No. 1 and No. 3 on either side eliminates these primary couples.

Secondary forces Similarly to the primary forces, the secondary forces of one piston when it is either at TDC or BDC is exactly counteracted

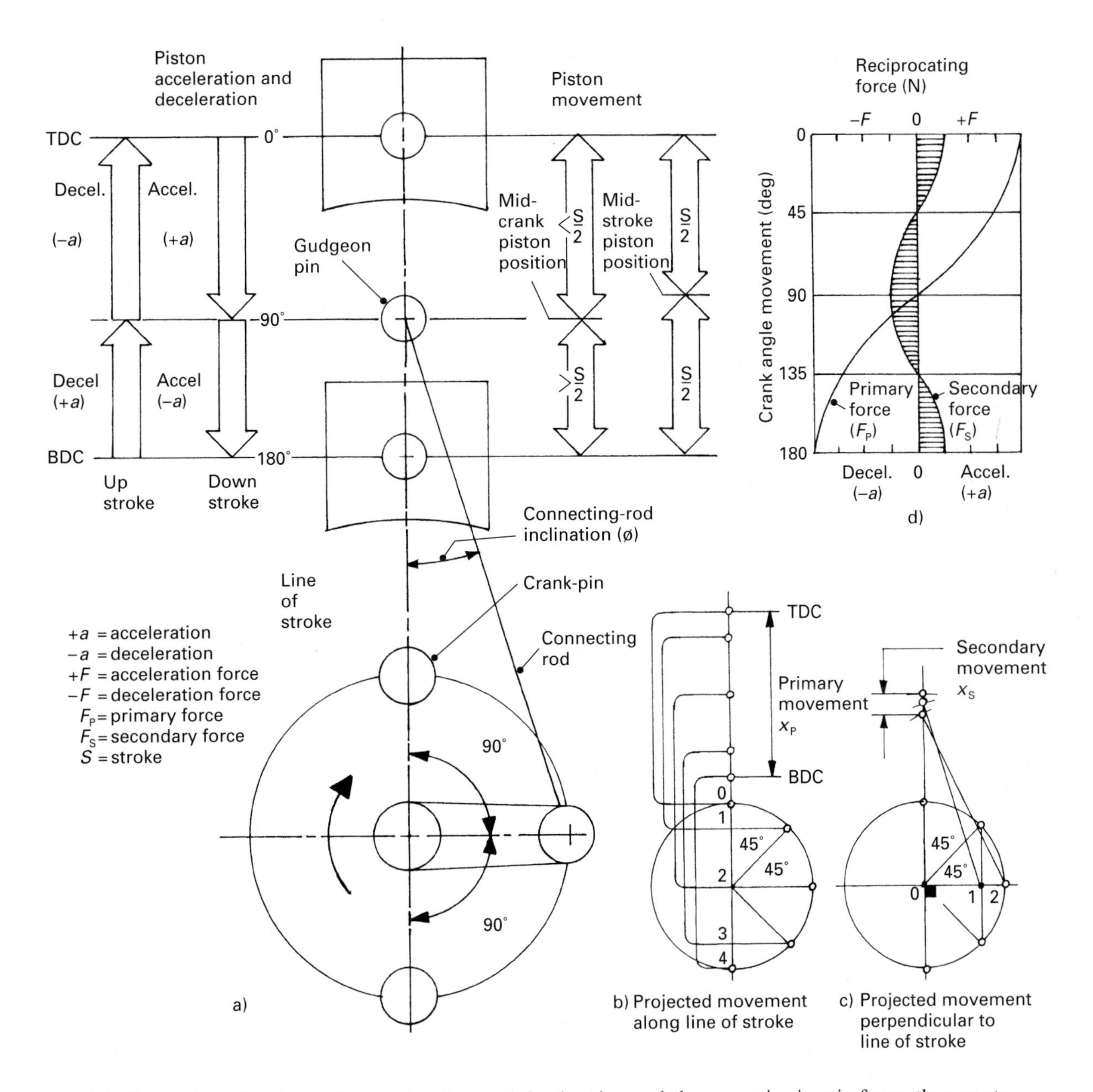

Fig. 12.1 Illustration of reciprocating acceleration and deceleration and the respective inertia forces they create

by the secondary forces of the two other pistons, that is $F_{s1} = F_{s2} + F_{s3}$ (also see graph in fig. 12.2) and again longitudinal couples are created due to the offset of these forces. However, these secondary couples do not necessarily have the same sense of direction as the primary couples, thus fig. 12.2(b) shows a clockwise primary and secondary couple but fig. 12.2(d) portrays an anticlockwise primary couple and clockwise secondary couple.

12.1.2 In-line four-cylinder engine balance (fig. 12.3)
The reciprocating balance of an in-line four-cylinder engine shows that the upward-acting primary forces of pistons No. 1 and No. 4 are exactly balanced by the downward acting primary forces of pistons No. 2 and No. 4 (see fig. 12.3). However the pistons at TDC have both the primary and secondary forces acting in the same upward direction, whereas the pistons at BDC have their secondary forces acting in opposition to the

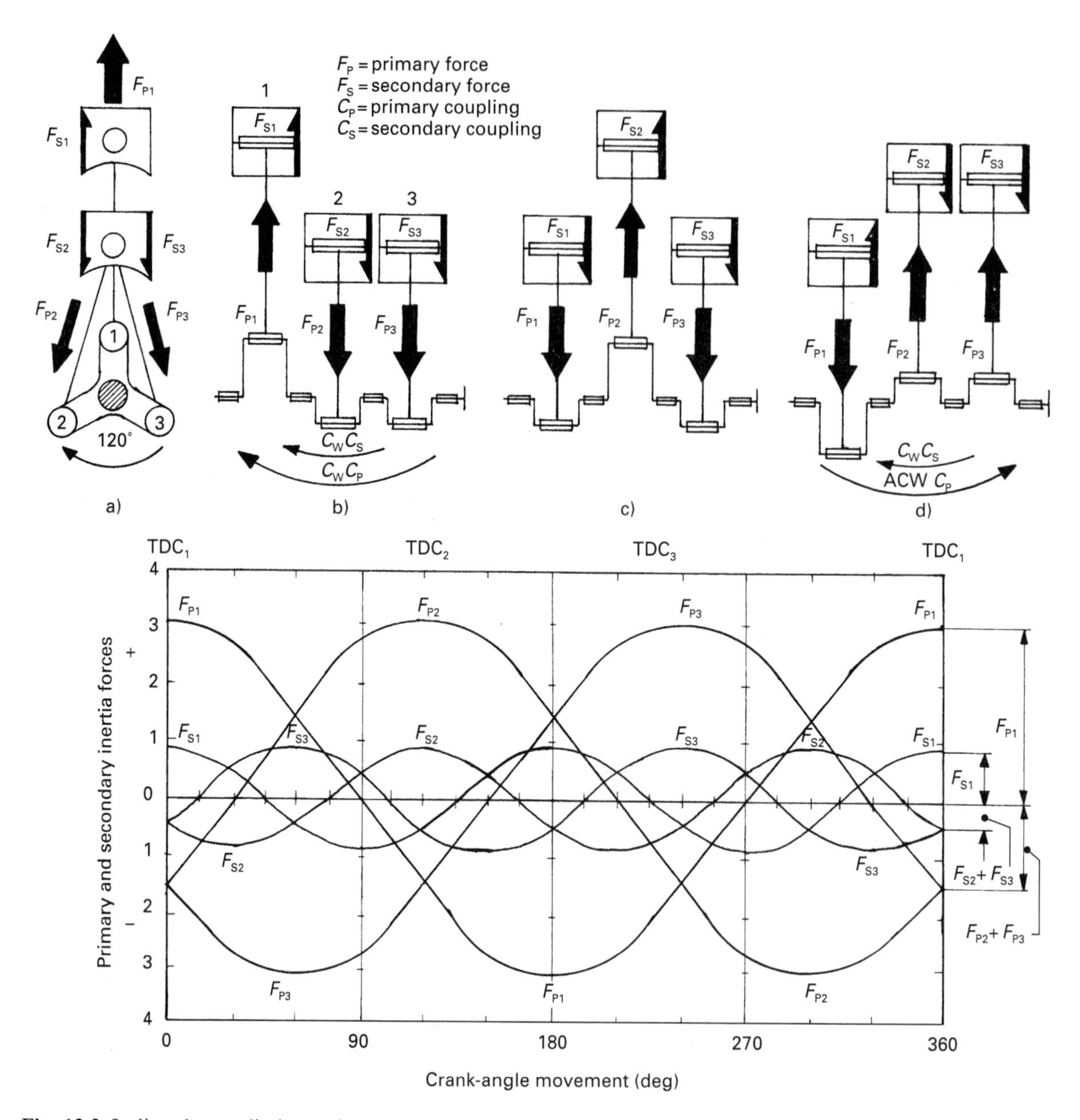

Fig. 12.2 In-line three-cylinder engine primary and secondary force balance

downward direction of the primary forces and therefore act in an upward direction. This results in an imbalance where the four secondary forces $4F_s$ acting together produce a vertical shake which becomes more noticeable with rising engine speed (see graph in fig. 12.3). There are primary couples C_p generated but these are absorbed by the rigidity of the crankshaft material. Secondary force imbalance is normally tolerated for in-line four-cylinder engines of up to two litres, but with larger in-line four-cylinder car engines, the increased magnitude of the secondary forces caused by the heavier reciprocating parts is not always acceptable. Some manufacturers now incorporate twin counter-shaft secondary force balancers.

12.1.3 90° twin-cylinder engine balance (fig. 12.4(a)–(d))

With the 90° twin engine the pair of connecting-rods share the common crankpin. Primary inertia

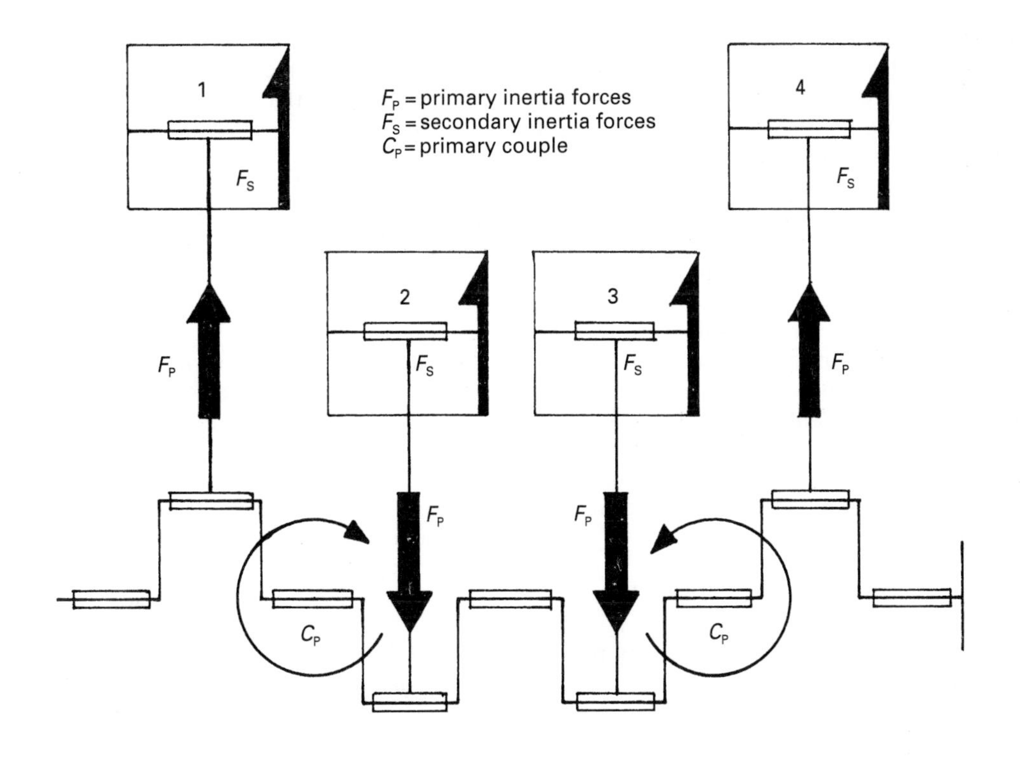

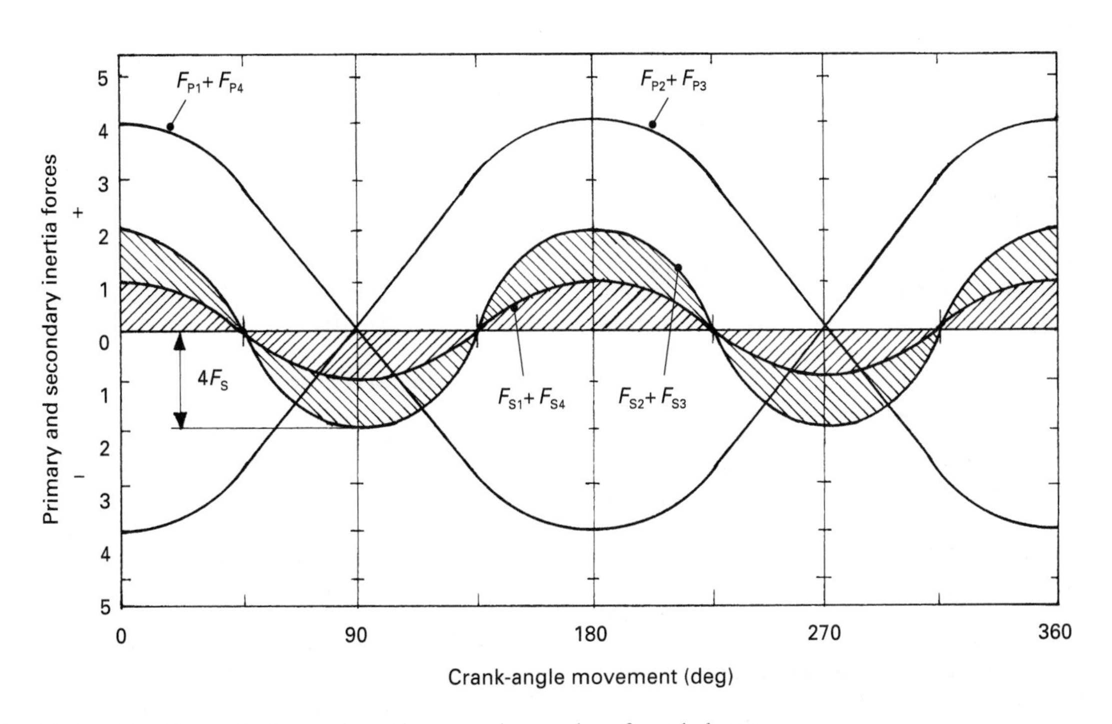

Fig. 12.3 In-line four-cylinder engine primary and secondary force balance

forces are completely out of balance if there are no counteracting balance weights. However by attaching counterweights to each web on either side of the crankpin on the opposite side of the crank, the primary inertia forces may be completely balanced (see graph in fig. 12.4). Unfortunately this configuration produces a secondary force imbalance which cannot easily be neutralised.

Primary inertia force The cyclic primary force balance using counterweights is now described as follows: 0° and 360° crank position (fig. 12.4(a)). With the crankpin in the TDC position relative to the left-hand cylinder, the No. 1 piston primary inertia force will be completely balanced by the counterweight positioned diametrically opposite the crankpin, whereas the right-hand piston will be in its mid-stroke position in which there is no primary force generated.

90° Crank position (fig. 12.4(b)) When the crankpin has rotated through 90° the right-hand No. 2 piston will be at TDC so that its primary inertia force is again counterbalanced by the counterweight, whereas the left-hand piston will be in the neutral mid-stroke no-primary-force position.

180° Crank position (fig. 12.4(c)) As the crankpin revolves through to the 180° position, the left-hand No. 1 piston will be at BDC so that again it is completely balanced by the counterweight while No. 2 piston will be in its zero-primary-force mid-stroke position.

270° Crank position (fig. 12.4(d)) A total crank angle movement of 270° brings No. 2 piston to BDC thus enabling the counterweight to neutralise its primary inertia force while the left-hand piston will again be in its mid-stroke zero-primary-force position.

Secondary force (fig. 12.4(a)–(d)) As can be seen from the triangle of forces in all the four extreme piston positions, the vertical components of the two secondary forces neutralise one another, but horizontal components produce a resultant secondary force imbalance of 1.414 F_s which is greater than either individual piston secondary force.

12.1.4 60° 'V' six-cylinder engine (fig. 12.5)

This six-cylinder engine has two cylinder banks, inclined at 60° to each other. Six cranks evenly spaced at 60° intervals make up the crankshaft configuration. The pistons of each cylinder bank are linked via connecting-rods to every second crankpin so that the crankpin interval for each cylinder bank is 120°. Hence the crankshaft can be considered to be two three-cylinder crankshafts merged into one (fig. 12.5). Consequently the 60° 'V' six-cylinder engine with its two banks of cylinders is in effect two in-line three-cylinder engines sharing a common crankshaft (see fig. 12.2(a)–(d)).

Primary and secondary inertia forces As with the in-line three-cylinder engine both primary and secondary forces are balanced, but there still exists a relatively large unbalanced primary couple and a much smaller secondary couple. However counterweights can be attached to the crankshaft to cancel the primary couples whereas the secondary couples are more difficult to neutralise and therefore are usually tolerated.

12.1.5 90° 'V' eight-cylinder engine balance with two-plane crankshaft (fig. 12.6)

This cylinder and crankshaft configuration has two cylinder banks of four cylinders inclined at 90° to each other. There are four cranks evenly spaced at 90° intervals. The first and last and the second and third cranks, form diametrically opposed pairs, are then arranged at right angles to each other; hence the term two-plane crankshaft (fig. 12.6).

A pair of connecting-rods, one for each cylinder bank are attached to a common crankpin so that the crankpin interval for each cylinder bank is also 90°. Consequently the 90° 'V' eight-cylinder engine with its two banks of cylinders can be considered as four 90° twin-cylinder engines sharing one crankshaft but without counterweights (see fig. 12.4(a)–(d)).

Primary and secondary inertia forces Conveniently the secondary resultant force horizontal imbalance which exists with each pair of pistons sharing one crankpin, as with the 90° 'V' twin-cylinder engine (see fig. 12.4(a)–(d)) are counterbalanced by a second pair of pistons spaced 90° apart and linked to a second crankpin positioned diametrically opposite to the first crankpin (fig. 12.6). These pairs of crankpins are set 180° apart with each crankpin linked to pairs of pistons with their line of stroke set at 90° to each other and provide both primary and secondary force balance. Because the primary inertia forces act on

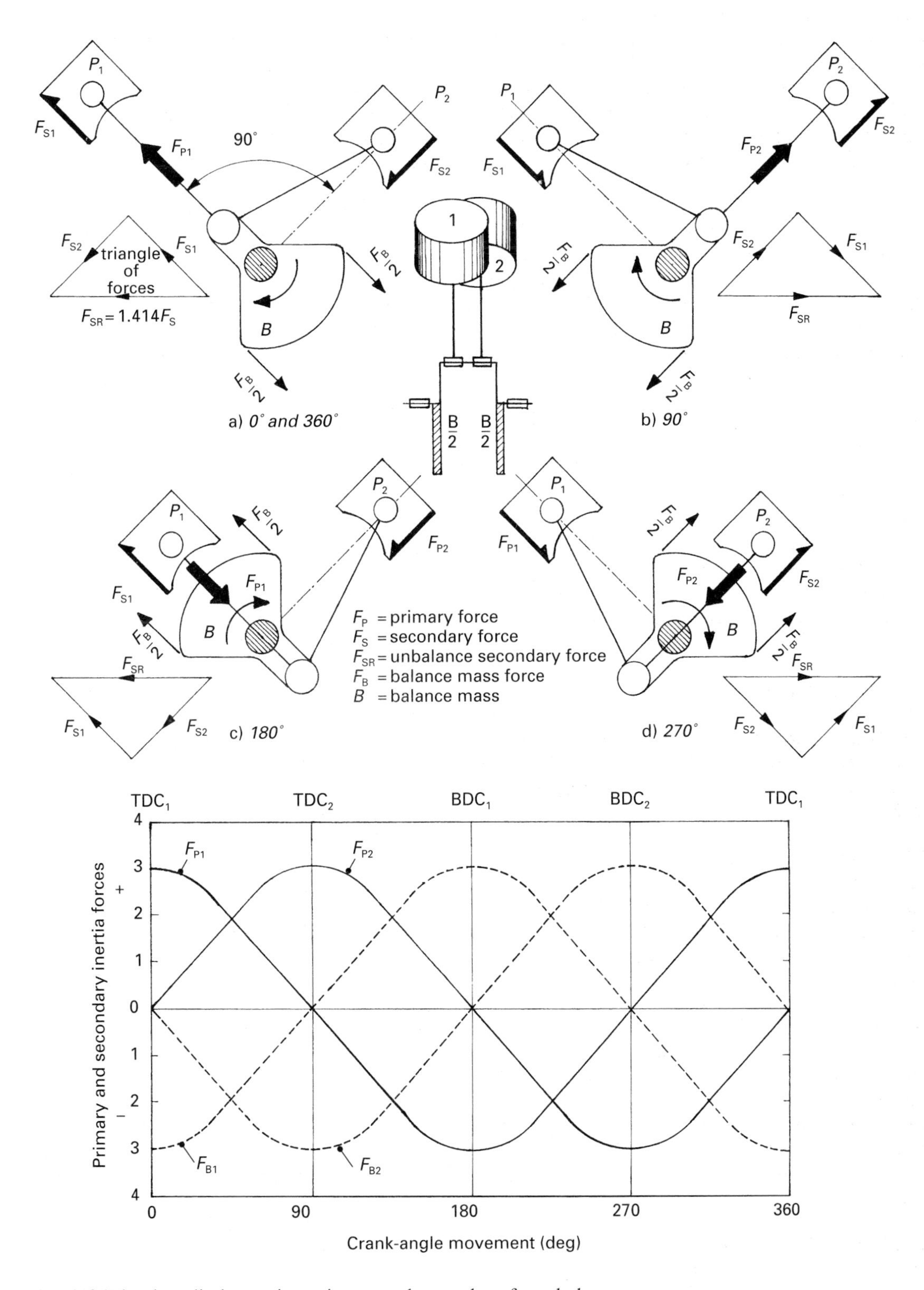

Fig. 12.4 90° 'V' twin-cylinder engine primary and secondary force balance

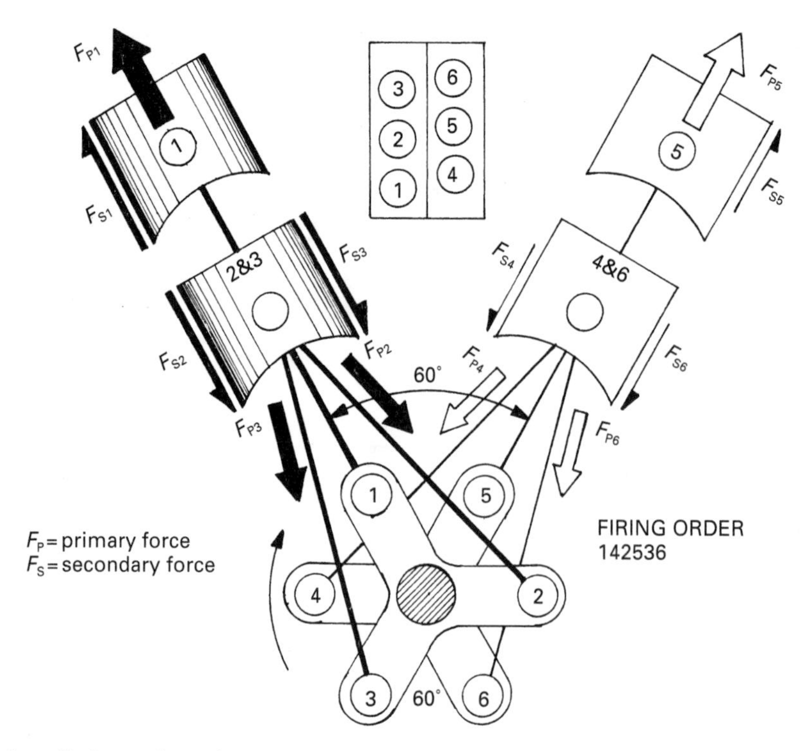

Fig. 12.5 60° 'V' six-cylinder engine primary and secondary force balance

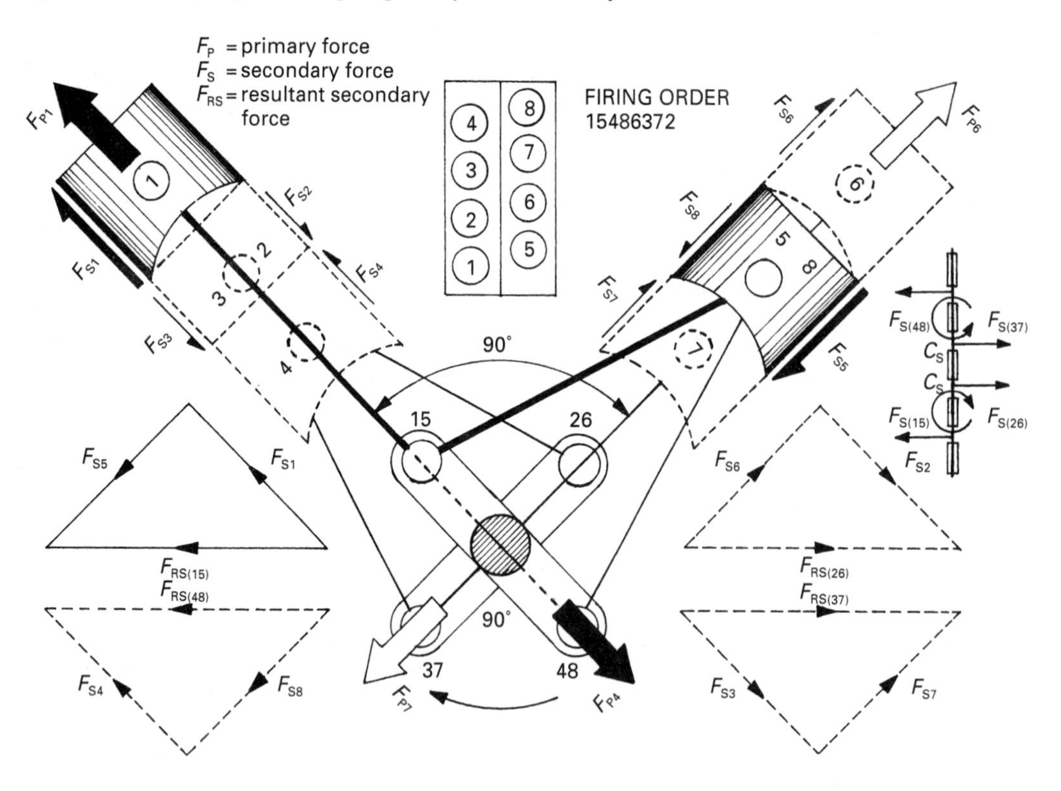

Fig. 12.6 90° 'V' eight-cylinder engine (two plane crankshaft) primary and secondary force balance

the crankpin in four different planes, primary couples are created, but fortunately these couples can easily be neutralised by attaching counterweights at various points along the crankshaft. In contrast to the primary forces, the horizontal resultant secondary forces are arranged symmetrically along the crankshaft so that the couples generated are absorbed by the stiffness of the crankshaft (see inset in fig. 12.6).

12.2 Four-cylinder chain drive parallel twin countershaft secondary force balancer (figs 12.7(a)–(d) and 12.8)

The out-of-balance secondary forces can be balanced by using twin countershafts and their weights B_L and B_R are equivalent in magnitude to the secondary inertia forces of all four cylinders i.e. $F_s + F_s + F_s + F_s = 4\ F_s$.

The countershafts are mounted either side of the cylinder and they are made to revolve at twice the crankshaft speed by means of a 2:1 chain and sprocket wheel drive (fig. 12.8). Thus the balance countershafts rotate counter to each other (B_L rotates clockwise and B_R rotates anticlockwise) at twice the crankshaft speed and are so timed that they counteract the positive secondary force $+\ F_s$ at either dead centre (0° at TDC and 180° after TDC) and the negative secondary force $-\ F_s$ at mid-crank position (90° after TDC; see graph in fig. 12.7). Secondary balance is achieved (fig. 12.7(a)–(d)) by the weights facing in the opposite direction to these secondary forces when in the vertical plane, and then for both balance weights to face either inward (135° after TDC) or outward (45° after TDC) and thereby oppose each other when they have moved to one of the two horizontal positions.

12.3 Torsional crankshaft vibration

Torsional crankshaft vibration is the rapid back and forth angular twisting along the crankshaft's length relative to the flywheel which occurs at certain rotational speeds known as critical speeds. As the crankshaft runs through a speed range, it may experience a short period of roughness (rumble) caused by violent torsional oscillations, and once above or below this critical speed the crankshaft will rotate smoothly again. These critical oscillations are caused by the frequency of the power impulse torque imposed on the crankshaft (these impulse frequencies increase directly with the crankshaft rotational speed) coinciding with the natural frequency vibration of the crankshaft or a whole multiple of this natural frequency. Then a condition of resonance will exist. In practice the natural frequency of vibration of the crankshaft is well above its designed maximum rotational speed, that is, the power torque impulse frequencies, so that only the synchronisation speeds of certain whole multiples of the natural frequency are able to produce critical vibrations. At any of these major critical speeds, if the engine is allowed to operate closely to or at one of these speeds, then the crankshaft torsional vibration stresses may exceed their safe working values and therefore would quickly result in the fracturing of the crankshaft.

To prevent crankshaft damage due to torsional vibration, many engines prone to this type of vibration install torsional vibration damper units to the front end of the crankshaft. There are two basic types as follows:

1) tuned rubber torsional vibrational vibration damper;
2) viscous fluid torsional vibration damper.

12.3.1 Tuned rubber torsional vibration damper (figs 12.9 and 12.10)

This form of vibration damper (fig. 12.9) utilises a rubber band which is either compressed or bonded between an outer annular cast-iron ring and an inner hub attached to the front end of the crankshaft. When there is no vibration the annular ring supported by the rubber band rotates freely with the crankshaft and pulley-wheel hub (fig. 12.10(b)). If the rotating crankshaft is suddenly subjected to a torsional twist (fig. 12.10(a)) which accelerates the shaft in an anticlockwise direction the inertia of the annular ring mass will tend to hold back. Thus the sandwiched rubber band becomes subjected to elastic deformation in the form of a shear strain.

After the front end of the crankshaft has twisted itself to its limit in the anticlockwise direction it then changes direction (fig. 12.10(c)), whereas the momentum of the inertia ring mass prevents it from reversing its direction initially so that it overshoots and in doing so distorts the rubber band.

Consequently, the continued to and fro torsional oscillation of the front end of the crankshaft causes a proportional distortion within the rubber due to the movement of the inertia ring mass always lagging behind that of the crankshaft's angular twist. Thus the unwanted strain energy stored in the crankshaft by the torsional vibration is transferred to the rubber vibration

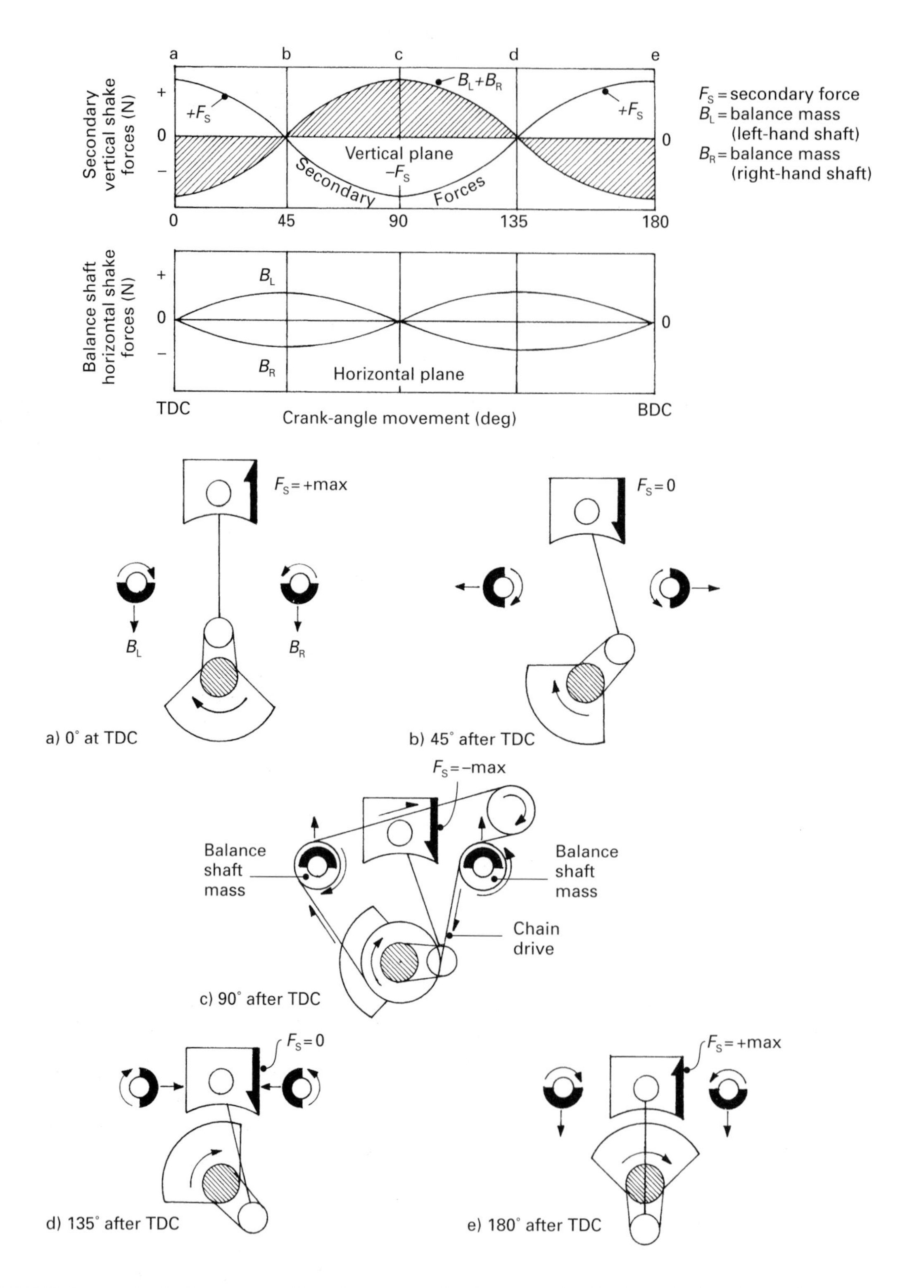

Fig. 12.7 Twin-countershaft secondary force balancer cycle of events suitable for an in-line four-cylinder engine

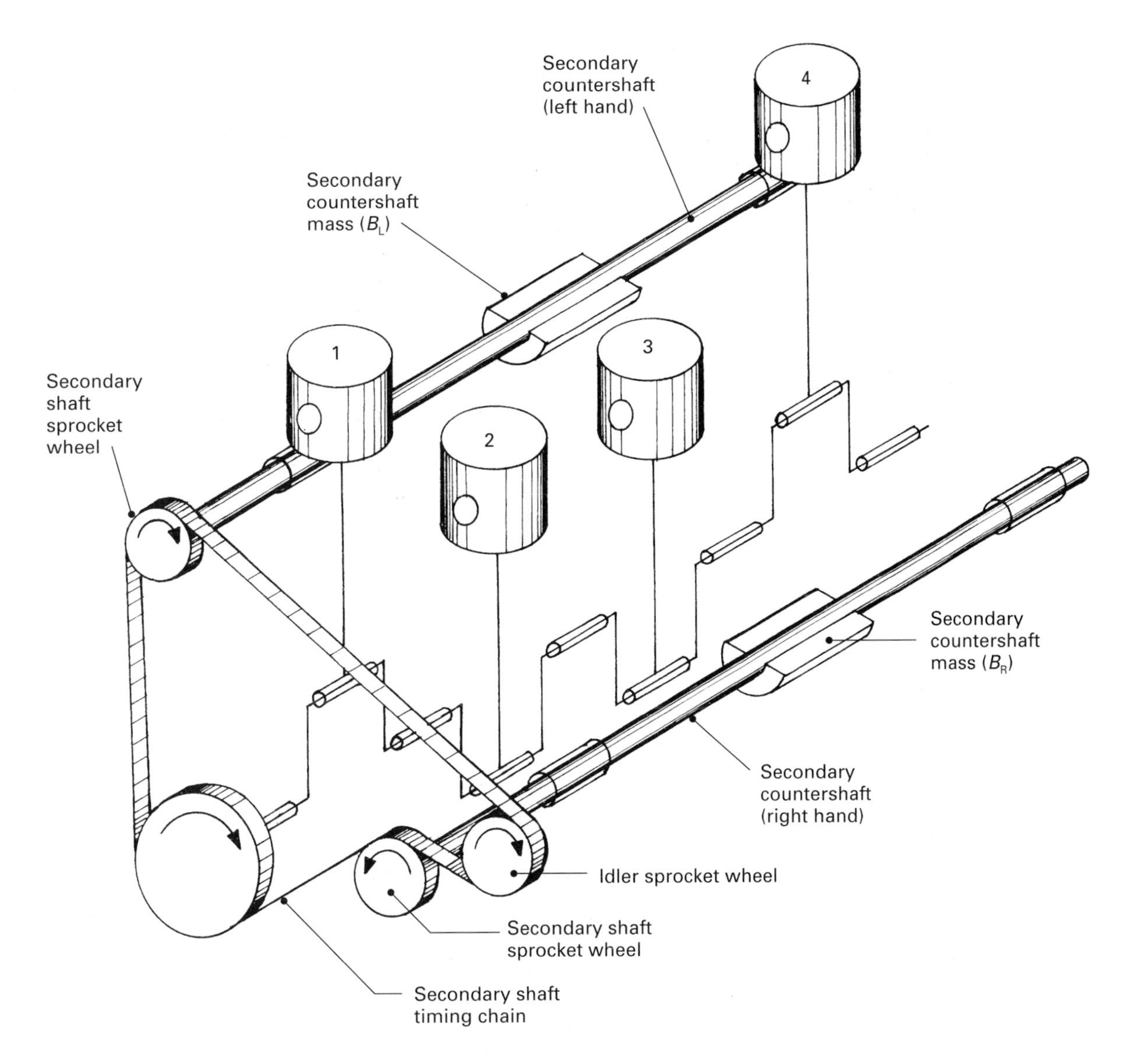

Fig. 12.8 Four-cylinder engine chain drive parallel twin and secondary countershaft balancer

damper unit where it is absorbed and dissipated by the out-of-phase movement of the inertia ring mass distorting the rubber band. Hence the energy lost from the crankshaft vibration is that gained by the strain energy imparted to the rubber and the internal friction it creates, after which this energy is dissipated in the form of heat to the surrounding air.

12.3.2 Viscous fluid torsional vibration damper (figs 12.11 and 12.12)

This consists of a rectangular sectioned annular ring (fig. 12.11) enclosed in a thin steel casing which is attached via a hub and pulley-wheel to the crankshaft at the furthest possible position from the flywheel, that is, at the crankshaft's front end. The clearance between the annular ring and the surrounding casing walls is filled with a viscous silicone fluid. This is able to maintain its properties through a wide temperature range and throughout the working life of the vibration damper unit.

If there is little or no torsional vibration the viscous resistance of the fluid drags the annular ring mass around with the casing and hub as the crankshaft rotates (fig. 12.12(b)) with very little or

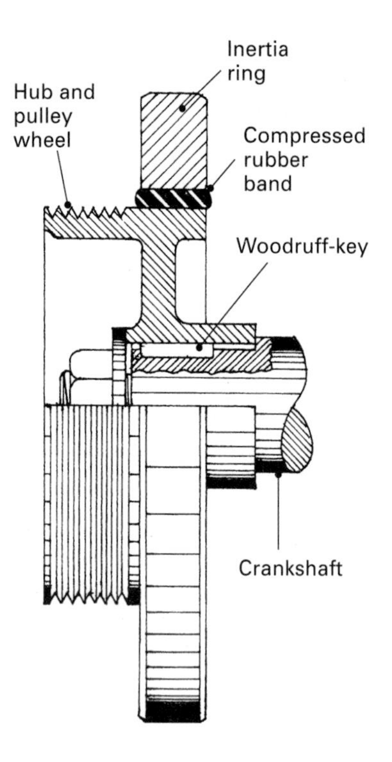

Fig. 12.9 Tuned rubber band vibration damper

even no slippage. However, if the front end of the crankshaft commences to oscillate about its axis relative to the flywheel rear end, in addition to rotating at some steady speed, then the inertia of the annular ring mass prevents it from oscillating with the casing due to its reluctance to change its direction of motion.

Thus if the crankshaft and hub suddenly accelerate in an anticlockwise direction (fig. 12.12(a)) the annular ring will continue to rotate at the mean speed of the crankshaft. It is only the viscous shear resistance of the fluid between the annular ring and casing walls that will cause the annular ring to be dragged at an increased speed in the same direction as the casing. Overall, however, there will be sufficient relative movement between the ring casing walls to produce a viscous shear within the fluid that absorbs and dissipates the internal strain energy building up within the crankshaft.

Once the crankshaft twist has reached its maximum amplitude in the anticlockwise direction the crankshaft reverses its twist and accelerates in the opposite clockwise direction (fig. 12.12(c)), this time moving against the momentum of the annular ring mass. Thus again the parallel walls of the annular ring and casing move relative to each other, and accordingly produce the viscous shear action needed to dissipate and damp down these critical torsional vibrations experienced by the crankshaft.

12.3.3 Crankshaft torsional vibration characteristics (fig. 12.13)

Torsional oscillation characteristics are shown in fig. 12.13. The full line curves illustrate the first

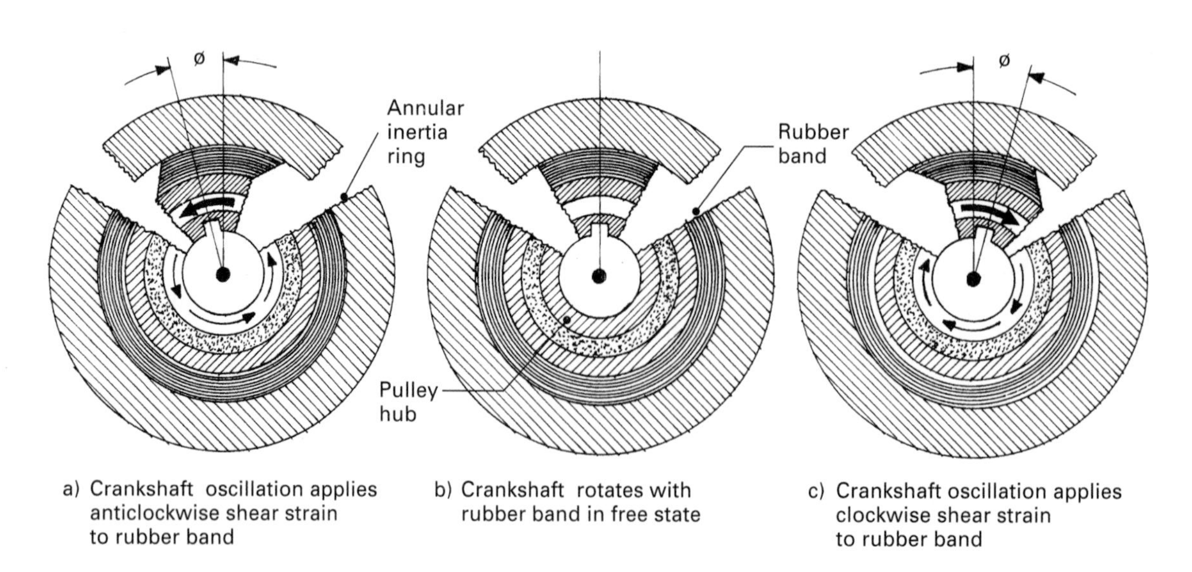

Fig. 12.10 Tuned rubber vibration damper principle

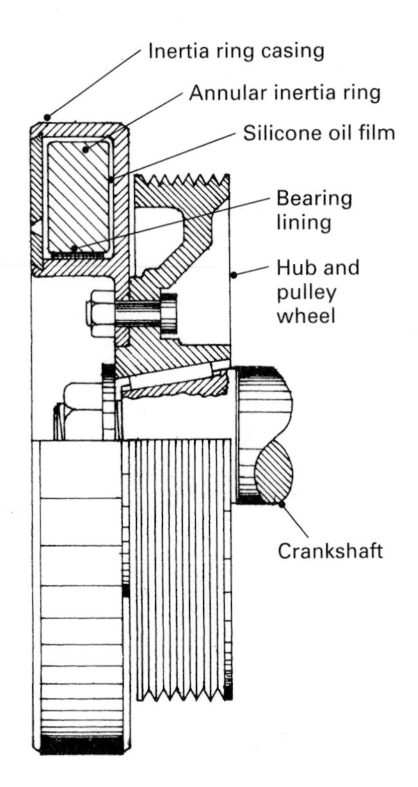

Fig. 12.11 Viscous fluid vibration damper

and second major order vibrations, whereas the broken line curves represent corresponding major order damped vibrations, which, as can be seen, have much lower peak amplitudes.

12.4 Force imposed on the piston

12.4.1 Side thrust and piston slap (fig. 12.14(a)–(c))
The triangle of forces making up the connnecting-rod upward reaction force F_c, the vertical downward gas pressure load F_g, and the horizontal side thrust F_s (fig. 12.14(a) and (c)) shows that the side thrust generated by the combustion load is many times larger than that produced by the compression gas load, i.e. (compression) $F_s < F_s$ (combustion).

When the crankshaft rotates, the reaction between the cylinder gas pressure and the inclined connecting-rod on its compression upstroke causes the piston to bear against the right-hand side of the cylinder bore (fig. 12.14(a)). As the piston approaches TDC the connecting-rod momentarily takes on an up-right position (fig. 12.14(b)). An instant later with now much higher firing stroke cylinder pressure, the piston snaps over to the left-hand side of the piston (fig. 12.14(c)). This impact of piston movement from one side to the other every time the piston reaches TDC produces a noise known as piston slap, its severity being proportional to the piston skirt side clearance and the rate of pressure rise in the cylinder due to combustion. Piston slap is usually at its worst when the engine is cold (piston skirt to cylinder wall side clearance is at a maximum), cylinder and piston wear is considerable and the ignition firing setting is aggressively advanced.

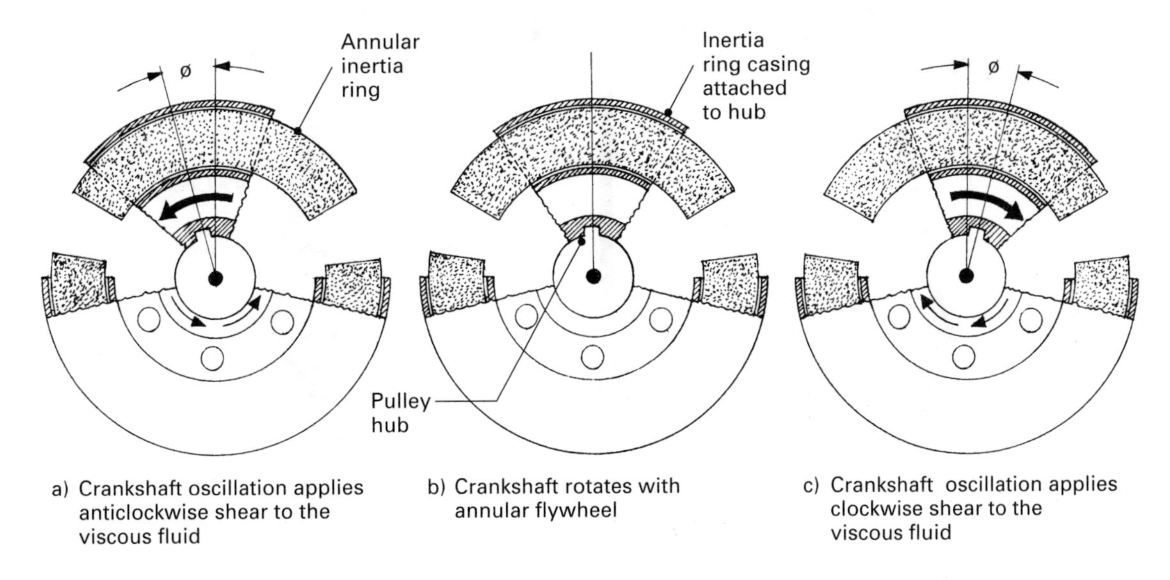

Fig. 12.12 Viscous fluid vibration damper principle

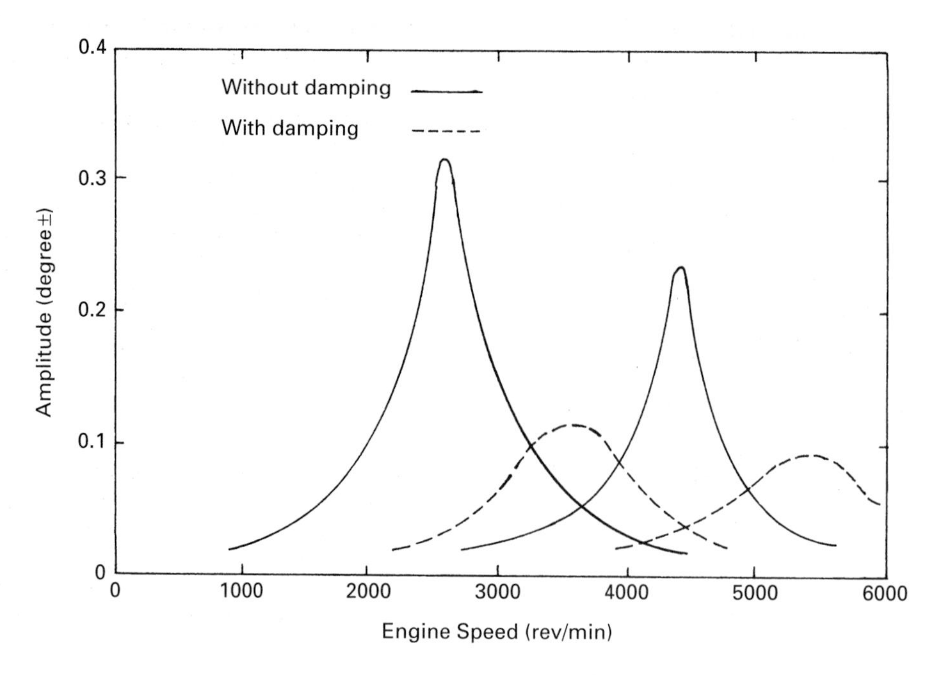

Fig. 12.13 Torsional vibration for a six-cylinder engine with and without damping

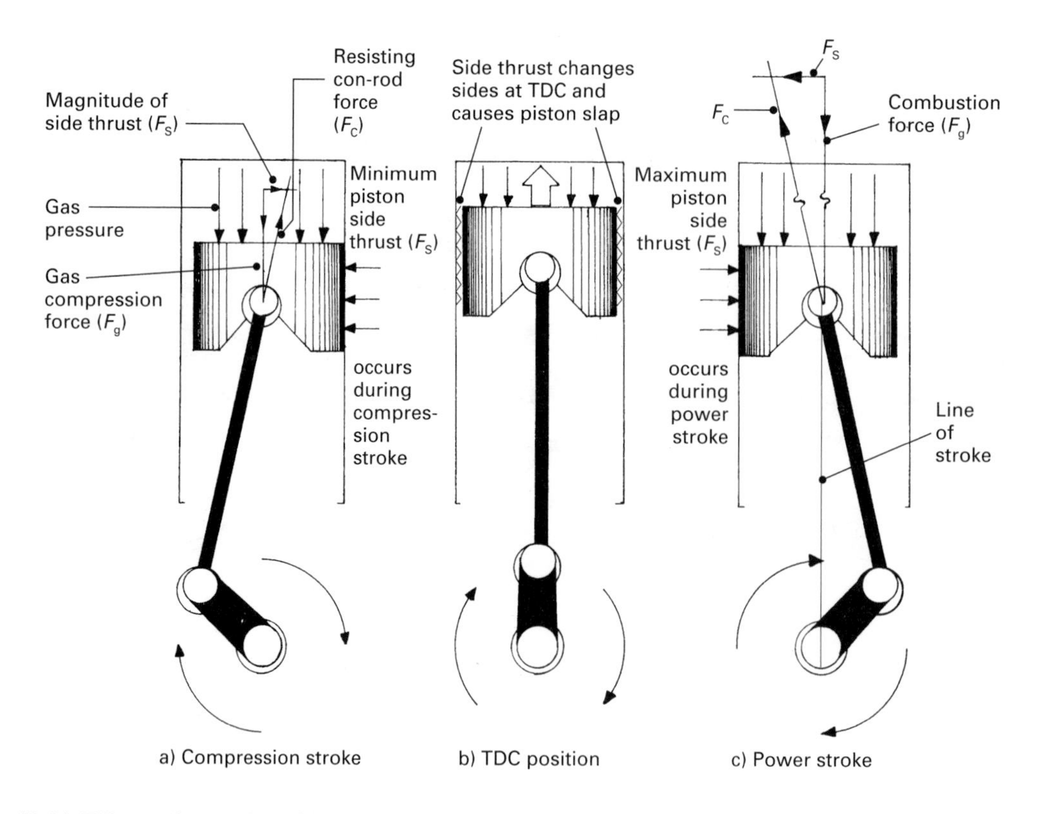

Fig. 12.14 Effects of central positioned gudgeon pin

12.4.2 Offset gudgeon pin (fig. 12.15(a)–(c))
Offsetting the gudgeon pin from the centre of the cylinder axis towards the side of maximum piston thrust (fig. 12.15(a)) causes the gas load acting on the piston crown between the gudgeon pin centre and the rim of the piston on the maximum and minimum thrust sides to be unequal. Accordingly this produces a clockwise torque which tilts the piston, allowing edge contact to be made on opposite facing walls simultaneously (fig. 12.15(b)). Subsequently the change over from minimum to maximum thrust side skirt to wall contact is transformed initially by piston tilt (fig. 12.15(b)), followed by the parallel alignment of the piston skirt to that of the rigid cylinder wall (fig. 12.15(c)). With very small gudgeon-pin offset of no more than 0.0125*d* where *d* is the piston diameter, there can be a useful reduction in piston to wall side impact and consequently reduced piston slap noise (fig. 12.15(c)). Pistons with offset gudgeon pins are usually marked on their crown with an arrow pointing towards the front of the engine, and must be assembled accordingly. Excess gudgeon pin offset would cause uneven piston and piston-ring to cylinder support, wear and poor piston-ring cylinder wall gas sealing.

12.5 Journal bearing load considerations

Gas pressure forces (fig. 12.16(a) and (b)) Gas pressure forces are created both on the pistons compression (contracting) upstroke (fig. 12.16(a)) and on the piston's combustion power downstroke (expansion) (fig. 12.16(b)). The peak combustion power stroke cylinder pressure may be three to five times greater than the maximum compression stroke pressure. In each case the connecting-rod is subjected to compressive stresses.

Inertia reciprocating forces (fig. 12.17(a) and (b)) Inertia forces are created by the reciprocating (to and fro) motion of the piston, gudgeon pin and approximately one-third of the connecting-rod masses. These forces increase with the square of the crankshaft rotary speed. Inertia forces pull the piston assembly away from the crankshaft big-end journal when the crankshaft is moving towards or away from the TDC position (fig. 12.17(a) and (b)) so that the connecting-rod is in

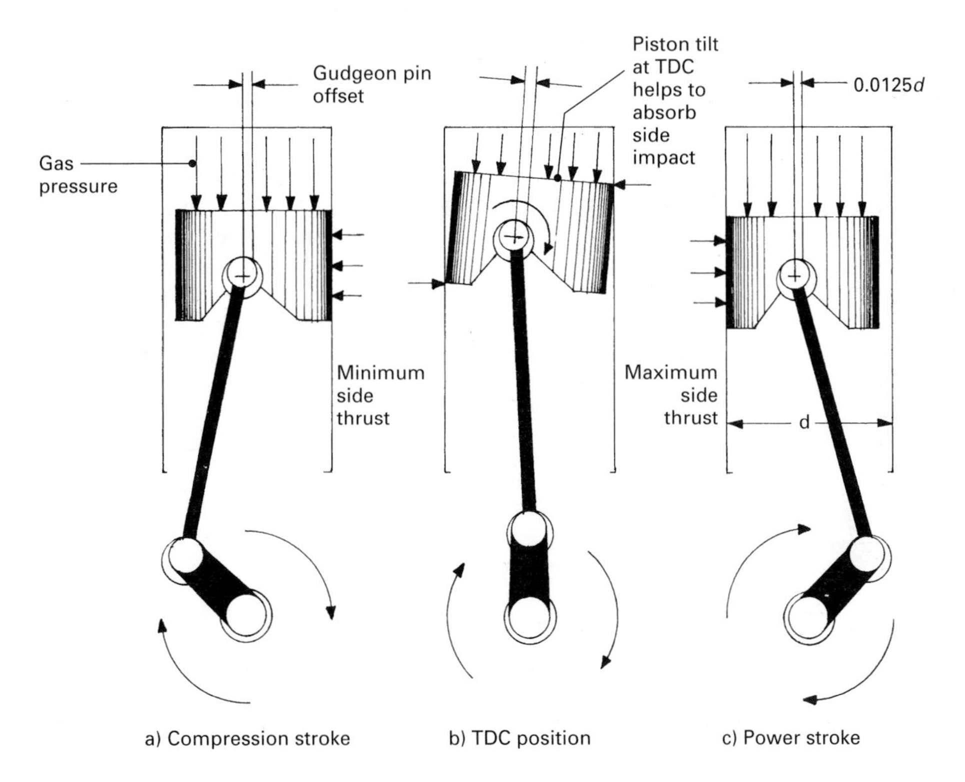

Fig. 12.15 Effects of offset gudgeon pin

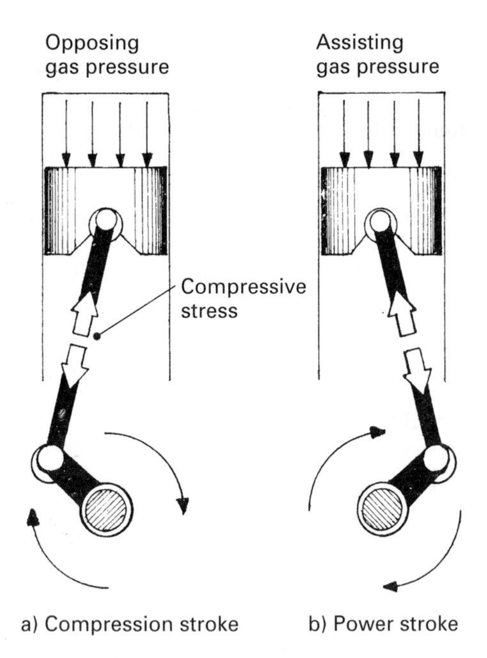

Fig. 12.16 Effects of gas pressure forces

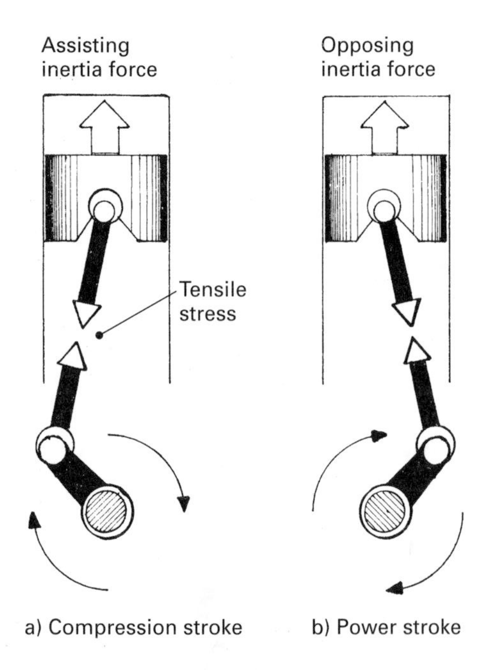

Fig. 12.17 Effects of reciprocating inertia forces

a state of tensile stress. Conversely, the piston assembly is pushed towards the big-end journal when the crankshaft is moving towards or away from the BDC region, thus causing the connecting-rod to be in a state of compressive stress.

12.5.1 Effects of combined gas and inertia forces on bearing load (fig. 12.18(a) and (b))

In the TDC crank angle region the inertia forces oppose the compression and power stroke gas pressure forces. Thus the inertia forces acting in opposition to the generated gas force will proportionally neutralise the gas load relative to their magnitude at any given crank-angle position, engine load and crankshaft speed. Consequently the big-end and main journal bearing load at any one instant are derived from the resultant gas and inertia force with respect to a given crank-angle position.

Gas and inertia forces can be represented by arrows and their individual magnitude by their length, as illustrated in fig. 12.18(a) and (b). The gas and inertia forces are shown with horizontal and vertical hatched lines (shaded lines) respectively, whereas the resultant of these forces is represented by the full black arrows. Figure 12.18(a) and (b) illustrates the situation for constant engine load (cylinder pressure) at low and high engine speeds. The resultant forces (load) at low speed (fig. 12.18(a)) pushes onto the big-end journal bearing, whereas at high engine speed (fig. 12.18(b)) the inertia forces exceed the gas pressure force (longer arrows), thereby creating a resultant force (load) that pulls against the big-end journal bearing.

12.5.2 Big-end journal bearing loads (fig. 12.19(a)–(c))

A deeper study of gas pressure and inertia forces and their resultant forces (bearing load) over a four-stroke cycle of events for similar engine loads but at three different engine speeds is shown in fig. 12.19(a)–(c). The shaded enclosed areas represent the resultant bearing load at any one crank-angle position. These graphs demonstrate how the gas and inertia forces interact at any one crank-angle position, according to engine load and speed on the big-end journal bearing.

Low engine speed (fig. 12.19(a)) At low engine speed the inertia forces are very small and do not alter significantly the steep rise in the dominating gas pressure load in the compression/power stroke region.

Medium engine speed (fig. 12.19(b)) At mid-engine speed the inertia forces considerably reduce the resultant big-end bearing load in the TDC region in the compression/power stroke region, whereas moderate generated inertia forces are virtually unopposed at TDC in the exhaust/induction stroke region.

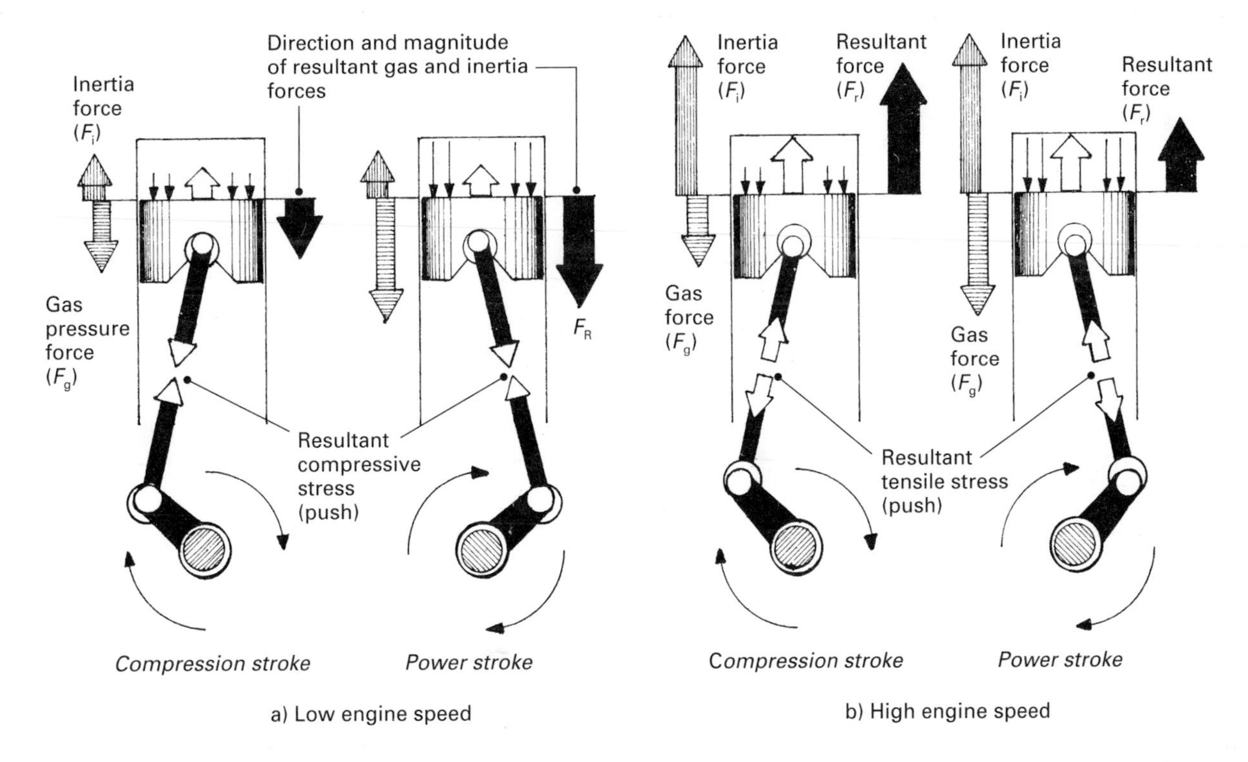

Fig. 12.18 Effects of engine speed on resultant gas and inertia forces (big-end bearing load)

High engine speed (fig. 12.19(c)) At very high engine speed the inertia forces have virtually cancelled out the gas pressure bearing load at TDC on the power stroke, whereas now a moderately large resultant force is created in the opposite sense at the end of the compression stroke. The large inertia force in the TDC exhaust/induction stroke region basically becomes the resultant load and dominates the big-end bearing load.

12.6 Engine and gearbox mountings

12.6.1 Dynamic and combustion disturbances

Most engines are balanced for both primary forces and couples, but secondary forces and couples do still exist with certain engine cylinder configurations which might give rise to vertical shake and horizontal rock. Flexible mountings are therefore interposed between the engine block and chassis or body members on to which the engine is attached to partially isolate the resulting vibrations from the body structure. There are other disturbing effects caused by combustion and inertia torsional reaction which cannot be eliminated easily, and must therefore also be absorbed by the elasticity and damping properties of the engine mounts. A torsional combustion reaction is imposed on the engine mass relating to the piston side-thrust caused by the combustion gas pressure periodic impulses and the inclined connecting-rod arising from the crank-throw's varying effective offset as the crankshaft rotates. A torsional inertia reaction is imposed on the engine block by the reciprocating mass of the piston and the connecting-rod which is restrained by the crankshaft and flywheel. Thus the oblique angle of the connecting-rod results in a side-thrust being exerted by the pistons against the cylinder walls as they reciprocate. This reaction is particularly prominent during overrun driving conditions.

12.6.2 Axis of oscillation (figs 12.20 and 12.21)

The engine and transmission unit should be suspended along an imaginary centre of rotation known as the principal axis. The principal axis passes through the centre of gravity of both the engine and transmission layout and is normally inclined between 10° and 30° to that of the crankshaft axis. Thus, in theory, the engine and transmission unit could be spun around this axis, then stopped in any angular position and released,

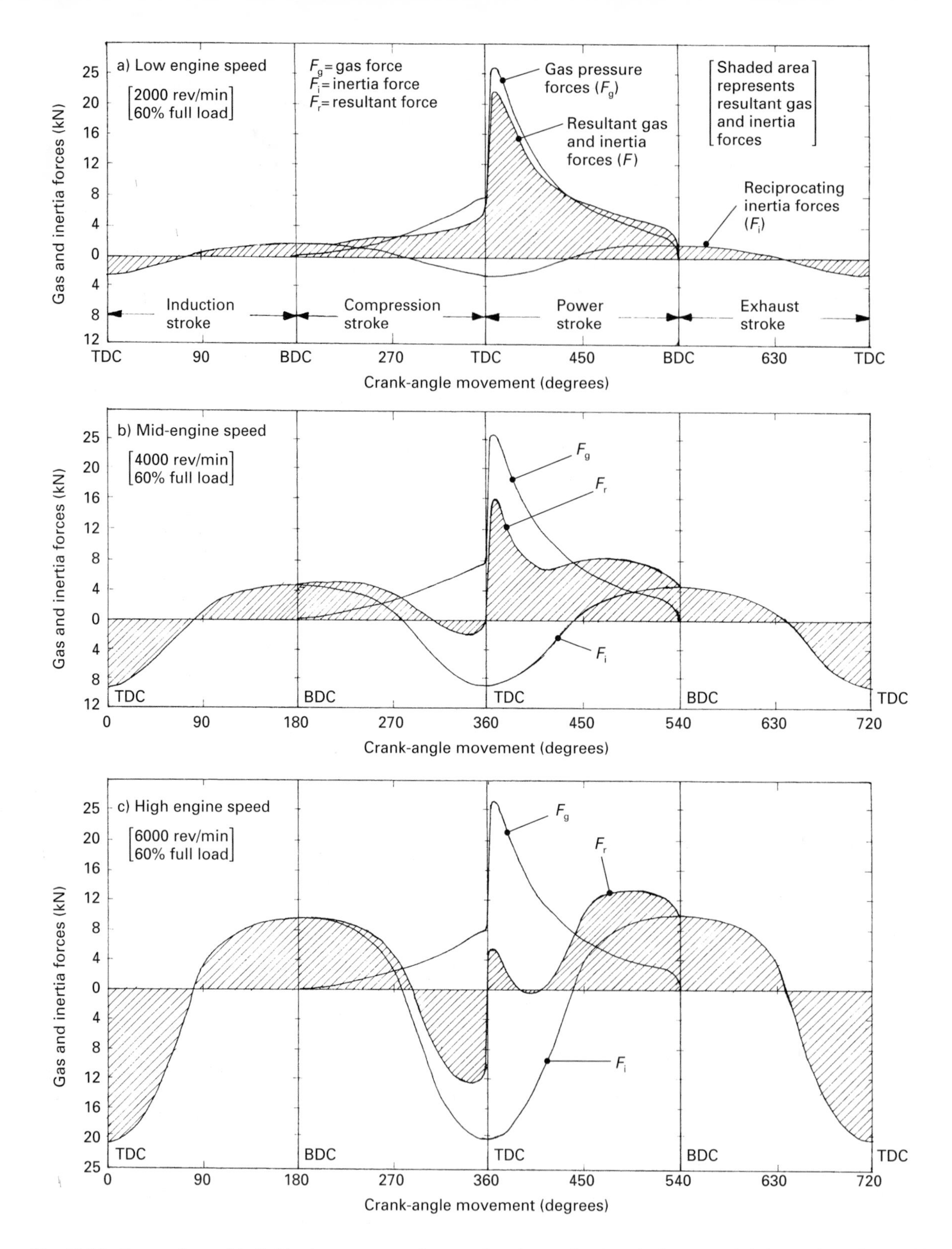

Fig. 12.19 Comparison of individual gas pressure force and reciprocating inertia forces and their resultant forces (big-end bearing load) at various engine speeds

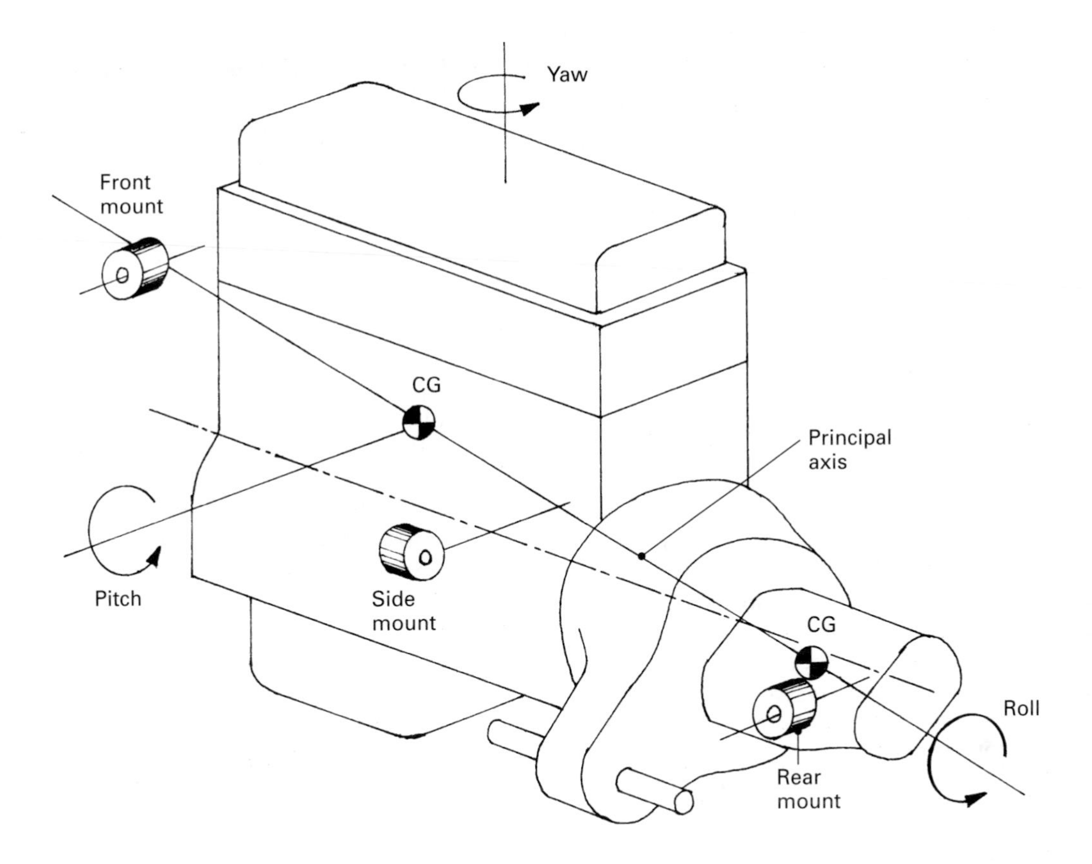

Fig. 12.20 Transverse mounted power unit with three-point support

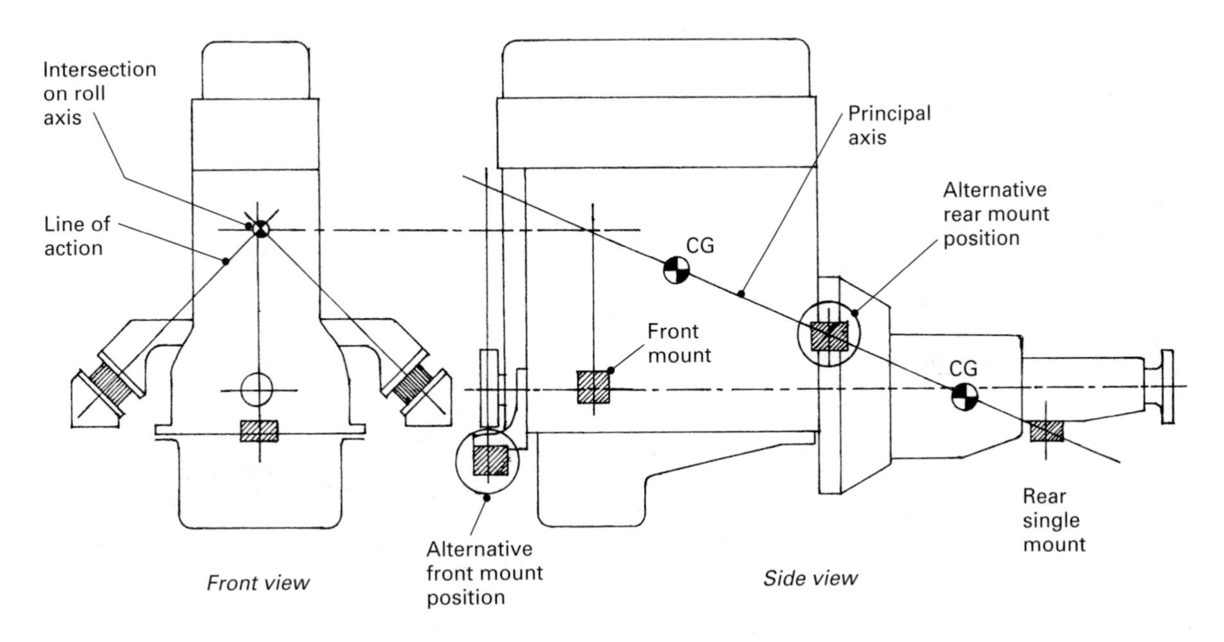

Fig. 12.21 Longitudinally mounted power unit with three- or four-point support

without it having any tendency to settle in one spot. Under these conditions the mass of the engine and transmission will not be subjected to a roll tendency when the vehicle is driven round a corner or bend in the road if the engine is mounted longitudinally, or when the vehicle is accelerated or braked in the case of a transverse-mounted engine.

12.6.3 Location of engine and transmission mountings (figs 12.20 and 12.21)

Flexible mounts for transverse engine layouts are normally positioned on brackets at the front and rear of the combined engine and transmission unit on the same plane as the centre of gravity roll axis (principal axis) (see fig. 12.20). These two mounts first support the static load of the engine and transmission, secondly they absorb both bump and rebound when the vehicle is driven over rough road surfaces, and thirdly they restrain the engine from pitching, as the body tends to roll when driving round sharp bends in the road. A third or even a fourth mount might be positioned on one side or even both sides of the engine to cope with the periodic power and inertia torque reaction impulses. Uneven power intervals or unequal power distribution between cylinders which result in vibrations, particularly when the engine is running light or idling, will be prevented from being transmitted from the engine to the chassis or body structure due to the interposing progressive stiffening elastic media.

With longitudinally positioned car engines a three-point mounting system may be used where there is a front mount support on either side of the engine, with a third mount positioned underneath the gearbox (see fig. 12.21). The front mounts are positioned below the principal axis but are inclined to each other in such a way that lines drawn perpendicular to the rubber blocks intersect at some point along the principal axis, this being equivalent of positioning the mounts flat on the same plane as the principal axis. On some heavy-duty commercial vehicles, the front mounts may be positioned further towards the front, aligning approximately with the crankshaft pulley (see encircled front mount). A rear mount at principal axis level may also be positioned on either side of the clutch bell housing (see encircled rear mount) to provide more substantial support in place of the single gearbox rear mount (see fig. 12.21).

12.6.4 Types of engine and transmission mountings

Single and double inclined wedge mounts (figs 12.22 and 12.23) Single and double inclined wedge mounts are used to suspend longitudinally positioned engine layouts. The single inclined mount (see fig. 12.22) consists of moulded rubber block bonded between two outer steel plates and held in position by inclined engine and chassis brackets. Sometimes a metal interleaf is inserted in the middle of the rubber moulding to restrain the rubber from becoming excessively distorted without raising the overall stiffness of the elastic media. When these mounts support the basic static downward load of the engine, the sandwiched inclined rubber block is initially subjected to a flexible shear stress, but with increased deflection the stress changes progressively from shear to a more rigid compressive stress. For heavy-duty applications the double inclined wedge (see fig. 12.23) provides more restraint on the fore and

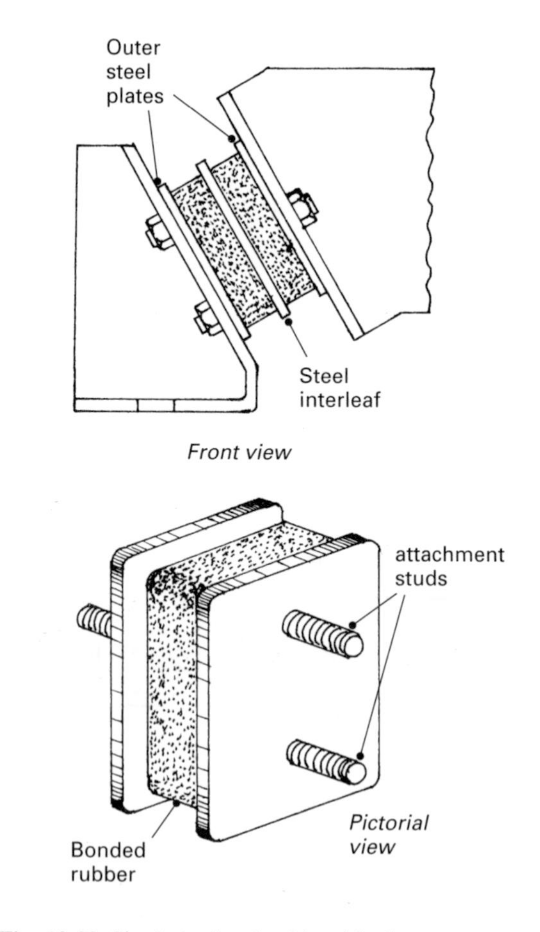

Fig. 12.22 Single inclined rubber block mount

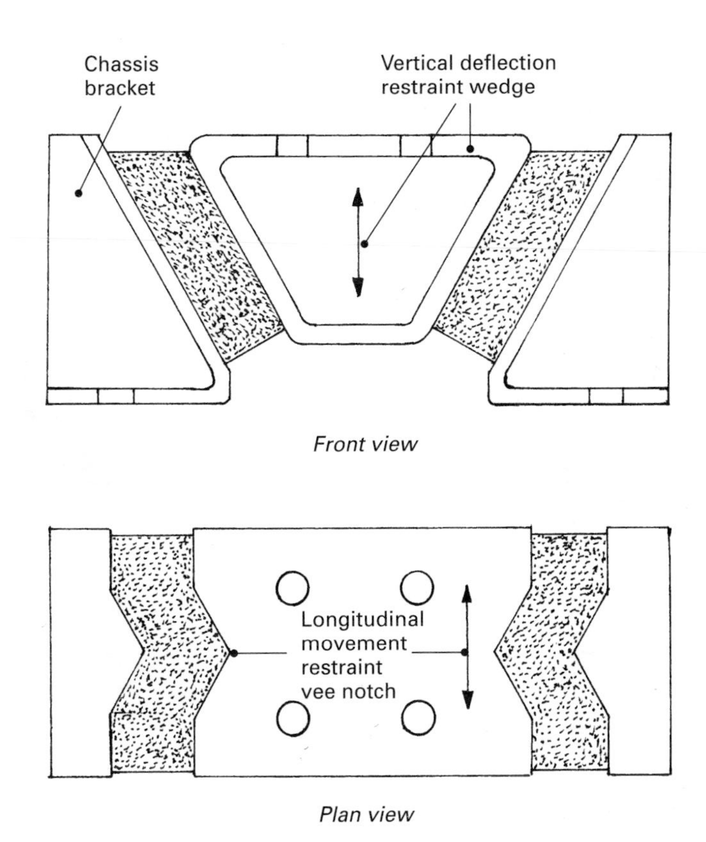

Fig. 12.23 Double inclined wedge rubber mount

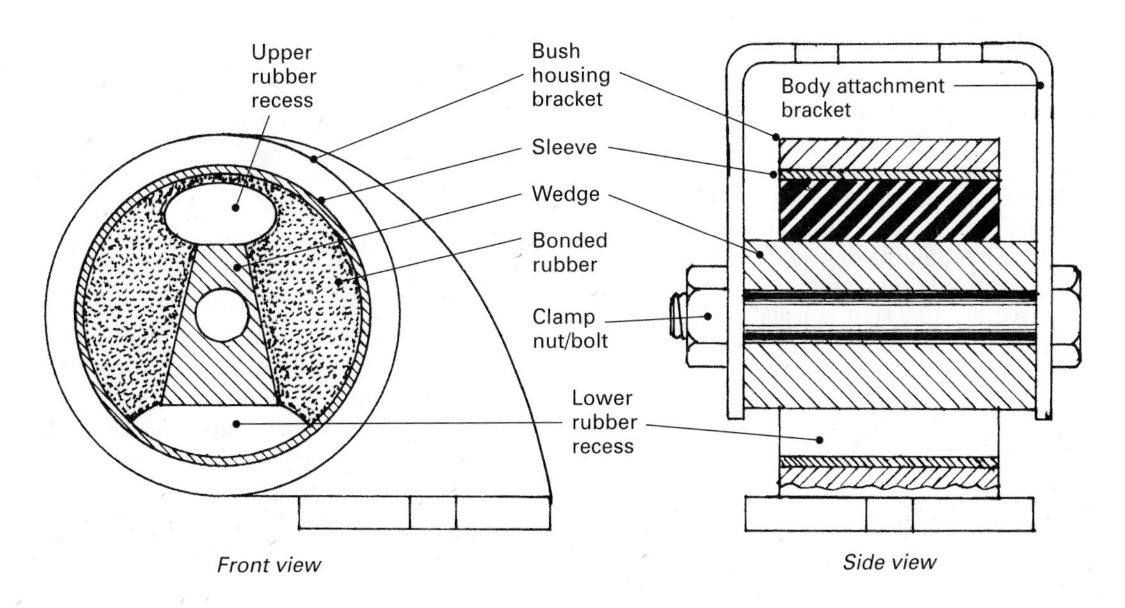

Fig. 12.24 Cylindrical bush vertical reaction power unit mount

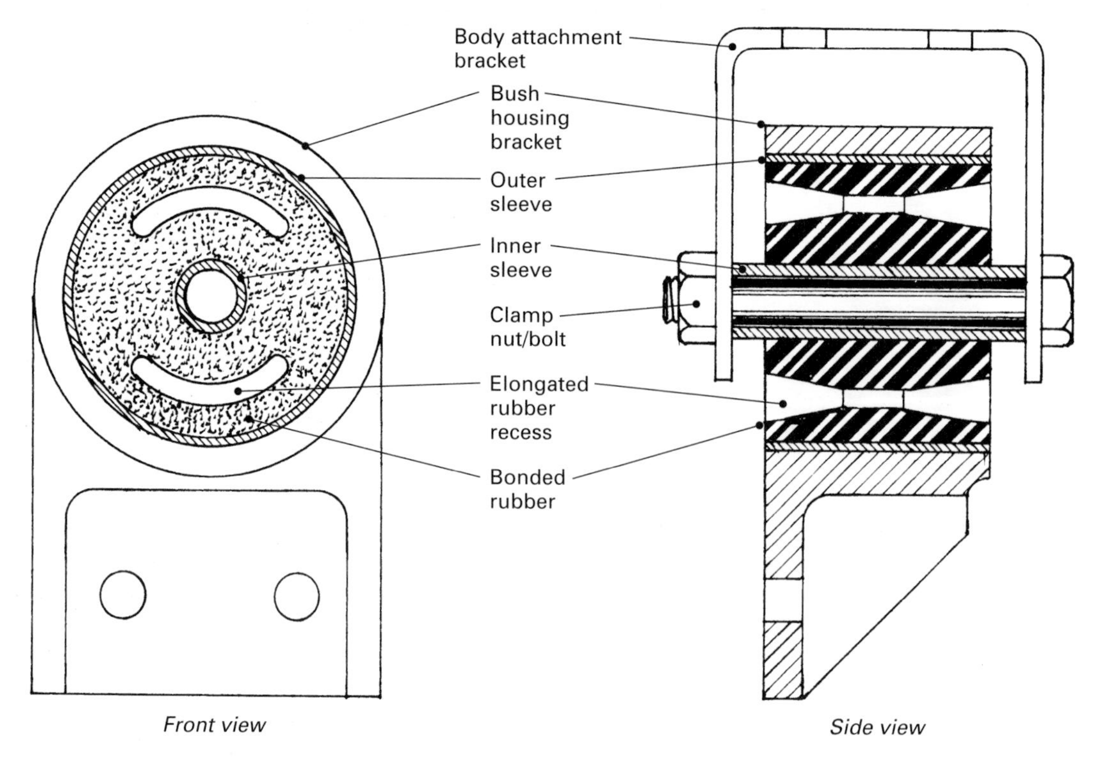

Fig. 12.25 Cylindrical bush roll reaction power unit mount

aft and lateral movement, in addition to the vertical progressive stiffening created by the wedge effect, by providing a transition from shear to compression stress as the distortion of the rubber increases.

Cylindrical bush mounts (figs 12.24 and 12.25) Cylindrical bush-type mounts are used extensively on transversely positioned engine layouts. With this type of mount a rubber bush is bonded to a sleeve which fits inside the cylindrical bore of each engine or gearbox support bracket. The shape of the rubber bush moulding depends upon its application. For the front and rear engine/transmission mounting (see fig. 12.24), it is usual for a wedge-shaped insert with a hole in the centre to be moulded in the middle of the bush, the hole being there to locate the bolt which attaches the engine/transmission bracket to the body support bracket. A small and large gap between the upper and lower rubber moulding and the cylindrical bore respectively, and the inclination of the sides of the wedge provides the required load/deflection characteristics. Thus, when installed, the wedge bears against the rubber, and as the vertical load is increased, the stressed rubber changes from a shear to a compressive mode to raise the elastic stiffness of the mounting. For the side engine mounts (see fig. 12.25), an inner sleeve is moulded in the centre of the bush to accommodate the bolt which clamps the engine/gearbox bracket to the body member support bracket. However, elongated rubber slots or recesses are moulded into the upper and lower regions of the bush. These cavities provide a relatively soft progressive vertical load/distortion characteristic of the bushing, but under extreme torque reaction these recesses close up, thus causing the rubber to go into a stiff compressive stressed state. These bushes also exert a constant stiff resistance in the horizontal fore and aft direction.

Note that transversely positioned engine and transmission unit layout mountings have to absorb not only the gearbox torque multiplication reaction, but also the final drive torque reaction which may increase the overall torque roll reaction fourfold.

13 Cylinder block and head construction

13.1 The cylinder block (fig. 13.1)

The cylinder-block assembly is the casting housing the cylinders, the crankshaft, and (depending on the design) the camshaft which controls the inlet and exhaust valves.

Within the cylinders, combustion produces rapid and periodic rises in temperature and pressure. These will induce circumferential and longitudinal tensile stresses – that is, around the cylinder and in the direction of the cylinder axis – see fig. 13.1. The reaction to the gas pressure is shown by the arrows tending to stretch longitudinally the set-bolts of the cylinder head and the main-bearing housing at the opposite ends of the cylinder block. Simultaneously the gas tries to expand outwards against the cylinder walls, so a plan section view of the cylinder walls would show a ring subjected to tensile circumferential stresses trying to expand the cylindrical sleeve. These induced stresses will be of a pulsating nature, so the cylinder will be continuously stretching and contracting while in operation.

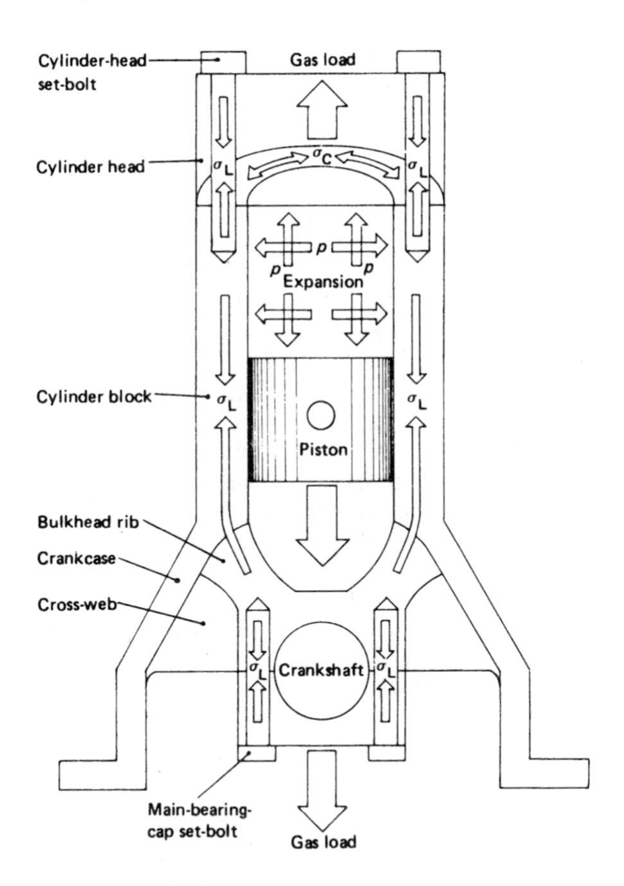

Fig. 13.1 Stress distribution in engine due to gas pressure (σ_L = longitudinal stress; σ_C = circumferential stress; p = gas pressure)

13.1.1 In-line cylinders

The in-line cylinder block assembly can have several variations. One is a separate cylinder head with a single monobloc casting forming an integral cylinder block and crankcase (fig. 13.2). In the example shown, the short crankcase (left-hand side) only comes down to the main bearing split level which is the traditional design for light and medium-duty petrol engines (fig. 13.2). However,

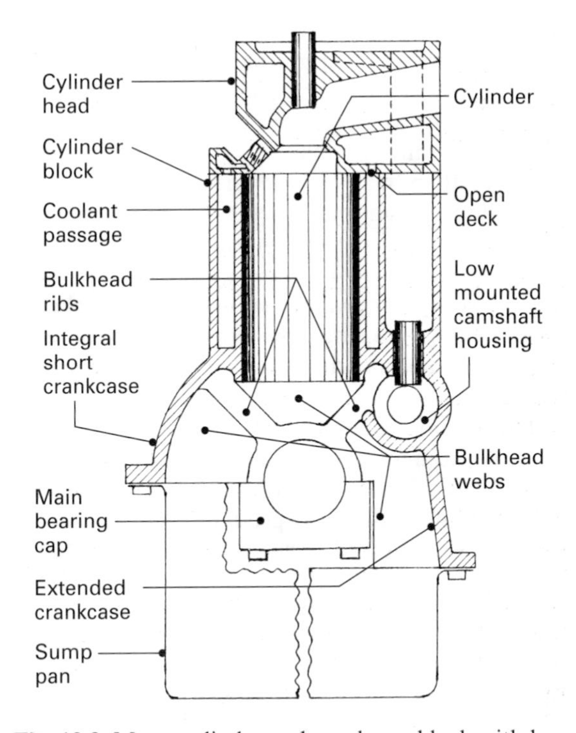

Fig. 13.2 Mono-cylinder and crankcase block with low mounted camshaft

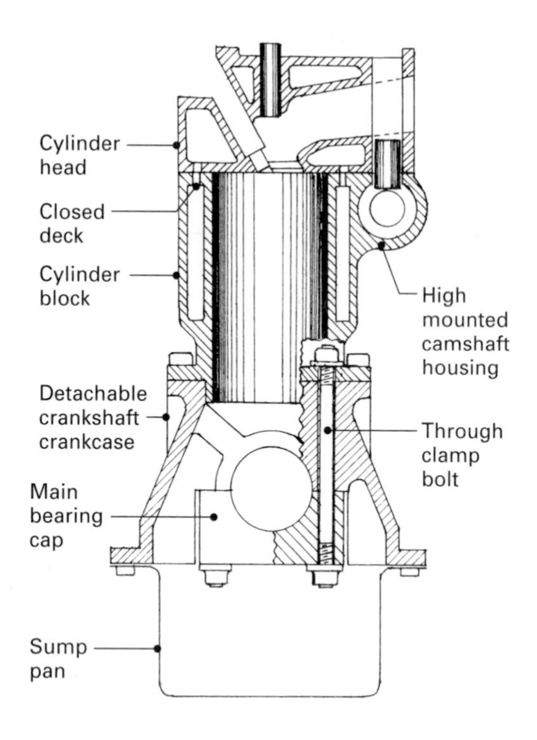

Fig. 13.3 Cylinder block with detachable crankcase with high mounted camshaft

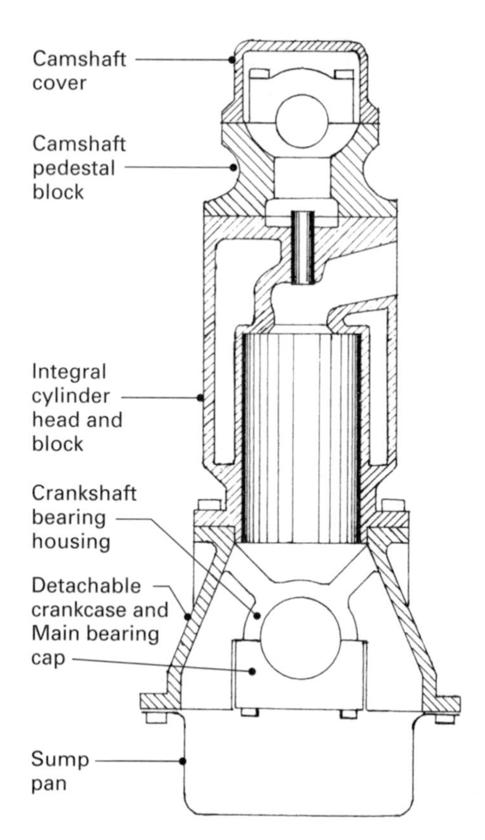

Fig. 13.4 Mono-head and cylinder block with detachable crankcase and overhead camshaft

to improve the rigidity of the crankcase, the extended crankcase (right-hand side) is now preferred for both petrol and diesel engines (fig. 13.2). Alternatively, for large diesel engines the cylinder block and crankcase may be separate castings (fig. 13.3), or there may be a separate crankcase with the cylinder head and block forming an integral single casting (fig. 13.4). A further combination (fig. 13.5) utilises a mono-cylinder and short crankcase block, which stiffens the crankcase and bearing housing, by extending them well below the main-bearing split level with a separate cast block.

The mono-cylinder block and crankcase (fig. 13.2) is the most popular arrangement for small and medium-sized engines since it is relatively easy to cast, is cheap to produce, and provides a very stiff combined structure. The detachable bolt-on crankcase (fig. 13.3) is used on some large diesel engines where, to save weight, an aluminium-alloy crankcase is bolted onto a cast-iron cylinder block. The combined head and cylinder block casting (fig. 13.4) with bolt-on crankcase has been used for heavy-duty diesel engine applications to minimise thermal distortion where the cylinder head meets the top of the cylinder bores, which is always a major consideration in design.

In a conventional engine, cylinder bores can be easily distorted as head bolts are tightened. However, this is totally avoided with the long through bolts (fig. 13.5) which sandwich together the upper camshaft bearing housing, lower camshaft bearing housing/cylinder head, cylinder/crankcase block, and the extended crankcase and main bearing housing. Furthermore the bolts, symmetrical distribution helps avoid the distortion to which aluminium engines can be prone by spreading the clamping load evenly.

Alternative cylinder configurations which may be preferred are horizontally opposed cylinders and vee-banked cylinders. These are described as follows.

13.1.2 Horizontal opposed cylinders (figs 13.6 and 13.7)

To enable the engine to be assembled and dismantled, horizontally opposed cylinders may have either a separate crankcase with banks of

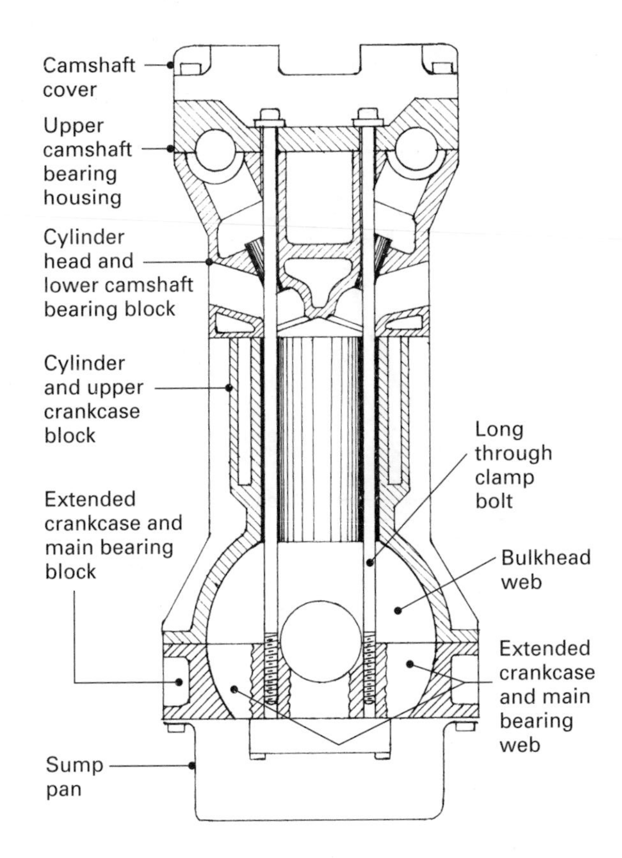

Fig. 13.5 Composite stacked – camshaft housing/cylinder head, cylinder/crankcase block, extended crankcase/main bearing housing with through clamp bolts (Rover 'K' series engines)

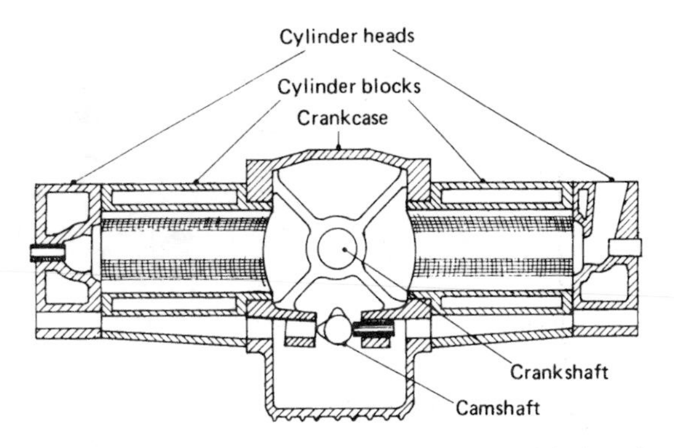

Fig. 13.6 Horizontally opposed cylinders with detachable crankcase

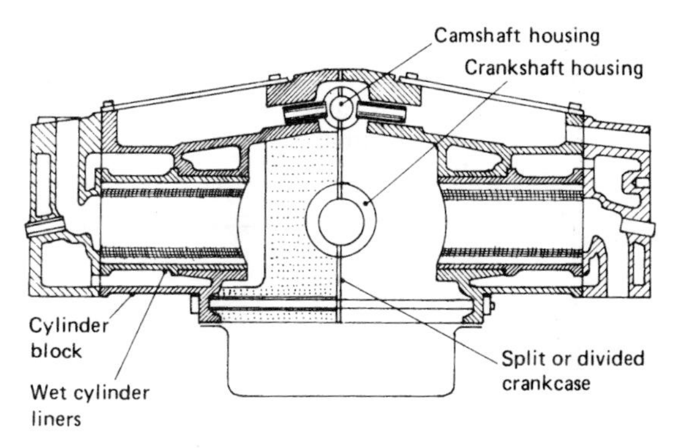

Fig. 13.7 Horizontally opposed cylinders with divided crankcase

two or three cylinders bolted on opposite sides (fig. 13.6) or two half-integral cylinder block and crankcase banks bolted together (fig. 13.7). There may be a central camshaft to actuate the valve push-rods, or twin camshafts, one for each bank, may be more appropriate for high performance.

13.1.3 Vee-banked cylinders (figs 13.8 and 13.9)
For engine cylinder capacities of 2.5 litres or above, the most compact and rigid arrangements use vee-banked cylinders. The most common angle between cylinder banks is 60° for four and six-cylinder engines (fig. 13.8), while 90° is preferred for eight-cylinder engines (fig. 13.9).

It is usual to have an integral cylinder and crankcase block (figs 13.8 and 13.9). To improve the rigidity of the main bearing housing, the integral crankcase walls may be extended below the main-bearing split level (fig. 13.9), or a separate crankcase/main bearing housing block can be bolted onto the underside of the cylinder/crankcase block (fig. 13.8). Large diesel engines using push-rod valve actuation normally have a central single camshaft housed in between the two cylinder banks. For high-performance petrol engines twin overhead camshafts are employed for each cylinder bank (fig. 13.8).

13.1.4 Coolant jacket (figs 13.2, 13.3, and 13.10)
Cast in the cylinder block are coolant passages which surround the cylinder walls circumferentially and lengthwise. The sides of the block thus form the walls of the coolant jacket for approximately the full depth of the cylinders. Near the bottom of the cylinders, the coolant passages end and the cylinder walls merge with the crankcase.

At the top of the cylinder, the coolant passages end either at the level of the block's joint face, which is then referred to as an 'open deck' (fig. 13.2), or just below the block's machined face, the joint surface then being known as a 'closed deck' (fig. 13.3). With a closed deck, the coolant circu-

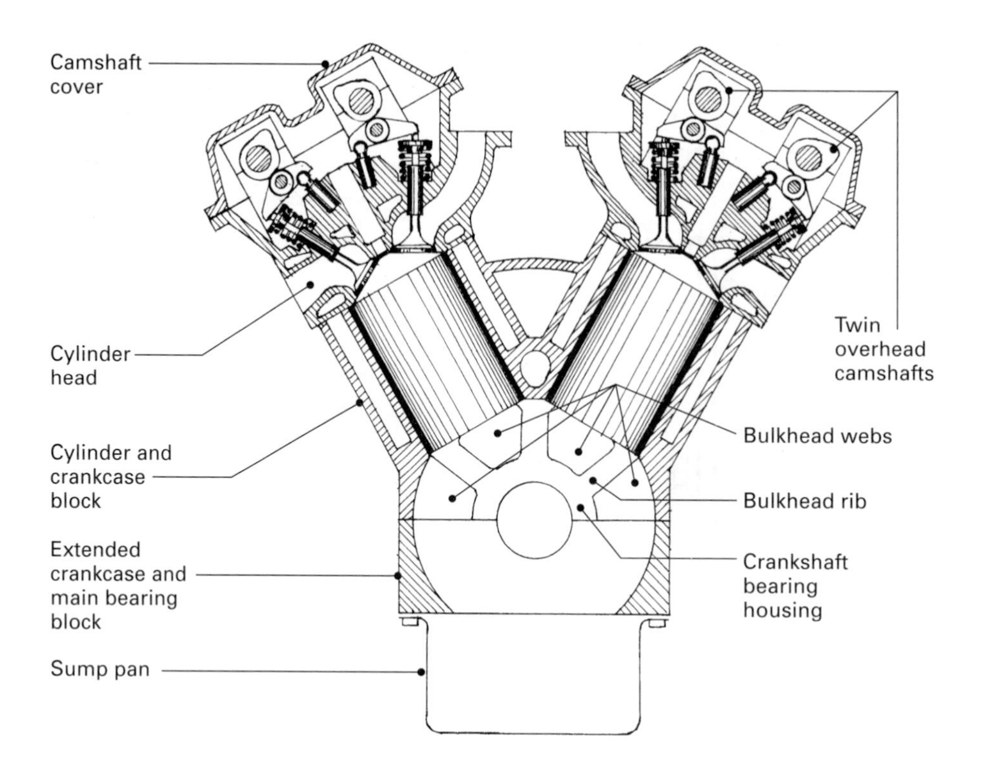

Fig. 13.8 60° Vee mono-cylinder and crankcase block with detachable extended crankcase/main bearing block and twin overhead camshafts

lation is provided by vertical drillings which communicate with corresponding holes in the cylinder head.

A closed deck is preferred to an open deck with respect to joint reliability, since the coolant-flow ducts between the head and the block can be drilled further away from the cylinder bore and there is normally more surface area to squeeze the gasket. On the other hand, it is easier to cast an open-deck cylinder block.

The cylinders are cast parallel and in a straight line. There may be either a small gap between the adjacent outside cylinder walls, to allow coolant to pass as necessary for heavy-duty engines; or, where space is limited, the cylinder walls may be 'siamesed' – that is, adjacent walls merge into a single continuous casting (fig. 13.10).

13.2 The crankcase (figs 13.2 to 13.11)

The function of the crankcase is to provide support for the individual main journals and bearings of the crankshaft and to rigidly maintain the alignment of the journal axes of rotation when they are subjected to longitudinal bending due to rotary and reciprocating inertia forces and the periodic torque impulses which tend to cause torsional distortion.

A tunnel-roof construction is provided which is partitioned-off by bulkhead cross-webs which mount and support the crankshaft main journals and bearings (fig. 13.11). This semicircular ceiling with spaced-out cross-webs provides a very stiff but relatively light crankcase construction.

Over the underslung crankshaft, the crankcase walls form a skirt which may either be separately attached to the cylinder block's lower deck (figs 13.4 and 13.5) or may merge into it as an integral casting (figs 13.1 and 13.2). The crankcase skirt may enclose the crankshaft from cylinder-block to crankshaft-axis level (fig. 13.2), but for extra rigidity the walls may extend well below the crankshaft (figs 13.3, 13.4, and 13.5) – this being preferred for both high-performance and heavy-duty engines.

To provide additional support to the cross-webs, ribs may run from the bottom of the cylinder block diagonally towards the main-bearing housings for all or just the front and rear cross-webs. With some aluminium-alloy integral cylin-

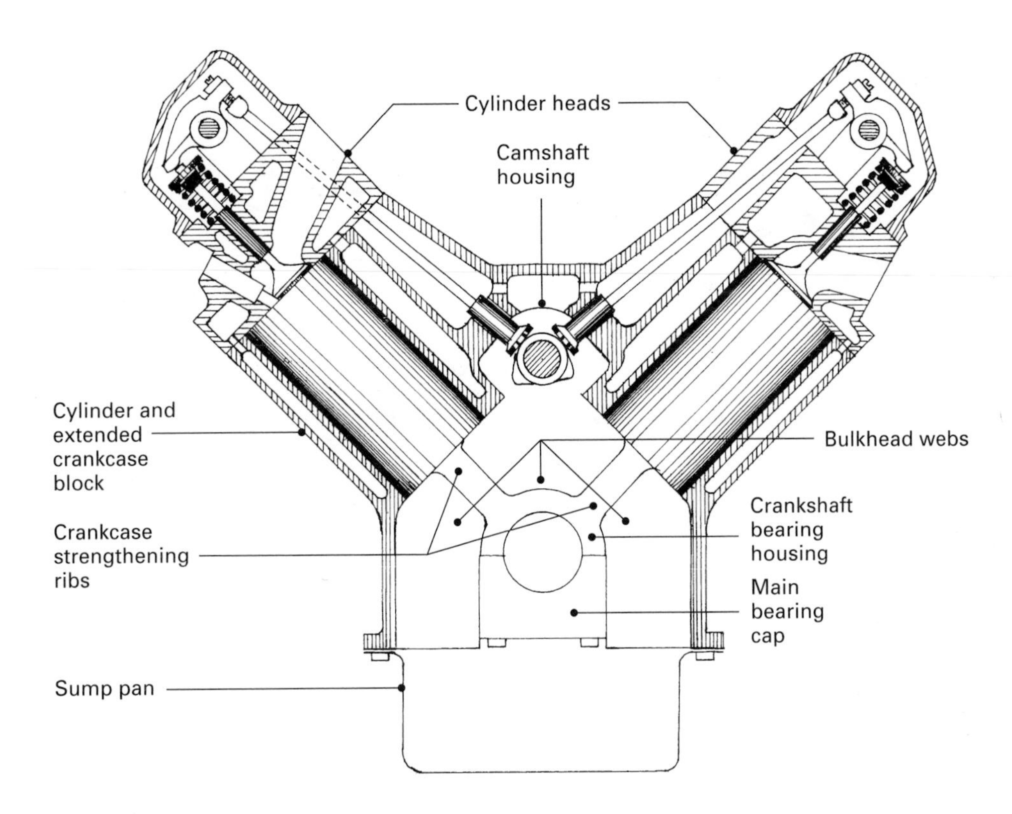

Fig. 13.9 90° Vee mono-cylinder/crankcase block with centrally mounted crankshaft

der-block-and-crankcase constructions, stiffening ribs are cast longitudinally and vertically downwards on the outsides of both the block and the crankcase walls.

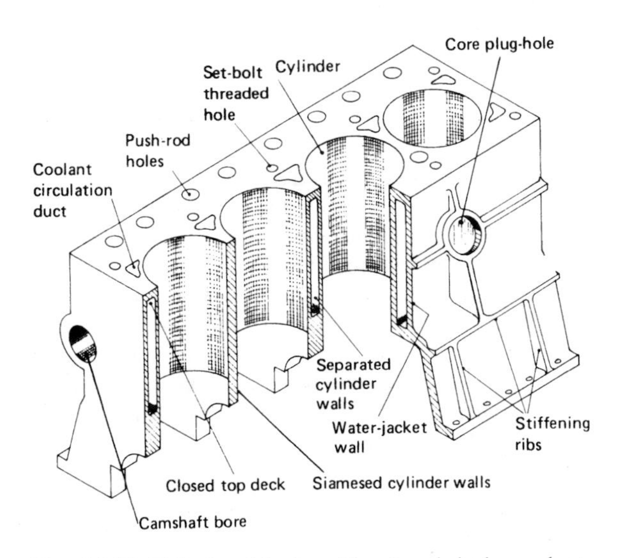

Fig. 13.10 Cylinder block with closed-deck coolant jackets, showing both separate and siamesed adjacent cylinder walls

The bottom of the crankcase walls are flanged both to strengthen the casing and to provide a machined joint face for the sump to be attached.

13.3 Camshaft location and support (figs 13.2 to 13.5, 13.12, and 13.13)

The function of the camshaft is to phase the opening and closing of each cylinder's inlet and exhaust cylinder-head poppet-valves relative to the crankshaft rotation.

Camshafts in the cylinder block are mounted parallel to the crankshaft and to one side of the cylinder (fig. 13.12) either low-down just above the crankshaft (fig. 13.2) or high a little below the cylinder head (fig. 13.3). Alternatively, the camshaft can be mounted centrally over the cylinder head on a pedestal support (figs 13.4, 13.5 and 13.13).

To support the camshaft there are usually three or five plain bush-type white-metal or tin–aluminium bearings which are a force fit in the cylinder-block or head-pedestal housing bores.

13.4 Cylinder-block materials

The cylinder block should be made from materials which, when cast in a monobloc form, have

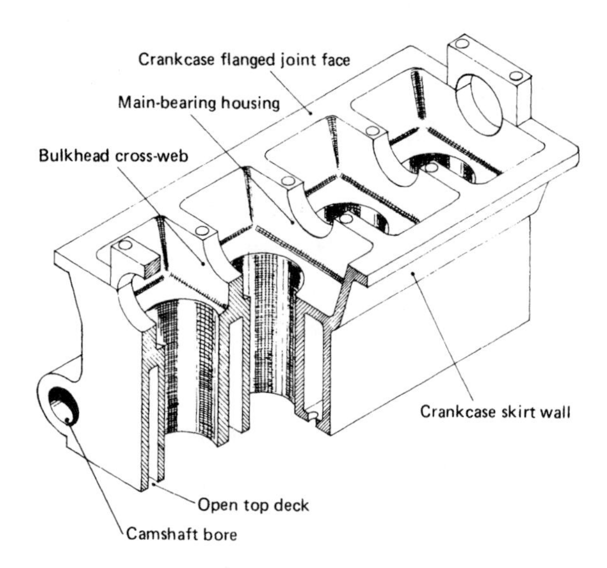

Fig. 13.11 Integral cylinder block and crankcase, showing both tunnel crankcase and open-deck water jackets

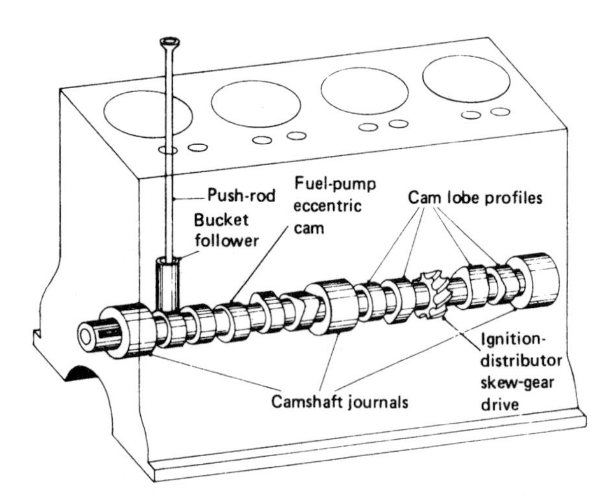

Fig. 13.12 Cylinder-block-mounted camshaft

adequate strength and rigidity in compression, bending, and torsion. This is essential so that the necessary support is provided against the gas loads and to the components which convert the reciprocating action of individual cylinder's mechanisms into a single rotary action along the crankshaft length.

The desirable properties of a cylinder-block material are as follows:

a) it should be relatively cheap,
b) it should readily produce castings with good impressions,

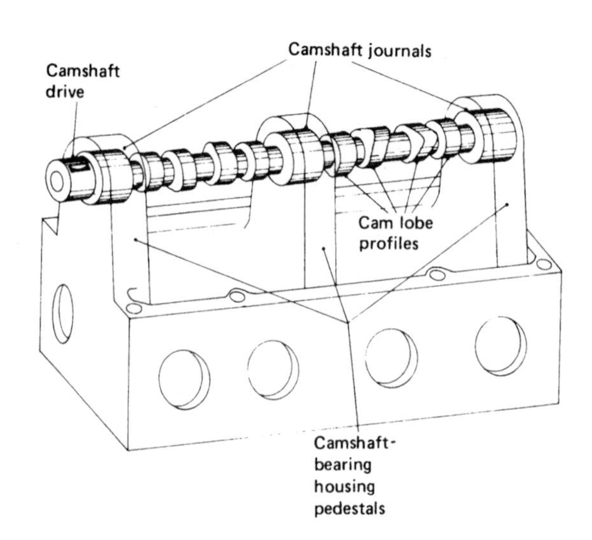

Fig. 13.13 Cylinder-head-mounted camshaft

c) it should be easily machined,
d) it should be rigid and strong enough in both bending and torsion,
e) it should have good abrasion resistance,
f) it should have good corrosion resistance,
g) it should have low thermal expansion,
h) it should have a high thermal conductivity,
i) it should retain its strength at high operating temperatures,
j) it should have a relatively low density.

Cast iron meets most of these requirements except that it has a low thermal conductivity and it is a comparatively heavy material.

Because of these limitations with cast iron, there has been a trend for petrol engines to adopt light aluminium alloys as alternative cylinder-block materials. Cylinder liners (see section 13.11) may be incorporated in cast-iron blocks as an option; but they are essential with the relatively soft light aluminium alloys, which cannot be used directly for wear-resisting cylinders. To compensate for the lower strength of the aluminium alloys, the alloy blocks are cast with thicker sections and additional support ribs, which brings their relative weight to about half that of the equivalent cast-iron blocks.

A typical cast iron would be a grey cast iron containing 3.5% carbon, 2.25% silicon, and 0.65% manganese. The carbon provides graphite to improve lubrication, with the silicon controlling the formation of a laminated structure known as pearlite which is mainly responsible for good wear resistance, while the manganese helps to strengthen and toughen the iron.

A common aluminium-alloy composition would be 11.5% silicon, 0.5% manganese, and 0.4% magnesium, with the balance aluminium. The high silicon content reduces expansion and improves castability, strength, and abrasion resistance, while the other two elements strengthen the aluminium structure. This alloy has good corrosion resistance, but it can absorb only moderate shock loads.

13.5 The cylinder head (figs 13.14 to 13.17)

The cylinder head is a casting which is assembled on top of the cylinder block, its functions being to house the inlet and exhaust poppet-valves and their respective ports, to house the spark-plug or injector location holes, and to form the upper face of the combustion chamber and take the combustion-pressure reaction. In addition, within the casting are coolant passages, cavities, and channels which surround the valve seats and ports and the sparkplug or injector bosses (fig. 13.14). These passages communicate with the coolant in the cylinder-block jacket through vertical ducts or drillings with corresponding contact faces in both the head and the block.

The mating faces of the cylinder head and block are ground flat so that, when a gasket is sandwiched in between and the head is bolted down, the gasket is squeezed together to form a gas-tight and liquid-tight joint.

13.5.1 Cylinder-head valve and port layouts (figs 13.15 to 13.17)

Both the inlet and the exhaust ports may emerge from the same side of the cylinder head, with the valves arranged side by side in a single row along the length of the head. This valve-and-port layout forms a loop-flow cylinder head (fig. 13.15). With this arrangement, both the inlet and the exhaust manifolds are on the same side, so that the induction can be pre-heated by the hot exhaust manifold to improve cold-running.

The same valve positioning can be used but with the inlet ports emerging on one side of the head and the exhaust ports on the other. This is known as an offset cross-flow head (fig. 13.16), and the advantages claimed are better breathing and lower exhaust-valve temperatures.

An alternative configuration, which is slightly more expensive but preferred for high-performance, positions the valves transversely across the cylinders, so that the inlet and exhaust valves form two separate rows along the cylinder head. The ports then emerge from the respective sides of the cylinder head to form an in-line cross-flow head (fig. 13.17). Generally the valves are inclined to each other in a hemispherical combustion chamber which permits larger valves to be used.

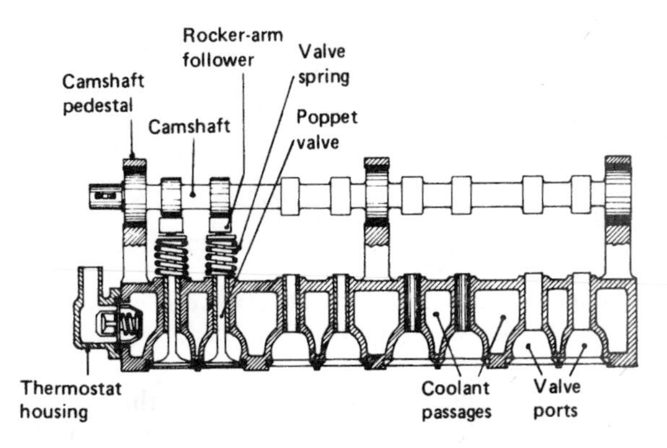

Fig. 13.14 Four-cylinder overhead-camshaft cylinder head

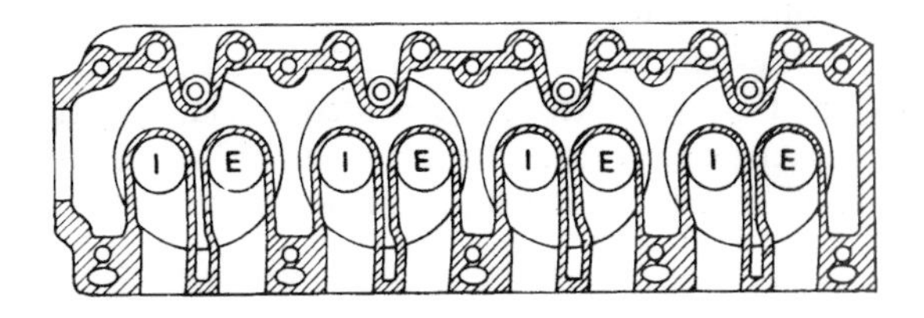

Fig. 13.15 Cylinder head with side-by-side valves and ports

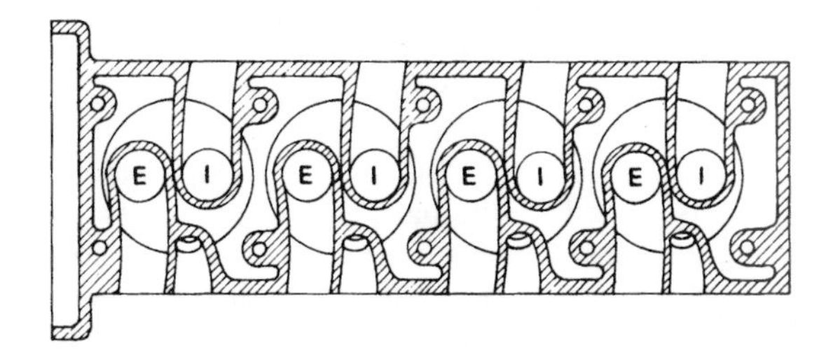

Fig. 13.16 Cylinder head with side-by-side valves and cross-flow ports (offset cross-flow head)

13.5.2 Thermostat housing (fig. 13.14)

To control the rate of coolant circulation to suit the amount of heat released by combustion, a thermostat valve is housed within and at one end of the cylinder head. It is usually situated in

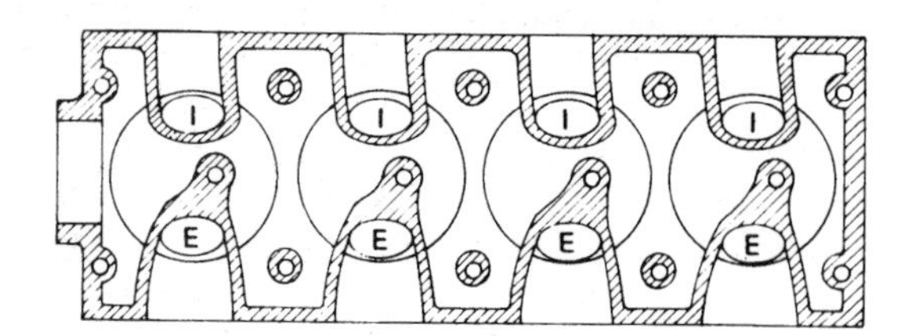

Fig. 13.17 Cylinder head with transverse valves and cross-flow ports (in-line cross-flow head)

the main coolant passage which takes the coolant from the head through to the top hose to the radiator, so that it can interrupt the flow of coolant before the engine has reached its working temperature (fig. 13.14). Once the operating temperature of the engine has been reached, the temperature-sensitive valve element will automatically open, thus permitting the heated coolant in the cylinder head to pass unrestricted to the radiator which dissipates the heat to the atmosphere.

13.6 Cylinder-head materials

The cylinder head should be made from a material which can readily be cast with complicated internal shapes both for the coolant passages and for the inlet and exhaust ports. The material's mechanical and thermal properties should be such that it is strong enough in compression to be clamped rigidly to the cylinder block by hold-down bolts or studs and is able to operate continuously under fluctuating gas pressures and temperatures. Generally the gas-pressure loads are not excessive for the available engineering materials, but the temperature gradients established across the thickness of metal between the combustion-chamber side and the coolant passages, between adjacent inlet and exhaust valve seats, and from the centre of the combustion chamber to the cylinder walls will produce an unevenness in the expansion and contraction of the metal in these regions. Consequently, thermal stresses will be created across the cylinder head, and these may eventually distort or even crack critical areas which are exposed to the heat of combustion.

The ideal cylinder-head material is the one which can limit the temperature of the cylinder-head combustion-chamber surface so that lubrication remains effective, combustible petrol-and-air mixtures do not overheat to cause detonation, hot spots are not established to promote pre-ignition, and high cyclic thermal stresses are avoided. Unfortunately under various operating conditions such as continuous full-load running on motorways or under part loads with weak mixtures and late ignition, surface temperatures will rise and local thermal stresses can easily reach dangerously high values if the heat cannot be adequately dissipated through the cylinder head.

The choice of materials is generally restricted to grey cast iron and aluminium alloys, but neither of these cast materials has anywhere near all the desirable properties.

The traditional cylinder-head cast iron meets most of the requirements, such as cheapness; good castability; good machinability; good corrosion resistance; adequate rigidity, strength, and hardness; and low thermal expansion. However, cast iron has the disadvantage of being heavy and having a low thermal conductivity.

The alternative material, aluminium alloy, provides slightly different merits: it has half the weight of equivalent cast-iron heads and its thermal conductivity is three times better than that of cast iron, so that the cooler-operating head allows higher compression-ratios to be used and there will be lower temperature gradients in the head so the likelihood of thermal distortion is reduced. The shortcomings of aluminium alloy are that it is more expensive; the corrosion resistance is not so good as cast iron's and can under certain circumstances lead to problems; it is much softer than cast iron and more care is needed during maintenance; it has a high thermal expansion, which can cause fretting between an aluminium-alloy head and a cast-iron cylinder block during starting and stopping conditions; and, finally, separate wear-resisting valve seats and guide inserts are essential.

The composition of the cast iron used is similar to that for the cylinder block (section 13.4), but slightly different aluminium alloys are preferred for the cylinder head. There are two commonly recommended, as follows:

i) 3.0% copper, 5% silicon, 0.5% manganese in a matrix of aluminium;
ii) 4.5% silicon, 0.5% manganese, 0.5% magnesium in a matrix of aluminium.

The additions of both copper and silicon reduce the thermal expansion and contraction and improve the fluidity and castability of aluminium. Copper added to aluminium hardens and strengthens the structure over a period of time (this is known as age-hardening), and silicon improves the abrasion resistance. Both manganese and magnesium improve the strength of the alloy. Unfortunately the corrosion resistance of the otherwise slightly superior alloy containing copper is inferior to that of the copper-free silicon–aluminium alloy.

13.7 Stud and set-screw threaded cylinder-block holes (fig. 13.18)

The cylinder head is assembled on the top deck of the cylinder block and is attached by either studs or set-screws which encircle the cylinder bores. The screwing down of the cylinder head to the block puts the cylinder head in compression and the studs or set-screws in tension, which tends to pull out and strain the metal around the threaded region on top of the cylinder block (fig. 13.18). To provide adequate support, therefore, the mass of the metal bosses surrounding the threaded holes should be as large as possible. To give sufficient stud or set-screw 'screw-in' joint strength, the depth of the threaded counterbore should be at least twice the diameter; and, to prevent local distortion on the surface of the deck when under tension, the threads cut in the block should start at least 0.3 times their diameter below the surface (fig. 13.18).

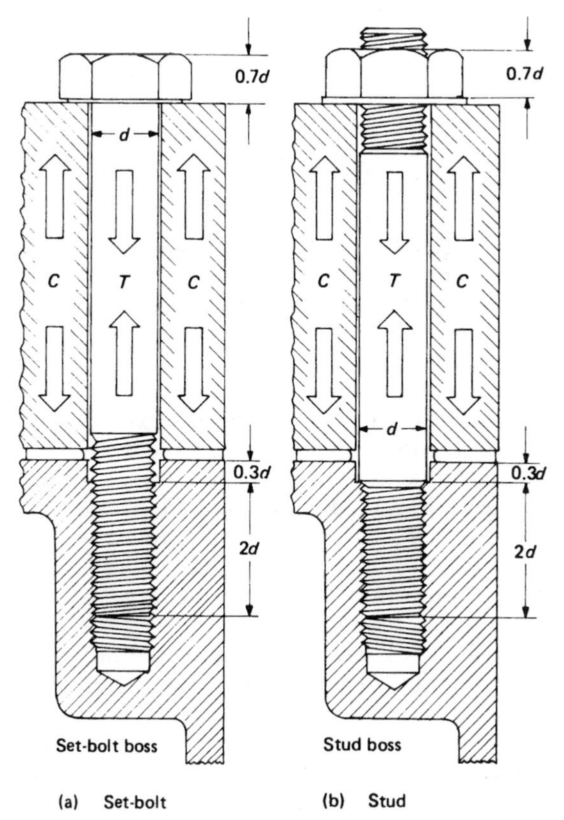

Fig. 13.18 Counterboring for cylinder head (d = set-bolt or stud diameter; T = tensile load, C = compressive load)

The hold-down-screw holes should be as close as possible to the bore, but if they are too near they will distort the top of the cylindrical bore out of roundness. Conversely, if the threaded holes are too far from the edges of the bores, the joint faces will tend to pull open during combustion and so their squeeze and sealing effectiveness will be reduced.

When using an aluminium-alloy cylinder head, it should always be held down by set-screws, otherwise any corrosion products formed between the studs and their respective holes in the cylinder head make it almost impossible to withdraw the head over studs which have been screwed into the block.

The minimum number of threaded hold-down holes in the top deck of the block is four, or in some cases five for engines with individual cylinder capacities up to about half a litre. Above this cylinder size for diesel engines, six, seven, or sometimes even eight or nine hold-down screws may be considered necessary to secure the head joint. There are generally two rows of threaded holes along each side of the cylinder-block top deck, and they are normally arranged with one hole in each far corner with the remainder spaced between pairs of cylinder bores. Thus, except for the ones at each end at the front or back, all the studs or set-screws share their compressive clamping effort between pairs of adjacent cylinder bores. Therefore each stud or set-screw influences the sealing of the top of the cylinder walls to the cylinder-head lower deck over approximately a quarter of the circumference of each adjacent cylinder.

13.7.1 Stud and set-bolt material

Good-quality hold-down studs or set-bolts may be made from manganese–molybdenum steel. A typical composition would be 0.35% carbon, 0.2% silicon, 1.6% manganese, 0.3% molybdenum, with the balance iron. After a heat treatment of hardening and tempering, the mechanical properties would be as follows:

Tensile strength	1000 N/mm^2
Yield strength	800 N/mm^2
Impact toughness	47 joules
Hardness	290 to 340 Brinell

The safe working strength for steel bar is usually taken as 80% of the yield strength of the steel, giving in this case a figure of 640 N/mm^2.

13.7.2 Stress distribution in a tightened cylinder-head set-screw joint (fig. 13.19)

Figure 13.19 shows a section of a cylinder-head set-screw hold-down joint, and the stresses are represented by the grid lines. Widely spread lines indicate low stresses, and closely spaced lines imply a high stress concentration. This illustration shows that within the set-screw there is considerable stress concentration at the shoulder of the set-screw head and where the set-screw enters the cylinder-block top deck. Within the threaded region of the block, at the joint interface, and where the shoulder of the set-screw head contacts the top of the cylinder head, the vertical and horizontal stress lines are close together. The horizontal crowded stress lines curving upwards as they merge into the threaded holes imply that the metal around the thread is being pulled away from the block towards the cylinder head. At the same time, as the set-screw is in tension, the horizontal stress lines around the cylinder-head hold-down set-screw hole curve and diverge both upwards and downwards away from the wall of the hole – this means that the stress is greatest near both the top and the bottom of the head and is least in the middle region.

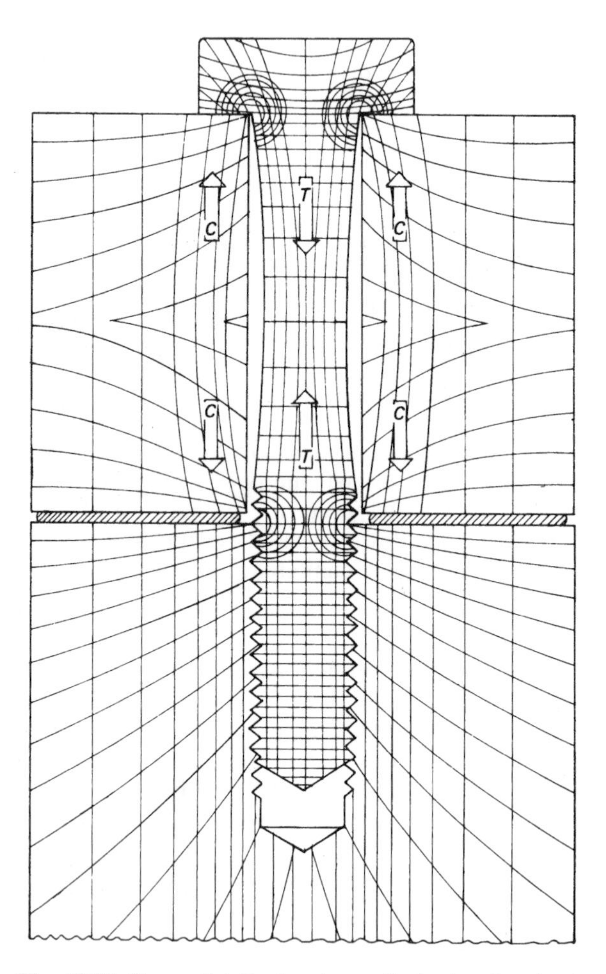

Fig. 13.19 Stress distribution in a cylinder head and set-screw (T = tensile load; C = compressive load)

The information provided by the stress grid lines within the cylinder-head clamping elements when a normal pre-tension is applied to the set-screw shows that a non-uniform complex stress distribution exists throughout the structure. Local stress concentrations such as may occur around the threaded joint may exceed the elastic limit of the set-screw material; so, although the tensile load may be far below the tensile strength of the steel, plastic strain may occur and eventually the static and pulsating loads may lead to fatigue failure.

13.8 The tightening-down of cylinder heads (figs 13.20 and 13.21)

The cylinder-head studs or set-screws are pretensioned by tightening them to a predetermined torque to achieve the following:

a) to provide sufficient compression between the joint and gasket faces so that adequate frictional force is established between the contact faces to prevent gasket leakage or blow-out at the joint gap;
b) to provide sufficient clamping load to compensate for differential thermal expansion and contraction of the head relative to the cylinder block when it heats up or cools, particularly with aluminium-alloy cylinder heads;
c) to provide sufficient pre-tensioning to compensate for the periodic fluctuating combustion pressure pushing the cylinder head away from the cylinder block, otherwise the gasket squeeze will be insufficient to prevent leakage;
d) to provide additional tensioning of the studs or set-screws to allow for any yielding of their shanks and threaded fixtures which might take place over a period of time;
e) to provide additional tensioning of the hold-down studs or set-screws to enable the joint faces to come further together as the gasket creeps and settles.

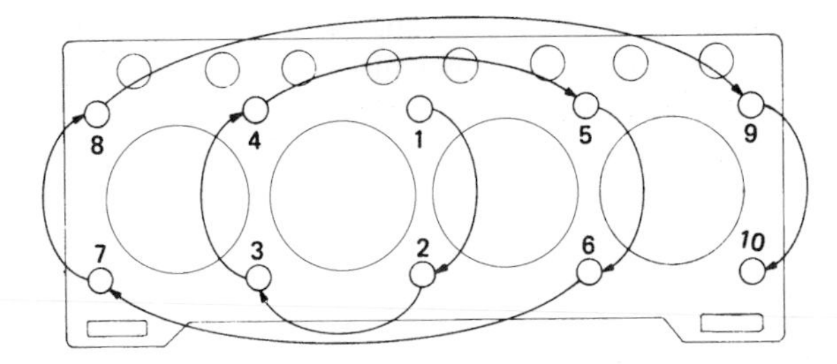

Fig. 13.20 Cylinder-head tightening sequence for four-cylinder petrol engine (Renault 18)

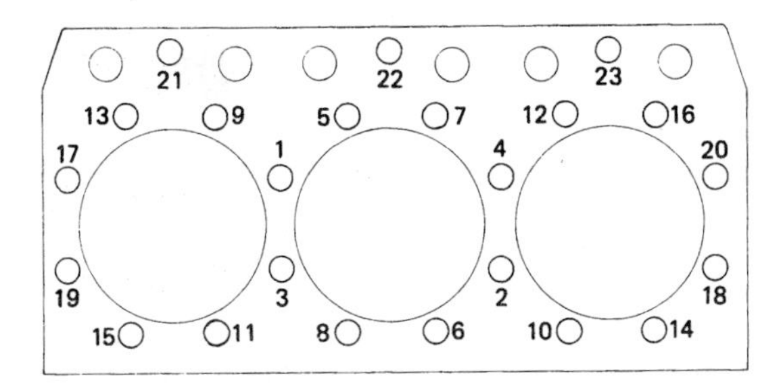

Fig. 13.21 Cylinder-head tightening sequence for six-cylinder diesel engine having twin heads (Rolls-Royce Eagle Mk III)

13.8.1 Procedure for replacing and tightening-down cylinder heads

Before fitting the cylinder head, scrape any carbon, corrosion products, and remains of the old gasket from both joint faces and examine the surfaces for cracks, local overheating (causing colouring of the metal), and distortion. Position the new gasket over the cylinder-block top deck, making sure that all the coolant and oil holes are correctly aligned.

The general procedure to adopt is to screw each set-screw or nut down fingertight in any order; then, starting from the middle of the head and working outwards in a spiral sequence, lightly tighten down the head. Finally, using a torque wrench and following the same tensioning sequence, tighten all the studs or set-screws to the recommended torque (see figs 13.20 and 13.21).

Unevenness in tightening the nuts or set-screws will deflect and distort the joint faces of the cylinder head and block and, if the deformation is excessive and the gasket compliance – that is, its ability to take up the space between the mating surfaces – is inadequate, permanent damage may result.

13.8.2 Considerations in tightening cylinder heads

The hold-down studs or set-screws should be strong enough to withstand the constant static tensile stress once they have been tightened and, at the same time, be adequately tough and fatigue-resistant to sustain the fluctuating tensile stresses caused by the cyclic combustion pressure.

To provide a continuous effective joint seal, the total clamping hold-down force around each cylinder bore should exceed at least 3 to 3.5 times the peak gas-pressure load.

As a general rule the studs or set-screws should never be tensioned beyond about 80% of their yield strength, so that they always operate well within their elastic range. However, localised stresses, such as are established in the thread-root region or near the set-screw head, may exceed these values and a small amount of plastic deformation may take place. The engine manufacturer usually takes this into account and includes a safety allowance when quoting tightening torques.

Excessive tightening will cause permanent strain in the studs or set-screws. With time, elastic recovery may be established, but the fatigue strength and safety factor will be reduced.

The drawing together of the joint faces when it is done uniformly will nevertheless produce an uneven distribution of pressure between adjacent studs or set-screws on the sandwiched gasket layers, these pressures obviously being greatest around the hold-down threaded holes and the least in between. The result of this varying pressure distribution is a variation of sealing capacity over the entire gasket joint. Furthermore, the compressive load on the gasket is constantly fluctuating due to the periodic gas loads, so the joint seal is continuously trying to adjust itself.

If tightening is begun from both ends, it will cause the gasket to extrude and pile the gasket material towards the middle. In addition there might be a degree of ripple formed in the gasket sheet. The net result will be uneven gasket thickness between the joint faces, so the head joint face will be distorted and strained to match the gasket profile.

Uneven tightening or varying gasket thickness around the cylinder bore may distort the top of the cylinder bore out of roundness, so that the piston rings do not seal effectively. This is particularly so for cylinder blocks which are fitted with liners that are designed to stand proud of the top deck face.

13.8.3 Torque wrench

A torque wrench is a lever normally attached to the head of a bolt or nut via a socket spanner. It includes some sort of measuring scale which permits the torque build-up to be observed or for the required torque limit to be initially set on the wrench. The torque applied to the bolts or nuts draw together and compress the components sandwiched in between, whereas the bolt or stud shanks are tensioned up to their pre-set clamping load necessary to seal the gasket joint.

Torque is the product of the perpendicular pulling or pushing force and the distance between the centre of the bolt or stud and the middle of the lever handle where the force is concentrated. Thus

torque = perpendicular force × effective length of lever Nm

$$T = F \times d \quad \text{Nm}$$

where F = force in newtons (N)
d = lever length (m)
T = torque applied to bolt or nut (Nm).

There are basically two types of torque wrench

1) the load-indicating torque wrench where the lever arm bends proportionally to the applied torque, and
2) the pre-setting torque wrench where the applied torque is permitted to increase up to the desired torque setting at which point the pre-set mechanism trips and makes a loud clicking sound.

Load-indicating-type torque wrench (fig. 13.22) The load-indicating-type wrench consists of a square socket drive adapter with a torque-arm and a rod-arm attached which are positioned parallel to each other. A handle is formed at the end of the torque-arm whereas the free end of the rod-arm acts as a pointer. A torque scale is attached at right angles to the torque-arm at the handle end of the arm. When the handle is pulled or pushed to the desired torque setting, the arm bends proportionally to the torque being applied to the square socket drive, whereas the pointer arm remains straight. Thus the deflection of the torque-arm will move the scale relative to the pointer. Hence the torque applied can be increased until the desired torque value is reached on the scale.

Pre-setting-type torque wrench (fig. 13.23) This type of wrench relies on pre-loading a compression spring housed in the handle of the torque

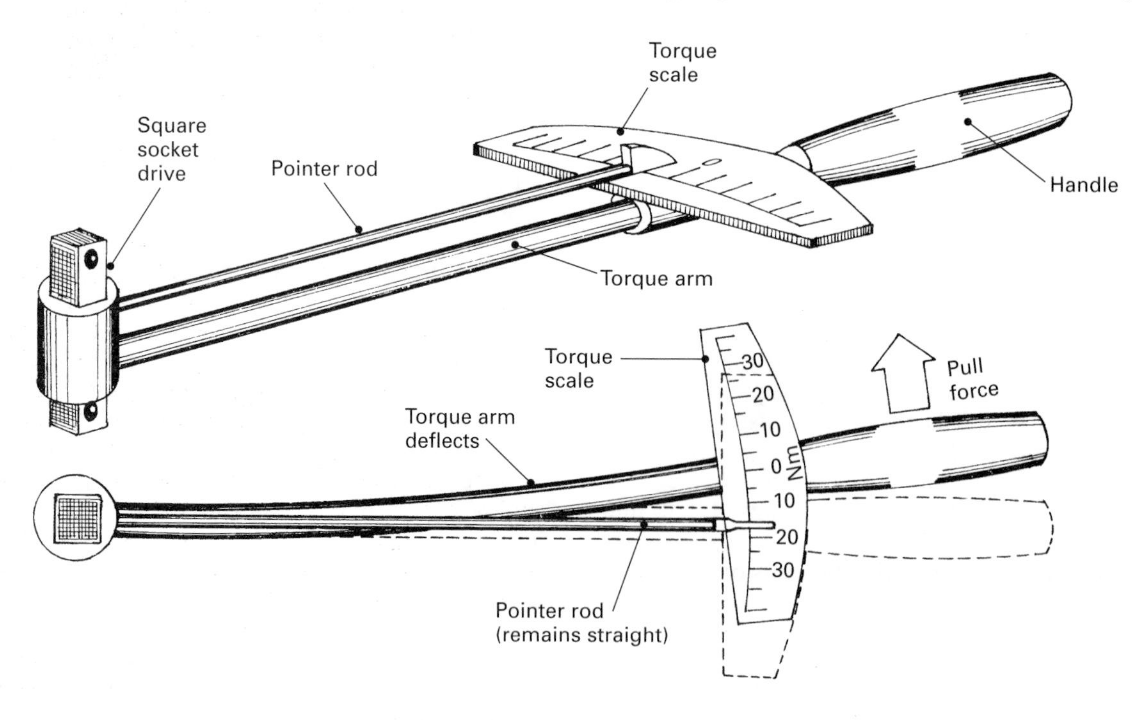

Fig. 13.22 Load-indicating torque wrench

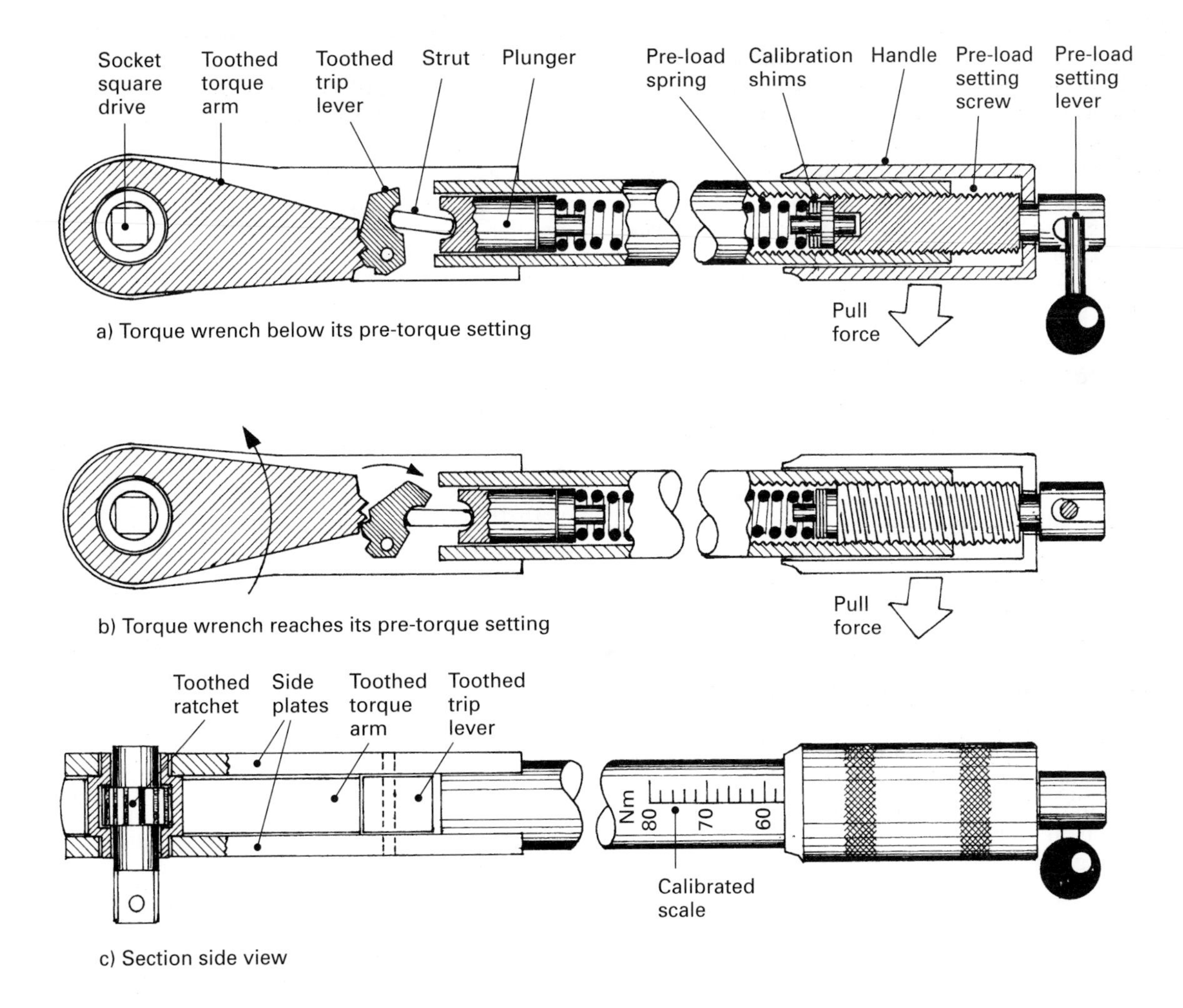

Fig. 13.23 Pre-setting torque wrench

wrench. The wrench consists of a tubular lever arm attached to the square socket drive side plates. Sandwiched between the side plates is a toothed torque-arm which is attached to the square socket drive via a toothed ratchet. The teeth of the torque-arm are made to mesh with teeth formed on one side of the pivoting trip-lever. The other side of this lever has a notch which in turn relays any angular movement of the trip-lever to the spring loaded plunger via a strut. The pre-load spring is trapped between the plunger and a pre-load setting screw. Increasing or decreasing the pre-torque is achieved by screwing in or out the pre-load screw via the pre-load setting lever located at the end of the handle.

When the handle is pulled round, the torque wrench rotates until the pre-setting torque is reached, at which point the toothed torque-arm overcomes the compression of the pre-load spring. The collapse of the pre-loading spring permits the torque-arm and the square socket drive to swivel relative to the side plates. This in turn rotates the trip-lever clockwise. It therefore forces the plunger via the strut to move further into its cylindrical bore against the pre-set thrust of the spring. At a given instant there will be a sudden relaxation felt in the resistance opposing the tightening of the bolt or nut and at the same time a distinct click will be heard. In this position the strut and the trip-lever/torque-arm teeth are aligned so that there is no more offset leverage tending to compress the pre-load spring. This signals that the torque wrench has reached its limiting tightening torque which has previously been set. The pre-load setting screw is screwed in or out to increase or decrease the limiting torque setting respectively. Calibration of the actual limiting torque setting to the scale on the cylindrical shank is achieved by altering the thickness of the shim pack so that the pre-set torque and the scale reading coincide. This calibration is carried out by the manufacturer but a calibration check should periodically be made.

13.8.4 Angular torque gauge (fig. 13.24)

This type of gauge is recommended by some manufacturers as a means of accurately measuring the final tightening of bolts or nuts used for cylinder-heads, big-ends, main-bearings and flywheel attachments.

This method of setting the final tightness of a bolt or nut is based on the axial and angular movement relationship between a male and female threaded joint. Once the bolt or nut slackness has been taken up, further rotation of either the male or female members of the threaded joint will stretch the shank of the bolt or stud in direct proportion to any relative angular movement. Hence the angular rotation of a bolt or nut provides a direct and accurate method of tensioning bolts and studs to the manufacturer's recommended values necessary to secure two or more components together.

The gauge consists of a square socket drive adapter onto which are mounted an upper cursor (pointer) transparent disc and a lower 360° protractor disc which has an adjustable reaction arm fixture attached to it. The upper transparent disc bears against a friction grip rubber washer, whereas the lower disc bears against a spring clip; both discs are separated by a wavy spring washer.

The lower protractor disc is prevented from turning by the adjustable reaction arm which is so adjusted that it comes into contact with any protruding component situated at cylinder head or camshaft level.

Initially the cursor pointer disc is twisted so that the pointer aligns with the zero on the protractor scale. When the ratchet lever or Tommy-bar is pulled round, the cursor pointer disc rotates with the square socket drive due to the friction grip washer. This then enables the angular movement of the socket drive to be read off on the protractor scale.

13.8.5 Procedure for tightening cylinder-head bolts (fig. 13.25)

A typical head tightening procedure would be as follows:

1) Tighten all the cylinder head bolts in the order shown to the first-stage specified torque using a torque wrench say 30 Nm.

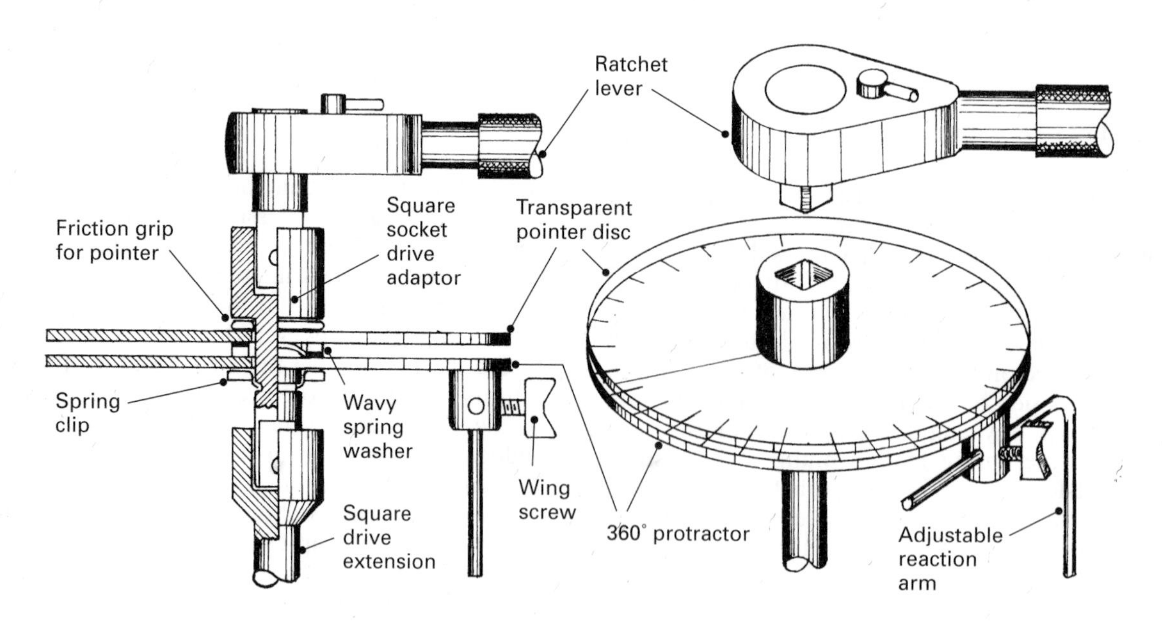

Fig. 13.24 Angular torque gauge

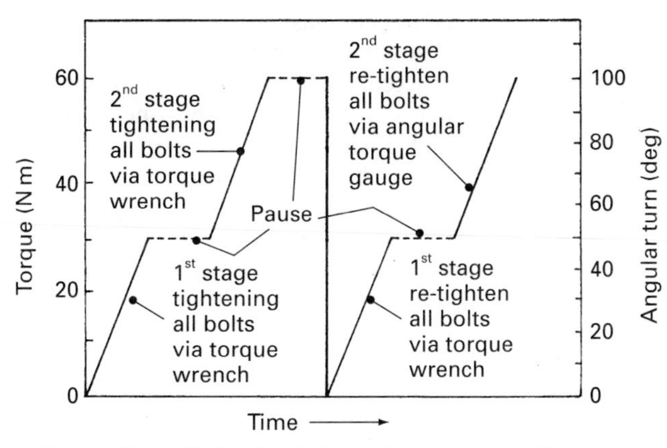

Fig. 13.25 Cylinder-head tightening sequence chart

2) Repeat this procedure but tighten the bolts to the second stage specified torque setting, say 60 N m.
3) In the reverse order loosen all the bolts completely.
4) Retighten all the bolts in the order recommended to the first stage specified torque setting, that is, 30 N m.
5) Using the angular torque gauge further tighten the bolts by turning them through the angle specified say 50–55° clockwise.

13.9 Cylinder-head gaskets

The mating surfaces of the cylinder head and block are machined and ground flat and smooth to within the working accuracy of the machines used. A magnified examination of the surface would show two main forms of irregularity:

i) waviness, where the surface profile consists of many very small humps and depressions caused by deflection and vibration of the machine's spindle or cutter:
ii) roughness which is superimposed on the wavy profile, consisting of an infinite number of tiny steep hills and valleys caused by the tearing action of the machining and grinding processes.

Therefore, where the mating faces are placed together, contact will be made only with the crests and peaks of these surface irregularities.

Severe unevenness of either joint face may be due to large machining tolerances, surface roughness, distortion, lack of uniformity of stud or bolt loading over the head, and localised operating-pressure fluctuations.

The function of the cylinder-head gasket is to fill the spaces between the joint faces so that a gas-tight and liquid-tight seal is maintained under all operating conditions.

The cylinder-head joint elements are classified into two types:

i) elongated – these are the bolts, studs, and their threads;
ii) compressed – these are the gasket, cylinder head, cylinder-block upper deck region, and cylinder liners and their flanges if fitted.

To maintain a good seal over long operating periods, all elongated elements must be subjected to elastic strain only. The working threads of the set-bolts and their holes in a cylinder block are 'elongated' and may consequently deform plastically. If, after a while, the bolts can be retorqued, this deformation may be completely compensated for. Without retorquing, a loss of sealing efficiency is inevitable.

13.9.1 Cylinder-head-gasket requirements

a) Provision for various apertures in the gasket:

   i) cylinders and combustion chambers;
   ii) lubrication passages;
   iii) coolant ducts;
   iv) holes for studs and bolts;
   v) holes for valve push-rods.

b) Resistance to chemical attack by

   i) fuels used and the products of combustion;
   ii) coolants and their additives for anticorrosion and antifreeze properties;
   iii) engine oils with their additives and contaminated lubricants.

c) Resistance to operating conditions due to

   i) widely fluctuating and pulsating internal pressures, including those experienced under detonation conditions;
   ii) widely fluctuating and pulsating temperatures and exposure to flame.

d) Stability under compression conditions, to avoid excessive settling:

   i) compression due to stud or bolt loading;
   ii) compression due to thermal expansion and contraction.

e) Essential considerations:

   i) adequate shear and tensile strengths, particularly if there are narrow sections;

ii) adequate provision for the cooling of the cylinder head and for minimising the effects of differential thermal expansion;
iii) maintenance of a gasket-thickness tolerance, as this will affect the compression-ratio;
iv) a gasket of simple construction, easy to assemble, and not readily damaged.

13.9.2 Types of cylinder-head gaskets (fig. 13.26)

Gasket seating conditions The gasket thickness and hardness must be chosen to match the degree of unevenness of either joint face due to large tolerances, distortion, surface roughness, or other factors such as lack of uniformity of bolt or stud loading.

Figure 13.26(a) shows the optimum gasket thickness, where the surface roughness has been completely filled in by both the elastic and plastic yielding of the softer sealing material. Figure 13.26(b) illustrates either a very thin or a very hard gasket material which, for the given tightening torque, will not yield and deform sufficiently to fill all the cavities on the surface. There will therefore be a certain amount of leakage due to gas or liquid moving in between and around each cone-shaped high spot.

Mono-metal sheet gaskets Provided that the cylinder head and block structure is very rigid, the finish of the machined faces is of a high standard, and the hold-down stud or bolt torque is high, gaskets can be made from single pieces of a sheet metal which is much softer than the joint faces, such as copper or very-low-carbon steel. These conditions are usually difficult to meet, so other forms of construction and combinations of materials are generally used.

Copper–asbestos gaskets (fig. 13.26(c)) The traditional gasket was a copper–asbestos–copper sandwich. This is soft and so fills the gaps more easily than steel. In practice, after the copper seal has been made by compression around the cylinder bore, it work-hardens due to the cyclic combustion impulses causing relative movement between the head and the cylinder block and subsequently does not have the resilience to reseal the joint gap under heavy fluctuating loads. Copper is also relatively expensive, so there has been a considerable decline in the use of this type.

Laminated steel-sheet gaskets (fig. 13.26(d)) Thin laminated soft-steel-sheet gaskets are used on rigidly constructed diesel engines. They consist of about six sheets held together by one of the outer sheets which forms a bead ring around the cylinder-bore holes. The soft steel readily yields to the shape of the joint faces, and any bulging or internal distortion of the gasket material is compensated for by the relative slipping of the laminated layers.

Corrugated steel-sheet gaskets (fig. 13.26(e)) Corrugated or embossed gaskets are very cheap and are suitable for relatively rigid joint faces with good surface finish.

The gasket is made from a single steel sheet. The cylinder bores, coolant passages, and push-rod and oil holes are initially stamped out, and crimped rings which can be of various depth are then formed around all these apertures. When the hold-down studs or bolts are tightened, the resulting elastic and local plastic strain of the steel ensures yielding to the joint faces. The greater the depth of the crimping, the higher will be the gasket pressure between the joint faces and the more effective will be the seal around the bore. Generally the depth of the crimped depressions is less for the coolant apertures.

Both sides of the gasket are coated with about 0.04 mm thick heat-resistant varnish. This eases the initial spreading and settling of the gasket as it is tightened down.

Asbestos-coated steel-sheet gaskets (fig. 13.26(f)) One form of composite gasket uses a single steel sheet which is coated with graphitised asbestos compound on each side. This material has good heat resistance and provides the necessary compliance between the rigid head and block mating surfaces. Separate ring beadings are moulded around each bore hole. These soft-faced gaskets yield readily to the contour of the joint faces to form a highly effective seal, and the bead protects the asbestos from direct flame exposure.

Laminated steel-and-asbestos gaskets (fig. 13.26(g)) This type of gasket uses alternate layers of steel and graphitised asbestos sandwiched together, with the outer steel sheet bent over and around each bore hole to form a reinforcement ring beading. With this construction, the steel sheet will seal the rigid cylinder-block face and the asbestos layer will yield to the more flexible cylinder-head face.

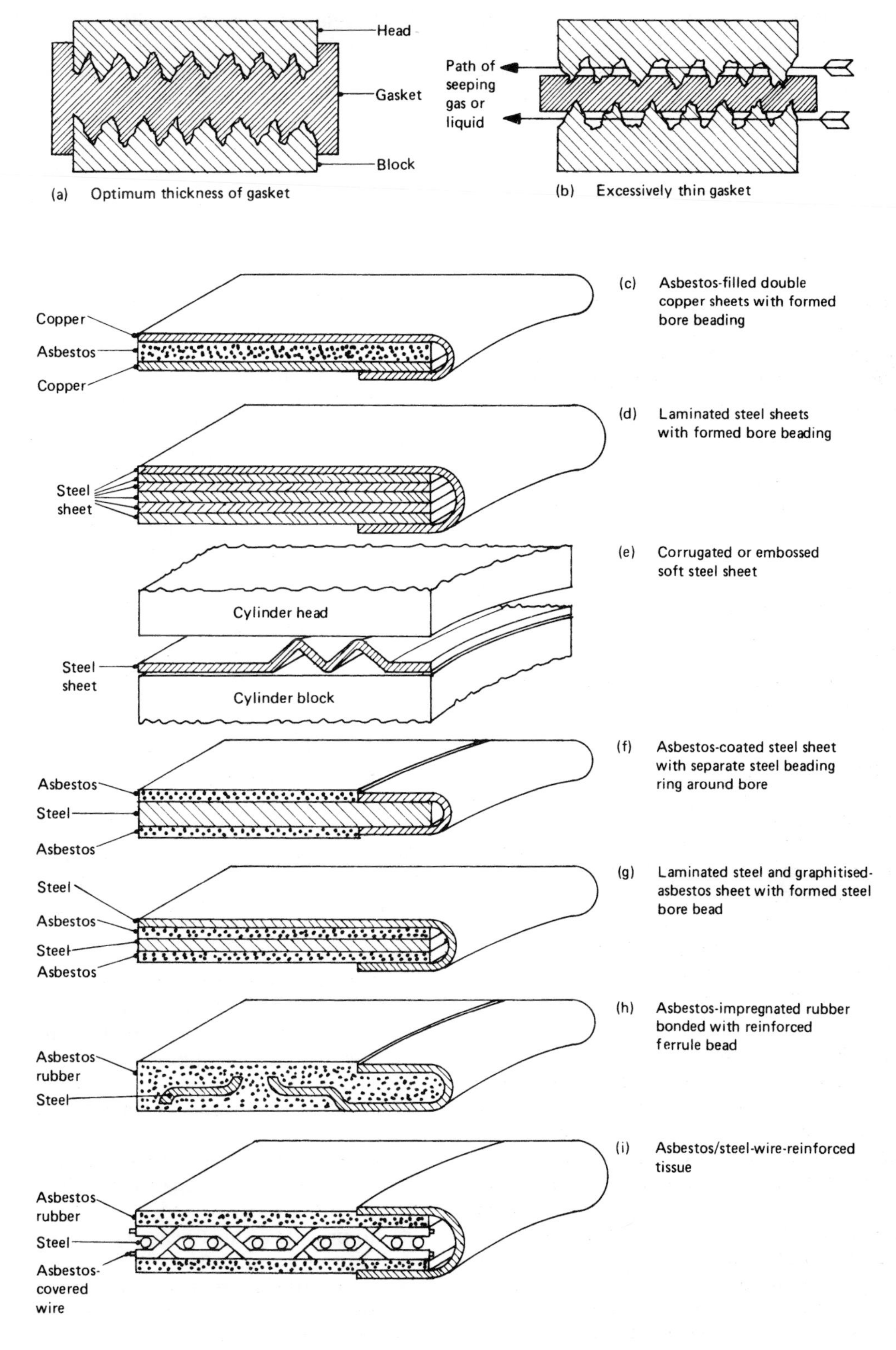

Fig. 13.26 Cylinder-head gaskets

Asbestos–rubber bonded gaskets (fig. 13.26(h)) These gaskets consist of an asbestos-impregnated rubber composition which has strengthening steel strips moulded into the soft material. In addition there are reinforcement ferrule beads formed around each bore hole. These gaskets tend to conform exceptionally well to unevenness of joint-face contours.

Tissue asbestos–metal gaskets (fig. 13.26(i)) The centre region of these tissue asbestos–metal gaskets is of interwoven asbestos-fibre-covered wire reinforcement which is held in between two asbestos-impregnated rubber sheets. In addition, steel reinforcement cylinder-bore ring beadings are incorporated to protect the asbestos matrix from the burning products of combustion. Although other types of gasket are less expensive, the asbestos–metal tissue type is superior with regard to both joining efficiency and the avoidance of distortion of the surrounding structure.

13.9.3 Considerations in selecting gasket materials

a) Gasket design should be able to cope with and compensate for relatively large differences in dimensions caused by machining tolerances on engine components and any flexing which might result under operating conditions.
b) Gaskets are subjected to plastic as well as elastic deformation irrespective of whether they are made from soft asbestos or hard steel materials.
c) Gaskets must be plastically deformed to give the necessary conformability to the slightest misalignment of the mating faces due to distortion and surface-texture waviness and roughness.
d) If the seal is to be effective, gaskets must expand and contract elastically (as do the other compressed components) when the cylinder head and block tend to separate and close due to the cyclic combustion pressure under operating conditions.
e) Gaskets must be vibration-resistant - that is, not suffer from fatigue failure – so that the seal around the cylinder bore is not harmed by excessive vibration amplitude caused by periodic impulses.
f) Gaskets must have good creep or settling resistance, to avoid loss of tightness. Therefore the compression stresses must be within the elastic range of the gasket material, and the hold-down bolts or studs too must not be overloaded either initially or in subsequent service. In practice some creep is unavoidable, even with the best materials. However, this is unimportant provided that the pre-load on the hold-down bolts or studs is sufficient to maintain the pressure necessary for maintenance of the seal.
g) Effective gasket sealing must be obtained with the minimum possible tightening torque in the hold-down nuts or bolts. Tightening should be necessary only once, and any gasket creep and settling down after initial assembly must not impair the quality of the seal.
h) Gasket collapse due to combustion pressure – referred to as 'blow-out' – will be minimised by having adequate friction between the surfaces of the gaskets and the faces that are to be sealed. Therefore, unless otherwise stated, gaskets should generally be installed in the dry condition.
i) Gasket metals all tend to age-harden due to vibration and variation of temperature. This will to some extent influence their sealing performance over a period of time.
j) Gaskets should be as thin as possible, to provide the maximum support for the head and block structure so that there is less likelihood of distortion occurring. However, components which are not so rigid or are machined with wide tolerances must through necessity have thicker sealing material.

13.10 Crankcase sump and baffle plates (fig. 13.27)

The container underneath the crankcase is known as the sump, and its functions are as follows:

a) to store the engine's lubrication oil clear of the rotating crankshaft in readiness for circulation within the lubrication system;
b) to collect the return oil draining from the sides of the crankcase walls and any oil ejected directly from the journal bearings;
c) to provide an easy path and a centralised storage area for any contaminants such as liquid fuel, condensed water, combustion products blown past the piston ring, and worn metal particles;
d) to provide a short recovery period for the hot churned-up and possibly aerated oil, enabling it to settle before it is recirculated throughout the lubrication system;
e) to provide a degree of intercooling between the hot oil inside and the air stream outside.

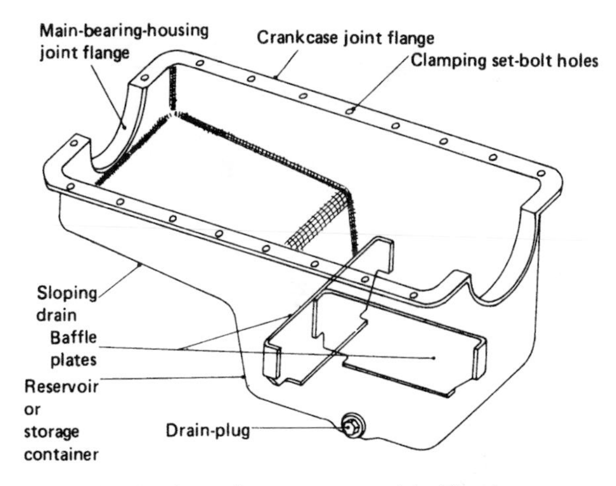

Fig. 13.27 Steel crankcase sump and baffle plates

The sump may be constructed from a single sheet-steel pressing (fig. 13.27) or it may be an aluminium-alloy casting with cooling fins and sometimes strengthening ribs. Both the pressed-out and the cast structures have a flanged joint face which aligns with a corresponding joint face on the underside of the crankcase. A soft flexible gasket sandwiched in between seals the joint, which is tightened down by set-screws.

The sump may be formed with a shallow downward slope at one end which merges into a relatively deep but narrow-walled reservoir region. The draining oil will gravitate towards the deep end, where it will completely submerge the lower part of the lubrication system's pick-up pipe and strainer. A drainplug is positioned at the lowest position in the bottom of the sump, so that used oil can easily be emptied out.

Generally the sump is not designed to contribute to the crankcase rigidity, but with some transverse front-wheel-drive engines the final-drive housing and sump form an integral unit. This is then normally an aluminium-alloy casting which is designed to withstand transmission loads and therefore must have structural stiffness support.

Cast aluminium alloy is much better than pressed steel in dissipating heat, and it does not promote resonant (vibration) noise – particularly if it has strengthened ribs on the outside. Conversely, the pressed-steel sump can be dented without any serious damage being caused, but the cast structure would not tolerate any impact and would crack.

Baffle plates are sometimes installed inside the sump to prevent oil surge due to the bouncing, rolling, and pitching of the vehicle, or to guide and prevent splashing about of the stray jets of oil escaping from the journals and bearings of both the crankshaft and the camshaft. For special applications a horizontal gauze sheet supported in a steel frame may be positioned just below the sump flange joint, the object being to strain out draining contaminants created by combustion-gas blow-by and to minimise oil splashing.

13.11 Cylinder-bore liners

Most petrol engines with cast-iron cylinders use a grey cast iron which has the desired casting and machining qualities and also possesses adequate mechanical properties such as strength, toughness, hardness, and wear resistance. Unfortunately, a single grade of cast iron when cast as a monoconstruction cannot have all the optimum individual mechanical properties, and separate cylinder liners are therefore sometimes preferred where prolonged cylinder life outweighs the extra cost, such as is the case for heavy-duty diesel engines.

With the trend for lighter power units to save fuel, there has been much research and development into the metallurgy of aluminium alloys and into methods of producing high-quality low-pressure-die-cast aluminium-alloy cylinder heads and blocks at reasonable cost. These alloys are generally much too soft to be used directly as cylinder-bore walls, so separate cylinder liners are essential.

When separate cylinder liners are to be used, they can be made from lightly alloyed cast iron, centrifugally cast into a cylindrical sleeve, machined, and then heat-treated to produce the optimum wear-resisting properties for their application. Hence the expected working life can be considerably extended.

These liners can be divided into two classes:

i) those which come into continuous direct contact with the bore-hole walls of the cylinder block – these are known as 'dry' liners;
ii) those which are supported only at each end in the cylinder block and are elsewhere in direct contact with the engine coolant – these are known as 'wet' liners.

13.12 Dry cylinder liners

Dry liner sleeves are provided in the following circumstances:

a) If the cylinder block is made from aluminium alloy, a stronger and much harder wear-resistant material is necessary for the cylinder-bore walls.

b) For some heavy-duty operating conditions, the normal wear resistance of a cast-iron cylinder block can justifiably be improved by incorporating sleeves with superior properties.
c) If, to reduce overall length, the cylinder block is designed with siamesed adjacent cylinder bores, then only dry liners would be suitable.
d) Dry liners may be used to restore to its original size a cylinder block which has been rebored two or three times due to excessive wear.
e) If both bending rigidity and torsional rigidity are essential, a cylinder block with 'cast-in' coolant passages and cylinder bores fitted with dry liners is generally more robust than a block incorporating wet liners.

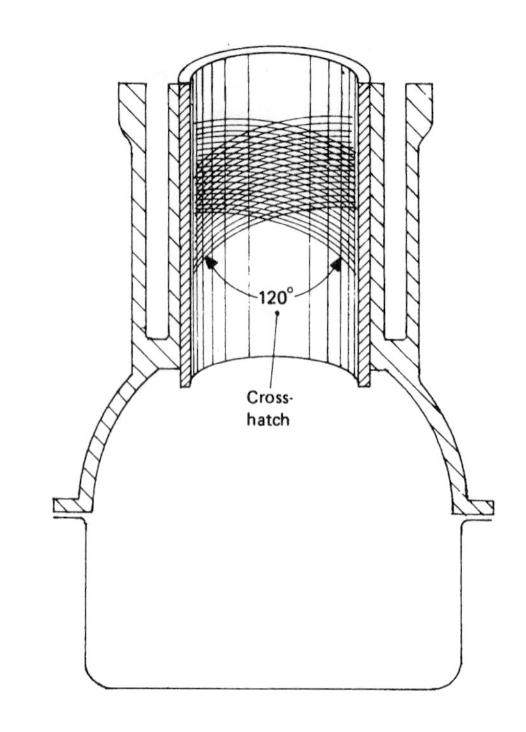

Fig. 13.28 Plain force-fit dry cylinder liner

13.12.1 Liner location fit in cylinder-block bores (figs 13.28 and 13.29)

There are three basic fits as follows:

i) cast-in fit,
ii) force (press) fit,
iii) slip fit.

Cast-in-fit liner With aluminium-alloy cylinder blocks using dry cylinder liners, the external cylindrical surface of the liner is machined with a helical groove running from top to bottom. These liners are usually preheated to 200°C and are then positioned in the cylinder-block casting dies before casting begins. After solidification, there will be a sound metallic bond between the aluminium-alloy block and the cast-iron sleeve.

Force-fit (press-fit) liner (fig. 13.28) The liner is a plain cylindrical sleeve which is held in position by the interference between its outer diameter and the bore-hole walls. The liner is located by drawing or pushing the sleeve into the cylinder block with considerable force, using suitable end-plates and guides and either a screw-and-nut draw-bar attachment or a hydraulic-press set-up.

Typical interference fits between the sleeve and the cast-iron cylinder block are as follows:

Bore diameter	*Interference*
75 to 100 mm	0.050 mm
100 to 150 mm	0.075 mm

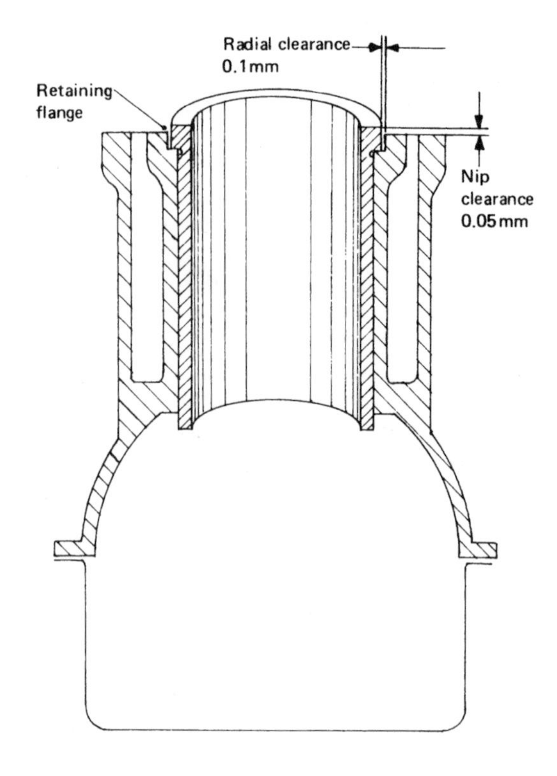

Fig. 13.29 Flanged slip-fit dry cylinder liner

Slip-fit liner (fig. 13.29) The liner is a cylindrical sleeve, flanged at one end to locate it and secure it in position. There is little or no interference between the liner and the block walls, and the liner is inserted by hand pressure – thus assembling these liners into the cylinder block can save considerable time. With these liners, the flange should project above the block face by 0.05 to 0.125 mm to prevent vertical movement relative to the block while in service.

Before fitting the new liner, remove rust, carbon, and any burrs from the cylinder-block cylinder walls and their counterbores, then check with an internal micrometer or any other similar gauge for diametrical distortion. When inserting a slip-fit liner, check the mating between the flange and the recess bore by blueing the sleeve top face, upturning the sleeve, and rubbing it against the counterbore face. Finally, check the sleeve bore for ovality with a micrometer gauge at two positions at right angles to each other at the top, middle, and bottom of the sleeve. If the difference at any of the positions checked exceeds 0.05 mm, rotate the sleeve through 90° in the cylinder block and recheck. If necessary, rearrange the sleeve again until the best position is found.

13.12.2 Machining a cylinder block to take a liner

When boring out the cylinder block to take the sleeve, the same care for alignment, circularity, straightness, diameter, and surface finish is necessary as when reboring a cylinder block which is to operate directly against its piston and rings.

First, dry liners have relatively thin walls and, when positioned in their holes, take up the contour of the finished wall profile. Any surface-roughness high spots or bulges in the bore will therefore quite easily distort the sleeve out of its true cylindrical contour so that the contact clearance between the piston skirt and the sleeve wall will vary and may even be too tight in places. Secondly, the heat from combustion is mainly extracted by conduction through the liner to the cylinder-block walls. If air pockets are produced by, say, ridge marks from a rough single-point cutting tool, local hot spots will form – causing distortion, rapid wear, and possibly piston seizure. The working tolerance for boring cylinder blocks is + 0.0000 to 0.0125 mm.

13.12.3 Liner-sleeve internal-bore allowance

Force-fit dry liners are normally supplied with an unfinished internal-bore diameter allowance of between 0.35 and 0.50 mm which has to be removed by boring and honing after the liners have been pressed into their respective cylinder-block bore holes.

Slip-fit dry liners may be supplied either as semi-pre-finished liners with an internal-bore allowance of 0.025 to 0.10 mm which after fitting is removed by honing or as pre-finished liners which have no internal-bore allowance and are therefore ready for use after they have been installed.

13.12.4 Liner internal-bore surface finish (fig. 13.28)

The liner bore surface should be honed leaving a surface controlled to 0.6 to 0.8 μm centre-line ('roughness') average with a cross-hatch angle of 120° (fig. 13.28). This produces an oil-retaining surface to provide optimum conditions for running-in new piston rings and cylinder bores ('ring bedding'), which is essential for both gas-sealing and oil control.

13.13 Wet cylinder liners

For motor-car petrol engines, wet cylinder liners may have some advantages if the cylinder block is made from aluminium alloy which has a high coefficient of expansion:

a) By isolating the bulk of the sleeve from the block, difficult expansion problems can be dealt with at one or two locating positions only.
b) The use of wet liners simplifies the casting of the cylinder block, be it cast iron or aluminium alloy, and it also allows a more suitable material and an appropriate heat treatment to be used to satisfy the structural requirements, rather than the cylinder-bore wear-resistance treatments which would normally be different.
c) The better outside surface finish and constant wall thickness of the liner improve both the thermal conductance and the uniformity of cylinder cooling around the piston-swept-volume region.

13.13.1 Liner location and support (figs 13.30 and 13.31)

With wet liners, the normal cylinder wall is eliminated, so the liner has to be more rigid than a dry liner. Wet liners are made a drop fit into the cylinder block at the top and near the bottom, or just at the bottom, the remainder of the sleeve being unsupported. To prevent leakage of the coolant, sealing O-rings are used.

Some liner sleeves have a flange at the top end which seats directly into a recess machined in the upper deck of the block, and occasionally a soft copper–asbestos or composite gasket fits underneath and between the flange and the block recess.

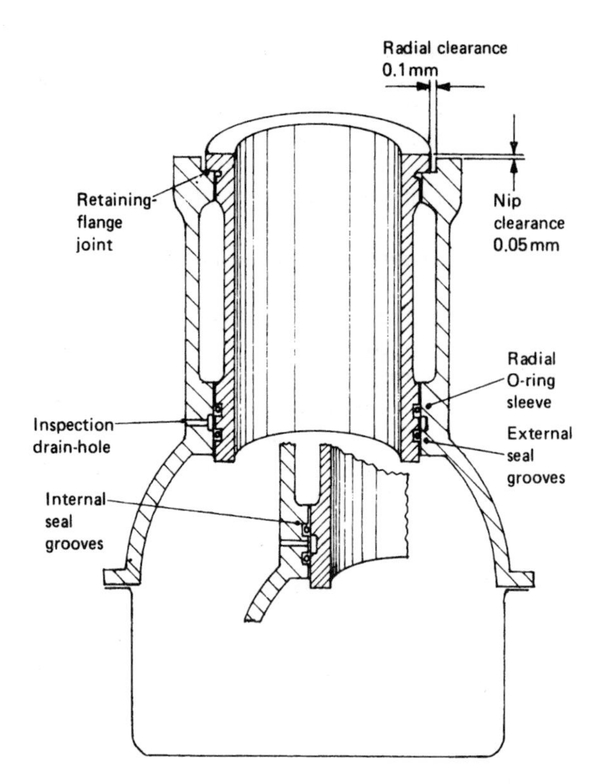

Fig. 13.30 Single sleeve support with open-deck wet cylinder liner

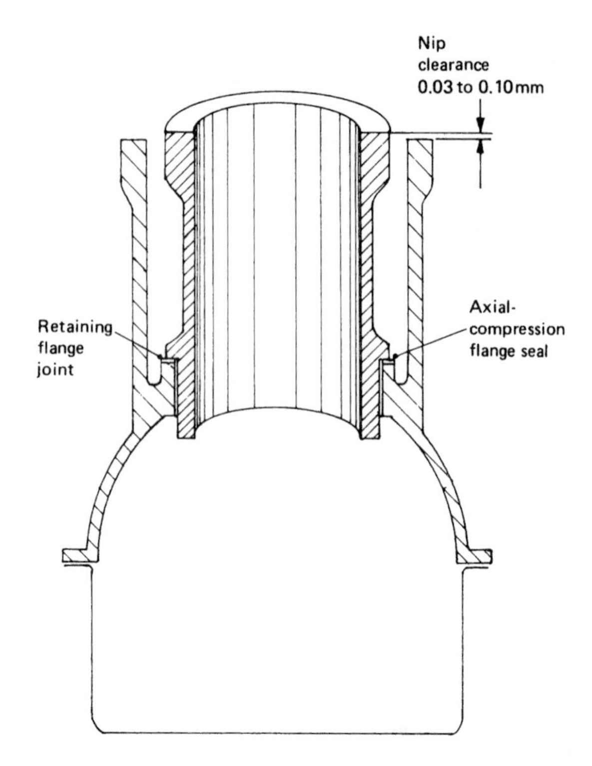

Fig. 13.31 Double sleeve support with closed-deck wet cylinder liner

To hold the sleeve in position, its flange is made to protrude above the block's top joint face by 0.05 mm for bores up to 100 mm diameter and by up to 0.175 mm for cylinder diameters ranging from 100 to 150 mm. A small clearance between the flange outside diameter and the internal recess diameter must also be allowed.

The bottom crankcase end of the liner is sealed by one or more rubber O-rings, usually fitted in grooves around the outside of the liner or in grooves machined into the internal bore which houses the liner (fig. 13.30). Sometimes an inspection drainhole is provided on the side of the block between the seals, so that any leakage through the seals will be immediately observed (fig. 13.30).

Another liner-sleeve support arrangement locates only the lower crankcase end of the liner, which is flanged so that it contacts a corresponding machined face in the block. A flat gasket is fitted between these two joint faces (fig. 13.31). The top of the liner sleeve has no side support but relies entirely on the vertical compression applied to the liner by the cylinder head and gasket when tightened down. To obtain the correct compressive support in the liner, its top face must project above the cylinder block's deck by something like 0.03 to 0.10 mm, depending on the diameter of the cylinder bore, and the tightening-down torque is of the utmost importance.

13.13.2 Installation procedure for wet liners

Remove the old gasket or/and sealing rings and clean the section of block that comes into contact with the liner, using a scraper and emery cloth until the metal is seen. Insert the new liner into the block without sealing rings or gaskets and turn it by hand to detect if there is any degree of tightness which could cause sleeve distortion.

The liner flange must be smooth and square in the counterbore, otherwise the flange will break off when the cylinder head is tightened down. This alignment may be checked by up-turning the sleeve, blueing the flange top, and turning it against the counterbore recess face, observing the bright patches which indicate high spots or uneven seating. Remove any burrs or dirt that might lift the flange, and measure the amount the liner flange stands proud of the block face, to ensure adequate clamping interference.

At this stage fit the sealing rings, taking care

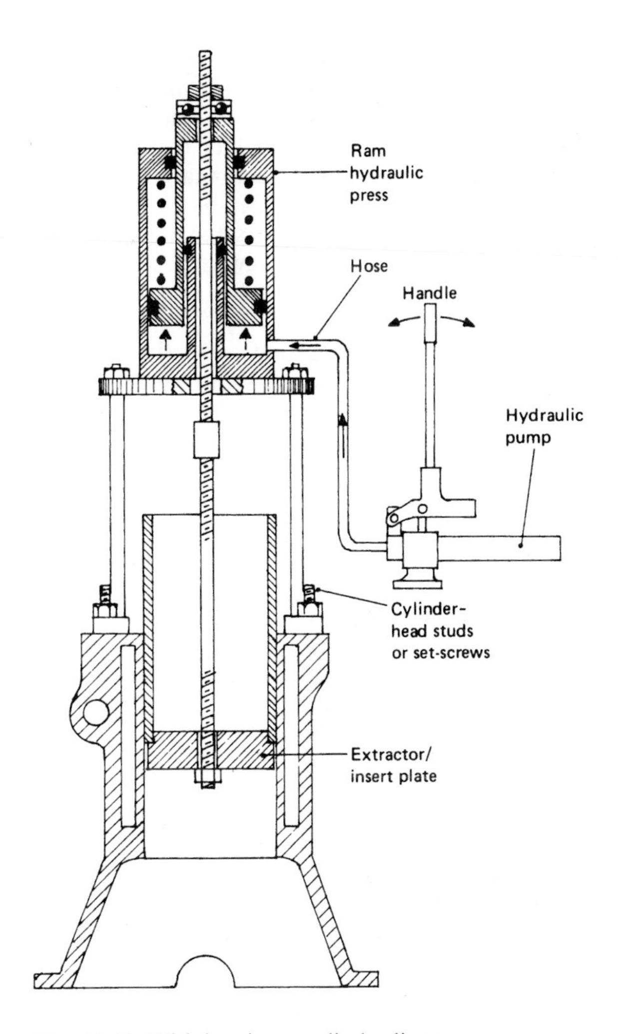

Fig. 13.32 Withdrawing a cylinder liner

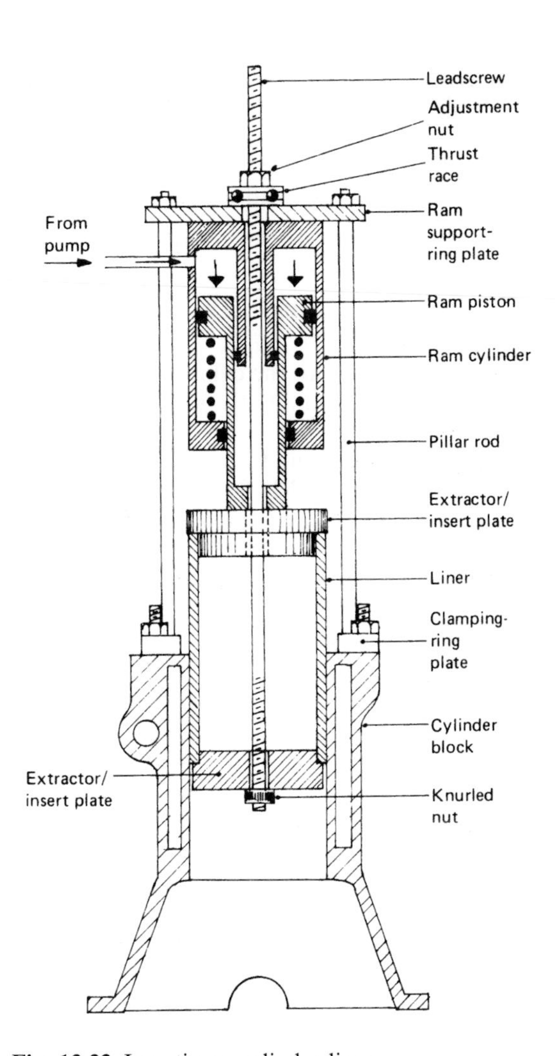

Fig. 13.33 Inserting a cylinder liner

not to overstretch or twist them. Applying a coating of 'Pressoline' or similar compound, work the liner sleeve into place as far as possible by hand, followed by lightly tapping home with a soft hammer – or a light press might be necessary for a large cylinder. Finally, recheck the sleeve cylinder bore for any misalignment or distortion.

13.14 Removal and replacement of press-fit liners

The procedures for removal and replacement of press-fit cylinder liners are as follows.

13.14.1 Removal of press-fit cylinder liners (fig. 13.32)

1 Place the clamping ring on top of the cylinder block over one bore and bolt it to the upper deck by means of either studs or set-screws. Screw the three or four short pillars into the clamping ring and then assemble the ram support-ring plate and the ram on top of these pillars.
2 Place the leadscrew through the hydraulic ram and cylinder and adjust the hexagon nut so that sufficient length protrudes at the bottom to allow the selected extractor/insert plate to be located and attached at the lower end of the liner by the knurled nut.
3 Connect the hose of the hydraulic pump to the power-press ram cylinder and check that the leadscrew is central and that both plates and the pillars are correctly aligned.
4 Operate the pump handle until the liner has been withdrawn the full stroke of the hydraulic ram.
5 Open the release valve on the pump and, as the

ram retracts, screw down the hexagon nut on the leadscrew to compensate for the initial travel of the ram.

6 Close the release valve and repeat pumping until the liner is fully withdrawn.

Repeat this procedure for each cylinder in turn.

13.14.2 Replacement of press-fit cylinder liners
(fig. 13.33)

1 Thoroughly clean out the cylinder bore and the shoulder recess and lightly smear them with clean engine oil.

2 With the clamping ring fastened to the cylinder block as in section 13.14.1, remove the shorter pillar legs and replace them with the longer ones provided.

3 Position the new liner over the cylinder bore and assemble the appropriate extractor/insert plate under the liner and the ram and thrust race on top of the liner. Insert the leadscrew through these items and the cylinder liner, and adjust the hexagon nut until sufficient length protrudes at the bottom of the liner to enable the lower extractor/insert plate to be secured with the knurled nut. Check that the liner is at right angles to the cylinder-block deck.

4 Begin to insert the liner by screwing down the hexagon nut on the leadscrew, and at the same time make sure that the liner is entering the cylinder squarely.

5 Continue inserting the liner with the hexagon nut until the frictional resistance increases. At this stage, draw the liner into the block by carefully operating the hydraulic-pump handle.

6 When the end of the ram stroke is reached, open the pump release valve and readjust the hexagon nut to take up the slack, then continue pumping until the liner has been pushed fully into position.

Repeat this procedure for all the cylinder liners.

14 Combustion and combustion chamber design

14.1 Spark ignition combustion process (fig. 14.1) The combustion process can be explained by the cylinder pressure variation on a crank-angle movement base indicator diagram. The combustion process may conveniently be divided into three phases or periods:

1) ignition delay;
2) rapid pressure rise or uncontrolled combustion;
3) after burning or late burning.

Ignition delay period (fig. 14.1) This phase is constant in time and therefore has a variable crank-angle movement with respect to engine speed. It is the period between the instant the timed spark passes between the sparking-plug electrodes and establishes a flame, to the point where there is sufficient burning to produce a noticeable rise in cylinder pressure above that of the compression pressure.

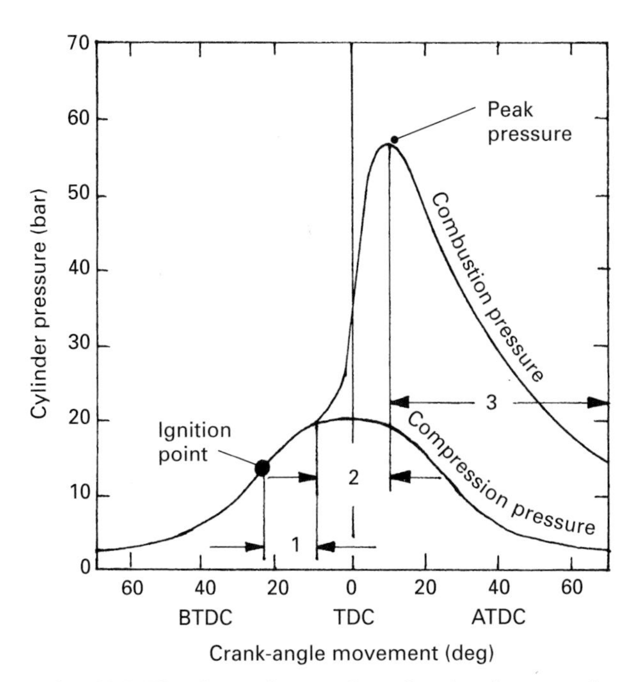

Fig. 14.1 The three phases of combustion in a spark ignition (petrol) engine

The length of this period is dependent upon a number of factors:

1) the temperature and energy released by each spark;
2) the nature and ignitablity of the fuel;
3) the initial temperature and pressure of the fresh charge in the cylinder;
4) the thoroughness of mixing of the air and fuel charge in the cylinder;
5) the strength of the air and fuel mixture in the cylinder.

Rapid pressure rise period (uncontrolled combustion) (fig. 14.1) This phase is approximately constant in crank-angle movement and covers the period between the initial pressure rise above that of the compression pressure to the point where it peaks in the cylinder. It is the period where the expanding flame front travels across the combustion chamber and releases the energy contained within the air/fuel mixture.

The duration of this period depends upon a number of factors:

1) the shape of the combustion chamber and position of the sparking-plug;
2) the thorough mixing of the air/fuel charge in the cylinder;
3) the strength of mixture within the cylinder;
4) the degree of induction and compression swirl;
5) the intensity of turbulence within the chamber. This tends to increase directly with engine speed and is the main factor in keeping this period constant in crank-angle movement.

After burning period (fig. 14.1) This phase covers the period from peak pressure to the point where the exhaust valve opens during the piston's power stroke. It is the period where the flame front has spread to the furthest chamber wall from the sparking-plug, but where something like a quarter of the fresh air/fuel mixture has yet to be burnt. However, it is more difficult for the flame to seek out the unburnt mixture pockets which will now be entangled with a blanket of burnt gases. As a result the remaining unburnt charge will burn at a much reduced rate. Consequently, due to the expanding cylinder volume and the slower speed of burning, the cylinder pressure will decrease rapidly.

14.1.1 Detonation (fig. 14.2)

Detonating combustion is identified by a ringing metallic knocking noise, sounding like a high-pitched ping. The term 'pinking' is commonly used to describe these audible uncontrolled shockwaves being repeatedly reflected from the walls of the combustion chamber. Detonation is mainly caused by the advancing flame front radiating heat, compressing and raising the temperature of the still unburned air/fuel mixture until the remaining combustible end-gas mixture ignites spontaneously. This results in the transmission of high velocity pressure waves through the chamber (fig. 14.2). These shockwaves considerably increase the heat transfer from the products of combustion to the surrounding chamber walls, thus causing local overheating which can do damage to the piston crown head gasket and electrodes and insulation of the sparking-plug. Furthermore, the very high and violent pressure waves, if allowed to persist, may destroy the oil film covering the cylinder wall piston and

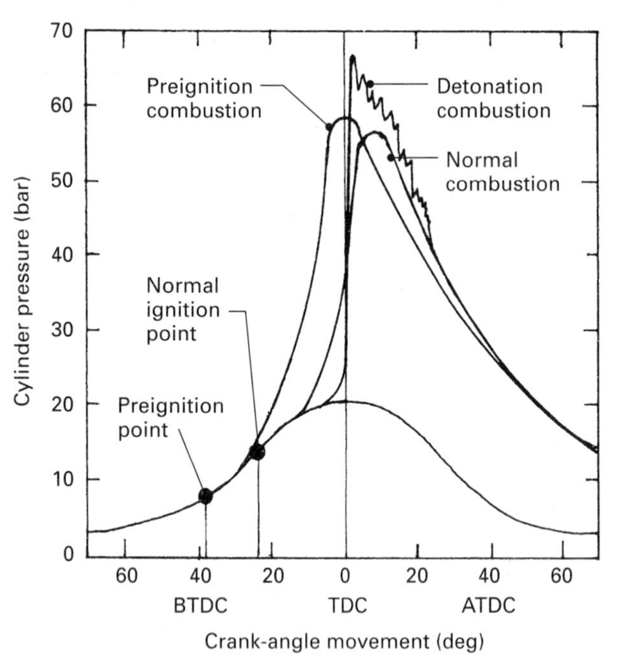

Fig. 14.2 Effect of cylinder pressure variation when preignition and detonation occurs

rings. This exposes the surfaces to corrosive erosion from the combustion products. It therefore promotes dry friction and rapidly increases cylinder and piston-ring wear. At the same time these very high repeating shockwaves tend to squeeze out the oil film separating the small and big-end rubbing pairs and this again increases bearing wear.

The influencing factors necessary to eliminate or minimise detonation are as follows:

1) maximise flame frontal area and reduce the distance the flame path travels;
2) maximise the speed of burning by increasing turbulence;
3) maximise the thorough mixing of the air/fuel mixture by promoting swirl;
4) maximise end-gas quenching without promoting the formation of hydrocarbons;
5) maximise the octane number of the fuel to match the engine compression-ratio;
6) retard ignition timing when necessary; however this does result in a slight loss of engine power.

Controlled and uncontrolled combustion (fig. 14.3(a) and (b)) Uncontrolled combustion occurs when the engine cylinder pressure and temperature rise to a dangerous level as the engine is accelerated under load. Thus the increased amount of fuel burnt under these brief conditions releases more heat energy than can be dissipated in the short time available. It therefore causes the advancing flame front to overheat the still remaining unburned end-gas until at some point these end-gases spontaneously combust.

Traditional two valve combustion chambers (fig. 14.3(a)) have the chamber walls surrounding the valve heads and a sparking-plug positioned off-centre. This provides mechanical squish and gives directional guidance to the incoming charge. As a result the chamber contour dictates a relatively narrow flame front which has to travel a considerable distance to the furthest point in the chamber. This type of chamber therefore has the recipe for promoting detonation every time the engine is subjected to heavy acceleration loads. In contrast a four-valve central sparking-plug pentroof combustion chamber (fig. 14.3(b)) provides a broad flame front and a relatively short flame travel. This chamber therefore promotes rapid flame spread. It therefore minimises the overheating of the end-gases and consequently

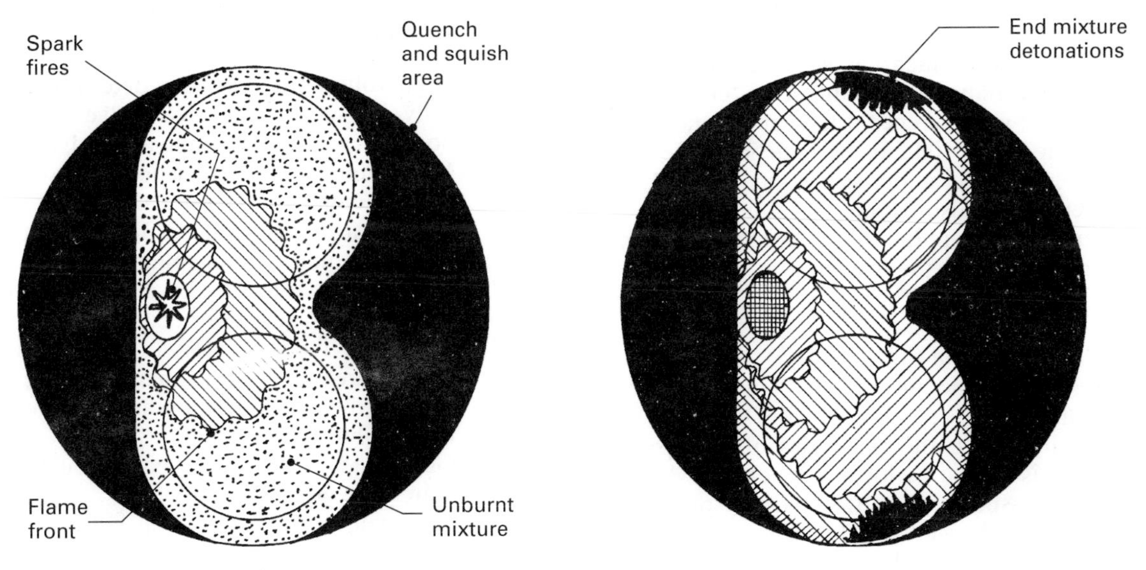

Flame front spread half complete *Flame front spread complete*

a) Uncontrolled combustion (detonation) in an inverted bath tube combustion chamber

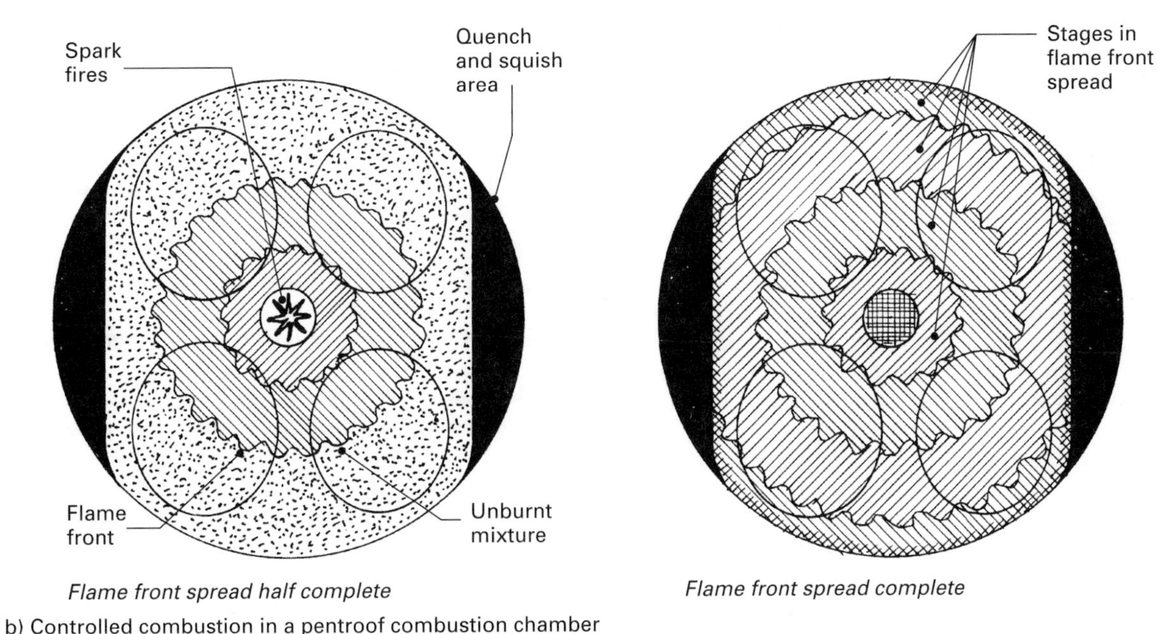

Flame front spread half complete *Flame front spread complete*

b) Controlled combustion in a pentroof combustion chamber

Fig. 14.3 Illustrating controlled and uncontrolled (detonation) combustion

suppresses any tendency to detonate under severe operating conditions.

14.1.2 Octane number

The octane number of petrol fuel is a measure of its ability to burn without knocking (pinking) in a spark-ignition engine. A special single cylinder variable compression-ratio engine run under standard conditions is used to measure the octane number of a commercial petrol under test and compares its knocking tendency with that of a blend of known octane number fuel. Two reference fuels are used, n-heptane which has poor ignition qualities and is given an octane number of zero, and iso-octane which has good ignition qualities and is given an octane number of 100. By blending the two reference fuels any octane value between 0 and 100 can be matched. The octane number of a fuel under test is then defined as the percentage of iso-octane in a mixture of iso-octane and n-heptane that shows identical knock characteristics to the test fuel.

14.1.3 Preignition (fig. 14.2)

Preignition is the uncontrolled ignition of an air/fuel mixture in the cylinder before the timed spark is fired between the electrodes of the sparking-plug. It is caused by local overheating which creates a hot-spot such as the sparking-plug central electrode, exhaust valve head, protruding sharp edges, e.g., head gasket or smouldering particles of carbon deposits. The local hot-spot will be produced towards the end of each compression stroke as the temperature of compression reaches the self-ignition temperature of the mixture. The combustion process that follows the premature ignition will be similar to that formed with a timed sparking-plug. However the danger of preignition is the early ignition (before the timed firing point) which provides more time for the burnt hot gases to remain in the cylinder before the exhaust valve opens so that more heat is transferred to the chamber walls. Therefore, as more heat is absorbed by the chamber walls, the hot-spot will reach the self-ignition temperature earlier and earlier during the compression stroke. Consequently the advancing premature point of ignition will also cause combustion to begin earlier, thus resulting in higher peak cylinder pressure and causing combustion to commence before TDC (see fig. 14.2). Negative work will therefore be done towards the end of the compression stroke in compressing the early stages of combustion, thus resulting in a loss of power. The real danger of preignition taking place in one or two cylinders of a multi-cylinder engine is that the good cylinders will continue to propel the vehicle, and the cylinders suffering from preignition will generate more heat and produce more negative work until permanent damage to the engine components occurs. Preignition can be caused by the excess of heat produced by detonation, late ignition timing, poorly mixed air/fuel charge, weak mixture, mis-matched sparking-plug heat range, and poor coolant circulation.

14.1.4 General principles of combustion chamber design for spark ignition engines

1) High volumetric efficiency determined basically by valve head size, valve lift, valve position, valve port shape and valve timing.
2) Maximum thermal efficiency by having the highest compression ratio that can be used with the available grade of fuel without instigating detonation and rough running.
3) Minimum heat loss by having the volume to surface area ratio at a maximum.
4) Minimum flame travel with the maximum frontal area determined mainly by the location of the sparking-plug and the shape of the combustion chamber.
5) Generated and controlled turbulence to match the engine speed.
6) Adequate compression squish to speed up the mixing of the charge but not to the extent that would leave overcooled quench areas in certain parts of the combustion chamber.
7) End-gas quench to prevent detonation, but not in excess as this will promote the formation of hydrocarbons.
8) Good exhaust gas and fresh charge scavenging of the spark-plug points.
9) Adequate exhaust valve cooling as this is the hottest region in the combustion chamber.
10) Adequate cooling of the spark-plug points to avoid the effects of preignition at the larger throttle openings.

14.1.5 Combustion chamber design terminology

Induction spiral swirl (fig. 14.11) Induction swirl is created by positioning the inlet port passage to one side and tangential to the cylinder axis. It therefore causes the fresh charge to enter the cylinder and rotate in a spiral fashion in a downward direction about the cylinder axis.

Tumble barrel roll swirl (figs 14.13 and 14.14) Tumble or barrel roll swirl is produced when there are twin parallel inlet valve port passages which cause the fresh charge to enter the cylinder and flow across and under the cylinder head in a downward rolling motion at right angles to the cylinder axis (transverse swirl). It there resembles a barrel roll or tumble dry spin motion.

Squish or compression swirl (fig. 14.6) Squish is the forced movement of the fresh charge created by the closing parallel gap formed between the flat underside of the cylinder head and the corresponding flat outer rim of the piston crown as it approaches TDC. The trapped charge is therefore squeezed into the combustion chamber space formed either in the piston crown or cylinder head where it then takes on a circular swirling motion.

Quench area (fig. 14.6) This normally refers to that part of the combustion chamber where the end gases are trapped in the narrow space formed between the cylinder-head and piston crown. Consequently the surrounding metal surface area, relative to this enclosed volume of charge, will be large and will therefore tend to overcool the last stages of combustion.

Turbulence (fig. 14.6) Turbulence is the randomly dispersed small irregular breakaways from the main flow stream of air or air and fuel mixture that takes on concentric spiral motions known as vortices or eddies. A high state of turbulence will agitate the mixing of air and fuel and therefore will considerably speed up the combustion process.

Helical inlet port (fig. 14.4) To raise the intensity of induction swirl in the cylinder, particularly with small direct injection diesel engines, the so-called helical port has been developed. With this concept fresh charge flows within the port above the valve seat and about its axis. It is then deflected by the passage walls which direct it in a spiral fashion around and downwards in the helix-shaped passage above the valve seat. The swirl generated inside this port is then discharged tangential from all directions around the annular opening formed between the valve head and its seat into the cylinder.

These ports promote a high level of cylinder swirl throughout the normal engine speed range but volumetric efficiency is impeded somewhat in the upper speed range.

14.1.6 Classification of spark ignition combustion chambers

Double-row side valve combustion chamber (T-head) (fig. 14.5) With this design the inlet and exhaust valves are arranged in separate rows on opposite sides of the cylinder. This cross-flow cylinder head arrangement draws the fresh charge up and in via the side inlet port and then expels

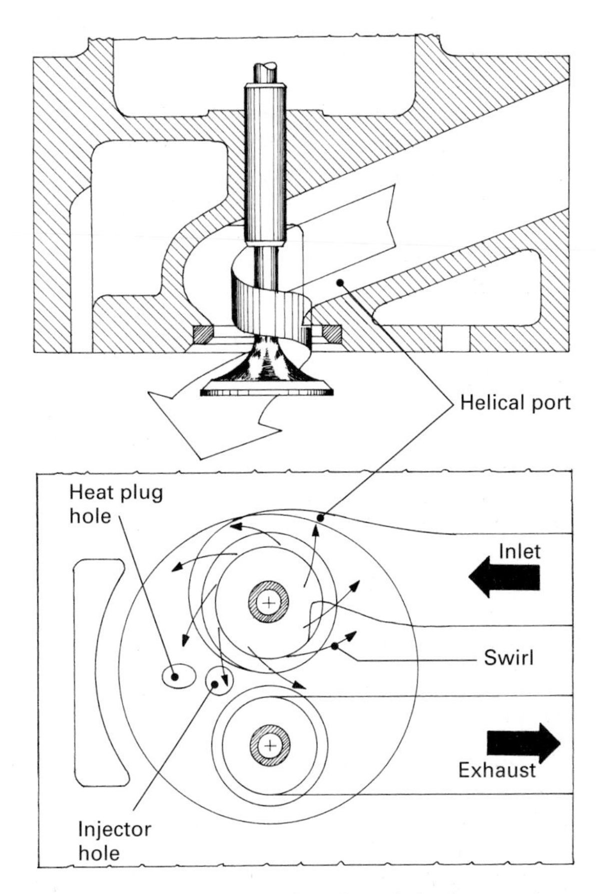

Fig. 14.4 Small diesel engine direct injection combustion chamber with helical induction port

the exhaust gases down and out of the side exhaust port. The combustion chamber space above each valve head through necessity tended to be large and therefore these engines were only suitable for very low compression ratios. Due to the very long flame path, the combustion process suffered from overheating of the end-gases and severe detonation. Having the valve and ports positioned on either side of the cylinder increased the width of the engine block and the cost of this twin camshaft engine.

Single-row side valve combustion chamber (L-head) (fig. 14.6) With this counter-flow cylinder head, both inlet and exhaust valve are actuated from a single camshaft located to one side of the cylinder. A plan view of the combustion chamber shows it to be a rounded triangular shape cast in the cylinder head with two corners of the chamber projecting over both the inlet and exhaust valves. The third corner of the chamber protrudes over the

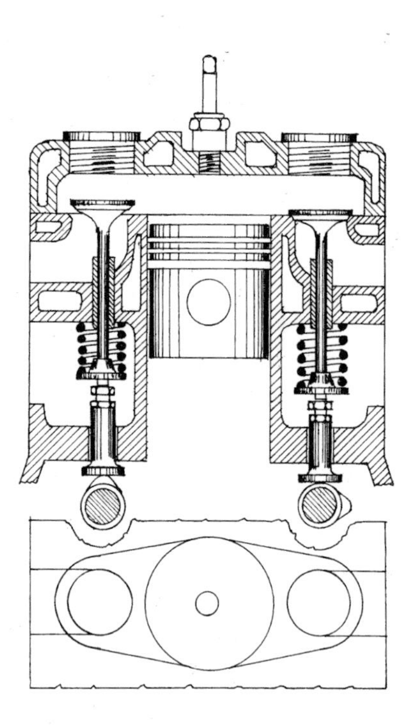

Fig. 14.5 Double-row side valve combustion chamber (T-head)

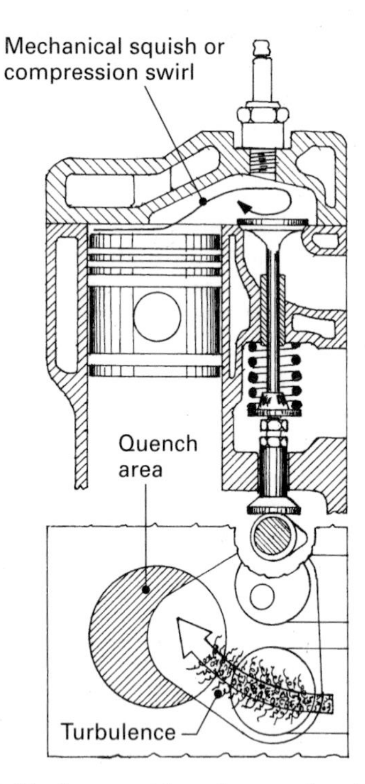

Fig. 14.6 Single-row side valve combustion chamber (L-head)

cylinder. The remaining flat region of the cylinder head also creates a large squish region as the piston approaches TDC. This squish produces a high amount of mechanical swirl which also promotes turbulence and speeds up the rate of burning. To reduce the delay period, that is, the start of combustion after the spark fires, the sparking-plug is located over the exhaust valve, the hottest region of the chamber. However, positioning the spark-plug over the exhaust valve does make for a long flame path even though a large flame front exists. These cylinder heads functioned reasonably well with compression ratios up to about 7.5:1, but scaling down of the clearance volume to increase the compression ratio does interfere with inlet breathing, exhaust gas scavenging and the combustion process.

Overhead inlet and side exhaust valve combustion chamber (F-head) (fig. 14.7) Chambers of this type form an elongated cavity in the cylinder head, overlapping both the cylinder and the space above the exhaust valve. Large inlet and exhaust valves and their ports can be utilised to promote good breathing of the fresh charge and the scavenging of the burnt gases. A large mechanical squish area is provided to considerably enhance the natural induction swirl created by the inlet port shape. However, the chamber cross-section in the region of the sparking-plug is relatively narrow. It thus restricts the flame front and at the same time the offset sparking-plug location produces an undesirable long flame path towards the centre of the cylinder. This chamber provides better engine performance in the mid to high-speed operating range than the single-row side valve cylinder head but is again restricted to an upper compression ratio limit of about 8:1.

Inverted bath tub combustion chamber (fig. 14.8) This shallow semi-heart-shaped chamber formed in the cylinder head provides a certain amount of directional control to the incoming fresh charge and the outgoing exhaust gases via the tongue or nose which partially separates both valves from each other. The flat rim area surrounding the combustion chamber with its widest point opposite the sparking-plug forms the squish zone to promote swirl. The spark-plug is positioned between the two valves but slightly offset from the centre of the chamber; thus this chamber contour provides a large flame front and a relatively short flame path.

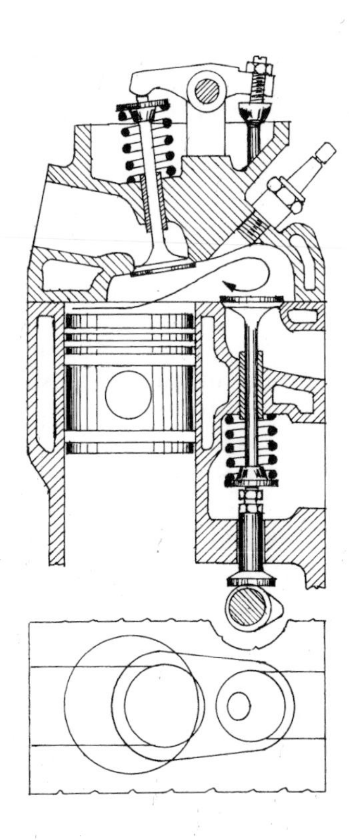

Fig. 14.7 Overhead inlet and side exhaust valve combustion chamber (F-head)

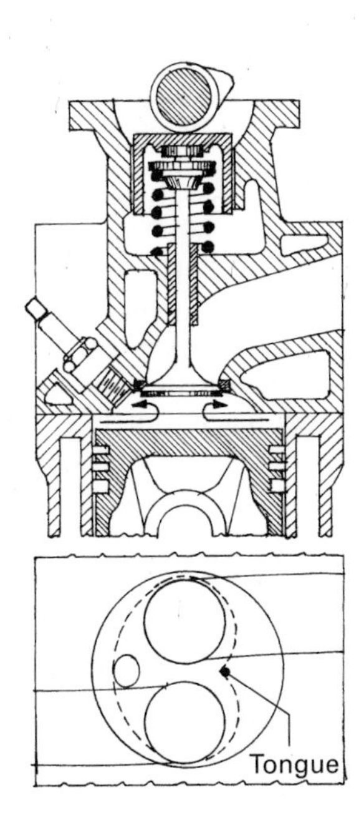

Fig. 14.8 Inverted bath-tub combustion chamber

Wedge-type combustion chamber (fig. 14.9) The inclination of the chamber roof provides a deep space near the sparking-plug which tapers to the wide flat squish zone opposite the sparking-plug. The burning process thus has an initially large frontal area and a relatively short flame path. The inclination of the valve stem reduces the curvature of both the inlet and exhaust ports and as such reduces the flow resistance of both the incoming and outgoing gases. A large quench and squish area is formed between the flat portion of the cylinder head and piston-crown on the opposite side to the sparking-plug. This configuration has the benefit of producing mechanical swirl partially at low speed but has the disadvantage of overcooling any charge trapped in this zone. It therefore tends to promote the formation of hydrocarbons.

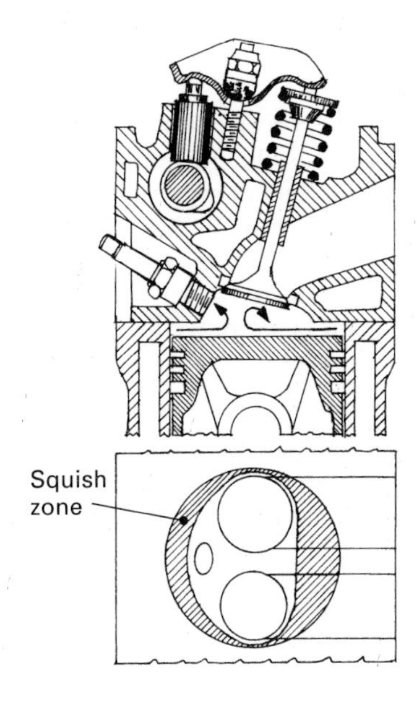

Fig. 14.9 Wedge-type combustion chamber

Bath tub combustion chamber (fig. 14.10) This design has a flat cylinder head similar to a diesel engine with the combustion chamber formed in the top of the piston crown. With the sparking-plug located over and to one side of the cavity, combustion takes place within the bowl formed in the piston crown; a rather narrow flame front exists, but the flame travel is relatively short. The flat annular region surrounding the crown cavity

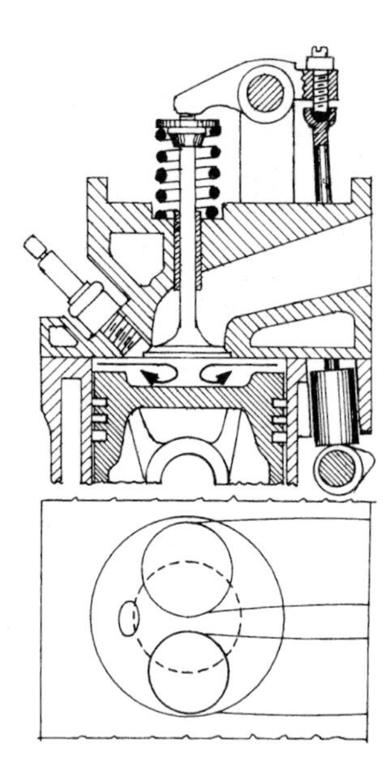

Fig. 14.10 Bath-tub combustion chamber

when the piston approaches TDB forms an effective squish area. Nevertheless this has to be balanced with the possible formation of hydrocarbons. The fresh air/fuel charge entering the cylinder is unrestricted but it has no directional guidance which could hinder the mixing process at certain engine speeds. Moderately high compression ratios can be employed without excessive detonation tendencies.

Hemispherical combustion chamber (fig. 14.11) A cross-flow chamber of this type has the inlet and exhaust valves inclined to each other as shown in the upper diagram in fig. 14.13. The inclination of the valve-stems permits larger valve-head diameters to be used compared to the common upright valve-stem layouts (see upper fig. 14.11). As a result of valve inclination higher volumetric efficiency can be achieved in the mid- to high-speed operating range than can be obtained with the conventional in-line upright valve arrangement. Unfortunately, there is not enough room to position the sparking-plug centrally between the valves so that they are slightly offset. Consequently, though there is a large flame front, the flame path can be long and a tendency to detonate can be difficult to control when compression ratios as high as 8.5 are employed. There is normally a small amount of squish provided on either side of the valve ports; nevertheless, more squish can be created if necessary by using a raised or domed piston-crown.

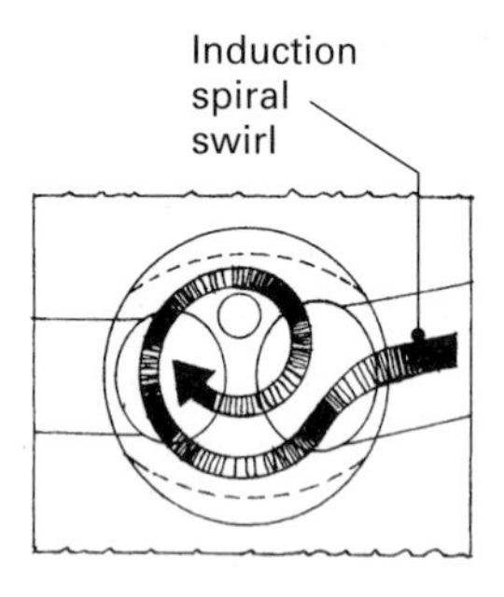

Fig. 14.11 Hemispherical combustion chamber

Triple-valve pentroof combustion chamber (fig. 14.12) Triple-valve pentroof combustion chambers have two rows of inclined valves in the cylinder head; one row supports twin inlet valves per cylinder whereas the other row has just one exhaust valve per cylinder. This valve consideration provides a semi-pentroof chamber shape with the exhaust valve offset, thus permitting the sparking-plug to be positioned more centrally than is possible with the two-valve hemispherical chamber A relatively large flame-frontal area and a short flame path is provided to speed up the burning process; also a moderate degree of squish provides the mechanical swirl required to improve the mixing of the air and petrol charge. Twin inlet valves have the benefits of vigorous mixing due to the induced barrel roll swirl. They also provide a larger flow path for the incoming fresh charge and are particularly effective in improving volumetric efficiency in the top operating speed range of the engine. The use of one large exhaust valve instead of two reduces the port surface area by roughly 30% compared to engines which have the same flow rate with two exhaust valves. This considerably reduces heat losses and allows the catalytic converter to reach full effectiveness in 60 seconds.

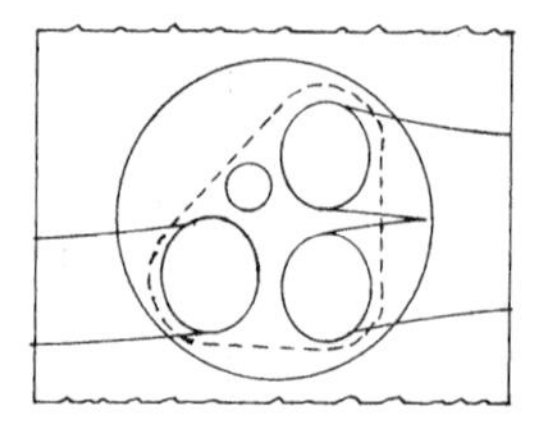

Fig. 14.12 Triple-valve pentroof combustion chamber

Quadruple-valve combustion chamber (figs 14.13 and 14.14) Pairs of inlet and exhaust valves inclined to each other are arranged in two rows; the sloping sides meet along an apex line to form a pent-shaped roof (fig. 14.13). This quadruple cross-flow cylinder head with twin inlet and exhaust valves provides a greatly improved flow path for the fresh and spent charge compared to an inclined two-valve cylinder head (fig. 14.14). In addition, there is now a centrally positioned sparking-plug which provides a large frontal area and a very short flame path that is very nearly equal in all directions. Accordingly it is possible to raise the engine's compression ratio to between 9:1 and 11:1 without instigating detonation. This then contributes to relatively high thermal efficiency over both the speed and load range of the engine. The twin path followed by the incoming air/fuel charge enters the pentroof chamber via both inlet ports, the charge flows under and across the roof, and it is then directed downwards in a semi-circular fashion to produce a tumble (barrel) roll swirl. Conversely, if a single inlet port is utilised, the incoming charge enters to one side of the chamber and is therefore directed in a circular path about the cylinder axis in a spiral fashion, hence this motion is referred to as spiral or whirlpool swirl. The high intensity of swirl is itself then exited into a turbulent swirl

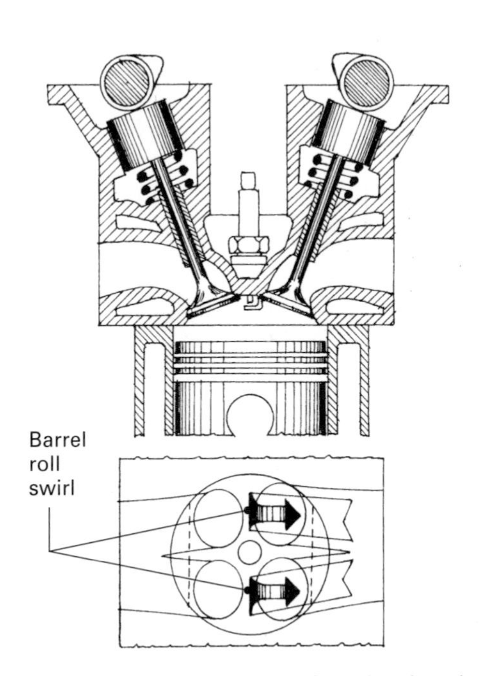

Fig. 14.13 Quadruple-valve pentroof combustion chamber

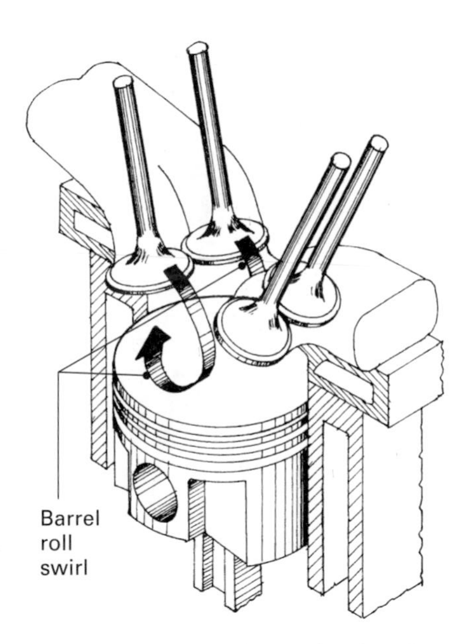

Fig. 14.14 Pictorial view of a quadruple-valve pentroof combustion chamber

which greatly enhances the thorough mixing of the air and petrol, this being a fundamental criterion necessary for consistent and stable burning of very lean mixtures. With the much enlarged twin inlet and exhaust port flow path, the expulsion and entry of the burnt and unburnt charge is made easier; this thus results in a much higher volumetric efficiency above the mid-speed range.

Twin-spark triple-valve combustion chamber (Mercedes) (fig. 14.15) Twin sparking-plugs reduce the flame front travel. They therefore give the engine a high lean-burn ability without detonation or mis-firing occurring and permit the retardation of the ignition timing of 5–10° crank-angle movement during start-up which results in higher exhaust temperature and therefore faster warm-up periods. The other major benefit of dual ignition is that it reduces the formation of hydrocarbons trapped in the piston lands by something of the order of 25%. The firing timing of the two spark-plugs are phased with the Mercedes-Benz engine; the second firing spark-plug trails the first firing sparking-plug by 16° crank-angle movement. The overall slower rate of burning reduces the noise of combustion by 3.3 dB (A) at the expense of a very small power loss. Each spark-plug takes it in turns to fire first so that the heat distribution on the piston crown is more uniform. Because of the more consistant and reliable burn-

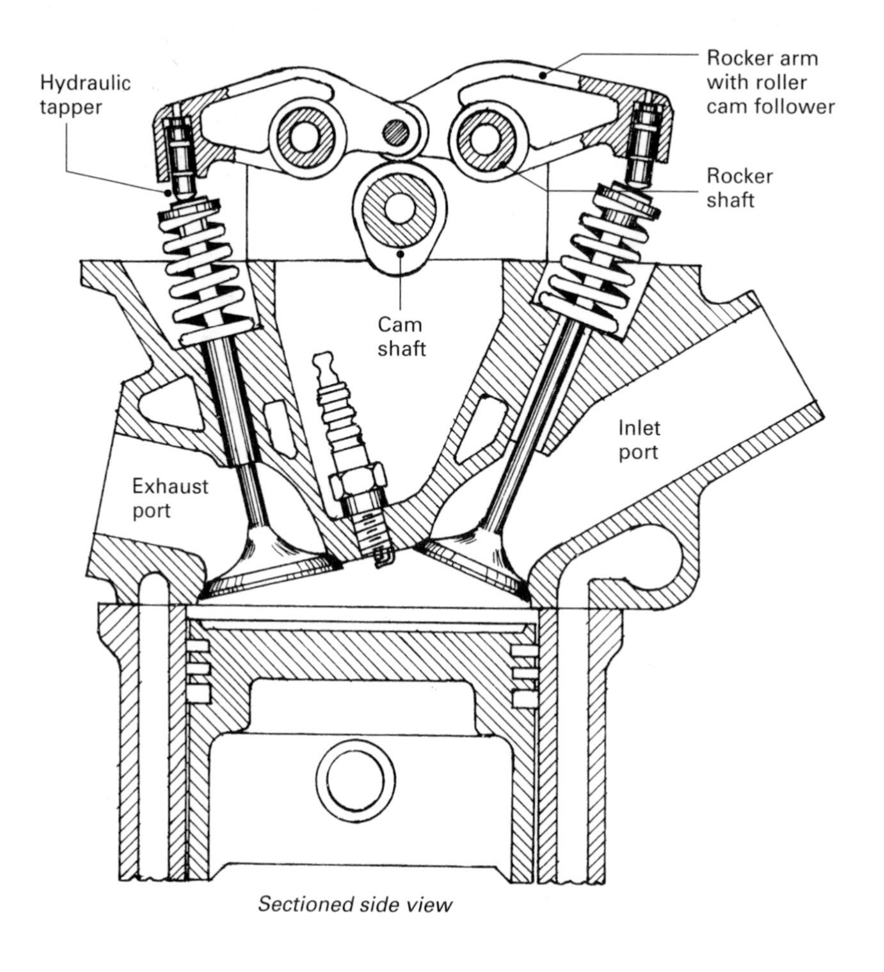

Sectioned side view

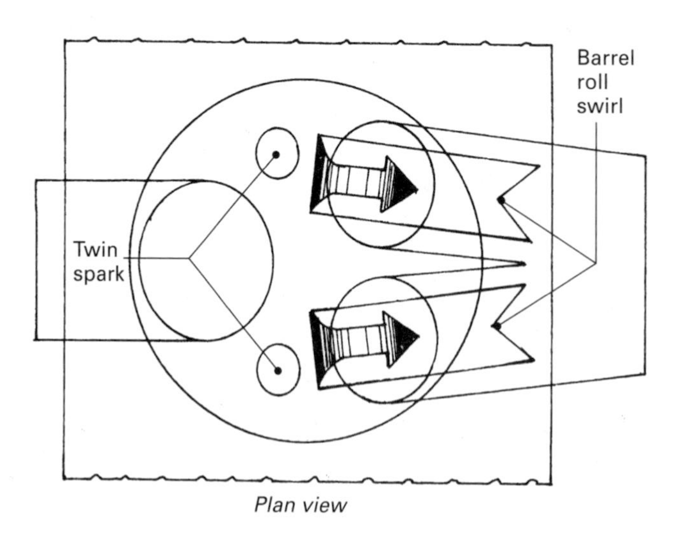

Plan view

Fig. 14.15 Triple-valve twin-spark pentroof combustion chamber (Mercedes)

ing with the dual-ignition, particularly with weak mixtures, a high degree of exhaust gas recirculation levels can be used to reduce combustion temperature and thus the emission of oxides of nitrogen will be minimised.

Direct-injection quadruple-valve petrol engine (Mitsubishi) (fig. 14.16(a)–(c)) This inclined four-valve cylinder-head with the central sparking-plug forms a pentroof combustion chamber; similarly the piston crown which forms the lower half of the chamber matches the pentroof profile. However it deviates from a semi-spherical cavity cast in the crown on the inlet-valve side of the chamber facing the sparking-plug and the fuel injector. This leaves the opposite tilted side of the crown to function as a large squish area. The twin intake ports flow vertically downwards into the roof of the chamber, whereas the twin exhaust ports exit horizontally.

Modes of operation

Idle and partial-load operating conditions (fig. 14.16(a)) Here the fuel is injected late towards the end of the compression stroke as this promotes a stratified form of air/fuel mixing. This enables ultralean air/fuel ratios of 30–40 to 1 to be burnt with large amounts of exhaust gas recirculation without mis-firing occurring. Exhaust gas recirculation is necessary to suppress the combustion temperature thereby, minimising the formation of oxides of nitrogen (NO_x).

Medium-load operating conditions (fig. 14.16(b)) Under these driving conditions fuel is injected early on the intake stroke so that the cylinder is homogeneously filled with a lean air/fuel mixture ratio ranging between 20 and 25 to 1.

Full-load operating conditions (fig. 14.16(c)) When full-load operating conditions prevail fuel is injected during the early part of the intake stroke. However, the homogeneous mixture is now at stoichiometric or slightly richer. It thus ranges from between 13 and 15 to 1. These enriched mixtures enhance maximum power, reduce the exhaust temperature and protect the three-way catalyst.

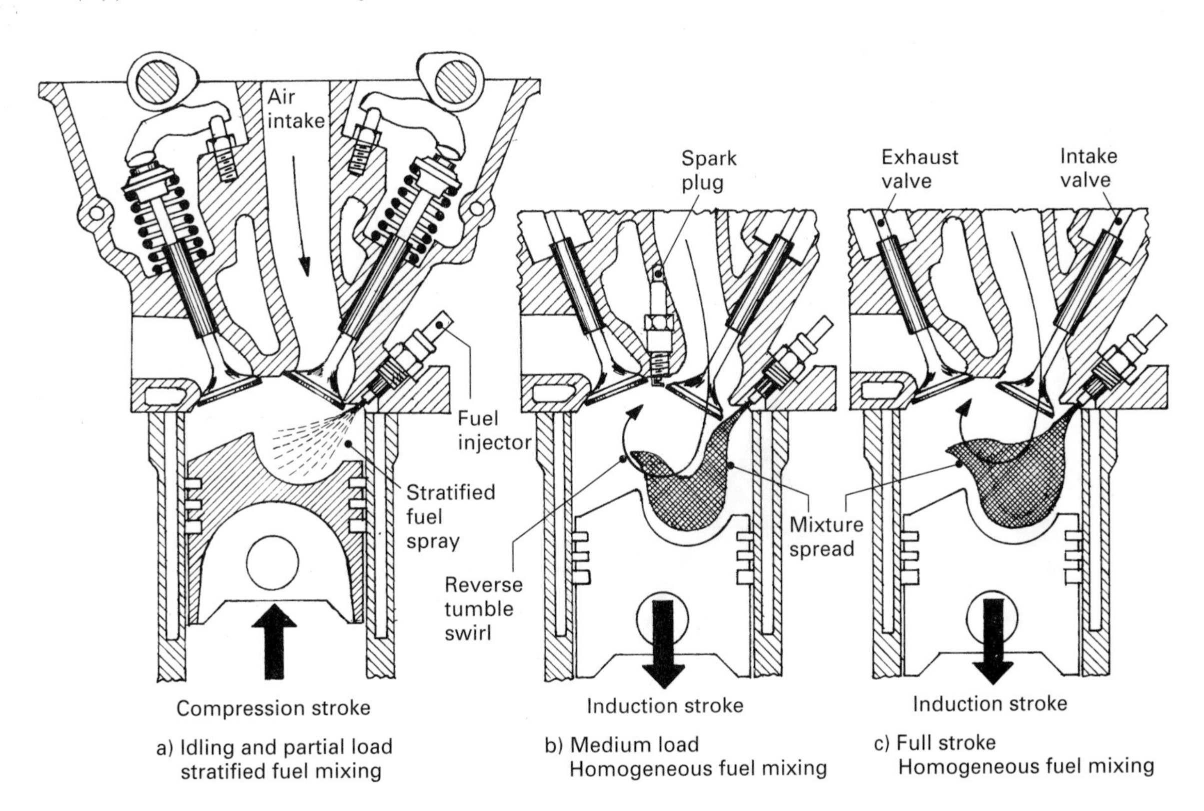

Fig. 14.16 Direct-injection quadruple-valve petrol engine (Mitsubishi GDI)

Fuel injector (figs 14.16(a) and 14.17(a) and (b)) A common rail fuel injection system is used which operates with a relatively low injection pressure of about 50 bar. To achieve the precise control of fuel entering the cylinder for the three modes of combustion an electromagnetic swirl type injector is positioned on the inlet-valve side of the combustion chamber with its spray formation directed towards the bowl-shaped cavity formed on inlet side of the inclined piston crown. A swirl injector was chosen to obtain a widely dispersed fuel spray to promote atomisation with relatively low injection pressure. This spray formation is sensitive to the air charge density. Thus injecting the fuel early on the induction stroke when the cylinder pressure is very near atmospheric produces an unrestricted widespread hollow cone structure ideal for homogeneous mixing (see fig. 14.17(b)). For injecting fuel late on the compression stroke as the piston approaches TDC, the pressure is in the region of 10 bar and produces a compact solid central core with faint cone-like haze due to the high drag forces opposing the penetration of the injected fuel (see fig. 14.17(a)).

To meet the wide ranging demands of the injector it is designed to perform in two states as follows:

1) For early injection on the induction stroke a widespread fuel spray is desired to produce a homogeneous mixing suitable for medium to full-load conditions.
2) For late injection towards the end of the compression stroke a compact spray is necessary so that the spray can be stratified within the cavity form on the piston's crown. This spray must also be highly atomised so that it can rapidly vaporise, thereby enabling it to sustain a high degree of exhaust gas recirculation before the mixture reaches the sparking-plug.

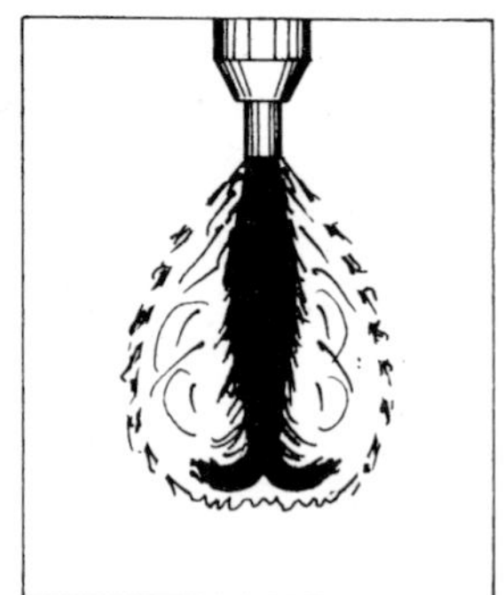

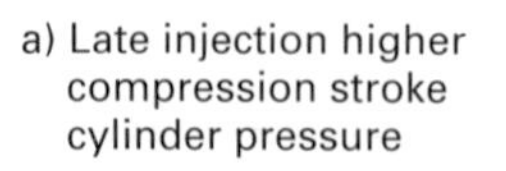

a) Late injection higher compression stroke cylinder pressure

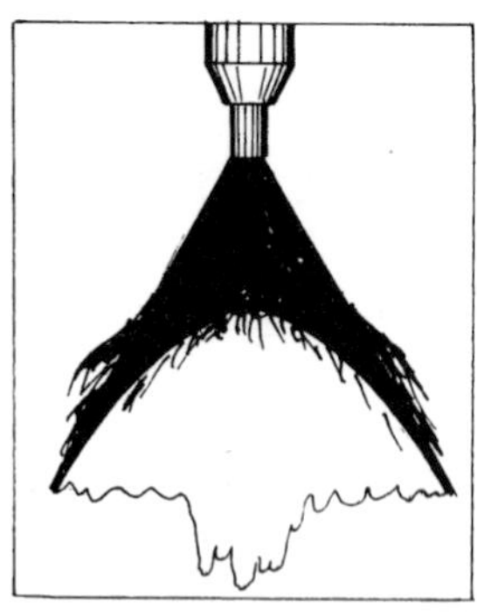

b) Early injection low induction stroke cylinder pressure

Fig. 14.17 Injection spray formation under low and moderate cylinder pressure conditions

Electronic control unit An electronic control unit monitors the constantly varying operating conditions of the engine and the demands of the driver and subsequently transmits electrical signal pulses to the injector to switch between the early and late injection modes and to adjust the corresponding air/fuel mixture strength to match these operating conditions.

The combustion process

Idle and partial-load combustion (figs 14.16(a) and 14.17(a)) During light-load operating conditions with the piston approaching TDC on its compression stroke, the piston cavity traps the air charge and moves it towards the centrally positioned sparking-plug in the roof of the chamber. The air movement within the piston cavity is enhanced by the squish flow from the exhaust to the intake side, as the gap between the piston crown and the roof of the chamber decreases. Before the piston reaches TDC on its compression stroke, fuel is injected into the high-density air mass; consequently the compact fuel spray attempts to penetrate the dense surrounding air, however, the frictional drag resistance proceeds to break up the fuel particles into a fine layered structure; this mixing process is known as stratification. The majority of this mixture will be concentrated around the sparking-plug, so that when the sparking-plug fires, ignition and burning only exist locally where the mixture is relatively rich, whereas outside this zone only air exists so that no burning takes place. Thus it can be seen that the very weak overall air/fuel mixture ratio of 30 to 40 to 1 can be sustained since in effect a rich mixture is being burnt, that is, an air/fuel mixture ratio of about 14.5 to 1 immediately underneath the sparking-plug. Once combustion is established it sweeps down on the descending piston on its power stroke.

Medium to full-engine load combustion (figs 14.16(b) and (c) and 14.17(b)) As more power is demanded the open throttle allows the full impact of the downward flow of air as it enters the cylinder. Reverse tumble swirl follows which promotes the vigorous and thorough mixing of the fuel when early injection of the fuel occurs on the piston's downward induction stroke. Under these conditions the finely atomised mixture is homogeneous, that is, it is distributed evenly thoughout the cylinder and combustion chamber. The mixture strength needed to sustain uninterrupted combustion ranges between 20 and 25 to 1 for medium engine loads, but for full engine power stoichiometric or even richer mixture strengths are essential, that is, air/fuel ratios of between 13 and 15 to 1.

The finely atomised fuel increases the charge cooling and the precise control of the injection and ignition timing enables a 12 to 1 compression ratio to be used without causing any combustion roughness.

Engine performance comparison Engine torque characteristics with early and late fuel injection are shown in fig. 14.18. It is claimed that the direct injected petrol engine compared to a conventional port injected engine of the same size uses 25% less fuel during town driving, and compared to a direct injected diesel engine of the same cylinder capacity, it has a 7% better fuel economy, 85% more power and 12% increase in maximum torque.

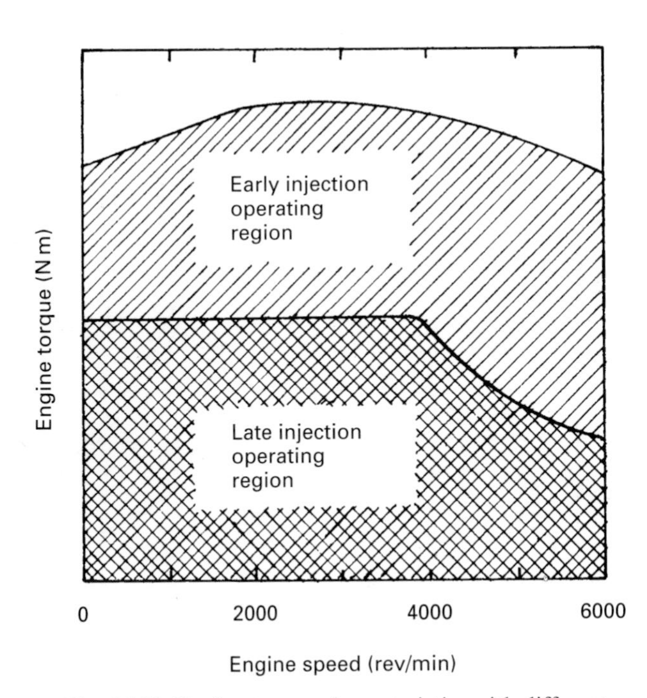

Fig. 14.18 Engine torque characteristics with different injection tuning modes

14.2 Compression ignition (diesel) combustion process (fig. 14.19)

The combustion process can be described in four phases or periods as follows:

1) ignition delay;
2) rapid pressure rise or uncontrolled combustion;
3) mechanical controlled combustion;
4) after burning or late burning.

Ignition delay period (fig. 14.19) The delay period is the time from when the instant fuel spray begins to enter the chamber to when early vaporised fuel droplets ignite and commence burning with the surroundings to cause the first noticeable pressure rise above that of the compression pressure in the cylinder. It thus includes the time for the early part of the fuel spray to penetrate into the hot compressed air mass followed by the time it takes for the hottest sites of the liquid/vapour cores to begin to burn.

The duration of this period depends upon the following factors:

1) the degree of fuel penetration and atomisation;
2) the pressure and temperature in the cylinder at the end of the compression stroke;

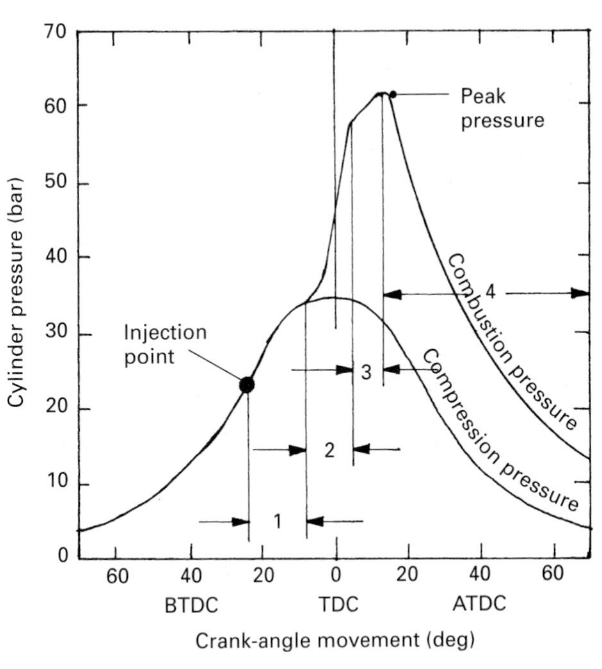

Fig. 14.19 The four phases of combustion in a compression ignition (diesel) engine

3) the degree of injection timing advance;
4) the amount of existing air turbulence;
5) the quality of the fuel, that is, its ability to ignite. This correlates to some degree to the Cetane number of the fuel.

Rapid pressure rise (uncontrolled combustion) period (fig. 14.19) This period follows on from the ignition and initial burning of the first fuel spray particles entering the chamber where the bulk of the penetrating spray droplets form vapour clouds, mix with the surrounding highly turbulent air swirl, and produce many nuclei flame sites which rapidly propagate burning throughout the agitated air/fuel mass. Rapid pressure rise continues until the majority of the unburnt fuel particles have been consumed.

The duration and steepness of this second period is mainly influenced by the following factors:

1) the degree of injection timing advance;
2) compression ratio and compression pressure and temperature;
3) the intensity of swirl and turbulence;
4) the duration of the ignition delay period.

Mechanical controlled burning period (fig. 14.19) This period follows on from the rapid pressure rise period and covers the part of combustion where the remaining fuel is partially surrounded and stifled by the burnt products of combustion. It therefore takes longer to react with the oxygen contained in the still-unused air charge. Consequently, the rate of burning and correspondingly the rate of pressure rise is much reduced. The duration of this period is greatly influenced by the quantity of fuel injected per cycle. Thus at no-load, the quantity of burnt products will be small and there will be a large excess of air so that the third mechanical controlled period will barely exist. Nevertheless, under full engine load the amount of fuel injected will be large which will produce a considerable amount of burnt products and proportionally less excess air. This results in an increase in the duration of the lower pressure rise rate period.

After burning or combustion termination period (fig. 14.19) This final period is represented by a rapid pressure reduction; fuel injection has already ceased but the late and slow burning of the last particles of the fuel to enter the cylinder is taking place, the completion of combustion being hindered by the barriers of burnt gases surrounding the few remaining and possibly isolated unburnt vapour specks.

14.2.1 Diesel knock (fig. 14.20)
Diesel knock is the loud noise produce by very high pressure rise rates in the cylinder. It normally occurs under idle and light-load operating conditions where the temperature in the cylinder is relatively low. The cause of this noise is mainly due to a prolonged ignition delay period. Ideally burning should take place progressively as the atomised fuel spray enters the chamber. Nevertheless an extended delay before ignition occurs means that a large proportion of the total fuel discharged will be in the chamber before burning actually commences. Thus a large amount of energy will be released by this unburnt fuel when combustion does finally begin. Subsequently a very rapid and steep pressure rise will take place over a very small crank-angle movement, resulting in a rough and noisy combustion process.

Factors which tend to extend the ignition delay period are as follows:

1) low compression ratios only marginally permit the self-ignition temperature of the fuel to

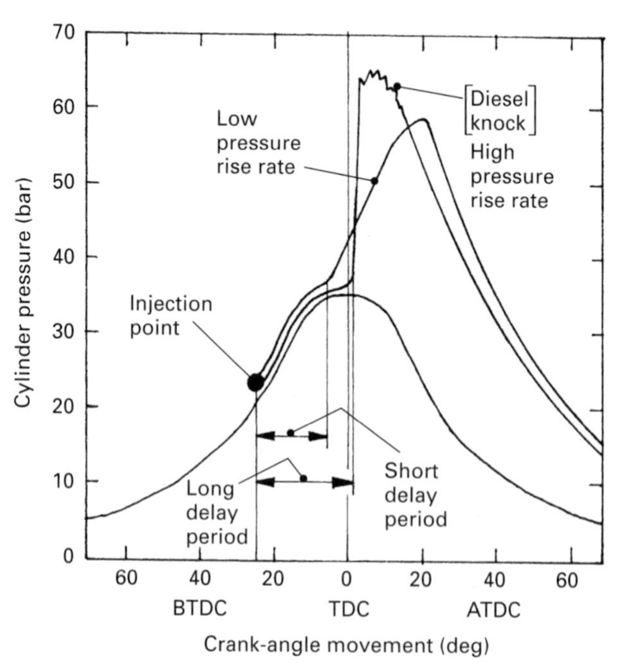

Fig. 14.20 Effect of cylinder pressure variation with short and long delay period

be reached as the piston approaches TDC on the compression stroke;

2) loss of compression pressure at TDC due to worn piston and rings or poorly seated poppet valves;
3) over-advanced fuel injection causes fuel spray to be discharged before the self-ignition temperature has been reached;
4) low air-intake temperature in cold weather reduces the peak compression temperature, thus making it more difficult for the air charge to reach the self-ignition temperature of the fuel;
5) poor fuel ignitability relating to a low Cetane number fuel;
6) low injector nozzle spring load producing a coarse fuel droplet discharge;
7) poorly atomised fuel spray due to worn needle to nozzle seat or partially blocked nozzle hole or holes.

14.2.2 Cetane number (fig. 14.20).

The cetane number of a diesel fuel is a measure of its ignition quality, that is, the speed with which it will ignite in an engine. It is measured in a test engine by comparing the performance of the fuel under test with that of blends of known cetane number fuels. The primary reference fuels are cetane which has a high ignition quality (short delay, fig. 14.20) and is given a cetane number of 100, and heptamethylnonane which has low ignition quality (long delay, fig. 14.20) and is given a cetane number of 15. Originally, the low ignition quality reference fuel was alpha-methylnapthalene which was given a cetane number of zero, but now heptamethylnonane which has a cetane number 15 has replaced it as a more stable compound. Diesel fuels for road vehicles these days have a cetane number of about 50.

14.2.3 Classification of compression ignition combustion chambers

Direct injection open quiescent quadruple-valve combustion chamber (fig. 14.21) These chambers are normally used for large (10–16 litre) low to medium-speed (up to 2500 rev/min) commercial vehicle diesel engines where low fuel consumption and high torque output is essential. A centrally positioned injector is mounted over a wide but shallow bowl formed in the cylinder head. Air movement is created by a moderate amount of induction swirl which directs the air in a spiral or whirlpool motion around the expanding cylinder space on the induction stroke, whereas inward compression swirl is achieved squeezing the air between the cylinder head and the relatively small piston-crown rim on its compression stroke. There are eight or more injector nozzle holes positioned at equal intervals around the spray nozzle and pointing towards the surrounding piston chamber walls. This combustion chamber design relies not so much on the intensity of air movement, but on the very high injection pressures (1000–1600 bar) that produce many penetrating and highly atomising spray outlets. Quiescent chamber design refers to the matching of the relatively low air speed movement to a large number of spray outlets which eject highly atomised fuel at very high pressures such that it maximises the possible amount of air and fuel mixing in the short time available. Large cylinder capacity direct combustion chamber diesel engines normally have compression ratios of the order 14–18:1 compared to 16–20:1 for smaller cylinder capacity direct combustion chamber engines.

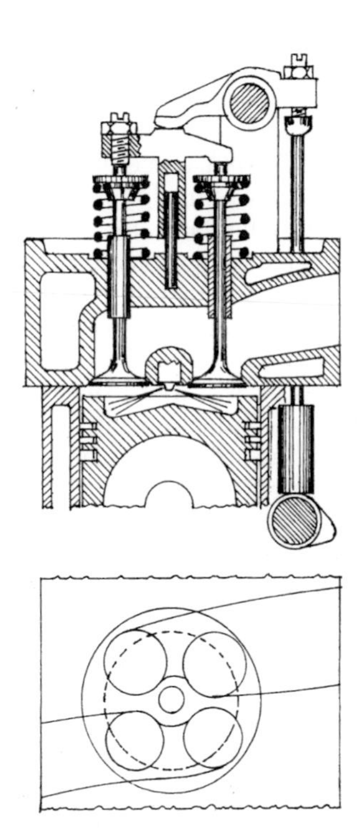

Fig. 14.21 Direct injection quiescent quadruple-valve combustion chamber

Direct injection semi-open deep bowel volumetric combustion chamber (fig. 14.22) Combustion chambers of this class are now utilised mainly for small to medium (2–8 litre)-sized engines which operate in the medium to high-speed range (up to 4500 rev/min) where low fuel consumption and high power output is required. Engine manufacturers have now refined these bowls in piston direct injection chambers for van application where some degree of combustion roughness can be tolerated at low speed, and for cars which have a high degree of noise insulation. Efficient combustion is achieved with the combination of induction spiral or whirlpool swirl produced by the helical-shaped induction port, and the compression squish inward swirl created by the large squish area formed between the flat parallel surfaces of the cylinder head and piston crown where the piston approaches TDC. To match the intensity of swirl fuel droplets are injected directly into the combustion chamber bowl at high penetrating pressures (800–1200 bar) that will atomise and thoroughly mix the fuel with the surrounding highly mobile air mass. The intense heat created by the highly compressed and turbulent air initiates local burning around the atomised fuel spray. The clouds of burning fuel and air mixture then expand and spread beyond the confined bowl, into the cylinder space formed between the cylinder head and piston crown.

Indirect injection with swirl-type combustion chamber (fig. 14.23) Indirect combustion chambers are used on small (1.5–3.5 litre) diesel car engines which can run with a clear exhaust up to 5000 rev/min and where a smooth low-speed low-noise level engine is essential. With this indirect combustion chamber design air is displaced from the main chamber above the ascending piston into a separate semi-spherical chamber (swirl chamber) via a passage cast in a stainless steel regenerative member that forms the underside of the swirl chamber. Thus not only does the air mass passing through the restricted throat or neck passage require a high degree of swirl and turbulence, it also becomes very hot. Just before TDC is reached on the compression stroke, fuel is injected into the dense highly mobile and turbulent air mass, and the fuel droplet spray is then rapidly

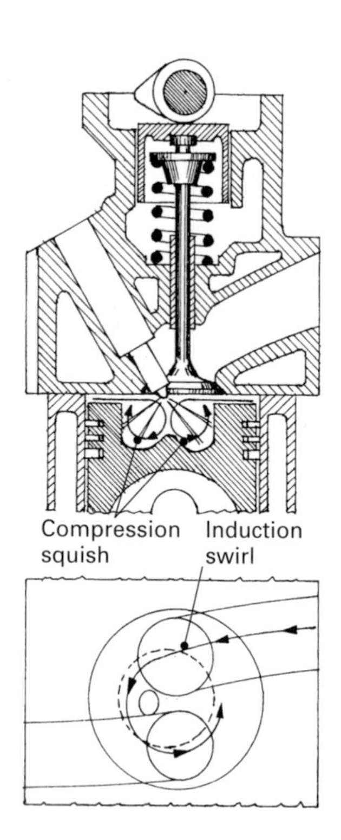

Fig. 14.22 Direct injection deep bowl volumetric combustion chamber

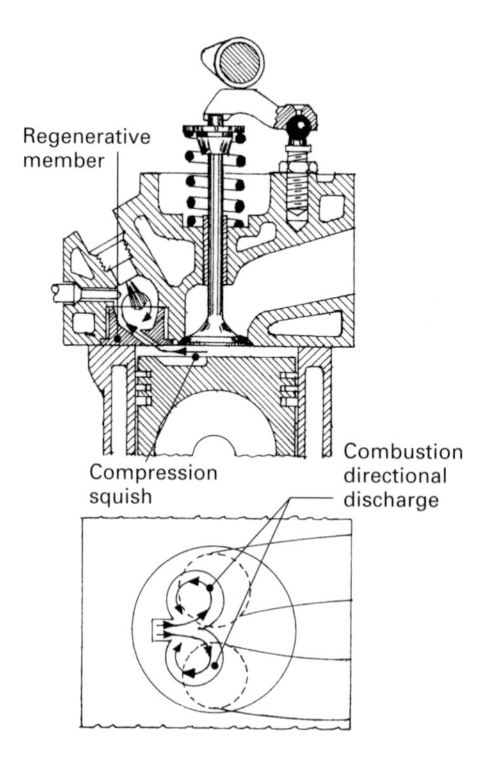

Fig. 14.23 Indirect injection with swirl combustion chamber

and thoroughly mixed; the heated fuel vapour now self-ignites, burns, and raises the gas pressure in the small swirl chamber. The burning mixture is then swept back through the narrow throat where it divides into two streams guided by the two cavities cast in the piston crown. The ability of the burning and unburnt mixture being projected back into the main cylinder chamber at a very high velocity in a double whirl motion contributes greatly to the thorough mixing and completion of the burning process in the short time available.

The disadvantage of any indirect combustion chamber design is that there are considerable heat losses from the swirl chamber and negative work done in pumping in fresh charge and expelling burning charge through the narrow communicating passage. Consequently power is lost (5 – 10%) and fuel consumption is higher (10–15%) compared to an equivalent direct injection combustion chamber. To compensate for slightly inferior cold starting, the indirect chamber has heater-plugs located horizontally in the side of the swirl chamber. However some direct combustion chambers are now fitting heater-plugs directed downwards over the combustion chamber piston-bowl. Indirect combustion chambers have greater heat losses than those of the direct combustion chamber and therefore operate with higher compression ratios of the order of 20–25:1 compared to 15–20:1 for the direct combustion chamber.

14.2.4 Comparison of the spark-ignition and compression-ignition charge filling and combustion process

With the spark-ignition engine a mixture of air and fuel is induced into the cylinder on the inlet stroke. After the spark fires at the end of the compression stroke and a flame is established, burning spreads throughout the combustion chamber seeking out the homogeneous still unburnt mixture (evenly distributed mixture). Conversely, with the compression ignition engine, air only is induced into the cylinder on the inlet stroke, after which a highly pressurised and atomised fuel spray is injected into the combustion chamber towards the end of the compression stroke, producing a heterogeneous mixture (unevenly distributed mixture throughout the chamber). These atomised droplets of liquid fuel absorb heat, vaporise and self-ignite once the threshold temperature (ignition temperature) of the fuel is reached at several focal points. At this instant these focal specks of burning fuel propagate through the mix of vapour and the immediately surrounding air, but do not spread to the outer region of the cylinder since this outer space is only occupied by the air charge. Under light loads burning will only concentrate over a small central region of the chamber, but as more fuel is injected with increased engine load, fuel penetrates further outwards, almost reaching the chamber walls. Accordingly, burning will likewise expand to match the fuel spray pattern.

15 Piston and connecting-rod assemblies

15.1 Friction and heat distribution of the piston assembly

The whole piston assembly absorbs something like 50 to 60% of the mechanical losses of the engine. For a typical piston with three rings, the first compression ring accounts for 60% of the friction work, the second compression ring for 30%, and the third oil-control ring for only 10%.

The energy from combustion heats the crown of the piston, and this heat has to be dissipated by way of the ring zone and skirt. Approximately 50 to 60% of the crown heat energy is transferred from the piston to the rings and then to the cylinder walls. The remaining heat-flow distribution is of the order of 20% through the ring lands and 20 to 30% through the skirt, 5% of this heat being carried away by the gas and oil but most being conducted through the cylinder walls.

15.2 Piston materials

The materials that pistons are made from should meet certain requirements such as good castability; high hot strength; high strength-to-mass ratio; good resistance to surface abrasion, to reduce skirt and ring-groove wear; good thermal conductivity, to keep down piston temperatures; and a relatively low thermal expansion, so that the piston-to-cylinder clearance can be kept to a minimum. Some of these properties will now be considered.

15.2.1 Mass considerations (fig. 15.1)

For high speeds, the reciprocating forces created by the pistons reversing their direction of motion must be as small as possible. This has made it necessary to turn to lighter materials than the cast iron and steel which were used on early slow-speed engines.

The obvious choice of the light metals was aluminium, which has a relative density of 2.6, compared with 7.8 for cast iron. Thus for a given volume, aluminium is one third of the mass of cast iron. This would reduce the mass of the piston in proportion, but, to maintain the rigidness of cast iron, the sections of the aluminium structures will be larger, offsetting the advantage to some extent. Aluminium is always alloyed with small amounts of other elements such as copper or silicon, the relative densities of these being 8.9 and 2.3 respectively. This will considerably improve the strength-to-mass ratio of the pistons, but will only marginally alter the mass compared to a piston made of pure aluminium.

Figure 15.1 shows a family of curves of piston mass against cylinder bore size. These clearly indicate how piston mass increases with diameter and how the piston metal or alloy influences the mass. At first sight it might be thought that magnesium or a magnesium alloy would be the ideal materal; however, due to their poor abrasion resistance, these are limited to car racing, where new pistons are fitted after each race meeting.

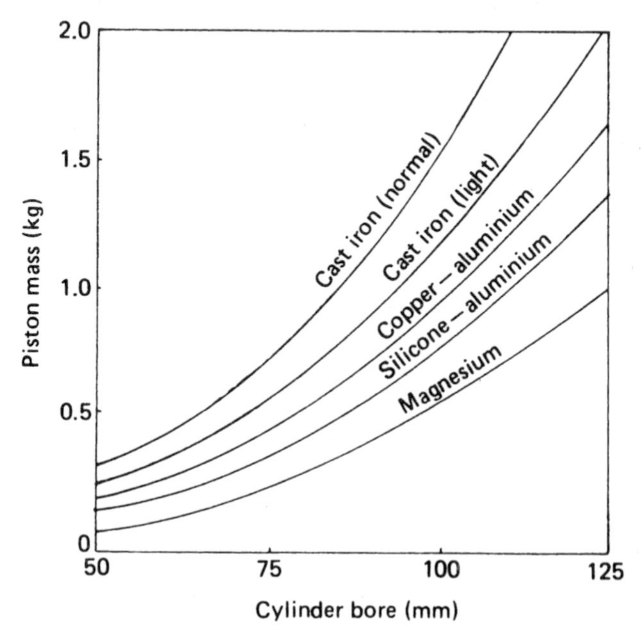

Fig. 15.1 Piston diameter and mass relationship for different materials

15.2.2 Strength and wear considerations (fig. 15.2)
Pure aluminium is not strong enough for use as a piston material, as it has a low tensile strength – about 92 to 124 N/mm^2 at room temperature, falling off progressively to about 31 N/mm^2 at 300°C, which is roughly the operating temperature in the centre of the piston crown. Furthermore, the soft aluminium has very little resistance to wear and scores readily. To overcome these limitations, small percentages of other elements such as copper, nickel, silicon, magnesium, and manganese may be alloyed with the aluminium, singularly or in various combinations. These elements produce not only improved strength over the operating temperature range, but also improved resistance to abrasion, this being mainly due to the elements forming hard particles within the aluminium.

Figure 15.2 shows the hot strengths of pure aluminium, of Y-alloy having 4% copper and 2.5% nickel, of 12%-silicon alloy, and of 22%-silicon alloy. At 100°C the Y-alloy is the strongest and the 22%-silicon alloy the weakest, with the 12%-silicon alloy in between. With increased temperature their hot strength decreases, but the rate of decline of the 22%-silicon alloy is less than that of the other two, thus at about 280°C its hot strength is superior to the other two alloys.

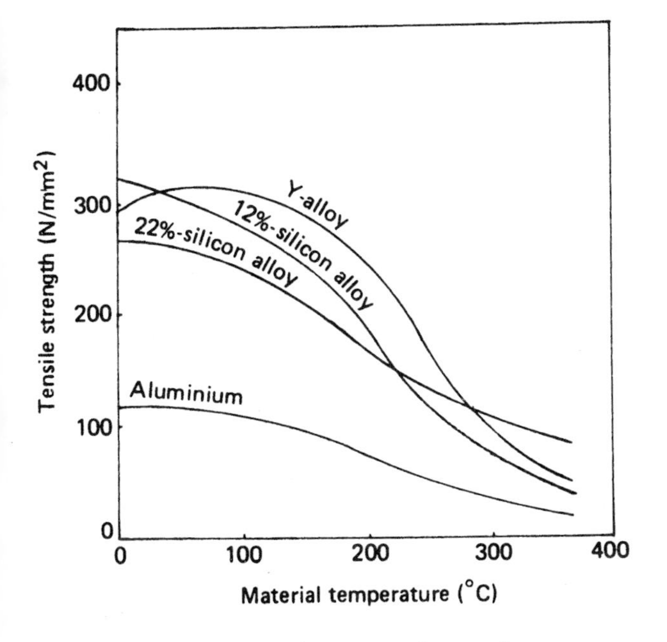

Fig. 15.2 Piston-material strength at various temperatures

15.2.3 Heat-conduction considerations (fig. 15.3)
Aluminium is a much better conductor of heat than cast iron. Considering silver as 100%, aluminium and cast iron have relative conductivities of 38% and 11.9% respectively. As the aluminium can conduct 3.2 times more heat away in a given period and alloy pistons have thicker sections than cast-iron pistons, heat transfer is superior with these light pistons. The better heat dissipation of aluminium-alloy pistons compared to cast-iron pistons greatly reduces the maximum piston-crown operating temperature, which is normally in the region of 250 to 300°C for alloy pistons and 400 to 500°C for cast iron.

Figure 15.3 shows how the piston's operating temperature increases as the engine speed rises and that the centre of the crown is the hottest region of the piston.

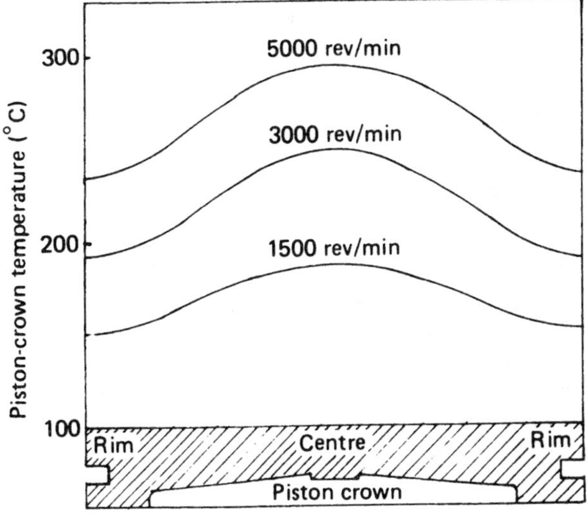

Fig. 15.3 Piston crown temperature at various engine speeds

15.2.4 Expansion considerations (figs 15.4 and 15.5)
One of the major disadvantages of aluminium as the base metal of a piston alloy is its high coefficient of linear expansion – in the region of 0.000 022 1 per °C, compared with 0.000 011 7 per °C for cast iron. This shows that the expansion of aluminium is almost twice that of cast iron, therefore extra clearance between the piston and cylinder at room temperature has to be provided, otherwise the piston would become tight and seize under operating conditions. However,

this clearance usually gives rise to piston slap when the engine is cold and consequently rapid wear. The development of low-expansion aluminium alloys has helped to reduce this problem, and their expansion properties are now discussed.

The best-known early aluminium-based alloy – the Y-alloy – has a coefficient of linear expansion of 0.000 024 5 per °C over a temperature range from 20 to 300°C. Most pistons are now made from silicon–aluminium alloys, there being two grades: with 12% silicon and with 22% silicon, these having thermal expansions of 0.000 021 and 0.000 017 5 per °C respectively. It can be seen that, as the silicon content increases, the thermal expansion is reduced, so that the cold clearance can be made smaller. The reductions in expansion relative to the Y-alloy are 11% and 40% for the 12%- and 22%-silicon alloys respectively, but the latter alloy still has a thermal expansion 50% higher than for cast iron.

Figure 15.4 shows the variation in expansion between cast iron and aluminium for various piston sizes at a mean temperature of 250°C. The increasing width of the shaded area indicates the greater expansion differences as piston diameter increases.

Figure 15.5 compares the expansion of cast iron and aluminium for a 75 mm diameter piston over a temperature range of 20 to 300°C. The shaded area can be considered as the difference between

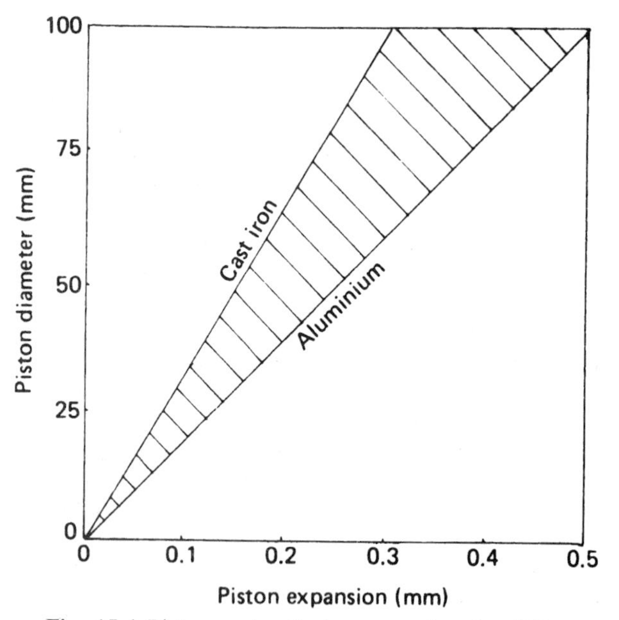

Fig. 15.4 Piston and cylinder expansion for different diameters

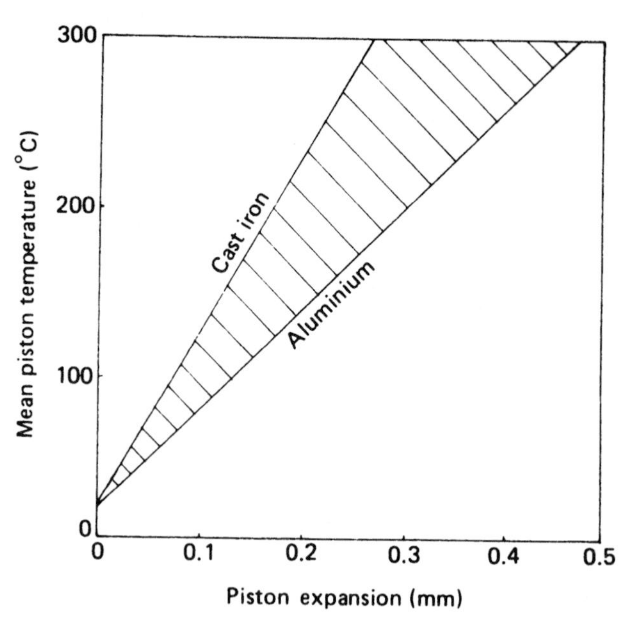

Fig. 15.5 Piston and cylinder expansion over a temperature range

the expansion of a cast-iron cylinder and that of an aluminium-alloy piston. In selecting a working clearance for a given mean working temperature, this difference will give the minimum working clearance when the engine is cold.

15.3 Piston nomenclature and design considerations (figs 15.6 and 15.7)

The highly worked piston has many features which influence its operating performance. These will now be identified and considered in some depth.

15.3.1 Ring-belt lands (fig. 15.6)

Several grooves are cut in and around the top of the piston to locate and house the piston rings (see section 15.6). The metal bands left between the grooves are known as 'lands', and their function is to support squarely the rings against the generated gas pressure and to guide them so they may flex freely in a radial direction.

The zone in which the rings and lands are grouped together is referred to as the ring-belt, and located in this belt are normally two compression rings and one oil-control ring. Sometimes for heavy-duty diesel applications there may be a third compression ring above the gudgeon-pin boss and a second oil-control ring situated near the bottom of the skirt.

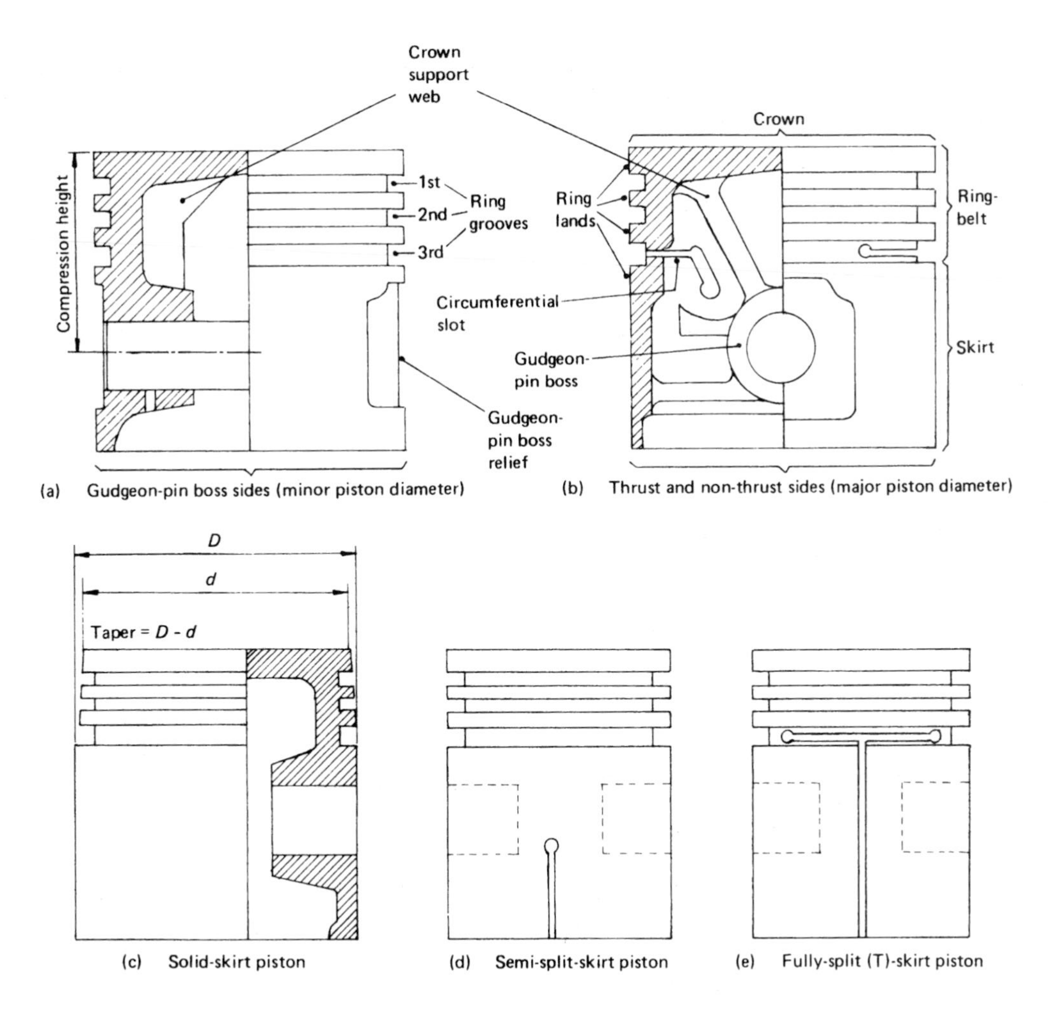

Fig. 15.6 Piston nomenclature

15.3.2 Skirt (fig. 15.6)

The piston skirt is that portion of the piston which continues below the ring-belt. Its function is to form a cross-head guide capable of absorbing the gas-pressure side-thrust created by the oblique angle made by the connecting-rod relative to the cylinder axis. The skirt should be internally structured to support the gudgeon-pin boss and of sufficient length to resist tilting of the piston under load, but it is not designed to support the piston crown against compressive loads.

Pistons are designed to operate with very small skirt clearances and, to prevent seizure under heavy loads, some petrol-engine piston skirts are made flexible so that their radial profile can adjust to the varying running conditions.

Some of the early pistons were provided with a vertical split from the bottom of the piston to the underside of the ring-belt on the same side of the skirt as the crankpin when the piston has passed TDC on its down-stroke (known as the non-thrust side). Some even had intersecting circumferential slots cut in the oil-control-ring groove above the gudgeon-pin bosses (fig. 15.6(e)) – such designs being known as fully-split-skirt pistons. If the operating temperature become very high and the working clearance marginal, then the skirt would be free to expand and close the split – in other words, the split provided a means of relief if the piston became tight due to overheating, particularly when initially 'bedding in'. The disadvantage of having a split is that it reduces the skirt's rigidity, so that the skirt tends to collapse inwards without elastic recovery. Thus the outcome will be a permanently reduced piston diameter, with a consequent increase in the piston slap, noise, and wear which the split was originally designed to cure.

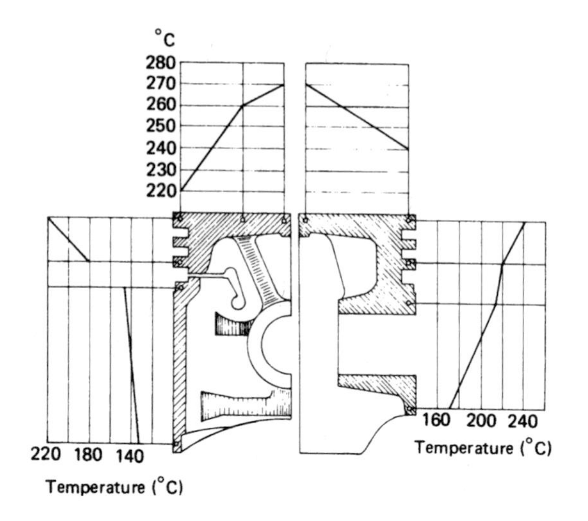

Fig. 15.7 Piston temperature gradients across the piston's crown and along its skirt

Skirts with splits which go only about half-way up are known as semi-split skirts. These are usually preferred as a compromise, and they also have blunting holes drilled at the end of the split to reduce the stress concentration created due to the split's notch effect (fig. 15.6(d)).

High-performance or heavy-duty pistons do not have any part of the piston skirt split. These are thus known as solid-skirt pistons.

15.3.3 Piston webs (fig. 15.6)

Webs are cast inside the piston between the crown and the gudgeon-pin bosses to act as struts. The compressive gas loads can then be transmitted direct from the crown to the gudgeon-pin bosses, and these forces are then transferred by way of the gudgeon-pin to the connecting-rod. Unfortunately the thick web sections form heat paths from the crown to the gudgeon-pin bosses which can lead to expansion problems if they are not carefully designed.

15.3.4 Gudgeon-pin bosses (fig. 15.6)

The combustion gas load is transmitted from the piston crown to the connecting-rod via struts and radial webs formed inside the piston on the gudgeon-pin axis. The bosses mate with the gudgeon-pin and so form a bearing surface for the rocking motion of the connecting-rod. The thick-sectioned webs also act as heat paths and divert most of the heat from the piston crown to the gudgeon-pin bosses and their adjacent skirt regions.

15.3.5 Thermal slots (figs 15.6 and 15.7)

In order to make the clearance across the thrust (gudgeon-pin) axis when cold very similar to the clearance when operating at working temperature, circumferential thermal slots are cut on the thrust and non-thrust sides of the piston between the ring-belt and the skirt. These horizontal slots extend about 90° on each side of the piston. The air gap formed tends to isolate the thrust and non-thrust sides of the skirt from the rest of the piston and so acts as a physical heat barrier. Heat will thus be redirected from the crown, straight to the gudgeon-pin bosses, instead of flowing directly to the close-fitting thrust and non-thrust sides of the piston skirt.

In some designs, the circumferential slots are situated in the oil-control-ring groove and end in inclined slots extending downwards into the upper part of the skirt. These elongated holes or slots provide additional heat barriers and so reduce even more the amount of heat reaching the thrust and non-thrust working faces of the skirt. This configuration of slots with drooping ends makes the skirt very flexible in the upper gudgeon-pin-boss region (which is the hottest) and yet the lower portion of the skirt will have the maximum support.

Figure 15.7 shows sections of the piston across thrust/non-thrust and gudgeon-pin sides. It can be seen that the skirt temperature gradient below the thermal slot on the thrust side is lower than that on the gudgeon-pin face side, so that small skirt-to-cylinder-wall clearances can be used. Considering the temperature gradient across the crown, the temperature decline between the centre of the crown and the point where the web connects the crown to the gudgeon-pin boss is less steep, because a large proportion of heat flows along this path to the gudgeon-pin boss and the skirt. Beyond the web to the crown outer edge there is a steeper drop-off in temperature, so the ring-belt is cooler on the thrust and non-thrust sides than on the gudgeon-pin sides.

15.3.6 Gudgeon-pin boss relief (fig. 15.6)

With a large amount of heat being transferred from the piston crown to the cylinder walls by way of the gudgeon-pin bosses, additional clearance must be provided in the skirt adjacent to these bosses. In these circumstances, the boss end-faces forming part of the skirt are sometimes recessed to provide a permanent relief, the reduction in skirt rubbing surface area being insignificant on the gudgeon-pin-boss skirt sides.

15.3.7 Skirt ovality (fig. 15.8)
Piston skirts are often ground to an elliptical or oval shape, having the smallest diameter across the gudgeon-pin bosses (fig. 15.8). The larger clearance at the gudgeon-pin bosses compensates for the majority of the heat flow tending to follow the main mass of metal of the gudgeon-pin webs and bosses. At working temperatures the oval shape will take on a circular form and hence match the section of the bore.

The amount of ovality is a maximum at the gudgeon-pin-boss level but is reduced progressively towards the bottom of the skirt, since the lower regions of the skirt operate at lower temperatures.

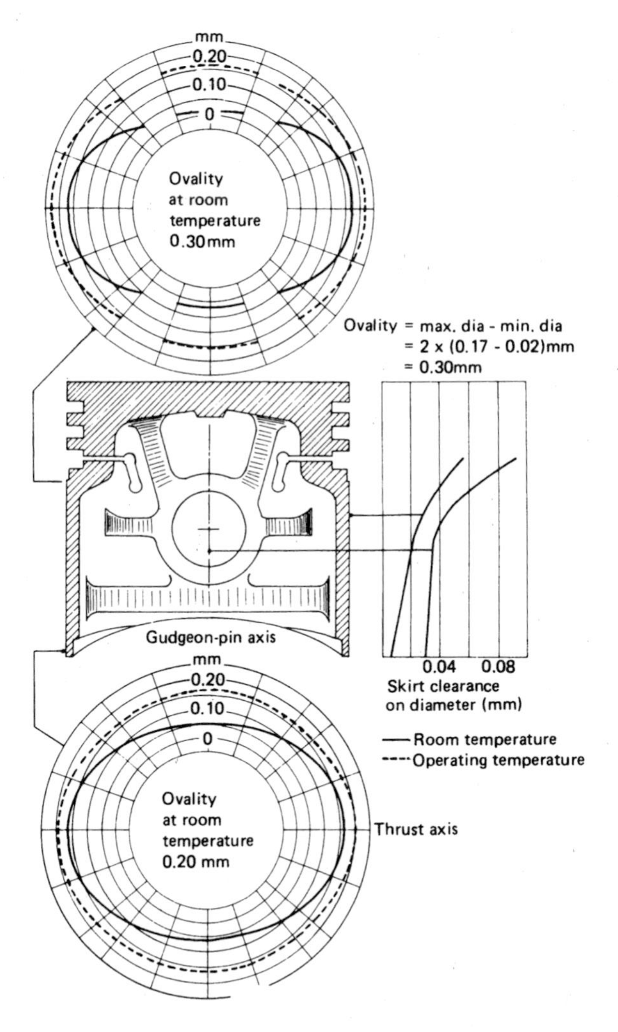

Fig. 15.8 Piston crown and skirt profiles, hot and cold

15.3.8 Piston taper (figs 15.6(c) and 15.8)
Piston temperature and hence expansion varies along the piston, being greatest at the piston crown and at a minimum at the bottom of the skirt. To compensate for this unequal expansion, the piston is tapered along its axis, the crown end having the smallest diameter (fig. 15.6(c)). The degree of taper will vary along the piston length – it usually being much greater in the ring-belt. The clearance between the piston skirt and the cylinder wall is about 0.05% at the bottom of the skirt and about 0.10% at the top of the skirt (fig. 15.8).

15.3.9 Compression height (fig. 15.6)
This dimension is the distance between the gudgeon-pin centre axis and the top of the piston crown. Engine manufacturers may offer different compression-ratios for the same basic engine and, to achieve this, a series of pistons may be made which are identical except for their compression heights.

15.4 Bi-metal strut piston (fig. 15.9)
This type of piston has a pair of steel inserts or struts extending over approximately 90° of the circumference on the thrust and non-thrust sides of the skirt. The outer face of each strut is cast with the aluminium alloy, but the inside face is separated from the piston ring-land region by a parallel gap originating in the oil-control-ring groove (fig. 15.9(b) and (c)). These gaps interrupt the heat flow from the crown to the thrust and non-thrust sides of the skirt and so minimise the average temperature which these sides normally attain.

The extension of these struts downwards provides a strengthening support to the skirt, yet there remains a degree of flexibility between the upper part of the piston and the gudgeon-pin bosses (which withstand the compressive gas loads) and the skirt (which supports the side-thrust) without creating concentrated stresses at the skirt-to-boss junctions which could result in fatigue cracking.

The principle of this design is to cast steel inserts into the skirt. These have a lower expansion rate than the bulk of the piston material, so they prevent the skirt from expanding as much as if it were made only from aluminium alloy.

During the casting solidification process, the inserts and the aluminium alloy become as one, and the steel strut prevents the alloy from con-

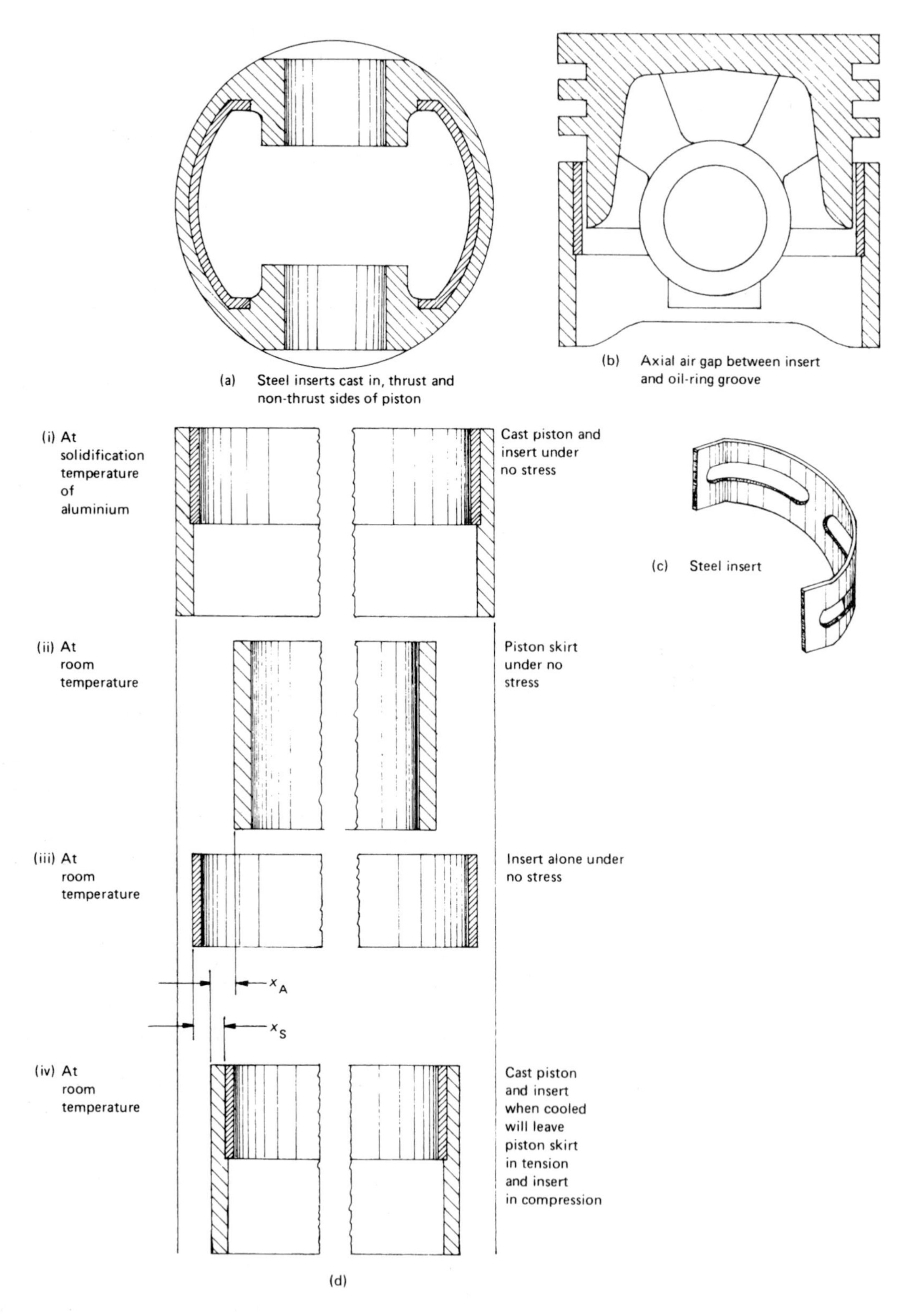

Fig. 15.9 Bi-metal strut piston

tracting its full amount when cooling, rather than restraining its expansion when it heats up.

Figure 15.9(d)(i) shows the size of the insert and skirt during casting at the solidification temperature of the aluminium alloy. Figure 15.9(d)(ii) and (iii) respectively show the undistorted but shrunk sizes of the skirt and the inserts if cooled separately to room temperature. Figure 15.9(d)(iv) shows the actual diameter of the skirt when the inserts have been bonded to the aluminium alloy and the bimetal combination has cooled to room temperature – it can be seen that the resultant size is somewhere mid-way between the sizes of the individual components in their free-state condition at room temperature.

A closer examination would show the steel insert is being compressed by x_s and the aluminium alloy is being stretched by x_A. Under operating conditions, the piston will become hot but there will be very little skirt expansion since most of the heat goes into removing the existing internal stresses set up by the contraction after solidification of the alloy. When the process is reversed and the piston cools, the different expansion rates of the struts and the skirt will again establish internal residual stresses which will limit the skirt's contraction.

In some of the earlier pistons which had steel inserts, struts were cast through the gudgeon-pin bosses, instead of on the thrust and non-thrust skirt sides. These designs certainly reduced the overall expansion, but the arrangement described above provides even better expansion control so that smaller skirt-to-cylinder-wall clearances can be employed.

15.5 Piston-ring nomenclature (fig. 15.10)

Ring diameter This is the diameter of the cylinder bore in which the ring is to operate.

Radial thickness This is the shortest distance between the outer and inner circumferential faces of the ring.

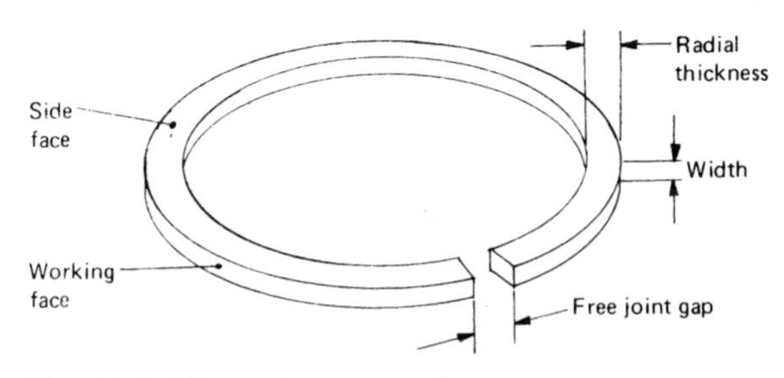

Fig. 15.10 Piston-ring nomenclature

Ring width This is the distance between the top and bottom side faces of the ring.

Side faces These are the flat parallel upper and lower faces of the ring which contact the sides of the ring groove.

Working face This is the outer circumferential surface which contacts the cylinder wall.

Free joint gap This is the circumferential distance between the two open ends of the ring in the free state.

Fitted gap This is the circumferential distance between the two open ends of the ring when it is placed in its bore.

Tangential load This is the tangential force applied between the two open ends of the ring which is necessary to close the free joint gap to its fitted clearance.

Cylinder-wall pressure This is the radial outward force per unit area of contact, assuming the pressure to be equally distributed around the ring.

15.5.1 Factors affecting ring performance

Radial thickness The cylinder-wall pressure exerted by a ring varies approximately as the cube of the ring's thickness – a small change in thickness will considerably alter the ring tangential load and hence the wall pressure. The radial thickness is usually related to the ring diameter d and should be at least $d/24$.

Width For a given cylinder-wall pressure, an increase in ring width increases the radial load on the cylinder wall, thus excessive width is undesirable. For a given tangential loading, narrow rings will bed in to the cylinder more quickly than wide ones. The narrow rings will also reduce ring flutter and resultant blow-by of combustion products at high engine speeds.

Free joint gap With a large free joint gap, the stress in assembling a ring over its piston will be low, but the stress in the fitted working piston will be high. With a small free joint gap the situation is reversed. In general, the most satisfactory compromise is a nominal free gap of $3\frac{1}{2}$ times the radial thickness.

Radial load and pressure The radial outward load exerted by the ring will depend upon a number of factors such as radial thickness, width, free joint gap, and the strength or modulus of elasticity of the ring material.

For a given radial load, a large working-face area will reduce the wall pressure and increase the thickness of the oil-film formed between the ring and the cylinder wall, but a small working-face area will do the opposite. Thus narrow rings increase the wall pressure and reduce the oil-film thickness. Similarly, a large radial thickness will considerably increase the radial load and wall pressure and might squeeze the oil film right down to no more than a few molecules thick (known as boundary-lubrication conditions). Bevelling the working faces reduces the wall contact area and increases the radial pressure if more oil control is necessary for a given application. Therefore the dimensions of the ring cross-section determine the degree of gas-sealing and adequate lubrication of the upper cylinder walls.

15.6 Piston-ring action

Piston rings may be divided into two broad groups:

i) compression rings, whose function is to seal the space between the piston and the cylinder wall so that compressed charge or gas cannot escape;
ii) oil-control (scraper) rings, whose main purpose is to control the amount of lubricant passing up to the top of the cylinder walls, although they also to some extent perform as compression rings.

15.6.1 Compression-ring action (fig. 15.11)

The piston ring is designed to expand radially outwards when fitted in its groove, so there must not be any interference between the ring side faces and the ring groove. In its free state, the ring is slightly larger than the cylinder bore so, when it is closed up in the cylinder, it will tend to spring outwards to apply pressure on to the cylinder wall; however, it is the gas pressure acting behind the piston ring which supplies most of the radial sealing force (fig. 15.11).

On the upward compression stroke, the compressed charge will move between the groove and the ring side faces, pass behind the ring, and press it against the cylinder wall, at the same time pushing down the ring against the lower ring groove of

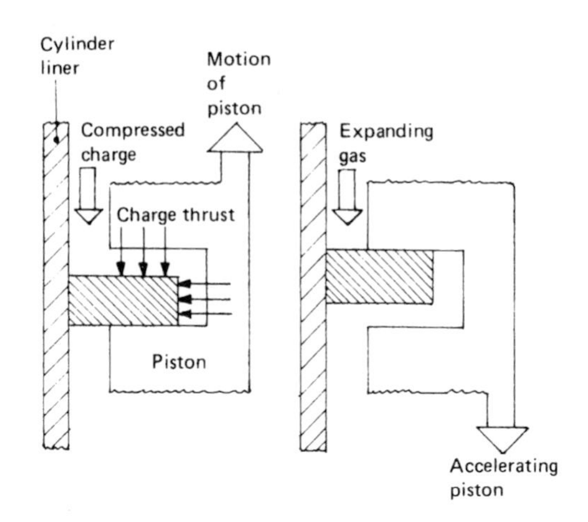

Fig. 15.11 Compression-ring action

the piston. This provides a very effective compressive seal without leakage, as long as the surface finish of the groove, ring, and cylinder wall is of sufficient standard.

On the downward power stroke a similar situation exists, but if the piston acceleration is greater than that of the ring, then the upper groove and ring side faces will be held firmly together to form the seal.

15.6.2 Oil-control-ring action (fig. 15.12)

During crankshaft rotation, more oil than is needed for lubrication will be splashed from the big-end bearings on to the cylinder walls. The oil-control scraper ring performs two functions: first it regulates the amount of oil passing to the combustion-chamber zone of the cylinder, and secondly it distributes a film of oil over the whole cylinder surface to lubricate the compression rings, the skirt, and the upper cylinder region.

The operating principle of the simple bevelled scraper ring will now be described. These rings are fitted with the bevelled working face pointing towards the cylinder head (fig. 15.12).

On the piston's upward stroke, the lower side face of the ring will bear hard against the adjacent ring groove and the bevelled working face of the ring will tend to slide over the oil, at the same time scraping a proportion of the oil ahead of itself. Excess oil will then accumulate in the clearance space formed between the groove and its ring, until it overflows through a row of drillings made in the back of the groove and finally drains to the sump.

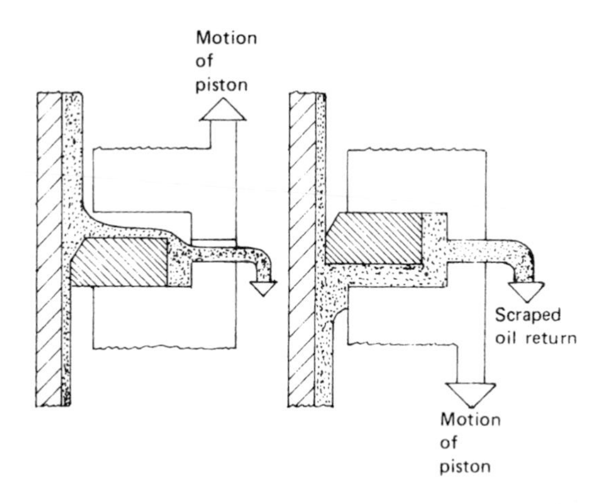

Fig. 15.12 Oil-control-ring action

When the piston reverses its direction and moves down its stroke, the ring will snap over to the top of the ring groove. The sharp edge of the working face of the ring will now scrape the oil down the bore and, at the same time, surplus oil will pass between the ring and its groove and out of the relief holes to drain off. These holes thus ensure that during this scraping movement there is no build-up of oil pressure which, with the consequent inward flexing of the ring, would push the oil up instead of down the cylinder.

15.6.3 Types of compression rings (fig. 15.13)
The most common types of compression rings are described below.

Chromium-faced plain or inlaid (fig. 15.13(a)) This is a single-piece ring tensioned by heat forming to produce the desired pressure pattern necessary for effective sealing of the cylinder bore. A hard chrome deposit on the ring working face extends the operating life of both the ring and the cylinder bore. These rings also influence the thickness of the oil film and hence oil control.

Plain inlaid (fig. 15.13(b)) Inlaid rings are used where it is desirable to retain a rectangular ring section. The well or groove on the working face is filled by depositing or spraying of chrome, molybdenum, or other suitable anti-friction materials.

Taper-faced (fig. 15.13(c)) A small amount of taper – 1° to $1\frac{1}{2}$° across the working face of the ring – provides excellent bedding qualities, as the small area of contact establishes a high radial pressure on the cylinder walls. The line contact

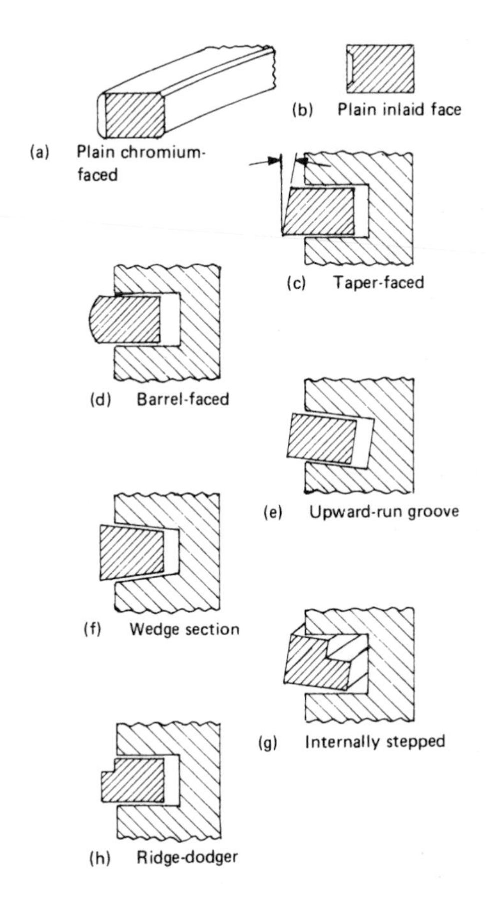

Fig. 15.13 Compression rings

of the taper also provides an effective scraping action on the down-stroke, so that it also regulates the oil supply.

Barrel-faced (fig. 15.13(d)) These curved face profiles provide a constant face radius. The degree of curvature is very similar to the amount of taper normally existing and is very critical for controlled and accelerated initial bedding. Sometimes these barrel faces are copper-plated to prevent excessive friction when bedding, which might cause scuffing.

Upward-run groove (fig. 15.13(e)) For special applications the ring groove may be slightly upward inclined, at 5°, to produce a bottom-edge line contact for rapid bedding while at the same time providing good ring side support without it becoming strained. These rings thus seal the gases at an early stage and maintain an adequate degree of oil control.

Wedge section (fig. 15.13(f)) For certain high-performance and heavy-duty engines, the problem of rings sticking in their grooves is sometimes minimised by tapering the side faces of the ring and groove to an included angle of 15°. The outward slope of the groove sides tends to ease fretting and sticking between the side faces of the rubbing pair as the ring flexes in its groove. This relative movement between the ring and the groove is mainly due to the change of piston tilt in its cylinder during every stroke and the continuous radial movement of the ring in its groove so that it conforms to the varying cylinder-wall profile.

Internally stepped (fig. 15.13(g)) The removal of a small square section from the top inside edge of the ring causes the ring to twist, thus producing a dished effect when the ring is assembled in its groove. The result of this internal relief is to form a taper-periphery working face which will give bottom-edge contact with the cylinder wall. This improves the ring bedding and provides a degree of scraping for oil control on the down-stroke.

Ring-dodger (fig. 15.13(h)) These rings are designed to clear the wear ridge formed at the top of the cylinder bore when replacement rings are fitted. Changing only the rings may be justified when the cylinders are loosing compression or consuming oil but the cylinder wear is relatively small.

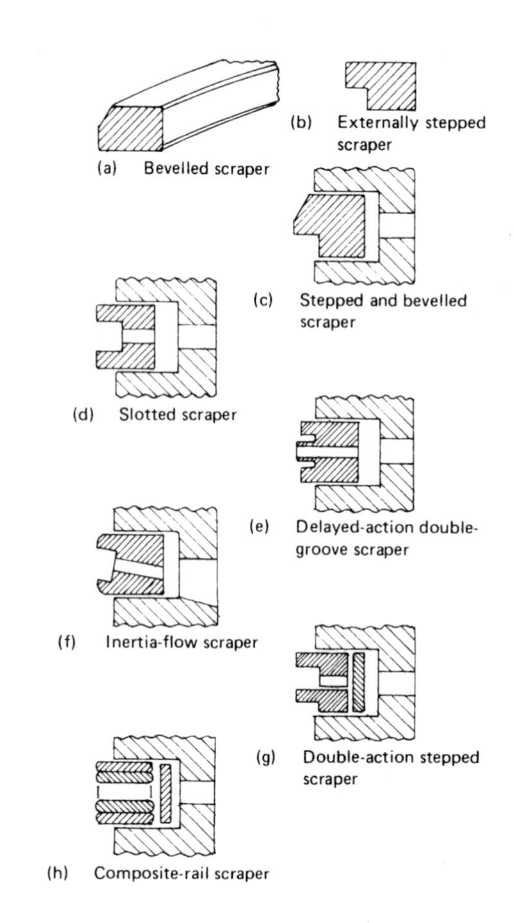

Fig. 15.14 Oil-control rings

15.6.4 Types of oil-control rings (fig. 15.14)

There are various kinds of oil-scraper rings to suit a whole range of different operating conditions, as described below.

Bevelled scraper (fig. 15.14(a)) This is a single ring with a narrow bearing working face. It should be installed with the bevelled side towards the piston head (see section 15.6.2 and fig. 15.12). With most scraper rings, oil relief holes must be provided at the back of the grooves, so that accumulated surplus oil can escape to the inside of the piston and hence drain to the sump.

Externally stepped scraper (fig. 15.14(b)) When the number of compression rings has to be reduced, a stepped scraper can be used to act both as a compression ring and as an oil-control scraper ring. The reduced width of the working face increases the pressure against the cylinder walls and thus provides improved sealing and scraping without excessively stressing the ring, as would be the case if these had been achieved by increasing the radial thickness.

Stepped and bevelled scraper (fig. 15.14(c)) These rings operate in the same way as both the stepped and the bevelled scraper rings, but the bevel reduces the working-face width so that for a given radial flexing load there will be an increase in pressure between the ring and the cylinder wall. For certain types of engine which need large amounts of oil to be splashed on to the cylinder walls, these rings provide a more effective means of regulating the upward movement of the oil.

Slotted scraper (fig. 15.14(d)) These very adaptable rings are formed from a rectangular section with a recess or groove round the centre of the working face. Elongated oil slots are also provided between the rear of this central groove and the back of the ring. The narrow ring lands so formed both provide a relatively high radial pressure against the cylinder walls and in effect establish a two-stage scraping action. If a greater degree of oil scraping is desired, the top edge of each ring land can be supplied bevelled.

Delayed-action double-groove scraper (fig. 15.14(e)) During the early stage of running, more oil may be necessary to lubricate the walls and piston assembly, but less oil is required as the cylinder walls become bedded. These slotted and grooved oil-control rings have a central land which projects further out than the other two lands when new; but, as the ring settles down, the central land wears away until eventually both of the outer lands control the scraping action.

Inertia-flow scraper (fig. 15.14(f)) On the upstroke, with this arrangement, oil flows freely past the top land of the ring and is scraped by the bottom land, so that oil moves through the drain-holes in the ring to the back of the groove where it can escape out of the channels formed in the piston section. During the down-stroke, the top land scrapes oil away into the drain-holes. The degree of oil control of this ring does not rely on the extremely high radial loadings but is greatly influenced by the inertia of the oil flow.

Double-action stepped scraper (fig. 15.14(g)) If there is ovality or heavy wear, most rings will not be able to fully conform to the cylinder walls. Better control can then be obtained by using twin scraper rings in a single groove. These two-piece rings have narrow lands with stepped segments to form oil slots. Each individual land is free to follow the contour of the bore at all positions of the stroke. A backing expander is sometimes used to increase the radial wall pressure for even more efficient oil control.

Composite-rail scraper (fig. 15.14(h)) These multi-rail scrapers are designed for use in worn cylinder bores where there may be bore ovality and cylinder taper or any other irregularities found in the cylinder walls. Only deep scores can prevent these rings controlling the oil supply.

The ring assembly consists of a number of steel rails having radiused edges which are hard-chromed. A crimped spring expands the rails against both sides of the groove – this eliminates the oil pumping which occurs between the ring and groove with a conventional ring – and an expander-ring spring behind the rails pushes them radially outwards so that they conform to the cylinder-wall profile. The rounded rail edges provide a smooth wiping action against the cylinder wall which squeezes the oil down the bore without scraping the walls and causing bore wear.

15.7 Piston-ring materials and methods of manufacture

Piston rings are made from high-quality cast iron having about 3.4% carbon with small alloy additions of up to 0.5% chromium and up to 0.25% molybdenum and nickel. The best microstructure is one which has uniformly distributed flake graphite in a pearlite matrix, but free ferrite tends to promote wear and scuffing.

Piston rings may be manufactured in various ways, four of which will now be briefly described.

i) Originally, piston rings were made from large sand-cast tubes or pots from which individual rings were parted-off to the correct width and then split. Residual compressive stresses were then created in the ring by hammering or peening round the inside of the ring, the magnitude and distance apart of the impacts being varied so that an outward-sprung free-gap profile was formed. When the deformed ring was placed in its bore, it became symmetrically circular, and at the same time exerted a radial outward thrust against the cylinder walls.

   More consistent homogeneous castings are obtained by casting the pots centrifugally, and, instead of hammering the inside of the ring, a heat-forming treatment is generally preferred.
ii) Rings may be individually cast and, after being machined and split, are then 'heat formed'. This is the process of opening the split ring on to a spacer and heating it until all the induced stresses created when opening the ring are removed. After cooling, the ring will retain its stretched shape but without being strained, and only when it is fitted in the bore will it exert an outward force.
iii) For large-scale production, rings may be cast directly to the normal-free-gap ring profile. The circumferential profile is obtained from pattern plates milled to the calculated shape for one bore diameter, so a whole series of pattern plates is necessary. After being

machined, split, and gapped, these rings will take up a truly circular shape and at the same time produce a uniform loading around the cylinder wall.

iv) Alternatively, rings may be cast to a truly circular shape and then 'cam-turned' to the open or free-state working-face shape. The rings are then split and gapped so that when they are squeezed into their bore they conform exactly to the cylinder's profile and exert the required radial outward pressure against the walls.

15.7.1 Phosphate-coated rings

Piston rings may suffer from scuffing and rapid wear while initially bedding down until the ring face and cylinder wall become seated.

To reduce the scoring and tearing of the surfaces while running in, the rings are generally treated with a porous phosphate coating. This surface layer protects the ring faces against generating very high local surface temperatures with the consequent intermittent cold welding and shearing of the contacting high spots.

These coatings are formed by immersing the ring in a bath of phosphoric acid and manganese. This produces an etched surface and deposits a layer of iron manganese phosphate all over the surface of the ring, which reduces the rubbing friction. The porous surface produced provides cavities for the worn particles and also acts as an oil reservoir which remains even when the coating has worn away.

After treatment, the rings are immersed in a hot bath of oil which is readily absorbed into the surface.

15.7.2 Chromium-faced rings

The introduction of chromium-plated compression rings has not only increased ring life but has also considerably reduced bore wear. It has been shown that ring life has been more than doubled under heavy-load operating conditions and increased up to fourfold where loads are light.

Chromium-plated working faces are usually used on the first compression ring, which is subjected to the highest working temperatures and the corrosive products of combustion. Sometimes, for heavy-duty conditions, additional rings are chromium-plated and they may also be plated on both side faces.

It is believed that the chrome face reduces the amount of abrasive particles becoming embedded on the ring surfaces and, because there is very little ring-face wear, there will be almost no debris to wear the piston groove and the cylinder wall. At the same time, very minute quantities of metallic chromium are transferred from the ring face to the cylinder walls during running. This tends to protect the relatively soft cast-iron cylinder-wall surfaces from abrasive wear and corrosive attack.

15.8 Piston and piston-ring working clearances

The procedure for measuring piston-ring clearances in the piston grooves and in the cylinder can be carried out under the following headings:

a) piston-ring side clearance,
b) piston-ring joint butting clearance (fitted-gap clearance)
c) piston-skirt-to-bore clearance.

15.8.1 Piston-ring side clearance (fig. 15.15)

The piston rings must be fitted with groove side clearance, this being the gap between the ring and land side faces. With insufficient clearance, the expansion of the lands will be sufficient to wedge the rings in their grooves – stopping the flexing and rotary movement required during running. If the condition prevails, the gas-sealing properties of the ring will be destroyed, causing blow-by which in turn leads to the destruction of the oil film followed by overheating and seizure. On the other hand, a loose fit of the ring in its groove will cause the ring to flutter. This hammers the ring against the groove faces, hence producing rapid groove wear. Excessive side clearance may also

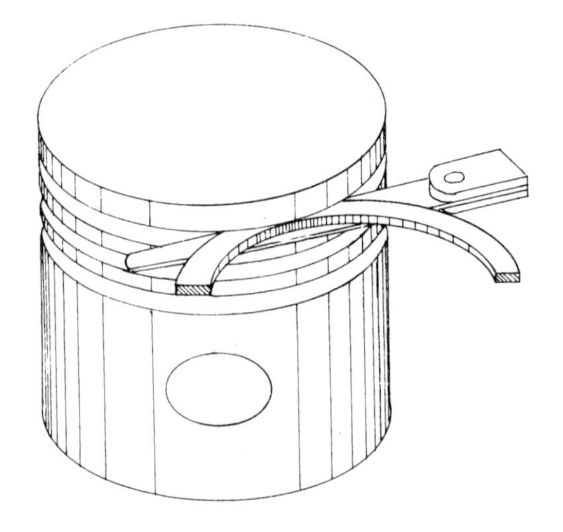

Fig. 15.15 Piston-ring side clearance

create a pumping action of the oil into the combustion zone.

Ring side clearance may be checked by removing the ring from the piston and rolling it around the outside of the piston in its groove. A suitable size of feeler gauge is then slipped between the ring and the groove to measure the clearance.

Typical minimum ring side clearances for pistons between 6 and 12 cm in diameter are as follows:

compression ring –	petrol engine	0.050 mm
	diesel engine	0.060 mm
oil-control ring –	petrol engine	0.040 mm
	diesel engine	0.040 mm

15.8.2 Piston-ring joint butting clearance (fig. 15.16)
A clearance must be allowed at the piston-ring joint to compensate for the expansion which takes place from cold to hot working temperatures.

Insufficient clearance will cause the ring ends to butt, expanding the ring against the cylinder walls with such a high outward pressure that the oil film will be sheared away, followed by semi-dry friction and overheating. In the worst case the brittle rings will tend to buckle until they eventually fracture. However, rings with large gaps may give rise to loss of compression, with the consequent blow-by over-heating effects.

Ring joint or fitted-gap clearance is checked by pushing the ring vertically down the bore, twisting at right angles, and then placing the piston on top of it to square it up. Different-size feeler gauges are then slipped inside the gap until a slight drag is felt.

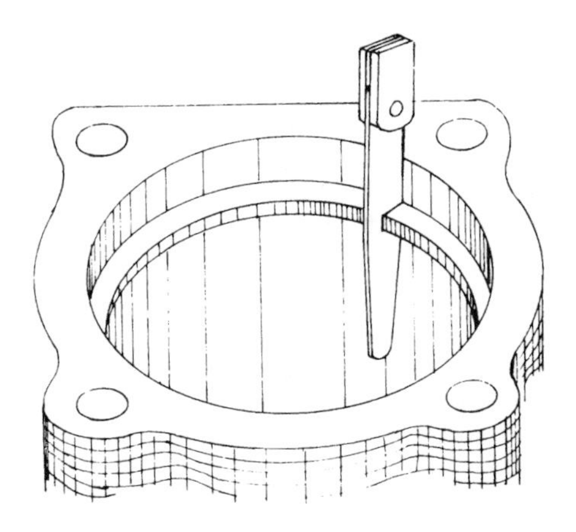

Fig. 15.16 Piston-ring joint butting clearance

Typical minimum ring joint clearances are as follows:

water-cooled four-stroke engines	0.03 mm per cm diameter
air-cooled four-stroke engines	0.04 mm per cm diameter

15.8.3 Piston-skirt-to-bore clearance (fig. 15.17)
The correct clearance between the skirt and the cylinder wall is necessary to eliminate piston slap when the engine is cold without the skirt seizing under heavy driving conditions.

To check this clearance, first insert the piston pull-scale feeler blade into the cylinder bore for its full length at right angles to the gudgeon-pin axis and then slide the corresponding piston, crown first, into the bore so that it traps the feeler blade between the cylinder wall and the piston skirt at its largest diameter. Then hold the piston stationary in the cylinder bore, without applying any side-thrust, and withdraw the feeler with a steady pull on the feeler-gauge spring scale, observing the reading required to remove the blade. A typical pull force will be between 20 and 35 N and the skirt clearance will be of the order of 0.03 to 0.10 mm, depending on the type of piston and its diameter.

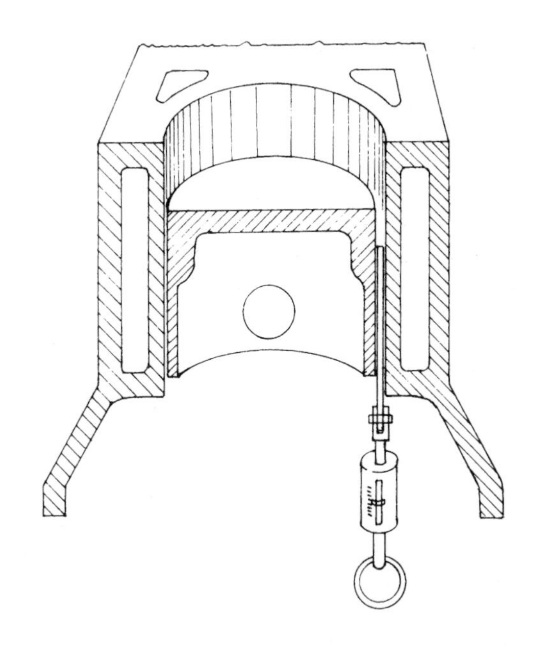

Fig. 15.17 Piston skirt-to-bore clearance

Typical minimum skirt clearances are as follows:

aluminium solid-skirt piston	0.010 mm per cm diameter
aluminium solid skirt with thermal slot piston	0.008 mm per cm diameter
aluminium split-skirt piston	0.005 mm per cm diameter

15.9 Piston and connecting-rod gudgeon-pin joints (fig. 15.18)
The piston and connecting-rod are coupled together by a gudgeon-pin which is supported in holes bored in the piston at right angles to the piston axis at about mid-height position, the centre portion of the gudgeon-pin passing through the connecting-rod small-end eye. This hinged joint provides a direct transference of gas thrust from the piston to the connecting-rod and at the same time permits the rod to pivot relative to the cylinder axis with an oscillating motion.

The temperature of the gudgeon-pin bosses in the piston is about 120 to 150°C for both petrol and diesel engines, and the temperature rise due to friction between the pin and the boss bores is something like a further 20 to 30°C, so the gudgeon-pin operates at a temperature of the order of 140 to 180°C.

The oscillatory movement of the connecting-rod during operating conditions tends to squeeze the oil film alternately from one side of the pin to the other under semi-boundary-lubrication conditions. This is in contrast to the rotating crankshaft journals, which operate under full fluid lubrication (see section 19.2.7).

Gudgeon-pins are usually made from low-carbon case-hardened steel of composition 0.15% carbon, 0.3% silicon, 0.55% manganese, and the remainder iron. This steel is carburised at a temperature between 880 and 930°C, refined at 870 to 900°C and oil quenched, then hardened by quenching from 760 to 780°C. Finally it is tempered at a temperature which must not exceed 160°C.

To obtain the longest service life from gudgeon-pins, they are generally lapped to a surface finish of 0.08 to 0.16 μm. A coarser finish will produce stress-raisers which will lead to fatigue failure and will also tend to pick up the softer bearing metal which the pin is rubbing against, but a smoother finish will prevent the oil clinging and wetting the pin's cylindrical working face.

Gudgeon-pin operating clearances are usually about 0.0075 mm, this being a critical factor for quiet running and long life with sparsely lubricated highly loaded rubbing pairs.

15.9.1 Gudgeon-pin fixtures
The methods of locating and securing the gudgeon-pin in position can be divided broadly into two types of fixture:

i) semi-floating,
ii) fully-floating.

Semi-floating pinch-bolt small-end-clamped gudgeon-pin (fig. 15.18(a)) With this fixture, the central portion of the pin has a full or partially formed circumferential groove. When the connecting-rod small-end is centrally aligned to this groove, the pinch-bolt draws the ends of the split small-end walls together until they clamp the pin tight. Thus relative movement takes place only between the gudgeon-pin and the piston bosses. This method of fastening the rod to the pin enables a narrow small-end to be used so that the width of the rubbing surface between the piston and the gudgeon-pin boss can be large.

Semi-floating force-fit small-end-clamped gudgeon-pin (fig. 15.18(b)) With this arrangement, the connecting-rod small-end faces are polished with emery cloth and heated evenly with an oxy-acetylene torch (230 to 320°C) until a pale-straw to dark-blue oxide colour appears on the bright surface around the eye. The gudgeon-pin is then forced through both the piston and the small-end eye until it is centrally positioned. The small-end then cools and shrinks tight over the pin. Again the relative rubbing movement will be between the pin and the piston bosses.

Semi-floating piston-boss-clamped gudgeon-pin (fig. 15.18(c)) This approach clamps the gudgeon-pin to one of the piston bosses and is used when the bearing properties of the piston material are not suitable for heavy-duty continuous oscillatory rubbing. The connecting-rod small-end is lined with an interference-fit phosphor-bronze plain bush bearing. This bush locates the gudgeon-pin and provides it with a low-friction surface. When tightening the tapered locking bolt, care must be taken not to strip the thread in the relatively soft alloy.

Fully floating gudgeon-pin with end-pads (fig. 15.18(d)) Gudgeon-pins which are allowed to float both in their piston bosses and in the small-

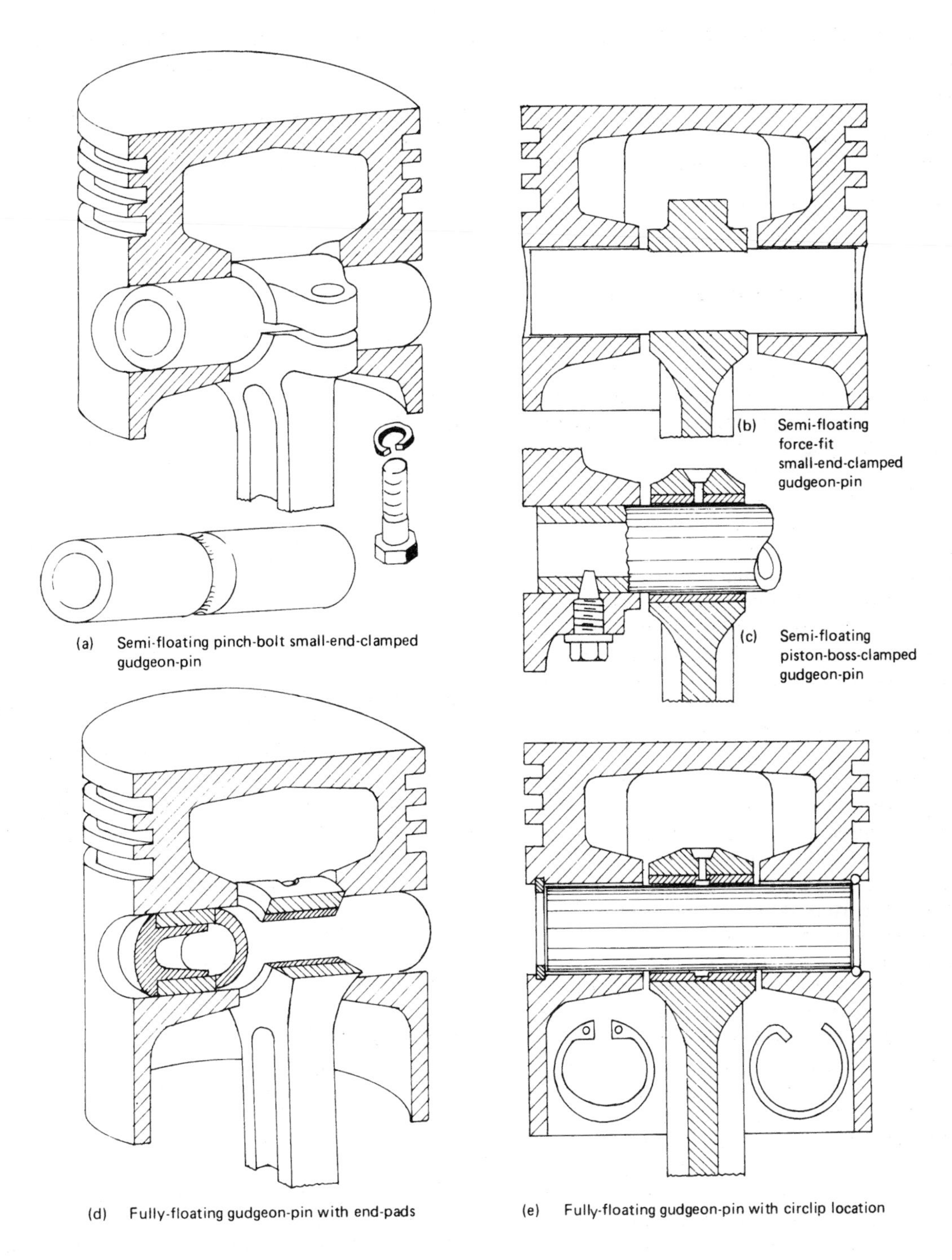

Fig. 15.18 Piston and connecting-rod joints

end eye must not be allowed to touch the cylinder directly, otherwise their very hard outer edges might score the walls. One method of preventing scuffing is to use spherical end-pads made from aluminium, brass, or bronze to act as buffers between the walls and the pin. During operation, the gudgeon-pin is free to revolve both in the small-end and in the piston boss, which tends to improve lubrication.

Fully floating gudgeon-pin with circlip location (fig. 15.18(e)) Fully floating gudgeon-pins extend the bearing-surface area to both of the piston-boss bores and the small-end bronze bush bearing. Engines with small connecting-rod-to-crankthrow ratios and large bore-to-stroke ratios have both large pivoting angular movement and heavy thrust loads on the piston skirt. Thus, under heavy-duty conditions, semi-floating pins may freeze in their bearings. The double swivel action of the fully floating pin reduces this tendency.

The gudgeon-pin is restrained from sliding from side to side by circlips which are positively located in internal circumferential grooves formed near the outer end of each gudgeon-pin-boss bore. Two types of circlip may be used: the heavy rectangular-section Seeger circlip or the lighter and cheaper but not so secure circular-section wire circlip.

15.10 Connecting-rod design, construction and materials (fig. 15.19(a) and (b))

Theoretically, the connecting-rod should be made as long as possible, to reduce the oblique angle made by the rod relative to the cylinder axis. Reducing this oscillatory angular movement of the rod about the small-end decreases the piston side-thrust and improves the reciprocating balance of the engine. Present-day practice is to reduce the overall height of the engine, and this has been achieved by reducing the length of the rod from 4.0 to 4.5 times the crankthrow down to 3.2 to 3.8 times, without the running quality deteriorating or shortening the engine's life.

The piston reciprocating-inertia loads produce both tensile and compressive stresses in the connecting-rod, but the combustion gas load is purely a compressive one of considerable magnitude at all speeds when the engine is operating under load or is accelerating.

To prevent buckling, the rod shank is made in an 'H' section – that is, with a central web and two end flanges (fig. 15.19(a)). This offers the greatest opposition to bending for a given weight of connecting-rod and at the same time provides adequate resistance to twist between the gudgeon-pin and the big-end crankpin axes.

The shank section is made to blend smoothly into both the big- and small-end-bearing hole bosses. The rigidity of the cap is improved by a reinforcing rib formed between the bolt or stud hole bosses – this helps to counteract the very large stresses and strains imposed on the cap half of the big-end by the inertia forces at high engine speed.

The big-end housing has to be adequately rigid and the bolt or stud hole bosses sufficiently hard on the surface to prevent the bolt (or stud) and nut bearing surfaces being forced out of parallelism. If the bolt head or nut wears a groove or deforms its housing seat when tightened or when subjected to dynamic loads, the uneven contact will introduce reversing bending stresses into the bolt or stud shank which will eventually cause fatigue failure.

Some connecting-rods have balancing lugs incorporated at the two extremes of the rod, so that metal may be removed from either end to obtain the desired small-end to big-end weight ratio and also to match the weights of individual connecting-rods.

Lubrication of the small-end by a hole drilled along the shank should be so arranged that the hole intersects the big-end bore circumferentially, to one side of the mid-position – see the insert in fig. 15.19(a). If this hole is too near to the central axis of the rod, it reduces the effective bearing area when at TDC; also, the inertia of the column of oil in the shank drilling will tend to pull oil upwards when the rod can no longer follow at the end of its stroke, therefore the highly loaded region between the journal and the big-end bearing will be starved of oil when it is most needed.

Most connecting-rods are made from steel forgings, but malleable or spheroidal-graphite iron castings or sinter forgings are beginning to be used for small to medium-sized petrol engines. A popular material used for both rod forgings and their clamping bolts or studs is manganese–molybdenum steel, its composition being 0.35% carbon, 1.5% manganese, 0.3% molybdenum, and the remainder iron.

15.10.1 Connecting-rod big-end-cap location and retention (fig. 15.19)

The bore of the connecting-rod big-end must be truly cylindrical, but this is not as simple to achieve as it might seem at first sight. In most cases, caps are fastened to the shank part of the rod by normally two or, for some heavy-duty applications, four studs and nuts or bolts and nuts. The problem comes about due to the inability to accurately position the stud or bolt holes on each side of the rod shank, and because there is always some clearance between the sides of the studs or bolts and the cap hole walls. Also, there is usually a small amount of eccentricity of the stud

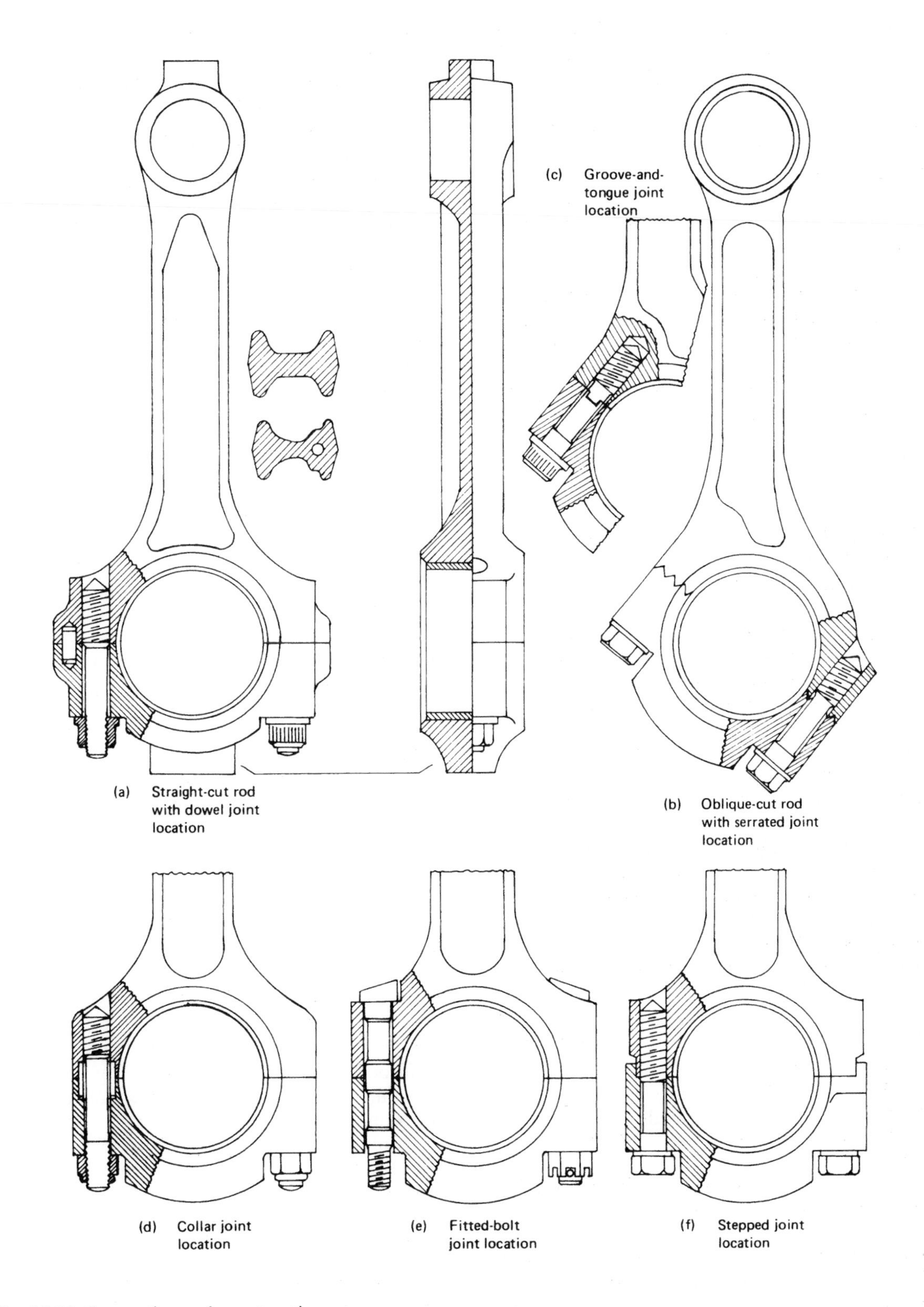

Fig. 15.19 Connecting-rod construction

and bolt created by the threads. The slightest misalignment between the shank and cap housings for any of these reasons will 'step' the two half bores and thus impose uneven loading on the shell bearing liners.

For the bolts or studs to provide the maximum support to the bearing liner, they should be located as near to the bore wall edge as possible without actually interfering with the fitted shell. Thus, under dynamic load conditions, the joint face near the bore will not be able to periodically pull apart and distort the cylindrical bore profile, as can happen with poorly designed big-end joint fixtures.

Big-end housing bore alignment may be reliably achieved by a number of methods, as follows.

Dowel-located joint (fig. 15.19(a)) This approach uses pieces of rod or tube forced into small blind holes drilled in the joint faces of the shank and cap. These locators are generally positioned furthest from the housing bore, so that the bolts can be as close as possible to the split housing bore.

Collar-located joint (fig. 15.19(d)) Location of the two half-housing bores is obtained by tubes or collars which slip over the studs or bolts with a clearance gap and are countersunk into each half of the bore housing at the split joint face.

Fitted-bolt joint (fig. 15.19(e)) Fitted bolts have transition fits at their extreme ends and in the centre region where they intersect the split joint. It is this close fit at the joint faces which centralises the two half-housing bores. The bolts have reduced diameters along their length except at the ends and in the middle joint zone – this form of bolt profile tends to reduce the stress concentration at the bolt head and in the threaded nut region.

Stepped joint (fig. 15.19(f)) Connecting-rod caps may be machined with lips at the end of the split joint face. These projections slot tightly over adjacent flats machined on the shank perpendicular to the joint face; thus they provide positive alignment of the two bore half-housings.

Oblique-cut rod with either serrated or groove-and-tongue joints For the least big-end bore distortion, the bore housing should generally be split perpendicular to the connecting-rod centre line, forming what are referred to as straight-cut caps. For some applications, connecting-rods have the split joint cut at an oblique angle to the cylinder axis, so that the big-end part of the rod will pass through the cylinder bore, thus allowing the piston-and-rod assembly to be withdrawn through the top of the bore.

The oblique-cut cap imposes additional loading on the clamping bolts and joint, so it is desirable to provide some form of extra support at the joint face. The two methods most commonly used are the serrated joint and the groove-and-tongue joint.

a) *Serrated joint* (fig. 15.19(b)) This joint has V-shaped grooves cut across both joint faces so that, when the two halves are clamped together, they tend to form a firm multi-wedge type of joint which does not have relative movement when tightened. Unfortunately, the drawing together of these V-shaped serrations is uneven and very elastic until fully clamped – this can make it difficult to introduce the designed joint compressive load when applying torque to the bolts or set-screws.
b) *Groove-and-tongue joint* (fig. 15.19(c)) Here the joint faces are machined so that a rectangular projection is formed on the shank joint faces and corresponding slots are made on the cap joint faces so that when the two halves are clamped together they form what is known as a groove-and-tongue lock-joint.

15.11 Connecting-rod shell liner bearings (fig. 15.20) Split half-shell thin-wall bearings are incorporated in the big-end-housing bore of the connecting-rod. The two halves operate under different conditions. The rod-shank half-shell is mainly loaded by the high gas pressure created by combustion over a very short period of time; whereas the rod-cap half-shell experiences the inertia loads from both rotating and reciprocating portions of the connecting-rod, these forces tending to remain for a much longer period of time.

To obtain the maximum service life from these bearings, it is very important to provide positive means of locating, aligning, and securing the shells in their housing bore; therefore the various fitting provisions will now be described and discussed.

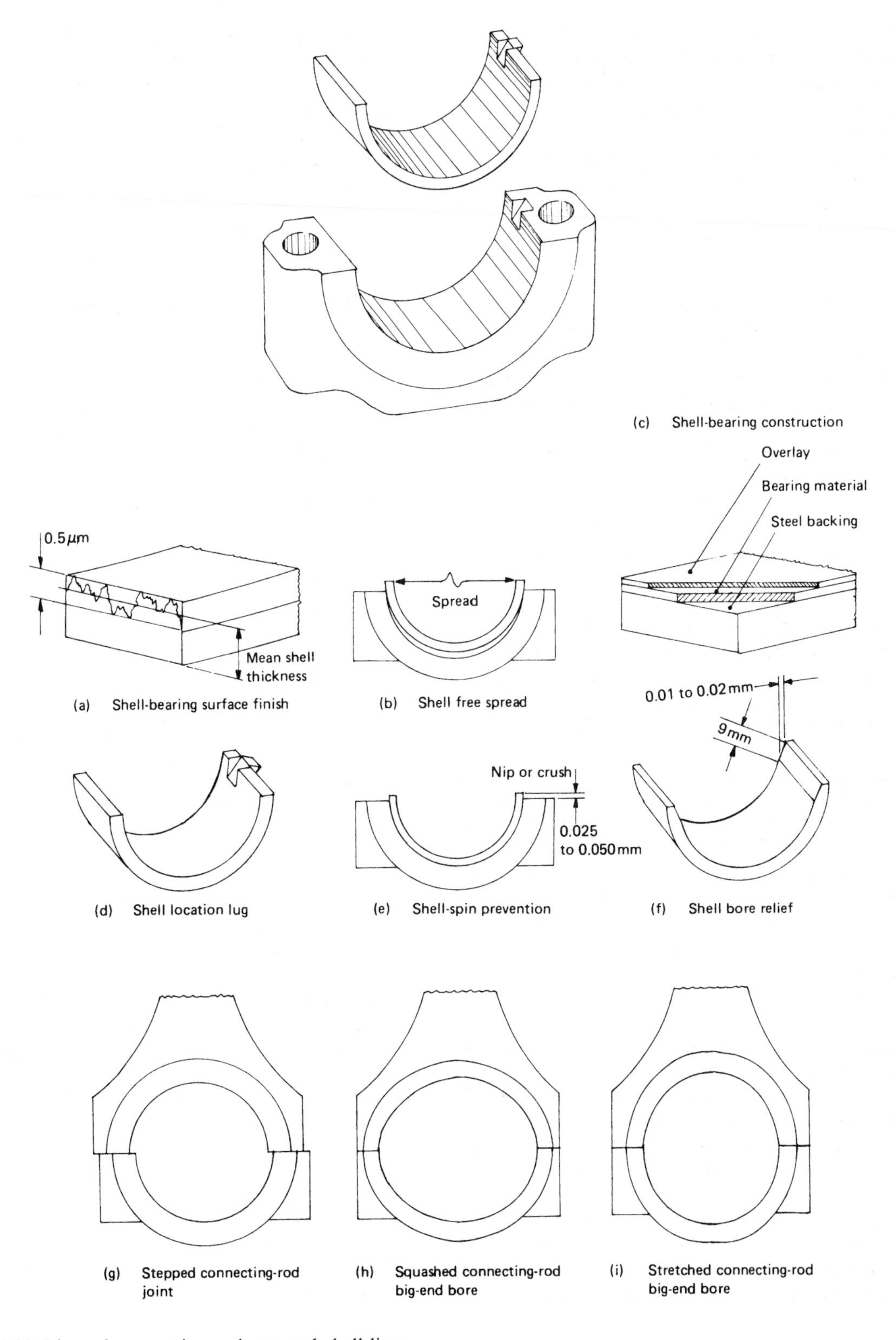

Fig. 15.20 Big-end connecting-rod cap and shell liner

15.11.1 Surface finish (fig. 15.20(a))

When magnified, all smooth surfaces resemble a series of peak-and-valley irregularities, the quality of surface finish being measured by their average height which is known as the surface roughness.

The quality of surface finish will influence the performance of the bearing:

a) smooth contact surfaces will initially reduce both the rate of wear and the 'running-in' time;
b) smooth contact surfaces will wear less rapidly than rough surfaces over their operational life;
c) smooth surfaces improve fatigue resistance, since it is the irregularities which tend to initiate fatigue cracks;
d) close working tolerances between journal and bearing can be achieved only if the surface roughness does not exceed the specified tolerance limits;
e) a smooth surface finish between the back of the bearing lining (shell) and the housing bore improves heat transfer – a rough finish will create air pockets which prevent the heat being dissipated, causing the bearing to run hot.

15.11.2 Shell free spread (fig 15.20(b))

Liner half-shells are slightly spread open in their free state – more for thin liners than for thick ones – so that they can be sprung or snapped into position against their respective bore half-housing walls.

This initial spread of the half-shells ensures that the steel backing of the shells is tensioned radially outwards, thus complete circumferential contact with the bore walls will be established and consequently the working-surface profile of the bearing will be truly cylindrical.

15.11.3 Shell location lug (fig. 15.20(d))

The liner half-shells have a pressed-out lug at one joint face which engages with a notch cut in the housing bore for locating the shell only. The lug and notch provide a rapid and effective means of aligning the shell centrally in its housing bore. The lug is not intended to act as a stop to prevent the shells spinning in the bore.

Initially, before fitting the shells into the big-end, the bore of the housing and the steel backing of the bearings should be wiped dry. This reduces the possibility of shell spin.

15.11.4 Shell-spin prevention (fig. 15.20(e))

Liner half-shells are prevented from rotating by the frictional force created between the shell backing and the bore-housing walls. To achieve this frictional grip, the half-shells protrude above the joint face by something like 0.025 to 0.05 mm when pushed into their bores. When the two half-joint faces are pulled together the excess overlap of the shell is crushed or nipped, so that the complete liner becomes an interference fit in the big-end housing. It is this which produces the radial outward force and the frictional resistance which prevents shell spin.

15.11.5 Liner-shell bore relief (fig 15.20(f))

Bore relief is obtained by having a slight taper at each end of the two liner half-shells on the inner circumferential bearing surface. This back-off relief prevents the internal faces of the bearing bulging inwards or becoming burred at the joint when the shank and cap are compressed tightly together. If the bearing loading is unavoidably high at the joint face – as, for example, when a connecting-rod is split at an oblique angle to its shank – the relief should be reduced or in certain cases omitted.

15.11.6 Roundness of the housing bore (fig. 15.20(g) to (i))

The liner must be cylindrical in shape after assembly in the big-end housing, but out-of-roundness distortion of the housing-bore profile can be due to a number of reasons, as follows:

Stepped cap (fig. 15.20(g)) This condition may be caused by

a) a shifted cap-and-shank split joint, due to a damaged joint-location device;
b) reversing the cap relative to the shank of a matched split joint;
c) stretched cap bolts or enlarged cap-bolt holes, which tend to create a loose fit between the bolts and their cap hole walls;
d) insufficient or uneven torquing of the cap bolts or studs – this can cause a shift of the joint faces under operating conditions;
e) interference between the recessed shoulder of the cap and the side of the torquing socket when fitted over the nut or bolt head – this can alter the relative alignment of the two bore half-housings.

Squashed-out bore (fig. 15.20(h)) This may be caused by

a) overtorquing the big-ends, which crushes and spreads the split-joint faces;
b) the cap having been filed down in an attempt to reduce journal-to-big-end bearing clearance;
c) insufficient shims having been sandwiched between the cap and shank joint-face housings if specified for certain applications.

Stretched-out bore (fig. 15.20(i)) Alternating reciprocating-inertia loading and flexing at very high speeds, if excessive, may cause the big-end bore to become permanently elongated in the direction of the rod length. A well proportioned big-end housing, good design, and the selection of appropriate materials with correct heat treatment can avoid this fault.

15.11.7 Big-end cap tightness

The bolts or studs should be torqued down so that the strain in their axial length imposes compressive stresses in the split-joint region of the two housings. The magnitude of tightening down – that is, tensioning the studs or bolts – should be increased until the bearing housing becomes truly cylindrical and the frictional force created by the compressive stresses is sufficient to prevent fretting between the joint faces when the big-end is subjected to both combustion gas loads and reciprocating-inertia loads. The roundness of the bore cannot be easily measured, nor can the frictional grip at the joint face, but the torque required to achieve these conditions can be predetermined, and it is this value which is quoted by the manufacturer for reassembling the big-end.

Figure 15.21(a) shows the recommended applied torque for bolts or studs of different nominal diameter. This clearly shows that the required holding torque increases with the square of the diameter; therefore a small change in shank diameter necessitates a considerable change in torque for large bolts or studs, but small-diameter bolts or studs are very sensitive to the magnitude of the applied torque.

Figure 15.21(b) shows the relationship between the tightening torque and the clamping tension produced in the bolts or studs for three different nominal diameters. The recommended torque for a safe working strength (taken as 80% of the yield strength of the steel) is shown by the circle at the end of each curve. Thus small bolts or studs must be very carefully tightened to avoid overstraining,

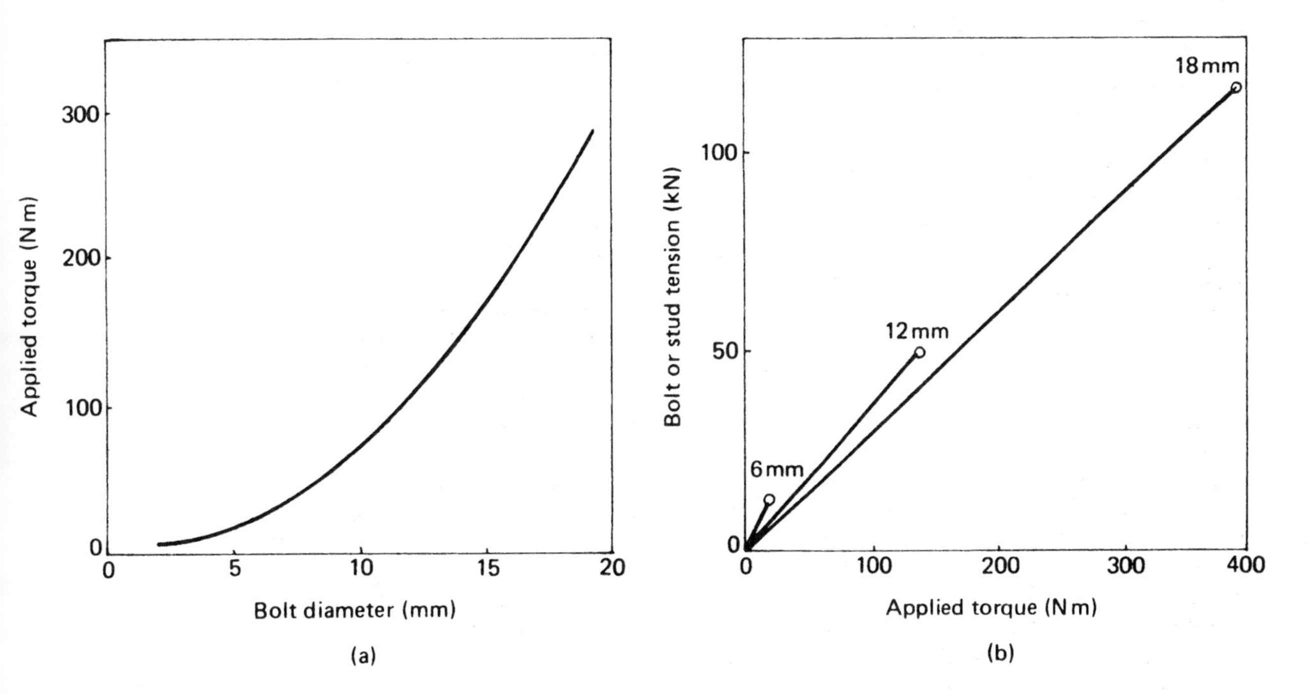

Fig. 15.21 Relationship between diameter, applied torque, and axial tension of bolts and studs

but large bolts and studs are not so critical. It should be observed that the increase or decrease in torque is directly proportional to the tension caused along the shank and between the screw threads. This means that the twisting leverage is also a direct measure of bolt and big-end-boss tensile and compressive loading.

If the nuts or bolts are overtightened, the big-end bore at the joint can become overcompressed and may distort the cylindrical profile of the bore and, in particular, of the shell-bearing inner face which is designed to rub against the crankpin journal with extremely small clearances. In addition, overloading the bolts and nuts or the studs and shank may stress them beyond their yield strength so that they become overstrained and may even strip their threads.

Conversely, if the joint is insufficiently tightened, the joint faces near the bore tend to elastically distort open under dynamic inertia loading, so that the bore becomes out of round and also relative movement between the two half-housings may be initiated. Also, the frictional force between the male and female threads of the studs and shank or the bolts and nuts provided by the torque tensioning may be inadequate to prevent the studs or nuts from slackening and becoming unscrewed if no locking device is provided. The practice of relying entirely on the generated friction between the meshing threads to lock them together – without a locking device – has become widespread for small and medium-sized engines.

15.12 Plain journal bearings (fig. 15.20(c))

Shell bearing liners consist of a low-carbon-steel cylindrical backing strip on to which the principal bearing material is cast or roll-bonded. Such a combination is known as a bi-metal bearing. The intermetallic bonding between the bearing alloy and the steel backing supports the bearing material and prevents it from progressively spreading outwards whenever it is subjected to the oil-film pressure generated by both the gas and the inertia loads.

It is essential for some bearing alloys to have a very thin overlay made from a soft anti-friction alloy. This may be electroplated on to the surface of the bearing matrix, and the composite bearing is known as a tri-metal bearing.

15.12.1 The selection of plain-journal-bearing materials

The properties required in an ideal bearing material may be summarised as follows.

High fatigue strength To resist the high fluctuating pressure in the lubricant film due to the periodic reciprocating-inertia and gas loads, a high fatigue strength is necessary.

High melting point To resist damage by high-temperature lubricant films and the reduction of yield strength of bearing alloys at elevated temperatures, a high melting point and hot strength are desirable. Note that oil temperatures in big-end bearings can reach values of 150°C.

High resistance to corrosion Corrosion is the destructive attack of a metal surface by a direct chemical or electrochemical reaction with its environment which tends to become more active as the temperature increases. At elevated temperatures, therefore, the bearing surface must be able to resist attack from worked-out or 'degraded' oil which has altered its composition and structure to form acidic lubricants.

Adequate hardness The bearing material must be sufficiently hard to resist abrasive wear and cavitation erosion (caused by the relatively soft bearing material being bombarded by high-velocity oil) and to sustain static and dynamic loads, but without sacrificing conformability and embeddability.

Good conformability This is the ability to tolerate misalignment between the bearing and the crankshaft. In general, conformability is inversely related to bearing hardness, so it may conflict with the desired strength and hardness.

Good embeddability This is the ability of the bearing surface to absorb dirt particles being carried round by the lubricant and so prevent scoring of the journal under high loads.

Good compatibility This is the resistance to local welding or pick-up on to the steel journal from the bearing when loaded under boundary-lubrication conditions but with a rotational speed insufficient to provide a thick hydrodynamic oil film (see section 19.2.7).

15.12.2 Classification of plain-journal-bearing materials

Journal bearing materials fall into three broad groups:

i) lead- or tin-based white metal (Babbitt metal),
ii) copper-based alloys,
iii) aluminium-based alloys.

White (Babbitt)-metal bearings White metals – also known as Babbit metals – are basically either antimony–tin or antimony–lead alloys. They have excellent conformability, embeddability, compatibility, and corrosion resistance; but their fatigue strength is not high enough for the loads experienced in main and big-end bearings of present-day engines. A slight improvement in load-carrying capacity is achieved by reducing the thickness of these bearing materials, but, even so, they still do not meet the more demanding requirements of today's high-compression-ratio engines.

White metal is largely used for camshaft and thrust bearings. The lead-based alloy has slightly superior hot strength to the tin-based alloy. Typical compositions are as follows:

Lead-based white metal		*Tin-based white metal*	
Antimony	15%	Antimony	7.5%
Tin	1%	Lead	0.2%
Arsenic	1%	Arsenic	0.1%
Copper	0.5%	Copper	3.0%
Lead	remainder	Tin	remainder

Copper-based bearing alloys Copper-based alloys are harder and have a higher fatigue strength than the white-metal alloys, but this is usually at the expense of conformability, embeddability, compatibility, and corrosion resistance.

The copper-based bearing alloys fall into three main classes:

i) copper–lead 70%/30% alloys – used for low- to moderate-duty petrol engines;
ii) lead bronze with 24% lead, 1.5% tin, and the remainder copper – used for moderate- to high-duty petrol engines;
iii) lead bronze with 8% lead, 5% tin, and the remainder copper – used for heavy-duty naturally aspirated and turbo-charged diesel engines.

Increased loads can be accommodated by using these high-strength alloys, but this is obtained only with a corresponding increase in material hardness and corrosion. Therefore, to improve the poor conformability, embeddability, compatibility, and corrosion resistance, the use of a soft overlay is essential. These overlays are either lead–tin–copper or lead–indium alloys. The latter have better fatigue strength than the former; therefore slightly thicker lead–indium overlays can be used, which improves the conformability and embeddability of the complete bearing.

Typical overlay compositions are as follows: 10% tin, 1% copper, and the remainder lead; or 5 to 10% indium and the remainder lead. An average overlay thickness is 0.02 mm, but different thicknesses are available for various applications.

Aluminium-based bearing alloys Aluminium-based bearing alloys are suitable for medium-to-heavy-duty operating conditions with both petrol and diesel engines, so they cover a similar application range to the copper-based alloys.

These bearing alloys fall into three main classes:

i) 20% tin, 1% copper, and the remainder aluminium – these alloys are unplated and are recommended for moderate-duty petrol engines;
ii) 6% tin, 1% copper, 1% nickel, and the remainder aluminium – with a lead–tin overlay, these alloys, are used for moderate- to high-duty petrol and diesel engines;
iii) 11% silicon, 1% copper, and the remainder aluminium – with a lead–tin overlay of 0.025 mm standard thickness, these alloys are designed to be used in turbo-charged heavy-duty diesel engines.

The tin–aluminium alloys are roll-bonded to a steel backing using a pure-aluminium-foil bonding layer, but the silicon–aluminium alloys are directly cold-roll-bonded on to their steel liners.

The high-tin-content aluminium alloy does not require an overlay since it has all the necessary desirable bearing properties, but the low-tin and silicon alloys are harder and therefore are overlayed to supplement their bearing properties.

Aluminium-based alloys do not suffer from corrosion attack as do the copper-based bearing alloys.

16 Crankshaft construction

16.1 Crankshaft nomenclature (fig. 16.1)
It could be said that the crankshaft is the central link-up for the power produced by the individual cylinders. The function of the crankshaft is to harness and phase the individual cylinders' power impulses transmitted through the mechanism of the connecting-rod which converts the reciprocating motion of the piston to a rotary motion at the crankshaft.

Crank-throw This is the distance from the main-journal centres to the big-end-journal centres – in other words, it is the amount the cranked arms are offset from the centre of rotation of the crankshaft.

A large crank-throw increases both the leverage applied to the crankshaft and the piston's stroke. Conversely, a small crank-throw reduces both the crankshaft turning-effort and the distance the piston moves between the inner and outer dead centres.

Crank-webs The cranked arms of the shaft which provide the throws of the crankshaft are known as crank-webs, their purpose being to support the big-end crankpin. To withstand both the twisting and the bending effort which will be created within these webs, they must have adequate thickness and width, but their mass must not be excessive, otherwise their inertia will tend to wind and unwind the shaft during operation.

Main journals These are the parallel cylindrical portions of the crankshaft which are supported rigidly by the plain bearings mounted in the crankcase to provide the axis of rotation of the crankshaft. The diameter of the journals must be sufficient to provide torsional strength and, so as not to overload the plain bearings, the diameter

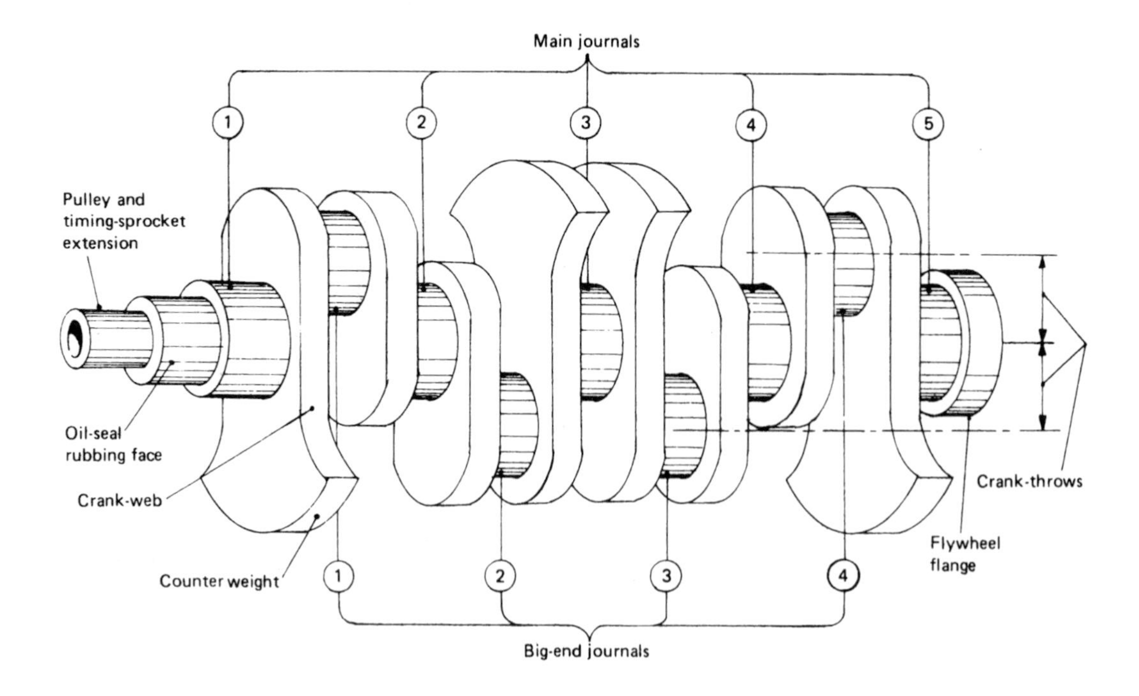

Fig. 16.1 Crankshaft nomenclature

and width of the journals must present an adequate projected bearing area (see fig. 16.4).

Big-end (crankpin) journals These journals provide cylindrical smooth surfaces for the connecting-rod big-end bearings to rub against, enabling the gas-pressure thrust to supply a continuous but varying turning-effort to the crankshaft.

16.2 Crankshaft proportions (figs 16.2 and 16.3)
To withstand the maximum cylinder pressure, the projected areas of both the big-end and the main-end journals – given by the product of the journal diameter and width – must be adequate (see fig. 16.4). To obtain a given projected area, either a small-diameter wide journal or a large-diameter narrow journal may be used.

Early engine design set out to minimise the surface speed of the journals by using wide small-diameter journals, but, with a better understanding of the mechanism of lubricated rubbing pairs and improved surface finish and materials, this is no longer considered necesssary, and the present trend is to increase the stiffness of the crankshaft and reduce its overall length by using narrow journals of large diameter.

For the necessary wall thickness and coolant passages, the closest the cylinder centres can be is something like 1.2 times the cylinder bore diameter for an engine which has a stroke equal to the bore (fig. 16.2). For a cylinder diameter D, the proportions of the crankshaft will then be as follows:

Cylinder bore diameter	D
Cylinder centre distance	$1.2D$
Big-end journal diameter	$0.65D$
Main-end journal diameter	$0.75D$
Big-end journal width	$0.35D$
Main-end journal width	$0.4D$
Web thickness	$0.25D$
Fillet radius of journal and webs	$0.04D$

The maximum diameter of the big-end for the connecting-rod assembly to pass through the cylinder is $0.65D$.

The big-end and main journal widths can also be expressed relative to their diameters – this gives big-end widths and main-end widths of 0.45 and 0.55 times their diameters respectively.

To prolong the fatigue life of the shaft, the fillet radius between journals and webs should be as

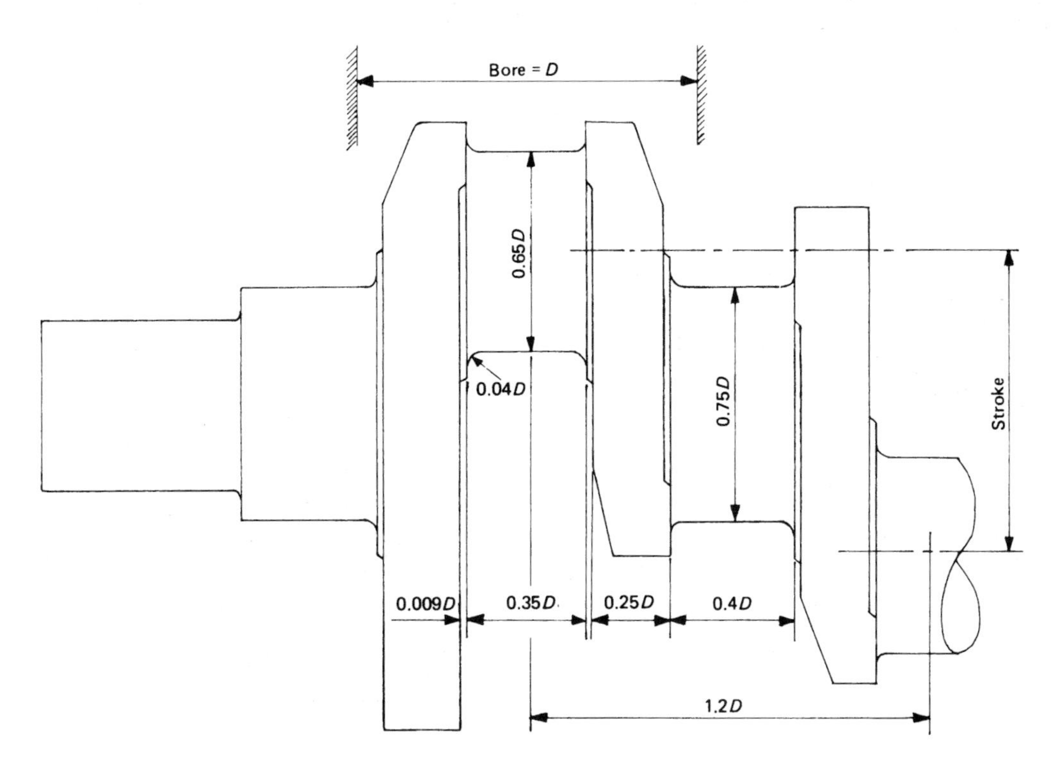

Fig. 16.2 Crankshaft proportions

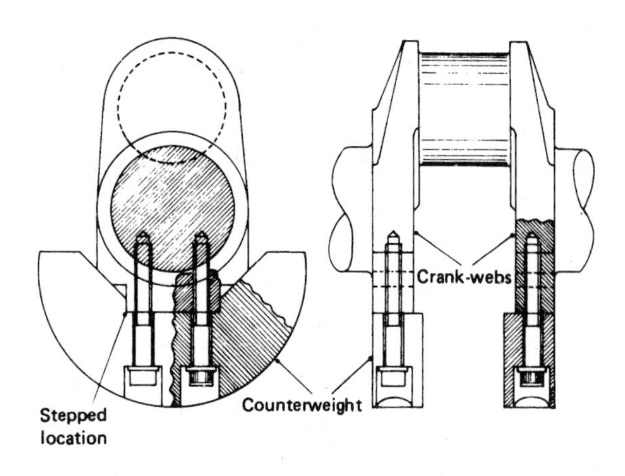

Fig. 16.3 Attached counterweights

large as possible and at least 5% of the journal diameter.

The degree of overlap between the diameters of the big-end crankpin and the main-end journal will depend on the length of stroke – that is, the crank-throw. A long stroke will have very little overlap, therefore thicker web sections are necessary, but a short-stroke engine will have considerable overlap which tends to strengthen the shaft.

Collars should be machined on the webs adjacent to the journals, so that the crankshaft and the bearings can be accurately aligned with the correct amount of side-float and, where necessary, to absorb the crankshaft end-float.

Most crankshaft proportions are designed so that the nominal stresses in the material under operating conditions do not exceed 20% of the tensile strength in bending and 15% in torsion.

Crankshaft journals should have a surface finish or roughness better than 0.5 μm after being ground, to minimise bearing wear.

16.3 Counterbalance weights (figs 16.1 and 16.3)
Crankshafts are usually designed with either integral (fig. 16.1) or attachable (fig. 16.3) counterweights. Their function is to counteract the centrifugal force created by each individual crankpin and its webs as the whole crankshaft is rotated about the main-journal axis. If no provision were made to oppose the outward pulls of the crankpin masses, they would tend to bend and distort the crankshaft so that excessive edge-loading would occur in the main bearings. To prevent the overloading of these bearings, each half crank-web is usually extended in the opposite direction to that of the crankpin, to counterbalance the effects of the crankpin.

Sometimes for large in-line and 'V' engines, bolt-on counterweights are used (fig. 16.3) as this makes it simpler to case or forge the crankshaft. One advantage of using detachable weights is that they can be made to slightly overlap the webs. With this increase in web width, more mass can be concentrated at a smaller radius from the axis of rotation. The great disadvantage of attaching weights to the web is that they have to be located and attached very accurately, as any error in assembly would result in an unbalanced crankshaft which could cause more damage than the weights were intended to prevent.

16.4 Crankshaft oil-hole drillings (fig. 16.4)
Oil is fed from the main oil gallery to each individual main journal and bearing. These bearings normally have a central circumferential groove facing the journal, so that oil will completely surround the central region of the journal surface.

Lubrication of the big-end journal is provided by diagonal drillings in the crankshaft which pass through the webs between the main and big-end journals (fig. 16.4). For effective lubrication of the big-end, these oil holes should emerge from the crankpin at about 30° on the leading side of the crank's TDC position. If the drilling were to intersect the surface in the middle of the projected area when the piston was at TDC, the high gas load would prevent an oil film being established between the rubbing surfaces. The position of the oil-passage exit from the main journal is not critical as it will be on the unloaded side during combustion and the central groove will distribute the oil in any case.

The drilled oil passages must not come close to

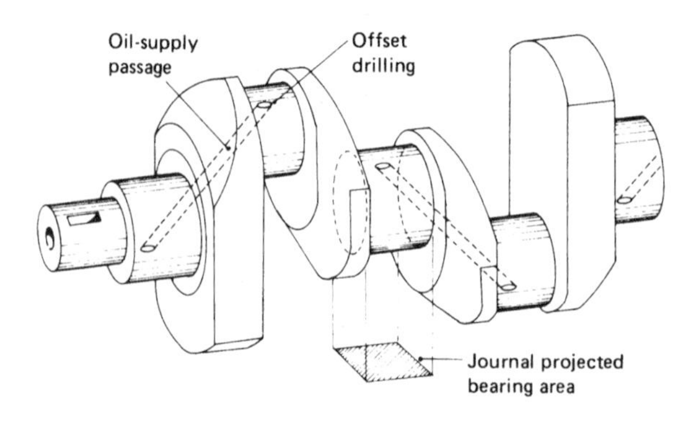

Fig. 16.4 Crankshaft with single diagonal oil drillings

the side walls of the webs or near the fillet junction between the journal and the webs, as this could cause high stress concentrations which could lead to fatigue failure. Furthermore, the emerging holes on the journal surfaces must be chamfered to reduce stress concentration; but the chamfering must not be excessive, otherwise it will destroy the formation of the oil film.

16.5 Fan-belt pulley-to-crankshaft attachment

(figs 16.5 and 16.6)

The pulley consists of the wheel which guides and suspends the fan belt and a central support known as the hub. The first main journal on the crankshaft has an extended parallel portion of slightly smaller diameter, and machined axially on the cylindrical surface of this are either one long or two short rectangular keyway slots. The Woodruff key which fits into the single long slot may have a flat floor, but the double-key flooring frequently is semicircular. The bores of the timing gear (which drives the camshaft) and the pulley hub both have internally machined slots or grooves spanning their full width and form a transition push fit with the shaft when assembled, but the driving torque is transferred from the shaft to the gear and pulley through the Woodruff key or keys (figs 16.5 and 16.6).

Finally, to prevent the components working loose due to vibration, a large set-bolt is screwed into the end of the crankshaft, either with a large washer or plate or direct, to clamp the assembly together.

The fan-belt pulley is designed to transmit power from the crankshaft to the various auxiliary components such as the water-pump, generator, power-steering and brake-fluid pump, and the cooling fan itself is fitted.

The cross-section of the belt is generally of trapezium shape with inclined side walls which correspond to the tapered sides of the pulley-wheel circumferential groove. In operation, the groove and belt walls are in parallel contact with each other, the drive being transmitted through frictional grip obtained by adjusting the tension in the belt and being reinforced by the wedge action of the tapered contact surfaces.

As the side walls of the belt wear, the belt section will move further into the groove and, provided the belt is periodically adjusted for tension, will continue to perform effectively. However, once the inside face of the belt touches the bottom of the pulley groove, the frictional force between

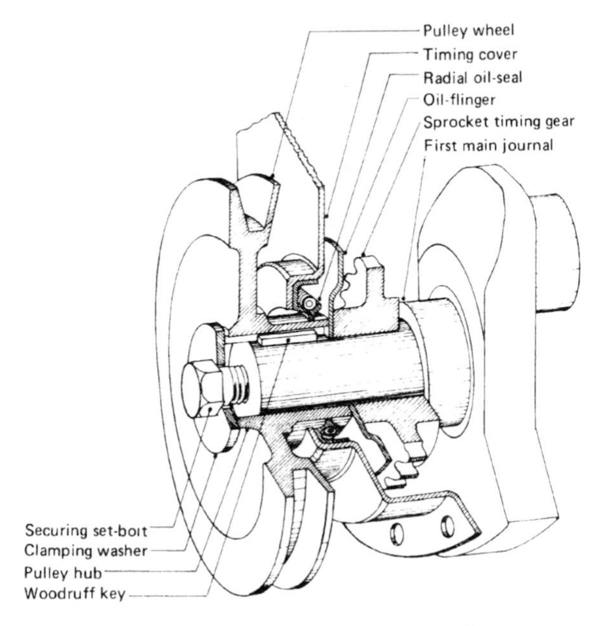

Fig. 16.5 Fan-belt pulley and oil-seal assembly

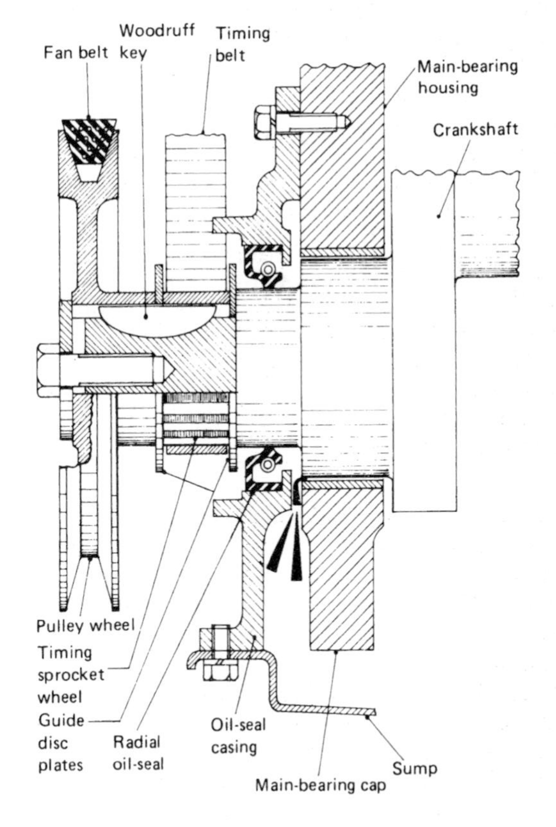

Fig. 16.6 Crankshaft pulley, timing sprocket, and oil-seal assembly

the tapered sides will be reduced and the belt grip will rapidly deteriorate.

Belts are mostly made from composite fabrics of rubber and nylon or terylene. These are reliable, but their life-span will depend upon the accurate alignment of the pulley-wheel system and the correct tension in the belt – a slack belt will cause slippage and wear, but an over-tightened belt will strain the fabric so that it disintegrates. Care must be taken when fitting new belts, as belt life will be considerably shortened by damaged side walls due to the screw-driver tearing the walls, overstraining the belt, or scratching or leaving burrs on the contact walls of the pulley.

16.6 Front and rear crankshaft oil-seals

Oil-seals must be provided to prevent oil escaping between the journals and bearings at both the front and the rear of the crankshaft. The three most common types of seals are as follows:

i) the radial-packing seal,
ii) the radial-lip seal,
iii) the spiral-thread clearance seal.

16.6.1 Crankshaft pulley-end oil-seals (figs 16.5 and 16.6, and 19.34)

The radial-packing and radial-lip seals are usually incorporated in the timing-chain or oil-pump housing surrounding the crankshaft's extended nose. The graphitised-asbestos packing or synthetic-rubber lip is made to rub against a sleeve which forms part of the pulley hub or the inner oil-pump member – see figs 16.5 and 16.6. In a very few arrangements, spiral-thread clearance seals are used, and in this case the helical 'spiral' is formed in the timing-case member and the plain cylindrical sleeve is generally an integral part of the pulley hub (see fig. 19.34(d)).

During operating conditions, most of the oil being squeezed out of the front main-end journal will be flung radially outwards as the oil spreads on to the rotating oil-flinger washer. Nevertheless, some oil will creep along the shaft and will meet the oil-seal packing or lip end on. This seal will prevent oil going beyond the casing, so it will then drain back to the sump. Likewise with the spiral-thread seal, the rotating movement of the shaft will tend to continuously screw back any excess oil, which is then returned to the sump by gravity.

16.6.2 Crankshaft flywheel-end oil-seals (figs 16.7 and 19.34a)

Oil leakage at the flywheel end of the crankshaft can be prevented by the radial-packing, radial-lip, or spiral-thread clearance seals (figs 16.7 and 19.34). The radial-packing or radial-lip seals are located by a press fit in either the rear bulkhead or the coverplate. If the spiral-thread clearance seal is adopted, it may be of either the external- or internal-thread type – the spiral thread is machined on to the crankshaft with the former and on to the inside surface of the cover-plate bore with the latter.

The axial movement of the escaping oil between the journal and the bearing is initially diverted by an oil-flinger or the side of the flywheel flange, which deflects it radially instead of allowing it to continue in an axial outward direction. The small quantity of oil which does still find its way outwards is then blocked off by the sealing action,

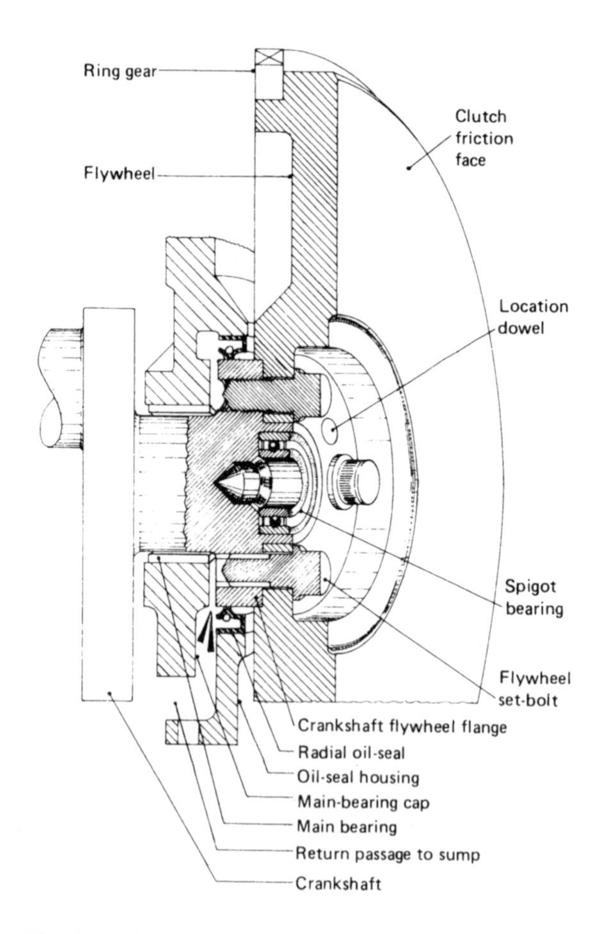

Fig. 16.7 Crankshaft-to-flywheel flanged joint and oil-seal

and the excess oil is then allowed to return to the sump.

16.7 Flywheel-to-crankshaft attachment (fig. 16.7)

Flywheel function Before the flywheel-to-crankshaft attachment is examined, an understanding of the various functions of the flywheel will emphasis the importance of having provision for a substantial joint.

The flywheel serves three main purposes:

i) to provide a solid disc plate which is rigid enough both to support the clutch assembly and to transmit the drive between the crankshaft and the first-motion shaft to the gearbox by means of the smooth annular friction face of the flywheel and the clutch driven-plate.
ii) to provide a carrier wheel for the large step-down ring gear when the engine is to be started. The small starter pinion gear is made to engage the flywheel ring gear – this gives something like a 10:1 speed reduction which will crank the engine at a speed of about 150 rev/min; conversely the torque imposed on the ring gear by the starter pinion gear will be the product of the pinion input torque and the gear ratio, thus providing the extra turning-effort necessary to initiate cranking.
iii) to provide angular momentum to the crankshaft so that it continues to rotate when the piston approaches inner or outer dead centre, and to store energy on the power stroke so that it will be given out on the three idle strokes, this being necessary to reduce crankshaft speed fluctuation throughout each cycle of operation.

Flywheel attachment (figs 16.7 and 16.8) The flywheel end of the crankshaft has a larger diameter than the main journals, to provide both a flanged face to support the flywheel and a circumferential rubbing surface for the rear crankshaft oil-seal. A shallow recess in the flywheel allows it to be aligned over the flanged end of the crankshaft, which provides concentric, i.e. radial, support for the joint (figs 16.7 and 16.8). To prevent relative angular movement between the crankshaft and the flywheel, due to the clearance between the bolts and their holes, one or two location dowels are countersunk in both mating faces. An alternative arrangement provides one of

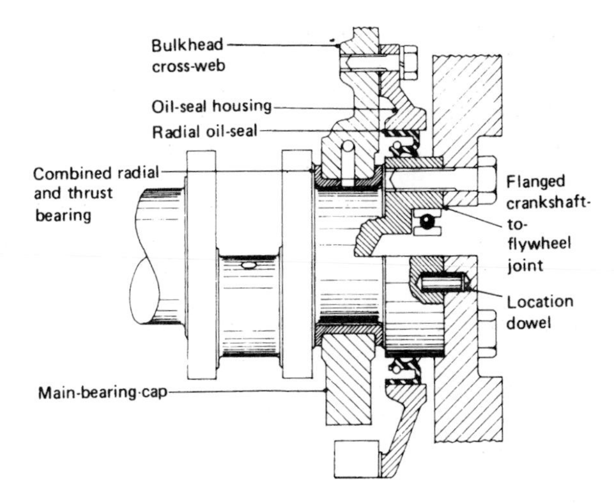

Fig. 16.8 Sectional view of main bearing with integral thrust flange, oil-seal, and flywheel joint

the clamping bolts with a shoulder which fits tight in its selected flywheel hole.

The drive torque is transmitted from the crankshaft to the flywheel by four or six bolts which clamp the flanged and recessed faces of the two components together. After the crankshaft has been either forged or cast to shape, each main journal has to be ground to size along a central axis which is established by two taper-ended countersunk holes drilled at each end. The shaft can then be supported between tapered centres while the necessary grinding is in progress.

A second larger hole using the same countersunk hole centre is bored part way in from the flywheel end of the crankshaft to house the outboard or straddle bearing provided to support the first-motion (spigot) shaft which carries the clutch driven-plate. The bearing may be a plain bush, needle rollers, or a ball-roller type.

When installing the flywheel, all the set-bolts should be initially lightly and evenly tightened, followed by torquing diametrically opposite set-bolts to the recommended tension. If this procedure is not adhered to, excessive flywheel wobble – known as 'run-out' – may result.

16.8 Crankshaft materials and heat treatments

Crankshafts are made from materials which can be readily shaped, machined, and heat-treated and which have desirable mechanical properties such as adequate strength, toughness, hardness, and high fatigue strength – and, of course, low cost. Traditionally these shafts were forged, which promoted all the necessary properties; but, with the evolution of the nodular cast irons and improvements in foundry techniques, cast crankshafts are now favoured for moderate loads, and only for heavy-duty applications do forged shafts predominate.

The selection of crankshaft materials and heat treatments for various applications is now discussed.

16.8.1 Manganese–molybdenum steel

This is a readily forged and machined relatively cheap popular steel used for moderate-duty petrol-engine crankshafts. It has an alloying composition of 0.38% carbon, 1.5% manganese, and 0.3% molybdenum.

This steel is heat-treated by quenching in oil from a temperature of 850°C, followed by tempering at 600°C, which will produce a surface hardness of about 250 Brinell number. The surface hardness makes it suitable for both tin–aluminium and lead–copper plated bearings.

The addition of manganese and the low carbon content improves the steel's toughness, strengthens the iron, and increases the depth of hardness. Molybdenum is added mainly to reduce temper-brittleness and mass effects which produce different cooling rates throughout the structure during heat treatment, so that the resultant steel retains a high degree of impact resistance.

16.8.2 1%-chromium–molybdenum steel

This forging steel is suitable for medium- to heavy-duty petrol- and diesel-engine crankshafts. It presents no difficulty in hot working, and afterwards can be easily machined to size. The alloying composition is 0.4% carbon, 1.2% chromium, and 0.3% molybdenum.

To produce the best mechanical properties, the steel is heat-treated by quenching in oil from a temperature of 850°C and then tempering at 680°C – this produces a surface hardness of about 280 Brinell number. If harder bearings are to be used, the journals can be flame or induction surface-hardened to 480 Brinell number and, for very heavy duty, a nitriding process can bring the surface to 700 diamond pyramid number (DPN). These journal surfaces are suitable for all tin–aluminium and bronze overlay bearings.

The function of alloying with chromium is to form hard carbides which considerably harden the steel. The chromium also strengthens the iron, but the disadvantage of this element is its tendency to promote grain growth and brittleness. The addition of a small amount of molybdenum stabilises the grain growth and improves the impact strength and the steel's machinability.

16.8.3 2½%-nickel–chromium–molybdenum steel

For heavy-duty diesel-engine applications, this steel is preferred – the alloying composition being 0.31% carbon, 2.5% nickel, 0.65% chromium, and 0.55% molybdenum. The steel is initially heat-treated by quenching in oil from a temperature of 830°C and then tempered at a suitable temperature not exceeding 660°C. A surface hardness in the region of 300 Brinell number will be obtained with no further heat treatment to either the core or the surface.

The effects of the alloying elements are as follows: nickel refines the grain, prevents grain growth during heat treatment, and both toughens and strengthens the core but tends to convert the carbon into its softer graphite form; conversely, chromium forms hard wear-resisting carbides, promotes grain growth, but prevents graphitisation. Thus it can be said that the nickel and chromium complement each other. The molybdenum is included to reduce any mass effects and brittleness and to improve the machinability of the steel.

This steel with its alloying elements is slightly more expensive than manganese–molybdenum and chromium–molybdenum steels, but the improved mechanical properties and less critical control of the heat treatment necessary might outweigh the cost disadvantage.

16.8.4 3%-chromium–molybdenum or 1½%-chromium–aluminium–molybdenum steels

These are forged steels used where hard high-fatigue-strength bearing materials are installed to withstand highly rated diesel-engine crankshaft loads. They are case-hardened by nitriding – absorbing nitrogen into their surface layers. The alloying compositions are as follows: 0.15% carbon, 3% chromium, and 0.5% molybdenum or 0.3% carbon, 1.5% chromium, 1.1% aluminium, and 0.2% molybdenum.

Initial heat treatment for both steels is oil quenching and tempering, at 920°C and 610°C or 890°C and 690°C respectively for the two steels. This is followed by nitriding the journals to produce the desired surface hardnesses.

The choice of these two steels depends on the engine applications. The 3%-chromium steel will have a relatively tough surface and hardness of 800 to 900 DPN; conversely, the 1½%-chromium-steel casing will tend to be slightly more brittle but with increased hardness, of the order of 1050 to 1100 DPN. Provided the nitriding is continued well into the journal fillets, the fatigue strength of these shafts will be increased by at least 30% compared to induction- and flame-surface-hardened shafts (see section 16.8.6).

16.8.5 Nodular cast irons

Due to their appearance and relative ductility, these cast irons are also known as spheroidal-graphite irons or ductile irons. They are grey cast irons having 3 to 4% carbon and 1.8 to 2.8% silicon but, instead of the usual formation of flake graphite, there are graphite nodules dispersed in a pearlite matrix. To achieve this improved structure, a very small quantity of about 0.02% residual cerium or 0.05% residual magnesium or even both is added to the melt, the result being the removal of the sulphur and the formation of many small spheroids in the as-cast material.

Nodular iron combines the advantages of grey cast iron – that is, low melting point, good fluidity and castability, excellent machinability, and wear resistance – with the mechanical properties of steel – relatively high strength, hardness, toughness, workability, and hardenability.

The surface hardness of as-cast nodular iron is greater than for steel of similar strength, their respective hardnesses being 250 to 300 and 200 to 250 Brinell number. Surface treatments such as flame or induction hardening can produce Brinell numbers of 550 to 580, and a form of nitriding can also be applied if necessary.

A large proportion of crankshafts for both petrol and diesel engines are now made from nodular cast iron in preference to the more expensive forged steel. To supplement the slightly inferior toughness and fatigue strength of these cast irons, larger sections and the maximum number of main journals are used.

16.8.6 Flame and induction surface-hardening processes

These methods harden the surface of the steel without the use of special compounds or gases. They can suitably be applied to steels which have sufficient carbon in them to respond directly to heat treatment – from 0.3 to 0.5% carbon. Hardness is achieved by rapidly applying heat to the surface only and then water quenching. As it is not necessary to heat the entire mass, the time required for hardening is greatly reduced and distortion of the journal is avoided.

Flame hardening is carried out by an oxy-acetylene flame which brings the surface layer to between 780 and 900°C, depending on the carbon-content equivalent of the different alloying elements in the steel. This is followed by a water-jet quenching operation. The actual timing for heating and cooling is critical, so it has to be predetermined and is mostly automatically controlled.

Induction hardening uses the same principle of imparting concentrated heat to the metal surface as does the flame-hardening process, but the danger of either overheating or burning the surface of the metal by a flame is avoided by inducing heat electrically into the surface to be hardened. In this process, an induction coil surrounds the journal. This coil carries a high-frequency current which induces circulating eddy currents in the journal surface, with a consequent raising of its temperature. Heat generated in the metal by induction is mostly confined to the outer surface of the journals – the higher the frequency of the current, the closer the heat is to the skin. Once the necessary temperature has been reached, the current is automatically switched off and the surface is simultaneously quenched by water jets which pass through holes in the induction block.

16.8.7 Nitriding surface-hardening process

Nitriding is a case-hardening process where the journals are exposed to an atmosphere of ammonia gas and heated to 500°C for a predetermined time, so that the nitrogen in the gas is absorbed into the surface layer.

Iron nitride does not confer hardness to any extent, and hence steels suitable for nitriding must contain alloying elements such as chromium, aluminium, and molybdenum which form hard nitrides. Aluminium forms stable aluminium nitrides which do not diffuse readily, and a shallow but intensely hard case is formed. Chromium contributes to the hardness, but chromium nitrides diffuse to a greater depth than aluminium nitrides and consequently hardness falls off more gradually from the surface to the core, thus reducing the risk of spalling (the tendency for the surface to flake off). The presence of molybdenum increases hardenability, gives grain refinement, and improves the toughness of the core.

With the nitriding process, the journals can be ground to their final size before treatment, as there is no quenching after nitriding which might cause distortion as is the case for other surface-hardening processes. The slow rate of penetration of the surface makes the cost of the process high –

for example, it takes 20 hours to produce a case depth of about 0.2 mm.

16.8.8 Carbonitriding surface-hardening process

'Tufftride' is the best-known salt-bath carbonitriding process. It involves immersion of the crankshaft in a bath of molten salts at a temperature of about 580°C for a relatively short cycle time of not more than two to three hours while both carbon and nitrogen dissociate from the salts and diffuse into the surface. Nitrogen is more soluble than carbon in iron, so it diffuses further into the material. Hard iron carbides and tough iron nitrides will be formed on the surface and greatly increase resistance to wear, galling (surface peeling), seizure, and corrosion. This outer skin is 6 to 16 μm deep and has a hardness varying from 400 to 1200 VPN, depending on the steel used. Underneath this outer layer, the remainder of the nitrogen goes into solid solution with the iron to strengthen it. This inner diffusion zone forms a barrier which prevents cracks from spreading and leading to fatigue failure.

This process produces similar properties to nitriding, but is much quicker and cheaper. However, the depth of hardness is usually less, which can be a problem if the shaft is to be reground. This surface-hardening treatment – sometimes known as soft nitriding – is becoming increasingly attractive for both steels and cast irons, and may replace other more expensive processes for a whole range of components which require surface hardness and corrosion resistance with plain carbon steels.

16.9 Crankshaft main-journal bearings (figs 16.8 to 16.10)

The crankshaft is underslung in the crankcase and is supported by main bearings housed in the cross-webs forming the bulkhead of the cylinder block. The main-bearing axis intersects the cylinder axis but is perpendicular to it. Half the bearing housing bore is machined out of the bulkhead web and the other half is formed in the bearing cap (figs 16.9 and 16.10).

The main-bearing caps are located and aligned to the bulkhead half-bore by methods similar to those used for connecting-rod big-end joints (section 15.10.1), the most common alignment devices being the dowel, collar, or stepped location joints.

Liner half-shell bearings are incorporated in between both halves of the housing bore. Their construction and fitting techniques are identical

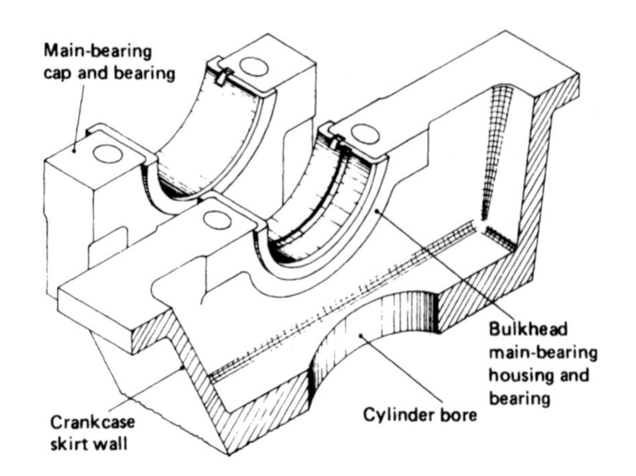

Fig. 16.9 Pictorial view of main bearing with integral thrust flange

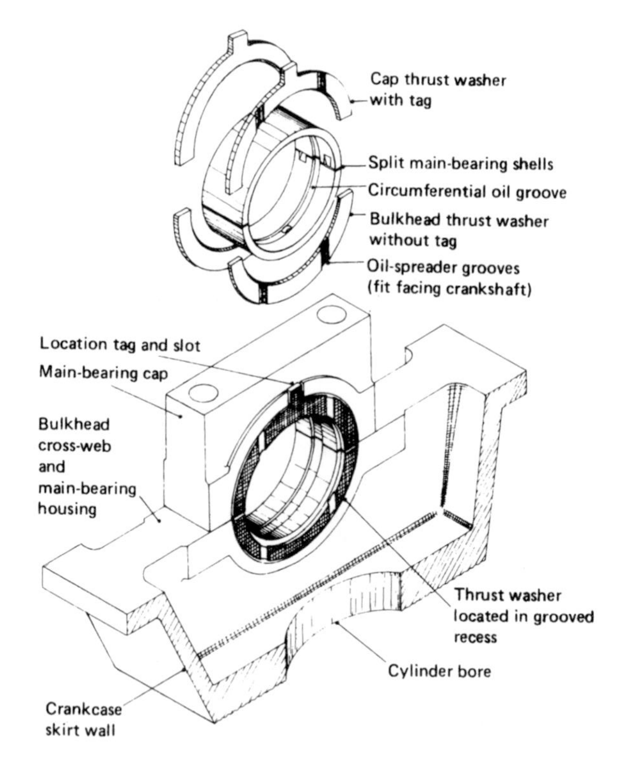

Fig. 16.10 Pictorial view of main bearing and separate thrust washers

to those described for the big-end in sections 15.11.1 to 15.11.7 under the following headings:

a) surface finish,
b) shell free spread,
c) shell location lug
d) liner-shell bore relief,

e) shell-spin prevention,
f) roundness of the housing bore
g) cap tightness.

The only major difference between the big-end and main-end shell bearings is that it is usual to have a central circumferential lubrication groove on the working face of the main bearings (fig. 16.10).

16.9.1 Crankshaft thrust bearings (figs 16.8 to 16.10)

The crankshaft is subjected to an additional load not experienced by the big-end journals and bearings – that is, end-thrust. End-thrust may be defined as the axial load transferred from the crankshaft to the main-bearing housing.

The axial thrust may be generated intermittently or continuously due to a number of causes:

a) Intermittent disengagement of the clutch will tend to push the flywheel against the crankcase.
b) Continuous torque conversion in some automatic transmissions tends to pull the flywheel away from the crankcase.
c) A helical valve-timing gear train when either accelerating or decelerating during operation conditions will push or pull the crankshaft axially one way or the other. Further loads may be imposed on the gear train when driving the various auxiliary components such as the injector pump, oil-pump, air compressor, power-steering hydraulic pump, water-pump, supercharger, etc.
d) The helical-gear primary drive on front-wheel-drive cars with integral engine and transmission imposes an almost continuous load on the thrust washers.

Under certain operating conditions, the crankshaft will be subjected to bending loads which will deflect and bow the shaft. This imposes severe strains on the cylinder block and crankcase, so that the axial alignment will become distorted. Simultaneously, the single main-bearing cap subjected to the end-thrust will not be stiff enough to prevent deflection. Consequently the contact between the thrust washer and the side of the crankshaft will be non-parallel, providing practically only edge contact, therefore only the crankcase half of the thrust washer tends to really contribute effectively to axial-movement control.

Thrust washers are fitted on each side of only one main-bearing housing bore, so that this takes axial crankshaft thrust in both directions for the whole crankshaft.

Manufacturers generally do use an upper- and a lower-half thrust washer, but in effect only the crankcase half is necessary, as this bears the majority of the axial thrust.

Because the thrust washers and the crankshaft web have a parallel-ring face contact, no wedge-shaped oil film can be generated to separate these rubbing pairs, so lubrication will be of a very marginal nature under continuous axial loading.

There are two basic types of main-journal radial and end-thrust bearings:

i) plain thin-wall bearings with integral thrust flanges (figs 16.8 and 16.9),
ii) plain thin-wall bearings with separate thrust washers (fig. 16.10).

With the integral flanged bearings, the steel backing is bent at right angles on each outer edge to form a flanged thrust face. These bearings can be simply pressed into position, usually near the flywheel end of the crankshaft, but they are

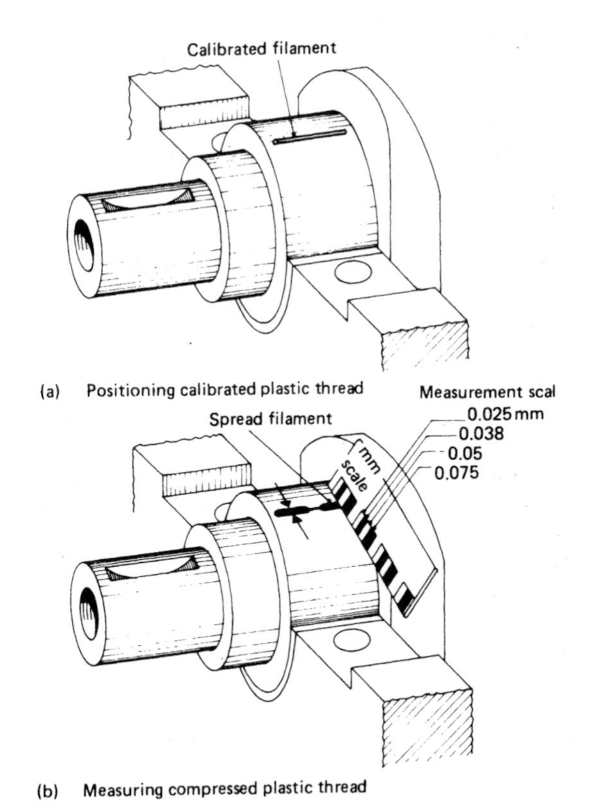

Fig. 16.11 Measurement of radial journal-to-bearing clearance

slightly more expensive to manufacture than the other type.

The separate-thrust-washer bearings are split rings which are located in grooves machined on both sides of a main-bearing bore housing. To prevent them from spinning, tags on each side of the cap half-washers align with slots machined on both sides of the main-bearing cap. No tags are provided on the crankcase half-washers.

After the bearings and thrust washers have been fitted, the end-float can be checked by inserting a feeler blade between the thrust washer and the crankshaft web. End-float should be within the specified limits, typical values being between 0.08 and 0.30 mm.

16.9.2 Measuring journal-to-bearing clearance (fig. 16.11)

Bearing wear can be accurately determined by the use of 'Plastigage' – this is a filament of plastic material of known cross-sectional area. Squeezing this filament between the journal and the bearing while tightening the bearing cap will flatten and spread it outwards, and the width of the compressed filament is then a measure of the thickness of plastic separating both surfaces, that is the diametrical journal-to-bearing clearance (fig. 16.11).

To obtain the correct reading, proceed as follows:

1 Remove the nuts or set-screws, the cap, and the bearing shells and wipe dry both the bearing and the journal surfaces.
2 Position a piece of Plastigage filament on top and across the full width of the journal.
3 Replace the bearing cap together with the bearing shell over the journal and evenly tighten down the nuts or set-screws to the specified torque.
4 Unscrew the nuts or set-screws and carefully remove the cap, but do not rotate the crankshaft while the plastic filament is in place. It will be observed that the plastic filament has been compressed in proportion to the journal-to-bearing clearance – the smaller the clearance, the greater will be the spread and vice versa.
5 Measure the filament width using the scale printed on the Plastigage pack. This gives a direct-conversion reading for filament thickness and hence journal-to-bearing clearance. Compare the measured reading with the manufacturer's clearance tolerance.

Note that the middle region of the filament is not squashed, since it aligns with the bearing's circumferential groove.

17 Valve timing diagrams, cam design and camshaft drives

17.1 Inlet and exhaust valve opening and closing periods

Consideration will now be given to improving the expulsion of the exhaust gases and the filling of the cylinder with fresh air/fuel mixture by virtue of early opening and late closing of both inlet and exhaust valves, and the compromise which must be made to optimise engine performance at both the lower and upper speed range, and at the same time to minimise the exhaust gas emission.

17.1.1 Inlet valve opens before TDC (fig. 17.1(a) and (b))

The inlet is made to open early before TDC towards the end of the exhaust stroke for a number of reasons:

1) To prevent excessive cam-follower shock loads and spring vibration the inlet valve is opened very gradually at first so that only a small amount of fresh charge can enter the cylinder for at least 10–20° camshaft movement.

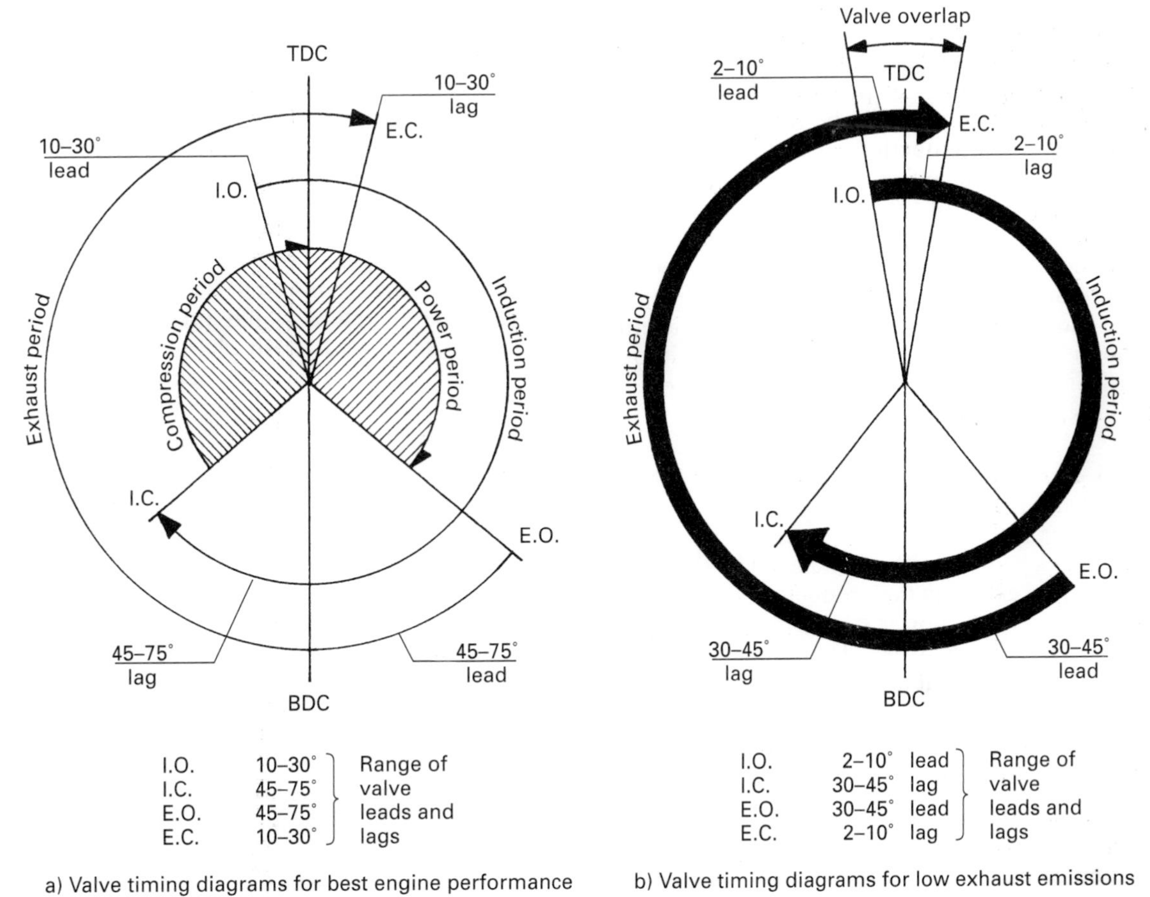

Fig. 17.1 Comparison of valve timing diagrams for both petrol and diesel engines

2) To compensate for the reluctance of the charge to enter the cylinder due to its inertia and its sluggishness to recommence moving every time the inlet valve closes and opens again.
3) To make use of the partial depression in the combustion chamber caused by the outgoing exhaust gases to commence drawing the fresh charge into the cylinder while the piston is in the TDC region and therefore cannot contribute to the inducement of the charge.

17.1.2 Inlet valve closes after BDC (fig. 17.1(a) and (b))

To maximise the air charge entering the cylinder, particularly at medium to high engine speed, the inlet valve is made to delay its closure to a point well beyond BDC. This extends the time for the incoming column of fresh charge to pile into the cylinder, thereby raising its pressure and density. Nevertheless, at low engine speed the momentum of the incoming charge is insufficient to oppose the upward moving piston at the beginning of its compression stroke just before the inlet valve closes. Subsequently a portion of the newly arrived charge will actually be pushed back and returned to the induction manifold. Thus the shortening of the effective compression stroke by the late closure of the inlet valve at low engine speed in effect reduces the nominal compression ratio and lowers the volumetric efficiency. Thus a compromise must be reached as to when the inlet valve should close so that a high volumetric efficiency is achieved at high engine speed without sacrificing too much volumetric efficiency loss in the lower speed range.

17.1.3 Exhaust valve opens before BDC (fig. 17.1(a) and (b))

The power lost in opening the valve early before BDC is reached is greatly outweighed by the gains in efficiently expelling the products of combustion near the end of the power stroke and at the beginning of the exhaust stroke. The power that could be produced by combustion towards the end of the power stroke if the exhaust valve was kept closed would be relatively small and ineffective for a number of reasons:

1) Towards the end of the power stroke the burning process slows down due to the spent portion of the charge suffocating and preventing the mixing and burning of the still unburnt charge.
2) The expanding cylinder volume on the power stroke causes a rapid reduction in combustion pressure.
3) As BDC is approached the effective crank-angle leverage quickly declines.

Taking these factors into account it is more advantageous to open the exhaust valve early, thus permitting the remaining kinetic energy of combustion to clear (blow-out) the cylinders before the completion of the power stroke, than to retain the products of combustion to the very end, and to rely solely on the piston on its exhaust stroke to sweep out the gases. Consequently the exhaust valve opening point in terms of crank-angle movement before BDC must be chosen so that the benefits gained in clearing out the exhaust gases and maximising the filling of the cylinder with fresh mixture is not disadvantaged by the power lost to the exhaust system over the mid-speed range of the engine.

17.1.4 Exhaust valve closes after TDC (fig. 17.1(a) and (b))

Efficient filling of the cylinder with fresh charge of air and fuel can only be achieved if all the burnt charge is completely removed at the end of each power stroke. A drawback to the complete expulsion of the exhaust gases every time the piston sweeps the cylinder on its exhaust stroke is that the gases occupying the combustion chamber cannot be positively removed at the end of the piston stroke and therefore would take up space that could be filled with fresh charge. This problem has largely been overcome by delaying the closure of the exhaust valve until after TDC so that the momentum of the outgoing exhaust gas column leaves a vacuum. At the same time fresh charge at the entrance of the early opening inlet valve will be induced to fill this void and simultaneously push any remaining exhaust gas out through its port.

17.1.5 Valve overlap (fig. 17.1(a) and (b))

This is the period where both inlet and exhaust valves are open at the same time in the TDC region. A large overlap encourages the removal of any remaining exhaust gases confined in the combustion chamber and at the same time providing an early start for the induction period. Nevertheless at low engine speed with a large exhaust lag there is sufficient time for fresh charge to actually be drawn into the exhaust port and manifold before the valve closes. As a result relatively large amounts of unburnt and partially burnt gases will pass out of the exhaust system under certain speed and throttle operating conditions. Conversely at low engine speed with a large inlet valve lead and a partially closed throttle, the depression in the induction manifold and ports may be greater than that in the combustion chamber and the exhaust port when the piston is the TDC region. Consequently some exhaust gases may be pulled back into the induction manifold thus diluting the fresh charge, upsetting the finely tuned mixture, and promoting slower burning and partial combustion. Because of the unacceptably high amounts of both hydrocarbon and carbon-monoxide which would be present in the exhaust system under the above conditions, there is a trend to sacrifice some engine performance for low emission by reducing somewhat the valve overlap.

17.2 Cam profile phases and valve opening and closing periods (fig. 17.2(a))

Cams are designed to control the open and close intervals of the inlet and exhaust poppet valves. The radial cam used for this purpose consists of a circular disc having a semi-oval triangular protrusion known as a lobe which extends for approximately 130° around the disc. Rotation of the cam causes its profile to slide against the smooth flat closed end of a cylindrical member known as a follower, which is itself restrained so that it can only move radially out or in, relative to the axis of the cam.

The cam profile has a follower lift or valve opening side and a corresponding follower fall or valve closing side. Both the lift and fall sides of the profile can be divided into three phases. These are

1) the cam ramp
2) the cam flank and
3) the cam nose.

Valve opening period (fig. 17.2(a)) First, there is a low rate rise ramp phase which joins the base circle to the cam-lobe. Secondly, there is a flank opening phase which accelerates the follower lift to its maximum speed position, known as the point of inflection, at which the flank concave curve meets the nose convex curve. Thirdly, there is the nose opening phase which decelerates the maximum follower lift speed to zero as the follower approaches its full lift position.

Valve closing period (fig. 17.2(a)) Conversely, the closing period starts with the nose closing phase causing the follower to accelerate from zero to its maximum speed under the exertion of the valve spring thrust. The flank closing phase then decelerates and the followers fall to very nearly a stop. Finally, the follower fall changes into a transitional low rate ramp closing phase to the point where it merges with the base circle dwell period. The follower then remains in its lowest position with the valve fully closed throughout the proceeding base circle dwell period.

17.2.1 Cam quietening ramp (fig. 17.2(a) and (b))

The quietening ramp lift contour joins the base circle to the start of the flank profile. The purpose of the ramp is to take up the follower to base circle clearance (tappet clearance) somewhere near its mid-contour position so that the follower's initial lift on the ramp commences at a very low minimum impact speed before moving onto the fast rise high impact flank portion of the cam profile so that the follower to lobe take-up occurs with the least shock. The upper and lower tappet clearance limits are chosen so that, providing tappet clearance adjustment is within this tolerance, contact of follower to cam lobe clearance take-up occurs somewhere along the quietening ramp portion of the cam. Note that if the tappet clearance is too small it may lead eventually to poor valve seating and loss of cylinder compression. Conversely a very large tappet clearance will cause follower-to-cam crash noise, as the follower contacting the cam after clearance take-up now takes place on the high lift flank portion of the cam.

17.2.2 Cam profile on a linear base (fig. 17.2(b))

Another way to illustrate the effective opening and closing periods of the inlet and exhaust valve is to plot the cam profile rise and fall vertically on a linear crank-angle base so that a better comparison of the cam rise and fall at any one instant can be made. The amount the cam profile lifts or lowers the follower per degree of camshaft rotation at any angular position of the camshaft relative to the follower is known as the rate of cam lift or fall. Thus it can be seen on the lift side of the profile that the ramp period has a very small constant rise rate whereas the flank's initial lift rate is small but quickly increases to a maximum at the point of inflection. Thereafter during the nose period the rate of cam lift steadily slows down to zero at the lobe's peak. An almost iden-

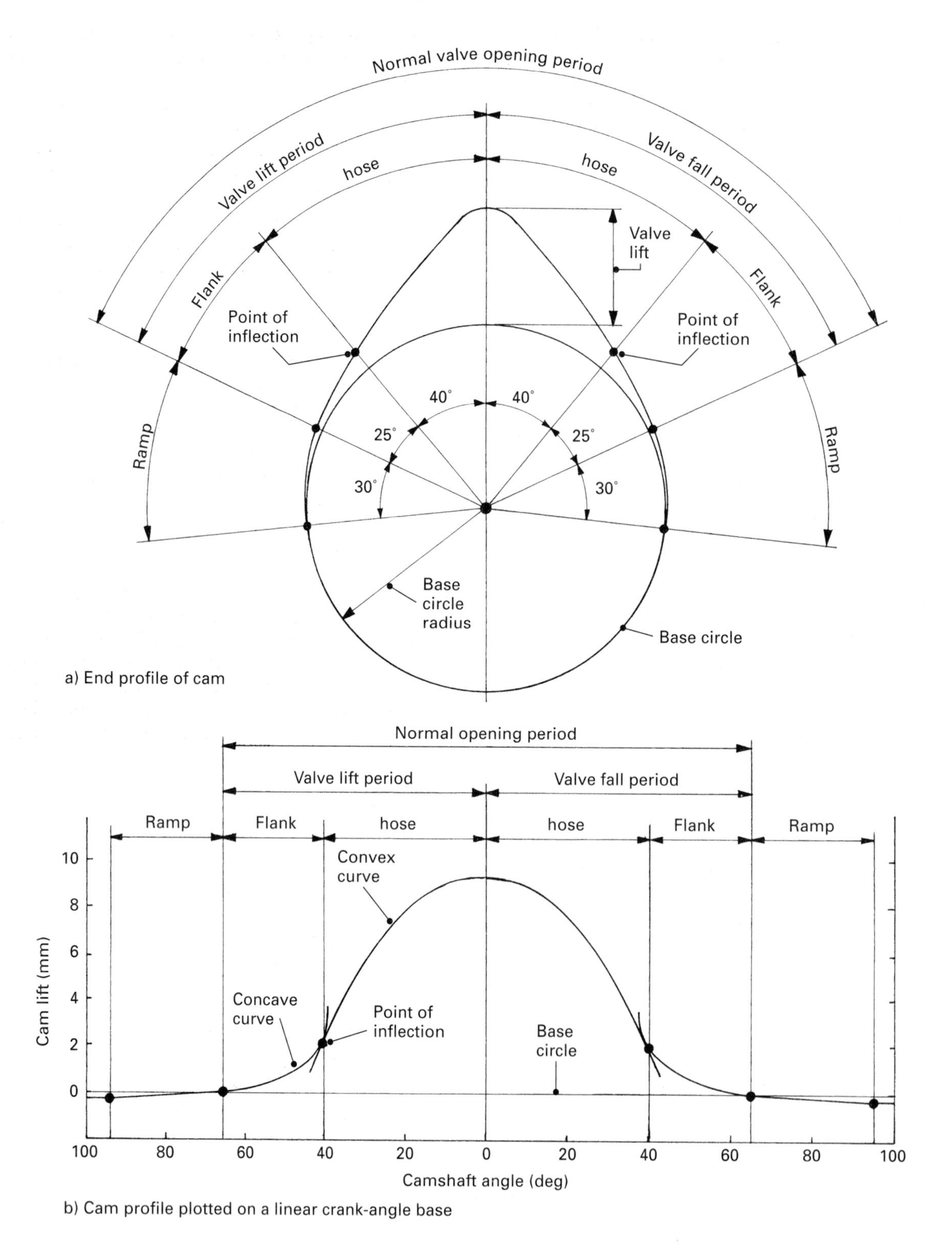

Fig. 17.2 Cam profile opening and closing phases

tical profile exists on the fall side of the cam lobe, but for design purposes the fall profile may be slightly modified. It is worth mentioning that the flank rise and fall period is cam controlled whereas the nose rise and fall period is spring controlled, that is, the spring keeps the follower hard against the cam profile.

17.3 Variable valve lift and timing control

Characteristics of fixed and variable valve lift and timing Variable valve opening period timing and lift is designed to overcome the limitation of having a fixed opening period and valve lift throughout the engine's speed range. The modern car petrol engine operates over a wide speed range, say from 1000 to 7000 rev/min. It therefore presents difficulties in optimising cylinder filling. In the low- to mid-speed range the engine performs better and the emission levels are lower if the intake valves open late, that is, after TDC on the piston's downstroke, and closes relatively early after BDC on its upstroke. Conversely, at high engine speed cylinder breathing is improved by opening the intake valve slightly earlier while the piston is in the TDC region and closing somewhat later after BDC on the piston's upstroke to make use of the ram momentum effect. However, with a fixed valve timing, a compromise is usually sought with the valve timing, so that the engine performance is optimised in the mid-speed range but at the expense of below optimum performance at the two extremes, that is, at low and high engine speeds.

The difficulties of matching the valve timing to suit the engine speed and load conditions can be partially overcome by having a set of cam lobe profiles for the low to medium speed range so that the air intake speed is kept well up, and at some predetermined operating conditions in the upper engine speed range, to switch to a cam profile which increases valve lift and enlarges the opening period of the valves as a means of maximising engine power.

17.3.1 Principle of dual valve lift and timing (fig 17.3(a) and (b), 17.4, 17.5 and 17.6(a)–(e))

The mechanism used in this example of a variable valve lift and timing arrangement only applies to the intake valves but a similar device can be used for the exhaust valves if desired. With this design there are pairs of intake and exhaust valves per cylinder. The two intake valves are operated by three cam-lobes while the two exhaust valves are actuated by two cam-lobes. To optimise the engine's performance in the low-speed range each intake valve has its own cam profile and thus its own opening period (see fig. 17.3(a)). The intake valve and cam which commences to open first is known as the primary valve and cam whereas the other valve and cam which starts to open slightly later is known as the secondary valve and cam.

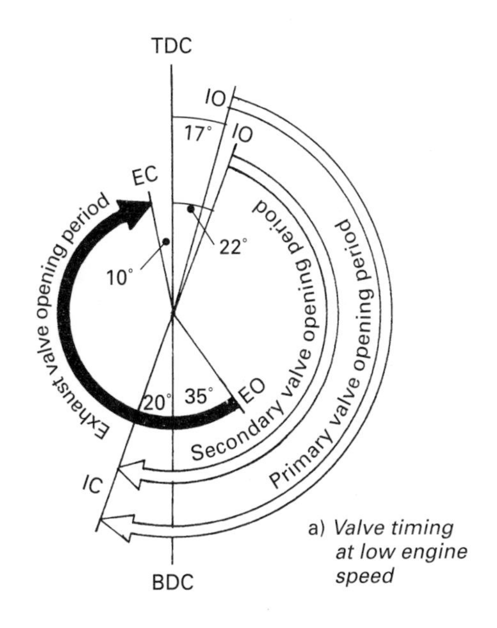

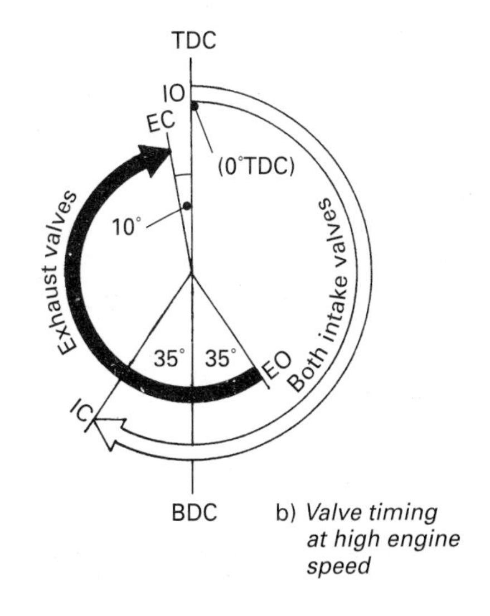

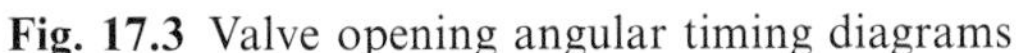

Fig. 17.3 Valve opening angular timing diagrams

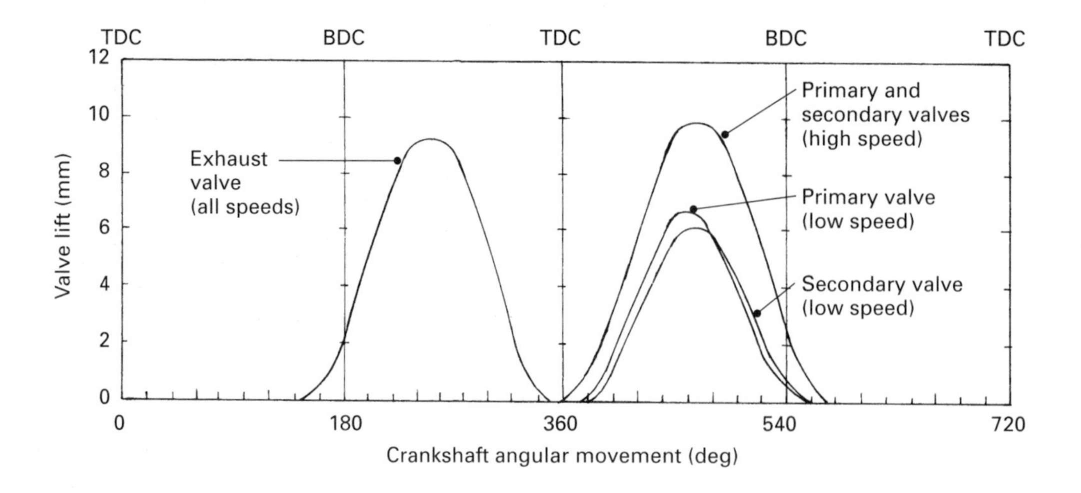

Fig. 17.4 Valve opening and left linear timing diagrams

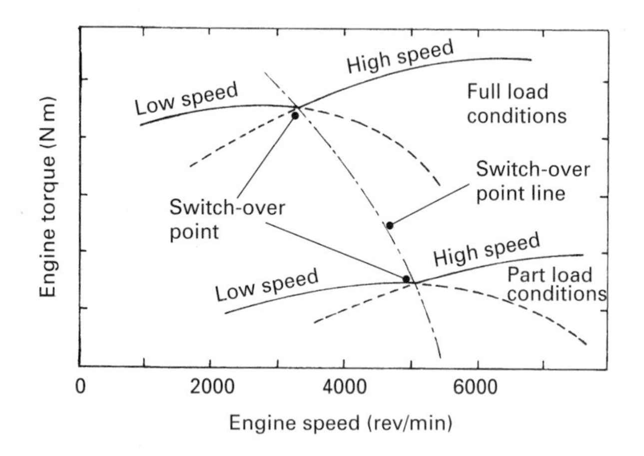

Fig. 17.5 Torque characteristics for dual valve timing

Both valves are made to close at the same time but if necessary the cams can be profiled so that one closes before the other. By phasing the opening of the primary and secondary valves, a spiral swirl can be produced in the cylinder so that vigorous mixing of the air and fuel charge can be achieved when the engine speed is still relatively low (see fig. 17.6(d)). In the upper speed range more demands are made on the quantity of charge entering each cylinder so that a third cam profile (mid-cam) providing an extended opening period and a higher valve lift is made to come into operation (see figs 17.3(b) and 17.4). This cam opens both valves simultaneously; it therefore alters the incoming charge movement from spiral to barrel roll swirl which is more suitable to high-speed filling and mixing conditions (see fig. 17.6(e)). The actual switch-over point speed where the cam-lobes change from low to high lift profile and an extended opening period depends upon the engine's coolant temperature, load, and vehicle speed. At full load the switch-over point occurs earlier than if the engine is operating under part-load conditions (see fig. 17.5).

Typical valve and timing specifications for the Honda 1.6 litre engine are shown as follows:

Valve lift at low engine speed

	intake		exhaust
	primary	secondary	
valve lift (mm)	6.5	6.0	9.2

Valve lift at high engine speed

	intake	exhaust
	both valves	
valve lift (mm)	9.7	9.2

Valve timing at low engine speed

intake open	primary secondary	17° ATDC 22° ATDC
intake closes	primary secondary	20° ABDC 20° ABDC
exhaust opens	both valves	35° BBDC
exhaust closes	both valves	10° BTDC

Valve timing at high engine speed

intake opens	both valves	0° TDC
intake closes	both valves	35° ABDC
exhaust opens	both valves	35° BBDC
exhaust closes	both valves	10° BTDC

Construction of a variable valve lift and timing rocker-arm device (Honda variable timing electronic control) (VTEC) (fig. 17.6(a)–(e)) The actuating valve mechanism consists of pairs of inclined intake and exhaust poppet valves. The intake and exhaust valves have their own rocker-arms which are mounted on corresponding twin intake and exhaust rocker-shafts. A single camshaft centrally located operates both the intake and exhaust valve opening periods. There are three intake and two exhaust cam-lobes formed on the camshaft per cylinder. Each of the two intake rocker-arms are separated by a mid-rocker-arm. This mid-rocker-arm freely pivots on the intake rocker-arm shaft; individual lost motion plungers and springs hold each of the mid-rocker-arms in contact with their corresponding mid-cam-lobes (see fig. 17.6(c)). Cylindrical holes are bored at right-angles to the rocker arms between the rocker-shaft hole and the contact-pad to receive the stopper and two synchronising pistons. The primary and secondary rocker-arms have blind holes, whereas the mid-rocker-arm cylindrical hole passes all the way though. Installed in these transverse bored holes are the stopper-piston, mid and end synchronising pistons. The spring-loaded stopper-piston pushes the mid and end synchronising pistons into place so that their exposed ends lie flush with the sides of the rocker-arms. A solenoid-valve and spool-valve control the engine's hydraulic pressure used to actuate the switch-over from the low speed cam-lobe profile to the high speed cam-lobe profile and vice versa. An electronic control unit fed by various sensors monitors the changing engine operating conditions. These variants are: engine coolant temperature, engine speed, engine load (measured in manifold depression), vehicle speed and oil pressure, and under suitable working conditions signals to the solenoid-valve when the cam-lobe changeover should take place.

Low to medium engine speed conditions (fig. 17.6(a)) At low engine speed the electronic control unit de-energises the solenoid winding. It therefore blocks the oil pressure reaching the head of the spool-valve and at the same time exhausts the oil trapped in this chamber. As the oil pressure decreases the stopper-piston spring pushes the synchronising-pistons across until the high-speed cam-lobe only controls the oscillating movement of the mid-rocker-arm. This means that the primary and secondary valves and their respective rocker-arms are now operated directly by their individual low-profile primary and secondary cam-lobes.With the primary valve opening leading the secondary valve opening by a small angular movement, the initial offset of the incoming air and fuel charge compels it to spiral around and down the cylinder bore as the piston descends (see fig. 17.6(d)).

Medium to high engine speed conditions (fig. 17.6(b)) At high engine speed the electronic control unit energises the solenoid-winding. This draws in the solenoid-armature, causing the valve to open and for the exhaust port to block. Oil now passes though to the end of the spool-valve. This forces it to shift over, and oil pump pressure is now applied to the end synchronising-piston chamber. Therefore at a given instant when all three rocker-arm piston bores align, the mid and end synchronising-pistons snap over, thereby making all three rocker-arms move in unison. As a result the opening and closing periods of both valves will now conform to the mid-high-lift cam-lobe profile. Since both valves open together the incoming charge flow pattern will be encouraged to produce a barrel-roll swirl (see fig 17.6(e)). In addition the high valve lift raises the cylinder filling volumetric efficiency.

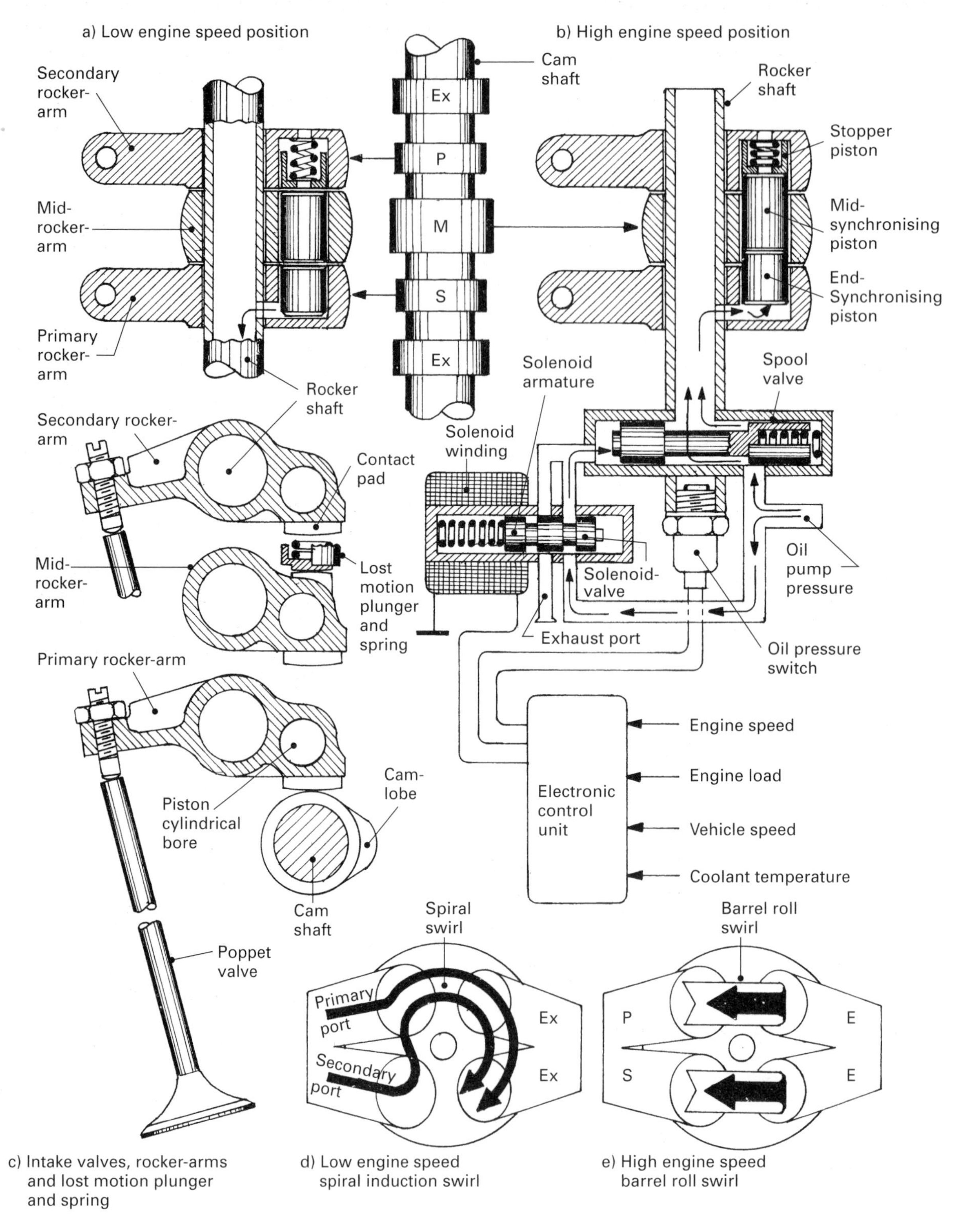

Fig. 17.6 Variable valve lift and timing rocker-arm and cam mechanism

17.4 Camshaft chain-belt and gear train drives

Crankshaft to camshaft drive (fig. 17.7(a)–(c)) With the four-stroke cycle engine, the cycle of events of the inlet and exhaust valve opening and closing is performed by the camshaft in one revolution, but the piston strokes – induction, compression, power, and exhaust – are completed in two crankshaft revolutions. Consequently, for the camshaft timing cycle to be in phase with the crankshaft angular movement, the camshaft has to turn at half crankshaft speed, that is, a 2:1 speed ratio.

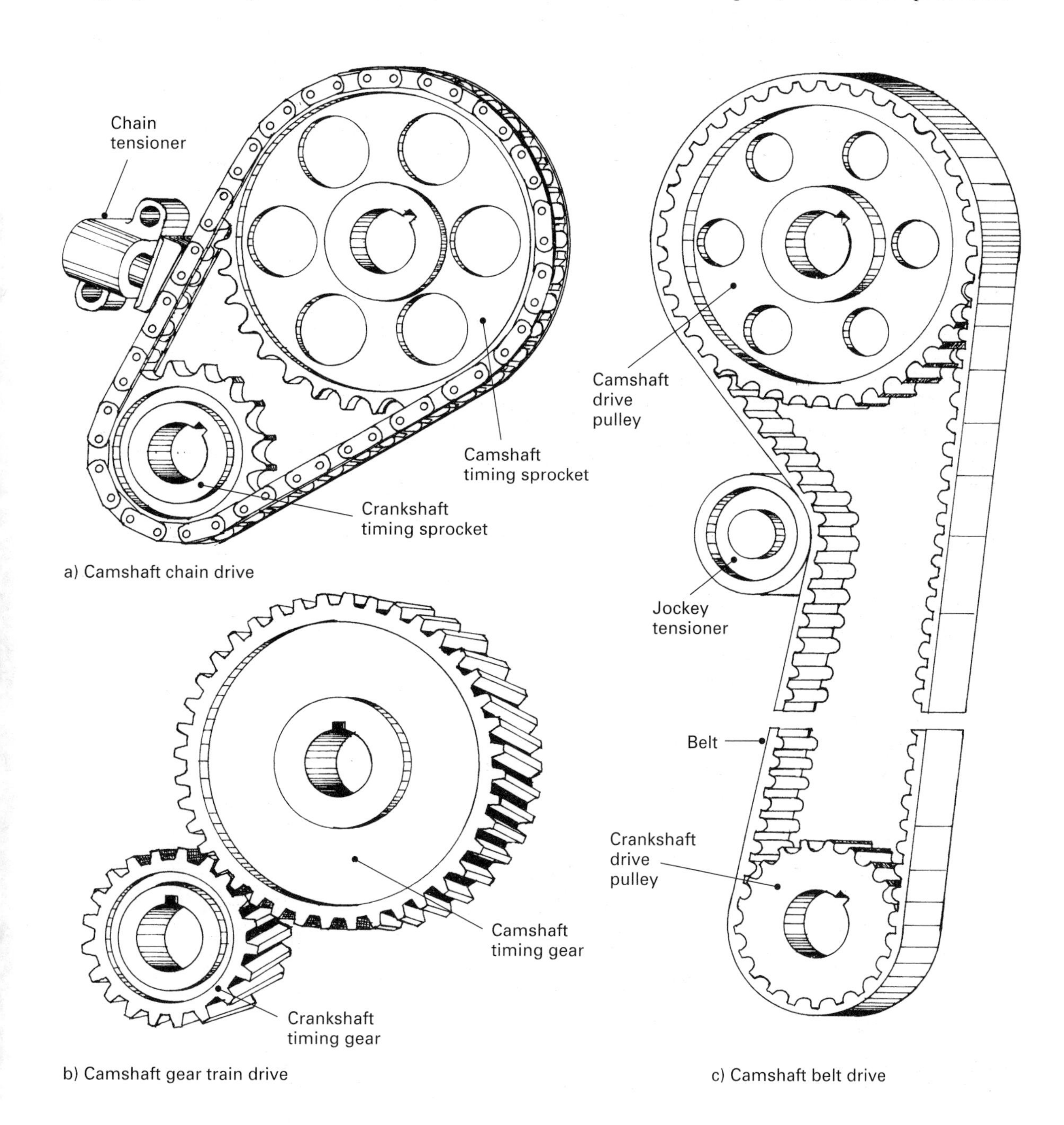

Fig. 17.7 Crankshaft to camshaft drive

The crankshaft to camshaft drive may be transmitted by three different methods: chain, belt, or gear (see fig. 17.7(a)–(c)).

17.4.1 Camshaft chain drive arrangement (fig. 17.8(a)–(d))
Chain drives are conveniently used where the span between the crankshaft and camshaft changes between hot and cold running, and where awkward positioning of driven shafts are made possible. To compensate for chain stretch and sprocket wheel wear, automatic chain tensioners are installed on the slack side of the chain between adjacent sprocket wheels. Where the chain span is large a vibration damper is sometimes placed adjacent to the taut side of the chain to absorb chain whip every time the engine decelerates.

Cylinder block mounted camshaft chain drive (fig. 17.7(a)) For low cylinder block mounted camshafts used for push-rod and rocker-arm valve actuation, a short chain with a two-to-one step-down crankshaft and camshaft sprocket-wheel pair (21/42 teeth) are commonly used with a mid-position chain tensioner to take up any free slack.

OHC with injection pump chain drive (diesel) (fig. 17.8(a)) This comprises a triangular chain layout with a large injection pump sprocket-wheel providing mid-span support on the taut side with a hinged-rail automatic tensioner on the slack side to take up free chain movement.

Twin OHC single-chain drive (petrol) (fig. 17.8(b)) Twin close mounted camshafts high above the crankshaft sprocket provides an isosceles triangle chain configuration. However, the crank to cam sprocket span is large so that support is given by a hinged-rail tensioner on the slack side and a single vibration damper strip on the taut side of the chain.

Twin OHC two-stage chain drive with rear-mounted camshaft drive sprockets (fig. 17.8(c)) The first-stage chain joins the small crankshaft sprocket to a large intermediate sprocket-wheel which simultaneously produces the two-to-one speed reduction and directly drives the inlet camshaft. However, the exhaust valve camshaft is indirectly driven by a pair of similar-sized sprocket-wheels attached to each camshaft at the opposite end (rear) to that of the intermediate sprocket-wheel (front of engine) and coupled together with a second chain. Chain slack and whip in the first stage is controlled by a hinged-rail automatic tensioner and a vibration damper strip respectively, whereas no tensioner is required for the short second-stage chain.

Twin OHC two-stage chain drive with intermediate sprocket-wheels (fig. 17.8(d)) A two-stage chain drive with a low mounted intermediate double

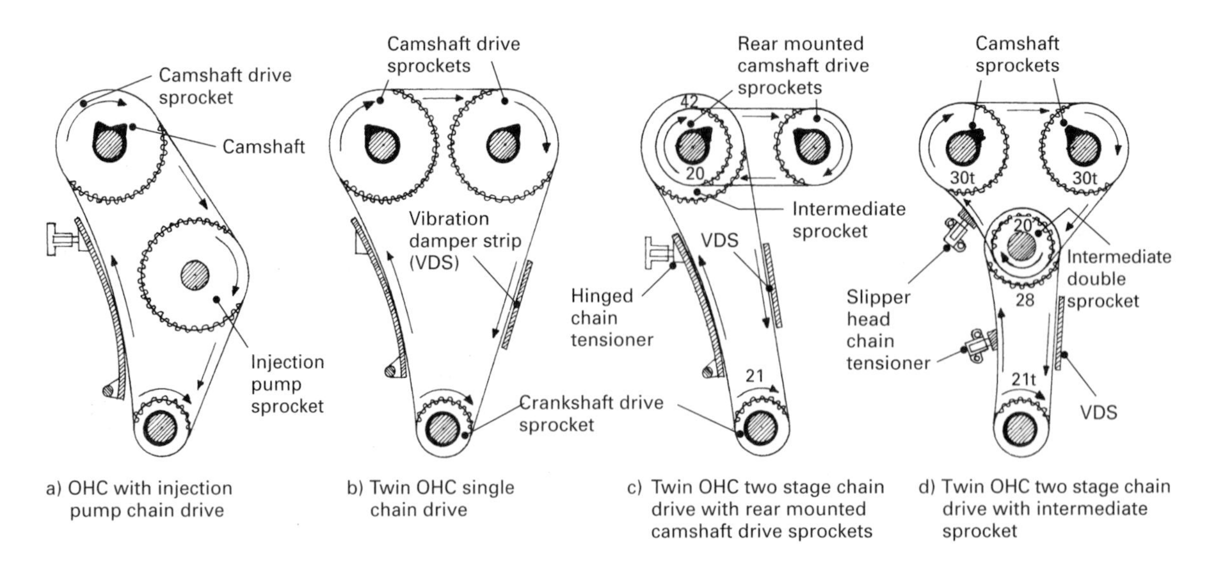

Fig. 17.8 Crankshaft chain drive arrangements

sprocket-wheel reduces the chain span between sprockets and enables the two-to-one speed reduction to be produced in two parts. Consequently the camshaft sprocket-wheel will be smaller, which is advantageous in making the cylinder head and camshaft assembly more compact, i.e. typically a 30-toothed sprocket-wheel instead of 42 teeth. In this arrangement each chain has a tensioner but only the first stage has a vibration damper strip.

Automatic chain tensioner To take up excess slack in the chain due to chain stretch and sprocket tooth wear, automatic chain tensioners are normally positioned between driver and driven pairs of sprocket-wheels. Two types are commonly adopted:

a) the slipper head tensioner suitable for short sprocket to sprocket wheel chain spans;
b) the hinged-rail tensioner which spreads its friction contact over a wider portion of the chain, making it partially suitable for large chain spans between sprocket wheels.

Slipper-head plunger and ratchet chain tensioner (fig. 17.9(a)) Tensioning of the chain is achieved by a spring-loaded plunger attached to a slipper-head supported in a cylindrical bore housing. Ratchet teeth are machined on one side of the plunger permitting the plunger ratchet teeth to slide over the spring pawl in its outward movement from the blanked cylinder as it progressively takes up new chain slack. However, the ratchet teeth set prevents any inward movement of the

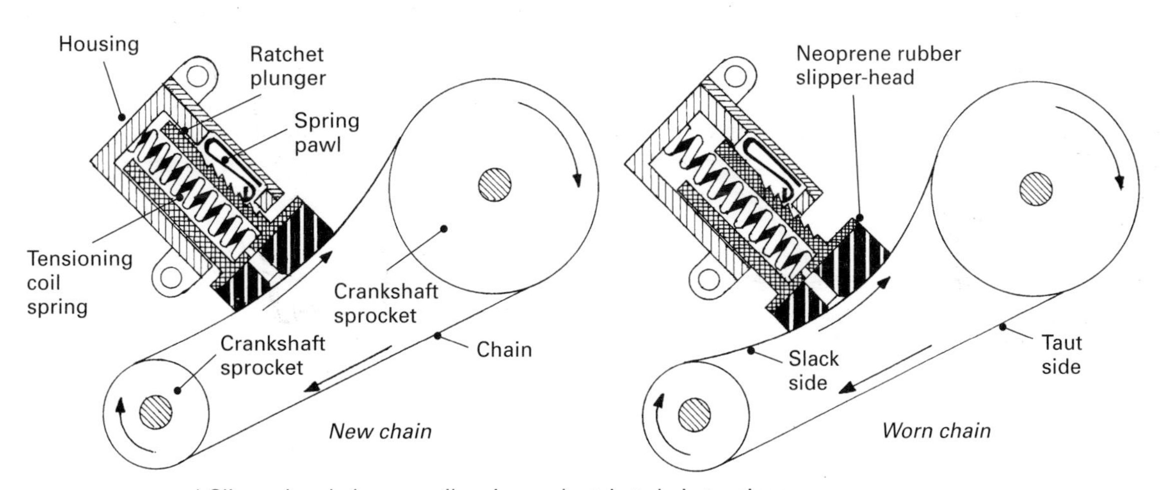

a) Slipper-head plunger coil spring and ratchet chain tensioner

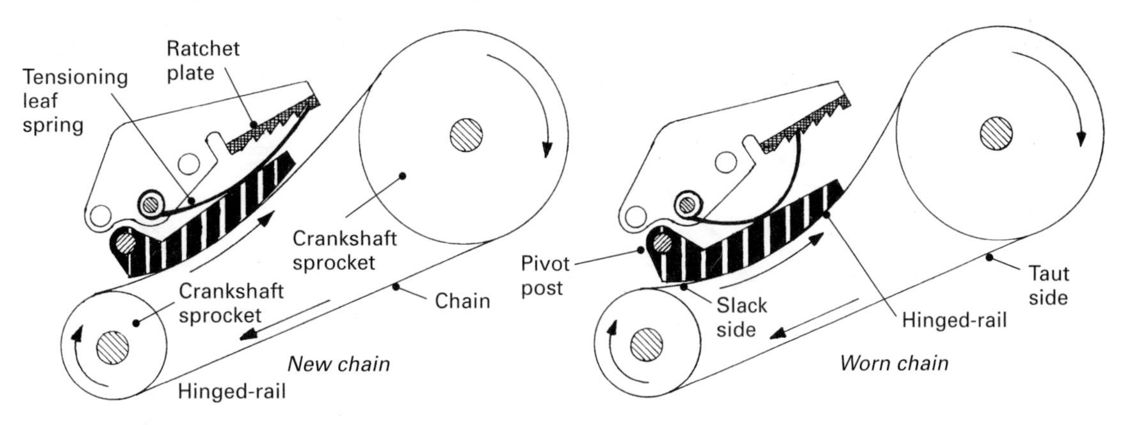

b) Hinged rail leaf spring and ratchet chain tensioner

Fig. 17.9 Automatic chain tensioners

plunger whenever the chain whips against the slipper-head. Sufficient free play in the assembled chain is provided by the gap between the active ratchet tooth shoulder and the spring pawl at any one time. In some automatic tensioner arrangements the outward spring load is assisted by oil pressure fed from the lubrication system. Oil is also then allowed to flow out from the slipper-head to lubricate its face contact with the chain.

Hinged-rail leaf spring and ratchet chain tensioner (fig. 17.9(b)) Chain slack is taken up with this tensioner design by a pivoting friction rail (hinged-arm) which is held against the chain by means of a bowed leaf spring constrained by a ratchet plate. The curled end of the spring is attached to a post, whereas the free end butts against the shoulder of the ratchet teeth. With a new chain and new sprocket-wheels the bowed spring will be partially flattened by the outward swing of the hinged-rail caused by the taut unworn chain, so that the free end of the spring butts against the shoulder of the furthest tooth from the spring post. As wear takes place the increased chain slackness permits the bowed spring to increase its curvature which correspondingly shortens its length so that the free end of the spring clicks into each tooth profile in progressive steps as chain slack increases. When whip against the hinged-rail occurs, the free end of the spring locks itself against the shoulder of the tooth it has been occupying. It therefore provides a stiff opposition to any outward movement of the rail caused by the impact of chain whip.

Roller chains (figs 17.10(a) and (b) and 17.11) Timing roller chains (figs 17.10 and 17.11 are made from pairs of numerous steel inner and outer side plate links. Large punched-out holes at each end of the inner plate tightly support steel

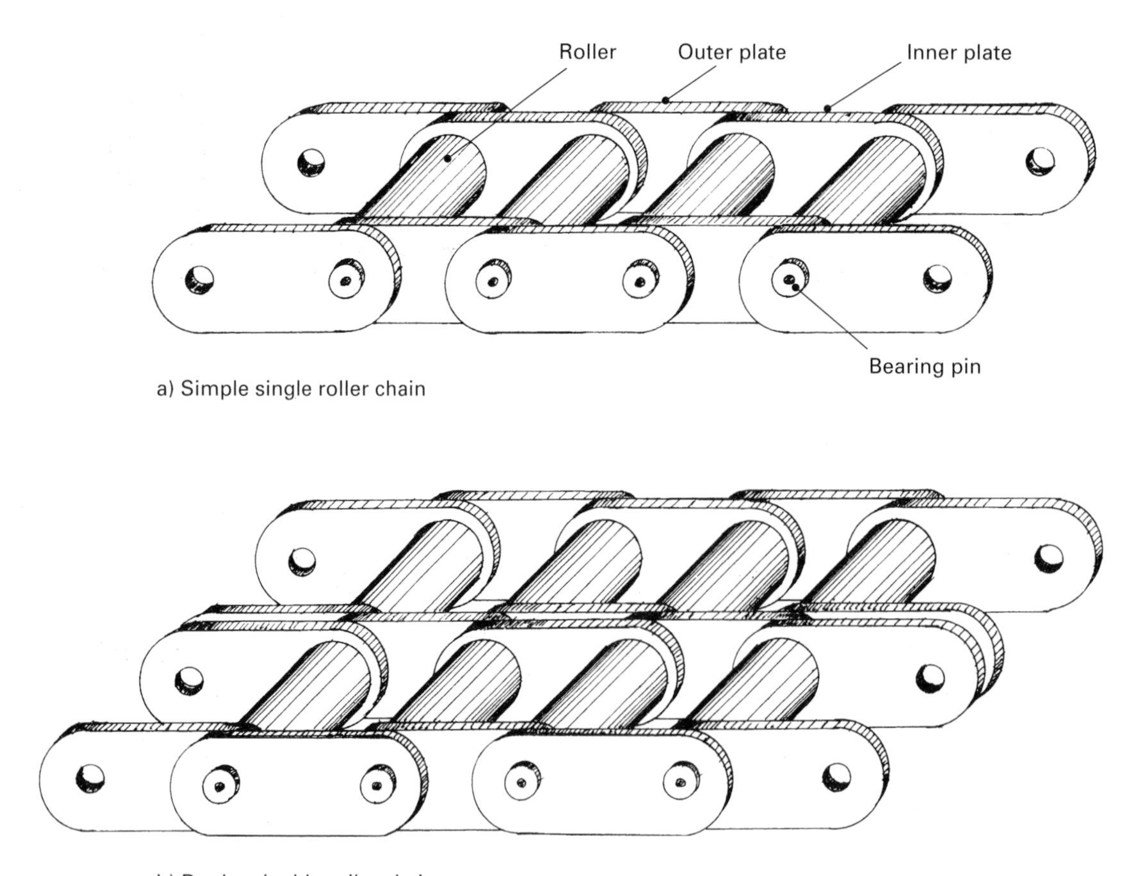

Fig. 17.10 Timing drive roller chain

cylindrical bearing–bushes, whereas the outer plates are held together via steel bearing-pins which slip through the inner bushes, but are a force fit in the smaller correspondingly outer side plate holes. The pin ends finally securing each hinged link assembly are swaged at their ends. To reduce friction wear between the chain and sprocket teeth, loose-fitting seamless rollers are initially assembled over each bush so that relative movement and load in the linked joints is shared between the pin-bush-roller and sprocket teeth.

Timing roller chains may be single (fig. 17.10(a)), double (fig. 17.10(b)) or even triple. However, for most purposes the single chain suffices, but for heavy duty or very heavy duty double or triple roller chains may be utilised respectively.

Wear occurs between each rubbing pair of surfaces, that is, the pin and inside of the bush, the outside of the bush and inside of the roller (fig. 17.11) and finally between the outer roller face and the sprocket tooth profiles (fig. 17.12(a) and (b)). High chain and sprocket wheel tooth wear

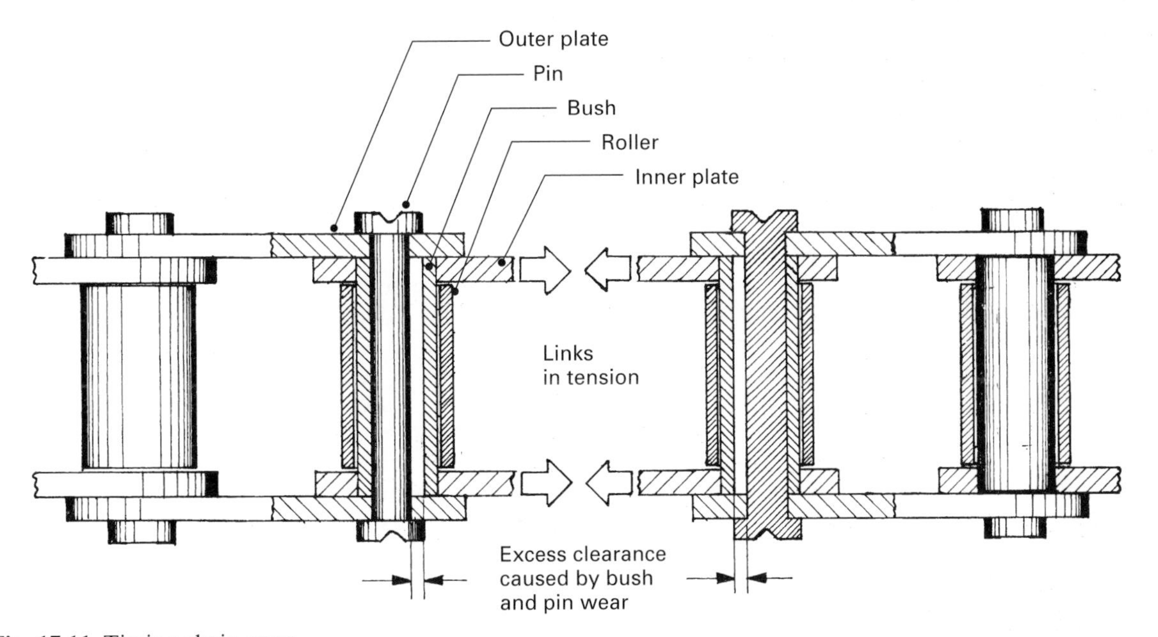

Fig. 17.11 Timing chain wear

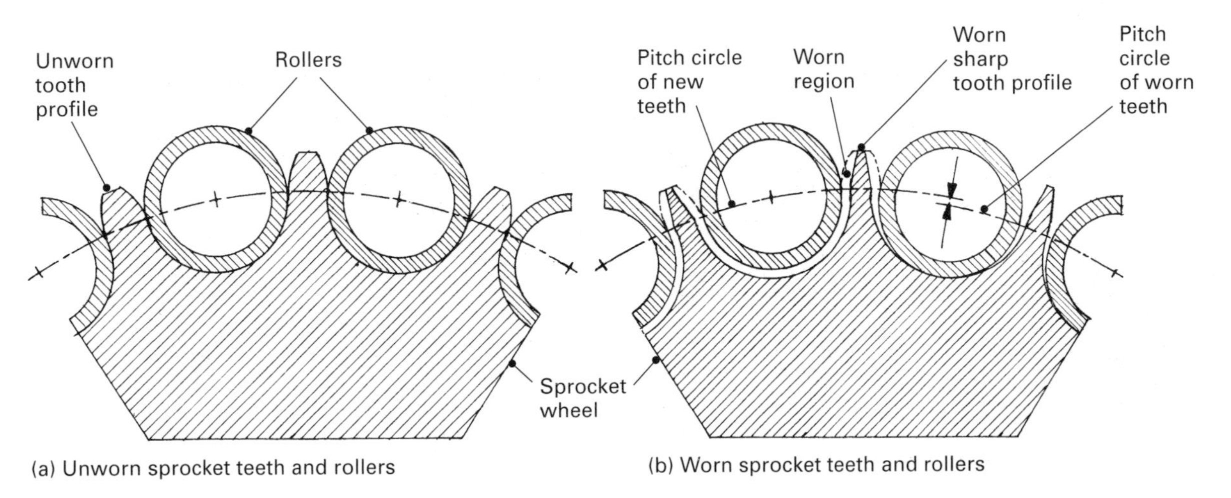

Fig. 17.12 Comparison of worn and unworn sprocket tooth profile

can usually be traced back to extended use of old and contaminated oil, insufficient lubrication, misalignment of sprocket wheels, incorrect tensioning of the chain, inadequate vibration damper support, excessive impact loads at very high engine speeds and, finally, poor design layouts of sprocket wheels and chain assembly (insufficient angle of chain wrap about sprocket wheels).

17.4.2 Overhead camshaft belt drive arrangements (fig. 17.13(a)–(d))

Timing belts transfer motion from the crankshaft to the camshaft by means of a pair of cog teeth pulley-wheels. The camshaft which rotates at half crankshaft speed therefore has twice as many cogs as the crankshaft pulley-wheel, i.e. something of the order 21/42 cogs. Positive motion from the driver to the driven pulley is provided via the internal and external cog teeth of the belt and pulleys respectively continuously interlocking with one and other (see fig. 17.7(c)). An adjustable jockey pulley is provided on the belt's non-drive side to prevent the belt cogs jumping over engaging pulley cogs due to the belt being slack. The jockey pulley-wheel rolls against the belt's smooth side and can be moved further towards or away from the belt to provide the correct working tension.

OHC and coolant pump belt drive (petrol) (fig. 17.13(a)) A triple cog pulley and jockey tensioner configuration uses its third cog pulley mounted low on the belt's non-drive side to drive the coolant pump, and just above it the jockey pulley provides belt tension and support but on the belt's opposite side. With this layout the belt passes from the outside of the cog coolant pulley to the inside of the smooth roller jockey pulley, thereby increasing the belt angle of wrap with both pump and jockey pulley.

OHC and injection pump belt drive (diesel) (fig. 17.13(b)) This comprises a triple cog pulley and jockey tensioner layout which has a large injection pump cog pulley mounted half-way down on the drive side of the belt with the jockey tensioner positioned high up on the non-drive side of the belt adjacent to the camshaft. Thus the belt receives additional support between the crankshaft and camshaft while also providing the drive for the fuel injection pump.

OHC coolant and injection pump belt drive (diesel) (fig. 17.13(c)) This is a quadruple cog pulley and twin jockey tensioner belt drive arrangement, which has a cog pulley driving the injection pump mounted between the crankshaft and camshaft on the drive side of the belt with the coolant pump pulley position low down on the non-drive side of the crankshaft pulley. A pair of jockey tensioners is used to increase the angle of belt wrap for both the injection pump and coolant pump drive pulleys, but only the jockey pulley mounted between the camshaft and coolant pump is adjustable. This configuration reduces the possibility of the injection pump cog pulley jumping over the engaging belt teeth under high cyclic impact loads.

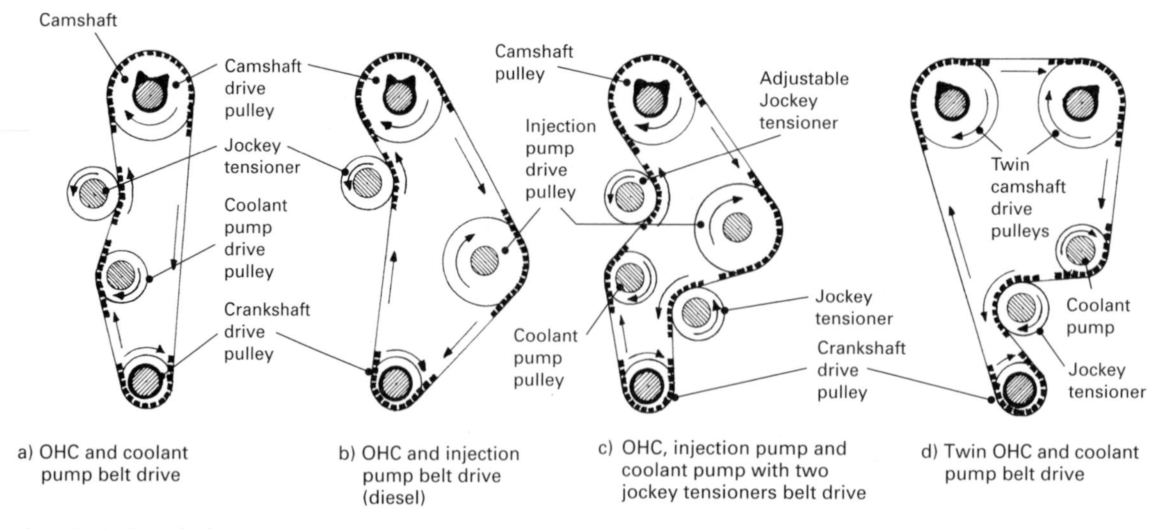

Fig. 17.13 Camshaft belt drive arrangements

Twin OHC and coolant pump belt drive (petrol) (fig. 17.13(d)) This is a quadruple cog pulley layout which has twin inlet and exhaust camshaft cog pulleys mounted side by side with the coolant pump and pulley mounted half-way down on the drive side of the belt. Belt tension and extra belt wrap is obtained for both the crankshaft and coolant pump pulleys by mounting the jockey tensioner deep in between them on the drive side of the belt.

Timing toothed belt (fig. 17.14) Timing belts are endless internally toothed straps, each tooth of which is identically profiled to match the external sprocket wheel teeth of the drive and driven pulley wheels (fig. 17.7(c)). The belt is made from a parallel layer of tensile load carrying glass fibre or similar material cords. These cords are strong, flexible and do not develop any permanent stretch (fig. 17.14). Moulded neoprene rubber teeth are bonded to the inside strands of the load-bearing cords to enable a continuous and positive grip to be maintained between the belt and sprocket pulley teeth. The synthetic rubber support backing is smooth on the outside and is bonded to the outward protruding profile of the strands. This keeps the continuously flexing layer of strands in position. Subsequently the tensile belt load will be equally divided between the cords. The smooth backing layer also provides the contact surface for the tensioning jockey idler pulley. To prolong the working life of the rubber teeth, a wear-resisting tough woven nylon fabric lining is bonded to the internal tooth of the belt.

Jockey wheel pulley tensioners (fig. 17.15(a)–(c)) An essential fixed belt tension is provided by adjustable jockey wheel tensioners. Belt tension is critical; therefore, great care must be used when assembling a new or worn belt. The jockey wheel pulley is generally supported on either needle or roller ball bearings mounted on a post which is itself attached to a hinged back plate or directly to the cylinder block.

Eccentric bush adjustment tensioner (fig. 17.15(a)) With the eccentric bush adjustment, the eccentric bush is twisted about its post by way of a double-pronged key until the correct belt tension is obtained. It is then secured by tightening the lock-nut screwed on the pivot post.

Pivot plate tensioner (fig. 17.15(b)) With the hinged plate assembly the plate is swung outwards about a pivot until the correct belt tension is obtained. It is then held in position by a lock-nut tightened over an elongated hole formed in the hinged plate

Compression spring preloaded plate tensioner (fig. 17.15(c)) A third method, employing again a hinged back-plate, uses a preloaded compression spring to automatically swing the hinged plate against the belt. The pre-determined spring thrust

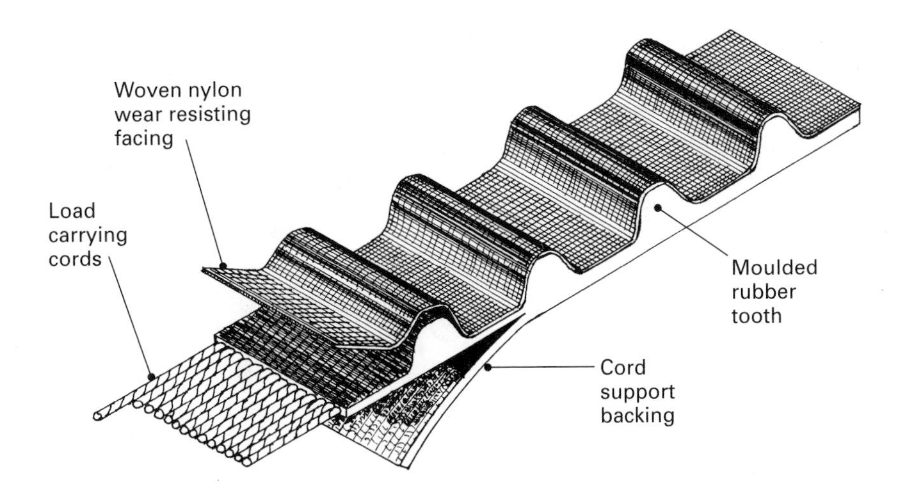

Fig. 17.14 Timing tooth belt drive

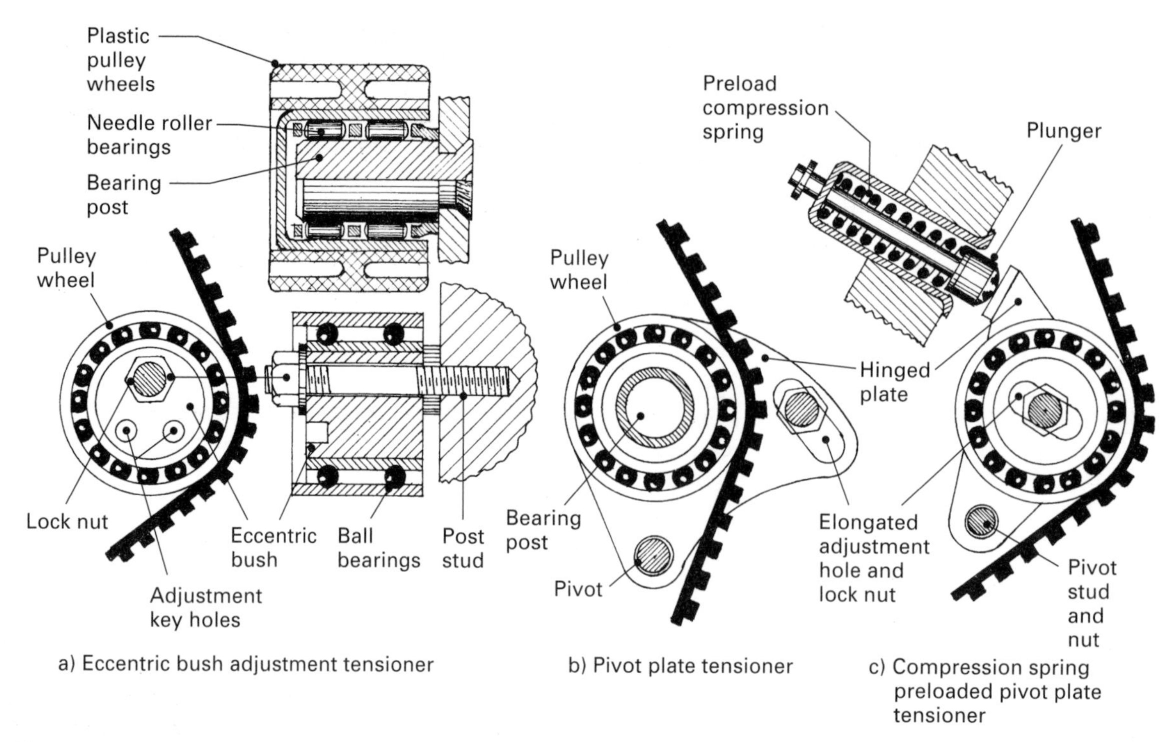

Fig. 17.15 Jockey idler pulley belt tensioners

thus correctly sets the belt tension. It is then fixed in this position via tightening the lock-nut overlapping the elongated hole.

Setting belt tension Adjustment of the belt can be carried out by loosening the jockey idler lock-nut and pushing it against the smooth face of the belt with a moderate thumb pressure, at the mid-point of the belt between the crankshaft and camshaft pulleys. The belt deflection should then be in the order of 6mm depending upon the manufacturers recommendation. The tensioner's position should be adjusted until the correct belt deflection is obtained before tightening the tensioner's lock-nut. Alternatively, adjust the position of the loosened tensioner until the belt can just be twisted 90° between finger and thumb half-way between the crankshaft and camshaft belt pulleys. At this point tighten the tensioner lock-nut. Neither of the above methods of determining the belt tension are sufficiently accurate; therefore timing belt tension gauges have been designed to measure the stiffness of the belt between adjacent pulley wheels (fig. 17.16) if the jockey pulley tensioner does not incorporate its own preload spring.

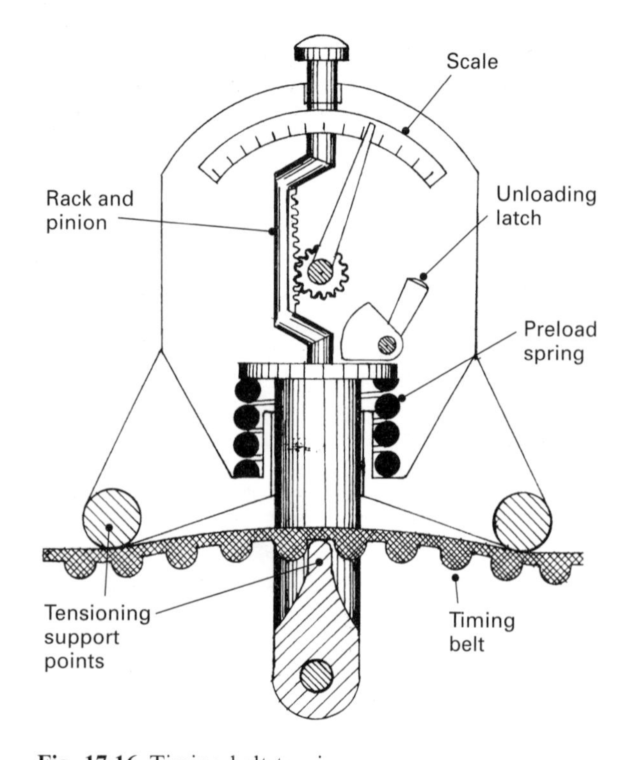

Fig. 17.16 Timing belt tension gauge

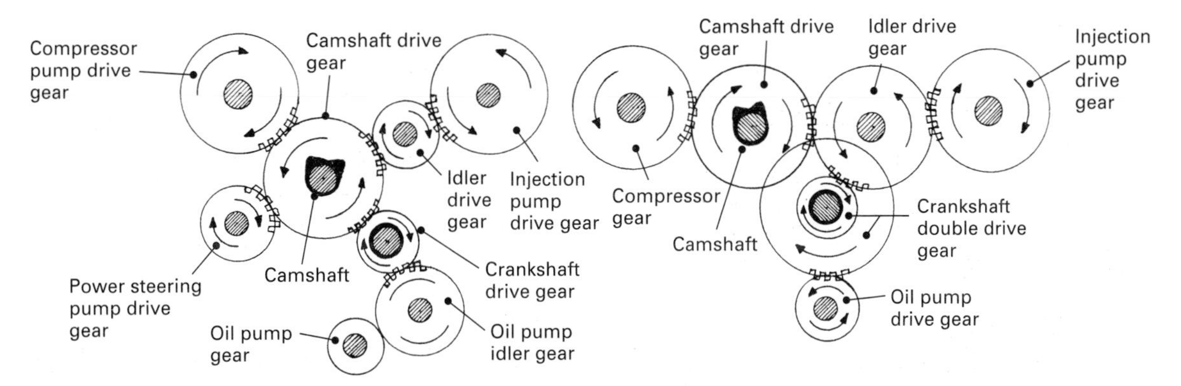

a) Direct driven camshaft with indirect idler driven injection pump

b) Indirect idler driven camshaft and injection pump

Fig. 17.17 Camshaft gear drive arrangements

Setting timing belt tension with gauge (fig. 17.16) To use the gauge, move the unloading latch to the no-load position. This pushes down the plunger and the central support contact against the stiffness of the calibrated spring. This enables the two upper and the central lower support contacts to be slipped across the belt in the mid-span position between adjacent pulley wheels. The unloading latch is then released permitting the calibrated compression spring now to exert its full upward force against the shoulder of the plunger. Rack movement will now rotate the pinion and attached pointer in proportion to the belt deflection indicated on the scale. The readings will vary according to the distance between adjacent pulley wheels and the width and length of belt. Therefore manufacturers will quote different scale deflection readings to suit the belt and pulley system's size and configuration. After setting the belt tension, the crankshaft pulley should be rotated two complete revolutions, enabling the belt to settle and accurately align itself between pulley wheels. Again slacken the jockey pulley lock-nut and reset the belt tension. Finally retighten the tensioner's adjustment lock-nut.

17.4.3 Camshaft gear train drive arrangements (fig. 17.17(a) and (b))

Timing gear trains are particularly suitable for medium to large diesel engine application where drive train loads can be considerable and reliability is essential. To ensure continuous tooth contact, reduce tooth profile fretting, pitting and wear, and to minimise increasing back-lash between meshing teeth, helical involute teeth are now always used.

Simple camshaft gear drive (petrol) (fig. 17.7(b)) Gear-wheel drives are mostly used where the camshaft is mounted low in the cylinder-block, and in its simplest form requires only a driver and driven two-to-one speed reduction pair of gear-wheels to mesh together.

Direct-drive camshaft with indirect idler-driven injection pump (fig. 17.17(a)) This timing gear layout is ideal for long push-rod and rocker-arm operated low mounted camshafts for both in-line and vee cylinder configurations. The offset injection pump requires an idler gear to reach it from the camshaft gear. However the compressor pump and power steering hydraulic pump being mounted on the same side as the camshaft are conveniently driven directly from it. The oil pump which is positioned below the cylinder-block is indirectly driven off the crankshaft via a large idler gear-wheel.

Indirect idler-driven camshaft and injection pump (fig. 17.17(b)) For high mounted camshafts, an intermediate idler gear-wheel is necessary to convey the drive to it. This idler gear also conveniently provides the drive for the offset injector pump, whereas the compressor pump mounted on the opposite side is driven directly from the camshaft. An additional large gear-wheel mounted on the crankshaft provides a speed step-up drive for the oil pump.

18 Poppet-valve operating mechanisms

18.1 The function of the valves and their arrangements

Improvements in engine performance owe a great deal to the steady but not so obvious improvements in the methods of inducing fresh charge and expelling spent gases from the cylinders.

To control the in and out movements of charge and exhaust gases in the cylinders relative to the piston positions in their bores, some sort of valve arrangement must be incorporated. On all modern engines, this is situated in the cylinder head. Various forms of valves have been used, such as sleeve, sliding, rotary, and poppet-type valves, but only the poppet-valve seems to have lasted the course.

A poppet-valve resembles a cylindrical stem with an enlarged mushroom disc at one end. The stem of the valve is situated in a guide hole made in the cylinder head centrally in a circular passage, and, if the valve stem is moved in and out, the valve disc head will open and close the ported passage leading to the cylinder.

Inlet and exhaust ports are both shaped so that they curve upwards and outwards emerging from one or both sides of the cylinder head. It is normal to have one inlet and one exhaust valve and port per cylinder, but twin inlet and exhaust valve-and-port layouts are preferred for some high-performance or large-capacity engines. In a very few cases, the use of twin inlets but only one exhaust valve is considered effective, based on the assumption that it is more difficult to induce the fresh charge to enter the cylinder than it is to expel the burnt gases.

Valves may be situated vertically or slightly inclined relative to the cylinder axis, so that the valve head conforms to the desired combustion-chamber contour.

18.2 Side camshaft with push-rod and rockers (figs 18.1 to 18.5)

This type of valve-operating mechanism is made up from the following components:

a) a camshaft
b) a cam follower (tappet),
c) a push-rod,
d) a rocker-arm,
e) a rocker-shaft,
f) a return-spring,
g) a poppet-valve.

The operating mechanism between the camshaft and the poppet-valve is known as the valve train.

18.2.1 Camshaft (figs 13.12 and 18.1)

The function of the camshaft is to provide a means of actuating the opening and controlling the period before closing both the inlet and the exhaust valves. A secondary function, but still a very important one, is to provide a drive for both the ignition distributor and the mechanical fuel-pump. The camshaft ensures that the cylinder's cycle of events occurs at the correct time

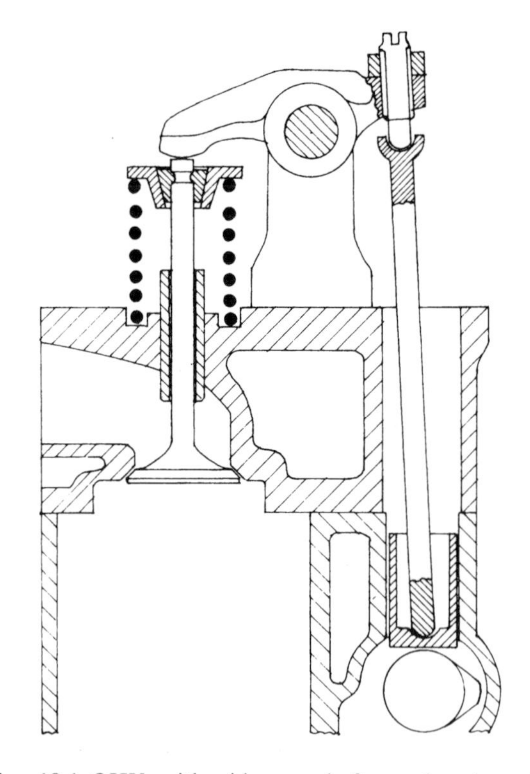

Fig. 18.1 OHV with side camshaft, push-rod, and rocker-arm-actuated valves

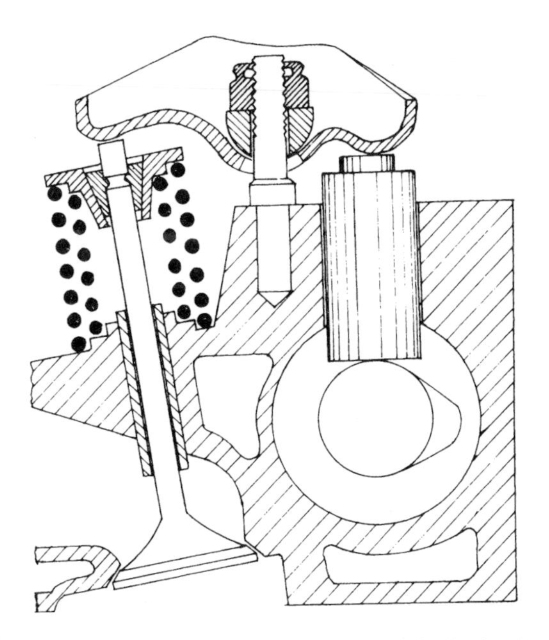

Fig. 18.2 OHV with high-mounted camshaft, bucket follower, and rocker-actuated valves

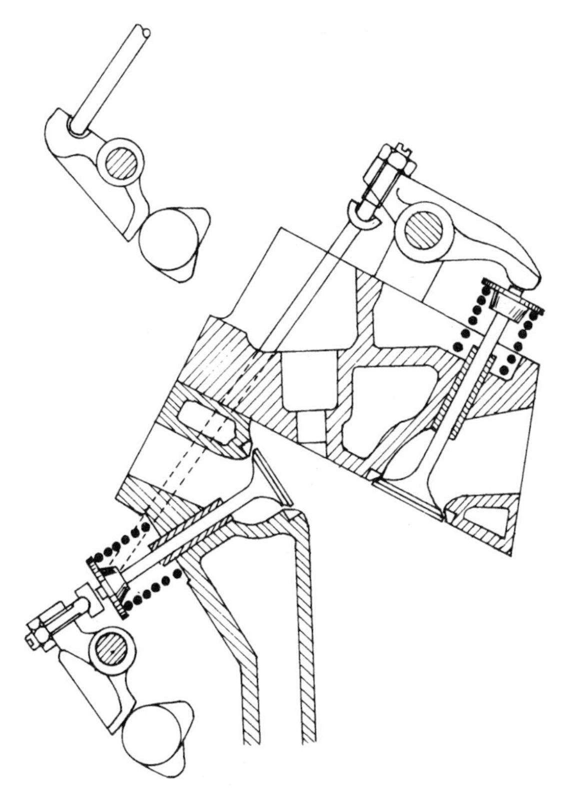

Fig. 18.3 Side-mounted camshaft with push-rod and rocker-arm-actuated OHV inlet and side-valve exhaust with direct-acting rocker-arms

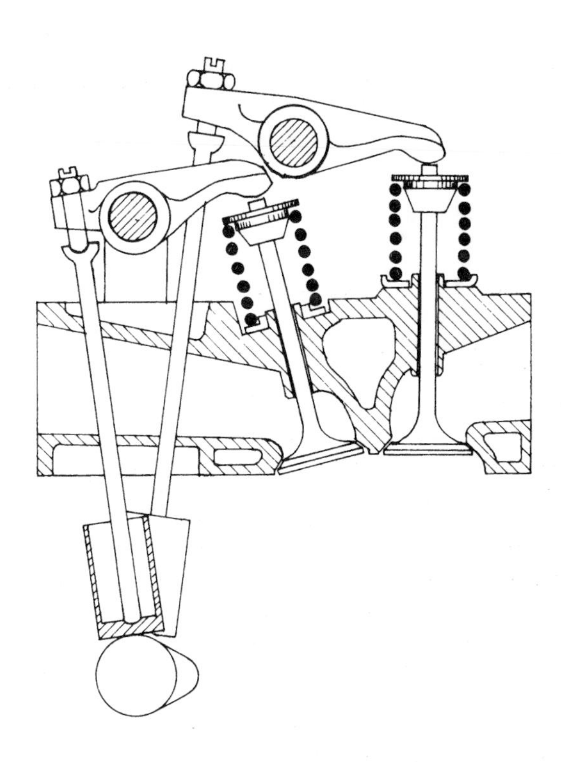

Fig. 18.4 OHV and side camshaft with push-rods, rocker-arms, and offset twin rocker-shafts

relative to the piston movement and simultaneously ensures that the individual cylinder's valve operations are in a sequence according to the firing-order.

The cam section is a base circle with an eccentric lobe. The radial distances between the centre of the base circle and the lobe profile are ground so that they relate precisely to the required valve lift and angular sequence of events.

Camshaft materials Chilled cast iron is now almost universally adopted as the standard material for camshafts. As might be expected, the necessary hardness is derived by placing metal chills (shaped pieces of cast iron) in the walls of the mould adjacent to where the cam lobes are to be, so that when the molten iron is poured it will freeze rapidly in these selected zones.

A typical cast-iron alloy used for camshafts might have a composition such as 3.3% carbon, 2% silicon, 0.65% manganese, 0.65% chromium, 0.25% molybdenum, with the balance iron.

The chilled zone is a complex composite consisting of approximately equal amounts of very hard and relatively brittle very tiny plates of iron carbide of hardness 900 to 1200 DPN and

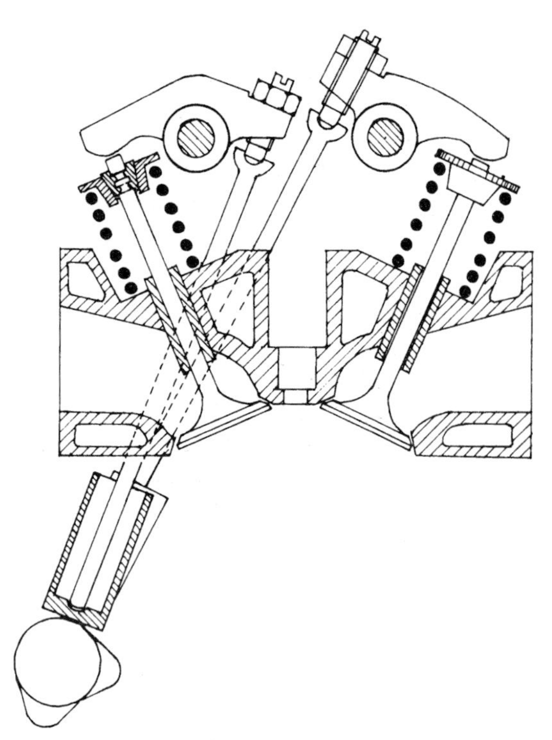

Fig. 18.5 Double-row valves and side camshaft with twin rocker-shafts and opposite-facing rocker-arms, push-rod actuated

relatively soft and ductile pearlite of hardness 200 to 300 DPN, giving a normal surface hardness of 450 to 600 DPN.

A section through a chilled-iron casting would show three distinct zones:

i) an outer layer which is the white-iron structure, containing a fine mixture of hard carbides and soft pearlite;
ii) a transition zone having a mottled appearance which contains both a carbide–pearlite mixture and graphitic grey iron together;
iii) a core region in which the normal soft and ductile graphitic grey-iron structure predominates.

18.2.2 Cam follower (tappet) (fig. 18.6)

The cam follower provides a method of converting the angular movement of the camshaft and its eccentric lobe into a reciprocating movement which is directly proportional to the amount the lobe profile deviates from the base circle.

There are two common forms of sliding followers, both being named after the shape they resemble: the mushroom follower and the bucket or barrel follower.

i) The mushroom follower has a relatively small-diameter solid cylindrical guide stem with a large disc- or mushroom-shaped head formed at one end to contact the cam profile. The other end is concave recessed to form a semi-ball-and-socket joint with the bottom of the convex push-rod end.
ii) The bucket or barrel follower resembles a hollow cylindrical sleeve with an enclosed bottom. Its top side has a concave recess to locate the push-rod semi-spherical end, and the underside is flat to convey the cam-lobe profile rise to the follower. Sometimes to both reduce weight and improve lubrication, helical slots are formed on the cylindrical walls.

With the mushroom-type follower, a shorter push-rod can be used which improves the rigidity of the valve-train system; but better side-thrust support is provided with the bucket follower.

To distribute wear over the bottom face of the follower, its centre line of stroke is slightly offset from the mid-width of the cam (fig. 18.6(c)). Thus, during operation, the follower will tend to revolve every time the lobe rubs against it.

Furthermore, the face of the follower is usually given a spherical radius of about 1 m, so that there is a very slight face curvature (fig. 18.6(a)). In fact the centre of the follower face will stand out about 0.08 mm and, to match this curvature, the cam lobes are ground with a taper something of the order of 3° to 4° (fig. 18.6(b)). This combination goes a long way towards reducing tappet and cam wear.

For some heavy-duty applications, roller followers may be justified. When these are used, the followers must be prevented from rotating, so a slot guide-screw is always provided (fig. 18.6(d)).

Cam-follower materials Similar to the camshaft, the follower is made from a chilled low-alloy cast iron consisting of iron, carbon, silicon, manganese, and chromium.

18.2.3 Push-rod (figs 18.6 and 18.7(b) and (c))

The push-rod is a strut whose function is to relay the to-and-fro cam-follower movement to one end of the pivoting rocker-arm. Both ends of the push-rod form part of a pair of semi-spherical ball-and-socket joints, to permit the rod to tilt slightly and revolve when the rocker-arm oscillates about its pivots. The bottom of the rod is convex and fits in a matching recess in the fol-

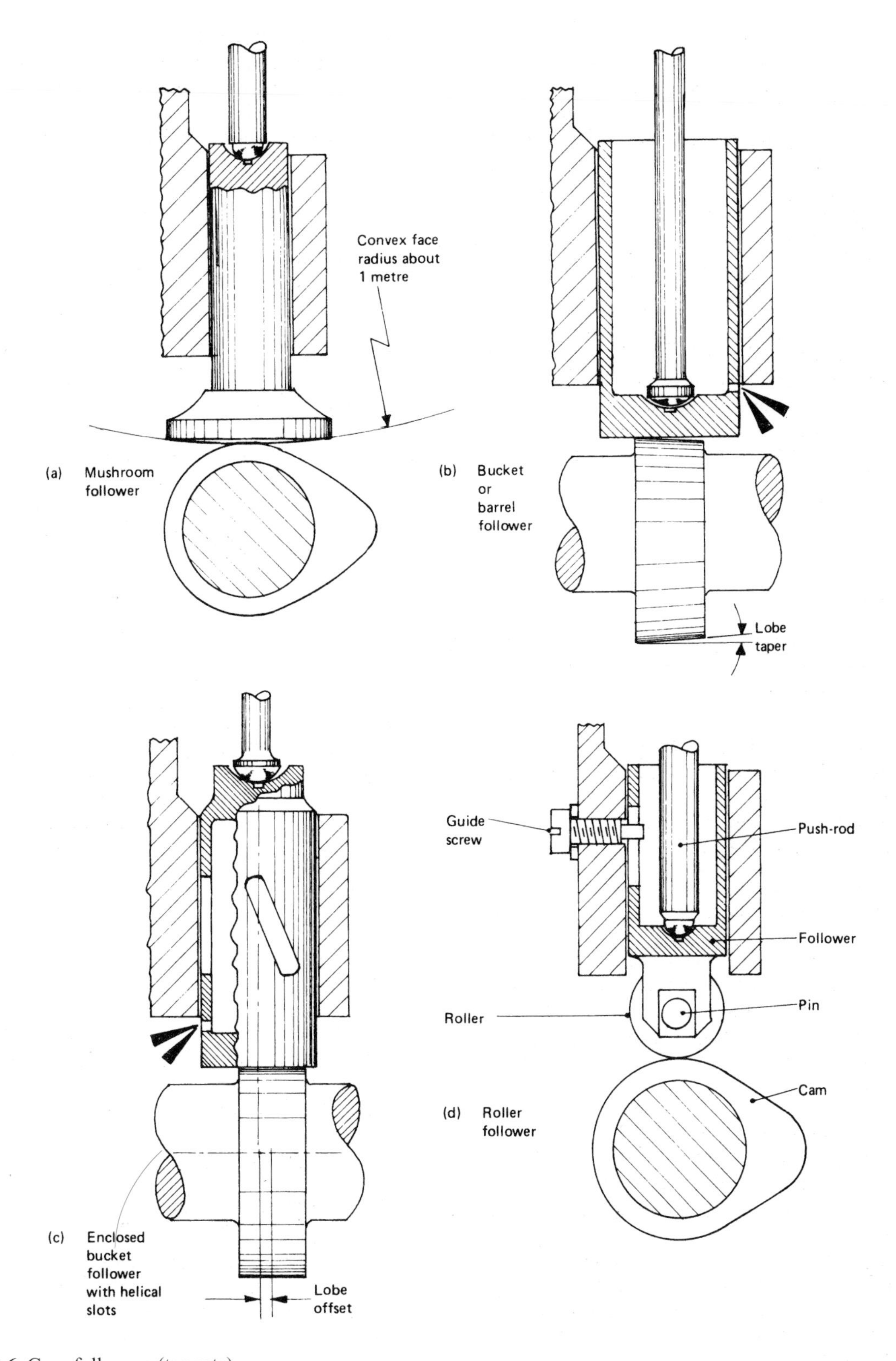

Fig. 18.6 Cam followers (tappets)

lower, while the top of the rod is expanded to support a concave-recess seat which will locate with the adjustable tappet screw on the end of the rocker-arm.

For medium-sized engines the push-rod may be solid (fig. 18.7(b)), but for large engines hollow rods which have hardened shaped end-pieces that are forced into the tubing are preferred (fig. 18.7(c)).

Push-rod materials Push-rods are usually made from carbon–manganese steel, a popular composition being 0.35% carbon, 0.2% silicon, 1.5% manganese, and the balance iron. The heat treatment given is to quench and harden the rods from a temperature of 840 to 870°C and then temper between 550 and 660°C – this should produce a hardness of 220 to 280 Brinell number. An alternative would be to use a steel with a higher carbon content such that it could be induction-hardened.

18.2.4 Rocker-shaft (fig. 18.8)

Rocker-shafts are machined from hollow steel tubing and are incorporated to provide a rigid pivot support for the rocker-arms.

These shafts are mounted and clamped on cast-iron or aluminium-alloy pedestals which are generally positioned between each pair of rocker-arms (fig. 18.8); thus a four-cylinder engine would have four pedestal support brackets.

Radial holes are drilled through the shaft to align with each rocker-arm for lubrication, and both ends of the shaft are plugged to prevent the oil from escaping. One of the support pedestals usually has a vertical drilling to supply the oil from the camshaft to the hollow rocker-shaft.

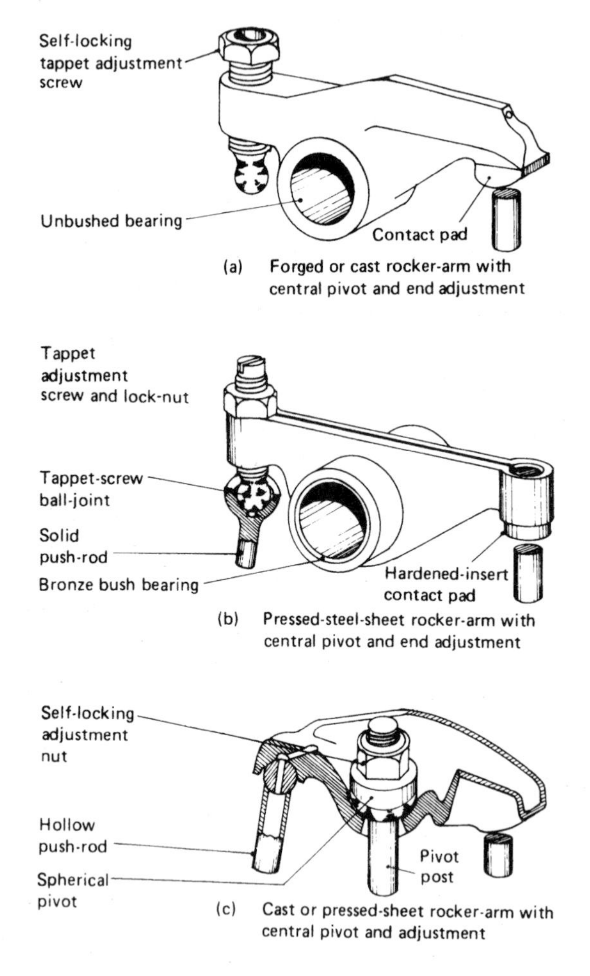

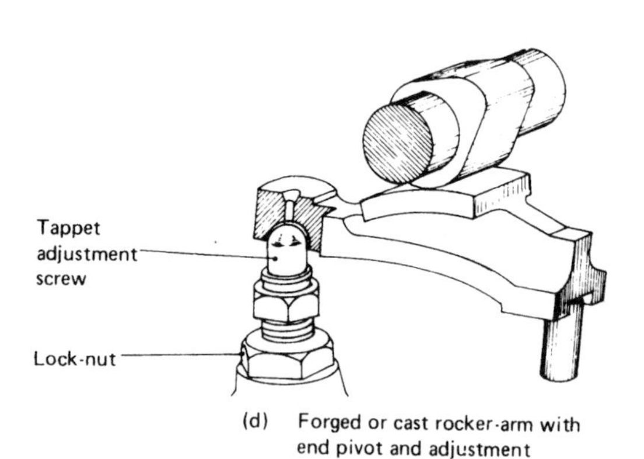

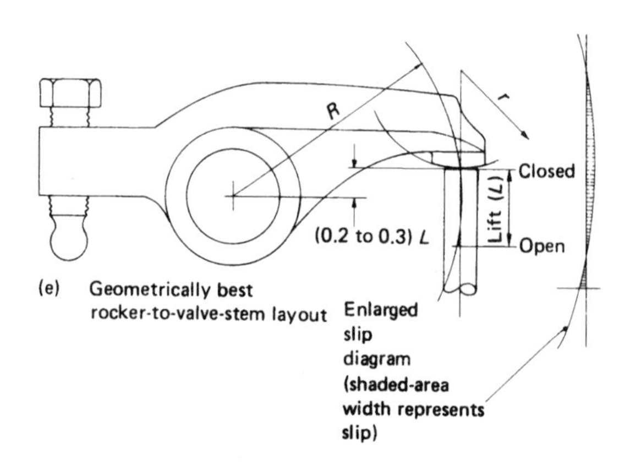

Fig. 18.7 Valve rocker-arms

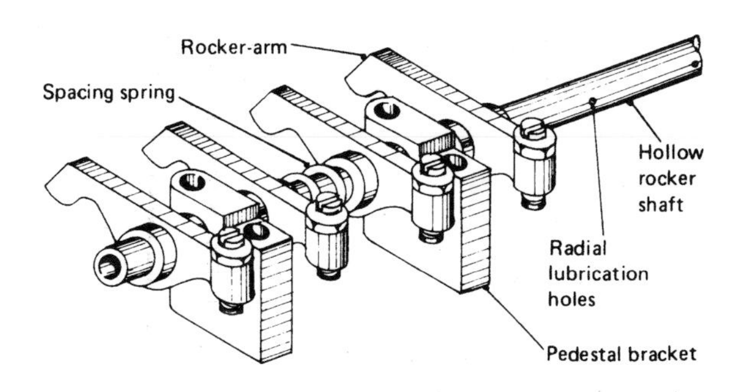

Fig. 18.8 Rocker-shaft assembly

This hole corresponds with an adjacent radial hole in the shaft. When reassembling the rockers and shaft, these two holes must align, otherwise the shaft will be starved of oil.

Rocker-shaft material These tubular shafts are made from carbon steel, a typical composition being 0.55% carbon, 0.2% silicon, 0.65% manganese, and the balance iron. After machining, the shaft is case-hardened to resist the oscillatory rubbing which takes place, either directly between the shaft and the rocker in petrol engines, or between the shaft and bronze bushes in heavy-duty diesel engines.

18.2.5 ***Rocker-arm*** (fig. 18.7(a) to (c))
By rocking or oscillating about its pivot, a rocker-arm relays the push-rod up-and-down movement to the stem of the poppet-valve. Thus this arm acts as a rocking beam.

The pivot – either a shaft or a spherical fulcrum post – is offset and positioned something like two-fifths of the way along from the push-rod end. Hence for a given cam lobe rise, the corresponding valve lift will be about 1.4 times greater, so that a lobe profile 40% smaller than would otherwise be necessary can be used. The actual rocker-arm pivot ratio and therefore the cam lobe size will vary somewhat, depending on design requirements.

Rocker-arm materials Rocker-arms may be manufactured from materials which can be cast, forged, or cold-pressed to shape. A selection of suitable materials is listed below, but other low-alloy steels may prove as durable or even better.

a) Malleable cast iron, induction-hardened at selected regions.
b) Forged medium-carbon steel, a typical composition being 0.55% carbon, 0.2% silicon, 0.65% manganese, and the balance iron. This can be hardened by quenching from 810 to 840°C and then tempered at a suitable temperature between 550 and 600°C.
c) Cold-pressed low-carbon steel, a typical composition being 0.2% carbon, 0.8% manganese, and the balance iron. Rockers using this method of construction have a hardened-steel contact pad attached at the valve-stem end.

18.3 Tappet clearance adjustment for push-rod mechanisms

Tappet clearance is necessary to allow for expansion and contraction of the valve and its operating mechanism.

There are two basic types of rocker-arm arrangement, which can be identified by the method of tappet adjustment:

i) push-rod-end adjustment,
ii) central-pivot adjustment.

18.3.1 ***Push-rod-end adjustment*** (fig. 18.7 (a) and (b))
The rockers are centrally pivoted on a rocker-shaft. At one end, the arm has a hardened face pad which is curved to smoothly contact the valve-stem tip. The other end of the arm has a threaded hole which secures an adjustable tappet screw with a lock-nut. The tip of this screw is formed into a hardened spherical ball which fits into a matching concave recess formed in the top of the push-rod.

Adjustment is achieved simply by slipping the correct size of feeler gauge between the valve-stem tip and the rocker pad, then slightly slackening the lock-nut and, using a screwdriver, turning the tappet screw to either increase or decrease the clearance. The correct clearance will be obtained when the feeler blade is just felt to grip as it is pulled across the valve-stem tip. Afterwards the lock-nut is tightened and the clearance rechecked.

Self-locking screws are sometimes adopted (fig. 18.7(a)). These have Dardelet-type threads – that is, the thread is formed with a shallow flat root instead of a conventional 'V' – but the tapped threads in the holes in the rocker-arms have a normal profile. It is thus the interference between the male and female threads which provides the self-lock.

18.3.2 Central-pivot adjustment (fig. 18.7(c))
This arrangement uses the hollow malleable-iron or pressed-steel type of rocker which has at one end a curved valve-tip contact face and at the other end a hardened spherical recess to receive directly the ball-ended push-rod. The rocker pivots on a case-hardened sintered-iron spherically faced fulcrum seat and is retained by a self-locking nut which is screwed on to a stud post pressed or screwed into the cylinder head. To prevent the rockers and push-rods swivelling about their central pivot, the push-rods are located and positioned by guide fork-plates mounted on the cylinder head.

The self-locking nut provides the adjustment for the tappet clearance. As before, the feeler blade is pushed between the valve-stem tip and the rocker face, and the central self-locking nut is turned either way until the correct clearance is observed by the degree of grip that the valve-stem tip and rocker impart to the feeler blade.

18.3.3 Camshaft positioning for push-rod tappet adjustment
Before the tappet clearance can be measured, the camshaft must be rotated until the follower is on the base of the cam and furthest from the cam-lobe nose.

There are three different methods of positioning the camshaft, as follows:

i) Rotate the crankshaft until the valve being adjusted is fully open; then turn further through one complete revolution to bring the follower on to the base of the cam. The camshaft moves at half crankshaft speed, so that to move the cam-lobe nose from its top position to the bottom, i.e. 180° camshaft movement, requires a corresponding crankshaft movement of twice as much, i.e. 360°.
ii) Consider a four-cylinder engine and remember that pistons 1 and 4 and 2 and 3 move in pairs. Crank the engine until the valves over cylinder 4 are on the 'rock' – that is, both valves are open. Cylinder 1 is then at the beginning of power, so adjust both of its valves. Again, crank the engine until the valves over cylinder 1 are on the 'rock' and adjust both valves of cylinder 4. Repeat this procedure for cylinders 2 and 3.

   Camshaft alignment procedure is as follows:

   Check valves 1 and 2 with valves 7 and 8 on the rock.
   Check valves 3 and 4 with valves 5 and 6 on the rock.
   Check valves 5 and 6 with valves 3 and 4 on the rock.
   Check valves 7 and 8 with valves 1 and 2 on the rock.
iii) Again consider a four-cylinder engine and observe that pistons 1 and 4 and 2 and 3 move together. When an inlet or exhaust valve of one cylinder is fully opened, the corresponding valve of the other paired cylinder is fully closed.

   Camshaft alignment procedure is as follows:

   Check valve 1 with valve 8 fully open.
   Check valve 2 with valve 7 fully open.
   Check valve 3 with valve 6 fully open.
   Check valve 4 with valve 5 fully open.
   Check valve 5 with valve 4 fully open.
   Check valve 6 with valve 3 fully open.
   Check valve 7 with valve 2 fully open.
   Check valve 8 with valve 1 fully open.

18.4 Overhead camshaft (OHC) (figs 13.13 and 18.9 to 18.15)
This type of valve operating mechanism is assembled from the following components:

a) a camshaft;
b) a cam follower – either
   i) a pivoted rocker-arm, or
   ii) a sliding inverted bucket;
c) a return-spring;
d) a poppet-valve.

18.4.1 Camshaft (figs 13.13 and 18.9 to 18.15)
These camshafts are arranged to be directly over the valve stems if inverted-bucket sliding followers are used (fig 18.9) or slightly to one side of the valve stems when pivoting rocker-arms are used (fig. 18.10).

Camshafts are supported by bearings between each pair of valves for high-performance or one bearing between every second pair for normal duty. The bearings may be mounted directly on to the cylinder head if rocker followers are incorporated, but a separate pedestal situated over the cylinder head is generally preferred when inverted-bucket followers are used. With rocker followers, plain white-metal or tin–aluminium bush bearings are mostly used, therefore the cam journal-bearing diameter has to be larger

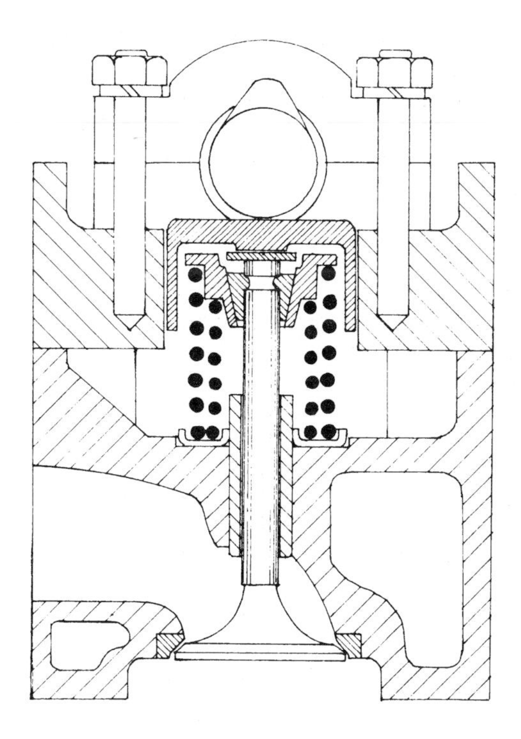

Fig. 18.9 OHC with direct-acting inverted-bucket followers

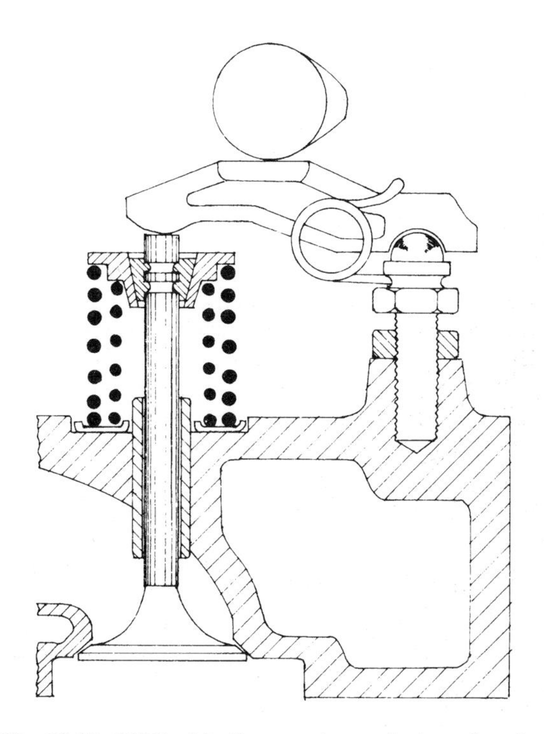

Fig. 18.10 OHC with direct-acting end-pivoted rocker-arms

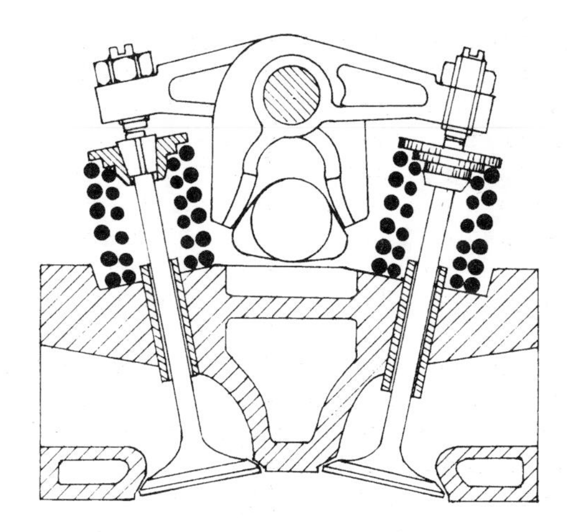

Fig. 18.11 Double-row-valves OHC with bell-cranked direct-acting rocker-arms

than the cam lobes, so that the camshaft can enter and pass through the bearing bores. When the bucket follower is used with a separate pedestal support, split half-shell white-metal or tin–aluminium bearings and caps can be used and the camshafts are made with smaller-diameter journals.

Single or twin camshafts may be adopted, depending on the type of combustion chamber and valve positioning which is to be used.

With a single overhead camshaft and bucket follower, the inlet and exhaust valves must be arranged in a single row. They may be either vertically positioned to conform to an inverted-bath-tub type of combustion chamber (fig. 18.9), or they can be inclined to suit a wedge-type chamber (fig. 18.2).

If a single camshaft is to actuate pivoted rocker followers, then the valves can be arranged either in one straight line (fig. 18.10) or in two rows with both inlet and exhaust valves inclined slightly to their cylinder's centre line to form a shallow hemispherical combustion chamber (figs 18.11 and 18.12).

Using twin overhead camshafts enables both the inlet and the exhaust valves to be considerably more inclined to the cylinder's axis so that a cross-flow cylinder head with a much deeper hemispherical combustion chamber can be designed (figs 18.14 and 18.15).

With all overhead-camshaft arrangements, considerably more lubricating oil is essential to flood the cam profiles than is necessary for mechanisms using push-rods, rocker-shafts, and rocker-arms.

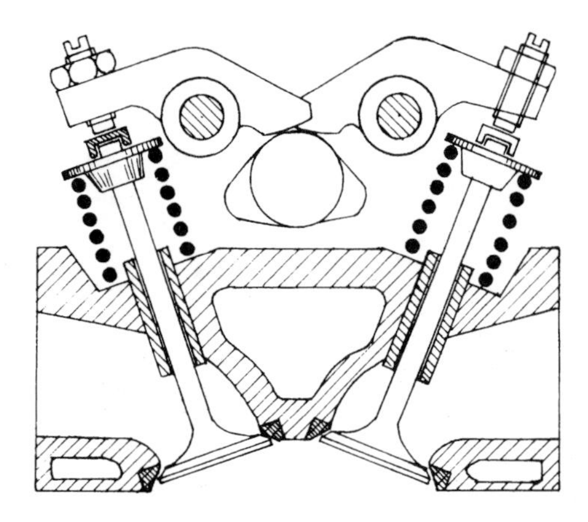

Fig. 18.12 Double-row-valves OHC with twin rocker shafts and direct-acting rocker-arms

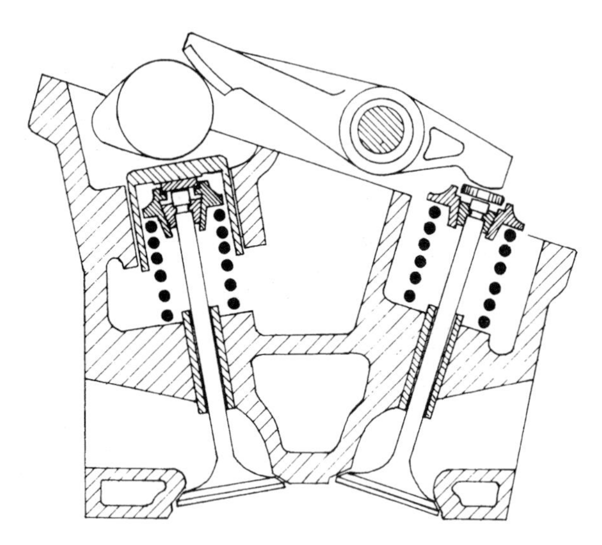

Fig. 18.13 Double-row-valves OHC with inverted-bucket inlet follower and rocker-actuated exhaust valves

18.4.2 Pivoted-rocker-arm followers

The cam-lobe rise and fall is relayed from the camshaft to the valve stem by way of a rocker-arm or beam. This may either be pivoted at one end to the cylinder head, with its free end bearing against the valve-stem tip, or it may be pivoted somewhere in between the camshaft and the valve stem.

End-pivot rocker (fig. 18.10) The end pivot consists of a spherical-head screw-post which fits into a concave recess underneath the rocker-arm. Adjustment is achieved by slackening the lock-nut and screwing the post in or out according to

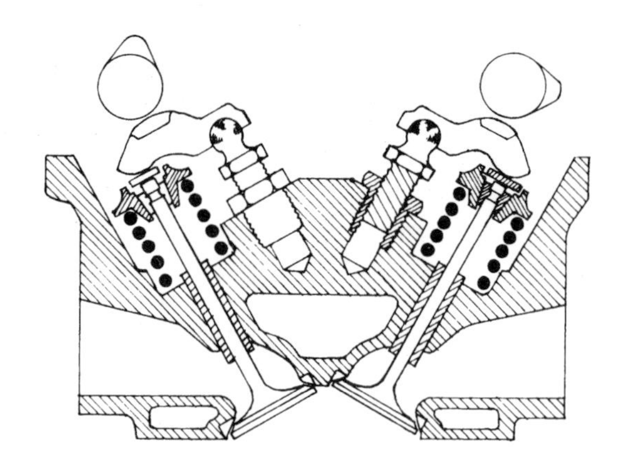

Fig. 18.14 Twin OHC with direct-acting end-pivoted rocker-arms

the required clearance between the valve-stem tip and the rocker contact face. To prevent the rocker-arm tilting from side to side while operating, a wide curved face pad is formed on the rocker-arm where it touches the cam profile (fig. 18.10).

Single rocker-shaft with central-pivot rocker (fig. 18.11) If the inlet and exhaust valves are to be inclined to the vertical and there is to be only one camshaft and rocker-shaft, then the rocker-arm can be bell-cranked as shown in fig. 18.11. At the camshaft end, the underside of the rocker-arm is given a convex shape to form the interface against the cam lobe profile, and a tappet screw and lock-nut are provided at the valve-stem end for adjustment.

Twin rocker-shafts with central-pivot rocker To obtain a greater degree of inclination for both the inlet and exhaust valves while still relying on only one camshaft, twin rocker-shafts can be incorporated so that the rocker-arms pivot on either side of the camshaft (fig. 18.12). This enables shorter and lighter rocker-arms to be installed and, at the same time, makes it easier to increase valve tilt so that a more favourable combustion-chamber configuration can be adopted. Again, simple tappet-screw adjustment is provided at the end of the valve stems.

Combined sliding bucket and central-pivot rocker (fig. 18.13) This unusual arrangement enables a single camshaft to operate a double row of inclined valves. Direct-acting inverted-bucket followers are used for the inlet valves, and centrally

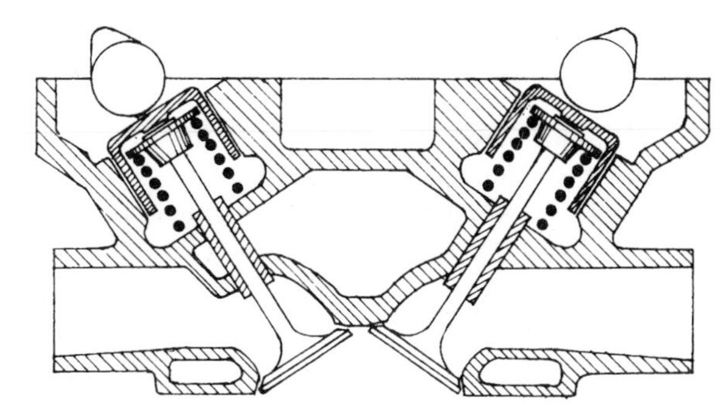

Fig. 18.15 Twin OHC with direct-acting inverted-bucket followers

pivoted rocker-arms actuate the exhaust valves (fig. 18.13). In this example the tappet clearances for both the bucket and the rocker-arm followers are adjusted by different thicknesses of shim pads. These are fitted between the valve stems and either the underside of the bucket followers for the inlet valves or the rocker curved-pad face for the exhaust valves.

18.4.3 Sliding inverted-bucket follower

To prevent the valve stem being subjected to side-thrust, a bucket-type follower is placed directly in between the cam profile and the valve stem. This construction provides the shortest and most direct transfer of movement from the cam to the valve stem, so there is no opportunity for any intermediate member to excessively compress or buckle under high inertia loads or to suffer from high-thermal-expansion problems.

Unfortunately, since there is no relay leverage multiplication from the cam to the valve, the valve lift will be exactly the same as the cam rise, so a larger cam lobe will be necessary than with the pivot-rocker types.

18.5 Tappet adjustment for direct-acting mechanisms

Adjustment of the free-play to allow for expansion and contraction movement between the camshaft and the valve-stem tip is obtained by either shim pads or a screw.

18.5.1 Shim-pad adjustment (fig. 18.16(a) and (b))

Adjustment is provided by a selection of shim-pad thicknesses in steps of 0.05 mm. These pads may be positioned as follows:

a) between the cam base circle and the adjacent recess ground into the top of the bucket follower (fig. 18.16(a));

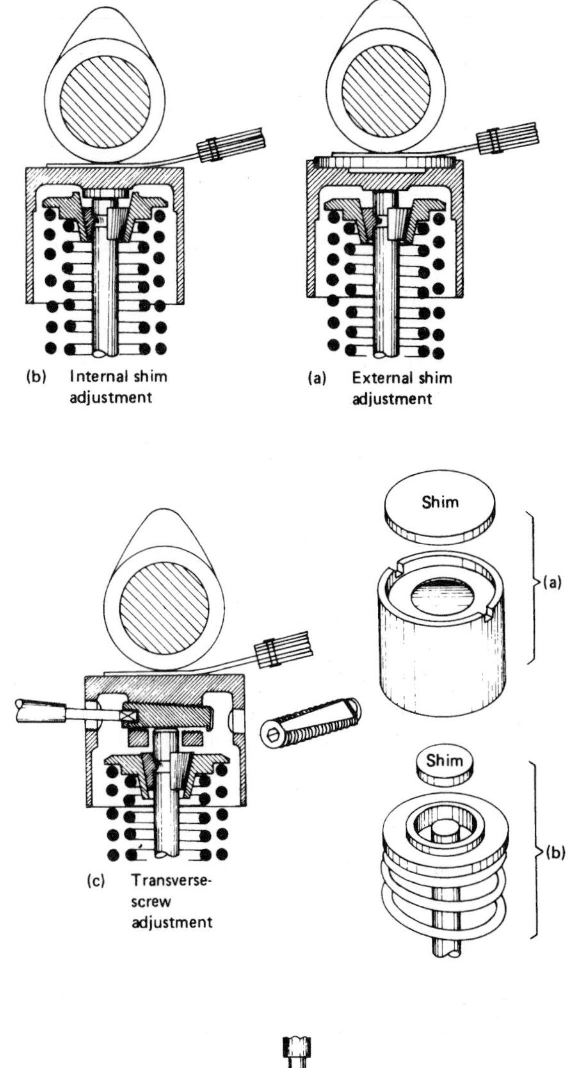

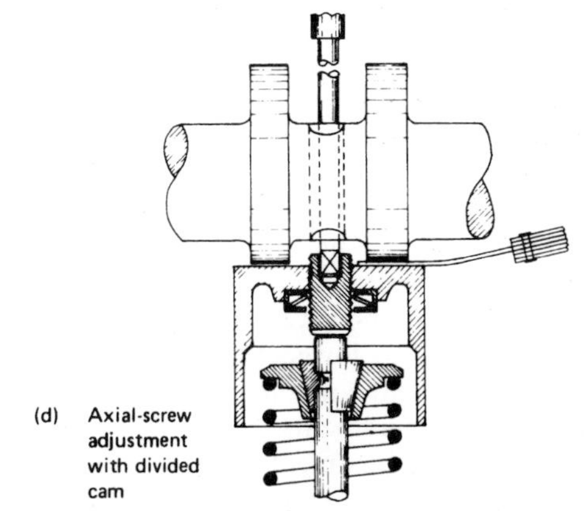

Fig. 18.16 Overhead-camshaft tappet adjustment

b) between the valve-stem tip and the flat ground on the underside of the bucket follower, the shim being maintained in the central position by the wall of the collet retainer hole (fig. 18.16(b)).

Shim-pad adjustment procedure Crank the engine until the cam lobe of the tappet being checked is pointing upwards. Check its clearance by pushing a suitable feeler gauge between the base circle of the cam and the top face of the follower and record the measured thickness on paper. Repeat this procedure for each individual valve assembly. Write down the amount each clearance gap deviates from the manufacturer's specified clearance.

If the shim pads are on the top of the follower (fig. 18.16(a)), compress the valve and spring and slip out the existing pad. Measure the pad's thickness and add or subtract to this the difference between the specified clearance and the measured gap. Replace the old shim with a selected shim pad with the corrected thickness. Continue this procedure for all the valves.

When the pads are between the stem and the follower (fig. 18.16(b)), then the camshaft has to be removed and all the tappet shim pads have to be measured and corrected in turn before the camshaft is replaced. After reassembling the camshaft, again check all clearances.

18.5.2 Tappet-screw adjustment

Adjustment may be by either transverse or axially arranged tappet screws situated in the inverted-bucket follower.

Transverse tappet screw (fig. 18.16(c)) With this adjustment device, the tip of the valve stem bears on a flat milled on a tappet screw in a transverse hole tilted at 5.5° to the horizontal. The angle between the flat and the axis of the screw is made the same as the screw inclination, so that the contact between the valve stem and the screw flat will be parallel. When an Allen key is inserted in a hexagonal socket in one side of the screw and is turned through one revolution, the axial travel of the taper raises or lowers the flat by 0.0075 mm so that adjustment of the valve clearance is altered in these increments.

Checking the adjustment is carried out in the usual way by rotating the engine until the cam-lobe nose is furthest away from the follower. The feeler blade is then pushed between the cam and the follower, and the tappet screw is adjusted until the recommended clearance is obtained.

Axial tappet screw (fig. 18.16(d)) An alternative arrangement has an axial tappet screw situated in the centre of the inverted-bucket follower. When setting the tappet clearance, a radial hole through the camshaft will align with the tappet screw so that an extended Allen key can be pushed into the hexagonal-shaped socket hole. To provide a permanent adjustment setting, a friction-grip device is used to lock the screw and to prevent it rotating under engine vibration. It should be observed that this method of adjustment is only possible because of the divided cam lobes, so this elaborate approach is not generally used.

18.6 Valve-actuating-gear requirements and considerations

The aim of good valve-train design has been to reduce the inertia of the valve-actuating mechanism as much as possible and to improve the rigidity of the whole valve system. Both of these improve the ability of the poppet-valve assembly to copy and follow closely the corresponding cam-profile lift movement over the whole engine speed range. If this is achieved, the timing of the valves will respond exactly to the designed periods of opening and closing, and the maximum time for exhausting and filling the cylinders will thus be available.

The traditional view that the inertia of a push-rod and rocker-arm is much greater than that of an overhead-camshaft valve gear is not justified; in fact there is very little difference between them, and in some cases the reverse is true.

With the overhead camshaft operating directly against an inverted-bucket follower, there is no multiplication of cam lift from the cam to the valve stem – in other words the cam lift must be equal to the valve lift; so, for a required valve lift, the cam base circle and lobe profile have to be proportionally larger than for the push-rod-and-rocker type of valve train. At the same time, because of the large cam profile, there are higher profile-to-follower velocities and relative rubbing velocities to contend with, so a substantial inverted-bucket follower is necessary to minimise side-thrust reaction caused by the cam action.

The main benefit of the direct-acting overhead camshaft is its greater rigidity, which enables the valve to be opened and closed quicker with much less vibration and undesirable oscillation. With the push-rod-and-rocker mechanism, both these components can be regarded as springs which are compressed every time the valve is opened by the cam lobe. The amount of this temporary elastic deformation will depend somewhat on the length of the push-rod, the degree of bending which will occur in the rocker-arm, and the magnitude of the continuously changing acceleration and deceleration of the camshaft.

Taking the rigidity of the overhead camshaft and the springiness of the push-rod-and-rocker layout into consideration, a valve operated by an overhead camshaft can follow a high-lift high-acceleration cam profile much more accurately and smoothly than can a push-rod-and-rocker system. The consequence of more precise valve timing is that optimum breathing and exhausting will be able to take place.

18.7 Advantages and disadvantages of the various valve mechanisms

18.7.1 Side camshaft with push-rod and rocker (fig. 18.1)

Advantages

a) A relatively uncomplicated short timing chain or simple gear train can be employed (see fig. 14.27(a) and (b)).
b) The rocker-arm leverage enables a degree of cam-profile lift multiplication to be imparted to the valve stem so that a smaller cam lobe can be used.
c) Adjustment and maintenance is usually simple and does not require the dismantling of any working engine components.

Disadvantages

a) When accelerating or operating at high engine speeds, the push-rod-and-rocker assembly is not able to relay to the valve the exact cam-profile lift, due partly to the elasticity of the system and to the resulting vibrations.
b) Larger tappet clearances are necessary to compensate for the very long valve-train mechanism expanding and contracting in service.
c) The increased number of contacting interface joints in the system tends to cause both more wear and more noise.

18.7.2 Overhead camshaft with sliding inverted bucket follower (fig. 18.9)

Advantages

a) This arrangement provides the most compact cam-to-valve mechanism which is also extremely rigid and direct, so that the valve movement responds very faithfully to the designed input cam-profile lift.
b) The valve stems are not subjected to side-thrust, so very little wear takes place provided that lubrication is adequate.
c) Tappet clearances are generally small and, once adjusted, tend not to alter over very long periods.

Disadvantages

a) A much more elaborate drive from the crankshaft to the camshaft is necessary.
b) Lubrication has to be positive and more accurately controlled and directed than for other systems.
c) Adjustment of tappets is considerably more difficult than for any other form of actuating mechanism used.

18.7.3 Overhead camshaft with pivoted rocker (fig. 18.10)

Advantages

a) The incorporation of the rocker-arm follower will provide a leverage ratio, so that a smaller cam profile can be adopted.
b) The inertia created by the pivot-arm follower is less than that of the sliding-bucket follower.
c) A single overhead camshaft can be made to actuate two separate rows of inlet and exhaust valves.
d) Adjustment of the tappet is generally straightforward.

Disadvantages

a) Motion from the cam to the valve is basically trying to bend the rocker-arm, so the stiffness of the system will not match that of the direct-acting bucket-follower arrangement.
b) Contact between the valve and rocker will impart a degree of side-thrust to the valve stem and guide.
c) There is an extra pivot joint in addition to the

other two contact interfaces to cause wear and noise.

d) These configurations tend to have very sensitive lubrication requirements.

18.8 The poppet-valve (figs 18.17 and 18.18)

The most highly loaded part of the valve is its head, which is subjected to uneven impact stresses across its diameter when the valve comes down on its seat and to thermal stresses due to the temperature variation over the head and between the head and the stem. Because of the shape of the valve, the maximum stress concentration appears on the surface of the valve's conical seat and in the region where the thickness of section changes from the head to the stem.

While operating under load, the temperature in the centre of the exhaust valve may be between 750 and 850°C (fig. 18.17) and between 450 and 550°C in the inlet valve. Since the mechanical and thermal stresses are cyclic, when failure occurs it is generally of a fatigue nature.

At these high working temperatures and under exposure to both dynamic inertia loads and the products of combustion, the surfaces of the valve may tend to oxidise and to corrode faster if the operating conditions are not suitably controlled. While hammering away in service, the valve seat will work-harden, and any carbon deposit between the seats will cause severe stress concentrations. At the same time, any gas leakage between the seats will cause local overheating which may initiate mechanical collapse, distortion or warping, and finally disintegration or burning of the valve's conical seat.

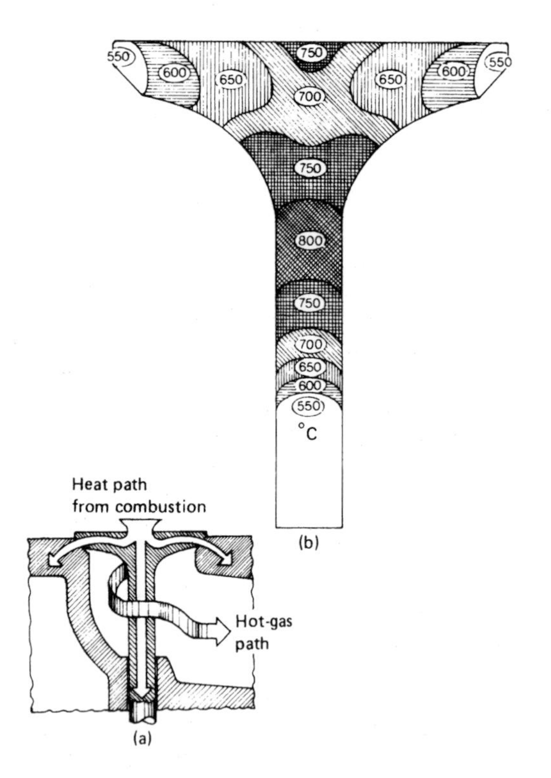

Fig. 18.17 Heat path and temperature distribution of an exhaust valve

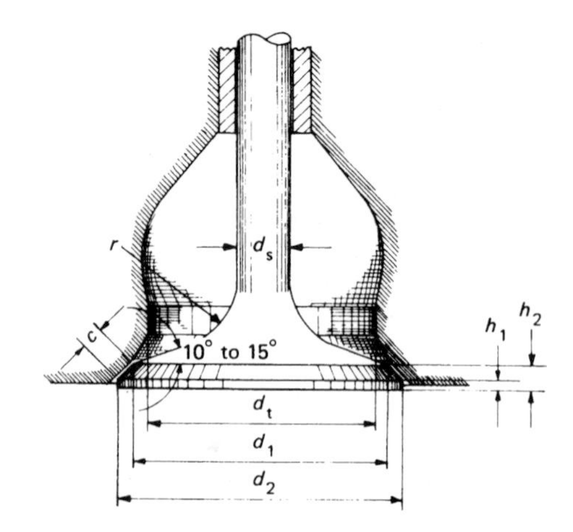

Fig. 18.18 Poppet-valve dimensions

The life expectancy of the valve can be improved if consideration is given to the following:

a) the selection of suitable valve and seat materials to withstand the high operating temperature, dynamic stresses, and the corrosive environment and to have good wear resistance under all operating conditions;

b) the shape of the valve should be in time-proven geometrical proportions so that the exhaust gases can flow with very little resistance between the valve and the seat and around the exposed portion of the stem without the stem absorbing excess heat, and yet have a section adequate to accommodate the stresses created due to the repeated impact loads.

Typical valve dimensions relative to the throat diameter (d_t) are as follows (fig. 18.18):

Maximum cone diameter d_2 $= 1.05d_t$ to $1.15d_t$

Minimum cone diameter $d1$ $= 0.95d_t$ to $1.0d_t$

Conical seat
width c $= 0.10d_t$ to $0.12d_t$
Parallel thickness
of head h_1 $= 0.025d_t$ to $0.045d_t$
Parallel and taper
thickness of head $h_2 = 0.10d_t$ to $0.14d_t$
Diameter of valve stem:
inlet $d_s = 0.18d_t$ to $0.24d_t$
exhaust $d_s = 0.22d_t$ to $0.28d_t$

The change from the disc head to the stem is initially given an angle of 10° to 15° to the horizontal followed with a small radius r.

The stem diameter of the exhaust valve should be 10 to 15% greater than that of the inlet valve, to improve the heat transfer to the cylinder head.

For the maximum amount of fresh charge or exhaust gas to flow between the valve head and its seat, the annular valve-opening area must equal the valve or port throat area. This occurs when the valve lift is equal to about one quarter of the valve head diameter. If the valve lift is less than the optimum, the engine's volumetric efficiency will be restricted; if it is much more, the inertia of the valve operating mechanism will be excessive, which will cause noise and rapid wear.

18.8.1 Valve and seat conical angles (fig. 18.19)

Valve-seat conical angles are normally 45° or 30° (fig. 18.19). As the angle is reduced for a given valve lift, the effective flow area around the valve can be increased, but the closing pressure on the seat for a given spring stiffness is reduced. Experience shows that good results can be achieved with either 45° or 30° conical valve-seat angle for the inlet valve but only a 45° seat angle for the exhaust valve, as this improves its heat dissipation.

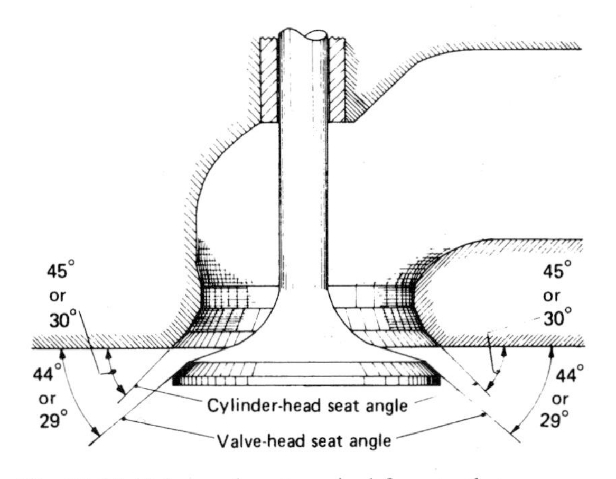

Fig. 18.19 Valve and seat conical face angles

The poppet-valve head and its seat in the cylinder head are ground to slightly different cone angles. The valve angle to the plane of the head is normally ½° to 1° less than the seat angle in the cylinder head. The difference in angles will result in a knife-edge contact around the outer edge of the valve and its seat which will rapidly bed the two together. This both protects the contact seating faces from combustion when closed and, at high engine speeds, allows the valve head to dish on impact so that it spreads its load over a larger surface area, thus minimising the contact stresses. Effective contact seating width should not be too wide, as this reduces the sealing pressure between contact faces and its ability to clear the products of combustion from the faces; nor should the contact faces be too narrow, as this reduces the heat path from the valve to its seat when it is closed.

18.9 Poppet-valve operating conditions

The operating conditions of the stem and of the head of a poppet-valve are quite different and therefore will be treated separately.

18.9.1 Valve stem and tip (fig. 18.7(e))

The valve stem operates in a guide sleeve which may reach a temperature of 400°C at the mouth of the exhaust port, progressively decreasing to something like 200°C at the valve-spring end. Under these conditions, the valve stem is subjected to a to-and-fro sliding motion and in certain cases to an additional rotary oscillating one. Therefore the stem and guide materials must be compatible so that they function effectively with very little wear for long periods under mostly boundary-lubrication conditions. By contrast, the valve-stem tip must be sufficiently hard not to be deformed by the tappet impact and in some designs must be able to resist any scuffing due to relative rubbing between the valve-stem tip and the rocker arm pad as it pivots – see fig. 18.7(e), where the shaded width represents the degree of relative movement between the rocker and the valve-stem tip.

18.9.2 Valve head and neck (18.17(a))

When the exhaust valve is closed, its crown is directly exposed to the combustion process. As the valve opens, the escaping and still burning gases flow around the rim of the head and underneath, encircling the valve's neck (fig. 18.17(a)). They then move through the port passages to the exhaust system.

If the valve is in the closed position, most of the heat is transferred to the valve seat and the coolant circulating in the cylinder head; but when the valve opens, the valve stem is the only path for the heat to travel. This explains why the hottest part of the valve will be its neck – that is, the transition between the head and stem – the next hottest the centre region of the crown down to its neck, and the coolest zone will be around the valve-head rim and the parallel region of the stem as it enters its guide (fig. 18.17(b)). Thus the valve-head temperatures may be as low as 500°C around the head rim, rising to 800°C at the neck under full-load conditions – and under abnormal conditions the maximum temperature may reach 900°C.

18.9.3 Valve-head loading

In operation, the conditions are such that the valve head and neck are subjected to longitudinal stresses due to the return-spring load and the valve assembly's inertia response which tends to momentarily deflect and cup the disc head at any instant of seating. At the same time, the large temperature gradient from the centre of the head to its rim and from the crown to the parallel portion of the stem creates large thermal hoop (circumferential) stresses in its head.

The strength of the material can cope with the combination of these two stresses provided that the contact seats are not distorted, or carbon particles are not trapped in between; but, if there is any unevenness around the circumferential seat interfaces, severe local stress concentrations may result which will quickly produce valve failure.

18.9.4 Valve requirements

The desired mechanical properties for an exhaust valve can be summarised as

a) sufficient strength and hardness to resist stretching and rubbing wear of the valve stem;
b) adequate hot strength and hardness to resist cupping of the head and rapid seat wear;
c) good fatigue resistance, to combat the repeated cyclic stress loading;
d) good creep resistance, to prevent the head from permanently deforming when operating at high temperatures and under alternating loads;
e) good corrosion and oxidation resistance at high operating temperatures and stresses and in an active corrosive atmosphere;
f) a moderate coefficient of thermal expansion, to limit the thermal stresses due to the large temperature gradient over the head;
g) good thermal conductivity from the valve's head, so that the heat from combustion can be dissipated readily.

18.9.5 Valve considerations

Valve temperatures decrease with increase in compression-ratio. This is because a greater part of the explosive energy is then converted to kinetic energy and less to heat. Arranging the inlet and exhaust valve ports alternately, i.e. I–E–I–E–I–E–I–E, instead of with like valves adjacent to each other, i.e. E–I–I–E–E–I–I–E, allows the valves to operate cooler, but this layout complicates the design of the induction and exhaust manifolds if they are both to emerge from the same side.

Sometimes the valve stem is tapered so that its diameter is slightly reduced at the valve-neck end, to compensate for the increased heat at the valve-port end.

The surface finish on the stem is important: if it is too rough, wear on the guide will increase; if it is too smooth, the lubrication film will be lost. The best surface finish has been found to be 0.5 µm. To improve the wear resistance, the stem surface may be given a very thin electrodeposited layer or 'flash' of chromium.

18.10 Poppet-valve materials

There are three basic alloy materials which generally meet the demands for exhaust valves, which may have to operate at temperatures in the region of 800 to 900°C:

i) the silicon–chromium steels,
ii) the austenitic chromium–nickel steels,
iii) the nimonic nickel-base alloys.

Early exhaust valves were made from what is known as a silicon–chromium steel, a popular composition being 0.8% carbon, 0.4% manganese, 1.3% nickel, 2% silicon, 20% chromium, and the balance iron. This steel could function well at a temperature of 650°C.

With the continuous development and improvement of engine performance, improved materials with better fatigue, creep, and corrosion resistance became necessary, so attention was focused on the austenitic steels. It was found that increasing the nickel content to 12% considerably improved the hot corrosion resistance and allowed the valve to operate at slightly higher temperatures. This then created the demand for the austenitic chromium–nickel steels such as the '21–12', which contains 0.25% carbon, 1.5% manganese, 1% silicon, 12% nickel, and 21% chromium.

A later improved austenitic chromium–nickel steel is the '21–4N'. This gives a higher hardness both when cold and hot and a greater degree of work-hardening. The composition of the 21–4N is 0.5% carbon, 0.25% silicon, 9% manganese, 21% chromium, 4% nickel, 0.4% nitrogen, and the balance iron. Basically, compared with the 21–12 steel, nickel is interchangeable with manganese but it needs to be in larger proportions, and nitrogen has been included. It is because of the large amounts of chromium and manganese that the steel will absorb the nitrogen, which, after solution treatment of the steel, has the effect of improving the wear resistance when subjected to high temperatures and loads.

With the trend towards leaner air–fuel mixtures under partial load, higher working temperatures will result and even more durable materials are necessary. The major contenders for high- and heavy-duty engines are the nimonic alloys, such as '80A' which has a composition of 0.05% carbon, 1% manganese, 0.6% silicon, 20% chromium, 2% cobalt, 2.5% titanium, 1.2% aluminium, 5% iron, and the remainder nickel. These nickel-base alloys have a higher hot strength and hardness and better fatigue and corrosion resistance than the austenitic steels, but they are of course more expensive. The expense can be partly overcome by having a two-piece valve with a nickel-base-alloy head being joined to a steel stem by friction welding.

The service life of both the austenitic steels and the nimonic alloys can now span to 150 000 km, compared with the traditional 22 000 km between relapping of the valve interfaces.

Inlet valves operate at temperatures of about 500°C and do not need such highly alloyed valve materials. A typical inlet-valve material is the silicon–chromium steel which has a composition of 0.4% carbon, 0.5% nickel, 0.5% manganese, 3.5% silicon, 8% chromium, and the balance iron.

For heavy-duty applications, exhaust-valve head-rim seats may be faced by a hard alloy such as stellite. This cobalt-base alloy can have a composition of 1.8% carbon, 9% tungsten, 29% chromium, and the remainder cobalt. Other surface treatments such as aluminising give beneficial protection to the valve head, but an aluminised facing must on no account be lapped, as this will destroy its protection properties.

18.11 Valve guides (fig. 18.20)

To support and guide the sliding action of the valve stem, so that the disc-shaped head with its conical seat is maintained in a central position to the valve seat while opening and closing, a guide hole in the cylinder head is necessary. Such holes, known as valve guides, may for normal duty take the form of directly drilled and reamed holes in cast-iron cylinder heads or, for aluminium-alloy cylinder heads or for heavy-duty cast-iron heads, separate sleeve or bush guides may be pressed into prepared holes.

The guide sleeve or bush may be made from good-quality pearlitic cast iron, to minimise stem-to-guide wear, or from bronze to improve the heat flow to the cylinder-head coolant passages for some high-performance applications. Sometimes shouldered bushes are provided to position the guide in the cylinder head and to provide additional support for the protruding extended sleeve, but the plain sleeve is adequate for most applications. When fitted into the cylinder head, the guide is usually made to protrude slightly above the spring-seat, to prevent too much oil draining down the stem.

The length of the guiding portion of the sleeve should be 8 to 10 times the stem diameter, to provide adequate support, and the external diameter of the bush sleeve should be something within 1.4 to 1.6 times the stem diameter. The

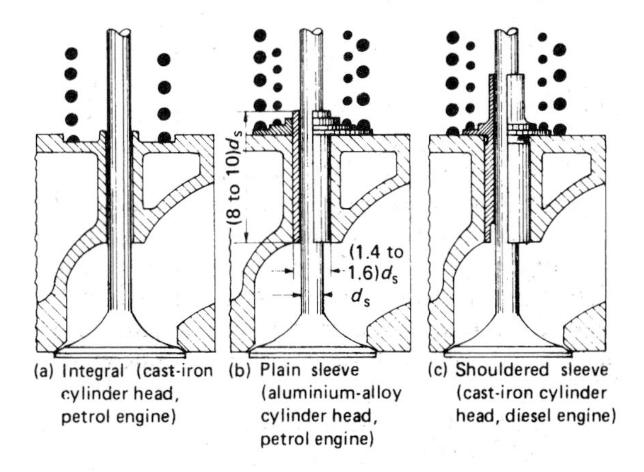

Fig. 18.20 Valve guides

clearance between the valve stem and the guiding surface of the sleeve varies from 0.02 to 0.05 mm in inlet valves and from 0.04 to 0.07 mm for the hotter-operating exhaust valves, these valves depending to some extent on the valve-stem and guide materials and their respective operating temperatures.

Worn integral guide holes can be reamed, and new valves with oversized stems are made for replacing the old ones. If existing valve guides are worn, they can be simply removed and new guides installed.

The valve-to-stem clearance must be sufficient to allow lubrication but not so large as to allow the stem to rock and so 'bell mouth' the valve guide. The rate of wear will increase as the clearance increases, this being due to the wear becoming localised. As wear takes place, the contact between the stem and the guide is not so effective, so there will be a rise in the valve's mean operating temperature. Lubrication may also deteriorate as a result of gum formation.

If valve guides are to be fitted, they should generally not protrude into the exhaust port, as this tends to raise the operating temperature of the exhaust valve's head.

18.12 Valve-seat insert rings (fig. 18.21)

For heavy-duty applications with cast-iron cylinder heads, or where the cylinder head is made from aluminium alloy, valve inserts must be incorporated both to withstand the high operating temperatures and the corrosive atmosphere around the valve seat porting and to provide a material with improved impact strength and hardness so that it is able to resist wear and therefore will normally require no servicing between major overhauls.

These inserts resemble a rectangular-section ring with a conical seat formed on one of the inside edges. The dimensions must be such that the insert is rigid enough to absorb the continuous hammering and to provide a sufficiently large mass to dissipate the heat from the poppet-valve head to the cylinder-head coolant system.

To meet these requirements, the radial thickness of the insert wall should be at least 0.10 to 0.14 times the throat diameter, and the external diameter of the insert should be within 1.2 to 1.3 times the throat diameter. A typical height for an insert is 0.15 to 0.25 times the throat diameter (fig. 18.21(a)).

Valve-seat inserts are force fits in recesses machined into the cylinder head. The interference in cast-iron heads is of the order of 0.0003 times

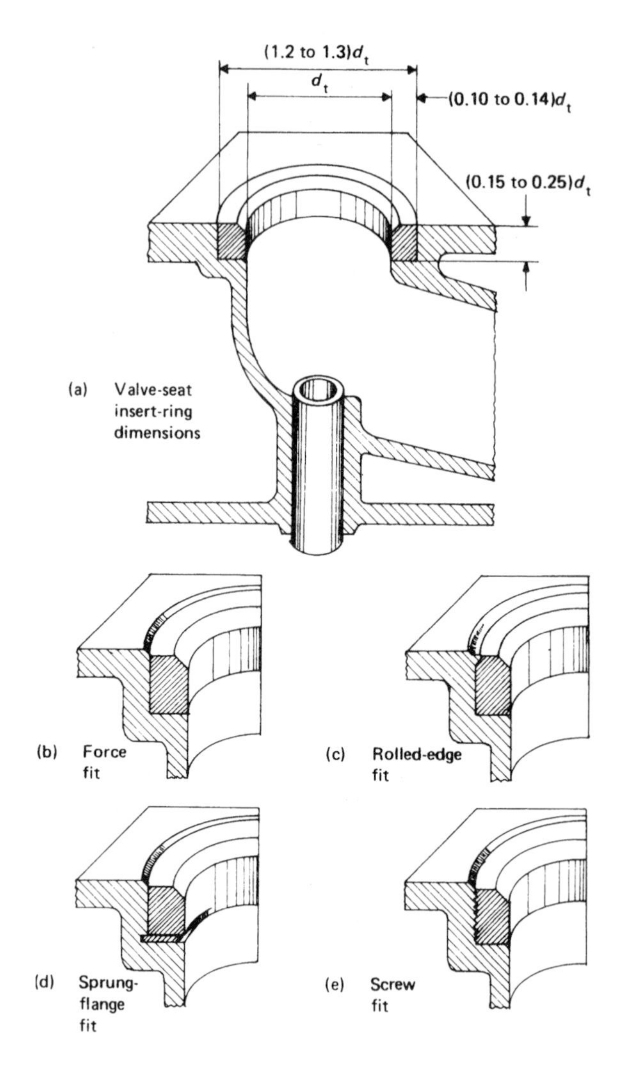

Fig. 18.21 Valve-seat insert rings

the exhaust-valve stem diameter. Typical interference fits are

Cast-iron cylinder head	0.019 mm per mm of outside diameter
Aluminium-alloy cylinder head	0.025 mm per mm of outside diameter

To facilitate the fitting of these rings, they are normally shrunk in liquid oxygen down to 180°C and then readily pressed into position. If this facility is not available, the cylinder head must be heated in boiling water for half an hour and the insert then be drawn into place as quickly as possible.

Besides the force- or shrunk-fit method of securing the insert rings in their recessed bores (fig. 18.21(b)), there are extra precautions which are sometimes taken:

Rolled-edge fit Here the insert ring is forced into its recessed bore, and the edge of the cylinder-head face is then rolled over to fill up the space provided by the chamfered outer top edge of the insert (fig. 18.21(c)).

Sprung-flange fit With this approach, circumferential slots are cut on opposite sides at the bottom of the ring and the lower portion of the insert ring is then distorted outwards. When forcing this type of insert into its bore, this lower part of the ring will spring out into a groove in the recess and so lock the insert permanently into position (fig. 18.21(d)).

Screw fit For certain aluminium-alloy cylinder heads, a screw-type joint is preferred, to provide a more positive grip in the cylinder head (fig. 18.21(e)). This compensates for the large differential expansion which usually exists between the insert and its recessed bore.

18.12.1 Seat-insert materials

These materials are selected for moderate to high duty with cast-iron or aluminium-alloy cylinder heads or for very-heavy-duty applications such as high-boost turbocharged diesel engines.

The following is a selected list of suitable insert materials:

a) For cast-iron cylinder heads for moderate to high duty, a low-alloy pearlitic cast iron would be suitable. A typical composition would be 3% carbon, 2% silicon, 0.4% phosphorus, 0.9% molybdenum, 1% chromium, and the balance iron. This would have a hardness of 270 to 300 Brinell number.
b) For an aluminium cylinder head for moderate to high duty, a high-nickel–copper austenitic grey cast iron which has both a high coefficient of expansion and good corrosion resistance would be used. A popular composition is 2.8% carbon, 2% silicon, 0.45% phosphorus, 1.8% chromium, 15% nickel, 7% copper, and the balance iron. This has a hardness of 160 to 240 Brinell number.
c) A cast-iron cylinder head for very-heavy-duty and high-temperture operation might be made from a high-chromium alloy cast iron containing 1.8% carbon, 1.8% silicon, 0.4% molybdenum, and 14% chromium. This would have a hardness of 270 to 320 Brinell number, or as high as 350 to 600 Brinell number if heat-treated.

18.13 Valve-spring retention (fig. 18.22)

The return-spring (or springs) holds the valve in the closed position until actuated by the camshaft. It acts against a spring-retaining plate attached to the end of the valve stem (fig. 18.22). This retaining plate has a central tapered hole which, when in position, fits over a shallow circular groove machined near the end of the stem. Wedged in between the spring-plate's conical wall and the grooved portion of the stem are two tapered half-collets which are locked to the stem groove by internal protruding circular notches or ribs. The continuous tension of the spring against the spring-plate maintains the collets' grip on both the stem and the spring-plate.

The underside of the spring-plate is normally stepped to locate either one or two springs. Similarly a second plate is sometimes placed over the valve stem and guide against the top of the cylinder head to hold the bottom end of the spring(s) in position.

18.14 Valve-rotators and spring retention

Valve-rotators may have a non-positive action which enables the valve to revolve at random when it is opening and closing or a positive action which directly rotates the valve as it operates. In either case the aim is to improve valve seating by clearing the seat faces of carbon particles which might otherwise become attached to them.

18.14.1 Non-positive valve-rotators

There are two common methods used: one approach uses loose-fitting collets, while the other adopts a thimble which fits over the valve-stem tip.

Split-collet valve-rotator (fig. 18.23(a)) This method uses specially designed collets which permit the valves to rotate under certain conditions. There are three annular grooves of semicircular section in the stem of the valve, and three semicircular ribs on the inner surface of each collet. Unlike conventional collets, these press against each other so that there is a small radial clearance between the valve and the collets. The vibration of

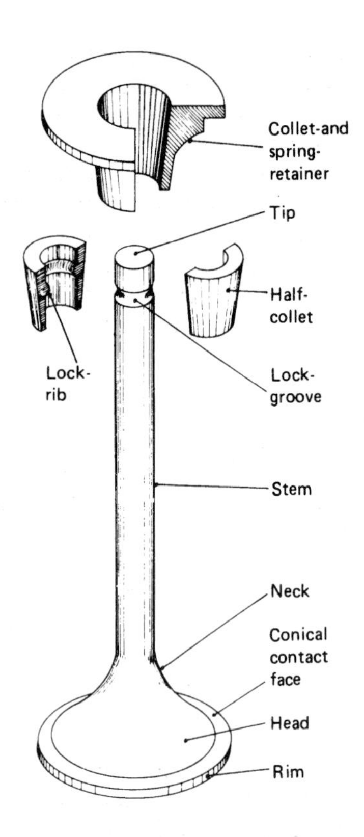

Fig. 18.22 Poppet-valve spring retention and terminology

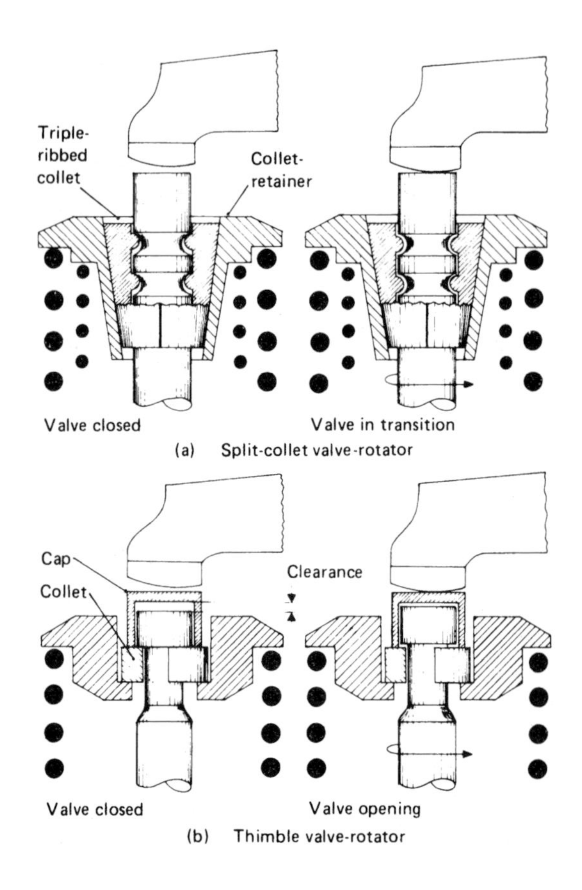

Fig. 18.23 Non-positive valve rotators

the valve gear is then sufficient to rotate the valve at between 15 and 25 rev/min at engine speeds greater than 1500 rev/min.

The collets are made from low-carbon-steel strip which is extruded to the final section. A length is cropped and pressed to the final curvature and then case-hardened.

Thimble valve-rotator (fig. 18.23(b)) The conventional valve is secured by a valve spring-retainer, and the spring always maintains pressure between the valve-stem end-groove and the tapered collets which secure the retainer. In consequence, the valve itself does not usually turn round relative to its seating. In order to extend valve and seat life between overhaul periods, it has been found that an arrangement which permits the valve to gradually reposition itself relative to its seat is advantageous. This is because the valve and seat will then be continuously making contact at different interfaces, so that any uneven wear or uneven heat distribution will be shared out.

With the thimble rotator arrangement, a steel cap fits over the end of the valve stem and rests on two semicircular collets which fit in the valve-stem groove below. The valve spring presses the retainer against these collets and so keeps the valve shut. When, however, the valve is to be opened, the rocker-arm presses down the cap which in turn bears against the two collets, and then moves the valve spring and retainer downwards. Since the spring pressure is now taken by the cap, the valve is freed from spring pressure – it is still moved downwards, because the closed end of the cap then abuts the valve-stem end, but it is free to turn. There is no automatic provision for turning the valve relative to its seating, but it is found in practice that the valve does tend to rotate in operation.

18.14.2 Positive valve-rotation ('Rotocap') (fig. 18.24)

This device has a ball-retaining plate with six ramped grooves (only four are shown) for balls

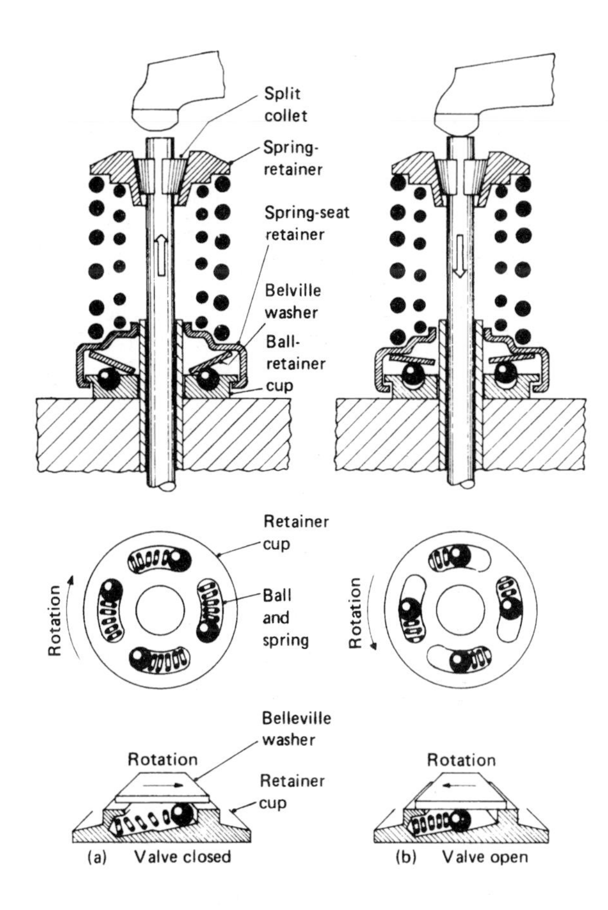

Fig. 18.24 Positive valve rotator ('Rotocap')

to roll along. Each of these balls is pushed to one side by a small spring. A Belleville-type dished spring washer fits over these balls to form an upper race. This is supported on its outer edge by a spring-seat retainer which holds the whole assembly together and also provides a seat for the helical-coil valve springs (fig. 18.24).

When the valve is in the closed position, the dished washer is supended between the spring-seat retainer and the ball-retainer. Under these conditions the balls are free to move to the top of the ramp and so abut against the end of the groove.

As the valve starts to open, the dished spring washer deflects as the compression load on the valve spring increases. The outer edge of the dished washer will bear against the spring-seat retainer as before, but the inner part of the washer will now bear against the six balls. As this load is taken by the balls, it pushes them down their ramps which are so shaped that, as contact with the washer is maintained, the spring-seat retainer is rotated and hence the poppet-valve is rotated by the same amount.

When the valve closes again, the washer returns to its original position between the spring-seat retainer and the ball-retainer. This releases the load on the balls so that the small bias springs will now be able to push the balls up their ramps and, in doing so, will turn the spring-seat retainer and the valve assembly back to its starting position.

The cycle of the balls rolling up and down the ramps will thus cause the valve head to partly revolve around the valve seat and so sweep clear any particles which may have attached themselves to either of the contact faces.

18.15 Valve compression return-springs (fig. 18.25)
The function of the valve return-spring or springs is to ensure that the valve lift or fall follows precisely the corresponding cam-profile motion imparted to the follower under both acceleration and deceleration operating conditions.

These springs are normally of the helical-coil type (fig. 18.25(a)), working under load in compression. When the free length of the spring is deflected – that is, compressed during loading – each part of the spring wire is twisted, so the material is therefore subjected to torsional stress.

The stiffness of a helical spring – that is, its ability to resist deflection when a load is applied – is proportional to the fourth power of the spring-wire diameter and is inversely proportional both to the cube of the mean spring coil diameter and to the number of active coils, all other conditions being equal, i.e.

$$x \propto \frac{d^4}{D^3 N}$$

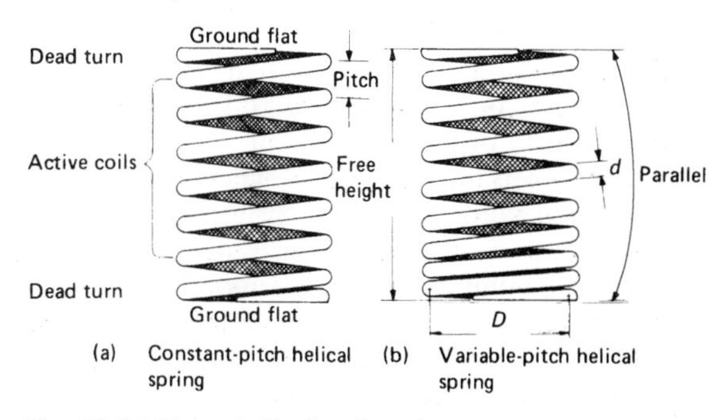

Fig. 18.25 Valve helical-coil springs

where x = spring deflection
d = spring-wire diameter
D = spring coil diameter
and N = number of active coils

This implies that doubling the wire diameter will increase the spring strength sixteenfold, halving the coil diameter will increase its strength eightfold, and halving the number of active coils will increase its strength only twofold.

The active coils are those coils which deflect when a coil spring is loaded. End turns or part turns on a compression spring which do not take part in deflection are referred to as dead coils. The dead coils at both ends of the spring are ground flat and perpendicular to the spring axis, with the two ends arranged diametrically opposed. This prevents the spring length bowing under compression – a combination of the essential torsional stresses with any unwanted bending stresses will considerably reduce the fatigue life of the spring.

The spacing distance between adjacent active coils is known as the pitch of the spring.

In the valve-closed position, the valve spring is loaded sufficiently to maintain the valve on its seat and to overcome the valve-train inertia forces so that the valve movement will closely follow the cam-profile motion at all speeds. If the spring is not to be overloaded, the spring load should not vary more than fourfold from the valve-closed to the fully open position. For the same reason, there should not be less than 4.5 active coils – however, too many coils will reduce the stiffness of the spring, so that spring surge is more likely to occur.

The valve and spring should be able to open slightly beyond the normal fully open position, to prevent the coils crashing together if overspeeding causes valve bounce.

18.15.1 Spring surge

Spring surge may best be explained by understanding what happens to a spring when it is periodically compressed and released rapidly and resonance occurs.

Resonance is the condition when the natural frequency of vibration of the spring (or multiples of the frequency – that is, other frequencies which are a whole number times the fundamental (natural) frequency) synchronises with disturbing vibrations generated by the action of the cam against its follower.

If resonance occurs, the coils at the stationary end of the spring completely or almost completely close together during the initial stages of valve lift, due to their inertia, and successive coils away from this end close up to a lesser extent. During the final stages of valve opening, the coils furthest from the cam follower close up most. A collapse of spring resistance so that some of the coils temporarily lose their pitch and move closer together sweeps or surges from one end of the spring to the other. When this spring surging takes place, the natural closing action of the coil spring will be out of control and the valve movement will therefore no longer correspond to the designed cam-profile rise, dwell, and fall.

18.15.2 Variable-pitch springs

One method of reducing spring surge is to have a variable-pitch spring in which the pitch between adjacent coils from the valve-stem tip to the stationary valve guide and seat is progressively made smaller. As the valve opens, the coils will then be progressively compressed solid, starting from the close-pitch cylinder-head end; hence the number of active coils is being reduced. Conversely, as the valve closes, the number of active coils increases. This variation of the number of active coils during opening and closing of the valve produces a variable spring rate and a constantly changing natural frequency which helps to discourage both spring resonance and surge.

18.15.3 Double springs

The use of double nests of springs enables high spring loads to be applied to the valve for a given valve-spring space. However, since to exert the same total force the stiffness of each spring in the nest will be lower than that of a single spring, the natural frequencies of the nested springs will also be lower (since a spring's natural frequency is proportional to the square root of its stiffness) and so surge will be encouraged.

One possible merit of using pairs of springs is that, under favourable conditions, resonance in one spring may be partially damped out by the non-resonant action of the other. There is, of course, also the safety aspect – if one spring breaks, the other will continue to operate, so the valve will not drop through into the cylinder and do immense damage.

18.15.4 Pre-stressing

Springs are usually pre-stressed by scragging – a process where the spring is compressed until the stress in the outside fibres of the wire is greater than the yield stress of the material, so that plastic deformation of the outside steel fibres occurs. The result of this is to induce residual stresses which raise the yield strength of the steel.

18.15.5 Shot-peening

All springs are given a short-peening treatment which has the greatest influence upon the performance of the spring and considerably improves the fatigue strength of the steel. The spring is bombarded all over with round particles of hardened steel (shot) at high velocity. This cold-works the surface of the wire coils and so produces compressive stresses in the outside fibres of the material. These residual compressive stresses help to prevent tensile stresses being established on the surface of the wire – if there were any imperfection in the outside fibres of the wire, such tensile stresses would initiate crack propagation leading to fatigue failure.

18.15.6 Valve-spring materials

Valve springs are made from either a plain high-carbon steel or a low-alloy chromium–vanadium steel.

The high-carbon steel contains 0.4 to 0.8% carbon, 0.3% silicon, and 1.0% manganese. The steel wire is ground all over to a good surface finish, followed by cold drawing to impart the necessary high-tensile properties. The wire is then coiled into the spring shape and is finally given a blueing – that is, a stress-relieving treatment.

The chromium–vanadium steel contains 0.4 to 0.5% carbon, 0.2% silicon, 0.6% manganese, 1.0 to 1.5% chromium, and a minimum of 0.15% vanadium. The steel wire is supplied in the softened state. It is surface-ground and then coiled to form. Finally it is heat-treated by hardening and tempering.

Carbon steels are easier to manufacture and they are not so notch-sensitive (susceptible to fracture due to surface imperfections such as roughness) as low-alloy steels, but the alloy steels do have good fatigue properties and do not suffer from relaxation.

Relaxation occurs when the spring – which could be operating at average temperatures of about 150°C or even more – under cyclic conditions of stress, plastically deflects so that it does not recover its original free length when an external load is released. Therefore over a period of time, the force exerted by the spring is reduced.

18.16 Automatic tappet clearance adjustment via a hydraulic element

The hydraulic tappet clearance adjuster performs the following functions:

1) changes in valve train actuating mechanism slackness due to expansion and contraction are automatically taken up;
2) relative large slack clearances due to worn components are tolerated and compensated for by the automatic adjustment;
3) initial tappet clearance adjustment is not generally necessary after distributing the cylinder head;
4) the automatic take-up of slack clearance ensures that the designed valve timing opening and closing periods are closely maintained.

18.16.1 Overhead camshaft end pivot rocker-arm with hydraulic valve clearance compensation pivot post element (fig. 18.26(a) and (b))

This type of zero tappet clearance mechanism utilises the rocker-arm end pivot as the hydraulic automatic compensating element. The hydraulic pivot end-post has a barrel-shaped body which slips into a cylindrical hole formed in the cylinder head; likewise a sliding fit plunger is positioned inside the pivot-post. Both the pivot-post and the plunger have grooves machined at mid-length on their outer faces with pairs of radial holes to permit the passage of oil from the oil gallery into the plunger's supply chamber. A ball-check valve and retainer is held against the underside of the plunger by a slack spring, the valve's function being to separate the supply chamber from the pressure chamber when the cam lobe is pushing down on the rocker-arm and opening the adjoining passage when the rocker-arm is in contact with the cam's base circle and the valve spring load has been removed. At all times the slack spring tries to extend and push the plunger outwards against the underside of the rocker-arm, thereby taking up free clearance slack that may exist.

Upstroke (valve opening) (fig. 18.26(b)) When the cam rotates through the base circle/rocker-arm contact period the poppet valve moves to its fully closed position, thereby relieving the load from the rocker-arm. The slack spring is now able to expand by pushing up the plunger until all the slack between the cam, rocker-arm and poppet valve stem has been taken up. Immediately the

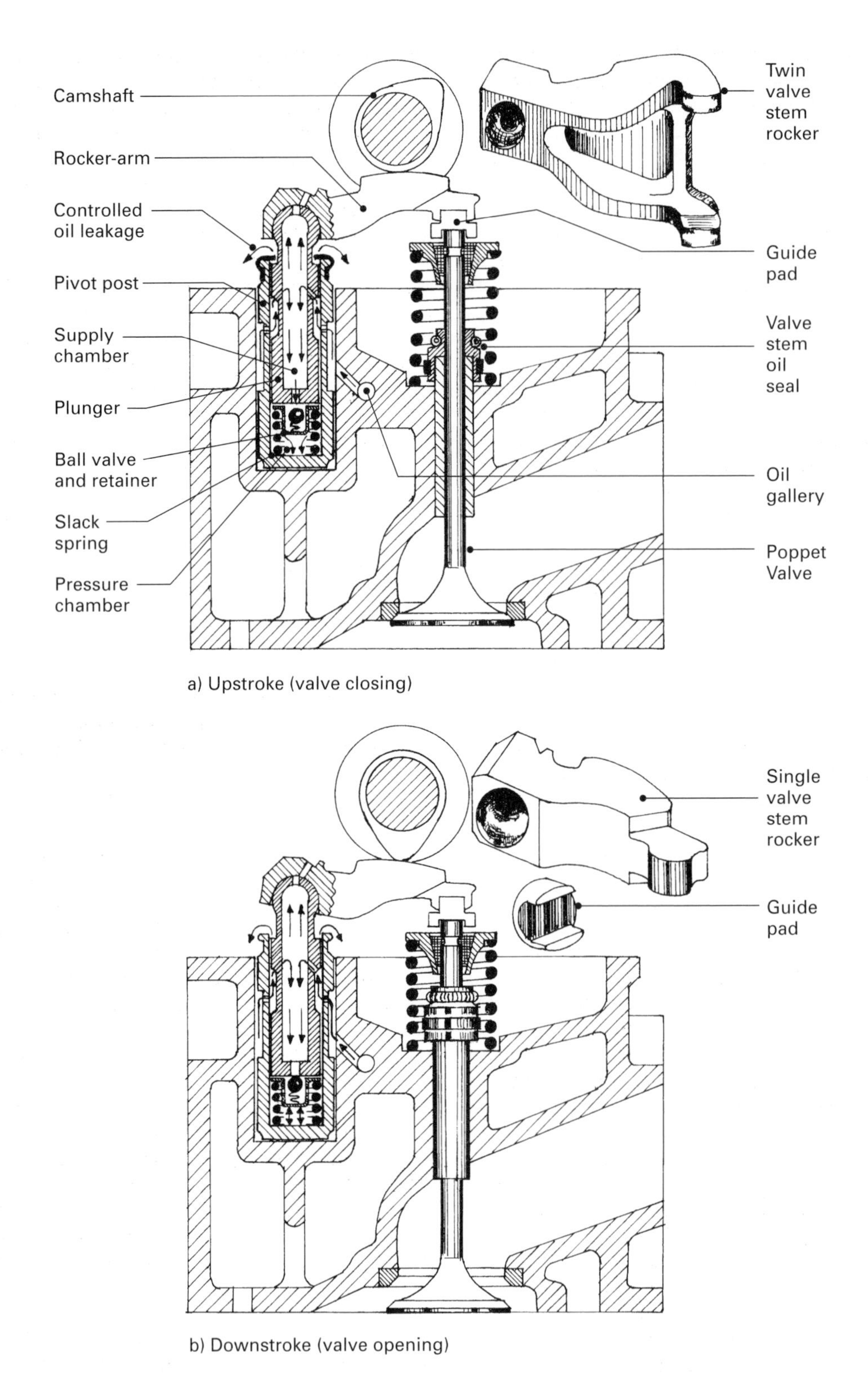

Fig. 18.26 Overhead camshaft end pivoted rocker follower with hydraulic valve clearance compensation pivot post element

pressure in the now enlarged pressure chamber drops to a stage where the supply chamber pressure exceeds that in the pressure chamber. At this point the ball-check valve snaps open. Oil will now transfer either to the pressure chamber to compensate for the contraction of a cold poppet valve and the leakage between the pivot post and plunger, or be expelled from the pressure chamber due to the expansion of a hot poppet valve under heavy load operating conditions.

Downstroke (valve opening) (fig. 18.27(b)) As the cam continues to rotate, the flank of the cam contacts the rocker-arm and commences to push it and the rocker-arm down. At this point the plunger is forced to move further into the pivot-post bore. This sudden reduction in the pressure chamber space now causes a rapid pressure rise in the chamber which snaps closed the ball-check valve.The trapped pressure chamber oil now becomes an incompressible rigid column permitting the cam to open the poppet valve without any loss of movement but at the same time ensures that the valve remains fully closed with the minimum of free movement during the base circle to rocker-arm contact period.

18.16.2 Overhead camshaft central pivoted rocker-arm with hydraulic valve clearance compensation tappet follower element (fig. 18.27(a) and (b)) The automatic slack clearance is taken up similarly to the hydraulic pivot-post (fig. 18.26(a) and (b)) valve train; however in this design the automatic hydraulic element is built into a bucket-type tappet follower, the task of the follower being to convert the cam-lobe's profile eccentricity into a vertical lift and fall motion which is then conveyed to the poppet valve via the centrally pivoted rocker-arm.

Downstroke (valve closing) (fig. 18.27(a)) As the cam rotates the follower to cam contact changes from the closing flank to base circle. It therefore permits the poppet valve to close. At this point the compressive load is removed from the valve-operating mechanism, that is, the rocker-arm and the hydraulic tappet-follower. The slack spring is now able to expand, and to enlarge the pressure chamber volume. Instantly the oil pressure in the pressure chamber falls below that existing in the supply chamber. This pressure reduction causes the ball-check valve to snap open and so permits oil from the supply chamber to feed into the pressure chamber to replenish oil loss due to leakage between the plunger and tappet follower body and allows for expansion or contraction slack clearance changes. Finally it compensates for wear between rubbing parts of the valve train linkage. The actual length of the tappet follower is thus adjusted every time the valve closes so that zero slack clearance is maintained under all operating conditions.

Upstroke (valve opening) (fig. 18.27(b)) As the cam rotates it brings the opening flank into contact with the tappet follower and so pushes the follower body upwards. Immediately, the pressure chamber volume decreases, causing the chamber's pressure to increase above that of the oil supply chamber; the ball-check valve then snaps closed. Any further upward movement of the follower also moves the plunger an equal amount, since the trapped oil in the pressure chamber now acts as an incompressible solid. It thus causes the hydraulic tappet and plunger element to act as a rigid strut. Hence there is no lost motion in the transference of movement from the cam to the poppet valve.

18.16.3 Direct-acting overhead camshaft with hydraulic valve clearance compensation inverted bucket-type follower element (fig. 18.28(a) and (b))

Upstroke (valve closed) (fig. 18.28(a)) The cam rotates to the no-lift base circle/tappet follower contact period. The compressive load is now removed from the pressure chamber, thus allowing the slack spring to extend and take up the slack clearance. This is achieved by the thrust sleeve and tappet follower being pushed upwards against the cam; at the same time the plunger sleeve is pressed downwards onto the poppet valve stem. The suddenly enlarged volume in the pressure chamber causes the instant collapse of the pressure within the chamber and thus makes the check-valve snap open. Oil will now flow between the supply and pressure chambers to equalize pressure and recuperate oil leakage and thus compensate for any variation in slack clearance due to expansion and contraction of the poppet valve and cylinder-head assembly caused by temperature change.

Downstroke (valve opens) (fig. 18.28(b)) The cam rotates to where the lift flank portion of the cam commences to push down the tappet follower. Instantly the squeezing together of the thrust

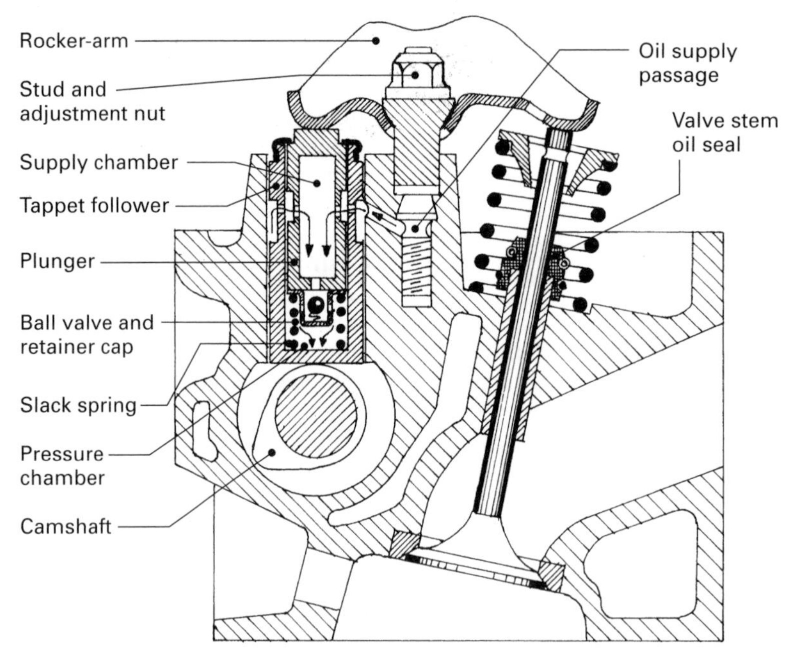

a) Downstroke (valve closing)

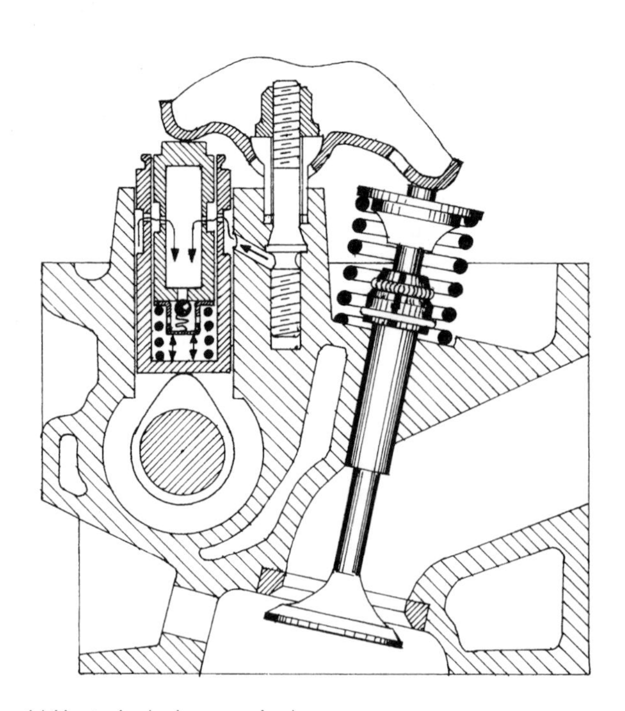

b) Upstroke (valve opening)

Fig. 18.27 Overhead camshaft central pivoted rocker-arm with hydraulic valve clearance compensation tappet follower element

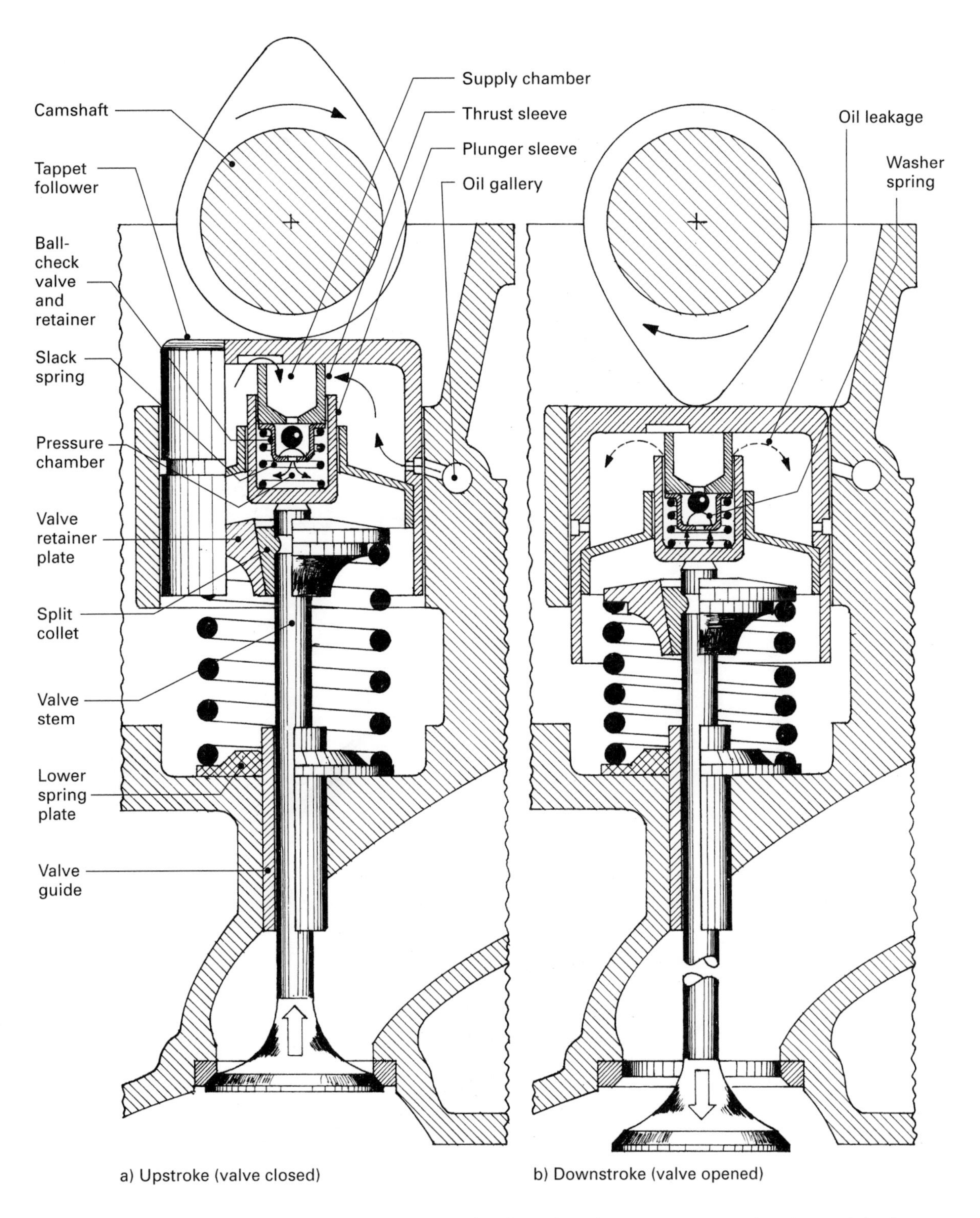

Fig. 18.28 Direct acting overhead camshaft with hydraulic valve clearance compensation inverted bucket-type follower element

and plunger sleeve reduces the pressure chamber space, raises the pressure within the chamber and quickly forces the ball-check valve into the closed position. The trapped oil now becomes an incompressible solid column; hence any further downward movement of the tappet follower produces an equal amount of poppet valve opening.

19 Engine lubrication system

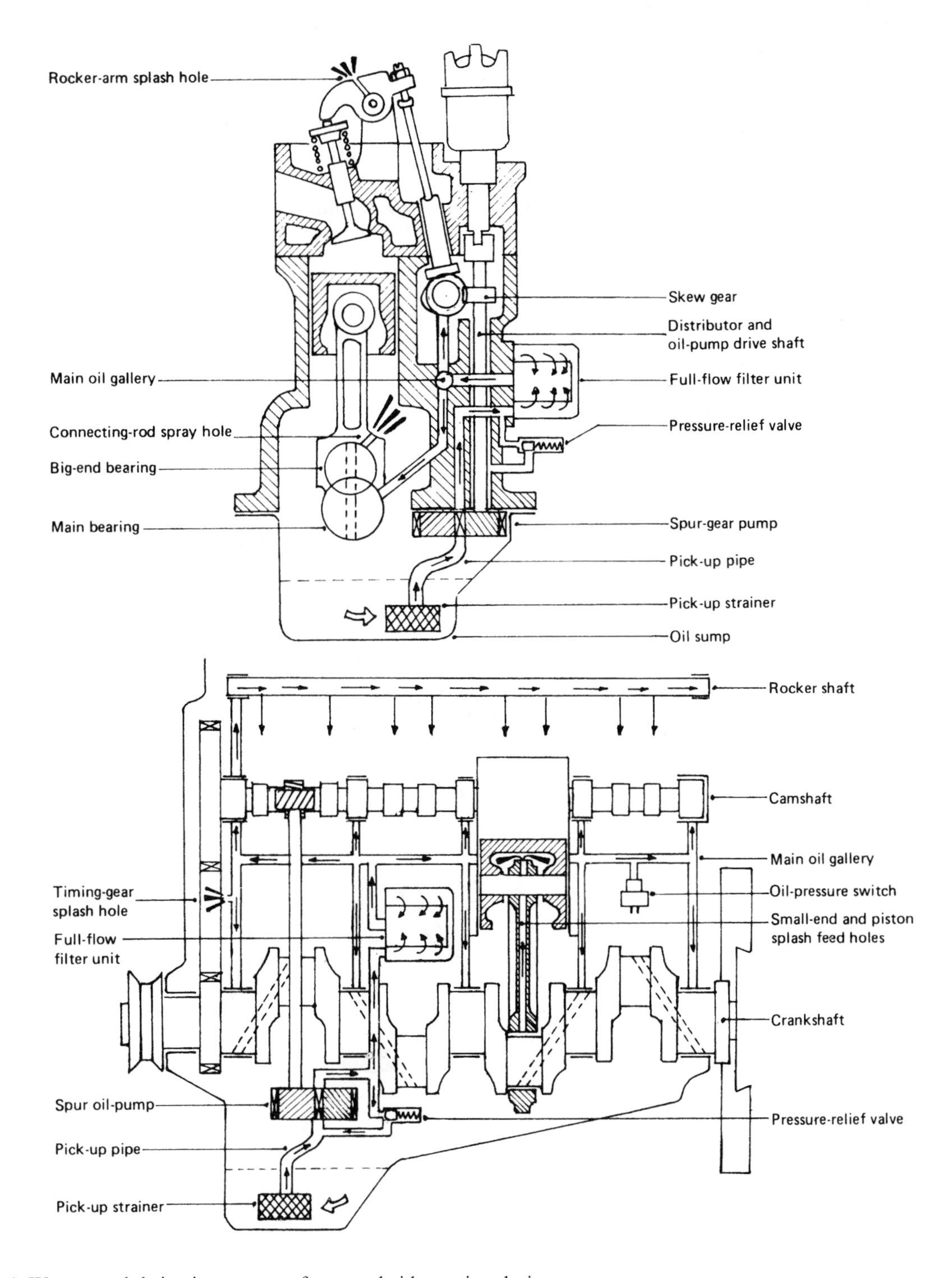

Fig. 19.1 Wet-sump lubrication system: front and side sectional views

19.1 Wet-sump lubrication system

Motor-vehicle engines today use what are known as 'forced-feed' lubrication systems, generally of the wet-sump type in which the sump acts as both an oil-drain return and a storage container. The forced feed comes from a rotary-type oil-pump which may be driven directly from the crankshaft or indirectly from the camshaft or any auxiliary shaft (fig. 19.1). When the camshaft rotates, oil from the sump is drawn through the submerged gauze strainer and pick-up pipe to the sump. The oil is then trapped, compressed, and discharged through a drilling to the lubrication system.

Control of the oil pressure is achieved by a pressure-relief valve situated on the output side of the pump. If the oil pressure becomes too high, the relief valve will open, bleeding any surplus oil back to the sump. This relief valve may be incorporated in the filter unit, the crankcase, or the pump housing (figs 19.1 to 19.3).

From the oil-pump, all the oil flows through drillings in the crankcase to a cylindrical full-flow filter unit. The oil circulates around the filter bowl, forces its way through the filter medium towards its centre, and flows out to the main oil gallery (figs 19.1 to 19.3).

19.1.1 Main oil gallery

The filtered oil flows from the filter unit to the main oil passage (known as the oil gallery) which lies parallel to the crankshaft. Most car- and commercial-engine manufacturers form the oil

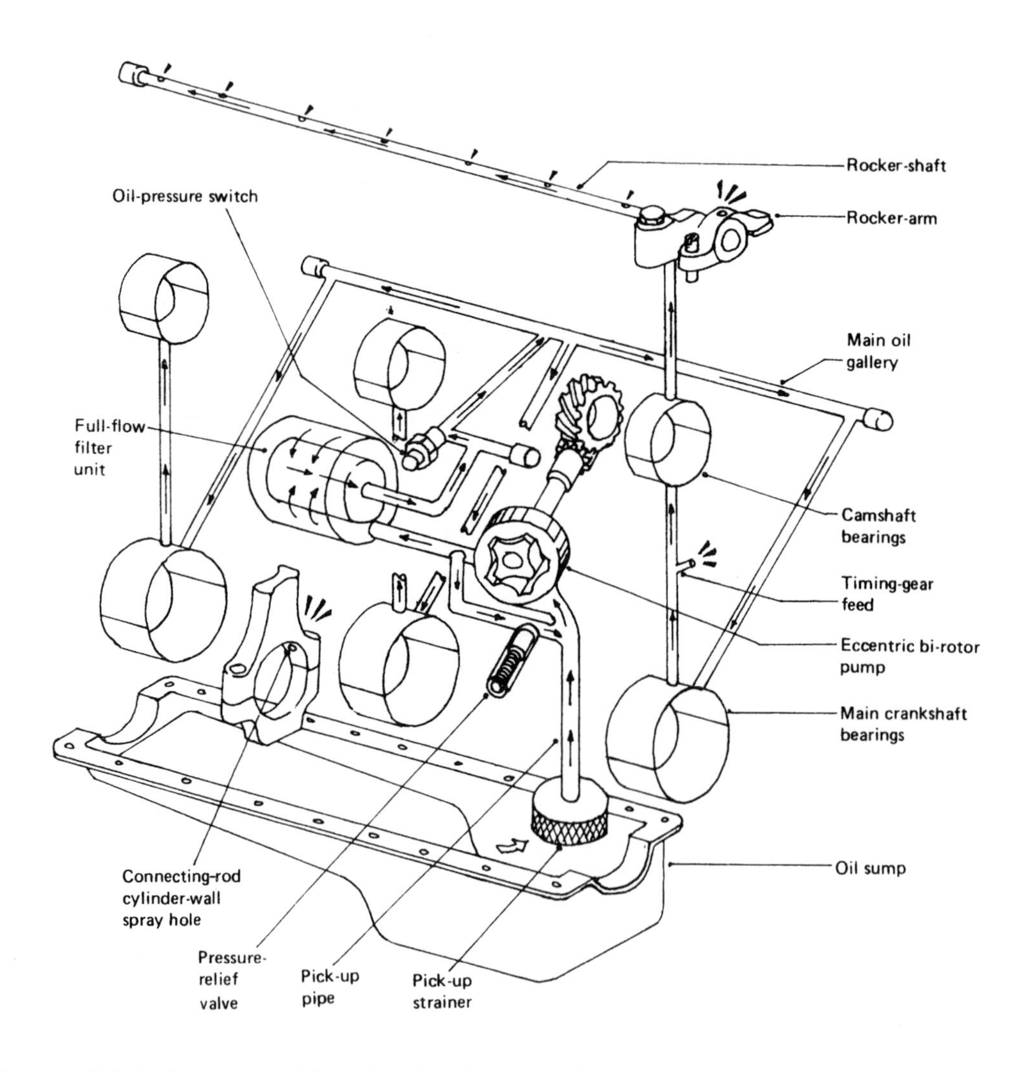

Fig. 19.2 Wet-sump lubrication system with push-rod-and-rocker valve mechanism

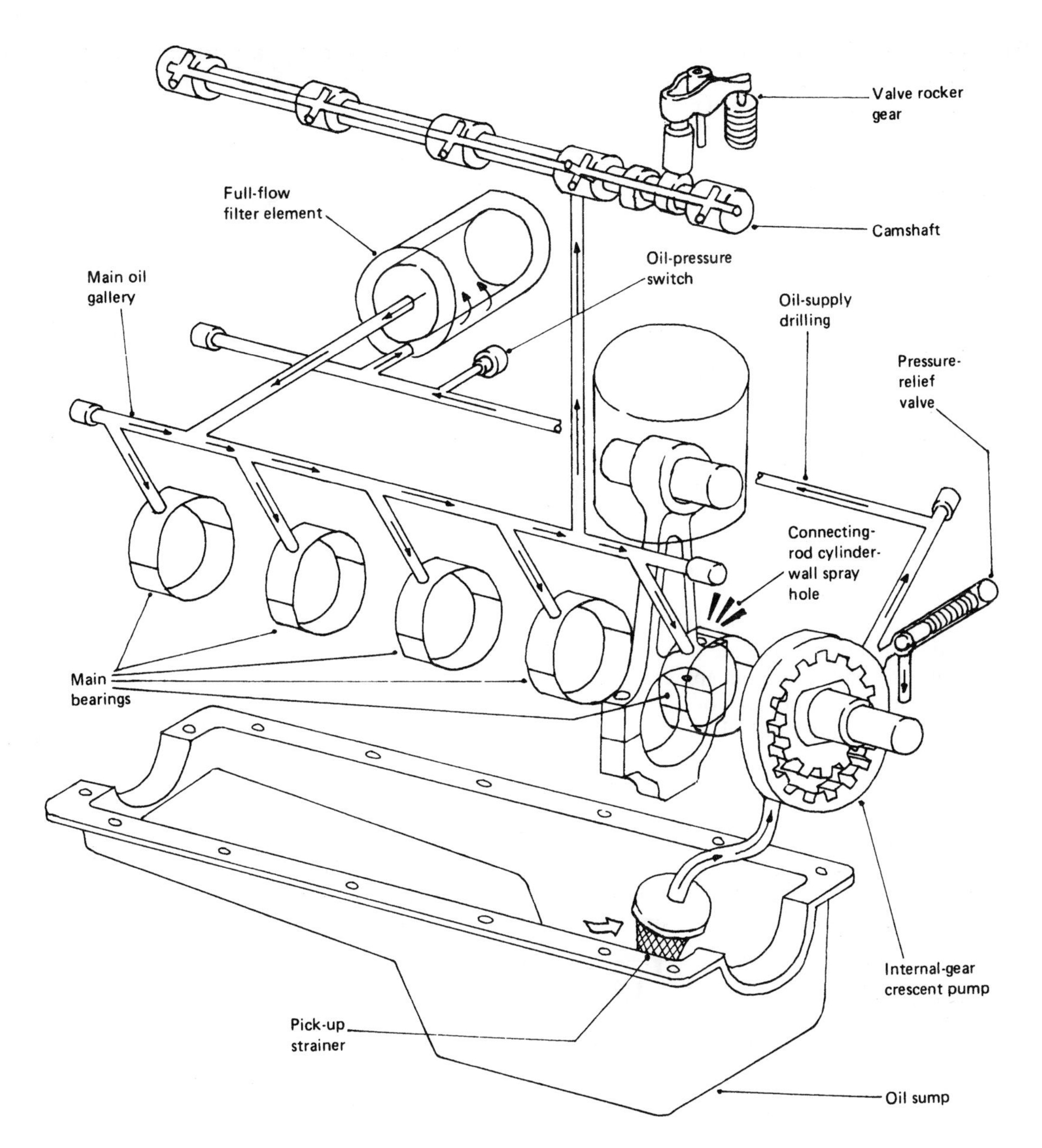

Fig. 19.3 Wet-sump lubrication system with overhead camshaft

gallery by drilling a hole in the crankcase the full length of the engine and plugging the ends.

By various branch cross-drillings in the crankcase, oil is distributed to the crankshaft main-journal bearings and in some cases to the camshaft bearings (figs 19.1 and 19.3). A few heavy commercial engines use a separate pipe situated underneath the main-bearing caps and bolted to them by pedestal brackets – drillings in these brackets connect the gallery-pipe oil to the main bearings.

19.1.2 Main- and big-end bearing lubrication

Cross-drillings in the crankcase from the main oil gallery provide an oil path to the main-journal bearings. A continuous oil feed to the big-end bearings from the oil grooves around the main-bearing liners is provided by diagonal drillings in the crankshaft. These pass from the main-bearing journal through the crankshaft web to the big-end crankpins (fig. 19.1).

19.1.3 Cylinder and piston lubrication (figs 19.4 to 19.7)

There are four separate techniques of cylinder and piston lubrication:

i) connecting-rod big-end side-clearance oil spray,
ii) connecting-rod big-end radial-hole oil spray,
iii) connecting-rod small-end radial-hole oil spray,
iv) crankcase fixed-jet oil spray.

Most modern engines today use one or a combination of these methods to achieve effective cylinder lubrication, any combination used depending mainly on the operating conditions expected from the engine.

Connecting-rod big-end side-clearance oil spray (fig. 19.4)) The simplest and most common method of cylinder and piston lubrication is by big-end side-clearance splash. Here the oil pressure of the lubrication system, the squeezing action between the connecting-rod and the big-end journal, and the amount of side clearance all combine to provide sufficient oil splash to the cylinder walls and the underside of the piston every time the crankshaft throw is in the TDC region.

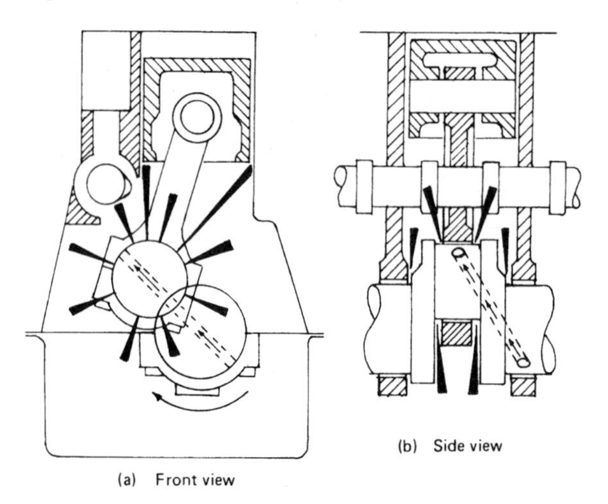

Fig. 19.4 Connecting-rod big-end side-clearance oil spray

Connecting-rod big-end radial-hole oil spray (fig. 19.5) Sometimes it is desirable to have a small radial drilling in each connecting-rod big-end which directs a squirt of oil to the thrust side of the cylinder bore once every revolution. The diameter of the hole and the angular positioning is a critical factor in this method of lubricating the cylinder.

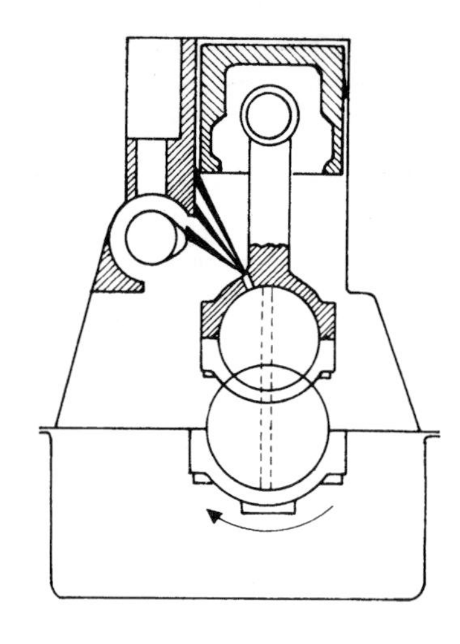

Fig. 19.5 Connecting-rod big-end radial-hole oil spray

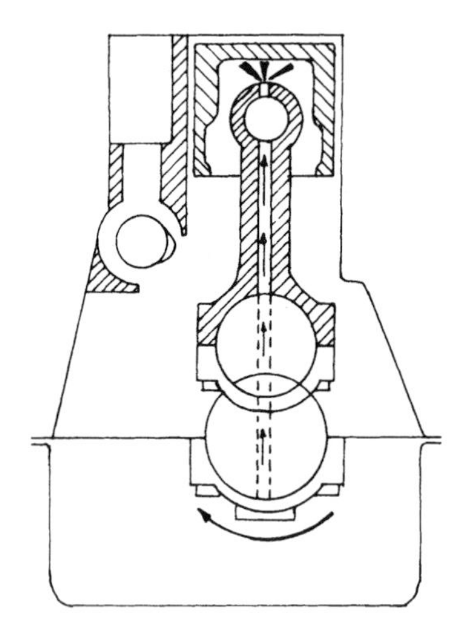

Fig. 19.6 Connecting-rod small-end radial-hole oil spray

Connecting-rod small-end radial-hole oil spray (figs 19.6 and 19.9)) For high-performance engines, the connecting-rod may have a drilling connecting the big-end to the small-end to provide a positive oil feed to the gudgeon-pin.

With heavy-duty diesel engines, the big-end bearings are grooved to provide a continuous flow of oil through the drilled connecting-rods to the small-end bearings. Two drillings in the small-end eyes sometimes supply jets of cooling oil to the ring-belt areas within the pistons.

Crankcase fixed-jet oil spray (fig. 19.7) In applications such as turbocharged heavy-duty diesel engines, a jet fixed in the crankcase projects upwards and provides a controlled and continuous spray of oil that cools and lubricates the underside of the piston. This method of oil supply is usually concerned more with reducing piston and ring temperature than with providing additional lubrication for the cylinder-and-piston combination.

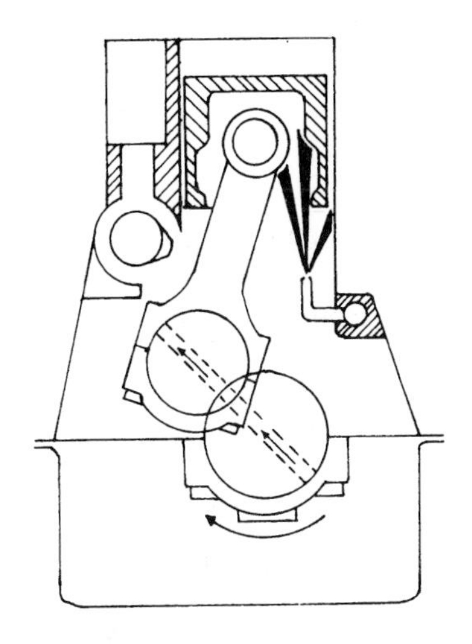

Fig. 19.7 Crankcase fixed-jet oil spray

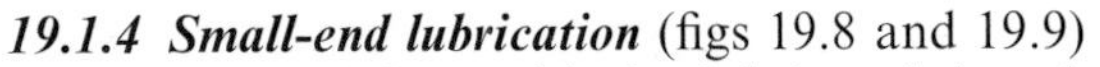

19.1.4 Small-end lubrication (figs 19.8 and 19.9)
The small-end is positively lubricated by the pumping action created by the piston ring scraping the oil from the cylinder bore on its downstroke. This oil will be pushed into the groove behind the lower piston oil-control ring, and from here it will flow along a drilled passage which intersects the piston gudgeon-pin-boss bores. As the small-end of the connecting-rod pivots, this oil will spread over the bearing surfaces. At right angles to the gudgeon-pin bosses, there are circumferential slots or drillings which allow some of the surplus oil to be splashed between the piston gudgeon-pin bosses and the small-end of the connecting-rod. This would be necessary if the gudgeon-pin was fully floating and there was no oil supply from the connecting-rod. In addition, there is a limited amount of splash from the big-end side clearance to complete the small-end lubrication.

19.1.5 Camshaft-bearing lubrication
Camshaft bearings are supplied with oil by basically four methods:

i) Oil is fed directly from the main oil gallery by individual cross-drillings in the crankcase to each camshaft bearing (fig. 19.1).
ii) A continuous supply of oil is obtained from drillings in the crankcase connecting the oil grooves in the crankshaft main bearings to the camshaft bearings (fig. 19.2).
iii) A single drilling passes oil from the main oil gallery to one of the internally grooved cam-

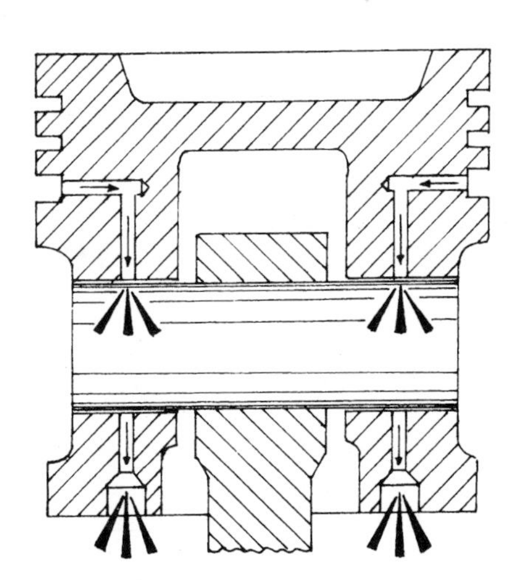

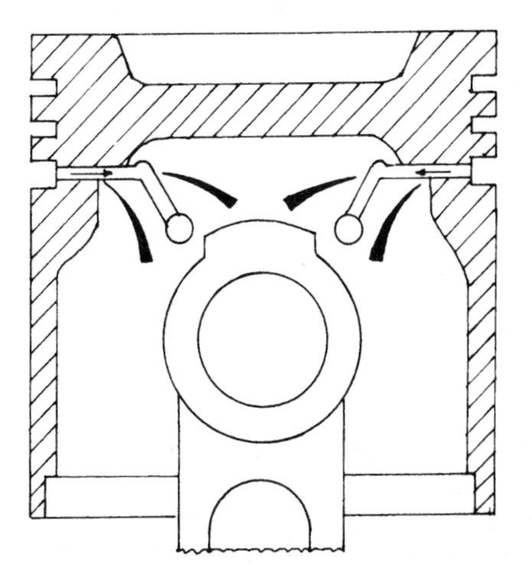

Fig. 19.8 Semi-floating gudgeon-pin with scraper-ring oil supply

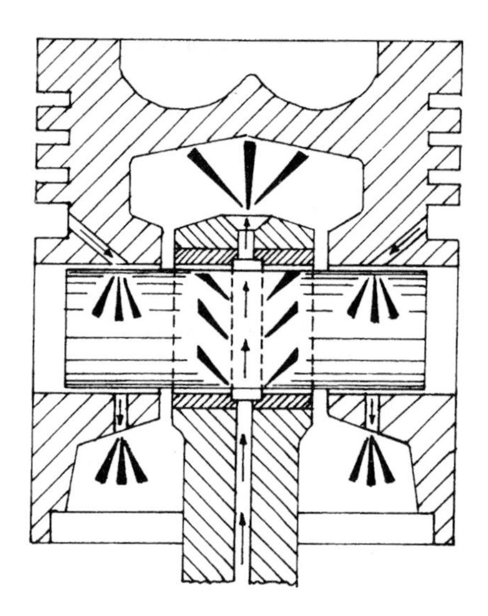

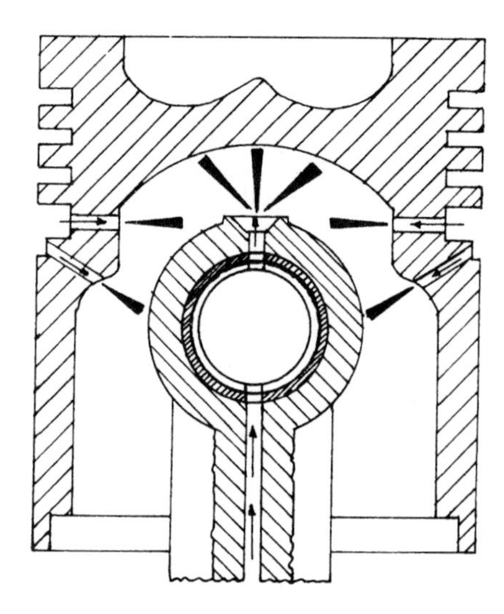

Fig. 19.9 Fully floating gudgeon-pin with scraper-ring and connecting-rod oil supply

shaft bearings. The oil can then enter a pair of radial cross-drillings into the hollow camshaft. A central axial oil passage through the camshaft feeds the other bearings via single radial cross-drillings (fig. 19.3).

iv) A separate camshaft oil passage is drilled into and along the length of the camshaft pedestal block. This drilling has intersecting holes connecting it to the various camshaft bearings (fig. 19.10).

19.1.6 Camshaft-lobe profile lubrication

Methods of lubrication of the camshaft lobes can broadly be divided into those for low- and those for high-mounted camshafts.

Low-mounted-camshaft lobe lubrication This depends mainly on:

a) big-end side clearance – this permits oil to be flung out and so splash the cam lobes each time the crank-throw aligns with the camshaft – that is, once every revolution of the crankshaft (fig. 19.4(a) and (b));
b) oil from the rockers draining and splashing on to the cam profiles (fig. 19.11);
c) oil mist which is created by the rotating crankshaft and the rocker and crankcase ventilation system.

High-mounted-camshaft lobe lubrication This depends on the type of valve-actuating mechanisms used:

a) With the direct-acting and centrally pivoted rocker-arm, the cam lobes are supplied with a cyclic splash from a drilling in the rocker-arm (fig. 19.12).

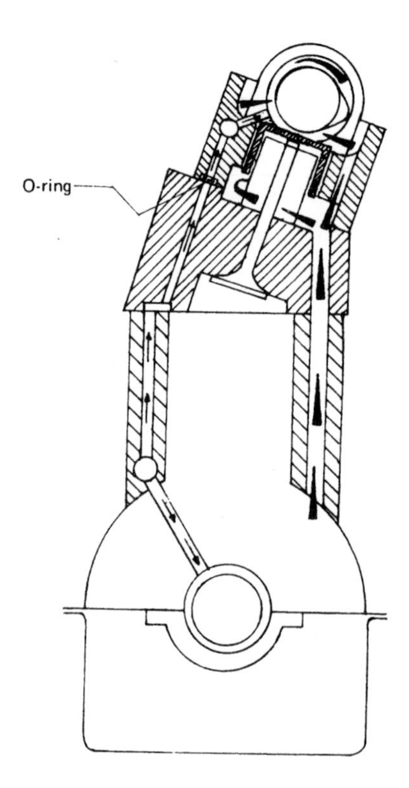

Fig. 19.10 OHC with directly actuated cylindrical follower

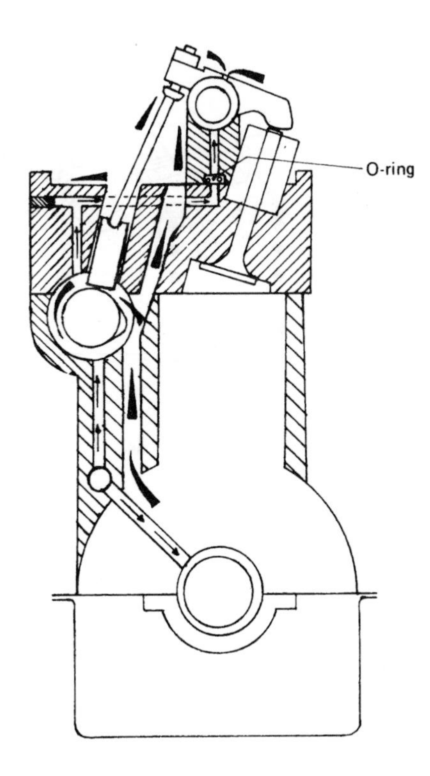

Fig. 19.11 High-mounted camshaft with push-rod-actuated rocker-arm

b) With the direct-acting and end-pivoted rocker-arm, the cam lobes are supplied by a spray directed on to the lobes. This is supplied by a pipe fixed between the camshaft-bearing pedestal supports (figs 19.13 and 19.18).
c) With direct-acting cylindrical followers, the cam lobes can be lubricated by three methods (fig. 19.17), as follows:

i) a simple oil-trough splash,
ii) a radial drilling intersecting the cam base circle,
iii) an oil spray coming from a drilled passage running the full length of the cylinder head.

19.1.7 Poppet-valve lubrication (figs 19.14 to 19.18)

The operation of the poppet-valve provides a reciprocating rubbing motion between the valve stem and its guide, and between the valve-tip/rocker-arm contact faces. One exception to the valve-tip rubbing action is provided by the direct-acting cylindrical cam follower, whose contact action is

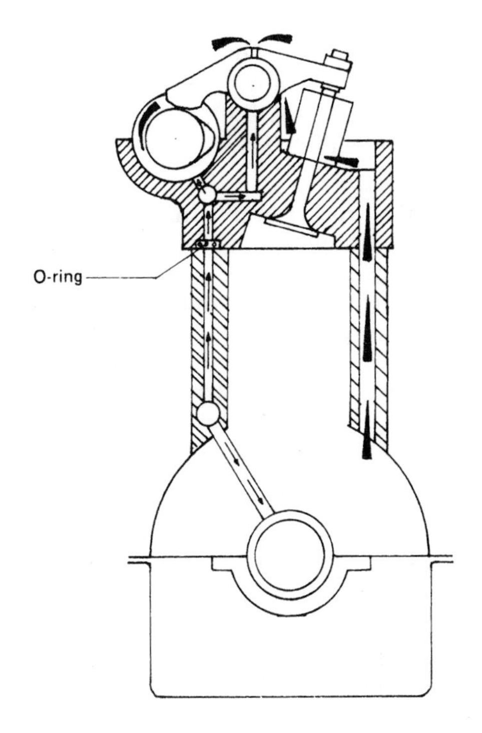

Fig. 19.12 OHC with centrally pivoted and end-actuated rocker-arm

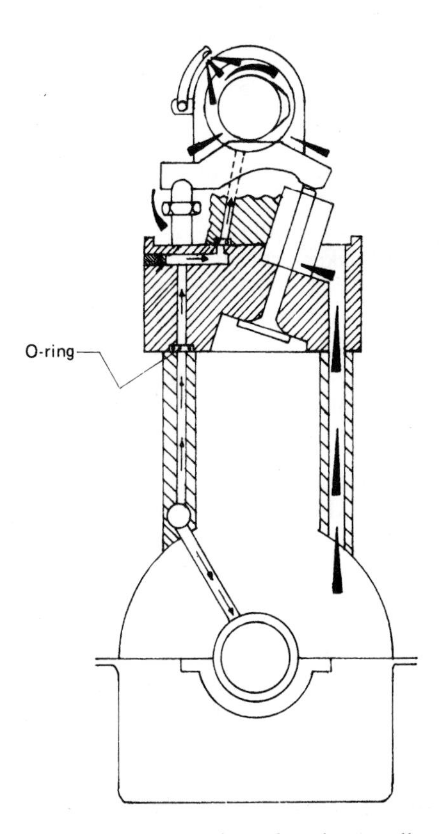

Fig. 19.13 OHC with end-pivoted and centrally actuated rocker-arm

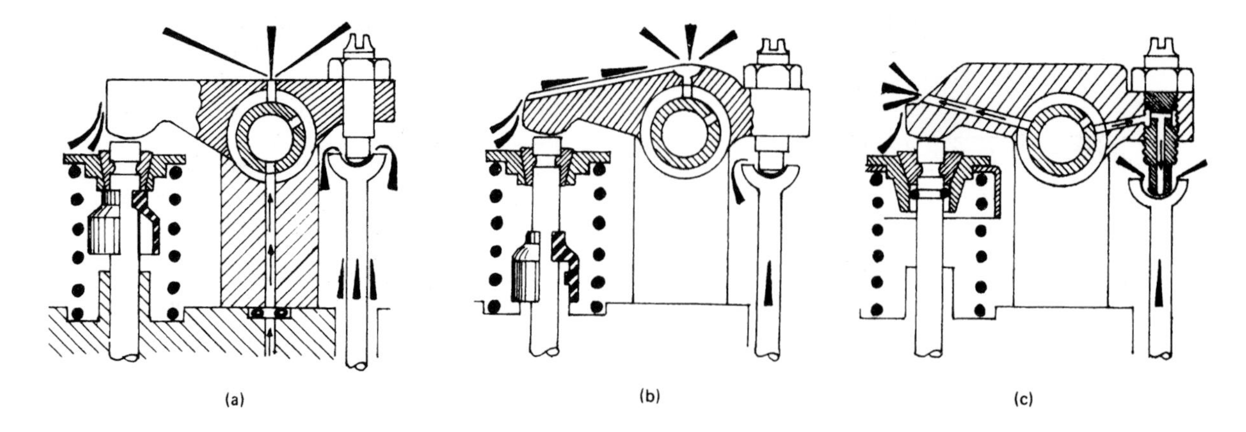

Fig. 19.14 Solid-rocker-arm oil hole and valve-stem oil-seal arrangements

a simple end-on contact, any side-thrust due to rotation of the cam lobe being absorbed by the follower alone.

The lubrication of the valve stem and tip is provided by splash and drainage of surplus oil from the rocker-arm and/or the camshaft lobes.

19.1.8 Valve rocker-arm-mechanism lubrication

The valve rocker-arm lubrication depends on the type of rocker-arm assembly.

Solid rocker-arm (fig. 19.14) These arms are supplied by an oil drilling or pipe running from one of the camshaft bearings to a hollow rocker-shaft which has radial holes aligning with each rocker-arm. The rocker-arm pivot hole may be bored and used directly over the shaft or it may be bronze-bushed with internal oil grooves.

Figure 19.14 shows three ways of feeding the oil to the valve stem-and-spring assembly and to the tappet and push-rod end.

i) The valve stem and the tappet assembly may be lubricated by a single vertical radial drilling in the middle of the rocker-arm (fig. 19.14(a)). Oil is squirted out in both directions as the arm rocks. This method generally supplies a sufficient quantity of oil for small engines.
ii) A more controlled method of lubricating the tappet and push-rod assembly is provided by a radial drilling in the rocker-arm at the pivot bore. The valve-stem end of the rocker-arm has an open grooved channel formed along the top of the rocker-arm where the surplus oil floods and drains down and over the valve and return-spring (fig. 19.14(b)). This system is preferred on some medium-sized petrol and diesel engines.
iii) An alternative for heavy-duty operation is provided by connecting the rocker-shaft feed to a hollow tappet screw so that oil will flow directly into the push-rod bowl-shaped seat and will then overflow and drain down the push-rod to lubricate the cam follower (fig. 19.14(c)). The valve-stem end of the rocker-arm has a horizontal hole drilled along it which feeds oil directly to the valve-and-spring assembly. One problem with this method of oil supply is that it may overlubricate the valve stem if no restriction is imposed on the oil supply to the rockers.

Steel-pressing rocker-arm with hollow push-rod (fig. 19.15) An oil drilling from one of the camshaft bearings supplies oil to the tappet-follower gallery drilling which lies parallel to the camshaft. Oil from this gallery flows around an annular groove in each tappet-follower body to ensure positive lubrication. A valve disc in the tappet (not shown) controls the flow of this oil through the hollow push-rod and to the rocker-arm and the valve.

Steel-pressing rocker-arm with central hollow stud (fig. 19.16) With this method, a passage from the first camshaft bearing supplies oil to the tappet-follower oil gallery drilled alongside the tappets and extending the full length of the cylinder head. Oil from the gallery flows around a recess machined on the tappet and to a short drilling which meets the central rocker-arm pivot-post stud. This stud is hollow and has a radial inter-

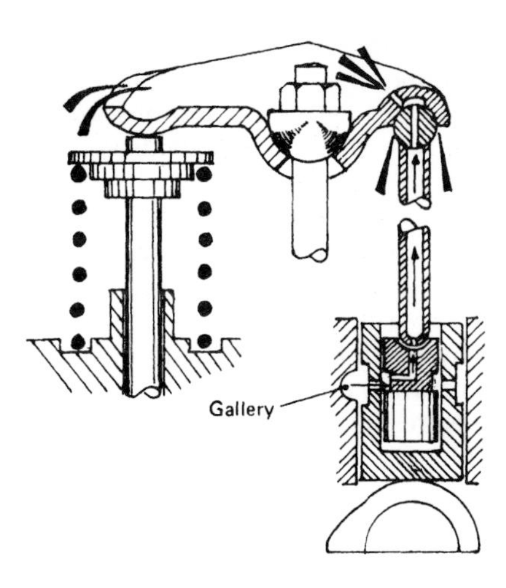

Fig. 19.15 Steel-pressing rocker-arm with hollow-push-rod oil supply

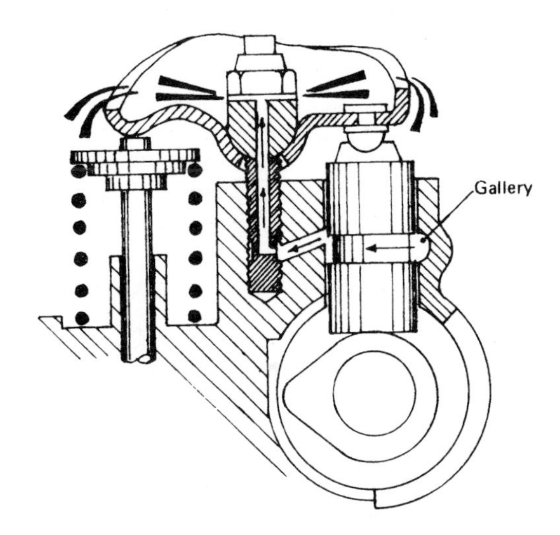

Fig. 19.16 Steel-pressing rocker-arm with central-hollow-stud oil supply

secting hole which connects the oil supply from the tappet gallery to the spherical rocker pivot. The oil then splashes and floods the rocker pressing, the overspill lubricating both the valve assembly and the top of the tappet follower.

19.1.9 Overhead-camshaft lubrication (figs 19.17 and 19.18)

Overhead camshafts may be actuated by either a direct-acting bucket follower or an indirect end-pivoted rocker-arm, and the methods of lubricating the cams are as follows:

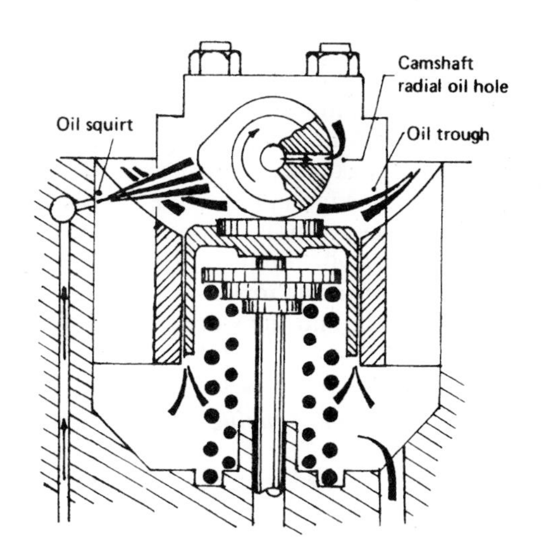

Fig. 19.17 OHC cylindrical direct-acting follower with a fixed-pedestal spray, a hollow camshaft with a radial oil hole, or simply a trough splash bath

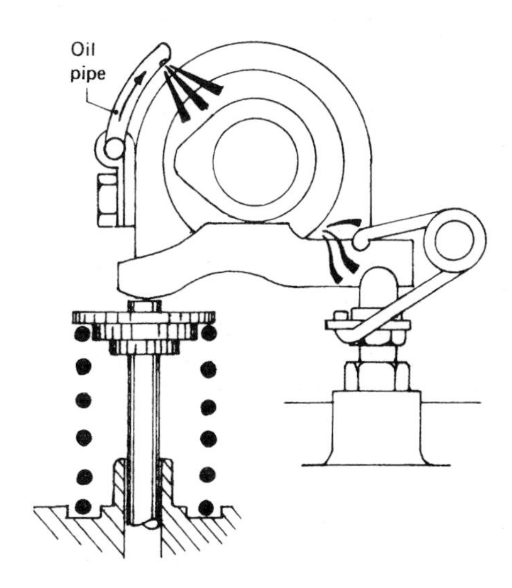

Fig. 19.18 OHC with end-pivoted rocker-arm and oil-pipe-supply spray

a) With the direct-acting camshaft arrangement (fig. 19.17), lubrication for the camshaft, follower, and valve stem is achieved either by a drilled hole along the centre of the camshaft axis with intersecting radial holes emerging on the base circle of the cam, or by a drilling in the pedestal casing parallel to the camshaft with projecting spray holes directed on to the cam profiles. In both cases the follower and the valve stem are lubricated by drainage of oil from the camshaft.

b) With the indirect end-pivoted rocker-arm arrangement (fig. 19.18), lubrication for the camshaft, rocker, and valve stem is obtained by an oil pipe mounted on the camshaft pedestal housing to spray oil on to the cam faces. The surplus oil draining from the camshaft also runs down the rocker-arm and lubricates its pivot joint and the valve tip and stem.

19.1.10 Lubrication of timing gears and chains (figs 19.1 and 19.2)

These highly worked components are usually lubricated by a small drilling which intersects the oil passage going from the main oil gallery to the first main bearing (fig. 19.1) or the passage from the first main bearing to the first camshaft bearing (fig. 19.2). Sometimes a small pipe from this drilling directs the oil on to the gears or chain. In addition, some manufacturers shape the sump to form a timing-gear oil trough, so that the draining oil will submerge the crankshaft gear – this then provides a continous upward oil splash to the rest of the camshaft drive.

19.1.11 Oil-pressure switch

A switch for an oil-pressure warning light is located at the end of a drilling intersecting the main oil gallery (figs 19.1 to 19.3). This switch is controlled by the ignition switch and operates a warning light when the oil pressure is zero or very low. At pressures above 0.3 to 0.5 bar, the light is extinguished.

19.1.12 Crankshaft oil passages (fig. 19.19)

Crankshaft oil passages are provided to feed oil from the main-journal bearing to the big-end journal.

Crankshaft with single diagonal oil passage (fig. 19.19(a)) The simplest form of oil passage is a diagonal drilling going from the main journal to

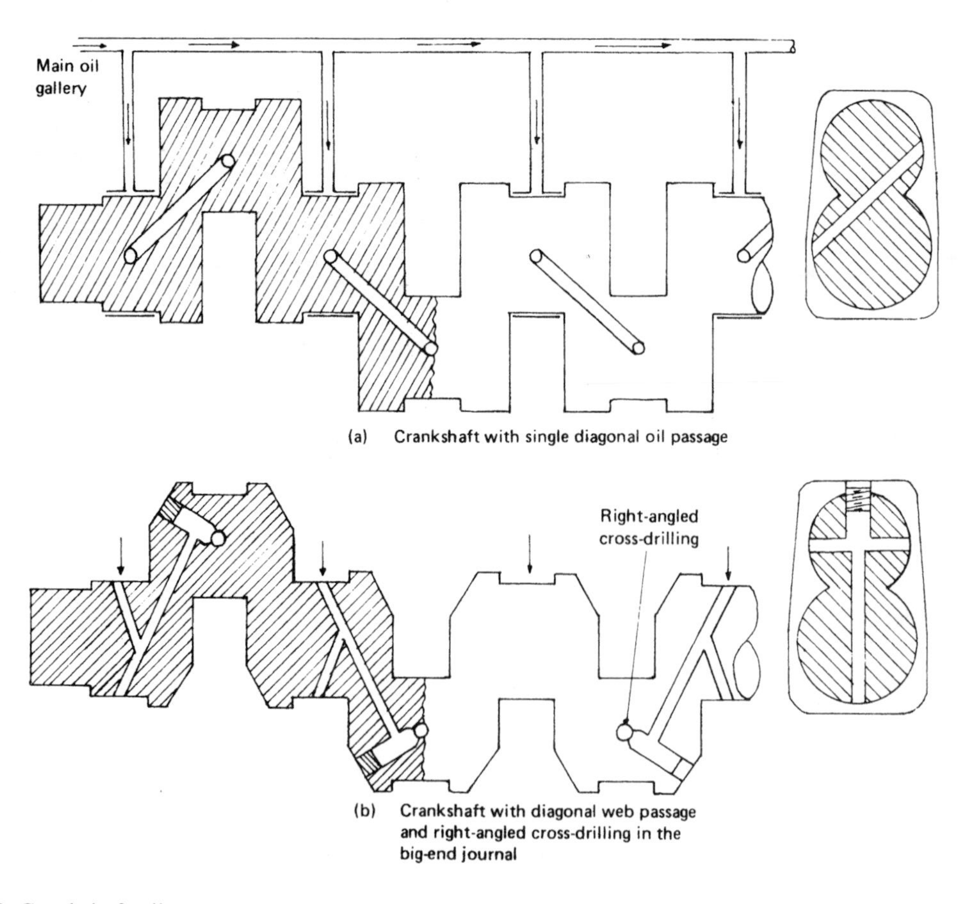

Fig. 19.19 Crankshaft oil passages

the big-end journal. It is usual to drill the diagonal hole at an angle to the crank-web centre-line. This is done so that, when the crank-pin is in the TDC position and the connecting-rod is pushing downwards due to combustion force, some oil will still be able to enter between the journal and the bearing. If the exit of the diagonal hole were exactly at the top of the big-end journal, in the TDC position oil would not be able to enter between the bearing and the journal. In addition, the effective projected bearing area would be reduced by the surface area occupied by the chamfered oil hole.

Crankshaft with diagonal web passage and right-angled cross-drilling in the big-end journal (fig. 19.19(b)) An improvement in oil delivery is achieved by having a cross-drilling going straight

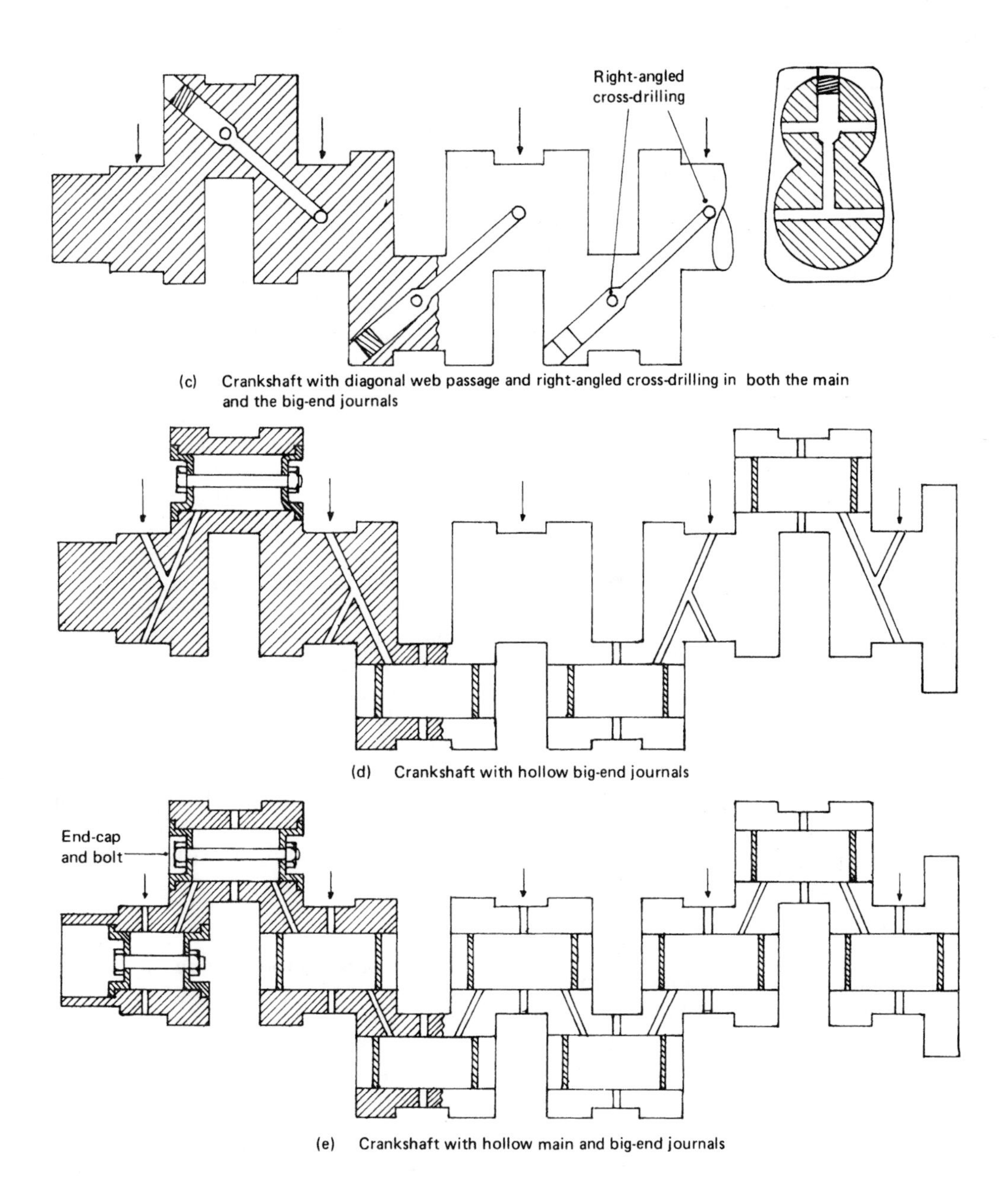

(c) Crankshaft with diagonal web passage and right-angled cross-drilling in both the main and the big-end journals

(d) Crankshaft with hollow big-end journals

(e) Crankshaft with hollow main and big-end journals

Fig. 19.19 (*continued*) Crankshaft oil passages

through the big-end journal and having a diagonal hole drilled from the main-bearing journal to intersect the big-end cross-drilling. A second hole is also drilled diametrically opposite the diagonal hole's entry in the main journal, so that, no matter whether the bearing is loaded at either the top or the bottom of the stroke, the other side of the bearing will permit oil to enter.

Crankshaft with cross-drillings in both the main and the big-end journals connected by diagonal passages (fig. 19.19(c)) An alternative is provided by having a right-angled cross-drilling in both the main and the big-end journals, these being joined by a diagonal web drilling. With these passages, oil will be able to enter the main journal and bearing at right angles to the connecting-rod centre line when the crankshaft is near either end of the stroke. This approach tends to provide a continous supply of oil to the big-end bearing over a complete cycle of operation.

Crankshaft with hollow big-end journals (fig. 19.19(d)) Large engines sometimes have hollow big-end crankpins which are enclosed by end-caps with a nut and bolt to hold them in position. A diagonal hole is drilled from the main journal to the hollow big-end pin, and a second hole drilled into the main journal intersects the diagonal hole. This layout stores oil in the hollow big-end and provides unrestricted oil to the big-end bearings.

Crankshaft with hollow main and big-end journals (fig. 19.19(e)) Some very large engines have both hollow main and big-end journals, and diagonal drillings connect each internal crankpin cavity with the next so that oil can flow from one end of the crankshaft to the other. The cross-drillings allow oil to pass between the journal and the bearing in most crank positions with an adequate supply of stored oil.

19.2 The mechanics of friction and lubrication

To appreciate how movement resistance and wear may be reduced and component life be extended, we shall consider friction and lubrication in some depth.

Components moving relative to each other will form either reciprocating or rotating rubbing pairs. The rubbing or sliding motion can exist under the following conditions:

a) dry solid friction,
b) boundary lubrication (section 19.2.6),
c) hydrodynamic lubrication (section 19.2.7).

19.2.1 Friction

Friction is defined as the resistance to motion when any two solid surfaces are pressed together and slide over each other. The magnitude of the frictional force depends on the surface roughness, the nature of the sliding materials, and the pressure between the rubbing pair.

19.2.2 Static friction (fig. 19.20(a))

The applied force required to make one solid surface just begin to move in contact with another will be greater than that required to continue sliding. The frictional force of rest is known as static friction, and it is this force which must be overcome when a crankshaft is initially cranked.

19.2.3 Kinetic or dynamic friction (fig. 19.20(a))

Once sliding has begun between two solid faces, the force needed to maintain uniform rubbing movement will be less than that required to overcome the starting resistance. The frictional force during motion may be referred to as kinetic or dynamic friction. It is this condition which exists between a journal shaft and bearing when running with very little lubrication.

19.2.4 Stick–slip friction (figs 19.20(b))

The sliding between some surfaces is not a continuous process but proceeds in a succession of jerks – hence the term stick–slip motion. This is due to one or other of the sliding surfaces having some elastic 'give', which is normally the case. The frictional resistance builds up to a maximum during the stick phase, when static friction has to be overcome, and then falls rapidly during the slip phase when the lower kinetic friction is acting. Stick–slip friction can exist if there is a failure of lubrication between crankshaft journals and bearings.

19.2.5 Coefficient of friction (fig. 19.20(c) and (d))

The resistance offered by a solid block of material sliding over another solid surface is known as the frictional force and may be caused by a combination of the following (fig. 19.20(d)):

a) by the interlocking and finally shearing of the weaker high spots on both surfaces (all surfaces – even the most highly finished –

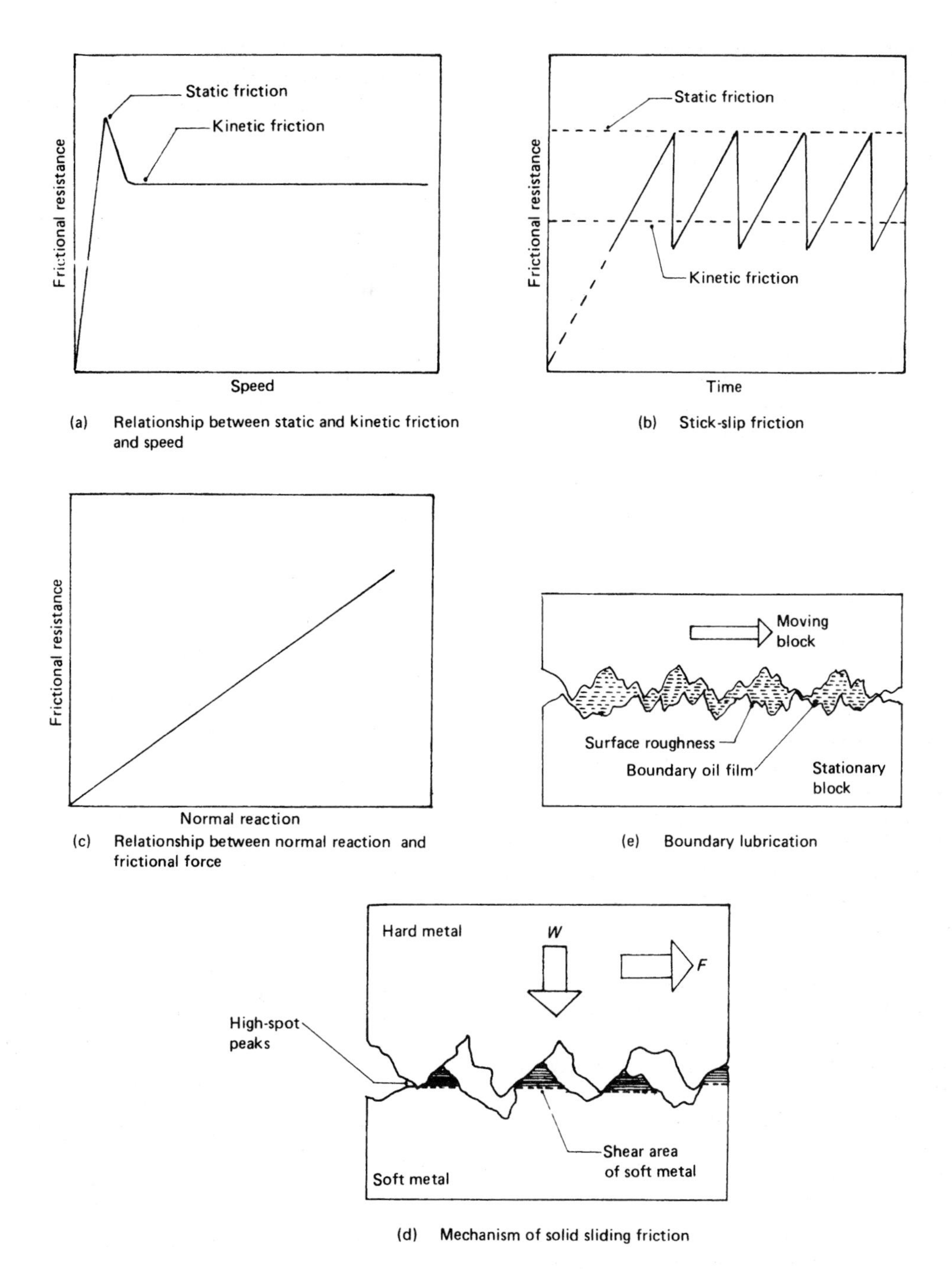

Fig. 19.20 Friction

have such high spots, due to their residual roughness);

b) by both elastic (recoverable) and plastic (permanent) distortion of strong and tough raised surface peaks;

c) by the lifting up and over, that is the moving apart, of the surface irregularities of very strong interface materials.

For any pair of solid rubbing surfaces, the magnitude of this frictional force has been found to be proportional to the normal reaction between the surfaces (fig. 19.20(c)). This means that the ratio of the frictional resistance to the force pressing between the solid sliding faces is a constant, and this constant is given the name coefficient of friction and the symbol μ (the Greek letter mu);

$$\text{i.e.} \quad \text{coefficient of friction} = \frac{\text{frictional force}}{\text{normal reaction force}}$$

$$\text{or} \quad \mu = \frac{F}{W}$$

Since it is a ratio of quantities both measured in newtons, the coefficient of friction has no units.

Typical values of μ for different rubbing metals under both dry and lubricated conditions are given in Table 19.1.

It should be observed that dry friction gives high coefficients of friction, and lubricating sliding surfaces establishes much lower coefficients of friction. To reduce wear, engine designers try to keep the coefficients of friction as low as possible for reciprocating and rotating rubbing surfaces, such as piston to cylinder, cam profile to tappet follower, and crankshaft journal to bearings.

19.2.6 Boundary lubrication (fig. 19.20(e))
This is an intermediate condition between dry friction and fluid hydrodynamic lubrication discussed below. It is basically a state where the lubricant fills the cavities which exist between all friction surfaces. The friction which occurs is due to the shearing of the lubricant film itself and to the metal-to-metal contact of the high spots of the sliding faces.

Good boundary lubrication depends on chemical affinity between at least some of the molecules in the oil and the metal surface to be lubricated. 'Chemical affinity' means chemical reactivity, and an oil which has been refined to the maximum degree of chemical stability is therefore unlikely to be able to provide very powerful boundary lubrication. Crankcase oils must be sufficiently stable chemically to have a long life in service before breaking down due to oxidation, which is in effect chemical degradation due to reaction with atmospheric oxygen. Special blending components are therefore used to achieve a compromise between the right degree of chemical reactivity and a tendency towards rapid oxidation.

Boundary lubrication takes place between cylinder walls and piston assemblies, valve stems and their guides, and journal bearings when rotating from standstill.

Table 19.1 Typical values of coefficient of static friction

Metals	Lubrication state	Coefficient of static friction
Hard steel on hard steel	Dry in air	0.6
	Oiled	0.14–0.2
Cast iron on steel	Dry in air	0.4
	Oiled	0.1–0.2
Phosphor bronze on steel	Dry in air	0.35
	Oiled	0.15–0.2
Copper–lead alloy on steel	Dry in air	0.2
	Oiled	0.1
White metal on steel	Dry in air	0.7
	Oiled	0.1

19.2.7 Hydrodynamic lubrication (fig. 19.21(a))
Imagine a journal shaft and bearing immersed in oil. The journal is always smaller in diameter than the bearing in which it rotates, so that a degree of convergence exists in the space between the two when the journal is loaded. The position where the shaft and bearing are at their closest is known as the point of nearest approach.

The formation of hydrodynamic journal lubrication takes place in four stages:

i) *Stationary or static friction* (fig. 19.21(c)(i)) When the shaft is stationary or revolving very slowly, there will be intimate contact between the shaft and the bearing at the base.
ii) *Boundary lubrication* (fig. 19.21(c)(ii)) When the shaft starts to rotate, it will climb up the bearing plane in a direction opposite to the direction of the rotation until the limiting frictional force is reached.
iii) *Semi-hydrodynamic lubrication* (fig. 19.21(c)(iii) As the journal's speed increases, it drags with it a clinging layer of oil, and another boundary layer of oil clings to the stationary bearing surface. Between these two layers, oil is moving in the same direction as the journal surface movement. Oil will thus be dragged or pulled into the thin end of the clearance space to form a converging wedge of oil. This oil film will eventually become strong enough to support and separate the shaft from the bearing. Thicker or higher-viscosity oils will be able to form stronger oil films and support heavier loads.
iv) *Hydrodynamic lubrication* (fig. 19.21(c)(iv)) The oil film formed between the two faces will increase in thickness with rising speed

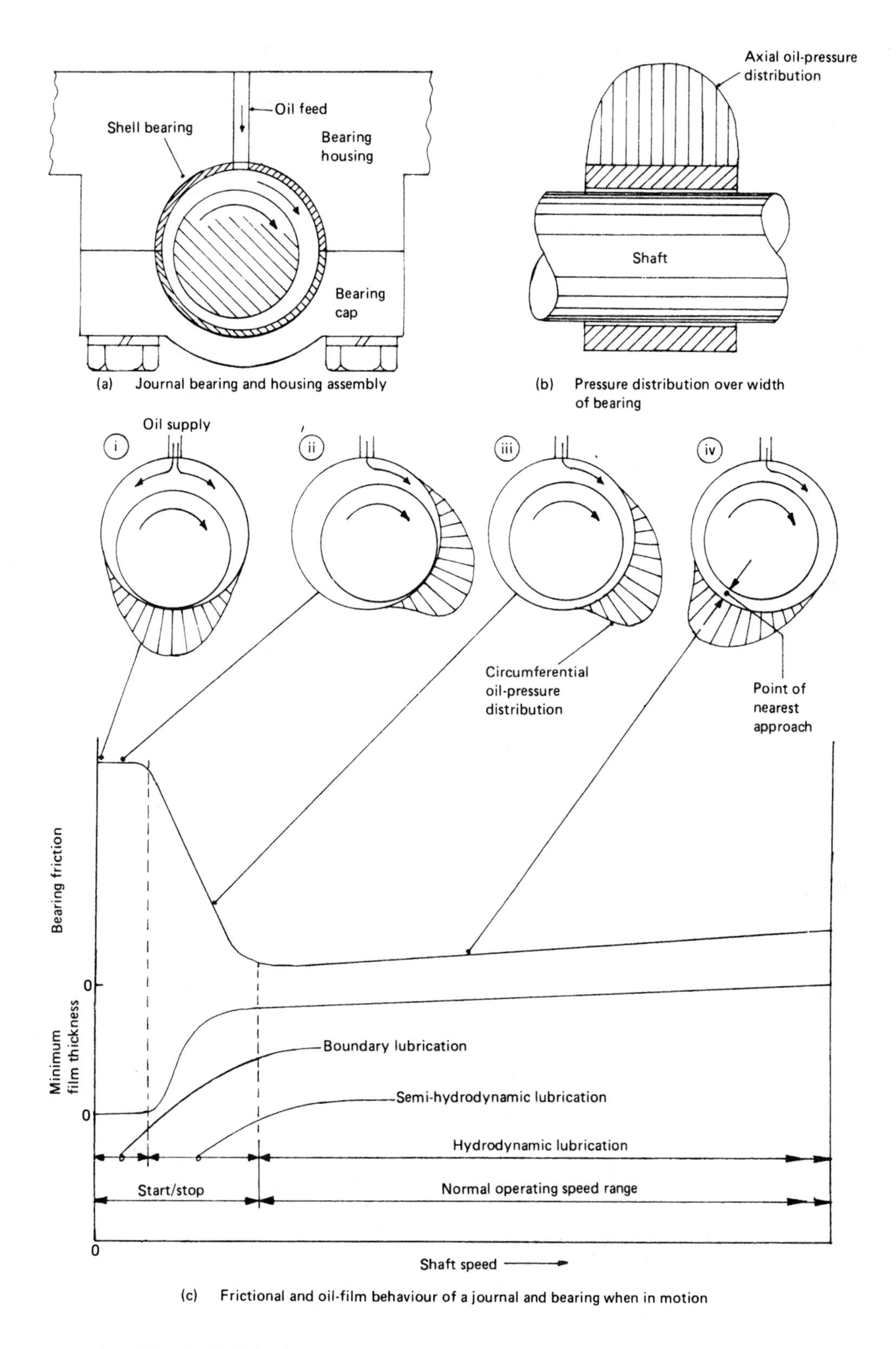

Fig. 19.21 Journal and bearing lubrication

and will push the shaft axis in the same direction as the direction of rotation, to the opposite side. In practice, a stable mean position will be established and the shaft axis will 'dance' about this point with fluctuating loads.

Hydrodynamic lubrication is possible only because most lubricants form strong attachments to the metal surfaces, and relative motion is achieved by the internal shearing of molecular layers within the oil-wedge film itself. In addition to crankshaft and camshaft journal lubrication, hydrodynamic lubrication can often develop on the middle areas of cylinder walls, where piston speeds are highest.

19.2.8 Journal and bearing frictional resistance, oil-film thickness, and speed (fig. 19.21(c))

The graph in fig. 19.21(c) illustrates the progressive frictional changes, from an initially high static-friction condition to a rapidly falling boundary-lubrication phase. This is then followed by the establishment of a stabilising semi-hydrodynamic lubrication condition. Finally, the formation of a purely hydrodynamic-lubrication oil film results in a slight increase in frictional resistance with respect to speed, this being basically due to the internal shearing of molecular oil layers over each other.

The oil-film thickness will vary over the shaft's speed range, being zero when the shaft is stationary or in the boundary-lubrication state but building up rapidly in the semi-hydrodynamic lubrication range. Then in the full hydrodynamic region the increase in oil-film thickness is very gradual.

19.2.9 Oil pressure and its distribution (figs 19.21(b) and (c))

The oil-film converging wedge will generate its own internal oil pressure to support the shaft. Figure 19.21(c)(i) to (iv) show the pressure distribution adjacent to the points of nearest approach. The pressure in these hydrodynamic oil films will be a maximum at the point of nearest approach, with a progressive decrease on each side due to the increasing clearance space.

With regard to the previous description of the formation of hydrodynamic lubrication, it should be observed that the oil supply does not necessarily have to be externally pressurised to separate the rubbing surfaces of the journal and bearing. An example of unpressurised lubrication was the case when big-ends were scoop-fed (fig. 19.22) and the main bearings were fed purely by gravity. Pressurisation does, however, increase the rate of oil circulation and extracts heat from the system, which is desirable with the present generation of high-output engines.

The pressure distribution along the bearing width is plotted in fig. 19.21(b). The shape produced is a parabola, the pressure being a maximum mid-way and dropping off on each side towards the exposed bearing edges.

19.3 High-pressure oil-pumps

There are four basic types of rotary-operating oil-pumps for pressure-feed lubrication systems:

i) external-spur-gear pump,
ii) internal-gear crescent pump,
iii) eccentric bi-rotor pump,
iv) sliding-vane eccentric pump.

These pumps are usually selected according to their ease and convenience of being driven and positioned. Other factors of importance will

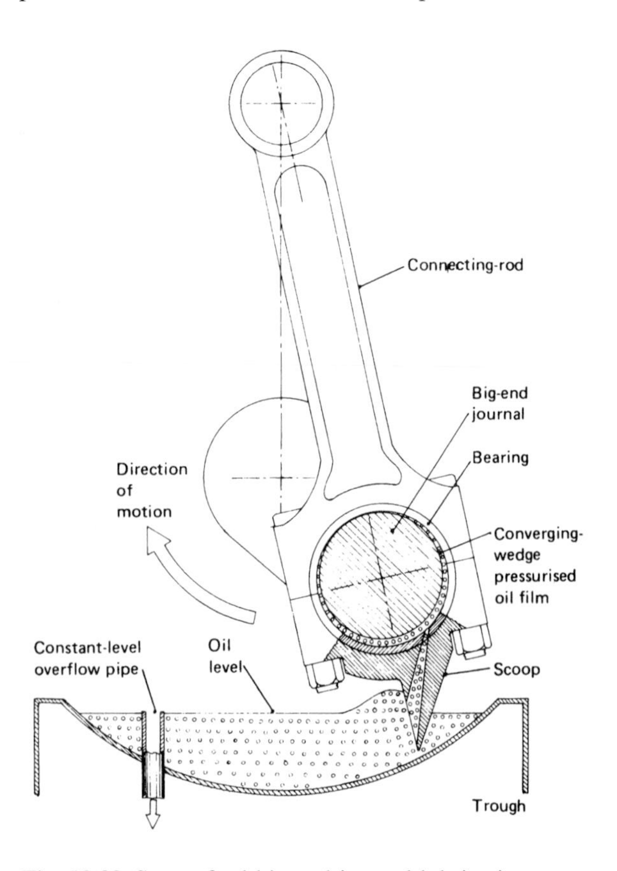

Fig. 19.22 Scoop-feed big-end journal lubrication

include expected pump life, oil flow-rate capacity, priming time, the ability to build up pressure at low speeds, and the ability to deliver oil under higher-pressure conditions continuously at high engine speeds.

19.3.1 External-spur-gear pump (fig. 19.23)
This pump consists of two identical meshing spur gearwheels housed in the pump body. The driving gearwheel is either shrunk on to the oil-pump drive shaft or is positively retained by a keyway. The drive shaft rotates in a bearing bore machined directly in the pump housing, and the driven gear revolves on a bearing post mounted within this housing.

The rotation of the gears creates a low-pressure area on the inlet suction side which draws in oil. As the wheels rotate, the oil filling the spaces between the gear teeth is sealed off by the housing walls. These trapped cells of oil are then moved around the periphery of each gearwheel, in opposite directions in the two gearwheels, to the discharge outlet port. The continuous displacement of oil to the outlet port will exceed the volume of the lubrication system passages, so that the excess discharged oil goes into pressurising the oil and in increasing the rate of oil circulation.

The pump components may be checked for correct clearance and wear using feeler gauges:

a) Measure the clearance between the gear tips and the pump body by inserting feeler blades beween them. This clearance should not exceed 0.2 mm.
b) Measure the backlash between the meshing gears by inserting feeler blades between them. This should be between 0.1 and 0.2 mm.
c) Measure the end-float between the end-plate and the gears by placing a straightedge across the open face of the pump casing, then check the clearance between this and the gear faces using feeler blades. This clearance should not exceed 0.1 mm.

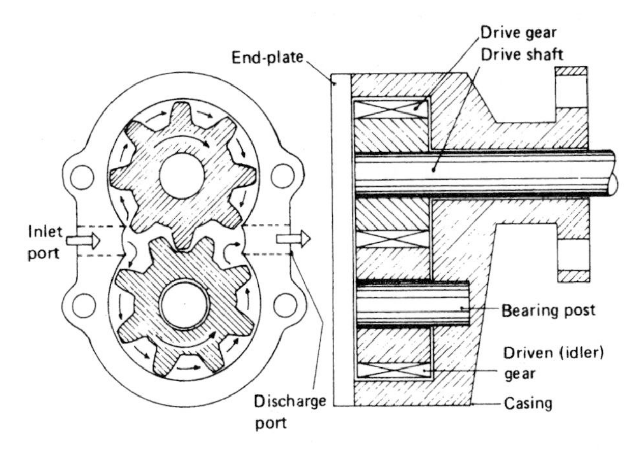

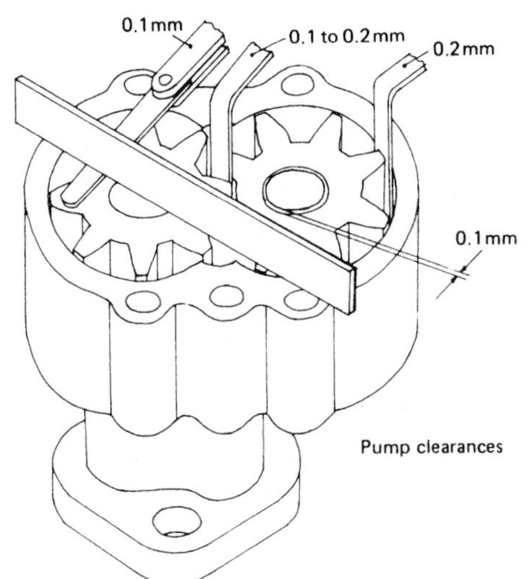

Fig. 19.23 Spur external-gear pump

19.3.2 Internal-gear crescent pump (fig. 19.24)
This pump consists of an internal-spur ring gear which runs outside but in mesh with a driving external-spur gear, so that its axis of rotation is eccentric to that of the driving gear. Due to their eccentricity, there is a space between the external and internal gears which is occupied by a fixed spacer block known as the crescent whose function is to separate the inlet and output port areas.

The driving gear is driven either by a separate shaft or is keyed and slips over an extension of the front crankshaft main journal. The outer-gear axis of rotation is maintained entirely by the pump casing wall. The housing of the pump around a crankshaft journal provides a very compact pump unit which is also capable of providing large flow outputs at relatively low crankshaft speeds.

The rotation of the gears creates a low-pressure area at the inlet suction end of the crescent which draws in oil. As the gearwheels rotate, oil will be trapped between the teeth of the inner driver gear and the inside crescent side wall, and between teeth of the outer gear and the outside crescent side wall. These teeth will then carry this oil round to the other end of the crescent, where it

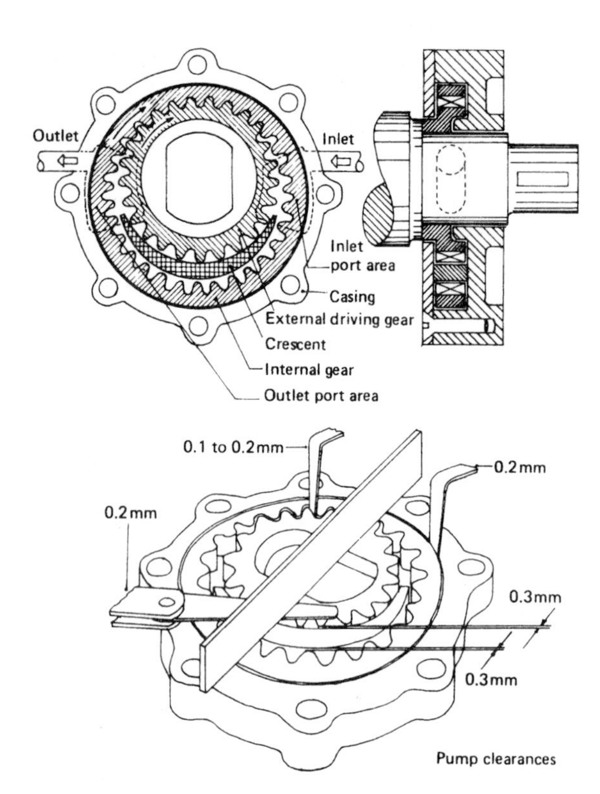

Fig. 19.24 Internal-gear crescent pump

will then be discharged by both sets of teeth into the outlet port chamber.

Each tooth space will hold a definite quantity of oil which will be moved round from the inlet to the outlet end of the crescent. The continuous supply of oil to the outlet side will increase the quantity of oil being discharged, but, once the space between the gear teeth has been filled with oil, the extra oil squeezed out from the teeth gaps will go into increasing the oil pressure. The mean oil pressure and the rate of circulation will then depend on the amount of oil escaping from the lubrication system's bearings.

The pump components may be checked for correct clearance and wear using feeler gauges:

a) Measure the clearances between the gear tooth tips and the crescent wall for each gear. These should not exceed 0.3 mm when both gears are in position with the pump body.
b) Measure the backlash between the meshing gears. This should be between 0.1 and 0.2 mm.
c) Measure the clearance between the outer gear and the body. This should not exceed 0.2 mm.
d) Measure the gear end-float clearance by placing a straightedge across the open face of the pump casing and slipping feeler blades between it and the gear side faces. This clearance should not exceed 0.2 mm.

19.3.3 Eccentric bi-rotor pump (fig. 19.25)

This pump consists of an inner and an outer rotor housed in the pump body, the outer rotor being eccentric to the inner. The inner rotor is pressed on to the oil-pump shaft and is positively retained by serrations. This rotor has four lobes which mesh with five segments in the outer rotor. Rotation of the inner rotor thus causes the outer rotor to revolve also, but at a speed which is slower by the ratio of the number of lobes to segments.

Initially oil is drawn, via the inlet port, into the space between the inner and outer rotors. Owing to their eccentricity and difference in size, the inner rotor drives the outer rotor at a slightly lower speed and the gap between the lobes increases, which induces oil to fill up this space. Eventually the space between the rotor lobes moves beyond the inlet port, thus trapping the oil, which is then carried around between the rotor lobes and segments. With further rotation, this effective cavity formed will decrease in volume and will eventually be exposed to the delivery port, so that the oil will be discharged under pressure to the oil filter.

The pump acts by continuous repetition of this process. Oil flowing into the space between the rotors from the inlet port is carried round between the rotor lobes and segments and then, as the space decreases, is forced through the pump discharge port.

The pump components may be checked for correct clearance and wear using feeler gauges:

a) Measure the clearance between the rotor lobe tip and the segment. This should be checked in two different positions and should not exceed 0.2 mm.
b) Measure the clearance between the outer rotor and the body by inserting feeler blades between them. This should not exceed 0.25 mm.
c) Measure the end-float clearance between the end-plate and the rotors by placing a straightedge across the open face of the pump casing and inserting feeler blades between it and the rotors, as shown in fig. 19.24 for the internal-

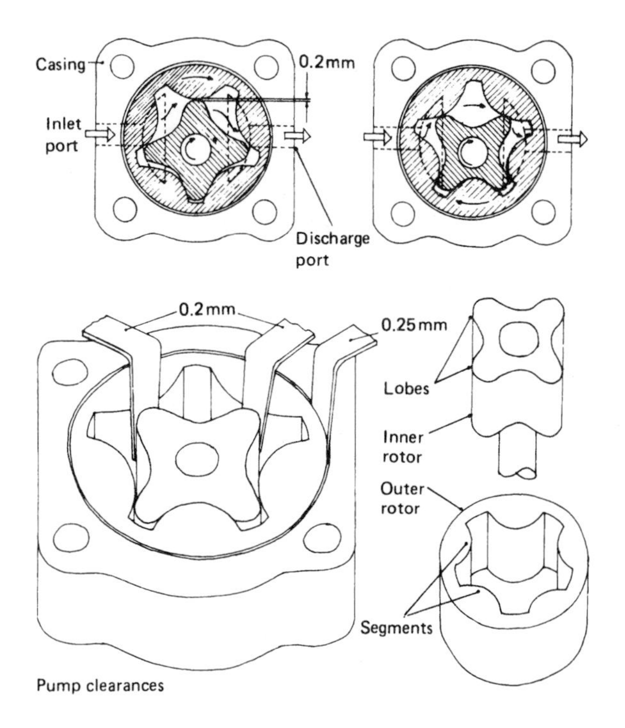

Fig. 19.25 Eccentric bi-rotor pump

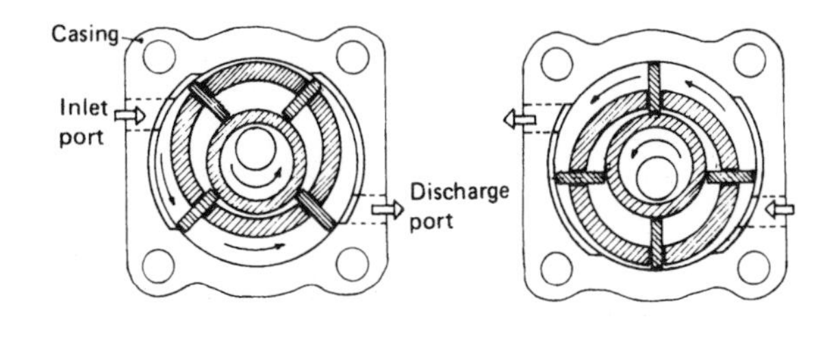

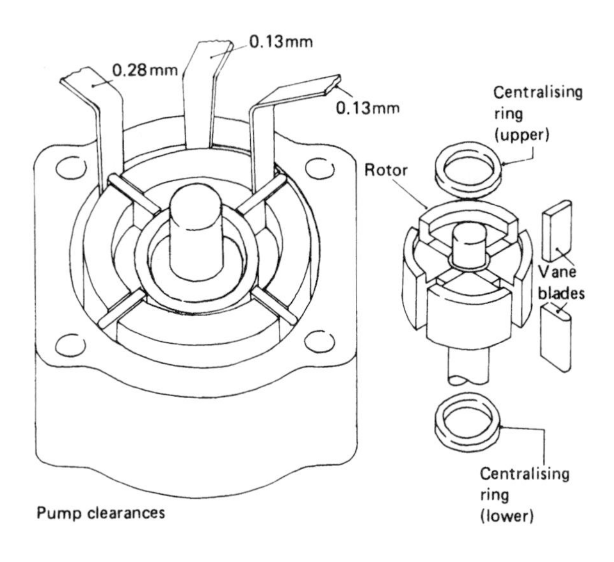

Fig. 19.26 Sliding-vane pump

gear crescent pump. This clearance should not exceed 0.2 mm.

19.3.4 Sliding-vane eccentric pump (fig. 19.26)
This pump consists of a rotor housed eccentrically in a cylinder bore machined in the pump body. The rotor is pressed on to the oil-pump shaft and is positively retained by a pin. Four sliding vanes are located in grooves machined in the periphery of this rotor and are positioned by centralising rings on each side of the rotor. The vanes are held against the pump-body wall by centrifugal force while the pump is operating.

As the rotor revolves, the vanes pass over the inlet port formed in the side of the pump body. The space between the vanes increases, due to the eccentricity of the rotor shaft to the casing wall, and oil will be drawn into the space between the rotor and the pump-body wall. This oil is then carried round between the vanes beyond the inlet port, where the space between the rotor and the pump bore decreases. Eventually the oil is forced out through the discharge port to the oil filter and oil galleries. The excessive quantity of oil being displaced will further increase the oil pressure in the engine's lubrication passages.

The pump components may be checked for correct clearance and wear using feeler gauges:

a) Measure the clearance between the rotor and the body by inserting feeler blades between them. This should not exceed 0.13 mm.
b) Measure the clearance between the vane and the body by inserting feeler blades between them. This should not exceed 0.28 mm.
c) Measure the clearance between the vane and the rotor groove by inserting feeler blades between them. This should not exceed 0.13 mm.
d) Measure the rotor and vane end-float by placing a straightedge across the open face of the pump body and then slipping feeler blades between it and the rotor and vanes. This clearance should not exceed 0.13 mm.

19.3.5 Oil-pump drive arrangements (fig. 19.27)
Crankshaft oil-pump skew drive is used when high-mounted camshafts are employed. A short

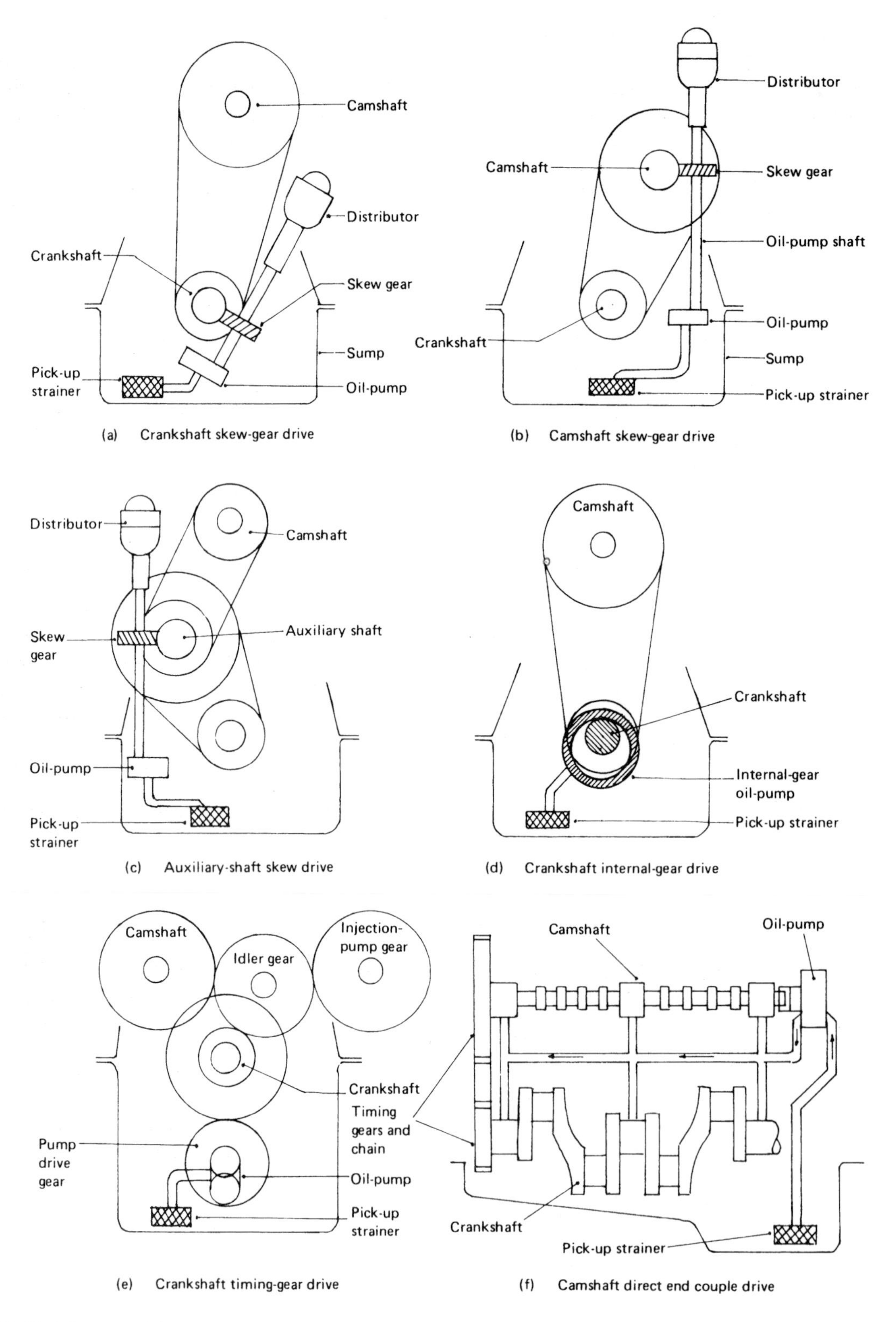

Fig. 19.27 Oil-pump drive arrangements

shaft with a skew gear meshes with a gear mounted on the front of the crankshaft and drives both the ignition distributor and the oil-pump with a 2:1 gear reduction (fig. 19.27(a)).

Camshaft oil-pump skew drive is used when a low-mounted camshaft is employed. A long shaft with a skew gear meshes with a gear machined directly on the camshaft. There is a 1: 1 gear ratio, and this shaft drives both the ignition distributor and the oil-pump (fig. 19.27(b)).

If a double-stage timing chain is used with a high-mounted camshaft and an auxiliary shaft (sometimes known as a jack shaft), it is often easier for the manufacturer to take the oil-pump drive from this shaft. Again it has a dual function of driving both the distributor and the pump (fig. 19.27(c)).

With small compact engines with high-mounted camshafts, an internal-gear crescent pump is most suitable. This pump occupies very little space and just sits over a keyed external gear on the crankshaft which has the dual function of driving the pump and generating part of the pumping action (fig. 19.27(d)).

For some transverse-mounted engines with low-mounted camshafts, a pump driven off a coupling situated at the end of the camshaft is preferred. This dispenses with a separate drive shaft and again is compact (fig. 19.27(f)).

For medium- and large-sized commercial engines, a pump driven directly from the crankshaft timing gear is preferred (fig. 19.27(e)). These forms of drive are more substantial and are necessary with large-output pumps.

19.4 Pressure-relief-valve control (figs 19.28 to 19.30)

The oil pressure in the lubrication system depends on the output from the oil-pump, which is directly proportional to the pump's speed. Pressure is built-up on the output side of the pump by the resistance caused by the fit of the bearing shells on the crankshaft journals – the smaller the bearing clearance, the higher the oil pressure. It is obvious that some sort of pressure-relief valve must be included in the system to prevent oil pipes from bursting, gaskets blowing, or the oil-pump drive gear stripping. Surplus oil vented through this valve will be returned to the sump. When the oil is cold and therefore thick, little oil flows through the bearings, even though the pressure may be up to the predetermined value, because of the relief valve returning oil back to the sump.

19.4.1 Types of pressure-relief valve

A simple relief valve may consist of a spring-loaded disc, ball, or plunger (fig. 19.28). When the working oil pressure has been reached in the oil gallery, the oil pressure will be sufficient to push back the spring so that the valve opens and bleeds off oil back to the sump. The actual setting pressure can be varied by altering the stiffness of the spring by means of an adjustable screw or by shims placed between the spring and the end-cap.

The disc-type valve is mostly used for filter bypass valves, the ball-type valve is more popular when the valve is incorporated in the pump housing for small and medium-sized engines, and the

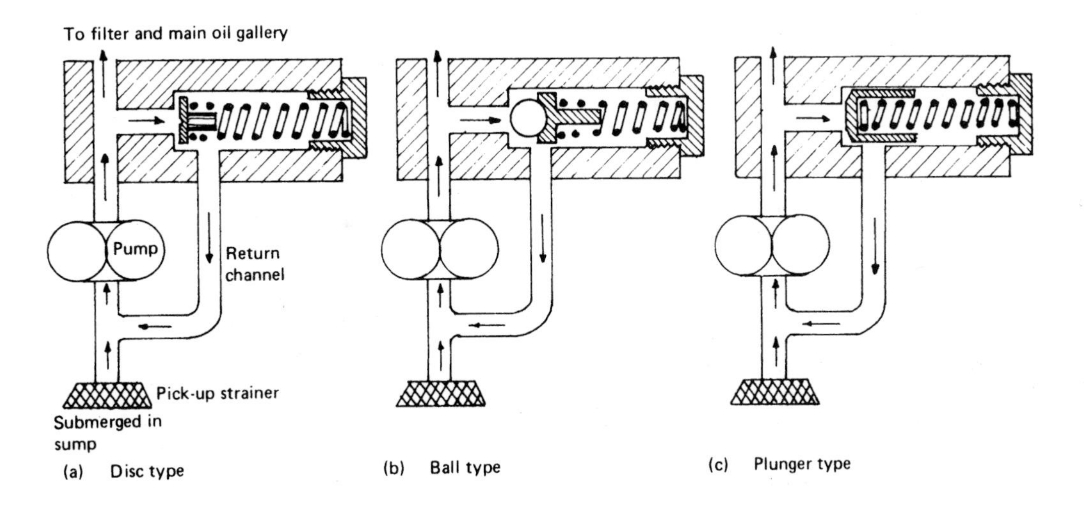

Fig. 19.28 Types of pressure-relief valves

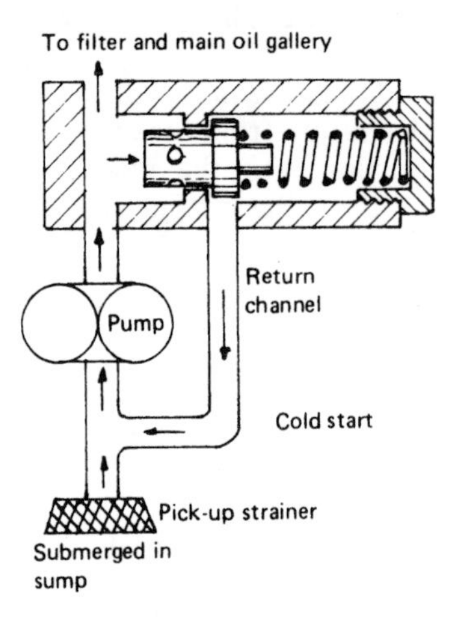

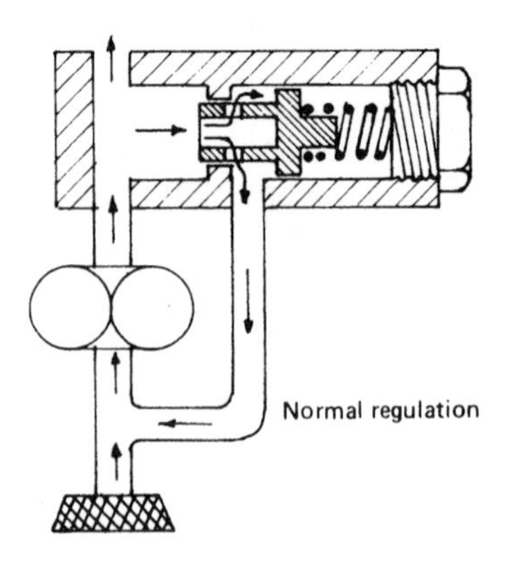

Fig. 19.29 Plunger relief valve with radial holes

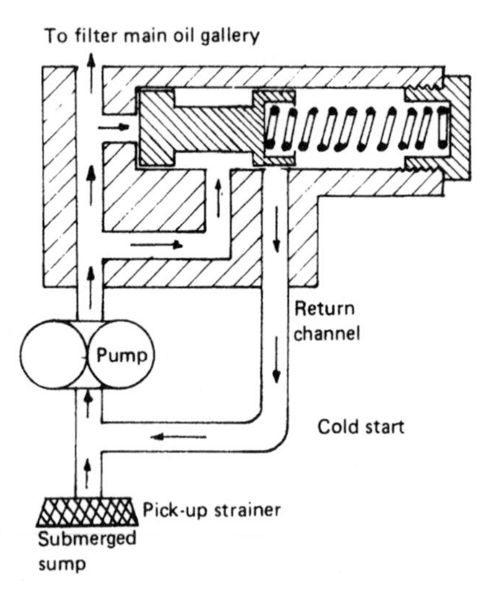

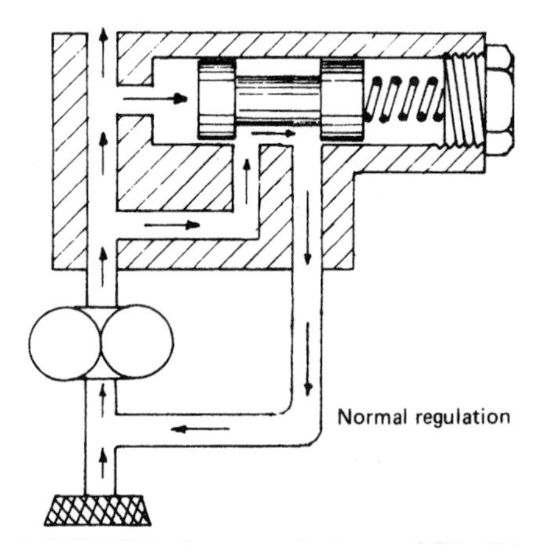

Fig. 19.30 Plunger relief valve with waisted middle

plunger-type valve is preferred on medium and large engines where better control is desired.

19.4.2 Plunger relief valve with radial holes

An improved plunger pressure-relief valve is shown in fig. 19.29. Here the plunger is hollow with radial holes. When the spring is compressed, the plunger moves back and exposes these radial holes to the low-pressure side. The oil will then flow out of these radial holes with the oil pressure still acting on the frontal area of the plunger. This provides a stable bleed-back of oil to the sump.

19.4.3 Plunger relief valve with waisted middle

A more sophisticated arrangement, which is used on some large diesel engines, is shown in fig. 19.30. Here the plunger has a waisted middle region which is exposed to the pressurised oil. The front face of the plunger is subjected to oil-gallery pressure at all times. When the predetermined pressure has been reached and the spring collapses, the oil will escape around the waisted plunger and its control edge back to the sump.

With this form of valve layout, the sensing face of the plunger is in direct communication with the oil gallery and is completely independent of the leak-off circuit. It is claimed that this method provides a sensitive and accurate way of regulating the rate of oil circulation.

19.5 Oil filtration

Oil filters are incorporated into the lubrication system to remove finely divided solid matter from the oil. Solid matter can be airborne and can get into the engine with the induction air and through the crankcase breather. Some abrasive matter originates within the engine as a result of combustion products blowing past the piston rings – this produces wear and corrosion which can be reduced to a very slow rate but in practice cannot be prevented completely.

It has been found that the most damaging abrasive dust is in the diameter range from 5 to 15 micrometres. Dust finer than this would not be able to enter the oil film, and coarser dust would not be able to bridge the oil film, so both finer and coarser dust will do less damage.

Oil filtration takes place in two stages:

i) a metal-gauze strainer surrounding the intake pick-up pipe prevents sludge and the larger particles of carbon and grit getting into the oil stream,
ii) a finer filter located on the outside of the engine removes the finer grit, dust, and metal particles.

The externally mounted filters may be classified by how they are connected into the lubrication oil-ways. There are two methods:

a) between the oil-pump and the main oil gallery – this is known as a full-flow filter system, since all the oil from the pump passes through it;
b) between the main oil gallery and the sump – this is known as a bypass filter system, since only a proportion of the oil from the pump passes through it.

19.5.1 Oil-filter materials (fig. 19.31(a))

The process of filtration is achieved by having a porous material which lets through the oil with very little resistance but prevents the suspended unwanted small solid particles from entering the oilways leading to the engine's bearings.

The unwanted particles are prevented from going through the filter medium either because they are larger than the pores or because, if they are smaller, they become trapped at changes of direction within the pores, which are honeycombed and tunnel in all directions.

The maximum particle size which can penetrate through the filter unit will depend upon the filter-medium material. As a rough guide, Table 19.2 lists the order of particle size, in micrometres, which can be transmitted.

Table 19.2 Characteristics of filter media

Filter medium	Maximum particle size transmitted (μm)	
Wire gauze	40–60	Surface filters
Wire-wound strainers	20–30	
Cloth	40–50	Depth filters
Felt	20–30	
Paper	10–20	
Cotton linters	5–10	
Plastic-impregnated paper	2–8	
Sintered metal	1–5	

Primary filters on the inlet side of the oil-pump normally use coarse materials and are generally known as surface filters, since the particles are retained on the outer surface of the filter medium.

Secondary filters on the outlet pressurised side of the oil-pump are generally known as depth filters, since the filtering process is progressive as the lubricant passes through the thickness of the filter medium. Unfortunately, as the filter medium becomes finer, the resistance to fluid flow generally increases, so that some of the better filtering media are suitable only for bypass filter systems.

The useful life of a filter ends when the filter becomes so clogged with particles and sludge that too much resistance is created to forcing the oil through it. To extend the period of use between filter replacements, the filter medium has its surface area increased by being wound in a star or corrugated fashion (fig. 19.31(a)). Coarser but thicker filter media can sometimes achieve equal results to fine thin-sheeted filter media, with less resistance to flow and relatively long life. Sometimes acetate blocks are used in addition to the corrugated-paper filter, to give supplementary fine filtering.

With a new filter cartridge, the flow resistance under normal duty lies between 0.2 and 0.5 bar. Generally a rise in resistance up to about 2 bar due to the clogging of the filter element is permissible.

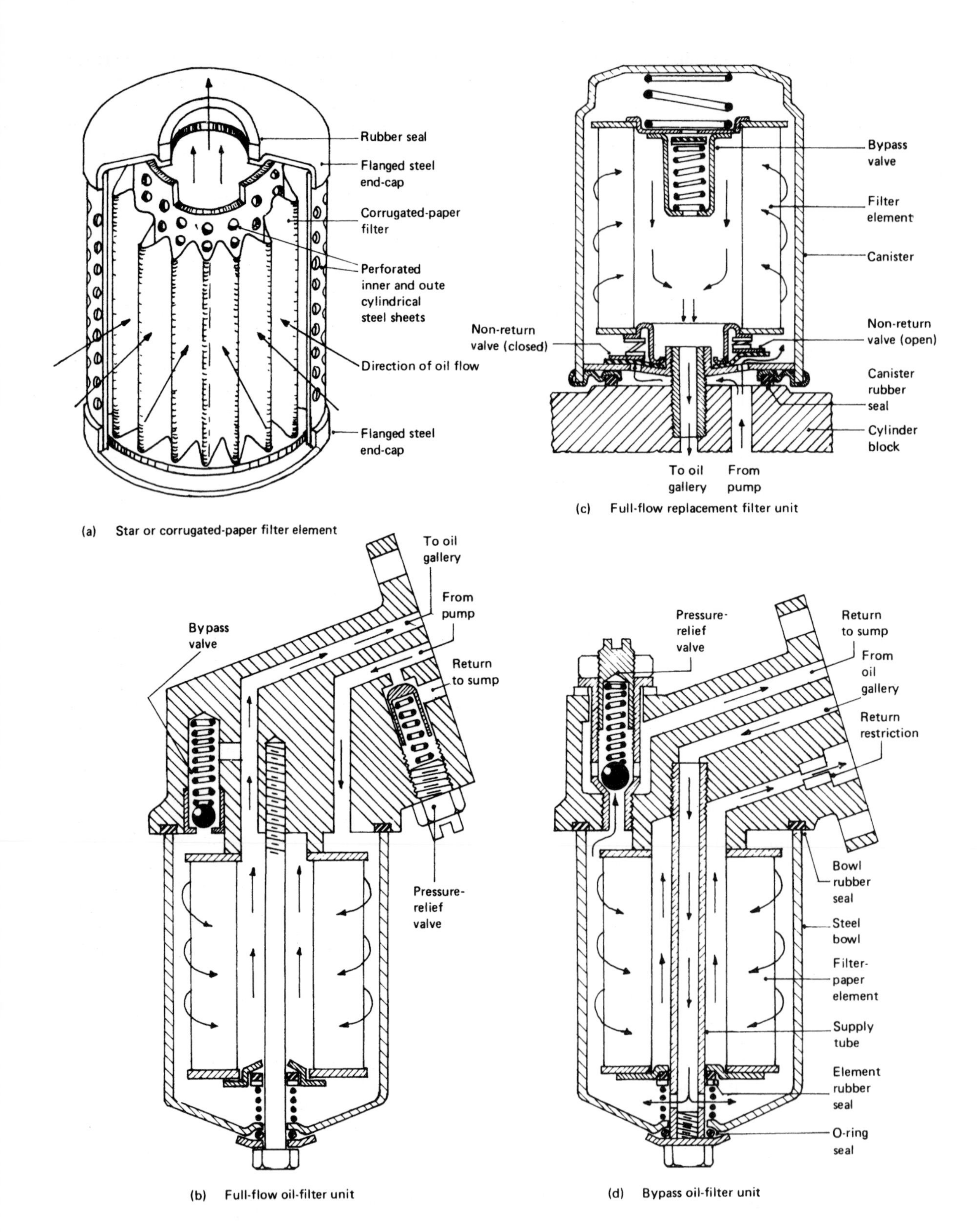

Fig. 19.31 Filter units

19.5.2 Full-flow oil-filter units (figs. 19.31(b) and (c))

Oil supplied by the pump enters the filter bowl, circulates, and forces its way through the filter-element material towards its centre (fig. 19.31(b)). The oil then passes out to the main oil gallery. If the filter element becomes choked, an emergency bypass valve opens and bypasses the filter element, so that unfiltered oil goes direct to the main oil gallery. The pressure setting of this valve is only 25 to 30% of that of the main pressure-relief valve – this ensures that blockage of the filter will not result in all the oil being dumped back to the sump and so starving the bearings of lubricant.

Inverted and horizontal-mounted filter units of the disposable-cartridge type (fig. 19.31(c)) incorporate a non-return diaphragm valve which allows oil to pass into the bowl under pressure but prevents the existing oil in the filter bowl draining back when the engine stops. This reduces priming time when the engine is restarted. A bypass disc-valve situated in the top of the filter element allows oil to pass unfiltered to the oil gallery if the pressure drop across the filter becomes excessive.

The filter-unit head sometimes houses a pressure-relief valve which will open and return oil back to the sump once the oil-gallery pressure has reached the blow-off setting (fig. 19.31(b)).

Filter-element materials in full-flow units are usually coarser or larger than in bypass units, to prevent excessive resistance to the flow of oil. This means that they are not so effective in preventing very small solid particles from entering the system.

19.5.3 Bypass-flow oil-filter units (fig. 19.31(d))

In these units, oil from the pump is forced along the inlet passage in the filter-head unit and down through the centre of a steel tube whence it flows from radial holes in the tube wall into the filter bowl, circulating the filter element (fig. 19.31(d)). It then penetrates the star-wound filter medium (fig. 19.31(a)), and the filtered oil then makes its way up the centre of the element and out through the oil restrictor. About 5 to 10% of the total oil circulating from the pump is metered by the oil-return restrictor orifice, then bypassed back to the sump.

When the safe working oil pressure is reached, the pressure-relief-valve spring and ball are pushed back to return surplus oil to the sump. Adjustment is achieved by screwing in to increase the pressure and out to decrease the pressure.

Generally the filter material used with bypass filters is very fine, since the filter does not interfere with the circulation in the lubrication system.

19.6 Engine oil-leakage prevention

The lubrication system runs between various engine components which have been assembled together. The joints and junctions of parts of the engine which involve oilways and oil storage must be made leakproof. The sealing of these components may be divided into two kinds:

i) static seals are used for components which are joined together with machined pressure-face joints or one machined face and a steel pressing,
ii) dynamic seals are required for components which have rotary motion relative to the engine-block assembly and have an external drive.

19.6.1 Static seals

A static seal is positioned between either two rigid components or one rigid and one semi-flexible component, so that, when they are bolted together, the seal deforms to occupy any available space existing between the components due to unwanted but unavoidable manufacturing imperfections such as surface roughness, waviness, distortion, and misalignment. Static seals are used to prevent leakage between the crankcase and the sump; between the cylinder block and covers for the timing gears and chain rockers, camshafts, and the cylinder head; and sometimes between the cylinder head and the cylinder block.

Cork, paper, or paper compounds are used for the common seals between the sump and the crankcase, the rocker cover and the cylinder head, and the timing cover and the cylinder block.

The sealing of the sump-to-crankcase joint at the flywheel end of the engine has always been a problem. Figure 19.32(a) shows one approach, where the main-bearing cap (with an extended width to house the crankshaft seal) has a rectangular groove machined round the outside to support a cork gasket strip. The cork is pressed round the semicircle of the grooved recess. The crankcase-to-sump flange gasket can also be seen lying on top.

Figure 19.32(b)shows the method used when the crankcase has extended side-members, so that the rear main-bearing cap appears to be sub-

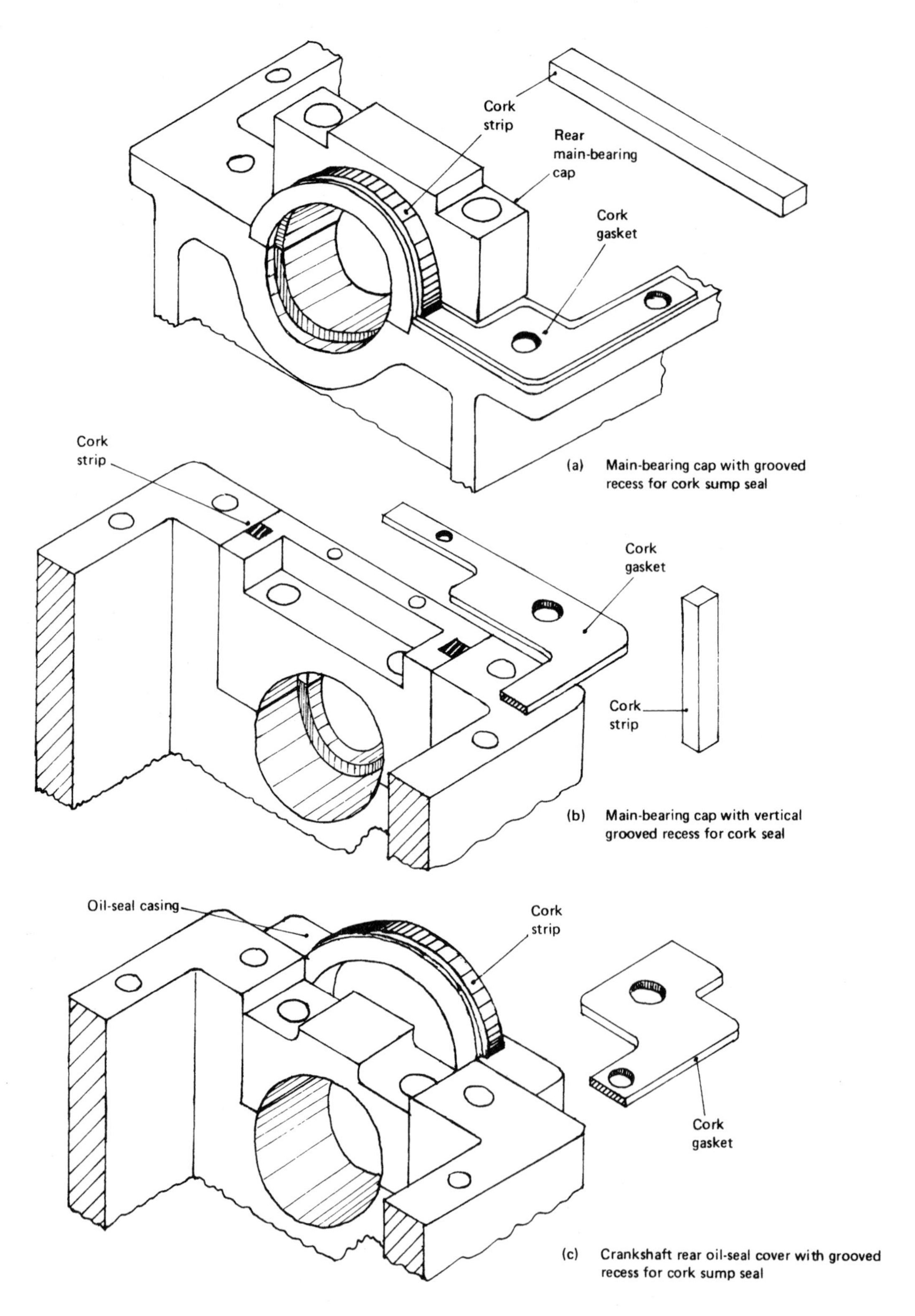

Fig. 19.32 Sump flywheel-end gasket sealing methods

merged in the crankcase. Sealing between the rear main-bearing cap and the crankcase side walls is achieved by a cork or asbestos strip which fits into straight grooves recessed on both sides of the bearing cap. The sump gasket can then simply sit over the crankcase flanged face and the main-bearing cap which lies flush with it.

Figure 19.32(c) shows a method of sealing the sump-to-crankcase joint when again the crankcase side walls are extended. In this case, the crankshaft rear oil-seal uses a separate housing or casing bolted on to the back of the crankcase. This casing has an external groove formed round the exposed semicircle of the casing. Again, a cork or asbestos strip is pressed around and into the groove, and the flat sump gaskets will lie on either side on the crankcase flanged face.

Oilways between the cylinder head and block, the camshaft block and the cylinder head, the rocker-shaft pedestal and the cylinder head, and external and internal pipes screwed into the crankcase may be sealed by synthetic-rubber O-rings (figs 19.10 to 19.13).

In some cases, oil sealing between the cylinder head and the cylinder block will also make use of the common cylinder-head gasket. This may be made from crimped steel, copper–asbestos, or steel–asbestos.

19.6.2 Dynamic seals

These seals prevent oil leaking past components which have relative rotary movement. Such seals take the form of contact radial or axial seals or spiral-thread clearance seals. They are used to prevent leakage of oil from both the front and rear of the crankshaft.

Spiral-thread clearance seal (figs 19.33(a) and (b) and 19.34(d)) This type of seal consists of a flanged rim or washer acting as an oil-flinger and a spiral thread machined on either the outside of the rotating shaft or the inside of the stationary member. When the shaft revolves, surplus oil which has splashed on to the crankshaft will be deflected by the oil-flinger and will drain down and back to the sump. A small amount of oil will still continue to spread over the oil-flinger towards the outer end of the crankshaft. This oil will then meet the spiral thread, which will screw it back towards the oil-flinger. Oil will then accumulate at the back of the oil-flinger shoulder where it will be flung out radially inside the cover-plate recess, to drain away. Most of the oil will be returned by the leading thread edge, and subsequently the remaining oil which does pass the first thread will be scraped back by the second thread followed by the third thread and so on.

Dynamic radial or axial contact seal (figs 19.33(c) and (d) and 19.34(a) to (c)) The radial-packing oil-seal (figs 19.33(c) and 19.34(c)) simply consists of asbestos rope or compound pressed into a trapezium-shaped recess in the crankcase or timing cover. The initial oversize of the packing section will force it against the shaft, so that it will continuously rub against the revolving shaft, even after it has settled down. Oil will in the first instance be deflected and flung out by the oil-flinger shoulder, but a reduced amount will find its way along the shaft or pulley sleeve to the packing. This oil will 'wet' the interface of the packing and the cylindrical surface and this will have the effect of creating a molecular oil barrier against any further outward movement of oil from the shaft.

The radial-lip oil-seal (figs 19.33(d) and 19.34(a)) consists of a flexible synthetic-rubber web moulded into a steel flanged sleeve which is itself a force fit in the cover-plate housing or timing cover. The lip-seal edge is held lightly in contact with the shaft and, to increase its resilience, has a circumferential garter spring applying some tension around the outside of the rubber just over the seal's lip. Any eccentricity of the shaft or the seal-housing bore will be absorbed by the flexing of the portion of the rubber web parallel to the shaft. These seals are usually single-lip, but double-lip seals are sometimes preferred on large engines, and if the latter are used it is usually recommended that grease is packed between the two lips when installing new seals. The inner lip's job is to stop the oil getting out, and the outer lip prevents the entry of dirt and dust from the atmosphere.

The axial-lip seal (fig. 19.34(b)) is similar to the radial seal except that the lip face bears against a flat smooth face machined on a washer disc. Light contact is maintained between the seal lip and the disc face by a coil spring whose tension has been predetermined for its application – this has been shown to be a critical factor in effective sealing. This method of dynamic sealing is generally preferred only on very large engines, where the space required and the pre-loading of the lip pressure are justified.

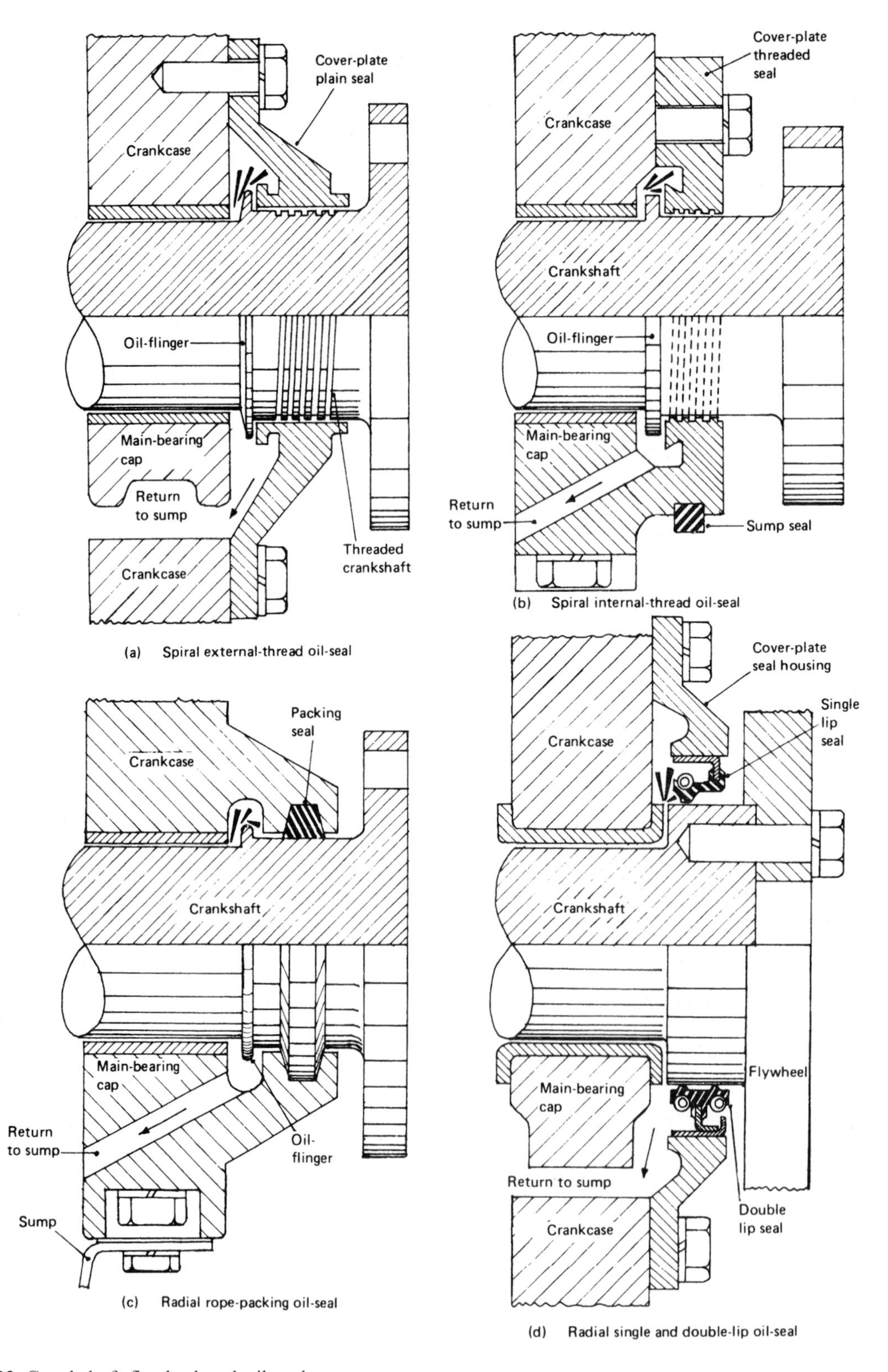

Fig. 19.33 Crankshaft flywheel-end oil-seals

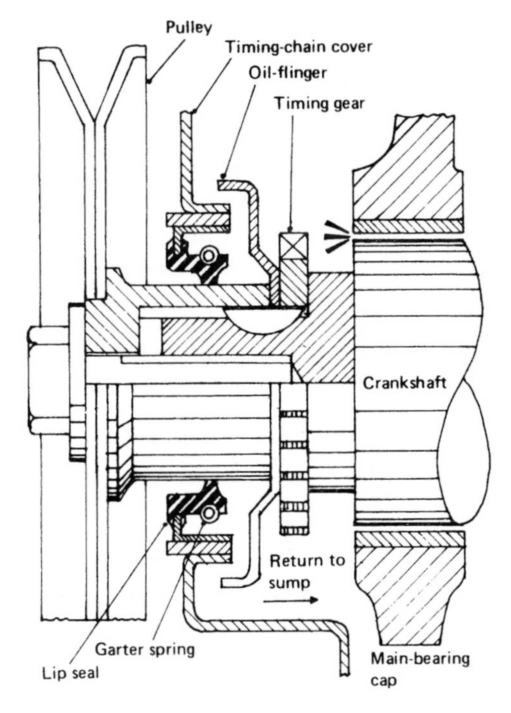

(a) Crankshaft front radial-lip seal

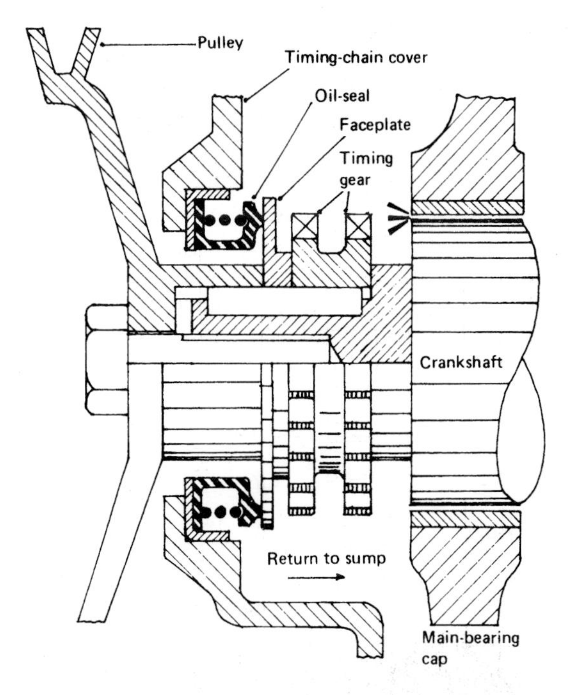

(b) Crankshaft front axial-lip seal

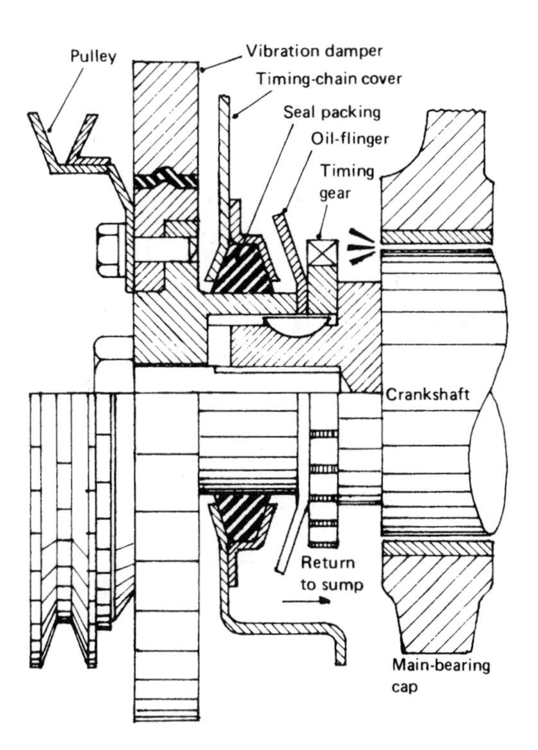

(c) Crankshaft front rope-packing seal

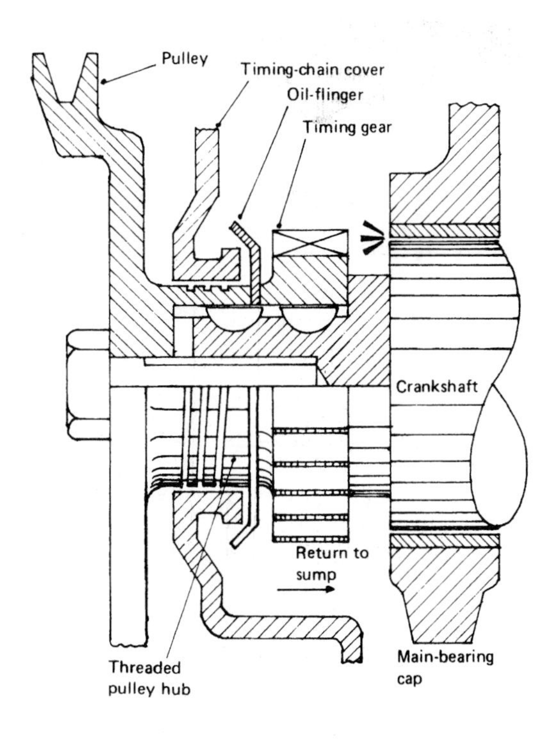

(d) Crankshaft front spiral-thread clearance seal

Fig. 19.34 Crankshaft pulley-end oil-seals

19.7 Dry sump lubrication system with liquid-to-liquid oil cooler (fig. 19.35)

Engines can be subjected to abnormal operating conditions, such as racing-car engines running continuously at high engine speeds, large diesel stationary engines working continuously at high brake loads, supercharged engines, engines used on earth-moving equipment which experience repeated bursts of heavy loading, and engines installed in vehicles which are designed to travel over rough ground and are subjected to large amounts of longitudinal and transverse tilt. Dry sump lubrication systems generally incorporate a pair of oil pumps:

1) a large low-pressure scavenging pump;
2) a smaller high-pressure circulation pump.

19.7.1 Scavenging pump (fig. 19.35)

This pump collects the hot aerated and expanded oil which drips into the sump from the deliberate spillage of oil from the crankshaft and camshaft journal bearings, oil spray that is directed onto the cylinder bore walls, the underside of the pistons, the faces of the camshaft lobes and the tim-

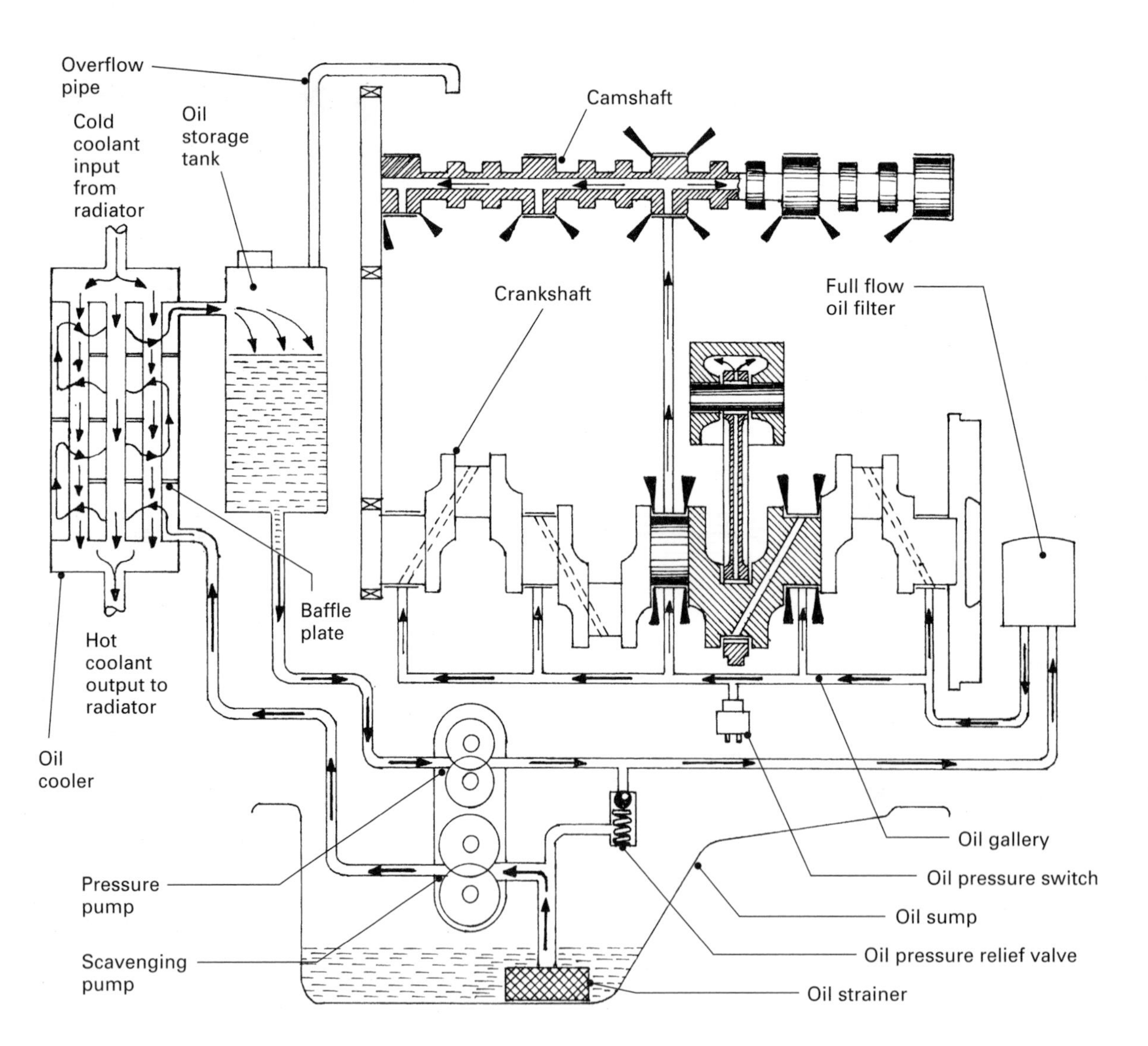

Fig. 19.35 Dry sump lubrication system with liquid-to-liquid oil cooler

ing gears, and dumps it into the storage tank by way of the liquid-to-liquid oil cooler.

19.7.2 High-pressure pump (fig. 19.35)
This pump draws cooled oil from the storage tank and delivers it under pressure via the full-flow oil filter to the main oil gallery. It is then fed to the individual main bearings, through diagonal crankshaft drillings and to the big and small end bearings of the connecting-rod. Oil is also supplied to the camshaft where it is distributed through a central drilling to each support bearing.

19.7.3 Oil cooler (liquid-to-liquid heat exchanger) (fig. 19.35)
This is a heat exchanger which extracts heat from the hot lubricating oil and conveys it through the circulating coolant to the radiator where it is dispersed into the surrounding atmosphere. This liquid-to-liquid heat exchanger consists of a stack of copper tubes supported by sealed end-plates and enclosed in an aluminium-alloy casing. The engine coolant flows through the tubes whereas the lubrication oil is guided by baffle-plates to pass several times over and across the tubes from one to the other, working its way from the entrance to the exit in the opposite direction to the flow of the coolant. Thus, as the oil passes one way and then the other over the tubes, heat will be transferred by conduction through the walls of the tubes. It is then absorbed by the coolant which then delivers it to the engine's cooling system radiator.

19.7.4 Oil storage tank (fig. 19.35)
After leaving the oil cooler, the de-aerated cooler oil lubricant enters the steep storage tank to provide a ready supply of cool oil at gravity head pressure to the inlet side of the high pressure feed pump. Excess tilting or pitching of the engine will not prevent this oil supply being delivered to the high pressure feed pump. Thus at all times high-pressure lubricating oil will be circulated to the main oil gallery and to the various parts that require lubricating.

19.8 Crankcase emission control

19.8.1 Piston ring blow-by
Piston rings can have a very near but never a perfect circular shape that matches the cylinder wall contour, if it is to exert the radial outward expansion force necessary to make contact between the rings and cylinder wall. In addition, the joint gap clearance will always be a source of gas leakage. Consequently, there is always a small amount of compression/combustion gas loss which escapes past the compression and oil control piston rings and so enters the crankcase. These gases may be burnt products of combustion carbon dioxide or they may be partially burnt products such as hydrocarbons, carbon monoxide and steam. Piston blow-by increases with engine speed and in particular when the piston and piston-ring and cylinder bore are worn.

The effects of piston blow-by are to increase the concentration of combustible air/fuel mixture and may cause an explosion in the crankcase. The gas and fume vapours may also condense and contaminate the lubrication oil and finally these gases will pollute the atmosphere if expelled from the crankcase. The normal technique to remove these unwanted gases is to pass them back into the engine cylinder so that they are burnt with the freshly supplied charge.

19.8.2 Positive crankcase ventilation (PVC) system (fig. 19.36(a) and (b))
The arrangement shown in fig. 19.36(a) and (b) is a camshaft-cover to the induction manifold circulation system.

Normal engine speeds (fig. 19.36(a)) Air is drawn from the air filter to the camshaft-cover via a tee-junction pipe and rubber hoses, and from the tee-junction pipe to the crankcase via the oil and mist separator through the vertical rubber hoses (fig. 19.36(a)). Thus air is then circulated around the crankcase. It collects the blow-by gases, and is then drawn up the two vertical camshaft drain passages. The air, gas and fumes under the left-hand camshaft-cover then pass through the tee-junction pipe to the right-hand camshaft-cover. The accumulated gases are then drawn into the induction manifold via the ventilation valve and pipe due to the depression created by the throttled air charge entering the cylinders (throttle valve not shown). These gas and vapour fumes are then drawn into the cylinders on each of the induction strokes so that in effect the engine is consuming its own blow-by gases and fumes. Some of these fumes will contribute to the combustion process and will then pass through the normal exit of the exhaust port to the exhaust system, where they may be given further after-burning treatment before being expelled into the atmosphere.

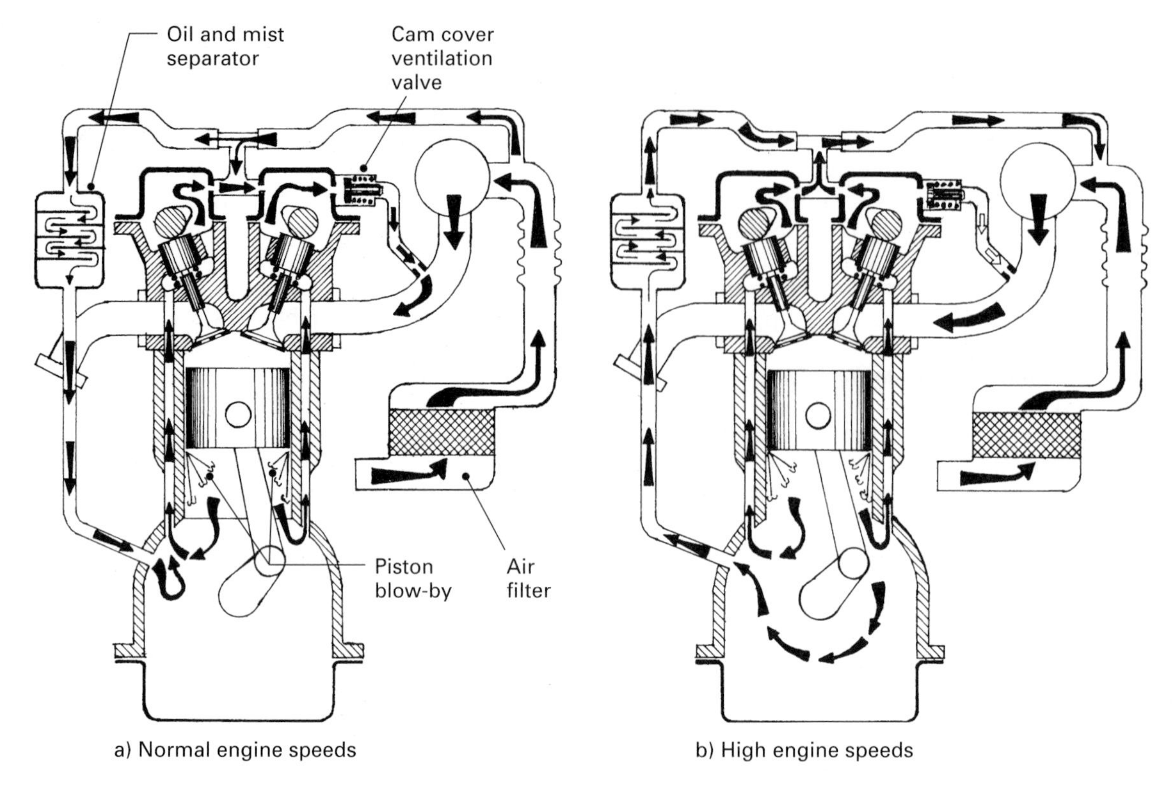

Fig. 19.36 Positive crankcase ventilation (PCV) system cam cover to induction manifold circulation

High engine speed (fig. 19.36(b)) At very high engine speeds (fig. 19.36(b)) the crankcase to induction manifold flow path is insufficient to accommodate the amount of blow-by due to the decrease in manifold depression when the throttle is practically wide open, as this prevents the ventilation valve fully opening. Hence the direction of flow in the oil and mist separator changes from cam cover to crankcase, to crankcase to air filter, and the flow from both cam covers now diverts from the ventilation valve to the air filter via the tee-junction pipe where the gases are again drawn into the induction manifold. Thus the blow-by gases are now drawn into the combustion chamber by way of the camshaft-covers and the oil mist separator passage-ways via the tee-junction pipe.

19.9 Low oil pressure switch and warning light circuit (fig. 19.37(a) and (b))

The low oil pressure switch which is screwed into an oilway in the cylinder block is designed to switch on the oil warning light bulb on the instrument panel whenever the oil pressure in the engine's lubrication system is below some predetermined setting. Thus when the engine is stationary and the ignition switch is turned on, a red oil warning light will be visible. If the engine is in good condition, the light should be extinguished once the engine starts and is running at idling speed and above. If it does not switch off, or if it flickers continuously after the engine has warmed up, then there is the danger of low oil pressure. Check the oil level in the sump and if need be investigate further.

The oil pressure switch is made of two parts: a threaded steel body and an insulation moulding. Sandwiched between these two halves is a rubberised diaphragm and a brass contact pressing. One side of the diaphragm is exposed to the lubrication oil via a central oil hole, whereas the chamber on the other side of the diaphragm contains a brass cupped disc moving contact which is held in the closed position by a return spring. This spring also serves as a conductor to the central terminal lead to the warning light bulb. With the engine stationary and the ignition switch turned on, the

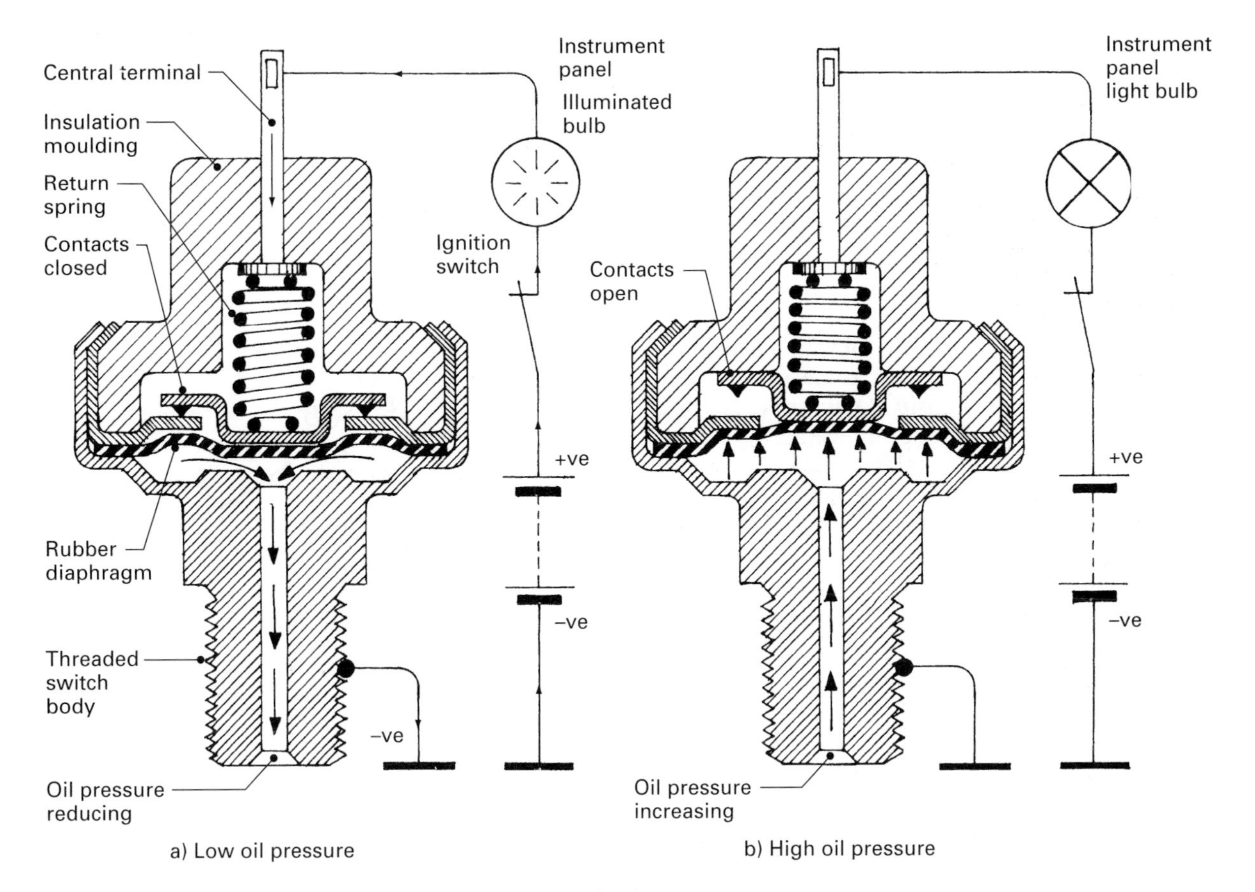

Fig. 19.37 Low oil pressure switch and warning light circuit

return spring closes the contacts, thus completing the warning light circuit. The bulb therefore should light up (see fig. 19.37(a)). When the engine is cranked and it starts up, the oil pump quickly builds up oil pressure. This pressure will be experienced by the underside of the diaphragm and so exerts an upthrust which pushes the moving contact clear of the fixed contact (see fig. 19.37(b)). The interrupted circuit will immediately switch off the warning light, indicating that the lubricating system oil pressure has reached some minimum value set by the manufacturers. This could typically be 0.5 to 2.0 bar when idling.

If the warning light does not come on when the ignition switch is turned on, short the warning light switch terminal to earth. If the light now comes on, suspect a faulty switch and replace it. If the warning light comes on continuously when the engine is running at all speeds, then substitute it for a switch that is known to work. If the light now goes out, replace the faulty switch. Finally if the light flickers or goes out only when the engine is cold or when revving up, check the lubrication system's oil pressure by removing the switch and screw a test gauge in its place.

19.10 Properties and selection of engine lubricants

Before considering the properties of an engine oil, a clear knowledge of the functions of the lubricant is necessary. Engine oils have to perform four major functions:

i) to lubricate the rubbing rotating and reciprocating components,
ii) to seal the piston rings and grooves against combustion leakage,
iii) to remove both the heat due to combustion which has been absorbed by the metal components and any heat generated in the bearings,
iv) to prevent corrosion of the internal components in the engine.

19.10.1 Viscosity

This is the ability of molecules of oil to cling together and so drag adjacent layers of oil when relative motion exists between a pair of solid surfaces separated by an oil film. Viscosity is thus a measure of an oil's ability to resist internal shearing, and it should be high enough to resist the oil being squeezed out from between two bearing faces when subjected to compressive loads.

19.10.2 Oiliness

Oiliness is that property of an oil which makes it a more effective lubricant than some other liquid which has the same viscosity. Oiliness is responsible for the boundary layer of molecules which can adhere or cling to a metal surface and provide lubrication after the bulk of the oil has been displaced or squeezed out. Under certain conditions of lubrication, one oil may reduce the friction in a bearing more than a similar oil of the same viscosity applied in the same way. The oil that effects the greater reduction in friction is said to be 'oilier'.

19.10.3 SAE classification of oils

Oils were originally identified by a vague grading such as 'light', 'medium', or 'heavy', but the viscosity of any grade differed from one producer to another. This difficulty was overcome by a special system of oil-viscosity classification devised by the Society of Automobile Engineers in the USA and known as the SAE system of classification. These SAE specifications are shown in Table 19.3.

The SAE numbers give no indication of quality, beyond providing a very rough description of the viscosity of a lubricant at a stated temperature.

The suffix 'W' is used to designate grades whose

Table 19.3 SAE crankcase oil specification

SAE Viscosity number	Viscosity range (Saybolt universal seconds (SUS))			
	At −17.8°C (0°F)		At 99°C (210°F)	
	Min.	Max.	Min.	Max.
5W	–	4 000		–
10W	6 000	12 000	–	–
20W	12 000	48 000	–	–
20	–	–	45	58
30	–	–	58	70
40	–	–	70	85
50	–	–	85	110

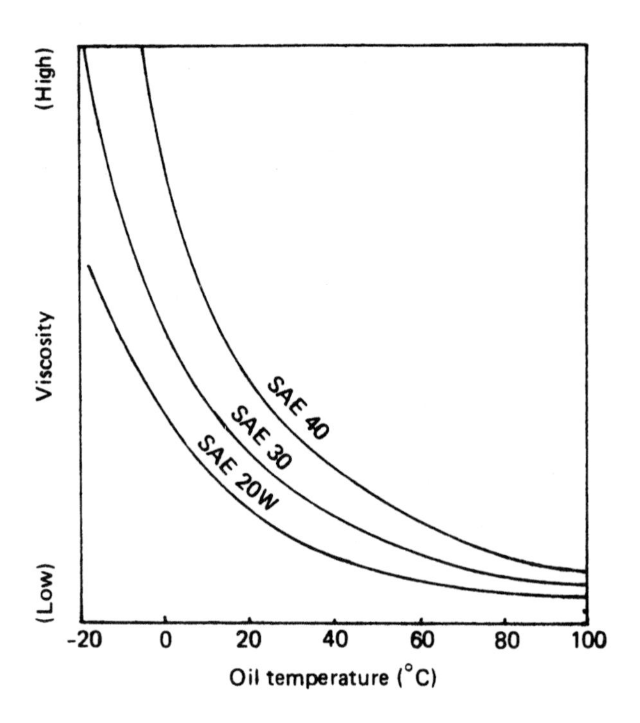

Fig. 19.38 Flow characteristics of different SAE oils

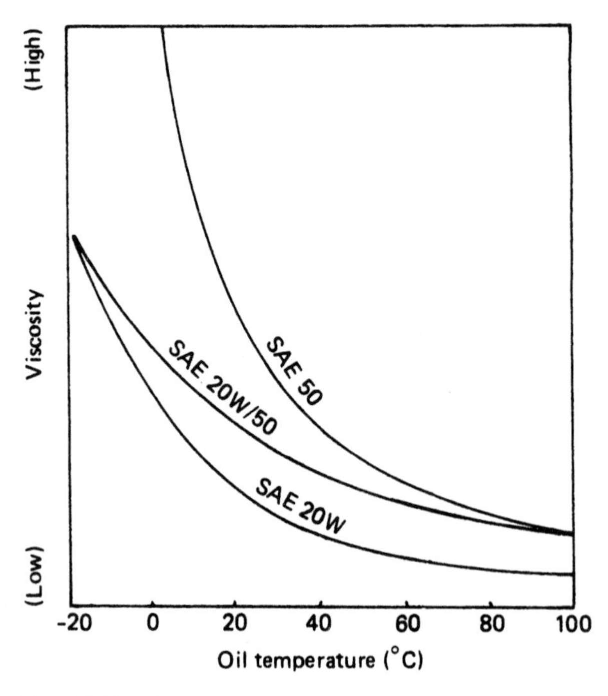

Fig. 19.39 Flow comparison of straight and multi-grade oils

viscosity limits are set at −17.8°C instead of 99°C. The lower temperature has been chosen in these cases to ensure better standardisation of cold-starting performance. Oils designated by 'W' are suitable for both winter and summer use.

Figure 19.38 shows the viscosity characteristics over a temperature range for three different SAE grades. An oil designated 20W/50 meets the requirement of SAE 20W at −17.8°C and the requirements of SAE50 at 99°C and is referred to as a multigrade oil (fig. 19.39).

19.10.4 Viscosity index, VI

A stated viscosity grade takes into account only the viscosity at one temperature – the way in which the oil alters in viscosity with temperature is not taken into consideration. Because of this, it is possible for an oil having a tendency to become very thick at low temperature and an oil that changes much less in viscosity for an equal change in temperature to be classified by the same SAE number. It is therefore necessary when considering oil quality – as distinct from viscosity only – to know the viscosity at two temperatures.

Viscosity index, VI, is a measure of a lubricant's consistency of viscosity with temperature change. A high viscosity index indicates that there should be little change in viscosity with a reasonable temperature rise. The advantage of a high viscosity index is that a low-viscosity oil may be chosen for an engine, so that the cranking resistance when starting the engine from cold is low but there is only a small loss of viscosity or 'body' at working temperatures.

Lubricants refined from petroleum have viscosity indices of 100 to 110, but with additives the viscosity indices have risen to 120 to 130. These additives consist essentially of very large inert molecules which can interfere with flow and so create an apparent viscosity which is far greater than that of the base oil. Since the additives' contribution to apparent viscosity does not vary so much with temperature as does the viscosity of the base oil, the effect is to flatten the viscosity–temperature curve. Viscosity improvers do not always show satisfactory chemical and physical stability, nor do they contribute anything but viscosity to the lubricating properties of the blend. Proper use of viscosity improvers is, however, valuable in practice.

19.10.5 Viscosity-index improvers (fig. 19.40)

Multi-grade oils do not actually thicken when they get hot, but they do not suffer the viscosity drop with temperature rise that would normally occur with lubricants refined only from petroleum.

With ordinary petroleum lubricants, the oil consists mainly of hydrocarbon molecules which

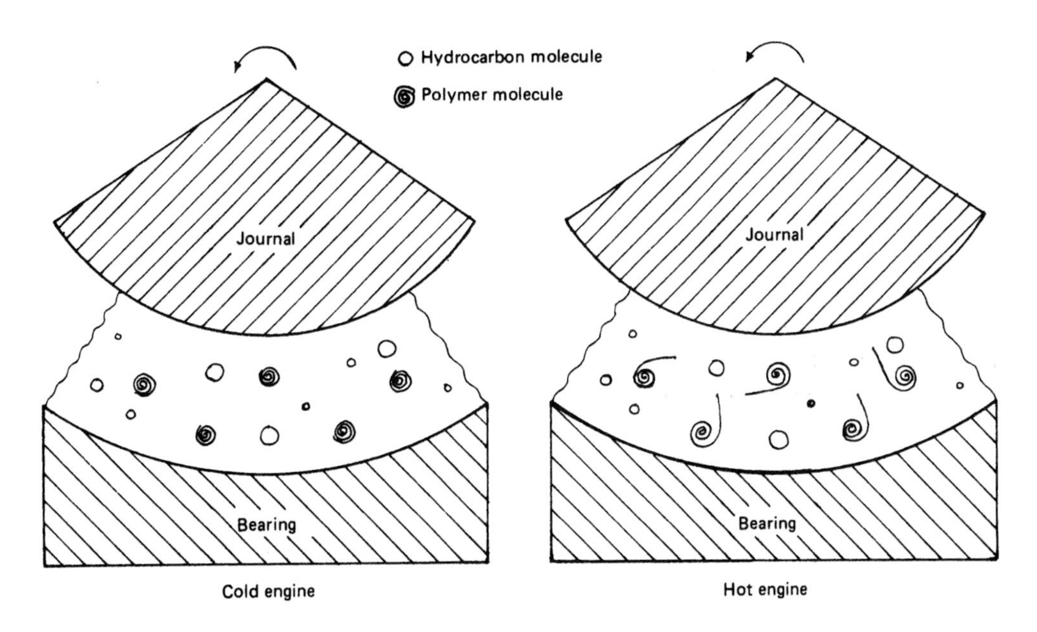

Fig. 19.40 Action of viscosity-index improvers

are in constant motion. When heated, the internal energy of the molecules increases so that they vibrate at higher frequencies, moving faster and further apart. This means that the internal friction of one molecule moving past another is reduced, so the lubricant becomes thinner.

To maintain the viscosity with rising temperature, the acceleration of the molecules must be reduced. At present, this has been achieved by adding a polymer such as one from the acriloid plastic family. These molecules are of the giant long-chain form and, when cold, are closely coiled-up, so that they occupy only slightly a larger space than any of the larger hydrocarbon molecules. The amount of polymer used is small and, when the liquid is cold, these polymer molecules are free to move through the liquid just like the bulk of the hydrocarbon molecules.

As the temperature rises, the polymer molecules unwind so that they occupy more space and restrict the acceleration of the hydrocarbon molecules through the liquid. The tendency of the hydrocarbon liquid to become thinner is thus compensated for by the increased resistance offered by the enlarged polymer molecules. This will prevent any large reduction in the oil viscosity as the temperature rises.

19.10.6 Engine-oil additives

To improve the various properties of lubricants, additives are included in the base oil. Their functions are as follows:

Detergents These improve engine cleanliness in the piston ring-belt and other hot regions. They prevent coagulation of deposit-forming gums, residues, etc. and hold them in suspension in the oil.

Dispersants These control cold-sludging and carry finely divided combustion products in coloidal suspension in the oil.

Oxidation and corrosion inhibitors These reduce the rate of oil oxidation and control alloy-bearing corrosion.

Anti-wear agents These protect rubbing components against wear, especially valve-train mechanisms where there is usually only boundary lubrication.

Oiliness agents These decrease friction of heavily loaded rotating and reciprocating components.

Anti-foaming agents By reducing the tendency to foam, these control oil foaming due to the churning and continuous movement of the oil.

Pour depressants These lower the solidification temperatures of oils.

Viscosity-index improvers These improve viscosity–temperature properties, control oil consumption, and make cold starting easier.

19.11 Correct maintenance of oil level

The oil level of the engine must be checked with the vehicle standing on a level surface. Withdraw the dipstick and wipe it with a clean rag or paper. Insert the dipstick fully into the dipstick tube and again withdraw it. The mark made by the oil on the lower end of the dipstick will indicate the oil level, which should lie between two scribed lines – if it does not, bring the level to the 'full' mark by pouring oil in the filler-cap hole.

Overfilling the sump may result in the following:

a) Excessive crankcase pressure build-up, which may damage the front or rear crankshaft dynamic oil-seal.
b) Static seals such as cork gaskets may be ruptured so that oil will leak out. Once a gasket is damaged, the movement of the vehicle will splash oil in all directions, so that leakage will continue even when the oil level is corrected.
c) Surplus oil may be splashed by the crankshaft against the cylinder walls, causing excessive oil consumption and blue exhaust smoke.
d) Excessive oil reaching the piston ring grooves will tend to carburise and may produce sticking rings and/or engine pinking.

Underfilling the sump may result in the following:

i) Too low an oil level may expose the pick-up pipe and gauze strainer so that air will be drawn into the lubrication system along with the oil. This may produce boundary-lubrication conditions instead of hydrodynamic lubrication.
ii) On long road bends or slopes, the oil may move to one side or end and so expose the pick-up pipe and gauze strainer. This may produce intermittent lubrication.
iii) The reduced amount of oil circulating may become overheated and so destroy the lubricant's properties and so increase and accelerate engine wear.

iv) The useful lubricating life of the oil may be shortened without the owner being aware that the oil needs to be replaced.

19.12 Oil contamination and replacement intervals

In time, oil in the lubrication system will become contaminated by soft and hard particles which can be abrasive and cause wear betwen bearings and journals and other sliding parts.

19.12.1 External contamination

The hard particles may come from dust induced into the air-intake-system crankcase breather, metallic particles worn away from moving parts, and casting sand still remaining in the engine after manufacture. Small pieces of dirt may enter the inside of the engine when changing the oil filter, removing the rocker cover when checking tappet clearances, or any other servicing which requires the removal of engine components and exposes the inside of the engine, e.g. removal of the cylinder head, timing gears and chain, distributor drive, etc.

19.12.2 Combustion contamination

Other particles may originate from the process of combustion, where the burning fuel forms carbonaceous particles, acidic material, and moisture. These come into the crankcase by piston-ring blow-by, and when mixed with the lubricant will tend to form sludge and deposits. If sufficient amounts of sludge and deposits exist, they will interfere with the filtering process so that abrasive particles will continue to circulate and so progressively damage the bearings.

In diesel engines, the sulphur in the fuel when burning with oxygen produces sulphurous acids. Small amounts of these acids may drain down the cylinder walls, mix with the oil, and attack the bearing materials.

19.12.3 Internal contamination

In addition to external forms of contamination, contaminants may be formed within the engine itself. The oil will oxidise over long periods of time, due to the high working temperature of the oil in the piston region. Oxidation is the process of the oil combining with the oxygen in the air at elevated temperatures – the higher the oil temperature, the more rapid will be the rate of oxidation. Oxidation causes thickening of the oil and formation of varnishes sometimes known as lacquer deposits. These deposits may form on the piston rings and in the piston grooves, eventually causing the rings to stick so that compression pressure will be lost. The net result will be a loss of power and an increase in fuel consumption. Some of the products of oil oxidation will be acidic and may corrode journal bearing materials.

19.12.4 Operating conditions

The amounts of sludge and deposits to a large extent depend on the mode of vehicle operation. Long-journey motoring keeps the engine working over a narrow and designed temperature range – this produces efficient combustion, so that by-products from burning and water formation are at a minimum. Short-journey running, or incorrectly regulated cooling systems, produces high levels of unburnt hydrocarbon products and moisture and crankcase condensation which will dilute the oil and destroy the lubricating properties of the oil.

19.12.5 Oil and filter replacement periods

Lubricating oils and filters should be changed regularly at the recommended intervals. The replacement of the oil and filters may be based on distance travelled or on a time-period scale. Vehicles used for journey operations always rely on distance intervals between oil changes, but stationary and marine engines and off-road vehicles will use an hour-meter which measures the number of revolutions the engine turns per hour. The frequency of oil and filter replacements should also take into account the driving or running conditions, such as summer or winter, dry or wet, dusty or clean, stop–start or continuous running, heavy- or light-duty working, etc. If the conditions are abnormal, the oil and filters should be changed over a short distance or time-span.

For industrial and marine engines, the oil should be changed about every 400 hours and the filters at either 400 or 800 hours, depending on the type of filter. With off-road vehicles, a typical change interval for oil will be reduced to 250 hours. The replacement time for both oil and filters should be shortened for heavy-duty continuous work.

With road journey-motoring applications, oil should be changed very 10 000 km (6000 miles) and filters every second oil change for cars, but commerical vehicles may need a change of filter at each oil change.

20 Cooling systems

20.1 Engine heat distribution and the necessity for a cooling system

The energy released from the combustion of fuel in the cylinder is dissipated in roughly three ways: 35 to 45% heat energy doing useful work on the piston, 30 to 40% heat expelled with the exhaust gases, and 22 to 28% heat carried away by heat transference. Thus approximately 25% of the heat generated must be transmitted from the enclosed cylinder through the cylinder walls and head to the surrounding atmosphere.

If the heat-flow rate through the metal is low, the temperature of the inner surfaces will rise to a point where the heat destroys the lubricating properties of the oil film on the cylinder walls. Simultaneously, thermal stresses will be established which may distort the cylinders and head and drastically reduce the working strength and wearing properties of the aluminium-alloy pistons.

20.1.1 Operating-temperature conditions of the engine (fig. 20.1)

To control the rate of heat removal from the cylinders, either a direct or an indirect air-cooling system is provided so that the temperature around the cylinder walls and combustion chamber remains relatively constant over the engine's normal speed and load range. The object of the cooling system, therefore, is to extract and transfer heat from the engine to the atmosphere at a rate which matches the rate at which energy is liberated by the burning of the fuel charge.

Unfortunately, different parts of the engine operate at different temperatures (fig. 20.1), so some regions inside the enclosed cylinder are more prone to overheating than others. A broad guide to the mean operating temperatures of the gas charge and the various zones in the cylinder is as follows:

Intake air	30 to 60°C
Peak-combustion gas	2000 to 2400°C
Exhaust gas	700 to 900°C
Cylinder wall near cylinder head	160 to 220°C
Cylinder wall near crankcase	100 to 150°C
Centre of cylinder head	200 to 250°C
Centre of piston crown	250 to 300°C
Cylinder-block coolant	80 to 100°C

20.1.2 Methods of heat transfer

The transfer of heat can take place only when there is a difference of temperature – heat will then flow from the hotter to the colder substance,

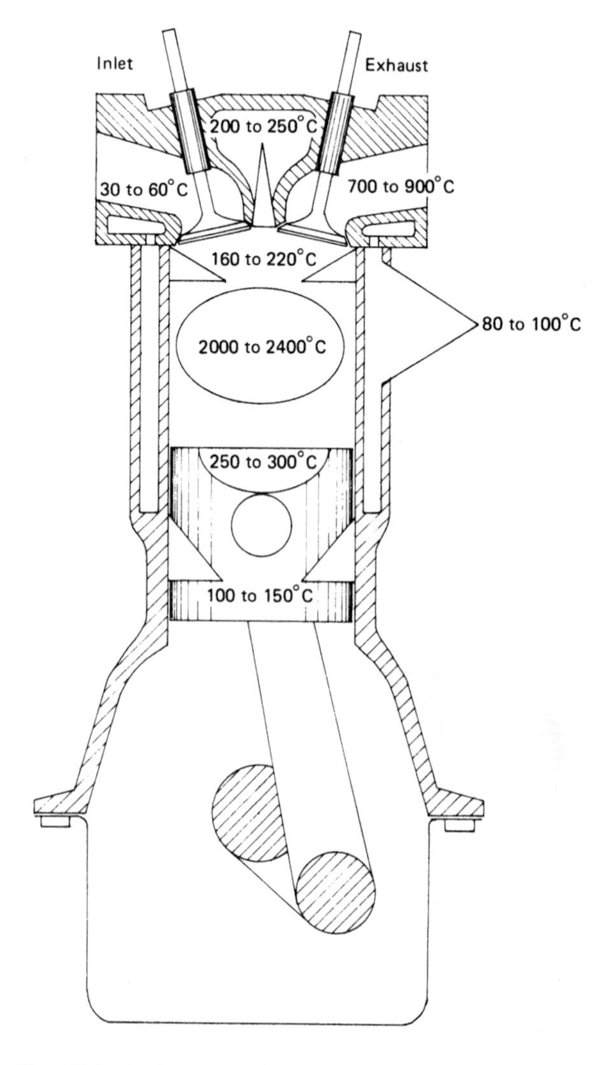

Fig. 20.1 Engine operating-temperature ranges

be it solid, liquid, or gas. The three methods of transmitting heat are conduction, convection, and radiation.

Conduction Conduction is brought about when heat is transferred from particle to particle throughout a body without any visible sign of movement. This type of heat flow is most effective in solids, but it can also occur at a much lower rate in liquids.

Convection Convection is established when heat is carried bodily by circulating currents moving particles through either a liquid or a gas. Natural or free convection currents are created entirely by changes in density due to differences in temperature at various levels in the liquid. Heat causes a fluid – be it liquid or gas – to expand. This makes the warm fluid less dense than the cooler one, so the lighter particles will rise and the heavier ones will sink – consequently a circulating current will be established. Forced convection may be achieved by a pump or fan which creates positive relative movement of the fluid over the stationary heated surface.

Radiation Radiation does not use a material medium to transmit the heat – in fact all substances, whether solid, liquid, or gaseous, emit energy by wave motion which radiates in all directions in straight lines with the speed of light. The emissive power of a radiating body is directly proportional to the fourth power of its absolute temperature, so the slightest increase in temperature can considerably increase the heat transfer by radiation.

20.2 Types of cooling systems

There are two basic methods of removing the heat from the engine:

i) direct air-cooling, where cool circulating air is made to come in contact with the exposed and enlarged external surfaces of the cylinder and head and thereby dissipate their heat to the surrounding air;
ii) indirect cooling (liquid cooling), where a liquid coolant is used to transmit the heat from the cylinders and head to a heat exchanger, usually referred to as the radiator. Movement of air through this radiator then extracts and dissipates the unwanted heat to the surroundings.

20.2.1 Direct air-cooling system (fig. 20.2)

If direct air-cooling is to be used, the surface area of the outside walls of both the cylinder and the head must somehow be enlarged to anything from five to fifteen times the plain cylindrical surface area if sufficient heat is to be transferred from the cylinder to the surrounding atmosphere.

Radial ribs or fins are used to increase the external surface area of the cylinder. The length of these fins will be greatest where the cylinder is hottest – near the cylinder head – and will progressively reduce towards the cooler-operating crankcase. Ideally the fins should have concave or hollow sides of parabolic form, the curvature of both sides meeting to form a very sharp outer edge in a razor fashion – this shape would give the maximum heat dissipation for a given weight of finned metal. Unfortunately this shape is difficult to cast, so a straight tapered fin with rounded outer edges is generally adopted as a compromise (fig. 20.2(a)).

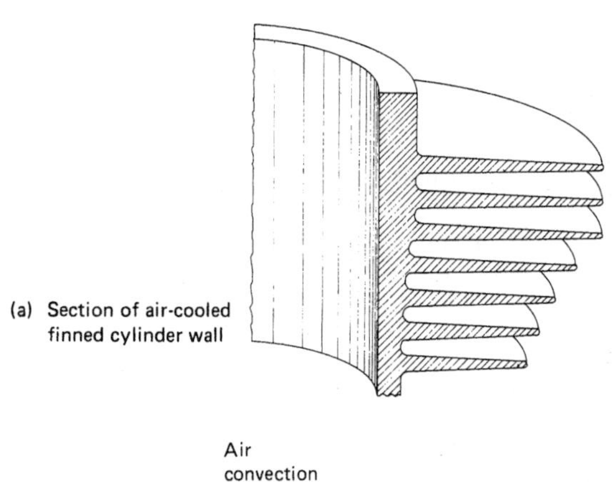

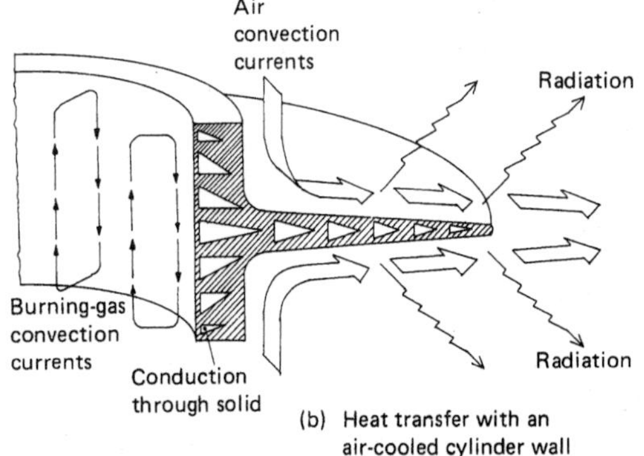

Fig. 20.2 Air-cooled finned cylinder wall

20.2.2 Heat transfer in a direct air-cooled engine system (fig. 20.2(b))

Heat from combustion is received by the cylinder walls, the combustion chamber, and the piston crown by both direct radiation and convection. The convection is created by the turbulent movement of the gases scrubbing against the metal surfaces so that fresh particles of the hot gas come into contact with each part of the surface to replace those that have given up part of their heat (fig. 20.2(b)).

Heat is conducted through the metal walls from the inner to the outer surface, due to the difference in temperature between these surfaces. Movement of air over the fins causes miniature whirling or eddying motion, resulting in surface friction which leaves a stagnant-air boundary layer close to the metal. The transfer of heat through this still air is mainly by conduction. Next to this stationary layer is a streamlined air flow which carries the heat away by convection, since fresh air particles continually pass over the air boundary layer. In addition, there will be some direct radiation from the cylinder fins to the surroundings, the magnitude of this radiation depending mainly on the working temperature of the cylinder.

20.2.3 Description of an air-cooled system (figs 20.3 and 20.4)

Air-cooled engines mounted on a motorcycle frame are usually exposed to the surrounding atmosphere. They can therefore rely entirely on the natural air stream caused by the forward movement to circulate air around the cylinders, head, and crankcase.

With multi-cylinder engines enclosed under a bonnet or boot, a more positive method of cooling is necessary. Controlled air-cooling is usually achieved by incorporating a fan which blows fresh air over the external finned surfaces of the engine. To improve the effectiveness of the blown air, the sides of the finned cylinders and heads are enclosed by a sheet-metal or plastic covering known as the cowling (fig. 20.3). The shape of the cowling guides the forced convection current around all the cylinders and provides a direct exit after the air has extracted and absorbed the heat from the engine.

Some engine configurations, such as the flat four-cylinder engines, employ baffles to improve the air distribution between cylinders and to direct additional air to critical components such as the oil-cooler (fig. 20.4(a)).

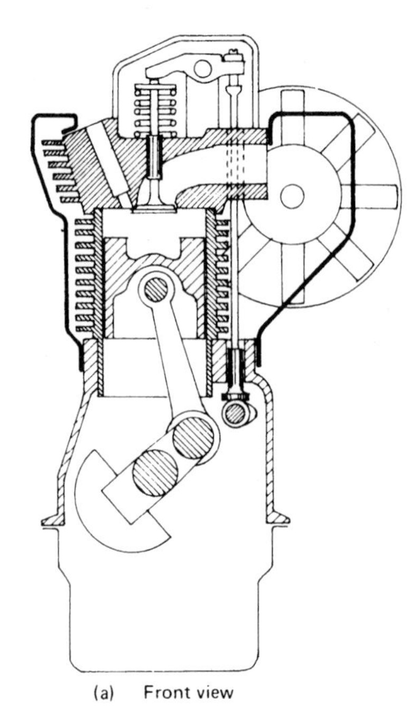

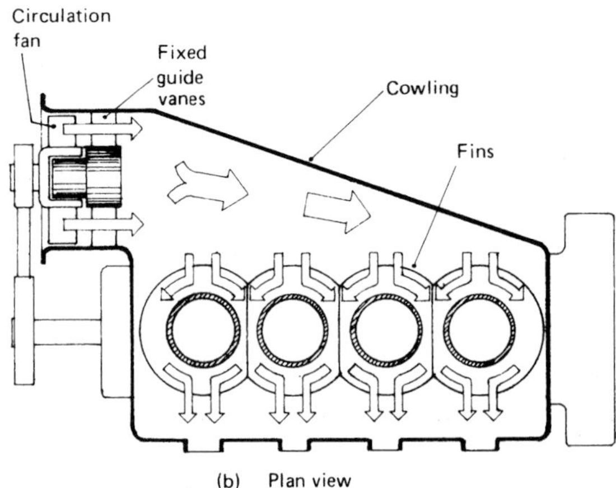

Fig. 20.3 Air-cooling sytem for an in-line four-cylinder engine

Fan blower (fig. 20.4) There are two classes of fan: the radial-flow type (fig. 20.4(c)), where the air is flung outwards by centrifugal force, and the axial-flow type (fig. 20.4(d)), where the air is pushed along parallel to the axis of the fan spindle. The radial-flow fan is more compact for a given output but tends to be noisy, whereas the axial-flow fan is more consistent and reliable when discharging large quantities of air; hence the former finds favour with small engines, while the latter is preferred for heavy-duty high-output engines.

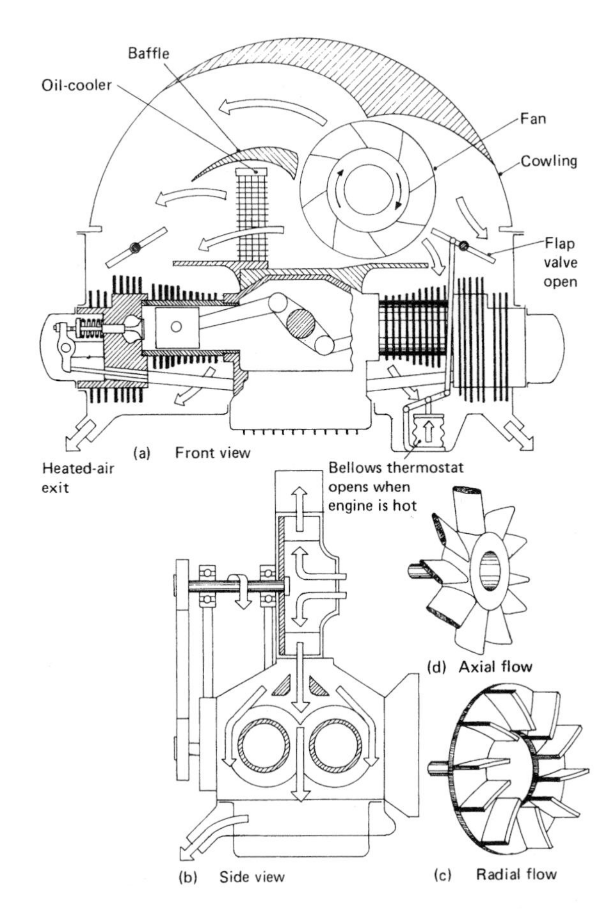

Fig. 20.4 Air-cooling system for a horizontally opposed four-cylinder engine

Fan discharge control (fig. 20.5) The amount of blown air circulating between the cowling and the cylinders may be regulated by a throttle ring situated on the inlet side of the fan, its function being to vary the fan's effective inlet-passage exposed area to suit the operating conditions of the engine (fig. 20.5). This can be automatically achieved by incorporating a thermostat in a hot working region of the engine so that it senses the temperature change. When the engine is cold, the thermostat will actuate either a linkage or a hydraulic servo connected to the throttle ring to restrict the flow of air to the fan. When the engine is hot, the restriction will be removed, thus permitting more air to circulate.

20.3 Heat transfer in an indirect liquid-cooled engine system (fig. 20.6)

The heat released from the burning of the atomised mixture of air and fuel is transferred in all

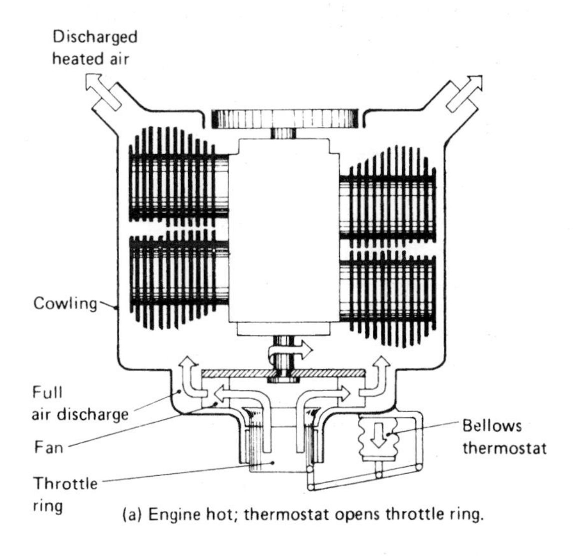

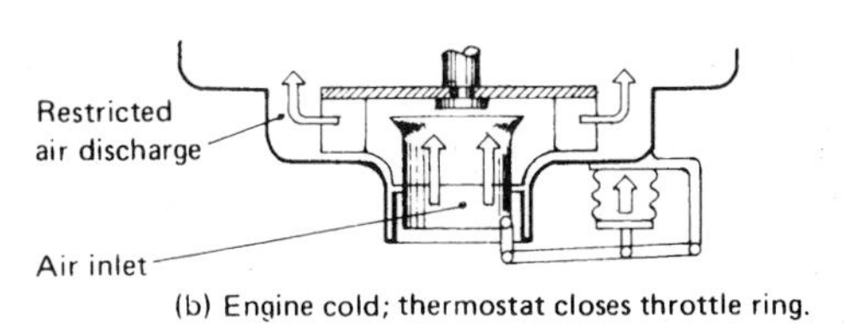

Fig. 20.5 Air-cooled engine with fan discharge control

directions to the metal walls of the combustion chambers, cylinders, and pistons – by direct radiation, by convection currents of gas rubbing against a stationary gas film which always forms on any metal surface, and then by conduction through this stagnant boundary layer of gas and a film of oil to the actual metal walls.

Due to the difference in temperature between the inner and outer cylinder walls, heat will be conducted through both the metal and any layer of scale formed. It is then further conducted through a thin stagnant boundary layer of liquid to the bulk of the coolant liquid in the passages around the cylinders.

20.3.1 Thermo-syphon liquid-cooling system (fig. 20.7)

To dissipate the heat given to the liquid, a radiator is provided, its function being to transfer the heat from the liquid to the surrounding atmosphere. The top of the radiator is connected by a hose to the cylinder head where the heated liquid has risen and collects, and the bottom of the radiator is joined by a hose to the lower region of the cylinder-block coolant passages (fig. 20.7).

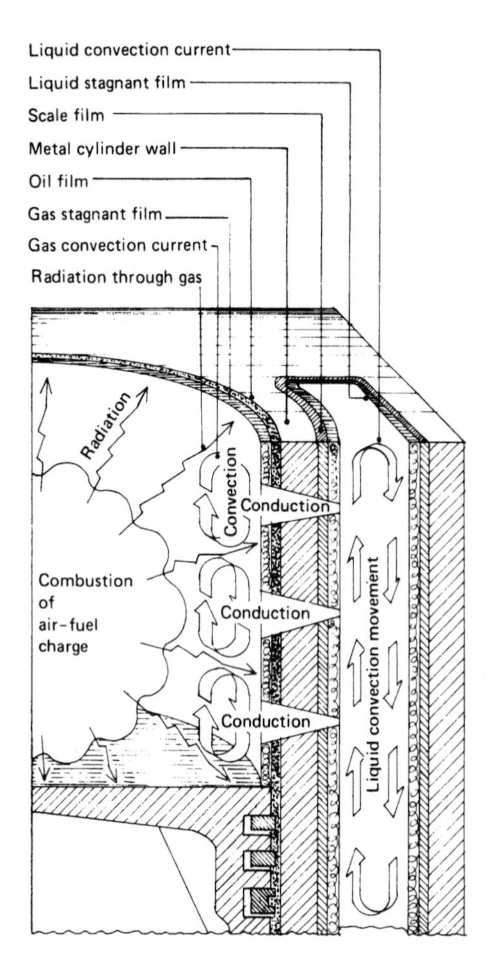

Fig. 20.6 Heat transfer with a liquid-cooled cylinder wall

As the liquid around the cylinders and head receives heat, it expands and becomes less dense relative to liquid which is not in contact with the hot metal walls; therefore the lighter hot coolant will rise to the highest point in the system, which is the header tank over the radiator tubes. At the same time, the liquid in the radiator tubes will be cooled by the air stream passing around the tubes and over the fins, consequently the density of the liquid in the tubes will increase, so that the cooled liquid sinks to the bottom. Since this cool liquid at the lowest point in the radiator is at the same level as the passages in the cylinder block, and they are joined by a hose, this heavier cooled liquid will replace the hot and less dense liquid in contact with the cylinder walls. Thus, once the engine has warmed up, a convection current flows between the engine and the radiator and so forms

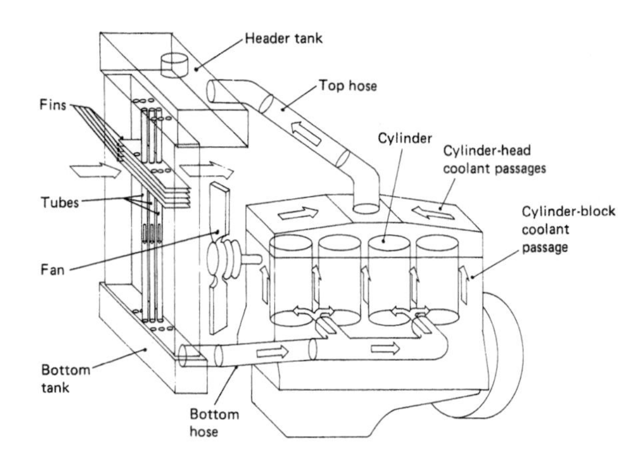

Fig. 20.7 Thermo-syphon liquid-cooling system

an enclosed circulating loop known as the thermo-syphon cooling system.

20.3.2 Description of a liquid-cooled system (fig. 20.7)

Radiator The radiator transfers the heat absorbed by the liquid coolant to the surrounding air, be it static or moving through the radiator matrix. The simple radiator is flat and rectangular in shape and consists of columns of spaced-out copper or aluminium-alloy tubes held in position at the ends by an upper header tank and a bottom tank. These tubes maximise the external surface area of metal exposed to the air stream. Attached to these tubes are layers of horizontal copper or aluminium sheets known as fins. These considerably improve the effectiveness of air-convection heat dissipation.

The upper header tank acts as a reservoir for the coolant and distributes the collected hot liquid evenly among the vertical tubes. The bottom tank collects the cooled liquid coolant from each tube and passes it through a single outlet back to the engine's coolant passages surrounding the cylinders.

Flexible hosing Upper and lower flexible fabric-reinforced rubber hoses connect the radiator to the engine cylinder-head and cylinder-block coolant jackets. The flexible hoses are necessary to absorb the relative movement between the radiator, which is bolted to the body, and the suspended engine which tends to vibrate while operating. Metal clips are slipped over the ends of the hoses and are tightened to hold the hoses in position and to prevent coolant leakage.

Coolant jackets To enable liquid to surround the cylinders, a space is provided between the cylinders and the sides of the block, thus creating external cylinder-wall surfaces so that heat can be transferred from the cylinder walls to the liquid coolant. An outer casing wall, known as the coolant jacket, is then formed to retain the liquid circulating around the cylinders. Similarly, coolant passages are cast inside the cylinder head so that the coolant can flow around the combustion-chamber walls, the inlet and exhaust ports and their valve seats, and the spark-plug or injector holes.

Fan To supplement the normal air movement across the radiator core tubes when the vehicle is stationary or moving very slowly, a fan is positioned on the exit side of the air stream passing through the radiator – that is, between the radiator and the engine. The fan-blade pitch and direction of rotation create a small depression on the fan side of the radiator matrix; therefore the air on the inlet side of the radiator will tend to force its way through the radiator core to try to cancel out this variation of pressure. The resultant air movement will thus provide a continuous air stream over the tubes and fins to dissipate the heat being circulated by the coolant.

20.3.3 Limitations of the thermo-syphon cooling system

a) Under certain conditions (such as pulling under load at low vehicle speed), unless a very large radiator and very large engine coolant passages are employed, the rate of coolant circulation created by the convection current cannot match the rate of transfer of heat from the cylinder walls to the coolant.
b) For adequate heat transfer, the radiator header tank should be situated at a higher level than the cylinder head. This would be impossible with modern body styles.
c) Without coolant-circulation control, the engine tends to be overcooled and very rarely reaches the optimum operating temperature, even over long running periods.
d) The large quantity of coolant enclosed in the cooling system prolongs the engine's warm-up period.
e) The large header tank, used to compensate for the low rate of coolant circulation, tends to overheat and so coolant is lost through evaporation.

20.3.4 Forced-convection pump circulation (fig. 20.8)

To speed up the rate of coolant circulation so that more heat can be removed in a given time, the basic thermo-syphon system is slightly modified by incorporating a centrifugal pump in series with the engine-coolant lower return-hose (fig. 20.8).

With this forced-convection-current system, coolant can be made to flow not only upwards but also along the full length of the cylinder-block coolant passages, so that the cooled liquid can be more uniformly shared out between the in-line cylinders. This helps to prevent uneven cooling or overheating of individual cylinders. Increasing the flow rate enables the radiator to work more efficiently, so it can be scaled down in size, and it is also unnecessary to mount the radiator any higher than the engine cylinder head, unlike when relying on natural coolant circulation.

The pump-assisted cooling system can make the engine operate more efficiently if it does not overcool the system – the system should function on the concept that it is there only to prevent critical zones in the engine from overheating. Furthermore, large volumes of liquid circulating round the engine should not hinder the cylinder and head quickly reaching their working temperatures. Controlling engine warm-up has partially been achieved by placing a thermostat valve in series with the top hose. When the engine is cold, the valve is closed – this prevents the bulk of the liquid circulating. Once normal working conditions are reached, however, the thermostat automatically senses the desired working temperature and opens the valve, and bulk circulation then begins.

To prevent excessive pressure build-up in the engine coolant passages, a bypass pipe circulates about one tenth (or even less) of the liquid directly between the cylinder-head thermostat housing and the inlet side of the pump. This also eliminates local boiling of trapped coolant due to lack of circulation.

20.3.5 Coolant circulating pump (fig. 20.9)

Description and construction of pump The pump consists of a fan-belt-driven spindle mounted on a pair of spaced-out ball-races which are supported in a sleeve fitting inside the pump cast-iron or

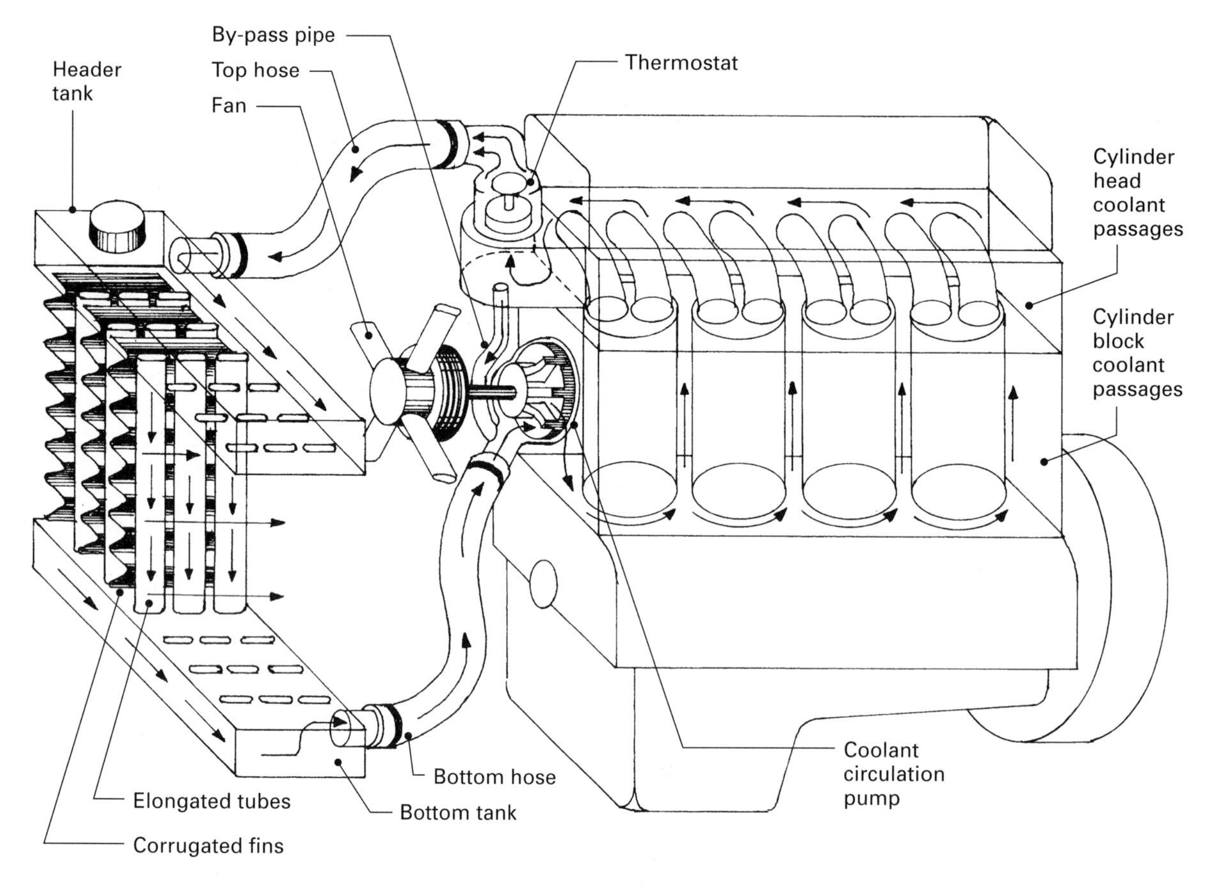

Fig. 20.8 Vertical down-flow pump-assisted liquid cooling system

aluminium-alloy casing (fig. 20.9(a)). One end of the spindle supports a pressed-on pulley wheel, and the other end has the impeller wheel pressed on to it. A friction-face ring-seal assembly is situated between the impeller and the casing to prevent coolant escaping between the rotating and stationary members. An annular-shaped inlet chamber faces and encircles the inner edges of the impeller vanes to maximise the supply of coolant to the impeller.

Pump operation The impeller casting resembles a disc with a number of curved (sometimes straight) blades known as vanes (fig. 20.9(b)). These blades or vanes break up the liquid coolant into cells so that, when the impeller rotates, the cells of coolant will also rotate. Centrifugal force will then force the liquid outwards through the diverging cell passages formed by these guide vanes. The coolant is then flung out at the periphery at a linear speed and force depending on the speed of rotation of the spindle.

On leaving the outer portion of the impeller tangentially, the coolant has maximised its energy of motion. This is then converted to pressure energy by means of the discharge involute, which is a spiral-shaped passage cast in the casing, its cross-sectional area increasing as it approaches the outlet port. Thus the enlarging involute collecting passage reduces the speed of the coolant coming out from the tips of the impeller vanes, which has the effect of pressurising the discharged coolant and so provides the means for increasing the rate of circulation throughout the system.

Impeller-spindle coolant seal Coolant leakage between the impeller spindle and the casing is prevented by a friction ring (fig. 20.9(a)) made from phenolic-resin-based plastics; graphites; carbon impregnated with synthetic resin; or carbon–ceramic, which exhibits much better thermal and chemical stability and emergency-operating characteristics such as required when starting or when engine overheating is causing increased rubbing

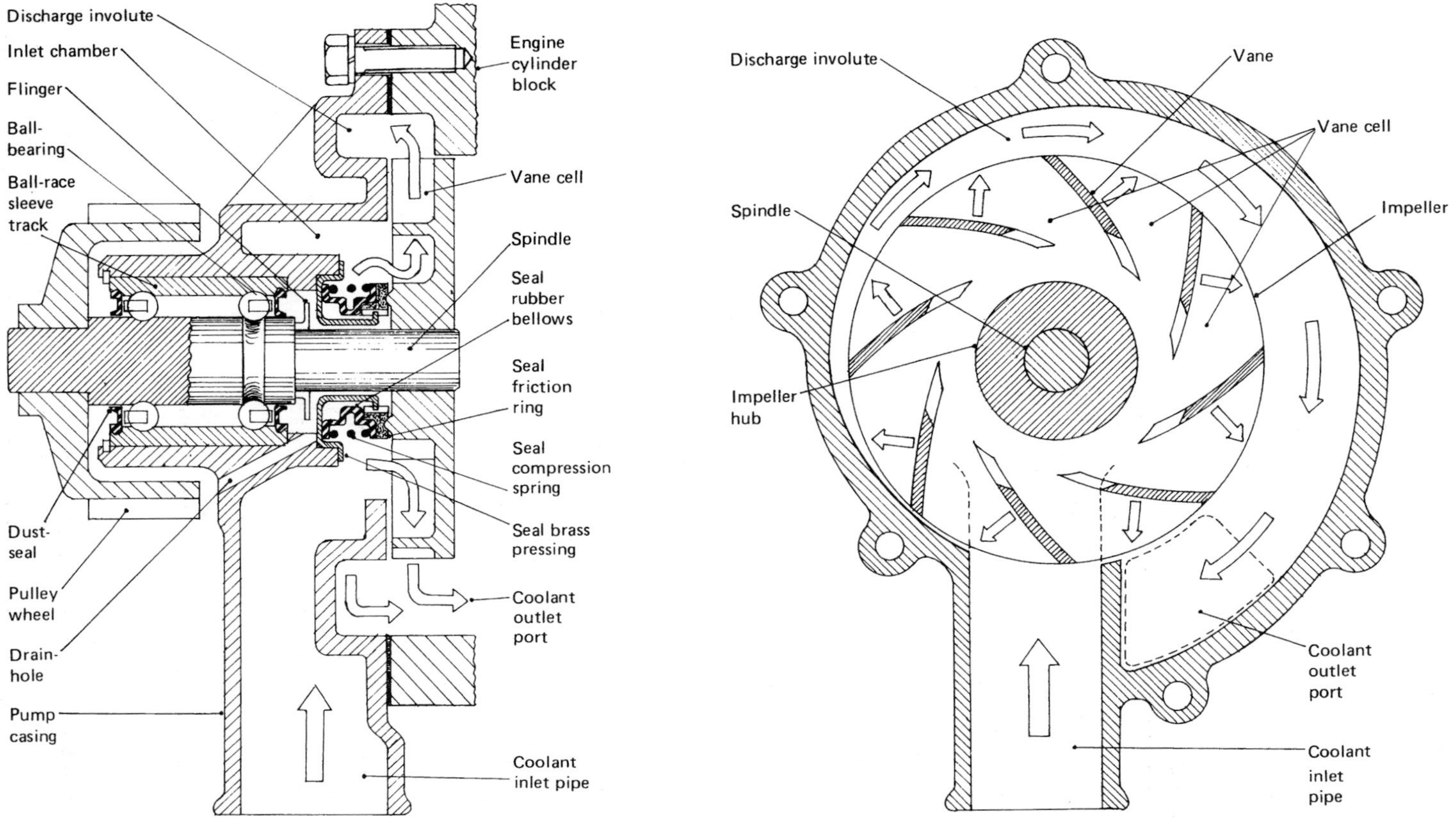

Fig. 20.9 Coolant pump

wear. The friction-face ring bears against a machined face of the impeller hub and is prevented from rotating by two lugs protruding from the brass pressing which are located in two grooves formed on the internal circumference of the ring. The ring is pressed into the rubber bellows which supports it, and a compression spring expands the bellows so that the friction-ring end pushes the ring face against the impeller hub while the other end reacts against the brass pressing to seal off the coolant.

Pressed over the spindle between the rear ball-race and the seal assembly is a coolant flinger washer, its purpose being to protect the ball-bearings from the coolant by deflecting and throwing out any surplus coolant which may pass between the friction-seal faces if there is an excessive pressure build-up in the cooling system or by natural seepage of coolant over a long period of time if the engine has not been operating. Any escaping coolant will then be able to pass out through the drain-hole. Additional synthetic-rubber ring seals are pressed in at each end of the ball-race outer sleeve to prevent dirt and water entering the bearings.

Pump faults Common coolant-pump faults may be due to the following:

a) undertensioned fan belt causing pulley slippage;
b) overtensioned fan belt causing rapid ball-bearing wear;
c) bearings pre-packed with grease have run dry and have overheated, or have become contaminated by coolant so that the grease film has been destroyed;
d) leaking seal, due to an overheated cooling system generating excessive pressure or to a badly assembled seal causing uneven contact between the friction ring and the impeller hub;
e) very weak water and antifreeze solution may cause excessive corrosion and erosion of the impeller;
f) very hard water mixed with antifreeze solution may cause the impeller to become encrusted with fur or scale to an extent which impedes the flow of coolant;
g) incorect thickness of the casing gasket or uneven tightening of the casing to the cylinder block may cause distortion and even cracking of the cast casing.

20.4 Comparison of air- and liquid-cooling systems

20.4.1 Air-cooling

Advantages

a) Air-cooled engines operate extremely well in both hot and cold climates.
b) Air-cooled engines can operate at higher working temperatures than equivalent liquid-cooled engines.
c) Air-cooled engines rapidly reach their working temperature from cold.
d) Air-cooled engines are marginally lighter than similar-sized liquid-cooled engines.
e) Air-cooled engines have no coolant-leakage or freezing problems.

Disadvantages

a) A relatively large amount of power is required to drive the cooling fan.
b) The large quantities of intake air passing into the cooling system can make the engine noisy.
c) The cooling fins can under certain conditions vibrate and amplify noise.
d) The pitch between cylinder centres has to be greater than in liquid-cooled engines, to permit the fins to extend between cylinders.
e) Each cylinder has to be individually cast, whereas a rigid monobloc construction is used by liquid-cooled engines.
f) To supplement the air-cooling, an oil heat exchanger is frequently necessary to prevent overheating of the lubricant.
g) The guide cowling and baffles surrounding the cylinders may hinder maintenance.

20.4.2 Liquid cooling

Advantages

a) Liquid-cooled engines provide greater temperature-uniformity around the cylinders, so there is less distortion compared with air-cooled engines.
b) The combined power consumption of the coolant pump and the fan in liquid-cooled units is far less than that of the air-cooled-engine fan.
c) The liquid-cooled engine cylinders are situated closer together, providing a very rigid and compact unit compared to the air-cooled engine.

d) Mechanical noise from the engine is damped by both the coolant and the jackets.
e) Liquid-cooled units are more reliable for heavy-duty work than air-cooled engines.
f) Hot coolant can readily be piped to heat the interior of the vehicle.

Disadvantages

a) Liquid-coolant joints are subject to leakage.
b) Precautions must be taken to prevent coolant freezing.
c) Liquid-cooled units take longer to warm up than do air-cooled engines.
d) The maximum liquid-coolant temperature is limited to its boiling point, whereas air-cooled engines can operate at slightly higher temperatures.
e) The coolant passages tend to scale, and the hoses and radiator tubes deteriorate with time.

20.5 Thermostat-controlled cooling systems

The function of the thermostat is to regulate heat dissipation by controlling the rate of coolant flow through the radiator. Engines are designed to operate most efficiently over a narrow temperature range (usually between 80 and 100°C) – it is therefore desirable that this temperature should be reached as soon as possible after starting and then be maintained.

The reasons for rapidly reaching the predetermined mean operating temperature and keeping it consistent within limits are as follows:

a) The joints of the various components are continually moving relative to each other until a steady mean temperature is reached. This movement can cause severe differential stresses and gasket leakage.
b) The working clearances of rubbing components such as the piston rings and cylinder, the journals and bearings, the valves and guides, tappet clearances, etc. will not be ideal until the engine is fully warmed up, so both noise and wear will be accelerated during the heating-up period.
c) Large amounts of condensed corrosive vapour in the cold crankcase may contaminate the lubricating system.
d) An improved uniform air–fuel mixture will be formed at the optimum working temperature, so more useful work will be done during combustion.
e) The optimum operating temperature will maintain the lubricating oil at the correct viscosity, so that the rubbing parts have the best lubricant protection with the minimum oil-film shear effort, which would otherwise consume power.
f) The vehicle's interior heater will operate efficiently only when the engine's designed operating temperature has been reached.

20.5.1 Types of thermostat-valve units

There are two basic forms of thermostat: one relies on the expansion of a volatile fluid enclosed in a bellows, and the other type depends on the expansion of plastic wax to compress a rubber boot. Both arrangements sense the change in coolant temperature and automatically open or close the thermostat valve so that it can regulate the coolant circulation accordingly.

20.5.2 Operation of the bellows-type thermostat (fig. 20.10)

The thermostat is usually situated in front of the cylinder head in a coolant-outlet housing; therefore all the coolant flowing through the top hose from the engine to the radiator has to pass through the thermostat-valve assembly.

The thermostat consists of a brass flexible bellows partially filled with a volatile fluid such as acetone, alcohol, or ether which boils between 65 and 85°C. A poppet-valve is attached to one end, and the other end is mounted in a brass frame which fits into the coolant-passage housing.

While the engine is cold, the bellows lobes will contract together so that the poppet-valve is closed. Under these conditions, coolant cannot circulate between the engine and the radiator. However, a restricted amount of circulation may take place in the engine through a bypass passage or pipe, to prevent differential expansion and the formation of steam at critical points in the cylinder and head.

Once the engine has warmed up, the heat will act on the bellows and the liquid will rapidly expand, pushing apart the individual bellows lobes so that the valve will begin to open. If the temperature continues to rise, the substance inside the bellows will change its state into a gas, and further expansion of the bellows will result until the valve is fully open.

The initial opening temperature – known as the 'crack' temperature – should be within 2 to 3°C of the value marked on the thermostat flange, but

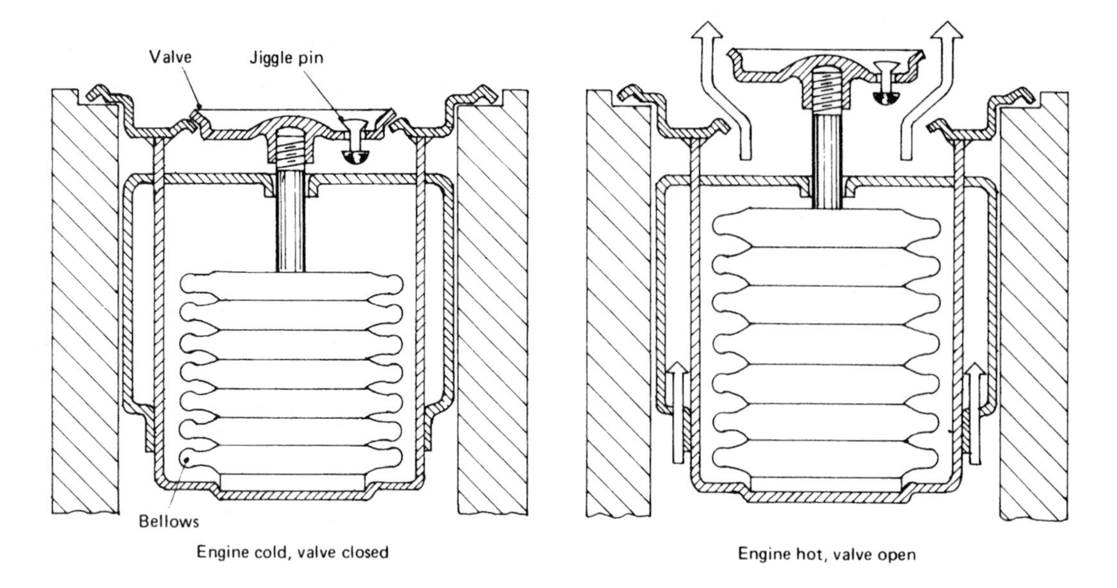

Fig. 20.10 Bellows-type thermostat

the fully open temperature may differ by as much as 14°C.

20.5.3 Operation of the wax-pellet-type thermostat (fig. 20.11)

The wax-element thermostat is situated in the engine's coolant-outlet housing and is usually held in position by a flanged-elbow coolant pipe which is indirectly connected to the radiator by the top hose.

This unit consists of a brass capsule with a stainless-steel thrust-pin situated in the centre. Around the pin is a synthetic-rubber boot which is sealed at the bottom and bears tightly around the thrust-pin. Within the annular space formed between the capsule case and the rubber boot, about 0.9 grams of microcrystalline paraffin-type wax is injected. This wax has a high coefficient of volumetric thermal expansion (0.28% per °C) and is generally mixed with copper powder to improve the thermal conductivity so that the capsule is more sensitive to changes in temperature. The thrust-pin is supported by a brass flanged frame which is mounted in a recess formed in the cylinder-head housing, and the central hole in this frame provides the valve seat. The open end of the capsule has a dome-shaped valve head attached which fits centrally in the thrust-pin frame. A return-spring is located between the outer frame and the underside of the valve head and provides a closing bias at all times.

When the temperature of the capsule is raised, the wax expands and squeezes in the rubber boot. This forces the capsule to move away from the fixed thrust-pin with a thrust not less than 150 N and hence opens the coolant passage. The shapes of the valve and seat are such that the coolant flow is partially throttled with small valve openings but towards the fully open position there is very little flow restriction.

The crack temperature is usually stamped on the base of the capsule, and the valve should start to open within 3°C of this value. The valve will continue to open progressively to the fully open position within a further 12 to 14°C of this temperature.

Figure 20.12 shows a thermostat tester. This is used to heat the thermostat up to operating temperature while observing the crack temperature when the valve starts to open and the temperature at which the valve reaches its fully open position.

When the temperature of the coolant flowing past the thermostat drops below the normal predetermined opening value, the valve will automatically close again, so a cycle of opening and closing will be repeated as the operating conditions change.

When the cooling system has to be filled, it normally needs to be self-bleeding. For this purpose, a small hole can be incorporated in the

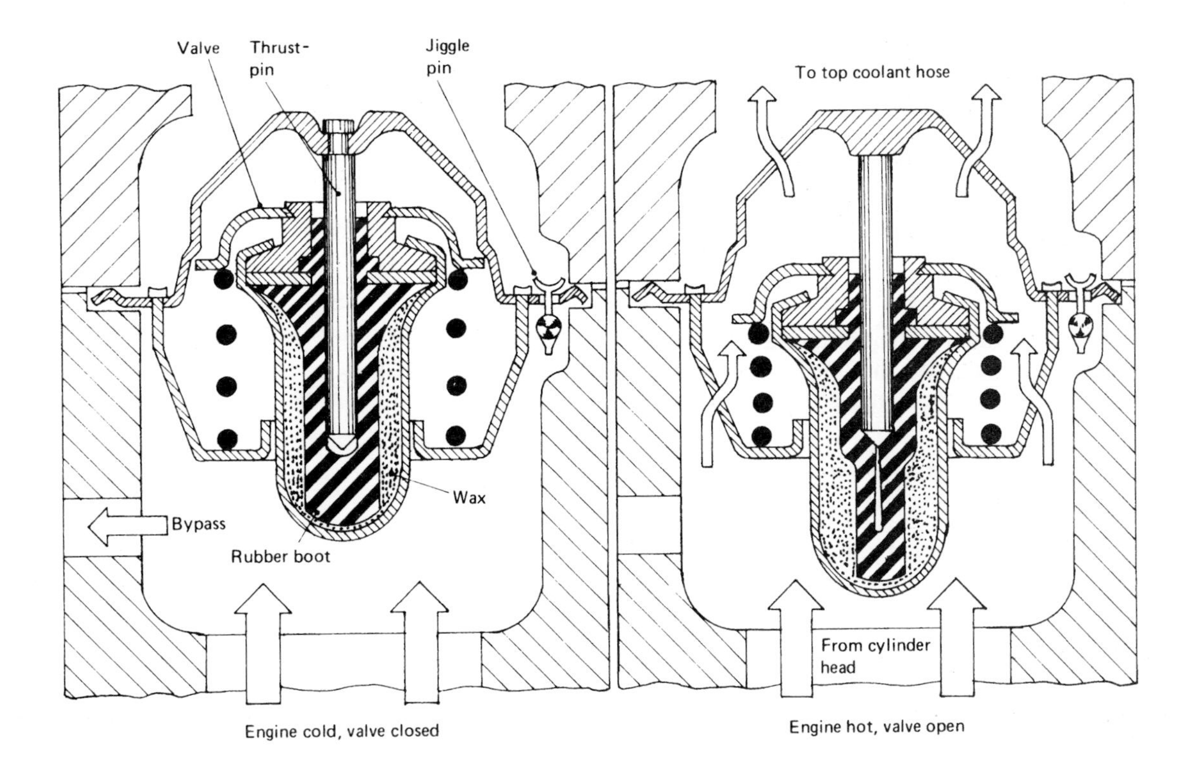

Fig. 20.11 Wax-pellet-type thermostat

thermostat flange. Alternatively, a slightly larger hole which is closed by a jiggle valve can be used. This valve will jiggle open and closed with varying cylinder-jacket coolant pressure when the main coolant circulation is prevented by the thermostat valve being closed.

A bypass passage must usually also be provided so that, when the thermostat valve is closed, a small quantity of coolant is able to circulate within the engine. This will avoid the possibility of hot spots forming in the cylinder head due to trapped and overheated coolant.

Both bellows- and wax-type thermostats tend to fail in the closed position, which obviously then causes overheating.

The wax-type thermostat has generally replaced the bellows type because it operates reliably within its opening and closing temperature range in pressurised cooling systems without suffering from failure due to pressure surges.

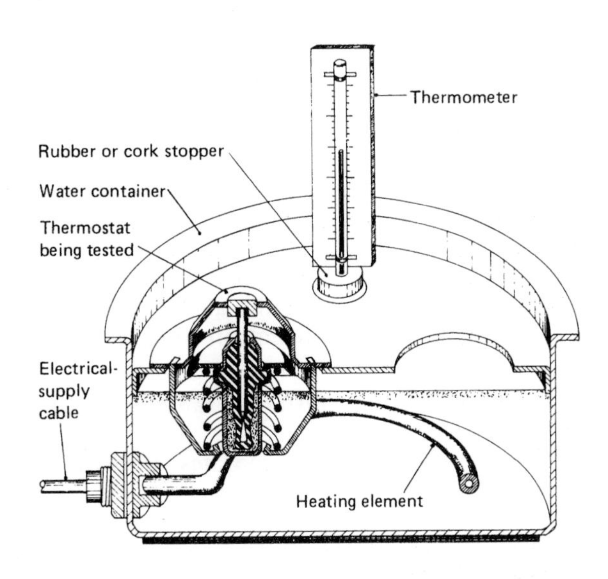

Fig. 20.12 Thermostat tester

20.6 Pressure radiator caps

If a radiator filler cap is of the vented type, steam vapour will escape and so frequent topping up with fresh coolant will be necessary. Conversely, if the cap forms an air-tight seal, the pressure in the cooling system will under certain operating conditions (such as climbing a steep hill) build up until the weakest point in the system gives way, causing loss of coolant. Typical leakage points would be dried-out hoses, soldered joints, corroded radiator tubes, etc. A properly designed pressure cap provides both pressure and vacuum relief under all operating conditions, and so avoids damage to any part of the system.

There are two main forms of pressure-cap construction: the simple open type, used mostly by car manufacturers, and the closed type generally preferred on commercial vehicles as it is slightly more robust. The closed-type cap has two relief valves which are independent of the cap filler-neck flange seal; whereas with the open type the pressure-relief valve and the flange seal are combined.

20.6.1 Reasons for using pressurised cooling systems (figs 20.13 and 20.14)

a) Pressurising the coolant allows it to operate at higher engine working temperatures without losses, so that the preparation of the air–fuel mixture for the combustion process will be improved and optimum lubrication-viscosity conditions will reduce frictional work. Figure 20.12 illustrates how the boiling point of water can be raised by increasing the filler-cap blow-off-pressure setting.

b) The rate of heat dissipation is proportional to the difference between the coolant temperature and the outside surrounding temperature. Therefore higher radiator operating temperatures increase the heat-transfer efficiency. It has been found that the cooling efficiency improves by 3.5% per 0.1 bar increase in coolant operating pressure.

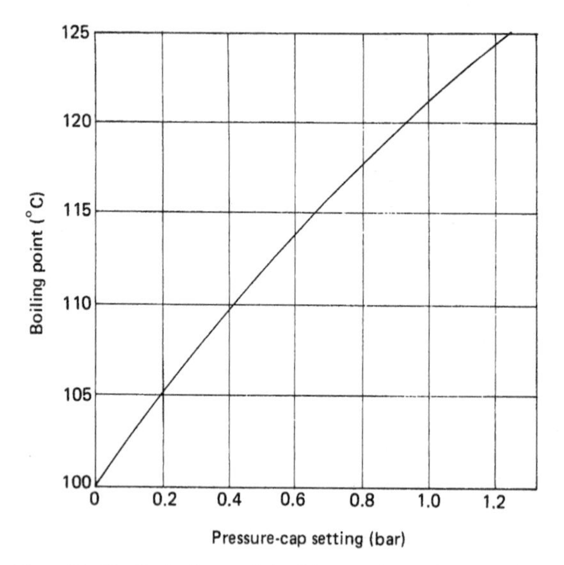

Fig. 20.13 Variation of boiling point of water with pressure

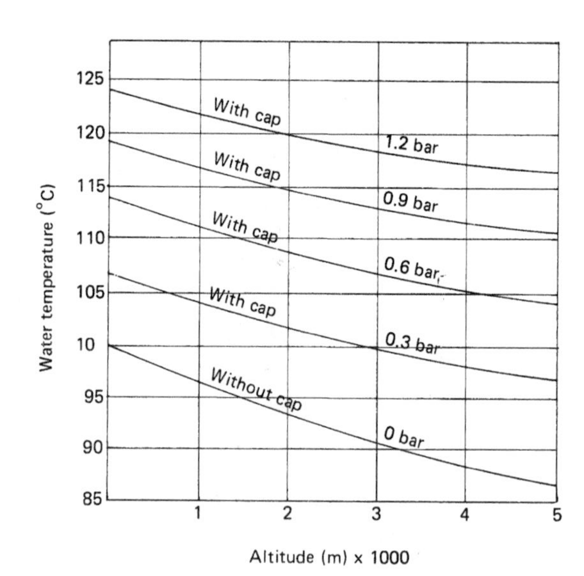

Fig. 20.14 Variation of boiling point of water with altitude

c) The higher operating temperature improves the transmission of heat to the atmosphere, therefore a smaller radiator can be used.

d) Sealing the radiator filler neck with a pressure cap prevents loss of coolant due to evaporation or surging and prolongs the intervals between topping-up.

e) Air pressure reduces with altitude, so, with an open cooling system, the coolant will boil as the vehicle climbs. A closed pressurised system will not sense the change in atmospheric pressure, so it tends not to overheat while driving over high terrains.

Figure 20.14 shows how the boiling point of water is lowered due to the reduction in atmospheric pressure as the altitude increases. It can also be seen that the boiling point of water can be raised proportionally over the altitude range by using filler caps with different pressure settings.

20.6.2 Operation of a pressurised radiator cap (figs 20.15 and 20.16)

The pressurised cap incorporates three main features:

i) a pressure-relief valve,
ii) a vacuum-relief valve,
iii) a safety removal device.

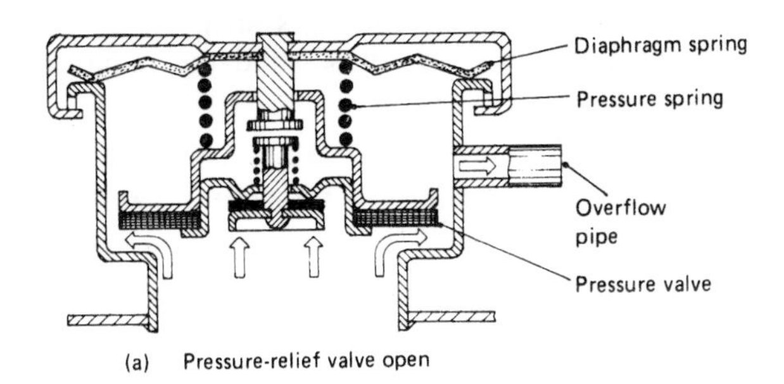

(a) Pressure-relief valve open

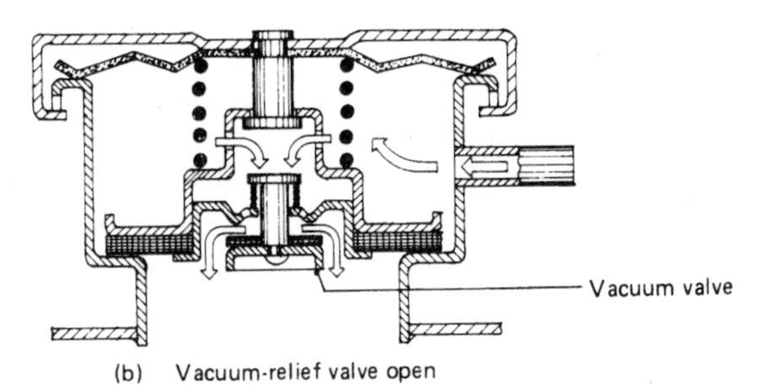

(b) Vacuum-relief valve open

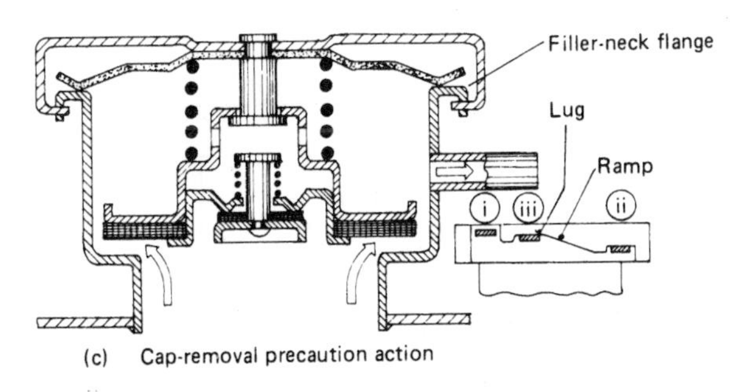

(c) Cap-removal precaution action

Fig. 20.15 Open-type pressure cap

Pressure-relief valve (figs 20.15(a) and 20.16(a)) During operating conditions, the temperature of the cooling system will rise and the coolant will expand. Furthermore, when the engine is working hard, the liquid flowing over the hottest zones in the cooling system will start to boil and vaporise. Since the expanded liquid and vapour cannot escape from the sealed system, the coolant pressure will be increased until the upthrust of the coolant and vapour equals the spring load which is keeping the disc-valve closed. With further pressure rise the valve will be pushed back, and excess coolant will initially fill the cap neck chamber before it escapes out of the overflow pipe to the atmosphere.

Under normal driving conditions the vapour pressure of the coolant will not exceed the valve opening pressure, so there will be no loss of coolant, or at most very little.

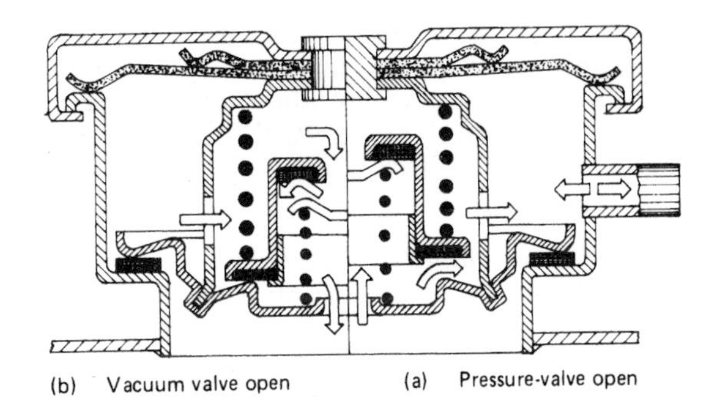

Fig. 20.16 Closed-type pressure cap

Vacuum-relief valve (figs 20.15(b) and 20.16(b)) Over a period of time some coolant may become lost to the atmosphere. When the engine cools and the liquid coolant contracts, a depression may then be created within the closed system. If this slight vacuum is not relieved, the atmospheric pressure may push in and flatten the rubber hoses and in extreme cases may be able to crush corroded radiator tubes. To prevent the likelihood of the hoses collapsing, a small vacuum disc-valve is mounted in the centre of the cap's pressure-relief valve and is pushed open by the difference of pressure existing between its two sides. Atmospheric air will then enter the normally sealed system to fill any vacuum.

This valve is generally set to open if the pressure difference across the valve exceeds about 0.06 bar.

Safety removal device (fig. 20.15(c)) The cap is secured to the filler neck by quick-acting bayonet-type locking lugs which are spring-loaded to ensure shakeproof fixing. Figure 20.15(c) shows the bayonet lugs in position:

i) ready for removal,
ii) partially removed, and
iii) fully locked.

When the cap is initially being removed, it should be partially twisted so that it engages a safety notch. This allows the cap to lift on its own, and the vertical movement will be limited by the bayonet lugs which contact the filler-flange ramps. The diaphragm locking spring will prevent the exit of steam from the filler-neck hole, so its only way out is through the overflow pipe. As a safety precaution against possible scalding, pause for a short while to allow the steam to escape. Further rotation of the cap will then release the cap from the filler-neck flange.

20.6.3 Cap operating pressure

The radiator cap blow-off pressure to a great extent regulates the boiling point of the coolant. In fact, for every 0.1 bar the system's pressure is raised, the boiling point is increased by between 1.5 and 2.5°C.

Cooling systems are designed to operate at different sealing pressures set by the cap's pressure-relief valve, ranging from 0.3 to 1.3 bar, the upper limit for a pressurised cooling system being the bursting pressure of the hoses and the radiator tubes.

20.7 Cooling systems incorporating an expansion tank (figs 20.17 and 20.18)

20.7.1 The advantages of incorporating an expansion tank

a) It eliminates periodic cap removal and topping up with fresh coolant.
b) It prevents the loss of excess coolant due to expansion.
c) It reduces the deterioration of the antifreeze

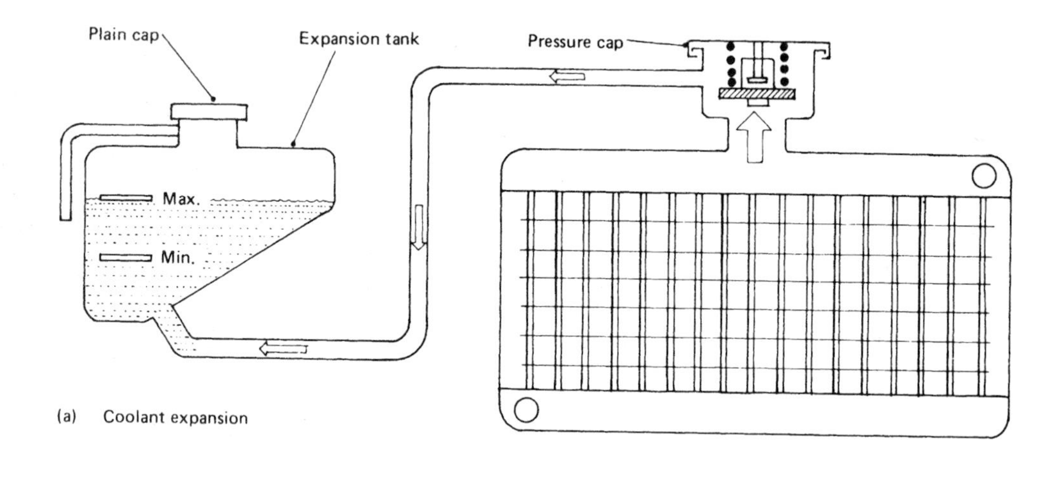

(a) Coolant expansion

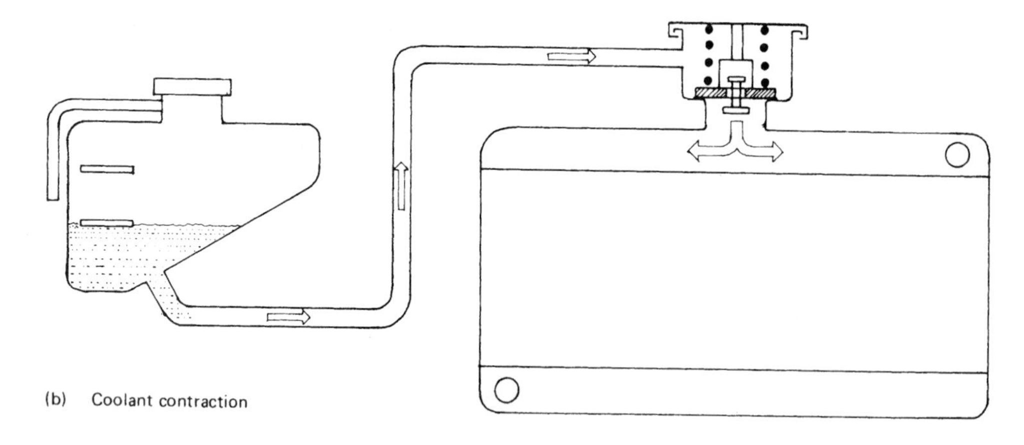

(b) Coolant contraction

Fig. 20.17 Vented expansion tank

and corrosion of the coolant jackets, as air is excluded from the system.

d) It enables smaller radiator header tanks to be used and permits the lowering of the radiator height relative to the engine.

20.7.2 Operation of an expansion-tank system

The sealed cooling system maintains the system completely full of coolant. Provision for expansion and contraction of the coolant is provided by incorporating an expansion tank into which excess hot coolant can pass and from which it can return as the coolant in the system contracts on cooling.

There are two basic layouts:

i) with a vented expansion tank,
ii) with a pressurised expansion tank.

20.7.3 Vented expansion tank (fig. 20.17)

This arrangement has the pressure cap directly on top of the radiator filler tube, and the vent pipe from the filler-neck chamber is connected to the bottom of an overflow plastic bottle known as the expansion tank. The top of this container is vented to the atmosphere, and a small quantity of coolant should be put into this tank when the system is initially filled.

Expansion of the hot coolant will force open the pressure valve and displace coolant to the expansion tank. When the system's coolant cools and contracts, the depression formed in the system draws the stored coolant in the expansion tank back into the sealed system. With this system, a rubber seal must be placed between the filler-cap diaphragm spring and the filler flange, to prevent air being drawn into the radiator.

20.7.4 Pressurised expansion tank (fig. 20.18)

A more popular layout has the pressure cap fitted on to the expansion tank, which is connected to the top of the radiator by a small-bore pipe. The usual filler hole on the radiator is blanked-off by a plug. The radiator is completely filled with coolant, and a small amount of coolant is also initially put into the expansion tank.

As the system warms up, coolant expands and overflows into the expansion tank which then becomes pressurised. When the temperature in the system drops, the coolant contracts and the pressurised expansion tank will now force coolant back into the main system in order to keep the system completely filled at all times.

The pressure cap on the expansion tank will operate just as if it were fitted directly on to the radiator filler neck.

20.8 Antifreeze coolant solutions (fig. 20.19)

Originally all liquid-cooled engines used water as the coolant medium to transfer the heat by convection currents from the engine to the radiator. Water has the fundamental properties necessary for a liquid coolant, such as a high specific heat capacity, a relatively high boiling point, a low viscosity, and ready availability. A major inherent limitation was that water freezes when the temperature drops below 0°C, so the water had to be drained overnight during the winter months.

Freezing implies that the coolant state changes

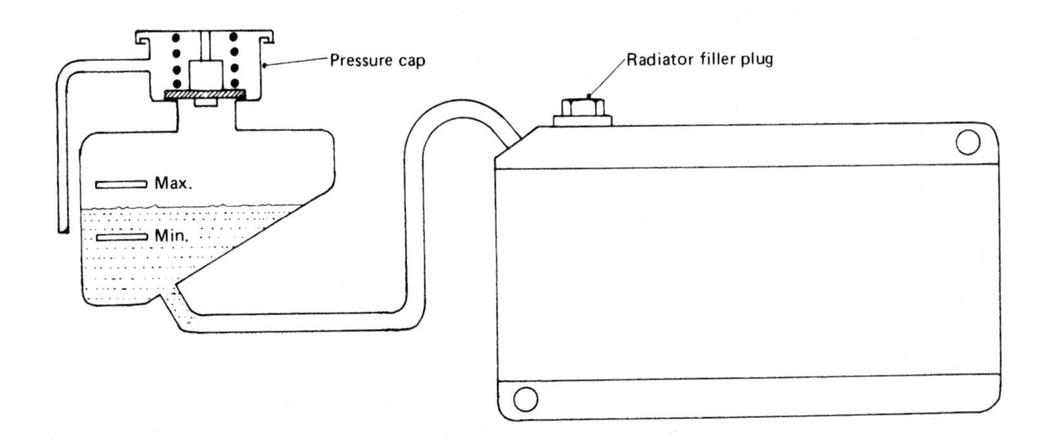

Fig. 20.18 Pressurised expansion tank

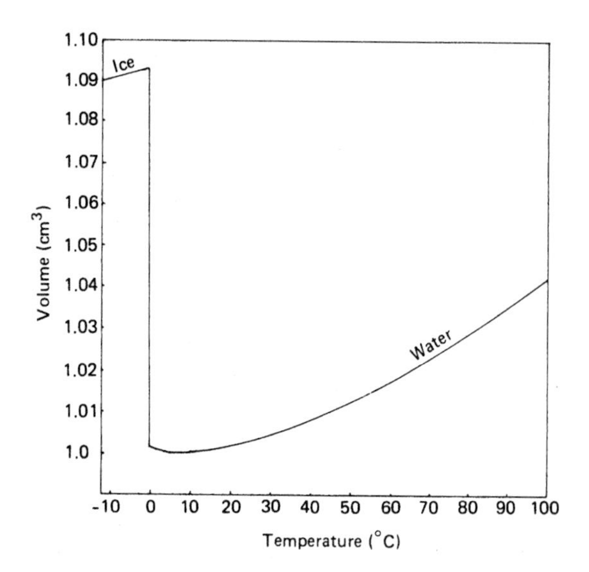

Fig. 20.19 Volume variation with temperature of 1 gram of water

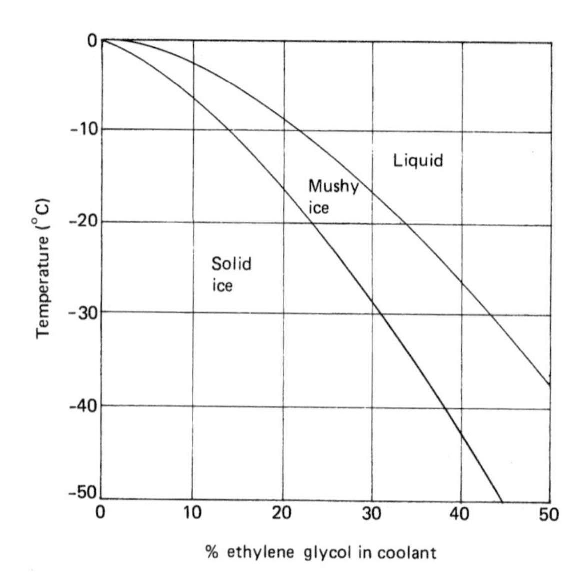

Fig. 20.20 Variation of freezing range with different strengths of antifreeze

from a liquid to crystalline ice which, in its free state, occupies approximately a 10% greater volume – see fig. 20.19. Unfortunately if the water in the cylinder-block and head passages were to freeze, it would rigidly solidify and exert pressure in all directions against the metal walls in its endeavour to expand. If all the water were converted to ice, the bursting pressure created by the trapped ice would be sufficient to rupture or crack the block and head at weak points and would certainly split open the radiator tubes.

To prevent damage to the engine and radiator during winter weather, suitable liquids or compound substances which go into solution with water are added to the water to lower the freezing temperature of the coolant below the temperature of the winter climate. These solutions are known as antifreeze solutions.

An antifreeze compound which greatly depresses the freezing point of water and remains in the system is calcium chloride. With added chromates such as sodium chromate, this also prevents corrosion of most metals used except aluminium.

Low-boiling-point liquids such as alcohol – usually ethyl alcohol or methyl alcohol – can be used, but they tend to be flammable and therefore are not preferred. The most popular antifreezes use the high-boiling-point glycols – usually ethylene glycol (boiling point 195°C) – which lower the freezing point of water in proportion to the solution strength. These solutions do not suddenly convert from liquid to solid at one temperature but gradually change in the form of mushy soft ice over a range of temperatures (fig. 20.20).

Typical solution strengths for ethylene glycol and degrees of protection are shown in Table 20.1.

Table 20.1 Typical ethylene-glycol solution strengths

Ethylene-glycol/water solution strength	Provides coolant circulation down to	Prevents ice damage down to
Normal UK winter (25%): 1 part antifreeze 3 parts water	−12°C	−25°C
Severe UK winter (33%): 1 part antifreeze 2 parts water	−20°C	−35°C
For all year round (50%): 1 part antifreeze 1 part water corrosion protection	−37°C	−50°C

20.8.1 Water suitability

The engine coolant water should have an acceptable mineral content and be of a known softness as shown below:

Chlorides (max.)	40 parts per million (p.p.m.)
Sulphates (max.)	100 p.p.m.
Total dissolved solids (max.)	340 p.p.m.
Total hardness (max.)	170 p.p.m.

20.8.2 Corrosion inhibitors

These are chemical compounds which form less than 10% of the antifreeze. They provide corrosion protection, acidity control of the coolant, and water-softening to avoid the formation of scale caused by mineral deposits.

Corrosion inhibitors protect the metallic surfaces of the cooling system against corrosion attack. The more commonly used cooling-system corrosion inhibitors are chromates, borates, nitrates, nitrites, benzotriazole, and tolyltriazole. The chromates, such as sodium chromate and potassium dichromate which are good corrosion inhibitors, are not compatible with glycols and they are therefore restricted to use in hot climates where only water coolants are necessary. The other inhibitor chemicals can be used with water and permanent antifreeze solutions.

20.8.3 Desirable properties of a good antifreeze

a) It should be able to go into solution with water to form a diluted water–antifreeze coolant mixture.
b) It should prevent freezing of the coolant down to the lowest operating-temperature conditions.
c) It should have as high a specific heat capacity as possible (ethylene glycol's is two thirds that of water), so that it can maximise the coolant's heat-dissipation capacity.
d) It should have a boiling point above that of water, so that the coolant can operate at higher more efficient temperatures. The boiling points of 30%, 40%, and 50% ethylene-glycol/water solutions are 104°C, 106°C, and 109°C respectively.
e) It should be non-flammable and have a flash-point well above the maximum possible operating temperature of the coolant. Ethylene glycol ignites at a temperature of about 125°C.
f) It should not have a high viscosity at operating temperatures, as this might prevent effective coolant circulation. Ethylene glycol is more viscous than water but it does circulate adequately.
g) It should not attack natural or synthetic radiator hoses. Ethylene-glycol solutions tend to shrink coolant-soaked hoses, so the clips should be retightened a few days after fitting new hoses.
h) It should be stable over a long period of time (two years), and it should not decompose to form corrosive products. Ethylene glycol in its pure state is corrosive to tin – lead solder alloys and to some extent to copper and aluminium.
i) It should be readily mixed with a corrosion inhibitor which prevents excessive corroding of non-ferrous components and retards the formation of scale from water-borne minerals.

20.8.4 Measuring an antifreeze-solution strength (fig. 20.21)

Coolant may be lost from the system for such reasons as leakages; replacement of hoses, gaskets, thermostats, or water-pump; or overheating

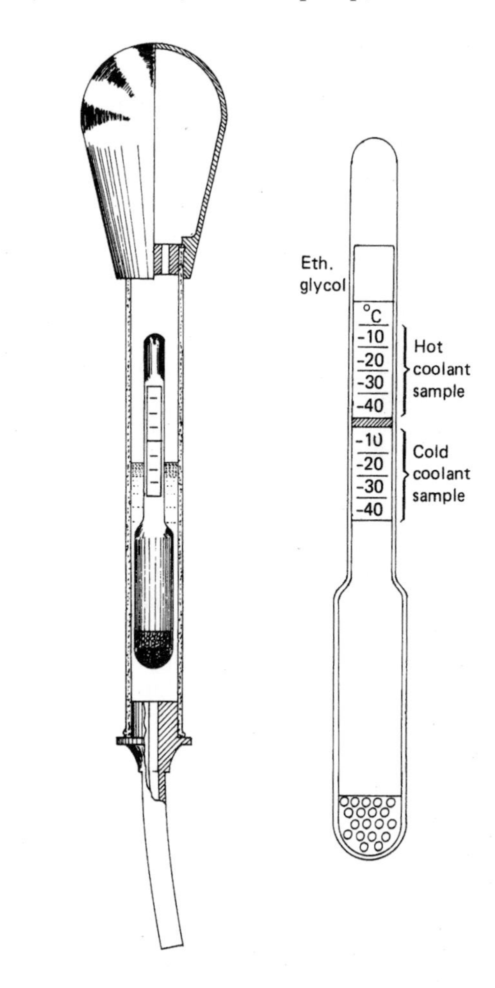

Fig. 20.21 Antifreeze hydrometer

due to fan-belt failure. The system then tends to be topped up with water for convenience, thus the antifreeze solution will become diluted, and this should at some stage be corrected.

The density of water and that of an antifreeze liquid are different, so, when water and antifreeze are mixed together, the resultant relative density will lie in between these two extremes, the actual value being directly proportional to the percentages of water and antifreeze present. This means that coolant antifreeze strength can be determined directly with a hydrometer gauge which will measure the relative density of a sample drawn from either the radiator or the expansion tank.

A hydrometer consists of a large glass-bore tube with a rubber suction ball at one end and a rubber withdrawal tube at the other (fig. 20.21). Inside the glass tube is a glass bobbin-type float whose large-diameter end is partially filled with lead shot for calibration purposes and whose small-diameter portion has one or more graduated scales. Sometimes there is one scale for methanol antifreeze solution, but there is always one for the most commonly used solution – ethylene glycol. Using the rubber suction ball, a sample of antifreeze solution is sucked into the glass tube, and the amount the float sinks relative to the liquid level is a measure of the liquid's relative density. However, the scales have been calibrated to convert this information into the freezing point of the sample of coolant withdrawn from the cooling system.

A further consideration is that the relative density varies considerably with the temperature of the solution, so two different-coloured scales are normally provided: one for hot samples (60 to 70°C) taken from an engine which has been stopped for only a few minutes, and one for cold samples (20°C) when the engine has not warmed the system.

Care should be taken when topping up a cooling system with an antifreeze solution, since if it comes into contact with the vehicle's paintwork it will spoil the finish.

20.9 Engine core plugs (fig. 20.22)

When the cylinder block is to be cast, a sand mould is prepared to the external shape required. Similarly, collapsible internal sand cores to occupy the shape of the coolant passages must be positioned and supported in the main mould.

After the iron has been poured and has solidified, the core sand has to be broken up and removed from the coolant cavities formed within the cylinder block. This is made possible by extending some sand coring in cylindrical form to the outside walls of the block – that is, meeting up the core with the outer sand mould. This then enables the unwanted core sand to be withdrawn through these core holes. It is obvious that these holes must be plugged if the coolant is to be retained. It can thus be seen that the primary function for core holes in the cylinder block and head is to enable coolant cavities to be formed during metal solidification and to provide an exit for the core sand. A secondary function is the use of the core plug as a safety valve to relieve internal pressure within the cooling passages if the liquid freezes solid, but this is effective only directly next to a core hole and in fact provides very little protection for the majority of the coolant-passage walls.

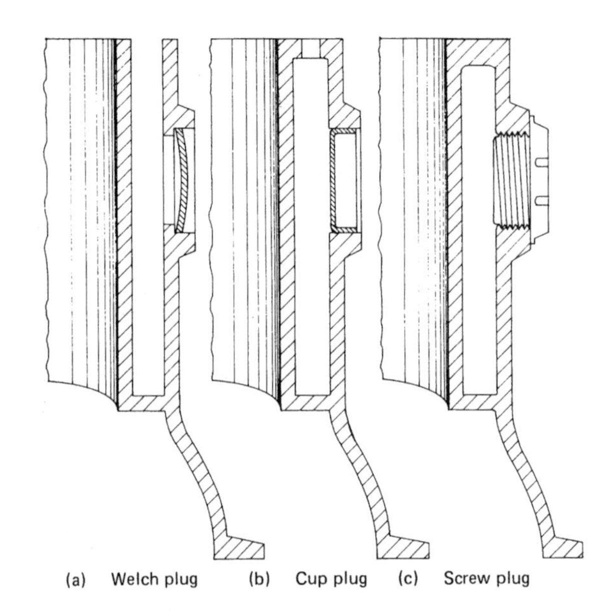

Fig. 20.22 Engine core plugs

There are three possible methods of blanking off the core holes: by welch plugs, cup plugs, and screw plugs.

Welch plug (fig. 20.22(a)) This resembles a dished disc. When fitted, it is pushed along the core hole hard against a shoulder formed in the jacket and then given several blows in the centre with a drift so that it spreads radially outwards to seal the hole. This type of plug can be conveniently installed in small core holes.

Cup plug (fig. 20.22)(b)) This is a deep-drawn low-carbon-steel pressing. When fitted, it has an interference fit and should be evenly tapped into position. This form of plug is adopted mostly for medium and large core holes.

Screw plug (fig. 20.22(c)) This is basically a steel or brass large-diameter stubby set-bolt which may have an external or internal hexagonal-shape head or countersunk holes so that it can be tightened or removed. This blanking plug is used only on a few large diesel engines where a permanent seal is necessary.

20.10 Maintenance requirements of the cooling system and components

Routine checks and servicing necessary for a cooling system are as follows:

a) Every 20 000 km, check that the fan belt is in good condition and correctly tensioned.
b) Check the condition of radiator and heater hoses annually, and tighten the clips if necessary.
c) If coolant is being lost, check the condition of the pressure-cap rubber seal and that it seats correctly.
d) Check the coolant-pump and thermostat-housing flanges for leakages annually, and tighten them both if necessary.
e) If the cooling system has been disturbed, check the strength of the antifreeeze solution using a hydrometer, and correct it if necessary.
f) If there is a grease nipple fitted, lubricate the impeller-shaft bearings every 20 000 km.
g) Flush out the cooling system every two years and refill with fresh antifreeze and water.
h) If the engine tends to overheat, remove the thermostat and check its opening temperature.

20.11 Cross-flow cooling system with liquid-to-liquid oil cooler

20.11.1 Cross-flow radiator (fig. 20.23)

With a cross-flow radiator there is very little thermo-syphon assistance and coolant is made to move through horizontal copper or aluminium-alloy radiator-tubes by the centrifugal force created by the rotating coolant (circulation) pump impeller driven by the crankshaft via the fan-belt or timing-belt. The cross-flow radiator has end tanks (chambers) that replace the top and bottom tanks of the vertical flow radiator. It therefore reduces the overall height of the radiator, permitting the car to have a streamlined low-profile bonnet (fig. 20.23). The radiator matrix consists of three rows of horizontal stacked tubes which have elongated cross-sections with copper or aluminium-alloy ribbon folded in triangular fashion, filling up the free spaces between adjacent tubes. The tubes are supported by input and output end tanks, respectively, their function being to collect the hot coolant coming from the cylinder-head and to discharge the somewhat cooled coolant to the input side of the circulation-pump.

20.11.2 Thermostat and pass flow control unit (fig. 20.23)

A cylinder-block input flow controlled thermostat is used in the system shown (fig. 20.23). Thus when the engine is cold the closed thermostat-valve blocks off the radiator return flow to the coolant (circulation) pump and, at the same time, opens the by-pass valve. This permits a small quantity of coolant to circulate between the relatively cool input to the coolant-pump and the hot exit from the cylinder-head. It therefore prevents the trapped coolant from overheating. When the engine has warmed up, the thermostat-valve opens and the by-pass valve closes; coolant circulation is now diverted from the by-pass pipe so that it has to flow through the radiator-tubes on its way to the input side of the coolant-pump via the thermostat-valve passageways.

20.11.3 Interior heater (fig. 20.23)

It can also be seen that the coolant is circulated through the interior heater matrix from the cylinder-head hot output to the upper chamber in the thermostat unit. It then passes to the input side of the coolant-pump (fig. 20.23). When the engine is cold and the thermostat valve is closed, the by-pass pipe and heater pipes control the only coolant that is returned to the input side of the pump. As a result the majority of the restricted coolant circulation is passed through the heater matrix tubes so that the warm-up time for the interior heater is at a minimum.

20.11.4 Oil-coolers (heat exchangers) (fig. 20.23)

The lubricating oil absorbs heat from the oil splashed onto the cylinder walls and pistons, from oil flowing over the hot cylinder-head and the heat generated between the rotating camshaft and crankshaft journal bearings. High-performance,

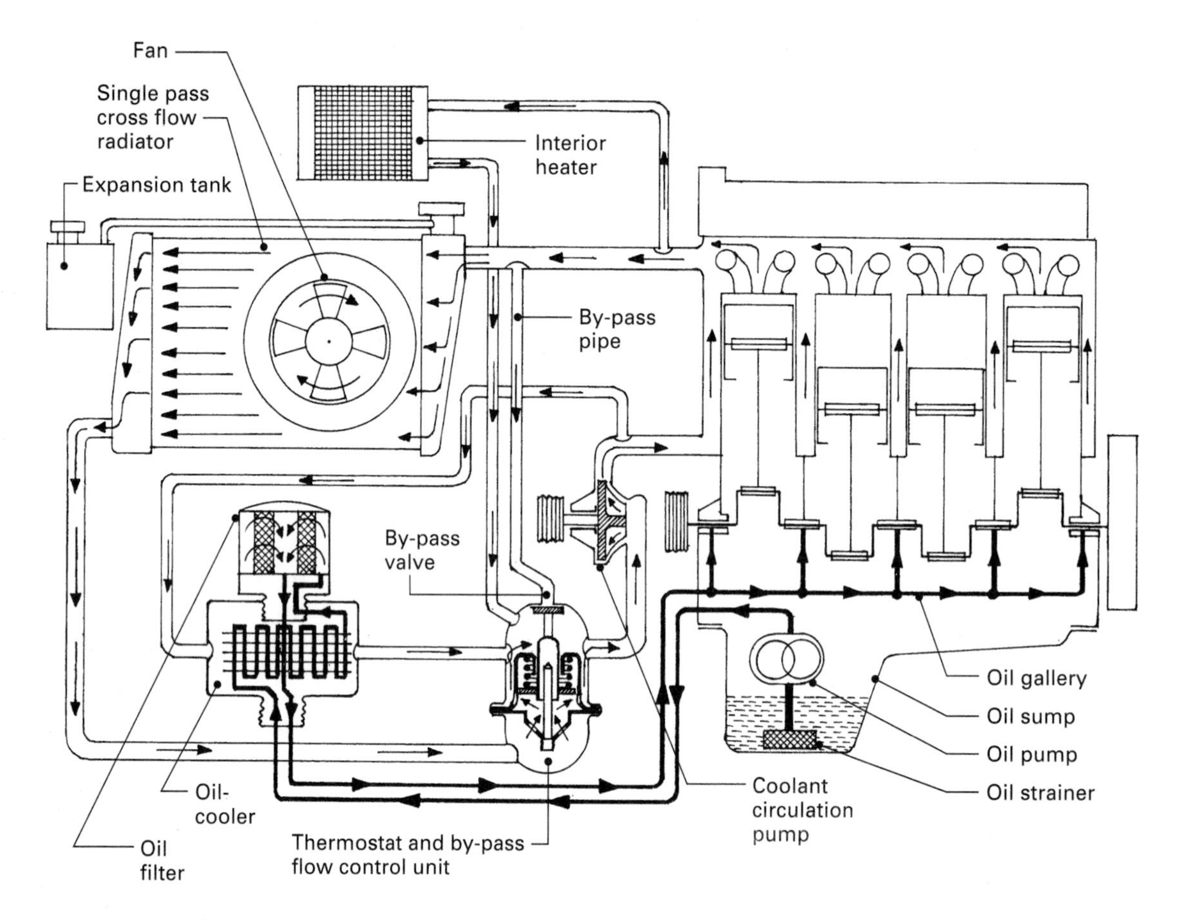

Fig. 20.23 Cross-flow cooling system with liquid-to-liquid oil cooler

heavy-duty or supercharged engines which work under heavy loads or at very high engine speeds may find the cooling system inadequate, so that the oil tends to operate at higher temperatures. Under these conditions it is sometimes necessary to pass the circulating lubricant through an oil-cooler to lower its mean temperature and to preserve the oil's effective life.

Oil-coolers are heat-exchangers, that is, they transfer heat from the hot lubricating oil to a cooler medium such as an air stream or the coolant from the engine's cooling system.

There are two basic types of oil-coolers:

1) liquid to air and
2) liquid to liquid.

Liquid to liquid This type of heat-exchanger functions like that of the engine cooling system radiator. It passes the hot lubricant oil through finned tubes which are directly exposed to the surrounding air stream caused by the moving vehicle or the engine's cooling system fan. Heat is continuously being transferred by conduction through the metal walls of each tube to the surrounding atmosphere via the much enlarged surface area exposure of the fins.

Liquid to liquid (fig. 20.23) This heat-exchanger transfers heat from one liquid, the hot oil, to another liquid, the cooler coolant circulating in the engine's cooling system. Both liquids are separated by many tubes. The heat is therefore transferred by conduction through the metal walls of each tube and then by convection current by circulating coolant to the radiator.

Combined oil cooler and filter unit (fig. 20.23) Hot lubricating oil is pumped from the engine's lubricating system's oil-pump outlet to one end of the oil cooler. It then passes through the cooling tubes, after which the cooled oil is discharged to the engine's main oil gallery where it divides to feed the crankshaft main bearings. The efficiency of the heat-exchanger is improved by making the oil flow around and across the tubes, from one side to the other, moving backwards and forwards several times before passing out to the full-flow oil filter (the zig-zag movement of the oil is achieved by incorporating baffle-plates to divide the oil chamber into a number of passes; (see fig. 20.23). The combined oil cooler and filter unit shown in fig. 20.23 screws onto the side of the cylinder-block in a similar way to that of a separate oil filter unit, except in this case there are input and output external coolant hoses to be connected.

20.12 Air temperature sensing viscous fan coupling (fig. 20.24(a) and (b))

Viscous fan-drives enable the fan speed to be matched to the engine's cooling needs under a wide range of driving conditions. The viscous fan-coupling therefore prevents overcooling of the engine, and the unnecessary use of the fan at high speed reduces the power consumption and

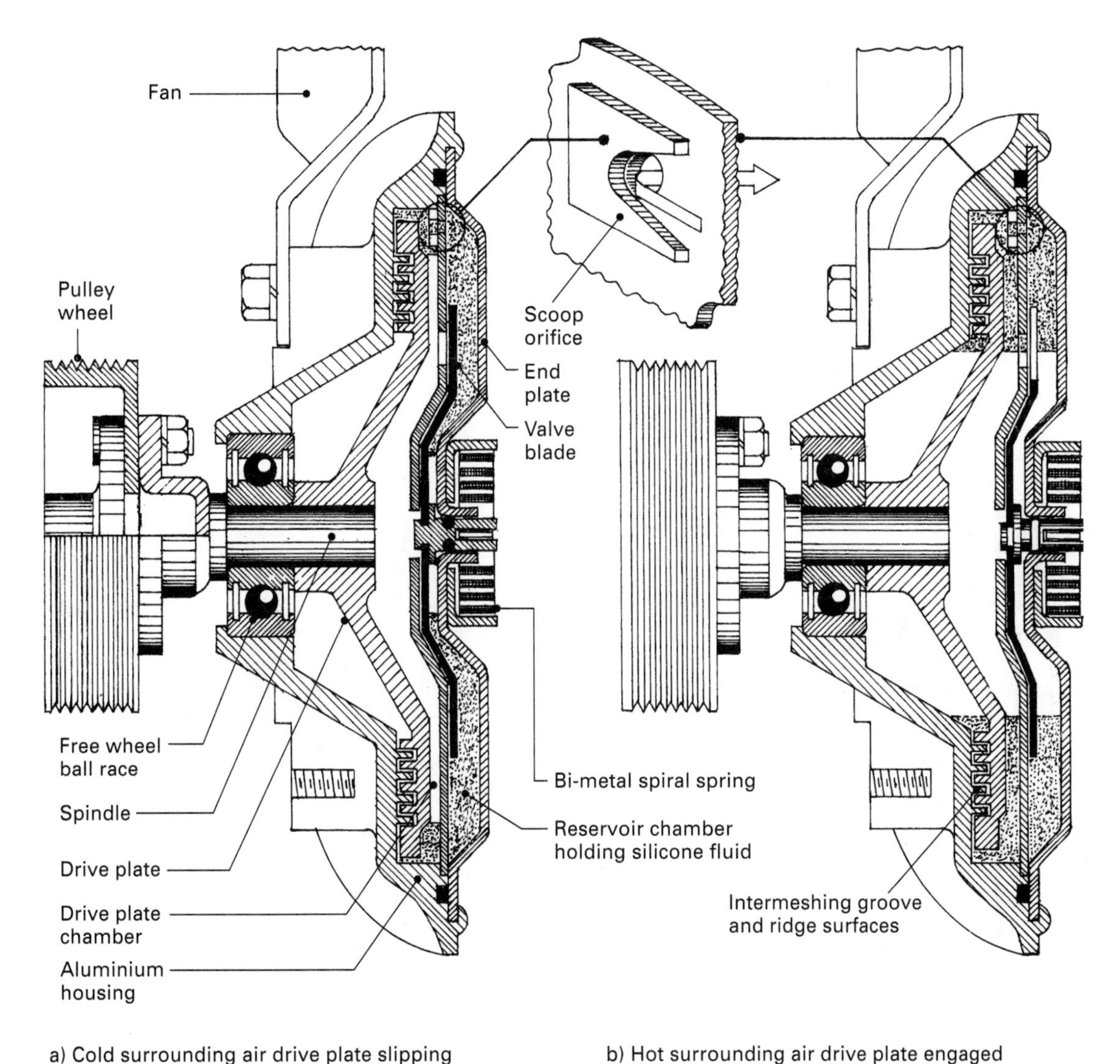

Fig. 20.24 Air temperature sensing viscous fan coupling

the noise usually associated with the churning of air.

Construction (fig. 20.24(a) and (b)) The coupling comprises an aluminium housing with cast cooling-fins, a steel separator-plate and a steel end-plate assembly, mounted on a single-row ball-race. Enclosed between the aluminium housing and the separator-plate is a disc-shaped driven-plate, whereas the space between the separator-plate and end-plate acts as a reservoir for the silicone fluid. A flanged spindle bolted on the belt pulley-wheel supports both the ball-race and the driven-plate. Attached to the outside of the end-plate is a bi-metal spiral spring which is sensitive to the ambient temperature of the surrounding air. This controls the opening and closing of the valve-blade. The separator-plate has a valve-orifice on the inside of the driven-plate annular friction surface and a scoop-orifice near the periphery of the plate. The fan-blades are themselves bolted to the aluminium housing.

20.12.1 Engine cold (figs 20.24(a) and 20.25(b))
When the engine is cold the bi-metal spiral spring will be closely coiled, thus moving the valve-blade to the closed position (fig. 20.25(b)). With the pulley, spindle, and aluminium housing rotating, the silicone fluid in the drive-plate chamber will be subjected to centrifugal force; hence the radial outward pressure on the fluid forces it to flow through the scoop-orifice into the reservoir-chamber on the opposite side of the separator-plate until the drive-plate chamber is practically empty. As a result, the clearance between the drive-plate and side members will interrupt the viscous drag so that the fan and housing assembly will free-wheel while the belt pulley-wheel and flanged spindle are driven at engine speed.

20.12.2 Engine hot (figs 20.24(b) and 20.25(b))
As the air temperature surrounding the viscous coupling rises, the bi-metal spiral spring tends to expand and uncoil, thereby twisting the valve-blade to the open position (fig. 20.25(a)). The stored silicone fluid in the reservoir-chamber is now able to return via the valve-orifice to the drive-plate chamber. Thus, as the chamber fills, the submerged fluid surface area between the drive-plate and the side member enlarges to increase the viscous drag between the coupling members. Consequently, the torque imposed on

Fluid flow
Orifice
Valve blade
Bi-metal spiral spring
Valve movement limiter
Separator plate

a) Engine hot valve orifice open

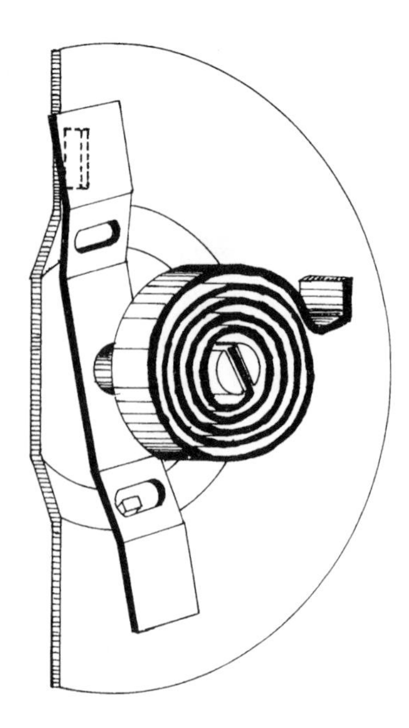

b) Engine cold valve orifice closed

Fig. 20.25 Operation of the bimetal spring for a viscous fan coupling

the outer housing and the fan assembly increases its rotational speed according to the fan air resistance and the speed of the belt pulley-wheel driven by the crankshaft-pulley via the fan-belt.

Conversely, when the vehicle is moving fast, the increased air stream moving through the radiator and over the engine cools the viscous-coupling assembly so that the bi-metal spiral spring tightly coils and causes the valve-blade to close off the orifice. Hence, the fluid circulation is blocked so that the majority of the fluid in the drive-plate chamber will be pumped, via the scoop-orifice, into the reservoir-chamber. Again the lack of viscous drag allows the outer housing and fan assembly to slip an amount proportional to the loss of surface area exposed to the fluid. These fan-speed to fan-drive speed characteristics are shown in fig. 20.26.

20.13 Cooling fan relay and thermal switch circuit (fig. 20.27(a) and (b))

The electric cooling fan system consists of a permanent magnet two-pole motor (similar to those used for windscreen wiper motors), fan relay and a thermal switch unit which is screwed into the radiator upper, lower or end tanks. The thermal switch unit is a heat sensitive switch which controls the 'on' and 'off ' running phases of the fan.

When the engine temperature reaches some predetermined value (typically 92 to 98°C), the thermal switch bi-metal disc suddenly distorts (buckles). This permits the relay-pin to move further into the capsule and at the same time allows the leaf-spring contact to close with the fixed contact (see fig. 20.27(b)). A small current now flows from the battery's negative terminal through the shunt relay winding to the negative earth via the closed thermal switch. As a result the winding's magnetic force field pulls down the spring-armature sufficiently to close the relay contacts. The supply current now passes through the relay frame, spring-armature, contacts and fuse. It then flows through the fan motor armature-windings and is then directed back to the battery's negative terminal via the earth return. The energised motor now commences to rotate at its self-regulating speed. As the engine cools the thermal switch temperature decreases until just below the contact closing temperature (typically 86 to 92°C); the bi-metal disc snaps back to its flat state at which point the thermal switch contacts are

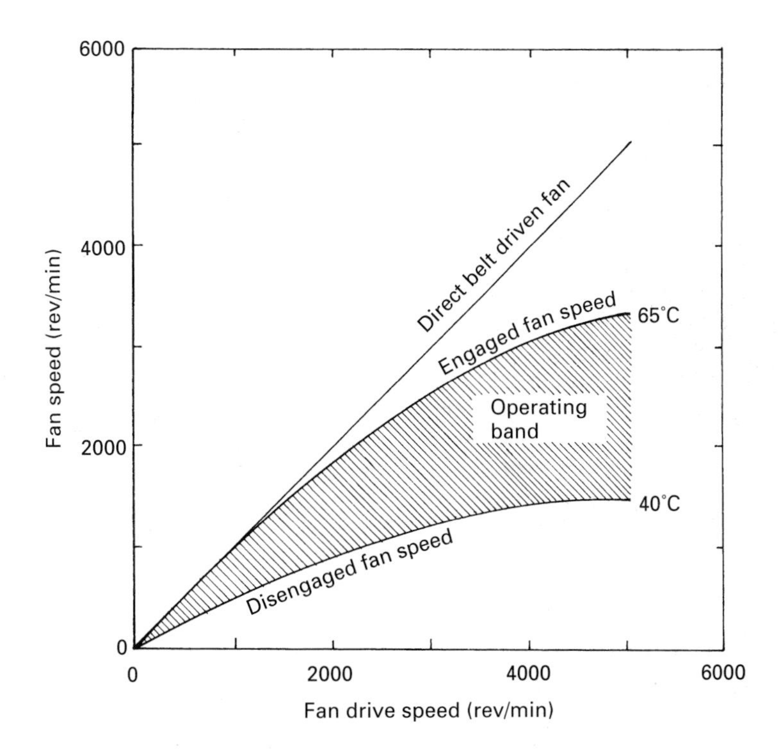

Fig. 20.26 Viscous fan drive with air temperature sensing speed characteristics

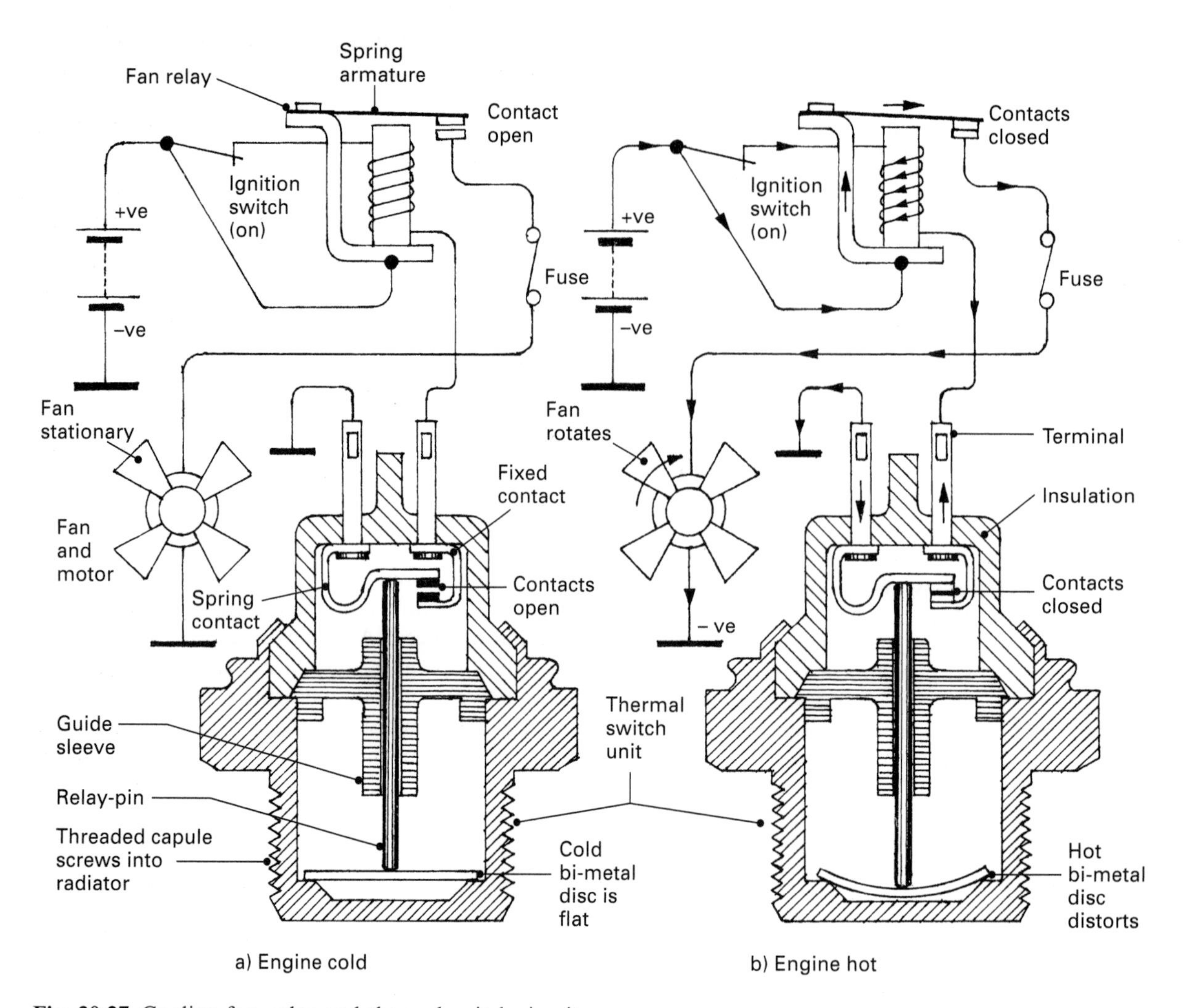

Fig. 20.27 Cooling fan, relay and thermal switch circuit

forced to open again, thereby switching 'off' the motor (see fig. 20.27(a)).

By these means the engine can operate at relatively high temperatures and only under extreme load and hot weather conditions, driving at very low speed in traffic or driving uphill in low gear will the fan automatically commence to rotate. In other words the fan's 'on' and 'off' periods are made to closely match the engine's heating and cooling cycle.

20.14 Interior heating and ventilating system

20.14.1 Heating and ventilating system layout (fig. 20.28)

To maintain the comfort and alertness of the driver and passengers, the interior of the driver and passenger compartment should be ventilated and heat must be made available in cold weather to warm the interior of the compartment. Air entering the compartment may be caused either by the air stream created by the moving vehicle or it can be forced in by an electrically driven blower. Cold air from the outside of the vehicle can be directed into the interior compartment through an enclosed plastic casing known as the heat distribution box. The actual amount of air entering and where it is diverted is controlled by the hot–cold deflector flap and an interior-screen deflector flap respectively.

Heat is supplied by the engine's cooling system. Hot liquid coolant is piped to a heat-exchanger and the cool air entering the heat distribution box freely flows through or is pumped. The air passes through the gaps between the tubes and fins of the heat-exchanger matrix. Thus heat is transferred from the hot circulating coolant to the cooler air stream entering the inside of the driver–passenger compartment via the heat distributor box. The amount of heat entering the compartment is controlled by the position of the hot–cold deflector flap, whereas the heat distribution between the footwell and the windscreen is controlled by the position of the interior-screen deflector flap.

20.14.2 Heating and ventilating system electrical circuit (fig. 20.28)

A direct-current permanent magnet two-brush unidirectional electric motor drives the squirrel-cage centrifugal-type blower. The speed of the motor and blower and therefore its pumping ability is controlled by varying the applied voltage. Adding the wire-wound resistors in series into the earth return circuit reduces the voltage applied to the electric motor. Thus the highest speed setting (4) is obtained when the electric motor return lead is connected directly to earth whereas the slowest speed setting (1) is switched in when the electric motor return lead is connected to earth via all three series connected resistors.

To reduce the current load on the ignition switch terminals the supply circuit between the battery and the electric motor is connected via the electric motor relay. When the ignition switch is turned on, current flows through the solenoid winding. This magnetises the winding core and pulls down the leaf-spring armature, thereby closing the contacts which join the battery to the positive brush of the electric motor. In some electric motor circuits a low current draw does not justify a relay connected in series. However, with other systems where a larger motor is used, a relay is switched in series with the battery supply and the electric motor only when the full supply voltage is applied to the electric motor in the high speed setting position.

20.15 Air conditioning

Air-conditioning lowers the temperature of the incoming air before it enters the inside of the vehicle. It therefore cools and dehumidifies the air space in the driver-passenger compartment and when required rapidly demists the insides of the surrounding windows. As a result an improvement in comfort, alertness and a reduction in driving fatigue is achieved for both driver and passengers.

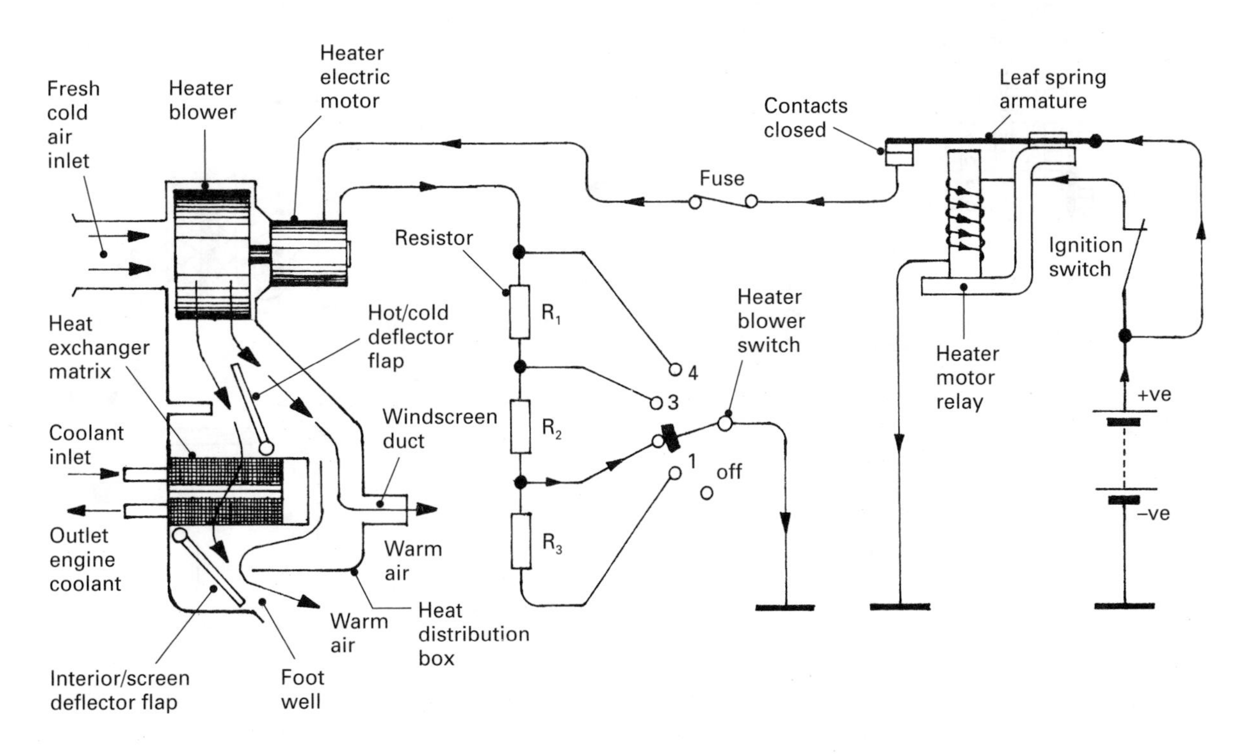

Fig. 20.28 Interior heater system and electrical circuit

The air-conditioning system (see fig. 20.29) operates on similar principles to a refrigerator. The major components of the system are as follows:

1) compressor
2) evaporator
3) condenser
4) receiver-drier
5) expansion valve
6) blower
7) dual high and low pressure switch.

20.15.1 Compressor

Engine-driven compressors are used to draw the low-pressure and low-temperature refrigerant gas from the evaporator, to compress it, and to pump it under increased pressure and temperature

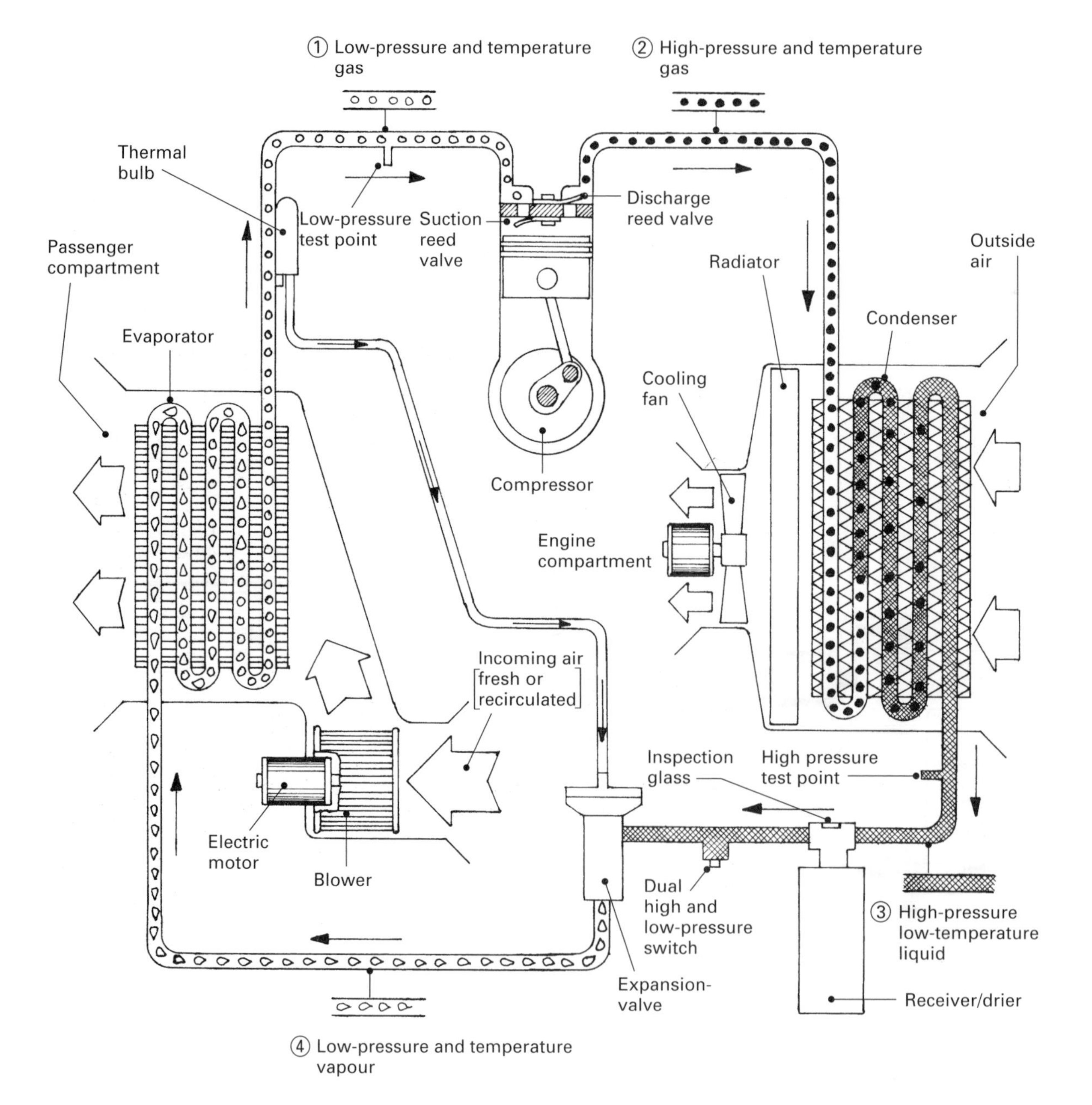

Fig. 20.29 Air-conditioning system

through the condenser to the expansion-valve which acts as a throttle restriction and is thus responsible for the pressure build-up.

There are basically two types of compressors, the vane-rotary-type compressor and the reciprocating-type compressor. The latter type can be subdivided into the crankshaft driven (fig. 20.29) and swash-plate drive type (fig. 20.30). Both piston-type compressors use disc or reed-type suction and discharge valves. At present the majority of air-conditioning systems use the swash-plate double-acting piston-type compressor; therefore only this type will be considered in detail.

Double-acting axial piston swash-plate drive compressor (fig. 20.30) The double-acting aluminium piston is really two opposing pistons spaced apart but cast as a single component. Each pair of

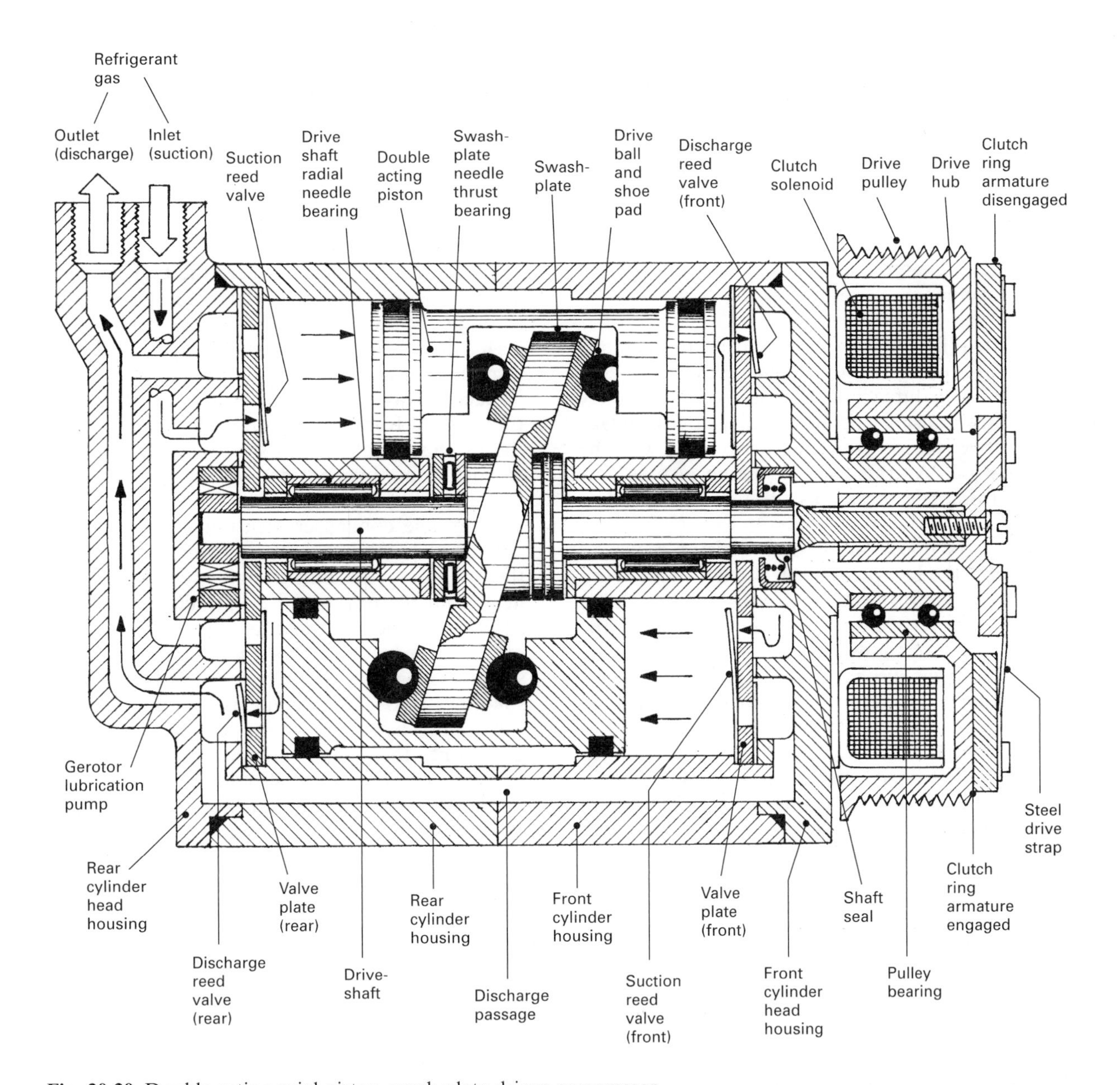

Fig. 20.30 Double-acting axial piston swash-plate driven compressor

pistons is positioned over the swash-plate and receives its drive through two balls and their shoe-pads. Each piston has a single nylon ring seal. There are three sets of double-acting pistons arranged at 120° intervals around the drive-shaft. The forged drive-shaft forming part of the swash-plate is supported by radial needle bearings mounted on either side of the swash-plate. Needle thrust bearings are mounted on both sides of the swash-plate to sustain the axial end-load reactions. Attached via splines to the front end of the drive-shaft is a hub which supports an annular armature ring through four drive straps, whereas the rear end of the drive-shaft is coupled to a gerotor-type lubrication pump which feeds the front and rear drive-shaft and swash-plate radial and end thrust needle bearings.

Valve plates with the suction and discharge reed valves fastened (riveted) to them are positioned at the front and rear end of the cylinders. Passages within the front and rear cylinder housings connect the discharge and suction chambers at both ends together.

Operation (fig. 20.30) When the electromagnetic solenoid is energised, the magnetic pull draws the annular armature ring against the drive pulley face thus causing the frictional grip to engage and drive the swash-plate drive-shaft via the splined hub. All three double-acting pistons now move from side to side as the tilted swash-plate rotates.

When the upper double-acting piston moves to the right, the low pressure in the left-hand cylinder opens the rear suction reed-valve thus drawing in the refrigerant gas; simultaneously the pressure in the right-hand cylinder rises, forcing open the front discharge reed-valve. Conversely, the lower double-acting piston which is moving in the opposite direction to the upper pistons causes the discharge-valve to open as the pressure builds up in the left-hand cylinder, and a pressure drop in its right-hand lower cylinder at the same time opens its suction-valve. A complete cycle of suction and compression is completed every revolution of the drive-shaft whereas there is a suction and compression cycle commencing and finishing every 120° of the drive-shaft's rotation.

20.15.2 Evaporator (fig. 20.29)

This is a heat exchanger and may be constructed similarly to that of the engine's cooling radiator where many tubes made from either copper or aluminium are joined at their ends to side tanks which form the inlet and outlet passages. Each tube is separated by a fin ribbon made from similar metal as the tubing and folded concertina fashion. This maximises the heat transference from the air to the refrigerant. Alternatively the evaporator may be constructed similarly to that of the condenser shown in fig. 20.29. The refrigerant vapour at low pressure and temperature flows through the tube or tubes, absorbing the heat from the warm air passing over the tubes so that the air leaving the evaporator is much cooler than when it first entered. It is the latent heat of evaporation of the refrigerant which extracts heat from the surrounding air via the fins (see fig. 20.33). The heat is then transferred by conduction through the metal walls of the tube or tubing to the refrigerant. The hot refrigerant is then expelled from the evaporator as it is drawn to the suction side of the compressor.

20.15.3 Condenser (fig. 20.29)

This is a heat exchanger where the high pressure and temperature of the refrigerant gas flowing through the tubes changes its state and is converted into a liquid. The evaporator consists of a single copper or aluminium tube which is folded in such a way that equal parallel lengths of the tube zigzag from side to side. The gaps between the tubing are filled with sheets of flat or corrugated fins, as this considerably increases the surface air of conducting metal exposed to the air stream. The condenser transfers the latent heat contained by the refrigerant from the inside of the tubing to the outside air stream. Thus the heat from the refrigerant is extracted and is dissipated by the surrounding air stream passing over and around the tube matrix. The hot pressurised refrigerant discharged from the compressor flows through the condenser tubing which is surrounded by the continuous displacement of air drawn into the engine's compartment by the ram effect of the moving vehicle and the engine's cooling system fan. This relatively cool air stream extracts heat from the refrigerant, causing it to give up its latent heat in its gas state at entry. It then progressively changes to a liquid state towards its exit from the condenser tubing (see fig. 20.33).

20.15.4 Receiver-drier (fig. 20.31)

Refrigerant from the condenser flows through the receiver-drier. Here water moisture is separated from the refrigerant by a filter screen and drier desiccant crystals. This prevents partial freezing

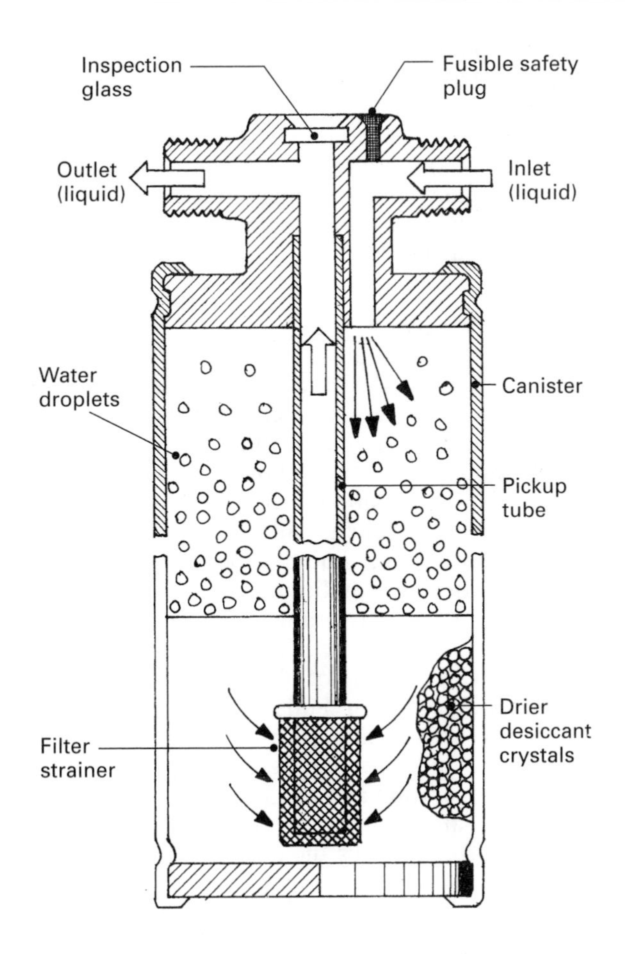

Fig. 20.31 Receiver-drier

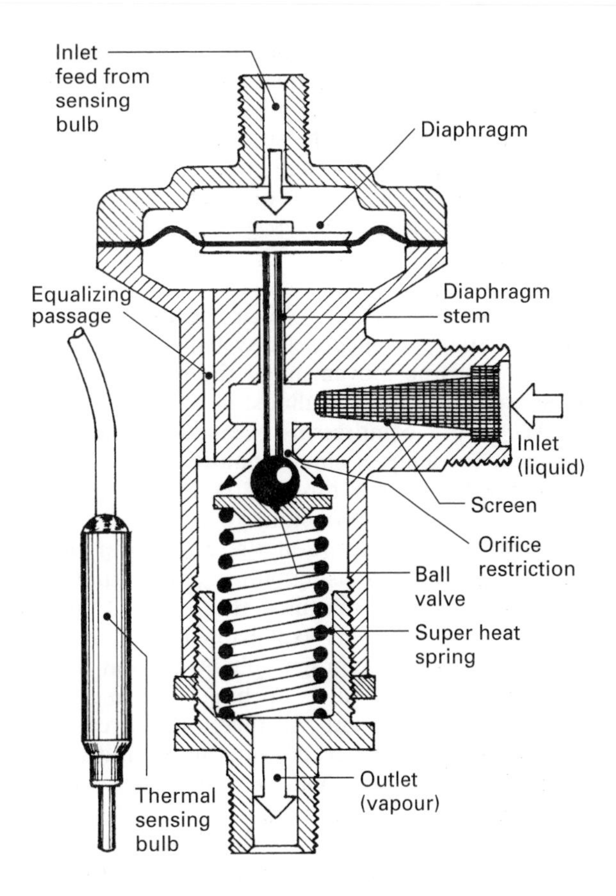

Fig. 20.32 Thermostatic expansion valve

and blockage of the expansion valve during the system's start-up when there is a tendency for water droplets to form. The receiver-drier also functions as a storage space for refrigerant so that it can compensate for very small refrigerant losses which occur over prolonged periods of time.

The receiver-drier consists of a cylindrical casing partially filled with desiccant crystals which absorb any water mist or droplets entering the container and remove any debris suspended in the liquid as the refrigerant flows from the condenser to the expansion valve. A central pick-up tube with a filter screen attached at its lower end diverts the dry waterless refrigerant to the expansion valve.

A glass inspection aperture is installed in the top of the receiver-drier canister to confirm the amount of charged refrigerant circulating through the pipelines. If the temperature of the refrigerant rises above 105°C a fusible plug positioned in the head of the receiver-drier melts, thus permitting the refrigerant to be discharged to the atmosphere.

20.15.5 Thermostatic expansion valve (fig. 20.32)
The expansion valve regulates the flow of refrigerant passing through the system. It therefore meters and controls the movement of liquid refrigerant reaching the evaporator. A lower flow rate reduces the refrigerant pressure and raises its temperature; conversely increasing the flow raises the refrigerant pressure and reduces its temperature.

A thermal sensing bulb filled with either carbon dioxide or refrigerant is fastened to the outlet tube of the evaporator. It senses the outlet temperature of the refrigerant which causes the fluid in the bulb to either expand or contract proportionally to the rise or fall in the refrigerant temperature. Thus any fluid expansion is relayed to the expansion valve upper diaphragm chamber. This pushes down the valve stem and accordingly increases the ball-valve opening so that more refrigerant may pass to the evaporator. As the refrigerant passes through the ball-valve restriction it produces a pressure drop on the outlet side. It therefore permits the refrigerant to expand and to convert to a vapour.

Operation (fig. 20.32) When the compressor is not rotating, the hydraulic system is pressurised but there is no movement of the refrigerant and the expansion valve is almost closed.

When the compressor is engaged and commences to pump, a small quantity of refrigerant is permitted to flow through the expansion valve thus causing the refrigerant to circulate. As the refrigerant temperature at the evaporator outlet rises, the thermal sensing bulb's temperature likewise increases. This causes the fluid within the bulb to expand. This expansion is then relayed to the upper diaphragm chamber and hence increases the ball-valve opening. The increased flow of the refrigerant will therefore decrease the temperature of the evaporator outlet and correspondingly the thermal sensing bulb. Consequently the fluid within the sensing bulb contracts. Thus it permits the superheat spring to partially close the ball-valve, thereby reducing the flow rate of the refrigerant. Note that a very high refrigerant temperature at the evaporator exit means that its flow rate is insufficient to match the heat absorption needed to cool the warm air moving around the tubing and through the evaporator before it enters the passenger compartment. Accordingly, the expansion valve opening must be increased sufficiently to maintain the refrigerant at a low enough temperature. A passageway between the spring chamber and the underside of the diaphragm makes the ball-valve more sensitive to the changing temperature conditions. It can thus be seen that the expansion valve automatically adjusts the flow rate to match the changing air temperature conditions.

20.15.6 Blower (fig. 20.29)

This is really a low-pressure air pump driven by an electric motor and is usually of the squirrel cage type. It draws the cool air surrounding the evaporator tube and fin matrix into the passenger compartment, and by selecting one of the blower speed settings, the interior temperature of the passenger compartment can be effectively regulated.

20.15.7 Dual pressure switch (fig. 20.29)

The refrigerant system is protected against excessively high or low pressure by the dual pressure switch. If the refrigerant pressure rises above, or falls below, the pre-set values, the dual pressure switch contacts open to interrupt the electromagnetic clutch circuit which thus brings the compressor to a standstill.

20.15.8 Refrigerant

A refrigerant fluid absorbs heat as it changes its state from a liquid to a gas and rejects heat as it reverses this process, and changes its state back from a gas to a liquid (see fig. 20.33). A refrigerant should have the following properties: be clean, non-toxic, non-corrosive, non-flammable, non-explosive, odourless and reliable when being cycled through its liquid–vapour states under normal working conditions. In addition it is essential for the refrigerant to have a low boiling point and a large latent heat of evaporation, that is, it must be capable of absorbing large quantities of heat when it boils and changes into a liquid. A low boiling point for the refrigerant allows the air-conditioning system to function at low temperature and in a pressure range within the evaporator which provides an efficient heat transfer process. Thus the quantity of refrigerant

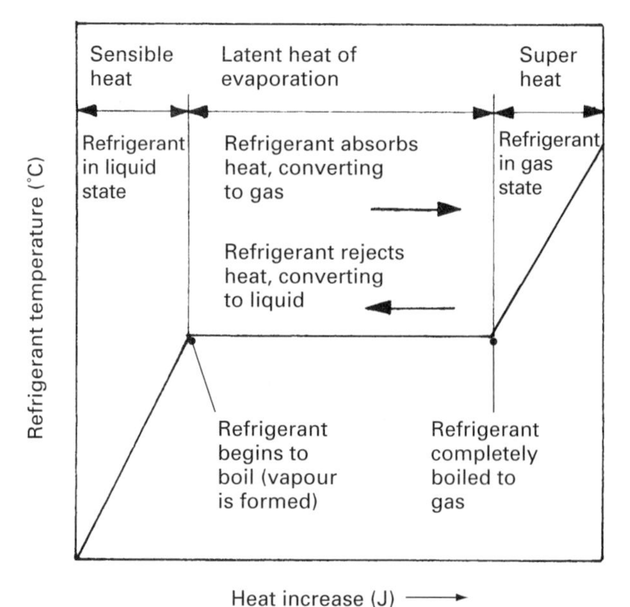

Fig. 20.33 Illustrative relationship between the refrigerant's temperature and heat content during a change of state

in the system should be relatively small and the energy consumed in circulating the refrigerant minimal. Thus the refrigerant R-12 when exposed to atmospheric pressure boils at −6.67°C and will evaporate from a liquid to a vapour when subjected to only 7% of the heat required by water to convert to steam. The refrigerant freon 12 which is a chlorofluorocarbon (CFC) known as R-12 was used extensively but it has now been replaced with tetrafluoroethane, coded R-134a, which is not so damaging to the Earth's ozone layer when released into the atmosphere. Note that refrigerants R-12 and R-134a must not be mixed as they are completely incompatible.

The following precautions should be taken when working with exposed refrigerant:

1) Always wear goggles to protect the eyes from splashed refrigerant as this could damage the eyes and even cause blindness.
2) Never let liquid refrigerant touch the skin as this causes severe frostbite.
3) Never discharge refrigerant from the air-conditioning system in a confined space because the refrigerant vapour contains no oxygen and therefore could cause breathing difficulties.
4) Never expose the refrigerant liquid or vapour to an open flame as this will produce a phosgene gas which is harmful if inhaled.

20.15.9 Vapour compression refrigerator cycle (fig. 20.34)

The air-conditioning system operates similarly to the refrigerator cycle where the evaporation and condensation processes take place when the refrigerant fluid is receiving and rejecting latent heat. When the engine is running and the magnetic clutch engages the compressor, the refrigerant is forced to circulate between the evaporator and the condenser via the compressor, receiver-drier and the expansion-valve (see fig. 20.29). The expansion-valve opening acts as an orifice and therefore creates a pressure build-up between the discharge side of the compressor and the expansion valve inlet, whereas there is a pressure reduction on the outlet side of the expansion-valve leading to the evaporator inlet. Consider the pressure-heat cycle shown in fig. 20.34.

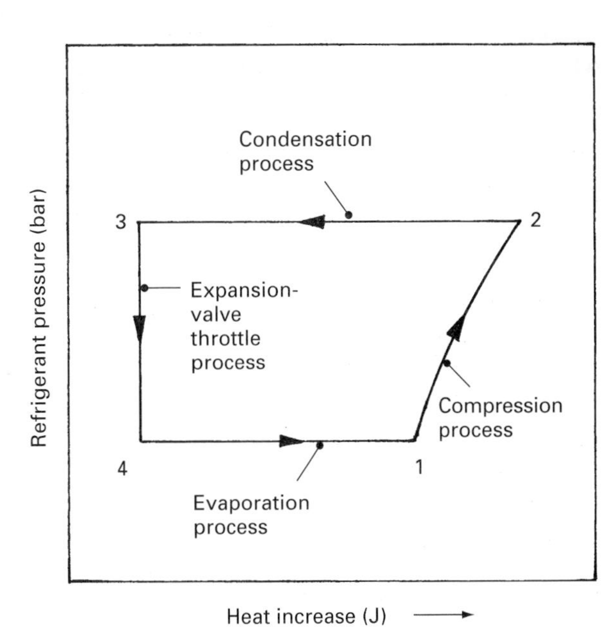

Fig. 20.34 Vapour-compression refrigerator cycle

Phases 1–2 (fig. 20.34) The refrigerant passing through the evaporator picks up heat from the warm air flowing over the tubing as it moves from the outside to the inside of the driver–passenger compartment. This heat coming from the warm air stream is partially used up in boiling the refrigerant (this being the latent heat of evaporation) and the remainder goes into further raising the temperature of the refrigerant, that is, superheating the refrigerant before it leaves the evaporator. However, this superheat should be kept to a minimum. It should be just sufficient to prevent any liquid entering the compressor and contaminating its lubrication. Remember the refrigerant boils at something like −7°C, so that the surrounding movement of warm air easily converts the refrigerant liquid-vapour into a gaseous state by absorbing heat.

The compressor draws in the low-pressure (0.5–2.0 bar) refrigerant gas at slightly superheated temperature and discharges the gas at a higher pressure (5–20 bar) and temperature.

Phase 2–3 (fig. 20.34) The refrigerant enters the condenser at high pressure and temperature. It then passes through the tubing. Heat is extracted by the cool air stream from the outside of the vehicle flowing over and around the condenser tubing; hence the refrigerant gas condenses to a liquid state. It therefore leaves the condenser still at high pressure but at a much reduced temperature.

Phase 3–4 (fig. 20.34) The high-pressure low-temperature liquid refrigerant then passes through the receiver-drier which removes any water and debris on its way to the expansion-valve. The expansion-valve restriction regulates the refrigerant flow to the evaporator. It therefore causes a pressure drop across the inlet and exit of the expansion-valve followed by a fall in temperature. Subsequently the refrigerant flowing to the evaporator now exists in a liquid-vapour state.

Phase 4–1 (fig. 20.34) The low-pressure and temperature liquid-vapour refrigerant flows through the evaporator tubing and in the process absorbs heat from the warm air blown from the outside of the vehicle into the driver–passenger compartment. As a result the heated low-pressure refrigerant quickly boils to a gaseous state as it passes through the warm evaporator tubing. At the same time the heat transference from the warm air coming in from outside the vehicle to the refrigerant via the evaporator fins and tubing walls lowers the temperature of the air stream before it enters the driver–passenger compartment.

21 Petrol-engine carburation fuel system

21.1 Layout of a petrol-engine fuel system (fig. 21.1)
A complete fuel system for a carburetted engine may be considered to include the following:

a) a fuel tank which stores the petrol and has a fuel-gauge sensor unit incorporated to indicate the amount of petrol in the tank;
b) a fuel lift or feed pump which transfers the petrol from the tank to the carburettor;
c) a fuel filter which prevents sediment, grit, and any contaminating particles which could block the petrol jets passing into the carburettor;
d) an air-silencer and filter unit which quietens the fast-moving air intake and prevents dirt from entering the engine;
e) a carburettor which merges air and petrol together so that they are mixed in the correct proportions and the petrol is finely atomised;
f) an induction manifold which collects the prepared air–fuel mixture and distributes it to the various inlet ports in the cylinder head;
g) supply and return pipelines which relay the petrol from the tank to the carburettor and return any excess petrol to the tank, thus providing a recirculating circuit free of vapour locks – petrol vapour accumulating in warm weather and preventing liquid flow.

21.1.1 Fuel tank (fig. 21.1)
The fuel tank, which is generally flat and semi-rectangular in shape with rounded corners, may be made from pressed steel (which is normally internally coated with a lead–tin alloy to protect it against corrosion), expanded synthetic-rubber compound, or flame-resistant plastic. Internally the tank normally has baffle plates arranged to prevent petrol surge when the vehicle accelerates, rolls, or pitches.

Holes are cut into the sides of the tank to locate the fuel-gauge sensor unit, the supply and return pipes, and the wire-gauze or nylon-mesh primary filter. A vent pipe is fitted to relieve the vapour pressure and to allow air to enter at atmospheric pressure. The filler tube usually has both the vent

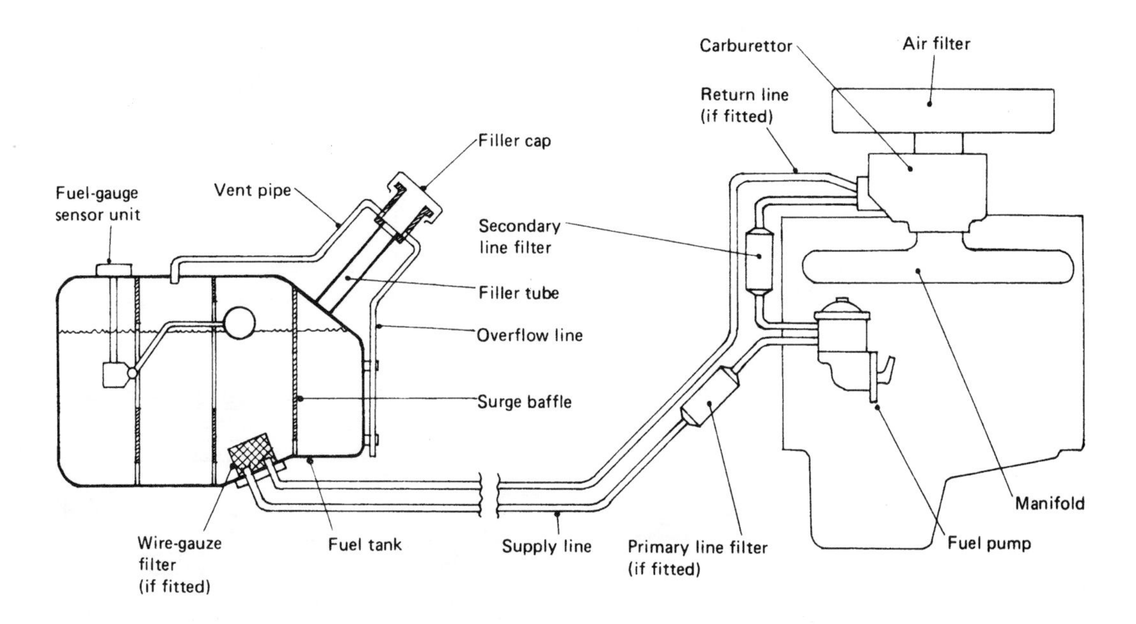

Fig. 21.1 Carburettor-type fuel system

pipe and the overflow pipe tapped into it near the filler-cap end. Non-vent filler caps are usually used, so that petrol vapour does not escape from this point, its only outlet being through the combined vent and overflow pipeline whose outlet is situated low down.

21.1.2 Position of the fuel tank (fig. 21.2)

Front-mounted engines normally have the fuel tank situated in the rear boot, either in the floor-pan for estate cars or over the rear axle for saloon cars, the latter position being preferred for safety reasons.

In contrast, rear-mounted engines have their fuel tanks in the front boot, either over the bulkhead or flat across the boot floor-pan, the latter providing more boot space but being more exposed to danger in a head-on crash.

21.1.3 Fuel pipes (fig. 21.1)

The supply and return lines may be made from steel or plastic piping and are secured by clips at several points along their length underneath the floor-pan between the fuel tank and the engine unit. Plain or braid-covered flexible rubber pipe-hoses are used to connect the more rigid pipes located under the floor to the feed pump and carburettor, which are bolted to the engine and therefore subjected to side-by-side rocking and vibration relative to the car body.

The fuel is delivered by the feed pump to the carburettor, where some of the fuel maintains the fuel level in the float chamber. When the float chamber is full, the float closes the needle valve, and fuel surplus to the engine's requirements is returned to the fuel tank by the return pipe. This pipe has a smaller bore than the supply pipe, as the fuel return flow is controlled by a restrictor orifice in the carburettor-connector return outlet, to prevent too much bypassing of fuel. The continuous flow of fuel through the system maintains the fuel temperature at an even level and ensures that vapour locks do not occur.

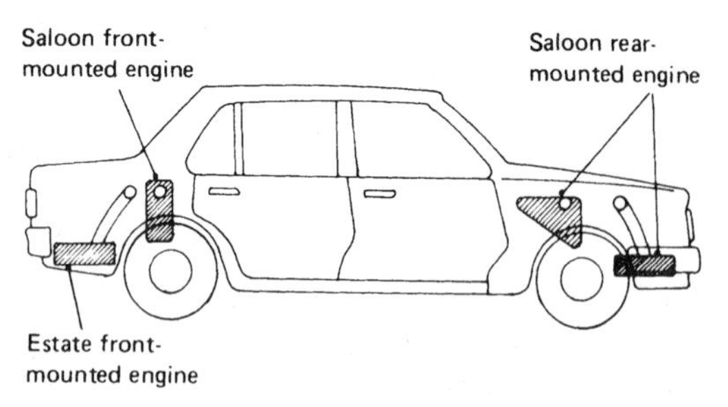

Fig. 21.2 Positioning of fuel tank

21.1.4 Petrol fuel-supply filters (figs 21.1 and 21.3)

To prevent dirt and fluff entering the feed pump, a relatively coarse strainer is necessary on the suction supply side of the pump (fig. 21.1). This primary filter may take the form of a cylindrical or bowl-shaped wire-gauze or nylon-mesh screen mounted in the fuel tank, an enclosed nylon-screen capsule connected in series in the suction line, or a nylon-mesh disc screen fitted inside the feed-pump inlet chamber.

On the pressure side of the feed pump, a secondary filter may be incorporated (fig. 21.1). This may be of the cylindrical capsule type with either a fine nylon meshing or a pleated-paper filter medium plumbed into the pipeline (fig. 21.3). Alternatively, just a simple wire thimble, or a nylon-mesh filter, or sintered permeable-bronze filters located inside the inlet port of the carburettor float chamber may be used.

21.2 Petrol feed pumps

Diaphragm mechanically operated feed pumps consist of a spring-loaded flexible diaphragm, usually made from laminated synthetic rubber and nylon fabric, sandwiched between an upper valve-chamber housing and a lower flanged pull-rod housing which is attached to the engine. Built

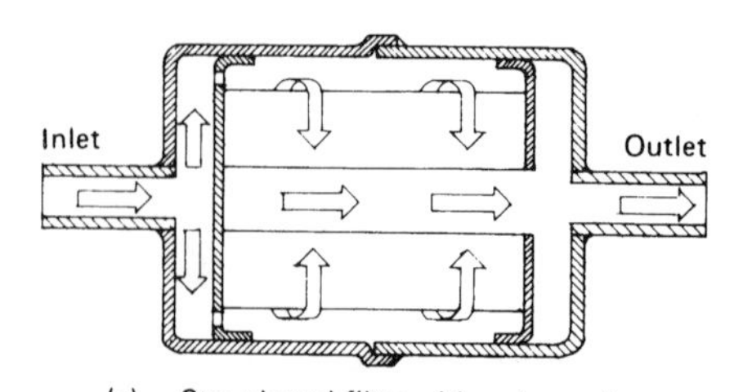

(a) Star-pleated filter with nylon casing

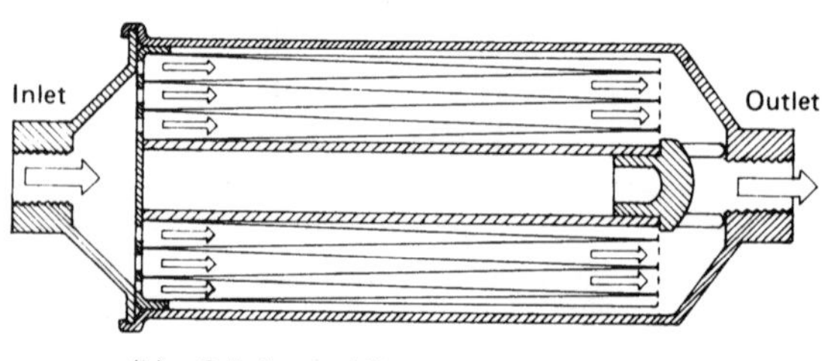

(b) Spiral-wound filter with aluminium casing

Fig. 21.3 Petrol-pipeline secondary paper filters

into the upper chamber are a pair of inlet and outlet valves which may take the following forms:

a) solid nylon discs,
b) rubber washers,
c) integral rubber lip and flap.

An eccentric lobe driven from the camshaft or a separate jack-shaft can operate either a bell-crank lever (fig. 21.4) or a push-rod (fig. 21.5) to provide the necessary inward diaphragm stroke movement, the outward displacement being derived from a diaphragm return-spring.

A separate detachable cap fits on top of the valve chamber to form an inlet chamber, sometimes with a nylon filter mesh included. The function of this chamber is to trap any sediment or water and to retain a small quantity of petrol above the diaphragm so that the valve-and-diaphragm assembly remains primed for long periods when the engine and pump stand idle.

Two types of mechanically driven petrol pumps are now considered:

i) bell-crank lever-arm operated,
ii) push-rod operated.

21.2.1 Diaphragm-type lever-operated petrol pump (SU)

Induction phase (fig. 21.4(a)) When the camshaft is rotated, the largest part of the lobe profile moves against the lever-arm so that this rocks about its pivot. The diaphragm will then be pulled away from the petrol chamber it is sealing; consequently the pressure of the enlarged chamber will drop below that of the atmosphere, and this then draws together and closes the outlet-valve lips. The result of this depression will be that the atmospheric pressure pressing down on the fuel level in the petrol tank will force petrol along and up the supply pipe and through the inlet port, to push open the inlet-valve flap and fill the expanding chamber space.

Discharge phase (fig. 21.4(b)) With further rotation of the camshaft, the smallest part of the lobe profile moves next to the lever-arm. This allows the lever return-spring to return the arm against the face of the cam without interfering with the position of the diaphragm, due to the elongated slot in the diaphragm stem. At the same time, the diaphragm return-spring is free to push at its own rate. This squeezes the petrol, which pushes back and closes the inlet-valve flap and forces open the outlet-valve lips. Petrol will now be transferred from the diaphragm chamber through the outlet port and then to the carburettor float chamber.

21.2.2 Diaphragm-type push-rod-operated petrol pump (AC)

Induction phase (fig. 21.5(a)) As the cam rotates, the lobe moves to its lowest position. The powerful pull-rod spring will now compress the weaker diaphragm return-spring, pulling the diaphragm outwards and away from the diaphragm chamber it encloses. The increased space in the diaphragm chamber will create a depression so that atmospheric pressure acting on the fuel level in the petrol tank will push petrol through the supply pipe and inlet port, to force open the inlet valve and fill the space in the diaphragm chamber above the diaphragm.

Discharge phase (fig. 21.5(b)) Continued rotation of the camshaft moves the lobe to its highest position, compressing the pull-rod spring so that both the push-rod and the pull-rod are forced inwards towards the diaphragm. The diaphragm return-spring, no longer compressed by the stronger pull-rod spring, squeezes the petrol inwards towards the discharge chamber. The pressurised petrol now pushes the inlet-valve disc back on to its seat and forces open the outlet valve. Petrol will then be displaced through the outlet port to the carburettor float chamber.

Due to the large amount of free play between the junction of the hollow diaphragm support and the shouldered pull-rod stem, the discharge rate is entirely independent of the push-rod movement – it is governed only by the stiffness of the diaphragm return-spring and the opposition pressure built up in the carburettor-to-pump pipeline.

21.2.3 Petrol-feed-pump pressure regulation

An operating-pressure range which is normally between 0.2 and 0.35 bar should be maintained in the petrol pipeline between the pump and the carburettor, the exact pressure range being determined by the stiffness of the diaphragm return-spring.

Once the engine has started and the working pipeline pressure has been reached, the spring controlling the return pumping stroke of the diaphragm remains partially or fully compressed, even though the push-rod or lever-arm continues faithfully to follow the eccentric cam-profile

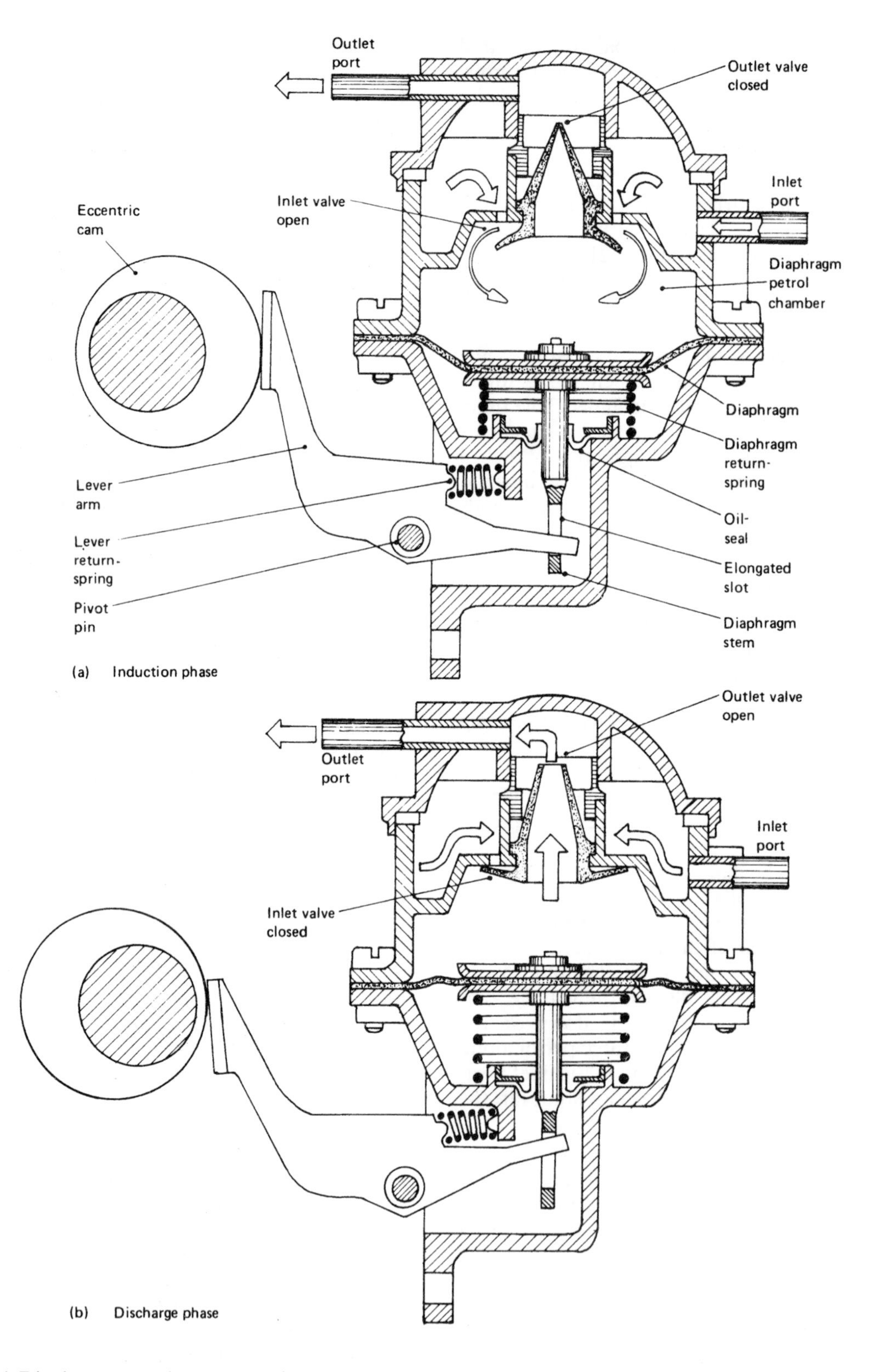

Fig. 21.4 Diaphragm-type lever-actuated petrol pump (SU)

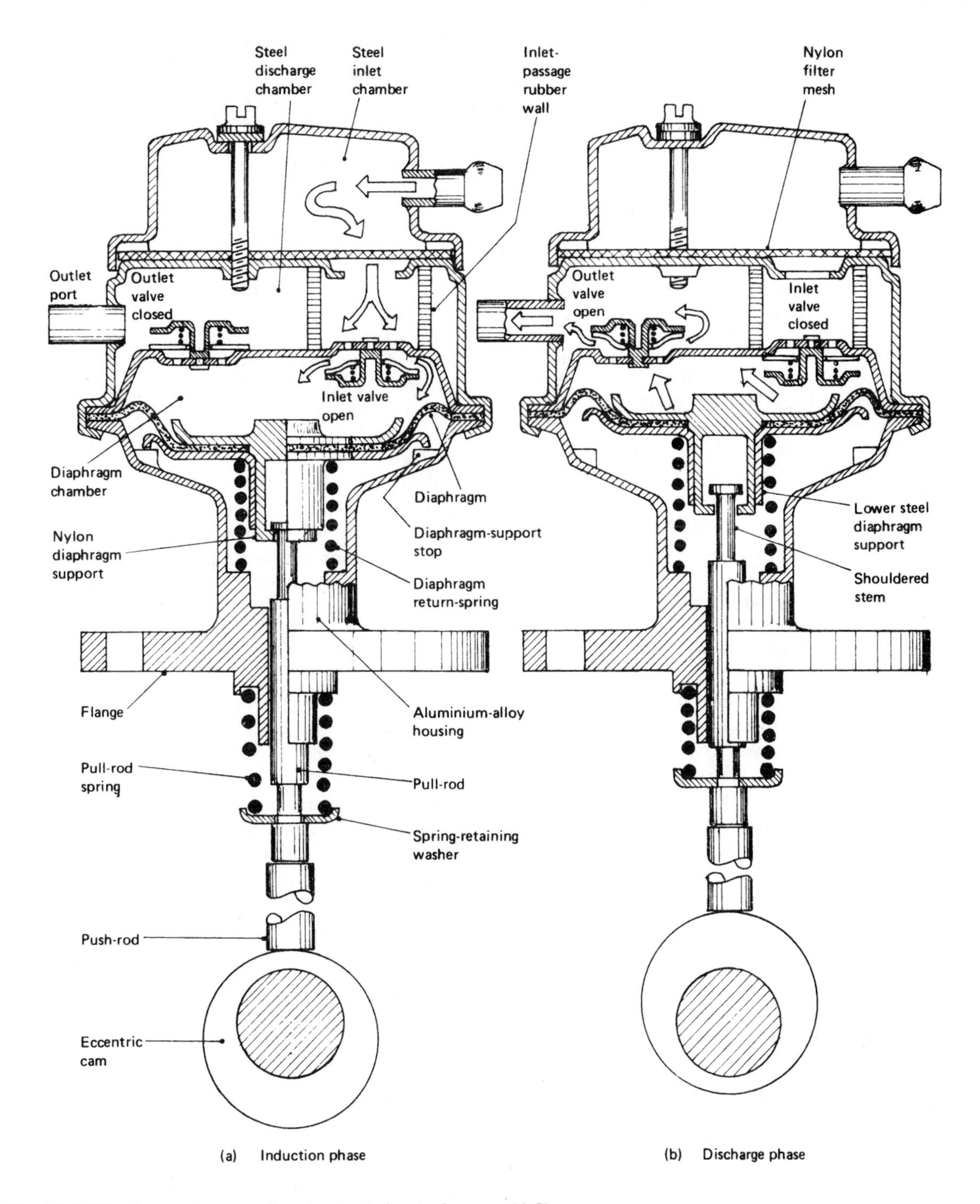

Fig. 21.5 Diaphragm-type push-rod-actuated petrol pump (AC)

movement. The restricted diaphragm movement thus decreases the rate of petrol discharge from the pump until the demands of the engine use up some of the petrol in the float chamber. The petrol pressure in the pipe-line will then fall below the predetermined minimum set by the strength of the diaphragm return-spring, and this again allows the diaphragm return-spring to start moving the diaphragm and hence displacing petrol into the pipeline. It can therefore be said that the pumping movement of the push-rod or lever-arm is constant, but the pumping stroke of the diaphragm is variable, it being governed by the amount of fuel consumed by the engine at different engine-speed and load conditions.

Controlled petrol pipeline pressure between the feed pump and the carburettor is important for efficient functioning of the carburettor under all operating conditions – if the pressure varies greatly, the following could occur:

a) very low pressure will not be able to keep up with the flow rate demanded by the engine at high-speed or heavy-load conditions,
b) excessively high pressure will tend to force open the carburettor float-chamber needle valve and sink the float so that the petrol level rises, consequently flooding the venturi (choke) tube and causing stalling of the engine.

21.3 Carburation

The function of the carburettor is to mix the fuel with the incoming air in the correct proportions to form a mixture which is combustible under engine operating conditions.

Petrol is a liquid fuel derived from crude petroleum whose major constituent elements are carbon (85 to 90%) and hydrogen (10 to 14%) but which also contains three minor elements: sulphur (0.2 to 3%), nitrogen (usually below 0.1% but sometimes up to 2%), and oxygen (up to 1.5%).

Refined petrol is composed of molecules consisting essentially of the elements carbon and hydrogen and so known as hydrocarbons. The proportions of these elements in the fuel are approximately 86% carbon and 14% hydrogen by mass. For chemically complete (stoichiometric) combustion under ideal conditions, a mixture of 15 parts of air to 1 part of petrol by mass is necessary or, as a percentage, 93.75% air and 6.25% petrol. Comparing these proportions on a volumetric vapour basis, this would be equal to 60 parts of air to 1 part of petrol or, in percentage terms, 98.4% air and 1.6% petrol.

The mixing of petrol with the air is achieved by inducing the liquid petrol into a rapidly moving air stream which will suspend and break up the liquid into very tiny droplets. This process is known as 'atomising' the fuel.

Air and petrol mixture strengths According to chemical combination requirements, the air–fuel mixture which gives complete combustion is composed of about 15 parts of air to 1 part petrol by mass. Thus a chemically correct or optimum mixture may be assumed to have an air–fuel ratio of 15 to 1.

Rich mixtures – that is, mixtures containing more than the optimum amount of petrol – usually produce more power than optimum and weak mixtures, the engine power generally being at its maximum when the mixture is about 15 to 20% rich, that is with an air–fuel ratio of between 12 to 1 and 13 to 1. The exhaust products of these rich mixtures normally have an excess of carbon monoxide and are visually observed as dark cloudy exhaust smoke. Prolonged running with very rich mixtures will result in sooting up of the combustion chamber and of the spark-plug electrodes.

Weak mixtures – that is, mixtures containing less than the optimum amount of petrol – usually produce less power than optimum and rich ones, but fuel economy is normally much better than that for the other conditions. For minimum fuel consumption, the mixture can be 15 to 20% weak – that is, with an air–fuel ratio of between 17 to 1 and 18 to 1. Burning is generally slow, and misfiring, overheating, and incomplete combustion will result if sufficient ignition timing advance is not provided to compensate for this prolonged combustion period.

Personal safety precautions Petrols give off flammable vapour at atmospheric temperatures and so should not be brought into contact with naked flames, electrical sparks, or excess heat – otherwise these petrol vapours mixed with air are liable to explode.

Exhaust gases contain poisonous fumes which should not be inhaled; therefore engines should be operated only in the open, not in confined garage spaces unless exhaust extractors are fitted to the ends of the exhaust-pipe system.

21.3.1 Parallel carburettor tube (fig. 21.6)

The principle of the carburettor may best be understood by considering a vertical parallel tube (fig. 21.6) indirectly connecting to the inlet port on the side of the cylinder head. Imagine the engine being rapidly rotated so that every time the piston moves from TDC to BDC on the induction stroke a fresh charge of air will pass through this tube and the inlet port on its way into the cylinder to fill up the space swept out by the movement of the piston. Because the diameter of the tube is

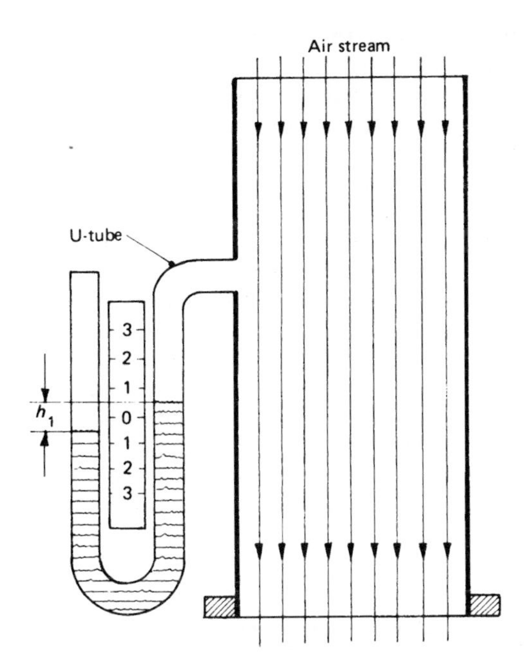

Fig. 21.6 Parallel pipe and U-tube gauge

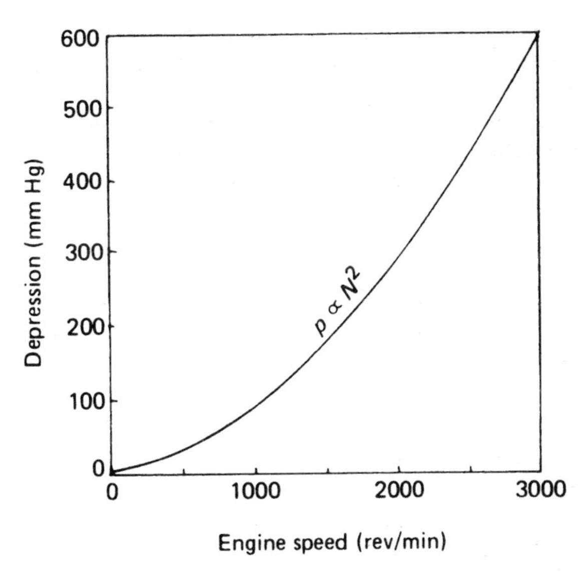

Fig. 21.7 Relationship between engine speed and induction depression

much smaller than that of the cylinder bore, the air will move through the tube at a speed several times that of the piston, to try to keep the expanding cylinder full of air.

The gas molecules in the rapidly moving air column in this tube will be stretched further apart than those on the outside in the atmosphere, so the pressure will be lower. Therefore, if a U-tube water-gauge is tapped into the side of the vertical tube, the atmospheric pressure acting on the water surface of the open end of the U-tube will push and displace the water towards the fast-moving column of air inside the vertical tube. The degree of depression created by the air movement can be observed and measured by the difference in height h of the two water levels. It has been found that the depression in the vertical tube increases in proportion to the square of the air speed; it therefore similarly relates to engine speed – see fig. 21.7.

21.3.2 Venturi carburettor tube (choke tube)

To increase the drop in pressure inside the vertical tube even further, the internal diameter of the tube may be reduced progressively over a short distance (fig. 21.8). This restriction – known as the venturi – will accelerate the molecules and increase their distance apart to a greater extent in the narrow neck of the tube, so that the quantity of air passing through can keep up with the rest of the air flow. The net result will be a concentrated drop in pressure at the tube's narrowest point, shown by the enlarged difference in the water levels h between the two halves of the U-tube gauge – see figs 21.6 and 21.8, the gauge readings being h_1 and h_2.

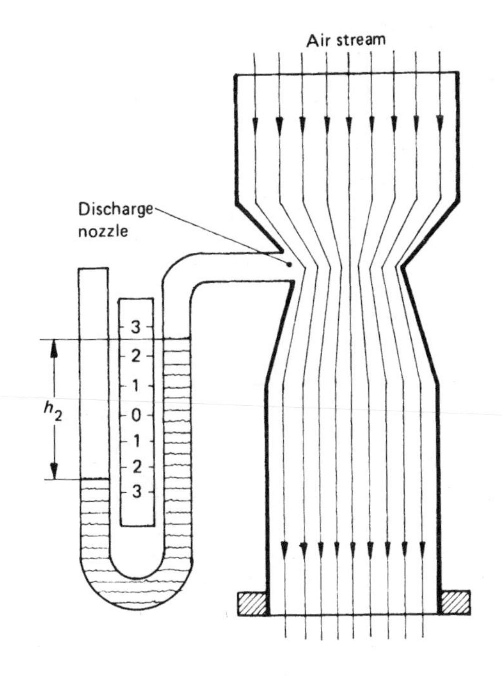

Fig. 21.8 Venturi pipe and U-tube gauge

This necked restriction is sometimes referred to as the choke tube, but it has also been named after the Italian engineer G.B. Venturi (1746–1822) in recognition of his studies on the flow of fluids. For optimum performance, the tube should have an intake angle of 30° and an output angle of 7° relative to the tube axis, with a smooth surface finish. The profile is chosen to concentrate a greater degree of depression in the region of the discharge nozzle. If fuel replaces the water in the U-tube, this could be pushed through the discharge nozzle into the venturi tube at a relatively low air speed, which would not be possible with the parallel-type tube.

21.3.3 Single-jet fixed-choke carburettor (fig. 21.9)

In vertical section, the basic carburettor is a modified U-tube – see fig. 21.9. Here the atmospheric side of the bent tube has been enlarged so that it can act as a reservoir and house the float-and-valve assembly. The other end intersects the venturi and so forms a spout and discharge nozzle. Dividing the two sides of the U-tube carburettor at the base of the bend is a restriction orifice known as the petrol jet, its function being to meter the amount of petrol flowing into the venturi.

Fuel mixing will take place when the speed of the air stream in the venturi (choke) tube is high enough to cause a sufficiently large pressure drop relative to the atmosphere in the region of the discharge nozzle. Thus, as the cranking speed of the engine is increased, a point will be reached when the air pressure near the discharge nozzle is sufficiently low for the atmospheric pressure acting on the petrol in the float chamber to push petrol into the spout so that it overflows from the nozzle into the venturi. This liquid petrol entering the venturi will be immediately torn apart into small filaments that break up and contract into various sizes of droplets. Therefore the petrol will be finely distributed continuously throughout the air stream as it is drawn into the engine cylinders.

21.3.4 Butterfly throttle valve (figs 21.9 and 21.10)

To control the speed and load output of the engine, a butterfly throttle valve is placed on the downstream side of the venturi. The spindle of the valve is connected to the accelerator pedal by relay-rods or cable and levers. The quantity of

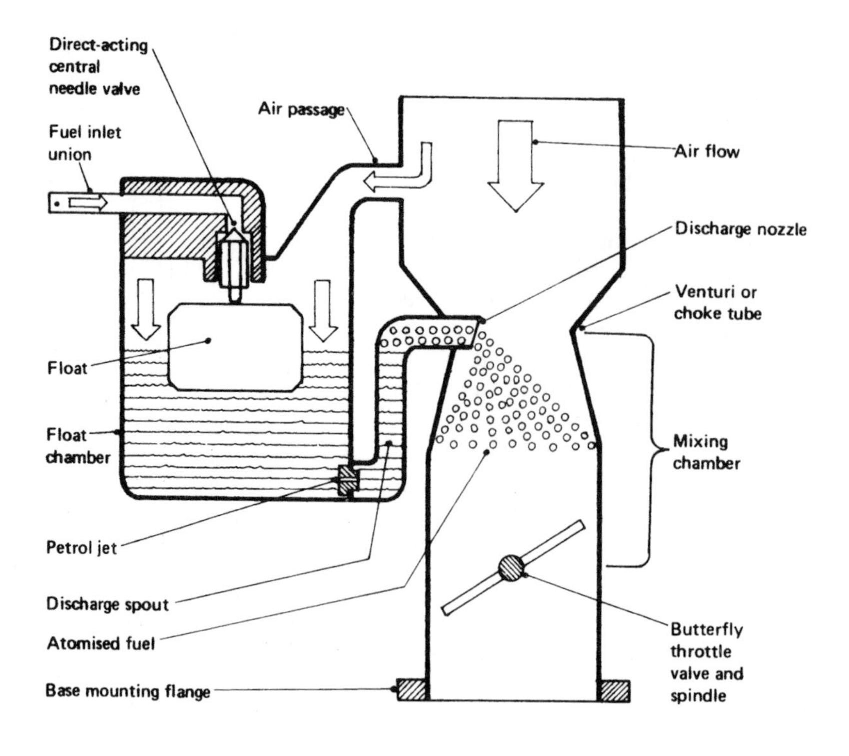

Fig. 21.9 Fixed-choke-tube single-jet carburettor

charge entering the engine can be varied by the degree of depression acting at the discharge nozzle, which depends on the angular position of the throttle-spindle opening the butterfly valve.

Figure 21.10(a) shows the carburettor with a series of U-tube gauges stationed at four positions from top to bottom as follows:

i) inlet side of venturi,
ii) on the venturi,
iii) venturi side of throttle
iv) engine side of throttle.

Looking at the U-tube gauges and remembering that pressure drop is proportional to the difference in water level h, the stations illustrate the following:

i) There is very little depression here as this is the widest point of the tube so the air does not move very fast in this region.
ii) The depression has been made to concentrate at this narrowest point in the tube, since the air has to move faster in this zone.
iii) The depression is reduced, as the tube has widened slightly in this region and so slows down the air stream.
iv) The depression is a maximum here, since the air stream is considerably restricted by the throttle.

If these depressions are plotted opposite the venturi-tube profile (fig. 21.10(b) and (c)), the horizontal width of the shaded area at any of these points indicates roughly the magnitude of the depression at that point.

The magnitude of the depression along the tube bore varies with engine speed and the throttle opening. A small throttle opening in mid-speed range will produce a large depression on the engine side of the throttle but will permit only a small amount of depression to exist on the intake side of the throttle (fig. 21.10(b)). With a large throttle opening in mid-speed range, the depression is now reduced on the engine side of the

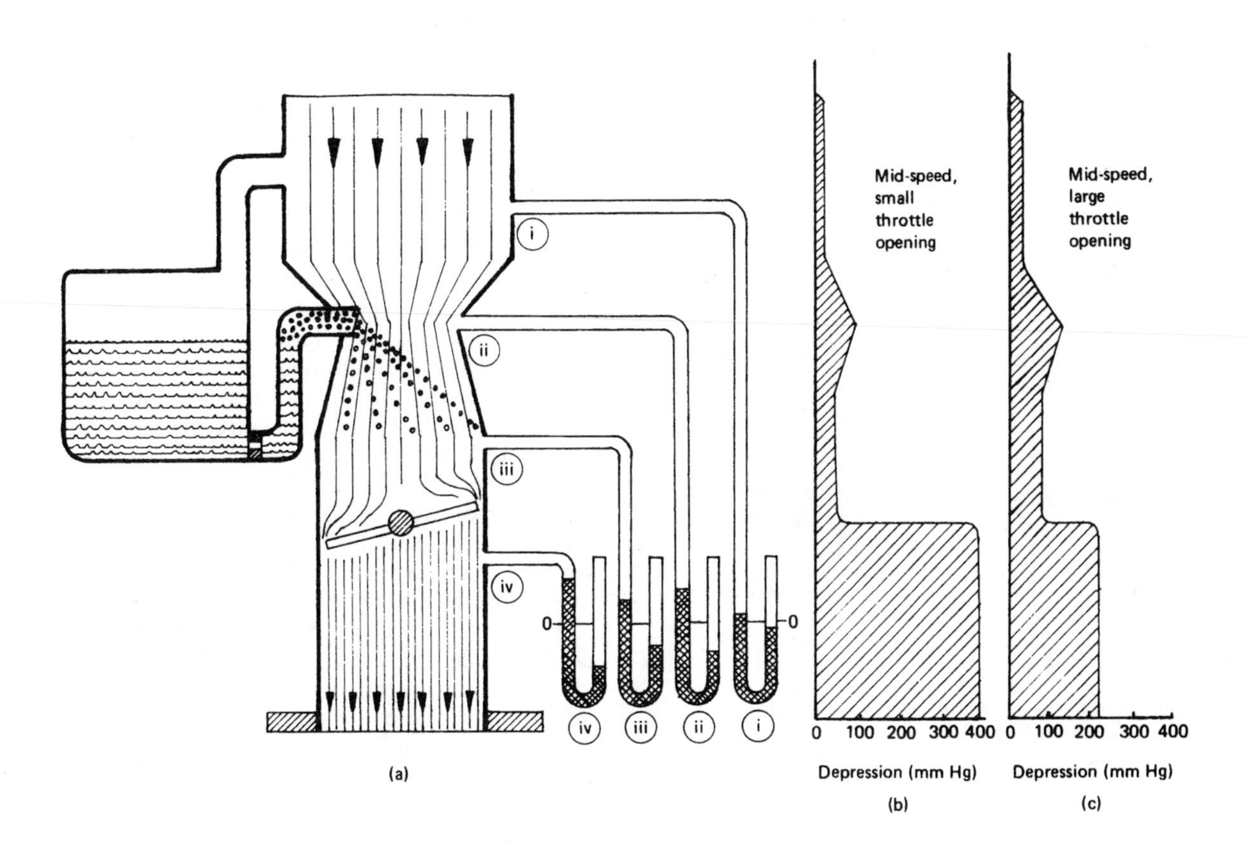

Fig. 21.10 Effect of throttle opening on induction depression

throttle and can concentrate further back in the venturi region (fig. 21.10(c)).

The conclusions to be drawn from this comparison are that the throttle opening regulates the degree of depression at the venturi and hence the amount of fuel spray which can be forced into the tube.

21.3.5 Carburettor float chamber (figs 21.9, 21.11, 21.14, and 21.15)
The function of the float chamber is to provide a reservoir of petrol of constant depth under steady-running conditions from which fuel can be instantly and consistently metered out to satisfy the demands of the engine and road conditions.

Within the reservoir chamber is a float which senses the changing petrol level. Floats are now made from either expanded synthetic rubber or nylon plastic, but the early floats were made from deep-drawn brass sheet and sometimes tended to leak along the joint seams.

The needle valve resembles a cylindrical stem with a conical tip which provides a self-centralising taper seat, a positive seal, and a progressive opening-and-closing action. The valve stem and tip may be simply a solid steel parallel pin with a tapered tip (fig. 21.11(a)) or a two-piece construction with a rubber-seat tip housed in the end of the fluted metal stem (fig. 21.11(b)). A further version is a three-piece valve with a rubber-seat tip and a spring-loaded ball situated in the body of the fluted stem (fig. 21.11(c)).

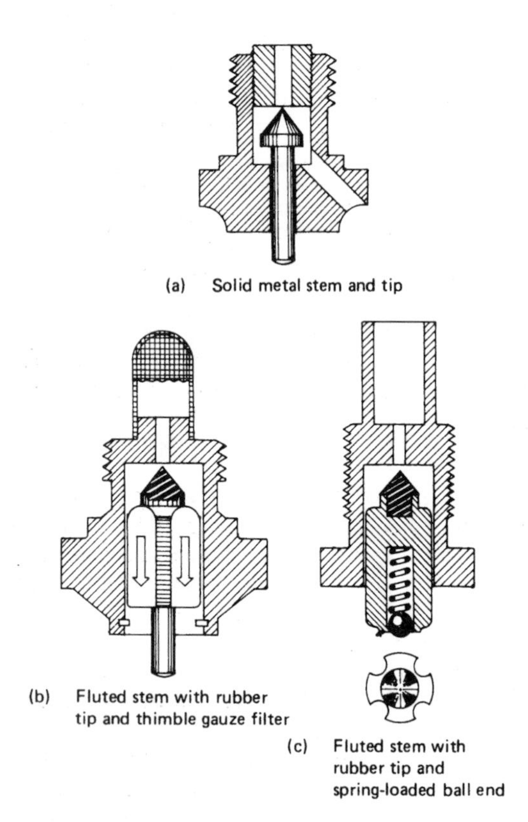

Fig. 21.11 Float-chamber needle valves

Both the rubber and spring-loaded type of valves are able to maintain a liquid-tight seal under normal engine-vibration conditions – under certain resonant-vibration conditions, this may not be possible with the single-piece needle valve, due to needle chatter.

Early float chambers had direct-acting centrally positioned needle valves (fig. 21.9), but with most present-day arrangements the needle valve is situated to one side of the float and is actuated by a hinged lever. There are two basic fuel-feed arrangements:

i) top feed, where the valve is positioned between the float and the pivot (see fig. 21.15);
ii) bottom feed, where the pivot is positioned between the float and the valve (see fig. 21.14).

In both cases the closing needle-valve force is provided by the combination of the upthrust of the hollow float which is partially submerged in the fuel and the leverage of the hinged arm.

Increasing the leverage or the size of the float will enable either a higher petrol-pump pipeline pressure or a larger needle-valve orifice to be employed – this improves the flow capacity of the valve without the side-effects of sinking the float and causing flooding at the discharge nozzle due to an excessively high steady-flow (static) petrol-level height. The actual petrol-level height in the float chamber is set at about 3 to 5 mm below the discharge-nozzle lower edge and generally may be adjusted by bending the hinged lever, but the actual setting procedure must be carried out as recommended by the manufacturer to avoid upsetting other parts of the system.

The level of fuel in the float chamber is thus controlled by the rise and fall of the float, which closes or opens the needle valve to cut-off or admit petrol from the pump as required. When the engine demands are large, such as when accelerating, the valve will remain just sufficiently open to keep the petrol at its correct height; but if the demands are small, as when idling or when the car is driven by its own inertia on over-run, the float level will remain high and so will close the valve against the petrol-pump line pressure until the level of petrol in the chamber has dropped sufficiently to actuate the float and so open the valve.

To prevent particles of dirt mixing with the petrol, the top of the float chamber is exposed to the pressure on the intake side of the venturi tube, which is filtered.

21.3.6 Limitation of the single-jet carburettor (figs 21.12 and 21.13)

The quantity of air consumed by an engine in unit time is directly proportional to the engine speed, but – due to the inertia of liquid flow – the rate at which petrol is drawn out of the discharge nozzle into the air stream will increase almost with the square of the engine speed;

i.e. $m_a \propto N$
and $m_p \propto N^2$
where m_a = mass flow rate of air
m_p = mass flow rate of petrol
and N = engine speed in rev/min.

This is illustrated in fig. 21.12, which shows the relationship between the percentage of petrol in the mixture and the engine speed. The point where the two curves cross gives the percentage of petrol by mass (6.25%) which can be completely burnt by the oxygen in the air charge.

If a jet size is chosen to give the correct mixture

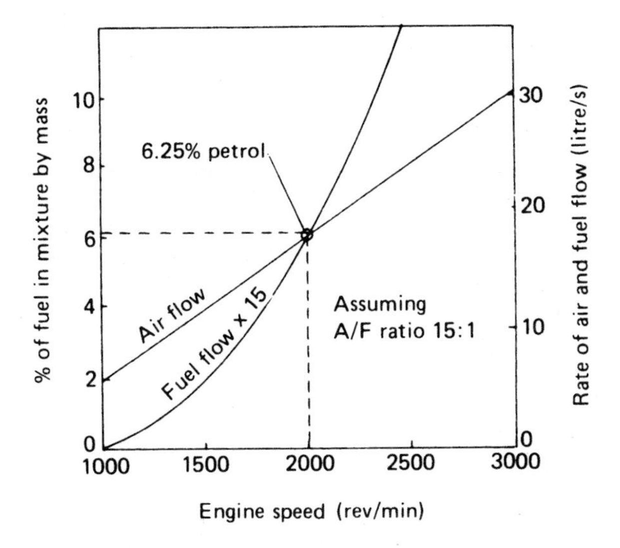

Fig. 21.12 Relationship between engine speed and fuel and air flows for a 1-litre engine

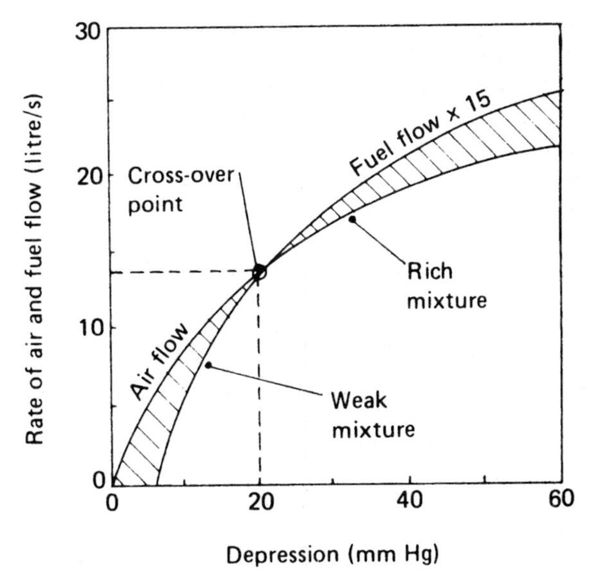

Fig. 21.13 Relationship between induction depression and fuel and air flows for a simple single-jet carburettor

at one predetermined fixed speed, then, at speeds below this, insufficient petrol will be forced into the air stream, producing a weak mixture and loss of power. With higher speeds, the quantity of petrol induced into the air stream will increase at a greater rate than the increase of air consumption, so that an over-rich mixture will be produced.

Figure 21.13 shows how the rates of air and petrol flow vary with different depressions established in the choke tube. This again shows a cross-over point – a depression below this point will provide only a weak mixture, but above it the mixture tends to become rich. It can thus be concluded that a petrol-jet orifice can be selected for optimum performance over only a narrow speed range.

21.3.7 Capacity-well compensation (fig. 21.14)

When the engine is developing only sufficient power to overcome the rotating friction losses – a condition known as idling – the level of the fuel will be the same in the float chamber, capacity well, and discharge spout.

During initial acceleration when the throttle is opened, the pressure drop created at the venturi will draw fuel from the discharge nozzle at a far greater rate than can be supplied by the petrol jet alone, but the capacity well will provide the extra fuel demanded. As it helps meet the increased demands for a rich accelerating mixture, the capa-

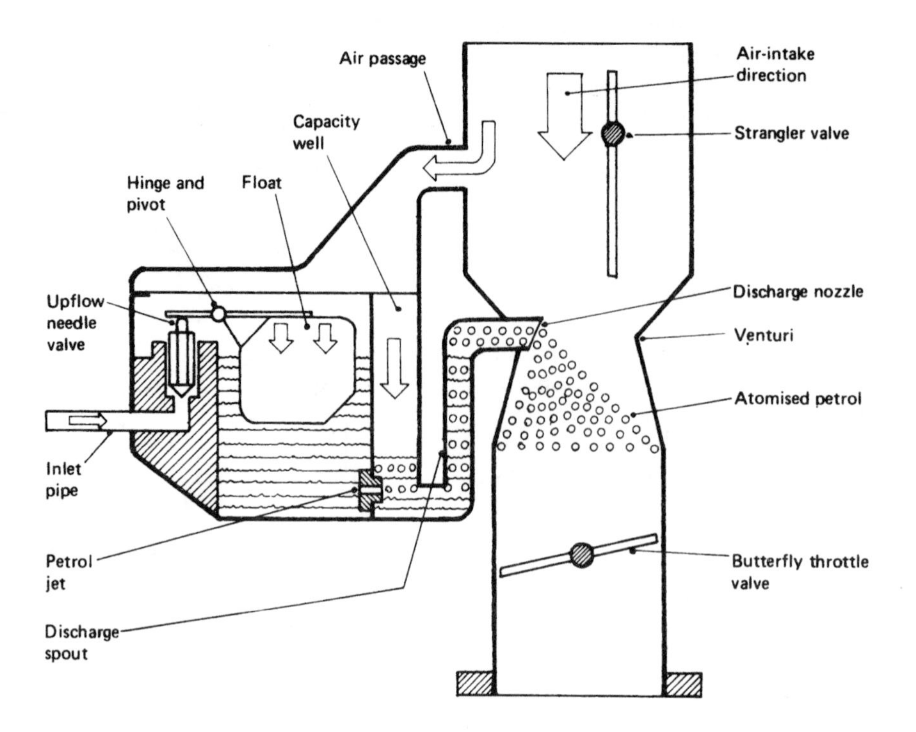

Fig. 21.14 Capacity-well-compensation carburretor

city-well fuel level will very quickly drop until the well is emptied (fig. 21.14). An emulsion of petrol droplets suspended in air will then be formed at the base of the well which will prevent any more enriching of the mixture and will improve the mixing and hence the quality of the mixture. Thus any further increase in speed or depression in the venturi will only result in a constant amount of fuel flow from the compensation petrol jet, since the air-passage bleed relieves some of the depression created across the petrol jet. Therefore the mixture strength when the capacity well empties will depend to a large extent on the petrol-jet orifice size and the degree of air bleed.

The limitations with this simple form of mixture correction are that it tends to overcompensate under large throttle openings and that it is not flexible enough under varying operating conditions.

21.3.8 Air-bleed and capacity-well compensation (fig. 21.15)

This system uses one fuel jet and an emulsion tube suspended at the top end and situated in the centre of the capacity well. Along the length of this tube are a number of small holes, and the top of the tube is exposed to the atmosphere by an air-correction jet which carefully controls the amount of air entering the tube (fig. 21.15).

Under no-load conditions, the fuel level in the well will be the same as in the float chamber. With initial throttle opening and depression, the fuel in the well will be consumed, thus providing an enriched mixture.

As the level of fuel in the well drops, it exposes the uppermost of the emulsion-tube holes. This allows more air to enter the capacity well and mix with the fuel, thus preventing any tendency towards undue richness.

The petrol level in the centre of the emulsion tube drops in proportion to engine fuel demands and so increases the amount of air bleed. Conversely, the emulsified-mixture level around the tube in the well tends to stabilise, and in fact is continuously drawn up the walls of the well to be discharged out of the nozzle into the venturi.

The degree of mixture-strength correction is determined to a considerable extent by the size of the air-correction jet. A very large air jet will tend to overcorrect a mixture-enriching tendency, but a very small air jet will not provide sufficient

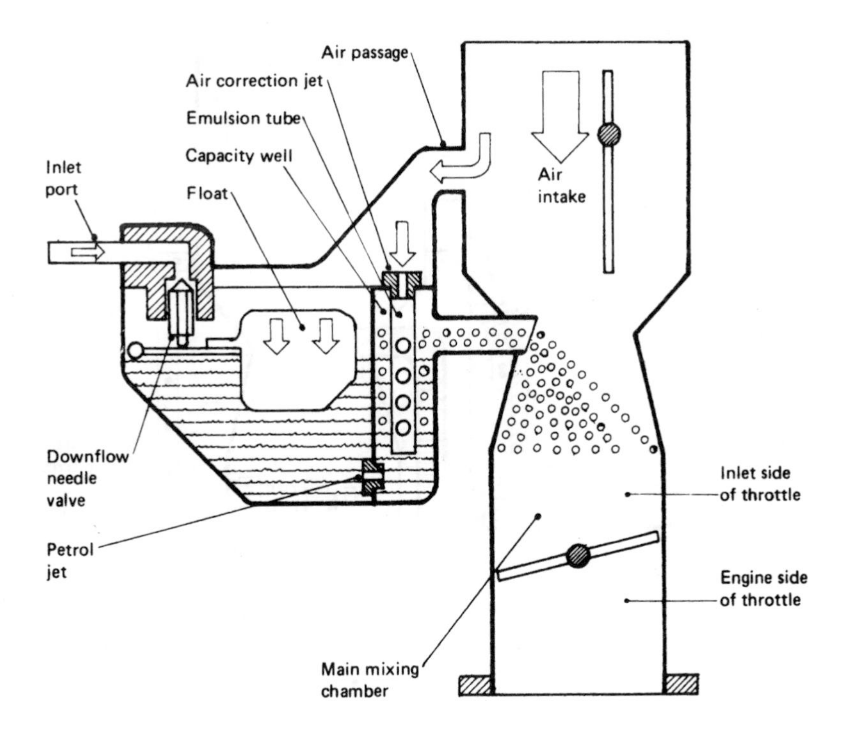

Fig. 21.15 Air-bleed and capacity-well-compensation carburettor

air bleed to correct the enriching mixture as the engine speed rises.

21.3.9 Idle running conditions

To bring the engine speed down to idling, the throttle valve must be closed against a slow-running adjustment stop (fig. 21.16). The result of this is that the depression in the venturi side of the throttle will be insufficient to raise fuel from the float-chamber level to the discharge-nozzle lower edge; therefore a separate supply of fuel must be supplied to the engine side of the throttle during idling. There are two approaches to supplying this separate air–fuel mixture for idling conditions:

i) air-screw control,
ii) volume-screw control.

Air-screw control (fig. 21.16) With this system, petrol is drawn from an idling well, through the idle petrol jet across to the air-control screw, and down to the idle duct intersecting the venturi-tube mixing chamber on the engine side of the throttle. The setting of the air-control screw provides a controlled degree of air bleeding so that an emulsion is produced to provide a pre-mix before the fuel enters the main intake stream. The reduced depression on the idle petrol jet allows the manufacturer to use relatively large orifices so that sufficient fuel will be drawn out.

To avoid a 'flat spot' when the engine accelerates from idle to main running conditions, due to the temporary weakening of the sources of the mixture supply as it changes from the idle duct to the main discharge nozzle, a progressive duct is provided just on the venturi side of the throttle valve. When the throttle starts to open, the progressive duct will immediately be subjected to the engine's depression so that an additional quantity of mixture will enter the venturi tube on the engine side of the throttle during this transitional stage.

Volume-screw control (fig. 21.17) This system has a fixed idle petrol jet and air bleed so that a metered amount of both air and fuel will be drawn into the idling passage to form an emulsion. This mixture will then be drawn past the volume-control screw into the main air stream on the engine side of the throttle, the volume of the pre-mixed supply being controlled by the positioning of the volume-control screw. When the throttle is opened slightly, the second duct –

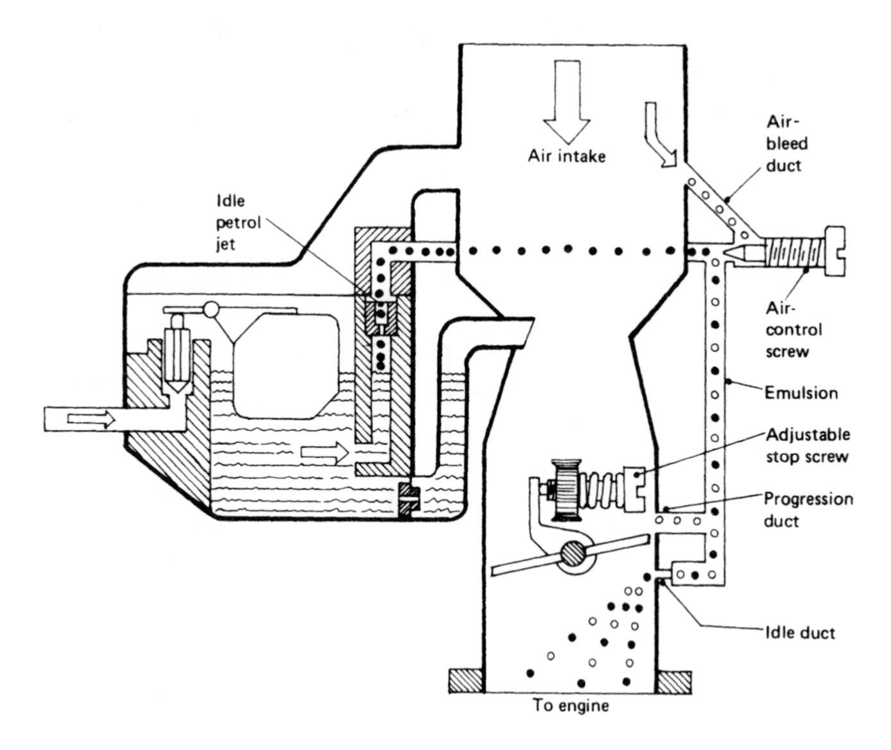

Fig. 21.16 Slow-running air control

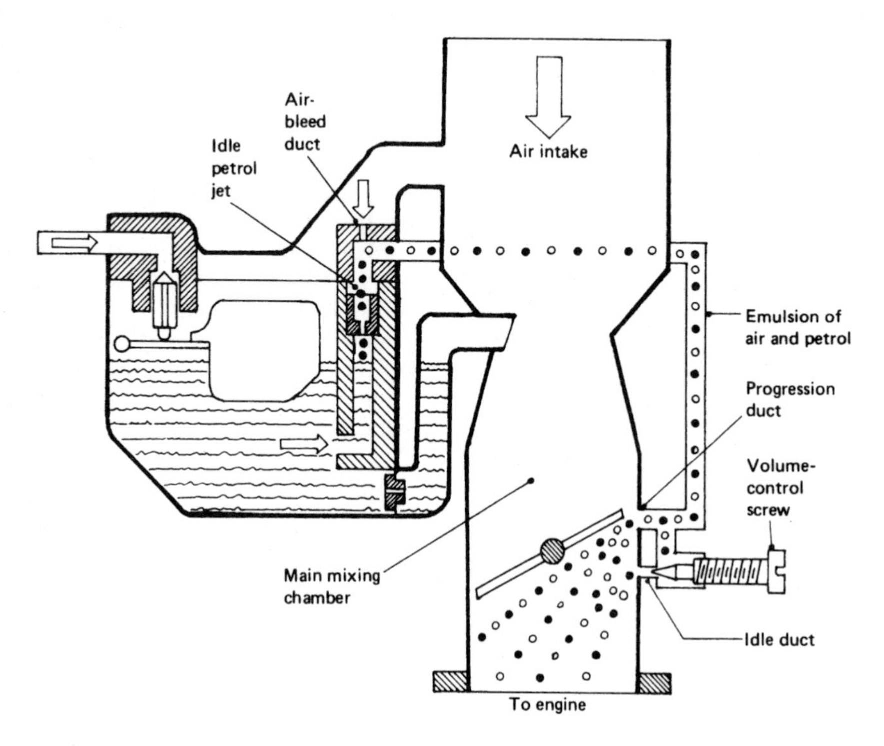

Fig. 21.17 Slow-running volume control

initially sealed-off by the closed butterfly throttle valve – is exposed to the engine depression and an additional quantity of emulsion mix will enter the main system. This gives a progressive smooth transition from idling conditions to the main venturi discharge as in the previous air-screw control system.

21.3.10 Strangler butterfly valve (figs 21.14 and 21.15)
When the engine is cold, only the low-boiling-temperature 'lighter' fractions of fuel will vaporise, so a richer mixture is necessary. The strangler butterfly valve is situated on the air-intake side of the venturi tube and has an interconnecting linkage joining it to the throttle spindle. To enrich the mixture, the strangler valve is partially closed; at the same time, the interconnecting linkage opens the throttle valve by approximately one third. This reduces the air supply and increases the depression in the mixing-chamber region, so that an increased amount of petrol spray will be forced into the mixing chamber.

21.4 Constant choke single-barrel carburettor (fig. 21.18)

21.4.1 Idle and progression system (fig. 21.18)
With the throttle closed, virtually all the induction vacuum will be on the engine side of the butterfly valve and therefore there will be insufficient depression in the mixing chamber formed between the inner venturi and the throttle valve to draw fuel from the discharge nozzle. Thus an idle system which by-passes the venturi side of the throttle and discharges directly into the chamber on the engine side of the throttle is utilised. Fuel from the float-chamber is drawn from the main petrol jet to the idle petrol jet tube where it meets air passing through the idle air jet. The emulsified mixture then passes radially outward from the idle tube where it then passes up, across and

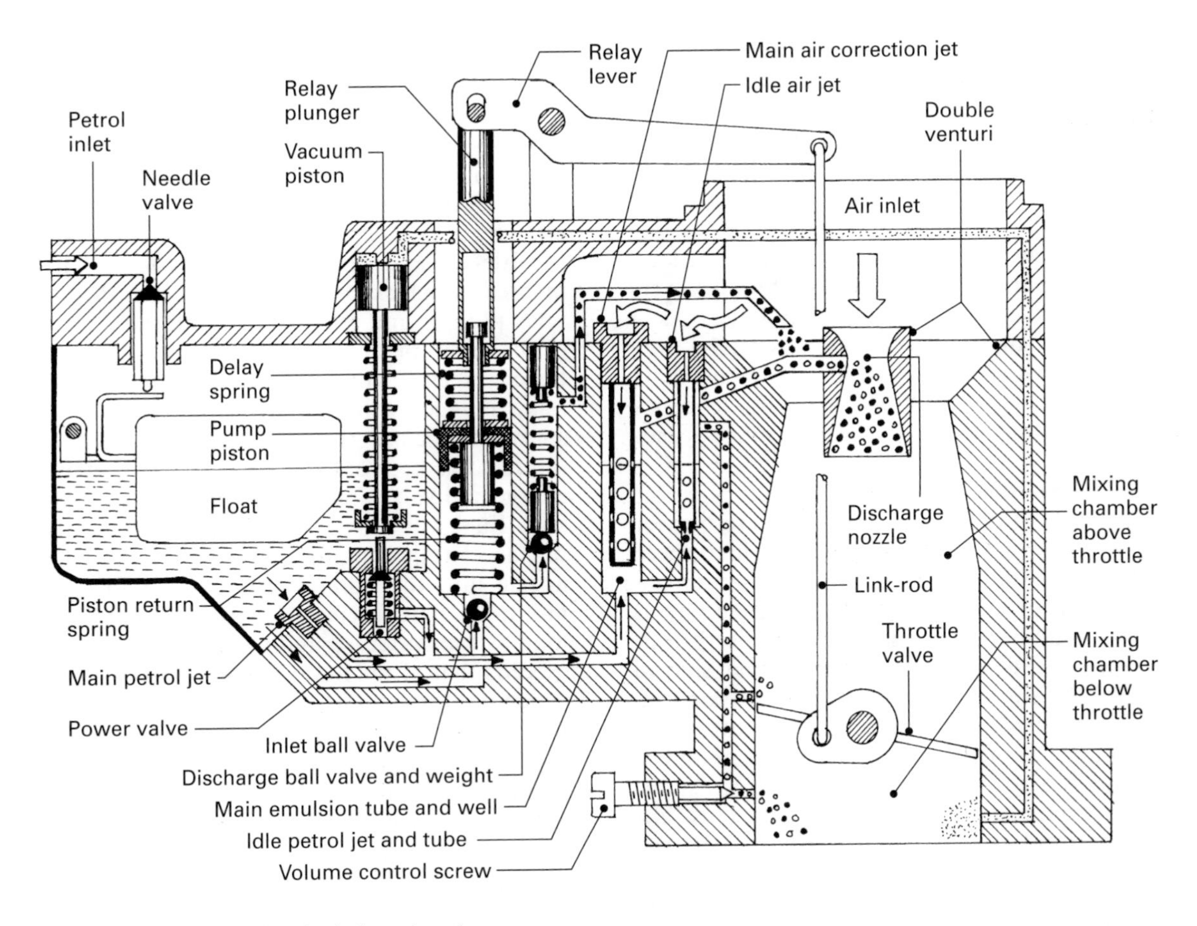

Fig. 21.18 Constant choke single-barrel carburettor

down the passage to the volume control screw which regulates the quantity of mixture feeding the chamber below the throttle. As the throttle valve spindle begins to twist open, the left edge of the butterfly valve moves above the progressive duct and exposes it to the depression from the engine. Hence a proportion of the emulsified mixture by-passes the volume control screw duct and discharges straight into the lower mixing chamber. Consequently, more emulsified mixture enters the lower chamber to increase the quantity of mixture supplied to the engine.

21.4.2 Main mixture control system (fig. 21.18)
When the throttle valve is partially opened a proportion of the vacuum created in the induction manifold will be transferred from the engine's side of the throttle to the mixing chamber and inner venturi above the throttle. Petrol will now be drawn from the petrol jet to the main emulsion-tube well, where it meets air also being attracted into the emulsion well via the air correction jet and the central passage into the emulsion-tube. The emulsion of air and petrol formed in the capacity well then passes to the discharge passage and is then discharged into the inner venturi where it mixes with the incoming air stream.

With increased engine demands the throttle will be opened further. This raises the concentration of vacuum at the discharge nozzle. Thus the increased draw of air/fuel mixture lowers the level of fuel in the emulsion-tube. It therefore uncovers more of the emulsion-tube radial holes. As a result more air enters the emulsion-tube well and mixes with the fuel supplied by the main petrol jet. It thus counteracts the natural tendency of over enrichment with increasing engine speed.

21.4.3 Power valve system (fig. 21.18)
The purpose of the power-valve (jet) is to provide additional fuel under full load and for high speed operating conditions and to be able to automatically cut in or cut out this extra fuel supply as driving conditions change. With small throttle openings the depression on the engine side of the throttle is relatively high. A passage connecting the underside of the throttle to the top of the vacuum piston therefore pulls up the vacuum piston and rod to its highest position against the resisting tension of its return-spring.This lifts the piston-rod clear of the power-valve (jet) and hence permits the power-valve return-spring to close the valve. Nevertheless as the throttle-butterfly-valve opens towards full engine load conditions, the depression on the engine side of the throttle decreases, the vacuum piston return spring is now able to push down the piston and rod and at the same time pushes open the power-valve. Additional fuel is now supplied separately to that supplied through the main petrol jet to provide the mixture enrichment which is necessary to produce full engine power.

21.4.4 Piston-type acceleration pump system (fig. 21.18)
The purpose of the acceleration pump is to supply, in the form of a fine spray, additional fuel directly into the incoming air stream every time the car is being accelerated and the throttle is opened quickly. If there was no acceleration pump fitted, the necessary increase in the quantity of mixture discharged from the nozzle to match the rapid response to the increased amount of air entering the engine would be much too slow. Hence this would result in an acceleration flat spot, that is, there would be a long delay before the initially over-weak mixture supply strengthened sufficiently to increase the speed of the engine. When the throttle valve is closed, the piston return-spring pushes up the piston to the top of its stroke. The slight vacuum created below the piston, lifts the inlet ball-valve off its seat so that petrol from the float chamber is now able to enter and fill the cylinder space below the piston. As the throttle-valve is opened, the relay-lever will be pivoted anticlockwise so that it pushes down the relay plunger and relay-spring until the underside of the pump piston contacts the head of petrol in the chamber. Further downward movement of the relay plunger will initially compress the relay-spring and close the inlet ball-valve, but once the fuel has been sufficiently pressurised, the discharge ball-valve and weight will be lifted from its seat. Fuel will thus be permitted to escape from the acceleration pump's discharge passage and nozzle into the outer venturi in the form of a fine spray. The discharge of fuel will continue as long as the stiffer expanding relay-spring tension exceeds that of the lower piston's return-spring at which point the pump discharge ceases. The function of the relay-spring is to allow the progressive expansion of this spring to extend the time of fuel discharge during the acceleration period. In addition, the weight on top of the discharge ball-valve prevents fuel being drawn into the outer venturi from the acceleration pump discharge nozzle if the depression in this region should become high.

21.5 Compound-barrel differentially operated carburettor (fig. 21.19)

One major problem with the simple barrel carburettor is that the venturi-tube must be small enough to maintain an air speed that will support the air/fuel mixture discharge into the venturi, without the heavier particles of liquid fuel gravitating and collecting on the walls of the venturi, as this would upset the mixture distribution uniformity of the incoming charge. If the venturi is relatively small a good mixture distribution would be obtained in the low to medium-speed range, but at higher speeds it would severely restrict the flow of charge passing through the carburettor, so that the filling volumetric efficiency of the cylinders would drop rapidly. Conversely, if a large venturi is chosen to provide the desired air speed at higher engine speeds, then at much lower engine speeds the air speed would be much too slow to atomise and support the emulsion of air/fuel mixture entering the venturi and mixing chamber.

The low and high speed conflicting venturi requirements can be mainly overcome by having a pair of barrels. The primary of these two barrels operates alone up to medium speeds and the secondary barrel is phased in to assist the primary barrel when the engine load and speed conditions necessitate the additional supply of air. The characteristics of the compound barrel are shown in (fig. 21.20).

21.5.1 Differentially operated secondary-stage throttle linkage (fig. 21.19)

The interconnecting linkage consists of a primary throttle-lever attached rigidly to the primary throttle-spindle, a long primary relay-lever which pivots on the primary throttle-spindle but is not

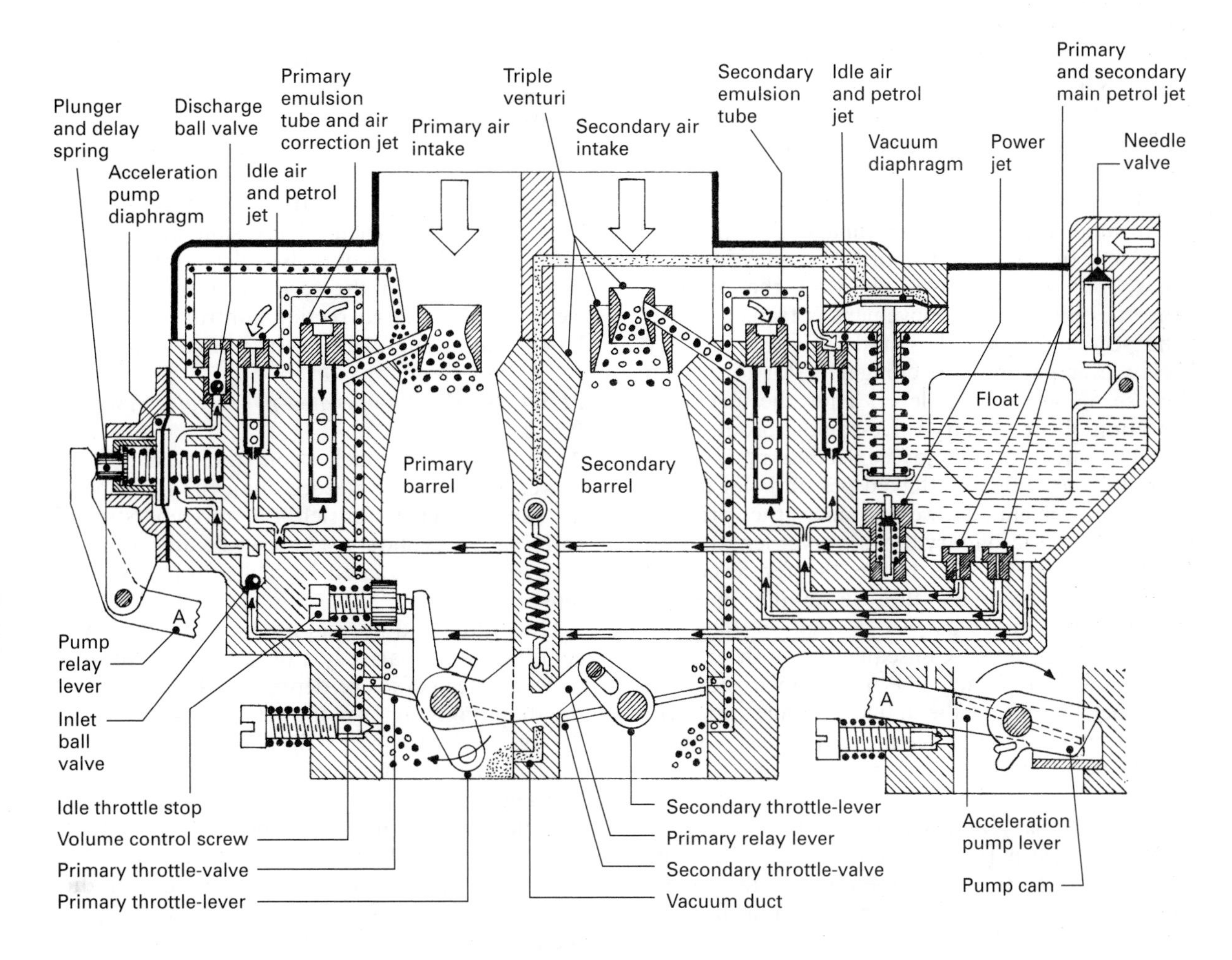

Fig. 21.19 Compound barrel differentially operated carburettor

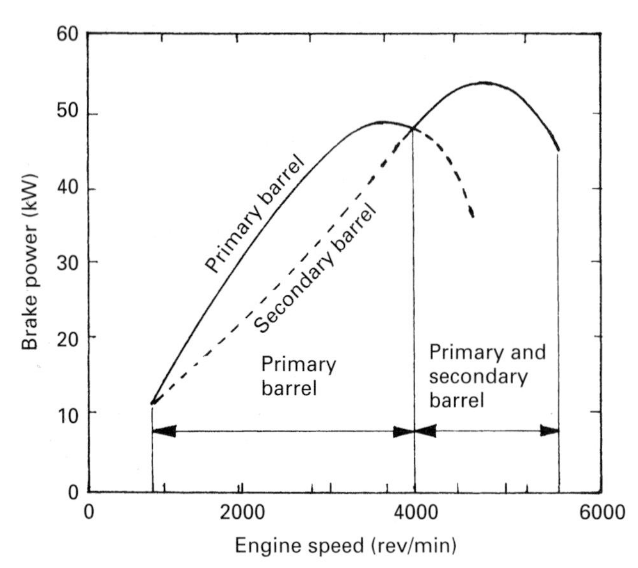

Fig. 21.20 Power characteristics for a 1.5 litre engine installed with a compound-barrel carburettor

fixed to it, and a much shorter secondary throttle-lever which is bolted to the secondary throttle-spindle. The throttle-lever only operates the primary throttle-spindle up to the 2/3 fully open position. Above this primary valve opening, a right-angled lug on the primary throttle-lever contacts the edge of the primary relay-lever so that both levers now rotate with the primary throttle-spindle. This movement is transferred to the secondary throttle-spindle via the primary relay-lever peg and the elongated slot formed in the secondary throttle-lever. Thus the clockwise rotation of the primary valve-spindle produces an anticlockwise motion to the secondary throttle-spindle due to the interlinked peg and slot downward movement. However, since the primary relay-lever to secondary throttle-lever length ratio is approximately 3 to 1, the shorter lever will rotate the secondary throttle-spindle approximately three times as fast as that of the long primary relay-lever. Consequently, both throttle valves will approach the fully open position together. Note that the actual leverage ratio will vary somewhat as the peg moves further along its guide slot.

21.5.2 Idle and progression system (fig. 21.19)
Petrol is drawn from the float-chamber via the main petrol jet to each of the idle jet-tubes where it meets incoming metered air. The emulsified air/petrol mixture is then drawn through passages to both barrel mixing chambers on the engine side of the throttle. However, the quantity of mixture entering the primary barrel can be adjusted by a volume control screw, whereas the idle discharge duct entering the secondary barrel has a fixed orifice. As the primary throttle begins to open, it exposes the progression duct to the engine side of the throttle, therefore permitting an additional amount of air/fuel emulsion to enter the primary barrel during the transition from idle to main mixture correction control.

21.5.3 Main mixture correction control (fig. 21.19)
Metered fuel from the main petrol jet flows to the primary emulsion-tube well where it meets metered air entering from the main air correction jet above. As the petrol is discharged into the inner venturi, the level of fuel in the emulsion-tube well falls. This uncovers bleed-holes in the lower half of the emulsion-tube so that both petrol and air will flow together to form an emulsion before it is discharged into the inner venturi. With increasing engine demands, the level of the fuel in the emulsion-tube well drops even further to expose more air bleed-holes. This therefore compensates for the over-enriching tendency of the flow of petrol from a fixed size of petrol jet. Once the secondary throttle commences to open there will be sufficient depression at the inner venturi of the secondary barrel to draw the air/fuel mixture from the secondary emulsion-tube well. The mixture correction control here is similar to that taking place in the primary barrel.

21.5.4 Diaphragm-operated power valve system (fig. 21.19)
As the engine approaches something like 80% full power, more fuel is needed than the main petrol jet can deliver to enrich the mixture strength entering the cylinders. This can be achieved with a power-valve actuated by a diaphragm vacuum sensor.

Under idle and normal mixture control driving conditions the partially closed primary throttle-valve creates sufficient depression below it to lift the diaphragm against the tension of the return-spring. This then keeps the diaphragm stem-end clear of the power-valve needle. Nevertheless, as the demands of the engine increase, the primary throttle-valve progressively opens until eventually there will be insufficient concentration of depression on the power-valve diaphragm vacuum duct to hold up the diaphragm and stem. At this point the diaphragm return-spring pulls down the diaphragm and stem until the power-valve needle is forced to open. Additional petrol will now be able to flow from the float-chamber directly to the primary emulsion-tube well to supplement the main petrol jet supply. When the engine speed and load is reduced, the primary throttle-valve will close somewhat. This re-establishes the depression below the throttle-valve so that once again the diaphragm will be raised to permit the power-valve needle to close.

21.5.5 Diaphragm-operated acceleration pump system (fig. 21.19)

The purpose of a petrol pump is to provide a separate pressurised fuel spray to enter the air stream to compensate for the sudden loss of depression in the venturi region when the throttle-valve is suddenly opened. Otherwise for a very short space of time there would be a weakening of the mixture strength resulting in sluggish acceleration. When the primary throttle is closed, the acceleration pump lever releases the pressure on the plunger and relay-spring, thus permitting the diaphragm return-spring to push the diaphragm outwards. The now enlarged petrol chamber creates a slight depression. This makes the discharge ball-valve fall down onto its seat, whereas the inlet ball-valve is lifted clear of its seat. Petrol now enters and fills up the enlarged diaphragm petrol chamber.

If the car is now accelerated, the primary throttle-valve and spindle rotate with the pump cam in the opening direction. At the same time, the cam pivots the accelerator pump relay-lever. As a result the plunger and relay-spring will force the diaphragm to the right. This compresses the trapped petrol, immediately the inlet ball-valve snaps closed, whereas the discharge ball-valve is jerked up to open the passage and to seal the air duct. Petrol will now be discharged into the outer primary venturi by way of the acceleration pump discharge nozzle. The function of the plunger and relay-spring is to prolong the duration of the acceleration period by allowing the spring to progressively expand as the petrol in the diaphragm chamber is consumed.

21.6 Constant-pressure or vacuum variable-choke carburettor (fig. 21.21)

The function of the carburettor is to provide a variable quantity of metered and atomised mixture to the cylinders under all running conditions. The constant-pressure carburettor has a variable-choke piston-type air valve situated at the entrance to the mixing chamber, and attached to the base of the air valve is a tapered needle which is submerged in a petrol-jet orifice immediately beneath it. The air valve has a central stem which guides its vertical movement, and the upper portion of the valve forms a piston which seals off a vacuum or suction chamber. The lower annular portion of the piston is subjected to atmospheric pressure, and the top side of the piston is exposed to a similar pressure to that of the mixing chamber by means of a small transfer hole formed in the air valve.

In its static position, the valve rests on a raised portion of the intake, which forms a venturi-jet bridge. A small air gap underneath it allows atmospheric air to flow through to the cylinders. A butterfly throttle valve is positioned on the downstream side, to control the quantity of mixture entering the engine.

21.6.1 Air-valve piston action (fig. 21.21)

With the engine at rest and no air flowing, the piston air valve is at the bottom of its stroke, with only a small air gap between the base and the top of the jet. When the engine is started, air flows through this gap, speeds up, and produces a pressure drop over the petrol-jet orifice. This drop in pressure is transmitted through the transfer hole to the upper side of the piston, with the result that the pressure in the vacuum chamber is near enough the same as that at the jet. The higher atmospheric pressure on the underside of the piston will create a pressure difference across the two sides, producing a resultant upthrust. The piston will thus be pushed up until a balanced position is reached when the upthrust is equal to the downward mass and spring force acting on the piston. This balance or equilibrium position will ensure that a constant pressure will always be maintained over the jet, whatever the engine speed and load conditions.

21.6.2 Mixture control (fig. 21.21)

Having established a constant depression at the jet, it will be seen that, with a constant-size jet, the same amount of fuel will be drawn into the air stream whatever the air speed, so that no mixture-strength correction can occur. To match an increased air flow, the effective annular petrol-jet orifice formed between the jet and the tapered needle will expand in proportion to the outward movement of the tapered needle as it rises with the air-valve piston. By choosing the correct needle profile, the restriction of fuel discharge can be made to correspond with the engine's

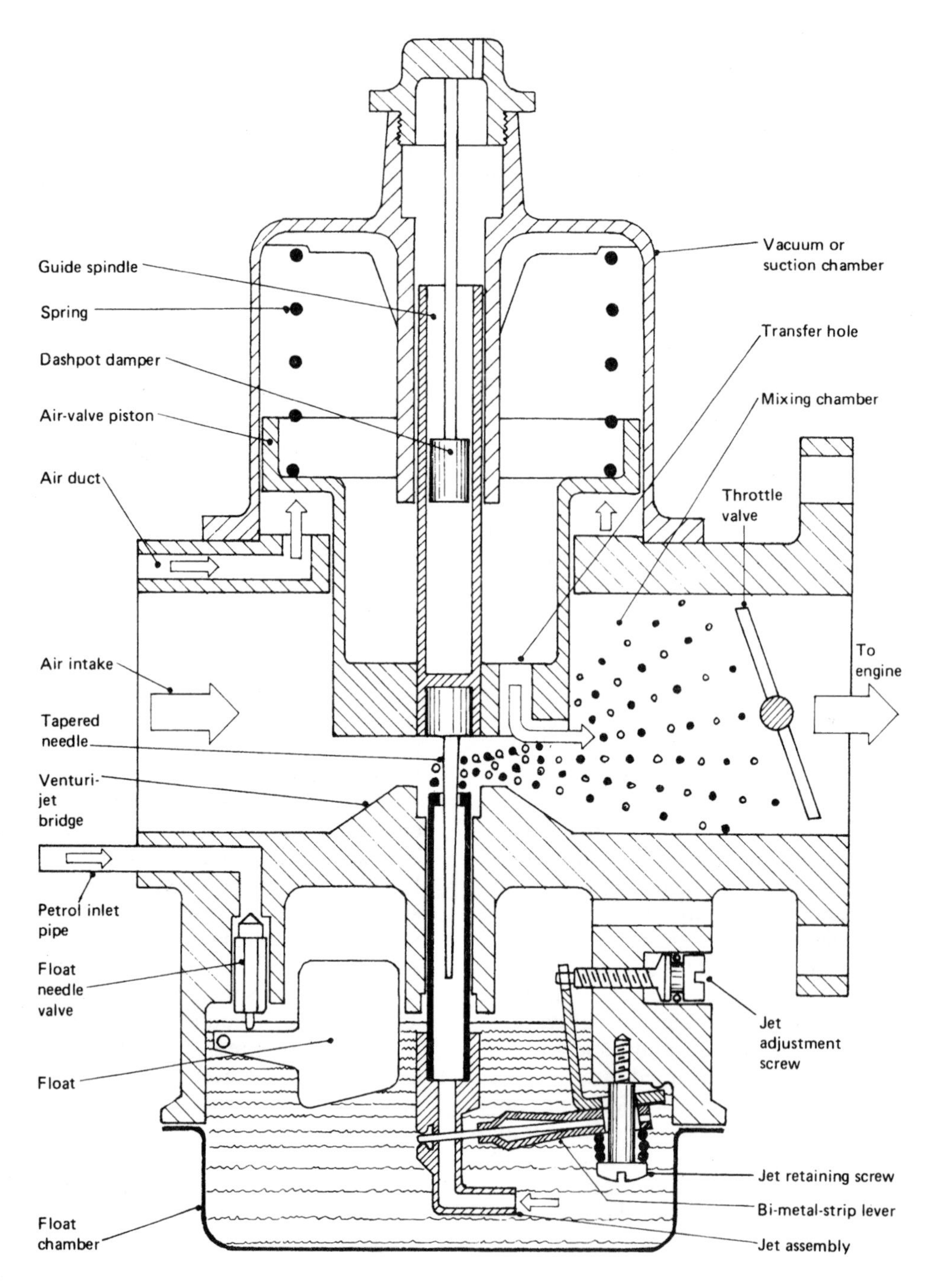

Fig. 21.21 Constant-vacuum variable-choke carburettor

air-consumption characteristics and so provide the best mixture ratio over the complete speed and load range without incorporating additional air or petrol jets.

If the piston lifts when the engine is running, it will allow more air to enter the cylinder, so that the depression at the jet bridge will be relieved immediately – thus there will be less fuel drawn from the jet, so the mixture strength will weaken. If the slow-running mixture is correct, lifting the piston slightly will increase the engine's speed, but lifting the piston further will result in the engine stalling due to the overweakening effect.

The mixture strength can be varied by altering the area of the jet orifice around the needle by an adjustment which moves the jet nearer or further from the needle, but it should be noted that any alteration will influence the discharge throughout the speed range. Figure 21.21 shows the jet-adjustment device in the form of a bi-metal-strip lever and a spring-loaded L-shaped lever – the height of the jet relative to the needle can be raised or lowered by screwing the horizontal set-screw out or in respectively.

The volume of mixture entering the engine during idling conditions is controlled by an idle-adjustment stop screw which limits the closing of the butterfly throttle valve.

21.6.3 Mixture temperature compensation (fig. 21.21)

The bi-metal strip is provided to compensate for the flow-discharge variation which occurs at different temperatures, due to the changing fuel viscosity. The bi-metal strip will curl upwards when the temperature rises, raising the needle and weakening the mixture. Conversely the strip will straighten out to a richer position when the temperature drops. Mixture correction over a wide temperature range is thus provided, consequently exhaust emission will be held to closer limits.

21.6.4 Cold-starting mixture control

To compensate for poor vaporisation when the engine is cold, a richer mixture must be provided. This is achieved simply by a hand-controlled cable which when pulled actuates a linkage which either (a) lowers the petrol jet so that the effective jet orifice is increased or (b) opens a separate cold-start metering valve which is exposed to the mixing chamber. Both approaches will result in considerably more fuel being drawn in at cranking speed. This excess fuel is progressively reduced by partially returning the control-cable knob towards the dashboard as the engine warms up.

21.7 Attitude of the choke tube

Carburettor choke tubes are broadly identified by three different directions of air draught:

i) up draught,
ii) down draught,
iii) horizontal draught.

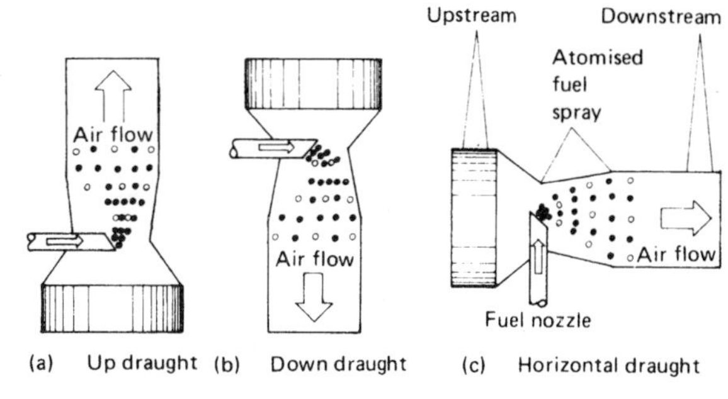

Fig. 21.22 Classification of air flow

Up draught (fig. 21.22(a)) This configuration makes the air flow move in opposition to the force of gravity, so that the heavier liquid fuel particles will not be drawn into the engine until they have been pre-heated and partially vaporised by the exhaust manifold. This draught is suitable for engines which burn paraffin-type fuels, such as tractor installations, but otherwise has no technical merit.

Down draught (fig. 21.22(b)) With this configuration the air movement is assisted by the force of gravity, so that better charge filling of the cylinder is obtained. When using this draught, adequate cold-start pre-heating of the liquid–air mixture must be provided, to avoid wet fuel draining into the cylinders, but without causing overheating of the charge once normal operating temperatures are established.

Horizontal draught (fig. 21.22(c)) Horizontal air draughts are neither aided nor hindered by the force of gravity, but the heavier liquid fuel will tend to collect on the floor of the choke tube whenever the air movement is very slow. This draught provides the most direct path for atomised charge from the carburettor venturi to the cylinders.

21.8 Induction and exhaust manifolds (figs 21.23 to 21.24)

Induction manifold Air and petrol mixture which has been atomised in the carburettor is conveyed to the inlet ports situated in the side of the cylinder head through pipes branching out from the base of the carburettor at the riser (figs 21.23 and 21.24). Such a layout – known as the induction manifold – may be provided by a single aluminium-alloy casting which has several passages sprouting out from a central inlet chamber.

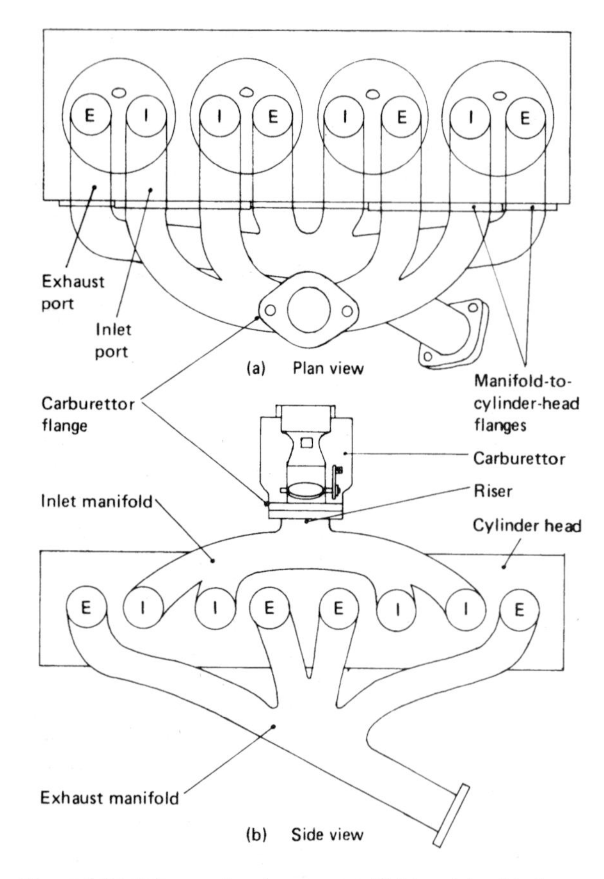

Fig. 21.23 Inlet and exhaust manifolds with side-by-side valves and ports

Exhaust manifold Conversely, exhaust gases escaping through the ports formed in the cylinder head are collected together by a cast-iron manifold in the form of flanged branch pipes bolted to the cylinder head (figs 21.23 and 21.24). These pipes merge at a common junction so that the exhaust gases can be passed through the exhaust-gas silencer system before being expelled to the atmosphere.

21.8.1 Pre-heating the induction manifold

To improve the atomisation of the mixture flowing from the carburettor through the induction manifold and into the engine cylinder, the manifold is sometimes provided by some means of pre-heating, such as by using heat from the exhaust manifold or the engine coolant system. Controlled heating of the fresh induced mixture charge will greatly assist in breaking up the heavier fractions of the petrol and any liquid droplets which may have collected on the floor of the

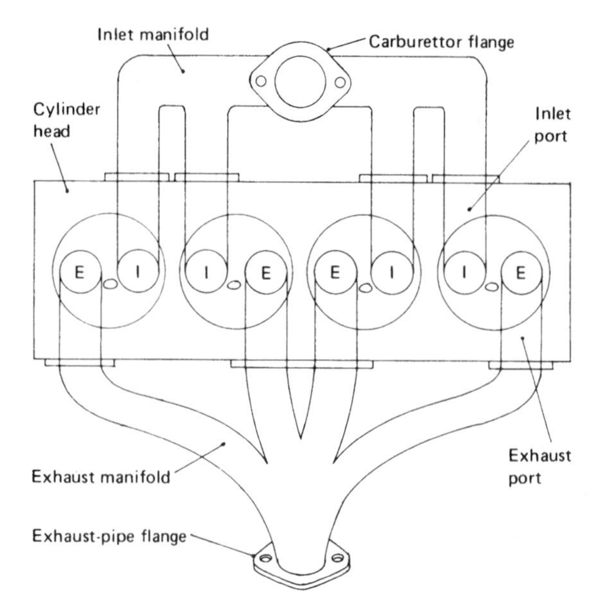

Fig. 21.24 Inlet and exhaust manifolds with side-by-side cross-flow ports

manifold passages. Unfortunately, if too much heat is supplied to the induction manifold, atomised mixture may actually be converted to vapour before it enters the cylinder, which would reduce the effective mass of charge induced per induction stroke.

Exhaust hot-spot heating When induction and exhaust manifolds are on the same side of the cylinder head (fig. 21.23), a degree of pre-heating of the air and petrol mixture can be obtained by designing the two manifolds so that they touch each other, the contact point being referred to as the 'hot spot' (fig. 21.25). This permits heat to be conducted from the hot exhaust to the much cooler induction manifold.

Coolant-circulation heating When manifolds are on opposite sides of the cylinder head (fig. 21.24), pre-heating from the cooling system can be achieved by surrounding the central induction riser with a jacket through which circulates the engine's heated coolant (fig. 21.26).

21.9 Multi-carburettors

To improve the distribution of the air and petrol charge so that all cylinders receive an equal quantity of a mixture of the same strength and thoroughness of atomisation, two or more carburettors are sometimes used. The advantages of

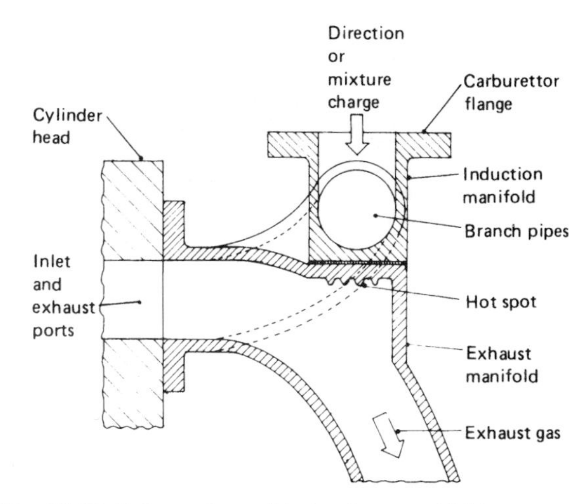

Fig. 21.25 Exhaust-heated induction manifold

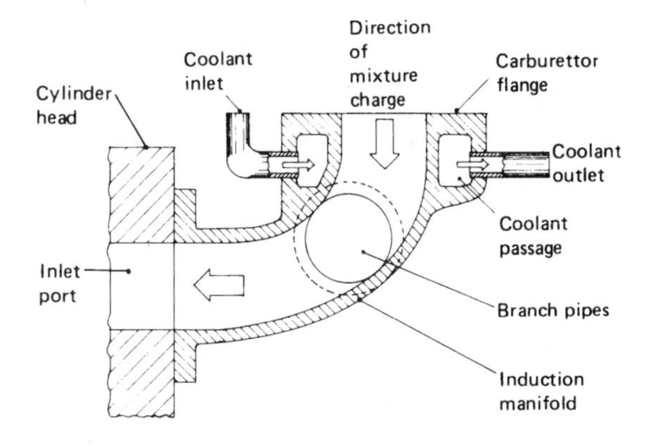

Fig. 21.26 Engine-coolant-heated induction manifold

using more than one carburettor are that fewer cylinders are fed by each carburettor, the mixture moves a shorter distance and more directly between the carburettor and the cylinders, and there is less induction interference between adjacent cylinders.

21.10 Air-intake silencers and cleaners

Air is a carrier of dust particles. As the engine consumes vast amounts of air – for example, a 2 litre engine consumes 5000 litres per minute of air at 5000 rev/min – air filters must be fitted to the induction air intake to prevent harmful abrasive grit, sand, dirt, and other particles from entering the cylinders or air jets and other vital parts.

21.10.1 Conditions of service and control

The air cleaner must clean effectively the whole of the air aspirated by the engine under full-throttle conditions but without causing large pressure drops across the filter medium. If a considerable pressure difference across the filter material is established, the engine may be starved of air under certain operating conditions. This will lower the engine's volumetric efficiency and reduce its performance, both by reducing engine power and by increasing fuel consumption.

In addition to filtering the air, cleaners are also designed to suppress the noise of the air through the induction system and the engine pulsations which produce objectionable hiss and roar. Most air cleaners therefore perform the function of both separating unwanted dust from the air entering the engine and silencing the mass of air rushing into the engine. Silencing the intake noise is effected by tuning the length and diameter of the tapered intake tube, in conjunction with employing a large-capacity air-resonance chamber where the air darts in all directions and is progressively dampened before it enters the intake system.

Regular air-cleaner maintenance is necessary, otherwise, if the cleaner is allowed to become choked with accumulated dust, the resulting restriction of air flow will cause sluggish performance and higher fuel consumption. In addition, under dusty operation conditions, unfiltered air can cause severe engine wear and damage to components such as cylinder bores, pistons and rings, and poppet-valve stems and their guides; and, if it can get into the crankcase, it will contaminate both the oil and the bearings and subsequently will score the crankshaft and camshaft journals. It should be borne in mind that sand is composed of silicates related to the silicon carbide used as a grinding-wheel material, which, as everybody knows, provides a very aggressive means of wearing or cutting away steel.

21.10.2 Paper-element dry air cleaners (figs 21.27 and 21.28)

These paper elements are constructed in two forms:

i) the annualar flat corrugated type (fig. 21.27)
ii) the cylindrical corrugated type (fig. 21.28)

In both designs the elements are constructed of impregnated specially prepared corrugated paper which vastly enlarges the surface area exposed to the ingoing air stream so that large quantities of air can pass through the medium with the minimum of restriction. With both annular and cylindrical corrugated elements, the ends of the

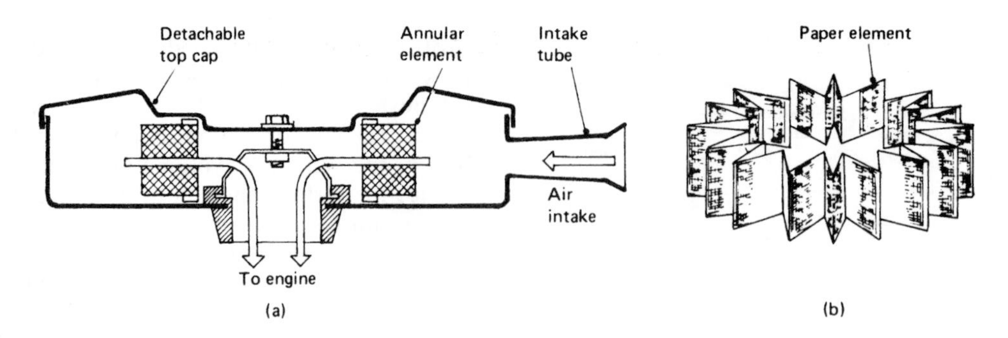

Fig. 21.27 Annular paper-element dry air cleaner

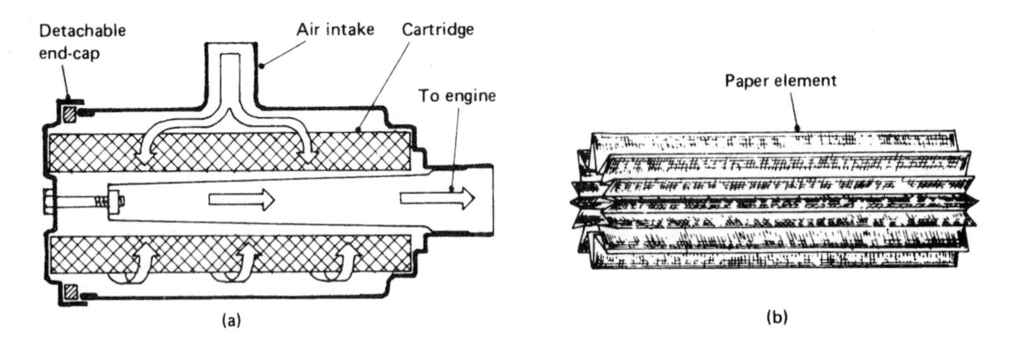

Fig. 21.28 Cylindrical-cartridge dry air cleaner

folds are held in position by rubber or plastic discs moulded to the paper (in addition, the cylindrical element is usually enclosed in a metal perforated frame). These replaceable filter elements are supported and clamped at both ends between the casing and its detachable cover cap.

The filter element is normally surrounded by a large resonance chamber which absorbs the momentum of the fast-moving ingoing column of air, and the interference caused by air turbulence in the chamber to a large extent damps out resonant noises associated with the pulsations created by the opening and closing of the engine's inlet valves.

These units are very light and more compact than the oil-bath cleaner units (see figs 23.28 and 23.29) for similar efficiencies, they can be mounted in any convenient position, and the elements are easy to change. Under normal operating conditions, element life is between 25 000 and 35 000 km.

22 Petrol engine fuel injection systems

22.1 Comparison of the various fuel supply systems

Fuel injection systems for petrol engines may be of the single-point injection SPI type which has its injector located at the entrance to the air intake just ahead of the throttle butterfly-valve. Air/fuel mixing and atomisation take place at low injection pressure upstream of the throttle. The mixture is then divided and distributed through the induction manifold to the various cylinders. The basic difference between the carburettor fuel system and the single-point injection SPI system is that the carburettor system relies on negative pressure created by the venturi to induce the metered fuel to enter the incoming air stream, whereas the single-point injection SPI fuel system injects fuel into an unrestricted air stream at something like 3 or 4 bar above atmospheric pressure. Conversely the multipoint injection MPI system injects fuel downstream of the throttle at the throat of each inlet-valve port just ahead of the inlet-valve so that each column of air charge is mixed with identical quantities of metered fuel.

The advantages and disadvantages of the single-point injection SPI and the multipoint injection MPI systems must be weighed against the short fuel mixing time available before the inlet-valve opens with the downstream port located at multipoint injectors, to the difficulties of providing a uniform distribution of air/fuel mixture to each port with the upstream single-point injection layout. Costs are obviously less with a single source of injection than with a multipoint source of fuel delivery.

22.2 Bosch KE-Jetronic multipoint petrol injection system (fig. 22.1)

This multipoint injection system injects fuel continuously into each cylinder-head inlet port. It is a mechanical-hydraulic injection system incorporating an electronic control unit (ECU) with electrical input sensors which monitor the engine's operating conditions and an electro-hydraulic output pressure actuator which controls air/fuel mixture correction.

22.2.1 Fuel pump (fig. 22.1)

The fuel pump unit is a roller-cell-type pump driven by a permanent magnet electric motor. The pump is composed of an eccentrically mounted rotor which has five radial slots machined around it (only four shown). Each slot contains a roller which moves outwards against the casing wall when the rotor revolves, thereby sealing the spaces formed between adjacent rollers. When the pump rotates, fuel is drawn through the inlet port to fill up the enlarged chambers or cells formed between each pair of adjacent rollers. As the rotor revolves the trapped fuel is progressively squeezed until it is exposed to the outlet-port. At this point the pressurised fuel is discharged to the accumulator at a pressure of over 5 bar.

22.2.2 Accumulator (fig. 22.1)

The accumulator serves two purposes: first it stores fuel when the engine is running and releases this excess fuel to maintain line pressure when the engine is restarted; secondly it smooths out any pressure fluctuation which may exist in the fuel discharge coming from the fuel pump.

22.2.3 Air flow sensor (fig. 22.1)

An air flow plate working on the suspended body principle is used to measure the amount of air entering the engine. The sensor is composed of a plate supported on a pivoting lever. The other end of this lever is weighted to balance the lever when in its free state. The sensor-plate is positioned in a specially shaped funnel which forms part of the air intake ahead of the throttle-valve.

Air drawn into the engine while it is running tends to lift the sensor-plate; the more the throttle is opened the greater will be the quantity of air entering the engine and the greater will be the upward pull on the sensor-plate. The movement of the sensor-plate which is a direct measure of the air flowing past it is relayed to the control-plunger which meters the quantity of fuel needed to match the air flow under different speed and load operating conditions.

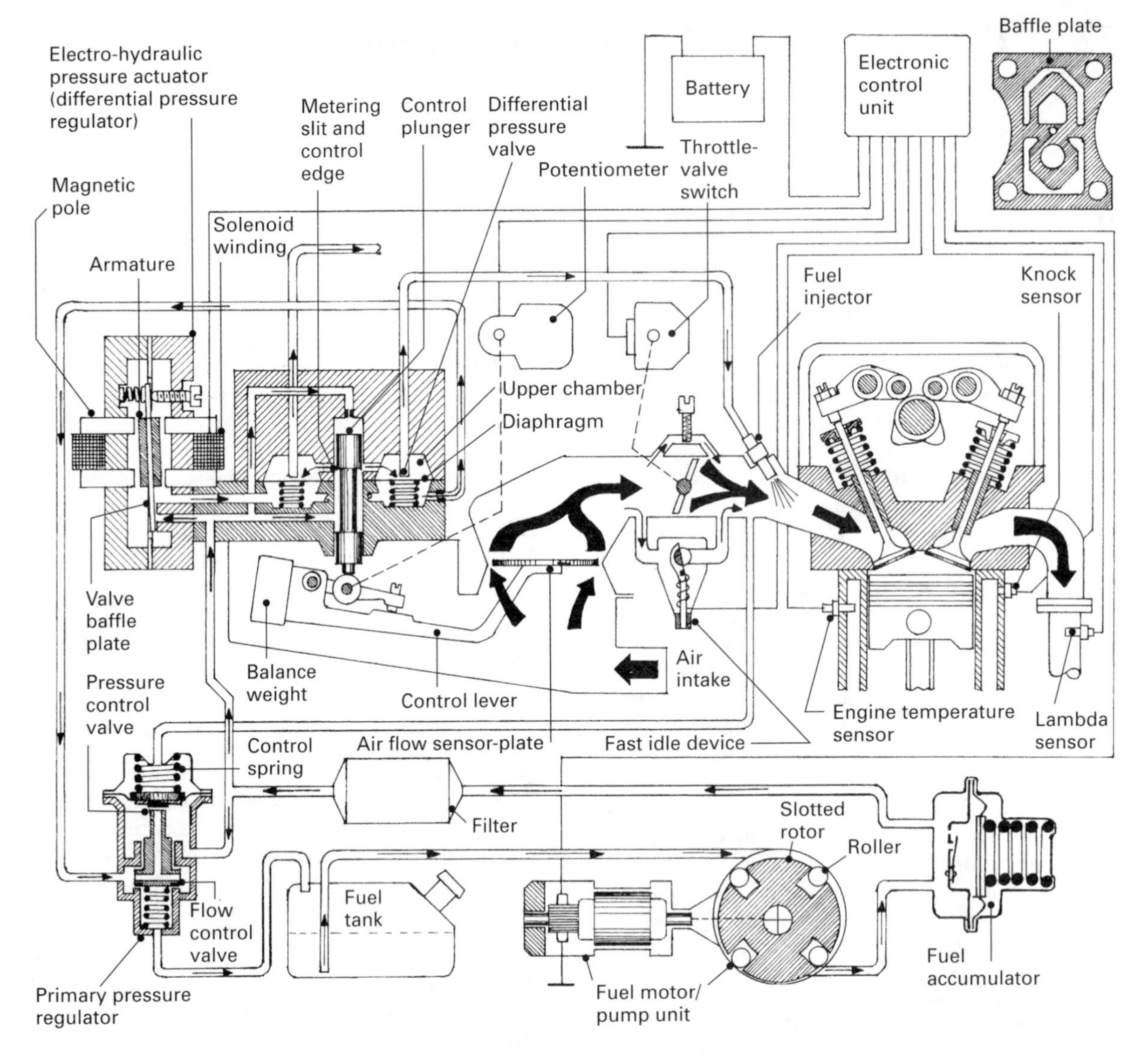

Fig. 22.1 Bosch KE-jetronic petrol injection system

22.2.4 Fuel distribution control plunger and barrel operation (figs 22.1 and 22.2)

The sensor-plate lift controls the quantity of fuel metered by the plunger and barrel slits before it is delivered to the individual cylinders. Fuel supplied from the pump flows through the accumulator and the filter to the barrel inlet port around the plunger central wasted region and out of the metering slits into the upper diaphragm chambers (fig. 22.1). There are as many slits and differential-valve upper chambers as there are fuel injectors and engine cylinders. When the engine speed is low the sensor-plate upward drag will be small so that only a very tiny portion of the metering slits will be open. As the demands of the engine increase with wider throttle openings, the greater will be the quantity of air entering the engine and proportionally greater will be the sensor-plate and plunger lift; consequently the larger will be the slit openings so that more fuel will be metered to enter the differential-valve upper chamber. Supply fuel is also conveyed to the barrel chamber above the piston crown via a restriction. The trapped fuel above the piston now acts as a damping device to smooth out air pressure pulsations

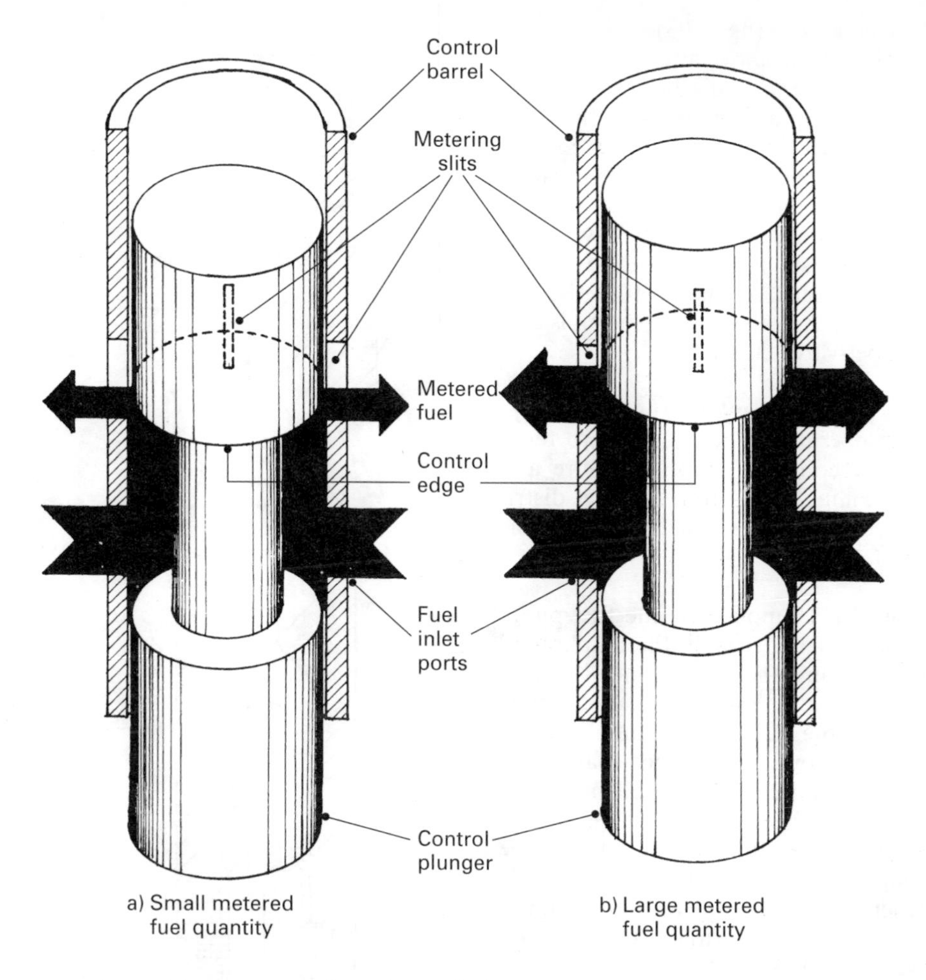

Fig. 22.2 Plunger and barrel fuel metering for a four cylinder engine

and to prevent the plunger overshooting when the throttle is suddenly opened.

22.2.5 Differential pressure valve (fig. 22.1)
When the barrel opening slit is small there will be a much larger resistance to flow across the slits than when the opening slit is large. Thus with a small opening slit there will be a large pressure difference between both sides of the slit, as opposed to a much smaller pressure difference when the slit is large. The consequence of the pressure difference varying across the slits when the quantity of metered fuel varies would mean that the fuel discharged from the metering-slits does not accurately relate to the same plate lift or fall. To overcome this pressure difference when both small and large quantities of fuel are metered, a spring-loaded diaphragm pressure valve is incorporated on the output side of each metering-slit. These differential pressure valves maintain a constant pressure difference between the upper and lower chamber of approximately 0.2 bar under most operating conditions. Thus when metered fuel flows into the upper chambers, the diaphragms deflect downwards until the predetermined differential pressure is reached, with the fuel discharging to the injectors via the central differential valve opening. If the quantity of fuel needed is small, the diaphragm deflection will also be small, so that only a small discharge will take place. However, if the fuel metered is large, the increased pressure in the upper chamber will push the diaphragm down further so that proportionately more fuel is delivered to the injectors. This device thus ensures a more accurate relationship between the quantity of air passing through to the engine and the corresponding metered fuel discharged to the injectors.

*22.2.6 **Fuel injector valve*** (fig. 22.3)
This type of valve opens above a pre-set pressure, normally about 3.3 bar, and discharges fuel continuously in the form of a spray into the throat of the cylinder-head inlet ports. Fuel pressure forces the poppet-valve to open against the resisting tension of the return spring. Once the valve is open, the floating poppet-valve oscillates at a high rate which assists the atomisation of the discharge spring. When the engine stops the line pressure drops until the valve return spring is able to pull the poppet-valve closed.

*22.2.7 **Primary pressure regulator*** (fig. 22.1)
The diaphragm-operated primary pressure regulator maintains a constant supply pressure and prevents fuel draining away from the fuel distributor/mixture control unit when the engine has stopped.

Engine running Fuel supplied by the fuel pump flows to the chamber underneath the diaphragm, causing it to lift against the upper control spring. At the same time, the lower weaker counter-spring pushes up the valve-body until it butts against the casing stop. Any further diaphragm lift due to excess supply pressure will move the diaphragm and hollow stem apart. Thus it permits surplus fuel to flow down the centre passage and out through the return-valve in the base of the casing, where it then returns to the fuel-tank. Fuel is also allowed to return to the fuel-tank from the lower chambers of the differential pressure valves.

Engine switched off If the engine is switched off the fuel pump stops pumping, and the pressure underneath the diaphragm drops, thus permitting the control spring to push down the diaphragm, first of all closing the hollow-stem at the top of the valve-body and secondly moving the valve-body further down against the upthrust of the weak counter-spring until the return valve seat at the bottom of the valve-body closes. Fuel will now be maintained under a slightly reduced pressure in the fuel distributor/mixture control unit, thereby preventing the formation of vapour-locks. This also enables the engine to easily start when hot.

*22.2.8 **Electro-hydraulic pressure actuator*** (fig. 22.1)
This solenoid-operated valve controls the pressure in the lower chamber of the differential pressure

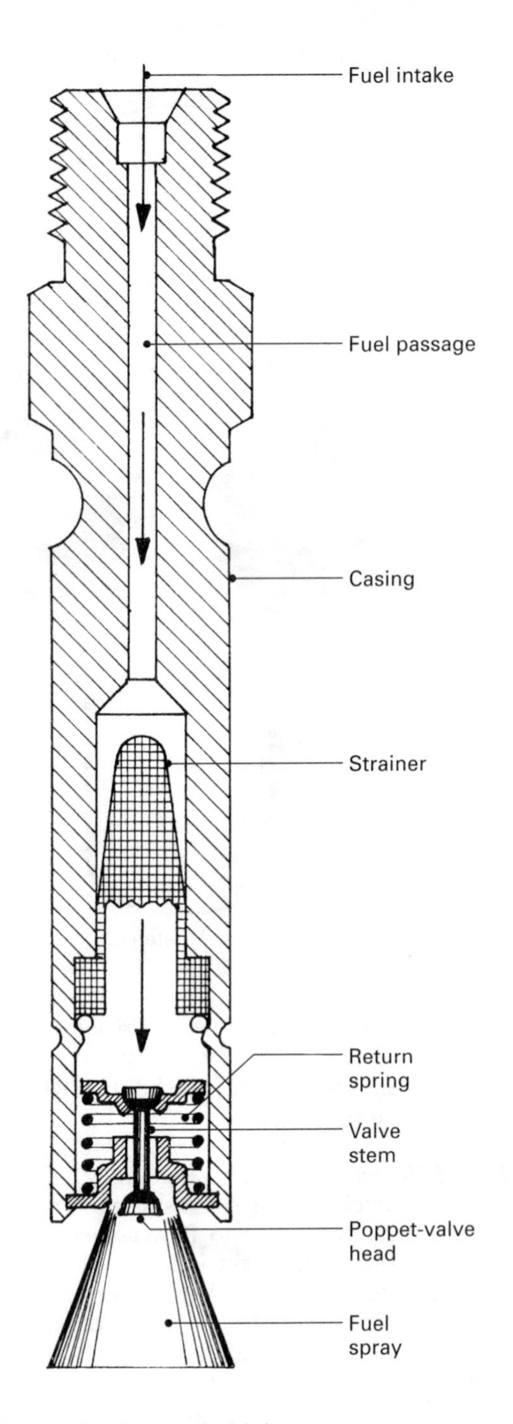

Fig. 22.3 Continuous fuel injector

valve so that it is effectively able to vary the quantity of fuel delivered to the injectors over the whole air flow operating range. It thus enables the air/fuel mixture correction to be made for a wide range of engine operating conditions such as start-up, warm-up, acceleration, full load, overrun, engine speed limitation etc.

Construction A flat armature piece is suspended on a thin spring steel baffle-plate between two double permanent magnetic poles which are separated by a solenoid winding from the electro-hydraulic pressure actuator. The rectangular baffle-plate has a 'U' slit stamped out at both ends so that the remaining narrow middle strap becomes a flexible pivot. An adjustment screw and spring provides the means of setting the correct preload tilt to the plate.

Actuator operation When the electromagnet is energised the armature tilts and tends to oppose the flow of fuel supplied at pump pressure, so that the higher the current passing to the solenoid winding, the greater will be the attraction force to close the baffle-plate valve.

Lean mixture A lean mixture is signalled by a small current so that fuel is able to flow to the lower differential-valve chamber almost unrestricted. The higher pressure in the lower chamber therefore applies an opposing force against the downward thrust on the diaphragm, exerted by the metered fuel in the upper chamber. Consequently less fuel will be discharged from each of the differential pressure valves to their respective injector.

Rich mixture A rich mixture is signalled by a large current which causes the armature and baffle-valve to tilt more, thus opposing and restricting the flow of fuel at supply pressure to the lower chamber. As a result the metering pressure in the upper chamber is able to overcome the upthrust of the reduced pressure in the lower chamber. It thus permits the diaphragm to move down more so that an increased amount of fuel will be discharged from each differential valve to the individual injectors.

22.2.9 Electronic control unit (fig. 22.1)

The purpose of the electronic control unit (ECU) is to receive information in the form of electrical signals from the monitoring engine sensors. It is then processed and compared with the programmed memory. This then provides a correction factor output signal which is fed to the electro-hydraulic pressure actuator (EHPA). It should be observed that the EHPA controls the pressure in the upper and lower chambers of the differential pressure valve. It thus determines the amount of fuel discharged to the fuel injectors via the differential pressure valve.

22.2.10 Air flow sensor potentiometer

This is a variable resistance attached to one end of the sensor-plate spindle, so that any movement of the sensor-plate produces a corresponding semi-circular sweep of the brushes over the resistance strip. The potentiometer brush movement position informs the electronic control unit of the relative position of the air flow sensor-plate as this is a measure of the power output of the engine. The direction the potentiometer brush twists relates to the increase or decrease of engine speed, whereas the speed at which it rotates indicates the magnitude of the engine's acceleration.

22.2.11 Throttle valve switch (fig. 22.1)

This switch unit has two contacts, an idle-contact and full-load contact. The cam that operates these contacts is attached to one end of the throttle-spindle so that the angular movement of the throttle rotates the cam the same amount. When the throttle moves to the closed idle position the idle-contacts close. This immediately sends a signal to the electronic control unit to enrich the normal part-load lean air/fuel mixture (note, a rich mixture is necessary when the throttle is closed at very low engine speed). With a wide-open throttle the full-load contacts close. This signals to the electronic control unit to enrich the air/fuel mixture to promote maximum power.

22.2.12 Start injector

A single solenoid-operated start injector is usually mounted somewhere central on the inlet manifold to provide an extra rich air/fuel mixture when the engine is starting from cold.

22.3 Multipoint electronic petrol injection (fig. 22.4)

This multipoint fuel injection system injects petrol intermittently into each cylinder head inlet port. It is an electronically controlled fuel injection system which relies on monitoring input sensors to keep the electronic control unit continuously updated with changes in the engine's operating parameters. This results in a processed and predetermined output response in the form of a pulsed signal to be relayed to each solenoid controlled injector.

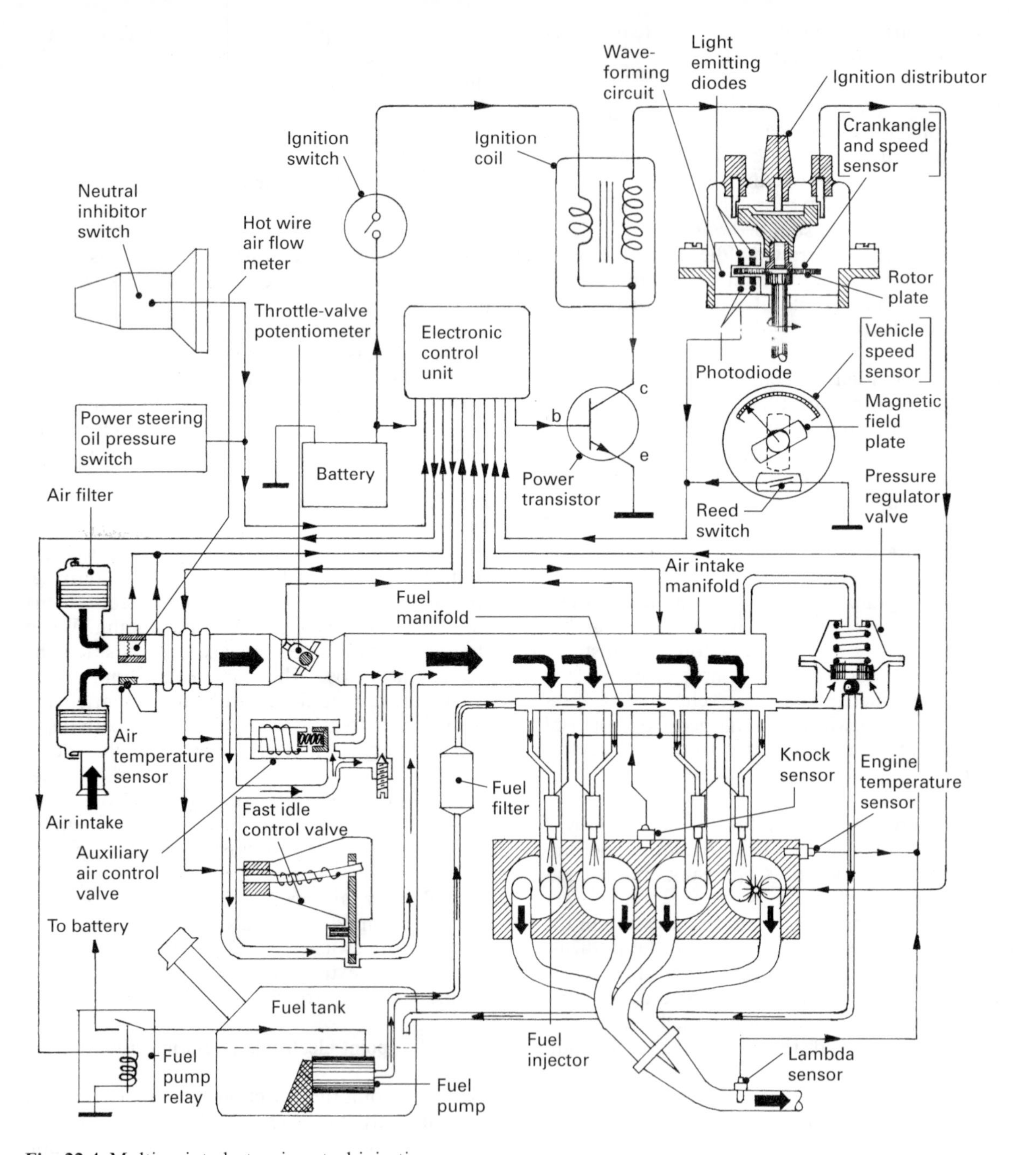

Fig. 22.4 Multi-point electronic petrol injection

22.3.1 Fuel supply system (fig. 22.5)

An electric roller-type pump supplies fuel from the fuel tank through a filter to the fuel manifold where it is distributed to each injector. The pressure of about 3 bar is maintained by the pressure regulator valve which returns excess fuel to the fuel tank. The continuous circulation of fuel prevents the fuel becoming overheated and forming vapour-locks which otherwise could cause difficult hot starting. The fuel volume capacity of the fuel manifold acts as an accumulator, thereby suppressing pressure fluctuations so that each

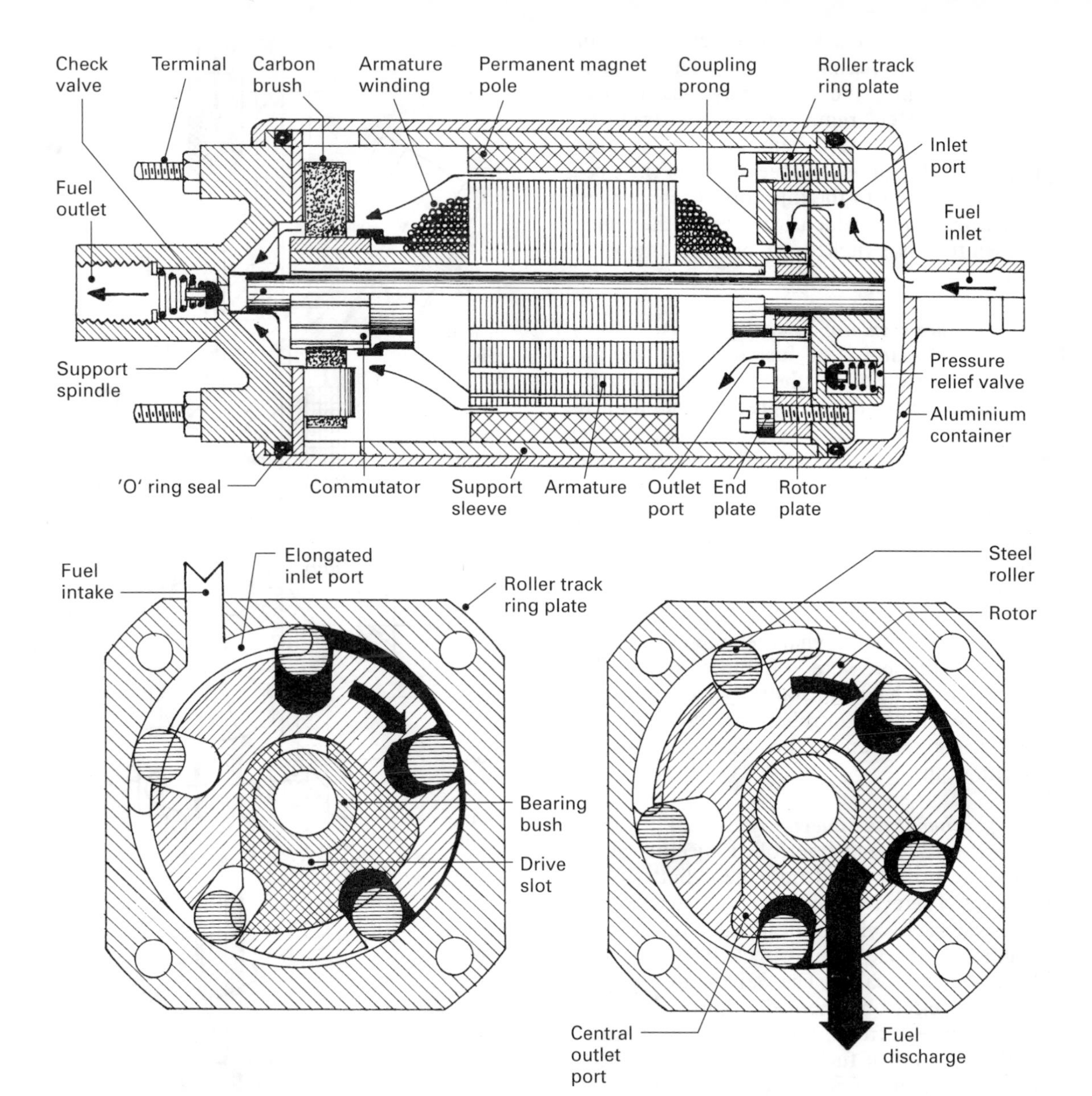

Fig. 22.5 Roller-cell-type pump

injector is subjected to the same constant fuel pressure.

The electric fuel pump unit is composed of two parts, an electric motor and a roller-type pump. The electric motor drives the pump whereas the pump delivers fuel under pressure to the individual injectors via the fuel manifold.

A roller-type pump consists of a cylindrical housing enclosing an eccentrically mounted rotor-plate. The rotor-plate usually has five radial slots machined around its circumference. Loose-fitting hollow rollers are trapped in each slot. This permits the centrifugal force to push the rollers outwards against the casing wall when the rotor is revolving. It therefore provides a seal between the casing and rotor-plate. The space created between pairs of adjacent rollers, known as cells, forms an increasing and decreasing volume which is connected to the inlet and outlet passages, respectively. Thus the enlarging cell volume draws in fuel from the fuel tank, whereas the cell volume squeezes and discharges the fuel through the electric motor and out to the injectors. There is normally a safety circuit which switches off the pump if the ignition is switched on but the engine is stalled.

*22.3.2 **Pressure regulator valve*** (fig. 22.4)
The injected quantity of fuel entering the air stream is controlled by the injection opening pulse duration with the fuel pressure maintained at a constant value. The function of the pressure regulator valve is therefore to hold the fuel pressure constant (in this case at 3 bar). This is achieved with a diaphragm-type pressure regulator valve which returns fuel to the fuel tank when the pump supply pressure reaches the opening pressure of the valve and control-spring. However, the pressure at the injector nozzle is subjected to the mixing chamber pressure variation; hence the pressure difference across the injector unit is held constant by exposing the pressure regulator control-spring chamber to the same mixing chamber pressure by an interconnecting pipe.

*22.3.3 **Multipoint fuel injector*** (fig. 22.6)
These are electronically controlled injector valves which spray metered fuel into the cylinder-head inlet ports. There is an injector situated at the entrance of each cylinder-head inlet-valve port, each injector nozzle being directed towards the neck of the inlet-valve. The valves are opened and closed electromagnetically with electrical pulses delivered from the electronic control unit. The amount of fuel discharged every injection is dependent upon the time the valve remains open, that is, it is controlled by the duration of the current pulse being transmitted from the electronic control unit. The fuel injector consists of an injector casing which houses the circular shaped solenoid, return-spring and armature. A nozzle body containing the needle-valve is secured to the lower end of the injector casing. The armature and needle-valve are attached and form the only moving member. Flutes forming part of the needle-valve serve both as a guide for the needle and as a passageway for the fuel to be discharged.

When the solenoid is de-energised the needle-valve is pressed against its seat by a helical spring. When the signal current pulse energises the solenoid, the armature which is attached to the needle-valve is lifted approximately 0.1 mm

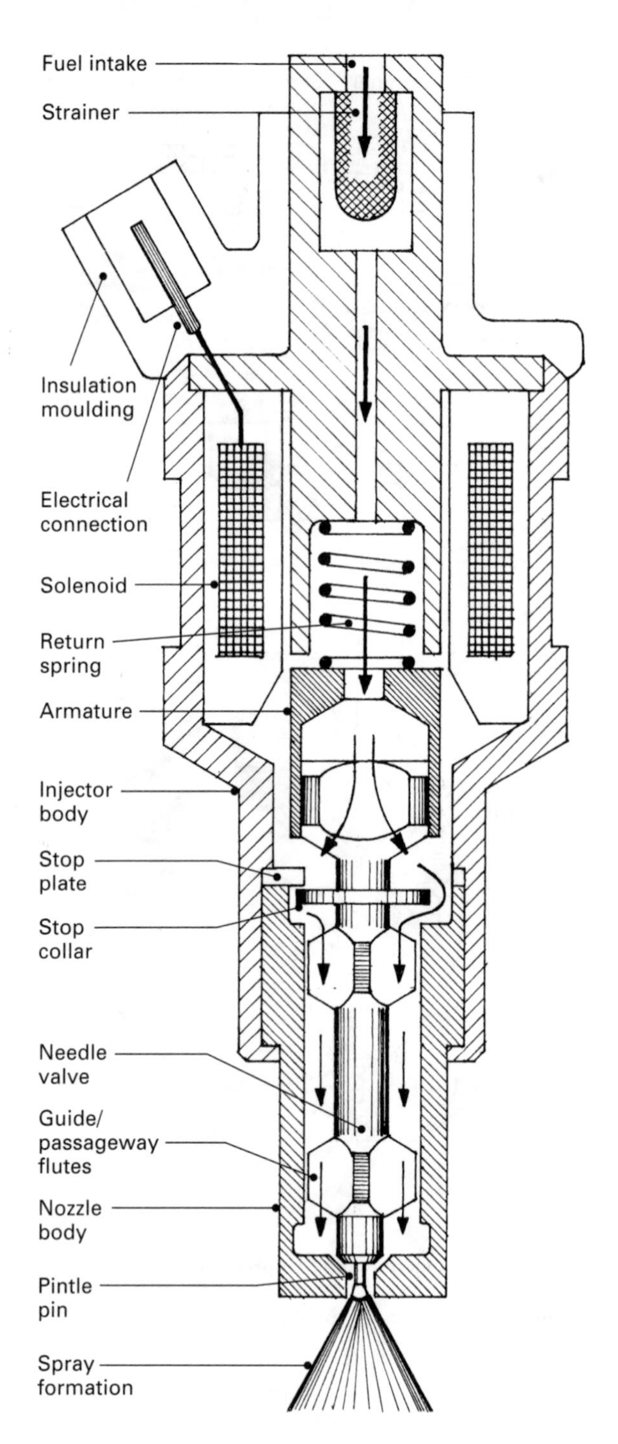

Fig. 22.6 Intermittent fuel solenoid-operated injector

at which point the needle stop-collar butts against the fixed stop-plate. Fuel now discharges from the now open annular orifice formed between the nozzle-body and the pintle-pin. The pintle-pin head at the top of the valve is specially ground to produce a conical atomised spray.

To vary the quantity of fuel discharge, the pull-in and release time of the valve can range from 1.0 to 2.0 milliseconds. An initial lift current of something like 5 amps opens the valve, followed by a much reduced holding current of about 1.5 amps which keeps the valve open for the remaining discharge time.

22.3.4 Fuel injection timing (fig. 22.7(a) and (b)) Two forms of fuel injection timing are in use with multipoint injection systems: these are simultaneous injection and sequential injection.

Simultaneous injection (fig. 22.7(a)) Fuel is injected into all the cylinder-head inlet-ports simultaneously twice each engine cycle, that is, a pulse signal of the same duration (width) is conveyed from the electronic control unit to each injector in the same crank-angle position and at the same time once every crankshaft rotation.

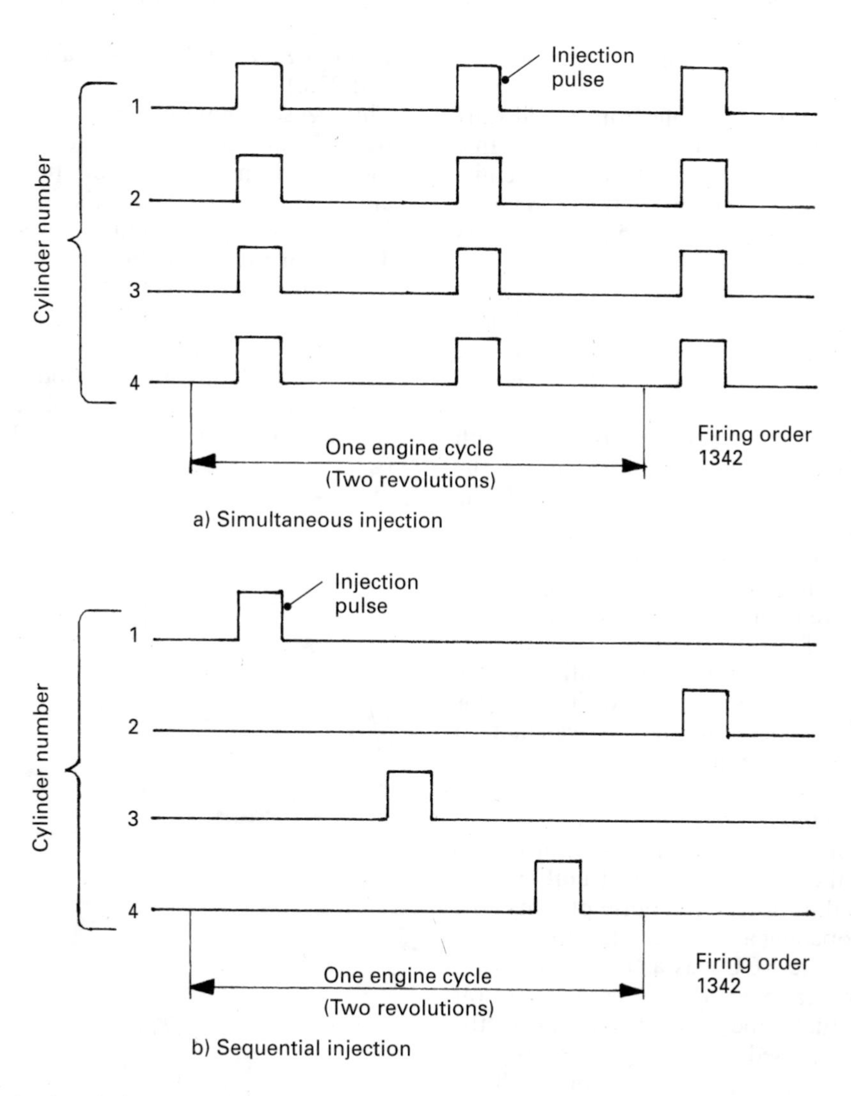

Fig. 22.7 Fuel injection timing

Sequential injection (fig. 22.7(b)) Fuel is injected into each cylinder-head inlet-port during each engine cycle in the firing-order sequence of the cylinders, that is, a pulse signal of the same duration (width) is transmitted from the electronic control unit to each injector in the same crank-angle position, in accordance to the firing-order of the cylinders once every second crankshaft revolution.

When the engine is being started and/or if the fail-safe system is operating, the system changes automatically to simultaneous fuel injection.

Fuel shut-off Fuel to each cylinder is cut off during deceleration or the operation of the engine at excessively high speeds.

22.3.5 Engine management system (fig. 22.4)

Engines today must be able to be efficiently operated under all driving conditions from a cold start to high performance demands, and at the same time the fuel consumption and the exhaust emissions must be kept to a minimum. If the engine is to operate effectively, various engine parameters must be monitored by appropriate sensors. These sensors rapidly transmit signals to a central source where they are collected, interpreted, and then computed. Outcome correction factors are then transmitted to both the fuel metering and the ignition timing systems. These objectives can best be achieved by integrating the fuel metering with the ignition timing, as this will optimise the engine performance. The coordination of the input sensor signals from the engine and the output correction commands which control both the quantity of fuel injection and the timing of the ignition spark are managed by the electronic control unit (ECU). Thus the function of the ECU is to process and compare the constantly varying input data with the programmed instructions stored in the memory and then to calculate a revised output pulsed signal.

22.3.6 Fuel injection control (fig. 22.4)

The quantity of fuel injected from the injector is controlled by the electronic control unit (ECU). Increasing the duration of the pulse prolongs the injector valve opening and so permits more fuel to be discharged. The ECU has a 'read only memory' (ROM) with a mapped-in program value for the amount of fuel to be injected. This means the duration of the pulsed output signal is program predetermined by the input signals from both the crank-angle and speed-sensor and the 'hot-wire' air-flow meter, that is, engine speed and air-flow dependent.

22.3.7 Ignition timing (fig. 22.4)

The ignition advance timing is controlled by the ECU in order to maintain optimum power and to provide the appropriate air/fuel ratio in response to the varying conditions the engine operates under. The ECU has a 'read only memory' (ROM) which stores mapped ignition timing data. The injection pulse duration and engine speed signals which are varying every moment are received by the ECU. This information is then assimilated by the computer which in turn transmits a responsive output current pulse to the ignition coil power transistor.

22.3.8 Crank angle and speed sensor (figs 22.4 and 22.8)

This sensor monitors engine speed and piston position, and sends signals to the ECU to control fuel injection and ignition timing. The crank-angle sensor is built into the ignition distributor. It has a rotor-plate and a wave-forming circuit. The rotor-plate has 360 slits for 1° signals and four slits for 180° signals. Light-emitting-diodes (LED) and photo-diodes are built in the wave-forming circuit. When the rotor-plate passes between the light-emitting diodes and the photo-diodes, the slits in the rotor-plate continually cut the light being transmitted to the photo-diode.

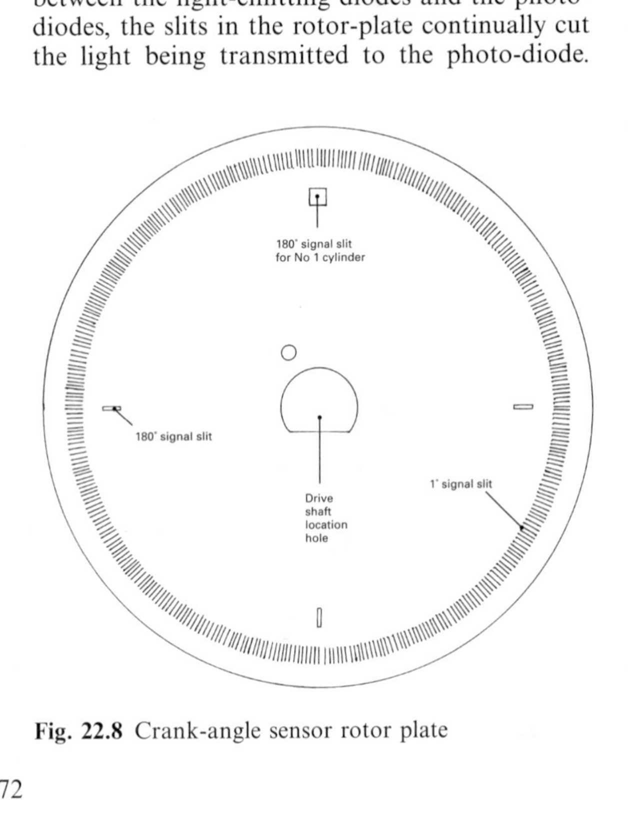

Fig. 22.8 Crank-angle sensor rotor plate

This generates rough-shaped pulses which are converted into on–off pulses by the wave-forming circuit. These pulses are then sent to the ECU.

22.3.9 Power transistor and ignition coil (fig. 22.4)
The power transistor is incorporated to amplify the ignition signal transmitted from the ECU before it passes to the primary winding of the ignition coil. A small current pulse signalled from the ECU is transmitted to the power transistor. This then switches on the transistor and energises the primary winding of the coil. When the pulsed signal ceases, the transistor circuit blocks. The following collapse of the primary circuit and electromagnetic field instantaneously induces a high-tension voltage into the secondary winding. This voltage is then transferred to the distributor cap and to the appropriate sparking-plug according to the engine's firing-order.

22.3.10 Detonation ignition retardation (figs 22.4 and 22.9)
The knock sensor allows the engine to operate with a higher compression ratio and optimum ignition advance and to burn a lean mixture under normal operating conditions and therefore permits the engine to produce more torque and power. However, in the event that the fuel does not match up to the compression ratio or the engines load becomes excessive, the detonation roughness of combustion will be detected by the knock sensor which instantly relays this information to the ECU. Temporary retardation of the ignition advance then takes place until normal driving conditions are restored.

Detonation sensor (fig. 22.9) The detonation sensor is bolted to the cylinder-block and senses the combustion knocking condition. The sensor is composed of an annular disc-shaped piezoelectric ceramic stacked between pairs of conducting and insulating washers and held under compression between the shoulder of a flanged sleeve and an absorption mass by a pre-loaded nut. The knock sensor functions on the 'piezoelectric effect' where mechanical forces (compression) acting on a crystal generate surface charges. A knocking vibration from the cylinder-block is applied as pressure to the piezoelectric ceramic element which then generates a proportionate alternating voltage at the surface of the crystal. This signal is then relayed to the ECU.

22.3.11 Vehicle speed sensor (fig. 22.4)
This sensor provides a vehicle speed signal to the ECU. The speed sensor consists of a reed-switch which is magnetically controlled and is installed in the speedometer unit. Its purpose is to transform the vehicle speed into a pulse signal.

22.3.12 Air flow meter (fig. 22.4)
The air flow meter measures the intake air flow rate. Measurements are made in such a manner that the ECU receives electrical output signals varied by the amount of heat emitting from the hot-wire placed in the stream of intake air. When this air flows into the intake manifold the heat generated from the hot-wire is taken away by the air stream. The amount of heat extracted from the hot-wire depends upon the quantity of air passing over it. Therefore, it is necessary to supply the hot-wire with more electric current in order to maintain the temperate of the hot-wire. Thus the change in current flow is monitored by the ECU as it is a direct measure of the amount of air passing through the intake manifold on its way to the engine cylinders.

22.3.13 Air temperature sensor (fig. 22.4)
The air temperature sensor monitors the incoming air temperature as this closely relates to the density of air passing into the engine. A change in electrical resistance, that is, signal voltage, is relayed to the ECU. This provides data which computes and adjusts the duration of the injector's opening pulse to match the density of the air passing into the cylinders. It therefore corrects and maintains the desired air/fuel mixture strength of the incoming charge.

22.3.14 Engine temperature sensor (fig. 22.4)
This sensor detects engine coolant temperature and transmits signals to the ECU. The temperature-sensing unit incorporates a thermistor whose electrical resistance decreases in response to the temperature rise. Thus the change in electrical current flowing through the thermistor is monitored by the ECU as it is a measure of the engine temperature.

22.3.15 Throttle sensor potentiometer (fig. 22.4)
The throttle sensor responds to the accelerator pedal movement. The sensor is a potentiometer (variable resistor) which transforms the throttle-valve opening or closing position into output voltages, and emits this voltage-signal to the ECU. A second inportant function of the throttle sensor is to detect how quickly the driver opens or closes

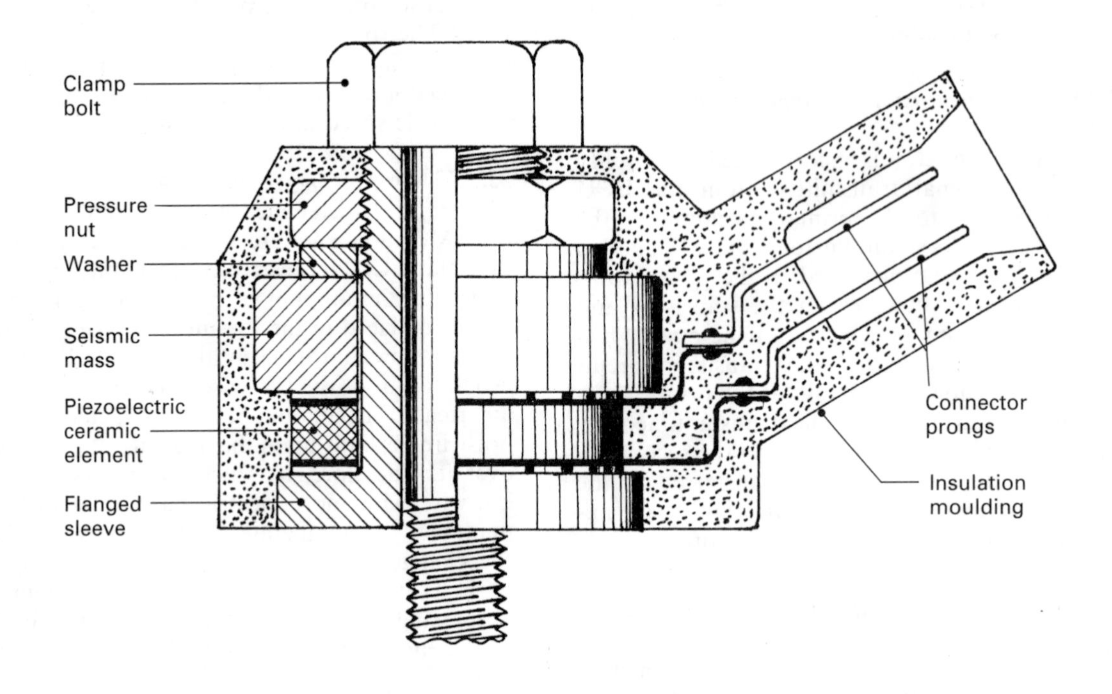

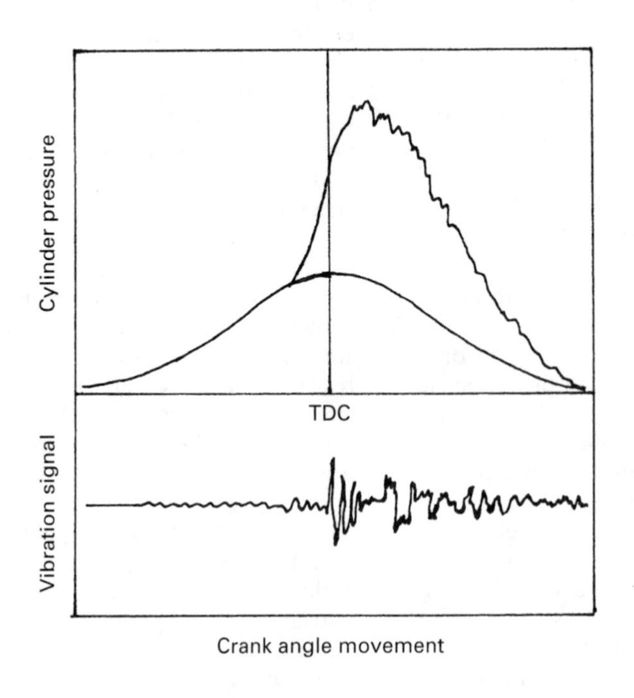

Fig. 22.9 Detonation sensor

the throttle-valve, as this is a measure of the expected acceleration or deceleration of the engine. The speed of throttle responses is thus also relayed in the form of a voltage-signal back to the ECU. The potentiometer brush position also controls fuel cut-off during engine overrun operating conditions.

22.3.16 Exhaust gas sensor (lambda sensor) (figs 22.4, 22.10 and 22.11)
This sensor is located in the exhaust pipe near the exhaust manifold where it monitors the oxygen content in exhaust gas (see fig. 22.10). The amount of oxygen in the exhaust system is monitored since it is a direct measure of the air/fuel mixture composition. The exhaust gas sensor element can be made of ceramic titania, the electric resistance of which drastically changes at the stoichiometric air/fuel ratio (see fig. 22.11). The titania sensor is wired in series with a comparater resistance built into the ECU, and a 1 volt supply is applied from the ECU between the free ends of both resistances (see fig. 22.10). Thus as the sensor's resistance changes due to the change in oxygen content in the exhaust gas, the voltage across the comparater resistance also changes. It is therefore automatically monitored by the ECU. Accordingly, the ECU adjusts the injector pulse duration to correct the air/fuel mixture strength as may be needed. To speed up the activation of the titania sensor when the engine is cold, a heater element is incorporated within the sensor unit.

22.3.17 Fast idle control valve (figs 22.4 and 22.12)
This device provides an air by-pass to supply excess air when the engine is cold. It thereby raises

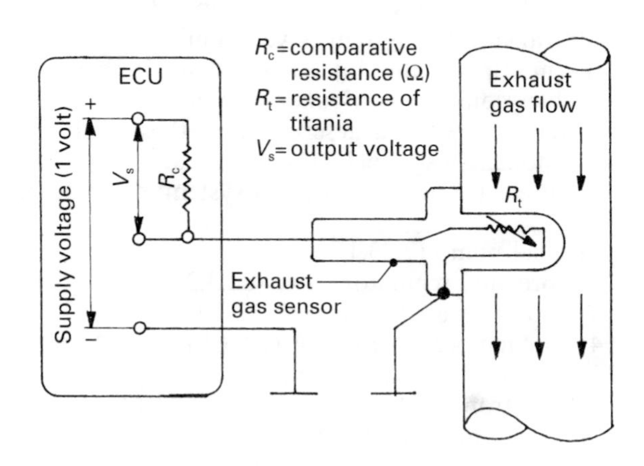

Fig. 22.10 Principle of the exhaust gas sensor

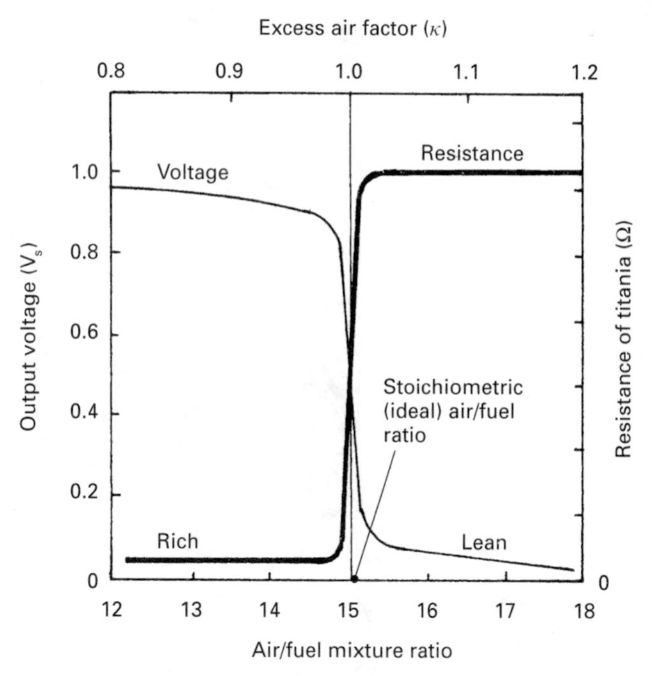

Fig. 22.11 Exhaust gas sensor resistance and voltage characteristics

the engine's speed from a slow hot idle to a fast cold idle during warm-up. A bi-metal heater and rotary shuttle-valve are built into the valve body. When the bi-metal temperature is low, the by-pass rotary port is closed. As the engine starts an electric current passes through the heater winding, and the bi-metal curved strip begins to straighten, causing the shutter to rotate and open the by-pass port. When the engine has reached its normal working temperature the bi-metal winding current ceases. This causes the bi-metal strip to cool and close the by-pass passageway. This brings the engine's speed back to the normal hot-idle speed setting.

22.3.18 Auxiliary air valve (fig. 22.4)
The auxiliary air control valve by-passes the throttle-valve to provide an additional air supply to the engine under idle speed conditions whenever extra load is applied to the engine. This slightly raises the idle speed, thus preventing the engine from stalling. The equipment that consumes power, particularly at idle speed, is as follows: the hydraulic pressure pump for the assisted power steering, torque converter drag when the automatic transmission is engaged, and the refrigerant pump for air conditioning when switched on. When the engine is at idle speed with no

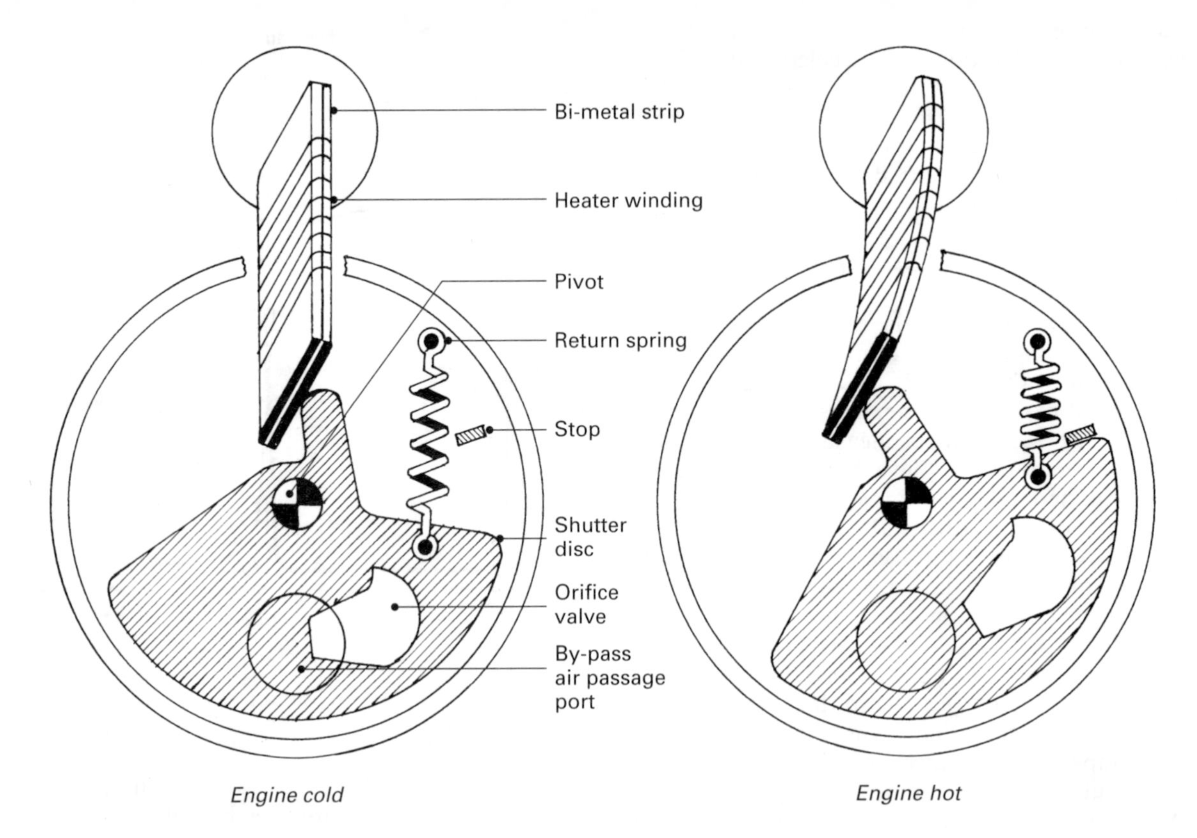

Fig. 22.12 Fast idle control valve

ancillary equipment operating and for speeds above idling, the auxiliary air control valve remains closed. However if an additional load is imposed on the engine at idle speed this is immediately signalled to the ECU which passes on an on–off pulse to the auxiliary air control valve solenoid. This energises the winding and pulls the armature/valve open. Additional air/fuel mixture now by-passes the throttle-valve to supplement the normal throttle-valve delivery. The duration of the 'on' voltage pulse determines the amount of mixture passing through the auxiliary air control valve passages. The correct hot idle speed is set with the adjustment screw when the engine has warmed up with all external engine load removed.

22.4 Single-point electronic petrol injection system (fig. 22.13)

This single-point injection system injects fuel intermittently through a single solenoid operated injector positioned centrally on the intake side of the throttle-valve. Air/fuel mixture is then distributed to each cylinder through the intake manifold similarly to a carburettor engine. Control is achieved by an ECU which receives input signals from various engine monitoring sensors. This data is then processed and compared with a memory program which then calculates corrections to the air/fuel mixture ratio. This revised information is then converted into output pulses to the central injector solenoid. The operation of this central injector system is very similar to that of the multipoint fuel injection system. It therefore enables common components to be used on both systems. A detailed description of these identical components listed below is given under the multipoint electronic fuel injection system:

1) fuel pump (22.3.1)
2) pressure regulator valve (22.3.2)
3) hot-wire air flow meter (22.3.12)
4) air temperature sensor (22.3.13)
5) engine temperature sensor (22.3.14)
6) electronic control unit ECU (22.3.13)
7) crank-angle and speed sensor (22.3.8)
8) vehicle speed sensor (22.3.11)
9) throttle-valve potentiometer (22.3.15)

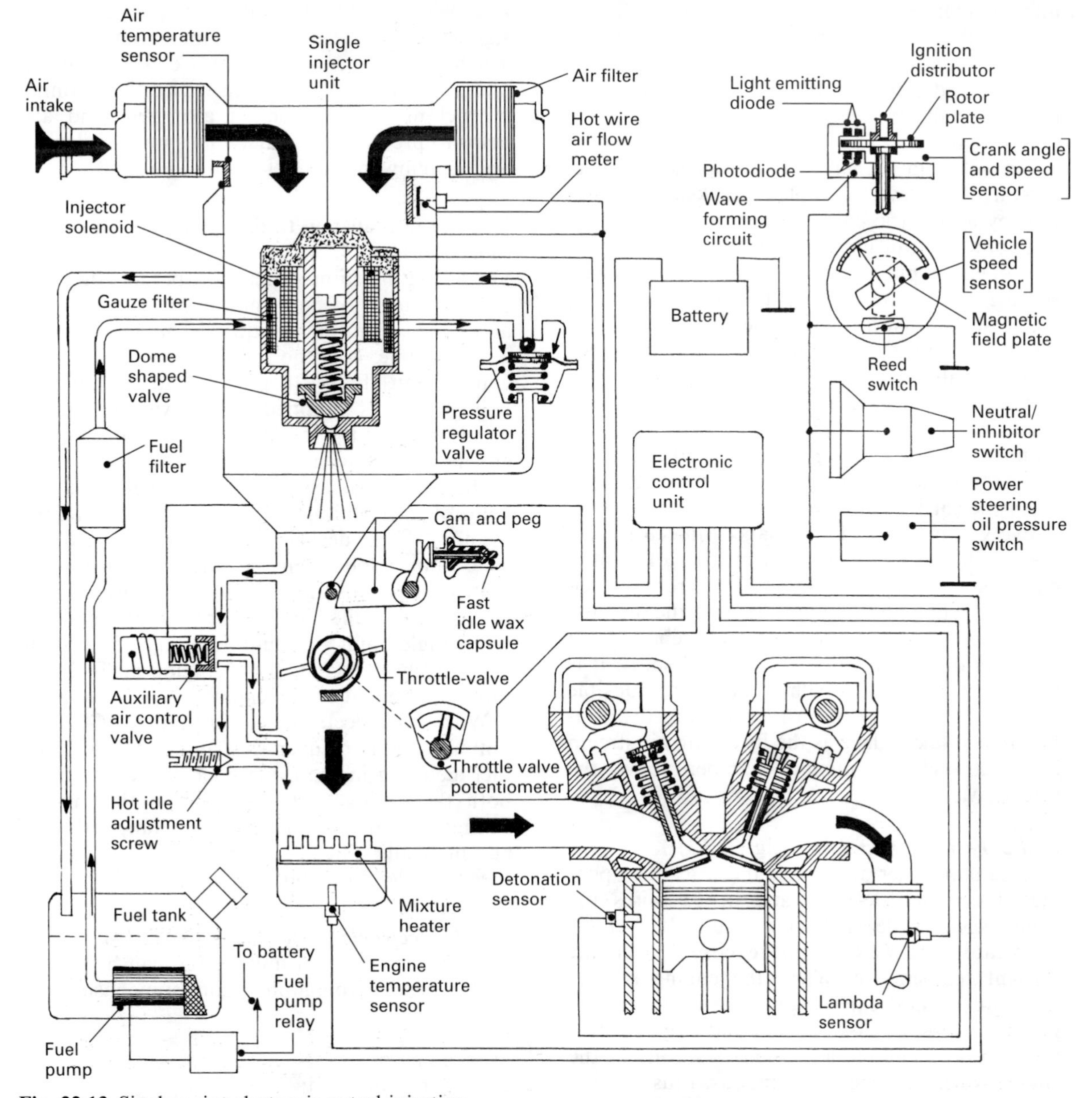

Fig. 22.13 Single-point electronic petrol injection

10) detonation knock sensor (22.3.10)
11) exhaust gas oxygen sensor (22.3.16)
12) auxiliary air control valve (22.3.18).

Other components which make up this type of fuel injection system are now described under the following headings:

1) solenoid-controlled central injector unit
2) fast idle cam device
3) mixture pre-heater.

22.4.1 Solenoid-controlled central injector unit (fig. 22.13)

This injector unit is located in the centre of the air intake passage. It contains a solenoid-operated low-inertia dome-shaped injector valve which is held on its seat when inoperative by a pre-tensioned return spring.

Fuel is supplied to the valve body by the centrifugal roller-type pump. Fuel enters and fills the valve-injector chamber via the ring-strip-gauze filter. It then passes through to the pressure regulator valve which when open returns the excess petrol to the fuel tank. Thus the continuous circulation and expulsion of petrol via the pressure regulator valve prevents the formation of vapour-lock which otherwise occurs.

The solenoid is energised and opens the injector-valve once every complete crankshaft revolution by an electric current pulse supplied by the electronic control unit (ECU). The frequency of this opening pulse which relates to the engine speed is fed to the ECU by the crank-angle and speed sensor. Nevertheless the amount of fuel discharge from the injector is dependent on the time the valve remains open. The duration of the valve opening period is determined by the data fed to the ECU from the various sensors which monitor the engine temperature, speed, load and acceleration/deceleration operating conditions when the car is being driven.

Every time the dome-shaped injector-valve opens, fuel escapes through a number of slanted discharge holes (not shown) where it then impinges with a swirling motion onto the inverted cup-shaped deflector walls. It then rebounds in the form of a highly atomised conical spray. It is this spray formation which is mainly responsible for the high quality of air/fuel mixture distribution which takes place as the incoming air stream flows through the mixing chamber above the throttle-valve.

22.4.2 Fast idle cam device (fig. 22.13)

This cam and thermo-capsule increases the opening of the throttle-valve and hence raises the engine speed while the engine is cold. It operates by means of a wax capsule exposed to engine coolant temperature. When the engine is cold the stainless-steel capsule pin is in its innermost position. This permits the cam and peg to rotate the throttle-spindle to a fast-idle position. As the engine warms up the wax expands, thus causing the rubber boot sealing the wax to distort and squeeze the pin outwards against the lever attached to the cam spindle. This rotates the cam anticlockwise, thereby permitting the peg to move to a lower point on the profile of the cam and so enables the throttle-valve to move to its slow hot-idle position.

22.4.3 Mixture pre-heater (fig. 22.13)

The electric mixture heater is positioned between the lower throttle mixing chamber and the intake manifold and is switched on by the ECU via a relay when the engine coolant temperature sensor signals a predetermined minimum temperature. The heater pre-heats the mixture entering the inlet manifold. It thus makes it easier for the incoming air/fuel mixture to remain atomised, and also helps to prevent any coarse fuel spray condensing and clinging to the inlet manifold walls.

22.5 An introduction to the electronic control unit (fig. 22.14)

For a fuel injection system to operate efficiently, a wide range of information must be known about the operating conditions of the engine at any one time. It is the function of the electronic control unit (ECU) to receive information from voltage signals supplied by the various sensors monitoring all the key operating variables of the engine. This information is then processed and evaluated and decisions are then made with the help of instructions stored in a programmed memory. These decisions are then converted into an output response in the form of a pulse current, which is instantaneously transmitted to the solenoid-controlled injector or injectors employed with single-point or multipoint injector systems respectively to match the changing air/fuel mixture requirement of the engine.

With some electronic control units (ECUs) the optimal ignition advance is integrated with the fuel metering in the same microprocessor; hence both the timing of the spark and the mixture strength at any moment is simultaneously and automatically adjusted to match all the possible speed and load combinations.

22.5.1 Electronic control unit (fig. 22.14)

An electronic control unit (ECU) can be considered to be a microcomputer consisting basically of three items

1) the microprocessor;
2) the memory and;
3) the input/output circuit.

The hardware of these items is made up from printed circuit boards with integrated circuits (ICs). Each integrated circuit has integrated semiconductor components on its silicon-chip. Individual transistors, diodes, capacitors, resistors, and coils also form part of the printed circuit board.

22.5.2 Power supply (fig. 22.14)

The power circuit reduces the battery/generator 12–14 volt output to a stable 5 volt supply

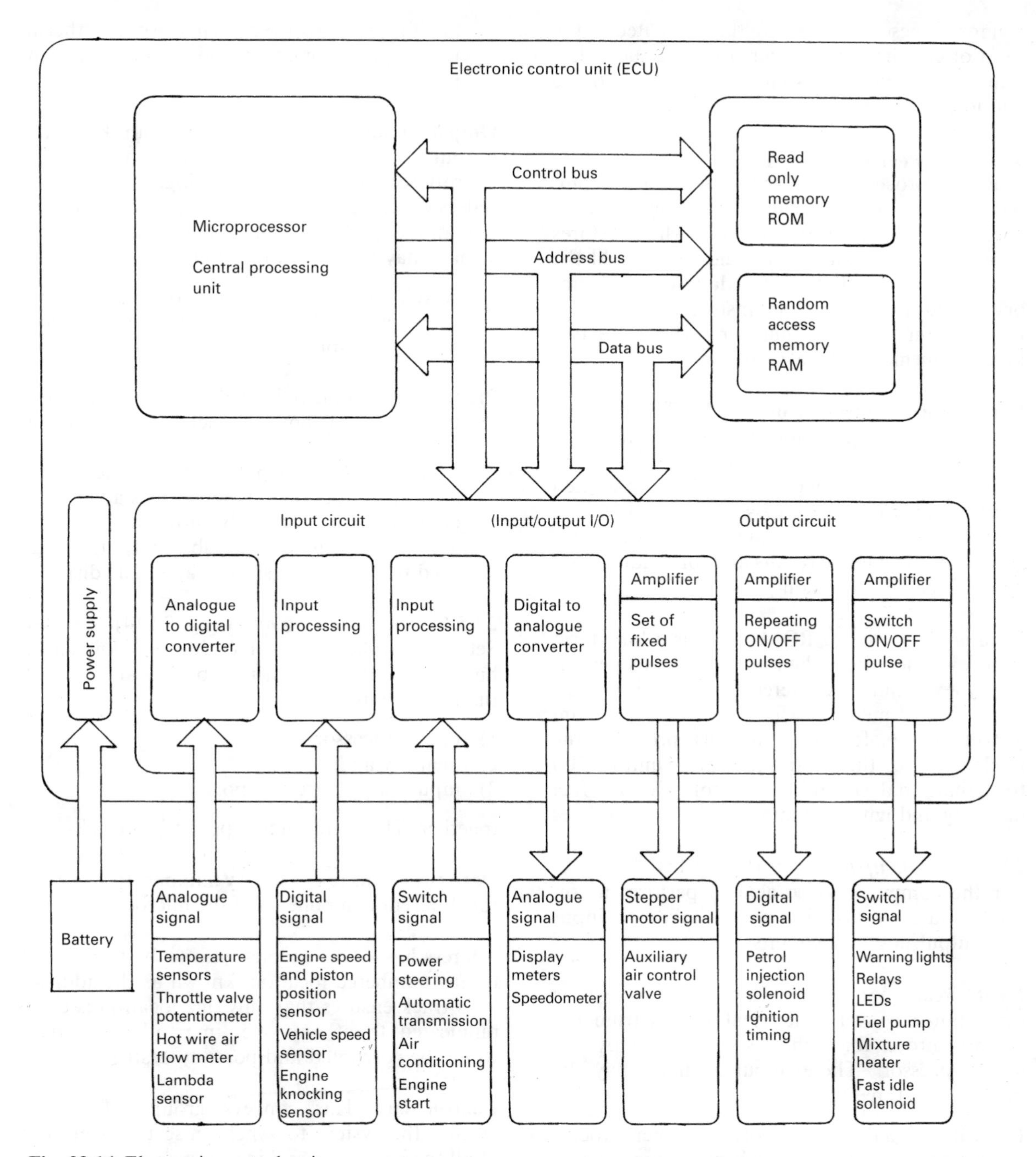

Fig. 22.14 Electronic control unit

which is the usual voltage input for most microprocessors.

22.5.3 Microprocessor (fig. 22.14)

Microprocessors are digital integrated circuits which can perform numerical calculations and make simple decisions using data in the form of binary numbers. The microcomputer receives information from the engine in the form of electrical signals from sensors and makes decisions judged on a program which is a sequence of predefined instructions stored permanently in the memory. These decisions are then executed in the form of command pulse signals to the fuel injection system and with some systems also to the ignition distributor.

22.5.4 Memory (fig. 22.14)

The microprocessor is the brain of the microcomputer, but in order to function it must be connected to a memory device which stores numerical data input results and which hold the instructions that tell it what to do next. The memory is used to store program instructions and data in the form of binary numbers. There are two kinds of memory in use as follows.

Random access memory or RAM This memory has a set of storage locations any of which can be accessed directly without having to work through from the first. Some memories can be both written to and read from, and access times are about the same for all locations. It is useful for temporary storage of incomplete results and for holding data before it passes to the output circuit.

Read only memory or ROM This is a long-term memory which holds data or instructions permanently and cannot be altered by the microprocessor. It is this memory which stores the program instructions (software) and thus provides the intelligence for the electronic control unit (ECU) to actuate and control the petrol injection fuel metering and ignition timing.

22.5.5 Input/output circuits (fig. 22.14)

For the computer to be able to perform useful control tasks it also has to be linked to input and output interface circuits.

Input circuit (fig. 22.14) This circuit, known as input ports, converts electrical signals from the sensors into a form that is recognised by the microprocessor. These input signals may be divided into three types:

1) switch signals which inform the microprocessor of some activity such as power-assisted steering under load, automatic transmission engagement, air-conditioning switched on, etc.
2) pulse signals which repeatedly switch on and off, the frequency or timing of the pulse is used to provide information such as engine and vehicle speed, ignition timing, etc.
3) smooth varying voltage signals which provide analogue signals from such sensors as thermistors, potentiometers, and hot-wire air flow meters.

Output circuit (fig. 22.14) This circuit, known as output ports, converts the decision made by the microprocessor binary digit logic into signal pulses which can activate the fuel injection metering and control ignition timing. These output signals may be divided into four types:

1) switch on/off pulse for operating such devices as warning-lights, relays, switches, mixture-heater, fuel-pump motor, and fast-idle solenoid;
2) repeated on/off pulse signals of varying duration to operate both the fuel-injector and coil-ignition;
3) a number of fixed pulses used to activate a stepper-motor which is sometimes adapted to control auxiliary air valve flow;
4) a calibrated smooth variable output voltage used to drive items such as analogue displays.

22.5.6 Interconnecting buses (fig. 22.14)

Sets of parallel tracks (conductors) known as buses are mounted on printed boards and connect the three items of hardware:

1) microprocessor
2) memory and
3) input/output circuits (ports)

together. There are three types of bus tracks:

Data bus This relays binary numbers to and from the different components of the microcomputer.

Address bus This enables data to be transferred to any numbered location, known as the address. This track ensures that only one location is communicated to at any one time, otherwise data would be corrupted and possibly destroyed.

Control bus This conveys groups of signals around the system to synchronise the operation of all the hardware, and in doing so permits (controls) only one memory location to discharge its content onto a database at any one time.

22.5.7 Input processing circuit (fig. 22.14)

Input signals are switch and pulse signals and therefore are already in digital form. However they have different signal voltages to the microcomputer. It is the purpose of the input processing circuit to reduce the input pulse signal voltage and filter out high-frequency noise.

22.5.8 Input analogue to digital converter (ADC) (fig. 22.14)
This is a device which converts the smooth varying voltage signal which is incompatible with digital systems into a binary number proportional to its amplitude, that is, to a digital input signal.

22.5.9 Output digital to analogue converter (fig. 22.14)
The purpose of this device is to convert digital data from the microcomputer into an equivalent analogue (smooth) voltage or current. It can be used to drive items such as analogue display meters, speedometers, etc.

22.5.10 Output amplifier (fig. 22.14)
The signals coming out from the microcomputer which operates at a voltage of 5 volts is generally too weak for direct activation of components. The output voltage may only be one or two volts or even less, and the output current may be of the order of a few milliamps. In contrast, the requirement for operating electrical components used on vehicles which function at battery/generator voltage of between 12 and 14 volts need energising currents of one to five amps or even higher. This problem is normally solved by integrating the battery supply with a power transistor which is activated by the small output current signal. This circuitry is generally referred to as the output amplifier.

23 Diesel-engine fuel injection systems

23.1 Layout of a diesel-engine fuel system (figs 23.1 and 23.2)
To develop an overall appreciation of how the diesel engine's fuel systems operate, it is best to describe individually the functioning of the various components of both the in-line and the distributor type of injection-pump layouts, as follows.

Fuel tank This stores the diesel fuel in readiness to meet the demands of the engine. It is usually positioned along the side of the vehicle's chassis, beyond the cab.

Primary filter This is often an optional component and may take the form of a coarse wire gauze which prevents large unwanted particles passing to the lift pump on its suction side. In some designs it may also use their differing densities to separate out water and heavy solid particles flowing with the fuel.

Lift pump This transfers fuel from the tank to the fuel-inlet connection on the injection pump, at a rate which varies with engine requirements. It is driven from an eccentric cam formed on either the injection pump or the engine camshaft.

Secondary filter This is connected in series with the lift pump on its pressure side, its function being to separate out very small particles of sand, dust, and grit and any other dirt which

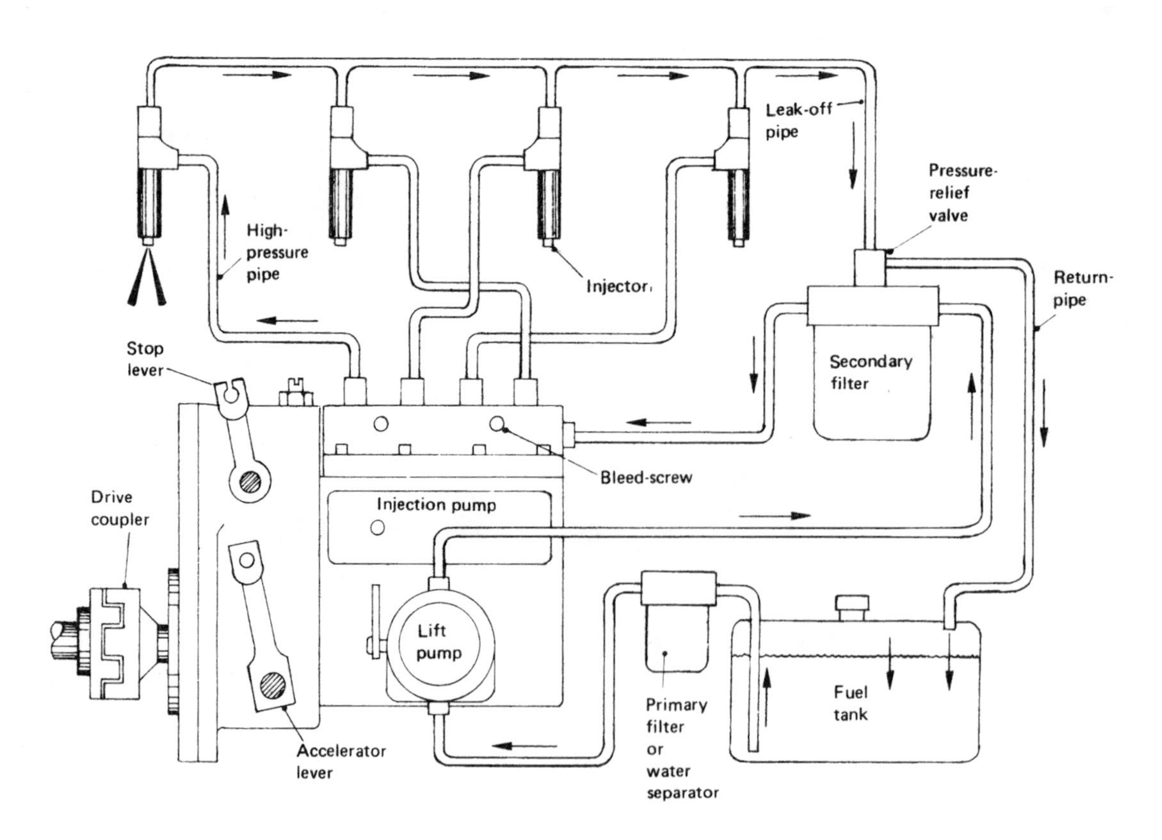

Fig. 23.1 In-line fuel-injection-pump system

may be mixed with the fuel, without restricting the flow of the fuel.

Pressure-relief valve or permanent-bleed orifice One or other of these devices is incorporated into the secondary-filter head.

The function of the pressure-relief valve is to regulate the maximum injection-pump-gallery fuel pressure and to provide a degree of fuel circulation. It also relieves the operating loads on certain types of lift pump.

The permanent-bleed orifice provides continuous fuel circulation and therefore both relieves the line pressure and greatly helps to purge the system of any air trapped, and so prevents air locks becoming established over a period of time.

Injection pump This meters and discharges under high generated pressure to each injector the exact amount of fuel in the correct sequence and at the right time to meet the demands set by the engine under all operating conditions. The unit is driven from the engine's timing gears and its output is controlled by the driver's accelerator pedal.

There are two basic kinds of injection pump: the in-line multi-pumping-element type, where each element is connected by a high-pressure pipe to its respective injector (fig. 23.1), and the distributor single-element type which performs the task of both pumping and distributing the fuel to each individual injector pipe (fig. 23.2).

High-pressure pipes These thick-walled steel pipes transmit the metered, timed, and pressurised fuel from either each injection-pump element or from the distributor head to the appropriate injectors.

Injector units The function of these injectors is to inject and atomise the fuel into the combustion chamber at the correct instant in the engine's operating cycle. They are mounted on the cylinder head and are triggered off by a predetermined pressure rise in the high-pressure pipelines.

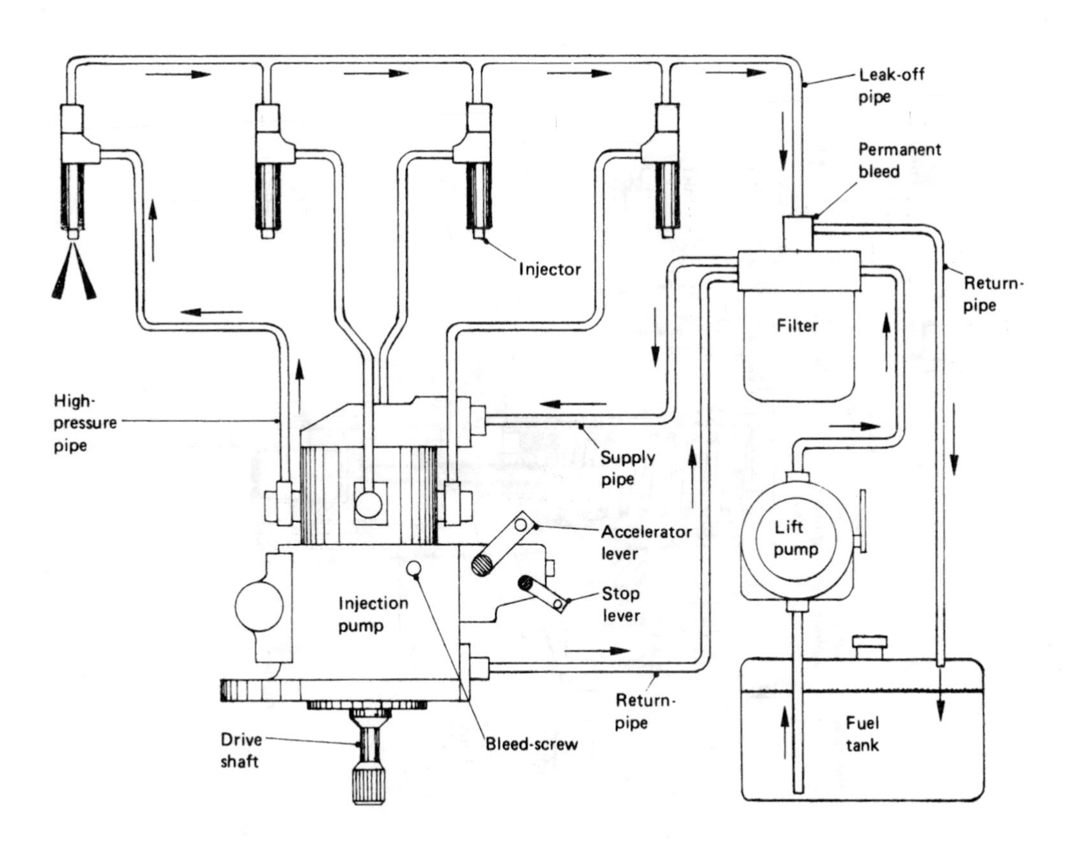

Fig. 23.2 Distributor-type-fuel-injection-pump system

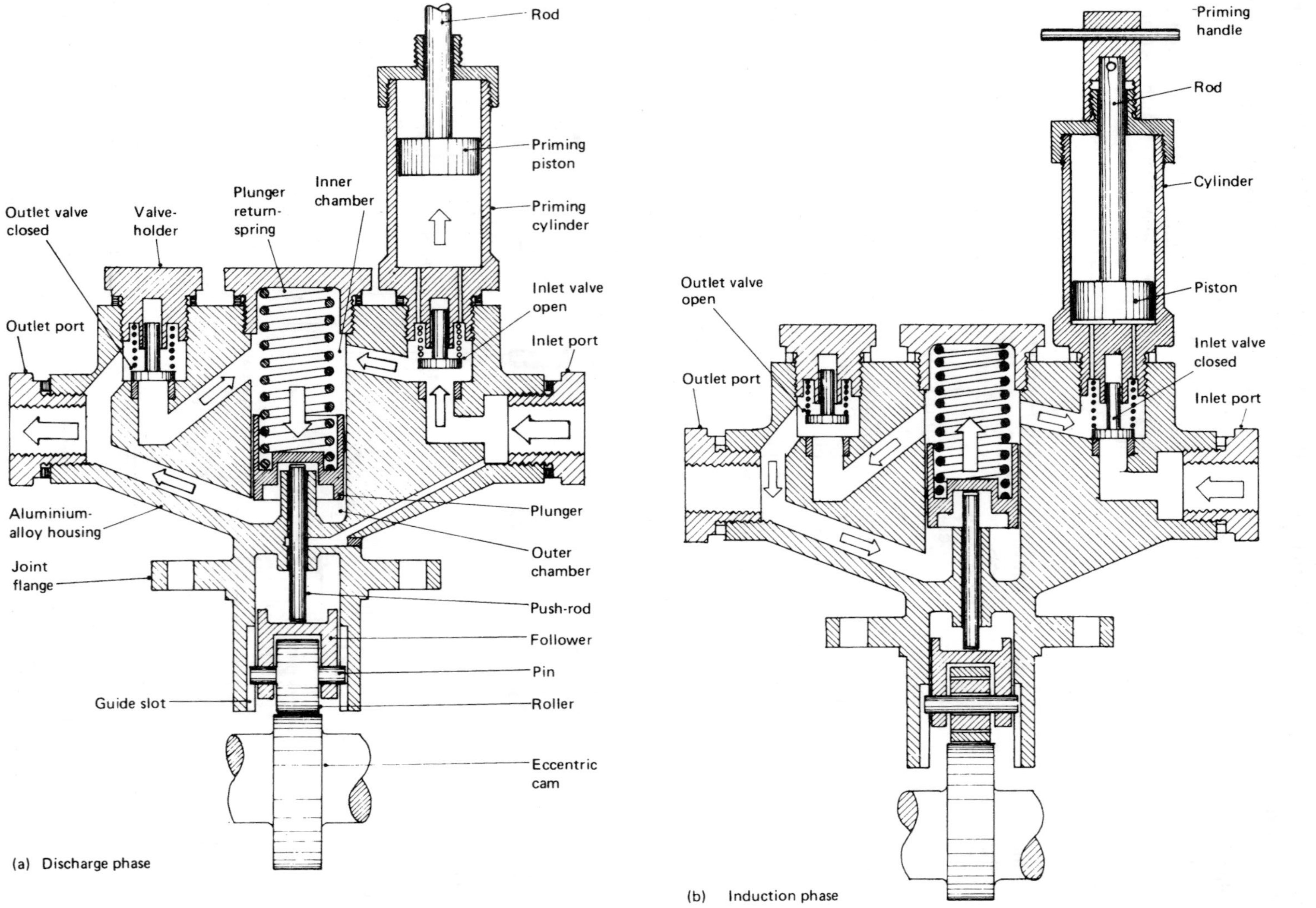

Fig. 23.3 Plunger-type diesel-fuel lift pump (CAV)

Leak-off pipes These allow carefully controlled fuel seepage between the injector nozzle and needle to provide lubrication for the sliding pair, the escaping fuel being passed back to the fuel tank via the pressure-relief valve or permanent-bleed device and the return-pipe.

Distributor-pump return pipe The internal lubrication of the single-element distributor pump relies entirely on diesel-fuel circulation, so this return-pipe is provided to complete the flow circuit (fig. 23.2).

23.2 Plunger-type fuel lift pump (fig. 23.3)

Inner-chamber filling and outer-chamber discharge phase (fig. 23.3(a)) When the cam lobe is in the minimum-lift position, the return-spring sweeps the plunger down, increasing the space in the inner cylinder chamber. The depression created by the plunger displacement closes the outlet valve and opens the inlet valve so that fuel from the tank will be forced through the inlet port and valve to fill up the inner chamber.

On the same inward stroke, but on the opposite side of the plunger, the fuel in the outer chamber will at the same time be pushed through the outlet port to supply fuel to the injection pump.

Fuel transfer from inner to outer chamber (fig. 23.3(b)) Continued rotation of the cam will start to lift the follower and plunger. The pressure build-up in the inner chamber will then close the inlet valve and open the outlet valve. Fuel will now be displaced from the inner chamber through the outlet valve to charge the outer chamber in readiness for the following pumping stroke.

Spring-controlled discharge phase (fig. 23.3(c)) Further rotation of the cam will move the cam lobe back to the minimum-lift position. The plunger return-spring alone will then return the plunger towards its outermost position and in doing so will displace fuel from the outer chamber through the outlet port to supply fuel to the injection pump.

Thus it can be seen that it is the stiffness of the plunger return-spring which determines the lift-pump line pressure. If the fuel in the system between the injection pump and the lift pump has not been consumed, the plunger spring will remain partially compressed until the demands of the engine will permit the spring to expand and displace fuel through the outlet port.

A plunger-type lift-pump circuit (fig. 23.4) generally includes a pressure-relief valve, built into the filter head, which controls fuel pressure at between 0.55 and 0.85 bar.

Prevention of fuel leakage past the push-rod Leakage of fuel along the push-rod and into the cambox housing is avoided by providing a return-path drilling between the inlet port and a small groove formed in the push-rod bore. Excess fuel will thus be drawn to the inlet port.

Hand-priming device (fig. 23.3) A hand-primer is incorporated into the feed system so that fuel can be transferred rapidly by hand-pumping from the fuel tank to the injection pump if the engine has been stationary for a considerable period, or if filters or part of the feed-system pipes have been dismantled and replaced.

The hand-primer consists of a single-acting plunger pump which draws fuel from the tank when the hand-plunger is pulled outwards on its intake stroke and delivers the fuel through the outlet port to the injection pump on the downward discharge stroke (fig. 23.3(a)). After priming, the plunger is held at the end of the delivery stroke by twisting the handle – this screws it to the cylinder so that it does not interfere with the cam-driven lift-pump action.

23.2.1 Combined piston-and-diaphragm-type lift pump (fig. 23.4)

For heavy-duty operating conditions, these combined piston-and-diaphragm-type pumps are preferred, the piston being provided to give additional support to the diaphragm. In the model shown in fig. 23.4, the inlet and outlet disc-valves are housed in the inlet and outlet ports, and a roller follower is placed between the cam and the push-rod to actuate the piston and the diaphragm, as this is more durable.

This type of pump operates in exactly the same way as the diaphragm-type petrol lift pump described in section 21.2.2 and fig. 21.5.

23.3 In-line injection pump (CAV 'Minimec')

23.3.1 Description and construction of pumping unit (fig. 23.4)

The in-line injector pump is a multi-pumping-element unit, there being one element for each engine cylinder. Each element consists of a plunger, a barrel, and a delivery valve which are arranged in a single row in a casing made in two

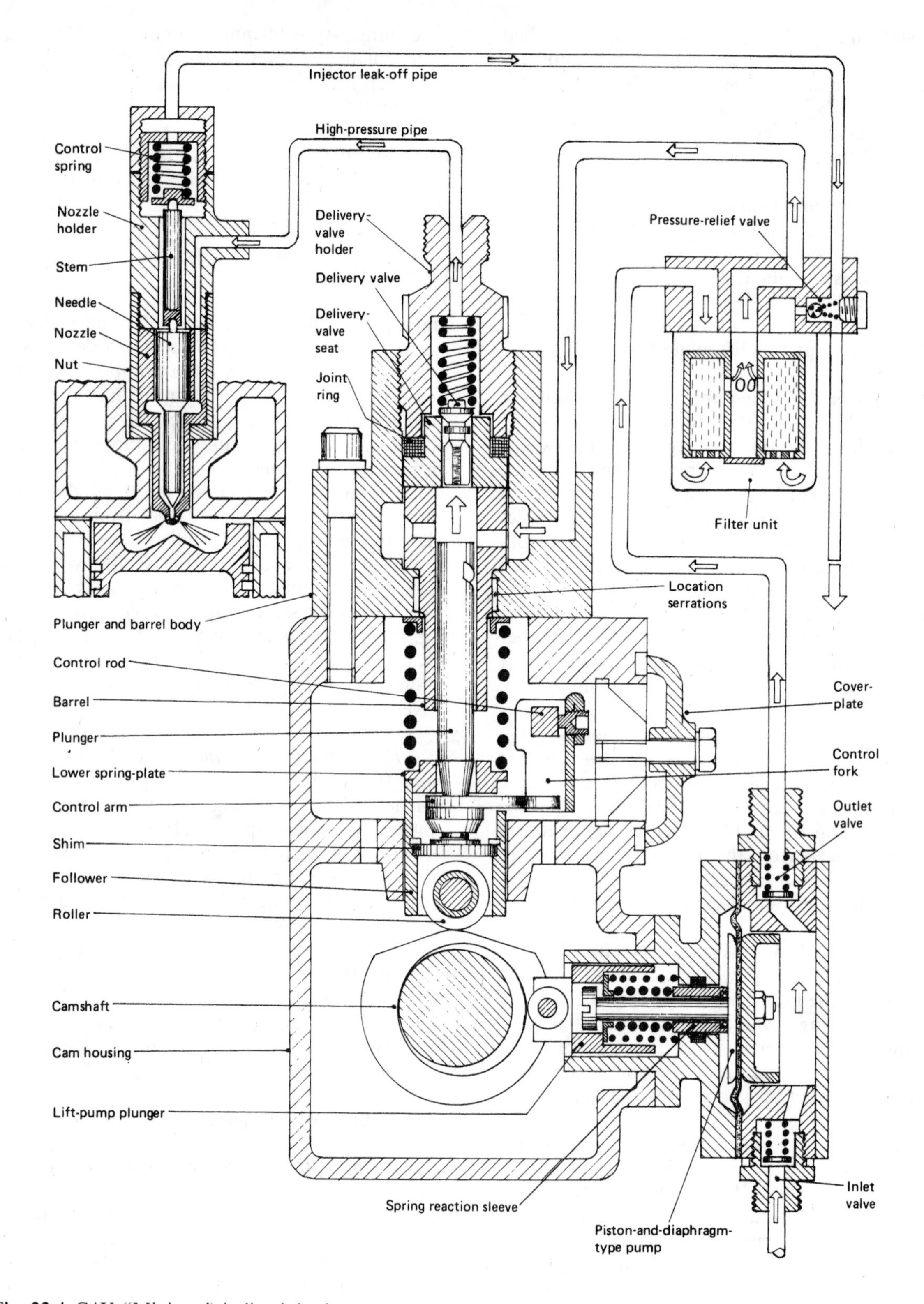

Fig. 23.4 CAV ('Minimec') in-line injection pump

parts – (i) a steel T-shaped body containing the fuel passages and pumping elements and (ii) an aluminium-alloy cambox and governor housing which contains the camshaft and bearings, cam followers, control rod, governor assembly, and operating linkage.

The function of the camshaft is to create the plunger movement for each pumping element so that it occurs in the firing order and at the correct instant in the engine's cycle of operation. The followers are provided to relay the cam-lobe lift to the plungers, and the governor is there to sense the changing engine speed and, in conjunction with the accelerator-lever position, automatically regulate the amount of fuel permitted to be injected into the combustion chamber.

23.3.2 Plunger-and-barrel pumping action

Filling of the barrel (fig. 23.5(a)) The plunger moves down the barrel and towards the end of its stroke uncovers both the inlet and spill ports, the inlet port being opened slightly before the spill port. Fuel at lift-pump pressure will be forced through both ports to fill up the space above the plunger (fig. 23.5).

Beginning of injection (fig. 23.5(b)) As the plunger begins its upstroke, it will first cut off the spill port and, as it continues to move up, it will push back a small quantity of fuel through the inlet port until the top edge of the plunger closes or cuts this port off from the space above the plunger (fig. 23.5(b)). The space above the plunger is completely filled with fuel, so the rising plunger will pressurise the trapped fuel and force open the delivery valve situated above the plunger. The continuing upward movement of the plunger will then transmit rising pressure through the existing fuel in the pipeline to the injector-nozzle needle (see fig. 23.4). Eventually this pressure build-up will lift the needle off its seat, and fuel will then be displaced along the pipeline to be sprayed out into the combustion chamber.

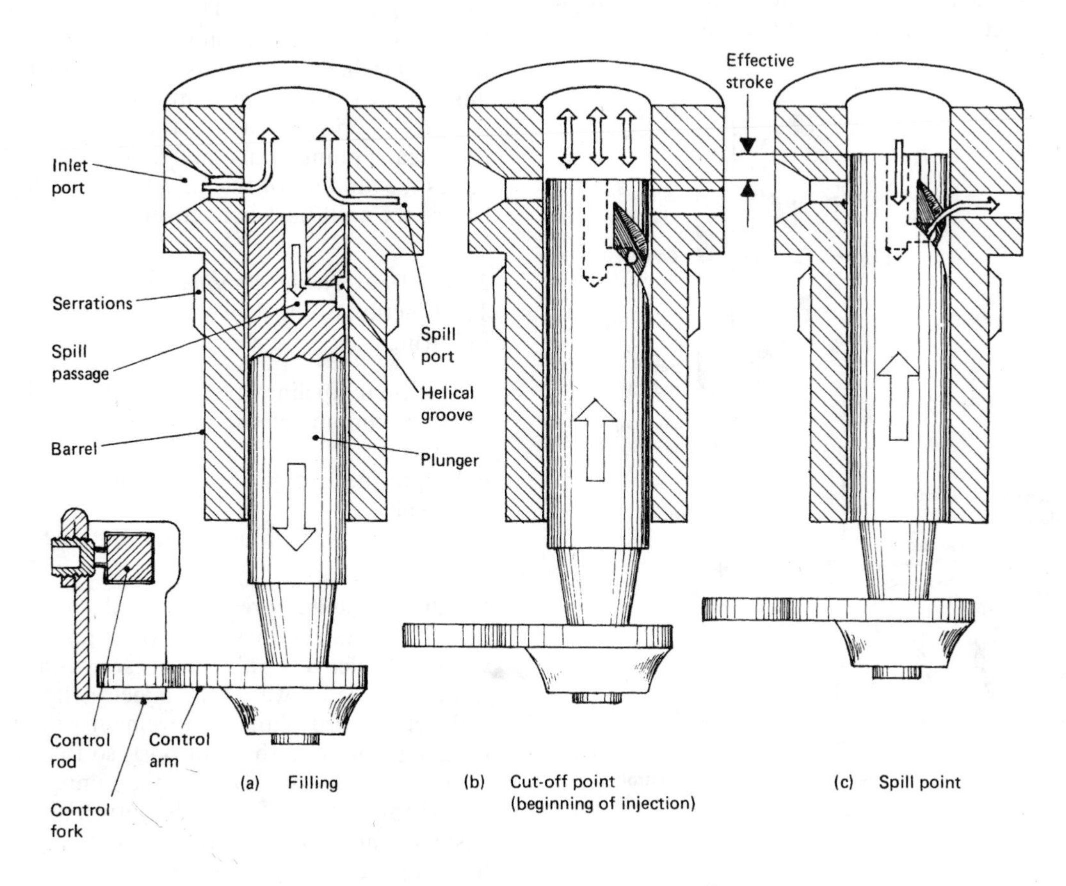

Fig. 23.5 Cycle of operation

End of injection (fig. 23.5(c)) Further upward movement will continue to force fuel past the delivery valve until the upper edge of the plunger helical groove is exposed to the barrel spill port (fig. 23.5(c)). The fuel pressure will then suddenly collapse as fuel escapes down the axial spill passage, across to the helical groove, and out through the spill port. The spring-loaded delivery valve will then shut, so that fuel delivery and injection will cease.

23.3.3 Output control

Description and construction of mechanism (fig. 23.6) The plunger stroke is controlled by the cam lobe and is always constant, but the part of it which actually pumps is variable. The point on the upward travel of the plunger at which the spill occurs can be altered by a device which allows the plunger to be partially rotated relative to the barrel. The mechanism used consists of an arm attached to the bottom of the plunger and engaging a box-shaped control-fork mounted on a square-sectioned control-rod (fig. 23.6).

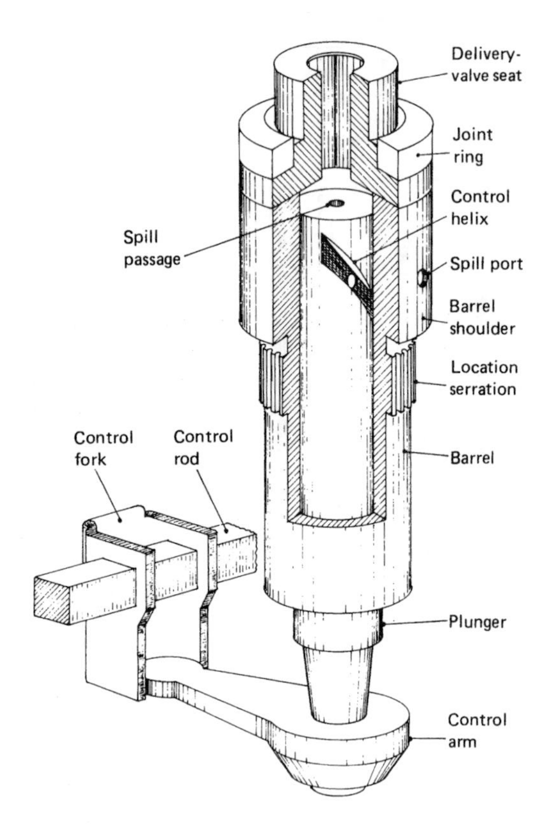

Fig. 23.6 Pumping elements

Control-arm and helix action (figs 23.7(b) and (c)) The partial rotation of the plunger in the barrel enables the position of the plunger helical spill groove to be varied relative to the fixed barrel spill port, thereby increasing or decreasing the effective pumping stroke of the plunger. For less delivery – say half-load (fig. 23.7(b)) – the helical groove aligns with the spill port earlier in the plunger upstroke, thus an early spill will reduce the amount of fuel injected. For a late spill, the effective plunger stroke will be lengthened to increase the delivered output – see fig. 23.7(c), which shows the full-load position.

Stop or no-load delivery position (fig. 23.7(a)) For no delivery – that is, the shut-off position – the plunger is rotated until the helical groove uncovers the spill port in the barrel for the entire stroke (fig. 23.7(a)) so that at no time can the fuel be trapped and compressed within the barrel.

Control-rod operation The control-rod is located to one side of the pumping elements and is positioned parallel to the camshaft. Its function is to link up all the plunger control-arms so that any to-and-fro movement of this control-rod will rotate each plunger an equal amout. It thus determines the amount of fuel injected by each plunger before the spill port is uncovered.

23.3.4 Delivery-valve action (fig. 23.8)

The function of the delivery valve is to provide the following:

a) residual pipeline pressure, so that each successive pumping stroke immediately actuates the injector;
b) rapid fuel cut-off, to eliminate injector-nozzle dribble;
c) positive continuous air purging or bleeding.

Residual pipeline pressure While the plunger rises on its injection stroke, the pressurised fuel will lift the delivery valve off its seat, enabling fuel to be displaced towards the injector (fig. 23.8(b)). Further upward plunger movement will align the helical groove and the spill port, so that injection ends. The delivery valve will then immediately be forced towards its seat by the return-spring. This ensures that the remaining fuel in the pipeline will be subjected to a residual pressure which is usually in the region of 30 bar.

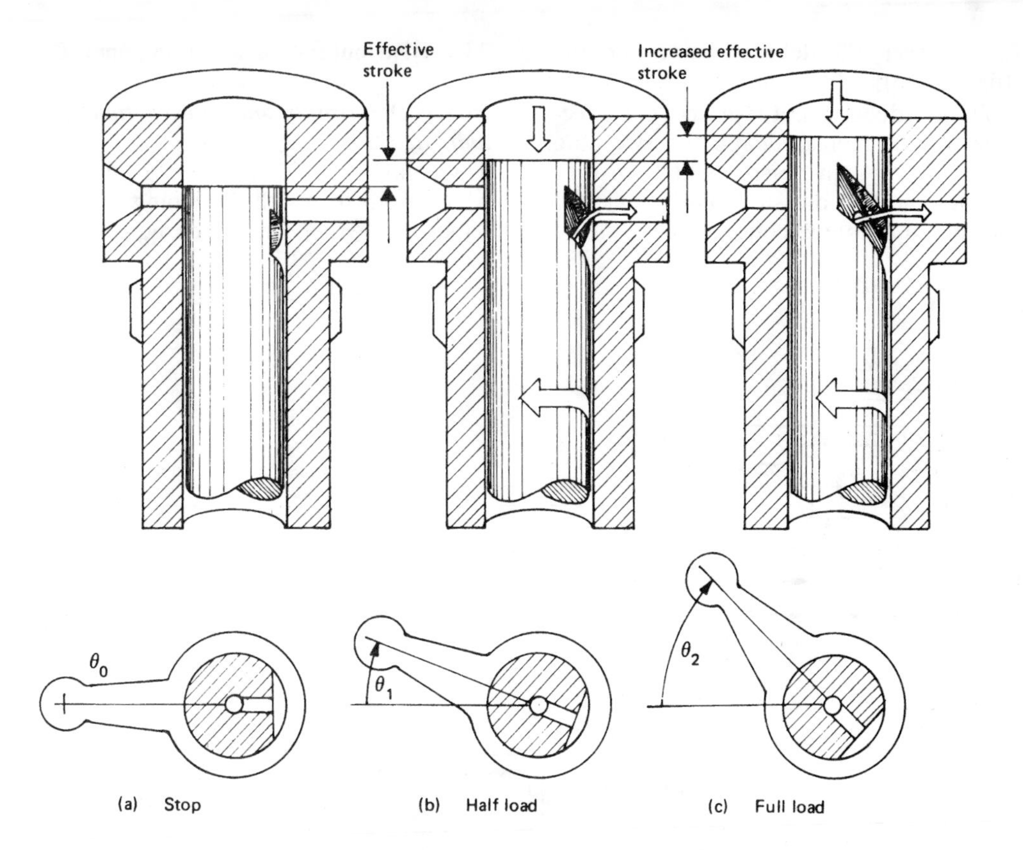

Fig. 23.7 Output control

Anti-dribble action In the process of closing the delivery valve, the piston portion of the valve initially cuts off the fuel pipeline above the valve from the barrel pumping chamber (fig. 23.8(c)) and then sweeps further down to rest on its seat (fig. 23.8(a)).

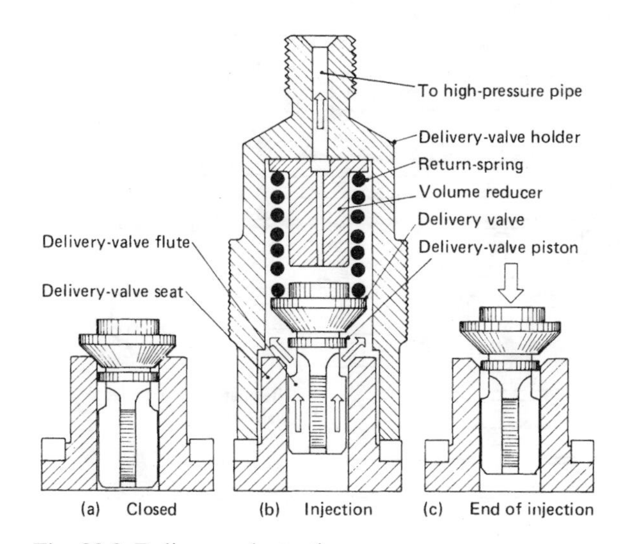

Fig. 23.8 Delivery-valve action

The plunger action of the close-fitting piston increases the total volume of the fuel trapped between the valve and the injector, and this causes a rapid pressure drop which unloads the injector line pressure to below the working pressure. Consequently the injector needle is able to snap on to its seat, thus providing a clean cut-off without nozzle dribble.

Automatic air bleeding An additional function of the delivery valve is to assist in working air out of the system, since the fuel volume between the delivery valve and the nozzle does not add to the pump clearance volume.

Volume reducer This is a cylindrical shouldered piece of low-carbon steel with a hole through the centre and a transverse slot across the enlarged shouldered end. It fits inside the delivery-valve holder and acts both as a spring-guide and to reduce the effective volume and compressibility of the fuel between the delivery valve and the injector (fig. 23.8(b)).

If the volume reducer is not replaced, the effective volume will be increased and the sweeping action of the delivery-valve piston will be insufficient to reduce the line pressure. The result will almost certainly promote injection dribble.

23.4 Distributor-type injection pump (CAV 'DPA')

23.4.1 Description and construction of pumping unit (fig. 23.9)

This pump unit consists of a single steel rotor which both forms the pumping element and distributes the fuel, and is designed to rotate in a

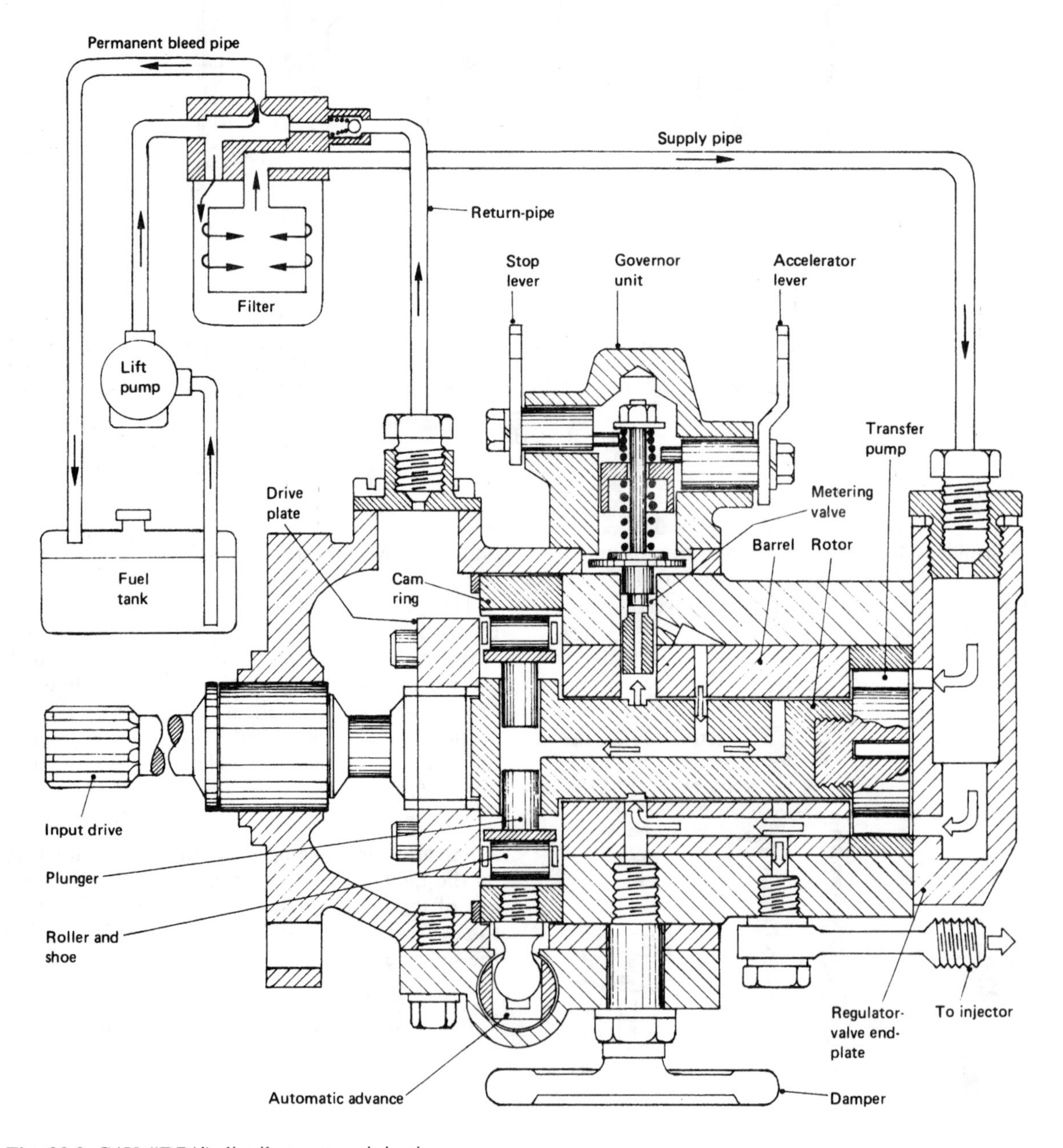

Fig. 23.9 CAV ('DPA') distributor-type injection pump

stationary steel cylindrical sleeve known as the hydraulic head.

Pumping section of rotor The pumping section of the rotor is flanged so that it is larger in diameter than the distributor section and it has a transverse cylinder bore which houses two opposed pumping plungers. A stationary internal-cam ring is carried in the pump housing and operates the plungers through rollers and shoes which are carried in slots machined into the circumference of the rotor flange. The cam ring has as many lobes as there are engine cylinders, and they are evenly spaced round the ring. The outward movement of the opposed plungers is controlled by metered fuel pressure.

Distributor section of rotor The distributor section of the rotor has a central passage which intersects the space between the pumping plungers. At the opposite end of the passage is a single radial distributor port. As the rotor revolves, this single distributor port aligns in turn with each of the discharge ports radially situated round the hydraulic head. These ports are connected to high-pressure pipes joining the pump to the individual injectors.

In between the two ends of the central rotor passage are equally spaced radial charge ports. There are as many of these as there are discharge ports, that is engine cylinders. As the rotor revolves, these charge ports align in turn with a single radial metering port drilled into the hydraulic head. This port supplies the metered fuel to the rotor for pressurising and distribution.

Fuel circuit The supply of fuel initially is provided by the lift pump which delivers it via the filter to the regulator-valve end-plate (fig. 23.9). Fuel first flows through the vane transfer pump which is screwed on to the end of the rotor – this increases the fuel pressure according to the speed of the rotor. The fuel then flows along a passage in the hydraulic head and around an annular groove to the metering valve. From here it flows to the metering port, where it enters the rotor.

23.4.2 Rotor and plunger pumping action (fig. 23.10)

The operating cycle may best be described by considering a single-cylinder injection pump in two phases – charging and injection – as follows.

Charging phase (fig. 23.10(a)) For a short period as the rotor revolves, the distribution port will be out of alignment with all the discharge ports, but one of the charge ports (only one shown) will align with the metering port in the hydraulic head. Fuel at metering pressure is thus able to flow into the central passage in the rotor, to force the plungers apart. In this position, both roller-and-shoe assemblies will be clear of all the cam lobes. The amount of plunger outward displacement is determined by the amount of fuel which can flow into the element for the short time the metering and charging ports are aligned.

Injection phase (fig. 23.10(b)) Continued rotation of the rotor will now misalign all charge ports with the metering port, but it will bring the distribution port into alignment with one of the discharge ports in the hydraulic head.

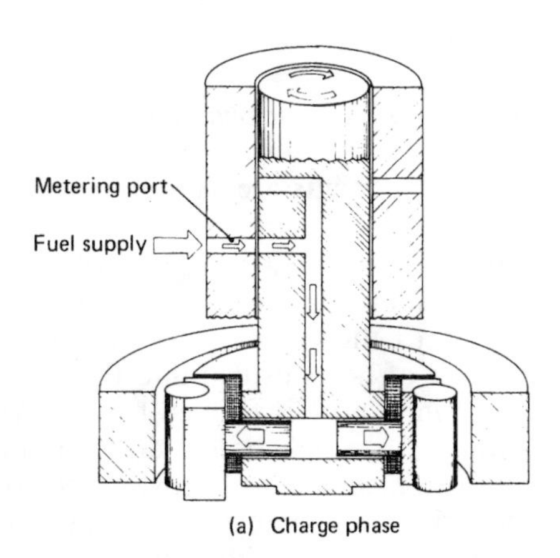

(a) Charge phase

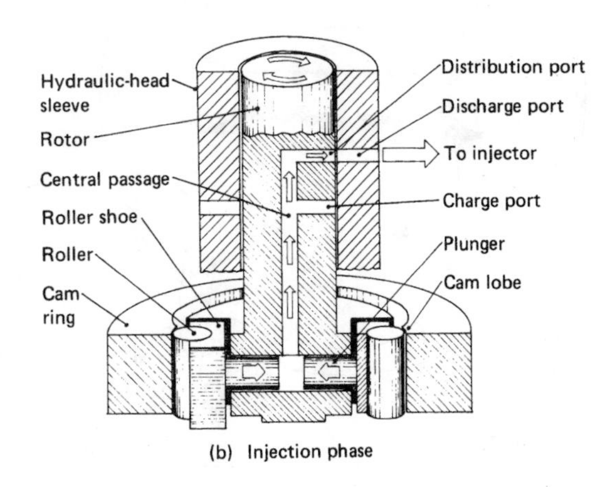

(b) Injection phase

Fig. 23.10 Cycle of operation

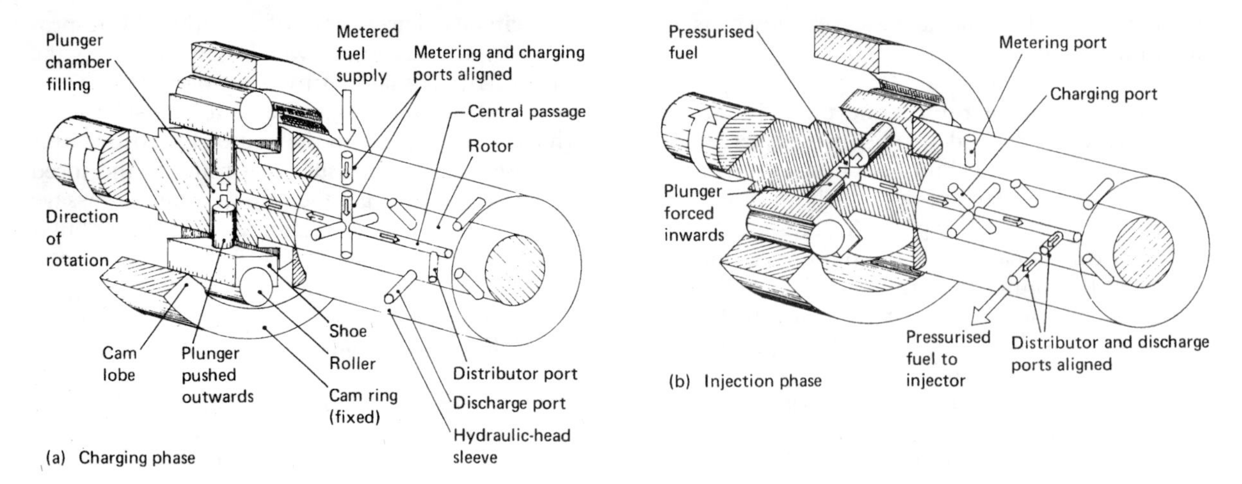

Fig. 23.11 Four cylinder DPA injection-pump cycle of operation

Simultaneously, both rollers will contact the flank of diametrically opposed cam lobes which force both plungers together. The generated high pressure in the fuel will now force it out through the discharge port and hence to an appropriate injector.

With a four-cylinder distributor (DPA) injection pump, the cycle of events is similar. There is still one pair of opposing plungers, one metering port, and one distributor port; but in this case there are four cam lobes, charge ports, and discharge ports. A charge or discharge event occurs at rotor intervals of 90°, as shown in fig. 23.11.

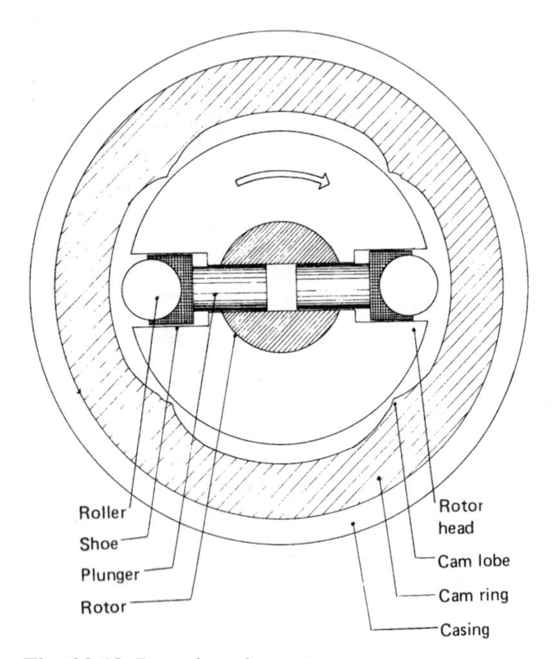

Fig. 23.12 Pumping elements

23.4.3 Cam-profile action (fig. 23.12)
Each cam lobe has two peaks, and injection ceases when the roller reaches the first peak, which is the higher. The valley between the two peaks ensures a rapid reduction in pressure in the injector pipe, preventing dribble and carbon formation at the injector nozzle at the end of the injector. The second peak prevents the pipeline pressure collapsing completely for long enough for the distribution port to be cut off from the discharge port. Residual pressure will then be maintained in the pipeline between the discharge port and the injector. Usually no delivery valve is necessary with this cam profile.

23.5 Injector unit

23.5.1 Description and construction (fig. 23.13)
The construction and function of the various injector-unit components are as follows:

Nozzle holder This is a forged-steel flanged housing which supports the various parts and is in turn bolted to the engine's cylinder head by means of a pair of holes drilled into the flange.

A threaded inlet port is formed to support an inlet adaptor, and a supply hole is drilled to intersect the pressure-face junction between the nozzle holder and the nozzle body. A central hole is drilled in the nozzle holder to house the spindle. The upper end of the holder has a large threaded

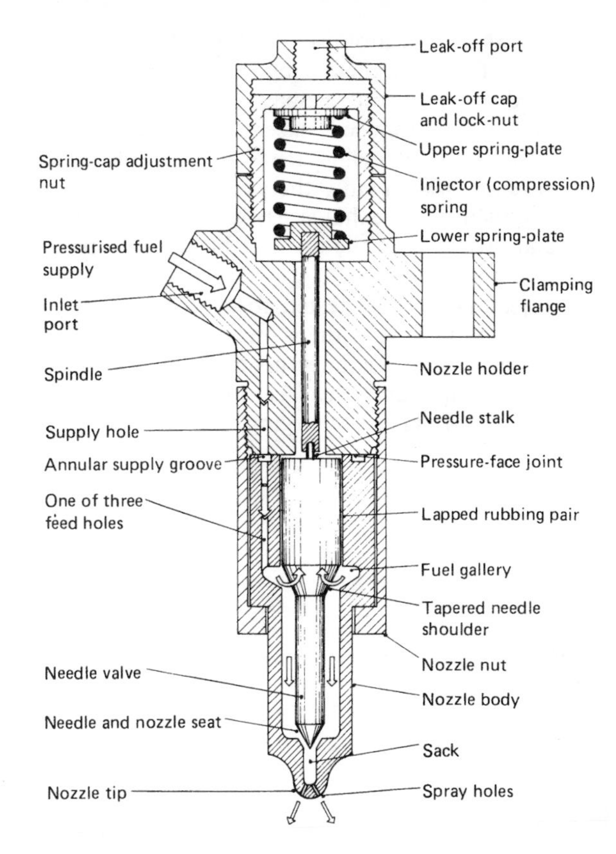

Fig. 23.13 Injector unit

countersunk hole made to support the spring-cap ajdustment nut.

Nozzle body This shouldered cylindrical and centrally bored steel body supports the needle valve. The bottom of the central blind hole is conically ground to form the nozzle-valve seat, and from three to five very small holes are drilled into the closed end of the nozzle tip to form the spraying nozzle. Half-way down the central hole is machined an internal groove cavity known as the fuel gallery, and three feed-holes equally spaced around the nozzle bore are drilled to intersect this fuel gallery. The pressure face of the nozzle body also has an annular groove formed so that fuel can circulate and supply the three feed-holes with fuel.

Needle valve This resembles a cylindrical spindle which has basically two sections of different diameters. The larger-diameter portion forms a lapped fit with the nozzle body. The smaller-diamter portion has a large clearance between it and the nozzle body, but its 60° inclusive-angle conical-ground end forms the needle-valve seat.

A small amount of leakage between the lapped nozzle body and the needle portion is provided to lubricate the rubbing surfaces. The taper shoulder joining the two different-diameter sections of the needle is ground to this shape purely to prevent stress concentrations fracturing the needle in this region under impact loading.

Spindle This relays the spring thrust to the needle valve from the spring situated between the spring-cap adjustment nut and the countersunk threaded hole in the nozzle holder. It therefore forces the needle valve on to the nozzle-body seat.

Injector spring This is a helical-coil compression spring positioned between the spring-cap adjustment nut and the spindle. It is located by both upper and lower spring-plates.

The function of the spring is to keep the needle valve closed until the designed opening fuel-line pressure is reached, and also to close the needle valve when this fuel-line pressure collapses. The actual opening fuel-line pressure can be varied by adjusting the initial degree of spring compression.

23.5.2 Injector action (figs 23.4 and 23.13)

Residual line pressure Existing pressurised fuel in the high-pressure pipeline enters the inlet port and moves through the supply and feed passages to fill up the annular space formed between the needle and the nozzle body. The pressure of this trapped annular column of fuel bears down on the floor of the nozzle at the lower end and presses upward on the tapered shoulder of the needle at the top end, the spring force at this stage being sufficiently large to maintain the needle tight on its seat.

Injection begins When the fuel pressure generated in the injection pump reaches some predetermined intensity, the delivery valve will open, permitting more fuel to enter the high-pressure pipe. This movement of fuel will increase the pressure underneath the needle's tapered shoulder until the equal and opposite reaction from the floor of the nozzle exceeds the spring load. At this point the needle will be lifted clear of its seat – fuel will thus be forced through the sack and out of the nozzle tip holes in the form of a finely atomised spray.

Injection ends Eventually the generated injection pressure will collapse, due either to the spill port of the in-line pump opening or to the rollers moving beyond the first cam-lobe peak in the case of the distributor pump. Immediately the spring load will exceed the opposing hydraulic residual pressure, and the needle valve will snap shut.

23.5.3 Fuel injection and atomising

The power output of a diesel engine depends upon the mass of atomised fuel injected into the cylinders, which is controlled by the amount of fuel displaced by each stroke of the injection-pump plungers. Finely atomised fuel droplets are essential for the minimum delay in ignition and controlled combustion propagation. Even though they may be atomised, excessive amounts of fuel injected into the cylinder will not be able to mix with enough air before the exhaust valves open – consequently the exhaust gas will be expelled into the atmosphere in the form of black and possibly sooty smoke.

Injectors may be tested for correct spray formation by removing them one at a time from the cylinder head and connecting each one to its high-pressure pipe and then running the engine at idling speed. The discharging fuel should appear

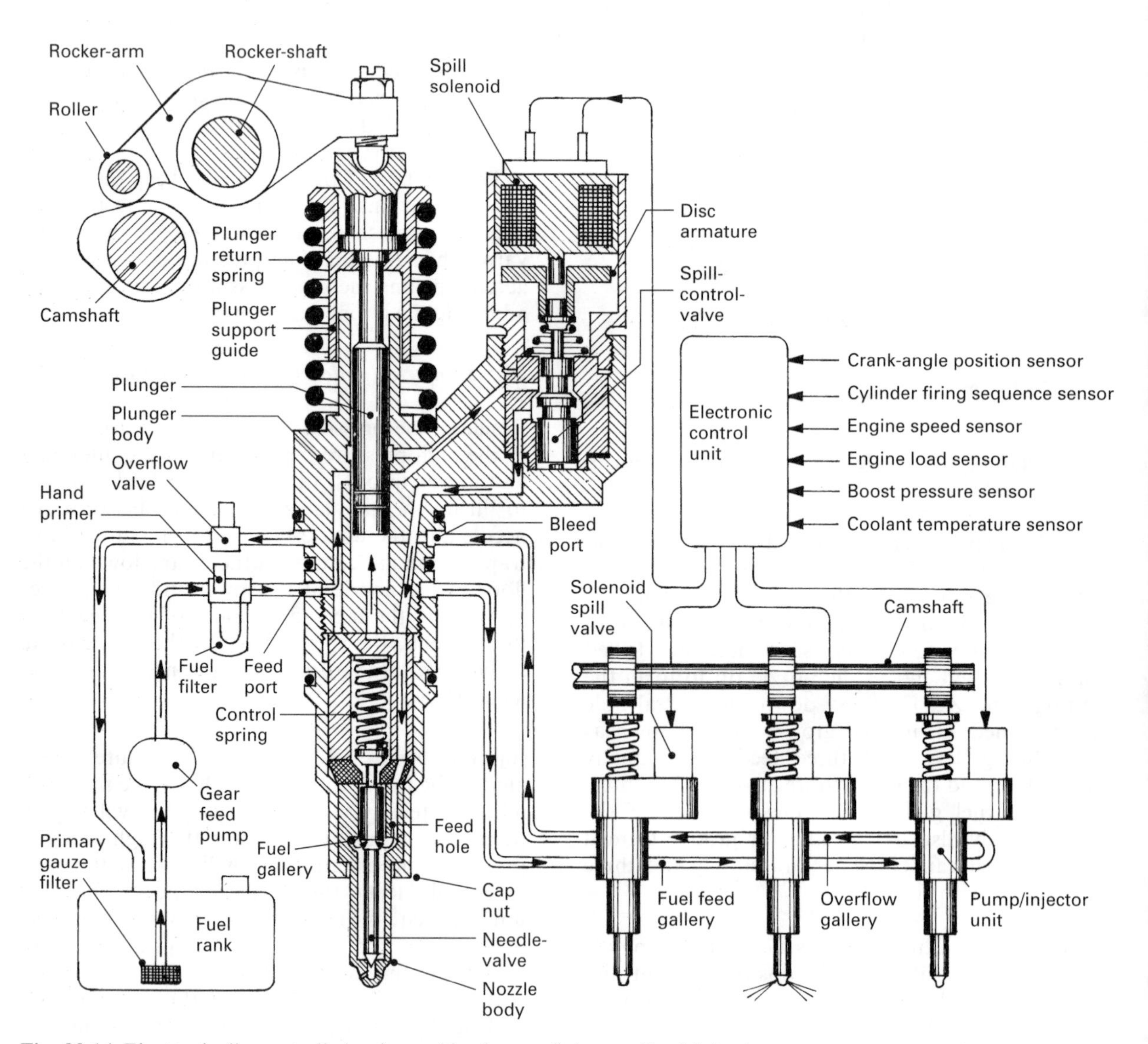

Fig. 23.14 Electronically controlled unit combined pump/injector diesel injection system

atomised, having the appearance of a fine mist without wet streaks.

23.5.4 Personal safety precautions when servicing diesel fuel-injection equipment

Diesel fuel may cause dermatitis if it comes into contact with the skin, so diesel equipment should be tested on the bench by using substitute fuel which is safe to handle.

Before testing diesel equipment in the workshop, make sure that the area is well ventilated and that an extractor fan is switched on to remove any spray or mist so that fumes are not inhaled. Do not inject fuel spray on to the hands, as high injection pressures will penetrate the skin and may cause poisoning.

23.6 Electronically controlled unit pump-injector diesel fuel injection system (fig. 23.14)

The central part of this fuel injection system is the combined pump and injector unit which is installed in the top of the cylinder-head vertically over the centre of the cylinder between the four valves. The upper part of the pump-injector unit houses the pump plunger whereas the lower part contains the screwed-on multi-hole injector. Two annular fuel supply and exit grooves are machined around the waist of the plunger body at a level just below the solenoid spill-valve. When the unit pump-injector is inserted in the cylinder-head these grooves align with internal fuel galleries which lie parallel to the length of the cylinder-head. Each side of these fuel grooves is a rubber 'O' ring groove. These grooves seal the fuel so that it cannot escape and the entry/exit fuel flow is kept separate. A single camshaft operates the unit pump-injector and both pairs of cylinder inlet and exhaust valves. This requires three cams per cylinder, a central injection pump cam and an inlet and exhaust cam on either side. A cam and rocker-arm actuates the pump plunger movement whilst a stout return spring maintains contact between the rocker-arm and cam-profile at all times.

Leak-off fuel from the needle-valve is fed back to the fuel feed port and gallery so that no external fuel return line is necessary.

Fuel is supplied by a gear-type feed pump driven by the engine timing gears. It supplies a relatively high fuel feed pressure continuously to the fuel feed gallery and unit pump-injector. An overflow-valve on the return pipeline from the overflow gallery regulates the fuel feed pressure to approximately 3.5 bar. A hand primer attached to the fuel filter housing is used to pump up the fuel and bleed the system when the engine is stationary.

23.6.1 Electronic control unit

The electronic control unit inputs signals from the various sensors such as crankshaft position, cylinder firing sequence, engine speed, accelerator position, boost pressure, air temperature, and engine coolant temperature. These signals are processed in the mapped computer program of the electronic control unit. The outcome is an electrical pulse to each individual unit pump-injector solenoid in the firing order sequence. These electrical pulses energise the respective solenoids at some predetermined crank-angle position before TDC and for a duration depending upon speed and load operating conditions. Individual performance between cylinders is balanced by evening out the amounts of fuel delivered per cylinder, particularly when the engine is running at idling speed.

The point of injection is advanced or retarded by altering the instant when the unit pump-injector solenoids are energised, whereas the amount of fuel injected is determined by the length of time the solenoids are energised.

A major advantage of the combined pump-injector unit is that the passageway between the pump pressure chamber and the injector nozzle is minimal. It thus permits very high injection pressures (up to nearly 2000 bar) to be generated, and very sharp fuel cut-off occurs at the end of injection so that fuel dribble at the nozzle nose is almost non-existent.

23.6.2 Phases of operation (fig. 23.14)

Filling phase (fig. 23.14) The rocker-arm roller has passed its highest lift position and is now moving towards the base circle no-lift dwell region of the cam. The plunger rises towards its highest outmost position with the spill control valve open. Fuel is drawn from the fuel tank via the gear-type feed pump. It is then pressurised and forced to flow though the filter to the lower fuel feed gallery in the cylinder-head. The fuel then passes through internal passages to the pressure chamber below the plunger. Filling continues until the plunger reaches its uppermost position. At this point the bleed port is exposed, and excess fuel then passes into the upper overflow fuel gallery on its return route to the fuel tank. This takes place via the overflow valve which is set to open at a pressure of 3.5 bar. This arrangement provides continuous fuel circulation, purges it of any air-locks and helps to cool the unit pump-injector.

Spill phase (fig. 23.14) The plunger reaches its outermost position and begins its downstroke with the spill control valve open. Fuel is now pushed out from the plunger pressure chamber through the open spill valve back to the lower fuel feed gallery. Spill continues on the plunger's downstroke as long as the spill valve remains open.

Injection phase (fig. 23.14) At a given instant on the plunger downstroke the spill-valve solenoid will be energised, thus causing the armature to be pulled towards the solenoid winding. It therefore closes the spill-control-valve. Immediately the fuel in the passageway between the plunger pressure chamber and the injector fuel gallery will be trapped. Thus any further downward movement of the plunger causes the incompressible fuel upthrust to lift the needle-valve off its seat. Consequently fuel will be discharged from the multi-hole nozzle into the cylinder. Injection continues on the plunger downstroke until the spill-control-valve opens again. The earlier or later the spill-control-valve closes the more advanced or retarded the point of injection will be with respect to the plunger on its downstroke.

Pressure drop phase (fig. 23.14) When the correct amount of fuel has been injected the spill solenoid will be de-energised, at which point the armature return-spring pushes open the spill-control-valve. Instantly the pressure in the plunger body and injector passageway drops below the nozzle's needle-valve opening pressure so that the injection phase ceases. Fuel flow will therefore be diverted back to the fuel gallery via the open spill-control-valve as long as the plunger continues to move on its downstroke.

The amount of fuel injected during each pumping stroke is determined by the time interval between the spill-valve closing and the instant it opens, that is, the time the spill-valve remains closed.

23.7 Electronically controlled unit pump diesel injection system (fig. 23.15)

This fuel injection system uses individual unit pumps driven from a low mounted camshaft in the cylinder-block which also operates the cylinder-head valves. Thus there are three cams per cylinder: a central pump unit cam with an intake and exhaust cam on either side. The injector, which is much slimmer than a unit pump injector, is installed in the centre of the cylinder-head. Its small diameter makes it particularly suitable for small cylinder diameter bores, that is, relatively small to medium-sized engines.

The pumps and injectors are interconnected by a relatively short high-pressure fuel line pipe so that precise injection timing and cut-off can be achieved with injection pressures of up to 1800 bar. However, so far pressures as high as 1600 bar have only been used, compared to the best in-line pumps which can now reach maximum pressures of 1100 bar.

The pump units are mounted on the camshaft side of the cylinder-block and fit into holes formed in the cylinder-block. These unit pumps consist of a pump housing and barrel which has a solenoid-operated spill-valve incorporated on one side above the level of the plunger. The lower part of the unit pump contains the plunger return-spring and roller-tappet follower which is driven directly off the engine's camshaft. A conventional low spring multi-hole injector is used. However two-stage injectors are becoming common with this type of fuel injection system.

23.7.1 Electronic control unit

An electronic control unit takes in messages in the form of electrical signals from a number of engine sensors. These include the following: acceleration pedal position, crank-angle position, engine speed, inlet manifold air pressure and temperature, and coolant temperature. It also receives signals from marks on the timing gears mounted on the crankshaft and camshaft to effect the correct phase synchronisation. The collective information is then compared with the values stored in the memory to give optimum performance, maximum fuel economy, minimum emission, low noise, smooth engine operation and good drivability. Output electrical pulses to the unit-pump solenoids are thus made to energise the closing of the spill-control-valve which in turn controls the injection timing and quantity of fuel injected.

23.7.2 Phases of operation (fig. 23.15)

Filling phase (fig. 23.15) The plunger moves down on its outstroke with the spill-control-valve open. Fuel is drawn from the fuel tank to the feed pump. It then flows at feed pump pressure to the upper feed gallery in the cylinder-head. The fuel

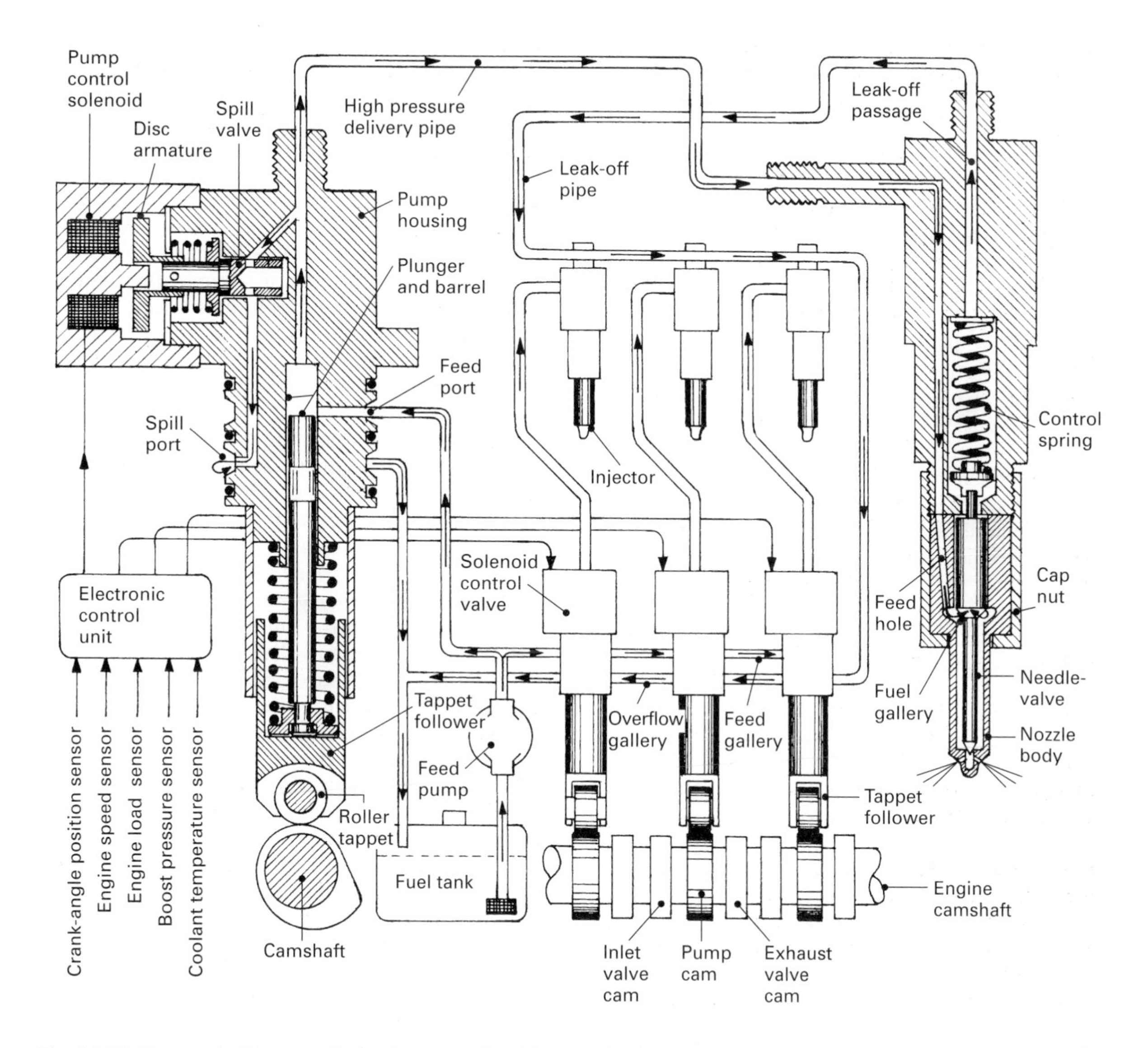

Fig. 23.15 Electronically controlled unit pump diesel injection system

then flows into the pressure chamber above the plunger via the feed-port. Filling continues until the plunger reverses its direction of motion and commences to move upwards, thereby blocking the feed-port from the pressure chamber.

Spill phase (fig. 23.15) With the plunger continuing to move on its upstroke, fuel will be pushed out from the pressure chamber to the spill-port and overflow gallery via the open spill-control-valve. Spill continues on the plunger's upstroke during the whole time the spill-control-valve remains open.

Injection phase (fig. 23.15) At a given moment on the plunger's upstroke the spill solenoid will be energised, thus causing the armature to be pulled towards the solenoid winding to the point when the spill-control-valve closes. At this point the fuel in the passageways between the plunger pressure chamber and the injector fuel gallery is trapped. Any further upward movement of the plunger therefore raises the fuel line pressure in the injector's fuel gallery to a point when it is sufficient to overcome the injector's control-spring downward thrust on the needle-valve. At this moment the needle-valve is rapidly lifted from its seat, and instantly fuel will be discharged from the multi-hole nozzle into the combustion chamber. Injection will continue on the plunger upstroke until the spill-control-valve opens again. Note that the instant at which the spill-valve closes is the point when fuel injection commences.

Pressure drop phase (fig. 23.15) When the measured quantity of fuel has been injected, the spill solenoid is de-energised causing the armature return-spring to open the spill-control-valve. Immediately the fuel pressure in the unit pump and the injector collapses. It thus permits the needle-valve to snap down onto its seat. Hence injection ceases and any further upward movement of the plunger will just displace fuel back to the feed gallery via the open spill-control-valve.

Note that the period when the spill-control-valve remains closed is equivalent to the injection duration which is also proportional to the quantity of fuel entering the cylinder per injection cycle.

23.8 Electronically controlled unit common-rail diesel injection system (fig. 23.16)

The common-rail injection system comprises basically three parts:

1) a high-pressure pump, in this example a self-contained in-line twin plunger pump with a solenoid-operated control-valve positioned on the side of the plunger and barrel;
2) a unit injector which includes a control plunger above the needle-valve and control-spring and a solenoid-operated injector control-valve mounted horizontally on top of the unit injector; and
3) a common-rail accumulator which is plumbed in between the high-pressure pump and the unit injector.

With this system of fuel injection, the pressures generated and injected are de-coupled, that is, they are not interconnected. The pressure produced by the high-pressure pump is transferred to the common-rail accumulator which stores the discharged fuel and maintains a constant fuel pressure in the common-rail fuel line of up to 1350 bar. This pressure is continuously available at the injector, in contrast to the conventional jerk injection pump which needs to repeatedly rebuild pressure for each injection cycle. The injection pressure generated is therefore independent of engine speed, and the quantity of fuel injected does not rely on a timed position of any one of the pumping plungers. Radial plungers are adopted for cars and in-line plunger pumps are preferred for commercial vehicles to generate the high pressure for the common-rail. The system's pressure is controlled by means of a pressure sensor. The unit-injector control-valve solenoid is activated by an electrical pulse supplied by the electronic control unit. The instant the solenoid is energised and the time it is applied determines the beginning and the duration of injection. Note the two factors determining the quantity of fuel injected per cycle is the constant pressure generated and the opening duration of the nozzle needle-valve.

Pilot injection, second and possibly a third stage of injection can be achieved by repeatedly activating the solenoid control-valve, whereas the injection rate can be modified by controlling the nozzle needle movement.

23.8.1 Electronic control unit

The electronic control unit is the brains of this fuel injection system. It contains a memory element, engine control microprocessor and a solenoid drive element. The engine control microprocessor receives and processes information signalled from the various sensors which monitor the engine variables such as crankshaft position, engine speed and load, boost pressure and temperature, coolant temperature, accelerator pedal position, common-rail pressure, etc. This data is then compared to the pre-stored memory. It is then computed and the resulting outcome is used to predict the necessary injection timing and the quantity of fuel delivered to the cylinder each power stroke for the prevailing operating conditions at any one time. This information is then converted into electrical signals which are then sent to the solenoid drive circuit; accordingly electrical pulses are relayed to each injector control solenoid in turn to activate the respective injectors.

23.8.2 High-pressure pump (fig. 23.16)

Fuel is drawn from the fuel-tank to the feed-pump. Here it is pressurised and forced to flow to the high-pressure pump inlet ports feeding both barrels. When each plunger in turn reaches its outermost position, fuel enters and fills the pressure chamber above the respective plunger. As the plunger on its upstroke cuts off the fuel supply, any further upward movement pressurises and displaces it so that it flows via the delivery-

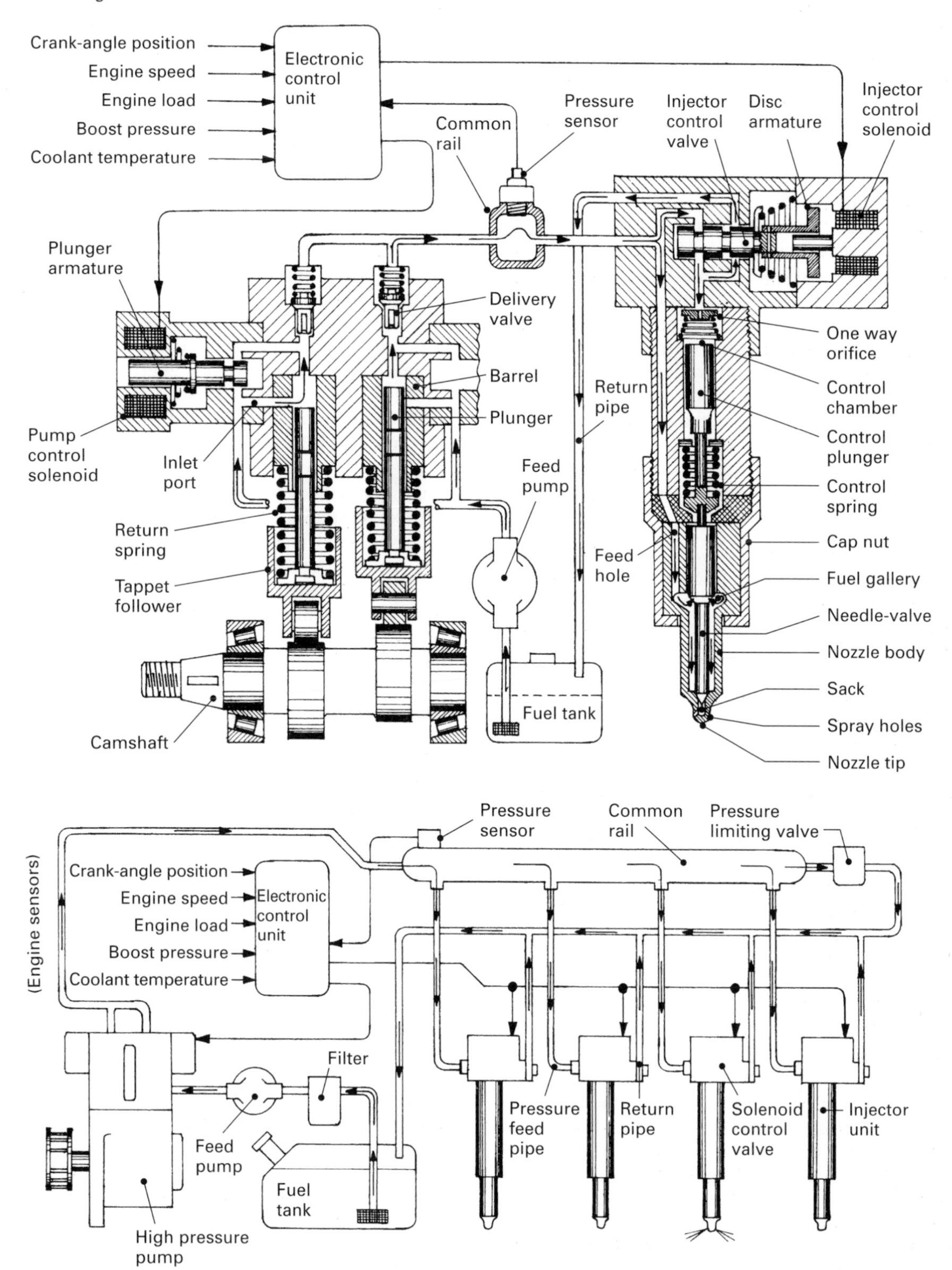

Fig. 23.16 Electronically controlled unit injector common-rail diesel injection system

valve to the common-rail accumulator. On the plunger's downstroke the delivery-valve snaps closed, thus trapping the fuel between the high-pressure pump and the unit-injector. As can be seen, the rotating camshaft continuously lifts and lowers the individual plungers so that a corresponding pumping and filling action takes place. The maximum pressure reached in the common-rail is monitored by the electronic control unit via the pressure sensor and at predetermined matched memory conditions an electric pulse is signalled to the pump control solenoid. Immediately the pump-control-valve opens and fuel thus spills back to the intake side of the barrel until the common-rail container pressure drops to the pressure assigned to the prevailing conditions, at which point the pump-control-valve closes and the cycle of events is repeated.

If the generated pressure exceeds some predetermined maximum a pressure limiting-valve opens and ejects the high-pressure supply fuel to the return pipeline.

23.8.3 Unit injector (fig. 23.16)

Inactive phase (fig. 23.16) High pressure fuel is continuously fed from the common-rail directly to the injector nozzle internal fuel gallery via the vertical passageway. During this inactive period the solenoid is de-energised. This causes the injector-control-valve to close the leak-off passage and to open the passage joining the common-rail to the control-plunger chamber. Consequently the pressure on top of the control-plunger holds the needle-valve in its closed position down on its seat.

Injection begins (fig. 23.16) At a given instant determined by the electronic control unit, an electric pulse energises the injector control solenoid. This pulls the control-valve towards the solenoid winding and closes the high-pressure fuel passage leading to the control-plunger and opens the leak-off passage. The fuel pressure above the control-plunger instantly collapses, whereas the fuel in the nozzle fuel gallery remains at the high constant common-rail pressure. Consequently the fuel pressure underneath the tapered needle shoulder is now able to lift the needle-valve clear of its seat against the downward resisting thrust of the control-spring. Fuel will thus be discharged through the sack and out of the nozzle holes in the form of a finely atomised spray.

Injection ends (fig. 23.16) When the calculated amount of fuel has been injected the injector control solenoid is de-energised. This enables the armature return-spring to very quickly push the injector control-valve to a position where the leak-off passage is closed and the high-pressure passage leading to the control-plunger is open. Immediately common-rail pressure is exerted on top of the control-plunger. This counteracts the common-rail pressure upthrust underneath the tapered shoulder of the needle-valve and permits the control-spring to snap down the needle-valve onto its seat, hence the injection phase ceases.

The quantity of fuel discharged into the combustion chamber per cycle is proportional to the time the injector solenoid is energised, that is, the duration of pressure application on top of the control-plunger. As a result it produces a hydraulic lock which prevents the needle-valve opening.

One-way orifice (fig. 23.16) The one-way orifice is incorporated above the control-plunger to provide an unrestricted pressure build-up to the plunger control chamber, but offers a high flow resistance during pressure release from this chamber. Subsequently this produces a gradual initial injection pressure rise and a sharp cut-off at the end of injection.

23.9 Cold start starting aids

23.9.1 Manifold heat operated thermostart (CAV)

Large direct injection engines usually use a manifold heater unit to preheat the cold air entering the cylinders, before attempting to start a cold engine, so that the cranking time before the engine fires can be reduced.

The preheating unit (see fig. 23.17) consists of a casing with a perforated flame sleeve pressed into an internal recess formed at the threaded end of the casing. The valve body is positioned in the centre of the hollowed casing, and is threaded at one end so that it can be secured to the casing via an insulation washer, permitting the diesel fuel supply pipe-union to be attached. An expansion-tube is secured to the valve-body at the opposite end; it contains a central valve-stem and a ball-valve which is trapped between the valve-stem and a fuel supply passage. Surrounding the expansion-tube is a heating coil. This element has emlarged loops extended to the open end of the flame-sleeve, earthed by a spot weld in the interior of the flame-sleeve. Conversely the opposite end

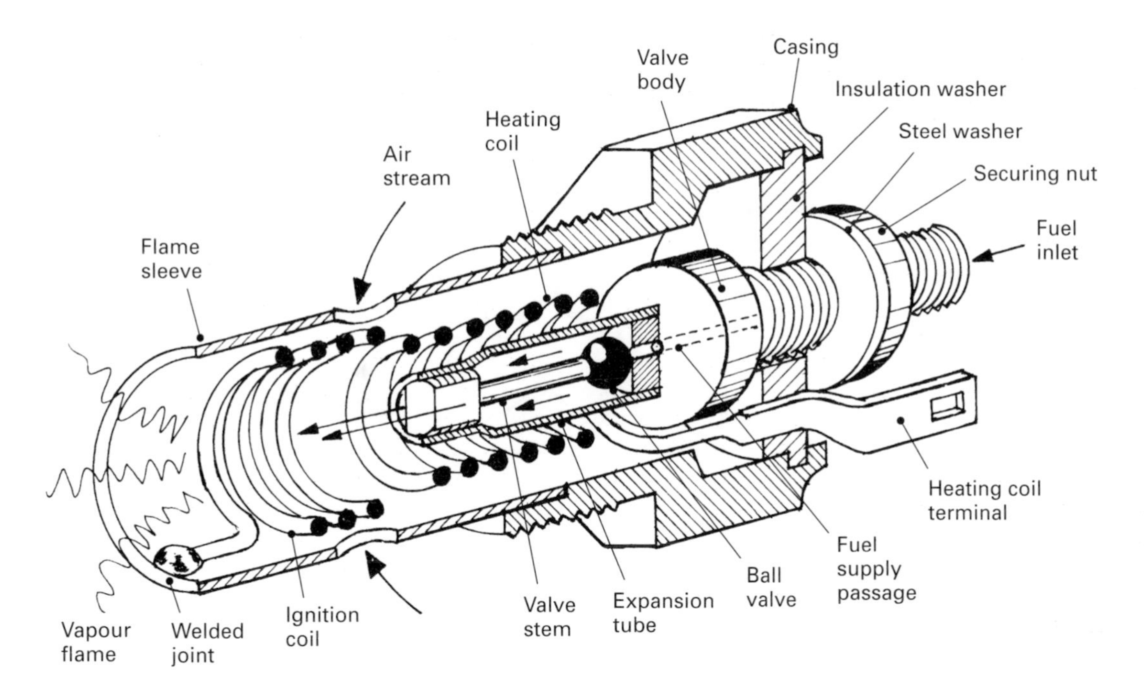

Fig. 23.17 Manifold heat operated thermostart (CAV)

of the heating coil is connected to the electrical supply terminal.

When the ignition switch is turned to the 'off position' the electrical heating coil current is interrupted so that the ball-valve will be pressed hard against the opening to the fuel supply passage by the valve-stem, and consequently no fuel can enter the expansion-tube.

Turning the ignition switch to the second position connects the heating-coil to the supply voltage, current than passes through the heating element coils, the heat generated then heats the expansion-tube and causes it to expand, whereas the relative cool valve-stem does not, as a result the ball-valve is released from its seat. Diesel fuel is now free to flow from the fuel supply around the ball-valve, along the expansion-tube and out through the two segmental shaped orifices.

When the ignition switch is turned to the third position the starter motor is energised, thus causing the engine to be cranked. The fuel droplets now meet the air stream passing through the large heater coil-loops, causing the heated fuel droplets to vaporise quickly, ignite and for the burning air-fuel mixture to be carried into the cylinders by the fast moving air charge.

The heated air charge entering the cylinders makes it easier for the injection fuel spray to ignite in the combustion chambers during cold weather conditions, this shortens the warm up time during which the combustion process is not operating at its most efficient and exhaust emission levels will be high.

23.9.2 Cold start glow plug (fig. 23.18)

To ensure reliability and make certain of ignition when cranking a cold engine and to start it when winter weather conditions prevail, all indirect combustion chambers and some direct combustion chambers for small and medium-sized diesel engines install glow-plugs which protrude directly into the combustion chamber. For indirect combustion chambers the glow-plug is positioned horizontally in the swirl chamber just below the inclined pintle injector nozzle (see fig. 23.18), whereas in the direct combustion chamber the glow-plug is installed at an inclined angle pointing downwards and protruding into the combustion chamber cavity formed in the pistons crown (see fig 23.18).

The glow-plug unit consists of a nickel-chrome wire filament covered in an inert refractory powder inside a stainless steel heat resisting sheath (heating tube), one end of the coil element is in contact with the internal spherical end of the sheath whilst the other end is joined to an

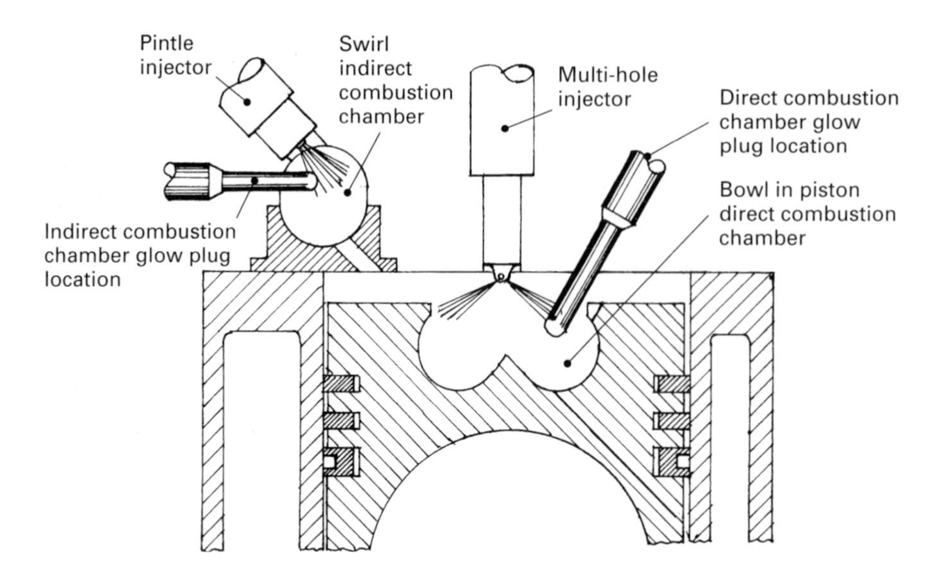

Fig. 23.18 Positioning of glow plug in direct and indirect combustion chambers

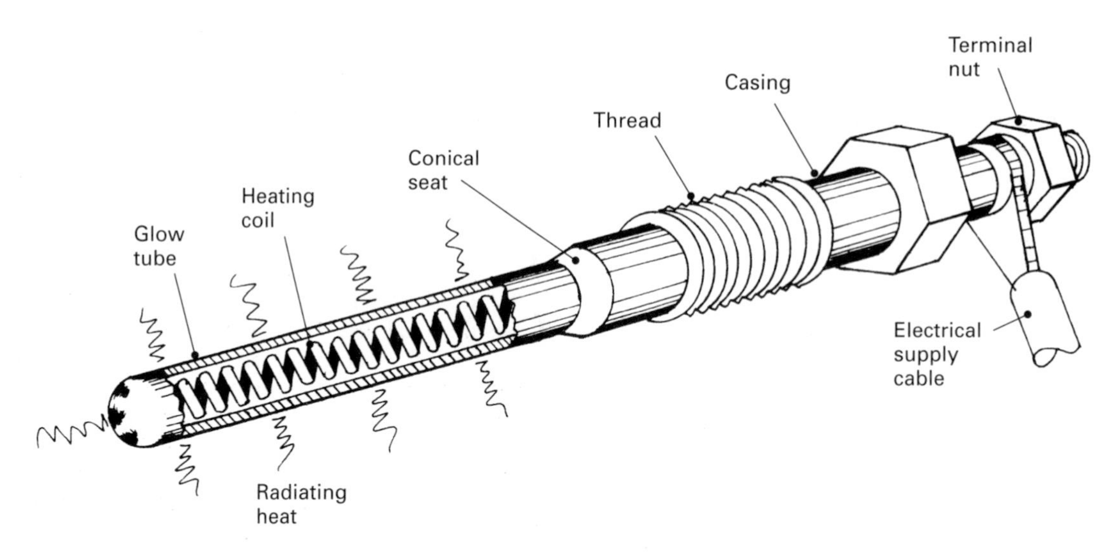

Fig. 23.19 Combustion chamber pencil type glow plug

extended connecting terminal which is insulated from the glow-plug casing (see fig. 23.19).

The glow-plug heating filament should be switched on between about 10 and 20 seconds before the engine is cranked to allow time for the wire heating coil to heat the metal sheath and for heat to transfer to the surrounding cooler air charge. Thus this additional heat provided by the glow-plug is added to the generated compression pressure heat so that the air temperature exceeds the threshold temperature needed for the injected fuel spray. The current consumption of each glow-plug heating filament is around 7.5 amperes when 11 volts is applied between the casing and terminal, this being equal to 30 or 45 amperes for a four or six cylinder engine respectively.

Operation (fig. 23.20) When starting the engine the ignition switch is turned from the 'off-position' (1) to the preheating position (2). A small current

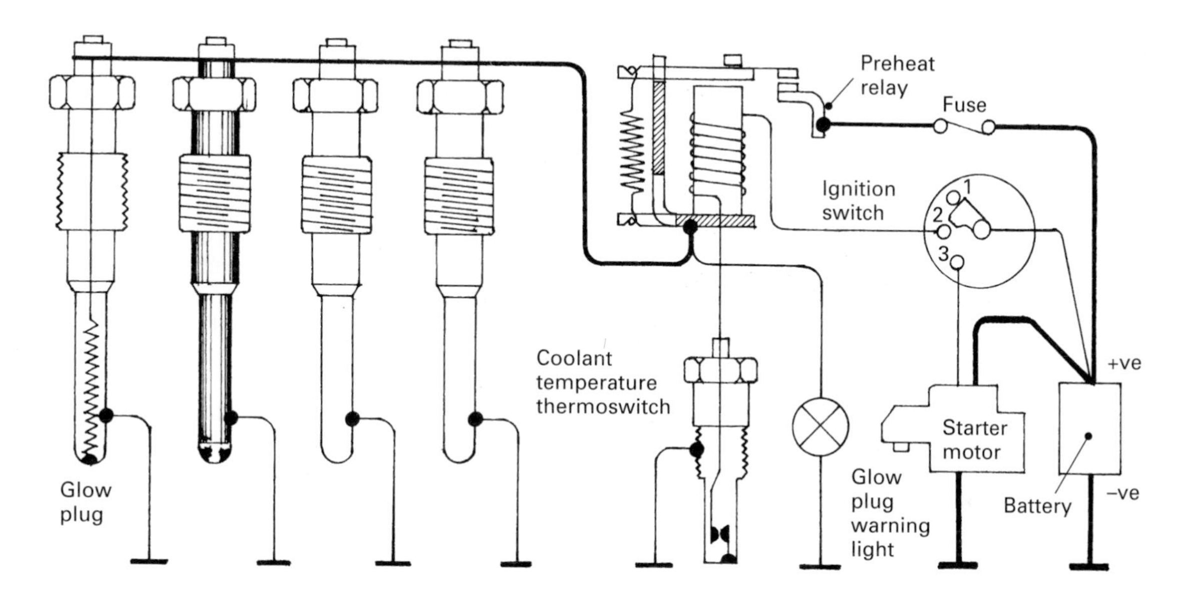

Fig. 23.20 Basic cold start glow plug preheated electrical circuit

now flows from the battery via the ignition switch contacts, relay solenoid winding, and coolant temperature thermo-switch to the negative earth-return. Immediately the electro-magnet field pulls down the armature-arm against the tension of the return spring which closes the relay contacts. A much larger current now passes from the batteries' positive terminal through the relay's closed contacts, and then to the glow-plugs' negative earth-return via each heating coil.

After a pause of about 20 seconds the ignition switch is turned further to position (3), and the starter solenoid now actuates the starter motor which then commences to crank the engine's crankshaft until the engine fires and starts to rotate the crankshaft under its own power.

While the ignition switch is in positions (2) and (3) the glow-plug warning light will be illuminated, informing the driver that the glow-plug circuitry is operative. Once the coolant temperature exceeds 40°C the coolant temperature thermo-switch contacts open, the current flow through the relay solenoid winding ceases, and the de-energised winding now releases the downward pull on the armature, thereby opening the relay contacts, resulting in the glow-plugs being switched off.

More sophisticated glow-plug circuits incorporate a preheat and postheat timer and a post-heat relay which allows the glow-plugs to continue to be switched on after the engine fires and is running under its own power.

23.10 Diesel-fuel filtration

Solid-particle contamination The amount of solid particles in liquid diesel fuel varies considerably in different parts of the world. In the United Kingdom an average figure of two parts per million by weight may be considered as normal, while in Africa the proportion may be as high as 230 parts per million. Not all the solid matter is sufficiently hard to damage the wearing surfaces of the injection equipment of course, but foreign matter will clog the filter elements. The contamination of the fuel in vehicles which are driven in normal traffic consists of organic components such as resins and asphalts, ageing products from diesel fuel, soot, and mineral compounds as well as rust, minute metal chips, and grit.

Abrasive concentration in fuel As little as 3 grams of natural abrasive in the 5 to 15 μm size range can wear out a six-cylinder in-line injection pump, and it makes little difference whether it is contained in a few litres or several hundred litres of fuel. If the lifespan of the injection pump is to be in the region of 150 000 km, and if a 10 litre engine is propelling a 20 tonne truck which has an average fuel consumption of 3 km/litre, this being a typical value, then it follows that the pump would be required to deal with 50 000 litres of fuel before it is worn out. This means, therefore, that the abrasive content passing through to the injector pump must be restricted to 0.06 grams per 1000 litres.

Water contamination Diesel fuel contains a concentration of molecularly bound (suspended) water up to 0.05%. In addition, free water is sometimes present due to storage and joining together of the suspended water molecules. The molecularly bound water does not itself cause any adverse effects during operation; however it is essential to separate the free water in the fuel, to avoid any residual sulphur in the liquid fuel combining with the water to form sulphuric acid which will then corrode the rubbing surfaces such as the plunger and barrel or the rotor and hydraulic head.

In a temperate climate, as much as 25 litres of water can accumulate in a 5000 litre tank by condensation from the air in a single year, or 0.5 litres of water in a week. In similar fashion, water may be deposited in fuel tanks or drawn in during rainy periods through tank vents by pressure fluctuation due to movement such as surge of the fuel within the tank. During cold-weather conditions, water in the fuel can also freeze in the fuel lines, causing partial fuel starvation to the engine.

Cold-weather wax formation Diesel fuel will freeze solid if cooled sufficiently. The fuel is a mixture of a number of component liquids which have a considerable range of freezing points, from as low as −28°C to probably as high as 48°C. There is no one freezing-temperature point from liquid to solid, but, as the temperature is lowered, the highest-freezing-point components tend to form tiny wax particles and these will grow in size as even lower-boiling fractions of the fuel solidify during long periods at low temperature.

For most diesel fuels in the United Kingdom, the wax particles start to form at temperatures in the range −12°C to −7°C. If a substantial quantity of these wax particles forms in the fuel system, it may be possible to start the engine satisfactorily but wax may accumulate in fuel filters and pipes during running and gradually strangle the flow of fuel to the engine until it loses power and finally fails.

Filter materials The particles required to be dealt with by the fuel filter may vary in size from 3 to about 40 μm. It can be assumed that particles larger than this could be held back by even the most rudimentary filters.

The most usual materials are wire gauze or nylon mesh for primary filters on the suction side of the lift pump and cotton, cloth, felt pads, and various types of paper or cellulose sheet for secondary filtration on the pressure side of the lift pump. Wire gauze and nylon mesh will hold back only the largest particles above 40 μm, cotton cloth above 25 μm, felt pads above 17 μm, paper and cellulose sheet will filter down to between 12 and 9 μm, and impregnated resin paper down to 4 μm. Most filter elements are now manufactured from resin-impregnated paper, and these elements will also filter up to a maximum water content of 3% from the fuel.

Under average working conditions, an element life of some 15 000 to 20 000 km or about 500 hours operation can be expected.

Properties of resin-impregnated paper filter The filter paper used nowadays has high filtering efficiency and is a good water separator due to a special impregnation process involving depositing on to the paper a surface-active precipitation agent which attracts petroleum liquids but repels water. When wetted with diesel fuel containing water, the flow resistance in the pores of the filter medium is lower for the diesel fuel than for water – this means that the fuel freely passes through the paper, but a considerable resistance is offered to the water. Eventually the small droplets of water do penetrate the paper, and they then combine together and, due to their density being higher than that of the fuel, gravitate to the bottom of the filter bowl, from where water can be drained off periodically.

The average abrasive-particle retention of these treated-paper filters is about 95%, or even better.

23.10.1 Secondary canister paper-filter units (fig. 23.21)

There are basically two types of paper filter-element construction, which can be identified by the direction and way in which the fuel passes through the filter medium:

i) cross-flow,
ii) up- or down-flow.

Cross-flow filters (fig. 23.22(a)) These filters have a pleated accordion-star type of construction. The pleats provide considerably more effective filter surface area than would be possible with a simple

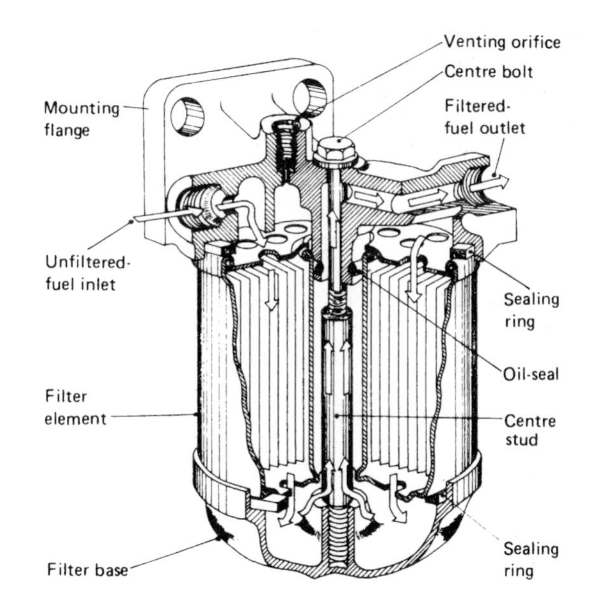

Fig. 23.21 Canister fuel-filter unit

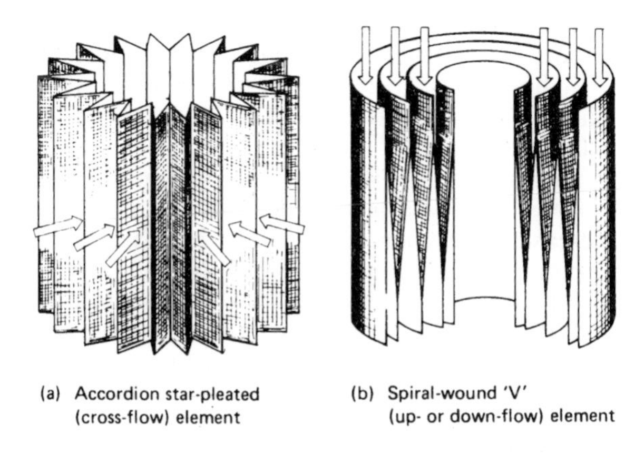

Fig. 23.22 Filter elements

cylindrical filter-paper sheet. With this construction, fuel flows radially across from the outside to the inside of the paper folds, which are sealed at the top and bottom by discs. The paper is usually resin-treated for high wet-strength. There is plenty of space between each pleat for the unwanted solid particles to be trapped without clogging the paper and reducing the fuel-flow-rate capacity.

Up- or down-flow filters (fig. 23.22(b)) These filters have a spiral 'V' type of construction – paper strips cemented top and bottom are wound around a tube to form a series of continuous V-shaped coils. A very large surface area is therefore exposed to the fuel – about 3500 cm^2 within the normal canister size. Fuel can flow either upwards or downwards through the paper folds, but usually a down-flow fuel path is preferred as it is easier to separate any water droplets from the fuel after filtration. The continuous V-shaped coils provide ample capacity for retained contaminants, while the 'crêped' paper permits unrestricted flow of filtered fuel between adjacent turns.

Most single-element cross-, up-, or down-flow filters will pass about 7000 litres before becoming choked. With a new element, the flow resistance under normal duty lies between 0.2 and 0.5 bar. Generally a rise in resistance to about 2.0 bar due to the soiling of the element is permissible.

23.10.2 Primary filter sedimenter (fig. 23.23)

A primary filter using gauze as a method of preventing dirt entering the suction side of the lift pump can become choked in cold weather, due to ice and wax formation, and will not in any case remove water which may be suspended in small droplets in the fuel. The primary-filter-sedimenter unit, on the other hand, offers no resistance to fuel flow at any time – its action depends on the difference in density between the relatively light fuel and the heavier solid abrasive particles and water, the unwanted matter settling at the bottom of the unit. To appreciate the principle of this unit, it should be understood that the heavier solid and liquid matter will remain in suspension with the fuel only when it is moving – if the fuel is

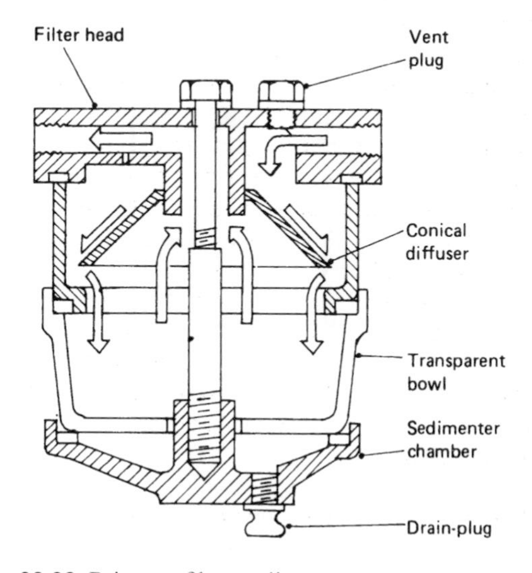

Fig. 23.23 Primary filter sedimenter

allowed to become static, the heavier matter quickly gravitates to the lowest position. This is known as sedimentation.

Initially the fuel containing droplets of water and solid particles enters into the top of the unit, flows around and over the conical diffuser funnel, and – because the space between the edge of the inverted funnel and the wall of the bowl is small – accelerates downwards. The higher-density impurities (water and solid particles) have more inertia, so they will continue in a downward direction for a longer time and distance. Because the inlet and outlets are kept as far apart as possible by the separator funnel in the comparatively large bowl, the movement of the fuel through the bowl will be relatively slow. Consequently the water droplets and heavier solid particles cannot be maintained in suspension and thus sink to the bottom, where they collect and can be drained off.

At low ambient temperatures, these units will also hold back wax crystals which have formed within the fuel, without causing the filter unit to become clogged.

23.10.3 Secondary filter agglomerator (fig. 23.24)
The secondary filter both prevents abrasive particles from flowing with the fuel from the lift pump to the injection pump and also separates out any water passing through the filter elements.

The mixture of fuel and fine water droplets passes through the lift pump and is emulsified by the pulsating action of the pump. It then flows

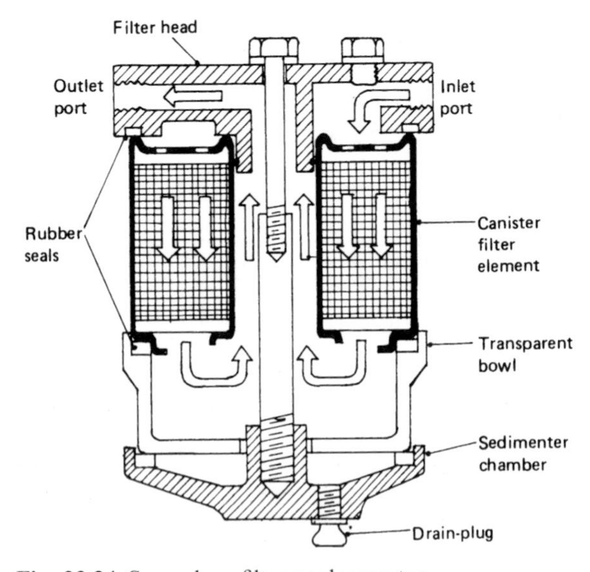

Fig. 23.24 Secondary filter agglomerator

into the filter head and down through the element, the fine pores of which retain the abrasive particles and solid contaminants but not the very fine water droplets, which are forced through these pores. As these minute water particles emerge on the other side of the paper-element surface, electrostatic attraction caused by the resin-impregnated paper pulls the individual water particles towards each other and thus they combine (agglomerate) to form larger droplets. The fuel and these larger water droplets then flow slowly downwards towards the relatively larger sedimenter chamber. The combined water droplets gravitate down to the base of the chamber, due to their density being greater than that of the fuel, and the lighter fuel then continues on its way up the centre of the filter to the outlet port. The accumulated water can be seen through the transparent chamber and periodically should be drained.

23.10.4 Twin-bowl filters (fig. 23.25)
The fuel flow in twin-bowl filters can be in series for dirty operating conditions or in parallel for high flow rates greater than 45 litres per hour, which is approximately the full-load demand of a 12 litre engine.

The use of a parallel-flow twin-bowl filter halves the flow rate through each element and increases the element life by more than three times as compared with that of a single-bowl filter used for the same application. If the bowls are connected in series it is necessary to change the second-stage filter element only after about the third change of the first filter element, but of course the unit has only the flow-rate capacity of a single filter unit.

23.11 Air-intake silencers and cleaners
Cleaning and filtering of the air is achieved by one or a combination of the following methods:

a) pleated paper,
b) wire or metallic mesh,
c) oil bath,
d) centrifuge.

23.11.1 Medium-duty dry air-cleaner (fig. 23.26)
This type of cleaner consists of a removable cover attached to the air-cleaner body which contains a replaceable paper filter element, with a row of plastic fins wrapped round it at the top end.

Air entering the upper part of the casing is directed to the fins on the element which give

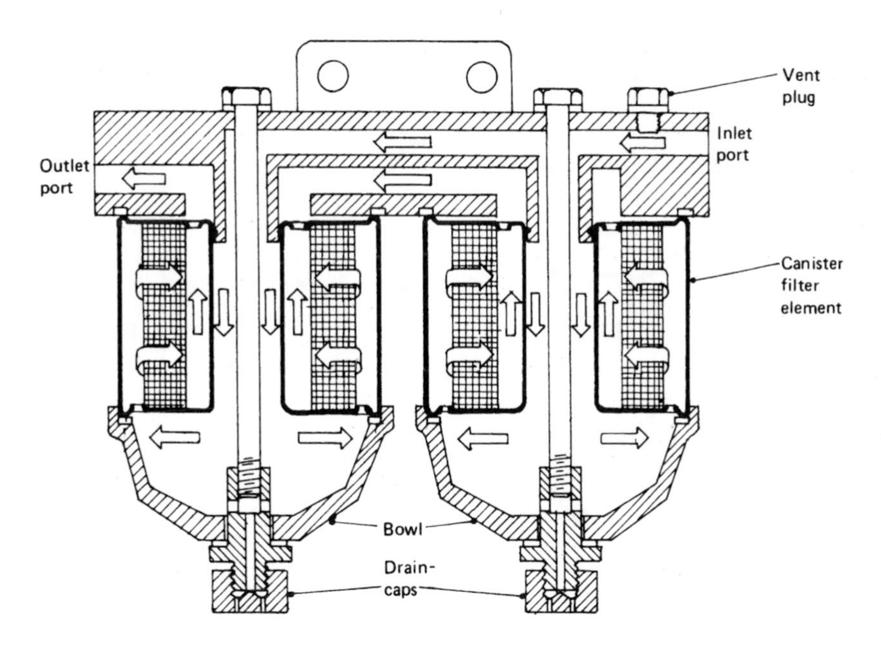

Fig. 23.25 Twin-bowl parallel-flow filter

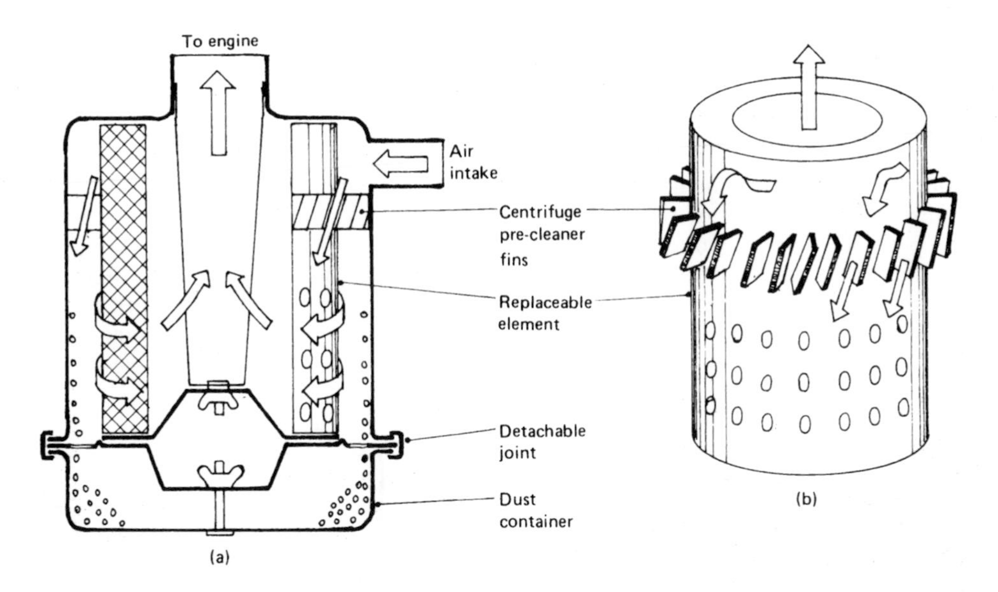

Fig. 23.26 Medium-duty dry air cleaner

the air a high rotational speed on its way down between the casing and element. This will separate a large proportion of dirt from the air by centrifugal action. This dirt will be thrown to the outside, where it flows down the inner wall of the casing and is ejected into the dust cup or container, which is baffled to prevent the re-entry of the dust. The pre-cleaned air then passes through the paper filter, which removes any remaining dust, before it enters the engine.

The dust container may be removed periodically and cleaned. The paper elements should be replaced every 50 000 km when operating under normal conditions.

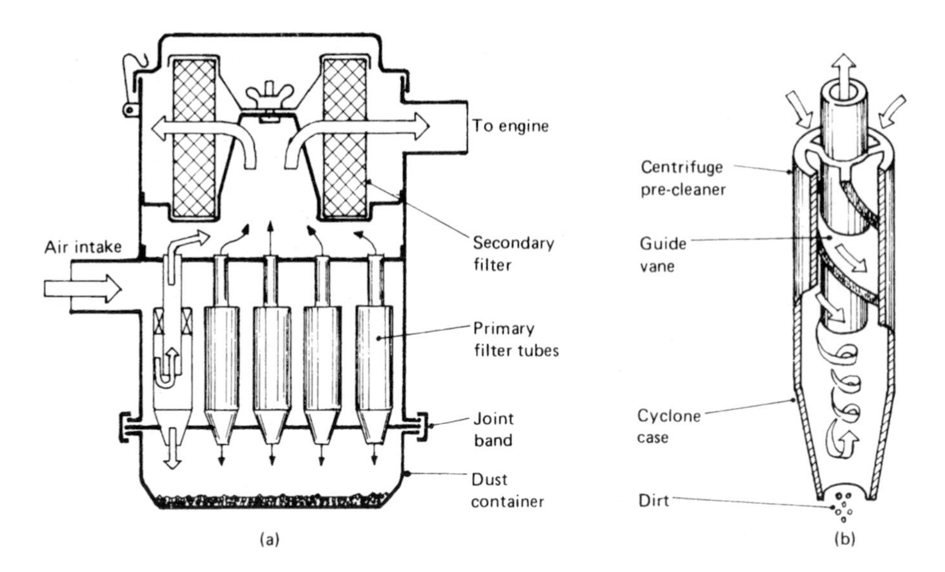

Fig. 23.27 Heavy-duty duo-dry air cleaner

23.11.2 Heavy-duty duo-dry cleaner (fig. 23.27)
This cleaner assembly consists of a bank of centrifugal cyclone pre-cleaners in series with a single annular-shaped replaceable impregnated paper filter element.

Air enters the cleaner case and is directed to the guide vanes of the cyclones. The vanes impart a swirling motion to the air and centrifuge the dust particles against the inner walls of the cyclones. Separate dust particles then gravitate to the dust container, so that approximately 95% of foreign particles are removed at this stage. This centrifugal action is fully effective at both high and low velocities.

The pre-cleaned air now reverses direction and penetrates the stagnant central core of the uncleaned-air spiral. It then passes up through the exit tube and enters the replaceable paper element. The air is filtered once more as it passes through the pleats of the paper before leaving the outlet port of the cleaner on its way to the engine. The overall efficiency for this design of two-stage filter is up to 99.9%.

23.11.3 Medium-duty oil-bath air-cleaner (fig. 23.28)
In this cleaner, air-cleaning is achieved in two stages. Unfiltered air first enters the air-inlet windows, from where it is directed downwards and around the annular space formed between the metallic mesh and the casing. The air then hits the oil-bath surface and is redirected towards the centre of the oil container. Due to their inertia, the heavier air-borne particles will continue to move down into the oil, but the lighter particles will still be suspended and carried by the air.

The pre-cleaned air will now rise, carrying with it oil mist. This oil mist will then condense as the air winds its way through the mesh, filtering any remaining dust. By the time this air has reached the top of the mesh, it will be dry and clean and it will then pass directly to the engine.

Periodically the oil container is removed, cleaned out, and refilled with fresh oil – a typical servicing interval would be 8000 km.

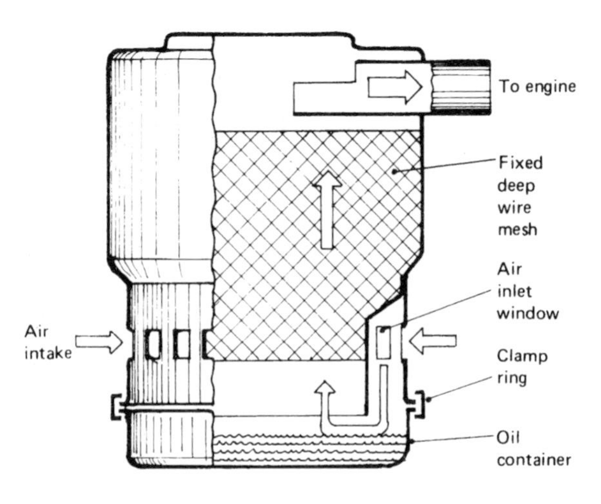

Fig. 23.28 Medium-duty oil-bath air cleaner

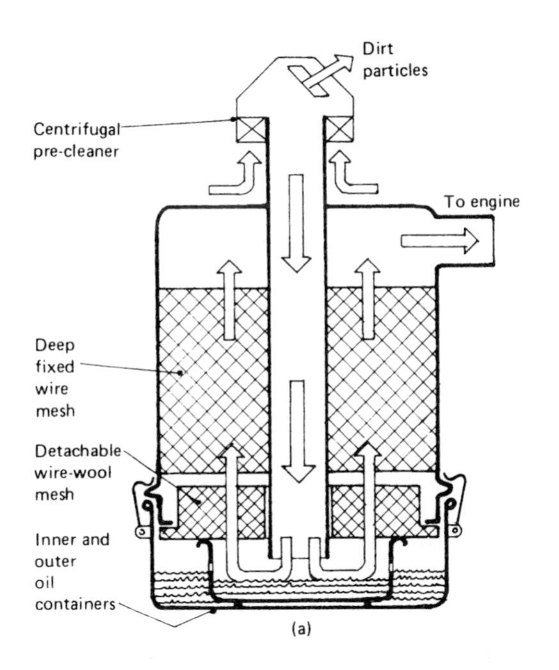

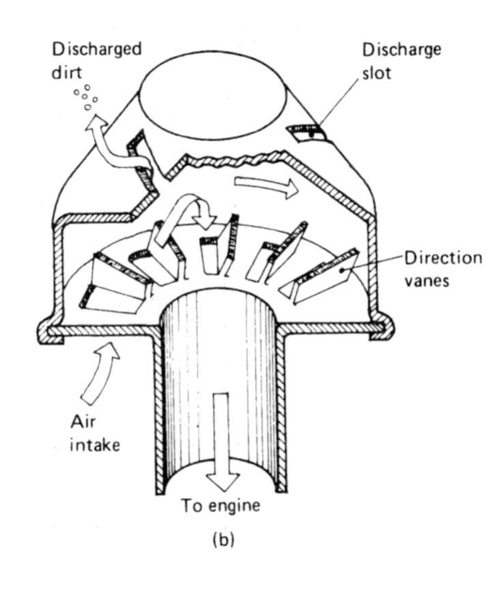

Fig. 23.29 Heavy-duty oil-bath air cleaner

23.11.4 Heavy-duty oil-bath air-cleaner (fig. 23.29)

These units incorporate a centrifugal pre-cleaner, an oil bath, a detachable shallow first-stage wire-wool-mesh element, and a deep second-stage fixed wire mesh.

Air enters the pre-cleaner, where a whirling motion is imparted to it and dirt particles are thrown out to the side of the casing by centrifugal force, being then discharged through specially designed slots (fig. 23.29(a)). Pre-cleaned air then accelerates down the centre of the cleaner (fig. 23.29(b)), impinges on the oil surface, and spreads radially outwards, forming an emulsion of air and oil mist. This then rises and enters the detachable wire mesh, which absorbs most of the dirty oil. The oil condenses and is returned to the inner and outer containers, where the dirt settles out from the oil, and the oil is then recirculated. The remaining relatively clean air and oil mist then enters the second-stage fixed mesh, which drys the air before it reaches the top and enters the engine. The detachable first-stage wire mesh reduces the dirt load of the second-stage wire mesh so that this rarely becomes clogged.

An inner container, which can be removed from the outer oil container, acts as a baffle in directing the oil-laden air to the filtering wire mesh and also controls the amount of oil in circulation and meters the oil to the filter element. The outer oil container supports the inner container and is a reservoir for oil and a settling chamber for dirt.

The inner and outer containers should be cleaned and filled with oil and the lower detachable wire mesh should be removed and replaced at regular intervals of about 8000 km, depending on operating conditions. Under severe operating conditions and where there are heavy dust concentrations, as in some African territories or under unusually severe conditions in the United Kingdom – such as in road construction, demolition work, sand and gravel pits. etc. – service attention to the cleaning may be necessary at least every 1200 km and in extreme cases every day.

24 Induction/exhaust manifold-exhaust silencer and emmision control

24.1 Induction wave ram cylinder charging

Every time the inlet-valve opens when the engine is running a column of air rushes through the induction tract passageways to fill the cylinders. At the same time the initial depression in the cylinder produces a negative pressure-wave which travels in the opposite direction through the column of air to where the inlet-tract entrance begins. When the negative pressure-wave reaches the open end of the tract, a corresponding positive pressure-wave is created which now travels in the reverse direction through the column of air towards the inlet-valve. If this positive pressure-wave can be so timed that it enters the cylinder towards the end of the induction period, then an increased amount of air will be crammed into the cylinder before the inlet-valve closes. Since it takes a definite time for the air to pass through the induction tract, varying the tract length alters the instant when the charge enters the cylinder at a given engine speed.

The science of induction wave-ram cylinder charging is therefore the process of matching the induction tract length to engine speed so that the incoming pressure-wave enters the cylinder at the same crank-angle position during the induction period throughout the engine speed range.

A variable induction-tract length is possible to achieve, but it is more feasible to pick two or three critical engine speeds where there is a downturn in volumetric efficiency, and then select tract lengths to optimise wave-ram charging at these speeds only.

24.1.1 Dual-stage induction manifold system with merging primary and secondary tracts (figs 24.1 and 24.2)

Dual-stage induction manifolds have a long and narrow cross-sectioned primary tract and a short

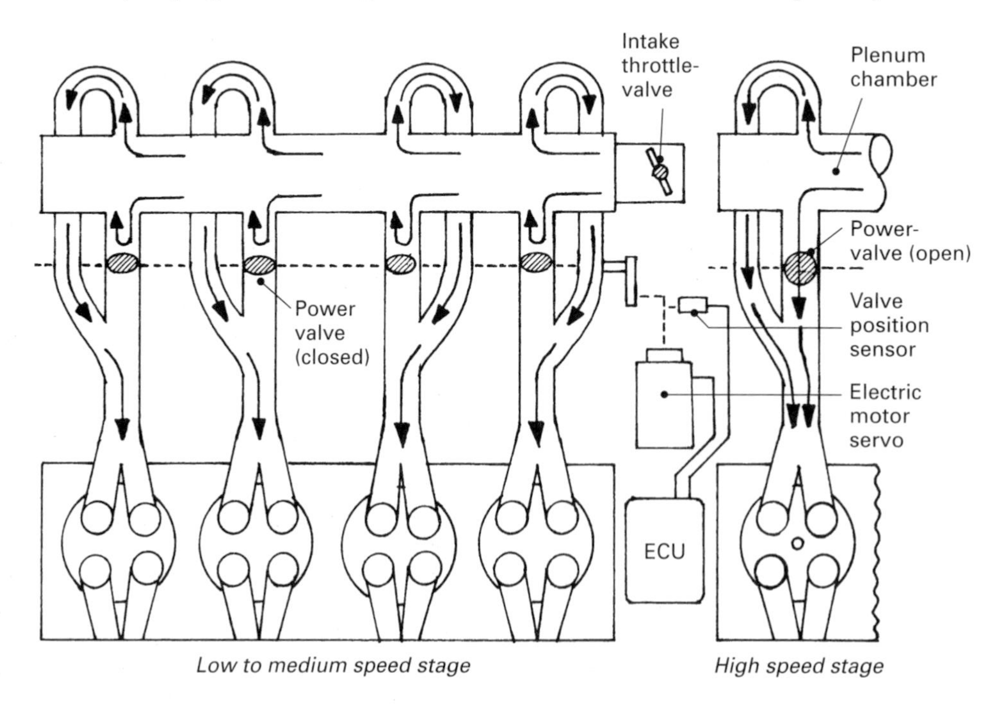

Fig. 24.1 Dual-stage induction manifold system with merging primary and secondary tracts

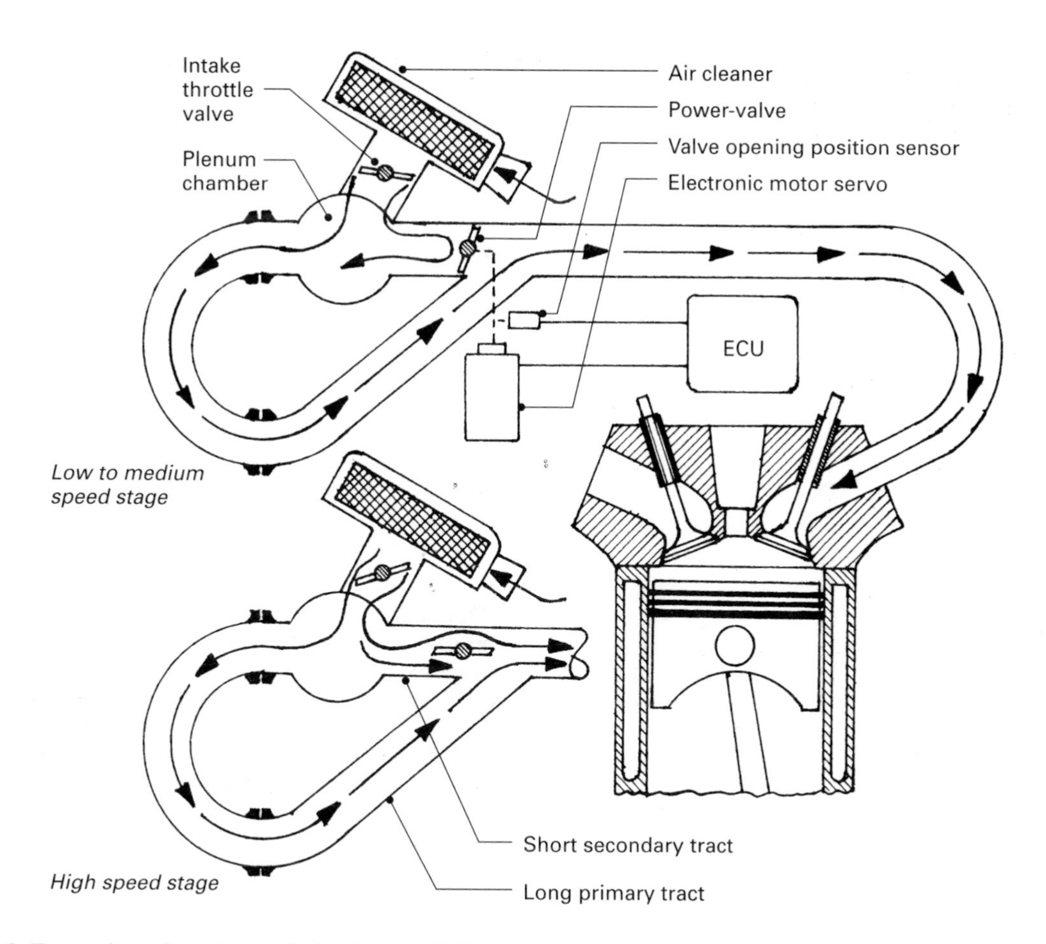

Fig. 24.2 Front view of dual-stage induction manifold system with merging primary and secondary tracts

and wide cross-sectioned secondary tract. Both tracts are fed from a common plenum-chamber but merge together some distance before the inlet-port.

At low to medium engine speeds, the power-valve in each secondary tract is closed so that the air stream flows through the long tract. Under these conditions the small bore long tract considerably increases the air velocity and correspondingly raises its momentum. Consequently an enhanced ram inertia crams air into the individual cylinders far in excess of that which would be possible with a shorter and larger bore passageway.

With increased engine speed the air velocity in the passageway will also increase. This raises the frictional wall resistance and lowers the charge density so that the primary tract finds it difficult to cope with the filling of the cylinders. As a result the inhaled mass per cylinder decreases and correspondingly the engine torque commences to fall.

To overcome the reducing volumetric efficiency at something like 2/3 maximum engine speed, the power-valves in each secondary tract are programmed to open. The relatively short and large cross-sectioned passageway of the secondary tract now enables the fresh charge to travel the shortest distance with the least resistance from the intake throttle to the cylinder-head inlet-ports. Thus there is a boost to the volumetric efficiency and accordingly engine torque in the upper speed range of the engine as shown in fig. 24.3.

24.1.2 Dual-stage induction manifold system with separate primary and secondary tracts (fig. 24.4)

The variation of a dual-stage induction manifold system differs from the merging primary and secondary tract in that the primary and secondary tract feed directly to separate inlet valve-ports. Thus at low to medium engine speeds with the secondary power valve closed, only the long and narrow bore primary tract supplies an air charge

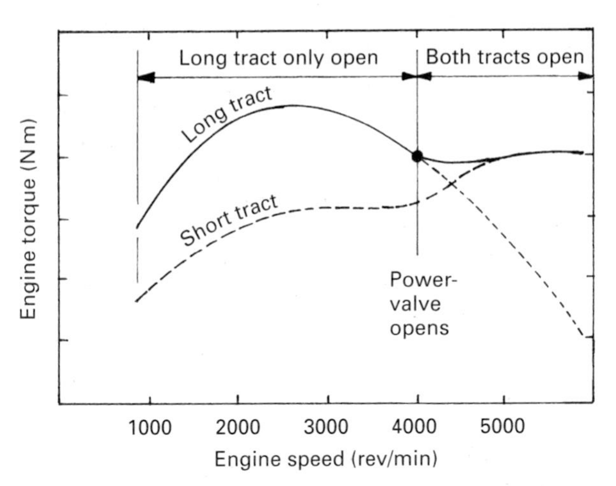

Fig. 24.3 Torque-speed characteristics for dual-stage induction manifold system

to one of the twin inlet-valves. The air charge (diesel engine) or air/fuel mixture (petrol engine) therefore enters the cylinders with considerable momentum at a tangent to the periphery of the cylinder. It thus imparts a vigorous swirl to the charge in a whirlpool or spiralling motion. The intensity of air movement provides exceptionally good mixing for the air/fuel mixture entering the cylinders or when fuel is injected directly into the cylinders as in the case of a diesel engine. It therefore promotes good low and medium-speed engine torque.

At higher engine speeds the small bore primary tract tends to throttle the air intake so that there would be a drop in cylinder volumetric efficiency. Consequently at some predetermined engine speed and load the short bore secondary tract power-valve progressively commences to open. The incoming air stream from the secondary tract supplements the primary supply and also modifies the air movement from a swirl to a tumble-roll in the cylinders. As a result the intensity of air and fuel mixing matches the engine speed, and considerably improves the cylinder filling process, thus ensuring ample engine torque at the top of the engine's speed range.

Operating the power-valve (figs 24.1, 24.2, 24.3 and 24.4) The power-valves are usually coupled together by a common spindle and can be operated by either a vacuum-diaphragm-actuator or an electric-motor-servo. The vacuum-diaphragm-actuator solenoid-controlled vacuum-valve see (fig. 24.4) is programmed by the electric control unit (ECU) to either open or close at a certain

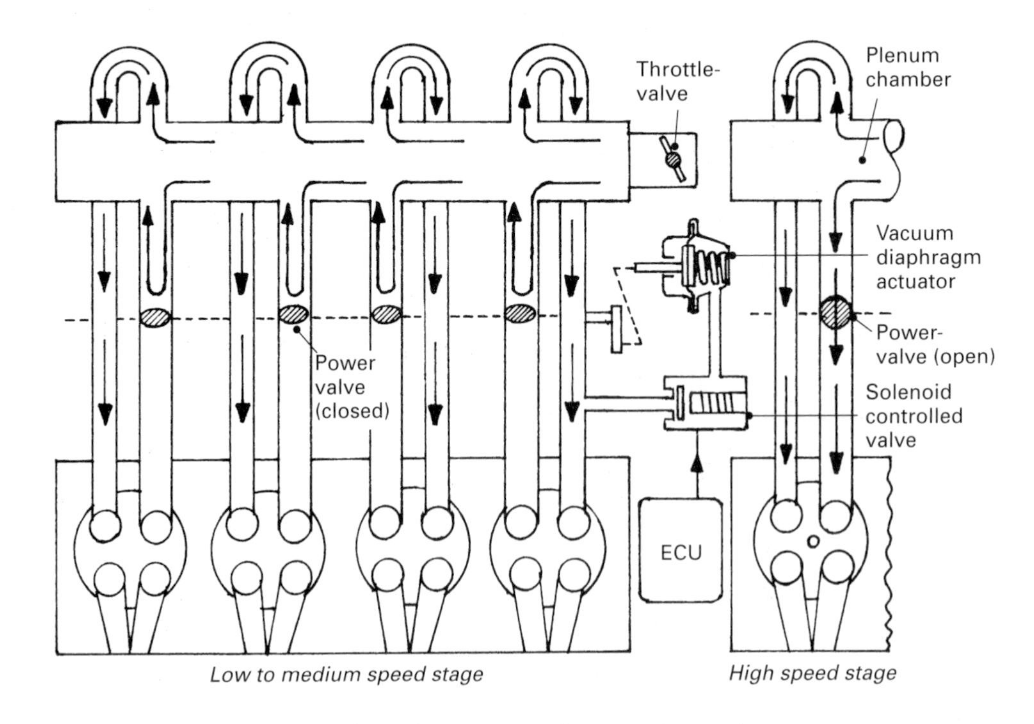

Fig. 24.4 Dual-stage induction manifold system with separate primary and secondary tracts

engine speed, whereas the electric-motor-servo with the aid of a power-valve angular-position sensor (see fig. 24.1) is programmed to progressively open or close the power-valve over some predetermined engine speed range.

24.1.3 Continuously adjusting variable-length induction manifold system (tickford design) (fig. 24.5)

This variable-length induction manifold comprises an outer cylindrical casing extending the full length of the cylinder-block with individual branch tracts which feed the inlet valve-ports, and a central mounted drum which has a scooped-out cavity stretching from one end to the other. The semi-circular space between the cylindrical casing and drum forms the variable tract length. Air is drawn in at one end of the casing and enters the drum's cavity which serves as a plenum-chamber.

At low engine speed the drum will be programmed so that the plenum-chamber will be at its furthest from the outlet port, thus a circular extended tract exists to promote a high degree of ram-inertia effects. It therefore causes more air charge to be packed into the cylinders per cycle.

As the engine speed increases, the servo-motor will be programmed to partially rotate the drum to match the tract length so that it optimises cylinder filling over the engines speed range.

At very high engine speeds the drum will have rotated to a position where the minimum circular-tract length is effective. It is claimed that this variable length induction-manifold boosts engine torque at both low and high engine speeds.

24.2 Supercharging

24.2.1 Fundamentals of supercharging

Supercharging is used to increase the air density of the air charge entering the cylinders so that there will be more oxygen available for every power-stroke than is possible with the conventional naturally aspirated engine which relies on induction depression to draw fresh charge into the cylinders.

Superchargers, which are also referred to as

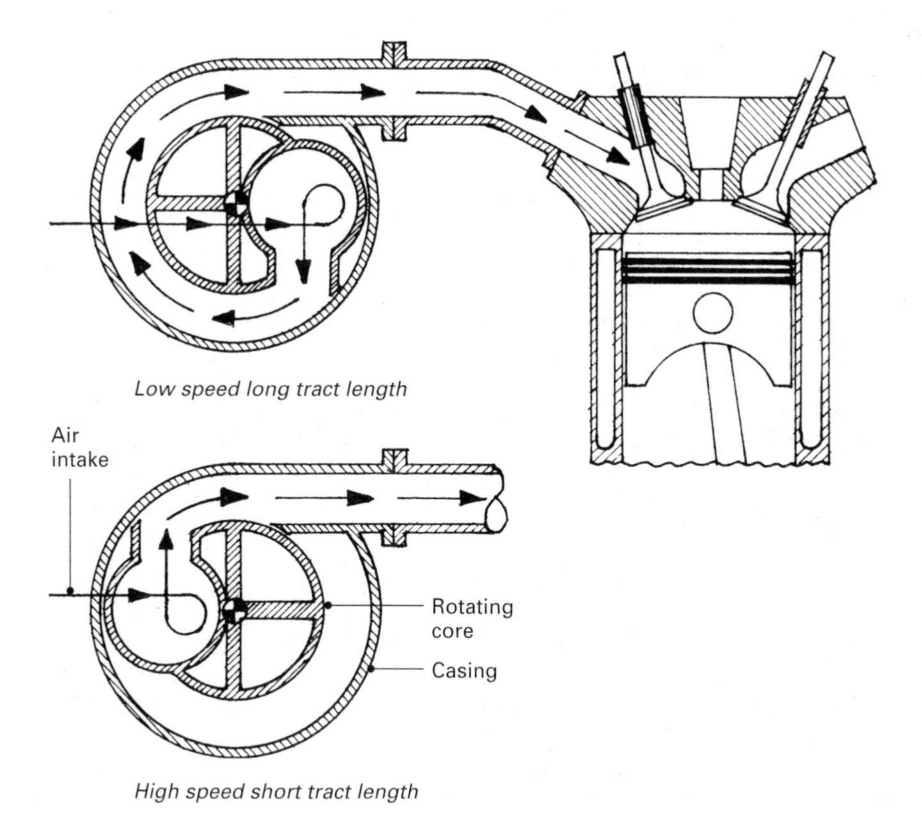

Fig. 24.5 Continuously adjusting variable-length induction manifold system (Tickford design)

compressors or blowers, are capable of raising the engine power output by as much as 25–50% above that of an equivalent naturally aspirated engine. Nevertheless there are some drawbacks: positively driven superchargers such as the Root's rotating lobe can consume as much as 15% of engine power and fuel consumption can suffer, whereas non-positive driven superchargers such as the turbo-charger consumes no power but is incapable of delivering boost power at low engine speeds.

Superchargers can be divided into two categories:

1) those which are driven positively by the engine via belt, chain, or gear-wheels and therefore consume engine power;
2) those which are non-positively driven but receive their energy from the expelled exhaust gases and do not consume engine power directly.

Supercharger classification	
Positive driven	Non-positive driven
Sliding vane Semi-articulating sliding vane Root's rotating lobe Male and female screw (Sprintex) 'G' Lader oscillating spiral displacer	Turbo charger Comprex pressure wave supercharger

Only four of these types of superchargers will be described in the following text. However before this, a brief mention of the meaning of boost pressure will be given.

Boost pressure and pressure ratio Supercharged engines pressurise the fuel charge above atmospheric pressure so that it is positively forced into the cylinder, whereas naturally aspirated engines induce charge into the cylinder by creating a negative pressure, that is, below atmospheric pressure. It is useful at this point to reflect on a few terms:

Atmospheric pressure Atmospheric pressure is the pressure exerted at sea level by the air and gas layer surrounding the Earth's surface.

Gauge pressure Gauge pressure is the pressure measured above atmospheric pressure and registers on a gauge.

Absolute pressure Absolute pressure is the combination of both atmospheric pressure and gauge pressure.

Boost pressure Boost pressure is the gauge pressure of the output delivery from the supercharger and quoted in bars.

A very common comparison used when comparing the state of engine tune is the ratio of absolute pressure to that of atmospheric pressure, this being known as the pressure ratio or the boost pressure ratio.

$$\text{Boost pressure} = \frac{\text{absolute pressure}}{\text{atmospheric pressure}}$$

$$= \frac{\text{boost pressure + atmospheric pressure}}{\text{atmospheric pressure}}$$

A general guide to the amount of supercharge is given as follows:

Intensity of charging	Boost pressure (bar)	Pressure ratio
naturally aspirated	0.0 and below	1.0 and below
low	0.0–0.5	1.0–1.5:1
medium	0.5–1.0	1.5–2.0:1
high	1.0 and above	2.0 and above

A typical maximum boost pressure ratio for a saloon car today would be in the region of 1.8:1.

24.2.2 Sliding-van supercharger

Construction (fig. 24.6(a) and (b)) This supercharger is composed of a drum mounted eccentrically within a cylindrical casing. Vane-blades fit into radial grooves formed around the periphery of the drum. The inlet port is situated where the gap between the drum and casing wall is large, whereas the discharge port is positioned where the drum and casing walls are together. Centrifugal force causes the blades to move outwards against the casing wall, thereby sealing the space between adjacent blades. Therefore it is necessary to have an oil drip feed into the casing to lubricate the blades and casing wall.

This type of blower is capable of delivering boost pressure at relatively low speed but it can only be for modest boost pressure in the mid to upper speed range.

Operating principle (fig. 24.6(a) and (b)) A rotating drum draws in air or an air/fuel mixture into the

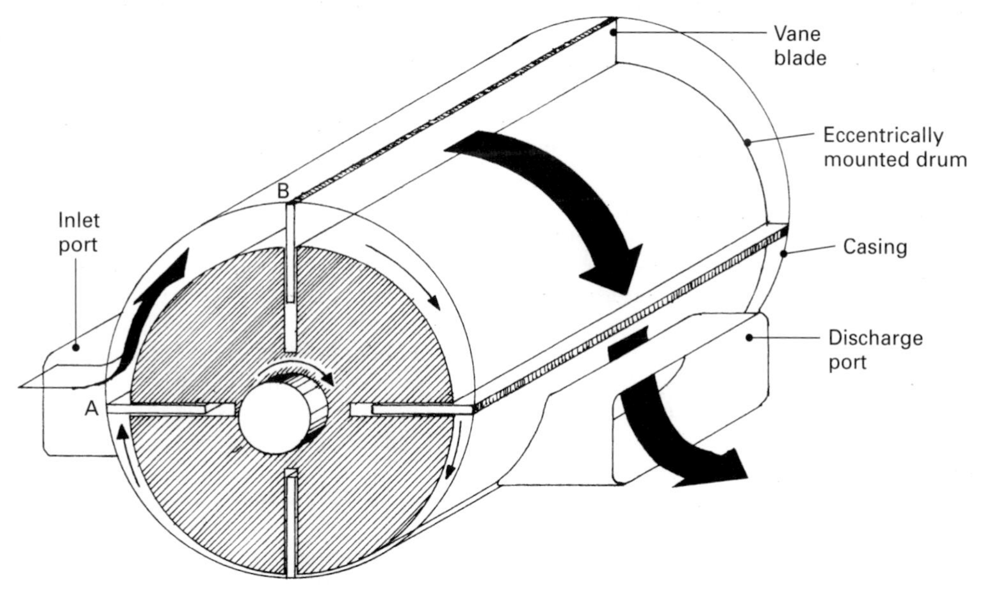

a) Rotating drum draws charge into the enlarging chamber space formed between adjacent vane blades (AB)

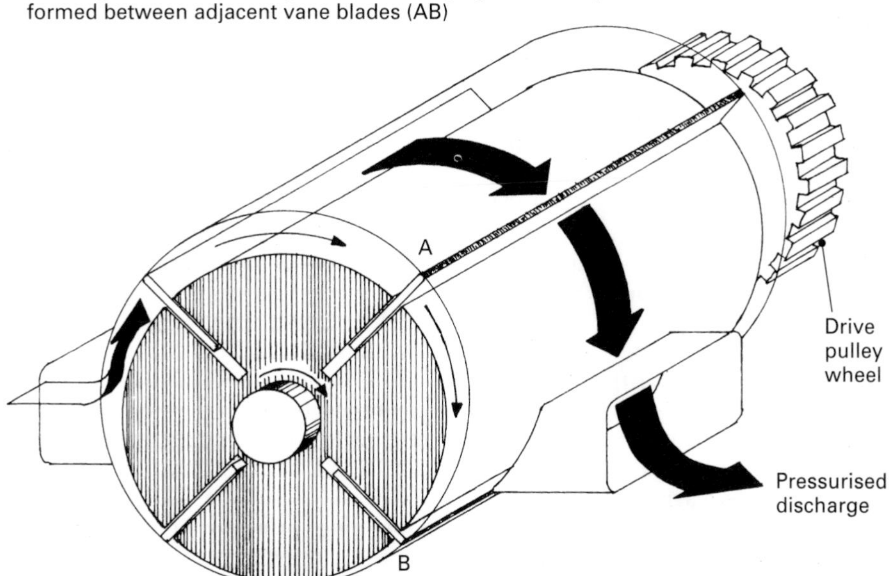

b) Rotating drum forces charge out of the reducing chamber space formed between adjacent vane blades (AB)

Fig. 24.6 Vane-type supercharger

expanding cell formed between the vane-blades A and B until blade A shuts off the inlet port (fig. 24.6(a)). Further rotation of the drum increases the volume of the sealed cell AB so that the pressure changes to a depression (the negative pressure phase does not contribute to the pumping action of the blower). This is followed by a reduction in the space formed between the drum and casing wall so that it compresses the charge, hence the charge pressure rises. Eventually the rotating blade B uncovers the discharge port so that the trapped charge in the contracting cell is squeezed out to the discharge port (fig. 24.6(b)). Initially flow-back of the stored charge in the induction manifold occurs, followed by equalisation of both cell pressure and reverse flow pressure. Any further movement of the drum now pushes the resultant charge through the outlet port and into the manifold. The charge contained in the manifold is then released to the individual cylinders according to the inlet valve opening sequence. With the eccentric drum and vane type of blower the charge is partially compressed between blade cells, but most boost pressure is caused by the cramming of more and more charge in the induction manifold, far in excess of that consumed by the cylinders.

24.2.3 Root's rotating lobe supercharger

Construction (fig. 24.7(a) and (b)) Two lobe rotors supported on ball or tapered-roller bearings are timed by a pair of equal-sized gear-wheels so that the lobes intermesh but never touch each other. Likewise there is an operating clearance between the ends and outer tips of the lobes and their flat and circular casing walls respectively.

Drive is transferred either by belt or gears to one of the rotors via a separate pulley-wheel or gear-wheel. For slow-speed engines there is usually a step-up gear-ratio, something like 1.0:1 to 1.5:1, whereas for a high-speed engine, there is a gear-ratio reduction in the order of 0.6:1 to 1.0:1 to limit the blower to a maximum safe operating speed. External lubrication of the bearings is necessary but no internal lubrication is required throughout the life of the blower.

Operation principle (fig. 24.7(a) and (b)) Rotating lobe rotors draw air or an air/fuel charge into the chamber formed between adjacent lobes AB until the trailing lobe A seals off the intake port, after which the trapped charge is carried around the periphery of the upper casing wall to the discharge port (see fig. 24.7(a)). Simultaneously charge is attracted into the chamber CD. The incoming charge in this chamber is then cut off from the intake port by the lobe C and then transferred around the circular lower casing wall to the outlet side of the casing. When the leading lobes B and D relative to their phasing uncover the discharge port, the compressed charge stored in the induction manifold rapidly flows back and mixes irreversibly with the unpressurised charge held in the lobe chambers AB and CD (see fig. 24.7(b)). Pressure equalisation between the charge already in the chamber and that which has rushed back from the manifold then takes place. Further rotation of the rotors then pushes both the recently arrived charge and the reverse flow fluctuating charge out into the induction manifold where it is stored until the sequence of cylinder-head valve opening transfers the charge to the various cylinders.

It should be observed that the rotating lobes only displace the charges from inlet to outlet port, and do not in any way compress it until it is ejected from the discharge port. Pressure build-up can only take place when the quantity of charge displaced into the induction manifold exceeds the amount of charge delivered to the cylinders. Thus the repeated discharge into the induction manifold with rising engine speed progressively increases the mass of stored charge and raises its pressure.

24.2.4 Screw-type (Sprintex) supercharger

Construction (fig. 24.8(a) and (b)) A pair of magnesium-alloy male and female screws have helical lobes that twist along their length. The screws are teflon-coated to protect them against corrosion. The configuration is such that the male screw has four convex-shaped lobes whereas the female screw has six concave spaces between its six lobes. Both screws are coupled together by a pair of timing gears which provide a gear ratio of 1.15:1 for the female and male screws respectively. This ensures that the concave and convex profiles of the two screws intermesh but they never touch each other. Likewise the outer periphery and ends of the screw operate with a small screw to casing clearance. Radial ball-bearings at each end of the screws support and maintain the correct working clearance between screws and casing. A typical engine to supercharger speed step-up would be 2.5:1 for a petrol engine.

Operating principle (fig. 24.8(a) and (b)) Air charge is drawn through the intake port to fill the expanding space created underneath the intermeshing male and female lobes AB as both screws rotate counter to each other. Further movement of the screw traps the air charge between adjacent lobes AB and CD and the cylindrical casing wall. These enclosed cells then move around the periphery of the casing towards the discharge port on the opposite side of the casing. The intermeshing male and female lobes AD on the discharge port side provide a merging cell with a contracting volume. At the same time the left and right-hand helical lobes form pairs of converging double diagonal vee-shaped cells along the length of

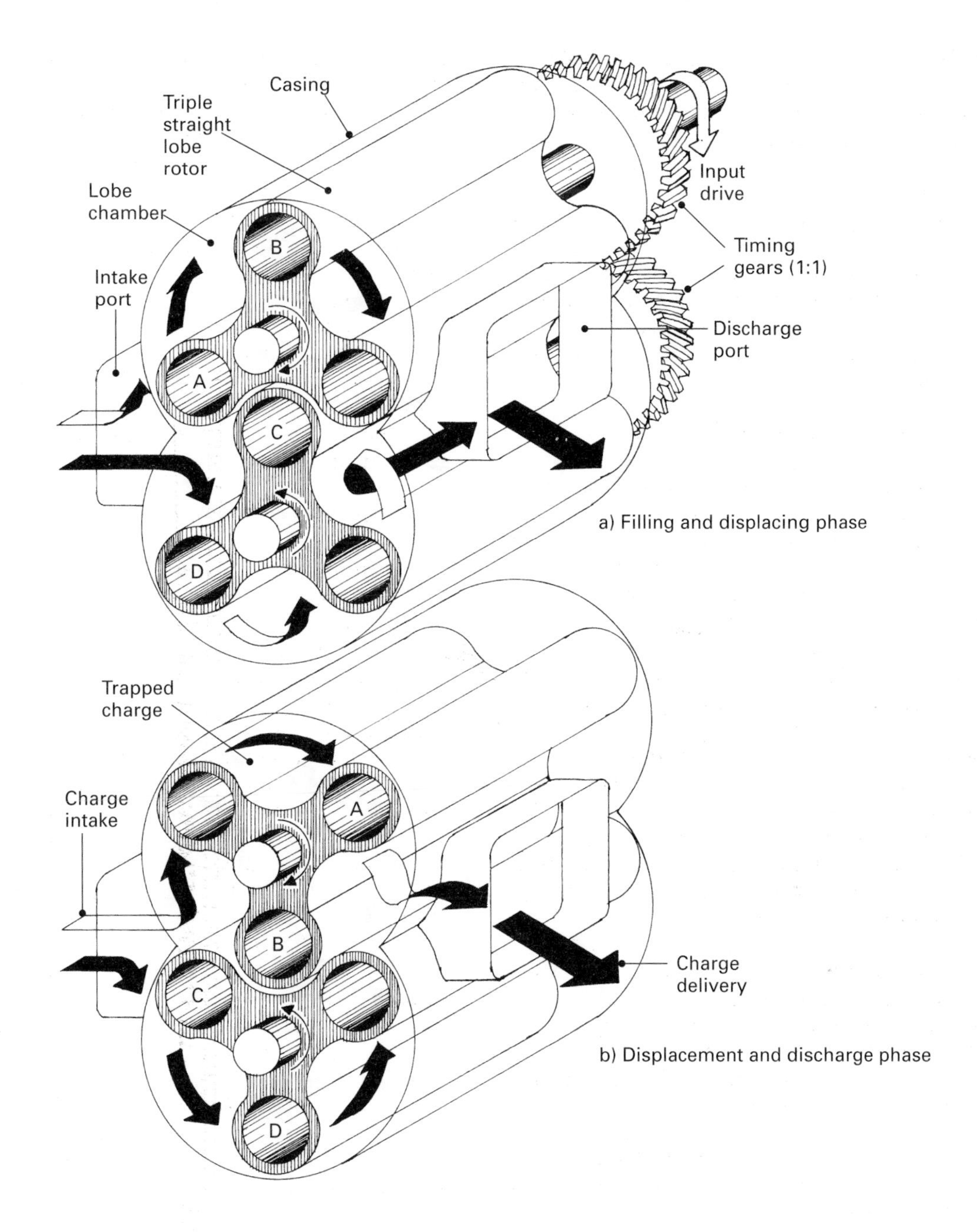

Fig. 24.7 Triple-lobe Root's-type supercharger

the screws. These lengthways converging cells squeeze axially the air charge trapped between the lobes as they approach the discharge port. Thus the pumping action is a combination of the circular displacement and axial compression of the air charge. These superchargers tend to consume very little power and operate at relatively lower temperatures so that intercooling is not always necessary, providing the pressure ratio is not in excess of 2:1.

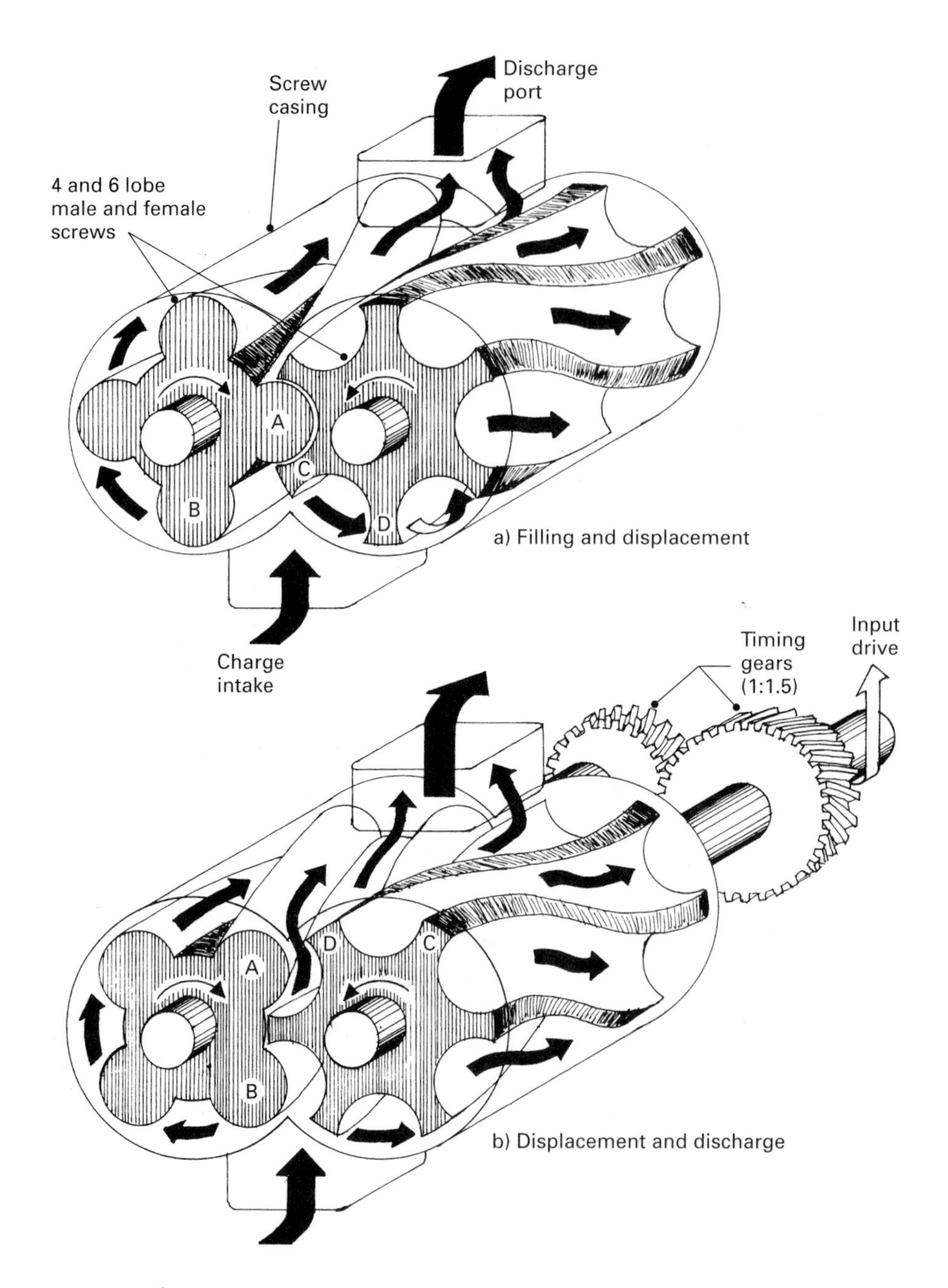

Fig. 24.8 Screw-type supercharger

24.2.5 Turbocharger

Construction (fig. 24.9) A turbocharger consists of a centrifugal compressor-wheel, and an exhaust gas turbine-wheel both being supported on a common spindle which is itself mounted on fully floating bearings, the compressor-wheel being enclosed in an aluminium-alloy housing, whereas the turbine-wheel which operates at much higher temperature is contained in a cast-iron housing.

Compressor (fig. 24.10(a)) The impeller-wheel is an aluminium-alloy casting, and has the configuration of a disc supported on a hub with about 12 or so radial blades which curve backwards to

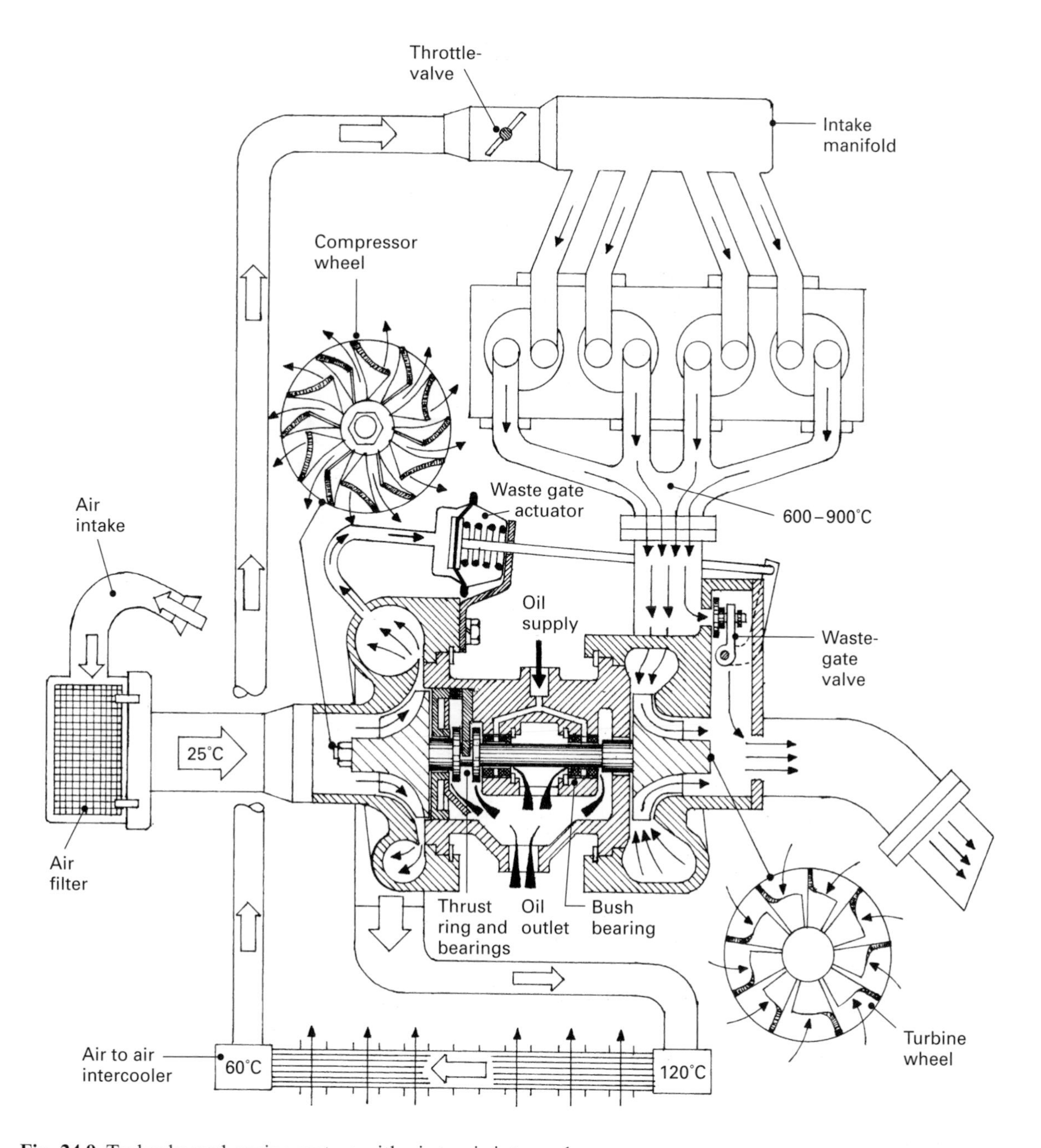

Fig. 24.9 Turbocharged engine system with air-to-air intercooler

the direction of rotation. A portion of every second or third blade near its root is bent forward as this tends to create a slight depression at the centre of the impeller-wheel when rotating. It thus assists the in-flow of air charge.

When the impeller-wheel is rotated, air is induced towards the hub of the impeller. It then enters the cells created between adjacent pairs of blades. The curved shape of the hub then changes the direction of flow from an axial to a radial one. This air is then subjected to centrifugal force which forces the air outwards towards the periphery of the wheel. The greater the speed of rotation of the impeller-wheel, the greater will be the outward thrust of air. The air then flows between parallel narrow walls (parallel wall diffuser) where

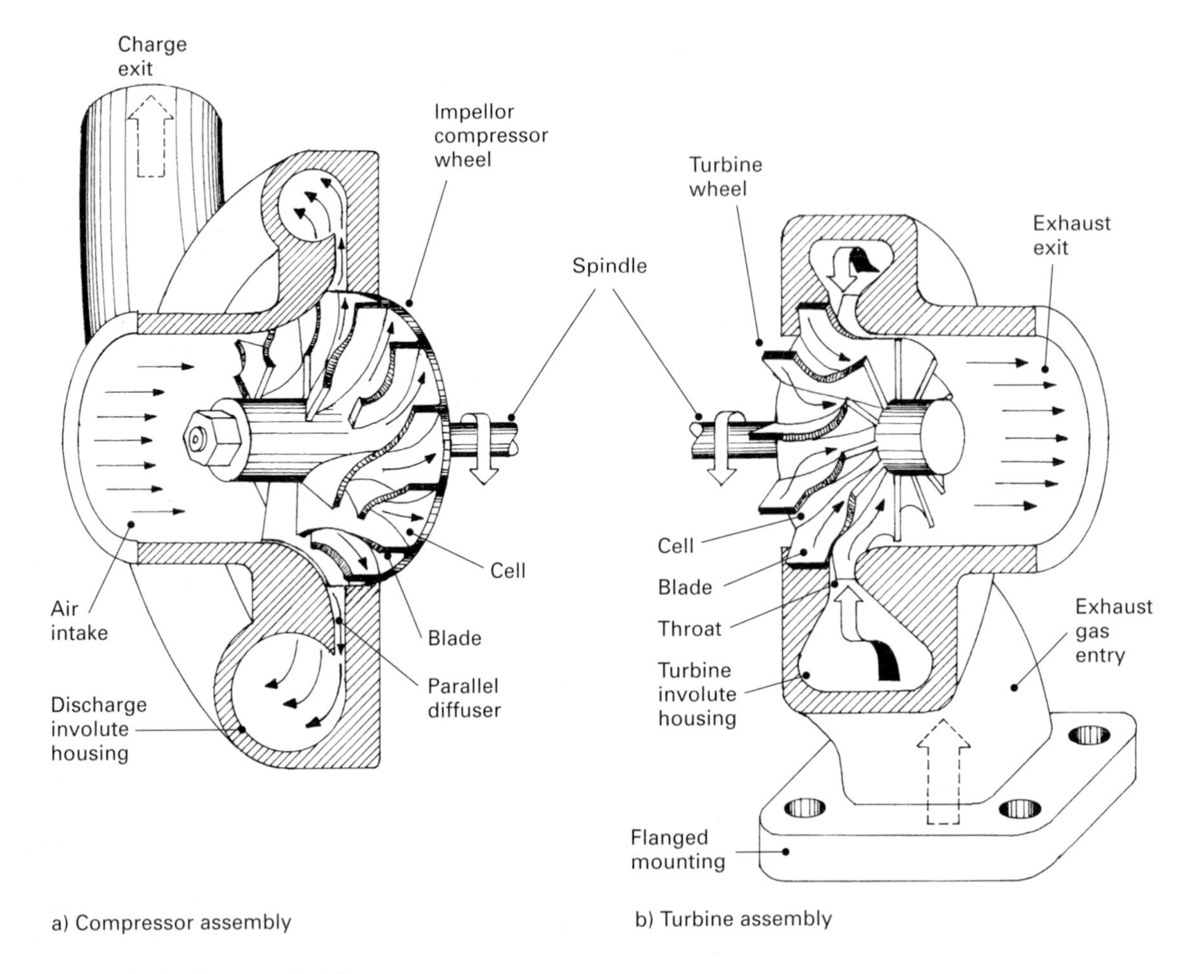

Fig. 24.10 Turbocharger principle

the velocity-energy is converted to pressure-energy. It then passes to, and is collected by, a discharge involute which encircles the parallel diffuser. This involute is in effect a passageway whose cross-section progressively enlarges towards its exit. The involute prevents the air leaving the diffuser from overcrowding the exit passageway, at the same time continuing the velocity to pressure energy conversion. The accumulative discharge from each impeller-cell is thus collected and then pushed out to the induction manifold.

Turbine (fig. 24.10(b)) The turbine-wheel is made from a heat-resisting nickel-based alloy which may have to operate at full engine load at temperatures up to 900°C. Alternative lightweight ceramic materials such as sintered silicon nitride are now available and are being used to a limited extent by some turbocharger manufacturers. The turbine-wheel takes the form of a curved hub which has its largest diameter at the spindle and its smallest diameter at the exhaust-gas discharge end. Cast around the hub are straight radial blades that curve backwards towards the outlet side of the turbine wheel.

When the engine is running, exhaust-gas from the exhaust-manifold enters the turbine involute housing. Its kinetic energy then drives it tangentially inwards through the throat of the turbine housing where it impinges against the blades. It is then diverted axially towards the exhaust exit. Thus the greater the quantity and the higher the pressure of the exhaust-gas released from the engine cylinders, the greater will be the kinetic-energy impact onto the turbine-blades and correspondingly the faster the turbine and impeller-wheel will rotate.

24.2.6 Boost-pressure control wastegate (fig. 24.9) Boost-pressure build-up characteristics with respect to engine speed usually show a very sluggish rise in boost-pressure at low engine speeds up to between 1500 and 2000 rev/min, followed by a steep pressure rise which must be limited if excessively high cylinder pressure which could damage the engine are to be avoided.

Boost-pressure control is generally achieved by a wastegate-valve which permits a proportion of the exhaust-gas expelled from the engine to by-pass the turbine-wheel and go directly to the exhaust system. Consequently when the wastegate-valve opens, energy which would drive the turbo faster is now diverted, such that the boost-pressure remains approximately constant above the speed at which the wastegate-valve opened.

The wastegate-valve is operated by a diaphragm actuator which in turn is controlled by output boost-pressure in the involute housing. This pressure is relayed to the diaphragm-actuator which at some predetermined pressure will overcome the stiffness of the return-spring to push open the wastegate-valve flap.

Lubrication (fig. 24.9) The rapid compressor/turbine-wheel response to the acceleration and deceleration and the life expectancy of the turbocharger relies considerably on the steady flow of clean oil supplied from the engine's lubrication system to the floating plain bearings and its free exit back to the engine's sump. The oil not only lubricates the spindle and bearing but it also cools the surrounding walls.

It is essential that oil of the recommended quality and viscosity is used, and that the oil is changed within the prescribed mileage and that there is no restriction in the oil inlet or the oil exit from the turbine unit.

24.2.7 Intercooling

Air delivered to the induction-manifold from the supercharger is subjected to reverse-flow pressure pulsation and turbulence and as a result a considerable amount of heat is generated. Now engine power is dependent principally upon the mass of air drawn into the cylinder per cycle, and increasing the charge pressure ratio alone will not permit the maximum quantity of charge to enter the cylinder. In fact the true measure of power potential is the density of the charge in the cylinder and this relates directly to the temperature of the air charge about to enter the cylinders. The lower the charge temperature at a given constant pressure the smaller will be its volume and hence more air charge is able to enter the cylinder.

It is estimated that where the output temperature from a supercharger may reach 120°C and this remains uncooled when delivered to the engine, then the power lost could be as high as 38%. Intercoolers which reduce the delivered air charge temperature to around 60°C can raise the engine power by as much as 20%. However the incorporation of an intercooler is only justified provided the boost-pressure ratio is 1.6:1 or more, as a lower operating pressure will not raise the charge temperature to such high values.

The function of the intercooler, or as the Americans call it the 'aftercooler', is to transfer heat from the compressed charge to another source which is at a lower temperature, such as the engine's cooling system, or directly to the surrounding atmosphere. Hence the intercooler is essentially a heat-exchanger. There are two types of heat-exchangers in common use. These are:

1) air to liquid and
2) air to air.

Air to liquid intercooler These intercoolers normally consist of many horizontal circular tubes through which coolant liquid from the engine cooling system passes. Copper cooling fins are closely stacked upright to the axis of the tube to speed up the heat transfer process. The compressed and heating air charge flows around and across these tubes. Heat tranfer then commences between the heated air and cooling liquid, by conduction through the copper-nickel alloy walls of the tubes and by convection-current as the coolant is circulated by the engine coolant-pump. These air to liquid intercoolers are capable of reducing a charge temperature of up to 150°C down to engine coolant temperature of about 85°C. Air to liquid heat-exchangers are usually attached to the engine and are so designed to form part of the induction manifold.

Air to air intercooler (fig. 24.9) With this type of heat-exchanger compressed charge passes through horizontally mounted elongated flat cross-section tubes. The external surface area of the tube is greatly increased by attaching fins made from corrugated copper or aluminium sheet between the horizontal row of tubes. The cooling atmospheric air flows between and across the tubes to increase the effectiveness of the cooling air flow.

The intercooler is usually placed in front of the engine radiator so that air will be drawn through the finned matrix by the engine fan or by the ram effect caused by the forward movement of the vehicle.

Heat-transfer takes place between the hot air charge and the much cooler atmospheric air stream by conduction through the copper or aluminium alloy walls of the tubes and then by convection current and radiation to the surrounding atmosphere.

Heat-exchangers of this kind are capable of reducing a charge temperature of 120°C to something like 60°C. Air to air heat-exchangers are not attached to the engine and are most effectively positioned in front of the radiator if space permits. The air to air intercooler is more popular than the air to liquid intercooler because it is able to bring down the air-charge temperature to much lower values of around 60°C as opposed to 85°C.

24.3 Exhaust manifold configurations

An exhaust manifold's purpose is to bring together the expelled exhaust gases from the numerous engine cylinders and to transfer them to the catalytic converter and silencer box via the exhaust system downpipe or pipes. Exhaust gas ejected from the engine cylinders is collected from each exhaust valve-port through branch passages which merge together to form a common passageway, hence the term manifold. The exhaust gas then flows to the downpipe/pipes through one, two and sometimes three flanged outlets.

The basic principle of good design is to keep the exhaust gases expelled from each cylinder into the branch-passage separate for as long as possible before permitting them to merge together. If the manifold branch-passages are very short the exhaust gas from individual cylinders is likely to reverse the direction of flow and move into adjacent branch-passages, thereby causing interference and impeding the flow of the gas. Equal branch-passage length is desirable but not always practical if identical scavenging from each cylinder is to be achieved. Long branch-passages enable the momentum of the column of exhaust gas moving along the flow-path to leave behind a depression in the exhaust port. If the exhaust valve closure is delayed, an extraction effect pulls the remaining unwanted gas out of the combustion chamber. It thereby permits the fresh charge to enter the cylinders unrestricted.

24.3.1 In-line four-cylinder exhaust manifold (fig. 24.11(a) and (b))

Conventional four cylinder manifolds have four branches of unequal length merging into a common central exit passage (see fig. 24.11(a)). This configuration can suffer from exhaust-gas interference between branch-passageways and unequal exhaust-port back-pressure, and with a limited exhaust-port depression pull effect to clear the exhaust gases from the combustion chambers.

A better but more expensive layout divides the manifold into two halves (see fig. 24.11(b)), 1–4 branches and likewise 2–3 branches converge, each converged exit being joined to one branch of the twin fork downpipe. In turn the twin fork downpipes converge into a single pipe a little way in front of the catalytic converter (see fig. 24.11(b)). The benefit of the step-by-step convergence of branch pipes and downpipes prolongs the interference of exhaust-gas discharge. Hence it minimises the gas back-pressure, and enhances the suction pulse extraction effect for each branch passage. Thus with a firing-order of 1342 the exhaust-gas expulsion sequence switches in turn between downpipe forks, that is, No. 1 branch discharges to the right-hand fork, No. 3 branch discharges to the left-hand fork, No. 4 branch discharges to the right-hand fork, and No. 2 branch discharges to the left-hand fork downpipe. This sequence being continuously repeated provides a 360° discharge interval between each downpipe.

24.3.2 In-line five-cylinder exhaust manifold (fig. 24.11(c))

The long branch-passageways in which branches 1–4 and 2–3 converge and branch 5 remains on its own (see fig. 24.11(b)) forms a very good compromise with this uneven cylinder number arrangement. Pairing branches 1–4 and 2–3 and keeping branch 5 separate provides extended irregular discharge intervals of 288°, 432° and 720°, respectively, from each of the triple downpipe forks. Branch to branch exhaust-gas interference is at a minimum and the extended discharge intervals, though uneven, permit a moderate degree of exhaust-gas depression extraction from the exhaust ports.

24.3.3 In-line six-cylinder exhaust manifold (fig. 24.11(d))

Dividing the manifold in two so that cylinders 1–2–3 discharge into one downpipe and cylinders 4–5–6 discharge into the other of the twin-fork downpipe

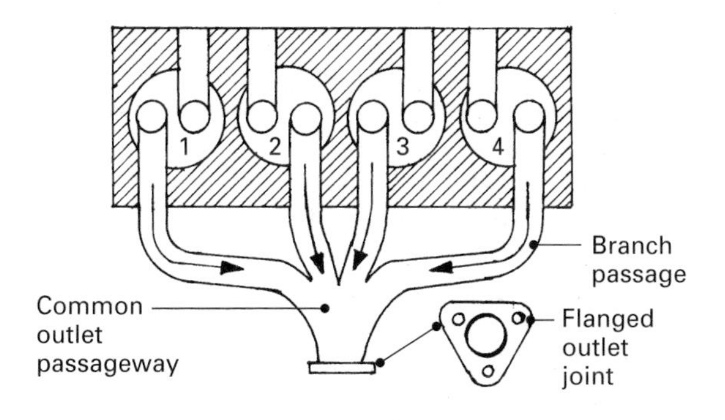

a) Four-cylinder engine with four-branch central downpipe exhaust manifold

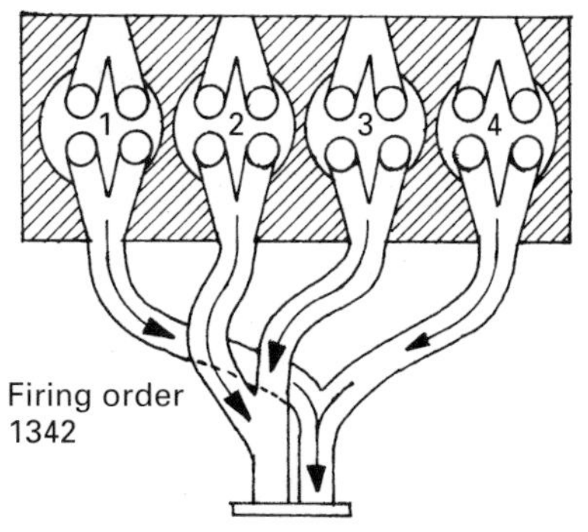

b) Four-cylinder engine with four-branch twin downpipe manifold

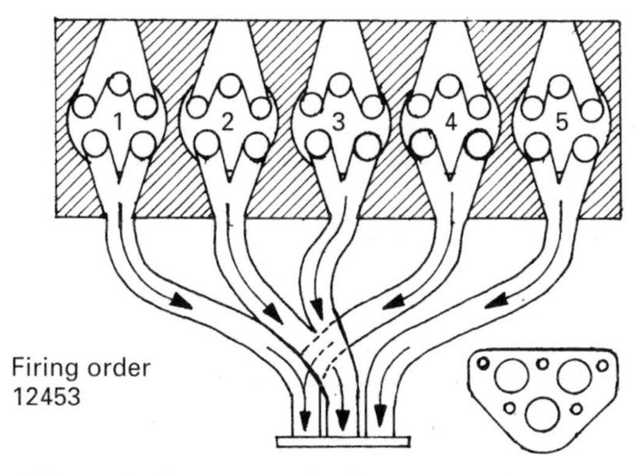

c) Five-cylinder engine with five-branch triple downpipe exhaust manifold

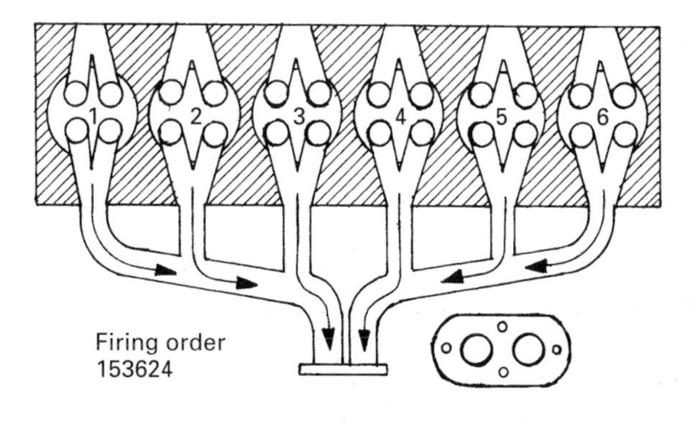

d) In-line six-cylinder engine with six-branch twin downpipe exhaust manifold

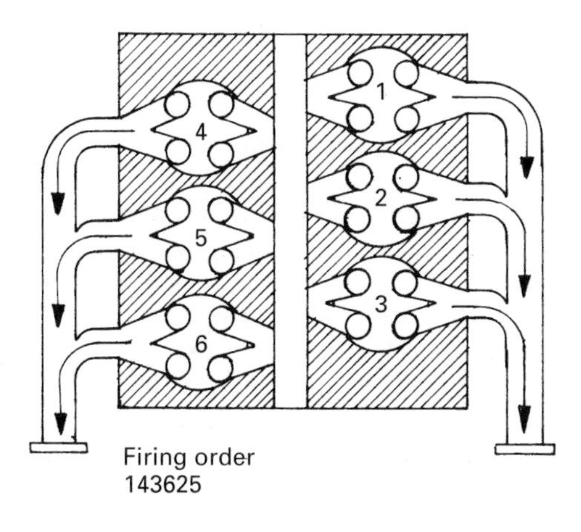

e) Vee-six cylinder engine with three-branch single downpipe exhaust manifold

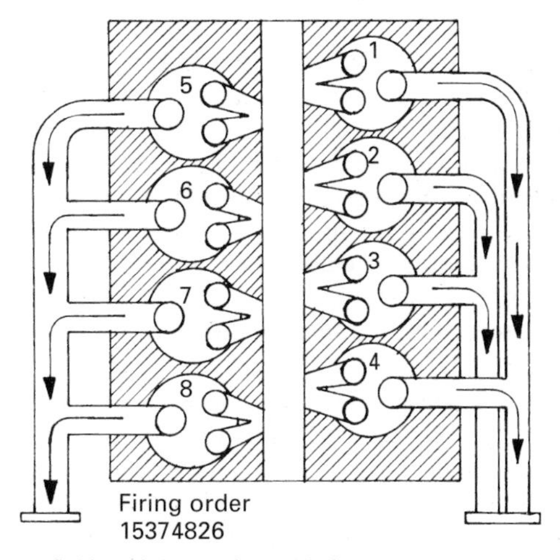

f) Vee-eight engine with four-branch single or twin downpipe exhaust manifold

Fig. 24.11 Exhaust manifold arrangements

layout, provides an equal exhaust expulsion interval between each half manifold exit of 240° crankshaft rotation. Adjacent branch-passage exhaust interference is minimal and exhaust-port depression extraction can be usefully utilised.

24.3.4 Vee-six-cylinder exhaust manifold (fig. 24.11(e))

This arrangement is in effect similar to the split manifold of the in-line six-cylinder engine; however, with a vee-cylinder banked engine each cylinder bank has its own three-branch manifold. Thus with a typical firing-order 143625, the exhaust discharge alternates from each cylinder bank, thereby providing a 240° crank-angle interval between discharges in each manifold. For normal performance the short branch-passages of equal length may have their passageways curved towards the rear, with each branch merging into a single passage, its exit being joined to a single downpipe at the rear of the engine. For greater engine performance, long branch pipes which take on a loop form of increasing radii towards the rear of the engine merge into a single downpipe via a conical collector.

24.3.5 Vee-eight engine exhaust manifold (fig. 24.11(f))

Each bank of a vee-eight cylinder engine can be considered as a separate four-cylinder engine, both banks being connected to the other by a common crankshaft and crankcase. An eight-cylinder engine has an exhaust discharge every 90° crankshaft rotation, and with an exhaust-manifold for each cylinder-bank and a single-plane crankshaft there should be an exhaust discharge interval for each downpipe of 180°. However, with the more popular two-plane crankshaft with a firing-order such as 15486372 there will be an uneven exhaust discharge interval from each cylinder-bank. For most purposes this uneven discharge interval is not considered serious enough to warrant an elaborate separate multi-loop manifold. Therefore, four short branch-passages of equal length merging into a common passageway, with one exit at the rear, usually suffice for each cylinder-bank, see fig. 24.11(f) left-hand bank, or a more effective but expensive arrangement as in fig. 24.11(f) right-hand bank.

24.4 Exhaust gas emission control

24.4.1 Exhaust gas composition (fig. 24.12)

When petrol burns to completion the exhaust gas products are basically water and carbon dioxide. However, the combustion in the engine's cylinders finds it difficult to completely burn the fuel; hence the exhaust gas composition now contains additional unwanted compounds of carbon monoxide (CO), hydrogen (H_2) and hydrocarbon (HC) and because of the high combustion temperature oxides of nitrogen (NO_x). Carbon dioxide (CO), carbon monoxide (CO) and oxygen (O_2) are measured as a volumetric percentage of the total exhaust gas composition while hydrocarbon (HC) and oxides of nitrogen (NO_x) are measured in parts per million because of the minute quantities involved. The exhaust gas composition is greatly influenced by the air/fuel mixture strength. Examining the exhaust gas composition versus air/fuel mixture ratio see (fig. 24.12) it can be seen that carbon dioxide (CO_2) is at its maximum at the stoichiometric ratio of approximately 14.75:1 and declines as the mixture strength moves either towards rich or weak. Conversely carbon monoxide (CO) is very near its minimum at the stoichiometric ratio but then rises steeply as the mixture strength enriches, while the oxygen (O) content is also almost at its minimum at the stoichiometric but increases steeply as the mixture ratio then moves to the lean side. Hydrocarbon (HC) is at a minimum in the region of optimum fuel economy 18:1 to 15:1 but then rises both as the mixture leans out and even more steeply as the mixture is enriched. Finally, very small quantities of oxide of nitrogen (NO_x) are seen to peak at an air/fuel ratio of approximately 16:1 and then drop rapidly as the mixture is either leaned or enriched. The high oxide of nitrogen (NO_x) concentration therefore conflicts with operating the engine just on the lean side of the stoichiometric ratio.

24.4.2 Exhaust gas emissions

Carbon dioxide (CO_2) Carbon dioxide is a product of combustion where the carbon in the fuel has been completely burnt. It is a colourless gas with a faint pungent odour and taste; however, it is not poisonous. Carbon dioxide when released from the exhaust pipe rises to the upper atmosphere and contributes to the formation of a layer around the Earth's atmosphere and is one of the primary causes of global warming.

Carbon monoxide (CO) Carbon monoxide is a product of combustion when the carbon in the fuel has not been completely burnt. It occurs with rich and poorly atomised mixtures and where the mixture distribution in the cylinder is

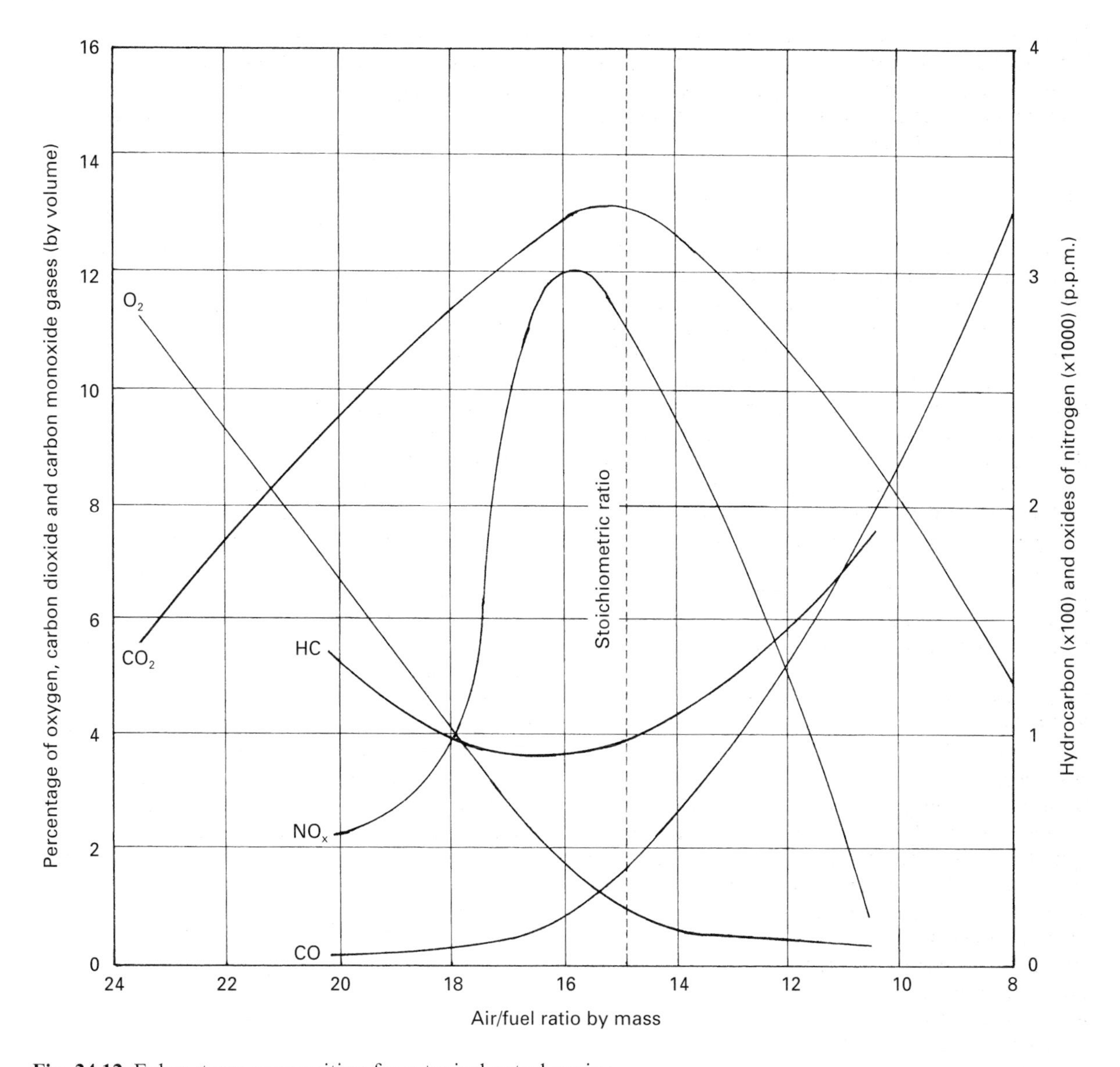

Fig. 24.12 Exhaust gas composition for a typical petrol engine

unequal. It is a colourless, odourless and tasteless gas which is poisonous when inhaled. Petrol engines operate in the region of the stoichiometric ratio, that is the theoretical quantity of air required for complete combustion of the fuel. The air/fuel mixture ratio may range between a rich and weak mixture that can vary from 10:1 to 20:1, respectively.

Hydrocarbon (HC) Hydrocarbon is a product of combustion and is the consequence of incomplete combustion of hydrocarbon fuel. Eye and throat irritation can occur. Most hydrocarbon compounds are harmless but there are some hydrocarbons which are cancer producing.

Oxides of nitrogen (NO_x) Oxides of nitrogen are the product of combustion and are formed when nitrogen combines at high combustion temperature. The first stage of oxidation is a colourless, odourless and tasteless gas; with further oxygen exposure it changes to a reddish brown colour. The latter is poisonous with a penetrating odour which can destroy lung tissue.

Lead (Pb) Lead is an antiknock additive and is introduced into some fuels in the form of tetraethyl-lead (TEL) or tetramethyl-lead (TML). Lead is a poisonous heavy metal. Exposure to leaded fuel emission can lead to absorption through the lungs and through the stomach and can affect the brain and nervous system.

Particulates Particulates are the product of combustion in diesel engines. Particulates are very tiny particles of solid carbon soot on which some hydrocarbon compounds have become absorbed. The soot is believed to form in regions of fuel spray which still exist after the gases have been cooled by the expansion process. Some of the organic hydrocarbons will also have cooled enough to condense and be absorbed by these tiny carbon particles. Traces of lubricating oil may also be found in the particulates. About 0.2–0.5% of the fuel mass may be emitted as particulates. These particulates can be inhaled and therefore form a health hazard as they may be retained in the lungs.

24.4.3 Exhaust gas catalytic converter

Construction of a three-way catalytic converter (fig. 24.13) Three-way catalytic converters have two basic components: an outer stainless-steel cylindrical or oval-shaped casing, within which is housed a honeycomb matrix core consisting of many parallel channel passageways through which the exhaust gas flows. Sandwiched between the honeycomb matrix and the casing is a layer of wire-mesh which protects the fragile matrix from both thermal expansion stresses and external denting.

The honeycomb passageways are made from a magnesium-silicate ceramic material which remains stable at exhaust gas expulsion temperatures which may reach 800°C to 900°C. A highly porous alumina (Al_2O_3) washcoat provides a rough surface finish, and covers the passageway walls for the purpose of increasing the effective surface area to something like 700 times. Furthermore the washcoat is impregnated by vapour deposition

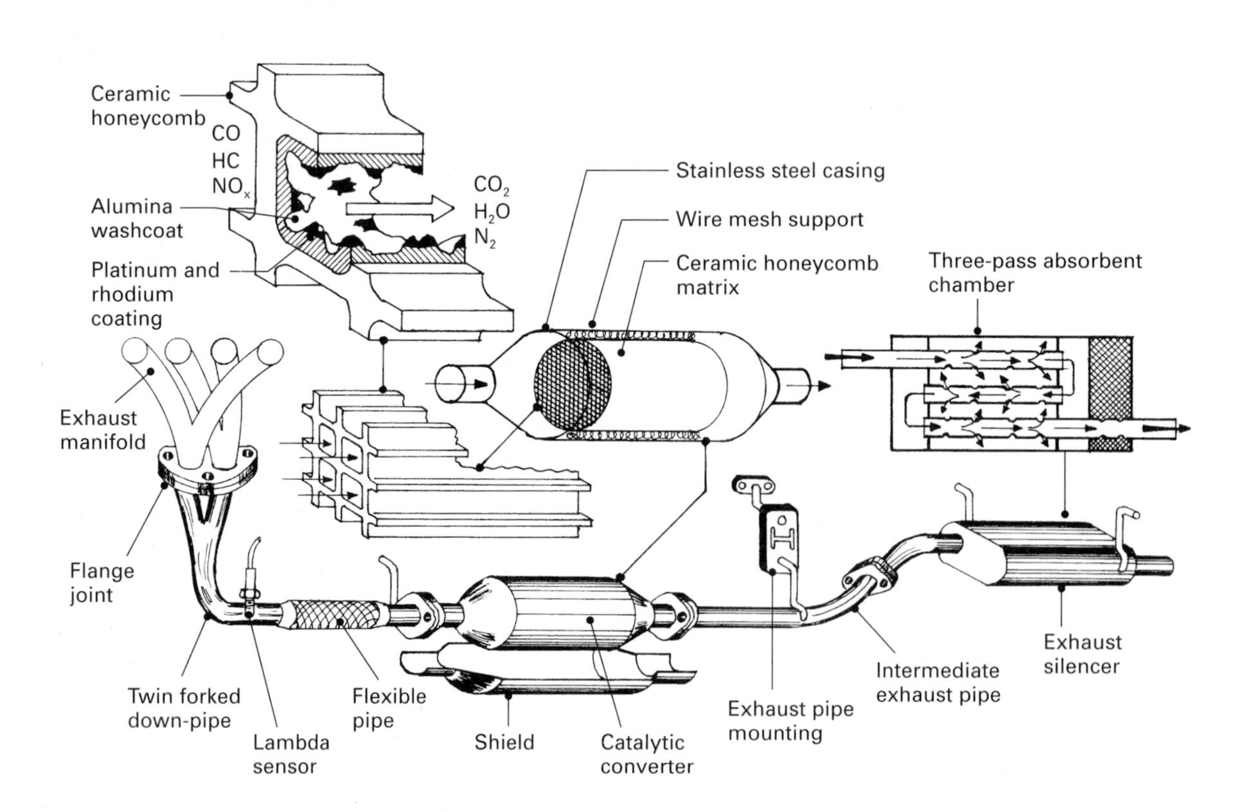

Fig. 24.13 Exhaust system with catalytic converter

with the noble metals platinum (Pt) and rhodium (Rh).

Three-way catalyst process (fig. 24.13) The three-way catalytic converter uses chemical reactions to convert the three main pollutants carbon monoxide (CO), hydrocarbon (HC), and the oxides of nitrogen (NO_x) in the exhaust gas to carbon dioxide (CO_2), steam (H_2O), and nitrogen (N_2).

The afterburn chemical reactions are speeded up by passing the exhaust gas through passageways which are exposed to a catalyst; a catalyst is a material that increases the rate of chemical activity without itself undergoing any permanent chemical changes. Platinum (Pt) and rhodium (Rh) are the most commonly used catalysts. The oxidation process (loss of electrons) of carbon monoxide (CO) and hydrocarbon (HC) is activated and speeded up in the presence of platinum (Pt) with a stoichiometric and slightly rich air/fuel mixture, whereas the reduction (separation of oxygen and nitrogen atoms) of oxides of nitrogen (NO_x) is activated in the presence of rhodium (Rh), again with similar air/fuel mixture strengths.

A gas flow temperature of above 300°C is necessary to produce sufficient chemical reaction from the oxidation of CO, HC and NO_x. Because the three pollutants CO, CH and NO_x are continuously being removed from the outgoing exhaust gas at the same time, the catalytic converter is known as a three-way catalyst. Abnormally high operating temperature bursts and poisoning of the catalyst's active material with contaminating lead and phosphorus causes loss in the catalytic effectiveness over a period of time.

24.5 Exhaust gas silencers (mufflers)

24.5.1 Silencing of exhaust gases

Exhaust silencers, or mufflers as they are known in the USA, are designed to suppress to an acceptable level the noise created by the exhaust gases expelled from the engine's cylinders. Exhaust noise is due to pressure-wave vibrations produced by the sudden and periodic release of gas energy through the exhaust ports. The exhaust system must suppress two basic types of noise:

Fundamental noise This noise results from the release of compressed products of combustion every time the exhaust valve is opened. The fundamental frequency of the exhaust noise depends on the engine's speed and the basic requirement of the silencer is to dissipate the energy of the resulting shock-waves.

Secondary noise Movement of gas within the exhaust silencer casing can be responsible for various crackles or rasps and can be the cause of the end-plates of the chambers or the chambers themselves rattling. Uneven firing pulses often exaggerate these noises. Cylinder-to-cylinder and stroke-to-stroke variation in combustion can also be responsible for promoting subharmonics, that is, sound-waves of different frequencies.

Approach to noise suppression Exhaust-gas noise suppression can be achieved by applying three different methods:

1) resonance chamber damping
2) reactive interference chamber damping
3) absorption material chamber damping.

Most modern exhaust silencers utilise a combination of two or even all three methods of reducing the level of exhaust-gas noise. The complete exhaust system should have the minimum of back pressure so that it will absorb no more than 4% of the engine's power and designers might well want the loss to be nearer 2 to 3%.

24.5.2 Resonant chamber silencer (fig. 24.14(a)–(h)) The cavity resonator method of sound suppression adopts the principle of a Helmholtz resonator (fig. 24.14(a)). The resonator chambers have different gas volume capacities and surround a central perforated straightthrough tube. The perforated narrow-neck holes along the tube connect the resonating-chambers to the passing through exhaust. As the exhaust-gas sound-waves travel along the straightthrough tube, the pulsating exhaust-gas flow at the entrance to each resonator-chamber creates a pressure variation within these chambers. This causes the exhaust gas trapped in the chambers to repeatedly expand and contract, that is, the gases vibrate. The vibrating exhaust gas mass in each chamber can be likened to a piston and spring which oscillates to and fro (see fig. 24.14(a)). The exhaust gas mass within each chamber has a natural frequency of vibration (similar to a piston mass and spring). This frequency depends upon the size of the chamber. The larger the chamber the softer the air spring and the lower will be its resonant frequency and vice versa. If the fundamental frequency of the exhaust gas pulsation

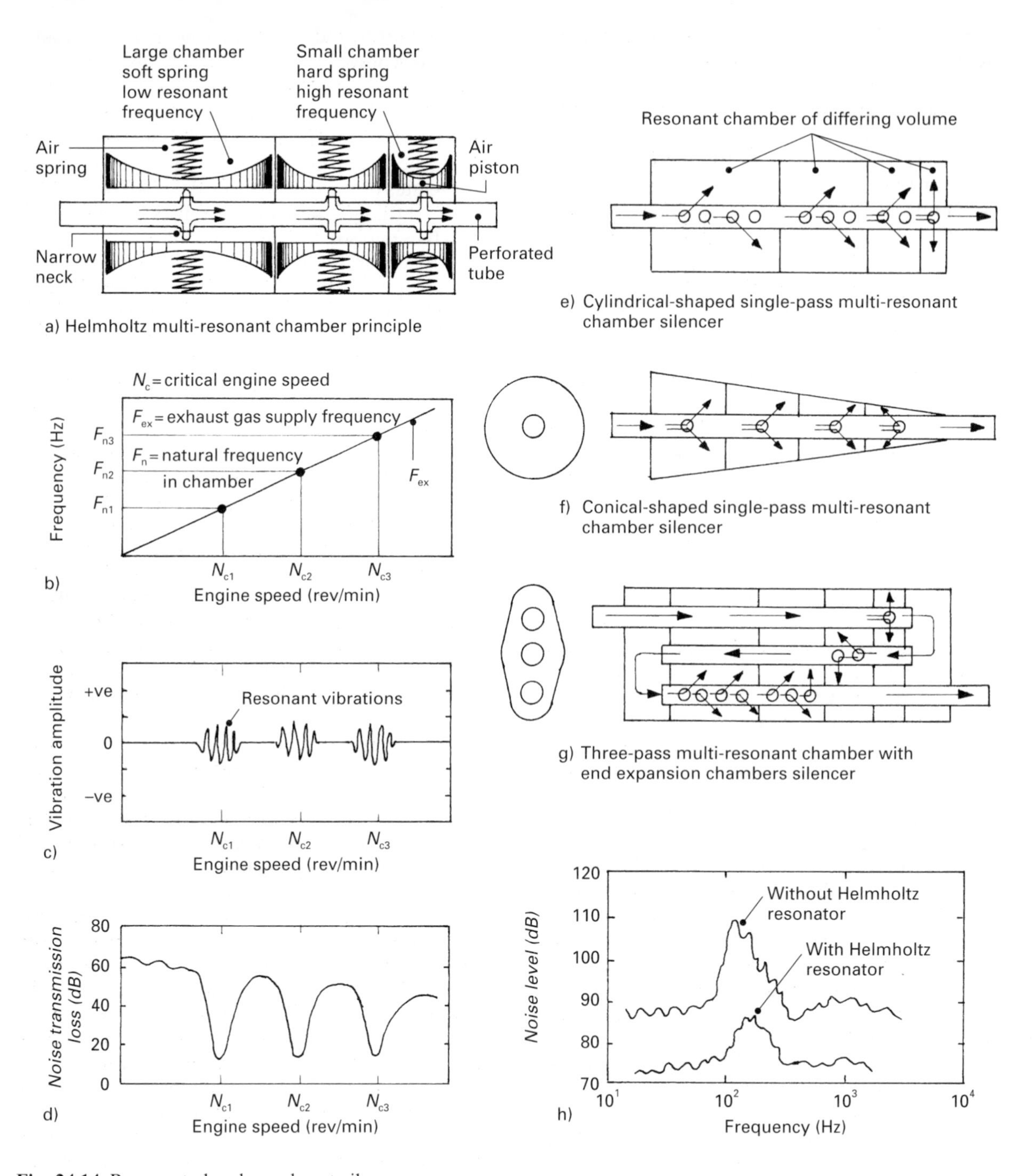

Fig. 24.14 Resonant chamber exhaust silencers

increases as it does with rising engine speed (fig. 24.14(b)), then it will eventually coincide with that of the exhaust-gas frequency (piston and spring effect) within each chamber in turn. At each of these critical frequencies, a resonance condition exists which promotes a large amplification of a chamber's vibrating gas (fig. 24.14(c)). This vibrating exhaust-gas absorbs the sound-energy passing the neck of the resonator at this particular frequency; consequently it considerably increases the noise transmission loss as the exhaust gas moves towards the pipe exit (fig. 24.14(d)). These silencers can be 'tuned' to resonate to a particular low frequency noise or boom which would normally be difficult to eliminate. Thus this approach is a very selective low-frequency narrow-band noise absorber. It is possible to combine several of these resonators in series on a piping system (fig. 24.14(e)) so that not only will each cancel out its own resonant frequency, but they can be made to overlap so that sound-energy is reduced (attenuated) over a wide range instead at sharply tuned points.

It is difficult to make the high-frequency resonating chambers small enough with the usual cylindrical construction, as the separating baffle-plates would be too close for frequencies above 600 Hz. One solution has been to make the multi-resonating chambers (fig. 24.14(f)) in the form of a frustum of a cone, and by varying the spacing of the baffle-plates, resonances of up to 1000 Hz have been achieved. To extend the frequency range, resonating frequency overlap, and to keep the overall length of the resonating silencer to a minimum two-pass or three-pass (fig. 24.14(g)) resonating silencers with end expansion-chambers are available. A comparison of exhaust gas noise without and with resonating chamber suppression is shown over a frequency range in fig. 24.14(h). Since these silencers are generally straightthrough and the resonators are really side branches they offer very little back-pressure to the expelled exhaust-gas and so consume very little engine power.

24.5.3 Reactive inference silencer (fig. 24.15(a)–(h))
The reactive interference method of sound suppression adopts the principle of a Quincke tube. This apparatus comprises a single inlet pipe dividing into two parallel branch sections of equal cross-sectional area; one branch length is fixed whereas the other branch length is variable. If the adjustable branch is set so that both branches are of equal length (fig. 24.15(a)), then the sound-waves passing through the branches travel equal distances to the outlet. Consequently the two sound-waves are in phase so that they merge to reinforce each other (fig. 24.15(b)).

Extending the adjustable branch means that the exhaust gas sound-waves have to travel a greater distance in this branch than in the fixed straight-through branch. If the extended path branch is made to be a half a wavelength longer than the other (fig. 24.15(e)), phased cancellation occurs where the branches meet, since the two waves come together 180° out-of-phase (fig. 24.15(f)), that is, both upper and lower half-waves cancel out. Thus the resultant interference is able to damp out unwanted sound-waves at selective frequencies.

The simplest form of reactive interference silencer is a straightthrough central tube with twin short parallel side branch tubes supported by two baffle-plates set slightly in from the end-caps to form separated expansion chambers, see fig. 24.15(c). The holes in the tubes permit some of the exhaust-gas to divert and short circuit the main passageways created by the tubes. Thus the relative distance travelled by the different sound-wave paths may be so arranged as to provide positive interference resulting in reactive damping of the input sound-wave noise. Various frequencies of exhaust-gas noise may thus be suppressed by this form of sound energy dissipation.

To extend the frequency range and effective damping of the expected sound frequency spectra, the exhaust gases can be made to double back, that is, to create a reverse gas flow. These silencer arrangements are classified as two, three, and double three-pass systems (fig. 24.15(d)–(h)) as opposed to the simple single-pass silencer as shown in fig. 24.15(c). As the number of passes increases the internal flow reversals also extract more acoustic energy, but correspondingly there will be an increase in exhaust-gas back-pressure.

24.5.4 Absorptive chamber silencer (fig. 24.16(a)–(f))
With the absorptive or dissipative method of sound suppression the sound-waves are dissipated by means of acoustically dead material which is able to absorb a large portion of the outgoing exhaust-gas. This material must be heat-resistant such as fibre-glass, long-fibre mineral-wool such as basalt or rock-wool, and wire-wool. These materials are soft, flexible, and provide a cellular structural mesh of interlocking pores. Such materials take in rapidly the sound-wave impulses and causes the material's fibres to move, thus converting sound energy into mechanical vibration and heat. The actual loss mechanism in the energy transfer are viscous flow losses caused by the wave propagation in and out of the porous mesh and frictional losses caused by the motion of the material's fibrous structure. The effectiveness (noise transmission loss) of the absorptive mesh increases with length of exposure of the exhaust-gas to the mineral-wool as it flows along the central tube (see fig. 24.16(b)). The pulsating gas flow

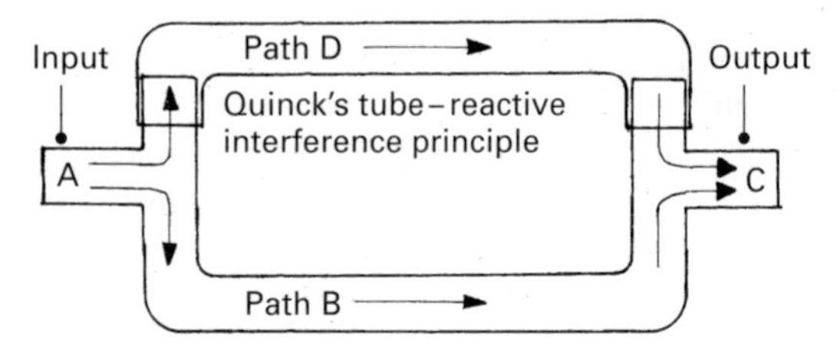

a) Twin exhaust piping with equal path length

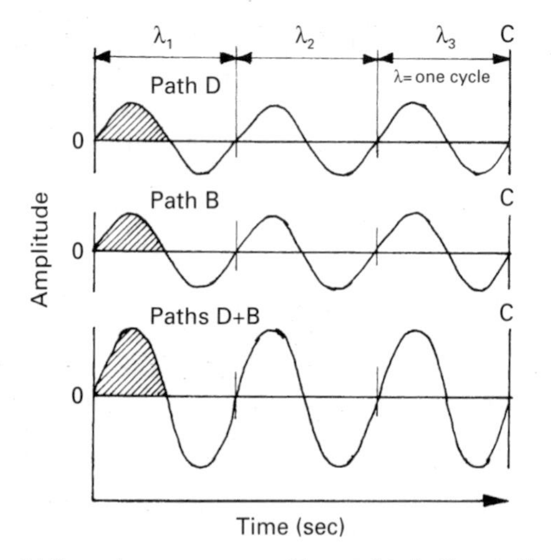

b) Sound wave curves with equal twin flow paths

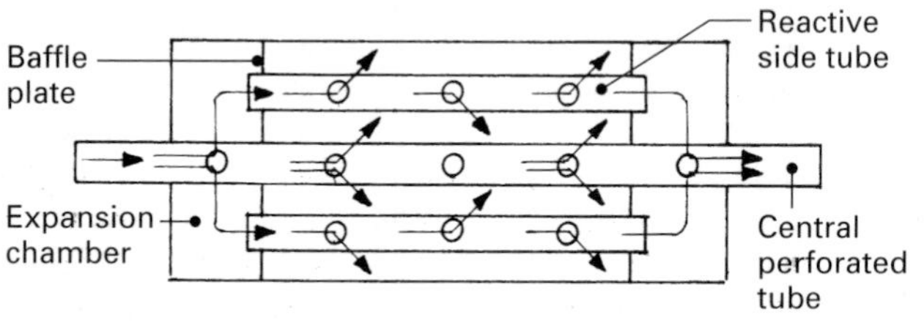

c) Single-pass reactive silencer (straight through)

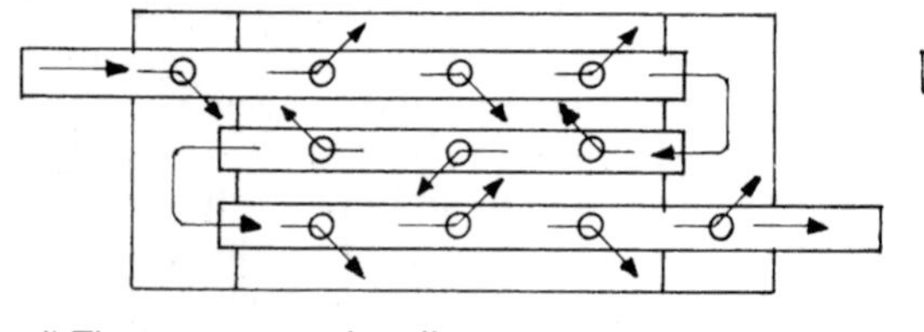

d) Three-pass reactive silencer

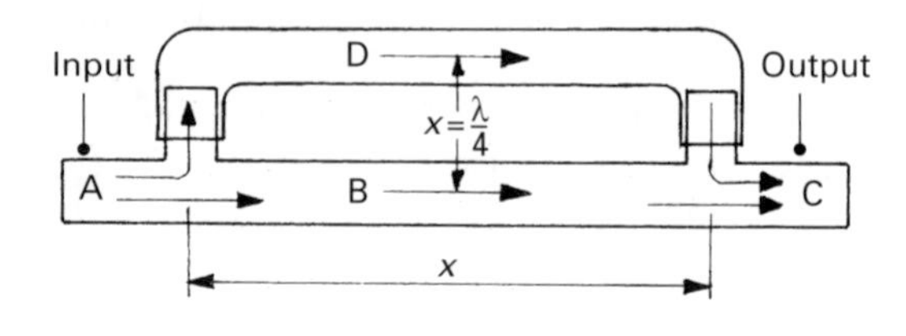

e) Twin exhaust piping with unequal path length

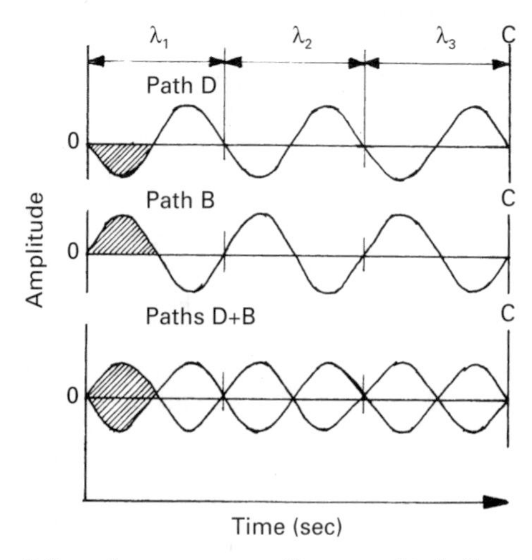

f) Sound wave curves with unequal twin flow paths

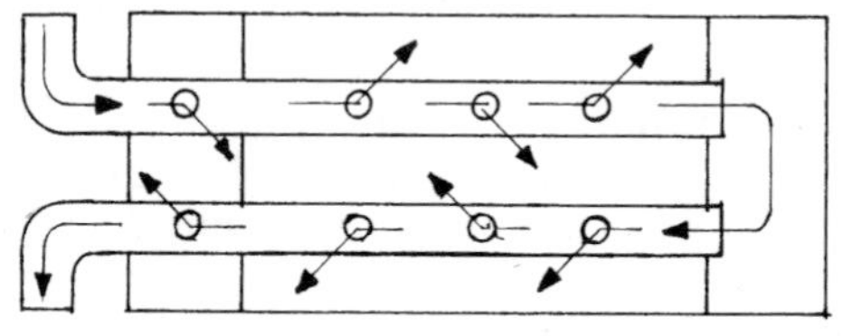

g) Double-pass reactive silencer

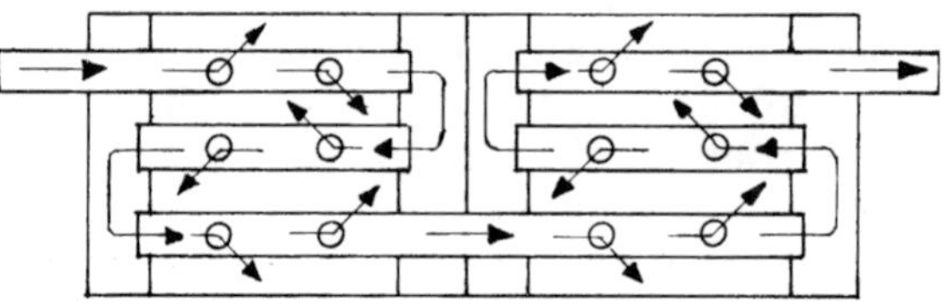

h) Double three-pass reactive silencers

Fig. 24.15 Reactive interference exhaust silencers

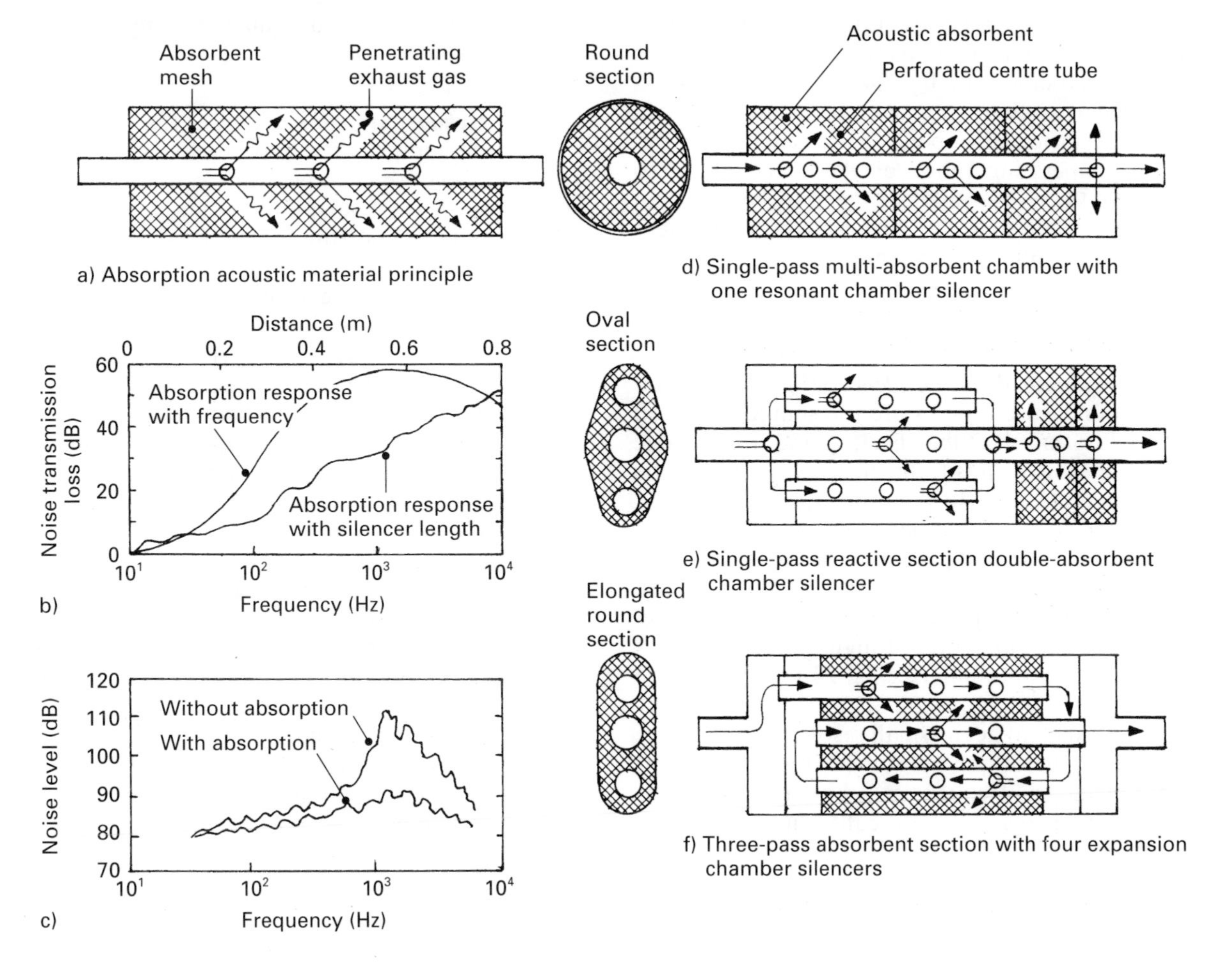

Fig. 24.16 Absorption exhaust silencers

in the central tube causes a rapid density variation through the holes and mesh on the surface of the tube so that sound-energy bombards the absorbent mesh and is thereby dissipated.The magnitude of the sound-energy is really very small so that the fibres do not become very warm due to the transformation, but the heat from the exhaust-gas will raise the temperature considerably. These absorptive silencers have relatively wide-band noise reduction characteristics but their performance drops off at the lower frequency range (fig. 24.16(b)). These silencers are very effective with such engines as the two-stroke diesel which makes a lot of high-frequency noise. The reason for the poor response at low frequency is that the larger wavelengths find it difficult to penetrate the absorbent material's interior so that only its bulk will be excited to vibrate. Nevertheless, if the incident sound-waves are of a higher frequency there is a lot of relative motion of the internal fibres and therefore good absorption.

The construction of the simple single-pass straightthrough absorptive silencer consists of mineral-wool or something similar packed into the annular space between a perforated or louvered central tube and the outer cylindrical steel-sheet casing (fig. 24.16(a)). To prevent conduction through the mineral-wool along the silencer's full length, the mineral-wool should be broken up by annular divisional baffle-plates (fig. 24.16(d)).

One disadvantage with absorptive materials is that their performance deteriorates with usage and so excess amounts of materials should be incorporated to compensate for the loss in effectiveness.

To improve the effectiveness of the absorption method of noise reduction over a wider and useful frequency range the absorption silencer is often combined with a single-pass reactive interference section as shown in fig. 24.16(e). A three-pass absorbent section may also be combined with end expansion-chambers to broaden the frequency operating range of the silencer and to improve the effectiveness of the absorbent section without increasing the silencer's overall length (see fig. 24.16(f)). Performance characteristics of an absorption-type silencer over an audible frequency range are shown in fig. 24.16(c).

24.5.5 Perforated tube for silencers (fig. 24.17(a)–(c))
Plain punched-out holes are the simplest method of perforating the tube sheeting and are suitable where the gas path is to be diverted from the tube or from a chamber to the tube (see fig. 24.17(a)).

Pierced or stabbed holes are particularly suited for tuned resonant-chamber sections as they extend the neck for the resonator (see fig. 24.17(b)).

'U' shaped stampings are punched out on steel sheet to form a louvred or tongue pattern which is then rolled into tubing (see fig. 24.17(c)). The size and number of louvres are aligned to the gas flow and are tuned to act as small resonators which are capable of smoothing out some of the subharmonics (secondary frequencies). Louvred tubing is used where there is difficulty in suppressing some of the unwanted sounds passing out from the exhaust.

24.5.6 Exhaust system enviroment (fig. 24.18)
At the downpipe where the exhaust system joins the manifold the temperature can reach 800°C (see fig. 24.18). Surrounding air flow reduces this rapidly to something like 500°C on the horizontal section to the catalytic converter. The converter casing can have an internal temperature of up to 700°C and a skin heat of 260°C. In the second or tail silencer the external skin can reach 220°C and even the tail-pipe perhaps 2–4 metres from the exhaust manifold may reach 190°C. In wet weather the hot metal is subjected to constant quenching and reheating, as water from puddles in the road is splashed up by the tyres. Inside the casing it suffers thermal shock and corrosion from the build-up of acidic condensation. The entire system is also subjected to vibration from the engine and movement of the vehicle itself.

Corrosion in the exhaust system results from the fact that part of the system does not reach the dew-point of the working gas which is about 95°C. Since the front portion of the exhaust system runs hotter than its rear end, the downpipe is less likely to corrode than the tail-pipe.

Currently most cars and vans are equipped with

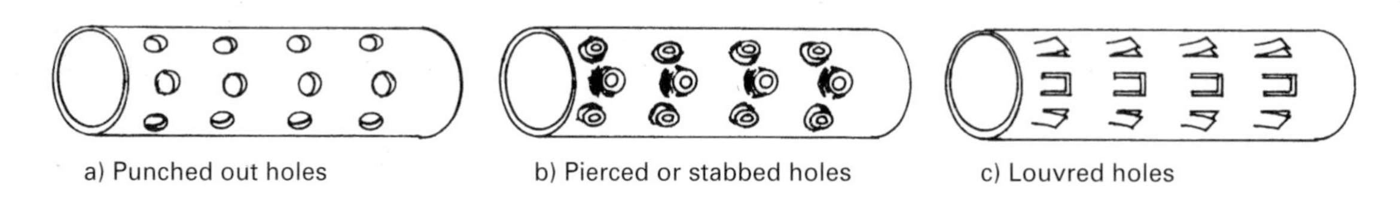

Fig. 24.17 Internal silencer box perforated tubing

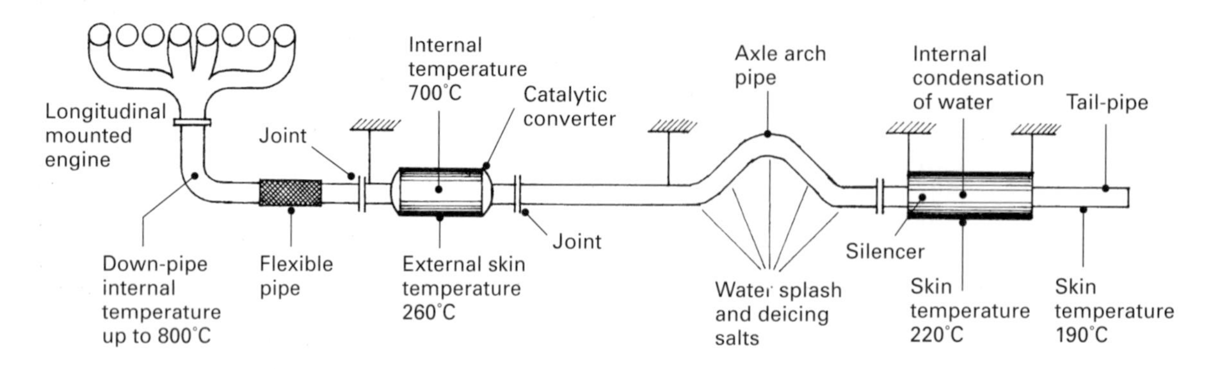

Fig. 24.18 Exhaust system operating temperature environment

mild steel (low carbon steel) downpipe and intermediate pipes. Stainless steel is used for the catalytic converter casing to safeguard the expensive and fragile ceramic honeycomb matrix, whereas the rear exhaust silencer shell and the tail-pipe is usually made from aluminised low-carbon steel as the cooler rear end of the exhaust system is more likely to corrode. Heavy trucks on long runs suffer less internal corrosion than cars which have more acidic gases and tend to have more cold start-stop motoring. Low-carbon steel silencers for commercial vehicles give a silencer life that is limited not by corrosion, but by physical considerations, such as damage due to fatigue resulting from poorly welded joints, resonant vibration, and incorrectly positioned inflexible and inadequate support mountings.

24.5.7 Aluminised low-carbon steel pipes and silencers

Aluminised low-carbon steel surfaces double the life of ordinary low-carbon steel sheeting used for piping and silencers when exposed to the hot exhaust gases, water spray and hot and cold operating conditions. Therefore they are commonly used for selective parts of the exhaust system. A brief account of the aluminising process for low-carbon steel components is given as follows. The tubing or silencer surface to be protected is grit-blasted and aluminium sprayed, usually to a thickness of 0.2 mm. It is then treated (covered) with a sealing composition which may be calcium-hydroxide, sodiumsilicate, or bitumastic paint to reduce oxidation during subsequent heating. It is then diffusion annealed in a furnace at 850°C for approximately half an hour. The final coating consists of a gradation of aluminium oxide. Such deposits will withstand oxidation for very long periods at temperatures up to 900°C. Above this temperature diffusion of iron into the aluminium becomes so rapid that the alloy layer becomes impoverished and the outer layer contains insufficient aluminium to provide further protection.

24.5.8 Stainless-steel pipes – catalytic-converters and silencers (fig. 24.19(a) and (b))

Steel surfaces exposed to the atmosphere slowly form a porous oxidised coating with age, that is, the surface corrodes (rusts) and consequently the steel loses weight, this weight loss being directly related to time (see fig. 24.19a). Conversely stainless steel which contains chromium and sometimes nickel, oxidises rapidly and tends to form a dense oxide coating. This dense oxide protects the surface from further oxidation and corrosion with respect to time as shown by the almost horizontal curve (see fig. 24.19(b)).

Stainless steel must have at least 10% chromium to offer significant resistance to exhaust gases and air corrosion at 600°C. With about 8% chromium the corrosion material weight loss rate is cut by 50% and with about 12% chromium the weight loss rate is only slightly better than for 10% chromium steel (see fig. 24.19(b)). Therefore there is very little advantage in increasing the percentage

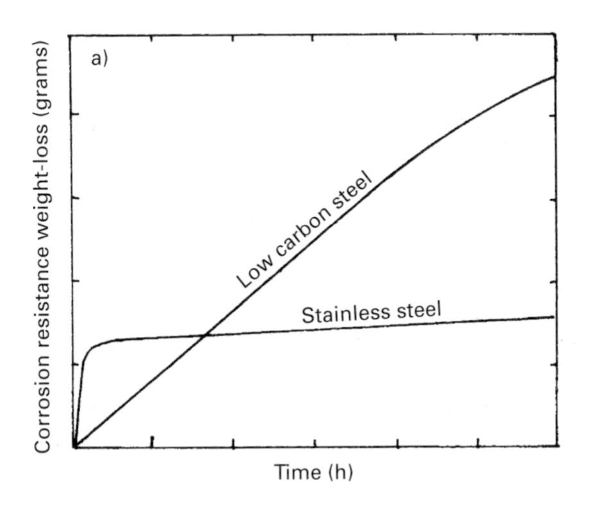

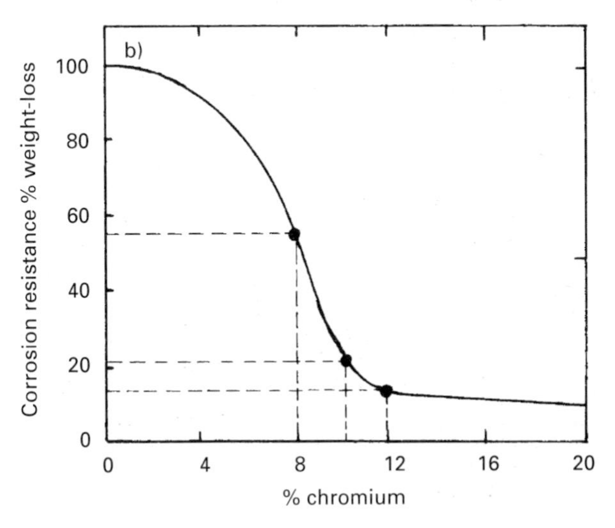

Fig. 24.19 Effect of chromium on the corrosion resistance of steel

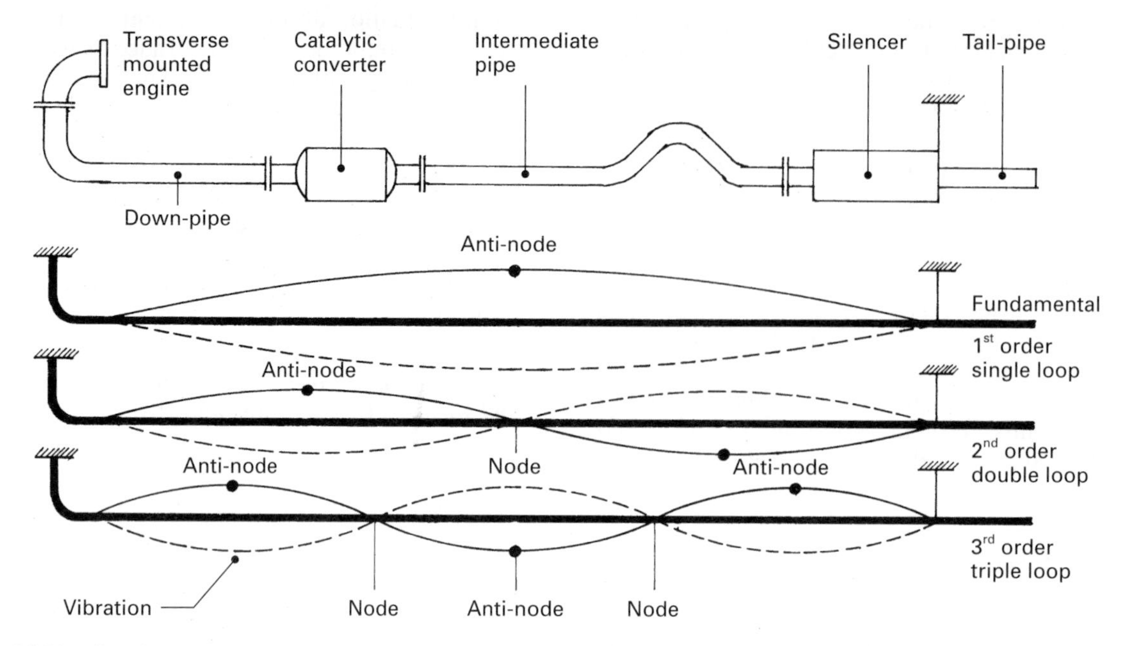

Fig. 24.20 Vibration resonance and exhaust system pipe mounting points

of chromium beyond about 14%. Stainless steel recommended for general use is BSC Hyform 409 muffle grade 12% chromium steel. This stainless steel can be formed in tooling used for low-carbon steel pipe. Shaping without cracking, and welding, bending, and flaring problems have been eliminated. Hyform 409 muffle grade pipe can be flared to a diameter 50% greater than that of the tube itself, while bending radii of 2–3 times the tube diameter can be made. This stainless steel can be welded by metal arc gas shielded (MAGS) and tungsten arc gas shielded (TAGS) and resistance welding techniques; no post-welding heat-treatment is necessary. The average life of an ordinary low-carbon steel silencer is about two years, aluminised low-carbon steel four years, whereas stainless steel silencers have a minimum life of at least six years and may even last the life of the vehicle.

24.5.9 Exhaust system vibration (fig. 24.20)

The exhaust system can be treated as a beam suspended between the rigid manifold end and the flexible tail-pipe end. When the engine is running at very low speeds some engines tend to suffer from severe vibration due to the longer power pulse intervals and cycle to cycle misfiring. This vibration can excite the exhaust system into some form of vibration between the supports. At very low engine speed a first-order single loop vibration is generated; at slightly higher speed it changes into a second-order double loop, and at even higher speed it converts to a third-order triple loop vibration. With higher speeds, still higher orders of vibration will take place.

24.5.10 Exhaust pipe and silencer support mountings (fig. 24.21(a)–(g))

Engines are flexibly suspended on the subframe or chassis structure of the vehicle, whereas the exhaust system is rigidly attached to the engine at its front-end, but it is also supported at various points along its length to the vehicle's understructure. To prevent the relative movement between the engine and the body structure straining and damaging the exhaust system, flexible support mountings are provided in the form of resilient rubberised-fabric straps, rubber-bonded blocks, silicon-rubber rings and moulded-rubber mounts (see fig. 24.21(a)–(g)). In addition, some exhaust pipe systems incorporate a flexible heat-resistant pipe section between the downpipe and catalytic converter to partially isolate the engine vibrations from the main exhaust system. Flexible exhaust system support-mountings attached to the body structure permit the piping system to align itself in its best possible unstrained position and at the same time prevent noise and vibration generated by the exhaust system being transmitted back to the vehicle structure. As can be

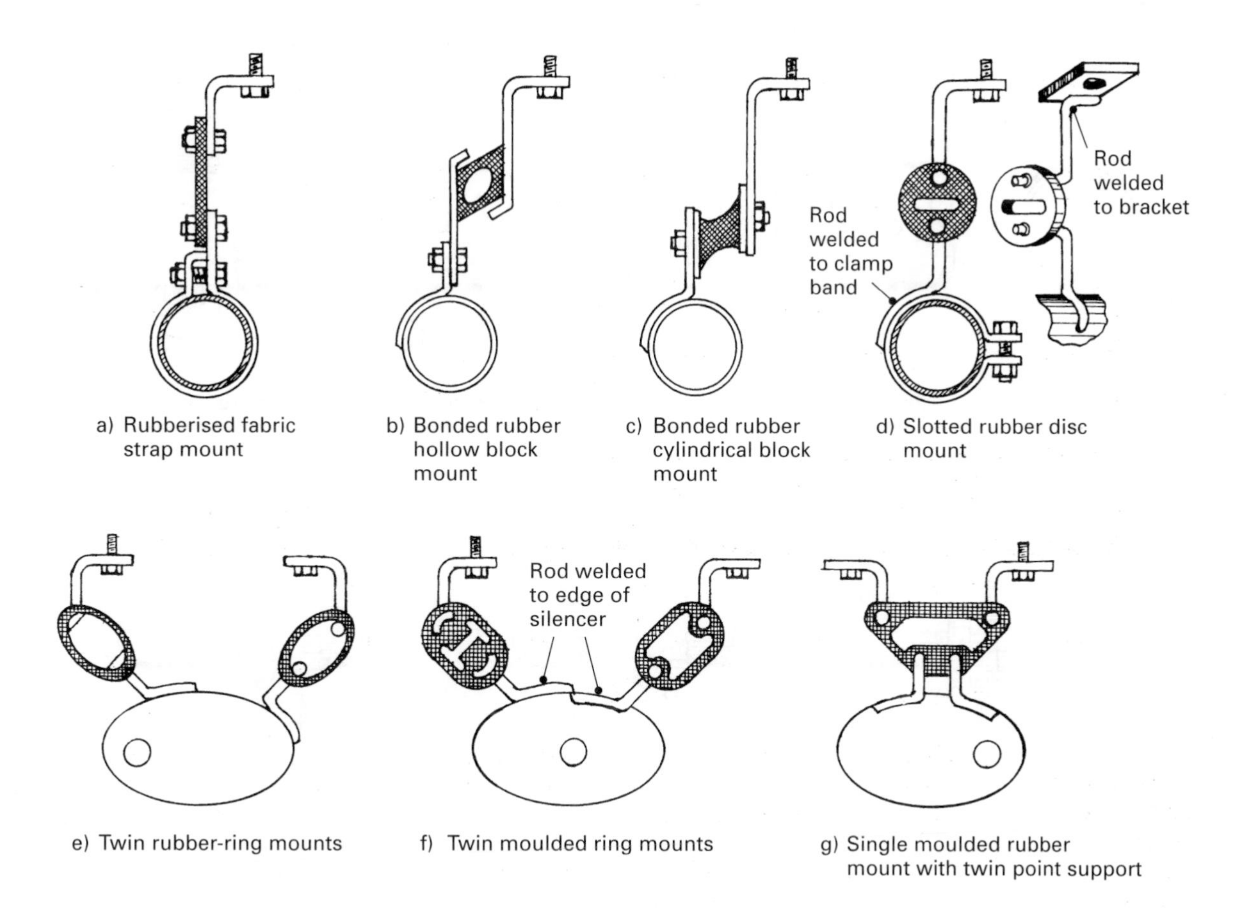

Fig. 24.21 Exhaust pipe and silencer mountings

seen in fig. 24.21(a)–(d) the exhaust pipes normally need only a single point support, whereas fig. 24.21(e)–(g) shows that the catalytic converter and silencer usually requires a two-point support to provide some degree of lateral constraint.

24.5.11 Exhaust pipe joints (fig. 24.22(a)–(f))
Pipe to pipe joints may be of the simple slip-over joint with either the pressed steel 'U' clamp (fig. 24.22(a)) or the preferred 'U' bolt clamp (fig. 24.22(b)). However these joints are usually difficult to separate after they have been in service for some time. A more substantial joint which provides a small amount of self-alignment on being assembled is the flared-pipe and ring with split-clamp (fig. 24.22(c)). This joint has been used extensively to join the downpipe to the manifold. A rigid flange and gasket joint (fig. 24.22(d)) is commonly used for both downpipe/manifold and for pipes being joined to catalytic converter and silencer boxes. The flared-pipe and ring flanged joint (see fig. 24.22(e)) is similar to the flared-pipe and ring with split clamp arrangement, but instead of the two half clamps, it relies on a pair of flange-plates to pull the joint tight. This type of joint clamping is sometimes preferred to the split-clamp when joining pipe ends together. An alternative to having each pipe end flared with a shaped compressive-ring in the centre is to have one pipe end flared with a nipple-piece fitted into the other pipe end (see fig. 24.22(f)). Again this permits a small amount of misalignment when the joint is drawn together.

24.6 Fuel tank evaporation control (fig. 24.23(a))
Petrol vapour from the fuel tank and induction manifold if released into the atmosphere is an unwanted pollutant which must be restrained. The amount of liquid fuel which can evaporate in a partially filled tank during warm and hot weather conditions can be considerable.

Evaporative emission of hydrocarbon pollutants

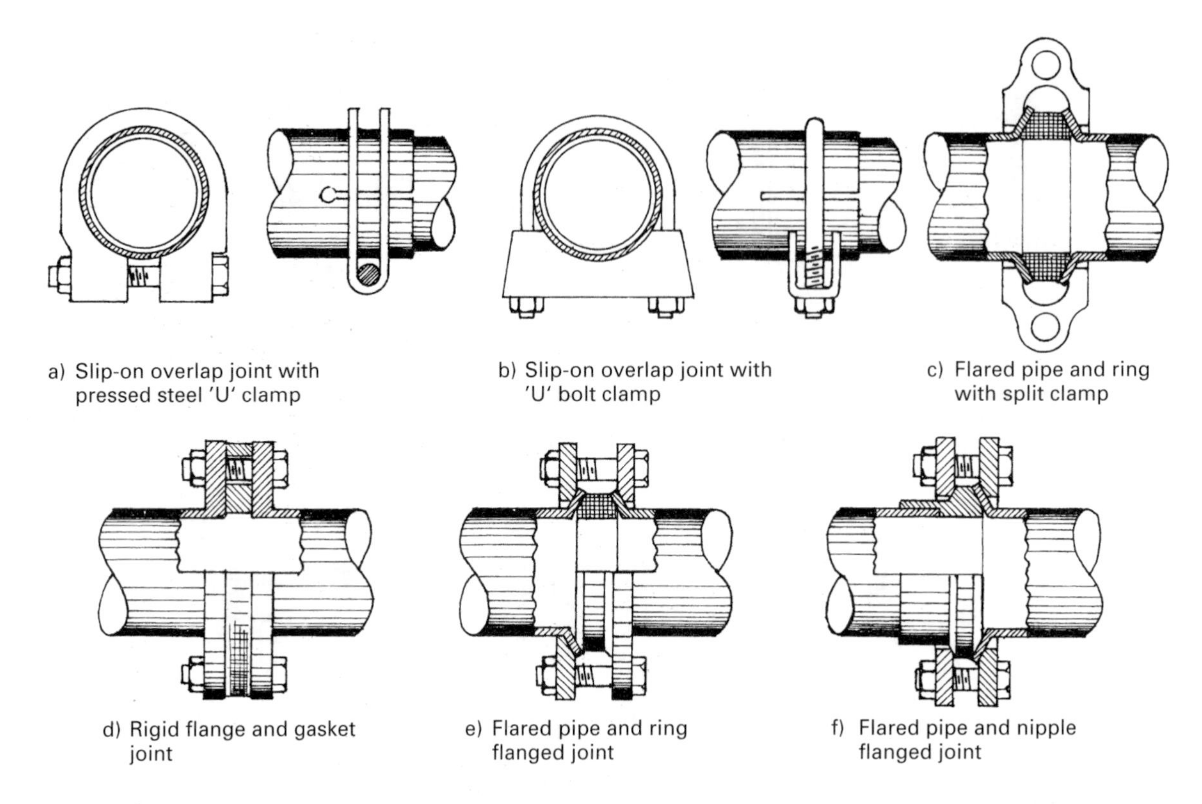

Fig. 24.22 Exhaust pipe joints

can be controlled when the engine is stationary by releasing the petrol vapour into a canister which contains activated charcoal, or carbon granules which absorb and store the vapour fumes.

Once the engine operates above idle speed conditions, the partially open throttle-valve subjects the vacuum cut-off diaphragm chamber to manifold depression, so causing the cut-off valve to open. Air is drawn from the bottom of the canister to the induction manifold. It thus purges the charcoal granules which have soaked up the vapours. Thus the air movement through the canister carries the stored vapour with it into the engine's cylinders via the induction manifold. It thus forms a small proportion of the air/fuel mixture and therefore is consumed as part of the combustion process.

When the engine runs at idle speed, only a small amount of stored vapour is permitted to flow into the induction manifold through the constant-purge orifice since the closed throttle-valve prevents vacuum getting to the cut-off valve diaphragm chamber.

Fuel tank filler-cap (fig. 24.23(b)) The purpose of the filler-cap is to seal the fuel-tank once it has been filled with petrol or diesel fuel. The filler-cap protects the fuel-tank against dust, dirt and unwanted litter entering the tank and at the same time it prevents fuel vapour above the liquid fuel level in warm weather escaping to the surrounding atmosphere. With a sealed fuel-tank, as the fuel is used and the level drops, the empty space above the fuel level changes from atmospheric pressure to a depression. To prevent the tank from collapsing as the vacuum builds up, a diaphragm-type vacuum-valve is incorporated into the filler-cap. This valve opens and permits air at atmospheric pressure to enter the tank every time the depression reaches some predetermined value.

Fuel check-valve (fig. 24.23(c)) This check-valve incorporates both a vacuum-relief-valve and an air-relief-flap-valve. The vacuum-relief flap collapses, permitting air coming from the base of the carbon canister to enter the fuel tank when-

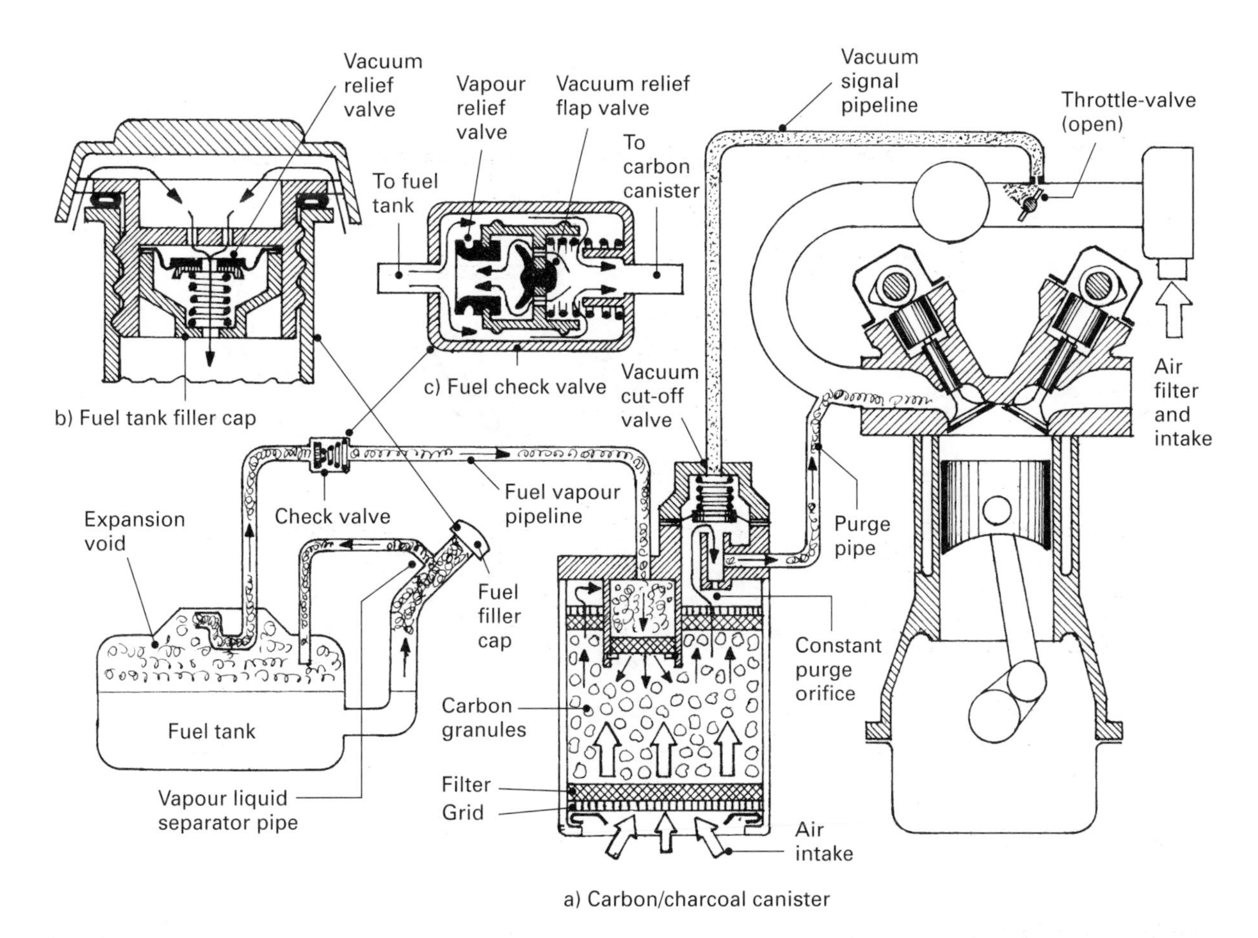

Fig. 24.23 Fuel tank evaporation control system

ever the depression in the fuel tank becomes excessive. Conversely the vapour-relief-valve pushes open whenever the vapour pressure in the fuel-tank becomes excessive. The vapour then passes to the induction-manifold via the carbon-canister.

24.7 Air intake temperature control (fig. 24.24(a)–(c))

Fuel injected into a cold air stream surrounded by cold metal walls would find it difficult to mix and atomise, and large particles of liquid fuel would tend to collect on the walls of the intake ports before being drawn into the cylinders. Consequently the quantity of air/fuel mixture distribution entering the cylinders will be poor resulting in erratic and incomplete combustion during the engine's warm-up period.

This intake temperature control is achieved by diverting pre-heated air surrounding the exhaust manifold when the engine is starting and warming up to the induction manifold. As the engine approaches normal operating temperature this diverted warm air passage is closed and at the same time the cold air intake passage opens to permit the cooler air to enter the induction manifold.

Operating conditions (fig. 24.24(a)) When the engine is cold or warming-up the bi-metal heat sensor closes the air-bleed-valve, thus permitting manifold vacuum to be conveyed to the vacuum actuator.

At idling and light-load conditions the high vacuum in the induction manifold fully lifts the diaphragm and flap-valve. This opens the hot air passage and closes the cold air intake passage so that only pre-heated air enters the induction manifold.

With higher engine speeds and load the wider opened throttle-valve reduces the manifold

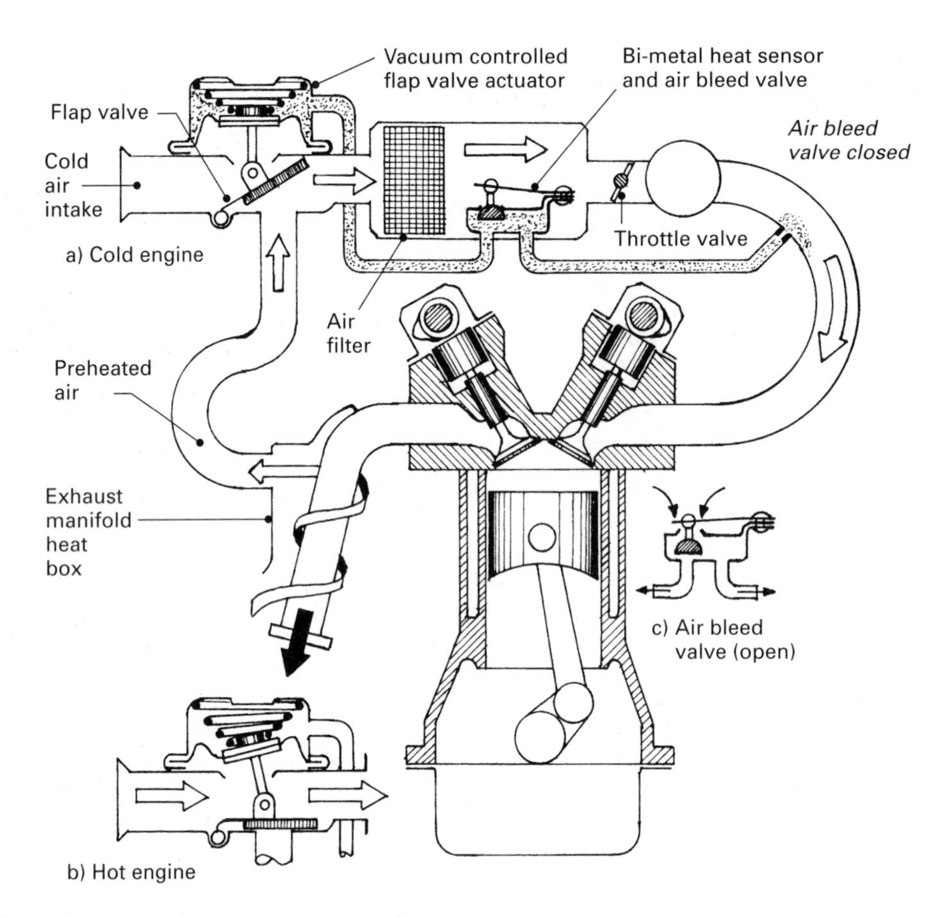

Fig. 24.24 Air intake automatic temperature control

vacuum so that the flap-valve only partially closes the cold-air intake passage; hence both hot and cold air will pass to the cylinders.

Once the engine reaches its normal working temperature the bi-metal heat sensor strip bends and opens the air bleed-valve (see fig. 24.24(c)). This prevents the manifold vacuum reaching the vacuum actuator. The diaphragm spring therefore takes over and pushes down the diaphragm. Consequently the flap-valve opens the cold-air intake passage and closes the hot-air passage (see fig. 24.24(b)). Only dense cold air now enters the cylinders as this is necessary to provide maximum engine performance.

24.8 Exhaust gas recirculation (fig. 24.25(a)–(c))
Oxides of nitrogen are formed by the reaction of oxygen and nitrogen at temperatures above 1370°C during the combustion process. One method of decreasing the amount of oxide of nitrogen produced during combustion is to recirculate a proportion of the exhaust-gas back into the cylinders. This has the effect of diluting the overall mixture, reducing the peak combustion temperatures reached, and correspondingly reduces the amount of oxide of nitrogen (NO_x) produced in the combustion process. The exhaust-gas recirculation system is designed to cut off the exhaust-gas supply to the induction-manifold at idle speed and very high engine speeds as this prevents unstable running at low speed and loss of engine performance at the top of the speed range.

Operation conditions (fig. 24.25(a)) At normal operating conditions the exhaust-gas pressure lifts up the air-valve diaphragm and closes the air-valve. At the same time the partially opened throttle-valve relays a vacuum to the exhaust gas recirculation (EGR) cut-off valve diaphragm-chamber. This pulls up and therefore opens the EGR cut-off tapered-valve. Exhaust-gas is therefore permitted

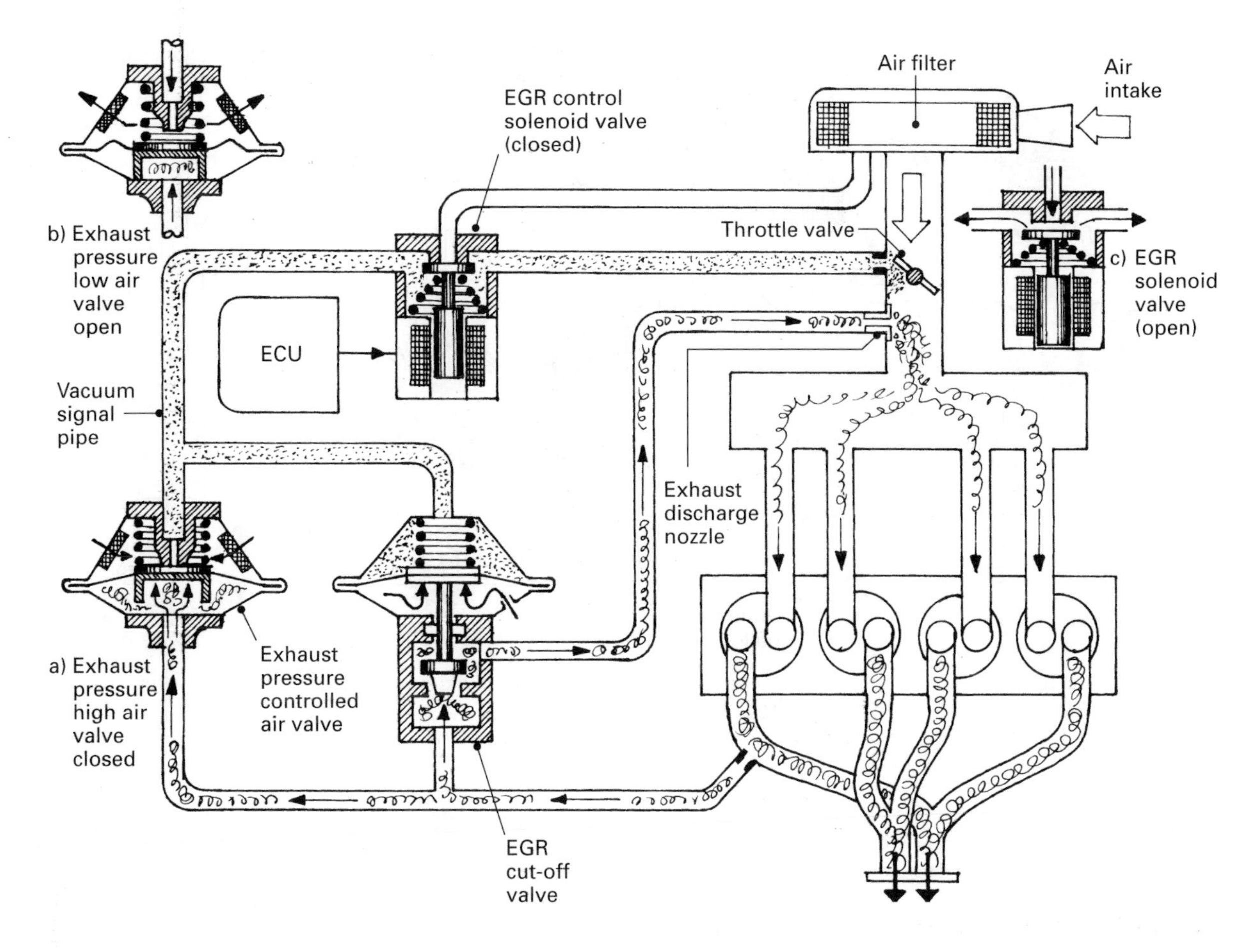

Fig. 24.25 Exhaust gas recirculation (EGR) system vacuum-operated

to flow from the exhaust-manifold to the induction-manifold via the discharge nozzle.

If the exhaust pressure is very low (see fig. 24.25(b)), the exhaust pressure operated air-valve remains open. This permits air to bleed into the EGR diaphragm-chamber so that the EGR tapered-valve remains closed. This prevents the exhaust-gas passing to the exhaust discharge-nozzle under these conditions.

At idle and very high engine speeds the electronic control unit (ECU) signals the solenoid-valve. This energises the solenoid and pulls open the air-valve (see fig. 24.25(c)). Air now enters the vacuum signal pipe and so prevents the ERG cut-off valve opening, again preventing exhaust-gases entering the induction-manifold.

24.9 Air injected exhaust system (fig. 24.26(a)–(b)) Secondary air injection into the exhaust ports towards the back of the exhaust-valve is an effective but nevertheless an expensive method used to reduce the carbon monoxide and hydrocarbon exhaust-gas emissions expelled with the burnt products of combustion. Air blown into the hot exhaust-gas mixes with the burnt and partially burnt gases. The products of combustion which still exist in the form of carbon monoxide and hydrocarbon are therefore able to react with oxgyen supplied by the air blast. This very nearly completes the combustion process so that mainly carbon dioxide, nitrogen and water are ejected from the exhaust system with only very small traces of carbon monoxide and hydrocarbon. The burning of the remaining carbon monoxide and carbon dioxide expelled from the cylinders in the downpipe raises the exhaust-gas temperature with the desired effect of heating the catalytic converter and speeding up the warm-up period when no pollutant conversion takes place. However, care in matching the air injection to the

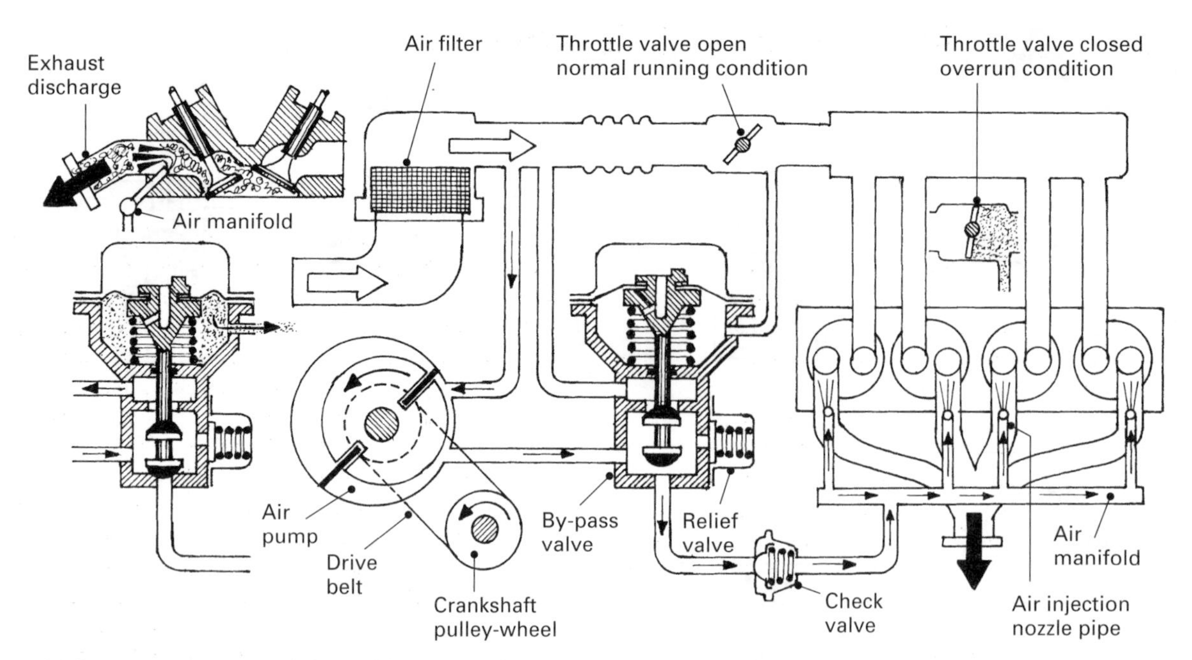

Fig. 24.26 Air injection with by-pass valve system

exhaust discharge is critical to prevent overheating and damaging the ceramic honeycomb matrix of the converter.

Antiback-fire prevention by-pass valve (fig. 24.26(a) and (b)) The sudden closure of the throttle-valve from normal steady running to one of deceleration or overrun operating conditions increases the vacuum in the induction-manifold. The consequences of this sudden closing of the throttle-valve is to monentarily enrich the air/fuel mixture in the manifold. This is due to the increased vacuum evaporating the fuel from the internal walls of the manifold and to the short delay before the air/fuel mixture is weakened. The enriched air/fuel mixture products of combustion come through the exhaust ports only partially burnt, and if this is permitted to mix with the air-blast, severe after-burning and back-firing in the exhaust system will take place. To prevent back-firing during deceleration a by-pass valve is incorporated whose function is to by-pass the air injection from the exhaust manifold and to divert it back to the filtered side of the air-cleaner/silencer.

Normal operating conditions (fig. 24.26(a)) Clean air from the air filter/silencer is drawn into the sliding-vane-type air-pump where it is compressed and discharged to the air-manifold and nozzle-pipes via the air by-pass valve. The air-pump is coupled via a pulley-wheel and belt to the crankshaft pulley. Its output therefore increases in proportion to the engine speed so that it roughly matches the amount of exhaust-gas expelled into the exhaust system at any one time.

Deceleration transition conditions (fig. 24.26(b)) When the accelerator-pedal is released to decelerate the engine, the vacuum on the engine side of the throttle-valve suddenly increases. Thus a vacuum is immediately relayed to the underside of the vacuum diaphragm. As a result, the diaphragm and spherical-valves will be pulled downward, closing the valve passage leading to the air-manifold, and at the same time opening the upper valve air return passage.

Thus air discharge from the pump, is returned to the intake side of the air-pump thereby preventing excessive after-burning and back-firing in the exhaust system. The air-line pipe between the by-pass valve and air-manifold contains a non-return check-valve to prevent flowback of the exhaust-gas into the pump during the by-pass period or following a pump drive failure.

25 Electrical wiring and lighting

25.1 Electron theory

All materials – whether solid, liquid, or gaseous – are made up of minute particles known as atoms. The simplest forms of matter are known as elements, and each element has atoms of only one type which is characteristic of the element concerned. More complex materials – known as compounds – are made up of molecules consisting of different types of atoms bound together.

An atom is made up of a small but relatively heavy nucleus containing positively charged protons together with neutral particles known as neutrons. The atom of each element has a different number of protons in its nucleus. Electrons – each having a charge which is equal and opposite to the charge on a proton – may be considered to orbit around the nucleus in much the same way as the planets orbit the sun (see fig. 25.1). There are as many negatively charged electrons in orbit as there are positively charged protons in the nucleus, so each atom is electrically neutral. The neutrons have no charge and just tend to cement or hold the nucleus together.

The proton and neutron particles have approximately equal masses, and practically the whole mass of the atom is concentrated in the nucleus – in fact the protons are about 1840 times heavier than the electrons.

Depending upon the atom, the electrons may orbit the nucleus in one or more concentric rings or shells. The simplest and lightest atom – that of the gas hydrogen – has only one shell containing one electron. The carbon atom, which is very light, has two shells and six electrons – two in the inner shell and four in the outer shell (fig. 25.1). Aluminium (a light metal) has three shells and thirteen electrons (fig. 25.2(b)); copper (a moderately heavy metal) has four shells and 29 electrons (fig. 25.2(a)); silver (a heavier metal) has five shells and 47 electrons (fig. 25.2(c)); while lead (a very heavy metal) has six shells and 82 electrons in its atom (fig. 25.2(d)).

There is a maximum number of electrons which can be accommodated in each shell, and a shell which is 'full' is relatively stable. An atom which has all its shells full is very stable, but most atoms have less than the maximum number of electrons in their outer shells and so are less stable and will tend to interact with other atoms so as to produce a more stable arrangement.

Electrons are held in orbit by a force of attraction similar to gravity which decreases with distance away from the nucleus. The electrons in the inner shells are therefore tightly bound to the nucleus, whereas the outer orbiting electrons are only weakly bound and may be dislodged from their orbits if sufficient energy is applied (usually

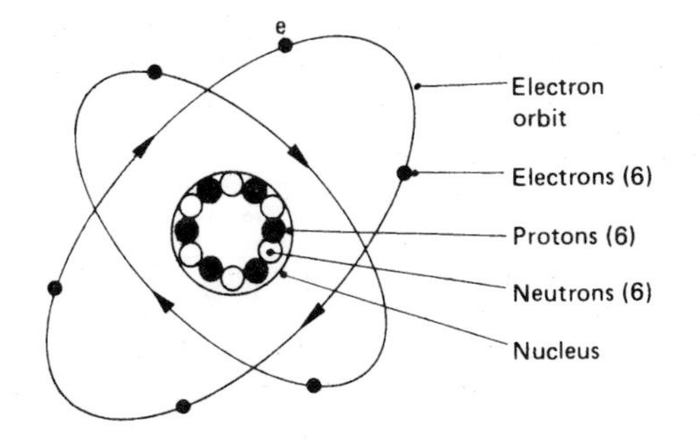

Fig. 25.1 Three-dimensional model of a carbon atom

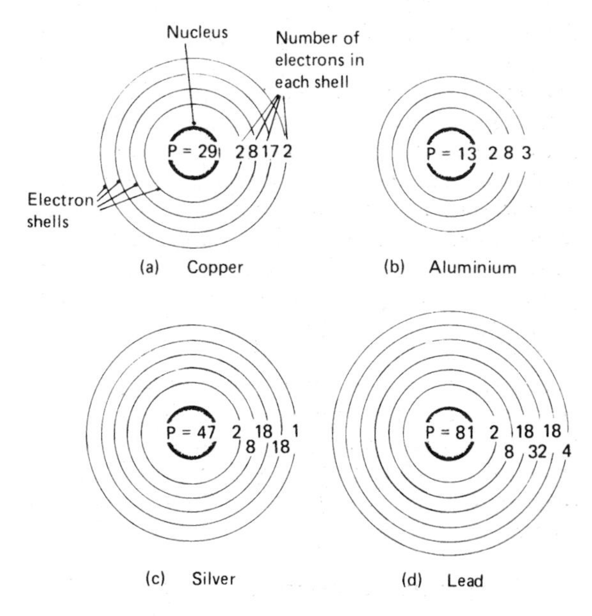

Fig. 25.2 Two-dimensional models of various atoms

in the form of heat). These outer electrons are therefore known as 'free' electrons.

Although some atoms may lose their outer electrons, others with an almost complete outer shell may effectively take in electrons to fill this shell. If an atom has lost or gained electrons, the number of electrons in orbit will no longer be equal to the number of protons in the nucleus and so the atom will no longer be electrically neutral but will have become a charged 'ion'. A loss of electrons results in a positively charged ion, as there are then more protons in the nucleus than electrons in orbit. A gain in electrons results in a negatively charged ion, as there are then more electrons in orbit than there are protons in the nucleus.

Particles with the same charge repel each other, whereas particles with opposite charges attract each other. Thus two similar positive or negative charged particles will tend to move away from each other, while oppositely charged positive and negative particles tend to drift together.

25.1.1 Electrical conductors (fig. 25.3)

A piece of metal conducting wire is composed of millions and millions of atoms spaced out in a three-dimensional pattern and bonded rigidly together by electrostatic attraction between atoms. The outer orbiting electrons in a metallic substance are easily detached from individual atoms and are then free to wander at random in between the now positively charged atoms (ions).

Electron movement in a definite direction can be achieved by applying a potential difference between the conductor ends by making one end of the conductor rich in electrons so that it becomes negatively charged and making the other end depleted of electrons so that it becomes positively charged. As would be expected, the previously randomly moving free electrons now are repelled from the negatively charged end but attracted by the positively charged end of the conductor – consequently there will be a steady drift of electrons through the conductor from the negative to the positively charged end as long as the potential difference is maintained.

Conductors are thus materials which have free electrons which can be made to flow in a given direction. The metals silver, copper, and aluminium are among the best conductors.

25.1.2 Electrical insulators (fig. 25.4)

Insulators are made from non-metals whose constituent atoms are held together by their outer shell electrons being shared between two or more atoms so that their negative charges are attracted to the positive charged nuclei of adjacent atoms. The electrons are tightly bound to the respective nuclei so that there are very few free electrons present and these are only very slightly displaced when a potential difference is applied across the substance. Insulators are thus substances which have outer-shell electrons which cannot drift from one atom to the next as is the case in metallic materials. Insulators are therefore non-metallic substances which have no free electrons drifting about, and examples of such materials are ceramics, glass, plastics, rubber, mica, paper, oil, etc.

No insulator is perfect, and some insulators have greater resistance to electron flow than others under different operating conditions such as dampness and temperature.

25.2 Electrical units

25.2.1 Current

The flow of electrons through a conductor caused by the application of a potential difference between its ends is known as an electric current (symbol I). The amount of current due to the drift of electrons through the conductor is large or small according to the number of electrons involved and the speed at which they move.

Before the electron theory was understood, it was assumed that current flowed from a positive to a negative potential, whereas the electrons actually drift in the opposite direction – from

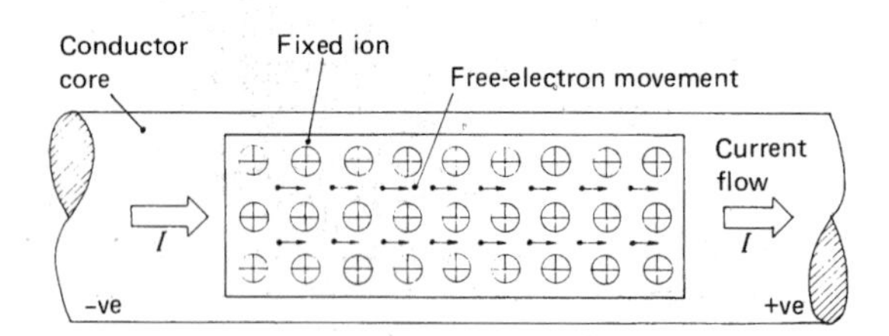

Fig. 25.3 Conductor

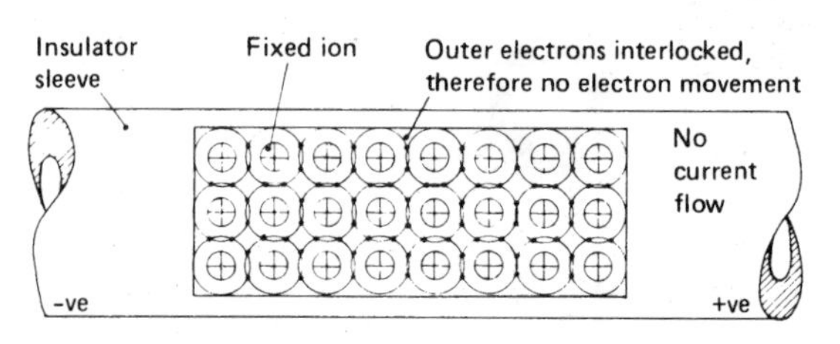

Fig. 25.4 Insulator

negative to positive. Nevertheless the original concept has become the standard convention for the directional movement of current – that is, current is by convention considered to flow from positive to negative.

The rate at which electricity flows in a circuit will depend on the number of electrons which pass a certain point in the circuit in one second. The unit of current is the ampere (symbol A), one ampere being equivalent to the flow of 0.63 million billion (0.63×10^{19}) electrons per second. Thus current is the rate of electron flow and is measured in amperes or electrons per second. The ampere may also be defined as the quantity of current forced through a conductor of resistance one ohm by a pressure of one volt.

25.2.2 Voltage

Voltage (symbol E or V) can be considered as the force or pressure that causes electrons to flow through a conductor.

A voltage is produced between two points when a negative charge (high electrical pressure) exists at one point and a positive charge (low electrical pressure) exists at the other point. The greater the excess of electrons at the negative potential and the greater the deficiency of electrons at the positive potential, the greater will be the difference in electrical potential or pressure or the greater will be the voltage.

The common sources of electricity are the generator and the battery. These can both be looked upon as electron pumps, since they will supply a continuous flow of electrons through a circuit such as the circuits used for the ignition system, the starter motor, and vehicle lighting.

The unit for measuring voltage is the volt (symbol V), which can be defined as the electrical pressure which on being steadily applied to a conductor of resistance one ohm will produce a current flow of one ampere.

25.2.3 Resistance

All conductors offer some resistance (symbol R) to the flow of current, and this is mainly due to three factors. First, each conductor atom resists the removal of an electron from its outer shell, as the electron is attracted by the protons in the nucleus. This attraction may be strong in a poor conductor, but is weak in a good conductor. Secondly, the movement of free electrons is impeded by the continuous collisions of electrons and atoms as the electrons move through the conductor. Thirdly, increased temperature raises the vibratory energy of the atoms so that more electrons will be dislodged from their orbits. An increase in the number of electrons flowing increases the number of collisions between electrons and atoms and as a result will cause internal friction, generate heat, and increase the resistance of the conductor. Raising the conductor's temperature has a similar effect, since this speeds up the vibrating movement of atoms within their framework so that the collision rate of electrons and atoms will rise, causing further hindrance to the flow of current.

The unit for measuring the electrical resistance of a conductor is the ohm (symbol Ω), which may be defined as the resistance which requires the application of one volt in order that a current of one ampere may flow.

25.2.4 Power

Energy is the ability to do work, and in mechanical terms work (W) is defined as the product of force (F) and the distance (s) moved in the direction of the force,

i.e. $W = F \times s$

The unit of energy is the joule, which is the work done when a force of one newton moves through one metre in the direction of the force.

Power (symbol P) is the rate of doing work and is measured in watts, where

1 watt = 1 joule per second

In electrical terms, power is given by the product of current and voltage,

$$\text{i.e. power (watts)} = \begin{matrix}\text{current}\\ \text{(amperes)}\end{matrix} \times \begin{matrix}\text{voltage}\\ \text{(volts)}\end{matrix}$$

or $$P = V \times I$$

Light-bulbs, generators, and starter motors are normally rated in watts to indicate the power they consume.

25.2.5 Electromotive force (e.m.f.)

An electrical force is applied to a conductor when one end of the conductor is given a high-potential negative charge and the other end is given a low-potential positive charge. Electrons then move from the end with the excess electrons to the opposite end which has a deficiency of electrons. The electrical force which enables a source of electricity such as a battery or a generator to create this movement of electrons is termed the electromotive force, or simply the e.m.f.

The electromotive force (e.m.f.) of a battery or generator is the open-circuit voltage measured across its terminals when no current is being supplied to any external circuit such as the ignition coil, starter, vehicle lights, etc.

25.2.6 Potential difference (p.d.)

Potential difference is the difference in electrical pressure or voltage required between two points in a circuit in order to make current flow between those two points.

When a battery supplies current to an external circuit such as the lights, there will be a directional flow of free electrons in the circuit. Some of the electrons collide, and these continuous electron collisions generate a degree of internal friction between the electrons which tends to oppose the movement of electrons both in the external circuit and within the battery itself. A small proportion of the battery's e.m.f. is therefore dissipated in maintaining electron flow within the battery, so the resultant voltage measured across the battery terminals when current is allowed to flow in an external circuit is less than when there is no current flow, and this reduced voltage is referred to as the potential difference (p.d.) across the battery. In other words, the p.d. across an electrical source is the closed-circuit voltage measured across the terminals of the electrical source (battery or generator) when current flows in the circuit.

25.2.7 Voltage drop (v.d.)

This is the voltage needed to overcome the internal resistance of the electrical source when current flows in an external circuit. If a large current discharge is required, for example in a starter-motor circuit, then the high rate of electron movement will increase the free-electron collision rate. Consequently much more voltage pressure is necessary to overcome the source's electrical resistance if adequate free-electron movement is to be obtained.

The voltage drop (v.d.) of a battery may be expressed as the difference between the open-circuit e.m.f. and the closed-circuit p.d. of the battery – the more the potential difference is reduced, the greater will be the voltage drop. For example if the battery voltage drops from an open-circuit normal 12 V down to 10 V when the starter is engaged, then the voltage drop (v.d.) will be 12 V – 10V = 2V.

The potential difference across a component in a circuit required to drive current through that component may also be referred to as the voltage drop across that component.

25.3 Basic vehicle wiring circuits (figs 25.5 to 25.8)

A vehicle's wiring system can be subdivided into a number of simple circuits each connected in series. These circuits each consist of the battery, the electrical component, its switch, and the following three wires or cables:

i) the feed wire – connecting one of the battery terminal posts to the switch;
ii) the switch wire – connecting the switch to the component;
iii) the return wire – connecting the component to the second battery terminal post, either directly (fig. 25.5) or indirectly via the frame of the vehicle (fig. 25.6).

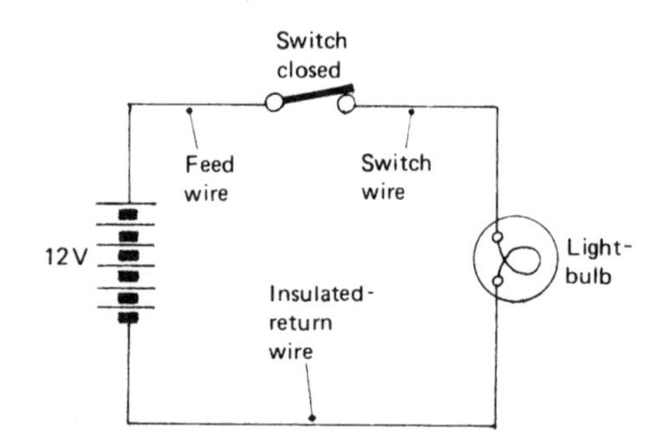

Fig. 25.5 Insulated-return circuit

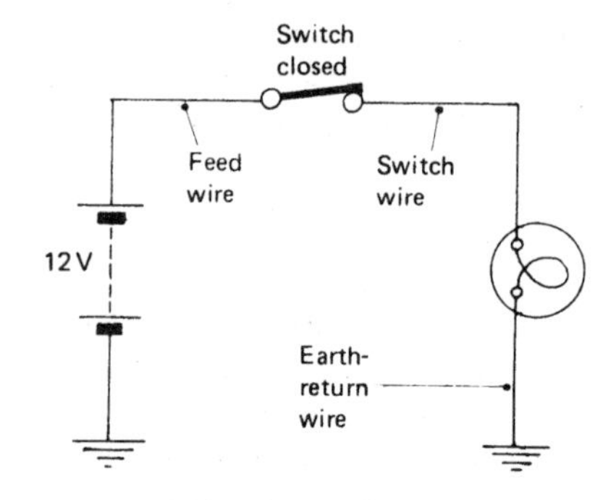

Fig. 25.6 Earth-return circuit

A more elaborate wiring circuit is formed when one switch controls several subcircuits which may have two or more components joined together in series or parallel (fig. 25.7). Overload fuses may also be included.

A typical vehicle lighting system is shown in fig. 25.8. Here it can be seen that some of the lighting

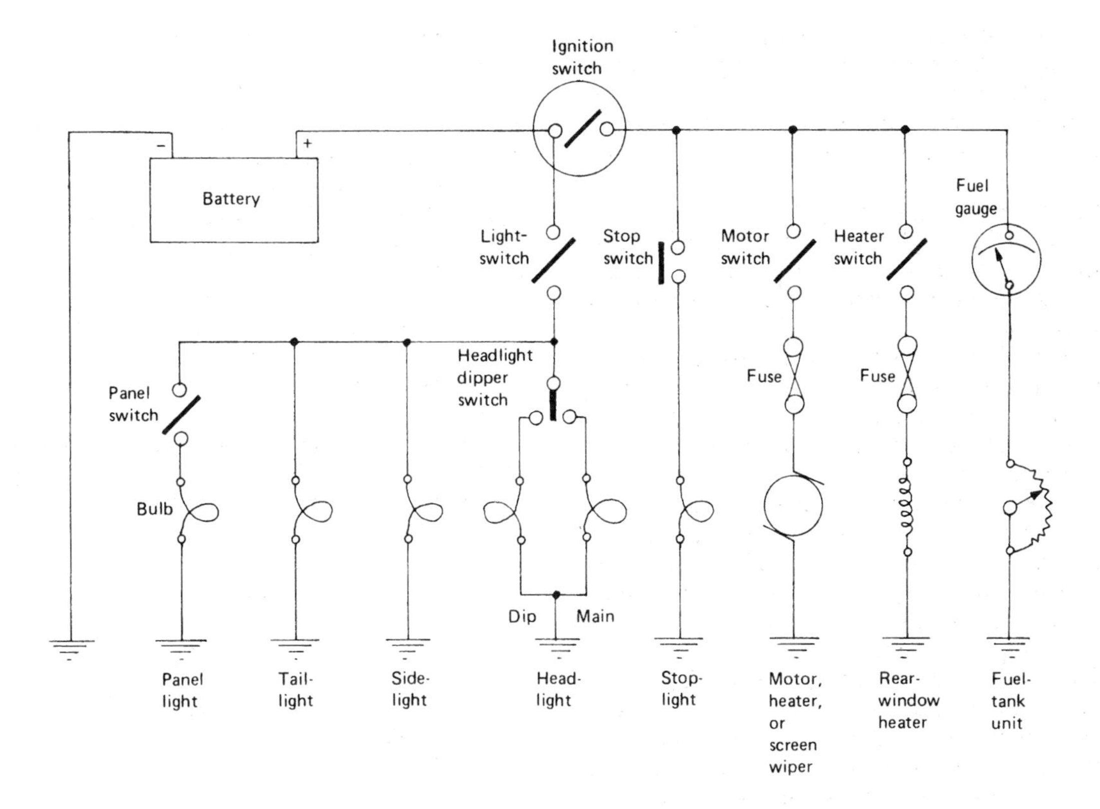

Fig. 25.7 Basic vehicle wiring circuit

functions independently of the ignition switch – such as the side-lights, tail-lights, and headlights – but other parts of the system, such as the stop lamp and the direction-indicator lights, will operate only when the ignition switch is turned on. It may also be noted that one switch may have two 'on' positions, so that it can switch circuits when necessary.

25.3.1 Earth return (fig. 25.6)

All electrical circuits require both a feed and a return conductor between the battery and the component it is to supply with electrical energy. If the vehicle has a metal structure, this can be used as one of the two conducting paths, which is then referred to as the earth return. A wire cable, known as the live feed cable, forms the other conductor. To complete the earth-return path, a short thick cable has one end bolted to the chassis structure while the other end is attached to one of the battery terminal posts – the electrical component then has to be earthed in a similar way.

For a complete vehicle's wiring system, only one battery-to-chassis conductor is necessary and any number of separate earth-return circuits can be similarly wired. Thus an earth-return system reduces and simplifies the amount of wiring and makes it easier to trace electrical faults.

25.3.2 Insulated return (fig. 25.5)

For some vehicle applications it may be necessary and safer to use a separate insulated-cable system for both the feed and the return conductors.

With separate feed and return cables, it is practically impossible for the cable conductors to short – that is, electrically contact each other – if chafed and touching any of the metal bodywork, since the body is not part of any of the electrical circuits and therefore is not live.

There are some vehicles, such as coaches and double-decker buses, which use large amounts of plastic panelling. For these it is not convenient to use an earth return – instead, an insulated return is more reliable and safer. From the safety aspect, an insulated return is essential for vehicles transporting highly flammable liquids and gases, where a spark could very easily set off an explosion or a fire.

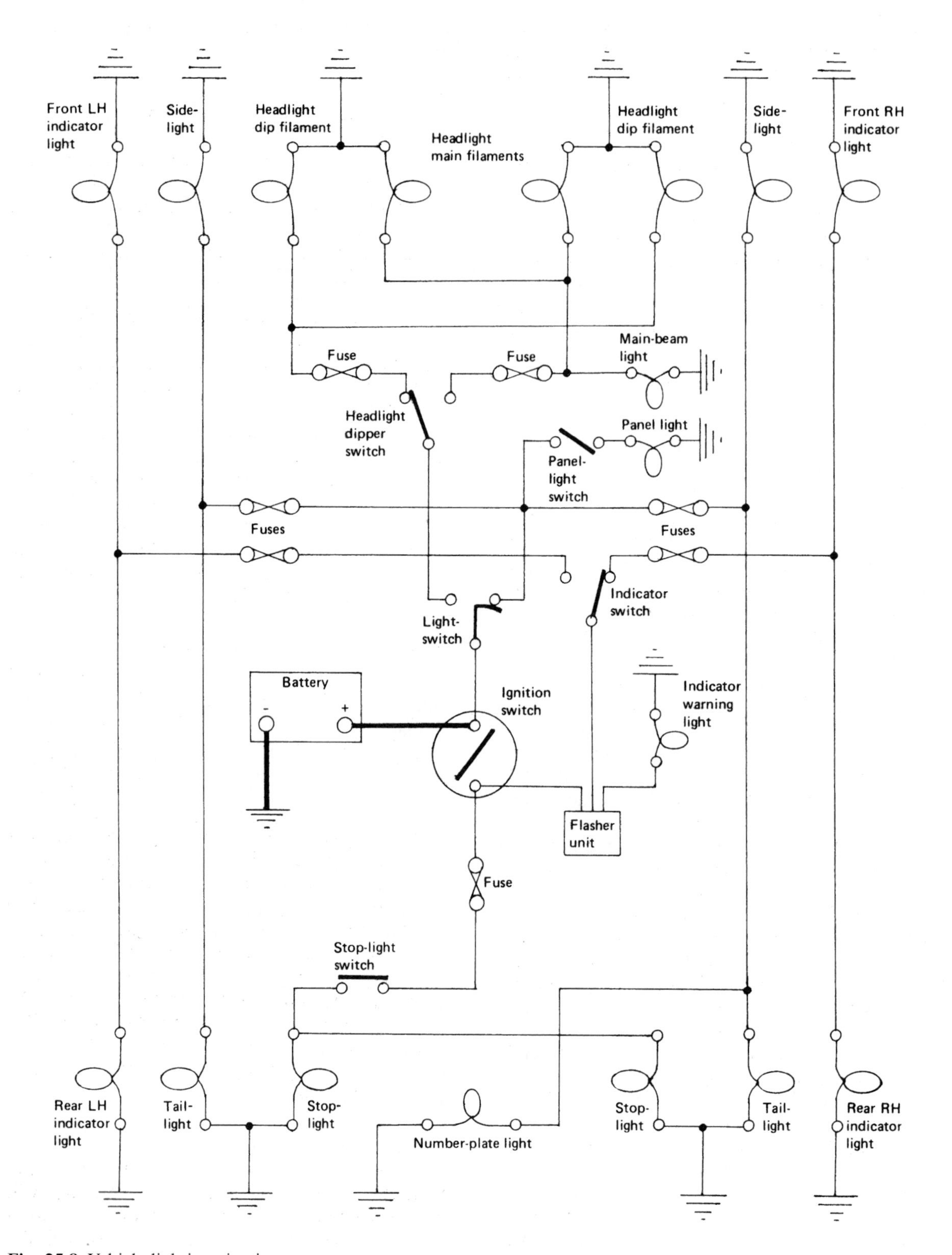

Fig. 25.8 Vehicle lighting circuit

The disadvantage of the insulated return is that the extra cable makes the overall wiring harness heavier, less flexible, and bulky – and of course raises the cost to some extent.

25.3.3 Open circuit

If a wiring circuit is not continuous, then the system is said to be open-circuit and no current will flow. This condition may be deliberate (such as when a switch breaks the circuit) or it may be unintentional, due to poor connections, partially broken or connected terminals, or fractured or burnt-out wire. Intermittent unwanted open circuits are usually difficult to trace and can cause damage to other electrical components, whereas permanent open circuits are normally easier to find.

25.3.4 Short circuit

This condition can occur in a wiring system if the insulation covering a live cable is chafed such that the bare wire underneath either touches some part of the metal earth return, such as the chassis, or is crossed with another exposed piece of wire so that a closed circuit with the battery in series results. The consequences may be sparking as the vehicle's body vibrates, and overheating at the short which may melt the insulation and expose more bare wire or other electrical connections. The continuous current flow will also rapidly discharge the battery. Eventually the wires may melt together and overload the shorted wiring so that it overheats again and burns through the wire, at the same time becoming a potential fire hazard.

Many wiring circuits on the vehicle are safeguarded by fuses so that, if the wiring does short and cause excessive current flow in the cable, the fuse will blow and prevent further current flow.

25.3.5 Electrical-circuit symbols (fig. 25.9)

Wiring diagrams are provided to enable electrical faults to be easily traced and diagnosed to specific parts of a circuit. To simplify the wiring diagrams, symbols are used to identify the various components, instruments, and wiring. Unfortunately different symbols have evolved to represent certain pieces of equipment making up the electrical system of a vehicle, and this does cause some confusion. However, a list of symbols commonly used by vehicle manufactureres has been included in fig. 25.9.

25.4 Cables

Cables are provided to carry electrical energy from one point to another with the minimum of energy loss.

Desirable properties of a cable are that it should

a) occupy very little space,
b) be easily manipulated to whatever shape is necessary,
c) be able to continuously accommodate a small degree of flexing,
d) maintain a minimum electrical resistance throughout the cable's life,
e) maintain an effective insulation covering throughout the cable's life.

The wire cable conductor is normally made from many strands of copper wire bunched together to form a core – this enables the cable to be considerably more flexible than if a single large-diameter strand were used. Another very important factor in favour of a multi-strand cable is that, if one or more strands is damaged and fails, then the remaining strands will share out the additional electrical load and still continue to conduct. Conversely, if a single large strand is adopted as the conductor, continuous bending of the solid wire will work-harden the copper, followed by fatigue failure.

25.4.1 Power capacity of a cable

In the selection of cable sizes, the most important factor is the power-carrying capacity of the cable. It should be remembered that power is given by the product of voltage and current, so that a 480 watt rated cable can carry either 480 W ÷ 240 V = 2 A at a house mains 240 V supply or 480 W ÷ 12 V = 40 A when a car battery applies 12 V.

For a cable conductor to carry a large current without a considerable voltage drop and without generating excessive heat, the conductor must have the minimum of electrical resistance. The resistance of a wire conductor is directly proportional to the length and inversely proportional to the cross-sectional area – thus doubling the wire length doubles the resistance, but doubling the cross-sectional area halves the resistance. This implies that, to keep the electrical resistance and hence the voltage drop to the minimum, the cable wire should be as thick and as short as possible.

25.4.2 Cable specification

Cable specification must stipulate the operating voltage, as the current-carrying capacity will greatly depend upon the applied voltage. The total core cross-sectional area mainly determines the continuous-current rating for wire cables, but

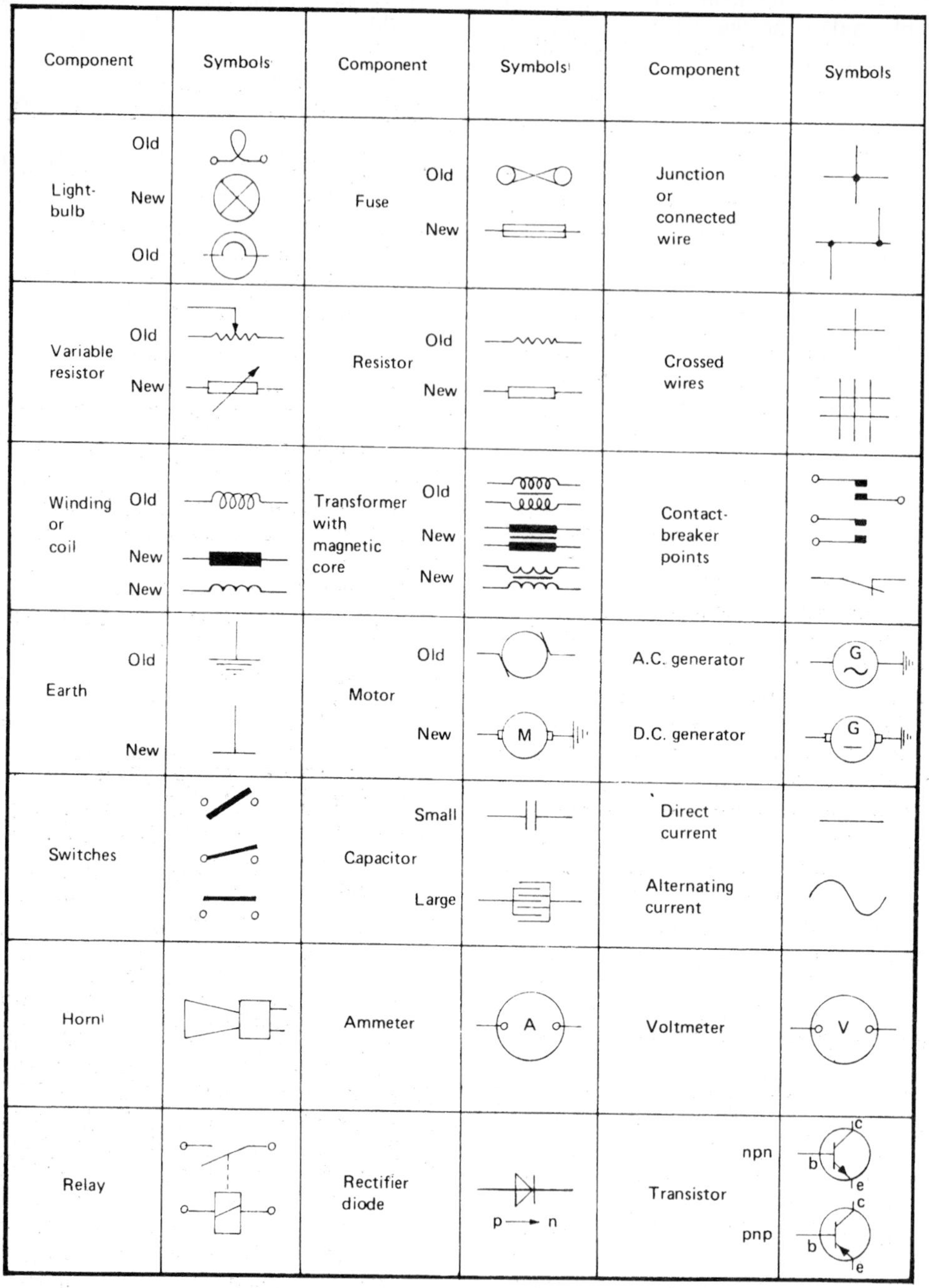

Component	Symbols	Component	Symbols	Component	Symbols
Light-bulb: Old / New / Old		Fuse: Old / New		Junction or connected wire	
Variable resistor: Old / New		Resistor: Old / New		Crossed wires	
Winding or coil: Old / New / New		Transformer with magnetic core: Old / New / New		Contact-breaker points	
Earth: Old / New		Motor: Old / New	M	A.C. generator D.C. generator	G G
Switches		Capacitor: Small / Large		Direct current Alternating current	
Horn		Ammeter	A	Voltmeter	V
Relay		Rectifier diode	p → n	Transistor: npn / pnp	c b e c b e

Fig. 25.9 Common electrical symbols

cable sizes are normally expressed by the number of strands making up the conductor core and the diameter of each strand – thus 28/0.30 mm implies there are 28 strands, each of 0.30 mm diameter.

Table 25.1 lists the most popular cable sizes, their electrical specifications, and motor-vehicle applications. This gives a general overall comparison, but consideration must be given to cable length and environmental conditions when choosing cable sizes. In general the total voltage drop in

Table 25.1 Cable specifications and applications for 12 volt circuits

Number and diameter in mm of strands	Nominal total c.s.a. in mm^2	Maximum continuous-current rating in amperes	Resistance in ohms per metre of conductor at 20°C	Voltage drop per metre of conductor at maximum rated current	Applications
14/0.25	0.7	6.00	0.0233	0.027	Low-tension ignition, side- and
14/0.30	1.0	8.75	0.0189	0.019	tail-lights, and body wiring
21/0.30	1.5	12.75	0.0141	0.013	
28/0.30	2.0	17.50	0.0094	0.009	Headlights
35/0.30	2.5	21.75	0.0073	0.008	Battery-feed and ammeter
44/0.30	3.0	27.50	0.0060	0.006	circuits
65/0.30	4.5	35.00	0.0041	0.004	Dynamo, ammeter, and
84/0.30	6.0	42.00	0.0031	0.003	alternator circuits
120/0.30	8.5	60.00	0.0022	0.0022	Heavy-duty dynamo, ammeter, and
80/0.40	10.0	70.00	0.0019	0.0018	alternator circuits (commercial vehicles, buses, coaches, etc.)
37/0.71	15.0	105.00	0.0011	0.0013	Car petrol engines } Light duty } Starter cable
266/0.30	20.0	135.00	0.0010	0.000762	Car/van (extra flexible) } Light duty } Starter cable
37/0.90	25.0	170.00	0.0008	0.00099	Small diesel engines } Light duty } Starter cable
61/0.90	40.0	300.00	0.0005	0.000462	Medium diesel engines } Heavy duty } Starter cable
61/1.13	60.0	415.00	0.0003	0.000293	Large diesel engines } Heavy duty } Starter cable

any one circuit should not exceed 5% of the battery voltage – that is, 0.6 volts in a 12 volt battery circuit – thus very long cables may have to use slightly higher rated wire to avoid an excessive voltage drop in their circuit. Where more than two cables of size 28/0.30 mm or over are bunched together, the maximum permissible current rating should be only 60% of the rating shown in Table 25.1 – this is because heat dissipation is less effective with bunched cables.

Cable conductors are normally stranded plain copper wire, but tinned copper wire is used for special purposes. The conductor core is normally insulated with high-grade p.v.c., which has good resistance to the effects of oil, water, and acid.

25.4.3 Ignition H.T. cable

Ignition cables operate under very high voltage surge conditions and therefore they require extra insulation covering compared with the normal low-tension cable. The conductor core may be either plain or tinned copper wire – alternatively, interference-suppression carbon-impregnated textile-core cable may be used. This TV and radio interference-suppression carbon-impregnated cable comes in two resistance ranges – for cables up to 300 mm in length, select a cable which has a core resistance of 25 000 to 34 000Ω/m; if the cable is to be longer than 300 mm, use a cable with a core resistance of 15 000 to 24 000 Ω/m.

Ignition cables are available in two sizes – 9/0.30 mm and 19/0.30 mm, that is with 9 or 19 strands of 0.30 mm diameter wire – with cross-sectional conductor areas of 0.65 and 1.35 mm^2 and approximate outside insulation diameters of 5 and 7 mm respectively.

The plain copper wire may be insulated and sheathed with just high-grade p.v.c. or with p.v.c. braided with strong cotton and treated with successive coats of glossy enamel to form a highly polished oil-, water-, and flame-resisting finish. Alternatively, to meet abnormal climatic and environmental conditions, a wire core of tinned copper may be sheathed with high-grade rubber and finally Hypalon or Vamac.

25.5 Cable colour code

To enable the various cables to be quickly identified when many cables are grouped together to form what is known as a wiring loom, the cable wire insulation is provided with a range of colours. British Standard AU 7 specifies a standard colour code for all electrical circuits likely to be installed on a vehicle. There are seven base colours used – brown, yellow, white, green, blue, red, and black – but these would not provide a sufficiently varied range of colours to cope with the number of wires branching out from the main cable harness. Therefore, to extend the colour range, additional colour combinations are made available by superimposing a thin line of a second tracer colour on to one of the base colours. These tracer-strip colours are brown, yellow, white, light green, blue, red, black, purple, pink, orange, and slate.

Base cable colours are associated with the following circuits:

Brown – main battery feed;
Yellow – generator circuits;
White – ignition circuits which do not require fusing;
Green – auxiliary circuits fed through the ignition switch and protected by fuses;
Blue – side-light and tail-light circuits;
Black – earth circuits.

Cables on wiring diagrams printed in black and white can be identified by single letters as follows:

B = black
U = blue
LG = light green
K = pink
O = orange
Y = yellow
N = brown
R = red
P = purple
G = green
S = slate
W = white

For a base colour with a second-colour tracer, the code is for example

NU = brown with blue tracer
GW = green with white tracer
RP = red with purple tracer

European vehicles using the DIN standard will code cables as follows:

BL = blue (*blau*)
BR = brown (*braun*)
GE = amber (*gelb*)
GR = grey (*grau*)
GN = green (*grün*)
RT = red (*rot*)
SW = black (*schwarz*)
WS = white (*weiss*)

The British Standard vehicle-wiring colour code is shown in Tables 25.2(a) and (b).

25.6 Cable connectors (fig. 25.10)

Connectors are used to join various parts of electrical circuits together. They may be used to connect cables to components from a more convenient point, to interrupt the main or sub-main loom, or to form a multi-junction point to spur off other cables.

Connectors are generally made from brass strip in a range of sizes to cope with different current demands. Depending on where they are to be used, they may be covered with p.v.c. or left uninsulated.

The cable core is attached to the connectors either by soldering or by crimping the connector over the wire strands. Both these methods of producing an electrical joint have proved satisfactory in service.

25.6.1 Ring and fork plain brass connectors (fig. 25.10(a) and (b))

These are generally used to connect cables to electrical components and are commonly seen as earth-return terminals. Additional electrical accessories when fitted to a vehicle frequently use these forms of connectors.

25.6.2 'Lucar' connectors (fig. 25.10(c)–(f))

These are rapid-connecting flat-type male and female terminal-to-cable or cable-to-cable connectors. Their main uses are connections to spade terminals on the ignition switch, lighting switches, voltage stabiliser, wiper and heater motors, etc. Depending on its polarity with respect to earth, the connector will be plastic-covered or bare – for example, an earth-return connector will be bare, and a connector which is live with respect to earth will be covered.

25.6.3 Snap-in bullet connectors (fig. 25.10(g) (i)–(vi))

The male and female connectors complement each other, and when pushed together the spherical top of the male unit locates into a recess in the female, providing a rigid secure connection. Junction connectors of this type are available in single, twin, three-way and four-way sizes, with either common or separate insulated internal contacts, depending on the application requirement. These junction connectors are popular for all types of lighting wiring circuits.

Table 25.2(a) British Standard AU 7 vehicle-wiring colour code

Colour		
Main	Tracer	Destination
Brown	–	Main battery feed
Brown	Blue	Control box (compensated voltage control only) to ignition and lighting switch (feed)
Brown	Red	Compression-ignition starting aid to switch. Main battery feed to double-pole ignition switch (a.c. alt. system)
Brown	Purple	Alternator regulator feed
Brown	Green	Dynamo 'F' to control box 'F'. Alternator field 'F' to control box 'F'
Brown	Light green	Screenwiper motor to switch
Brown	White	Ammeter to control box. Ammeter to main alternator terminal
Brown	Yellow	Dynamo 'D' to control box 'D' and ignition warning light. Alternator neutral point
Brown	Black	Alternator warning light, negative side
Brown	Pink	
Brown	Slate	
Brown	Orange	
Blue	–	Lighting switch (head) to dipper switch
Blue	Brown	
Blue	Red	Dipper switch to headlight dip beam. Headlight dip-beam fuse to right-hand headlight (when independently fused)
Blue	Purple	
Blue	Green	
Blue	Light green	Screenwiper motor to switch
Blue	White	Dipper switch to headlight main beam (subsidiary circuit – headlight flasher relay to headlight). Headlight main-beam fuse to right-hand headlight (when independently fused). Headlight mainbeam fuse to outboard headlights (when outboard headlights independently fused). Dipper switch to main-beam warning light
Blue	Yellow	Long-range driving switch to light
Blue	Black	
Blue	Pink	Headlight dip-beam fuse to left-hand headlight (when independently fused)
Blue	Slate	Headlight main-beam fuse to left-hand headlight or inboard headlights (when independently fused)
Blue	Orange	
Red	–	Side- and tail-light feed
Red	Brown	Variable-intensity panel lights (when used in addition to normal panel lights)
Red	Blue	
Red	Purple	Map-light switch to map light
Red	Green	Lighting switch to side- and tail-light fuse (when fused)
Red	Light green	Screenwiper motor to switch
Red	White	Panel-light switch to panel lights
Red	Yellow	Fog-light switch to fog-light
Red	Black	Parking-light switch to left-hand side-light
Red	Pink	
Red	Slate	
Red	Orange	Parking-light switch to right-hand side-light
Purple	–	Accessories fused direct from battery
Purple	Brown	Horn fuse to horn relay (when horn is fused separately)
Purple	Blue	
Purple	Red	Boot-light switch to boot light
Purple	Green	
Purple	Light green	
Purple	White	Interior light to switch (subsidiary circuit – door safety lights to switch)
Purple	Yellow	Horn to horn relay
Purple	Black	Horn or horn relay to horn push
Purple	Pink	
Purple	Slate	Aerial-lift motor to switch 'up'
Purple	Orange	Aerial-lift motor to switch 'down'
Green	–	Accessories fused via ignition switch (subsidiary circuit fuse A4 to hazard switch (terminal 6))
Green	Brown	Reverse-light to switch
Green	Blue	Water-temperature gauge to temperature unit
Green	Red	Left-hand flasher light
Green	Purple	Stop-light to stop-light switch
Green	Light green	Hazard flasher unit to hazard pilot light
Green	White	Right-hand flasher light
Green	Yellow	Heater motor to switch, single speed (or to 'slow' on two-speed motor)
Green	Black	Fuel gauge to fuel-tank unit or change-over switch

Table 25.2(b) British Standard AU 7 vehicle-wiring colour code (*continued*)

Colour			Colour		
Main	Tracer	Destination	Main	Tracer	Destination
Green	Pink	Choke solenoid to choke switch (when fused)	White	Black	Ignition coil C.B. to distributor contact-breaker. Rear heated window to switch or fuse T.A.C. ignition
Green	Slate	Heater motor to switch (or to 'fast') (on two-speed motor)	White	Pink	Radio from ignition switch
Green	Orange	Low-fuel-level warning light	White	Slate	Tachometer to ignition coil
Light green	–	Instrument voltage stabiliser to instruments	White	Orange	Hazard warning feed (to switch)
Light green	Brown	Flasher switch to flasher unit 'L'	Yellow	–	Overdrive
Light green	Blue	Flasher switch to left-hand flasher warning light	Yellow	Brown	Overdrive
Light green	Red	Fuel-tank change-over switch to right-hand tank unit	Yellow	Blue	Overdrive
Light green	Purple	Flasher unit 'F' to flasher warning light	Yellow	Red	Overdrive
Light green	Green		Yellow	Purple	Overdrive
Light green	White		Yellow	Green	Overdrive
Light green	Yellow	Flasher switch to right-hand flasher warning light	Yellow	Light green	Screenwiper motor to switch
Light green	Black	Screen-jet switch to screen-jet motor	Yellow	White	
Light green	Pink	Flasher unit 'L' to emergency switch (simultaneous flashing)	Yellow	Black	
Light green	Slate	Fuel-tank change-over switch to left-hand tank unit	Yellow	Pink	
Light green	Orange		Yellow	Slate	
White	–	Ignition control circuit (unfused) (Ignition switch to ballast resistor)	Yellow	Orange	
White	Brown	Oil-pressure switch to warning light or gauge	Black	–	All earth connections
White	Blue	Choke switch to choke solenoid (unfused). Rear-heater fuse unit to switch. Electronic ignition T.A.C. ignition unit to resistance.	Black	Brown	Tachometer generator to tachometer
White	Red	Solenoid starter switch to starter push or inhibitor switch	Black	Blue	Tachometer generator to tachometer
White	Purple	Fuel pump No. 1 or right-hand to change-over switch	Black	Red	Electric speedometer
White	Green	Fuel pump No. 2 or left-hand to change-over switch	Black	Purple	
White	Light green	Screenwiper motor to switch	Black	Green	Screenwiper switch to screenwiper (single-speed) relay to radiator fan motor
White	Yellow	Starter inhibitor switch to starter push. Ballast resistor to coil. Starter solenoid to coil	Black	Light green	Vacuum brake switch to warning light and/or buzzer
			Black	White	Brake-fluid-level warning light to switch and hand-brake switch
			Black	Yellow	Electric speedometer
			Black	Pink	
			Black	Slate	
			Black	Orange	Radiator fan motor to thermal switch
			Slate	–	Window lift
			Slate	Brown	Window lift
			Slate	Blue	Window lift
			Slate	Red	Window lift
			Slate	Purple	Window lift
			Slate	Green	Window lift
			Slate	Light green	Window lift
			Slate	White	Window lift
			Slate	Yellow	Window lift
			Slate	Black	Window lift
			Slate	Pink	Window lift
			Slate	Orange	Window lift

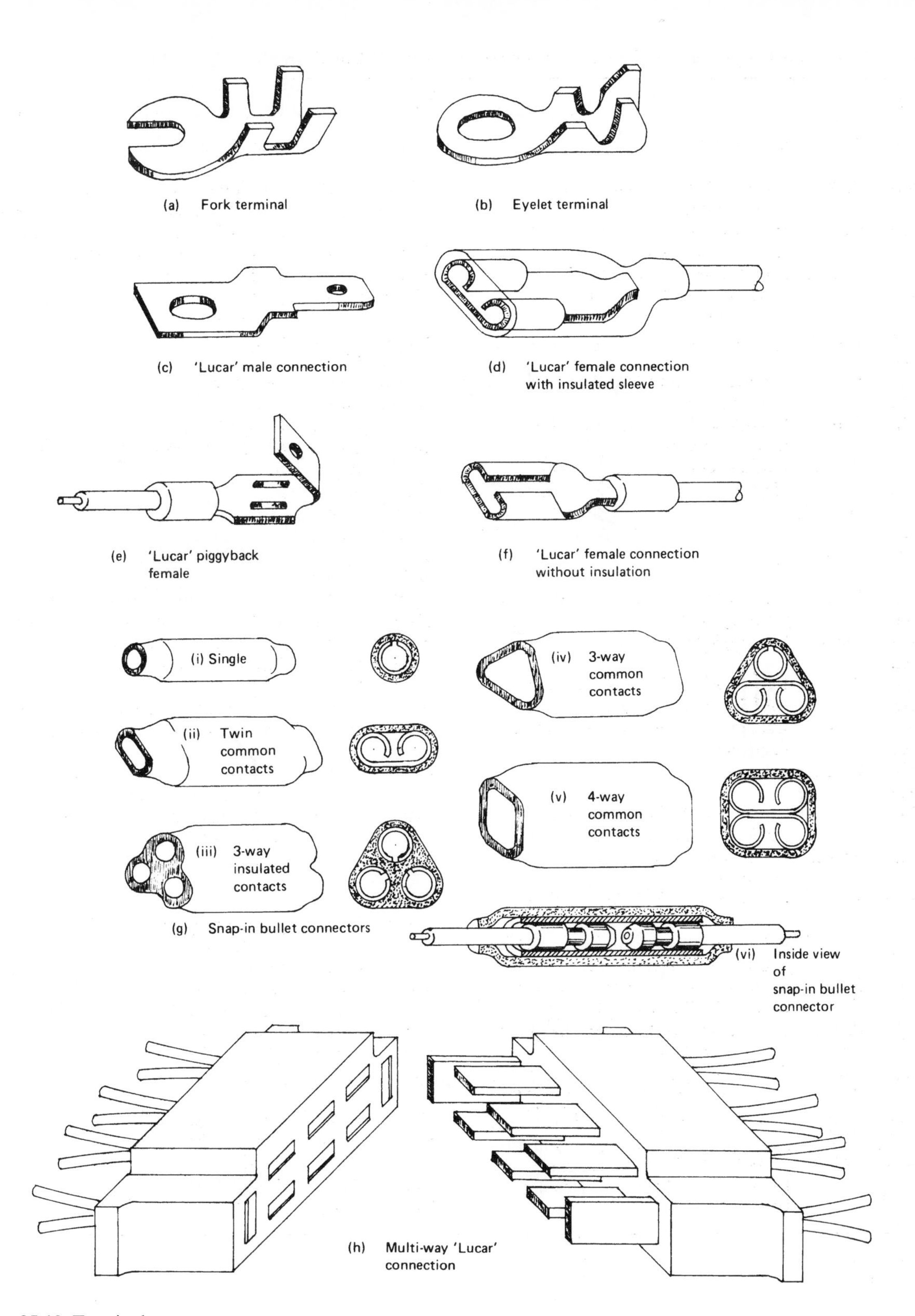

Fig. 25.10 Terminal connectors

25.6.4 Multi-way 'Lucar' connectors (fig. 25.10(h))
Connectors of this type are useful in connecting the various looms together in the bulkhead region or for connecting up sublooms such as the dip-switch and indicator units. One feature of these male and female spade connectors in a group form is that they cannot be incorrectly connected.

25.7 Printed circuits (fig. 25.11)
Where there are many small electrical components and circuits close together – such as on an instrument panel – the circuit wiring may be replaced by a single copper sheet which has been stamped or printed out to form a network of conducting strips which twist and bend in such

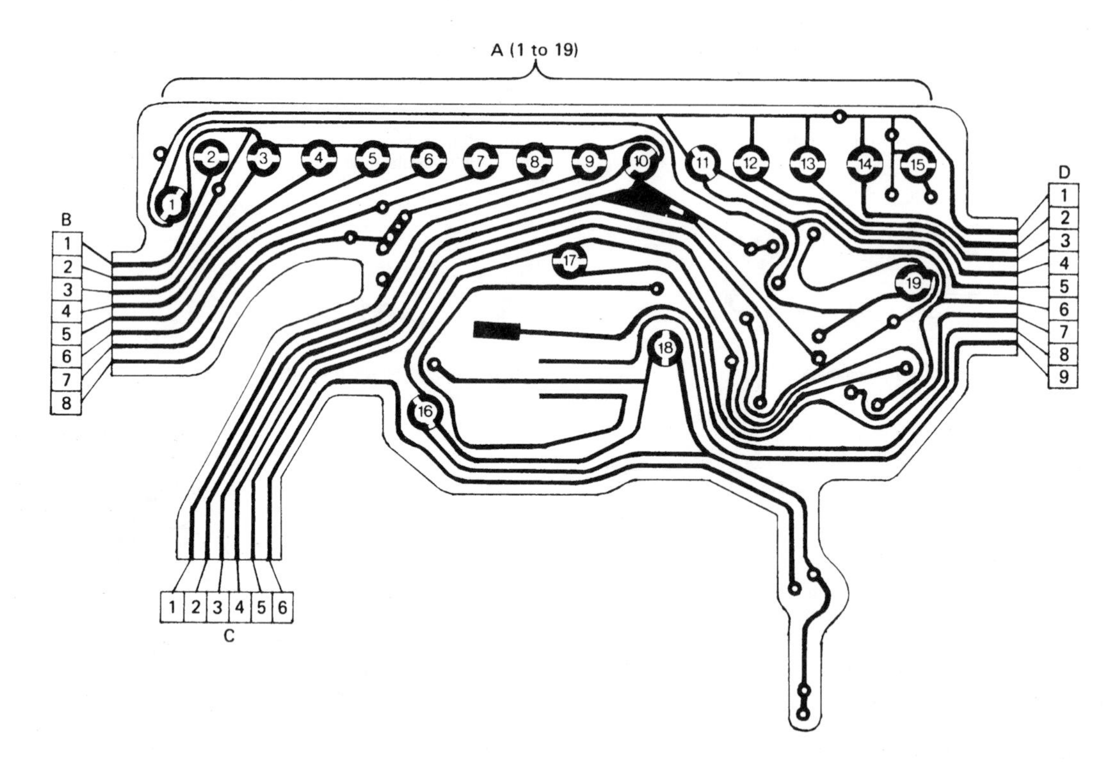

Fig. 25.11 Instrument-panel printed-circuit diagram

Light-bulb holders A
1 Speedometer light
2 Side, rear, and licence-plate lights
3 Headlight dipped-beam warning light
4 Headlight main-beam warning light
5 Rear-screen demister warning light
6 Front-fog-light 'on' warning light
7 Brake-circuit incident and hand-brake 'on' warning light
8 Oil-pressure warning light
9 Battery-charging warning light
10 Brake-pad-wear warning light
11 Hazard warning light
12 Rear-fog-light 'on' tell-tale
13 Choke 'on' warning light
14 Low-fuel warning light
15 Automatic transmission incident warning light
16 Clock light
17 Direction-indicator light
18 Coolant-temperature light
19 Fuel-gauge light

Connector B
1 Side, rear, and licence-plate warning light
2 + after ignition switch (not in use)
3 Headlight dipped-beam warning light
4 Headlight main-beam warning light
5 Rear-screen demister warning light
6 Front-fog-light 'on' warning light
7 Brake-circuit incident and hand-brake 'on' warning light
8 Oil-pressure warning light

Connector C
1 Battery-charging warning light
2 Brake-pad-wear warning light
3 + after ignition switch
4 Coolant-temperature-gauge light
5 Fuel-gauge light
6 Lighting

Connector D
1 + after ignition switch
2 Low-fuel warning light
3 Choke 'on' warning light
4 Rear-fog-light 'on' tell-tale
5 Hazard warning light
6 Earth
7 Direction indicators
8 Clock
9 Tachometer make and break

a way that they do not cross or interfere with each other.

These printed circuits, as they are called, provide a very compact, simplified, and reliable method of connecting up the numerous conducting circuits. For example, fig. 25.11 shows an instrument-panel printed circuit – the row of horizontal large circles represents the panel warning lights, and the small eyeholes are the terminal points which are joined to the various instruments such as the fuel gauge, temperature gauge, clock, tachometer, etc.

25.8 Fuses (figs 25.12 and 25.13)

Fuses are installed in series in an electrical circuit to prevent unpredicted excessive current flow damaging electrical components, melting cable insulation, or even burning out the cable wire strands.

Very large current surges can occur if an electrical component overheats and develops an internal short, or if a live feed wire short-circuits to earth. A fuse is the weak link in an electrical circuit. If current flowing in a circuit suddenly rises to the maximum rating of the fuse conductor, the temperature will instantly rise to the melting point of the fuse material and immediately a gap will be blown through the fuse, thus preventing any further flow of current in the circuit.

25.8.1 Fuse rating

Fuses are given two different ratings, which may be explained as follows:

Continuous-current rating This is the maximum current a fuse can carry continuously. Its value is normally half that of the fusing-current rating.

Fusing-current rating This is the peak current at which the fuse blows. The fusing-current rating is based on the fuse carrying its continuous-current rating for 5 minutes after which the rated fusing current is applied – the fuse should then blow within 10 seconds.

Manufacturers normally mark a current rating on the fuse, but it it is not always obvious if it refers to the continuous or the blow value. Lucas and CAV glass fuses (Table 25.3) are normally marked with the fusing rating, whereas the Continental ceramic-type fuses (Table 25.4) are identified by their continuous rating. Only replace fuses with their correct current rating – never put in a higher-rated fuse, as this may overload or damage part of the circuit. If replacement fuses

Table 25.3 Specification for glass fuses

Identifying colour (amperes)	Continuous-current rating (amperes)	Fusing-current rating (amperes)
Blue	1.0	2
Green	1.5	3
Red	2.5	5
Blue on green	4.0	8
Pale blue	5.0	10
Light blue	7.5	15
Blue on yellow	10.0	20
Pink	12.5	25
Green on white	15.0	30
White	17.5	35
Yellow	25.0	50

Table 25.4 Specification for ceramic fuses

Current rating* (A)	Identifying colour	Test fusing current and time	
		1.5 × rating	2.5 × rating
5	Yellow	Blows within 1 hour	Blows within 1 minute
8	White	Blows within 1 hour	Blows within 1 minute
16	Red	Blows within 1 hour	Blows within 1 minute
25	Blue	Blows within 1 hour	Blows within 1 minute

*Current rating is the continuous-current carrying capability.

Table 25.5 Fuse current rating for tinned copper wire

Current-carrying capacity (amperes)	S.W.G.	Wire diameter (mm)
5	35	0.21
10	29	0.34
15	25	0.51
20	23	0.61
30	21	0.81
38	19	1.02
65	17	1.42
78	15	1.83

are not immediately available, a single strand of tinned copper wire of appropriate rating may be tied between the fuse clips as a temporary measure – see Table 25.5.

It has now become common practice for nearly every individual electrical circuit to be fused, so that any fault which causes excessive current flow will put only one or two components out of action when the fuse blows. This makes it much easier to trace the problem.

Typical fuseboards for glass and ceramic fuses are shown in figs 25.12 and 25.13.

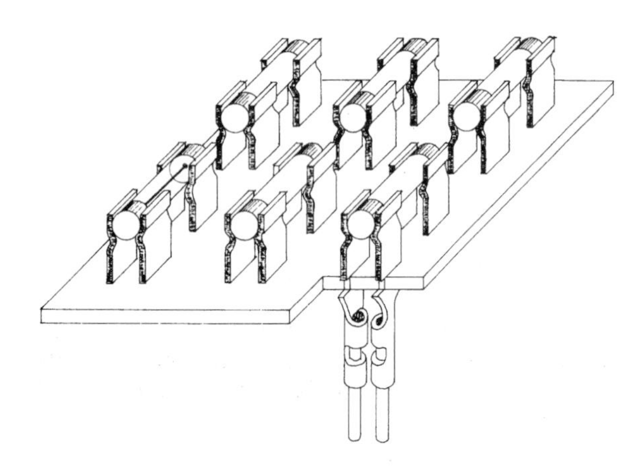

Fig. 25.12 Glass-fuse-holders and board

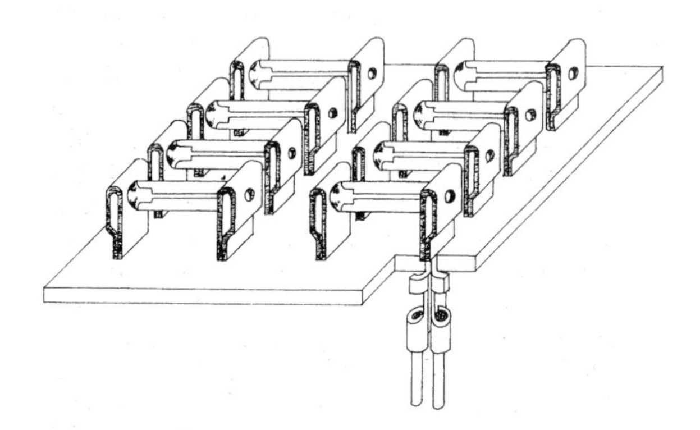

Fig. 25.13 Ceramic-fuse-holders and board

25.9 Light-bulb fundamentals

An electric light-bulb changes electrical energy into heat and light when current flows through its incandescent filament. The design must be such that the bulb can be operated for long periods at a high temperature without appreciable deterioration in the intensity of radiated light. Due to the low voltages used for vehicle electrical systems, the bulb filament coils can be relatively short and thick compared to the domestic light-bulb, so that they can withstand the shock and vibration normally experienced by the vehicle's bodywork when it is being driven.

Most light-bulbs consist of a tungsten-wire filament wound in the form of a helical coil which is suspended between two conducting and supporting lead-wires, the whole assembly being enclosed in a sealed glass bulb.

A simple current-conducting filament exposed to the atmosphere loses too much heat, so that less energy is available for producing a bright light for a given electrical power consumption. Furthermore, a filament operating at high temperatures in air will tend to oxidise and deteriorate rapidly.

25.9.1 Vacuum-filled bulbs

To improve the lighting effectiveness of the filament, it can be enclosed in a bulb from which the air has been removed. Such light-bulbs are known as vacuum bulbs. The vacuum prevents large amounts of heat being transferred from the filament to the internal surface of the bulb by convection and, because a vacuum does not support combustion, it prevents filament oxidation.

Unfortunately most metals vaporise at high operating temperatures. The actual critical temperature varies with the surrounding gas pressure – a vacuum lowers the vaporisation temperature, and an increased pressure raises it. Thus, if the operating temperature of the filament is very high, tungsten will vaporise and condense on to the internal surface of the bulb so that the latter becomes blackened, reducing the radiated light. In addition, the vaporisation will reduce the cross-sectional area of the filament, so that its electrical resistance is increased and its temperature is decreased, so again reducing the resulting light intensity. Vacuum-filled bulbs are therefore operated at temperatures limited to about 2200°C, although the melting point of tungsten is about 3500°C.

25.9.2 Inert-gas-filled bulbs

Higher operating temperatures for filaments can be achieved with inert-gas-filled bulbs. Gases such as nitrogen and argon which do not support burning are introduced into the bulb at various pressures. This allows the filament temperature to be raised to about 2600°C before excessive filament vaporisation and glass blackening take place.

Gas-filled bulbs consume about one third less power than vacuum-filled bulbs for the same light brilliancy.

Filament efficiency is further improved by winding the coil in a tight helix so that less surface area is exposed to convection and thus heat transfer to the inside of the glass bulb is minimised.

25.9.3 Optimum voltage versus bulb light output and life expectancy

Light-bulbs are designed to operate at the maximum generator charging voltage, which is usually set at 13.5 volts for a 12 volt battery system. If the voltage applied to the filament deviates from this optimum value (13.5 V), the light output and life expectancy will also change. Thus if the voltage rises 10% above the optimum 13.5 volts, due to incorrect generator regulator setting, the light output will increase to something like 145% of the optimum but the life expectancy will fall to only 28% of the optimum life. Conversely, if the operating voltage drops 10% lower than the designed optimum (13.5 V), the light output will be reduced to 67% and the life expectancy will be increased to something of the order of 440% of that expected with a continuous potential difference of 13.5 volts.

Generally, small bulbs used for side- and tail lights, instrument panels, etc. have an estimated life of about 200 hours if operated at optimum designed voltage. This value is halved for headlight bulbs, i.e. 100 hours.

25.10 Headlight reflectors (fig. 25.14)

The object of the headlight reflector is to direct the random light rays produced by the light-bulb into a beam of concentrated light by applying the laws of reflection.

A reflector is basically a layer of silver, chrome, or aluminium deposited on a smooth and polished surface such as brass or glass. The outer surface of this layer soon tarnishes in air, so, when a glass reflector is used, the surface in contact with the glass is made the reflector and, to provide further protection for the coating, its back face is usually painted with shellac varnish or something similar.

Consider a mirror reflector that 'caves in' – this is called a concave reflector. The centre point on the reflector is called the pole, and a line drawn perpendicular to the surface from the pole is known as the principal axis (fig. 25.14). If a light-source is moved along this line, a point will be found where the radiating light produces a reflected beam parallel to the principal axis. This point is known as the focal point, and its distance from the pole is known as the focal length.

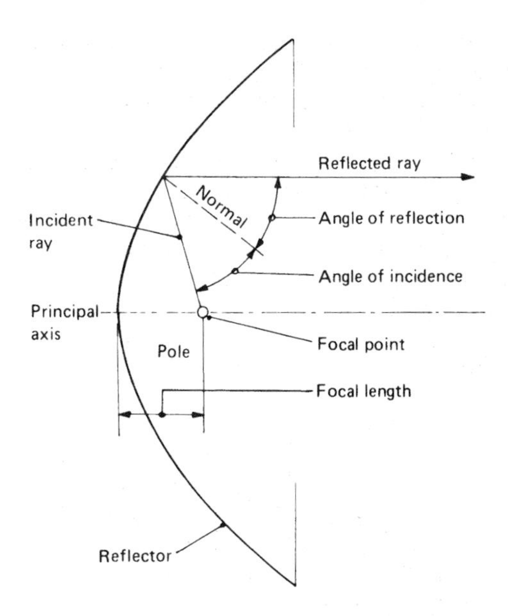

Fig. 25.14 Reflector theory

25.10.1 Parabolic reflector

If the concave reflector were large and truly spherical in section, there would be many focal points necessary for all the light rays to be reflected into a parallel beam; with only one focal point, therefore, some of the light would be reflected as a convergent beam (fig. 25.15(a)). Because of this, the concave spherical reflector is not suitable for headlights, and a parabolic reflector is used instead. (A parabola is a curve similar in shape to the curved path of a stone thrown forward into the air.) A parabolic reflector (fig. 25.15(b)) has the property of reflecting rays parallel to the principal axis when a light-bulb is placed at its focal point, no matter where the rays fall on the reflector. It therefore produces a bright parallel reflected beam of constant light intensity.

25.10.2 Deep or shallow reflectors

With a parabolic reflector, most of the light rays from the light-bulb are reflected and only a small amount of direct rays disperses as stray light (fig. 25.15(b)). The intensity of reflected light is strongest near the beam axis (except for light cut off by the bulb itself), dropping off towards the outer edge of the beam.

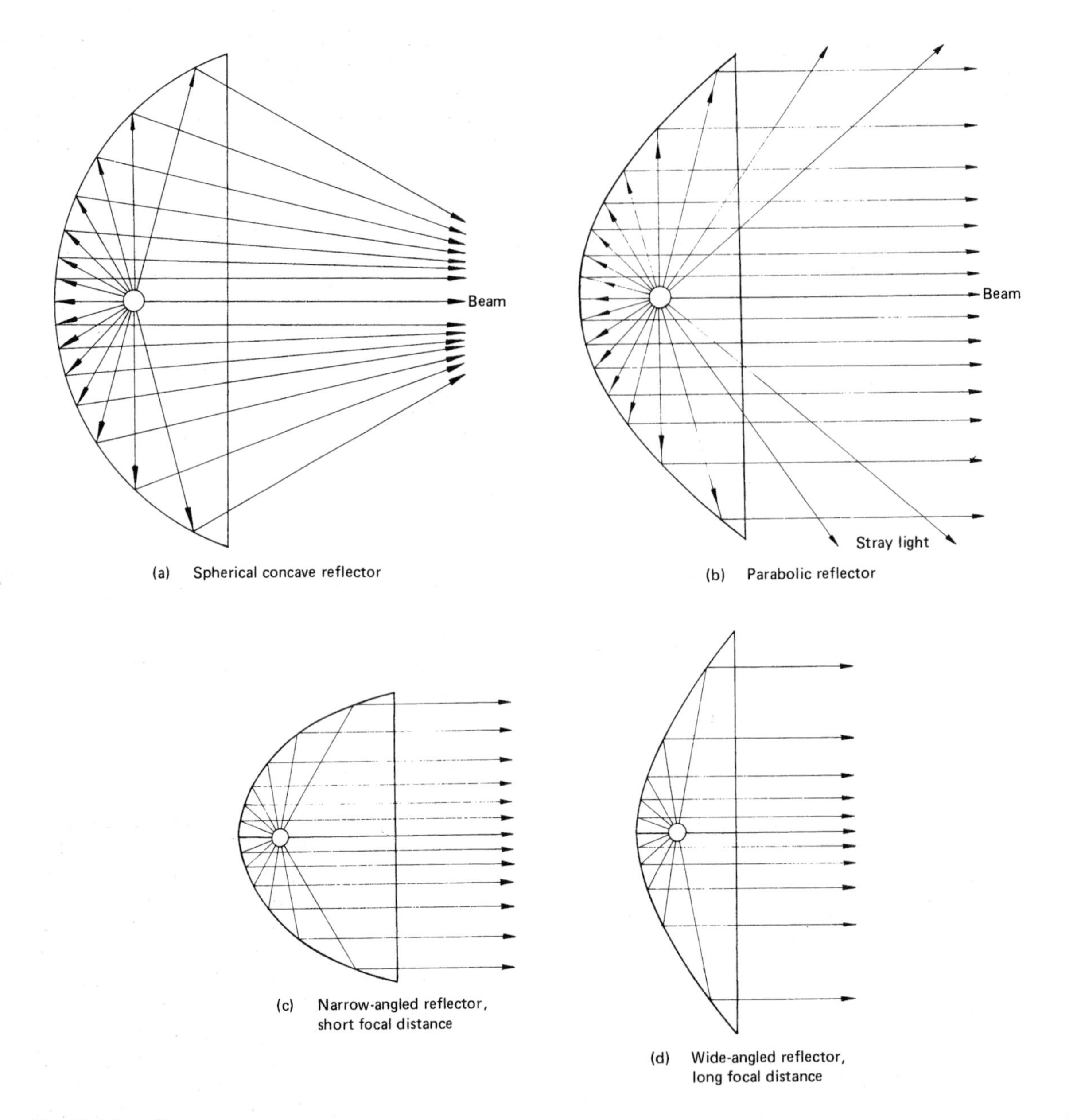

Fig. 25.15 Reflectors

The actual amount of reflected light on the outer fringe of the beam will depend to some extent on the depth and size of the reflector. A deep narrow-angled reflector gives a concentrated beam with little stray light, making it suitable for spotlights (fig. 25.15(c)). Conversely, a shallow wide-angle reflector increases the stray light and reduces the concentrated light (fig. 25.15(d)). These shallow reflectors – with sufficiently high filament wattage to intensify the beam – are adopted for headlights, the scattered light providing horizontal spread for lighting the side of the road.

25.10.3 Filament position relative to the focal point (fig. 25.16)

The position of the bulb relative to the reflector is important if the desired beam direction and shape are to be obtained.

Consider fig. 25.16(a) – here the light filament

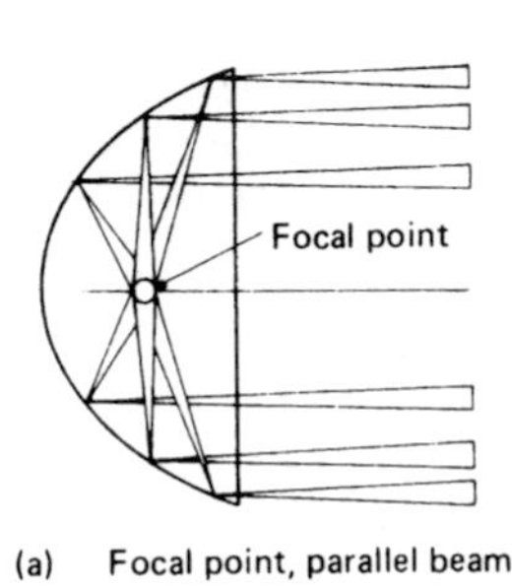

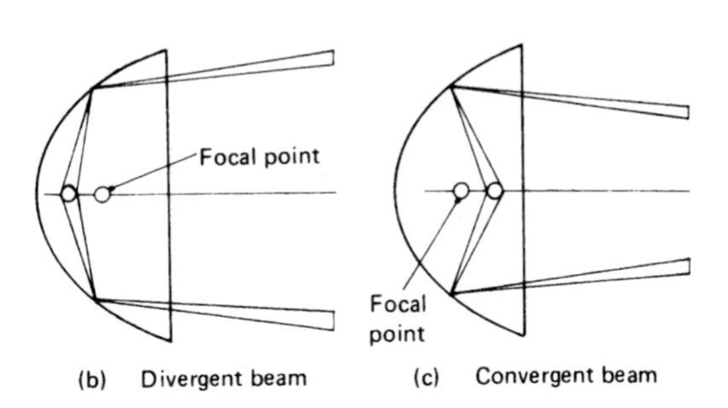

Fig. 25.16 Relationship between bulb position and beam formation

is at the focal point, so the reflected beam will be parallel to the principal axis. If the filament is between the focal point and the reflector (fig. 25.16(b)), the reflected beam will diverge – that is, spread outwards – along the principal axis. Alternatively, if the filament is positioned in front of the focal point (fig. 25.16(c)), the reflected beam will converge towards the principal axis.

25.11 Headlight arrangments

Headlight bulbs are generally of the double-filament (bi-focal) pre-focus type. They differ slightly in design as follows.

25.11.1 Offset main and dipped beams (fig. 25.17)
The main-beam and dipped-beam bulb filaments are offset respectively below and above the focal point, so that the top dipped beam will converge in a downward direction and the main beam will converge in an upward direction. The correct beam settings will be obtained when the reflector itself is tilted a very small amount downwards until the main beam is horizontal to the road – at the same time this increases the dipped-beam downward angle to illuminate the road surface a little way ahead of the vehicle.

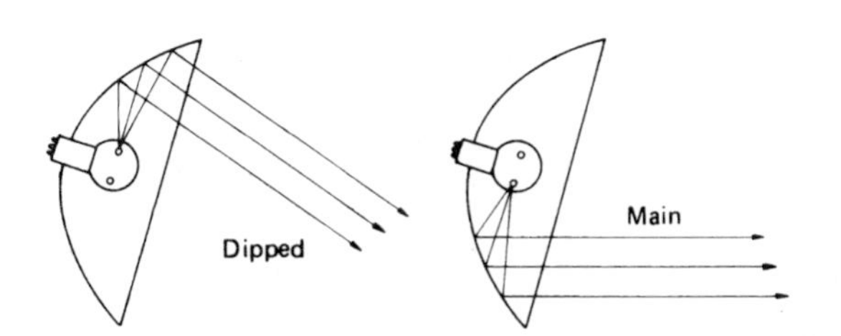

Fig. 25.17 Offset main and dipped beams

25.11.2 Offset dipped beam (fig. 25.18)
With this type of bulb the main filament is positioned at the focal point, but the dipped filament is displaced slightly above it. Since the main filament is fully focused, the reflected light will be parallel to the principal axis to provide a long concentrated beam, while the dipped filament being out of focus gives a downward but somewhat distorted beam which, however, is directed by the front lens to form a wide-spread beam on the road.

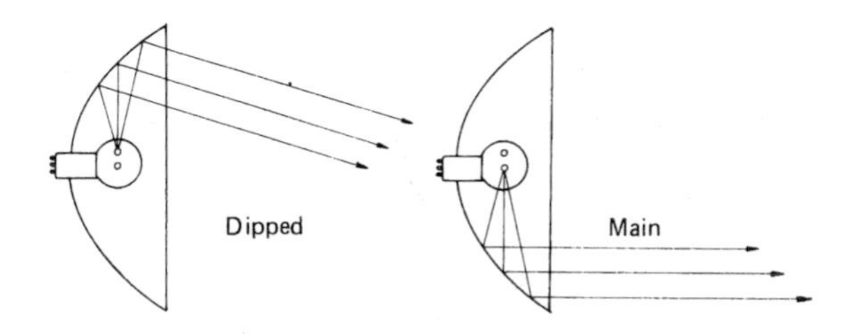

Fig. 25.18 Offset dipped beam

25.11.3 Offset shielded dipped beam (fig. 25.19)
This most popular double-filament bulb has the main filament positioned at the focal point of the reflector and the dipped filament brought forward of the focal point. Situated just below this dipped filament is a metal cup or shield. As with the two other types of bulb, the main filament provides a reflected far-reaching beam parallel to the principal axis. Conversely, the dipped filament produces an out-of-focus converging beam, the upper half pointing in a downward direction but the light rays from the lower half being prevented from

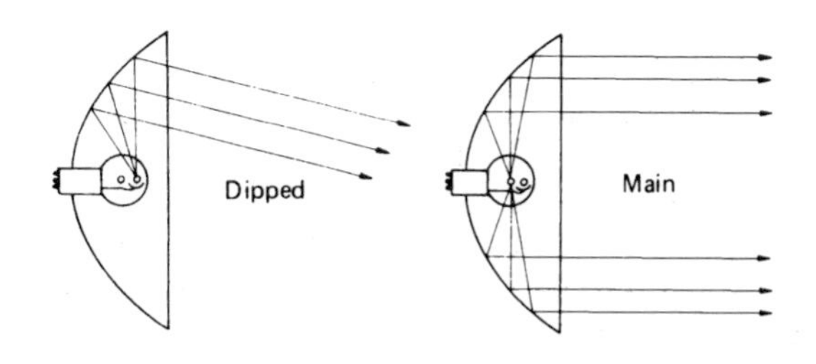

Fig. 25.19 Offset shielded dipped beam

striking the reflector due to the shield, so that none of these rays can cause dazzle by being reflected upwards.

25.11.4 Sealed-beam bi-focal unit (fig. 25.20)

Sealed-beam headlights are a further development from the separate pre-focused light-bulb and reflector–lens assembly. This design uses the reflector and lens to form a large all-glass sealed envelope filled with an inert gas to replace the conventional glass bulb. The two heating filaments are accurately mounted directly inside an aluminised glass reflector on to supporting wires which are located in the reflector by ceramic insulators.

The advantages of these units are precise filament focus and – more important – the prevention of entry of dirt, dust, grease, and moisture which causes the rapid deterioriation of the reflecting properties of the reflector. Due to the very large reflector area, filament-metal transfer due to evaporation of the tungsten does not blacken the glass lens. It is claimed that light intensity only drops off by at the most 3% of its original value during its normal four-year life expectancy, compared with a 40% reduction for the conventional double-filament (bi-focal) light-bulb. Tungsten–halogen sealed-beam headlights are also available, and the bulbs for these are discussed in detail in section 25.14. The major disadvantage is that the whole unit has to be replaced if one of the filaments fails.

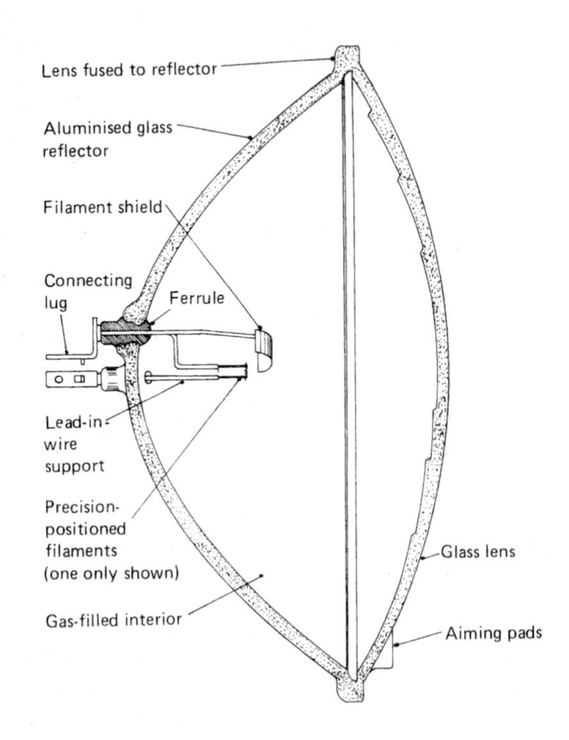

Fig. 25.20 Sealed-beam unit

25.12 Headlight cover lens

A good headlight should have a powerful far-reaching central beam, around which the light is distributed both horizontally and vertically in order to illuminate as great an area of the road surface as possible. These requirements cannot be met solely by the design of the reflector, but the beam formation can be considerably improved by passing the reflected light rays through a transparent block of lenses. It is the function of the lenses to partially redistribute the reflected light beam and stray rays so that a better overall road illumination is achieved with the minimum of glare.

One cause of glare which cannot be corrected by the reflector is due to the shape and size of the bulb filament – even if the filament is accurately focused, there is no absolute point source of light; therefore some reflected light rays will deviate and form both divergent and convergent rays which if not controlled will cause dazzle.

25.12.1 Block-prism lens

Lenses work on the principle of refraction – that is, the change in the direction of light rays when passing into or out of a transparent medium such as glass. The headlight front-cover glass lens is divided up into a large number of small rectangular zones, each zone being formed optically in the shape of a concave flute or a combination of flute and prism (fig. 25.21). The shape of these fluted prism sections is such that, when the roughly parallel beam passes through the glass, each individual lens element will redirect the light rays to obtain an improved overall light projection.

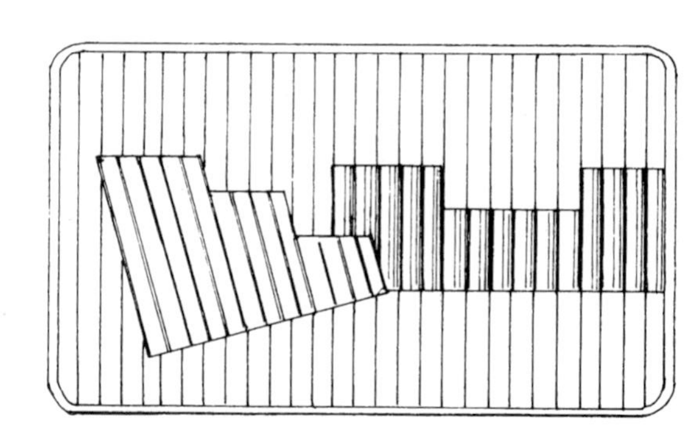

Fig. 25.21 Block-pattern headlight lens

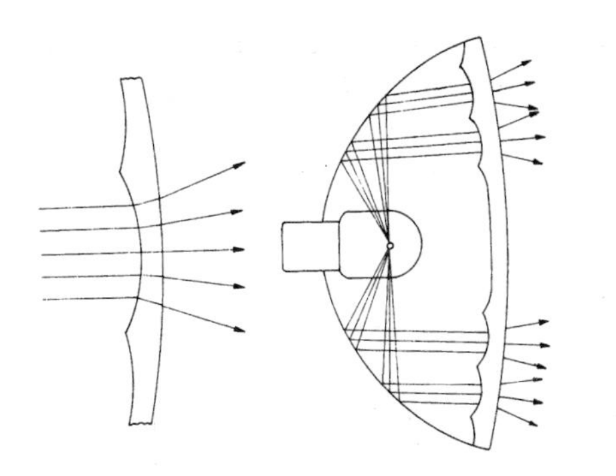

Fig. 25.22 Horizontal beam control

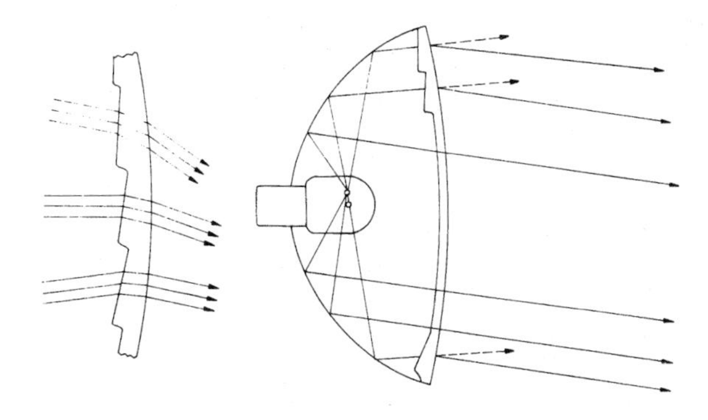

Fig. 25.23 Vertical beam control

The total lens-element pattern is designed to provide optimum road illumination for both the main and the dipped beam. Thus one group or block of prism elements will project the rays straight ahead for the main beam, while another block of lens elements may spread the rays out horizontally by the diverging concave-flute effect (fig. 25.22) and at the same time sharply bend downwards these rays to give diffused local lighting just in front of the vehicle by the prismatic-lens effect (fig. 25.23).

25.13 Classification of light-bulbs

25.13.1 Festoon (fig. 25.24(a))
The glass envelope has a tubular shape, with the filament stretched between brass caps cemented to the tube ends. This type of bulb was commonly used for number-plate and interior roof lighting.

25.13.2 Miniature Edison screw (MES) (fig. 25.24(b))
This is the well known flash-light bulb with a screw cap. It was frequently used with a ballast resistance wound round the body of the lamp, but is now seen only on early car models.

25.13.3 Miniature centre contact (MCC) (fig. 25.24(c))
This has a bayonet cap – that is, two locating pins project on either side of the cylindrical cap – of diameter about 9 mm. It has a single central contact, with the metal cap body forming the second contact, and an internal wire is included to support the V-shaped filament. It is made with various power ratings, e.g. 1, 2, 3, 4, and 5 watts.

25.13.4 Capless glass (fig. 25.24(d))
These bulbs have a semi-tubular glass envelope with a flattened end which provides the support for the protruding wires which are bent over to form two contacts. They are made with various power ratings up to 5 watts and are used for side-lights and parking, number-plate, and panel lights etc.

25.13.5 Single-contact small bayonet cap (fig. 25.24(e) and (f))
These bulbs have a bayonet-cap diameter of about 15 mm with either a small spherical or a large semi-spherical glass envelope enclosing a single filament. There is a single central contact, with the metal cap body itself forming the second contact. The small bulb is normally 6 watts and is used for commercial vehicle side- and tail-lights, while the larger 21 watt bulb is used for car and commercial-vehicle direction-indicator lights, hazard lights, reversing lights, and rear fog-lights.

25.13.6 Double-contact small bayonet cap (fig. 25.24(g))
These bulbs are similar in shape and size to the large semi-spherical single-contact 15 mm bayonet-cap bulb. There are two filaments, one end of each being connected to an end contact, while both of the other ends are joined together and earthed to the cap body to form the third contact. These caps have offset bayonet pins so that the two different-wattage filaments cannot be confused. One filament is used for the stop-light and the other for the tail-light, and they are rated at 21 and 5 watts ('21/5 W') respectively.

25.13.7 Pre-focus double filament (fig. 25.24(h))
Bulbs of this type are used only for headlights where it is very important for both filaments to be accurately positioned relative to the reflector, both along the principal axis and offset from it in

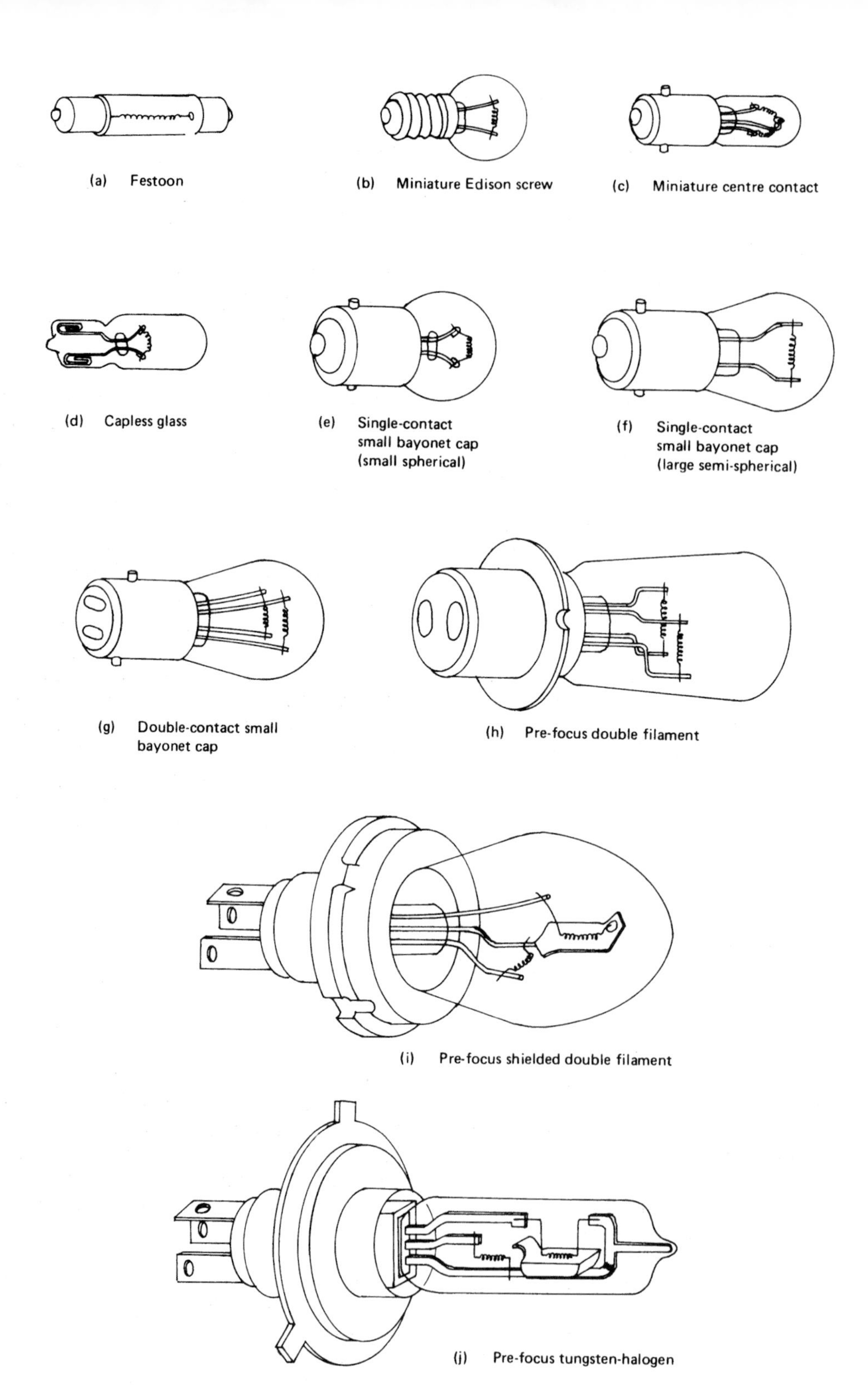

Fig. 25.24 Classification of light-bulbs

the correct plane. This positioning is achieved by a notched flange which is formed around the metal cap. When the bulb is fitted into the holder, the flange will be flush against a reference shoulder so that the main filament will be precisely on the focal point of the reflector; at the same time, the notch will be located by a projection which sets the bulb's angular position so that the dipped-beam filament is directly over the focal point. The normal ratings for these main/dip bulbs are 45/40, 50/40, and 55/40 watts.

25.13.8 Pre-focus shielded double filament (fig. 25.24(i))

This type of headlight bulb must also be accurately focused, with the main-beam filament positioned on the focal point and the dipped filament in front of this point with the shield positioned directly below the dipped filament. Again this positioning is provided by the grooved-shouldered bulb cap which can fit in only one position inside the holder.

With this bulb design, the cap body is insulated from both filaments and three separate terminals are provided, two filament end wires sharing one of these terminals. The wattage ratings of these bulbs are similar to those of the pre-focus offset-dip double-filament bulbs.

25.14 Tungsten–halogen light-bulbs (fig. 25.24(j))

When current passes through a filament, the tungsten heats up and glows and provides light. At the same time, the surface of the tungsten begins to melt and eventually boils and gives off vapour. This metal vapour will be transferred to the cooler internal surface of the glass, where it condenses to form a black film which prevents light passing through the glass.

It has been found that if the argon and nitrogen filler gases are replaced with an element from the halogen group – such as iodine, bromine, chlorine, or fluorine – metal-vapour condensation on to the inside surface of the bulb is prevented. The halogen gas produces a regenerative cycle – that is, the tungsten will still evaporate off the surface of the filament but, as it moves away by convection, it reacts with the halogen gas to form a heavy gaseous tungsten halide which will not condense on the cooler glass but moves towards the filament where, provided it is very hot, the tungsten will be deposited back on to the filament while the halogen returns to its previous state. As, due to the reaction with the halogen, the tungsten vapour is continually being removed and returned to the filament, the filament is able to operate at a higher temperature of around 2900°C which increases the filament's light output.

With halogen-filled bulbs, though, another problem has been created by the hot halogen gas attacking any cooler metal within the glass envelope, such as the other filament not in use and the metal shield which is always at a lower temperature than any current-conducting filament. This problem has been solved by replacing the pure halogen gas by dibromomethane gas which contains bromine, carbon, and hydrogen. This gas operates in the same way as halogen gas except that, on the cooler surfaces and as the bulb cools down after the operation, the hydrogen and bromine combine to form hydrogen bromide gas which is not aggressive towards metals. Two filaments can therefore be placed together in the same envelope and be switched on separately without the other one deteriorating.

Because of the higher operating temperature of tungsten–halogen bulbs, ordinary glass tends to crack, so a more expensive quartz glass is used which will tolerate large temperature changes without excessive internal strains in the glass. Quartz glass must *never* be fingered, because the natural oils from the fingertips result in hot spots forming on the glass during operation which will cause distortion and glass failure. If by accident the glass is touched by hand, it should be thoroughly cleaned with methylated spirit before being replaced.

The power ratings for the main- and dip-beam filaments are normally 60/55 watts.

25.15 Light-bulb location and attachments

Side-light and headlight holders and their securing arrangements will now be considered.

25.15.1 Side-lights (fig. 25.25)

Side-lights commonly use single-contact bayonet-type bulbs. These fittings consist of a cylindrical brass cap supporting a sealed glass bulb which encloses the filament and conducting wires. The bulb has two short pins projecting outwards on opposite sides of the metal cap.

The corresponding bulb-holder is in the form of a short brass sleeve which fits over the bulb cap and is surrounded by a supporting plastic moulding which also acts as an electrical insulator.

The bulb-holder sleeve has two parallel slots formed diametrically opposite each other, beginning at the sleeve's open end. At the far end, these slots then turn at right-angles for a short distance

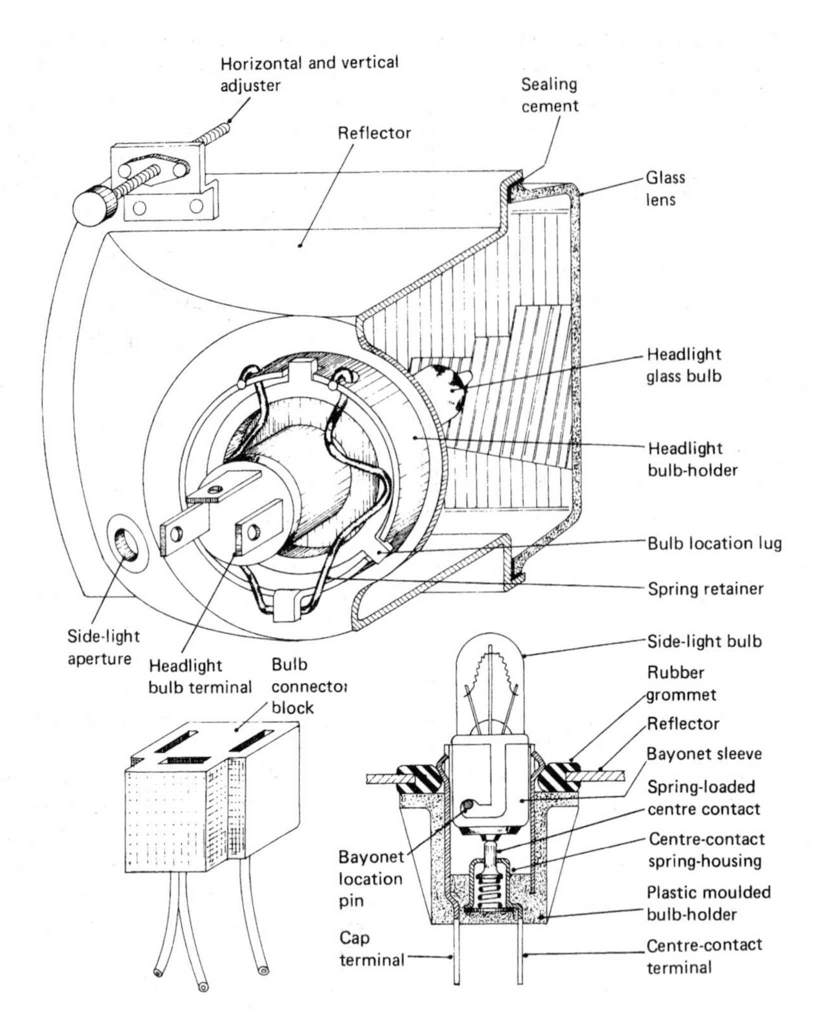

Fig. 25.25 Headlight and side-light assembly

and are then terminated by a circular hole so that they resemble the letter 'J'.

In the base of the holder is a brass cylindrical casing which houses a spring and acts as a guide for the flanged brass contact pin.

When inserting the bulb, the bayonet pins are first aligned with the slots, and the glass bulb is then pushed towards the end of the parallel slots against the compression of the spring. It is then twisted until the pins rest in the circular stop-holes and is then released. Thus the pins are trapped and so bear against the circular upper edges of the holes – this prevents them accidentally coming out due to any imposed vibration.

25.15.2 Headlights (fig. 25.25)

Halogen headlight bulbs have a metal cap with an enlarged circular flange which has three lugs projecting outwards from it. These lugs fit into cut-outs formed in the rear of the bulb-holder rim which forms part of the reflector.

Two of the lugs are spaced slightly closer to each other than to the third one, so that the cap will only fit one way into the holder. Thus, when positioned in the holder, the dip filament is either offset from the focal point towards the top or, if the dip filament is shielded, it faces upwards.

When replacing a blown bulb filament with a new bulb, pull off the wiring-harness plug from the rear of the light unit, remove the protective rubber or plastic cover (not shown), squeeze together from the middle both sides of the retainer clip, and swing the retainer clip outwards to release the bulb cap and withdraw the bulb.

Align the unequally spaced lugs of the new bulb with their corresponding cut-out slots around the rim of the holder without touching the glass part, as this might damage it. Push the bulb into position and secure the cap by pressing the two halves of the spring clip against the cap flange and clipping the spring under the lip formed at the base of the rim holder. Finally refit the rubber or plastic cover and the wiring-harness plug.

Tungsten-filament headlight bulbs are similarly fitted, but in their case the cap of the bulb has notches which align with projections formed in the bulb-holder.

25.16 Four-headlight system

A two-headlight system which has a pre-focus double-filament bulb in each reflector is relatively effective but to some extent is a compromise to satisfy both main- and dipped-beam conditions.

The light intensity for both operating conditions can be maximised by employing four headlights – that is, two headlights adjacent to each other on each side of the vehicle.

The two inner headlights have a single-filament bulb positioned at the focal point, while the outer headlights are each fitted with pre-focus double filaments, one being on the focal point and the other slightly offset.

When the main beam is switched on, both the inner headlights project their beams and additional off-focus beams will be provided by the outer-headlight offset filaments. Such a combination will give a far-searching beam provided by the inner headlights with a shallow broad beam provided by the outer headlights.

If now the switch is changed from main to dipped beam, the inner-headlight filaments and the offset outer-headlight filaments will be extinguished, while the outer-headlight filaments positioned on the focal point will now provide a concentrated dipped beam.

25.17 Headlight settings (fig. 25.26)

To obtain the best road visibility without dazzling on-coming vehicles, it is necessary to set the headlight reflectors relative to the vehicle body under certain specified conditions.

If there is no special alignment equipment available, the vehicle with its normal load should be driven on a level flooring to squarely face a smooth screen wall 7.62 m from the headlight lenses.

The correct headlight reflector setting gives main beams straight ahead with both beams parallel and horizontal to the ground.

This setting can be obtained by measuring the centre height of the headlight from the ground and marking a horizontal line across the screen at this height. Switch on the main beam, cover one of the lights, and screw the reflector vertical-adjustment screws in or out to tilt the reflector. Continue doing this until the centre spot of the beam aligns with the horizontal line. Repeat this procedure for the other headlight. The distance between headlight centres should then be measured across the lights and then checked on the

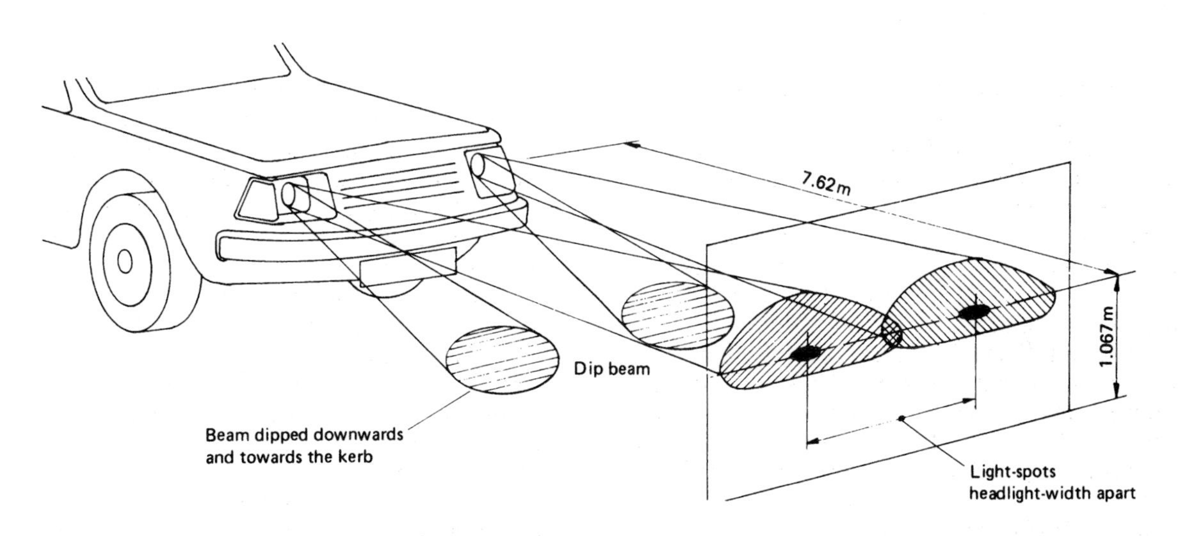

Fig. 25.26 Headlight beam setting

screen. If there is a discrepancy, screw in or out the horizontal-adjustment screws on the side of the lights to swing the reflectors round until the beams are parallel.

A final check is to switch the headlights to dip, stand at the screen, and bend down to an eye-level height of 1.067 m facing the lights. Under these conditions there should be no eye dazzle.

26 Coil ignition system

26.1 Ignition considerations

The combustible mixture of air and petrol is set alight by a spark occurring between two electrodes in the combustion chamber at the end of the compression stroke just before TDC. It is the function of the ignition system to periodically provide a spark of sufficient heat intensity to ignite the charge mixture at the predetermined position in the engine's cycle under all speed and load conditions.

Individual cylinders require a single spark every second revolution, so the frequency of firing for a four-cyclinder engine say at a maximum speed of 6000 rev/min will be $(6000/2) \times 4 = 12\ 000$ sparks per minute or 200 sparks per second. There is thus an extremely short interval between firing.

The voltage necessary to ionise the air between the electrodes so that the spark will bridge the air gap can vary from as little as 5000 volts when the gap is small, the engine is hot, the cylinder pressure is low, and a chemically correct air–fuel mixture is being burnt, to a value of 20 000 volts when the spark-plug electrodes are badly eroded, the air gap is large, the engine is cold, high cylinder pressures exist, and either very weak or very rich mixtures are induced into the cylinders.

26.2 Ignition-system equipment

The understanding of how the ignition system operates may be simplified by considering the functions of each component making up the complete system (fig. 26.1).

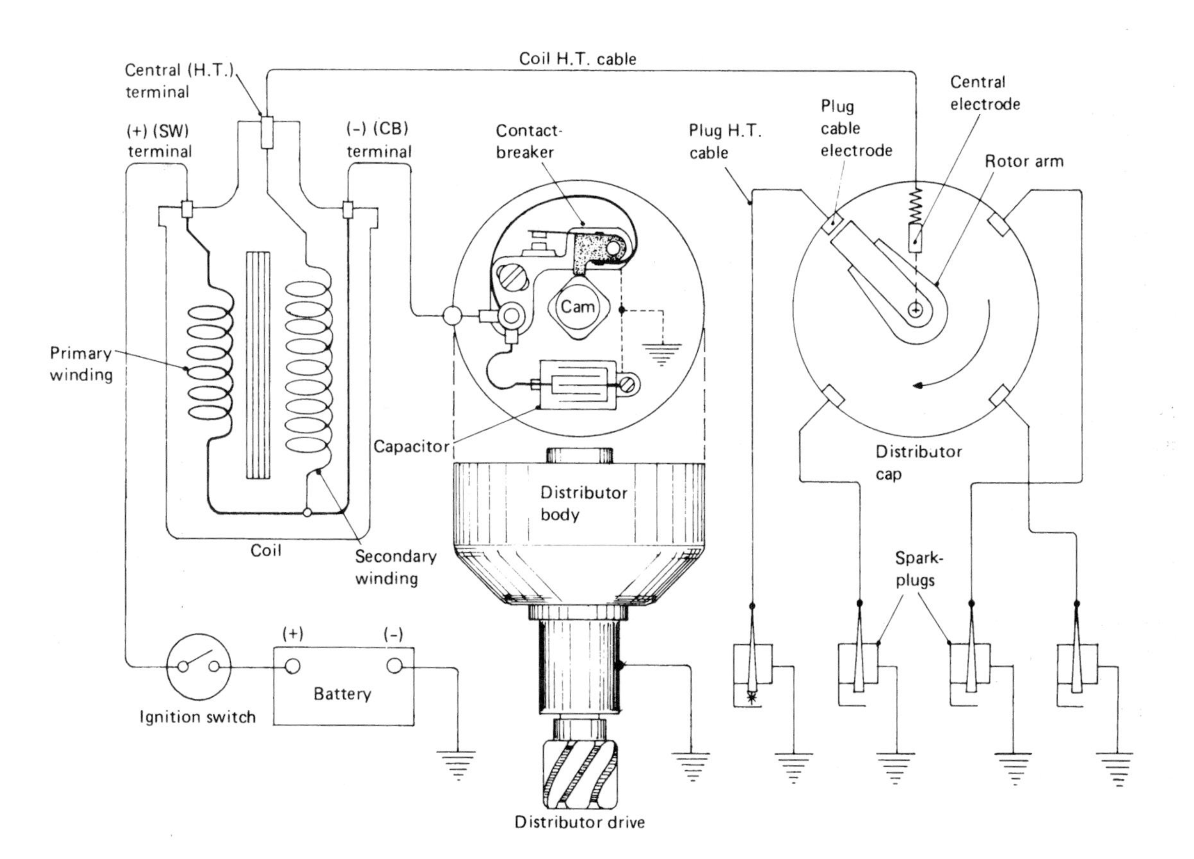

Fig. 26.1 Coil ignition system

Battery This is usually a 6 or 12 volt battery. It stores chemical energy which can be converted into electrical energy to supply the flow of current through the primary circuit of the ignition system immediately when required.

Ignition switch This switch is connected in series in the coil primary-winding circuit. It enables the driver to switch on or off the electrical supply available from the battery as required to operate the ignition system.

Ignition coil This is basically an electrical step-up transformer which stores the electrical energy and converts the relatively low battery voltage into a high-intensity voltage.

Contact-breaker points These contact points control the timing of the spark. They periodically interrupt the current flow in the primary winding of the coil so that a high-intensity voltage will be induced into the secondary winding of the coil.

Capacitor This component provides the conditions for a rapid electrical interruption in the primary winding when the contacts open, so that more electrical energy is induced into the secondary circuit in the form of a high-tension spark and less energy will be available to cause arcing and damage to the contact-point faces.

Distributor high-tension circuit This circuit distributes and conducts the high-tension voltage pulses generated in the coil to the individual sparkplugs in the correct order of firing and in phase with the engine's operating cycle.

Distributor centrifugal automatic-advance mechanism This device, situated in the base of the distributor body, senses any change in engine speed and automatically advances or retards the point of firing relative to the crankshaft angular position to suit the engine speed.

Distributor vacuum advance mechanism This device, situated on the side of the distributor body, senses any variation of manifold vacuum, which is a measure of engine load, and automatically alters the point of firing with respect to crankshaft angular position to suit the load.

Spark-plug This component supports and insulates the two electrodes in the combustion chamber and enables the gap between the electrode tips to be accurately set.

26.3 Fundamental electromagnetism definitions

Magnetism Any magnet has an area around it in which its magnetic effects may be observed, such as iron filings being attracted or compass needles being deflected. This area is called a magnetic field, and the magnetic effects are observed to act along magnetic lines of force.

A magnetic field is also created around a conductor when an electric current flows through it. The greater the current, the stronger will be the magnetic field. This magnetic field surrounding a wire conductor may be represented by concentric circles (fig. 26.2(a) and (b)). The direction in which the magnetic field acts around a conductor will be clockwise when the current flows into the paper and anticlockwise when the current is moving out from the paper (fig. 26.2(b)).

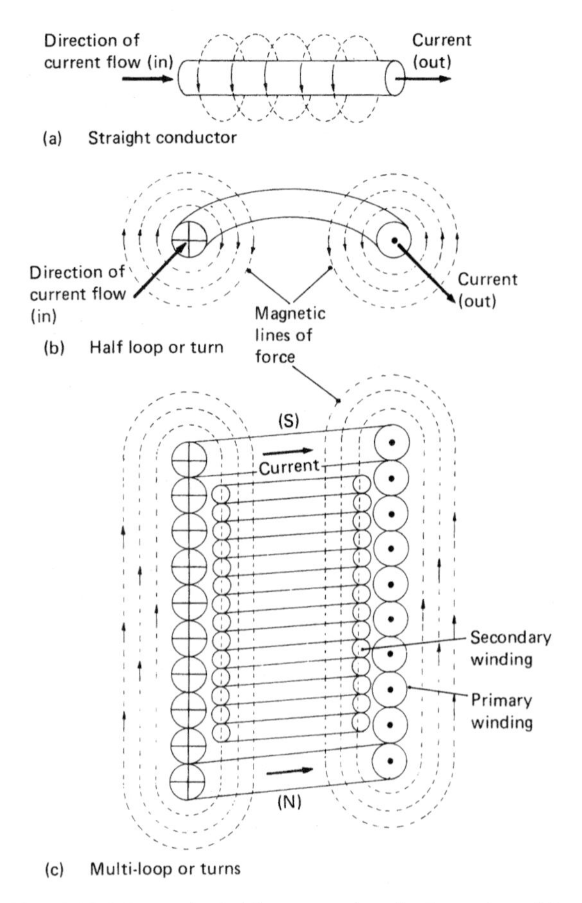

Fig. 26.2 Magnetic fields set up by single and multi-loop conductors

Electromagnetic induction A conductor situated in a magnetic field will have an e.m.f. induced in it if the conductor interacts with the magnetic field by either of the following methods:

a) when the conductor cuts or is cut by magnetic flux (lines of magnetic force);
b) when there is a change of lines of magnetic force passing through the conductor, i.e. if the magnetic flux (the strength of the magnetic field) increases or decreases.

This is known as electromagnetic induction.

The faster the conductor cuts or is cut by the magnetic flux, or alternatively the faster the change of magnetic flux linking with the conductor, the greater will be the e.m.f. induced.

Self-induction If a current of increasing or decreasing magnitude is passed through a winding, the magnetic flux due to the current will also expand or shrink respectively. This changing flux will link with the loops of the winding and will therefore induce an e.m.f. into the winding. Because the induced e.m.f. in the winding is due to changes in the winding's own magnetic flux, it is called a self-induced e.m.f.

Mutual induction When two windings are wound close together (fig. 26.2(c)) and current is passed through one winding (the primary winding), the magnetic flux produced around the primary-winding conductor will also link with the other (secondary) winding. If the current in the primary winding is increased or decreased, the changing magnetic flux which results will cut through the secondary winding, and an e.m.f. will thus be induced in this secondary winding. Both windings are then said to have mutual induction.

Voltage due to self-induction The magnitude of a voltage induced into a secondary winding depends upon the following:

a) the number of winding turns and the ratio of primary to secondary winding turns;
b) the strength of the magnetic field surrounding both windings, created by the circulating electric current;
c) the rate of magnetic field change – that is, the speed at which the magnetic lines of force cut through the winding conductors.

26.4 Ignition-coil construction (fig. 26.3)

The ignition-coil assembly consists of two separate coil windings – the primary and the secondary – each winding having many layers of insulated wire. Both of these windings are wound around a soft-iron core. Surrounding the outside of the windings is a soft-iron sheath, and the whole coil assembly is submerged in thin insulating mineral oil and housed in an aluminium case. The coil windings are supported and insulated by a porcelain base at the bottom of the case and are held in position at the top by a moulded plastic tower cap which has both the L.T. (low-tension) and H.T. (high-tension) terminals moulded into it.

Iron core The soft-iron core strengthens the magnetic field caused by current flowing through the primary windings but readily loses its magnetism when the current is interrupted. If the core was one large iron bar, the change of flux linkage would cause unwanted induced currents to flow within the core. Such currents whirl around like the eddies in water, and hence are called eddy currents. To minimise eddy currents, the core is made up of about thirty soft-iron strips, each 1 cm wide and 11 cm long, insulated by varnish from

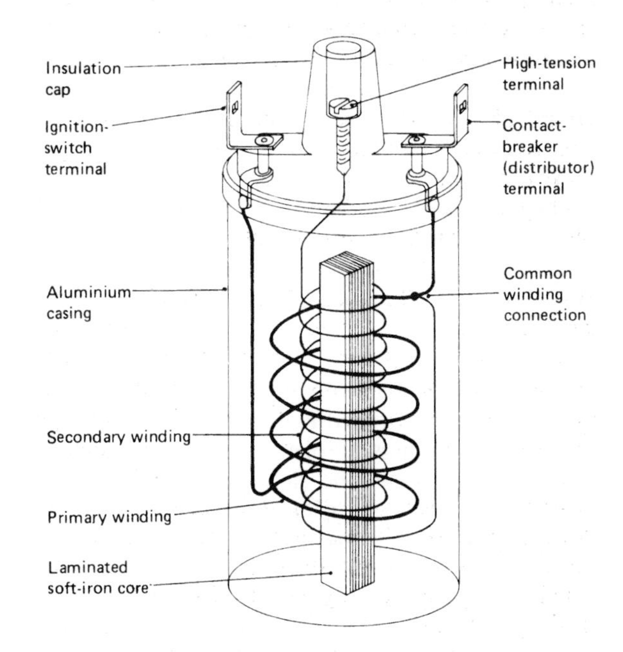

Fig. 26.3 Ignition-coil unit

each other and bundled together to form a laminated structure (fig. 26.3). The induced eddy currents in these strips will be much reduced compared to those induced in a single-piece core, so the energy losses due to these eddy currents will be much smaller.

Secondary winding Wound around the laminated iron core and insulated from it is the secondary winding, with anything from 15 000 to 30 000 turns of very thin enamelled copper wire. One end of this secondary winding is connected to the primary winding near the contact-breaker terminal, and the other end is joined to the central H.T. terminal (fig. 26.3).

Sometimes the iron core is connected in series in between the secondary winding and the central H.T. terminal – this method of connection reduces the chance of the winding insulation rupturing and electrically shorting with the core.

Primary winding This winding consists of heavy-gauge enamelled copper wire wound anything from 250 to 350 times around the outside of the secondary winding so that its total length provides a resistance of about 3 ohms. The turns ratio between the secondary and primary windings is from 50:1 to 150:1. Each layer of wire in both the primary and the secondary windings is separated from the next by insulating paper. One end of the primary winding is joined to the battery or switch terminal, and the other is connected to one end of the secondary winding and to the contact-breaker L.T. terminal (fig. 26.3).

The primary winding is deliberately wound over the secondary winding instead of the other way around so that the heat created by the primary current is easily transferred to the oil and casing.

Iron sheath A sheet of soft iron with a series of vertical slits is wrapped twice around the outside of the coil windings to form a sheath. This sheath reduces stray magnetic fields and thus energy losses from the coil, while the slits restrict the formation of eddy currents.

26.4.1 Description of the ignition system (figs 26.1 to 26.3)

The ignition coil forming part of the ignition system is basically a transformer energised by either a 6 volt or 12 volt battery which supplies direct current. Electromagnetic induction is made use of in the transformer to convert a low voltage to a high voltage.

The ignition-coil transformer consists of a primary winding (or coil) and a secondary winding wound on a soft-iron core which intensifies any magnetic field between them (fig. 26.3). When a current flows in the primary winding, a magnetic field is produced. If the primary current changes, then the strength of the magnetic field or flux will change too. The strength of the magnetic field will depend on the current flowing and the number of turns in the coil.

The secondary winding is situated in the magnetic field produced by the primary winding and so will have an e.m.f. induced in it by mutual induction – but only when the primary current and hence the magnetic field is changing.

The e.m.f.'s in the primary and secondary windings are related as follows:

$$\frac{E_2}{E_1} = \frac{N_2}{N_1}$$

where E_1 = primary e.m.f
E_2 = secondary e.m.f
N_1 = number of turns in primary winding
and N_2 = number of turns in secondary winding

Thus if the secondary winding has many more turns than the primary winding, then the secondary e.m.f. will be much greater than the primary e.m.f. Such a transformer is known as a step-up transformer. A typical step-up turns ratio would be $N_2/N_1 = 100$.

The ignition system can be considered to be made up of two separate circuits:

i) a low-tension (L.T.) primary circuit obtaining its electrical supply from the battery,
ii) a high-tension (H.T.) secondary circuit which provides the path for the periodically induced high-voltage pulse necessary to force a spark to jump the air gap between the spark-plug electrodes.

Primary circuit This low-tension circuit consists of the battery, the ignition switch, the distributor contact-breaker, the coil primary winding, and the earth return, all connected in series to form a closed loop when the contact points are closed (fig. 26.1).

The distributor contacts are made to periodically open and close by a rotating cam which is driven by the engine at half crankshaft speed.

Every time the points open, a high voltage will be induced into the secondary circuit.

Secondary circuit This high-tension circuit consists of the coil secondary winding, a cable from the coil to the distributor, a central cap electrode, a rotating rotor arm, outer distributor-cap electrodes so positioned as to form a circle, high-tension cables connecting the individual cap electrodes to their respective spark-plugs, and the earth-return path through the cylinder head (fig. 26.1).

Whenever there is an induced voltage surge in the secondary winding – that is, when the contact points just open – the rotor arm will be in line with one of the outer cap electrodes so that the voltage impulse will jump the clearance space between the rotor-arm tip and the segment-shaped electrode (fig. 26.4). This high-voltage

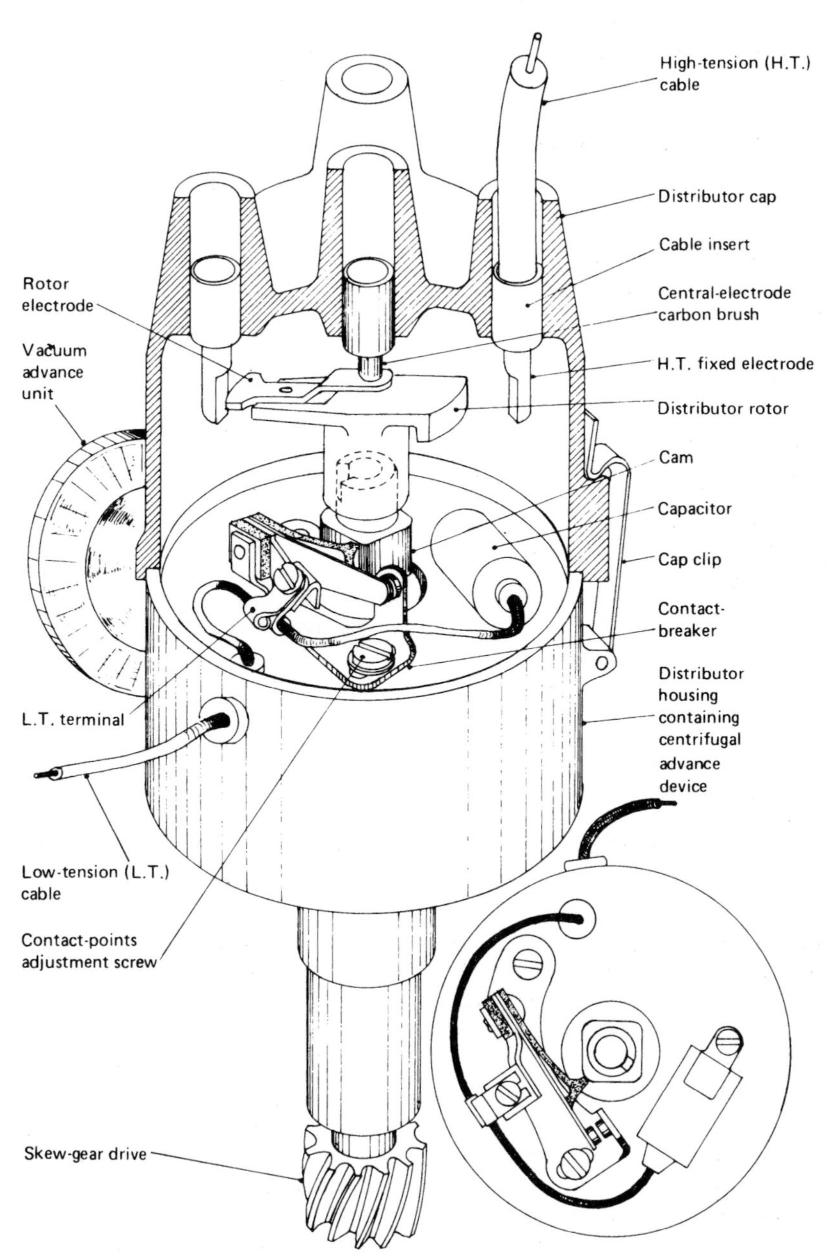

Fig. 26.4 Ignition distributor

surge will then be conveyed along the cap cable to the spark-plug, where it forces a spark to bridge the air gap existing between the central and side electrodes.

26.4.2 Operation of the ignition coil

If the ignition switch is closed and the contact-breaker points are together, current will flow from the battery through the primary winding and the earth-return path back to the battery. The primary winding is wound closely over the secondary winding so that, when current produced by the battery voltage circulates, it produces a magnetic field which interlinks both the primary and the secondary windings (fig. 26.2(c)).

Any change in this primary current will interact with the primary winding's own magnetic field and will induce an e.m.f. in the winding by the self-induction effect. Because the self-induced e.m.f. acts in a direction so as to oppose the changing current, it is often referred to as the back e.m.f.

When the rotating distributor cam opens the contact points, the primary current very rapidly falls to zero and the magnetic field also decays rapidly. Self-induction acts so as to oppose these changes, and a back e.m.f. is induced in the primary winding as the primary current falls to zero. As any e.m.f. induced in a conductor is proportional to the rate of change of magnetic flux linking with the conductor, the back e.m.f. will be very large, as the change producing it is very rapid. By transformer step-up action, an even larger e.m.f. is thus induced in the secondary winding and is fed to the spark-plug gap to produce a spark.

If, when the contacts open, a back e.m.f. of 200 volts is induced into the primary winding due to the magnetic flux cutting through the winding conductors, and the secondary winding has 100 times as many turns as the primary winding, then the secondary induced voltage will be 100×200 volts = 20 000 volts.

With further rotation of the distributor cam, the contact points are made to close by the tension of the return leaf spring. Current will then again flow through the primary windings to restore the magnetic field around the windings in readiness for the next point of firing. This cycle of growth and collapse of primary-winding current will continually repeat itself at a frequency proportional to engine speed.

26.5 Capacitor function and operation

A capacitor consists of two sets of conducting plates separated by insulating material. It may be made either from two or more strips of aluminium foil with insulating material interleaved between them or from strips of sheet insulating material with metallic layers deposited by evaporation on to one face of the sheet insulation itself. To reduce space, these strips are rolled up tightly and they are then impregnated with insulating compound and placed inside an insulated cylindrical aluminium casing (fig. 26.5). At each end of the roll, one set of conducting sheets (plates) protrudes from the others. One set of these exposed outer metal edges of the roll directly contacts the inside of the casing through a spring washer, and a metal disc mounted on an insulator button electrically contacts the end of the other set. The roll is then hermetically sealed in the container casing.

A short piece of wire with insulation covering is joined on to the metal disc. This wire receives the flow of current from the circuit of one set of foils or metallised strips, the conducting path for the other set being provided by the casing itself.

26.5.1 Capacitor capacitance

In operation, the capacitor plates are continually going through a cycle of charge and discharge. The charged state is when the contact-breaker points open and the plates are of different potential, one being very positive and the other very negative. The discharged state is the condition

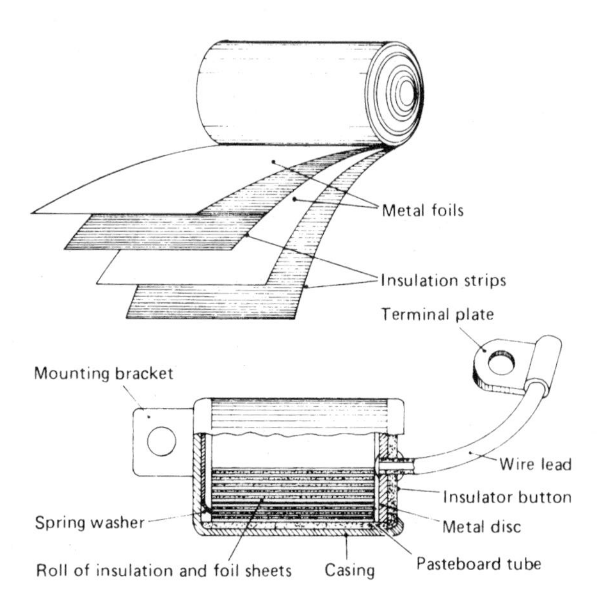

Fig. 26.5 Capacitor construction

existing when the contact-breaker points close and both plates rapidly reach the same polarity.

As the potential difference across the two plates increases, so will the amount of stored charge. The ability of a capacitor to hold a charge is known as its capacitance, and capacitance is measured in units known as farads. The farad is a large unit, so the smaller microfarad (μF) unit is used. Typical values for the capacitor range from 0.18 to 0.25 μF.

26.5.2 Function and operation of capacitor

While the ignition switch is closed and the contact-breaker points are together, current flows from the battery and through the primary winding, returning to earth to complete the circuit. When the contact-breaker points open, the current flow will be interrupted and the collapse of the magnetic field will generate a large surge of voltage in both the secondary and the primary windings.

If the capacitor were not connected across the contacts (fig. 26.1), the back e.m.f. in the primary winding would cause a spark to arc and bridge the gap made between the contact faces – that is, current would continue to flow in the primary winding for a very short time. Consequently the collapse of the magnetic field would be prolonged, so the existing lines of force would cut through the secondary winding at a relatively low speed and only a low voltage surge in the secondary circuit would therefore result.

With the capacitor connected in parallel with the contact-breaker points, the surge current in the primary winding when the contacts open finds an easier path by way of the capacitor plates. This diversion of the surge current to the capacitor has three effects:

i) the primary-current flow stops instantly, so that the lines of force will cut across the secondary windings at a very high rate – in fact, with the capacitor connected across the contacts, the electrical break in the primary circuit is some 20 times faster than if there were no capacitor present;
ii) arcing across the contact points will be minimised, so pitting of the point faces will proceed at a very slow rate;
iii) when the contacts close again, the charge stored in between the capacitor plates discharges into the primary winding and so helps to accelerate the build-up of a new magnetic field in the primary winding.

26.6 Distributor contact-breaker construction (figs 26.1 and 26.4)

The contact-breaker consists of two electrical points generally made from tungsten. One contact is mounted on a steel pressing which is fixed by a screw or screws to the distributor baseplate, thus making it adjustable. The other contact is mounted on a pivot arm which may oscillate on a pivot post attached to the steel pressing, and a curled leaf spring riveted to the pivot arm tends to hold the contacts together (fig. 26.1). Alternatively, a leaf-spring pivot arm may be used – this is rigidly attached to the fixed steel pressing of the earthed contact so that the oscillating motion is obtained by the flexing of the leaf-spring arm (fig. 26.4). In both cases, the two contacts are electrically insulated from each other when the points are apart.

26.6.1 Operation

The opening and closing of the contact points is achieved by the cam profile bearing against a fabric or nylon extension heel forming part of the pivot arm. There are as many cam lobes as there are cylinders, so four- and six-cylinder engines will have four and six cam lobes respectively. When the cam rotates, each lobe in succession will push against the pivot-arm heel, thus separating the points. When the heel is between two lobes, the contacts will be returned to the closed position by the spring tension.

26.6.2 Adjustment

Adjustment of the contact-point gap is achieved by cranking the engine until the pivot-arm heel aligns with the peak position of any lobe. Feeler gauges are then inserted between the contact faces, and the set-screw or -screws holding the fixed contact are then loosened so that the relative position of the plate can be altered to obtain the correct gap setting. Typical gaps are 0.4 and 0.6 mm.

An excessively large gap will tend to advance the point of ignition, while a very small gap will retard the point when firing occurs.

26.6.3 Contact-breaker dwell (fig. 26.6)

Contact-breaker point dwell refers to the time period from the points just closing to when they open again at a specified distributor camshaft speed of rotation. Thus, with rising engine speed the dwell time period will become smaller and smaller. Time is an inconvenient method of measuring dwell, and it is therefore usual to quote

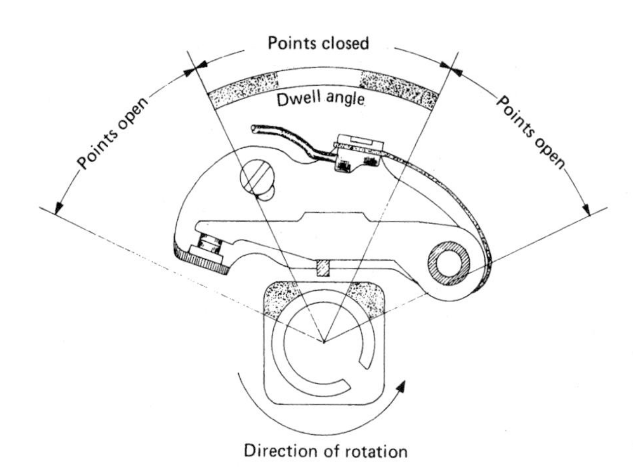

Fig. 26.6 Dwell angle is that portion of cam rotation when the contacts are closed

dwell in terms of the distributor-cam angular movement during which the contact-breaker points remain closed. This dwell angle remains approximately constant and is independent of camshaft speed.

Distributors are designed to provide the greatest dwell angle possible between point closing and opening, so that the ignition coil's primary-circuit current has sufficient time to establish a strong magnetic field before it is interrupted. This is most important with rising speed, as the time available for the growth of the magnetic field between firing impulses becomes critical.

The dwell angle can be varied by adjusting the contact-breaker point gap – the point gap is reduced to increase the dwell angle and is enlarged to decrease the dwell angle. Unfortunately, very large dwell angles reduce the point gap so much that excessive arcing may result.

Each manufacturer of distributors recommends a dwell angle to suit his own design of contact-breaker and cam lobe profile. Typical dwell angles are 48° to 52° for four-cylinder engines and 36° to 40° for six-cylinder engines.

26.7 Mechanical centrifugal advance device (figs 26.7 to 26.9)

26.7.1 Purpose of device

The time it takes to initiate combustion – that is, the period from the spark occurring to the instant when burning actually begins – is approximately constant and is known as the delay period. The significance of this constant time delay in burning is that with rising engine speed the corresponding crankshaft angular movement during this delay period will increase. For example, if the delay period spans an angular crankshaft movement of 6° at 1000 rev/min, the delay period will be increased to 12° at 2000 rev/min and to 18°, 24°, and 30° respectively at 3000, 4000, and 5000 rev/min.

The consequence of this lateness in burning in the cylinder with rising engine speed is that the gas-pressure thrust on the piston will not be at its optimum, nor will it occur at the most effective point during the power stroke.

To compensate for the inherent retarding point of ignition which occurs with rising engine speed, a flyweight advance mechanism is universally adopted. This device senses rotational-speed variations and alters the instant of firing in almost direct proportion to changes in engine speed.

26.7.2 Description of mechanism (figs 26.7 and 26.8)

The mechanism consists of a drive-plate driven at camshaft speed, which transmits similar rotary motion to the cam-plate and cam by means of a pair of steel flyweights and retraction springs (fig. 26.7).

These retraction springs may be of equal length and stiffness rate, or of unequal length and rate – in the latter case, the stronger of the two springs has an extended loop, making its overall length greater than that of the weaker one.

26.7.3 Operation of device (figs 26.7 and 26.8)

With the engine idling, there is very little centrifugal force on the flyweights, and the retraction springs will pull the flyweights into their most inward position. When the engine speed is increased, the centrifugal force on the flyweights increases and they swing outwards about their pivot pins (fig. 26.8(a) to (c)).

As the weights move outwards, their straight flanks roll partially around the reaction-plates. This twists the cam-plate and cam about their central axis relative to the drive-plate. The cam-plate will thus rotate slightly ahead of the distributor drive-shaft, and consequently the cam lobes will contact the moving contact-point heel earlier and so advance the contact-points' angular opening position. An alternative version of the flyweight mechanism is shown in fig. 26.9(a) and (b).

26.7.4 Maximum advance limit

The cam-plate has an extended sickle-shaped arm formed on it, the tip of which will contact the

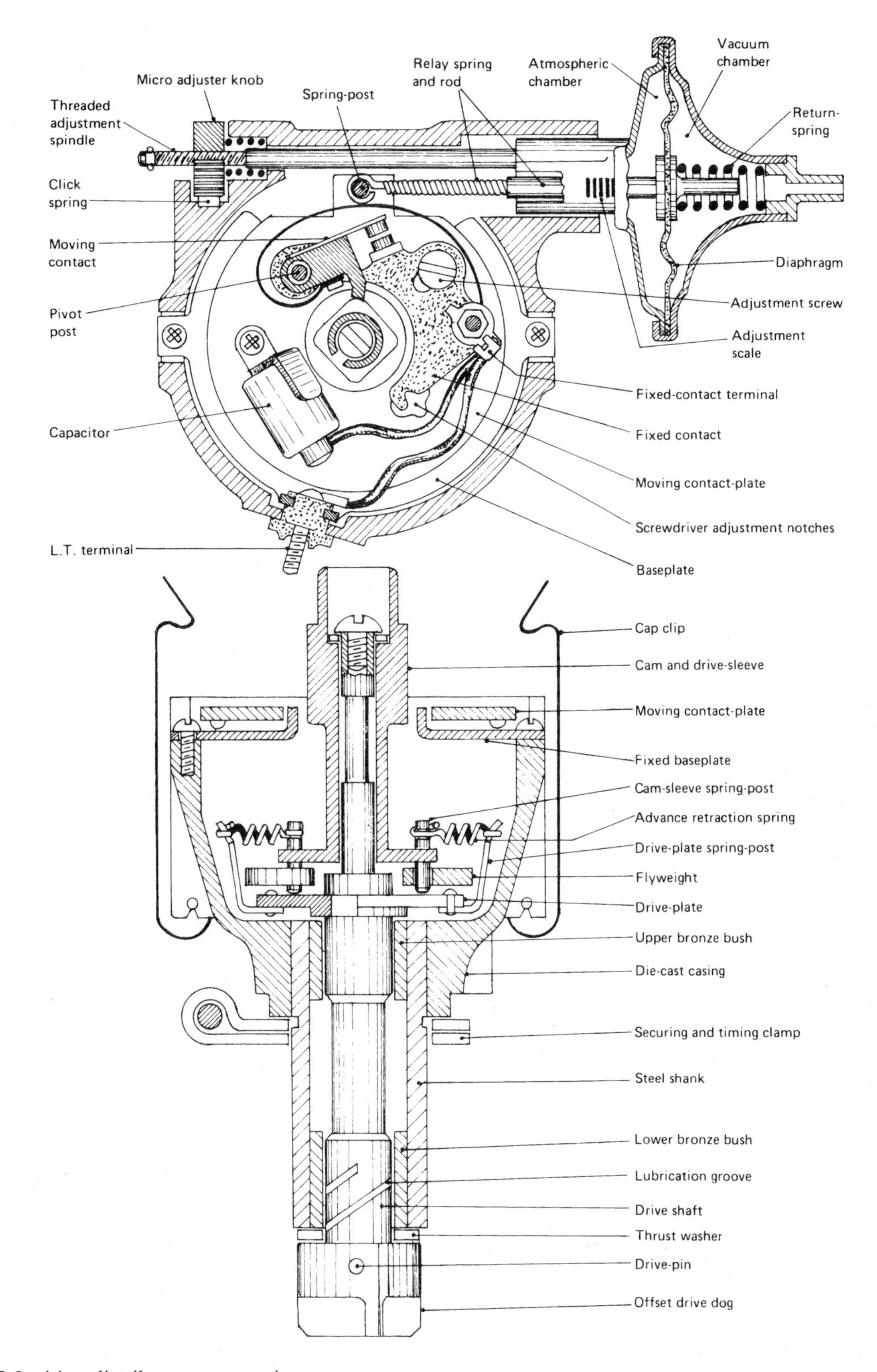

Fig. 26.7 Ignition distributor construction

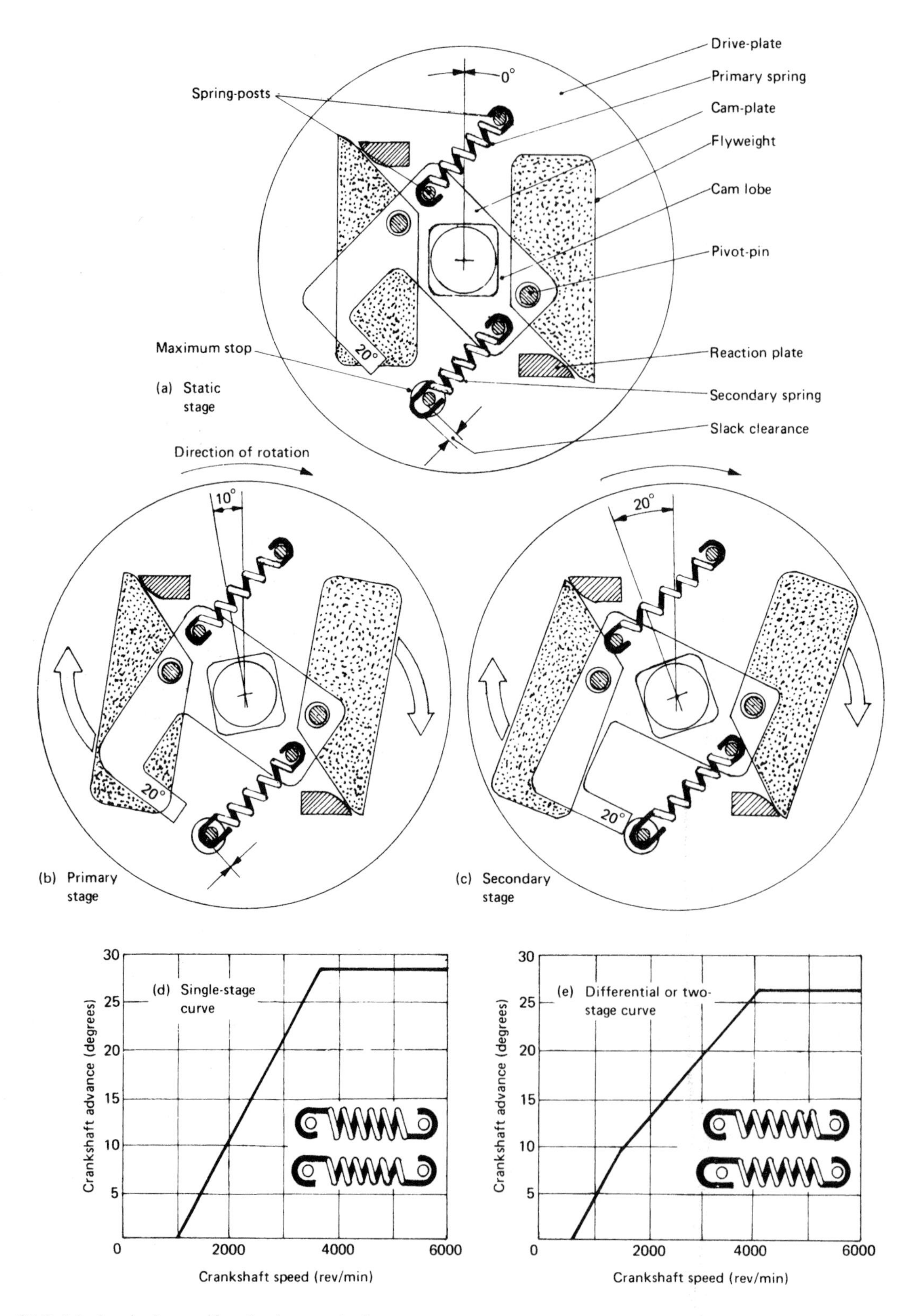

Fig. 26.8 Mechanical centrifugal advance device

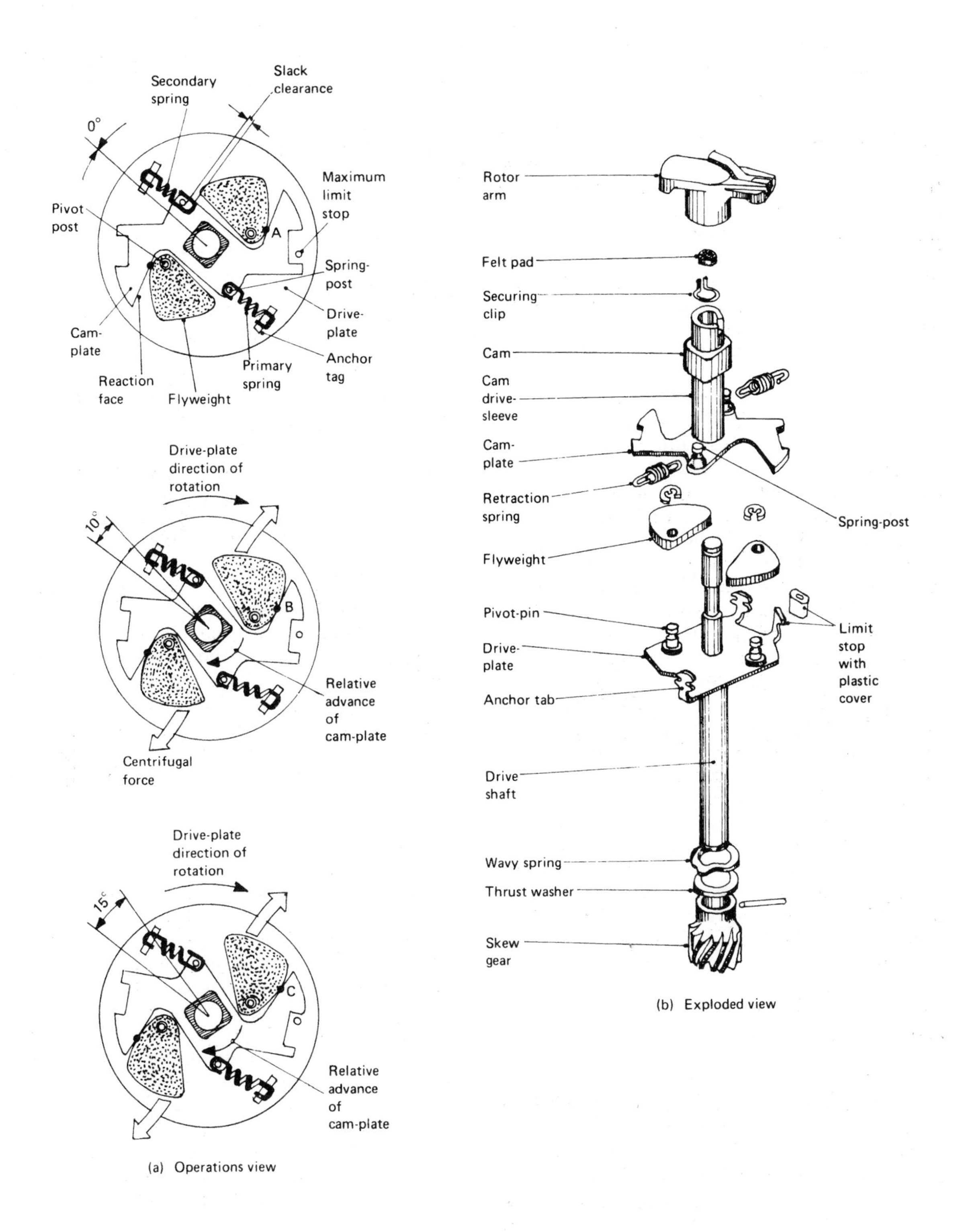

Fig. 26.9 Alternative method of mechanical-advance-device construction

enlarged stop-post when the maximum designed advance is reached. The length of this arm will be formed to give the correct limiting advance for its application, and this value will be stamped on the top face (fig. 26.8(a) to (c)).

26.7.5 Equal-spring-assembly action

When both springs have the same stiffness rate, they share out the restraining resistance offered over the complete speed and advance range, so that a linear curve of advance against speed will be obtained (fig. 26.8(d)). Once the extended arm of the cam-plate butts against its stop-post, there will be no further advance, even though the engine speed may continue to rise. This is shown by the horizontal portion of the advance curve.

26.7.6 Differential-spring-assembly action

Some engines prefer the spark to be fired just before TDC when being cranked for starting, but then benefit by a rapid advance while the engine is accelerated and begins to pull, followed by a more gradual advance once the transmission has been moved into a higher gear. To match the engine's optimum advance requirements, different-rate (differential) pairs of springs may be incorporated instead of the equal-rate ones.

The weaker (primary) spring first restrains the flyweights' throw-out alone, while at this stage the stronger (secondary) spring is inoperative as the slack provided by its elongated hook or loop has not yet been taken up. Eventually the movement of the cam-plate relative to the drive-plate will bring the secondary spring into action (fig. 26.8(b) and (c)), and the increased resistance offered as this stronger spring becomes tensioned will result in less distributor advance for a given increase in engine speed (fig. 26.8(e)). This secondary-spring stage is matched to the increased heat and turbulence generated within the combustion chamber which lead to slightly less advance being required with further rising engine speed.

26.8 Manifold vacuum advance device (figs 26.7 and 26.11)

26.8.1 Purpose of device

This device senses any variation of manifold vacuum (depression), which is an almost direct measure of engine load. It operates entirely independently of engine speed. Any change in manifold vacuum will alter the ignition setting – that is, it will move the point at which the spark occurs, advancing the spark with an increase in vacuum

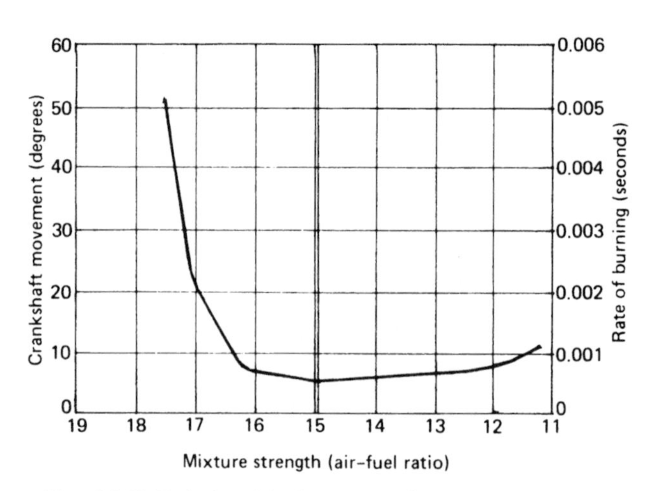

Fig. 26.10 Relationship between mixture strength and rate of burning expressed in both seconds and crank angle for an engine running at 2000 rev/min.

but retarding the point of firing when the vacuum decreases.

A variable ignition point is necessary to compensate for the different burning rates of various mixture strengths (fig. 26.10) and the reduction in charge density under part-throttle conditions. Weak mixtures take longer to burn than rich ones, so that under part-throttle, light-load, maximum-economy conditions when weak mixtures are burnt, additional ignition advance is desirable. This contrasts with wide-open-throttle full-load conditions, where rich mixtures are burnt and a much retarded point of firing is required to achieve optimum power or maximum acceleration.

26.8.2 Description of mechanism (fig. 26.7)

This mechanism consists of a contact-plate which can revolve relative to the baseplate about the distributor camshaft axis. It has the fixed and moving contact-breaker points screwed to it, and is coupled to a diaphragm by means of a relay spring. One side of this diaphragm is subjected to any manifold vacuum, and the other side is exposed to atmospheric pressure. A helical return-spring tends to expand the vacuum chamber if there is no vacuum present.

26.8.3 Operation of device (fig. 26.11(a) to (c))

When the vacuum chamber is connected to a depression in the manifold, a pressure difference is produced between the two sides of the diaphragm. Atmospheric pressure will then deflect the diaphragm, compressing the return-spring

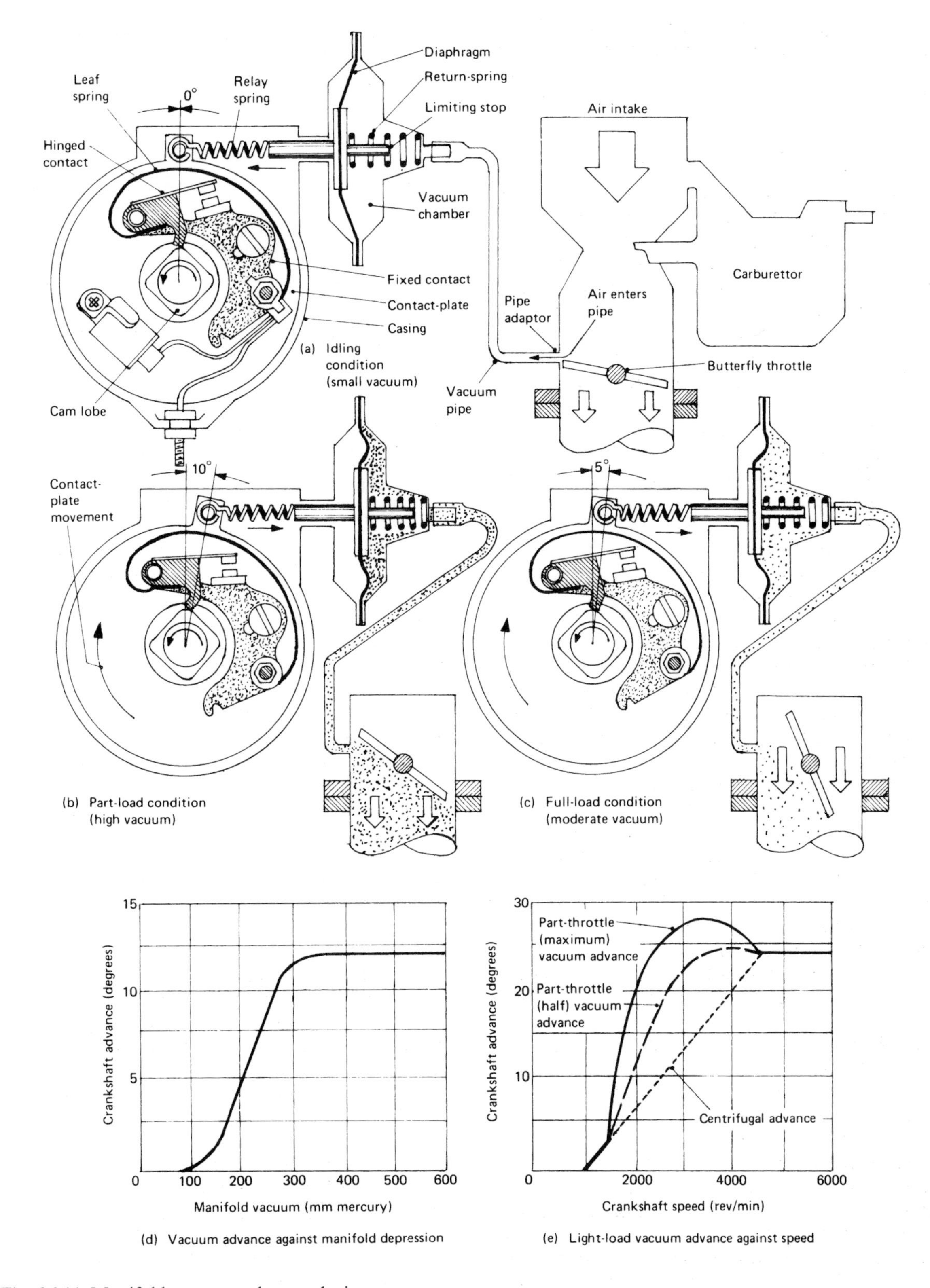

Fig. 26.11 Manifold vacuum advance device

and, by means of the relay spring and rod, twisting the contact-plate about its axis in a direction opposite to the direction of rotation of the camshaft drive. This has the effect of moving the pivoted contact heel so that it touches the cam lobe earlier – the contact points will thus open in advance of the normal static angular setting.

26.8.4 Low-speed idling (fig. 26.11(a))

With the engine idling, the butterfly throttle valve will seal off the manifold vacuum from the vacuum-pipe adaptor; hence both sides of the diaphragm will be subjected to atmospheric pressure so that the return-spring will fully retard the ignition setting.

26.8.5 Wide throttle opening under full load (fig. 26.11(c))

If the engine is pulling hard with wide-open throttle, there will be little vacuum created in the induction manifold. This is because the throttle restriction produced is small relative to the piston speed; thus there will be insufficient depression created in the vacuum pipe to overcome the spring, and therefore the point of ignition will not be advanced.

26.8.6 Part throttle opening under light loads (fig. 26.11(b))

When the engine is running under light conditions at relatively high speed (such as overrun), the throttle will only be partially open with a relatively high piston speed. A high degree of vacuum will be established in the induction manifold and in the vacuum-pipe adaptor, and this will overcome the tension of the diaphragm return-spring and so advance the ignition setting in proportion to the created depression (fig. 26.11(d)) until the diaphragm stop contacts the vacuum chamber's casing.

Notice that the vacuum advance is independent of and is superimposed on to the centrifugal advance effect and that it operates only while the engine load is light with moderately high engine speed. Above a certain engine speed, the pulling-load requirements rise; therefore the throttle has to be progressively opened and thus the vacuum advance declines from its peak value until it eventually merges with the centrifugal curve (fig. 26.11(e)).

26.8.7 Micrometer adjustment (fig. 26.7)

The micrometer adjuster provides a convenient method for altering the static ignition timing accurately for small increments of advance or retardation.

The micrometer adjustment is part of the vacuum unit which is mounted on a sliding sleeve supported by a boss cast into the side of the distributor casing. The moving contact-plate is connected to the vacuum diaphragm by a relay spring which passes through the centre of the diaphragm sleeve. A threaded spindle is cast in the end of the diaphragm sleeve, and the screwed end of the spindle protrudes through the other side of the casing and is held in position by a serrated and internally threaded disc or knob which partially fits into a shallow groove. Ignition timing is adjusted when the knob is rotated. Since the knob is restrained from end movement, due to the shallow groove, the screwed spindle instead slides the vacuum unit further in or out of the sleeve boss. The consequence is that the relay spring rotates the moving contact-plate, and this advances or retards the point of ignition.

A further refinement is provided by the scribed marks formed on the top face of the sleeve. Each space between these marks is equivalent to 4° crankshaft movement, and their position relative to the edge of the casing thus provides a reference point. When the ignition timing is initially set by rotation of the distributor body, the micrometer scale is first zeroed by screwing the vacuum unit towards the casing until the scribed divisions have been covered by the sleeve boss.

Note that the adjustment knob is prevented from rotating due to vibration by a small click spring which flexes against the rim serrations.

Manufacturers consider the micrometer adjustment device a luxury which is very rarely justified these days.

26.9 Ignition timing

26.9.1 Reason for early ignition setting

For most engines, the optimum instant for the spark to fire is not at TDC or a little after on the beginning of the power stroke but is instead near the end of the compression stroke just before TDC. The reason for this is that, once the spark has fired, it takes some time for burning of the air–fuel mixture to begin; so, in effect, the piston will have moved some way down its power stroke before burning has actually started if the spark fires exactly at TDC. Now the gas pressure generated by combustion depends on the extent to which the charge has been compressed relative to its original volume; so, if the resulting gas-pressure thrust acting on the piston is to be a maximum, the space between the cylinder head and the piston crown must be at a minimum when combustion actually begins.

26.9.2 Meaning of ignition timing

The position of the point of firing relative to TDC is known as the ignition timing and is expressed in degrees of crank-angle movement. It is set by positioning the distributor body relative to one of the cam lobes so that the contact points have just opened, at which instant the collapse of the magnetic flux around the ignition-coil windings will induce into the secondary winding a very high voltage of sufficient magnitude to force a spark to jump between the spark-plug electrodes.

26.9.3 Static ignition timing

Static ignition timing is the setting of the distributor contact points so that they open when one piston is almost at the end of its compression stroke at some crankshaft relative angular position before TDC as specified by the engine manufacturer. At the same time, the distributor camshaft drive must be positioned such that the rotor arm points to the cap electrode and high-tension cable corresponding to the cylinder which is about to begin its power stroke.

26.9.4 Static ignition timing procedure (fig. 26.12)

To simplify the positioning of piston 1 in its cylinder at the correct angular crankshaft position at the point of ignition (normally before TDC), most engine manufacturers provide a timing datum (a mark or marks) on the front cover of the timing gear, chain, or belt which can easily be aligned with a notch or notches on the inside of the crankshaft pulley. The ignition timing procedure described below applies to the use of pulley and timing-cover marks.

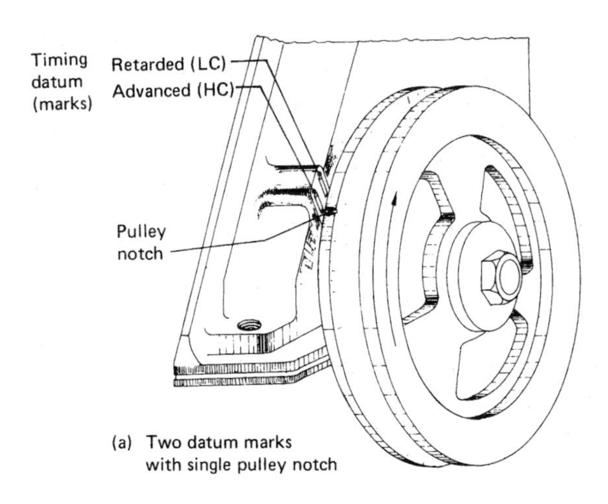

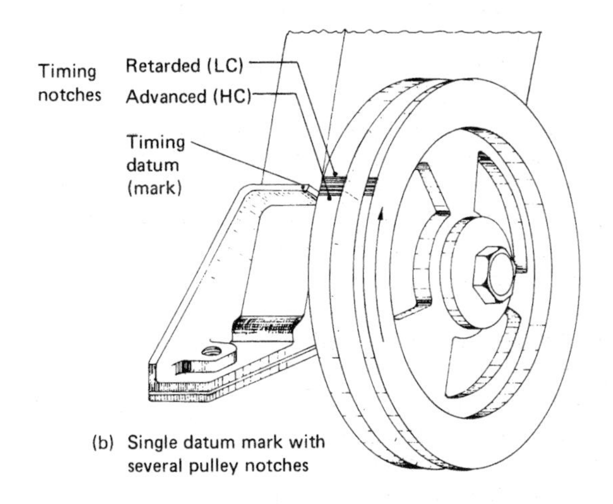

Fig. 26.12 Ignition pulley and timing-cover marks

1 (a) Remove all the spark-plugs, then rotate the engine by either pulling the car forward with top gear engaged or by pulling round the fan belt until piston 1 approaches TDC on its compression stroke. (This may be established by placing a finger over the spark-plug hole of cylinder 1 and feeling the pressure developed in the cylinder.)

(b) Alternatively, for an in-line engine, the compression stroke for cylinder 1 can be found by removing the rocker or cam cover and observing when the valves over the last piston are on the rock – that is, when the exhaust valve is just closing and the inlet valve is beginning to open. In this position, piston 1 will be in the region of TDC on its compression stroke. The crankshaft should then be turned backwards (anticlockwise) slightly and stopped.

2 Rotate the crankshaft forward (clockwise) until the notch on the rear of the crankshaft pulley is aligned with the appropriate timing mark on the timing cover. (Refer to the manufacturer's data, as there may be more than one mark for engines with different compression-ratios or different petrol octane ratings (fig. 26.12(a) and (b)).

3 Remove the distributor high-tension cap by pulling off the clips on each side.

4 Slacken off the distributor-body clamp bolt and rotate the body in the direction opposite to the direction of rotation of the rotor arm until the contact-breaker points are just opening. Observe which of the cap's high-tension segment-shaped electrodes the rotor arm is then aligned to.
5 Tighten the distributor-body clamp bolt and replace the distributor cap and clips.
6 The H.T. terminal for the cap electrode adjacent to the rotor arm will correspond to cylinder 1. Working round the cap in the direction of rotation of the rotor, connect the other sparkplug cables to the cap in the appropriate firing order – e.g. 1, 3, 4, 2.
7 Typical static ignition firing settings are from 4° to 12° before TDC.
8 A slight readjustment of the distributor may be necessary and should be carried out on the road if heavy pinking (high-pitched explosive sounds) occurs when accelerating in top gear on wide throttle opening from 30 to 70 km/h. The ignition should be retarded until a trace of pinking can only just be heard under these conditions of acceleration. This is done by slackening off the distributor-body clamp bolt and very slightly rotating the body in the direction of rotation of the rotor arm, then tightening the clamp.

25.10 Spark-plug function

A spark-plug is designed to be screwed into a hole formed in the cylinder head so that its electrodes protrude into the combustion chamber. The function of the spark-plug is to introduce the spark to ignite the prepared combustible charge at both high and low cylinder gas pressures and temperatures and in air–fuel mixtures of differing strengths. To achieve this under the varying operating conditions, the plug must conduct into the combustion chamber the high-voltage surge produced by the ignition system and discharge this in the form of a spark across the electrode gap. The heat energy then liberated must be able to start off the combustion process.

26.10.1 Operating environment

At the point of the cycle at the end of the compression stroke when the plug fires, the pressure inside the combustion chamber can be anything from 3 to 15 bar, depending on the compression ratio and the throttle opening, and the temperature within the chamber will be between 200 and 400°C. During peak combustion conditions, the pressure and temperature in the cylinder will be raised to about 70 bar and 2000°C respectively. To function effectively, the electrode and insulator temperatures must be somewhere in between the conditions at the end of compression and at peak combustion. It has been found that the best operating temperature for the insulator tip is in the range of 450 to 850°C. If the tip is too hot, rapid electrode wear and pre-ignition will result. If it is too cool, excessive fouling will occur, followed by misfiring.

26.10.2 Construction (fig. 26.13)

The spark-plug consists of three main components:

i) the insulator,
ii) the body or shell,
iii) the electrodes.

Insulator The function of the plug insulator is twofold:

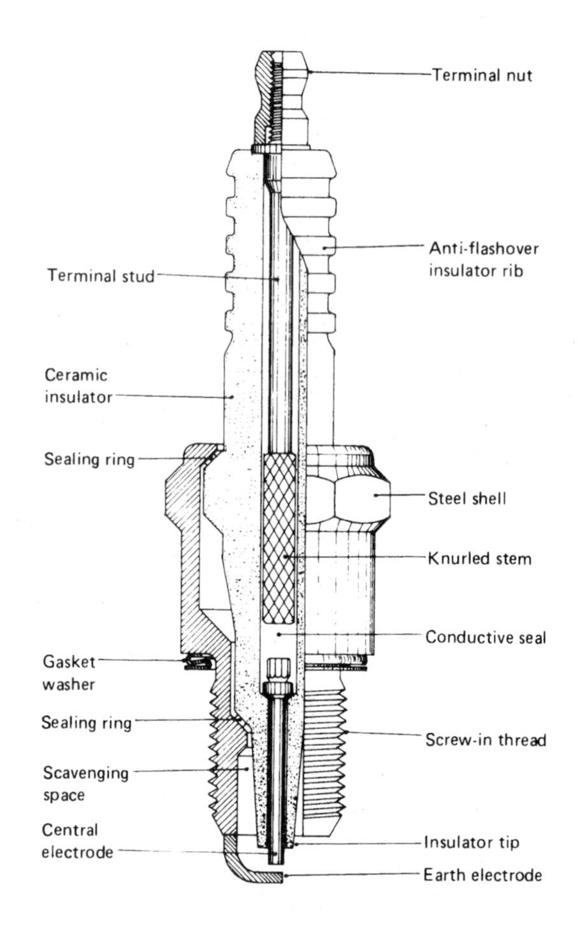

Fig. 26.13 Spark-plug

i) to separate the central electrode from the metal shell of the plug and so prevent the high-voltage surge supplied by the ignition coil leaking to earth within the plug shell under operating conditions;
ii) to control the main operating temperature of the central electrode by means of the insulating material's thermal conductivity, shape, and the length of its heat path.

A very popular insulator material is aluminium oxide, which is an alumina ceramic. This meets most of the insulator's requirements:

a) high electrical resistivity at operating termperatures;
b) high mechanical strength at operating temperatures;
c) high thermal-shock resistance;
d) high thermal conductivity;
e) low thermal expansion rate;
f) a durable surface to withstand abrasion, erosion, and chemical attack from the products of combustion.

Body shell The function of the body is to support and house both the insulator and the electrodes. It must be robust enough to secure the spark-plug in the cylinder head.

The plug body shell is normally made from high-quality low-carbon steel. The body is formed by impact extrusion – a process in which a small round bar is fed through a form of multi-punch press to shape the shell. The threads are then cold rolled, resulting in a low-friction profile. The body dimensions must be adequate to resist distortion by stretch and twist when subjected to excessive torque.

Electrodes The function of the electrodes is to provide an adjustable spark-gap between two electrical conducting points which are exposed to the combustion process. There are two electrodes: a central one supported and mounted in the insulator and a side or earth electrode which is welded to the plug body.

Electrodes are usually made from nickel alloy with small additions of chromium, manganese, and silicon to meet the following requirements:

a) good hot strength,
b) good electrical conductivity,
c) good thermal conductivity,
d) good spark-erosion resistance,
e) good chemical-corrosion resistance,
f) low coefficient of expansion.

Sealing To prevent gas from escaping between the insulator and the outside steel body and the insulator and the central electrode, permanent gas-tight seals are necessary. There are several different forms of seals in use, as follows:

a) solid copper rings,
b) composite compound rings,
c) dry powder,
d) metal powders fused into glass.

Desirable seal properties are

i) high thermal conductivity;
ii) high pressure and thermal-shock resistance;
iii) high oxidation and corrosion resistance;
iv) either good or poor electrical conductivity, depending on the spark-plug design.

An additional possible leakage point exists between the plug and the cylinder head. This may be sealed by either a flat seat and a ring gasket or by a taper conical seat which has a wedge action when tightened.

26.10.3 Spark-plug size (fig. 26.14)

The portion of the plug which screws into the cylinder head is identified by basically two dimensions:

i) the outside diameter across the screw thread. This is known as the thread diameter. Typical sizes are 10 mm, 14 mm, 18 mm, and 22 mm.

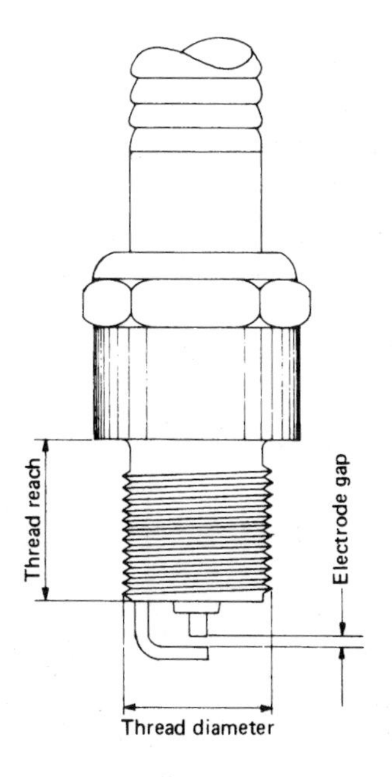

Fig. 26.14 Spark-plug dimensions

ii) the distance between the shoulder of the plug body and the end-face. This is known as the thread reach. Typical sizes are 9 mm, 11 mm, 13 mm, and 19 mm.

26.10.4 Spark-plug gap (fig. 26.14)
This refers to the shortest distance between the central electrode and earthed side electrode. Plug gap clearances range between 0.5 and 0.8 mm – a typical value would be 0.6 mm. Adjustment of the gap is achieved by bending the earthed electrode in or out until the contact faces grip a feeler gauge placed between them.

26.10.5 Fouled spark-plugs
If spark-plugs are operated for excessively long periods without removal of the deposits of carbon formed over the central electrode insulation during cold starting and under load, the high voltage may track to earth over the dry or sometimes wet deposits so that insufficient energy will be available to jump the spark-gap of the electrodes.

26.10.6 Spark-plug routine maintenance
Spark plugs are normally renewed every 15 000 km (10 000 miles) and they should be removed, inspected, cleaned, and adjusted at intervals of 7500 km (5000 miles).

1 Unscrew the spark-plug from the cylinder head, using a properly fitting socket or box spanner.
2 Examine the condition of the electrodes and the insulator for wear, cracks, damage, and colour.
3 Remove soft dirt and oil deposits from the outside of the plug and then sand-blast the hard deposits such as baked carbon and fuel residues using the special-purpose cleaning machine.
4 Bend open the earthed electrode slightly and, using a flat file, file square the opposing eroded electrode faces.
5 Adjust the gap with a feeler gauge by carefully bending the earthed electrode with the gap-setting tool.
6 Finally, initially screw in the plugs by hand and then, using a torque wrench, tighten the plugs according to the manufacturer's recommendations. Note do not overtighten.

26.11 Transistors (fig. 26.15(a)–(c))
A transistor is a semiconductor and is basically an electronic solid state switch device. Transistors are made from two types of semiconductor (silicon) material, one being positive P-type material which has a deficiency in mobile electrons and therefore

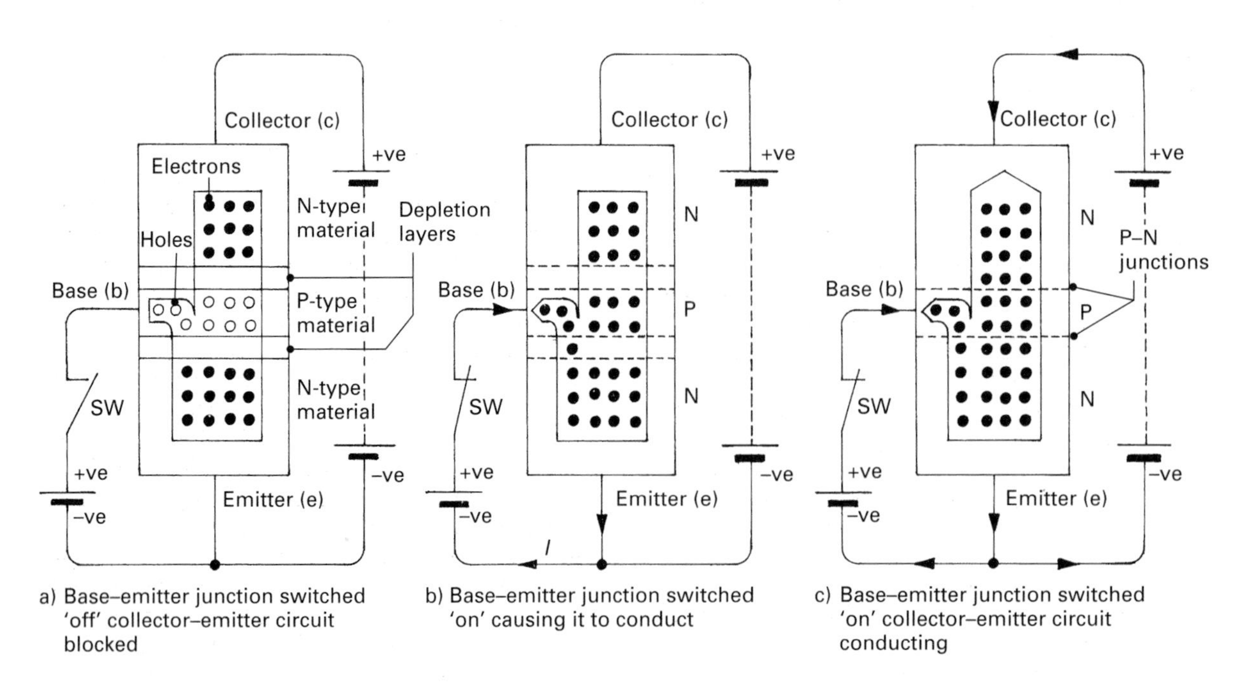

a) Base–emitter junction switched 'off' collector–emitter circuit blocked
b) Base–emitter junction switched 'on' causing it to conduct
c) Base–emitter junction switched 'on' collector–emitter circuit conducting

Fig. 26.15 Transistor action

for simplicity is said to be filled with holes, and the other is negative N-type material which has an excess of mobile electrons (see fig. 26.15(a)). Normally this wafer of positive material is sandwiched between two slightly thicker wafers of negative material. This construction is known as an NPN transistor. This combination forms two negative-positive (N-P) junctions.; there are three connections to this silicon chip: the middle connection which activates the switch is named the base 'b', while one of the two outer sandwich connections is known as the emitter 'e' since it emits electron charges, and the other side connection is called the collector 'c' since it collects electron charge carriers. At each P-N junction the electrons diffuse just inside the P-type material and likewise the holes diffuse into the boundaries of the N-type material. This narrow region therefore becomes neutral as there are no free moving electrons or holes. It is therefore known as the depletion layer.

Transistor action (fig. 26.15(a)–(c)) When the base-emitter (b-e) junction circuit switch is closed, the depletion barrier breaks down. Electrons then cross from the emitter N-type material into the base P-type material and at the same time electrons from the battery negative terminal will replace those lost by the emitter. This movement of electrons therefore forms the (b-e) junction circuit (see fig. 26.15(b)). Remember the convention for the direction of current flow is opposite to the movement of electrons. Very quickly the P-type material becomes saturated with electrons; consequently the depletion layer of the rest of the (b-e) junction and that of the (b-c) junction will instantly shrink and be replaced by free electrons (see fig. 26.15(c)). As a result the P-type material middle region of the semiconductor temporarily changes to the N-type material. Immediately electrons from the negative battery terminal which supply the emitter will be attracted to the collector due to the positive potential on this side (the positive terminal of the battery has a deficiency of electrons whereas the negative end has an excess of electrons). Hence current will flow though the (c-e) circuit as long as the (b-e) junction current continues to apply a small trigger current. When the (b-c) junction circuit is interrupted, the base instantly changes back to its previous P-type material with its non-conducting depletion layer. In other words the (c-e) circuit blocks the flow of current. This switches off the transistor (see fig. 26.15(a)). A simplified way of presenting the transistor and its connections is shown in fig. 26.16.

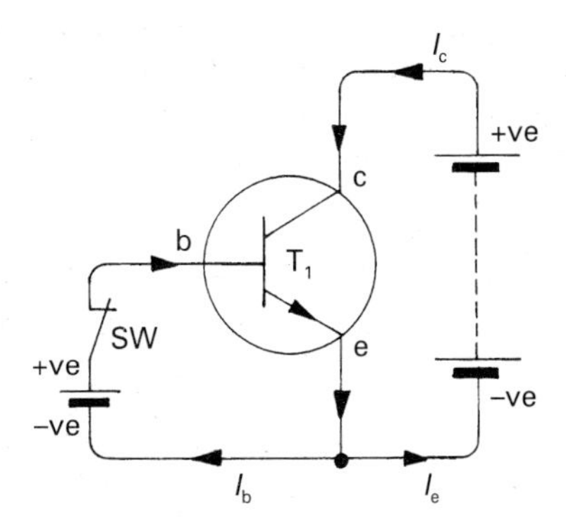

Fig. 26.16 Symbolic representation of a transistor circuit

Thus it can be seen that the transistor functions similarly to a relay switch where a small current controls the on/off switching for a circuit requiring a much larger current flow.

26.12 Electronic ignition control

26.12.1 Basic induction semiconductor ignition system

The basic components of an electronic solid state ignition system are the induction pulse generator driven by the distributor drive, a primary and secondary ignition coil, a transistor switch which makes and breaks the primary winding energising current and a conventional cap and rotor.

Induction pulse generator (fig. 26.17(a) and (b)) The induction pulse generator consists of a moving iron four-tooth reluctor and a four-projection permanent magnet stator with its induction winding. A magnetic flux field will be established around the permanent magnet and reluctor. As the rotating reluctor moves towards and away from the stator projections, the air gap decreases and then increases causing a strengthening and then a weakening of the magnetic flux field. This therefore induces an alternating voltage into the induction winding. This induced voltage rise produces a positive half-wave as the reluctor teeth approach the fixed stator projections, reaching a maximum just before the corresponding teeth and projections align with each other. The voltage then rapidly decreases, changes polarity, and generates a second half negative wave as the reluctor teeth move away from the stator projectors. This cycle of events being continuously repeated, the frequency of this generated voltage

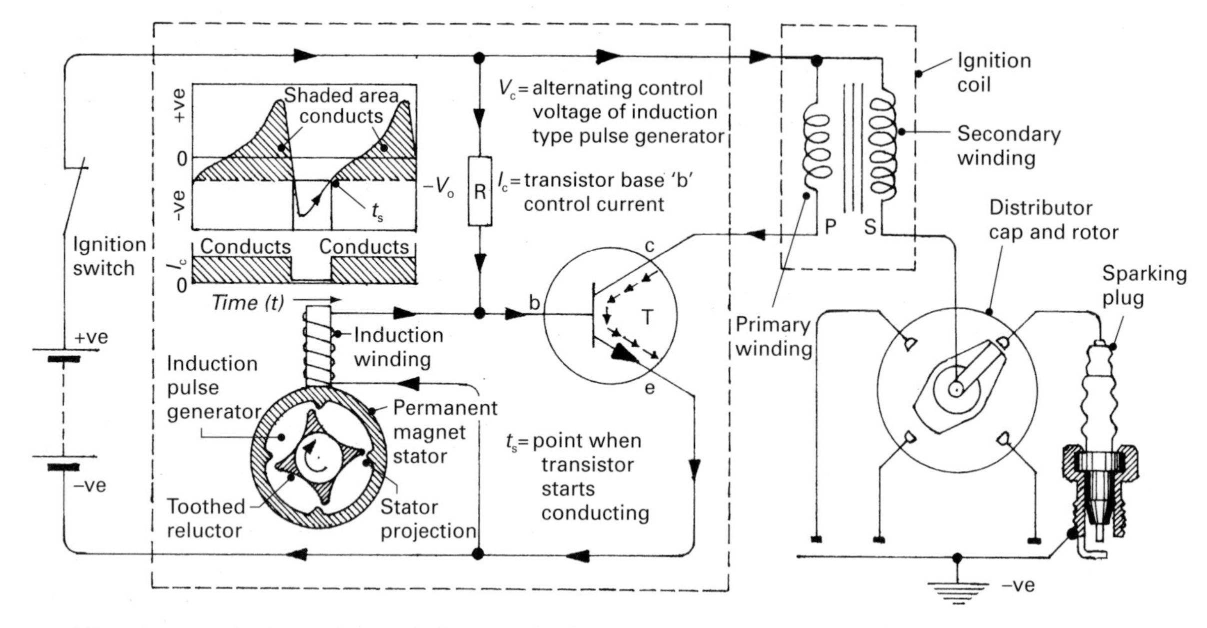

a) Transistor conducting – primary winding energised

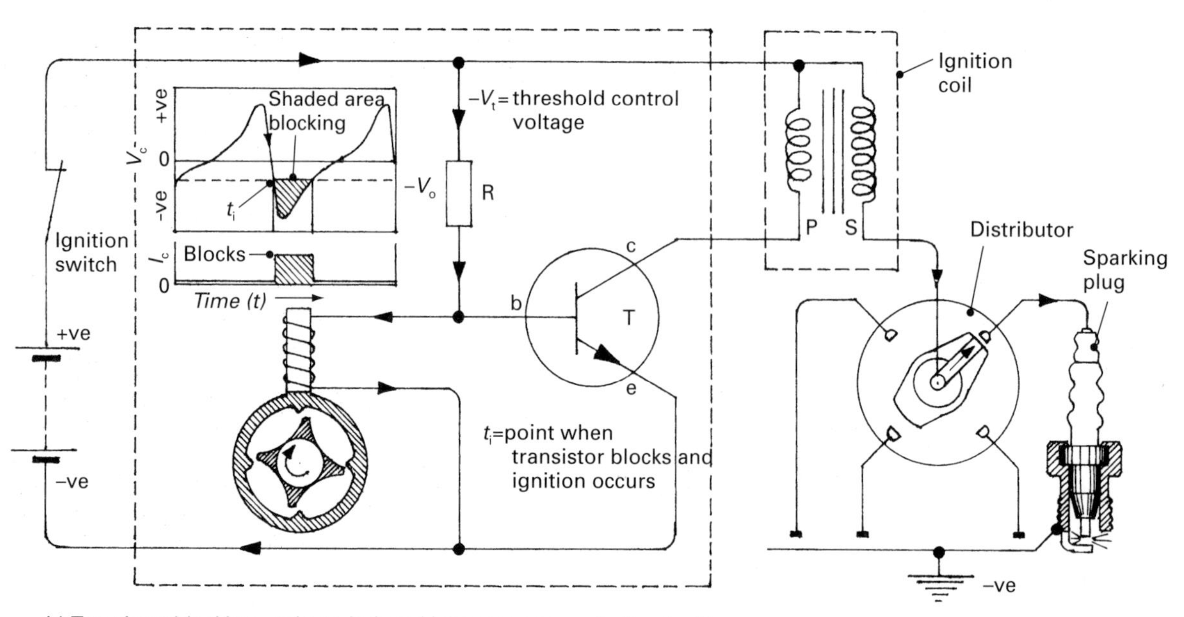

b) Transistor blocking – voltage induced into secondary winding-ignition occurs

Fig. 26.17 Very basic electronic ignition system

is proportional to the engine speed and the number of cylinders in the engine. For the latter there must be the same number of teeth/projections as there are cylinders.

Primary winding energised (fig. 26.17(a)) As the reluctor teeth move away from the stator projections the negative half-wave voltage decreases until the threshold control voltage $-V_0$ is reached at point t_s. This is the minimum voltage necessary when applied to the base 'b' of transistor T to switch on the (c-e) circuit. Current now flows from the battery positive terminal though the primary winding and back to the battery negative terminal via the conducting (c-e) circuit of the transistor. The energised primary winding quickly establishes a magnetic flux which interlinks both primary and secondary windings.

Primary winding circuit interrupts – ignition occurs (fig. 26.17(b)) As the revolving reluctor teeth again approach the stator projections the generated positive voltage reaches its peak and then rapidly declines to its negative half-wave. When the induced voltage reaches the negative threshold control voltage $-V_0$, at point t_i the base 'b' switches off the transistor causing the (c-e) circuit to block. Instantly the magnetic flux field surrounding the primary and secondary windings, collapses so that the flux lines of force cut through the windings thus inducing a large voltage into the secondary winding. The high-intensity voltage drives a very small current though the central high-tension lead to the appropriate sparking plug via the rotor and distributor-cap. It then ionises the air gap (produces its spark) on its return path to the negative earth.

26.12.2 Inductive semiconductor ignition with induction-type pulse generator

Pulse generator function (fig. 26.18(a)) The pulse generator (see fig. 26.18(a)) provides a periodic alternating voltage wave from those frequencies related to engine speed. Its output signal is then amplified by the first transistor in the module box before it is fed to the second output power transistor which switches on and off the current flowing though the ignition coil primary winding.

A constant reference voltage V_r is supplied by the battery to the induction winding of the generator. Its value fixes the mid-position between the alternating half-wave of the generated signal (see fig. 26.18(b)). The on and off state of both transistors is controlled by the negative wave of the signal current. The start of the primary current dwell (energising primary winding) t_s takes place on the gradual slope on the negative half-wave while the point of ignition t_i occurs on the steep slope of the negative wave (see fig. 26.18(b)).

Ignition switched on, engine stationary (fig. 26.18(a)–(d)) When the ignition is switched on, current from the battery positive terminal passes through the resistor R_2, the generator induction stator winding. It then goes to the negative ground via the diode D_2 (see fig. 26.18(a)). This provides the base 'b' of transistor T_1 with a positive potential. The transistor collector-emitter (c-e) circuit will therefore conduct. With the (c-e) circuit of T_1 switched on, the base 'b' of T_2 will be connected directly to the ground negative potential so that the (c-e) circuit of the power transistor T_2 blocks and the ignition coil primary winding will be on open circuit (see fig. 26.18(c) and (d)).

Generating a voltage signal (fig. 26.18(a) and (b)) As the engine is cranked, the reluctor trigger wheel (see fig. 26.18(a)) rotates with the distributor rotor shaft so that the trigger teeth will periodically pass by the corresponding stator teeth and so generate a fluctuating magnetic flux across the tips of the rotating and stationary teeth. Correspondingly an alternating voltage signal will be induced into the induction winding of the generator (see fig. 26.18(b)).

Beginning of primary current growth (fig. 26.18(a), (c), (d) and (e)) During the short time the trigger teeth align with the stator teeth, the generator magnetic flux strength will rapidly change so that the generated alternating signal voltage now opposes current coming from the battery positive terminal via R_2 at base 'b' of T_1 (see fig. 26.18(a)). This causes the base 'b' of T_1 to become negative thus making its (c-e) circuit block (switch off). Current from the positive battery terminal now flows though R_3 to the base 'b' of the power transistor T_2 and so energises the base-emitter (b-e) junction. Instantly its (c-e) circuit conducts (switches on) (see fig. 26.18(c) and (d)). Current will then flow through the ignition coil primary winding to the negative ground via the T_2 (c-e) circuit, permitting it to build up its magnetic flux within the ignition coils windings (see fig. 26.18(e)).

Ignition point (fig. 26.18(a), (c), (d) and (f)) Just as the reluctor trigger teeth pass the stator teeth, an abrupt change of alternating signal voltage polarity takes place, and instead of the positive voltage decreasing, it now commences to increase until the threshold trigger voltage V_t (switch on voltage) of T_1 is reached (see fig. 26.18(c) and (d)). At this point t_i ignition occurs, the (c-e) circuit of T_1 conducts and T_2 blocks (see fig. 26.18(a)) due to it now having a negative base 'b' (see figs. 26.18(c)

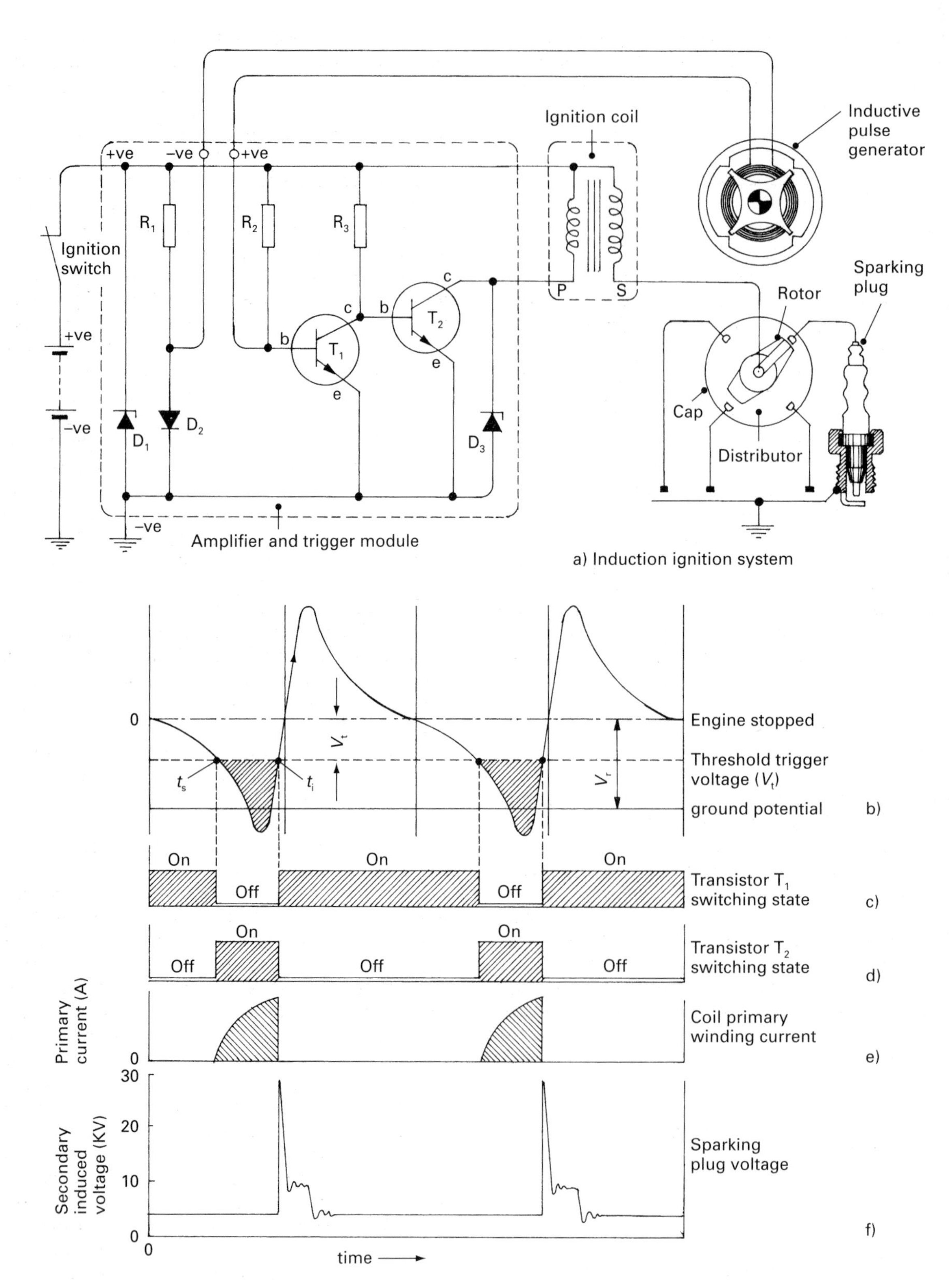

Fig. 26.18 Inductive semiconductor ignition with induction pulse generator

and (d)). The sudden interruption of the ignition coil's primary winding current flow causes its magnetic field to collapse, thereby inducing a very large voltage into the secondary winding (see fig. 26.18(f)).

26.12.3 Inductive semiconductor ignition with the Hall generator circuit

The Hall effect (fig. 26.19(a) and (b)) When a supply current I_s passes though a metal layer (such as a silicon semiconductor) shielded from any magnetic flux, the electron charge will be uniformly distributed thoughout the semiconductor layer. If a magnetic field 'B' is applied at right angles to the supply current I_s moving across the conductor, then an electromagnetic force acts on each electron perpendicular to the direction of both the magnetic field and current flow. This tends to push the electron charge towards one end so that there is a surplus of electrons in the top half and only a few in the bottom half of the semiconductor layer. As a result, a potential difference is created between the upper and lower terminals. Thus if the two ends of the semiconductor layer were connected together by an external conductor, electrons from the upper half of the semiconductor where there is an excess amount of electrons would migrate through the external conductors to the lower half of the layer where there is a deficiency of electrons. The movement of electrons in the external conductor is therefore trying to re-establish a more evenly balanced distribution of electrons through the semiconductor layer. The imbalance of electrons between the upper and lower ends of the semiconductor thus produces a potential difference, that is, a voltage between the opposing terminals known as the 'Hall voltage', and the deflection of the electrons caused by the magnetic field crossing the current flow is known as the 'Hall effect'.

If a magnetic diversion vane is placed between the south magnetic pole and the semiconductor layer, the magnetic flux field will be diverted from the semiconductor layer. This causes the electrons to rapidly realign themselves so that they are again uniformly spread thoughout the semiconductor layer. Thus since the potential difference between the two terminals is non-existent the Hall voltage will also be zero.

Hall-type generator (fig. 26.20) There are two basic parts to the generator: a stationary magnetic assembly with a tiny Hall integrated circuit attached to its side and a rotating vane rotor made in the form of a circular steel disc with downward bent lugs known as the vanes. The stationary member consists of a permanent N and S magnet mounted back to back on the inside of a vane rotor, and a semiconductor layer and a Hall integrated circuit on the outer side. A right-angled magnetic flux conductor is positioned over the permanent magnets on the inside and a similar but shorter one is positioned on the opposite side of the air gap over the semiconductor layer and the Hall integrated circuit. The whole stationary assembly is moulded into an insulated base. When the vanes move away from the air gap a magnetic flux field path is quickly established across the air gap. It also passes at right-angles through the semiconductor layer to generate a Hall voltage. As the vane passes through the air gap it screens and diverts the magnetic flux path from the semiconductor layer, and therefore switches off the Hall voltage. The width of each vane determines the dwell angle needed for the primary current to build up and there are as many vanes as there are engine cylinders. The Hall integrated circuit, consisting of transistors T_1 and T_2, amplifies the small generated Hall voltage and converts its trapezium-shaped switch on/off pulse to a rectangular-shaped one. It further amplifies the signal output voltage and current before it is passed to the output power transistor T_3 and T_4 of the trigger box which controls the opening and closing of the ignition coil's primary winding circuit.

Hall integrated circuit and trigger action

Primary winding circuit, completed flux established (fig. 26.21(a)–(f)) When the vane rotor revolves, one of the vanes enters the Hall generator air gap and thereby screens the Hall layer so that the flux abruptly decreases down to the cut-out threshold B_2 (see fig. 26.21(b)). Since the generated Hall voltage is now below the minimum required to switch on transistor T_1, both T_1 and T_2 cease to conduct (see fig. 26.21(a), (c) and (d)). Consequently, current is now available for the base 'b' of T_3 via resistor R_6. This switches on T_3 and allows current to flow via its (c-e) circuit to the base of T_4. Instantly the (c-e) circuit of T_4 starts to conduct (see fig. 26.21(a) and (d)). Current now flows from the positive terminal through the primary winding and then passes through the conducting (c-e) circuit of T_3 and T_4 back to the battery negative terminal to complete its circuit.

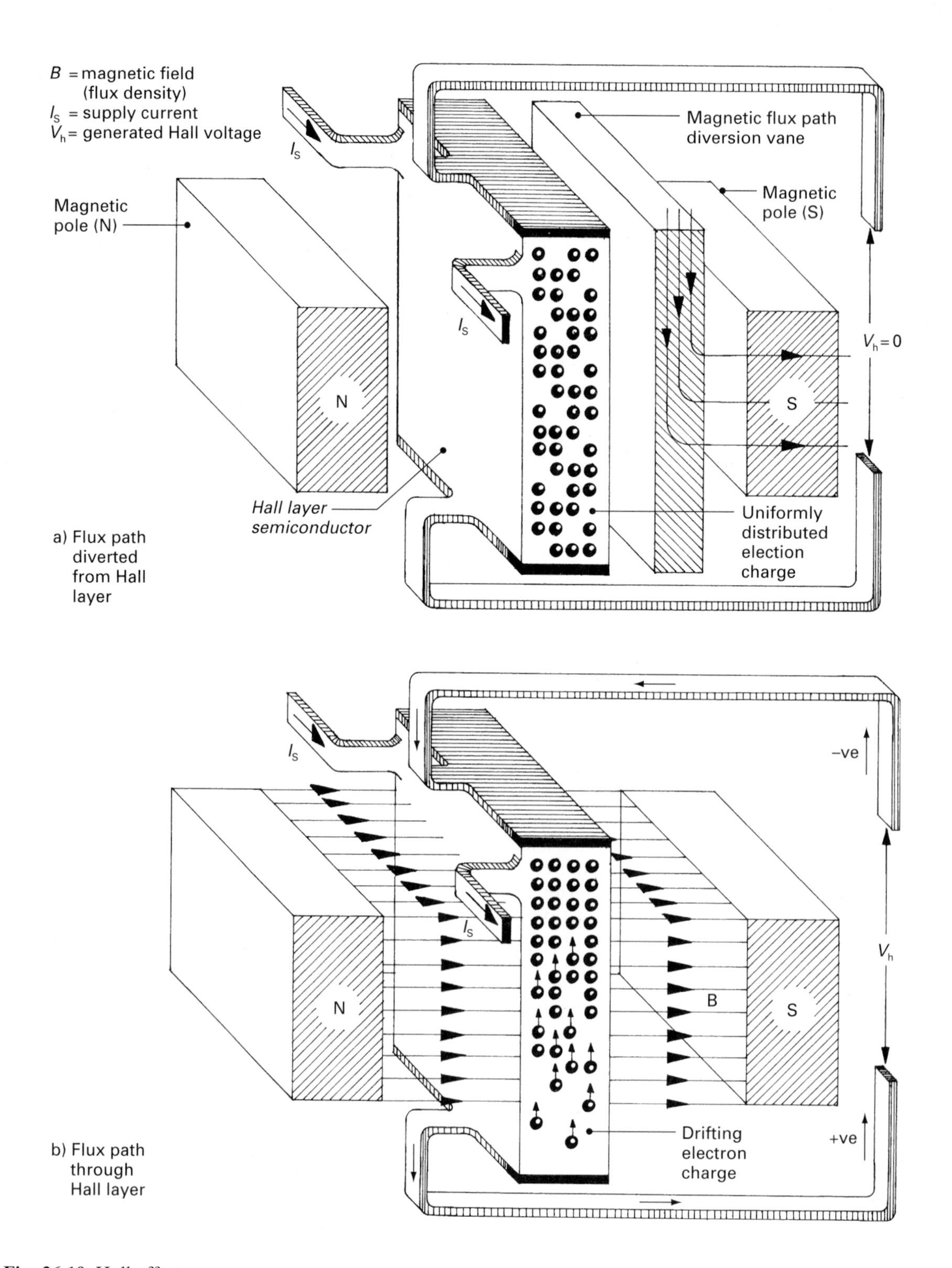

Fig. 26.19 Hall effect

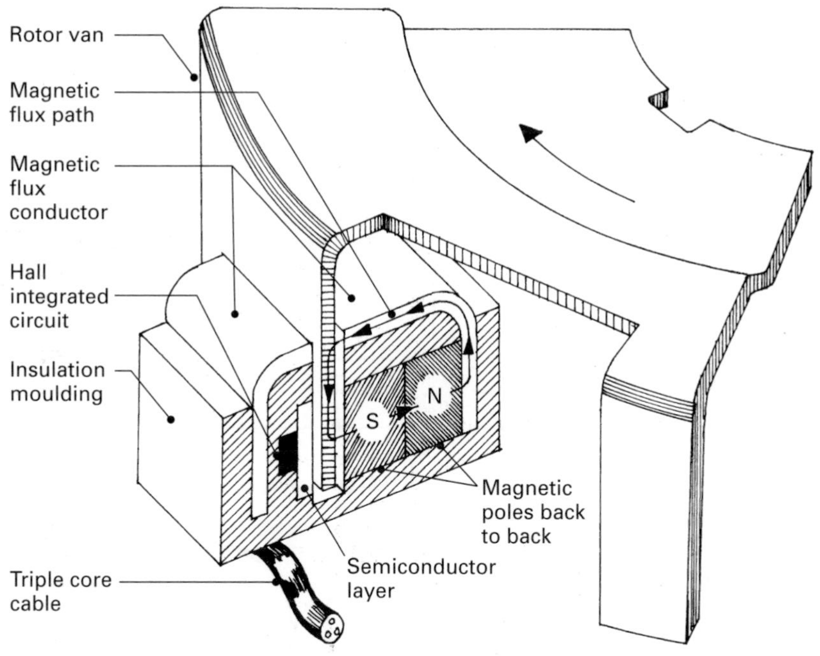

a) Flux path diverts from Hall layer

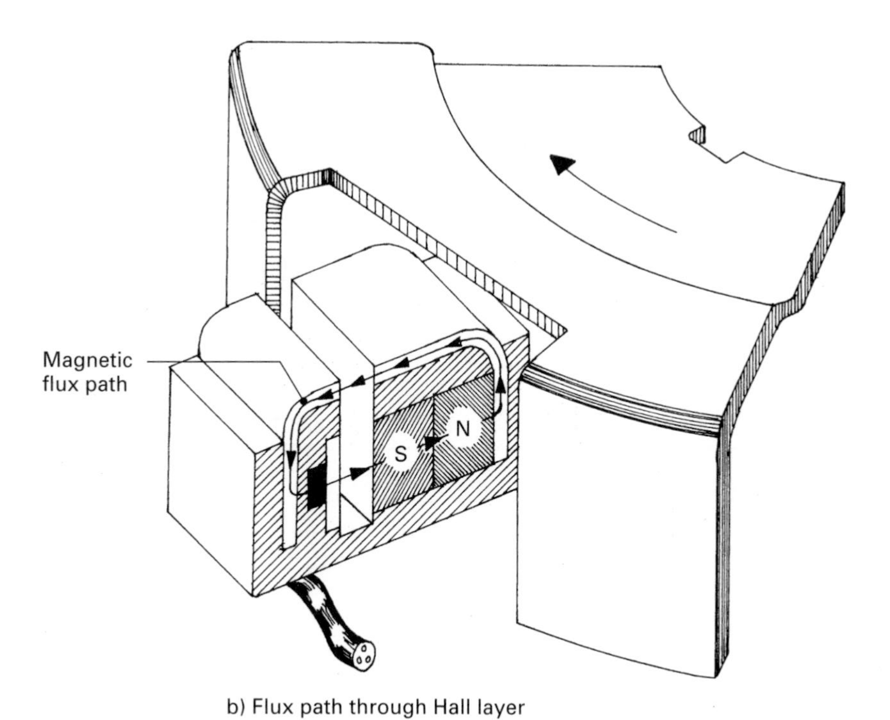

b) Flux path through Hall layer

Fig. 26.20 Hall generator trigger

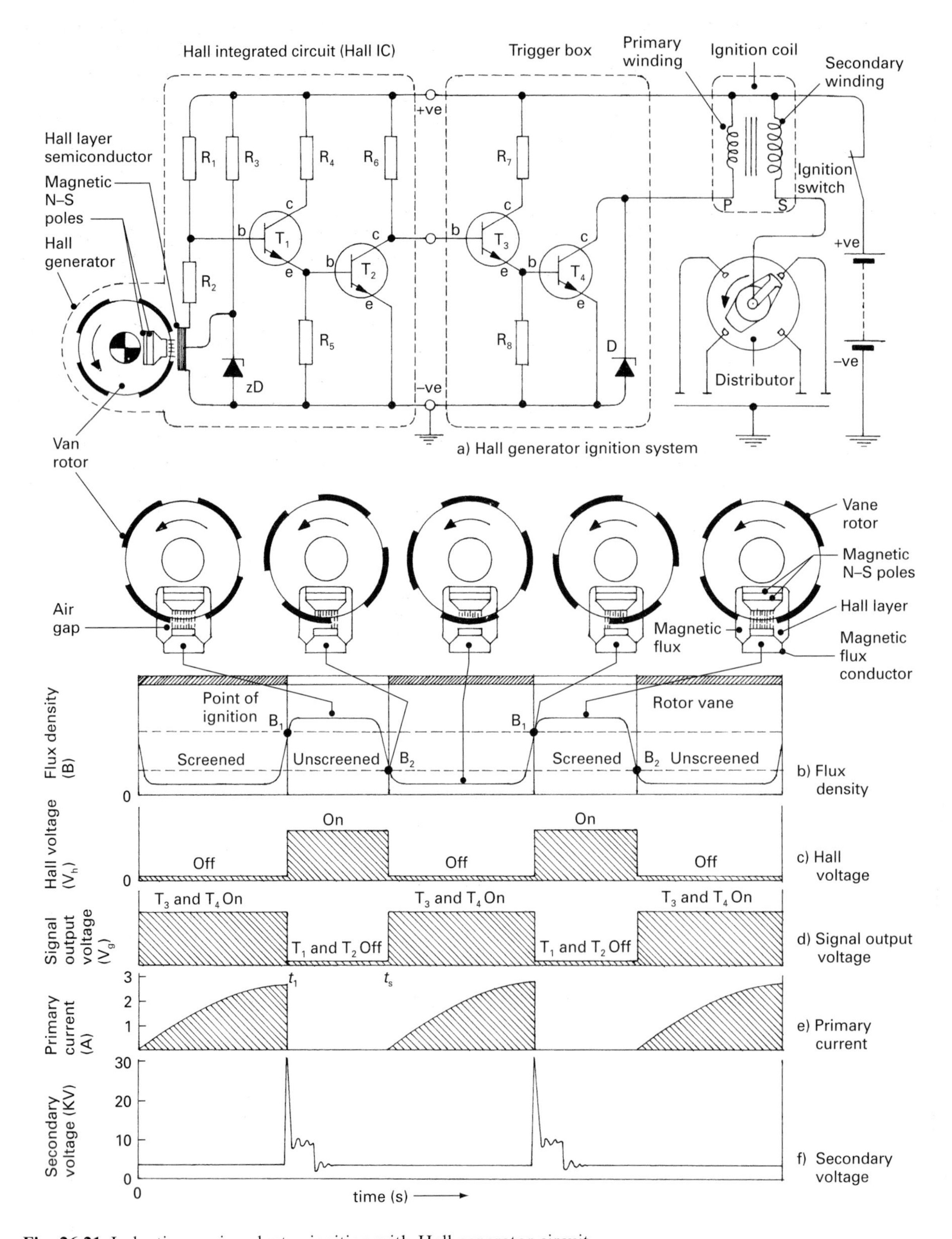

Fig. 26.21 Inductive semiconductor ignition with Hall generator circuit

Note that the conducting circuit of T_3 has a higher resistance than that of T_4 due to resistor R_8. Therefore the power transistor T_4 takes most of the primary current circuit load. The build-up of current in the primary winding therefore establishes a magnetic flux field around both primary and secondary ignition coil windings in readiness for the next primary circuit interruption and high-tension voltage inducement into the secondary winding (see fig. 26.21(e)). The two output transistors T_3 and T_4 form what is known as a Darlington pair; they are coupled together in such a way as to considerably improve and sharpen the switching action compared to a single output transistor for the make and break primary winding circuit of the ignition coil.

Primary winding circuit, interruption ignition occurs (fig. 26.21(a)–(f)) As the vane rotates the vane shield moves out of alignment with the air gap. It permits the magnetic flux to be re-established across the air gap. The flux density will therefore rise until a cut-in threshold 'B_1' is reached, this being the point of ignition (see fig. 26.21(b)). At this point, the Hall voltage feeding the base 'b' of transistor T_1 is sufficient to switch on its collector-emitter (c-e) circuit (see fig. 26.21(a), (c) and (d)). Immediately T_1 conducts and an amplified current flows to the base 'b' of T_2, thus also switching on its (c-e) circuit. At the same time transistors T_2 and T_4 switch to the blocked state because the current previously flowing through resistor R_6 which fed the base of T_3 now finds a lower resistor path through the transistor T_2, that is, the base of T_3 is earthed through the (c-e) circuit of T_2 (see fig. 26.21(a) and (c)). The abrupt interruption of the primary winding of the ignition coil causes its magnetic flux field to collapse and for the shrinking flux field (lines of force) to cut through the secondary winding. A very high voltage is instantly induced into the secondary winding (see fig. 26.21(f)) which causes a corresponding generated current to flow to the sparking-plug when it completes its circuit to earth by jumping the air gap and hence creating the ignition spark.

27 Batteries, generators and starter motors

27.1 The lead–acid battery

27.1.1 Storage-battery cell concept

The function of the battery is to store chemical energy which, when required, can readily be converted into electrical energy during the process of discharging.

The electrical energy is created when two electrochemically dissimilar substances are immersed in a liquid conducting media known as the electrolyte, forming an electrical cell which will produce a direct current (d.c.) voltage when the circuit is completed.

27.1.2 Primary and secondary cells

There are two classes of battery cells, known as primary and secondary types.

Primary cells can only be discharged once and are then discarded. These cells are used in torches, transistor radios, calculators, etc., but they are unsuitable for motor-vehicle purposes.

Secondary cells have a reversible action – this simply means that the chemical change which occurs inside the cell when delivering current can be reversed by applying a direct current to the cell terminals. These secondary cells can be repeatedly charged and discharged, thereby making them suitable for motor-vehicle applications. The most successful secondary cell in terms of reliability, life expectancy, and cost is the lead–acid variety, whose action depends on the chemical activity of lead and dilute sulphuric acid.

27.1.3 Simple lead–acid cell (fig. 27.1)

A lead–acid battery cell in its simplest form consists of two lead plates with terminals connected to them submerged in a container partially filled with a solution of dilute sulphuric acid (fig. 27.1). When the two terminals are connected to a direct-current source of electricity, current passes from one plate to the other, and after a period of time the positive plate (known as the anode) will form a dark brown lead peroxide (symbol PbO_2) on its surfaces, but the negative plate (known as the cathode) will just remain as grey metallic lead (Pb). The cell is then in a charged state.

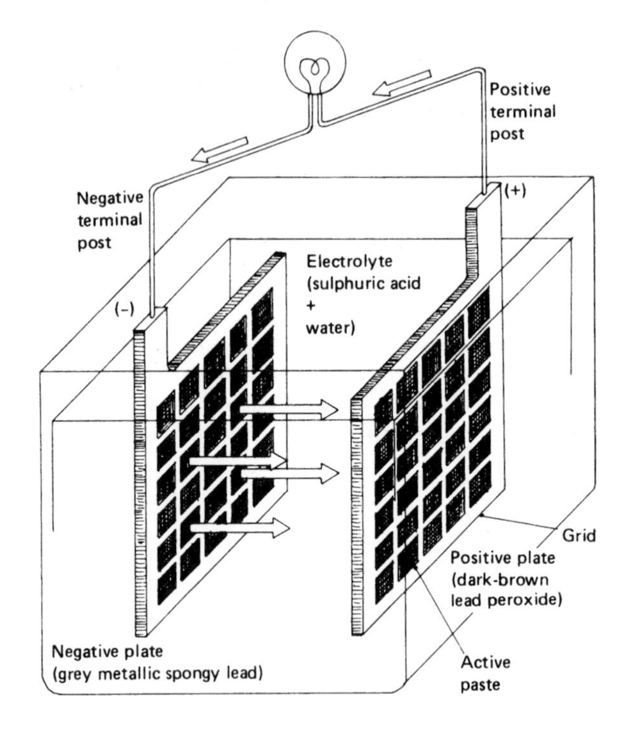

Fig. 27.1 Simple lead–acid cell

If the charging current is removed and the terminals are connected to a light-bulb, say, then current will flow from the positive to the negative terminal through the external circuit formed by the filament, which thus lights up until the chemical change is completely reversed and both plates have converted back to plain metallic lead.

27.1.4 Battery construction

Container (fig. 27.2) This is a moulded box divided into a number of compartments or cells – six for a 12 volt battery or three for a 6 volt unit. The material used must have high acid-proof and insulating properties and considerable mechanical strength. Polypropylene is commonly used, having replaced the earlier heavy bulky hard rubber compounds.

On the bottom of the cells are moulded ribs on

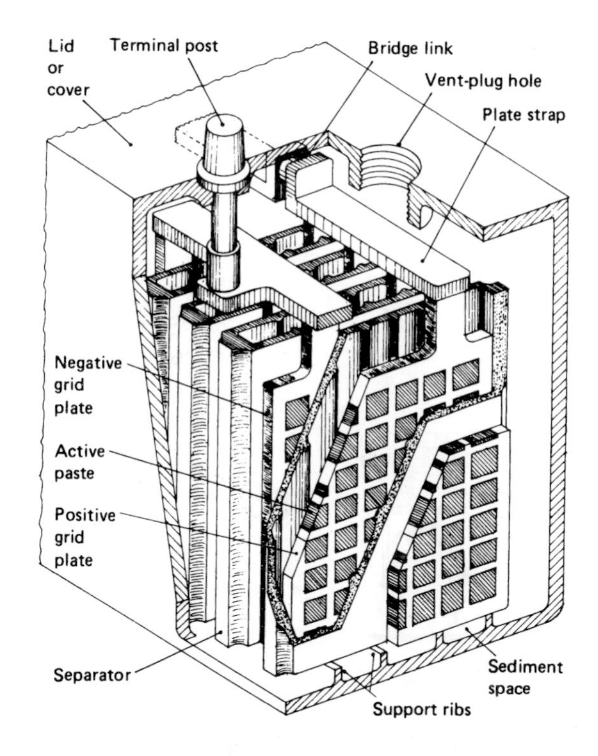

Fig. 27.2 Sectioned view of a single battery cell

which the plate assembly rests – these are provided so that any active material falling from the plates does not bridge across the bottom of the plates. A lid with vent plugs or multi-cell lift-up vent tops encloses the top of the cells, and the joint between this cover and the casing is generally heat-sealed to improve the rigidity of the battery box.

Plates (fig. 27.2) Plates are made in the form of a lattice grid which supports the active material and conducts the electricity to and from it. The lead used for making the grids is alloyed with small additions of antimony or calcium, to facilitate its casting or cold working and to improve its mechanical and electrical properties. The windows of both the positive and the negative grids are then filled with the same special preparation of lead oxide (PbO) paste. After a process known as forming in which the plates are electrically charged, the paste is converted to a special spongy form of grey lead (Pb) on the negative plate while the positive plate is filled with a brown paste of lead peroxide (lead dioxide, PbO_2).

Separators (fig. 27.2) To prevent short-circuiting between plates of opposite polarity, sheets known as separators are placed between the plates to prevent them from touching.

Separators in sheet form are made from non-conducting but porous material arranged to interfere as little as possible with the chemical action between the plates. They are usually grooved on the side facing the positive plates, to allow free circulation of the electrolyte. Several materials have been used for separators, including special types of wood, ebonite, glass wool, porous rubber, resin-bonded cellulose, and (possibly the most widely used) sintered polyvinyl chloride (p.v.c.).

Plate groups (fig. 27.3) Each cell has two groups of plates, one group being positive and the other negative, the two groups being interleaved with each other (fig. 27.3(a)). There is always an odd number of plates, the extra one being negative because of the greater chemical reaction around the positive plates.

The current-carrying capacity of a battery –

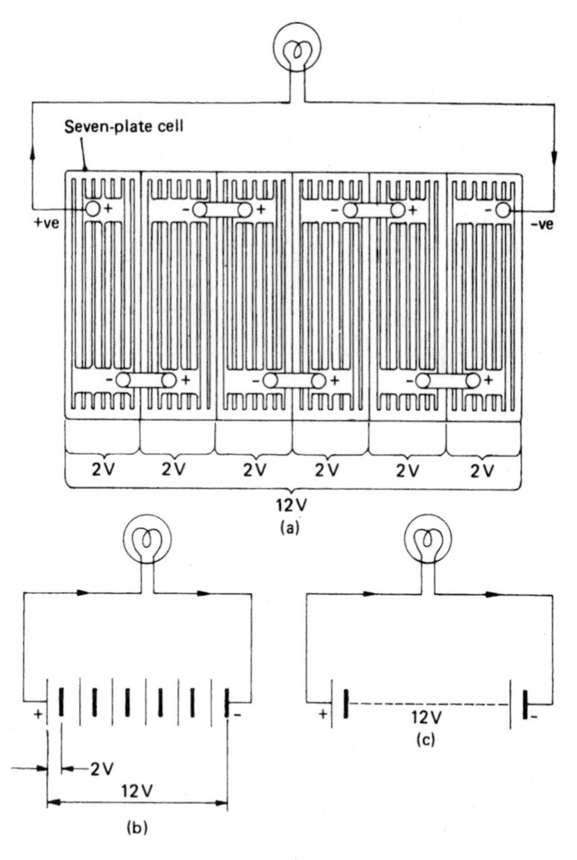

Fig. 27.3 Battery-cell plate grouping and symbolic representation

that is, the cells' ability to provide a continuous current supply – is determined mainly by the total active exposed areas of the two dissimilar (positive and negative) plates; hence the number of plates in each group depends upon the storage capacity for which the battery is designed.

A battery's discharge-rate capacity is usually expressed in ampere hours (A h) and is given by the current which can be supplied continuously before the voltage falls below a given value (usually 1.8 volts per cell) for a continuous discharge period which is usually either 10 hours or 20 hours. Thus a battery rated at 40 A h should be able to continuously provide 4 amperes for a period of 10 hours or 2 amperes for 20 hours.

Plates of the same polarity are interconnected by welding in position a common lead-alloy strap which in turn is intercell-linked with a bridge connector (fig. 27.2). Batteries normally used for cars have either seven or nine plates per cell – that is, three positive and four negative or four positive and five negative respectively.

Cells are said to be 'in series' when the positive pole of one cell is connected to the negative pole of the adjacent cell. If each cell has a nominal potential of two volts, three cells are required in series to make a 6 volt battery or six cells for a 12 volt battery. The two end cells of a battery have external terminal posts welded to the unconnected link straps so that the battery can be electrically connected to an external source, one post having a positive polarity and the other negative.

Battery cells are commonly represented by two parallel lines, the positive plates being identified by a long thin line and the negative plates by a short thick line (fig. 27.3(b)). A shorthand method of representing a group of battery cells is shown in fig. 27.3(c).

Terminal posts and lugs The two battery terminal posts and lugs act as junction points to enable the battery to be connected to the various wiring circuits used on the vehicle. These posts must be adequately proportioned and well supported by the battery lid to absorb the high current discharge when the starter motor is engaged, vehicle vibration, rough handling when being connected or disconnected, corrosion, and the constant pull of the heavy cable.

Figure 27.4 shows the various methods of attachment. For heavy-duty applications fig. 27.4(b) is possibly the best, but for light duty fig. 27.4(d) is frequently used. Figure 27.4(c), which does not need any spanner or screwdriver to remove it, has become very popular, whereas fig. 27.4(a) is very rarely used these days.

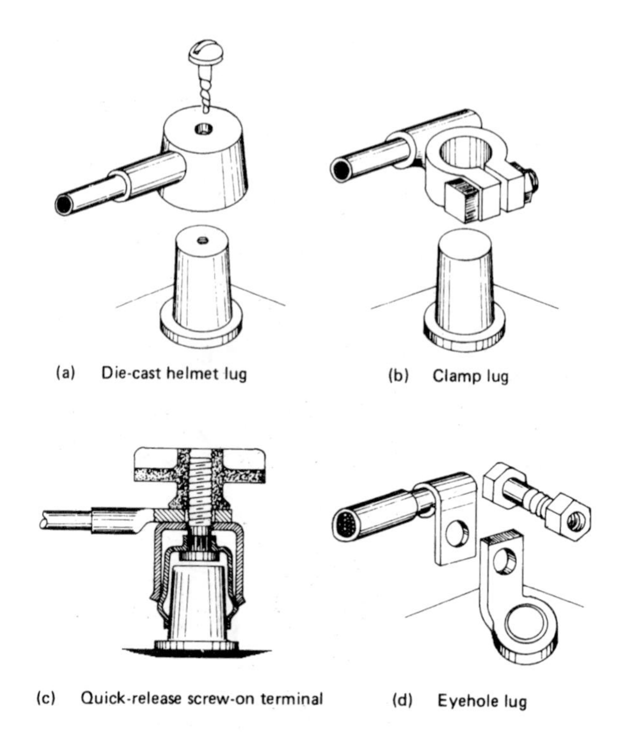

Fig. 27.4 Battery terminal posts and lugs

27.1.5 Discharge and recharge cycle of operation

Discharge When a battery is delivering electrical power, it is said to be discharging. The energy is produced by the acid in the electrolyte gradually combining with the active paste in the plate grids. This combination produces lead sulphate ($PbSO_4$) in both the negative and the positive plates. A cell is completely discharged when the chemical reaction is completed and both plates are entirely sulphated – as they are then composed of identical material, the terminal voltage will then be zero. In practice a battery should never be allowed to reach a completely discharged state, as this will cause damage to the plate grids and active paste and the sulphate becomes difficult to convert back to active paste.

Recharge The object of charging is to drive all the sulphate out of the plates and return it to the electrolyte. A direct current is passed through the cell in the opposite direction to that during discharge. During recharge, the lead sulphate ($PbSO_4$) in the positive plate is converted into lead peroxide (PbO_2) while the lead sulphate in the negative plate is converted into spongy lead (Pb). The electrolyte gradually becomes stronger as sulphate from the plates combines with the hydrogen (H_2) in the water to form sulphuric acid (H_2SO_4) until no more sulphate remains in the plates and all that absorbed by the plates during discharging has been returned to the electrolyte.

27.1.6 *Battery chargers*

The power supply from the mains is an alternating current (a.c.) in nature, i.e. the direction of the current is continuously changing, and it is at a very high potential – 220 to 240 volts. This form of power source is unsuitable for charging batteries, and it is the battery charger's role to modify the form of supply to meet the battery requirements.

Basically the battery charger performs three functions:

i) it incorporates a step-down transformer which reduces the voltage to suit the battery being charged,
ii) it incorporates a rectifier which converts the alternating current to a direct current so that the discharging chemical reaction within the battery cells is reversed,
iii) it automatically adjusts the charge rate to match the state of charge.

When connecting a battery to a battery-charger unit, connect the positive lead of the charger to the positive terminal post and the negative lead to the negative terminal post. Always remove the battery vent plugs to allow gas to escape, and check the electrolyte level before charging.

Batteries should never be left in a discharged state and, if they are only to be used occasionally, they should be periodically charged at the normal recharge rate by an independent source such as a portable battery charger to prevent excessive sulphate forming on the plates.

The recharge rate should be normally set to approximately one tenth of the battery's ampere-hour capacity – for example, if the battery has 40 ampere-hour capacity, the charger should be adjusted to provide a current flow of 4 amperes.

27.1.7 *Relative-density (specific-gravity) test* (fig. 27.5)

In a lead–acid battery cell, the chemical reaction takes place mainly between the active paste and a water-diluted sulphuric-acid solution known as the electrolyte. The sulphuric acid is initially diluted by mixing it with water until the electrolyte has a relative density (specific gravity) of approximately 1.28 – i.e. until the mass of a unit volume of electrolyte is 1.28 times the mass of the same volume of pure water.

If the cell is fully charged, the active pastes are free of sulphate so the electrolyte strength is at its highest. As the cell discharges, the active pastes react with the sulphuric acid in the electrolyte and lead sulphate is formed in the plates. This means that the sulphuric acid is gradually transferred from the electrolyte into the active paste and so the strength of the solution will be proportionally reduced. Now the relative density of the electrolyte varies directly with the solution strength, so the relative density can be used to estimate the state of charge of each battery cell. The relative density is measured using a hydrometer (fig. 27.5).

A hydrometer syringe is in the form of a glass tubular body with a soft rubber bulb at one end and a rubber sampler tube at the other end. Inside the glass body is a glass float which has a vertical relative-density scale which has been calibrated by means of the lead shot situated in the base of the float.

To use the hydrometer, immerse the sample tube in the battery-cell electrolyte, squeeze the rubber bulb and release it to withdraw a sample of electrolyte. Allow the float to rise in the fluid and then read off the scale at surface level. If the individual cell readings vary by more than 0.040, one or more cells may be faulty. Table 27.1 shows how the relative density is related to the state of charge.

The relative density of the electrolyte varies with temperature, and when reading its value from a hydrometer it is necessary to use a standard temperature as a reference. This standard is normally 15°C. For every 1.5°C above or below 15°C, add or subtract 0.001 respectively from the hydrometer reading.

Table 27.1 Relative density and state of charge

	Relative density	
State of charge	Normal	Tropical
Fully charged	1.28	1.23
Half charged	1.20	1.16
Discharged	1.12	1.08

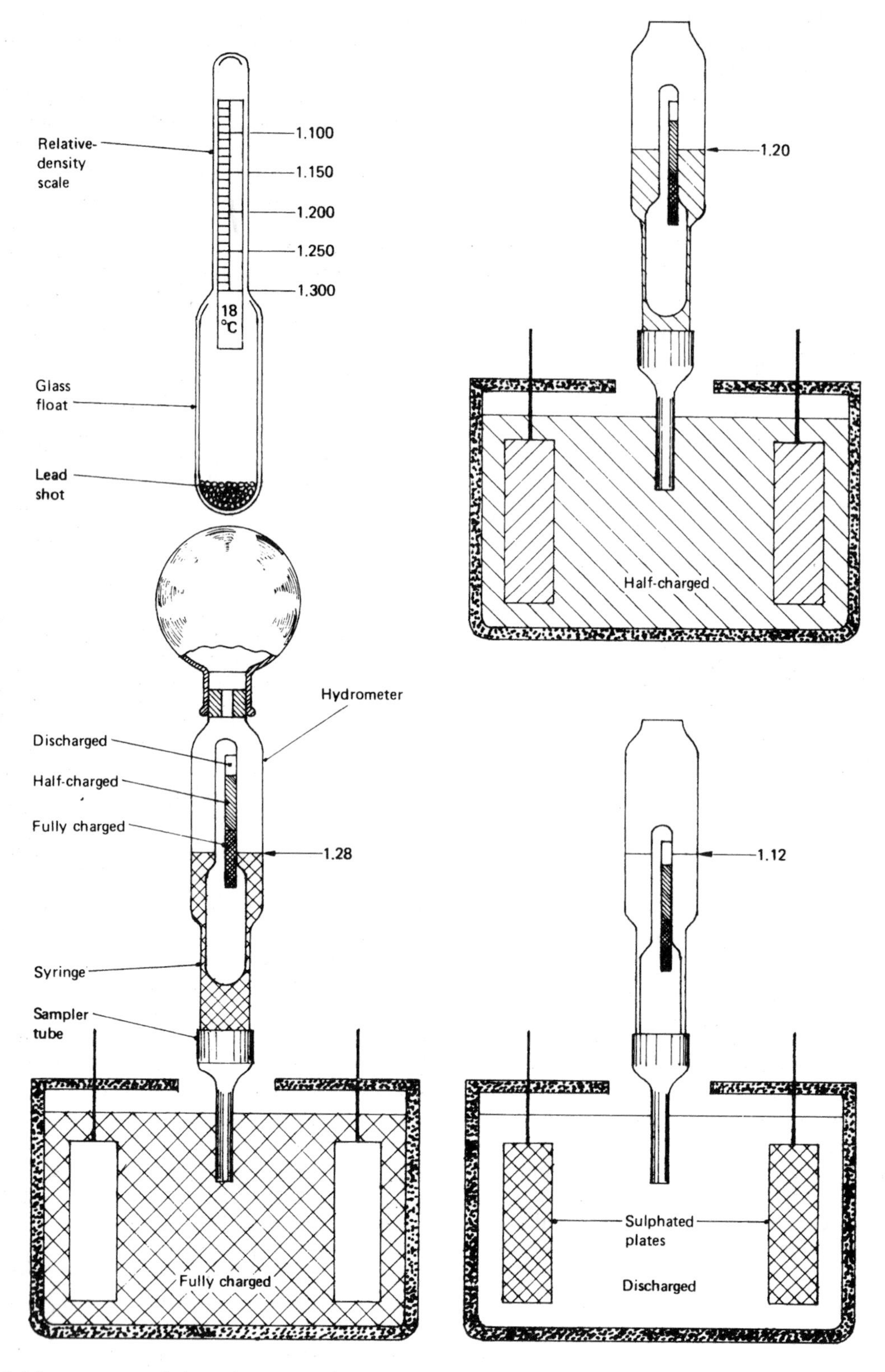

Fig. 27.5 Measurement of electrolyte strength

Example A hydrometer reading of 1.22 at a temperature of 21°C will indicate a corrected electrolyte relative-density reading of

$$1.22 + 0.001 \times \frac{21 - 15}{1.5} = 1.22 + 0.004$$
$$= 1.224$$

27.1.8 High-rate discharge test (fig. 27.6)
Measuring the relative density is the first step in determining a battery's state of charge, and the readings from the hydrometer scale will give a good indication of the state of charge, provided that the cells were originally filled with equal-strength solutions and no spillage has taken place. Unfortunately the relative-density readings cannot tell if some of the cell plates are shorting or have shed active paste and therefore will be unable to sustain a heavy current draw, such as when cranking the engine.

The high-rate discharge test applies an electrical load across the battery terminals by means of either a fixed resistor (which is usually made from an alloy such as manganese – eureka) or a variable resistor consisting of a carbon pile. Both of these materials are able to retain their resistance over a wide temperature range, which is an essential feature since the energy dissipated as heat during discharge will be considerable.

One version of a heavy-discharge tester is shown in fig. 27.6. It consists of a pile of carbon discs or plates clamped together with a nut and bolt. The bolt has a plastic knob moulded on to its head which enables the operator to adjust the degree of compression applied between the plates. The carbon pile is connected in series with an ammeter which measures the current discharge and a shunt resistor which is connected in parallel with the ammeter (this resistor is there to protect the ammeter from the high current flow), and a voltmeter is placed across the output terminals so that the voltage drop can be observed while the battery is discharging.

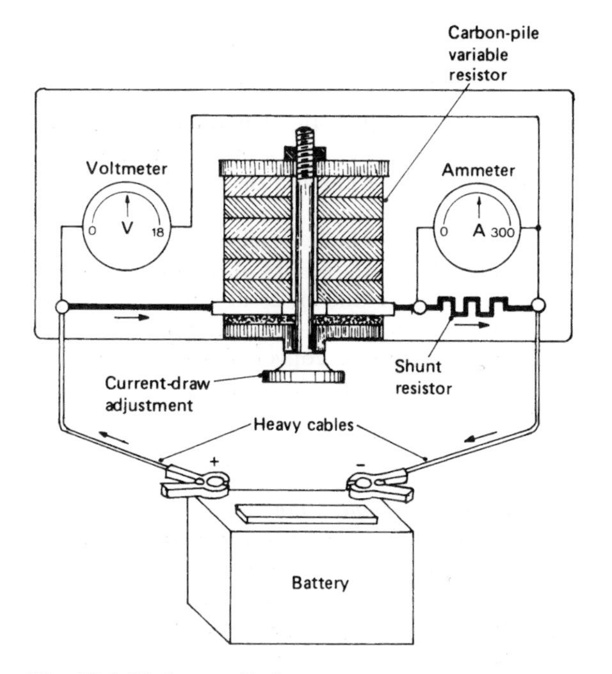

Fig. 27.6 High-rate discharge tester

To carry out the test, the carbon pile is first slackened by unscrewing the bolt one or two turns, then the heavy cable clips are clamped over the battery terminal posts. The adjustment knob is then rotated by hand so that the carbon pile is squeezed together, thus reducing the electrical resistance of the carbon and so permitting the battery to discharge through the carbon, the shunt resistor, and the ammeter.

Observe the ammeter scale and increase the discharge current until it reaches a value three times that of the ampere-hour (20 hour rate) capacity of the battery. For example, a 60 A h battery should have a discharge test current of 3 × 60 = 180 amperes. Hold this current for only 10 seconds and at the same time observe the voltmeter reading. A 12 volt battery in good condition should be able to maintain a voltage greater than 9.6 volts. If it drops just below this value, the battery should be recharged; but if the voltage rapidly drops below 6 volts it may be assumed that one or more cells are faulty. Note: this test should not be carried out on a battery which is known to be discharged.

27.1.9 Battery care and maintenance

Topping up Once every month for normal batteries or annually for 'maintenance-free' batteries, the level of the electrolyte in the cells must be inspected. The correct level generally is just above the top edges of the separators, which stand about 6 mm above the plates – this ensures that the plates are adequately covered. Only distilled water or de-ionised water must be added to replace any water lost from the electrolyte due to evaporation.

Battery installation When installing a battery, make sure that it rests level and does not rock. Tighten down the mounting bolts and brackets evenly – if they are overtight, the casing may distort and crack; conversely, if the battery is loose it may vibrate and cause internal cell damage.

Before connecting the cables to the battery, determine the polarity of each cable and only connect the positive cable to the positive battery terminal post and the negative cable to the negative post. If the cables are reversed, severe damage to the electrical system will result, such as diodes and transistor circuits being destroyed and field windings of both the starter motor and the generator being burnt out. It is best to disconnect the earth battery terminal first when removing a battery and reconnect it last when replacing one.

One or both of the battery cable terminals should be disconnected when any large component is being dismantled or removed from the vehicle or when electrical components such as the starter and generator are being removed. This prevents the possibility of a short circuit caused by a cable rubbing against a metal surface or by one of the cables becoming entangled and pinched between components.

Keep the battery clean and dry, and immediately wipe away water spillage. Any acid spillage on metal should be removed with household ammonia or baking soda and hot water, to minimise corrosion. After neutralising corroded metalwork, repaint these areas with acid-resisting paint.

Battery-terminal corrosion Corroded battery posts and terminals can prevent good electrical continuity. Corroded parts should therefore be covered for a short period with a rag soaked in hot water – this rapidly removes the products of corrosion. After cleaning the posts and terminal connectors, smear them with petroleum jelly or, if this is not available, grease. This provides a limited amount of protection against further corrosion.

Mixing electrolyte Whenever acid and water are to be mixed, add the acid slowly to the water while stirring with a wood or glass rod – *never* add the water to the acid, as this is dangerous. If the water is poured on to the acid, the water (being lighter) will not mix with the acid but will float on top of it. A chemical reaction between the acid and water will then generate heat so that the water boils, and steam and acid spray in the form of small explosions will then shower the operator.

Personal safety precautions Never smoke or allow a spark or naked flame to come close to the battery when it is being removed or installed, as the oxygen and hydrogen gases generated by charging can readily ignite and explode.

If acid is splashed on to any part of the body, the area affected should be immediately treated with sodium bicarbonate solution or, if this is not available, clean water.

27.2 Generators

A generator is a machine which converts mechanical energy to electrical energy by electromagnetic induction. Machines which generate electricity all make use of the basic principle that, when relative motion between a conductor and a magnetic field takes place, it creates an electromotive force within the conductor which produces a flow of current. The relative motion is obtained either by passing the conductor through the magnetic field or by moving the magnetic field across the conductor.

The direction of a magnetic field or its flux lines, as they are sometimes referred to, between the two ends of a permanent magnet is assumed to be from the north to the south pole. The direction in which the current flows depends on the direction of relative movement between the conductor and field – thus, if a conductor moves backwards and forwards across a magnetic field, the direction of current flow in the conductor will likewise alternate.

Generator machines are broadly classified under two headings:

i) Dynamos, in which the generated alternating current is rectified by a mechanical switch known as the commutator to give a direct-current (d.c.) output.
ii) Alternators, in which the generated current is continually reversing its direction of flow, hence these are known as alternating-current (a.c.) machines. The a.c. generated in this form of machine is converted to a direct current by means of diode rectifiers.

27.3 Alternators

27.3.1 Principle of operation

The principle of the alternator is illustrated by fig. 27.7(a) and (b), which show a single conductor loop passing through the pole-pieces of a soft-iron horseshoe-shaped yoke, its open ends forming the leads for the external circuit (in this case a light-bulb). Pivoting in between the pole-pieces is

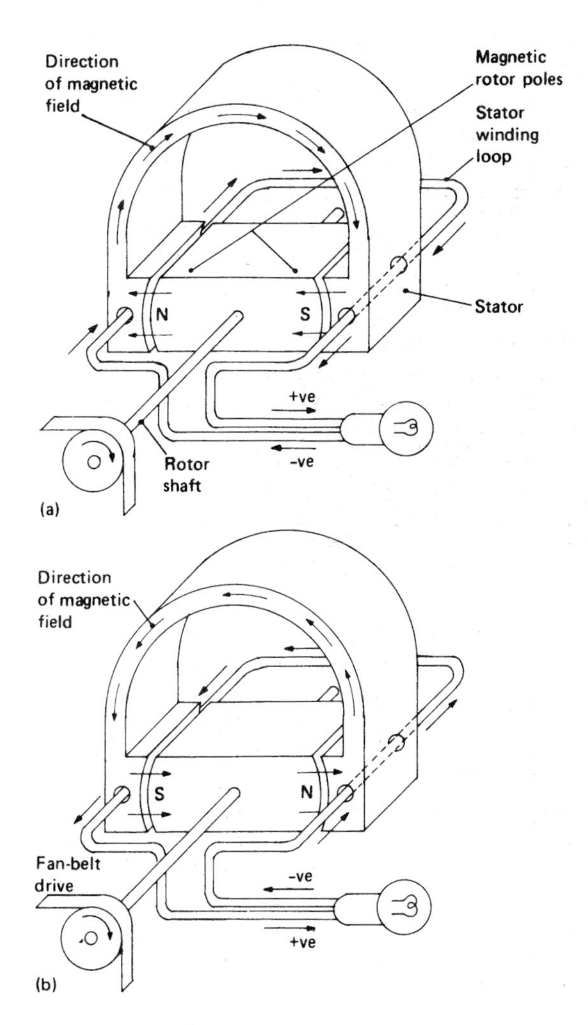

Fig. 27.7 Basic single-loop alternator

a permanent magnet, which establishes a magnetic field (lines of flux) around the yoke. When the fan belt drives the pulley wheel, the permanent magnet rotates on its axis so that the magnetic field in the yoke will be continually in a state of growth and decay as the orientation of the magnet changes relative to the pole-pieces. Consequently, as they expand or contract with the frequency of the magnet's rotation, the magnetic-flux lines will constantly be cutting through the two half-conductors of the loop.

Observing fig. 27.7(a) and (b), it can be seen that, whenever the two magnet poles are adjacent to the yoke poles, the maximum flux lines cut through the two half-conductors so that a flow of current will be created in the conductor loop by the induced electromotive force (e.m.f.).

Figure 27.7(a) shows the magnet rotating clockwise with its north pole on the left and the south pole on the right-hand side of the iron yoke, thereby making the lines of flux circulate round the yoke from the north to the south pole in a clockwise direction. At the same time, due to the movement of the magnet, the lines of flux will cut through the conductors and the induced voltage will produce a flow of current in the conductor loop in a clockwise direction.

If the magnet is now revolved a further half revolution (fig. 27.7(b)), the position of the magnet poles will have reversed, with the north pole now being on the right of the yoke and the south pole on the left-hand side. Consequently the direction of the lines of flux round the yoke will be in an anticlockwise direction. This reverses the generated flow of current so that it also moves in an anticlockwise direction.

Thus the rotation of the magnet causes the yoke poles to continually change their north and south polarities, hence the direction of the flux-path lines is constantly being reversed and therefore the current in the conductors is continuously changing between a maximum value in one direction and a maximum value in the opposite direction.

A current which repeatedly changes its direction of flow is known as an alternating current (a.c.), and with a two-pole magnet the change of direction occurs once every complete revolution of the magnet, the output produced by one complete revolution being known as an alternating-current cycle.

27.3.2 Construction

Practical alternators incorporate many conductor windings around a ring-shaped yoke, these commonly being referred to as the stator windings and the stator yoke (fig. 27.8(a)). Also, to reduce the voltage fluctuation even more, the rotor is made in two halves, each half having several segment poles of like polarity so that when they are fitted together they form a ring of alternating north and south poles (fig. 27.8(b)).

27.3.3 Generating an alternative voltage and current (fig. 27.9)

Imagine a single conductor loop with its two parallel ends embedded diametrically opposite each other in the cylindrical stator and a rotating bar-magnet with its two poles supported on a spindle. At any instant the bar-magnet flux path from N to S is completed via the air gaps and through the cylindrical stator (see fig. 27.9). A voltage is gen-

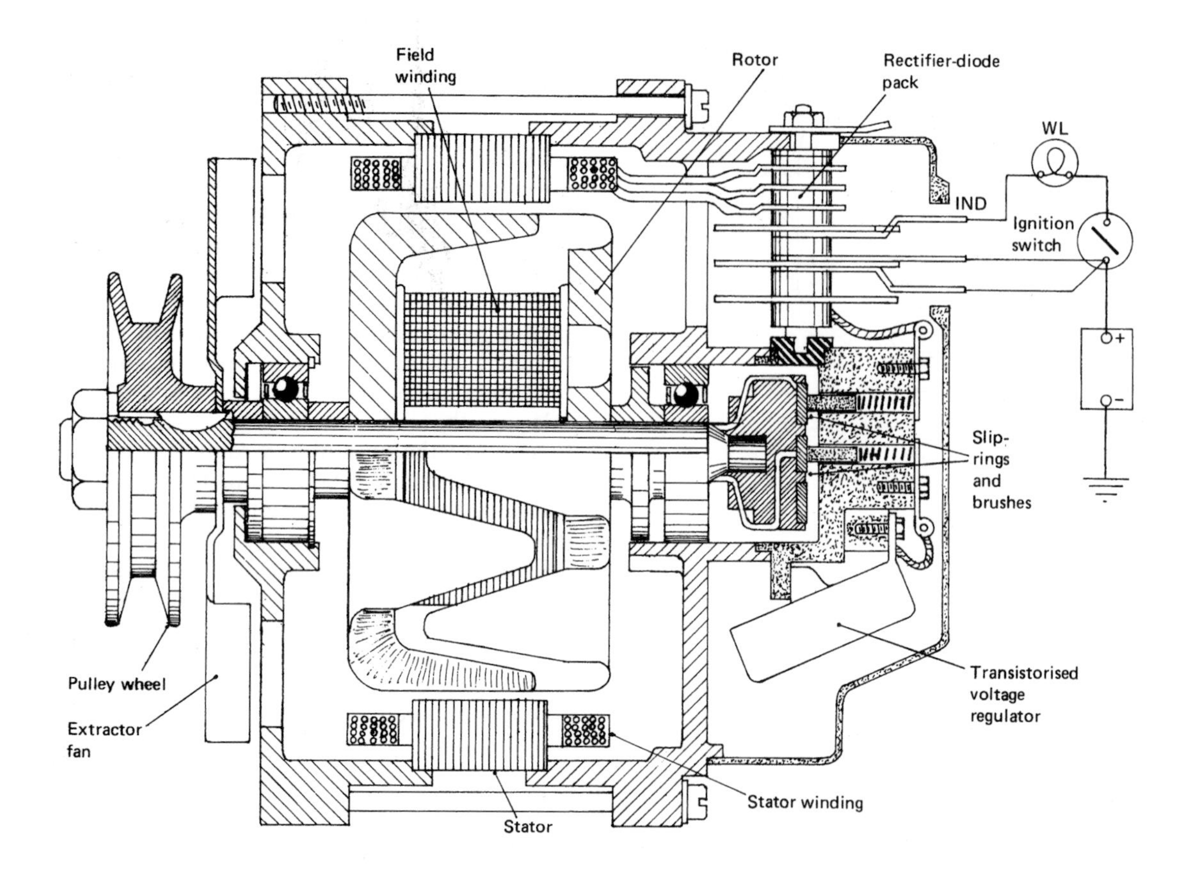

(a) Section view

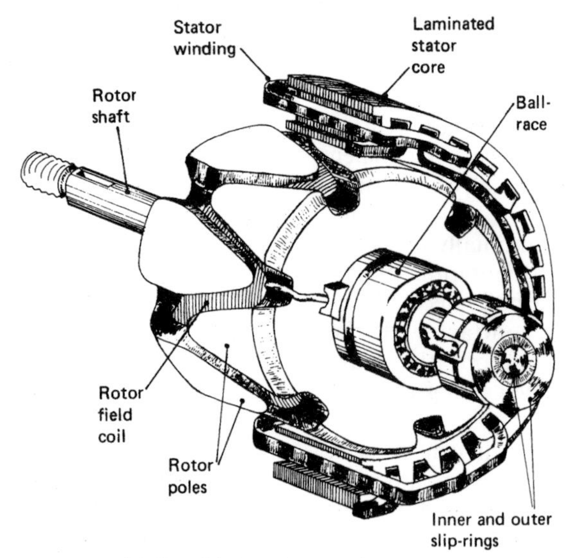

(b) Pictorial view of rotor and stator

Fig. 27.8 Alternator

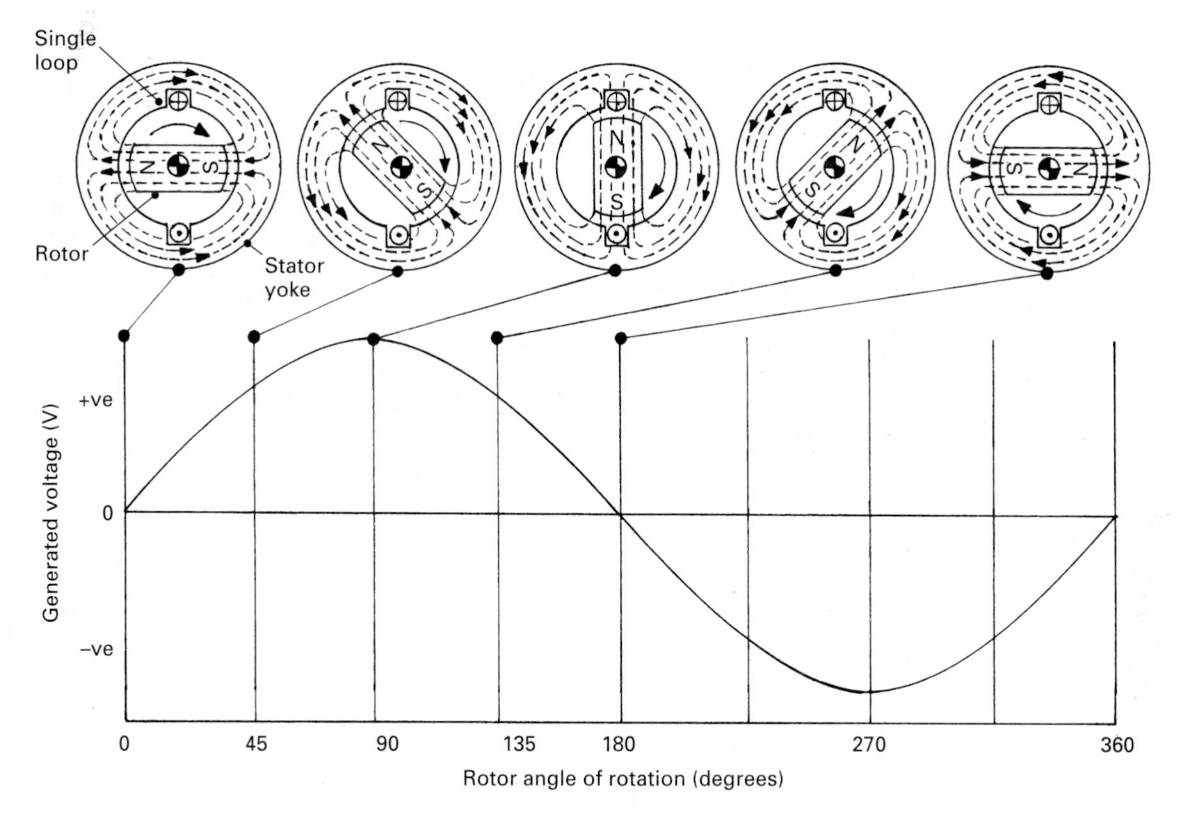

Fig. 27.9 Generating voltage cycle of a simple single-loop two-pole rotor alternator

erated in a conductor when magnetic lines of force are made to cut across or through the conductor. When the direction of the magnetic field is at right angles (zero angular position) to the top and bottom conductor ends, the magnetic flux (lines of force) path passes though the cylindrical stator but it does not cut across the conductor, consequently there is no voltage generated (see graph in fig. 27.9). As the rotor poles approach the upper and lower parallel loop conductors they progressively come under the influence of the field force and at the same time the field force commences to cut through the conductors. The maximum lines of force cutting through the conductor at any instant occur when the rotary poles are parallel (adjacent at 90° angular position) with the conductors; thus in this position the induced or generated voltage will also be at its peak. At 45° and 135° angular movement of the rotor, the lines of force cut the upper and lower parallel conductors at an inclination so that there will be fewer lines of force being cut per degree of rotation compared with the 90° angular position when the rotor poles align with the conductors. Accordingly the induced voltage will be less than the possible maximum. If the voltage produced is measured at, say, every 10° of rotor rotation for one complete revolution, it would be seen that the shape of the alternating voltage curve conforms very nearly to a sinusoidal wave.

27.3.4 Circuit rectification

The alternating current generated in the alternator would be quite adequate for illuminating light bulbs but it is unsuitable for recharging batteries, which require a direct current source of supply to convert the lead sulphated plates back to their active state of spongy lead or lead peroxide. This difficulty is overcome by converting the alternating current to a direct current by incorporating solid state rectifier diodes in series with the output leads.

These semiconductor rectifiers permit the generated current to pass through in one direction but block the flow of current on the other half-cycle when the direction is reversed. This is an inefficient method of rectification, as only one half-cycle provides useful power, and therefore full-wave rectification using what is known as a four-diode bridge circuit is commonly employed. Diodes will allow current to pass in one direction only and are represented by a triangle with a perpendicular line on its nose which indicates the direction in which current will be allowed to pass.

27.3.5 Single-phase bridge circuit full-wave rectification (fig. 27.10(a)–(c))

The diodes are so arranged that the alternating voltage generated in the alternator stator winding produces a likewise alternating current which when passed through the diode bridge circuit is rectified to produce a direct current suitable for charging the battery.

For the positive half-wave, current from the battery negative terminal passes through the lower right-hand negative power diode, through the upper and then the lower stator winding, and then passes through the upper left-hand positive power diode to the positive battery terminal (see fig. 27.10(a)). Thus the generated charging voltage opposes the battery's voltage potential and pumps current into the positive terminal and takes current out at the negative terminal.

When the rotor has revolved a half revolution, the direction of current flow reverses this being the negative half-cycle (see fig. 27.10(b)). Current again flows from the battery's negative terminal, but this time it passes through the lower left-hand negative power diode, through the lower and then the upper stator winding and out through the right-hand positive power diode to the positive terminal of the battery.

The generated alternative voltage, the two separately rectified half-cycles, and the combined rectified half-cycle are shown in fig. 27.10(c).

27.3.6 Three-phase alternator (fig. 27.11)

A major deficiency of the single-phase stator-winding alternator with full-wave rectification is that the resultant positive half-wave voltage generated still fluctuates between zero and its peak value every 90° rotation of the rotor. A three-phase alternator reduces this voltage variation by having two additional generated voltage cycles. These three similarly generated voltage half-waves are then so arranged that they are 120° out of phase with each other. Consequently, the three positive half-waves overlap each other so that the variation between maximum and minimum generated voltage with respect to the angular movement of the rotor is much reduced.

Stator and windings (fig.27.11) A basic three-phase alternator has three identical and separate windings X, Y, Z which are wound around the three evenly spaced (120° interval) stator poles which are attached to the cylindrical-shaped yoke. Instead of having six conducting wires, one for each end of a winding, the wires from one end are usually connected together to form what is known as a star connection, thereby reducing the external connecting wire leads to three. Each of these wire leads is connected between a positive and negative power diode of the rectifier's bridge circuit.

Rotor and winding assembly (fig. 27.11) A 12-claw pole rotor is commonly used and is illustrated in fig. 27.11. It consists of two opposing six-claw pole halves which when assembled on the rotor shaft intermesh, the bent-over fingers or claws from each half have similar polarities, so that the N and S poles alternate. Wound around the centre and between the half-pole pieces is an excitation field winding, its ends being connected to a pair of insulated copper slip-rings which are also mounted on the rotor shaft. Current is taken to and from the slip-rings via a pair of spring-loaded carbon brushes, one acting as an earth-return while the other is connected to the three exciter diodes and battery via the voltage regulator and the warning light.

27.3.7 Full-wave voltage and current rectification of a three-phase alternator (fig. 27.11)

By having three identical and independent windings X, Y, Z arranged at equal intervals around the stator, a rotating two-pole magnet rotor would produce three alternating sine-waves which are 120° out of phase with each other. Subsequently when one generated voltage wave is declining another one is increasing and since there is an overlap of 120° between each winding phase the resultant voltage is never permitted to drop to zero. As with the single-phase alternator, the three-phase alternator generates an alternating voltage and current which can be converted into a direct voltage and current by incorporating a three-phase full-wave diode bridge circuit.

27.3.8 Full-wave bridge circuit rectification for each winding phase (fig. 27.12(a)–(c))

X winding generating (fig. 27.12(a)) When the rotor pole aligns with the X winding, a voltage will be induced into the winding; first in one

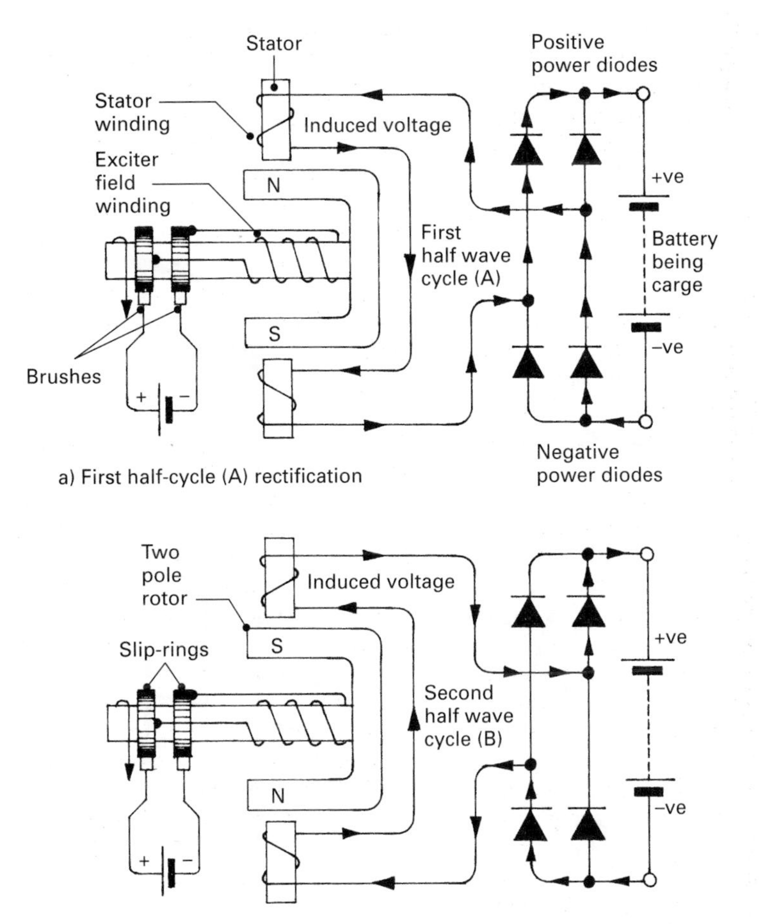

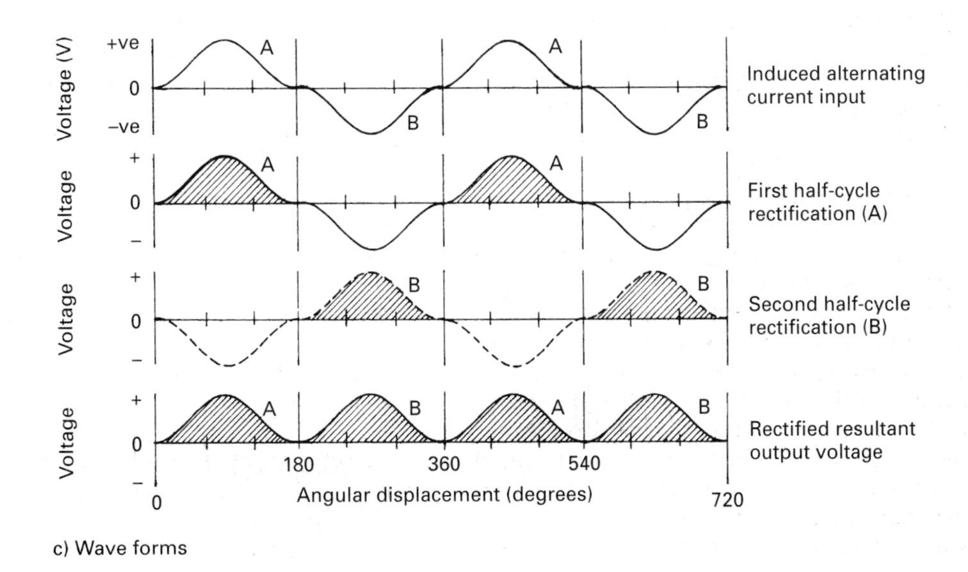

Fig. 27.10 Single-phase bridge circuit full wave rectification of positive and negative half-waves

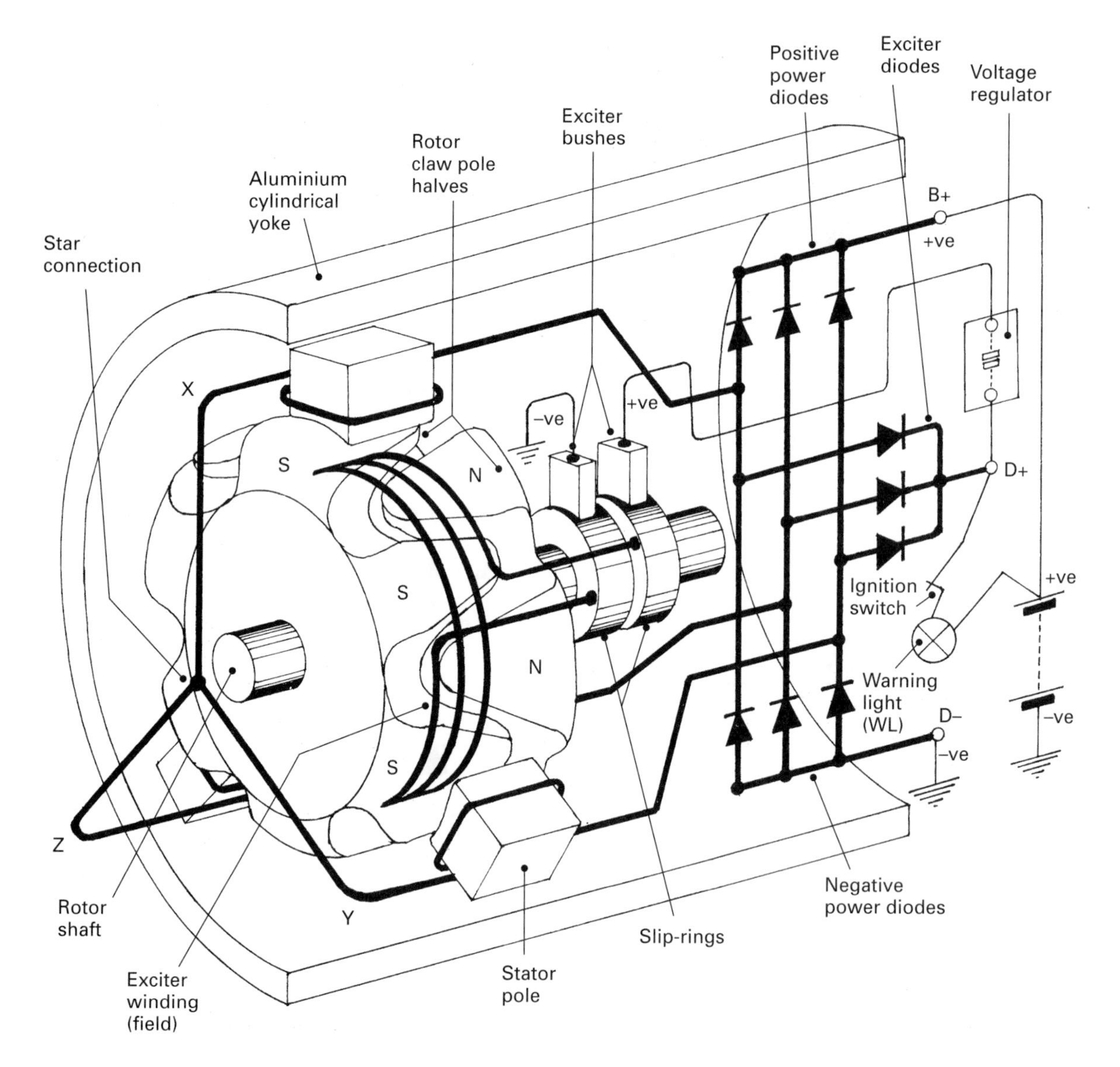

Fig. 27.11 Simplified construction of a three-phase claw-pole alternator with slip-rings and solid state voltage rectification

direction and then in the opposite direction. For the positive half-cycle current flows (full arrows) from the negative terminal through the lower middle negative power diode. It then passes though the Y and then the X windings, and finally passes out through the upper right-hand positive power diode before arriving at the positive terminal. Conversely when the current reverses its direction of flow, the negative half-cycle (broken arrows) moves from the negative terminal through the lower right-hand negative diode through the X and then the Y windings, and finally passes through the upper middle positive diode to the positive terminal. The upper graph in fig. 27.12(a) shows the complete alternator voltage sine-wave whereas the lower graph shows the rectified two half-waves.

Y winding generating (fig. 27.12(b)) As the rotor pole moves around to the Y winding, voltage will be induced into the winding and the corresponding positive half-cycle current will now flow (full arrows) from the negative terminal, through the lower left-hand negative power diode, through the

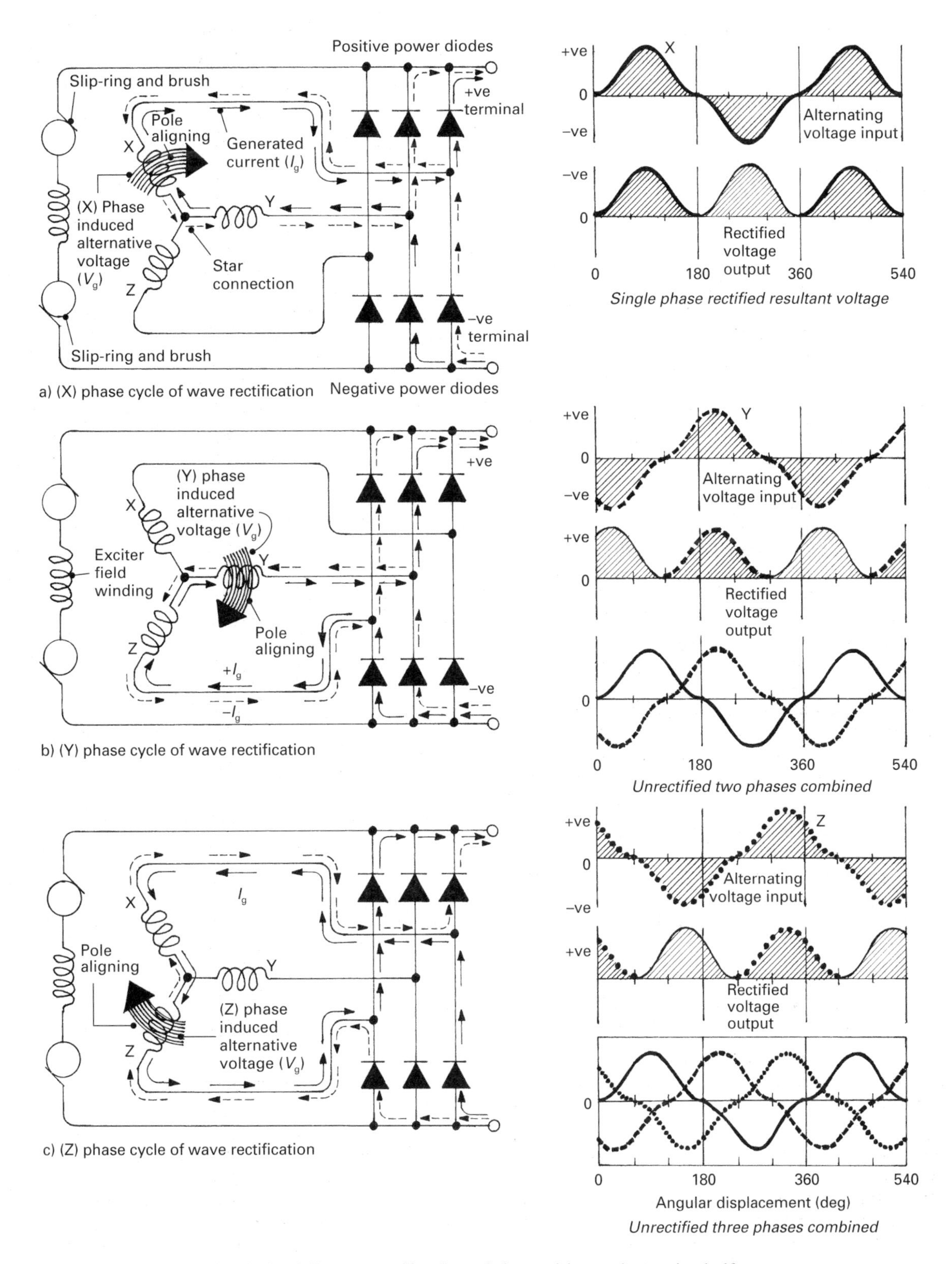

Fig. 27.12 Three phase circuit for full wave rectification of the positive and negative half-waves

Z and then the Y windings and out through the upper middle positive power diode to the positive terminal. The current then reverses its direction of flow, the negative half-cycle current (broken arrows) moves from the negative terminal though the lower middle negative power diode through the Y and then the Z windings, and finally it passes through the upper left-hand positive power diode to the positive terminal.

The alternating voltage sine-wave generated and the rectified negative half-cycle is identical to that for the X winding phase (see fig. 27.12(b)) but the Y winding voltage phase is 120° out of phase with that of the X winding. The lower graph in fig. 27.12(b) shows both the X and Y winding alternating voltage wave formations together on the same axis.

Z winding generating (fig. 27.12(c)) Further rotation of the rotor now brings the rotor pole adjacent with the Z winding. The induced voltage for the positive half-cycle produces a current (full arrows) which flows from the negative terminal, through the right-hand negative power diode, through the X and then the Z windings, and then comes out to the positive terminal via the upper left-hand negative power diode. During the reverse current flow, the negative half-cycle current (broken arrows) moves from the negative terminal through the lower left-hand negative power diode, through the Z and then the X windings where it then passes through the upper right-hand positive power diode on its way to the positive terminal. Again identical generated voltage sine-waves are produced and rectified (see graphs in fig. 27.12(c)), but the phasing of the Z winding is this time off-set 120 ° and 240° from the X and Y winding phases respectively. The lower graph in fig. 27.12(c) shows the X, Y and Z winding voltage

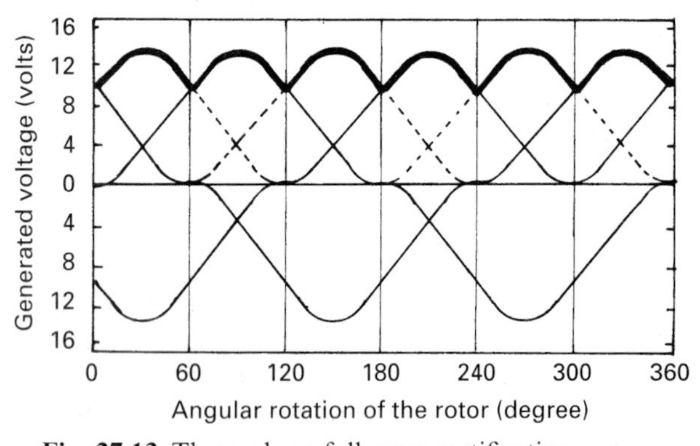

Fig. 27.13 Three-phase full-wave rectification

phases together on the same angular displacement axis. It can be see that the resultant voltage is much more uniform. Likewise the rectification of all three phases collectively produces a very consistent direct voltage supply (see fig. 27.13) with only a moderate small ripple in the voltage supply.

27.3.9 Alternator design considerations (fig. 27.14)
The 12-pole claw rotor which is commonly used requires a 36-slot stator. This means the angular distance or pitch between slots is 360/36 = 10°. The three windings are separately wound from side to side in a wave formation around the stator. Each phase winding is off-set by a slot pitch so that the parallel adjacent conductor repeating sequences will be X Y Z, X Y Z, etc. Thus if the slot pitch is 10°, then three slots, one for each winding, will span an angular width of 10 × 3 = 30°. Likewise the width of each rotor pole to match the angular spacing for three stator slots will be 360/12 = 30°. To complete one generated voltage positive and negative half-wave in one cycle, an adjacent pair of N and S poles will have to sweep past three stator slots which hold the X Y Z windings through an angle of 60°. Therefore in one revolution of the rotor there will be 360/60 = 6 alternating cycles completed. In contrast a two-pole rotor in one revolution will only complete one alternating voltage cycle.

The magnitude of a generated voltage is proportional to the flux density, the speed of cutting the lines of force, and the length of the conductors subjected to this flux field. Consequently to increase the voltage generated the length of each winding is extended by threading nine conductors though each slot (see fig. 27.15) instead of using a single loop as shown for simplicity in fig. 27.14. The flux field path between adjacent pairs of N and S rotor poles is completed by encircling each stator slot as shown in fig. 27.15. This arrangement therefore concentrates the flux field around the winding conductors so that the maximum number of flux lines of force cut through the conductors when the rotor is turning. A cross-sectional view of a modern alternator is shown in fig. 27.16.

27.3.10 Field winding excitation (figs 27.11 and 27.17)
The simple alternator with a permanent magnet rotor increases its generated voltage almost directly as the rotational speed of the rotor. This is because the magnitude of the induced voltage

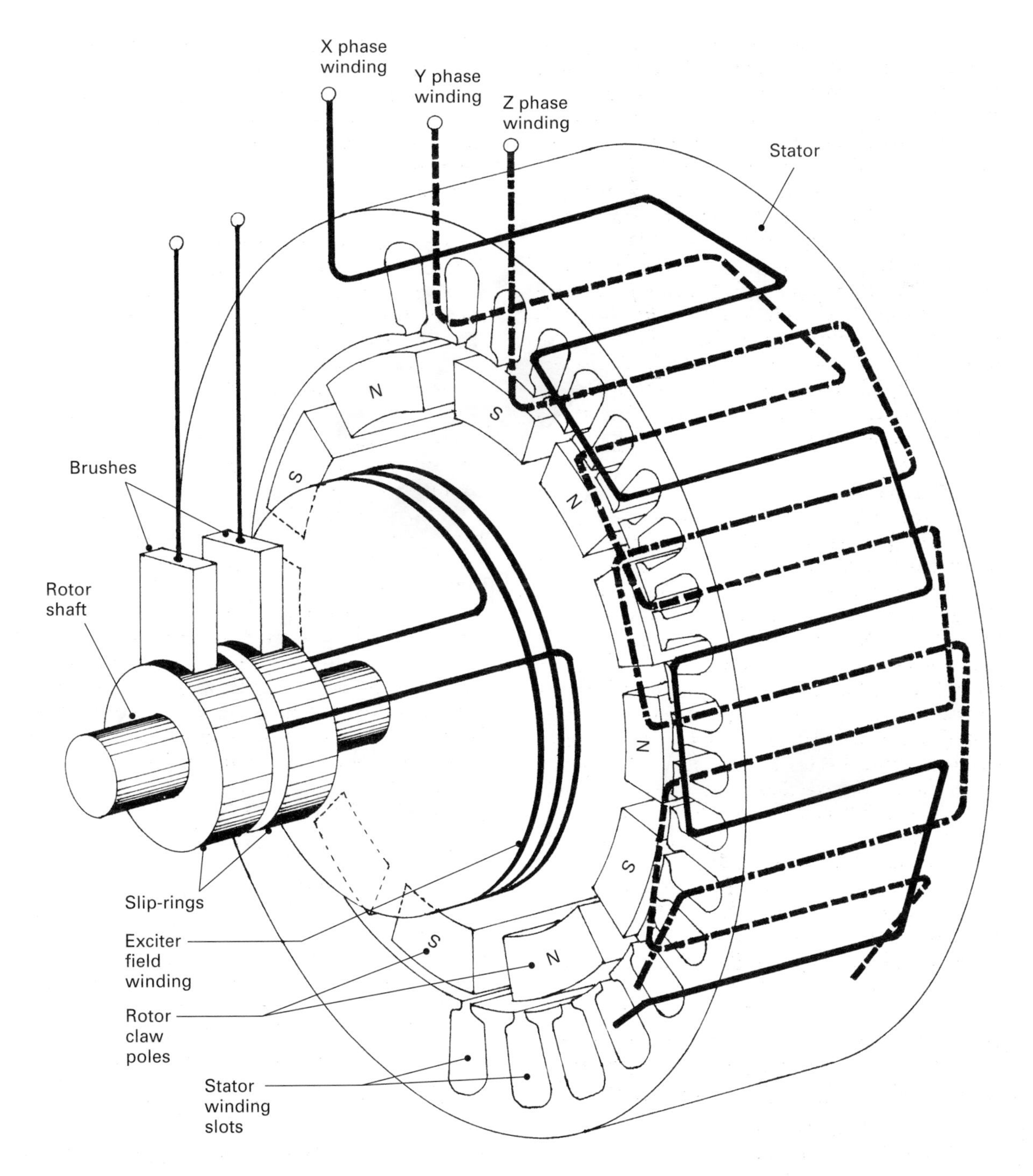

Fig. 27.14 Three-phase stator and rotor winding formation

into the stator windings is proportional to the flux density and to the rate at which the magnetic flux lines of force cut through the conducting windings. Thus with this type of alternator there would be no output control of the generated voltage so that eventually, if the speed was continually permitted to rise, it would overload and burn out the windings. In addition, the charging voltage applied to the battery and the voltage supplied to the rest of the vehicle's electrical equipment

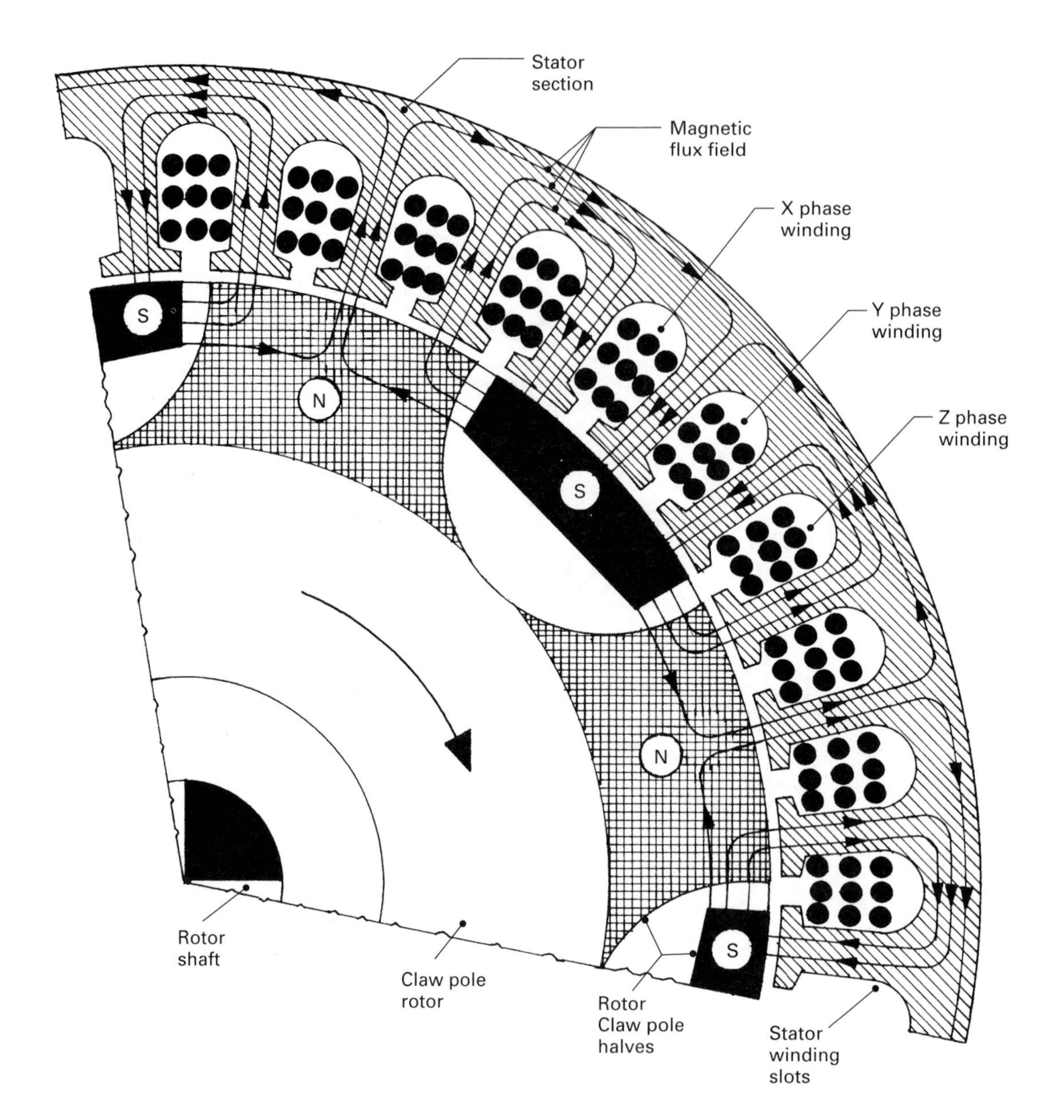

Fig. 27.15 Section and view of three-phase stator and windings with rotor poles

must be strictly controlled to avoid overcharging the battery, blowing the light bulbs and damaging other electrical components.

The output of the alternator can be regulated by replacing the permanent magnet rotor with a soft iron rotor and incorporating a field winding, sometimes known as the excitation winding, around the core of the rotor, this winding being energised initially by the battery and then by the three exciter diodes via a pair of slip-rings and brushes. When there is no current flowing through the field winding there is no magnetic flux field and therefore no voltage is generated when the rotor rotates. However if current is supplied to the field winding it excites, that is, it establishes a magnetic field around, the rotor poles so that lines of force will be formed between the rotor poles and the stator and its windings. If the current supplied is increased, the generated voltage will also rise due to a corresponding increase in the strength of the flux field. Conversely it will decrease as the flux field weakens due to a reduc-

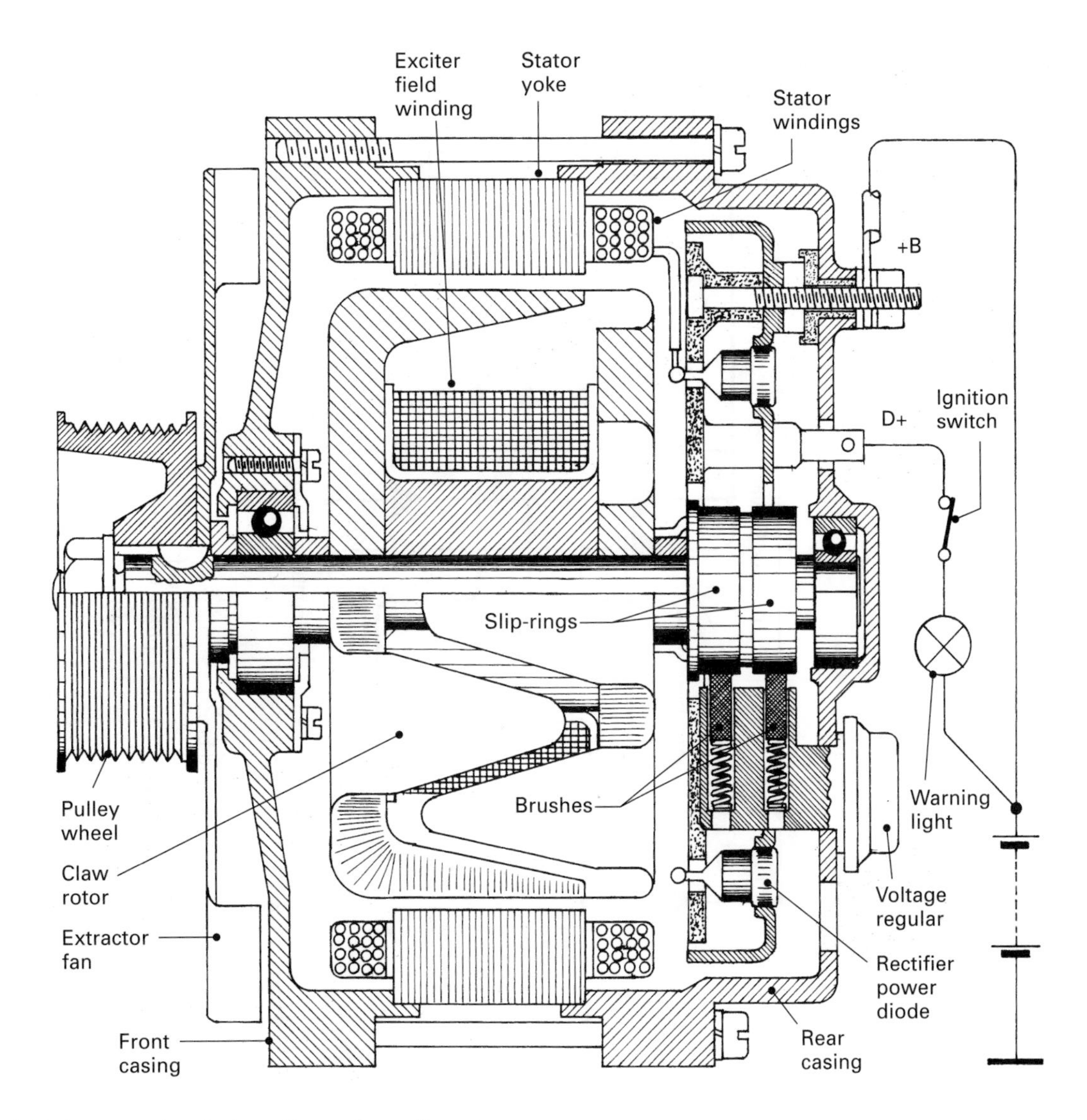

Fig. 27.16 Cross sectional view of a modern alternator

tion in current flow through the field winding. Thus it is the field current which controls the magnitude of the magnetic excitation of the rotor poles and hence the density of the flux path surrounding the stator windings.

In actual practice the field current is regulated by the make and break of the field winding circuit which is continuously interrupted by the voltage-regulator's power transistor switching action.

Initially to excite the field winding, pre-excitation current is supplied by the battery's positive terminal via the warning-light filament through the voltage-regulator to the positive slip-ring brush. It then passes through the field winding and is returned to the battery's negative terminal by way of the earthed slip-ring brush. Once the rotational speed of the rotor is sufficient for the alternator to generate its own current, the field current will be fed by the three exciter diodes.

27.3.11 Warning light (figs 27.11 and 27.17)
When the ignition switch is turned on current from the positive battery terminal flows through the warning-light bulb, the voltage-regulator, positive slip-ring brush, and through the field exciter winding where it then returns to the battery's negative terminal via the negative earth return slip-ring brush. The warning-light bulb

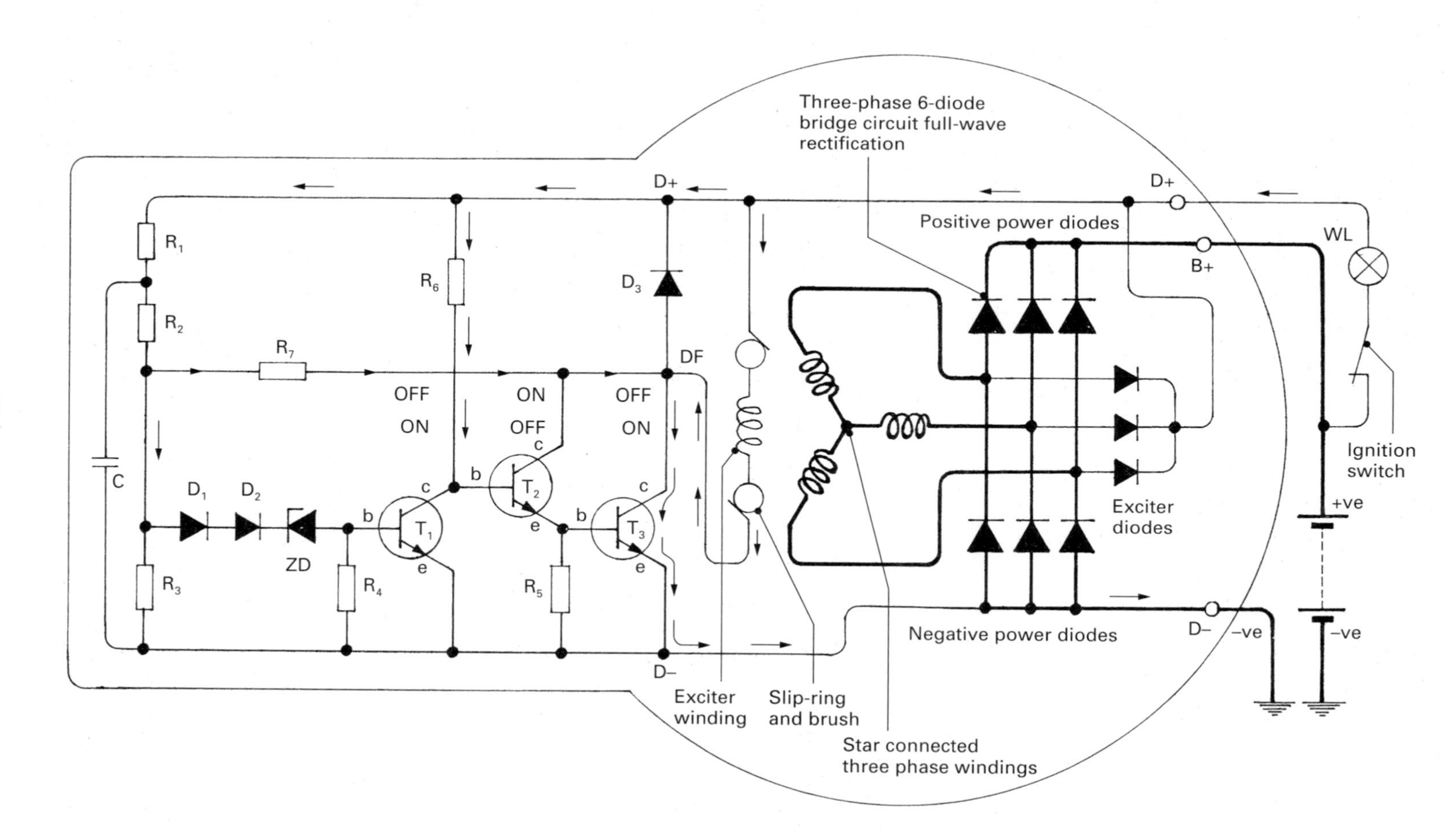

Fig. 27.17 Voltage regulator with 9 diodes and self-excited rotor field circuit (Bosch EE 14 V3)

filament resistance limits the circulation of the field current but this is sufficient to excite and start off the self-generating process of the alternator as the engine is being cranked.

Until the warning-light goes out, the battery provides the pre-excitation current to the field winding. Once the warning light is extinguished the alternator commences to charge the battery and to supply current to any of the other electrical components switched on. This normally occurs below idling speed as the engine is being started. The reason for the warning light to go out is that the alternator-generated voltage builds up and opposes the battery voltage until they are of equal potential, or the generated voltage is only one or two volts above that of the battery, so that there is very little current passing though the warning-light bulb to illuminate it. Thus the warning light should always be extinguished above idling speed if the alternator is operating correctly.

27.3.12 Reverse current switch off (figs 27.11 and 27.17)
When the engine is starting the battery is prevented from discharging through the three-phase stator-windings as the rectifying power diodes are polarised with respect to the battery current in the reverse direction. Thus the diodes conduct in the forward direction as indicated by the arrow symbol pointing towards the positive battery terminal, whereas in the reverse direction the diodes block. Consequently the generated current can flow from the alternator to the battery but the battery cannot be discharged by current being drained through the shunted stator windings which are connected across the battery terminals.

27.4 Voltage regulator

27.4.1 The need for voltage regulation
An alternator voltage-speed characteristic under no load operating conditions is a relatively linear one. Thus as its rotational speed increases so does its generated output voltage. Therefore it is necessary to regulate its output to protect its windings and to match its output supply to the battery's charging needs and the electrical equipment of the vehicle. Current limitation is achieved by the inherent stator-winding reaction. This limits the maximum current permissible for full load (at full load the majority of excitation ampere-turns are dissipated by the stator-winding reaction, and only a small proportion of the ampere-turns is effective). At high speed a weak magnetic flux is sufficient for generating the necessary terminal voltage. The voltage-regulator controls the voltage output of the alternator by switching 'on' and 'off' the excitation current flowing through the field winding circuit once the output voltage of the alternator reaches some predetermined maximum regulated voltage. At all speeds below the regulating voltage, the exciting current is switched on. However the alternator usually commences to generate its own voltage slightly below the engine's idling speed. The duration of the on-off switching phases is extremely short; it may be a few milliseconds. It depends upon the alternator speed and the external electrical load, that is, the battery's state of charge, the number of lamp bulbs, heaters or radio switched on and other electrical equipment installed in the vehicle which may be consuming current. A switching rate of 1000 cycles per second could be exceeded under certain operating conditions. When the alternator-generated voltage reaches the upper limit of the regulating voltage, the field excitation current flow is interrupted. This reduces the magnetic excitation of the alternator and thus the generated voltage. As the generator voltage drops to the lower regulating limit, the excitation current again commences to flow, the alternator magnetic excitation rises, as does the generating voltage, and the cycle of events repeats itself. Thus the rapid opening and closing of the field winding circuit controls the mean excitation current which in turn determines the generated output voltage of the alternator. This excitation field current normally does not exceed 3 A. If the speed is low and the load is relatively large the duration of the 'on' phase will be long and that of the 'off' phase will be short; conversely, if the speed is high and the electrical load is small the duration of the 'on' phase will be short, whereas the 'off' phase will be long.

27.4.2 Voltage regulator with rectifying diodes and self-excited rotor field circuit (Bosch EE 14 V3) (fig. 27.17)

Basic components (fig. 27.17) The voltage-regulator consists of a control transistor T_1 and a pair of output transistors T_2 and T_3 which control the field winding's magnetic excitation by periodically interrupting the field current flow. A unidirectional breakdown Zener-type diode Z_D is used to activate the control transistor at the set regulator voltage. Resistors R_1, R_2 and R_3 form a voltage divider which sets the proportion of generated voltage feed to the base of transistor T_1. Diodes D_1 and D_2 provide temperature compensation which adjusts the regulating to a limited extent to match the ambient temperature, whereas diode D_3 protects the power transistor T_3 from high voltage surges each time the field-winding current is interrupted. Capacitor C is incorporated to smooth out the rectified direct current voltage ripple.

Generated voltage build-up, output power transistor switched on (fig. 27.17) When the ignition is switched on, current flows from the positive battery terminal to the positive terminal of the regulator D_+, through resistances R_1, R_2 and R_3 and returns to the battery's negative terminal via the regulator terminal D_-. Similarly current flows from the exciter diodes via resistor R_6 to the base 'b' of transistor T_2. This switches on its collector-emitter (c-e) circuit, and in so doing feeds current to the base 'b' of transistor T_3 which also now switches on its collector-emitter (c-e) circuit. The excitation current now flows through the collector-emitter (c-e) circuit of T_3 and through the field winding. As a result it increases the magnetic excitation of the alternator, thereby raising the generated voltage; likewise the voltage across the voltage divider and the Zener diode Z_D will increase.

Generated voltage reaches its regulating output, power transistor switches off (fig. 27.17) When the generated voltage V_g across the Zener diode reaches its breakdown voltage V_{zd} the Zener diode commences to conduct in the reverse direction. Current now flows via terminal D_+, through resistors R_1 and R_2 and the Zener diode to the base-emitter (b-e) of transistor T_1. The collector-emitter (b-e) circuit of T_1 now conducts so that the base 'b' of transistor T_2 is earthed and its voltage reduced to zero. The collector-emitter (c-e) circuit of transistor T_2 therefore ceases to conduct. As a result of transistor T_2 blocking, the current fed to transistor T_3 ceases, causing it also to block. Instantly the excitation current is interrupted, thus reducing the magnetic excitation of the alternator, and likewise reducing the generated voltage. Once the generated voltage has decreased to below the Zener diode breakdown voltage (regulating voltage) transistor T_1 switches 'off' and transistors T_2 and T_3 switch 'on', which again permits the exciting magnetic field to build up. This cycle of opening and closing the exciting field circuit continuously repeats itself. Every time the field winding circuit is abruptly interrupted, the collapse of its magnetic flux field induces a high voltage in the field winding. This would be applied across the collector-emitter (c-e) circuit of transistor T_3 and would therefore completely destroy it. To overcome this problem diode D_3 is shunted across the field winding so that the induced voltage caused by the sudden opening of the field circuit provides an easier conducting path for this voltage to discharge into.

27.4.3 Checking alternator charging system

Check belt drive Check alternator drive belt tension and pulley-rotor shaft attachment.

Check battery Check the battery's state of charge. If the battery is discharged temporarily replace the battery with a charged one before carrying out electrical tests.

Check warning-light circuit (figs 27.11 and 27.17) Turn on ignition switch. The warning-light bulb should be illuminated. If the bulb does not light, disconnect the D_+ lead from the alternator and connect it via a test lead to a good earth point. The bulb should now light, but if it still does not light up suspect a blown bulb. This should therefore be replaced. Re-attach the D_+ lead to the alternator terminal. However, if the bulb does light up when the D_+ lead is earthed, suspect a defective voltage-regulator or exciter field winding and brushes. Remove and examine the alternator for defective field windings or brushes and if these are not at fault replace the voltage-regulator with a new one.

Check if alternator is charging (figs 27.11 and 27.17) If the warning-light bulb does come on with the ignition switch turned on, start the engine and observe the warning light as the engine speed rises to its idling speed. A correctly operating charging system will extinguish the warning light just before the engine reaches idling speed when being started. However, if the warning-light bulb is on but is dim, flickers or is bright suspect an internal fault in the alternator.

Check if rectified diodes are functioning (figs 27.11, 27.16 and 27.17) With the engine idling increase the engine speed to 1500 rev/min and then electrically load the alternator by switching on both the side and head lights, the heater blower motor and the rear window heater and observe the warning light. If the charging system is functioning correctly the warning light will not come on. However, if the warning light does show a dim light the alternator output is insufficient to cope with the electrical load and the following test should be made. Idle the engine and connect a voltmeter between the B_+ and D_- terminals. If the reading is below 1/2 volt, the rectifier diodes should be in good order. However, if the reading is above 1/2 volt, suspect a faulty diode or diodes or a bad connection in the diode bridge circuit. Remove, dismantle and examine alternator and test diodes for continuity.

Check regulator (figs 27.11, 27.16 and 27.17) Start the engine and increase its speed to 1500 rev/min. Apply electrical load by switching on all the lights, heater blower motor, rear window heater, radio, etc., and observe the warning light which should be extinguished. Connect a voltmeter between the B_+ terminal and a good earthing point and measure the voltage reading. This should be within a voltage range of 13–15 volts. If the reading is higher than 15.5 volts suspect a defective voltage-regulator and replace it; conversely, a voltage reading of below 13 volts indicates a faulty alternator. Remove the alternator, dismantle and test the continuity of the stator windings, field winding, slip-ring and brushes and rectifier diode pack.

27.4.4 Care in fitting alternators
When fitting or removing a battery or alternator, care must be taken to ensure that the leads are connected up the right way round first time – otherwise the alternator rectifier diodes may be damaged when the engine is started.

27.5 Dynamos
These machines operate on the same principle as the alternator but, instead of the magnetic field rotating with the conductor stationary as in the case of the alternator, the dynamo is so arranged that the magnetic field is stationary and the conductors are made to rotate.

A simple dynamo (fig. 27.18) might consist of a horseshoe magnet with two poles formed at its open end. Mounted between these poles on a support spindle is a single conductor loop with its ends joined to the two half split-ring segments. Two carbon brushes are arranged on either side of the ring segments to transfer the generated and rectified direct current from the split segments to the external circuit to be supplied with electricity.

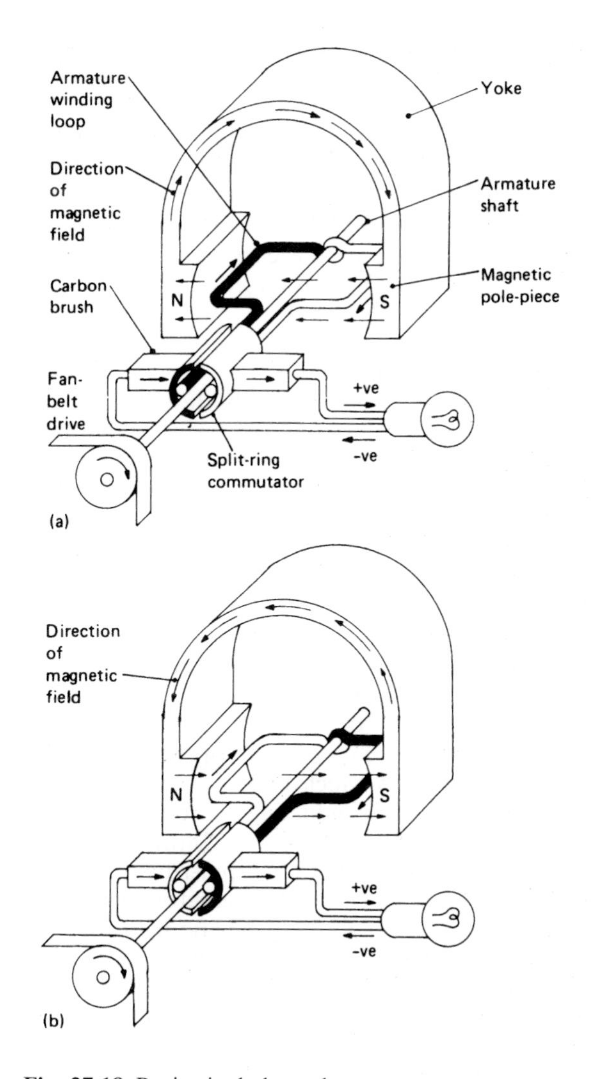

Fig. 27.18 Basic single-loop dynamo

27.5.1 Current generation
As can be seen from fig. 27.18(a), the rotating conductor loop is under the influence of the magnetic flux established by the magnet poles. As the two half-conductors cut through this flux, an electromotive force will be induced into the conductors, producing a flow of current.

It can be seen that the direction of current flow in the black half-conductor adjacent to the north pole is towards the rear of the loop, and that in the white half-conductor next to the south pole is towards the front. Current therefore flows in a clockwise direction through the loop, coming out at the right-hand split-ring and brush and returning by the left-hand brush and ring segment.

If the conducting loop rotates a further half-revolution (fig. 27.18(b)), the black half-conductor will be next to the south pole while the white half-conductor will be adjacent to the north pole. Thus the direction of the current flow in each half-conductor has been reversed.

27.5.2 Commutation

The reversal of current flow in both half-conductors as they are rotated cannot be avoided, but the resultant flow of current can be made unidirectional by a sort of switch which reverses the connections between the armature loops and the external circuit in time with the reversals of current in these loops. The name 'commutator' is given to the special form of switch used to commute or change the current flow from one kind into another – in this instance from an alternating current into a unidirectional one.

Commutation is obtained by connecting the two ends of the conductor loop to two half split-ring segments which make rubbing contact with a pair of conducting carbon brushes arranged on either side (fig. 27.18(a) and (b)). As the conductor loop and the split-ring segments rotate together, the two half-conductors will continuously pass through the magnetic fields of the north and south poles, and at the same time each half-segment changes its position and moves from one brush to another. As the direction of the current flow changes in the conductors, the segments will switch over and make contact with the opposite brush so that the direction of the current flowing to and from each individual brush remains the same. Thus it can be seen that the split-ring and brushes act as a mechanical switch to rectify the alternating current so that only a direct current is passed on to the brushes and any external circuit.

The generated direct current for a single-loop armature fluctuates from zero to a maximum and then back to zero every 180° of rotation, therefore armatures are made up from a number of evenly spaced loops and split-ring segments so that when the voltage is declining in one or more loops it is increasing in others. All the individual currents are continuously being collected by the brushes as they contact each pair of commutator segments in turn to produce a combined steady direct-current output (fig. 27.19(b)).

27.5.3 Output control (fig. 27.19(a))

With the simple dynamo in fig. 27.18 there is no control over the amount of voltage and current being generated as the armature speed is increased. Practical dynamos control their voltage and current output by varying the strength of the magnetic field established between the poles. This is achieved by changing the permanent-magnet poles for soft-iron ones and then winding coils known as the field windings around each pole-piece so that they become electromagnets (fig. 27.19(a)). Regulation is then provided by restricting and varying the amount of current passing through these windings.

27.5.4 Cut-out (fig. 27.20)

If the generator and battery were permanently connected together, the battery would discharge through the generator armature to earth when the generator was either stationary or when its generated output voltage was less than that of the battery. It is therefore necessary to provide some sort of switch which disconnects the battery from the generator and reconnects it when the generated voltage exceeds the battery voltage.

The switch arrangement used to open and close the battery-to-generator circuit is known as the cut-out. It consists of a pair of contacts operated by an electromagnet relay which senses any change of generated voltage and automatically breaks or joins up the battery to the generator circuit.

The cut-out unit consists of an electromagnet relay having two bobbin windings, one connected in shunt (parallel) with the generator output and the other in series with the battery and the generator (fig. 27.20). The series winding is open-circuited by the contacts, which are held open by spring tension. When the generator speed rises and its output voltage rises above that of the battery, the magnetism of the shunt winding becomes adequate to overcome spring tension and closes the contacts. Charging current now flows from the generator to the battery through the series winding, and extra magnetism will now be created to assist that of the shunt winding to hold the contacts together.

If the speed of the generator falls so that its voltage is less than the battery voltage, current will begin to flow in the reverse direction through the series winding – that is, from the battery to the generator. The magnetism of the series winding now opposes that of the shunt winding and partially cancels its magnetic field – this allows the

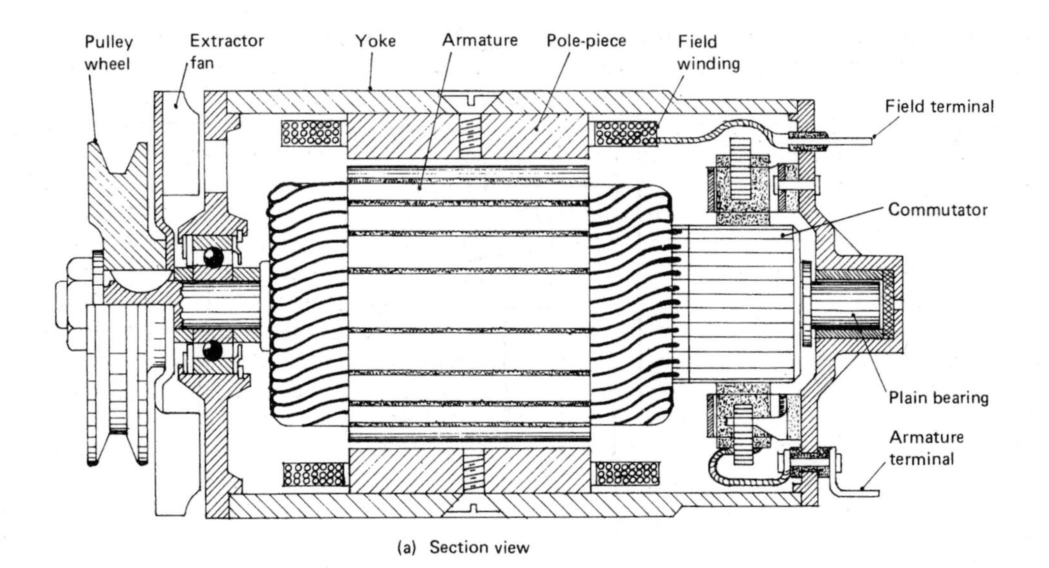

(a) Section view

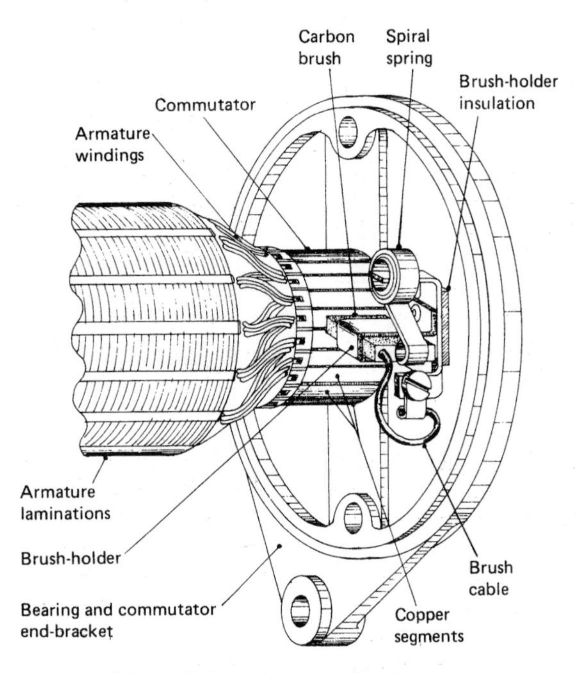

(b) Detail of commutator and brush-gear

Fig. 27.19 Dynamo

contacts to open under spring tension, thereby disconnecting the generator from the battery.

27.5.5 Output regulation

Generator output is regulated by limiting the current flowing through the field windings and hence reducing the magnetic field strength. This is achieved by putting a pair of vibrating contacts in series with the field-winding circuit – when generator output reaches the upper limit, the contacts will open and close with increasing frequency as the generator speed rises or the battery demands are high. To prevent excessive contact arcing when they open, a resistor is

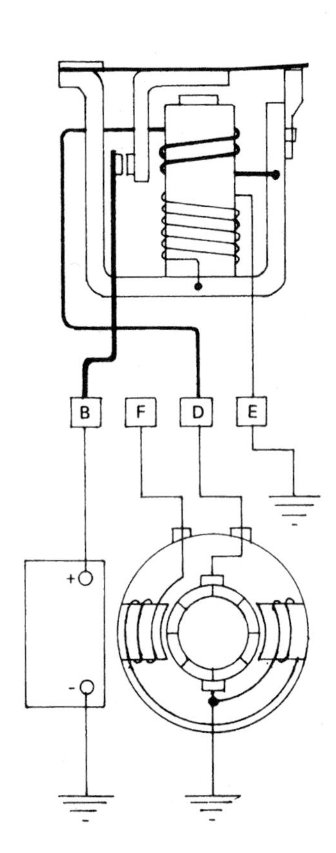

Fig. 27.20 Cut-out

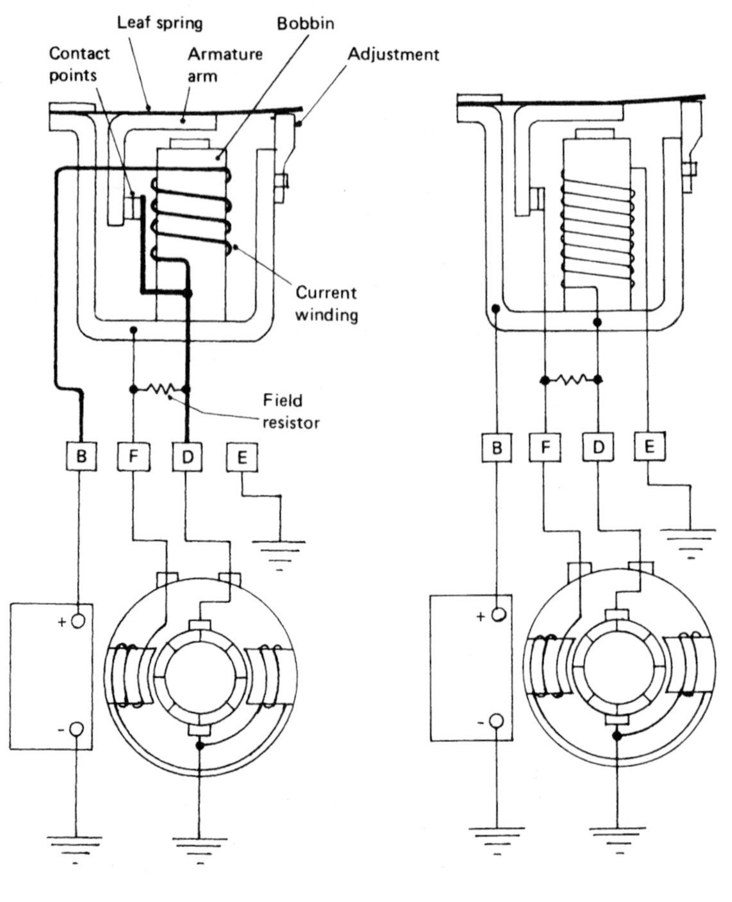

Fig. 27.21 Current regulator

Fig. 27.22 Voltage regulator

shunted across the points so that, when they do open, a reduced flow of current will pass through the resistor in preference to jumping the contacts.

27.5.6 Current control (fig. 27.21)

Current control is necessary to safeguard the generator windings against overloading due to excessive current demands for such things as a flat battery or perhaps a large lighting or heater load. Control is derived from a series current winding wound on to a relay bobbin (fig. 27.21), its function being to interrupt the current flowing through the generator field windings when the predetermined maximum charging rate has been reached.

With increased speed, generated current will flow from the generator (dynamo) to the battery by way of the series current winding. When the generated output reaches its maximum, the series bobbin winding is sufficiently energised to pull the L-shaped armature arm downwards so that the contacts open against the tension of the leaf spring. The generator field winding will then be on open circuit; consequently the magnetic field strength across the generator poles will be reduced and the output will fall. The energising current in the series bobbin winding will thus weaken and the regulator contacts will close again to allow the output to rise. This opening-and-closing cycle continues at between 60 and 100 times per second, thereby limiting the maximum output-current flow.

27.5.7 Voltage control (fig. 27.22)

Voltage control is required in addition to current control, to restrict the maximum generated voltage which would otherwise continue to increase with rising speed. If not checked, this voltage build-up would overload and blow such things as light-bulbs, heater elements, and other electrical equipment.

As the battery becomes charged, its terminal voltage rises – causing a general increase in line voltage between the generator and the battery. When the line voltage reaches its predetermined value, the voltage shunt winding wound on to a relay bobbin is sufficiently energised to pull downwards the armature arm (fig. 27.22), thus opening the contacts against the spring tension. This places the field resistor in series with the generator field winding and so reduces the field exciting current and hence the generator's magnetic field strength so that the generator output voltage will drop.

In practice the cut-out and the current and voltage regulators are not built as separate components but are combined as one control unit.

27.5.8 Reasons for preferring alternators to dynamos

Alternators are now preferred to dynamos, as alternators have no mechanical rectifying switches (the commutator and brush gear) to wear and because the charging current is carried by stationary windings rather than by rotating windings as is the case with dynamos. Also, the design characteristics of the alternator enable larger charging currents to be generated at low engine speeds compared with the dynamo.

27.6 Starter motors

A starter motor is a machine which converts the electrical energy stored in the battery into mechanical energy when required to crank the engine for starting purposes.

A simple electric motor consists of a horseshoe-shaped soft-iron yoke with field windings wound around each of the two pole-pieces (fig. 27.23). Between the pole-pieces is a rotating loop conductor known as the armature winding with its ends attached to each half-segment of a split-ring. When the battery supplies current to the starter, it flows from the positive battery terminal to the right-hand brush and segment, round the armature loop, and comes out of the left-hand segment and brush; it then circulates through the left- and right-hand yoke field windings and returns to the negative battery terminal.

27.6.1 Magnetic-torque production (figs 27.23 and 27.24)

The flow of current through the yoke field windings converts the yoke into an electromagnet so that a magnetic field or flux is established between the pole-pieces, travelling from the north to the south pole. Similarly, current flowing through the armature loop creates concentric rings of magnetic flux around the two half-conductor cores, and, as shown in fig. 27.24, this magnetic flux flows anti-clockwise around the left-hand conductor and clockwise around the right-hand conductor.

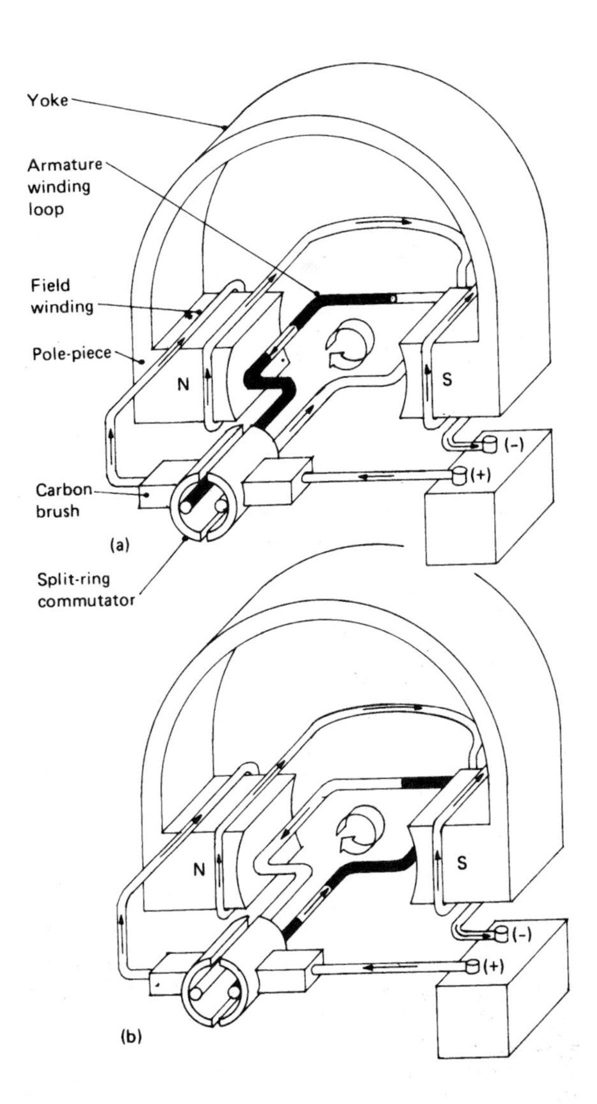

Fig. 27.23 Basic single-loop starter motor

Referring to fig. 27.24, it can be seen that where the magnetic lines of force between the yoke poles and those for the armature conductors flow in the same direction, the two sets of lines merge and strengthen each other. This is shown below the left-hand conductor and above the right-hand

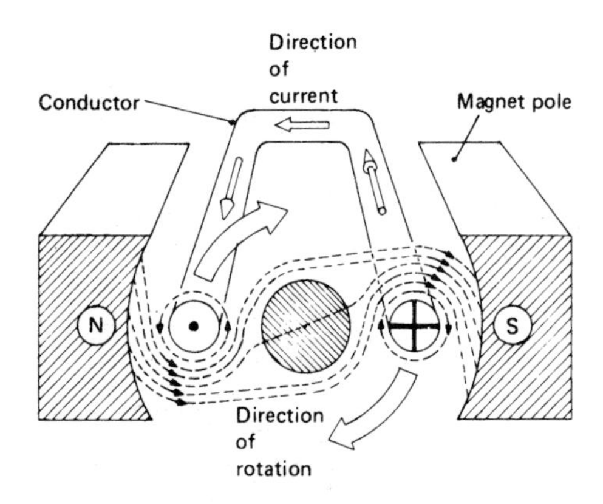

Fig. 27.24 Production of a magnetic torque

conductor. Conversely, where the yoke and armature magnetic fields flow in opposite directions, they neutralise each other and so the field strength in this region is very weak – above the left-hand conductor and below the right-hand conductor. The difference in the magnetic field strengths above and below each conductor will result in a net upward force being exerted on the left-hand conductor and a net downward force on the right-hand conductor which in effect apply a clockwise-rotating torque to the armature loop.

27.6.2 Commutation (fig. 27.23)

If the ends of the armature loop were connected directly to the battery, the magnetic-field interaction would rotate the loop so that the two half-conductors changed places with each other, and at the same time this would alter the direction of the current flowing in the conductor relative to the north and south poles of the yoke, with the result that the loop would then reverse its direction of rotation and turn backwards.

To make the armature loop rotate continuously in only one direction, the direction of current flowing in each half-loop must be reversed every half-revolution. This is known as commutation. Commutation is obtained by connecting the armature-loop ends to a split-ring (fig. 27.23) so that each half-ring segment contacts a different brush during every half-revolution of the armature; consequently the direction of the current flow within the armature loop will be repeatedly reversing.

27.6.3 Starter specification (fig. 27.25)

A typical starter motor has four field windings and poles and many armature conductor loops, so that a more powerful and uniform magnetic torque is produced. To reduce the current flowing in each copper–carbon brush, four evenly spaced brushes are normally incorporated, which may be of either the face type (fig. 27.26(a) and (b)) or the radial type (fig. 27.26(c) and (d)).

27.6.4 Inertia-drive starter motor (fig. 27.25)

The object of the inertia drive is to permit the starter motor to attain sufficient speed and hence power before it engages with the engine flywheel starter-ring gear.

Because starter motors are relatively small, they develop insufficient torque to rotate the engine directly; therefore the drive from the starter to the engine is transmitted by a toothed pinion attached to the end of the motor armature shaft and a ring gear which is fitted around the engine flywheel (fig. 27.27). To start the engine, the pinion has to be moved into mesh with the ring gear. A gear reduction between 10:1 and 12:1 will multiply the starter torque sufficiently to crank the engine at least at a minimum starting speed of about 100 rev/min.

The internally threaded pinion screws onto an externally threaded sleeve which is splined onto the armature shaft. When the starter is first energised, the shaft and sleeve suddenly rotate, but the pinion does not have time to revolve, due to its inertia; therefore the armature shaft's threaded sleeve will screw the pinion in towards the armature until it engages with the flywheel ring gear. Once the pinion and ring gear are in mesh, power will be transmitted to the stationary flywheel, consequently cranking the engine from rest to a speed at which the engine develops its own power. With further rising engine speed, the flywheel ring gear will run faster than the starter pinion, so that it now screws the pinion all the way back along the screwed sleeve until the pinion is thrown out of engagement with the ring gear.

As a further refinement to the drive, a strong compression spring is fitted at the end of the armature shaft to cushion the shock loading when the pinion initially meshes with the ring gear.

27.6.5 Starter solenoid switch (fig. 27.28)

The starter solenoid switch is provided for the following reasons:

a) to enable a relatively small current to control a very large current of maybe several hundred amperes.
b) to reduce the voltage drop in the starter circuit by enabling much shorter cables to be used.

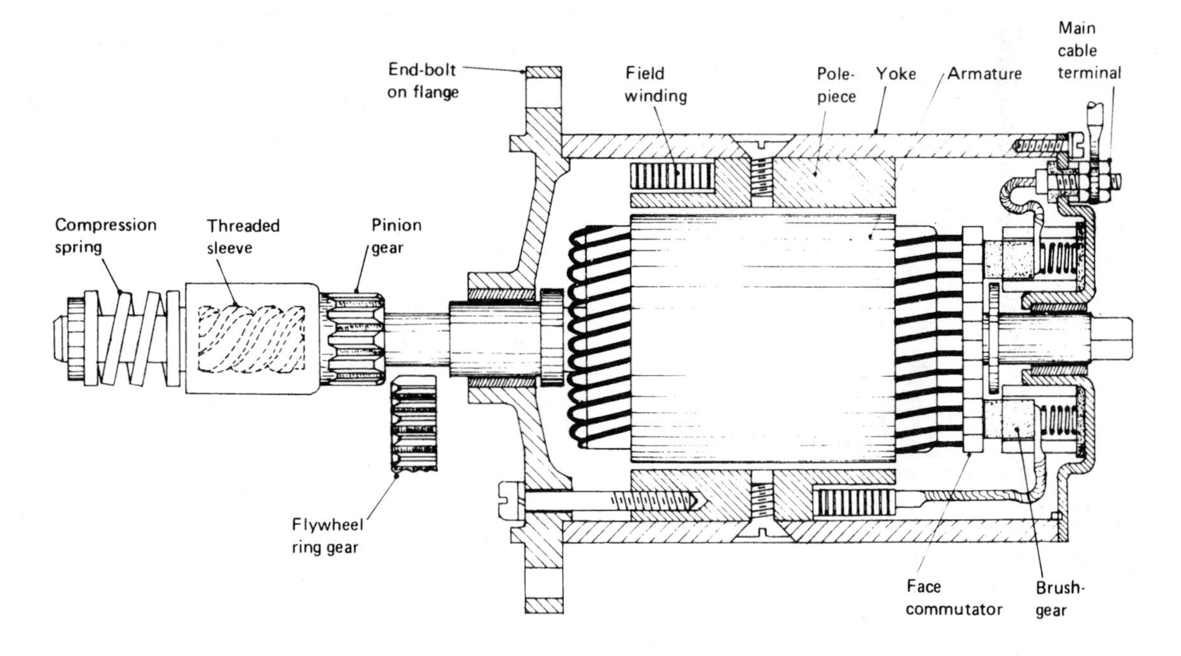

Fig. 27.25 Inertia starter motor

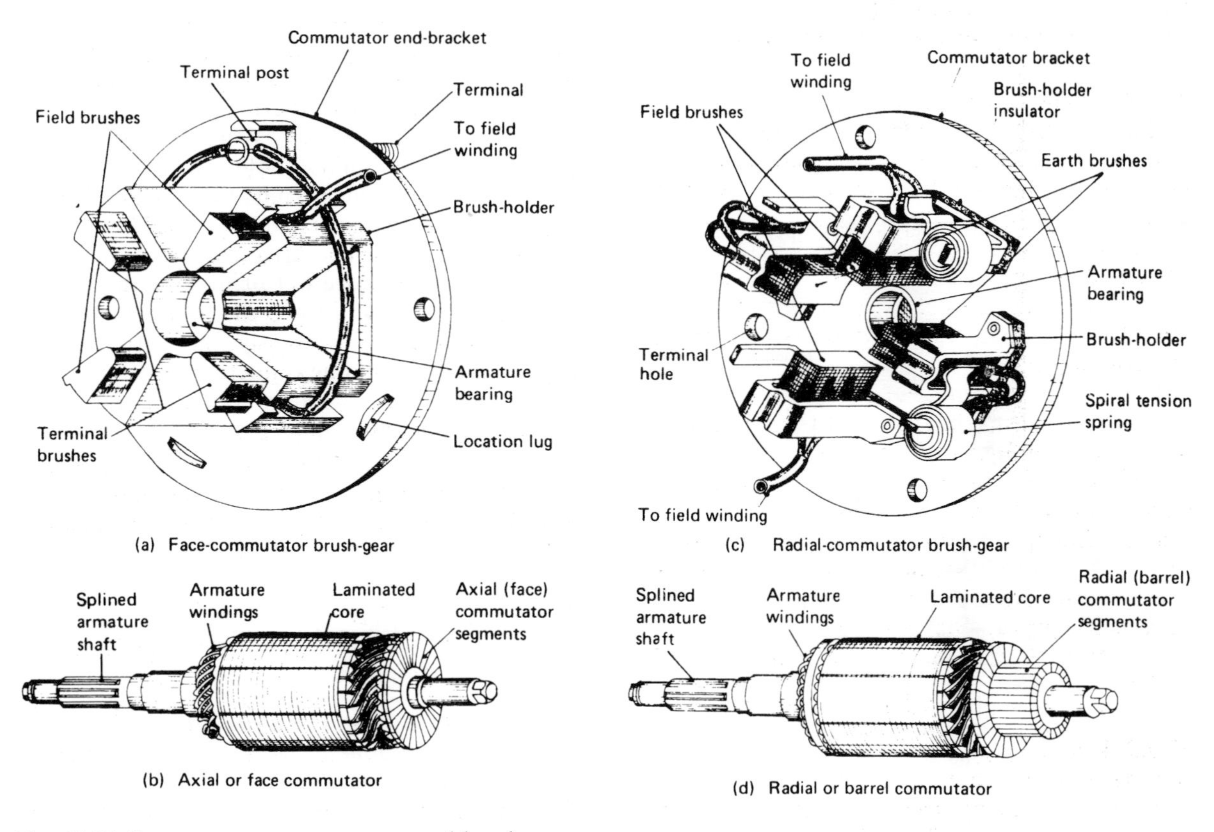

Fig. 27.26 Starter-motor commutators and brush-gear

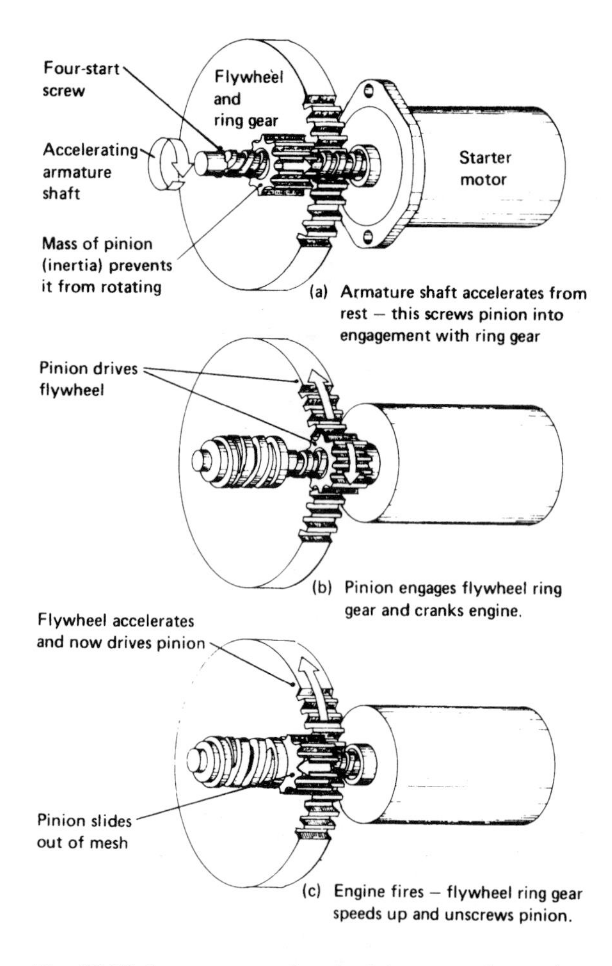

Fig. 27.27 Starter-motor inertia-drive operating cycle

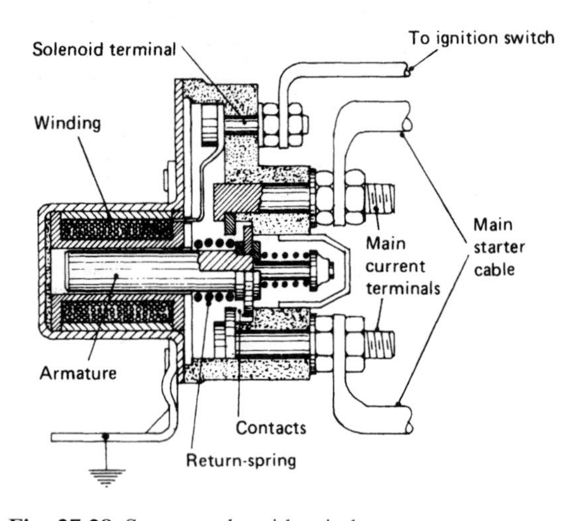

Fig. 27.28 Starter solenoid switch

The solenoid switch consists of an electromagnet winding with one end earthed to its casing and the other end joined to a small terminal (fig. 27.28). When the ignition-starter switch is turned on, a small current will energise a solenoid plunger and a moving contact. This bridges the gap between two fixed heavy-current contacts so that current flows from the battery through the cables and contacts directly to the starter motor.

Figure 27.29 shows the complete wiring circuits for an inertia starter motor.

27.6.6 Pre-engaged starter motor (fig. 27.30)

A pre-engaged starter solenoid circuit is designed to ensure that the armature-shaft does not revolve until the drive pinion is fully engaged with the flywheel ring-gear. Within the solenoid unit are two windings, a pull-in winding which pushes the drive pinion into mesh with the flywheel ring-gear when the engine is stationary, and a hold-in winding which keeps the pinion in mesh during engine cranking. The pull-in heavy-gauge low-resistance winding is earthed through the armature-winding and brushes, whereas the hold-in light-gauge high-resistance winding is directly earthed.

Phases of pinion engagement and torque application (fig. 27.31(a))

Ignition switch turned off When the ignition is turned off the pull-in and hold solenoid windings are de-energised. This permits the solenoid armature return spring to expand and push out the solenoid-armature. The drive pinion at the same time is forced to move out of mesh with the flywheel ring-gear by tilting the shaft fork-lever anticlockwise. Simultaneously the current flow to the armature-winding is interrupted, thus causing the armature to quickly spin to a standstill. The drive pinion and clutch assembly continues to move away from the ring-gear until the spinning drive pinion, clutch and brake-disc come into contact with the fixed brake-disc member. The friction generated by the disc rapidly brings the drive pinion assembly to a standstill so that another attempt to start the engine can be made quickly by the driver if the engine has failed to start.

Ignition switch and starter switch turned on/drive-pinion moving into mesh (fig. 27.31(b)) With the ignition switch closed the starter switch is turned to the start position. Current now flows through both of the solenoid pull-in and hold-in windings.

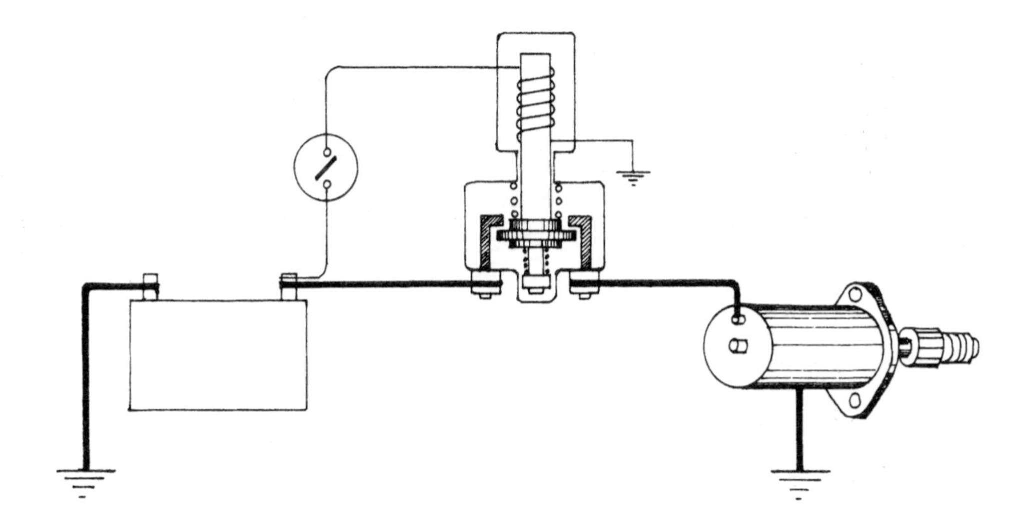

Fig. 27.29 Inertia-starter circuit

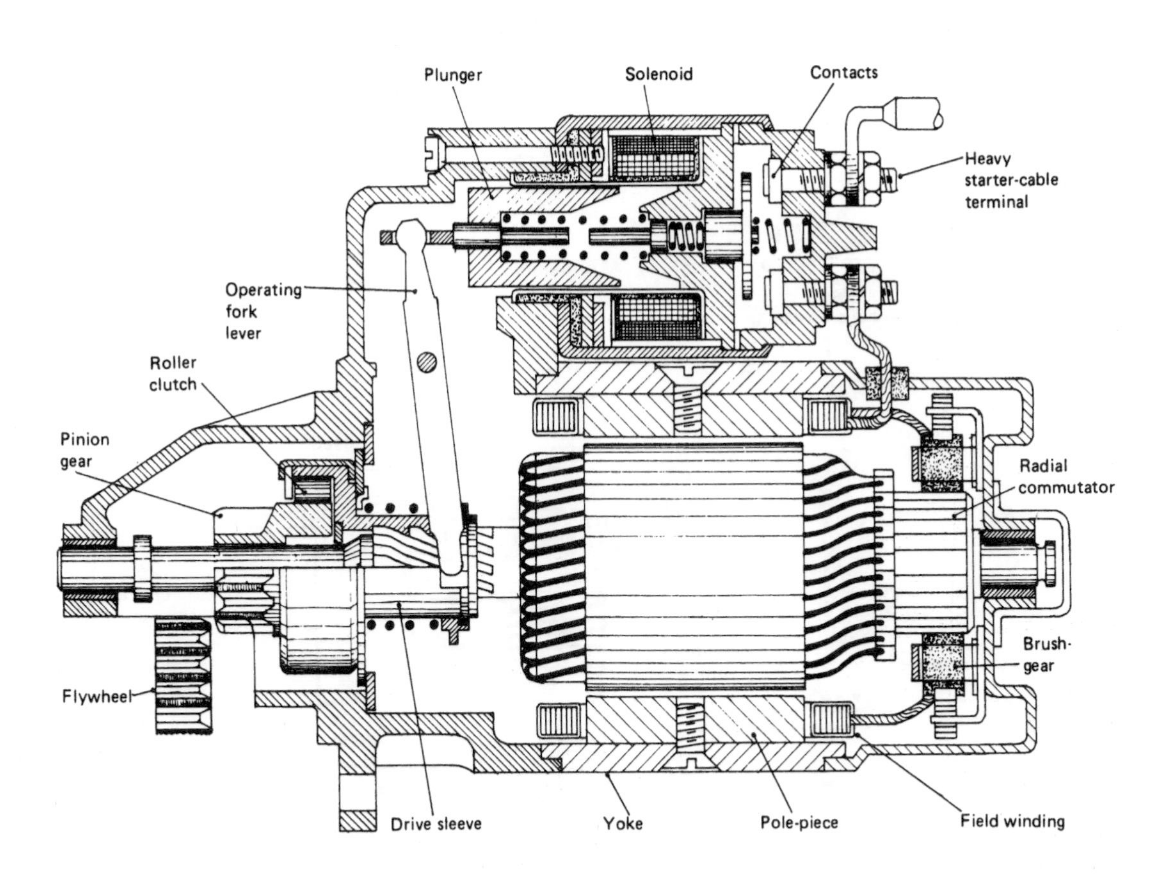

Fig. 27.30 Pre-engaged starter motor

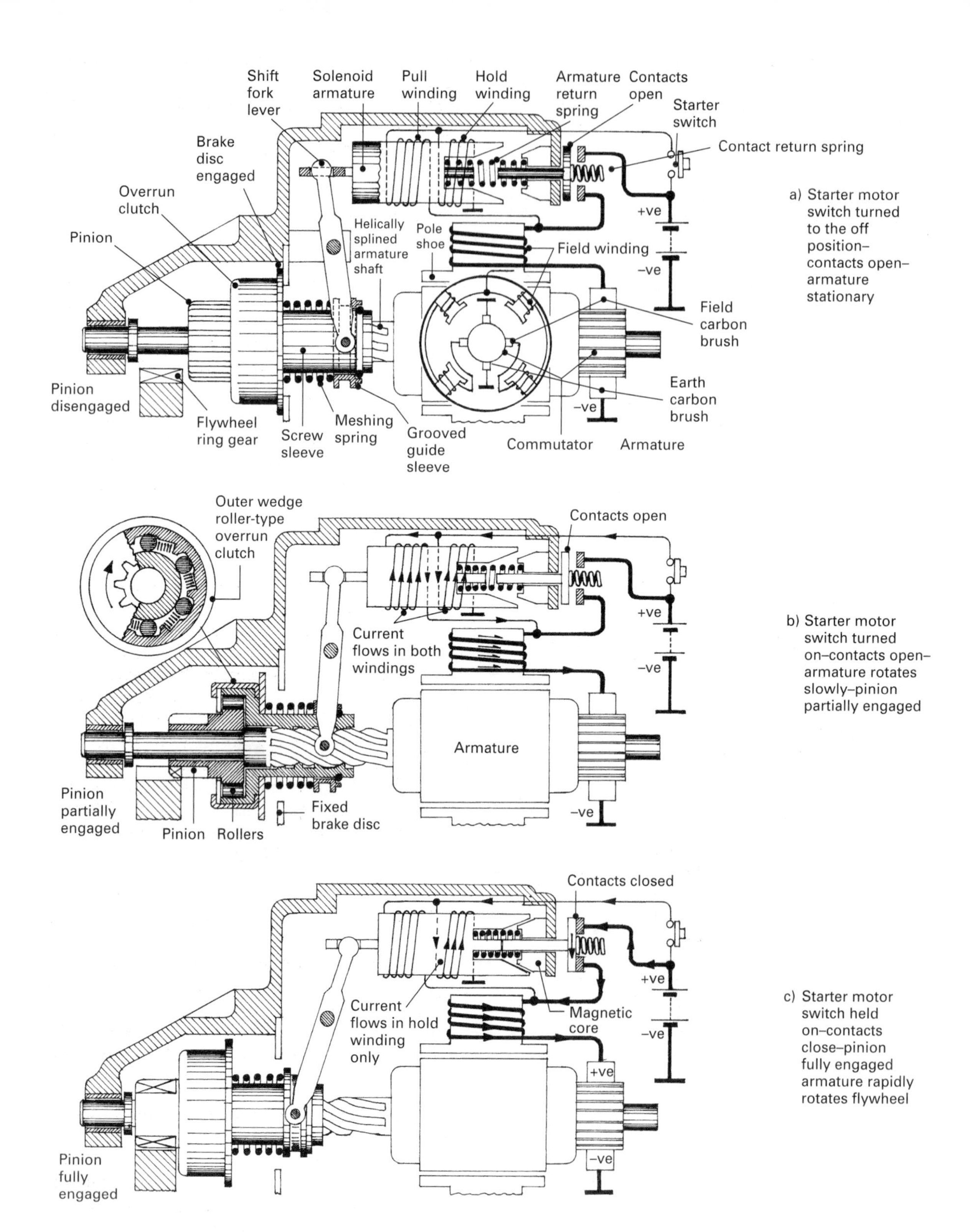

Fig. 27.31 Pre-engaged drive starter motor

The combined magnetic field of the pull and hold windings draws the solenoid-armature towards the magnetic core where the flux strength is greatest. This tilts the shaft fork-lever and brings the drive pinion into mesh with the flywheel ring-gear teeth spaces via its spiral twist movement on the armature-shaft. If tooth-to-tooth abutment occurs so that the pinion does not slide into mesh with the ring-gear, the grooved guide sleeve compresses the mesh spring, thus permitting the solenoid-armature to move fully over to the closed solenoid contacts. Current now flows through the armature-windings of the motor and the energised armature then commences to rotate. The pinion teeth and the spaces between the ring-gear teeth now align, thus enabling the drive-pinion to slip into mesh.

Ignition and starter still switched on/drive pinion fully meshed/armature rotates (fig. 27.31(c)) With the ignition switch closed the starter switch is turned to the start position. The solenoid-armature has been pulled fully over to the magnetic core, thereby closing the solenoid contacts. The closed contacts now short circuit the pull-in winding so that current ceases to flow through the winding. However the drive pinion is prevented from moving out of mesh by the still energised holding winding's magnetic field which is sufficient to keep the pinion in mesh. With the solenoid contacts closed, current flows from the positive battery's terminal through the field-winding, positive brush and armature-winding and to earth via the negative brush. Subsequently the energised field and armature windings produce the turning effort needed to rotate the armature-shaft.

When the starter switch is released the hold-winding is demagnetised and the spring-loaded solenoid-armature is pushed back to the off position. Consequently the contacts open. This causes the current flow to the field and armature windings to cease and the armature-shaft to spin to a standstill. At the same time the pinion is pulled out of mesh via the tilt of the shift fork-lever ready for the next start application.

Overrun roller clutch (fig. 27.31(b)) This consists of inner and outer ring members separated from each other via a number of cylindrical rollers. The outer ring member has cam ramp steps, one for each roller. These are formed around its internal circumference, whereas the inner ring member's external surface has a continuous circular track.

Springs are used to preload the rollers so that they roll up their respective ramps. In one direction of drive, the rollers wedge themselves between the outer and inner ring members, that is, when the outer ring member drives the inner ring member; conversely, when the inner ring member becomes the drive, the rollers release the wedge lock, thus allowing the inner and outer ring members to free-wheel.

Should the drive pinion remain in the engaged position with the flywheel ring-gear when the engine fires, starts, and picks up speed, then the drive-pinion which forms part of the clutch's inner ring member unlocks itself from the outer ring member which is attached to the armature-shaft by the rollers rolling down their respective ramps. As a result the armature is freed from the drive pinion and so prevents the armature-winding commutator and brush gear from possible damage due to running at excessively high speeds. Note that if the armature-shaft tends to rotate faster than the drive pinion the rollers are compelled to roll up their ramps, thereby locking both inner and outer ring members together.

27.6.7 Axial (sliding armature)-type starter motor (fig. 27.32(a)–(c))

This heavy-duty starter motor is designed so that the armature, shaft and drive pinion as a whole move forward axially to engage the flywheel ring-gear; hence it is known as an 'axial-type starter'.

Starter-to-flywheel tooth meshing takes place at low axial and rotational speeds, but once the drive pinion is fully engaged, the armature delivers its full drive torque with a corresponding speed rise. Thus the impact of the pinion engagement before the starter develops its full torque is minimised.

There are three field windings:

1) a main series winding which produces the most drive torque once the drive pinion is engaged,
2) a pull-in series winding which draws the armature assembly axially towards the flywheel ring-gear and
3) a hold-in shunt winding which keeps the pinion and armature assembly in the engaged position while the engine is being cranked.

An extra long armature commutator is used to ensure a continuous brush continuity when axial movement of the armature occurs. A solenoid-operated two-stage switch controls the pinion-engagement and power-applied phases. To prevent damage to the pinion and armature due to excessive overload or in the event of an engine

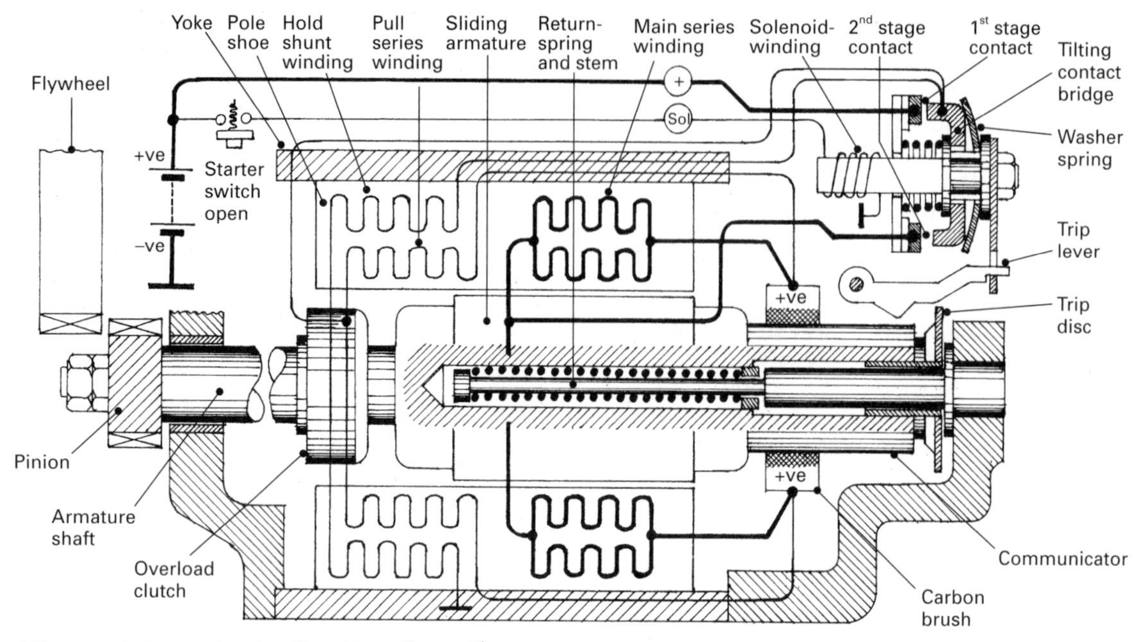

a) Starter switch turned to the off position–1st and 2nd contacts open–pinion disengaged

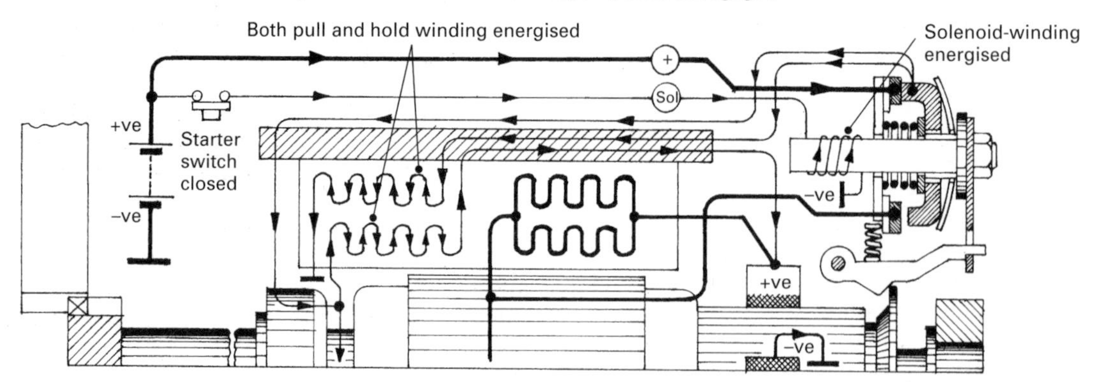

b) Starter motor switch turned on–1st contact closes–armature slides–pinion partially engaged–armature rotates slowly

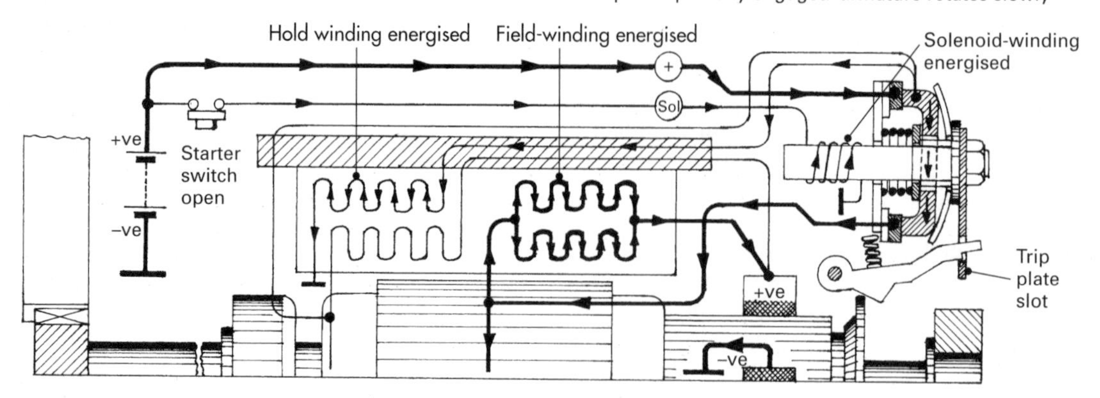

c) Starter motor switch held on–1st and 2nd contacts closed–pinion fully engaged–armature rapidly rotates flywheel

Fig. 27.32 Axial (sliding armature)-type starter motor (CAV)

back-fire an overload multi-plate clutch is incorporated in the drive between the armature and drive pinion. Its limiting torque is set above the lock torque of the starter but it will slip below the shear strength of the pinion teeth. In addition there is a freewheel device incorporated into the overload clutch to release the armature from the pinion if the latter has not disengaged itself from the flywheel ring-gear when the engine has started and is running on its own.

Phases of pinion engagement and torque application

Starter motor switch turned to the off position (fig. 27.32(a)) With the starter switch turned to the off position, all three windings are de-energised. This permits the central return-spring to expand, thereby driving the pinion out of mesh and the armature assembly away from the flywheel ring-gear. If the starter switch is released because the engine has not fired, allow the flywheel to come to rest before commencing to turn on the starter switch again. Once the engine has fired and is rotating on its own, immediately release the starter switch to avoid damage to the pinion and flywheel teeth.

Starter motor switch turned on/armature moves drive pinion into mesh (fig. 27.32(b)) When the starter switch is turned to the on position, current flows from the positive battery terminal through the solenoid-winding to the negative earth. The energised solenoid-winding draws the first-stage upper contacts together. Current now flows through both the pull and hold field windings. This produces a magnetic field force which slowly rotates the armature and simultaneously draws the armature and pinion into mesh with the flywheel ring-gear.

Starter motor switch still turned on/engaged drive pinion and armature rotate (fig. 27.32(c)) As the armature approaches the end of its axial engagement travel the trip-disc contacts and pushes up the trip-lever so that it clears the trip-plate slot, thus allowing the second stage lower spring-loaded contacts to snap closed. The second-stage contacts now short circuit the high-resistance series pull-winding through the lower resistance field (main series) winding; current through the pull-winding therefore ceases. Nevertheless the hold-winding is still able to keep the armature assembly and pinion in the engaged position. Current now passes via the tilting bridge contacts through the main field-winding and then by way of the negative brushes through the armature windings, finally returning to the battery's negative terminal via the earth-return positive brushes. This results in the maximum interaction of the field and armature winding's magnetic flux so that the full armature torque is exerted via the drive pinion to the engine's flywheel.

A free-wheel device built into the overload-clutch assembly automatically disengages the armature from the drive pinion if the engine fires and speeds up without the drive pinion being released. It therefore prevents the armature and commutator from centrifugal damage.

27.6.8 Co-axial (sliding-pinion)-type starter motor (fig. 27.33(a)–(c))

Co-axial or sliding-pinion-type starter motors are used on large capacity diesel engines and are constructed so that the drive pinion alone moves axially forward to engage the flywheel ring-gear, and at the same time the armature is made to revolve slowly to ease the alignment of the pinion teeth with the spaces provided between the ring-gear; hence this kind of drive is known as a 'co-axial'-type starter.

A series-parallel field-winding is wound around the four poles, and built into the drive pinion end of the starter are two solenoid windings: a pull series winding which draws the drive pinion into mesh with the flywheel ring-gear, a hold shunt winding which keeps the pinion engaged during the engine's cranking period, and a limiting resistor which reduces the current supply to the series field-winding during the time the pinion is moving into mesh.

Phases of pinion engagement and torque application

Starter motor switch turned to the off position (fig. 27.33(a)) With the starter switch turned to the off position, both first and second-stage contacts are in their open position. Consequently the field, pull and hold windings are de-energised. This therefore allows the pinion return-spring to slide the drive pinion inwards and away from the flywheel ring-gear.

Starter motor switch turned to the on position/drive pinion moves into mesh (fig. 27.33(b)) When the starter switch is turned to the on position, current flows from the battery's positive terminal through the overspeed-relay solenoid winding to earth.

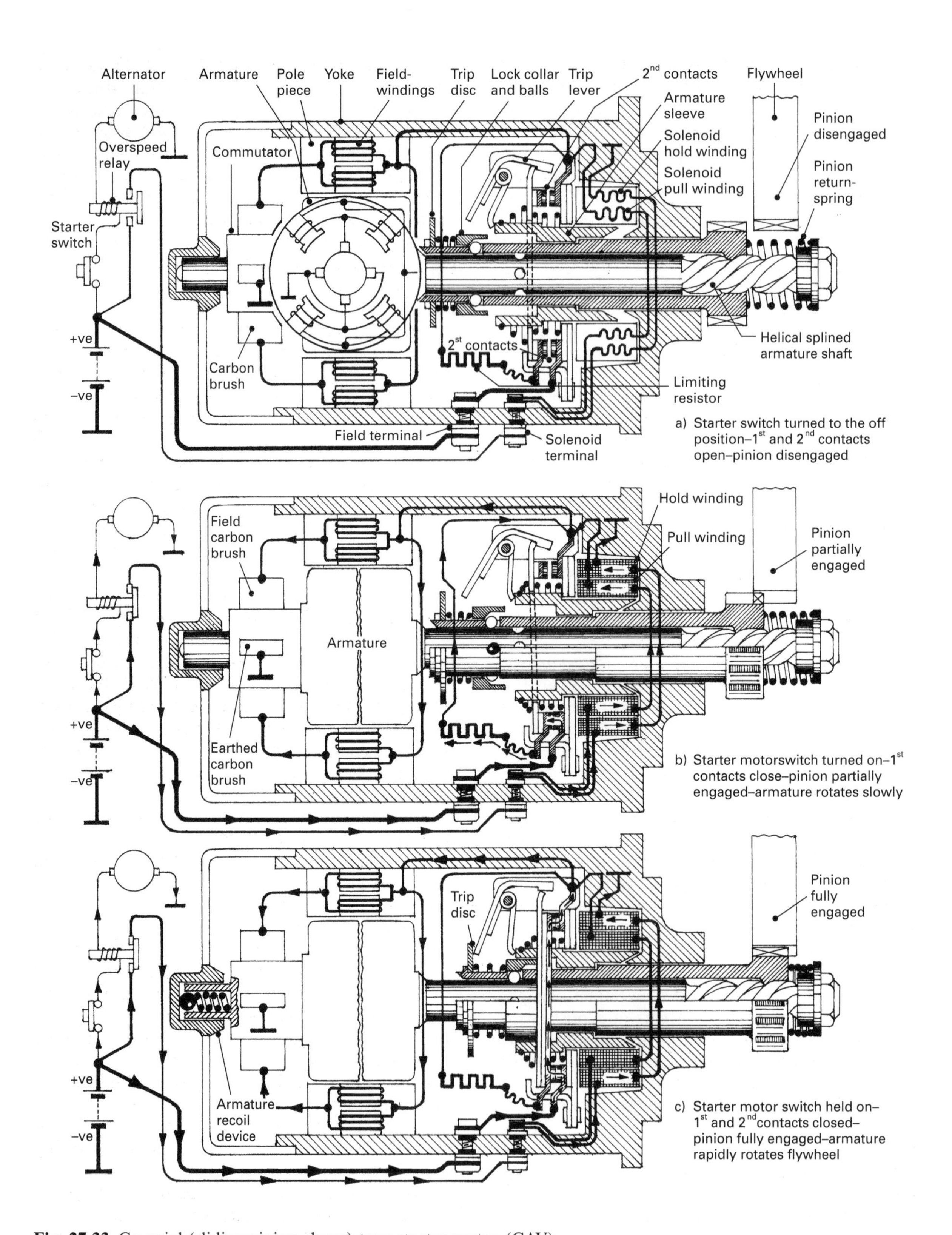

Fig. 27.33 Co-axial (sliding pinion sleeve)-type starter motor (CAV)

This causes its contacts to close. Current is now permitted to flow through the solenoid pull, field and armature windings to earth via both pairs of positive and negative brushes, and through the solenoid hold-winding straight to earth.

The magnetic field produced by the pull and hold windings now draws the armature sleeve into the solenoid core towards the flywheel ring-gear and at the same time causes the lower first-stage contacts to close. Current now flows across the first-stage contacts through the heavy-duty limiting resistor, field and armature windings and to earth via the pairs of positive and negative brushes. The restricted current flow through the field and armature windings due to the resistor's resistance partially energises them. This causes the interacting magnetic flux force to slowly rotate the armature. Consequently, the combined forward and rotating movement partially engages the pinion with the flywheel ring-gear. The pinion is then prevented from rotating by the engine's resistance to being cranked. However, the pinion is made to slide fully into mesh by the slowly rotating armature-shaft screwing the held pinion along its helical thread.

Stater motor switch still turned to the on position/ drive pinion engaged/armature rotates (fig. 27.32(c)) As the drive pinion reaches its fully engaged position, the trip-disc hits the trip-lever and consequently tilts the lever anticlockwise. It therefore releases the spring-loaded contact-bridge so that the second-stage contacts now close, thereby short-circuiting both the limiting resistor and pull winding. The battery now supplies its full current potential through both the field and armature windings. As a result, the full magnetic flux torque rotates the armature and subsequently cranks the engine via the flywheel ring-gear. At the same time as the pinion has moved to its fully engaged position, the four steel balls located around the pinion sleeve fall into recesses in the armature shaft. This allows the spring-loaded lock collar to slide over the balls, thus holding them in position. Thus the pinion will be held in the outward engaged position until the switch is turned to the off position.

Armature recoil device (fig. 27.33(c)) If the pinion should become worn and burrs are present the pinion teeth may not be able to mesh in the spaces between the flywheel ring-gear teeth. This can be overcome by a recoil ball and spring recessed into the armature-shaft at the commutator end. Every time the pinion moves outwards and butts the flywheel teeth the screw action between the armature-shaft and pinion will cause the armature to react and move away from the pinion by compressing the recoil spring, and at the same time the twist and impact motion caused by the helical screw thread when the starter switch is turned 'off' and 'on' is usually sufficient to free the pinion from its burrs.

Overspeed relay (fig. 27.33(a)–(c)) An overspeed relay is included in the starter wiring circuit to prevent accidental engagement of the starter when the engine is running above idling speed and to protect the starter from being driven by the flywheel.

The voltage applied to the relay winding when the engine is starting is that of the battery but when the engine fires and accelerates the alternator voltage rises and opposes the battery voltage; accordingly the actual voltage applied across the relay-winding is the difference between the battery and the alternator voltage. When the voltage applied across the relay-winding has been reduced to around 1 volt the overspeed relay contacts open. This interrupts the current flow to the starter motor solenoid field terminal. Hence the field and armature windings are de-energised and turning the starter switch to the on position will not energise the starter until the engine comes to rest.

Note that the relay winding can be earthed directly instead of passing though the alternator to earth. In this event, the starter motor will function as normal but there will be no overspeed protection.

27.6.9 Starter cable

When the starter motor is initially engaged, the turning-effort necessary to crank the engine (known as the lock torque) demands a peak current draw which may be three times the continuous current (see Table 25.1) – that is, anything from 300 to 600 amperes, depending on the size and type of engine. The cable transmitting starter-circuit current must therefore be capable of carrying a relatively large continuous current which may be trebled for very short periods during lock-torque conditions.

Most starter circuits use an earth-return system where the chassis or body structure acts as the return path for the starter-to-battery circuit. The advantage of this system is that only a relatively short connecting strap needs to be used. This strap is made from tin-coated copper strands woven together, with eyelet terminals at each end. One end is usually attached to one of the motor bell-housing clamping bolts, and the other end is bolted to the nearest convenient point on the chassis or steel member of the body.

28 Electrical auxiliary equipment

28.1 Instrument panel gauges and transmitter senders

28.1.1 Moving coil meter (fig. 28.1)

This very versatile instrument is ideally suitable for use as a voltmeter, ammeter or ohm-meter as it provides sensitive accurate readings and can be easily calibrated to measure a wide range of electrical units. Meters of this kind consist of a rectangular-shaped coil wound on a light aluminium frame made from insulated copper wire pivoted on steel needle points carried by hardened (jewelled) bearings and controlled by a pair of upper and lower non-ferrous hair springs which are coiled in opposite directions to neutralise expansion effects. Current is fed into and out of the coil by the spiral hairsprings, which also provide the controlling resistance torque. The coil is mounted between upper and lower insulated bearing-plates which are in turn supported on studs screwed into each of the two concave-shaped permanent magnet pole-pieces positioned on either side of the coil. A central soft iron core supported by two side plates is provided to intensify the magnetic field by reducing the length of air gap across which the magnetic field has to pass. The core also ensures that a radial magnetic field of uniform density is distributed around the semi-circular pole-pieces and the cylindrical core. Current to and from the coil is taken by way of terminal tags formed on the upper and lower insulated bearing-plates via soldered wire leads.

When a current flows through the coil it produces a magnetic force field which interacts with the permanent magnet's own force field, and depending in which direction the current passes through the winding produces a resultant force field which strengthens the field behind the (sectioned view) windings and weakens the field ahead. Consequently the vertical sides of the coil will be pulled in a clockwise direction in opposition to the hairspring's elastic resisting tension. Thus as more current passes through the winding, the greater will be the magnetic turning effect, and the greater will be the angular deflection of the pivoting coil. In other words, the coil and dial pointer are made to revolve in direct proportion to the current flowing through the coil.

Voltmeters, ammeters and ohm-meters all work on the same basic principle; the only difference is how the meter is connected to the circuit being tested. The voltmeter has a resistor connected in series to limit the voltage applied across the meter, the ammeter has a resistor connected in parallel with the meter so that most of the current bypasses the meter, and the ohm-meter has a small battery connected in series with the meter and with the component being tested. Multi-meters are moving coil instruments with a number of resistors which can be switched either in series or parallel with the meter to provide an extended range of volts, amps and ohmic readings (see fig. 28.2).

28.1.2 Temperature and fuel level transmitters

Temperature semiconductor transmitter (figs 28.3 and 28.4) These are usually made from a material whose ohmic resistance is temperature sensitive and is adapted so its resistance forms a measure of its surrounding temperature. Temperature sensors of this kind are known as thermistor semiconductor transmitters made from a material which has a negative temperature coefficient (ntc), that is, its resistance decreases with increasing temperature. This contrasts with the less popular positive temperature coefficient (ptc) material where the resistance increases with rising temperature (see fig. 28.3). The thermistor consists of powdered oxides of nickel, cobalt and manganese which are formed into small pellets. These pellet discs are installed at the base of a brass threaded capsule. A sheathed brass heat-sink, spring and insulated terminal connects one side of the pellet disc to the temperature gauge while the other side is earthed through the casing. When the temperature is low there are only a few mobile electrons contained in the pellet (see fig. 28.4(a)), so that very little current flows through the semiconductor disc; however, as the temperature rises, the concentration of mobile electrons quickly increases (see fig. 28.4(b)). It therefore

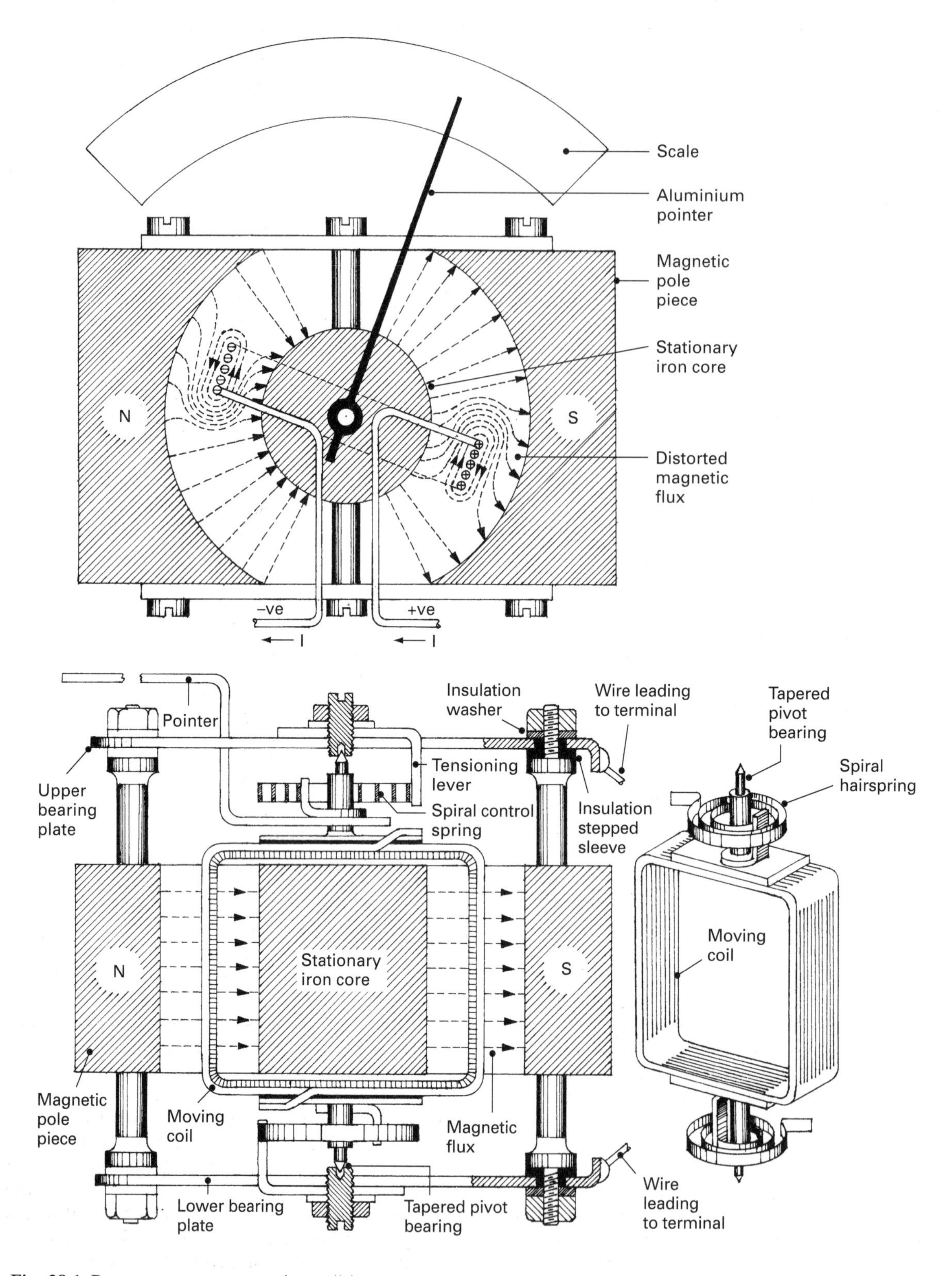

Fig. 28.1 Permanent magnet moving coil instrument

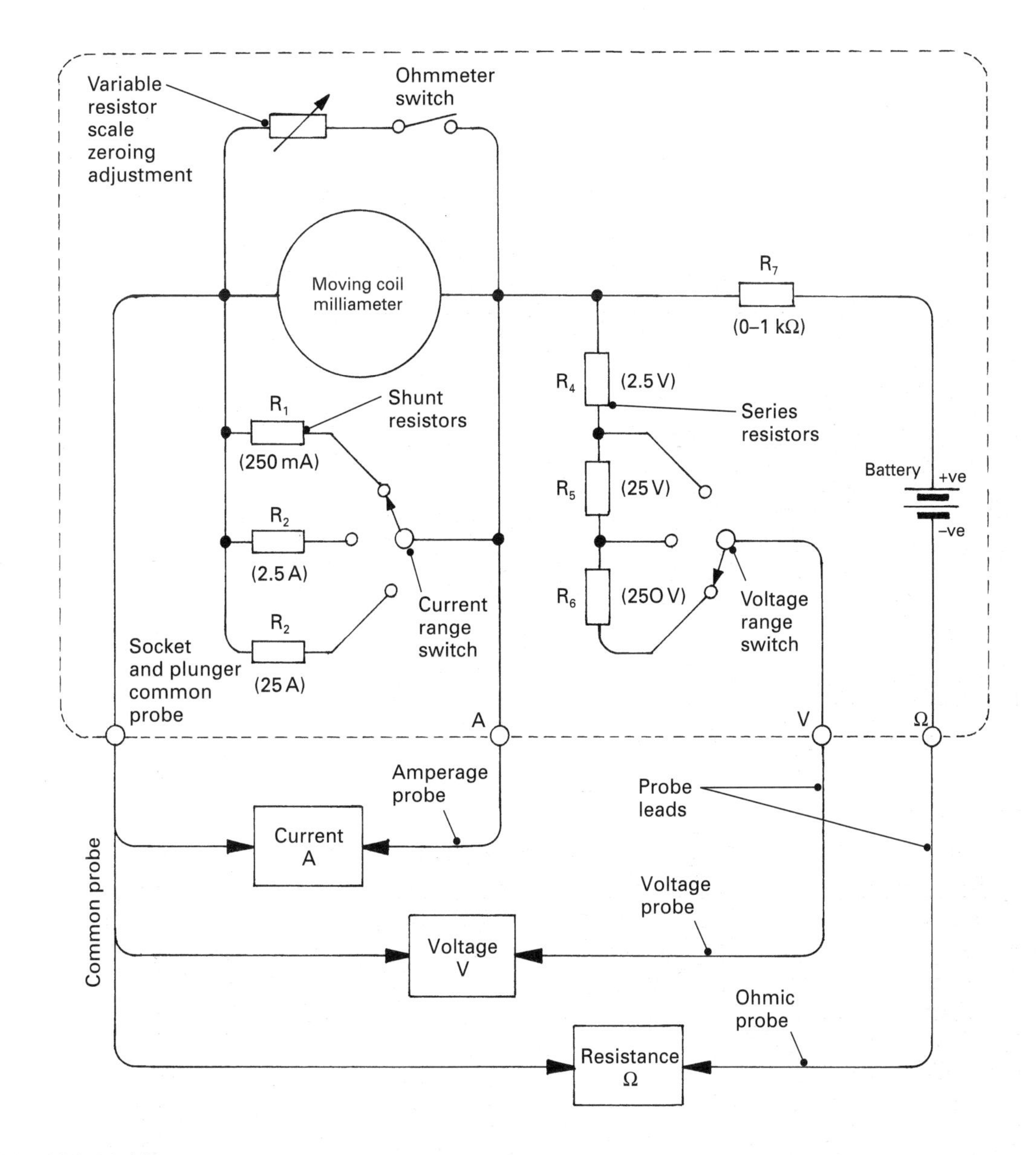

Fig. 28.2 Multimeter

supports a much higher current flow. A typical thermal transmitter resistance at 60°C and 90°C would be 750–850 Ω and 150–250 Ω respectively.

The temperature transmitter is screwed at some convenient point into a cooling passage in the cylinder-head to enable it to sense the changes in coolant temperature. When the engine is cold the thermistor will be high but as the coolant temperature rises there will be a corresponding (non-linear) decrease in resistance which is monitored in terms of an increase in current flow to the temperature gauge. Conversely, as the temperature decreases, the thermistor resistance increases so that less current is fed to the temperature gauge.

Fuel level variable resistor transmitter (figs 28.5 and 28.6) The most common type of fuel level transmitter is the rheostat variable resistor and float arrangement where the rise and fall of the fuel level height is monitored by a float attached to a pivoting swing-arm (see fig. 28.5). A brush or contact arm which also pivots with the swing-arm bears against a wire wound resistor or a resistor

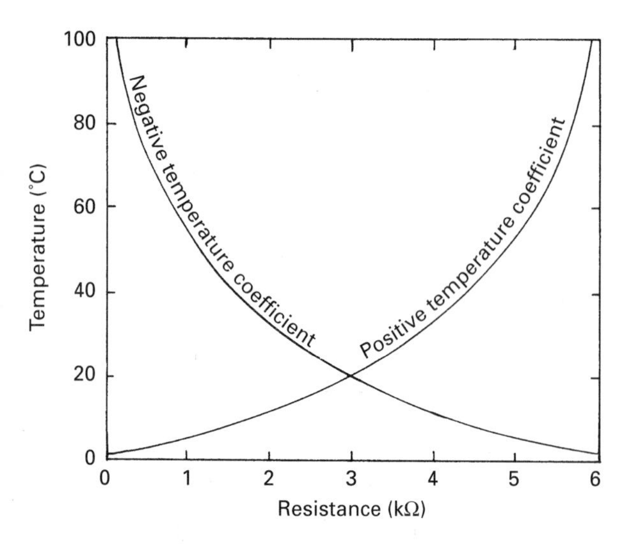

Fig. 28.3 Typical temperature response of both negative and positive temperature coefficient semiconductors

strip. The resistor's width is made to vary along its length so that its effective resistance for any position of the sliding contact matches the shape of the tank and hence its fuel content. A change in resistance and therefore current flow to the fuel gauge is obtained by connecting one end of the resistor to the fuel gauge and taking the sliding contact arm to earth at the pivot. As the fuel level moves up or down, the effective length of the resistor also changes. It therefore proportionally changes the current flow to the fuel gauge, this being a calibrated measure of the amount of fuel in the tank. Typical values for transmitter resistance and float lift from empty, half full and full would be 8Ω/110mm, 75Ω/50mm and 180Ω/0mm, respectively (see fig. 28.6).

28.1.3 Moving iron coolant temperature and fuel level gauges

Moving iron gauge (figs 28.4(a) and (b) and 28.5(a) and (b)) With this type of gauge two coils or windings wound on poles are positioned approximately at right angles to each other, and positioned at the axis of intersection of the two poles is a pivoting iron vane and dial pointer. The two coils are connected in series with each other, and the free ends of the coils are connected to the battery's positive and negative terminals via the ignition switch and an earth return. The link between the two coils forms a junction which is connected to a variable resistor transmitter, be it a fuel level or a coolant temperature sensor. When current flows through both coils the magnetic pull at their respective poles attracts the pivoting iron vane and it is the difference in the amount of current flowing through each of the two coils which determines the extent the iron vane and pointer swivel about their pivots.

This type of gauge is identified by the pointer instantly returning to the zero position when the ignition is switched 'off'.

Moving iron coolant temperature gauge and semiconductor transmitter (fig. 28.4(a) and (b)) When the ignition switch is turned on current flows from the battery's positive terminal through both coils to the battery's terminal via the earth return. A resistor is shunted across the deflection coil so that some of the current flow by-passes the deflection coil therefore making its magnetic attraction slightly less than that of the control coil.

When the coolant is cold the semiconductor transmitter resistance will be very high so that very little current flows to earth via the transmitter. Therefore the control coil will exert a greater pull on the iron vane than the resistor-shunted deflection coil. As the coolant temperature begins to increase, the transmitter resistance decreases, so that more current will flow via the transmitter to the negative earth than through the control coil. Consequently the deflection coil exerts a proportionally greater pull than that of the control coil which, in effect, is steadily being starved of current as the transmitter diverts more of the current flow from the control coil. Thus the angular rotation of the vane and pointer is a measure of the conductivity of the transmitter and hence the coolant temperature.

Moving iron fuel level gauge and variable resistor transmitter (fig. 28.5(a) and (b)) When the ignition switch is turned on current flows from the battery's positive terminal through both coils to the battery's negative terminal via the earth return. A resistor is shunted across the control coil so that some of the current flow by-passes the control coil, therefore making its magnetic attraction slightly less than that of the deflection coil when both coils receive the same current supply.

When the fuel tank is empty the fuel level transmitter resistor will be at its minimum so that most of the current flowing through the control coil is returned to the battery's negative earth via the transmitter and very little current passes through

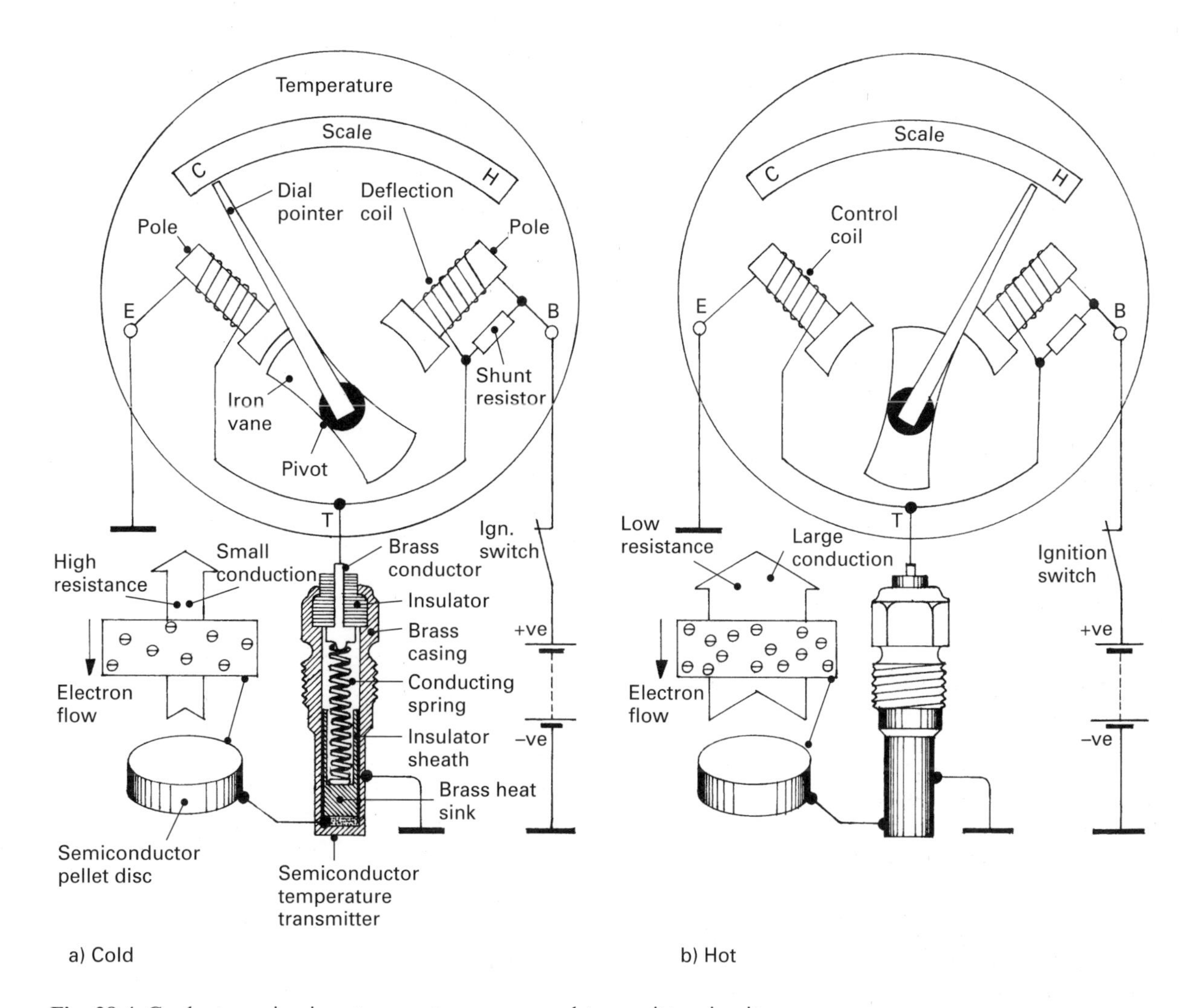

Fig. 28.4 Coolant moving iron temperature gauge and transmitter circuit

the deflection coil. Accordingly the magnetic pull exerted by the control coil attracts the vane and holds it and the pointer in the empty position.

As the fuel tank is filled, the rising float swivels the variable resistor towards its maximum resistance position. This causes more of the current passing through the control coil to divert through the deflection coil as opposed to going through the transmitter on its way to the battery's negative terminal. Consequently the deflection coil exerts a progressively greater magnetic pull than that of the resistor-shunted control coil which does not pass so much current. Hence the angular rotation of the vane and pointer is a measure of the resistance of the transmitter, which in effect corresponds to the tilt of the swing-arm and thus the fuel content.

28.1.4 Bi-metal thermal coolant temperature and fuel level gauges

Bi-metal thermal gauges (fig. 28.7(a) and (b)) This form of meter comprises a fixed fulcrum arm and a 'U' shaped bi-metal strip arm with an insulated heating winding wound around one leg. Both arms have their free ends bent in such a way as to support and guide a slotted aluminium pointer. The gauge is zeroed and calibrated by the manufacturer via a pair of sawtooth adjusters which support each of the two arms by rivet fixture points to the gauge's casing. The adjustm... such that when no current flows th... winding the pointer is in the... bi-metal arm. the left-hand side of the ... is wound around

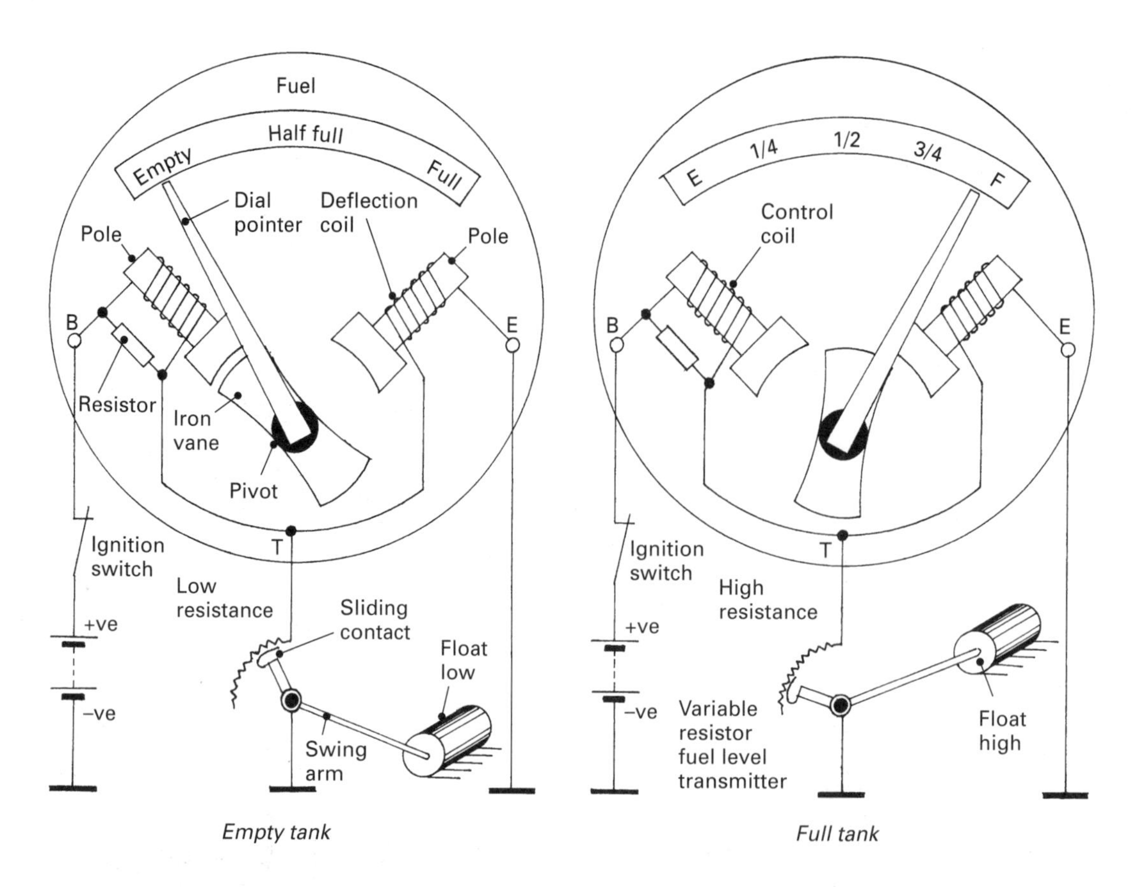

Fig. 28.5 Fuel tank content moving iron gauge and transmitter circuit

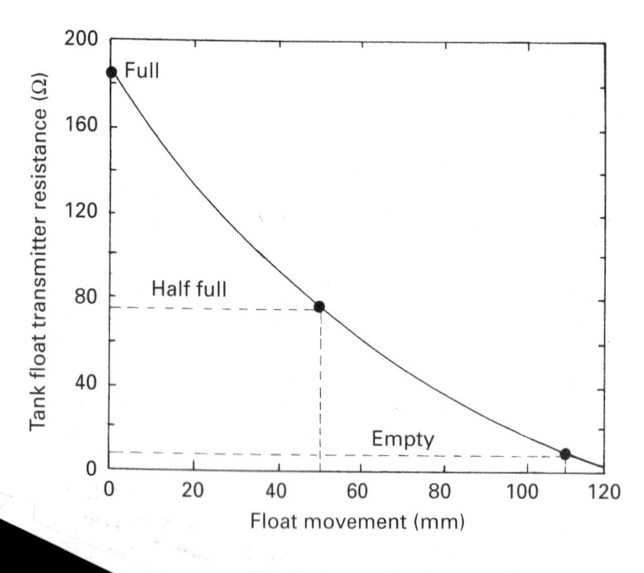

-hip between float transmitter move-

One end of the winding is connected to the positive battery's terminal via the voltage stabiliser and the other end is connected to either the fuel level transmitter (see fig. 28.8(a) and (b)) or the coolant temperature transmitter (see fig. 28.7(a) and (b)). When the ignition switch is turned on and current flows through the heating winding, the bi-metal strip distorts in proportion to the heat generated, and in so doing deflects the suspended pointer across the calibrated scale. Bi-metal thermal-type gauges for accurate readings require a constant voltage supply. It is therefore necessary to interpose a voltage stabiliser unit in series with the gauge. The bi-metal thermal gauge responds slowly to changing conditions due to its inherent thermal lag. It therefore does not sense fuel surge when braking or cornering.

Coolant temperature bi-metal thermal gauge and semiconductor transmitter (fig. 28.7(a) and (b)) With the ignition switch turned 'on' current flows

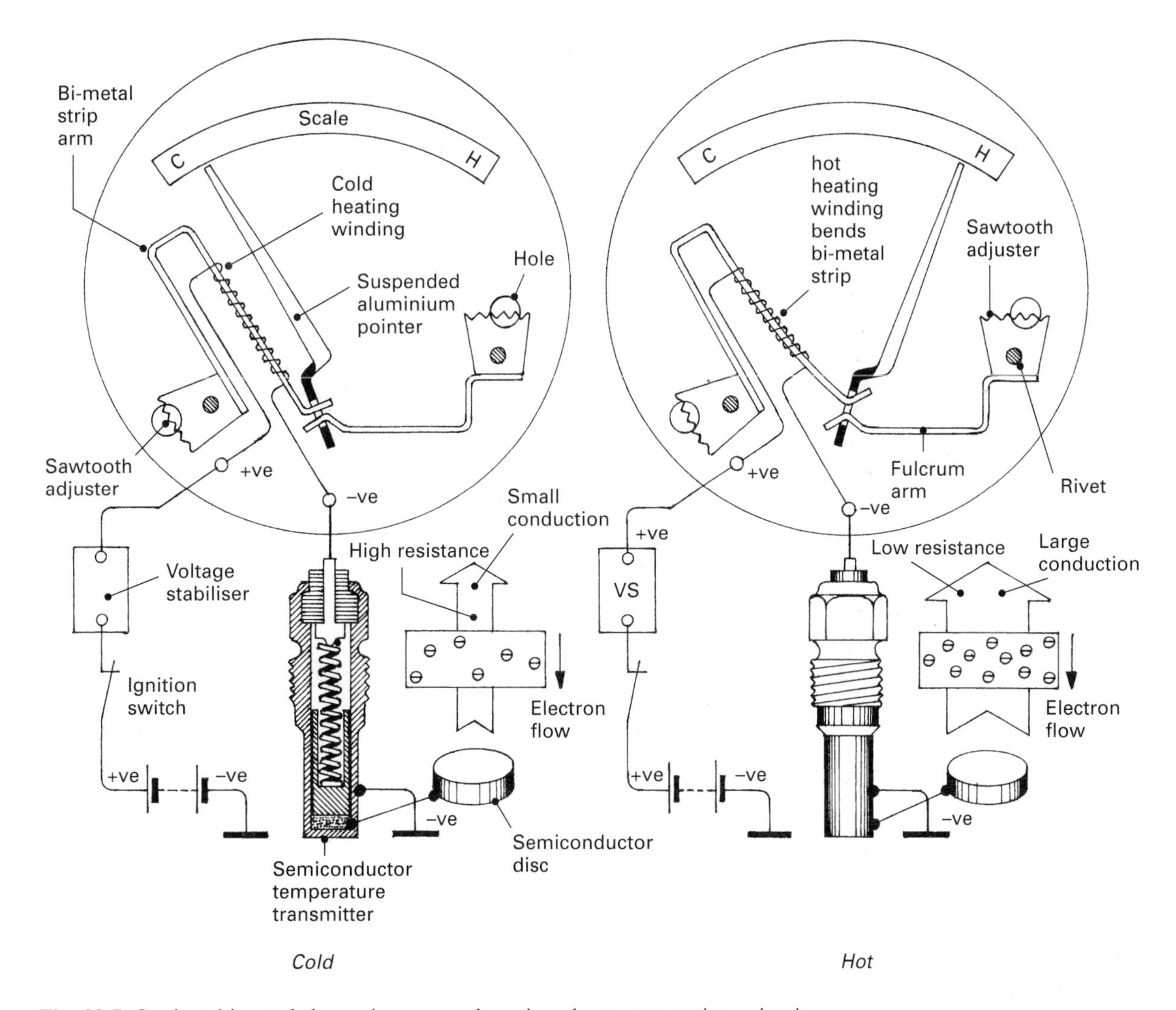

Fig. 28.7 Coolant bi-metal thermal gauge and semiconductor transmitter circuit

from the positive battery's terminal via the voltage stabiliser through the heating winding of the gauge and back to the battery's negative terminal via the temperature transmitter earth return. When the engine is cooled very little current can pass though the semiconductor pellet so that the winding's heating effect on the bi-metal strip is minimal and the corresponding pointer movement will be small. As the coolant temperature rises, the semiconductor pellet permits more current flow through it and through the bi-metal heating winding. Consequently the bi-metal strip will bend and correspondingly the pointer will deflect proportionally to the coolant temperature increase. Conversely as the temperature decreases the current flow through both the semiconductor pellet and the winding reduces. The bi-metal strip will therefore tend to straighten so that the pointer is brought back to its original cold position.

Fuel level bi-meter thermal gauge and variable resistor transmitter (fig. 28.8(a) and (b)) With the ignition switch turned on current will flow from the positive battery's terminal via the voltage stabiliser through the heating winding of the gauge and back to the battery's negative terminal via the float level transmitter earth return. When the fuel level is low the effective resistance of the fuel transmitter is high so that very little current can pass through the transmitter and bi-metal heating winding; hence the heating effect on the bi-metal strip is minimal.

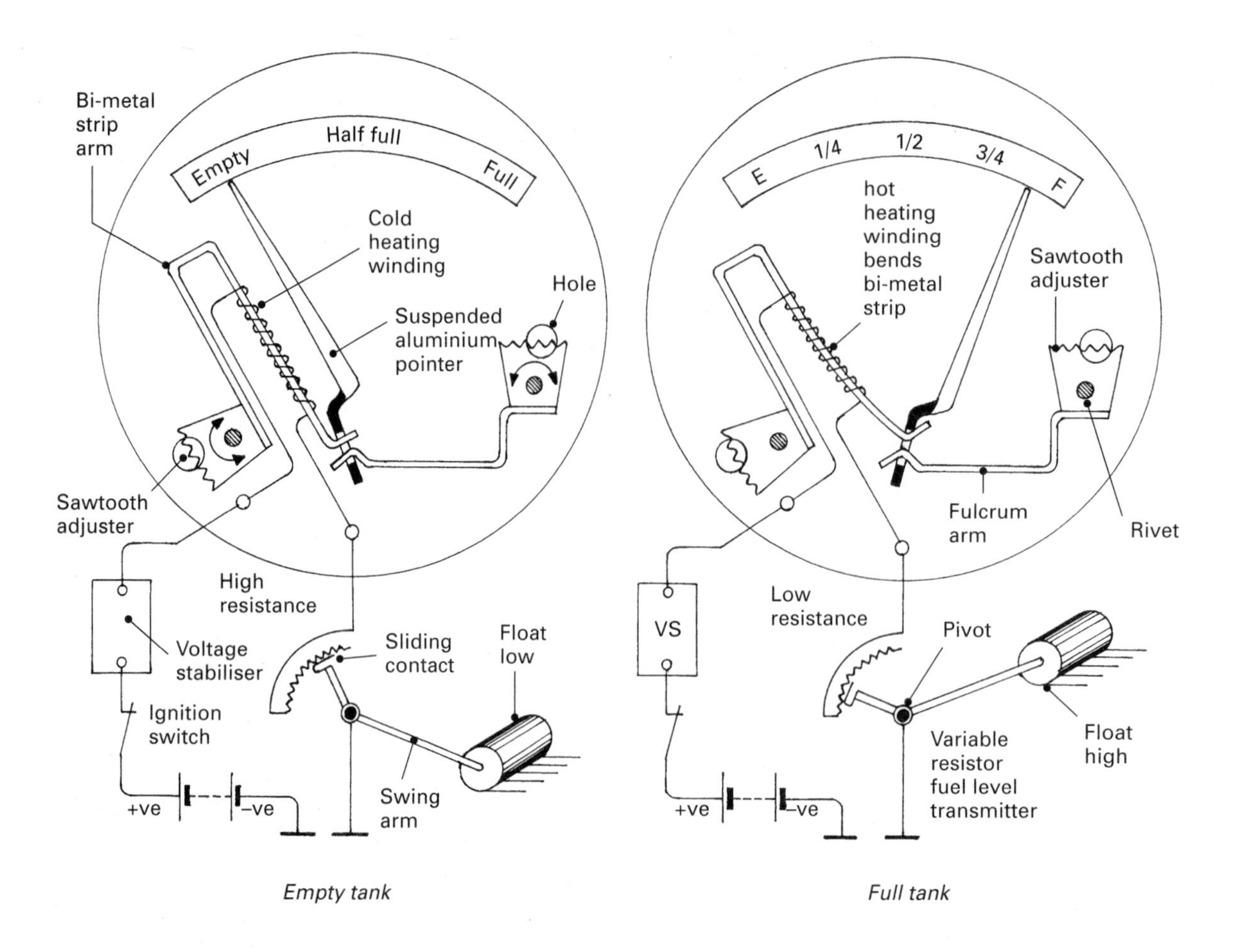

Fig. 28.8 Fuel tank content bi-metal thermal gauge and transmitter circuit

As the fuel tank fills up the float rises and swivels the swing-arm and the slide so that the effective resistance is reduced. Consequently, the bi-metal strip will bend and correspondingly cause the pointer to deflect proportionally to the increased amount of fuel pumped into the tank. Conversely, as the tank empties the current flow through both the transmitter resistor and the heating winding reduces. The bi-metal strip will tend therefore to straighten so that the pointer is brought back towards its empty position on the scale.

28.1.5 Voltage stabilisers

Certain types of instruments such as bi-metal thermal gauges rely on a constant voltage supply since the magnitude of the current flow to the gauges is affected by the voltage; otherwise it would not accurately reflect the changing input conditions from the sensing transmitter, be it coolant temperature or fuel level, etc. However, the voltage supply from the battery is consistently varying from may be 10.5–13.5 volts depending on whether the battery is discharged or is being charged by the alternator. It is therefore essential to interpose a voltage stabiliser between the battery and gauges. There are two main types of voltage stabilisers in use for gauges:

1) the bi-metal thermal stabiliser and
2) the Zener diode stabiliser.

Bi-metal thermal voltage stabiliser (fig. 28.9) The bi-metal voltage stabiliser consists of a 'U' shaped bi-metal leaf spring strip which is mounted at one end to the instrument terminal (I) while its free end becomes the moving contact and a fixed contact forming part of the battery terminal (B). An

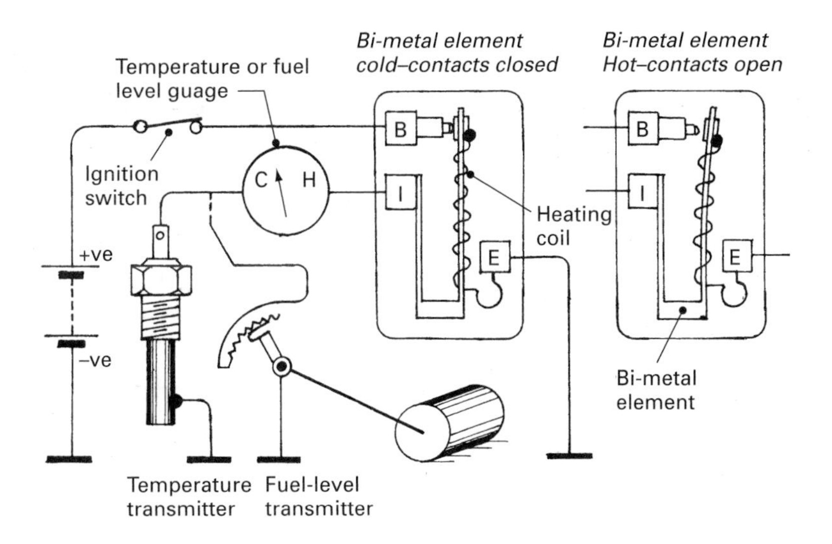

Fig. 28.9 Thermal voltage stabiliser

insulated heating coil is wound around the free arm of the bi-metal leaf spring, this coil being subjected to the battery's voltage supply.

When the ignition switch is turned on, current flows from the battery's positive terminal through the closed contacts. It then passes through the gauge and sensing transmitter to the negative earth via the bi-metal leaf spring and the instrument terminal (I). Current also flows through the heating coil; this causes the generated heat to distort and bend the bi-metal leaf spring away from the fixed contact so that the current is interrupted. After a very short pause the bi-metal leaf-spring cools, straightens out and once more closes the contacts. Current therefore again flows through to the gauge and sensing transmitter. Thus the continuous heating and cooling cycle of the bi-metal leaf spring breaks and makes the supply voltage output to the gauge and transmitter in the form of a square wave voltage pulse (see fig. 28.10), resulting in an average steady voltage input to the gauge of about 10 volts. The gauges used are not sensitive enough to react to the repeated voltage interruption, but they do respond fairly accurately to the average voltage output from the stabiliser.

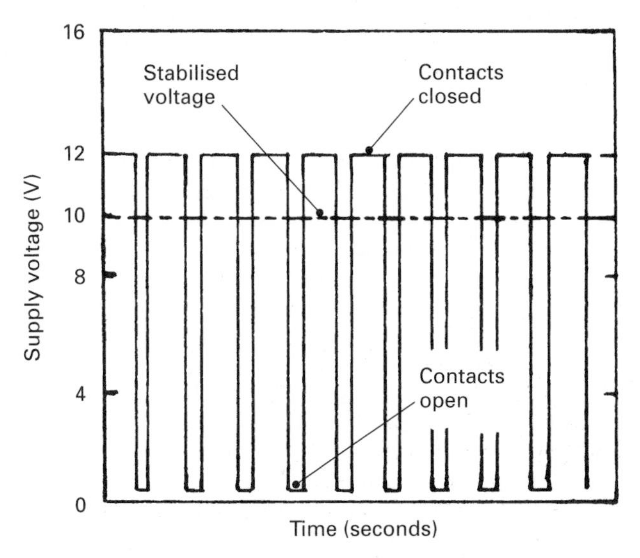

Fig. 28.10 Thermal stabiliser voltage pulse characteristics

Zener diode voltage stabiliser (fig. 28.11) This solid state voltage stabiliser is becoming more popular than that of the bi-metal thermal type since there are no moving parts, it is very cheap to manufacture and is proving to be more reliable.

A Zener diode voltage stabiliser consists of a resistor in series with the battery which reduces the voltage supply to the gauge to a value which can be consistently provided by the battery even if the battery is in a low state of charge. The Zener diode is wired in parallel with the gauge and transmitter, that is, it is shorted between the gauge and earth. When the voltage supply is below the Zener diode breakdown voltage V_z current will only flow through the diode in the forward direction, that is it blocks the current flow to earth. However, when the supply voltage exceeds the

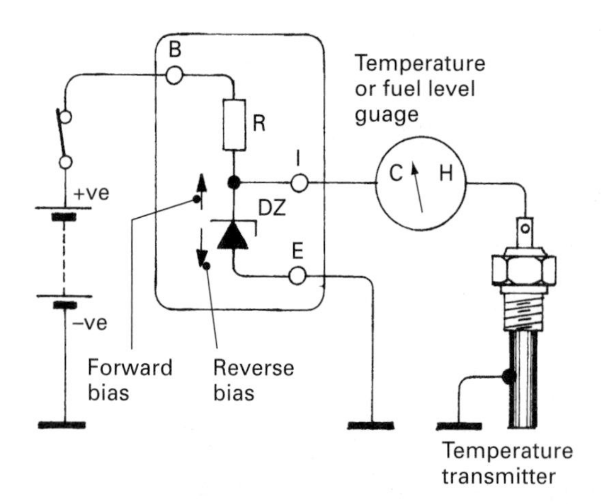

Fig. 28.11 Diode voltage stabiliser

breakdown voltage V_Z of the Zener diode, current suddenly commences to flow through the diode in the reverse direction (see fig. 28.12). Thus a proportion of the current supply by-passes the gauge and transmitter whenever the voltage tends to rise above the reference Zener voltage. It therefore keeps the supply output voltage to the gauge and transmitter constant.

The Zener diode voltages used for this purpose are either 6.2 volts or 9.1 volts depending upon the working voltage of the instrument panel gauges and transmitters.

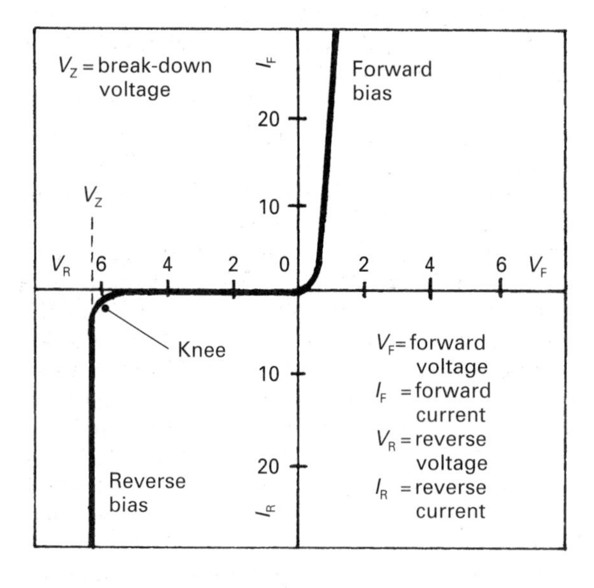

Fig. 28.12 Zener diode characteristics

28.1.6 Air-core cross-coil fuel level and coolant temperature gauges

Air-core cross-coil gauge (figs 28.13 and 28.14) This gauge, which can be used to measure fuel level or coolant temperature, consists basically of three coils wound across each other and connected in series. Two of the coils A and C are wound either side of a pivoting bar-magnet and dial pointer, and a third coil B is split so that each half of the coil is wound at right angles over coils A and C on either side of the pivoting magnet (see fig. 28.13). The bar-magnet is submerged in a viscous silicon fluid to damp out pointer fluctuations due to pulsating signal responses from the transmitter (see fig. 28.14) and a pull-off permanent magnet (not shown) is incorporated to return the moving bar-magnet and its pointer back to its zero position when the gauge is switch 'off'. All three coils are connected in series; coil A is connected to the battery's positive terminal via the ignition switch, while the free end of coil C forms the earth return to the battery's negative terminal, and interposed between coils A and C is the transmitter junction connection whose purpose is to provide a variable resistance proportional to the change in either fuel level in the tank or coolant temperature (see fig. 28.15). A ballast resistor R_b is connected in series with the coil A to step down the supply voltage and a second resistor R_c is shunted across coils A and B for calibration purposes.

Principle of operation (fig. 28.15) When the ignition switch is turned 'on' all three coils will be energised by the current flowing through the windings. A magnetic field will be produced in each winding and these all interact to produce a resultant field known as the magnetic axis along which the bar-magnet will be attracted and correspondingly the pointers will respond. The intensity of the magnetic field in each coil is proportional to the number of winding turns and the current passing through it, and since the number of winding turns is fixed, the only variable is the current flow, this being maintained very nearly constant in coils A and B but varying considerably in the split coil C see fig. 28.15(a) and (b)).

The magnitude and direction of these magnetic force fields can be shown in terms of arrow length and inclination. When the fuel tank is nearly empty there will be a high resistance in the fuel level transmitter. The current flow in coil C will

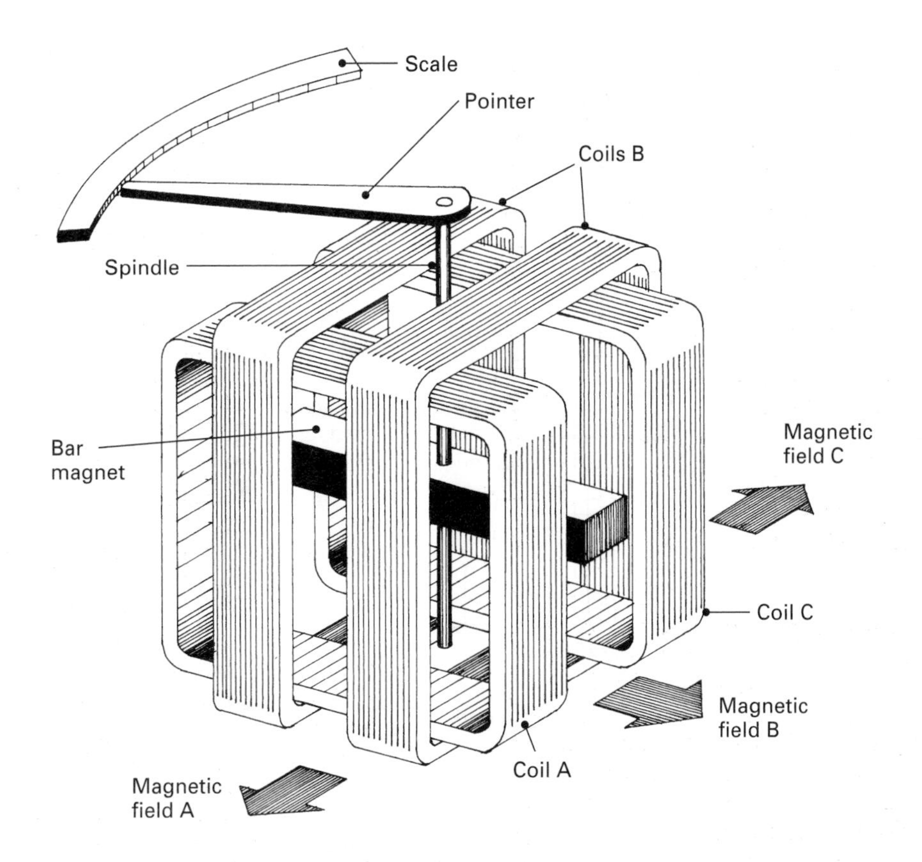

Fig. 28.13 Pictorial view of a air-core cross coil gauge

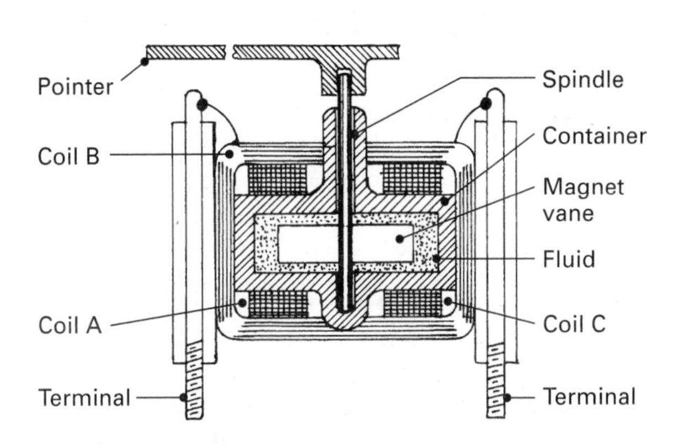

Fig. 28.14 Sectioned view of an air-core cross coil gauge

therefore be near its maximum and correspondingly so will its magnetic force field strength C_e as shown by the length of the horizontal arrow pointing towards the right. At the same time coils A and B which have a similar number of windings to each other produce their own magnetic force field. Coil A's force field acts in the opposite direction to coil C, whereas the split coil B acts perpendicular to both coils A and C. Subtracting the magnetic force field C_e from A_e and putting them into rectangular components produces a resultant magnetic force field R_e and its direction Θ_e is show in fig. 28.15(a).

As the float level moves towards full, the resistance of the fuel level transmitter decreases and therefore shunts a large proportion of current away from coil C. Correspondingly the magnetic force field in coil C decreases to C_f, whereas the intensity of the magnetic force field for coils A and B are still equal to each other but have increased very slightly due to the reduction in resistance of the current's return path through parallel circuits of coil C and the transmitter resistor. Thus subtracting the magnetic force field C_f from A_f produces the resultant magnetic force field R_f and its direction Θ_f which has now swung clockwise, that is, the bar-magnet will now have twisted through an angle of $\Theta_e + \Theta_f$ (see fig. 28.15(b)).

Note that the fuel level transmitter can be replaced by a semiconductor coolant temperature

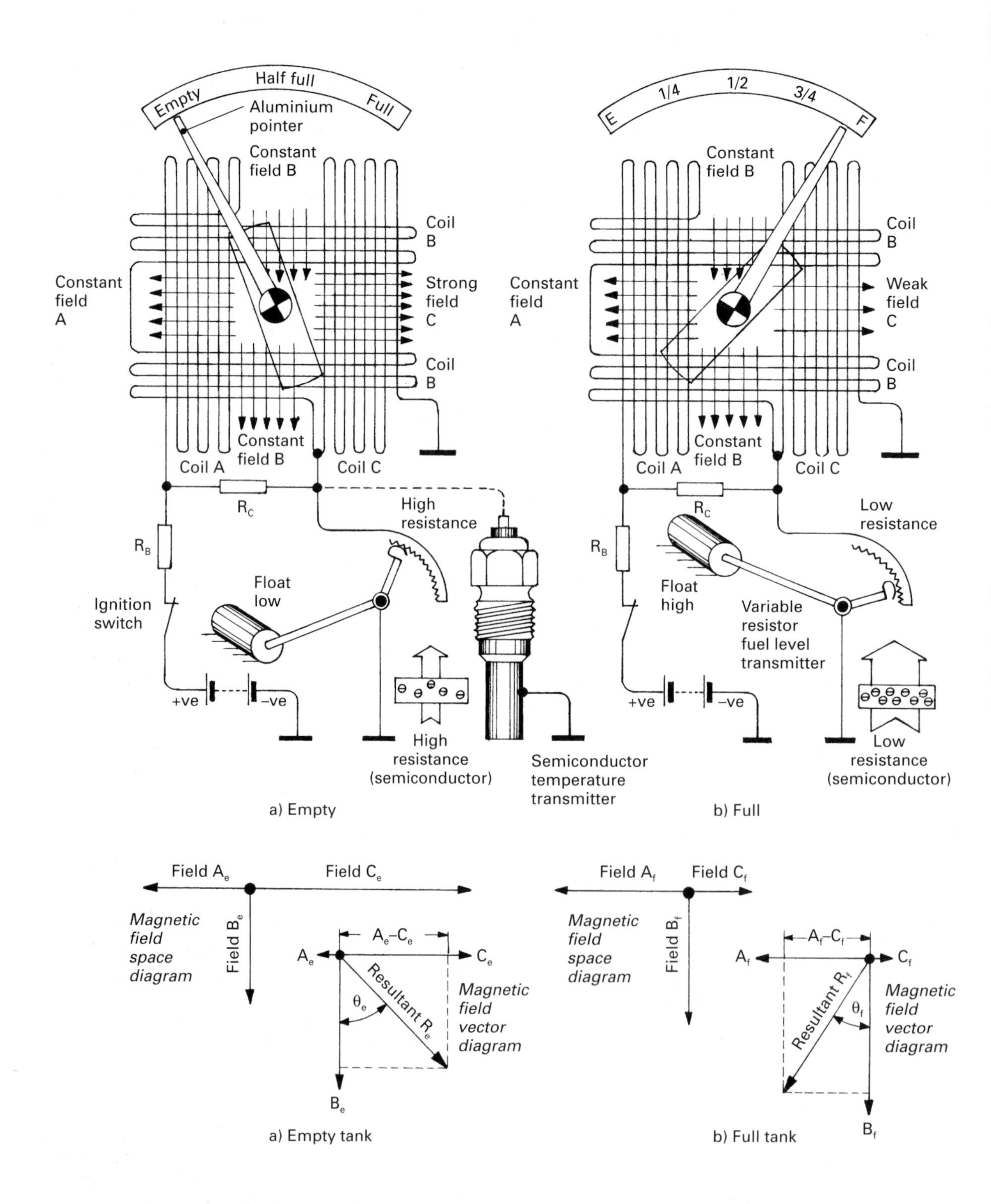

Fig. 28.15 Air-core cross coil fuel level gauge and variable resistor transmitter

transmitter to monitor the coolant temperature as the engine operating conditions change (see insert in fig. 28.15(a)).

28.2 Direction indicators

28.2.1 Thermal vane flasher unit (figs 28.16(a)–(e) and 28.17)
Such flasher units consist of a fixed contact mounted on a stud base and a ribbon strip held contact stretched across a spring steel vane which is thermally sensitive to its make and break switching (see fig. 28.16(a)–(e)).

The flasher unit is connected in series with the signal lights and therefore the switching frequency is regulated by the bulb load; consequently an incorrect bulb wattage rating will alter the number of times per minute the contacts open and close. A blown bulb reduces the current flow through the ribbon strip. This prevents the contacts opening so that the remaining serviceable signal bulb or bulbs and dash panel light illuminate but do not flash; similarly the normal clicking stops.

When the direction indicator switch is moved to either the right or left-hand side 'on' position, current flows via the flasher unit terminal B through the closed contacts, the metal ribbon and metal vane to the L terminal. It then passes to the selected signal light bulbs causing them to light up (see fig. 28.17). The heat generated by the current passing through the metal ribbon causes it to expand to a point where the tension of the distorted spring steel vane is reduced sufficiently to snap the vane and ribbon contacts upwards into its almost relaxed state against the stop arm, fig. 28.16(e). Instantly the contacts open, this interrupts the current flow and extinguishes the lights and at the same time the ribbon strip cools causing it to contract. Hence the vane bows once again make it deflect back to the closed contact position and current flow resumes to illuminate the signal lights, fig. 28.16(d). This make and break switching is automatic and continuous during the whole time the direction indicator switch is in one of the 'on' positions with the ignition switch also turned to the 'on' position.

Hazard warning lights (fig. 28.18) If the vehicle breaks down and is causing an obstruction, the hazard warning lights should be switched 'on'. The hazard warning lights are the same signal lights as the direction indicator lights and use the same flasher unit circuit. However they are wired up via the the hazard warning light switch to override both the ignition switch and the direction indicator switch (see fig. 28.18). The multipole hazard warning light switch is connected so that all the signal lights on both sides of the vehicle flash continuously with the engine switched off. An instrument panel hazard warning light is illuminated whenever the flasher lights are switched on so that the driver is made aware that they are operating when the vehicle is being driven.

28.2.2 Electronically controlled direction indicator (fig. 28.19(a) and (b))
In its simplest form the flash frequency is controlled by a stable free running solid state multivibrator with an output relay to switch the current supplied to the indicator light 'on' and 'off' (see fig. 28.19(a) and (b)).

When the direction indicator is switched either to the left or right 'on' position, both transistors T_1 and T_2 draw base current through resistors R_3 and R_4. However, owing to manufacturing tolerances one conducts more than the other. The higher transistor quickly saturates, that is, switches on, whereas the low conducting transistor cuts off. Each transistor switches automatically between its two states, that is, conducting and blocking. This switching 'on-off' cycle continuously repeats itself. As a result the output voltage can be taken from the collector c of either transistor but, in this case, the output is taken from transistor T_2 and is alternately high (the supply voltage) or low (zero volts). It therefore generates a continuous square pulse wave.

An output relay is still preferred as it can absorb high peak currents every time the signal lights are switched on, whereas a power transistor circuit capable of withstanding such high initial peaks would be much too expensive. A Zener diode Z_D is inserted between the feed terminals to stabilise the supply voltage and a capacitor C_1 is included to smooth out any voltage spikes coming in from the voltage supply source. A diode D is also included to protect the integrated circuit against incorrect polarity.

Operation

Transistor T_1 conducting T_2 blocked (fig. 28.19(a))
Assume T_2 was 'on' and has just switched 'off' while T_1 was switched 'off' and has just switched 'on'. Current now flows through to the base b of T_1 causing its collector-emitter (c-e) junction to

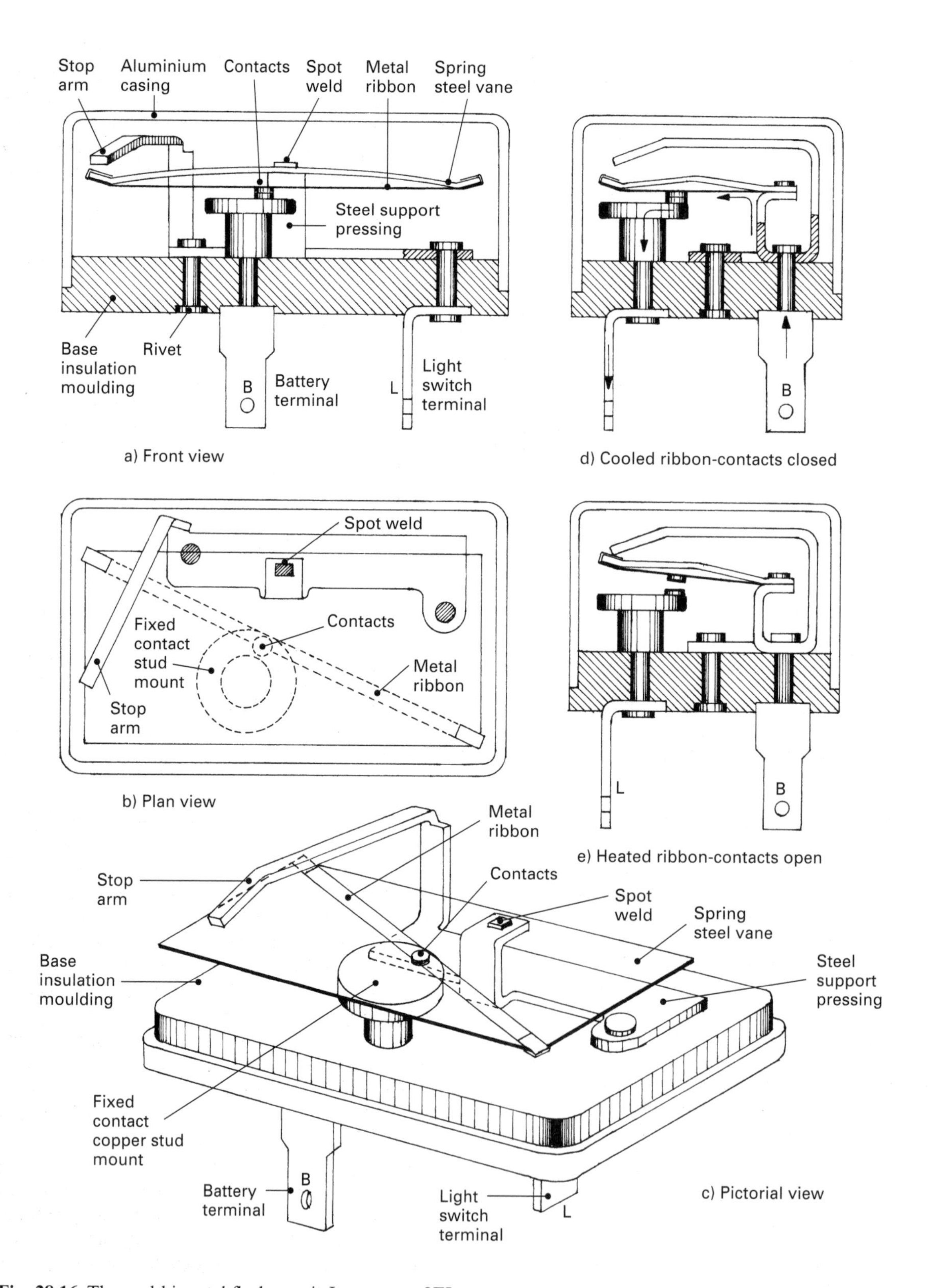

Fig. 28.16 Thermal bi-metal flasher unit Lucas type 8FL

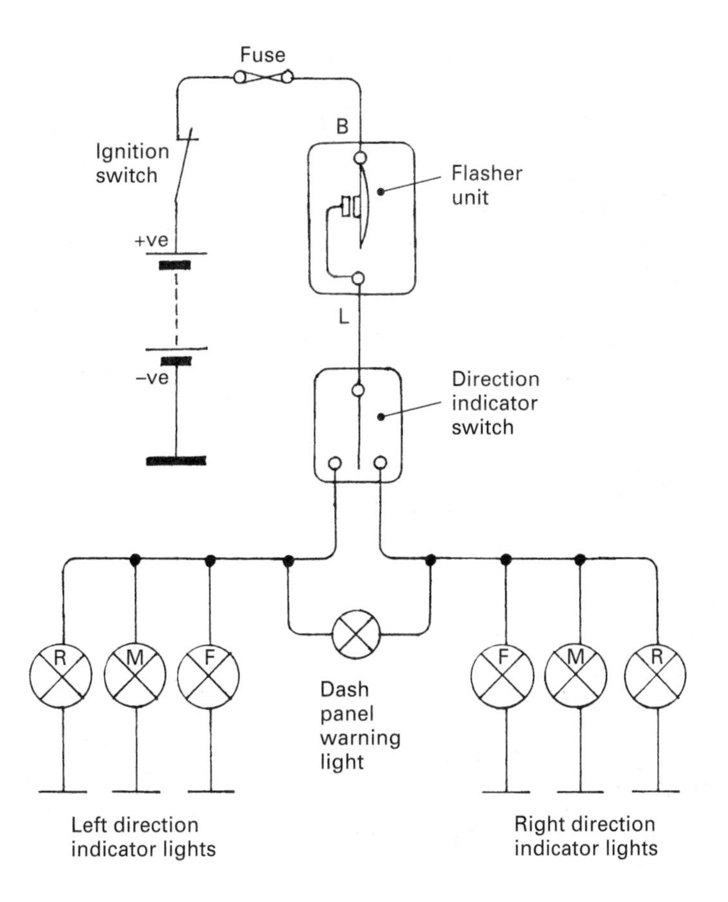

Fig. 28.17 Direction indicator lights and thermal flasher unit circuit

commence conducting. Instantly the voltage of the collector c of T_1 falls to zero but, since capacitor C_2 has not had time to discharge, it applies a negative voltage to the base b of T_2 thereby causing it to block (switch off).

Transistor T_1 blocks T_2 conducting (fig. 28.19(b)) Now capacitor C_2 starts to charge up via R_3 but, when it rises to 0.6 V (minimum base switch-on voltage) the base b of transistor T_2 quickly saturates and switches on. Meanwhile capacitor C_3 has been charged via R_6. Therefore when T_2 saturates (switches on) the collector c voltage of T_2 falls to zero. The charged capacitor C_3 now applies a negative voltage to the base b of T_1 causing it to switch off. With the collector voltage of T_2 now zero, capacitor C_3 commences to charge through R_4. However when its voltage reaches 0.6 V, so will the base b of T_1. It thus causes the transistor to saturate, that is, to switch on again. This circuit will continue to switch back and forth between the two conducting and blocking states at a rate controlled by the values of C_2, R_3 and C_3, R_6.

Every time transistor T_2 saturates and switches on, current passes from the emitter e of T_2 to the base b of T_3, thus switching on transistor T_3. Current now flows from the input positive supply through the relay shunt winding to the negative earth. The energised shunt winding's magnetic flux immediately pulls down the armature-arm, thereby closing the relay contacts. As a result current is now directed to either the left or right-hand side turn signal lights. An instant later transistor T_3 blocks and cuts off, the shunt winding's magnetic field collapses, and the relay contacts open. This interrupts the current flow to the turn signal lights and hence extinguishes the light, the whole cycle of events being repeated continuously until the indicator switch is moved to the 'off' position.

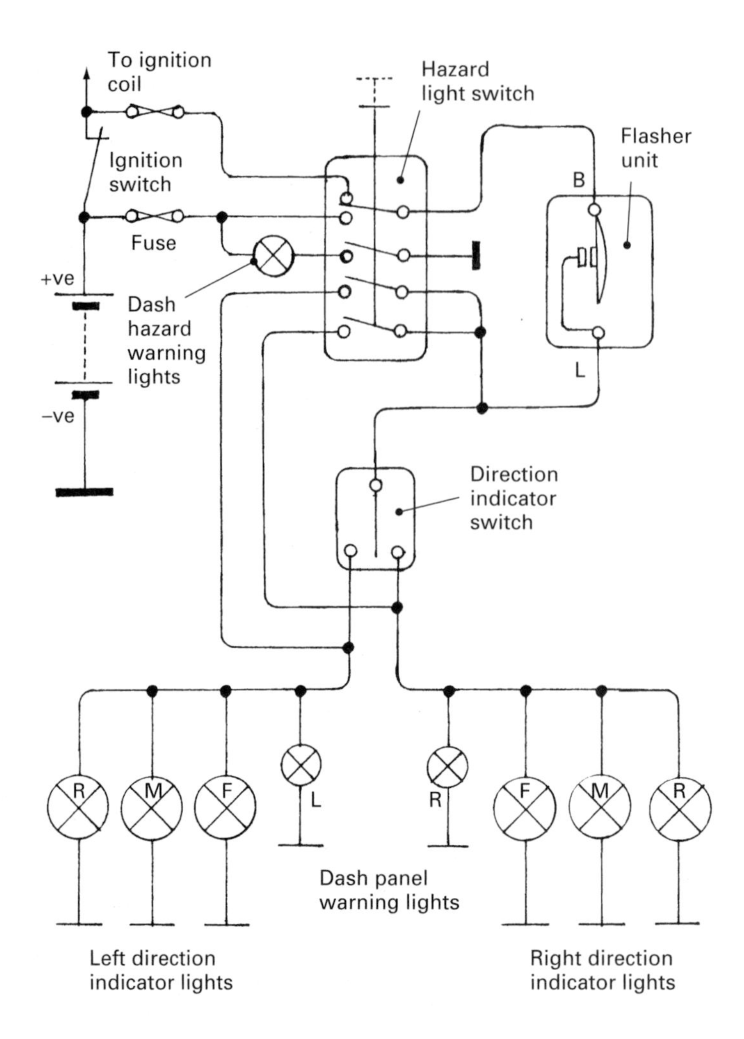

Fig. 28.18 Direction indicator lights and thermal flasher unit circuit with hazard warning light facility

28.3 Windscreen wiper systems (fig. 28.20 (a)–(d))
Wiper motors can either be of the two-pole shunt-wound field winding type or the more common permanent magnet two-pole design which dispenses with the field winding and uses a high-energy permanent magnet which consumes less current and is more reliable.

The shunt-field-winding-type motor may be single or two-speeds. The two-speed version incorporates a resistor which is switched in series with the shunt-field winding when low speed is selected, and is switched out when high speed is required (see fig. 28.20(a) and (b)).

The permanent magnet two-pole motor may be single or two-speed. The single-speed motor has two brushes situated on opposite sides of the commutator (see fig. 28.20(c)), whereas the two-speed wiper motor has three brushes. The third brush which is smaller than the other two is stepped and is positioned approximately 60° on the leading side of the negative brush (see fig. 28.20(d)). Low speed is obtained when current is supplied to the two brushes diametrically opposite to each other. The high wiper speed circuit is completed when the supply current is switched between the positive large brush and the negative narrow off-set third brush. The output rotational speed after the gear reduction is in the order of 45 and 65 rev/min for low and high speed respectively. Generally the wiper motor is wired up via a fuse on the ignition coil side of the ignition switch. This means that it cannot be energised when the ignition switch is turned off.

Parking the wiper blades in their lowest posi-

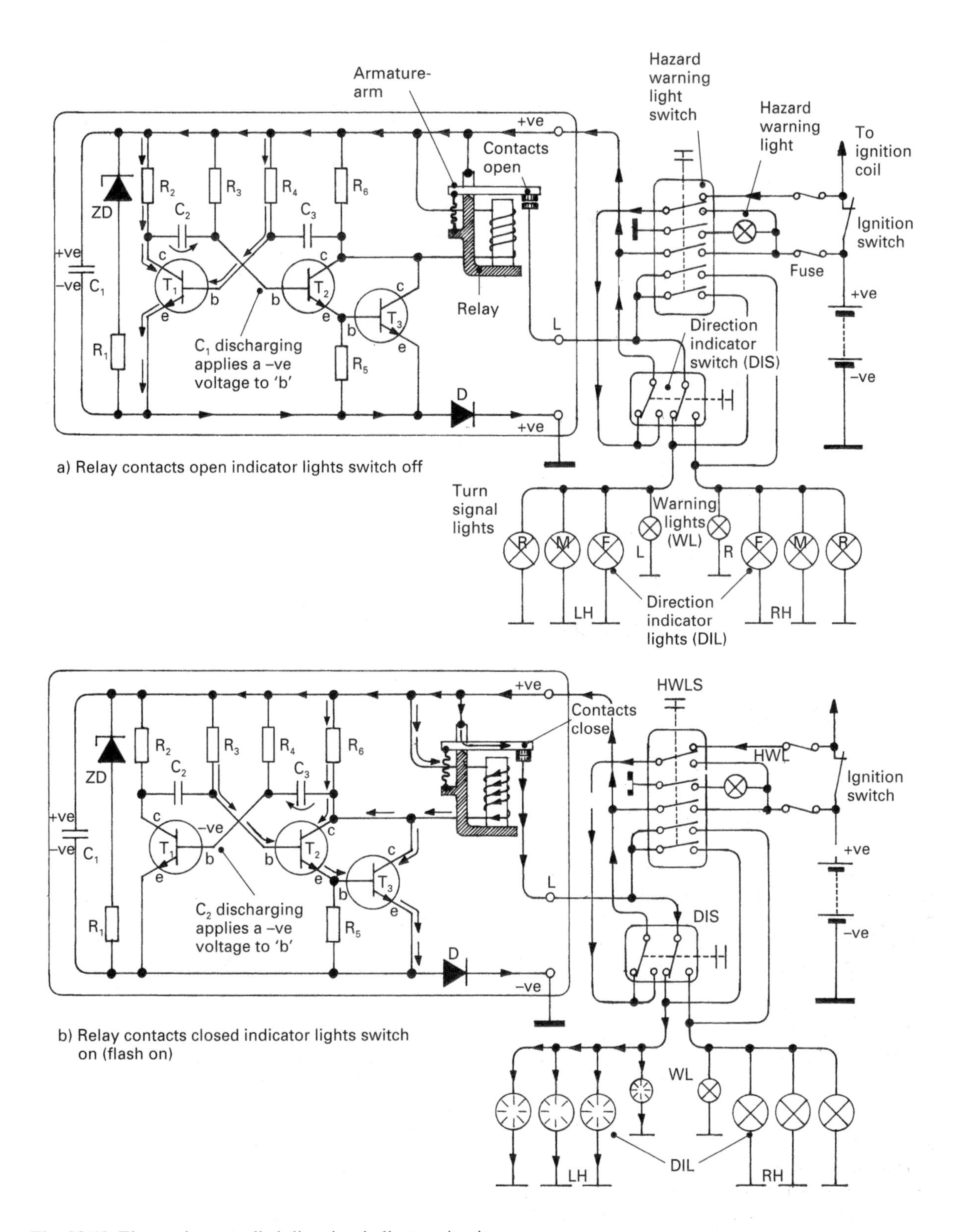

Fig. 28.19 Electronic controlled direction indicator circuit

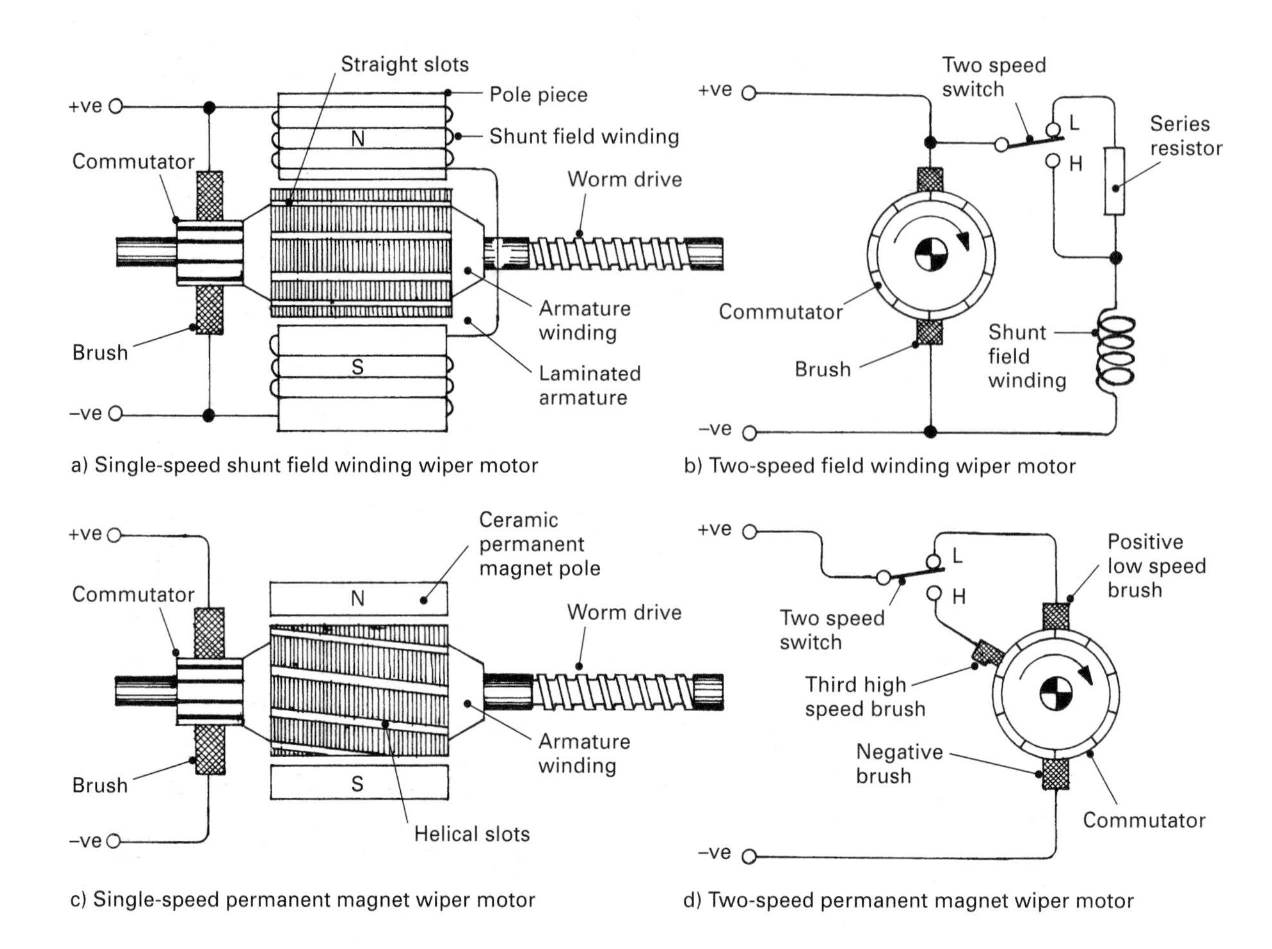

Fig. 28.20 Single and two-speed windscreen wiper motors

tion on the windscreen when the wiper motor switch is turned 'off' is achieved by a separate switching arrangement to be described later in the text. Overload protection of the wiper motor is usually obtained via a thermal overload switch which breaks the supply voltage circuit whenever the current drawn exceeds some pre-set value. Excessive current draw can be caused by the wiper blades becoming stuck or seizing, wiping on a dry screen, experiencing heavy rain, snow or ice. The contacts of this type of switch usually close once the switch has cooled. However, until the cause of the overload is removed, these contacts will continuously break and make.

28.3.1 Permanent magnet two-pole motor (fig. 28.21)

With this type of motor, the magnetic field or flux is produced by the high-energy ceramic permanent N and S pole magnets which are mounted inside the cylindrical yoke cover. To reduce eddy current losses in the armature, a laminated core is used. The armature core is supported on the armature-shaft and is mounted between the poles to form a conducting path between the poles for the magnetic field. When current is passed through the armature windings via the two carbon brushes, a separate magnetic field is established around the longitudinally located conducting wire of the windings. This second electromagnetic field created around the winding conductors interacts with the two pole permanent magnet field. Consequently a distorted resultant magnetic field is produced which exerts a net twisting effort on the winding conductors, causing the armature to rotate.

Connecting the ends of one of the armature winding loops to a battery via a pair of slip-rings

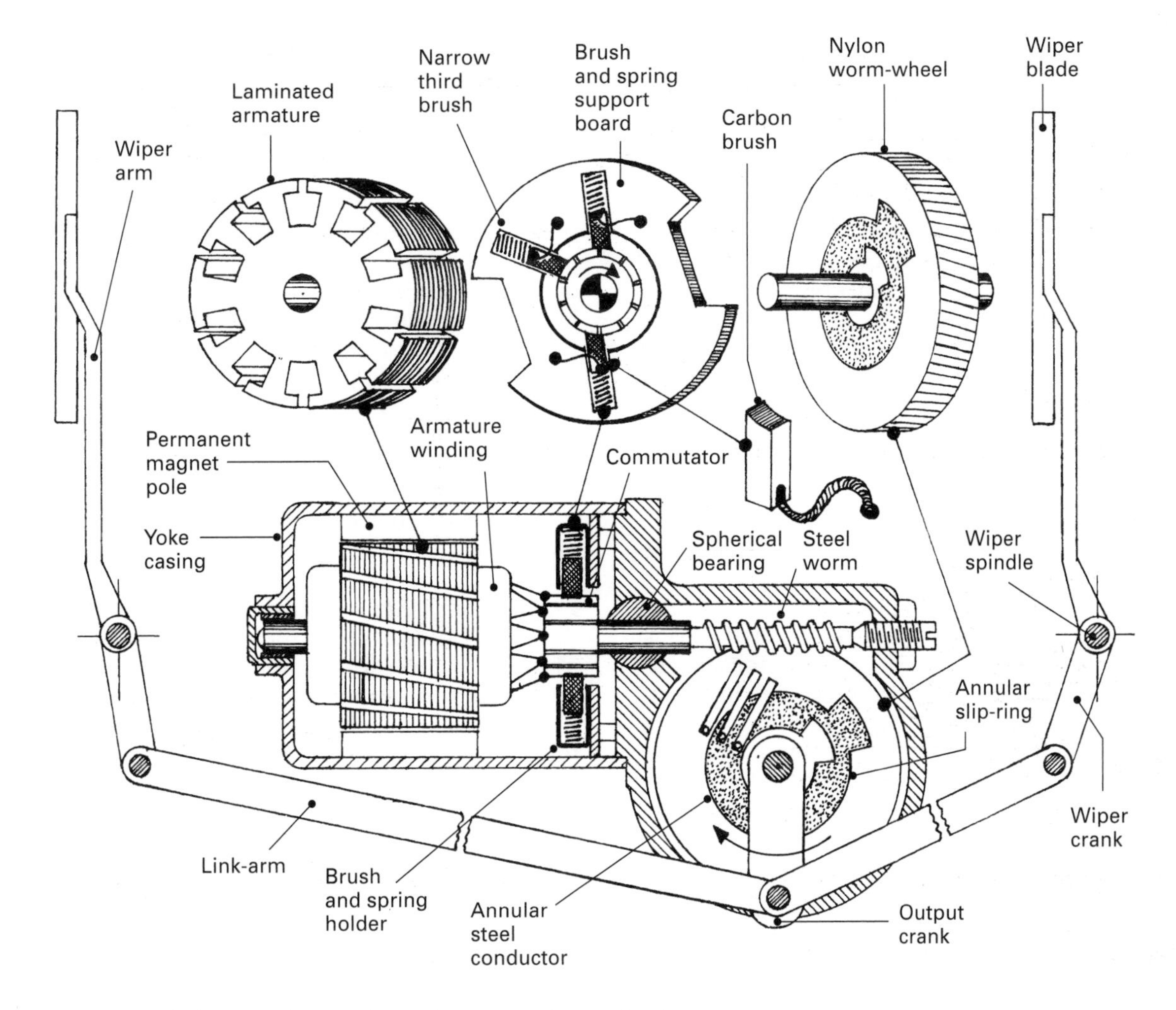

Fig. 28.21 Windscreen wiper and electric motor drive mechanism

and brushes would create a magnetic field interaction, causing the two half conductors of the loop to rotate a half revolution. However, the current flow through the loop will still be in the same direction when the two half conductors change places; consequently the magnetic field interaction will reverse the direction of the magnetic force field imposed on the conductor, thus causing the loop to rotate back to its original position. To overcome this problem, the direction of current flow to each winding loop can be continuously reversed every half revolution of the armature by connecting the winding loop ends to two segments of a split-ring. Current supplied to these segments via the brushes positioned diametrically opposite each other enables a different segment of the split-ring to contact each of the fixed brushes every half revolution of the armature. As a result, the direction of the current flow in the winding is repeatedly changing every 180° of armature rotation. Thus the permanent magnet poles and the winding loop conductor's magnetic field interact in such a way as to exert a force on each of the two conductors perpendicular to the magnetic field of the two pole-pieces. Thus the two forces act in the opposite sense to each other and therefore makes the loop rotate in one direction only.

The wiper motors which are normally used have a series of ten winding loops wound around and through the slots of the laminated armature These windings are connected also to ten segments of the multi split-ring known as the commutator. Thus instead of one magnetic pull pulse every half revolution of the armature shaft if a single loop winding were to be used, there are ten pull pulses, so that there is a magnetic torque pulse every 30° of armature rotation. This number of windings provides a relatively uniform and continuous mean torque suitable for driving the windscreen wiper blades.

Back EMF and the constant speed motor (fig. 28.23(b)) With the permanent magnet motor the flux field between the magnetic poles and the supply voltage to the armature windings are both constant. A generator is constructed like a motor, in as much as it has an armature and windings, commutator, brushes and pole-pieces which are permanent magnets or electromagnets, that is there is a field winding. Therefore, when a supply voltage is applied across the brushes, the armature is made to rotate and at the same time the rotating armature windings generate their own voltage. This voltage is known as the back emf (emf is short for electro-motive-force; the back emf = supply voltage − voltage drop in the armature winding circuit), and opposes and cancels out an equal amount of the supply voltage. The speed of the motor will increase until the build-up of the voltage generated by the motor itself reduces the net supply to a point where there is only sufficient voltage to overcome the output load without any further increase in armature speed. Hence this type of motor has a constant speed characteristic since there is very little change in speed over the normal working load range of the motor.

Windscreen wiper drive mechanism (fig. 28.21) Power is transferred from the armature-shaft to the output through a worm and wheel 50:1 gear reduction. Either a single or a two-start worm-gear is machined on one end of the armature shaft. This meshes with a 100 or 50 toothed nylon-worm wheel to match the worm starts respectively. Thus if the wiper motor crank-arm is designed to rotate at low and high speeds of 45 and 65 rev/min respectively, then the corresponding armature speeds will be $45 \times 50 = 2250$ rev/min and $65 \times 50 = 3250$ rev/min respectively for a 50:1 gear reduction.

The wiper motor may be central or off-set mounted depending upon the space available for the linkage. With the centrally positioned motor, see fig. 28.22(a), motion is transferred from the motor crank-arm to each wiper spindle via a push-pull link and crank-arm, whereas if the motor is mounted to one side of the windscreen, see fig. 28.22(b), the drive is taken through a push-pull link to the nearest wiper spindle crank-arm and then through a connecting arm to the second wiper spindle crank-arm. It should be observed that the motor crank-arm continuously rotates, whereas the wiper arm oscillates to and fro for every revolution of the motor crank-arm. The amount of oscillatory movement of the linkage mechanism and also the angle of wipe is determined by the length of the crank-arm. However the blade sweep is normally restricted to a maximum of 120°.

The large inertia and frictional resistance produced at the wiper reversal point means that great care must be given to the geometry of the wiper spindle and the motor crank spindle's relative position so that the wiper reversal point occurs near the ineffective crank-angle movement at the motor end.

28.3.2 Two-speed permanent magnet wiper motor (fig. 28.22(a) and (b))

These two-speed wiper motors are designed to operate at two distinct speeds namely, 45 and 65 wiping cycles per minute for low and high speed respectively. The armature torque produced is proportional to the product of the magnetic flux between the poles and the armature current, whereas the armature speed is proportional to the supply voltage and inversely proportional to the flux between the poles.

Low speed (fig. 28.22(b)) When the two brushes positioned diametrically opposite each other are connected to the voltage supply, the torque produced by the magnetic pull on the armature under stall conditions will be at a maximum. Once the static resistance of the wiper has been overcome the armature's speed rapidly rises to the steady low speed range while its output torque drops to something like 20% of its stall torque.

High speed (fig. 28.22(b)) The upper wiper speed range is obtained by incorporating a third brush which leads the negative brush by 60°. Therefore the supply voltage is applied between the third brush and the positive brush which is positioned furthest from it. This results in fewer winding loops between the 120° apart brushes and more loops on the opposite 240° apart side. Hence the series resistance of the loops trapped at any instant between the 120° brush spacing is lower than if the brushes were spaced 180° apart. Accordingly more current will flow through the

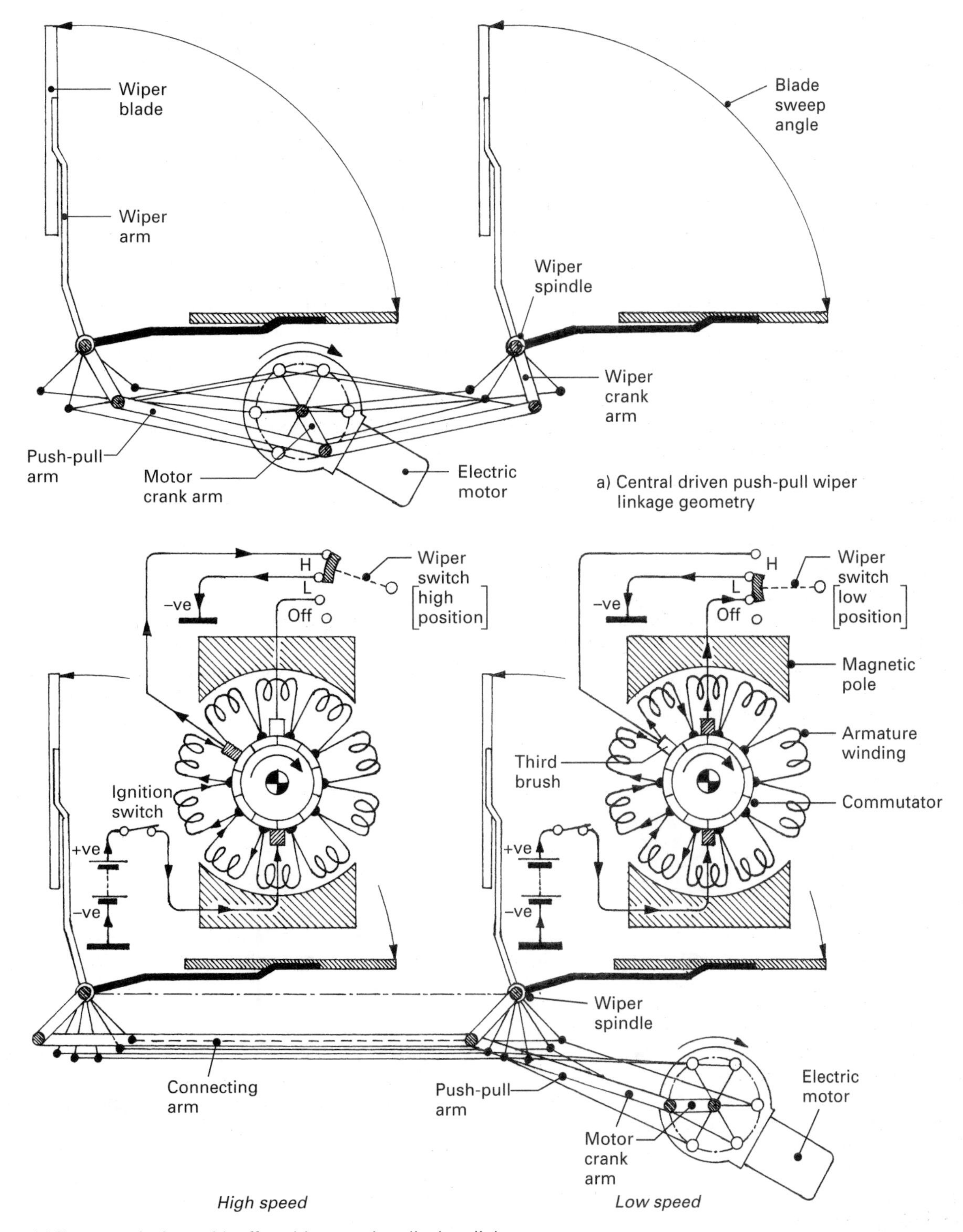

Fig. 28.22 Wiper linkage geometry

loops between the 120° spacing and proportionally the magnetic force field intensity in this region increases. Typically 1.5 amps is used for the low speed 180/180° brush spacing and 2.0 amps for the high speed 120/240° brush spacing when tested under light load with the wiper link arm disconnected. Consequently the armature will rotate faster with the 120° high magnetic field intensity brush spacing but, since the intensity of the magnetic force field is weaker for the remaining 240° angular displacement, its overall stall torque capacity is less than when the brushes are opposite each other as when in the low speed setting. Thus the high speed range should not be used in heavy snow or on a partially wet windscreen.

28.3.3 Two-speed wiper with cam-operated limiting and braking switch

The need for a parking switch mechanism (fig. 28.23(a) and (b)) With the basic wiper motor circuit, when the wiper switch is moved to park, the supply voltage is switched 'off' causing the wiper blade sweep to stop somewhere between its two extremes. The actual position when the blades come to rest will depend at what angular position the armature supply current is cut off, the inertia of the blade system and the friction drag of the blades on the windscreen. This problem is largely overcome by a limiting switch. Its contacts are made to open and interrupt the voltage supply at the same angular position of the armature whenever the wiper switch is moved to park. This occurs regardless of the position of the wiper blades when the wiper motor switch on the steering column is moved to the 'off ' position.

Regenerative or dynamic braking (fig. 28.23(b)) To rapidly bring the armature to a standstill when the voltage supply to the brushes is switched off, an additional pair of contacts to that of the limiting switch is shunted across the two brushes. This shorting of the brushes results in the armature winding's self-induced current flowing in the opposite direction to that of the supply current when it was switched on. This self-induced current therefore produces its own magnetic field which interacts with the magnetic field between the pole-pieces in such a way as to attempt to reverse the rotational direction; hence it speedily brings the armature to rest.

Wiper two-stage park switch (fig. 28.23(a) and (b)) Wiper parking is controlled by a cam and plunger-operated two-stage contact. The first stage cuts off the supply voltage to the armature winding whenever the wiper blades reach the parking edge of the windscreen, whereas the second-stage contacts short circuit the commutator brushes. This brakes the armature, thus rapidly bringing the blades to a stop in their park position.

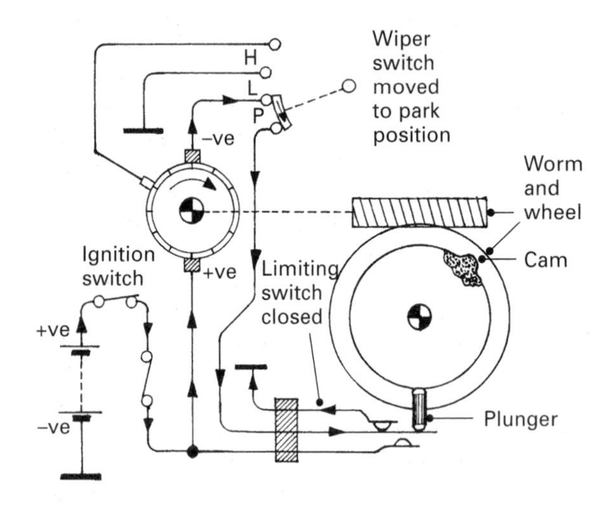

a) Wiper switch moved to off position, electric motor continues to rotate

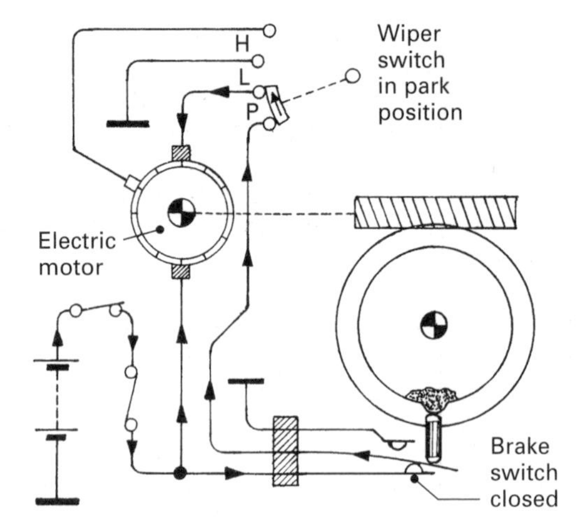

b) Wiper switch in off position, limiting switch opens and brake switch closes, electric motor stops

Fig. 28.23 Two-speed wiper with cam operated limiting and braking switch

Wiper switch moved to park position (fig. 28.23(a) and (b)) When the wiper switch is moved to the park position the negative earth return is completed by passing through the closed contacts of the limiting switch. This means that the armature continues to rotate (see fig. 28.23(a)) until the cam via the plunger on the side of the nylon worm-wheel pushes open the first-stage limiting contacts. The supply voltage to the armature winding is therefore instantly cut off. An instant later the cam via the plunger closes the second stage regenerative brake contacts (see fig. 28.23(b)). This causes the armature winding brushes on the commutator to be joined together (shunted), so that the regenerative brake action quickly brings the armature to a standstill.

Wiper switch moved to 'on' position When the wiper switch is turned to either the low or high-speed position, the limiting switch contacts are overridden by the wiper switch contacts which then connect the negative brush directly to earth. Immediately the supply voltage is connected to the armature windings via the brushes and commutator, the energised armature commences to rotate and to actuate the wiper blade sweep motion.

28.3.4 Two-speed wiper with slip-ring reed contact limiting and braking switch (fig. 28.24(a)–(d)) With this design wiper parking is controlled by three fixed copper reed contacts and are so positioned that they bear against a rotating annular shaped contact plate attached to the worm-wheel. The centre supply contact is connected to the wiper switch park contact. The left hand brake contact short circuits the commutator's two brushes every time the wiper blades arrive at their park position when the wiper switch is moved to park, whereas the right hand limiting contact acts as an earth-return with the wiper switch in the park position.

High speed (fig. 28.24(a)) With the steering-column-mounted wiper switch moved to high position and the ignition switch turned on, current flows from the battery to the positive brush through the armature-winding and out to the negative off-set third brush. The circuit is then completed by current passing to the wiper switch through the high 'H' contact to the earth-return via the switch shunt-bar. The energised armature now rotates and drives the oscillating wiper blades at a high speed.

Low speed (fig. 28.24(b)) With the wiper switch in the low position and the ignition switch turned 'on', current flows from the battery to the positive brush, through the armature-winding and out to the diametrically opposite negative brush. The current is then completed by current passing to the wiper switch, to the low 'L' contact to the earth-return contact via the switch shunt-bar. The energised armature now rotates and drives the oscillating wiper blades at a reduced speed.

Wiper switched to park – electronic motor continues to rotate (fig. 28.24(c)) With the wiper switch moved to park with the ignition switch still turned on, current continues to flow from the battery to the positive brush, through the armature and out to the opposite negative brush. The current then passes to the wiper switch from the low 'L' contact, through the shunt-bar to the park 'P' contact. It then passes to the middle supply contact. The circuit is then completed by current passing from the supply contact to the limiting contact earth-return via the rotating annular steel conductor. Hence the armature continues to rotate.

Wiper switch remains in park position – limiting switch contacts open and brake contacts closed – electric motor stops (fig. 28.24(d)) With the wiper switch in the park position and with the ignition switch still turned on, current continues to flow from the battery's positive terminal through the armature and wiper switch, to the middle supply contact, and to earth via the limiting contact brushing against the rotating annular conductor. However, as the worm-wheel continues to revolve, the annular conductor cut-out inner segmental gap aligns with the limiting contact. Immediately the earth return circuit is interrupted so that current ceases to be supplied to the armature. At the same time the brake contact aligns with the outer segment extension of the annular conductor. Consequently the positive and negative armature brushes become shunted together through the limiting and braking contacts bearing against the armature conductor. As a result the self-induced current in the armature windings (motor acting as a generator) flows in the opposite direction to that when the supply current was being delivered. It therefore produces an armature magnetic field which interacts with the pole magnetic field in such a way that it tends to reverse the rotational direction of the armature. The armature is thus rapidly brought to a standstill in the

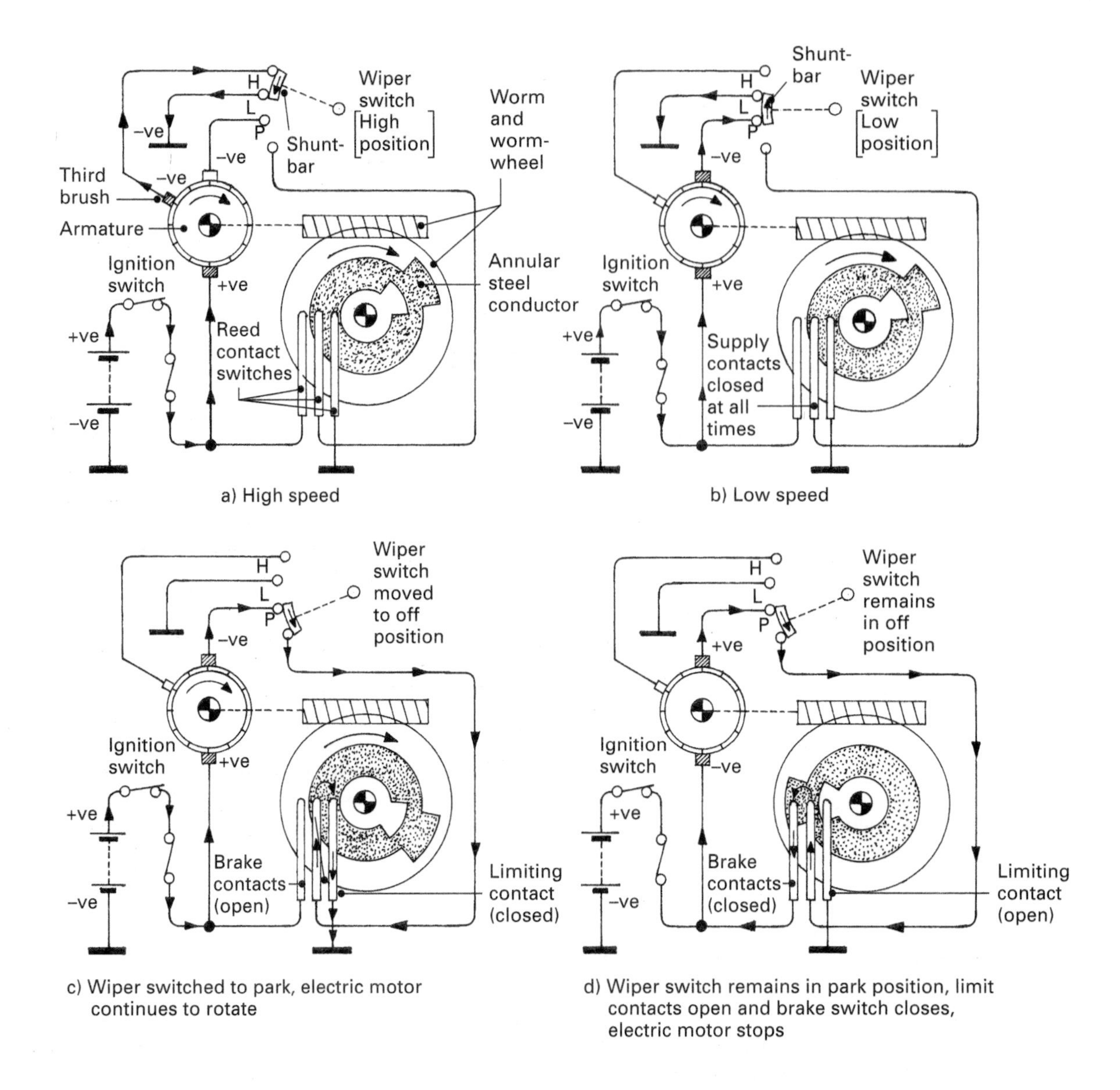

Fig. 28.24 Two-speed wiper with slip-ring reed contact limiting and braking switch

desired angular position, that is, to park the wiper blades against one of the edges of the windscreen.

28.3.5 Two-speed and intermittent windscreen wiper system (fig. 28.25(a)–(d))

Wiper switch moved to intermittent position (fig. 28.25(a) and (b)) When the wiper switch is moved to the intermittent position, current flows from the battery positive terminal through the armature windings via the commutator's low-speed negative and positive brushes to the intermittent terminal 'I' contacts of the steering column wiper switch, and then to the upper contact of the solenoid relay switch via its 'L' shaped frame (see fig. 28.25(a)). Current is then directed back to the steering column wiper switch to the inner earth-return terminals to complete the circuit to the negative side of the battery. The energised armature windings will now rotate the armature on its low-speed setting. It should be remembered that every time the wiper switch is moved to park, the armature comes to rest in the same angular position where the limiting contact aligns with the inner segment

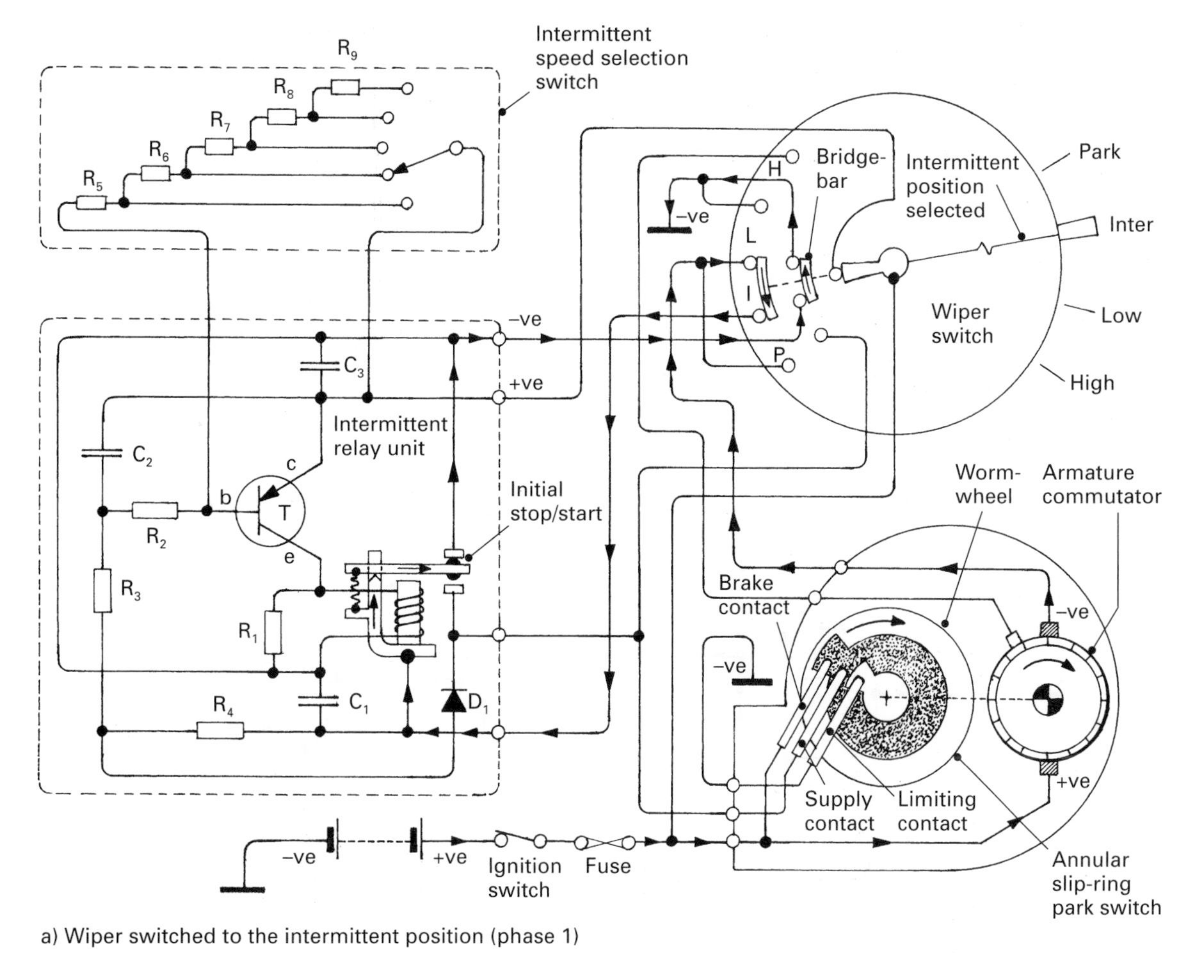

Fig. 28.25 Two-speed and intermittent windscreen wiper system

insulation gap, whereas the brake contact bears on the outer segment of the annular conductor.

Thus as soon as the armature commences to rotate, the limiting switch contact closes on the annular slip-ring to form an earth-return. The capacitor C_2 is now earthed through the resistor R_3 and diode D_1 via the closed limiting switch contact. It is therefore able to charge the capacitor C_2 and to apply a positive voltage between the base 'b' and emitter 'e' of the transistor T via the potential divider resistors R_6 (different resistors between R_5 and R_9 may be selected), R_3 and R_2 which instantly switches on the transistor T (see fig. 28.25(b)). Current will now pass between the collector 'c' and emitter 'e' of the transistor to energise the relay shunt-winding. Its armature-arm is thus pulled down, opening and closing the upper and lower contacts respectively. The earth-return for the capacitor C_2 is now switched from the steering column wiper switch to the limiting contact earth-return of the motor's slip-ring parking switch. As the motor armature revolves, the limiting contact earth-return opens the circuit, causing the motor armature to stop.

However the capacitor C_2 now commences to discharge current to the base 'b' of the transistor T, so that it remains switched on, that is, current continues to pass between the collector 'c' and the emitter 'e'. Thus the lower relay contacts remain closed. The discharge time of the C_2R_3 circuit is approximately the interval of intermittent wiper operation when the motor is switched off and the wiper blades are inactive. When the capacitor C_2 is discharged, the transistor T switches off, blocking current flowing to the relay winding. This causes the relay spring-loaded lower contact to

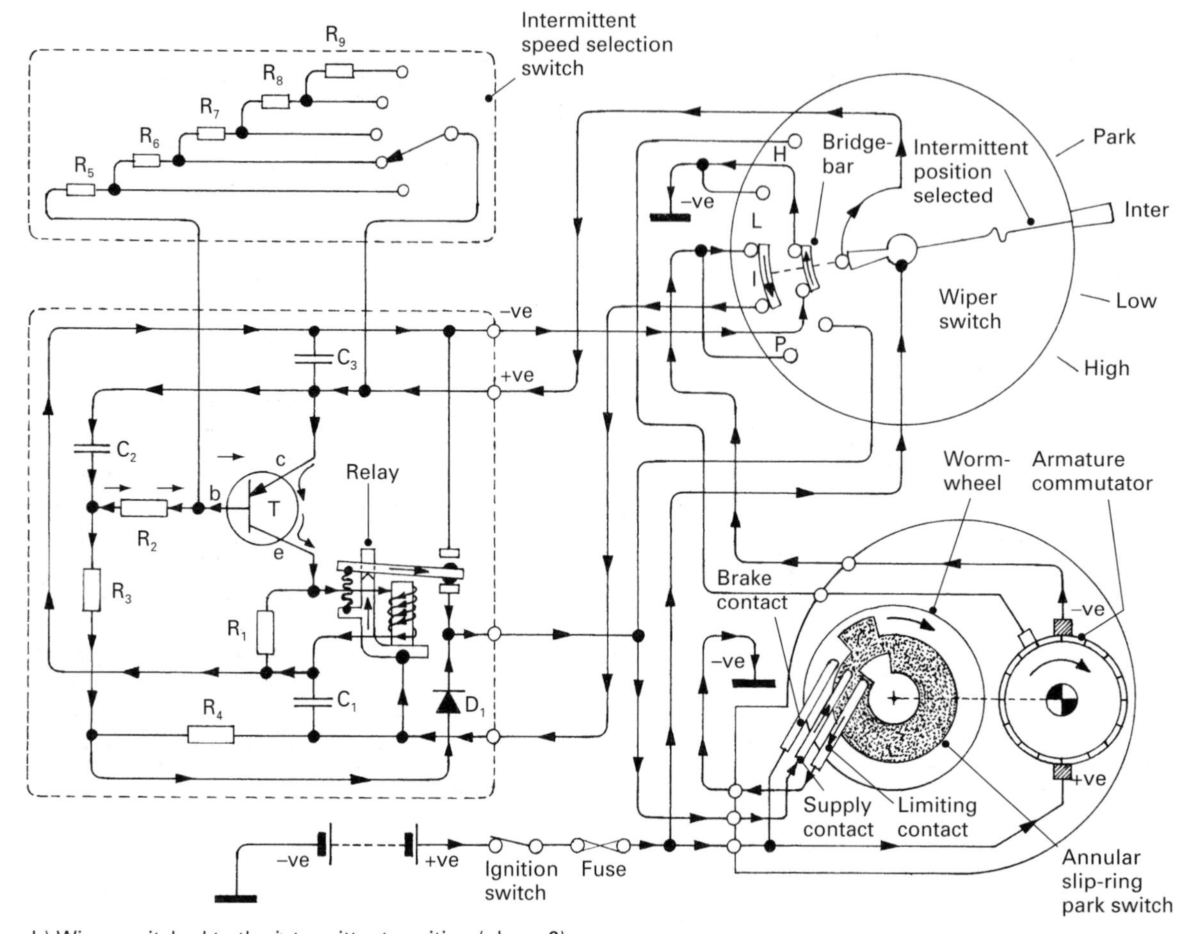

b) Wiper switched to the intermittent position (phase 2)

Fig. 28.25 Two-speed and intermittent windscreen wiper system (*continued*)

open and for the upper ones to close (see fig. 28.25(a)). Consequently the earth return for the negative armature brush is again completed through the steering column wiper switch. The re-energised armature once more commences to rotate and to drive the wiper mechanism.

The discharge time can be shortened or extended from roughly three seconds to twelve seconds by changing the voltage drop across the resistor connecting the positive supply voltage to the base 'b' of the transistor T, that is, different resistors from R_5 to R_9 can be switched in to vary the intermittent period (inactive time) of the wiper motor.

Low wiper speed (fig. 28.25(c)) With the steering column wiper switch moved to low speed position, current flows from the battery supply through the armature low speed positive and negative commutator brushes to the wiper switch low speed earth-return terminal 'L' via the bridge-bar. This completes the circuit from the negative armature brush to earth to energise and rotate the motor armature and to activate the wiper blades at the slower speed.

High wiper speed (fig. 28.25(d)) With the steering column wiper switch moved to high speed position, current flows from the battery supply through the armature positive and negative high speed third brush to the wiper switch high speed earth-return terminal 'H' via the bridge-bar. This completes the circuit from the negative third brush to earth to energise and rotate the motor

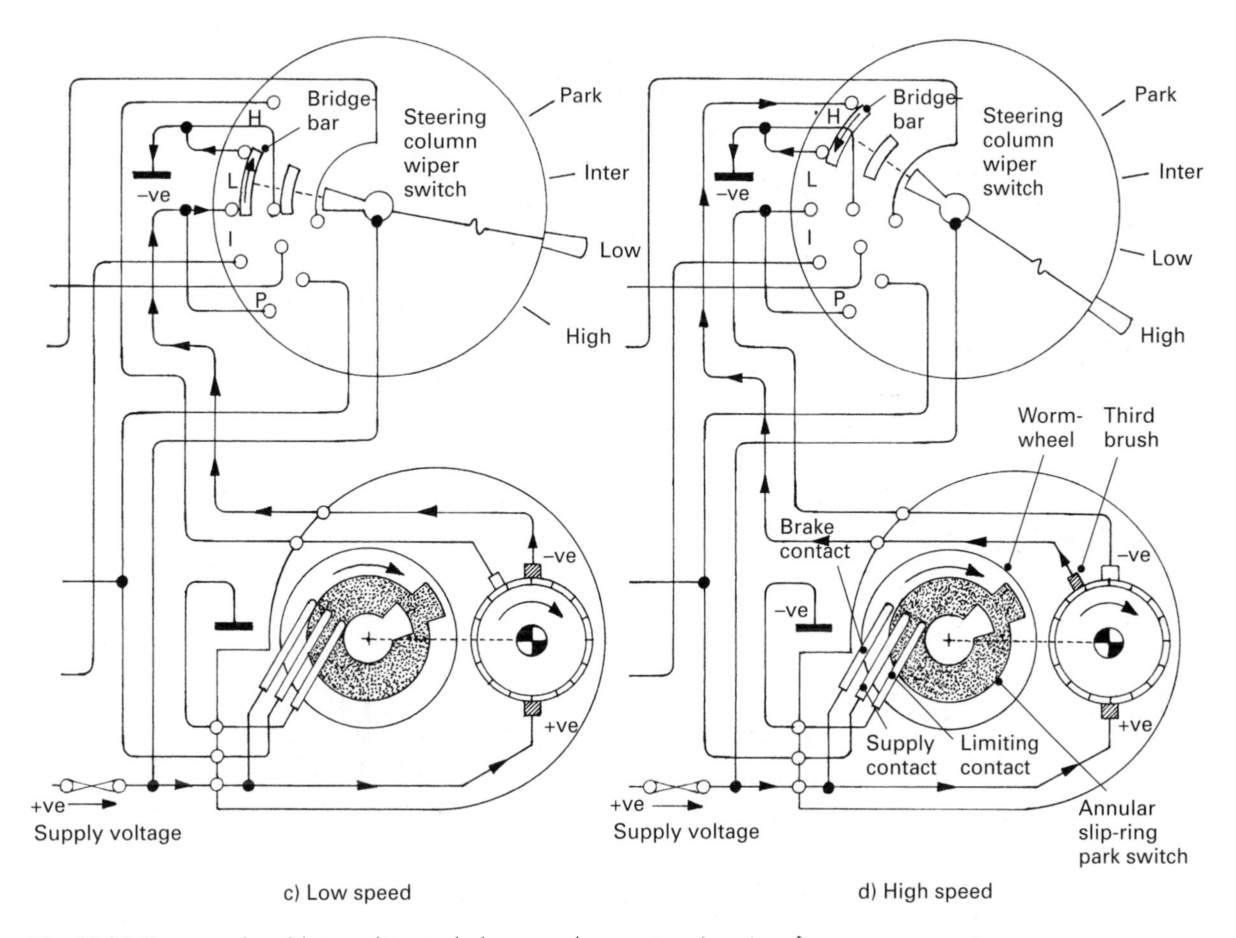

Fig. 28.25 Two-speed and interm ittent windscreen wiper system (*continued*)

armature, thereby activating the wiper blades at the high speed setting. This high speed should not be used if there is ice and snow on the windscreen.

Wiper parking (fig. 28.25(e)) With the steering column wiper switch moved to park position, current flows from the battery supply through the slow speed positive and negative commutator brushes to the wiper switch park terminal 'P' to the output terminal via the outer bridge-bar. Current then passes to the body of the relay and through its upper closed contacts. It is then directed back to the wiper switch to the inner park terminal. The inner bridge-bar links the circuit path again back to the middle supply contact which bears against the slip-ring. The earth-return circuit is then completed via the limiting contact. The armature continues to revolve until the inner segmental insulator gap in the slip-ring aligns with the limiting contact and the outer segment conductor aligns with the brake contact. At this point the earth-return path for the armature brush is interrupted and the slow speed negative and positive brushes are short circuited. This results first in the current supply to the armature to being cut off and secondly in the self-induced current (motor acts as a generator) momentarily reversing the resultant direction of pull by the magnetic interaction so that the armature comes to an abrupt halt.

28.4 Power window winders

Motor-driven window winders generally consist of a dual polarity direct-current motor with a built-in circuit breaker for each power window with its worm and worm-wheel gear reduction and a 'lift' and 'lower' mechanism which can take the following forms:

1) sector and pinion gear and lever
2) pulley wheel cable and reel
3) cable rack and pinion.

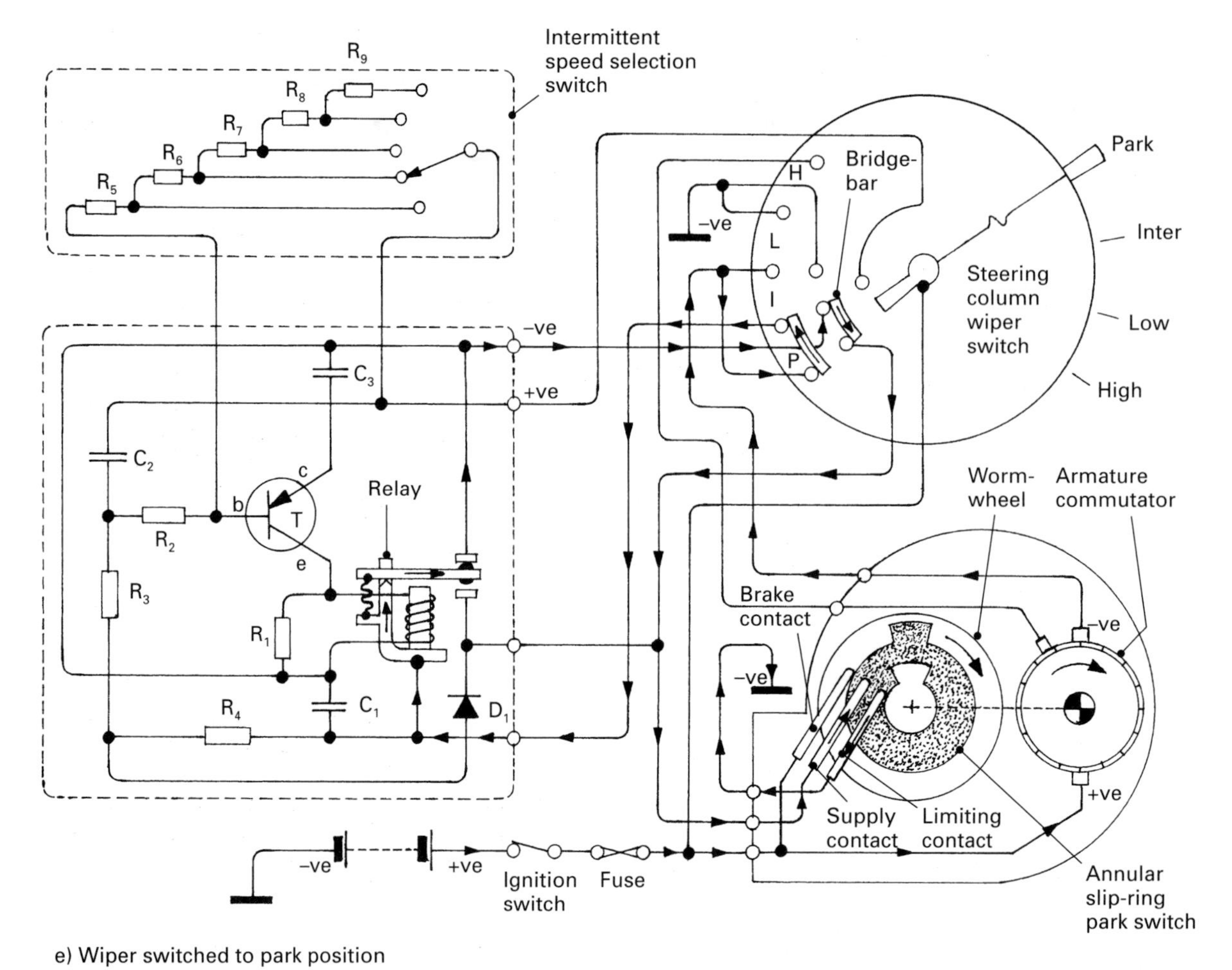

Fig. 28.25 Two-speed and intermittent windscreen wiper system (*continued*)

The individual power window winder can be operated by an up–down switch mounted in each door for passengers to operate and by a master switch panel mounted at some convenient position to allow the driver to operate each passenger door window. There is also a main switch which can be used to override the passengers door window switches, thus preventing children opening and closing the windows but at the same time still permiting the driver to operate his own door window switch.

Each motor has a circuit breaker incorporated to prevent overloading and damaging the motor and possibly the wiring circuit.

The up or down motion of the windows is achieved by switching the polarity of the motor, that is, changing the direction of current flow through the armature windings of the motor.

28.4.1 Sector arm and pinion power window mechanism (fig. 28.26)

The up and down movement of the window is controlled by a toothed sector arm which is permitted to swivel on its pivot when power is provided by a dual polarity motor through a two-stage gear reduction consisting of a worm and worm-wheel and a sector and pinion which provides the first- and second-stage reduction respectively. To balance the off-set upthrust on the window, a second arm is made to pivot half-way along the sector-arm in a diagonal position so that it contacts both the fixed and moving horizontal slides. The load reaction on the fixed slide

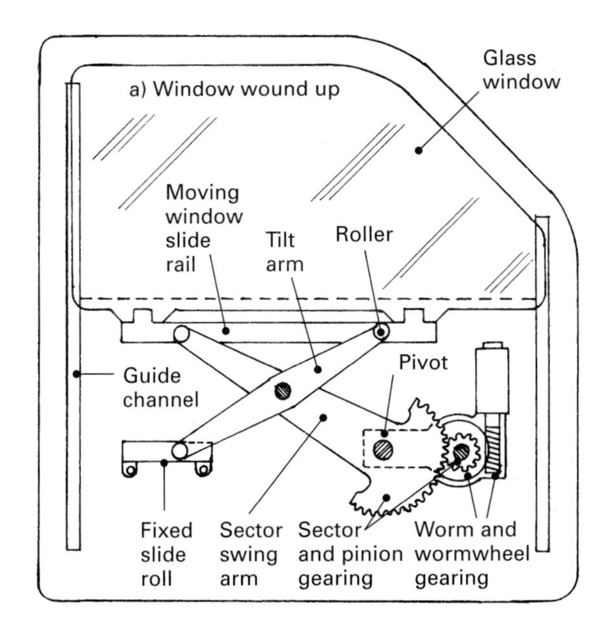

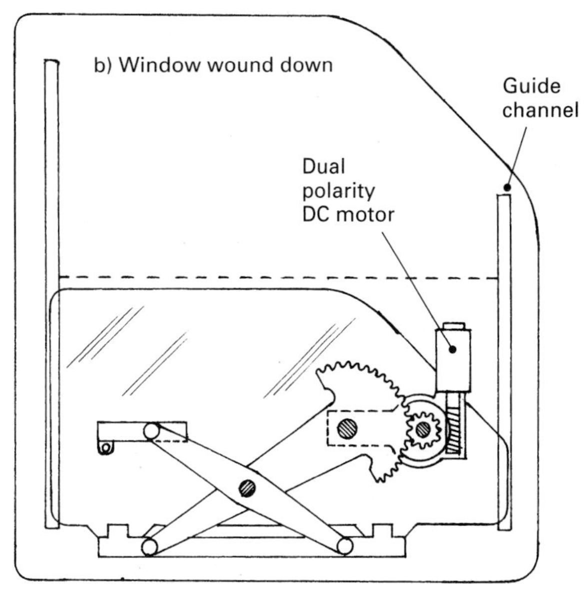

Fig. 28.26 Sector-arm and pinion power window mechanism

ensures that the other end of the tilt-arm exerts an equal up-thrust at the other end of the window slide-rail with that of the supporting sector-arm itself.

28.4.2 Pulley-reel and cable power window mechanism (fig. 28.27)

Power is transferred from the dual polarity motor to the window in two stages, first via the worm and worm-wheel reduction gear and second by the reel and pulley system which pulls the cable in either a clockwise or anticlockwise direction when the motor is energised and is made to rotate. The cable is wound around the reel several times to increase its grip on the reel.

The up and down movement of the window is controlled by a Bowden multi-strand cable supported between an upper and lower pulley-wheel. The cable is attached to the window via a slide-saddle which is restrained to a vertical movement by a guide-rail. The cable converts the rotary motion of the input drive at the cable-reel to a vertical slide-saddle motion.

The direction the window moves, be it up or down, is determined by switching the polarity of the armature brushes, that is, changing the direction of current flow through the armature winding.

Dual-polarity motor (fig. 28.28) This is a narrow or flat-shaped horseshoe ferrite permanent magnet two pole motor with an eight-slot armature and an eight-segment commutator with two brushes and a built-in circuit breaker. The bi-directional rotating armature is controlled by switching the polarity of the brushes, that is, reversing the direction of current flow through the armature windings.

Circuit breaker (fig. 28.28) A circuit breaker connected in series with the armature winding is incorporated into each dual polarity motor. Should the window door switch or master switch be held on for a prolonged period in the fully up or down position, or if something has become jammed between the window and door frame, then the motor will be subjected to stall or lock condition's causing a high rise in the current flow. Consequently, it would tend to overload the motor, thus endangering the continuity of the armature-winding, commutator and brushes. The overloading of the motor is prevented by interposing a circuit breaker in series with the armature winding so that if the motor should stall, the bi-metal contacts heat up and bend away from its adjacent fixed contact, thereby interrupting the supply current to the motor. Once the circuit breaker has cooled down the circuit will automatically reconnect and again switch on the motor.

The inserted graph in fig. 28.28 shows that if a

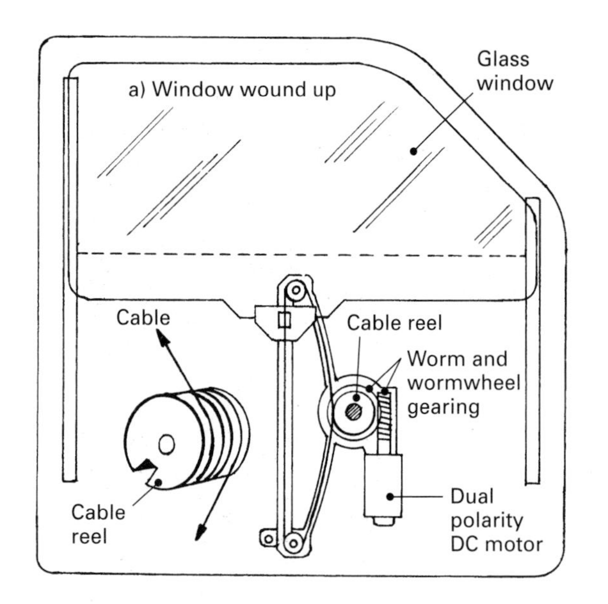

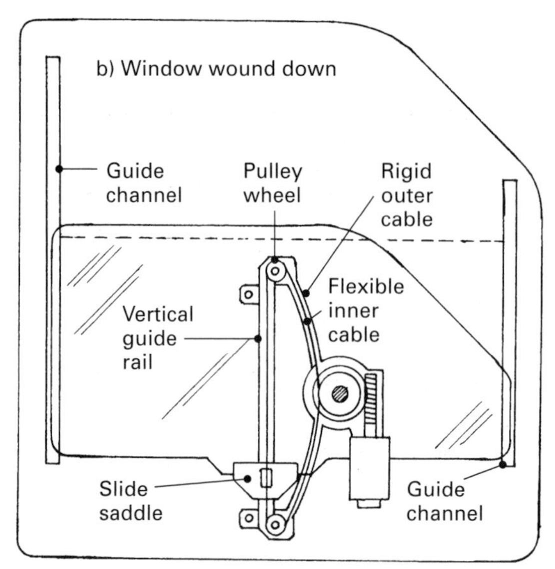

Fig. 28.27 Pulley-reel and cable power window mechanism

small current flows through the bi-metal contacts, the contacts would remain closed for a considerable time; conversely if a large current flows through the contact, the bi-metal strip contacts distort and break the circuit early. For example, if a current of 30 A passes through the contacts it would take roughly 8–20 seconds for them to open.

Power window switch (fig. 28.28) The power window switches are of the self-centring rocker type which control a pair of two-way contacts with a neutral position. When the switch is rocked to the down position, the upper set of contacts connects the battery's positive terminal feed to the motor while the lower contacts complete the current flow return circuit to the battery's negative terminal via the earth-return. This makes the motor rotate in the clockwise direction. Pressing the rocker in the opposite direction towards the up position changes the upper set of contacts to the earth-return and the lower contacts become the feed. The current flow to and from the motor is therefore reversed, that is, the motor polarity has been switched and the motor will rotate in the opposite anticlockwise direction.

28.4.3 Power window wiring circuit (fig. 28.28) Current is supplied to the individual window motors through the main fuse, power window relay and to the four 15 amp window fuses. It then passes to the individual motors via the window master switch and window door switches. When the ignition switch is turned to the 'on' position, voltage will be applied to the power relay window. Its magnetic pull will then close the contacts. If now the master window switch or a window door switch is pushed to either the window up or down position, from the positive battery terminal current will be supplied to the individual window motor through the main fuse, power window relay contacts, and individual motor fuse. It then passes to the motor via either the master switch or the door window switch depending upon which of the two switches have been activated.

Examples of clockwise (down winding) and anticlockwise (up winding) circuitry is now described.

Left-hand front window winding down (fig. 28.28) Current is supplied to the left-hand front window motor through the main fuse, window power relay, left front window motor fuse, the window door switch upper contact and circuit breaker before flowing in and out from the armature windings via the brushes to produce the magnetic torque. Current then passes to the left front window door switch and the corresponding master switch lower bridged contacts, and finally it passes to the negative earth via the main switch.

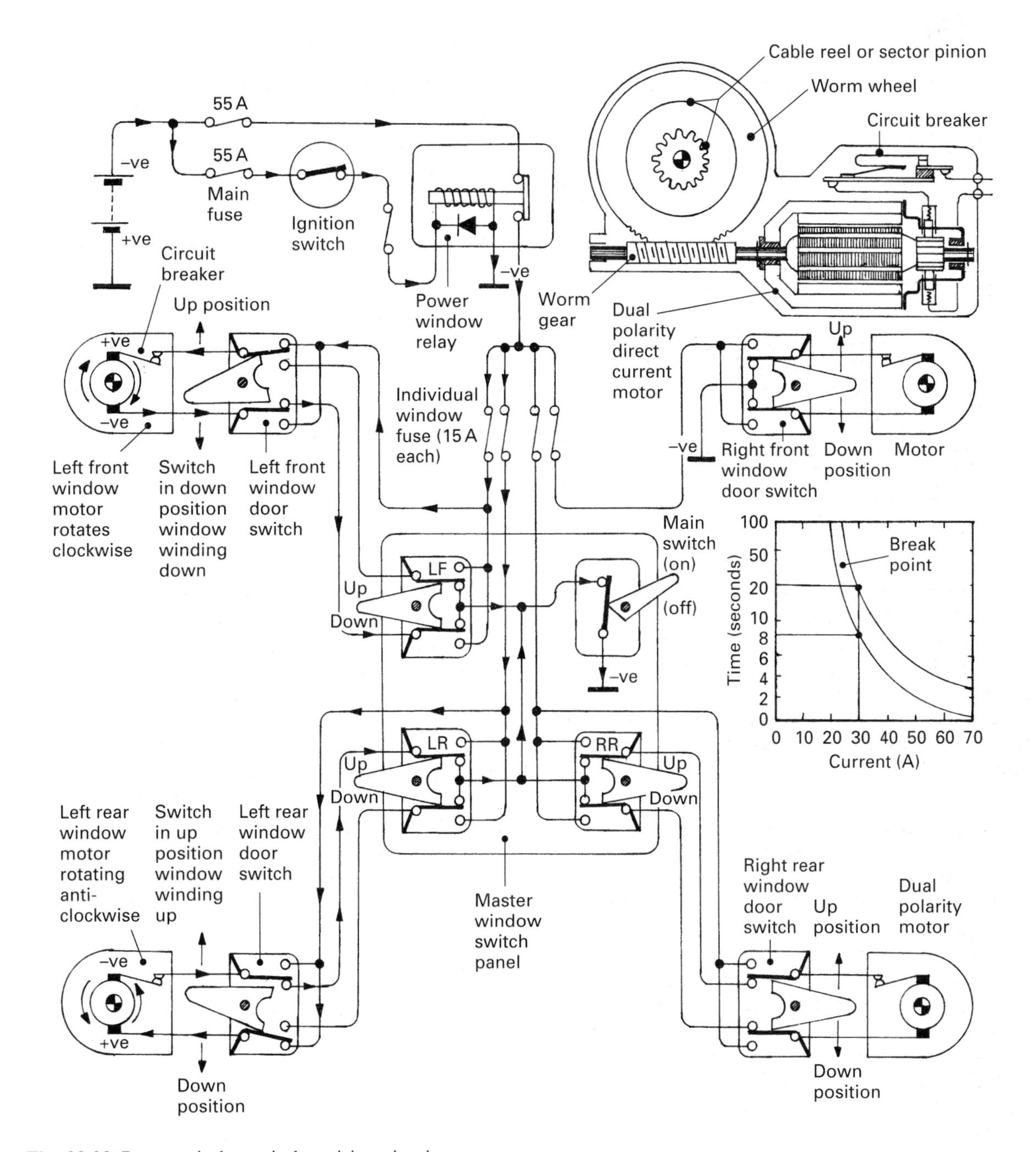

Fig. 28.28 Power window winder wiring circuit

As a result the motor armature is compelled to rotate clockwise.

Left-hand rear window winding up (fig. 28.28)
Current is supplied to the left-hand rear window motor through the main fuse, power window relay, left rear window, motor fuse, door switch and lower bridged contacts, before flowing in and out from the armature windings via the brushes to energies the motor. Current then passes via the circuit breaker to the left-rear window door switch and the corresponding master switch upper bridged contacts, and finally it passes to the negative earth via the main switch. Accordingly the

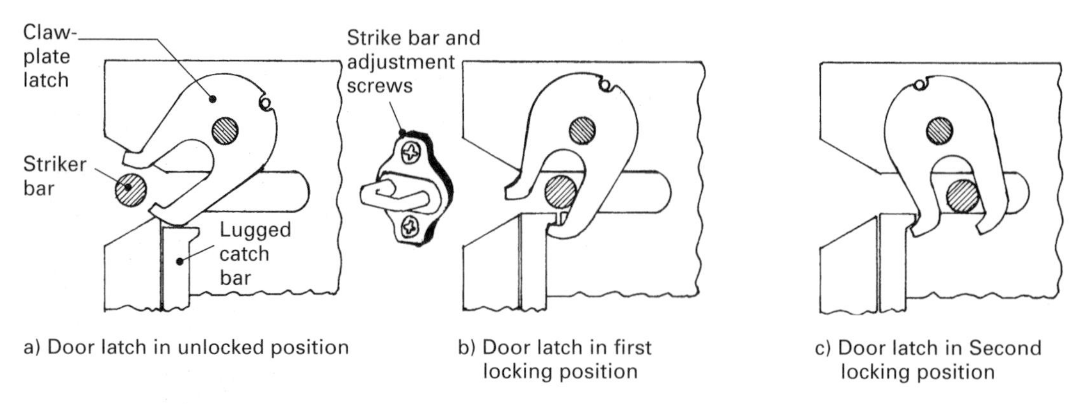

Fig. 28.29 Door latch operating phases

motor armature is compelled to rotate in an anticlockwise direction.

28.5 Central door locking

Central door locking enables the driver to lock or unlock the driver's door and the passenger doors simultaneously when the door key is inserted and turned in the driver's door. Sometimes the central lock-unlock control point facility is extended to the front passengers door and the boot lock can be included in the system. Central door lock systems can be activated by electro-pneumatic, solenoid-capacitor or by dual-polarity motor actuators. The majority of central door lock systems use a permanent magnet dual-polarity motor installed inside each door, which, when commanded, reverses its direction of rotation, thereby activating the locking and unlocking of the doors.

28.5.1 Claw-plate and bar-type latch door lock (fig. 28.29(a)–(c))

This door lock device is made in three parts:

1) a 'U' shaped bar striker screwed on the side of the door pillar;
2) a disc-plate with two claws protruding from its circular profile which is attached by a pivot to the door; this claw-plate latch is spring loaded so that it always returns to the open position; and
3) a spring-loaded vertical lugged catch bar which locks the closed door.

Operating sequence (fig. 28.29(a)–(c))

The striker bar aligns with a horizontal slot in the door which tapers outwards at the mouth. As the door closes the horizontal slot moves towards and over the striker bar until it hits the first claw. The latch-plate is then forced to twist anticlockwise to a point where the spring-loaded lugged catch bar clears the claw, thus enabling it to spring upwards. The lug on the catch bar therefore traps the claw-plate latch in its first locking position (see fig. 28.29(b)). Further movement of the door towards the fully closed position swivels round the claw-plate latch even more until the second claw clears the lugged catch bar and as before trips the spring-loaded catch bar. It thus causes the lug to hold the claw-plate latch in the second fully locked position (see fig. 28.29(c)). To release the door latch so that the door can open, the spring-loaded lugged catch bar is pulled down by the manual release slide, thus enabling the spring-loaded claw-plate latch to swivel clockwise towards the unlocked position (see fig. 28.29(a)).

28.5.2 Door actuator and latch mechanism (fig. 28.30(a) and (b))

The locks are operated by a dual-polarity permanent magnet two-pole motor which usually has an eight-slot armature and a corresponding eight-segment commutator. When the voltage supply positive and negative polarities are reversed the motor changes its direction of rotation. To reduce speed and increase the actuating turning effort of the motor a two-stage reduction gear is built into its output drive. A pinion attached to the armature-shaft drives an integral wheel and pinion gear. This pinion in turn meshes with a rack which converts the rotary motion of the motor into a linear two and fro movement. The rack with an extended rod at one end is made from a plastic moulding. The rod end is linked to a bell-crank lever and pawl via an elongated eye and pin joint.